CONN'S TRANSLATIONAL NEUROSCIENCE

CONN'S TRANSLATIONAL NEUROSCIENCE

Edited by

P. Michael Conn

Senior Vice President for Research,
Associate Provost and Professor of Internal Medicine and Cell Biology at
Texas Tech University Health Sciences Center, Lubbock, TX, United States

ELSEVIER

AMSTERDAM • BOSTON • HEIDELBERG • LONDON
NEW YORK • OXFORD • PARIS • SAN DIEGO
SAN FRANCISCO • SINGAPORE • SYDNEY • TOKYO

Academic Press is an imprint of Elsevier

Academic Press is an imprint of Elsevier
125 London Wall, London EC2Y 5AS, United Kingdom
525 B Street, Suite 1800, San Diego, CA 92101-4495, United States
50 Hampshire Street, 5th Floor, Cambridge, MA 02139, United States
The Boulevard, Langford Lane, Kidlington, Oxford OX5 1GB, United Kingdom

Library of Congress Cataloging-in-Publication Data
A catalog record for this book is available from the Library of Congress

British Library Cataloguing-in-Publication Data
A catalogue record for this book is available from the British Library

ISBN: 978-0-12-802381-5

For information on all Academic Press publications
visit our website at https://www.elsevier.com/

Publisher: Mara Conner
Acquisition Editor: Melanie Tucker
Editorial Project Manager: Kristi Anderson
Production Project Manager: Lucía Pérez
Designer: Matthew Limbert

Typeset by TNQ Books and Journals

Contents

List of Contributors

T. Abbruscato Texas Tech University Health Sciences Center, Amarillo, TX, United States

S. Al Aïn Université du Québec à Trois-Rivières, Trois-Rivières, QC, Canada

N. Alan University of Pittsburgh School of Medicine, Pittsburgh, PA, United States

A. Alfonso University of Illinois at Chicago, Chicago, IL, United States

S.T. Ali Texas Tech University Health Sciences Center, Lubbock, TX, United States

M. Avila Texas Tech University Health Sciences Center, Lubbock, TX, United States

B.S. Barker University of Virginia Health System, Charlottesville, VA, United States

D.M. Benedek Uniformed Services University of the Health Sciences, Bethesda, MD, United States

J.K. Bizley University College London, London, United Kingdom

E.A. Buffalo University of Washington School of Medicine, Seattle, WA, United States

P. Cabrera National Institutes of Health, Bethesda, MD, United States

M. Catani King's College London, London, United Kingdom

J.J Cheng Johns Hopkins Hospital, Baltimore, MD, United States

Y. Chen National Institutes of Health, Bethesda, MD, United States

J.H. Chien Johns Hopkins Hospital, Baltimore, MD, United States

K. Choi Uniformed Services University of the Health Sciences, Bethesda, MD, United States

A.O. Claudio Texas Tech University Health Sciences Center, Lubbock, TX, United States

A. Dabrowski Harvard Medical School, Boston, MA, United States

N. Dafny University of Texas McGovern Medical School Houston, TX, United States

M.S. Dawson King's College London, London, United Kingdom

J.C. DeToledo Texas Tech University Health Sciences Center, Lubbock, TX, United States

A.M. El-Dokla Texas Tech University Health Sciences Center, Lubbock, TX, United States

R.A. Feroze University of Pittsburgh School of Medicine, Pittsburgh, PA, United States

S. Franklin University of Kentucky, Lexington, KY, United States

J.A. Frasnelli Université du Québec à Trois-Rivières, Trois-Rivières, QC, Canada; CÉAMS, Research Center of the Hôpital du Sacré Cœur de Montréal, Montréal, QC, Canada

C.R. Goodlett Indiana University–Purdue University Indianapolis, IUPUI, Indianapolis, IN, United States

S. Gorantla Cleveland Clinic Foundation, Cleveland, OH, United States

A.T. Gulledge Geisel School of Medicine, Hanover, NH, United States

J.F. Hejtmancik National Institutes of Health, Bethesda, MD, United States

J.C. Hoyle The Ohio State University, Columbus, OH, United States

X.-Z. Hu Uniformed Services University of the Health Sciences, Bethesda, MD, United States

B. Hwang Texas Tech University Health Sciences Center, Lubbock, TX, United States

L. Jennes University of Kentucky, Lexington, KY, United States

C.E. Johanson Alpert Medical School at Brown University, Providence, RI, United States

L. Johnson Queensland University of Technology, Brisbane, Australia

E. Johnson-Venkatesh Harvard Medical School, Boston, MA, United States

P. Julayanont Texas Tech University Health Sciences Center, Lubbock, TX, United States

I.N. Karatsoreos Washington State University, Pullman, WA, United States

I. Kim Vanderbilt University Medical Center, Nashville, TN, United States

J. Kim Texas Tech University Health Sciences Center, Lubbock, TX, United States

V.E. Koliatsos Johns Hopkins University School of Medicine, Baltimore, MD, United States

K.R.R. Krishnan Rush University, Chicago, IL, United States

P. Laengvejkal Texas Tech University Health Sciences Center, Lubbock, TX, United States

M.A. Lane Drexel University, Philadelphia, PA, United States

F.A. Lenz Johns Hopkins Hospital, Baltimore, MD, United States

M.D. Lumpkin Georgetown University School of Medicine, Washington, DC, United States

O. M'Hamdi National Institutes of Health, Bethesda, MD, United States

J.D. Mainland Monell Chemical Senses Center, Philadelphia, PA, United States; University of Pennsylvania School of Medicine, Philadelphia, PA, United States

M.L.R. Meister University of Washington School of Medicine, Seattle, WA, United States

S. Melmed Cedars-Sinai Medical Center, Los Angeles, CA, United States

S.L. Miller The Geisel School of Medicine at Dartmouth College, Lebanon, NH, United States

G. Mittleman Ball State University, Muncie, IN, United States

P.T. Morgan Yale University, New Haven, CT, United States

J.M. Nickerson Emory Eye Center, Atlanta, GA, United States

I. Noorani University of Cambridge, Cambridge, United Kingdom

A. Okai Texas Tech University Health Sciences Center, Lubbock, TX, United States

A. Ozpinar University of Pittsburgh School of Medicine, Pittsburgh, PA, United States

M.K. Patel University of Virginia Health System, Charlottesville, VA, United States

L. Pessoa University of Maryland, College Park, MD, United States

H. Pollard Uniformed Services University of the Health Sciences, Bethesda, MD, United States

S.A. Prescott The Hospital for Sick Children and University of Toronto, Toronto, ON, Canada

M.K. Racke The Ohio State University, Columbus, OH, United States

S. Ratté The Hospital for Sick Children and University of Toronto, Toronto, ON, Canada

P.J. Reier University of Florida, Gainesville, FL, United States

G.C. Rosenfeld University of Texas McGovern Medical School Houston, TX, United States

J.V. Rundo Cleveland Clinic Foundation, Cleveland, OH, United States

D. Ruthirago Texas Tech University Health Sciences Center, Lubbock, TX, United States

J. Ryu Johns Hopkins University School of Medicine, Baltimore, MD, United States

R.F. Sekula Jr. University of Pittsburgh School of Medicine, Pittsburgh, PA, United States

K. Shah Texas Tech University Health Sciences Center, Amarillo, TX, United States

R. Silver Barnard College, New York, NY, United States; Columbia University, New York, NY, United States

C.H. Soubrane Pfizer Neuroscience and Pain Research Unit, Cambridge, United Kingdom

G.J. Stephens University of Reading, Reading, United Kingdom

E.B. Stevens Pfizer Neuroscience and Pain Research Unit, Cambridge, United Kingdom

R.S. Swenson Geisel School of Medicine, Hanover, NH, United States

S. Tantikittichaikul Texas Tech University Health Sciences Center, Lubbock, TX, United States

C. Trimmer Monell Chemical Senses Center, Philadelphia, PA, United States

H. Umemori Harvard Medical School, Boston, MA, United States

R. Ursano Uniformed Services University of the Health Sciences, Bethesda, MD, United States

D.M. Waitzman University of Connecticut Health Center, Farmington, CT, United States

D.M. Welch The Ohio State University, Columbus, OH, United States

H. Wilms Texas Tech University Health Sciences Center, Lubbock, TX, United States

C.-K. Wu Texas Tech University Health Sciences Center, Lubbock, TX, United States; University of California Irvine, School of Medicine, Orange, CA United States

G.H. Wynn Uniformed Services University of the Health Sciences, Bethesda, MD, United States

L. Xu Johns Hopkins University School of Medicine, Baltimore, MD, United States

H.H. Yeh The Geisel School of Medicine at Dartmouth College, Lebanon, NH, United States

G.T. Young Pfizer Neuroscience and Pain Research Unit, Cambridge, United Kingdom

L. Zhang Uniformed Services University of the Health Sciences, Bethesda, MD, United States

L.V. Zholudeva Drexel University, Philadelphia, PA, United States

N. Ziogas Johns Hopkins University School of Medicine, Baltimore, MD, United States

S. Živković University of Pittsburgh Medical Center, Pittsburgh, PA, United States

Preface

Organizing a book is a great opportunity to think about the progress of a field. What has come about in the recent past? What is currently important? In the case of the neurosciences, much has been accomplished since mid-1980s, and this volume includes topics that might not have appeared in a similar book written until 2016.

The authors have intentionally focused on material that is both current and essential to the neurosciences and, for brevity and to avoid redundancy, assumed that the reader is already versed in cell biology and biochemistry, as examples.

The authors were identified from individuals known to be excellent teachers and thought leaders in the neurosciences. I am especially pleased with the care that they have taken to prepare and structure their chapters and by their enthusiasm for this project.

As a special feature of the book, we have included short "translational" chapters, which were designed to bring the attention of those interested in disease states to the underlying etiology, described in the basic science component of the book. I am grateful to Dr. John Detoledo, Chairman of Neurology and his colleagues at Texas Tech University Health Sciences Center, for undertaking this component of the project.

Finally, I thank staff at Elsevier for helping to coordinate the process.

P. Michael Conn

CHAPTER

1

Cytology of the Central Nervous System

L. Jennes

University of Kentucky, Lexington, KY, United States

The central nervous system is formed by two main classes of cells, the neurons and the glial cells. Both cell types contain all the typical structures and organelles of regular cells but, in addition, they exhibit specific specializations that give these cells their unique functions. In the following section, we focus on these specializations and only briefly describe the general cell biological and morphological structures that are found in most cells.

NEURONS

Neurons are highly specialized, excitable cells that contain a cell body or soma which includes the *perikaryon* (the region around the nucleus) that is the central site of synthesis and degradation of proteins, lipids, and carbohydrates. Usually, a prominent nucleus resides in the center of the soma and is the site where chromosomal DNA and many associated proteins, needed for uncoiling of the DNA and transcription of DNA as well as synthesis of messenger and transfer RNAs, are located. Because neurons are typically very active in protein synthesis, the nuclei of many neurons appear only lightly stained when observed with the microscope since much of the DNA is uncoiled and does not trap histological dyes. This light portion of a nucleus is the transcriptionally active *euchromatin* while chromatin that is condensed and transcriptionally inactive and stains dark with specific dyes is called the *heterochromatin*. The appearances of the two forms of chromatin as light or dark are transient and reflect the degree of ongoing transcriptional activity.

Inside the nucleus is a prominent *nucleolus*, which is the site of ribosomal RNA (rRNA) synthesis. This structure contains a large number of proteins that facilitate rRNA synthesis and processing, as well as transport first to the nucleus and then into the cytoplasm. The nucleolus typically contains three main components that differ in their function and protein content: a fibrillary center that contains large amounts of RNA polymerase I; a dense fibrillary component that contains RNA processing proteins; and a granular component in which the proteins are located that are needed for the final rRNA processing (Fig. 1.1).

The nucleus is surrounded by a *nuclear envelope*, which is formed by two parallel membranes that are continuous with the rough endoplasmic reticulum (rER). The inner nuclear membrane is covered by the nuclear lamina, which consists mostly of lamin intermediate filaments and associated binding proteins. Thus, this structure

Conn's Translational Neuroscience
http://dx.doi.org/10.1016/B978-0-12-802381-5.00001-4

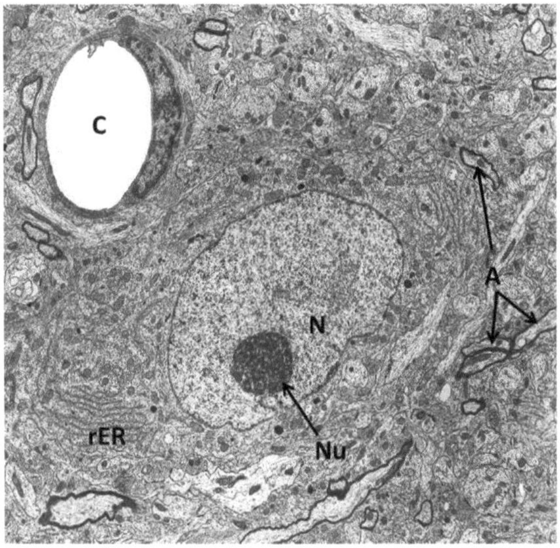

FIGURE 1.1 Electron micrograph of a neuron in the brain stem showing a prominent nucleus (N) and nucleolus (Nu), as well as extensive rough endoplasmic reticulum (rER). Myelinated axons are indicated by (A), and (C) indicates a capillary.

is fairly rigid and gives the nucleus its shape. In addition, the nuclear lamins participate in a variety of biological processes, including organization of the chromatin, localization of nuclear pores, and reorganization of the nucleus before and after mitosis.

The nuclear envelope contains many *nuclear pores*, which are sites for diffusion of small molecules and facilitated transport of larger molecules into and out of the nucleus. At these specialized sites, the inner and outer nuclear membranes fuse and form openings between the nuclear plasm and cytoplasm. A nuclear pore is a complex structure that consists of about 50 proteins (nucleoporins). These proteins form an octagonal structure as well as a central channel or plug. The nuclear pore allows substances of about 5000 Da or less to freely diffuse through the pore while larger structures require a nuclear localization signal sequence or patch of basic amino acids to be recognized by nuclear import receptors. The import receptors bind to the signal sequence and to nucleoporins and the complex moves slowly through the pore into the nucleus where the complex dissociates.

Continuous with the nuclear envelope is the *endoplasmic reticulum*, which is a system of membranous channels that form stacks of cisternae or flattened sacs. Much of the endoplasmic reticulum can be studded with ribosomes (*rER*) indicating that this is the site where proteins are assembled and modified either for export or to be incorporated into other organelles, such as the Golgi apparatus, endoplasmic reticulum, lysosomes, or plasma membrane. The sizes and numbers of the endoplasmic reticulum cisternae vary with the protein synthetic activity of the cell and are largest in cells that are highly active. Collectively, the stacks of the endoplasmic reticulum in neurons are referred to as *Nissl bodies* when seen in the light microscope. Proteins that are newly synthetized on ribosomes have a signal sequence that is recognized by a signal recognition particle and guided through the membrane into the lumen of the endoplasmic reticulum. Here, the signal sequence is removed from the newly formed protein and posttranslational modifications, such as glycosylation, are initiated. Proteins inside the cisternae are packaged into small transfer vesicles and transported from the rER to the forming or "cis" face of the Golgi apparatus, which is closest to the nucleus.

The *Golgi apparatus* consists of concave stacks of smooth, membrane-bound cisternae whose main functions are to modify and sort proteins. After the transfer vesicles from the rER fuse with the membrane of the cis-Golgi apparatus and have delivered their content into the lumen of the cisternae, the proteins are transported from the "cis"-face to the "trans"-face or the maturing face of the Golgi apparatus. During this transport, proteins are modified enzymatically by the addition of select oligosaccharides or by phosphorylation or sulfatation. Eventually, the proteins reach the trans-Golgi apparatus where they are packaged into a variety of vesicles and transported to their destinations, which include the plasma membrane, secretory vesicles, endosomes, or other membrane-bound organelles. Proteins that are destined for release are packaged in the Golgi apparatus into secretory vesicles. These vesicles fuse with the plasma membrane and spill their content into the extracellular compartment, for example, at an axon terminal. This process implies that the vesicular membrane is inserted into the plasma membrane and causes an increase in surface area. Thus, compensatory membrane retrieval by endocytosis occurs to prevent swelling of the terminal or increase in surface area.

In the cisternae of the trans-Golgi apparatus, proteins are sorted depending upon their final destination. Proteins that are destined for *lysosomes* receive mannose 6-phosphate (M6P) during their passage through the cis-Golgi apparatus and, upon arriving at the trans-Golgi apparatus are recognized and bound by specific M6P receptors. These receptors help packaging the lysosomal enzymes into small clathrin-coated vesicles, which pinch off the trans-Golgi network and become primary lysosomes after the clathrin coat detaches and the M6P-receptor proteins are recycled. Lysosomes contain about 40 different hydrolytic enzymes that function most efficiently at around pH 5.0. The membranes of lysosomes contain ATP-dependent H^+ pumps that control the acidic environment inside a lysosome. Should a lysosome be damaged, the acidic pH is neutralized immediately by the higher pH of the cytoplasm and autodigestion is prevented.

Newly formed lysosomes that have not yet participated in degrading activities are *primary lysosomes* and appear as small (~0.05–0.5 μm), clear vesicles. Lysosomes that are active in catalytic degradation of protein, carbohydrates, nucleic acids, or lipids are called *secondary lysosomes* and are characterized by a heterogeneous content. Lysosomes are involved in the degradation of extracellular material that is taken up by *endocytosis*. During this process, extracellular material is concentrated at the outside of the plasma membrane, which invaginates to form a pit and then pinches off at the cytoplasmic site to form an endocytic vesicle. The concentration of material at the outside of the plasma membrane may be receptor mediated. The endocytic vesicle fuses with a primary lysosome and degradation of the ingested material is initiated.

Depending upon the size of the endocytic vesicle, two types of processes are distinguished. *Phagocytosis* describes a process in which larger particles or substances are taken up by "phagosomes" which fuse with primary lysosomes to form secondary lysosomes. *Pinocytosis* describes a process by which fluids are taken up in small vesicles (~100 nm). Lysosomes are also important for the elimination of intracellular organelles that become nonfunctional due, for example, to aging. In this event, the endoplasmic reticulum surrounds this organelle to form an *autophagosome*. This autophagosome fuses with a lysosome, and catalytic degradation is initiated (Fig. 1.2).

Continuous with the rER is the *smooth endoplasmic reticulum* (*sER*), which is characterized by small stacks of parallel membranes that form a network of interconnected channels and cisternae. Ribosomes are absent in this organelle. While the sER can have a wide variety of functions, such as detoxification or participation in steroid synthesis, depending upon the cell type, the best studied function in neurons is related to the regulation of intracellular Ca^{2+} levels. The membrane of the sER contains large integral membrane proteins that function as receptors for inositol 1,4,5-trisphosphate (IP_3) as well as gated Ca^{2+} channels. When IP_3 binds to its receptor, the Ca^{2+} channel is opened and allows Ca^{2+} to move from the lumen of the sER to the cytoplasm, thus raising the cytoplasmic Ca^{2+} concentration very quickly. In addition, a second type of Ca^{2+} channel is present in the sER membrane; however, this protein is directly regulated by cytoplasmic Ca^{2+}. Thus, when cytoplasmic Ca^{2+} binds to and activates this channel, additional Ca^{2+} is released from the lumen of the sER into the cytoplasm.

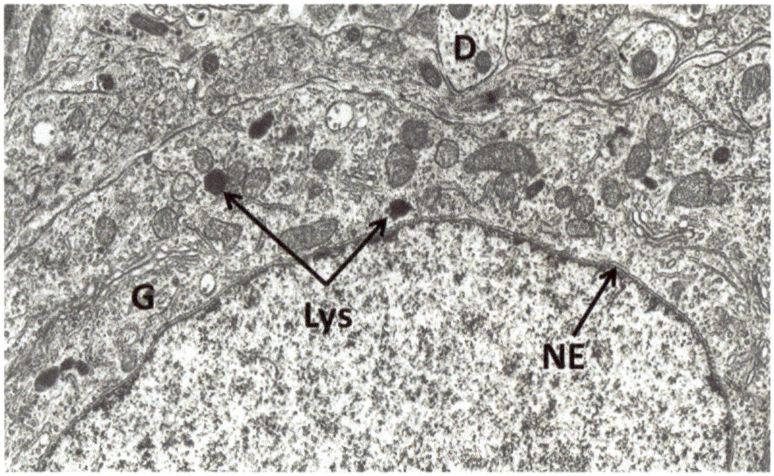

FIGURE 1.2 Higher magnification of the cell in Fig. 1.1 showing a Golgi apparatus (G), nuclear envelope (NE), as well as lysosomes (Lys). A dendrite is labeled (D).

The structure and function of *mitochondria* in neurons are similar to other cells; they are the sites of oxidative phosphorylation which results in the formation of adenosine triphosphate (ATP). ATP in turn, is used to transfer energy within a cell. Mitochondria have a distinct architecture. They are surrounded by two unit membranes, an inner and an outer membrane which are separated by a narrow space, the intermembranous space. The outer membrane is unfolded and contains the protein "porin" which forms aqueous channels through this membrane, allowing free diffusion of substances up to a MW of 5kDa. The inner membrane forms many folds or cristae which extend into the matrix and is highly specialized and far less permeable than the outer membrane due to the presence of high amounts of the phospholipid "cardiolipin." The different functions of mitochondria are strictly associated with separate compartments. The inner membrane contains proteins of the electron transport chain that produces ATP by oxidative phosphorylation. The matrix contains enzymes that metabolize pyruvate and fatty acids to generate acetyl coenzyme A which is further processed in the Krebs cycle. In addition, the matrix contains circular DNA, mitochondrial ribosomes, and transfer RNA (tRNA). While most mitochondrial proteins are encoded by genes in the nucleus, some proteins required for oxidative phosphorylation, are encoded by genes in the mitochondrial DNA. Mitochondria can divide by binary fission which occurs in a semi-autonomous fashion, independent of cellular/nuclear division.

The *cytoskeleton* plays an essential role in the maintenance of shape and function of a neuron and acts as a dynamic, internal framework as well as an extensive system to transport organelles and molecules throughout the cell. The cytoskeleton includes three families of filaments which differ in size, structure, mechanism of assembly/disassembly and functions: microfilaments, intermediate filaments (neurofilaments), and microtubules.

Microfilaments have a diameter of 7nm and are composed mostly of actin which can bind ATP and then polymerize into long polar and helical filaments. Although actin filaments are asymmetrical with a distinct "plus" (or "barbed") end and a "minus" end, they have the potential to grow on both ends by the addition of ATP-bound globular actin. However, growth is usually faster at the "plus" end. Polymerization of actin filaments usually involves the process of "treadmilling" in that actin monomers are added at the "plus" end while other actin monomers are removed from the "minus" end by depolymerization. This process is highly regulated and involves several families of binding proteins: "thymosin" concentrates actin monomers in a reserve pool and thus prevents their use in polymerization; "profilin" binds to actin monomers and promotes assembly into microfilaments by exchanging bound ADP with ATP; "gelsolin" cuts actin filaments and binds to the exposed "plus" end, thereby preventing growth at the newly formed barbed end; and "cofilin" causes depolymerization of actin filaments at the "minus" end.

Microfilaments are relatively short and concentrated in the portion of the cytoplasm next to the plasma membrane. Here, the microfilaments, together with a host of actin-binding proteins, form a dense, three-dimensional network which is important for the development of neurons, transport of organelles and molecules, as well as for the formation of synapses. Transport of organelles and larger molecules is accomplished by using the molecular motor protein "myosin V," which is a bifunctional protein whose head contains ATP and a binding site for actin while the tail binds to a family of adaptor proteins that connect the filament to the vesicle or organelle to be transported.

Neurofilaments belong to the family of intermediate filaments, which are about 10 nm in diameter. These filaments are chemically very stable and tough and represent the structural backbone of a neuron. Neurofilaments are formed by polymerization of monomers, which are about 46 nm long and characterized by a head-, rod-, and tail region from the N- to the C-terminal. Initially, two monomers form a dimer which, in turn, aligns in an antiparallel orientation with a second dimer to form a tetramer. Eight tetramers form a protofilament and eight protofilaments align in a rope-like fashion to form one filament. The formation and disassembly of neurofilaments are regulated by phosphorylation and result in a structure that is concentrated especially in axons to provide physical support.

Microtubules are long, dynamic, hollow structures with about 25 nm in diameter. Microtubules provide a way of transport throughout the entire length of a neuron and are very important for the maintenance of neuronal cell shape and movement of mitochondria and small membrane-bound vesicles from the perikaryon through the axon to the presynaptic terminal.

Microtubules are formed by polymerization of α- and β-tubulin heterodimers into protofilaments, 13 of which align longitudinally and form a polar tubule which has a distinct "plus" or "growing" end and a "minus" end where growth is slow. In axons, all microtubules are oriented in the same direction with the "minus" end toward the perikaryon and the "plus" end toward the axon terminal. The microtubules carry membrane-bound transport vesicles, mitochondria, nonvesicular proteins and synaptic vesicles from the perikaryon to the terminal by "anterograde" transport. This transport is accomplished by "kinesin" molecular motor proteins that bind to microtubules and can move, one tubulin heterodimer at a time toward the "plus" end of the tubule while carrying its cargo. The type of axonal transport involving motor proteins is fast and can move organelles about 50–100 cm/day. Cytoskeletal and cytosolic proteins are transported by "slow" axonal transport which travels at a rate of 0.2–2.5 mm/day only in an anterograde direction. Spent materials need to be transported back from the axon terminal to the cell body for disassembly or recycling. This process of "retrograde" transport is accomplished by the motor protein "cytoplasmic dynein," which moves on microtubules from the "plus" end in the axon terminal toward the "minus" end of the cell in the perikaryon.

DENDRITES AND AXONS

Neurons have the unique property to receive, process, integrate, and transmit information over sometimes very long distances. Specific features associated with different parts of a neuron accommodate these tasks: *dendrites* usually receive a signal which is then propagated along the perikaryon to the axon which carries the signal to its terminal. In most neurons, many dendrites extend from different regions of the perikaryon and taper off with increasing distance from the cell body. Dendrites can branch extensively and develop dendritic trees, which form the portion of a neuron that receives most of the synaptic input. Often, dendrites contain small protrusions or spines along their surface, which cause a significant increase in the accessible surface area available for synaptic innervation (Fig. 1.3). The cytoplasmic composition of dendrites, especially at their base, resembles the one of the perikaryon and contains most organelles, such as the rER or lysosomes but usually not a Golgi apparatus. Dendrites contain many microtubules and neurofilaments as well as microfilaments, which are particularly dense in dendritic spines where the microfilaments are involved in the control of spine size and shape. The sER reaches sometimes the base of dendritic spines, which may also contain polyribosomes indicating that local protein synthesis can occur. When a neuron receives synaptic signals at a dendrite, these signals are integrated and propagated to the perikaryon in the form of an electrical signal, the synaptic potential. If this potential reaches a certain threshold, an all or a none response is initiated at the initial portion of an axon, the *axon hillock*, and an action potential is generated that travels unidirectionally along the axon to its terminal. Here, the arriving action potential causes the release of neurotransmitter from the presynaptic terminal. The axon hillock is a highly specialized area in the perikaryon devoid of most organelles; however, its plasma membrane contains a very high density of voltage-gated ion channels that are needed to initiate an action potential. *Axons* are fairly thin structures with diameters between 1 and 20 μm that maintain a uniform caliber throughout their

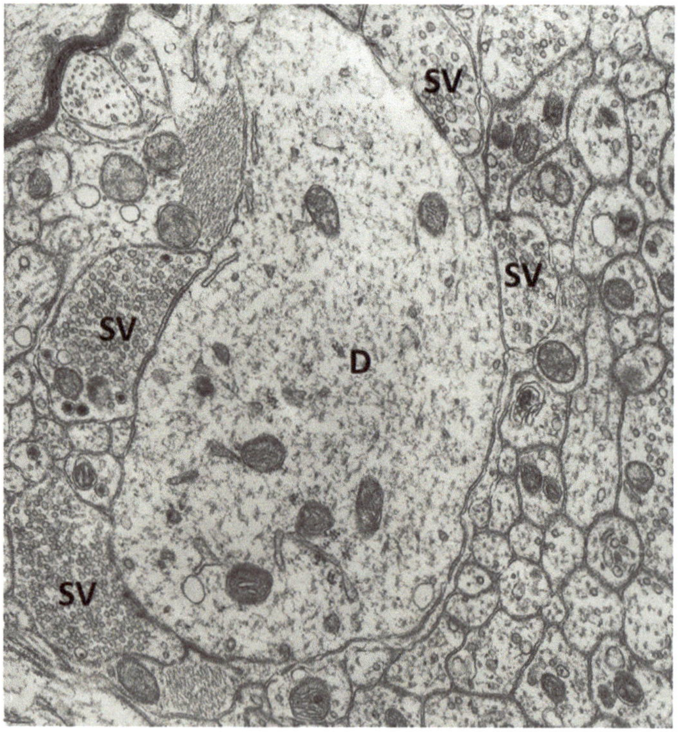

FIGURE 1.3 Electron micrograph of a dendrite (D) in the brain stem showing multiple sites of innervation by presynaptic terminals containing a large number of small, clear synaptic vesicles (SV).

extent. Axons vary greatly in length and can reach about 5 ft, they can branch into collaterals that arise at obtuse angles, and thus can innervate different targets. Axons contain large amounts of intermediate filaments for stabilization and microtubules and associated proteins to accommodate a high demand for transport of neurotransmitter and proteins to the presynaptic terminal. There is very little, if any protein synthesis occurring in axons and polysomes are usually absent. Thus, each organelle or protein used in an axon terminal is assembled in the perikaryon and then transported by anterograde mechanisms to the terminal.

Axons terminate in bulb-like structures, the *presynaptic terminal*. While there are significant variations in their appearance, synapses have several common features: the presynaptic portion contains mitochondria, components of the smooth endoplasmic reticulum and a large number of membrane-bound synaptic vesicles, which are concentrated near the presynaptic terminal membrane. This terminal membrane is coated on the axoplasmic site by electron-dense material that forms the presynaptic density. This region is greatly enriched in proteins that are involved in docking of the synaptic vesicles, fusion of the vesicular membrane with the presynaptic membrane, exocytosis of the vesicular neurotransmitter content, and removal of released neurotransmitter. The presynaptic terminal is separated by an extracellular synaptic cleft (about 20–40 nm wide) from the postsynaptic density of the cell that receives the synaptic input. This density contains very high concentrations of receptors to bind the released neurotransmitter, ion channels, and enzymes to degrade the released neurotransmitter (Fig. 1.4).

Synapses can be classified by the site of a neuron that is contacted for innervation:

- axo-somatic synapses are formed on the perikaryon of a neuron;
- axo-dendritic synapses describe an axon terminating on a dendrite;
- axo-spinous synapses are restricted to axon terminals on dendritic spines;
- axo-axonic synapses occur between two axons.

Chemical synapses transmit signals very rapidly to the adjacent postsynaptic site with a synaptic delay of about 0.3–5 ms. However, a second family of synapses exists that uses electrical current to transmit information. These are the *electrical synapses* or gap junctions; they can transmit a signal almost instantaneously without any delay by passing current directly from one cell to the adjacent cell and, in contrast to chemical synapse, they can act in a bidirectional fashion. Electrical synapses are formed by connexin integral membrane proteins, six monomers of which assemble into a hemi-channel that is called "connexon." This hemi-channel spans the neuronal plasma membrane and aligns with a connexon of an adjacent neuronal membrane to form a hydrophilic, intercellular channel. The channel has a pore of about 1.5 nm and allows small organic molecules, Ca^{2+} and cyclic adenosine monophosphate (cAMP) to freely move between the two cells. Usually, a large number of connexons concentrates in patches that can greatly vary in size. At these patches, the plasma membranes of adjacent neurons are separated by only 4 nm, which is much smaller than a 20 nm space of a synaptic cleft. Connexins are members of a large gene family that has about 20 members and are expressed in a tissue-specific manner.

Neurons can be classified by their appearance, complexity of their dendrites, and size. Most neurons fall into one of the following categories:

- *Unipolar neurons* have only one process and are found mostly in invertebrates.
- *Bipolar neurons* are usually oval in shape and contain two processes, a dendrite that receives signals usually from the periphery and an axon that propagates the signal to the central nervous system. Bipolar neurons

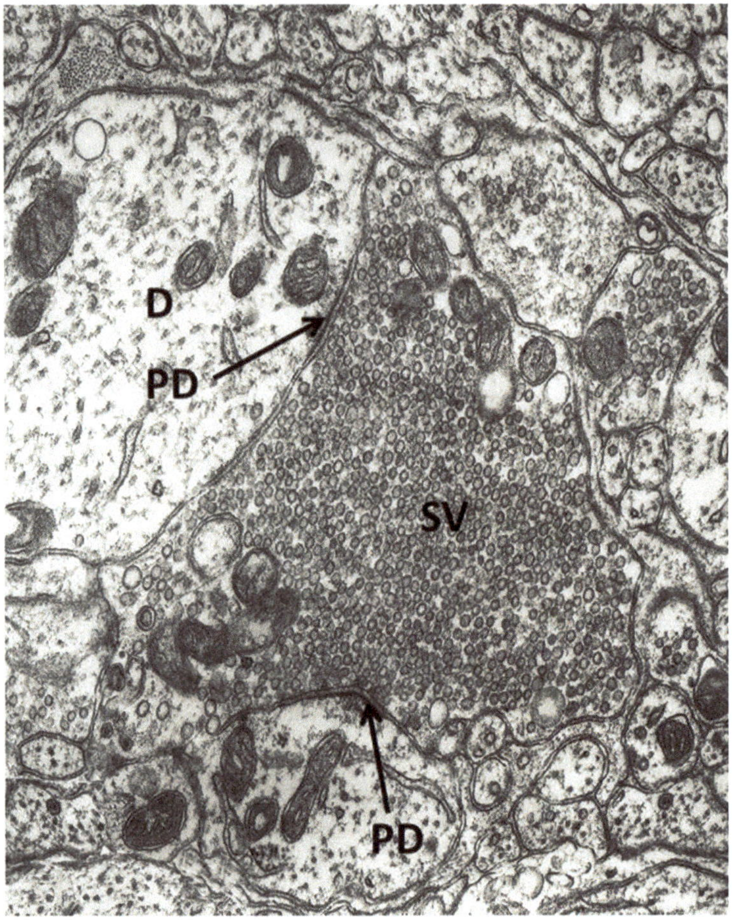

FIGURE 1.4 Electron micrograph showing a presynaptic axon terminal (SV) that innervates two dendrites (D). Both axo-dendritic synapses exhibit postsynaptic densities (PD), indicated by the arrows.

are found in sensory organs, such as the retina, olfactory epithelium, and the auditory system.

- *Pseudounipolar neurons* are variations of bipolar neurons in that they have two processes which fuse during their development into one short common axon. This axon splits into one branch that terminates in the periphery while the second branch terminates in the spinal cord. This way, stimuli from the periphery bypass the cell body and reach the axon terminal without delay. This type of neuron is found in sensory ganglia of cranial and spinal nerves.
- *Multipolar neurons* are characterized by many dendrites that can originate from different regions of a perikaryon. These neurons vary greatly in size, shape, and complexity of their dendritic tree, and they represent the most common type of neuron in the central nervous system.

NEUROGLIA

Neuroglia cells represent the most numerous cell family in the central nervous system with 5–10 glial cells per neuron or 350billion cells per brain. Glial cells retain their ability to divide; provide metabolic and structural support for neurons, and maintain conditions that allow adequate functioning of neurons. They are less electrically excitable than neurons and do not form chemical synapses. Based upon their appearance, function, and origin, four types of glial cells have been identified in the central nervous system: astrocytes, oligodendroglia, ependymal cells, and microglia.

Astrocytes are the most numerous type of glia and account for about one half of all cells in the brain. They are divided into two categories of glia, the protoplasmic astrocytes and the fibrous astrocytes. *Protoplasmic astrocytes* are present predominantly in gray matter and are characterized by many, often short, branched processes. Sometimes, these processes form sheets that can separate different neuronal cell processes or neurites and form close associations with synapses. Protoplasmic astrocytes are involved in many regulatory processes; they buffer K^+ and Cl^- ions, they regulate extracellular neurotransmitter concentrations through specific uptake mechanisms, and they play an important role in the formation of synapses by, for example, secreting substances that stimulate the development of synapses. In addition, some astrocytes release neurotrophic and gliotrophic factors, such as "glial-derived neurotrophic factor," which promotes survival of select neurons.

Fibrous astrocytes are present predominantly in white matter (ie, myelinated fiber tracts). These cells have long, thin processes that do not form sheets.

Both types of astrocytes form expansions at the end of their processes, the "astrocytic end feet." These structures completely cover the basement membrane of capillaries and small arterioles as well as the outer surface of the brain forming an insulating layer, the "glia limitans." Thus, astrocytes form a physical barrier which is important for the control of movement of small molecules into the brain.

Oligodendrocytes are relatively small glial cells that develop from precursor cells in the ventral ventricular zone during the late prenatal and early postnatal period. The main function of oligodendrocytes is to provide an insulating cover that surrounds individual axons and thereby allows an increase in the speed of conductance of electrical signals. Oligodendrocytes have several processes that can flatten into sheet-like structures and wrap many times around portions of individual axons. Each process of an oligodendrocyte covers a small segment of an axon, and a narrow gap or node of Ranvier of uncovered axon remains between adjacent segments. These nodes contain a large number of Na^- ion channels and are critical for the saltatory conduction of action potentials. Each oligodendrocyte can form several myelin sheets covering different segments of one axon, or they can provide myelin to several segments of different axons. Usually, axons are covered by myelin sheets from the initial segment to the terminal.

Ependymal cells form a simple, cuboidal epithelium that lines the inner surfaces of the brain, including the ventricles and the central canal in the spinal cord. Ependymal cells are linked to each other by desmosomes while the base of the cells is in contact with astrocytes. In most regions of the ventricular system, ependymal cells have kinocilia at their apical surface which are thought to be important for moving the cerebrospinal fluid throughout the ventricles and central canal of the spinal cord.

In certain regions of the lateral, third and fourth ventricles, the ependymal epithelium is continuous with the *choroid plexus* which consists of highly specialized and polar cells that form cellular sheets and float in the cerebrospinal fluid. The cells of the choroid plexus rest on a basement membrane, the cells are connected to each other by tight junctions and the lateral plasma membranes form extensive interlocking folds with adjacent choroid plexus cells. Underneath the choroid plexus cells lies a dense network of fenestrated capillaries that also lack tight junctions. This allows plasma filtrate to freely move from the blood into the extravascular region, however, due to the junctional complexes between adjacent choroid plexus cells, movement of solutes toward the apical ventricular surface is prevented. Instead, transport and exchange mechanisms move ions and micronutrients through the choroid plexus cells, which sets up an osmotic gradient that is neutralized by water that follows passively. The cerebrospinal fluid consists mostly of water and ions, some glucose and very small amounts of protein. The cerebrospinal fluid leaves the ventricular system at the foramen magnum and the aperture of Luschka of the fourth ventricle as well as at the terminal cisterna of the spinal cord to reach the subarachnoid space. Thus, both internal and external surfaces of the brain are bathed in cerebrospinal fluid.

Microglia are small cells with short processes that are found throughout the central nervous system. In contrast to the other families of glial cells, microglia are not derived from a neuroectodermal lineage. Instead, they originate from the bone marrow mesoderm and are derived from monocytes. Microglia enter the central nervous system early in development and remain inconspicuous for a long period of time during which they migrate through the brain, screening for immunologically relevant events. Thus, microglia participate in the management of inflammatory and immune responses in the brain. Microglia can become highly phagocytic after insults to the brain, such as stroke, when they remove cellular debris or other pathological substances, such as beta-amyloid. In addition, they are antigen-presenting cells and they release cytokines that attract other cells, including neutrophils and lymphocytes to move across the blood–brain barrier into the central nervous system.

Acknowledgment

I would like to thank Dr. Bruce Maley, Department of Anatomy and Neurobiology, University of Kentucky for the electron micrographs.

Further Reading

[1] Kandel ER, Schwartz JH, Jessell TM, Siegelbaum SA, Hudspeth AJ. Principles of neural science. 5th ed. New York: McGraw-Hill; 2013.

[2] Albers B, Johnson A, Lewis J, Morgan D, Raff M, Roberts K, et al. Molecular biology of the cell. 6th ed. New York: Garland Science; 2014.

[3] Purves D, Augustine GJ, Fitzpatrick D, Hall WC, LaMantia A-S, White LE. Neuroscience. 5th ed. Sunderland (MA): Sinauer Associates; 2012.

CHAPTER

2

Ion Channels

B.S. Barker[1], G.T. Young[2], C.H. Soubrane[2], G.J. Stephens[3], E.B. Stevens[2], M.K. Patel[1]

[1]University of Virginia Health System, Charlottesville, VA, United States; [2]Pfizer Neuroscience and Pain Research Unit, Cambridge, United Kingdom; [3]University of Reading, Reading, United Kingdom

ION CHANNELS

Ion channels are protein molecules that span across the cell membrane allowing the passage of ions from one side of the membrane to the other. They have an aqueous pore, which becomes accessible to ions after a conformational change in the protein structure causes the ion channel to open. Once opened, ion channels show selectivity for the class of ions that are allowed to pass through them. Some ion channels only allow a specific cation or anion to pass through; for example, sodium (Na^+) channels only allow Na^+ ions to pass through while potassium (K^+) channels only allow K^+ to pass through. Other ion channels allow passage of a broader group of ions, for example, *N*-methyl-D-aspartate receptors (NMDA) are a class of ion channels that allow Na^+, Ca^{2+}, and K^+ ions to flow through them.

Ion channels are divided into those that are opened by changes in membrane potential, *voltage-gated ion channels*, and ion channels that are opened by the binding of a ligand, such as a hormone or a neurotransmitter, *ligand-gated ion channels*. For both types of ion channels, opening will lead to the passage of ions down their respective electrochemical gradient, which typically occurs at very fast rates. In this chapter, we introduce both voltage-gated and ligand-gated ion channels that are abundantly expressed within the central nervous system (CNS) and the peripheral nervous system (PNS).

Voltage-Gated Ion Channels

Voltage-gated ion channels play a critical role in generating and facilitating the transmission of information, in the form of action potentials, within the CNS, PNS, and the cardiovascular system. In this section, we introduce three classes of ion channels widely expressed within the CNS and the PNS: sodium (Na^+) channels, calcium (Ca^{2+}) channels, and potassium (K^+) channels.

Na^+ Channels

In 1952, Hodgkin and Huxley published several papers describing the critical cellular components responsible for action potentials in the squid (*loligo*) giant axon. Cation replacement

Conn's *Translational Neuroscience*
http://dx.doi.org/10.1016/B978-0-12-802381-5.00002-6

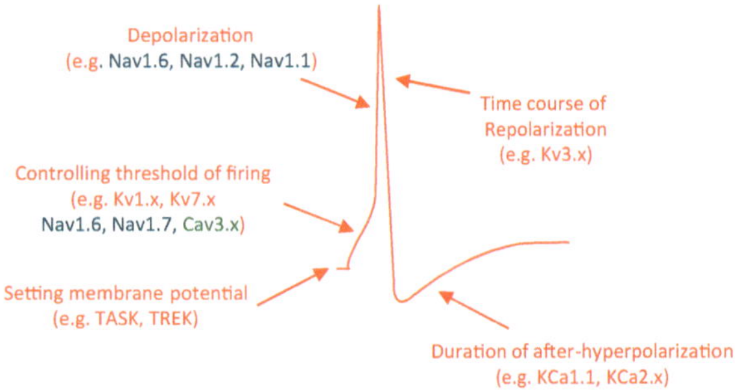

FIGURE 2.1 Functional role of ion channels in neuronal action potential generation.

experiments were used to demonstrate that the rapid upstroke of the action potential was caused by a brief, rapid increase in Na$^+$ ion conductance while K$^+$ channels were shown to be responsible for membrane repolarization (Fig. 2.1). While several studies were performed on Na$^+$ currents in tissue, it was not until 1980 that the Na$^+$ channel protein was identified using photo-affinity labeling by α-scorpion toxin. In 1984, researchers discovered the cDNA sequence and primary protein structure encoding the electric eel (*Electrophorus electricus*) Na$^+$ channel.

Na$^+$ Channel Structure and Function

The Na$^+$ channel α-subunit is a single polypeptide of about 1700–2000 amino acid residues with a molecular weight of about 260 kD consisting of four homologous domains (DI–DIV), each containing six α-helical transmembrane segments (S1–S6; Fig. 2.2). Both the C-terminus and N-terminus are located intracellularly. Although expression of the α-subunit alone is sufficient to form the ion-conducting channel, modulation of channel expression levels and gating is thought to involve one or more of the auxiliary β-subunits (β1–4). The biophysical properties of Na$^+$ channels, that is, the manner in which ion channels open and inactivate, vary among the different isoforms. Together with the differential pattern of expression of these isoforms within excitable cells, Na$^+$ channels play a major role in modulating neuronal activity in a very precise manner.

Selectivity and permeation of Na$^+$ ions is achieved by the segments connecting S5 to S6 from each domain, known as the P-loops. These reentrant loops contain four conserved amino acids—aspartate (D), glutamate (E), lysine (K), and alanine (A)—termed the DEKA motif, and they form the narrowest part of the channel. Studies on the crystal structure of the voltage-gated Na$^+$ channel from *Arcobactor butzleri* (Na_VAb) suggest that the pore of the Na$^+$ channel is composed of an extracellular funnel, a narrower selectivity filter, a central cavity, and an intracellular activation gate. The Na$^+$ ions are conducted through the channel in a partially hydrated form. The pore of the Na$^+$ channel is lined by the S6 segment and it is this segment that contains most of the binding sites for many of the Na$^+$ channel-blocking drugs.

Na$^+$ channels are voltage-gated, which means that they open as the membrane potential is depolarized. The S4 segments of each domain contain repeats of positively charged residues of either arginine (R) or lysine (K) separated by two hydrophobic residues. This orientation allows for the formation of an α-helix. It is these positively charged residues that form the voltage sensor, for not only the Na$^+$ channel, but also for Ca^{2+} channels and the K$^+$ channels as we shall see later in this chapter. The arrangement of the positive residues is such that they align along one side of the S4 segment, allowing each of the S4 segments to respond to changes in membrane voltage and move, either linearly or in a spiral motion, outward, thereby opening the channel. Movement of the S4 positive charges out of the membrane during activation of the channel can be detected and is referred to as gating current. Recent crystal structure studies identify the voltage-sensing domain (VSD) as including segments S1 through S4.

The outward movement of the S4 segments is coupled with movement of the fast inactivation gate, the cytoplasmic linker between DIII

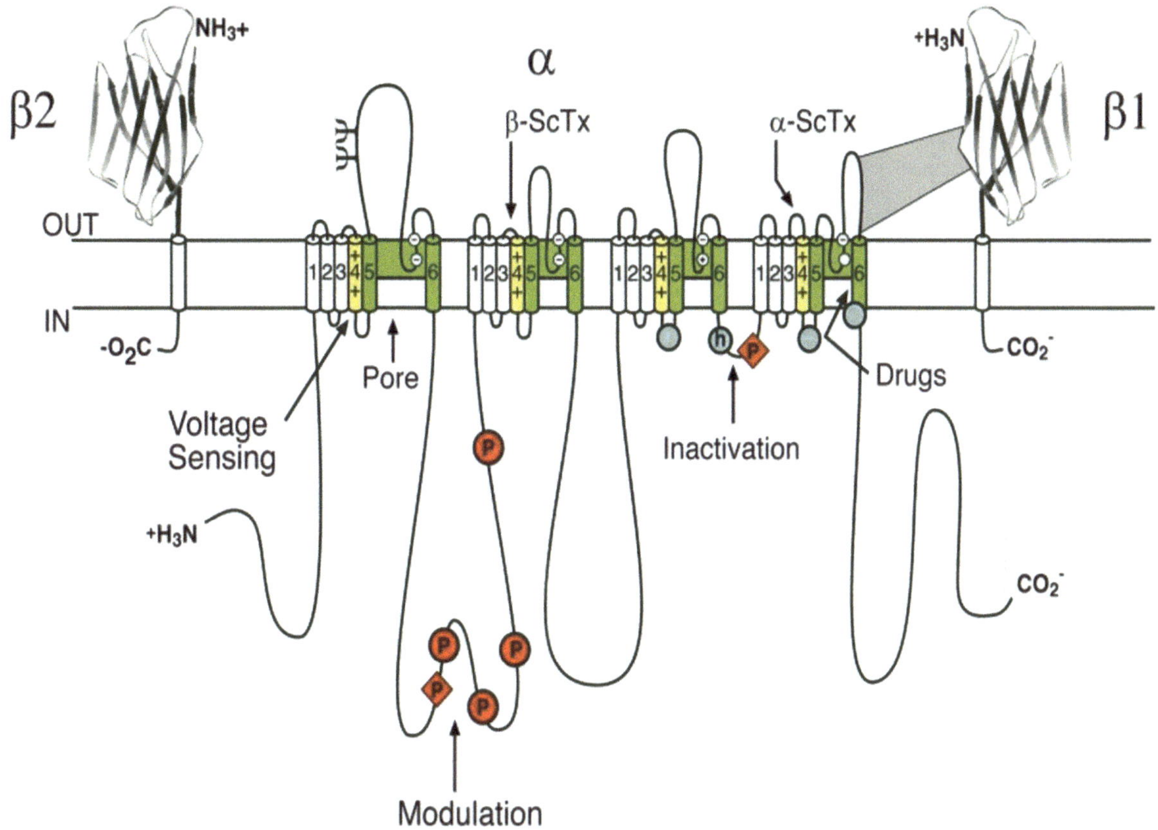

FIGURE 2.2 The primary structures of subunits of the voltage-gated Na^+ channel. Cylinders represent probable α-helical segments. Bold lines represent the polypeptide chains of each subunit with length approximately proportional to the number of amino acid residues in the brain sodium channel subtypes. The extracellular domains of the β1- and β2-subunits are shown as immunoglobulin-like folds. ψ, sites of probable N-linked glycosylation; *P* in *red circles*, sites of demonstrated protein phosphorylation by PKA (*circles*) and PKC (*diamonds*); green, pore-lining segments; white circles, the outer (EEDD) and inner (DEKA) rings of amino residues that form the ion selectivity filter and the tetrodotoxin binding site; *yellow*, S4 voltage sensors; *h* in *blue circle*, inactivation particle in the inactivation gate loop; blue circles, sites implicated in forming the inactivation gate receptor. Sites of binding of α- and β-scorpion toxins and a site of interaction between α- and β1-subunits are also shown. *Adapted from Catterall WA. From ionic currents to molecular mechanisms: the structure and function of voltage-gated sodium channels. Neuron 2000;26:13–25.*

S6 to DIV S1, known as the DIII–IV linker (Fig. 2.3). Fast inactivation of the Na^+ channel occurs within a 1–2ms time frame. A cluster of three hydrophobic amino acids, isoleucine (I), phenylalanine (F) and methionine (M), known as the IMF motif, within the DIII–IV linker is required for fast inactivation. Additional sites for fast inactivation include residues within DIV S6 and within the S4–S5 loop of both domains III and IV, which likely act as "docking" sites for the fast inactivation gate. The C-terminus of the Na^+ channel may also interact with the III–IV linker to stabilize the inactivation gate and prevent reopening of the channel. Fast inactivation of Na^+ channels is thought to contribute to action potential termination and regulation of the refractory period.

Under prolonged depolarization, Na^+ channels undergo a process known as slow inactivation. In contrast to fast inactivation, slow

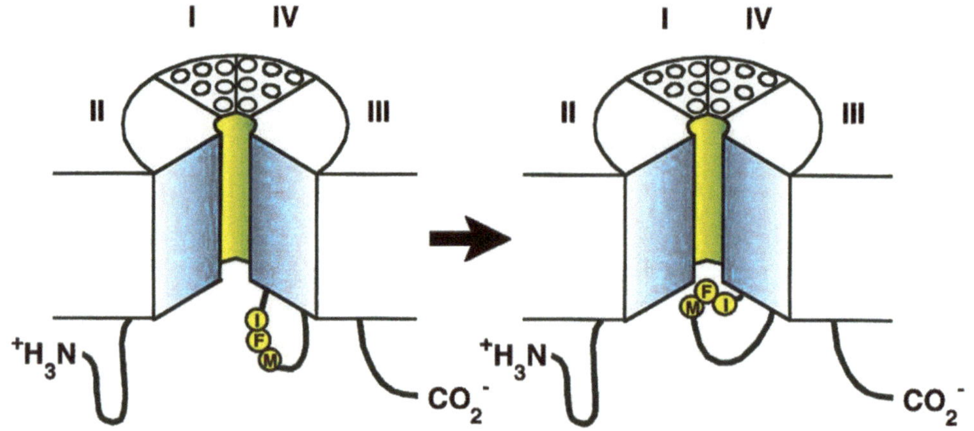

FIGURE 2.3 Mechanism of inactivation of Na^+ channels. The hinged-lid mechanism of Na^+ channel inactivation is illustrated. The intracellular loop connecting domains III and IV of the Na^+ channel is depicted as forming a hinged lid. The critical residue Phe-1489 (F) is shown occluding the intracellular mouth of the pore in the Na^+ channel during the inactivation process. *Adapted from Catterall WA. From ionic currents to molecular mechanisms: the structure and function of voltage-gated sodium channels. Neuron 2000;26:13–25.*

inactivation occurs on a much slower time scale of seconds and is a more complex process that does not involve the fast inactivation gate, but most likely involves the selectivity filter and the S6 segments. Slow inactivation is thought to contribute to overall membrane excitability by increasing action potential thresholds, thereby limiting action potential burst durations and importantly, limiting the propagation of action potentials within dendrites. Once the Na^+ channel has entered the inactivated state, it has to recover back to the closed state for it to open on the next depolarization. Recovery occurs during repolarization of the membrane and involves the movement of the S4 segments back into the membrane and relocation of the inactivation gate away from the pore.

Na^+ Channel Isoforms

To date, 10 α-isoforms (Na_v1.1–1.9 and Na_x) and four β-subunits (β1–4) have been cloned (Table 2.1). Of the 10 Na^+ channel isoforms, nine (Na_v1.1–1.9) have been cloned and functionally expressed. Na_x channels are structurally very different from the other members of the Na^+ channel family and are thought to play a role in sensing the sodium level in the brain. In the absence of subtype-selective Na^+ channel antagonists, the different isoforms have been classed according to their sensitivity to the puffer-fish toxin tetrodotoxin (TTX) such that Na_v1.1–1.4, 1.6, and 1.7 are classed as TTX-sensitive channels with IC_{50}s in the nM range while Na_v1.5, 1.8, and 1.9 are resistant to TTX with IC_{50}s in the mM range (Table 2.1).

The CNS expresses Na_v1.1, Na_v1.2, Na_v1.3, and Na_v1.6, and genes for all four isoforms are located on chromosome 2. Na_v1.1 is localized to the dendrites and cell bodies of excitatory neurons. Na_v1.1 is also expressed at the axon initial segments (AIS) of fast spiking parvalbumin-positive inhibitory neurons, and therefore, likely plays an important role in modulating network excitability via activation of inhibitory circuits. Mutations in Na_v1.1 have been associated with human epilepsy including Dravet syndrome and genetic epilepsy with febrile seizures plus (GEFS+). Na_v1.2 is expressed in cell bodies, dendrites, and axons of unmyelinated neurons. During development, Na_v1.2 is replaced by Na_v1.6 at nodes of Ranvier. Mutations in

TABLE 2.1 Mammalian Na^+ Channel Subunits

Type	Former Name	Genbank Number[a]	Gene Symbol	Chromosomal Location	Splice Variants	Primary Tissues
$Na_v1.1$	rat I	X03638 (r)	SCN1A	Mouse 2	$Na_v1.1a$	CNS
	HBSCI	X65362 (h)		Human 2q24		PNS
	GPBI	AF003372 (gp)				
	SCN1A					
$Na_v1.2$	rat II	X03639 (r)	SCN2A	Mouse 2	$Na_v1.2a$	CNS
	HBSCII	X61149 (r)		Human 2q23–24		
	HBA	X65361 (h)				
		M94055 (h)				
$Na_v1.3$	rat III	Y00766 (r)	SCN3A	Mouse 2	$Na_v1.3a$	CNS
				Human 2q24	$Na_v1.3b$	
$Na_v1.4$	SkM1, μ1	M26643 (r)	SCN4A	Mouse 11		Skeletal muscle
		M81758 (h)		Human 17q23–25		
$Na_v1.5$	SkM2	M27902 (r)	SCN5A	Mouse 9		Uninnervated skeletal muscle, heart CNS, PNS
	H1	M77235 (h)		Human 3p21		
$Na_v1.6$	NaCh6	L39018 (r)	SCN8A	Mouse 15	$Na_v1.6a$	
	PN4	AF049239 (r)		Human 12q13		
	Scn8a	AF049240 (r)				
	CerIII	U26707 (m)				
		AF049617 (m)				
		AF050736 (h)				
		AF225988 (h)				
		AF003373 (gp)				
$Na_v1.7$	NaS	U35238 (rb)	SCN9A	Mouse 2		PNS
	hNE-Na	X82835 (h)		Human 2q24		Schwann cells
	PN1	U79568 (r)				
		AF000368 (r)				
$Na_v1.8$	SNS	X92184 (r)	SCN10A	Mouse 9		DRG
	PN3	U53833 (r)		Human 3p22-24		
	NaNG	Y09108 (m)				
		U60590 (d)				

Continued

TABLE 2.1 Mammalian Na^+ Channel Subunits (Goldin et al., 2000)—cont'd

Type	Former Name	Genbank Number[a]	Gene Symbol	Chromosomal Location	Splice Variants	Primary Tissues
$Na_v1.9$	NaN	AF059030 (r)	SCN11A	Mouse 9	$Na_v1.9a$	PNS
	SNS2	AJ237852 (r)		Human 3p21–24		
	PN5	AF118044 (m)				
	NaT	AB031389 (m)				
	SCN12A	AF126739 (h)				
		AF188679 (h)				
		AF109737 (h)				
		AF150882 (h)				
Na_x	$Na_v2.1$	M91556 (h)	SCN7A	Mouse 2		Heart, uterus, skeletal muscle Astrocytes DRG
	Na-G	M96578 (r)	SCN6A[b]	Human 2q21–23		
	SCL11	Y09164 (r)				
	$Na_v2.3$	L36179 (m)				

[a] *The letter in parentheses after each accession number indicates the species of origin for the sequence, as follows:* h, *human;* r, *rat;* rb, *rabbit;* m, *mouse;* gp, *guinea pig;* d, *dog.*
[b] *This gene was originally assigned symbols* SCN6A *and* SCN7A, *which were mapped in human and mouse, respectively. The two most likely represent the same gene, and the* SCN6A *symbol will probably be deleted.*
Adapted from Goldin AL, Barchi RL, Caldwell JH, Hofmann F, Howe JR, Hunter JC. Nomenclature of voltage-gated sodium channels. Neuron 2000;28:365–8.

$Na_v1.2$ have also been associated with GEFS+ and benign familial neonatal-infantile seizures (BFNIS), a seizure in infants that persists during the first 12 months of life, but not after that initial developmental period. $Na_v1.3$ is considered to be an embryonic isoform since its expression is high during development. When expression is detected, it is thought to localize to somatic regions suggesting a role in integrating dendritic depolarization and neuronal excitability. $Na_v1.6$ is abundantly expressed in neurons of the CNS. It is concentrated at the AIS of both excitatory and inhibitory neurons and at nodes of Ranvier in myelinated neurons, where it mediates the initiation and propagation of action potentials. Mutations in $Na_v1.6$ are associated with epileptic encephalopathy type 13 (EIEE13). Patients with EIEE13 not only experience early onset of seizures (between 0 and 18 months of age) but also experience intellectual disability and developmental delay. $Na_v1.6$ is known to interact with β4-subunits to produce resurgent Na^+ channel currents in neurons, which play an important role in driving neurons to threshold potentials, thereby modulating burst firing.

Both $Na_v1.4$ and $Na_v1.5$ are considered the muscle Na^+ channel isoforms. $Na_v1.4$ is located on chromosome 17 and is highly expressed in adult skeletal muscle, where it plays a critical role in controlling muscle excitability, and thus, muscle contraction. $Na_v1.4$ channels are located along the sarcolemma and T-tubules. Mutations in this gene give rise to myotonia and periodic paralysis disorders. $Na_v1.5$ is located on chromosome 3 and is referred to as the cardiac isoform since it is the predominately expressed isoform in cardiac tissue. Mutations in $Na_v1.5$ account for two types of cardiac arrhythmias; Brugada syndrome (loss of channel function) and long QT 3 (LQT3; gain of channel function).

The PNS expresses a number of Na^+ channel isoforms including $Na_V1.1$, $Na_V1.6$, $Na_V1.7$, $Na_V1.8$, and $Na_V1.9$, however, only $Na_V1.7$–1.9 are specific to the PNS. $Na_V1.7$ is located on chromosome 2 while $Na_V1.8$ and $Na_V1.9$ are located on chromosome 3. $Na_V1.7$ is normally highly expressed in small diameter dorsal root ganglion (DRG) neurons and sympathetic ganglion neurons. $Na_V1.7$ channel currents are fast activating and inactivating, but slowly recover from inactivation. The channel is thought to play a role in establishing excitability of neurons making it a likely target for transmission of the pain sensation. The $Na_V1.8$ isoform is exclusively expressed in sensory neurons, most abundantly in the small diameter, nociceptive DRG neurons. Expression is detected along peripheral nerve axons and nerve terminals in the skin. $Na_V1.8$ channels generate a TTX-resistant Na^+ current that is slow-activating and slow-inactivating, but with fast recovery from inactivation kinetics. Biophysical properties and expression profiles make $Na_V1.8$ a possible target in neuropathic and inflammatory pain. The $Na_V1.9$ isoform is highly expressed in small diameter, nonpeptidergic neurons of the DRG under normal conditions. Expression has also been detected in the free nerve endings of skin. The isoform also mediates a TTX-resistant Na^+ current that is persistent in nature and has activation and inactivation curves that would allow it to be active at resting membrane potentials, thereby, generating a subthreshold depolarizing current. $Na_V1.3$ mediates a TTX-sensitive, fast-activating, and fast-inactivating Na^+ current in DRG neurons and has been shown to be greatly upregulated following nerve injury. $Na_V1.6$ is thought to be expressed at the nodes of Ranvier in PNS myelinated neurons and also unmyelinated neurons. $Na_V1.6$ channels produce resurgent Na^+ currents in DRG neurons and could also play a role in controlling burst firing in these neurons. The isoforms' role in pain is unclear since many studies show a downregulation in $Na_V1.6$ expression levels in animal models of pain.

Na^+ Channel β-subunits

Na^+ channels interact with two functional classes of auxiliary β-subunits, of which β1 and β2 are the original members. A third auxiliary subunit, β3, is structurally most similar to β1 sharing 57% sequence identity, and a fourth subunit, β4, is homologous to β2. The β-subunits—β2, β3, and β4—are localized to chromosome 11, with β1 being localized to chromosome 19. The β-subunits are about 30–40 kDa in size and are comprised of a large immunoglobulin-like extracellular domain at the N-terminus, a single transmembrane α-helix, and a short intracellular segment at the C-terminus (Fig. 2.2). β1 and β3 are thought to interact with the α-subunit in a noncovalent manner and modulate activation, inactivation, and recovery from inactivation gating parameters of Na^+ channels. Structure and function studies have identified the P-loop of domain IV of the Na^+ channel α-subunit as a likely interaction point for the extracellular domain of β1 and most likely β3, accounting for the fine tuning of Na^+ channel activation and inactivation gating. Disruptions in this interaction, as a result of mutations in either the α-subunit or the β-subunit, are known to result in changes in cellular excitability resulting in epileptic seizures. An interaction between the intracellular domain of the β-subunit and the Na^+ channel has also been suggested and could modulate channel inactivation and/or trafficking of the channel to membrane. β2 and β4 share 35% homology and are covalently linked to the α-subunit via a disulfide bridge. β2-subunit co-expression with Na^+ channels leads to alterations in the gating parameters of the channel. β4 is responsible for the generation of resurgent Na^+ currents. These unique currents are generated as a result of open channel block of the Na^+ channel α-subunit at depolarized voltages and the subsequent unblocking that occurs on repolarization, resulting in an inward Na^+ current that can drive neurons to firing threshold. Studies have identified the β4-subunit intracellular C-terminal domain as being responsible for the open channel block. Although most studies have focused on resurgent currents generated by β4-subunit interaction with $Na_v1.6$, it is likely that other Na^+ channel isoforms are capable of interacting with β4 to generate these currents. In addition to modulating channel gating parameters, β-subunits may also function as cell-adhesion molecules, interacting with the extracellular matrix and cytoskeleton proteins.

Na^+ Channels and Disease

In view of the fact that sodium channels play a critical role in initiating and maintaining the excitability on numerous cell types, it is not surprising that alterations in Na^+ channel function, as a result of a genetic mutation (channelopathy), or as a result of trauma (acquired channelopathy), can lead to disease states. Here we describe some diseases that result from abnormal Na^+ channel function.

Epilepsy

Epilepsy is a neurological disorder characterized by episodes of excessive synchronized neuronal excitability that manifests itself into the generation of epileptic seizures. To date, mutations in $Na_v1.1$, $Na_v1.2$, and $Na_v1.6$ have been associated with inherited epilepsy with the most number of mutations being associated with the $Na_v1.1$ isoform. Mutations can lead to either a gain in the function of the channel, where channels do not completely inactivate and generate a persistent Na^+ current (I_{NaP}), or a loss of function of the channel, where typically the cell membrane expression levels of the channel are reduced. Due to the expression patterns of $Na_v1.1$ (AIS of inhibitory neurons) and $Na_v1.6$ (AIS of excitatory and inhibitory neurons), loss- of-function mutations of $Na_v1.1$ and gain-of-function mutations in $Na_v1.6$ give rise to some of the most severe forms of epilepsy that are also associated with sudden unexpected death in epilepsy (SUDEP) in a number of patients. Since $Na_v1.2$ is transiently expressed within the AIS during early postnatal development, seizures associated with gain-of-function mutations manifest themselves during infancy and do not continue into adulthood.

Na^+ channels have also been implicated, in part, with increases in neuronal activity associated with acquired epilepsy. Acquired epilepsy can occur as result of traumatic brain injury, stroke, brain tumors, or infections, as well as other neurological disorders. Numerous studies have examined the role of Na^+ channels in establishing and facilitating the seizures associated with acquired epilepsies and have demonstrated not only increases in Na^+ channel expression but also alterations in the biophysical properties of the Na^+ channels predicted to lead to an increase in neuronal activity. In particular, increases in currents associated with $Na_v1.6$ have been demonstrated in animal models of acquired epilepsy.

Pain

DRG neurons express at least five of the nine Na^+ channel isoforms–$Na_v1.1$, $Na_v1.6$, $Na_v1.7$, $Na_v1.8$, and $Na_v1.9$. Numerous studies have focused on understanding the expression pattern of Na^+ channel isoforms after axonal injury or inflammation. Although $Na_v1.3$ is not expressed at high levels in adulthood, it has been shown to be upregulated in animal models of peripheral nerve injury and spinal cord injury. This upregulation suggests a role in increasing the neuronal hyperexcitability of primary afferents, central neurons, and pain sensation. Likewise, both $Na_v1.8$ and $Na_v1.9$, which are highly expressed in DRG neurons, have also been implicated in the development of pain. $Na_v1.8$ expression is upregulated in human patients with chronic neuropathic pain and in animal models of inflammatory pain. $Na_v1.8$ knock-down studies lead to an attenuation of pain sensation in neuropathic pain models. $Na_v1.9$ upregulation has been associated with inflammatory and diabetic neuropathy pain, but its role in the development of neuropathic pain is still unclear. Support for the role of $Na_V1.7$ in pain sensation is compelling. Patients with a disorder known as congenital indifference to pain (CIP) cannot recognize painful stimuli, but have normal autonomic and central function. Genetic analysis of these patients revealed a loss of function mutation in SCN9A, the gene for $Na_V1.7$. Furthermore, gain-of-function mutations of $Na_V1.7$ are associated with a heightened sense of pain. Inherited erythromelalgia (IEM) and paroxysmal extreme pain disorder (PEPD) are two severe pain disorders. IEM patients suffer from a very severe burning sensation, mainly in the lower extremities. PEPD patients suffer from extreme pain from the rectal, mandibular, and ocular areas. In view of the importance of $Na_V1.7$ in pain transmission, it is not surprising that current drug development strategies for pain management are heavily focused on the development of $Na_V1.7$-selective blockers.

Migraine

Mutations in $Na_V1.1$ have been associated with familial hemiplegic migraine (FHM) type 3. Both gain-of-function and loss-of-function mutations of $Na_V1.1$ can give rise to migraines, but underlying mechanisms remain unknown. Some patients experience a visual aura before the onset of the migraine and associated hemiparesis. In other cases, patients with FHM also have seizures. Interestingly, seizures in these patients can occur without prior onset of a migraine.

Pharmacological Block of Na^+ Channels

Almost all of the clinically used Na^+ channel antagonists exhibit state-dependent blocking characteristics. That is, at hyperpolarized potentials, block of the Na^+ current is less than that observed at more depolarized potentials. This voltage dependence of drug binding is well described by the modulated receptor hypothesis in which drugs have a much greater affinity for the inactivated states of the channel than for the resting states of the channel. This difference in affinity is not only thought to be an important mechanism of action for anticonvulsants and analgesics but also for reduced toxic side effects. Compounds with increased affinity for the inactivated state of the Na^+ channel would selectively suppress action potential bursts occurring during epileptic seizures or pain sensation without major effects on normal action potential generation. Affinity for the inactivated state would lead to a shift in inactivation kinetics, thereby reducing the voltage range of channel availability and activation, also referred to as the window current. This shift would reduce the proportion of Na^+ channels available to open and generate a depolarizing current. Furthermore, once bound to the inactivated state of the channel, recovery from the inactivated state would also be impaired leading to an accumulation of inactivated Na^+ channels and use-dependent block during repetitive action potential firing by hyperexcitable neurons. The binding site for many of the Na^+ channel blockers is thought to lie within the inner pore of the channel. Mutagenesis studies suggest interaction with the phenylalanine (F) residue within domain IVS6 is important for the inhibition of the channel, however, it is likely that additional sites also play a role in stabilizing drug binding including sites within DIVS6, DIIIS6, and DIS6, and residues within the selectivity pore.

Many of the currently available anticonvulsant drugs (AEDs) inhibit Na^+ channels, in part, as a mechanism of action. Commonly used drugs included phenytoin, lamotrigine, carbamazepine, and oxcarbazepine. These drugs preferentially bind to the inactivated state of Na^+ channels and have shown efficacy in the treatment of generalized tonic–clonic seizures and partial seizures. By virtue of the fact that AEDs target neuronally expressed Na^+ channels, they have also been used in the treatment of neuropathic pain. Local anesthetics are a group of compounds that also inhibit Na^+ channels and are used clinically to provide regional anesthesia and in the management of acute and chronic pain conditions. The compounds can be divided into those that are relatively short acting such as lidocaine and mepivacaine, or long lasting such as bupivacaine and ropivacaine.

Sodium Channel Toxins

Much information has been learned from the use of toxins that bind Na^+ channels. To date, eight distinct neurotoxin binding sites have been discovered. The various toxins can be classified as those that bind to and occlude the pore and those that alter the voltage dependence of gating. Toxins that bind and occlude the pore include tetrodotoxin (TTX), saxitoxin (STX), and μ-conotoxins. These toxins interact with site 1, delimited by the P-loops of DI–IV at the extracellular opening of the transmembrane pore. TTX deserves special mention since its use has been instrumental in not only classifying the different Na^+ channel isoforms but also in the initial discovery of the channel and the identification of residues within the selectivity filter.

Several toxins alter channel gating by binding to the Na^+ channel extracellularly or via interactions within the channel pore. Toxins that alter gating through an extracellular binding site include α-scorpion toxins and sea-anemone toxins, which interact with the extracellular loop of DIV S3–S4, and scorpion β toxins and spider β toxins, which interact with the extracellular loop of DII S3–S4. In general, these toxins either block inactivation by preventing movements in DIV S4, or enhance activation by stabilizing DIS4. Toxins that act by binding to sites within a transmembrane region of the Na^+ channel include veratridine, batrachotoxin (BTX), grayanotoxin, brevetoxin, and ciguatoxin. Brevetoxin and ciguatoxin act at DIS6 (site 5) and enhance Na^+ channel activity by causing a hyperpolarizing shift in activation and blocking inactivation. Veratridine, BTX, and grayanotoxin interact with DIS6, DIIS6, and DIVS6 (site 2) and cause persistent activation, block of inactivation, and a hyperpolarizing shift of the voltage dependence of activation.

Ca^{2+} Channels

Seminal work by Paul Fatt and Bernard Ginsborg first described the role of Ca^{2+} ions as an important second messenger in large crustacean muscle in 1958. It took further work in mammalian skeletal and cardiac muscle to identify action potentials that were carried by a Ca^{2+} influx independently of Na^+. These studies led to the discovery and the advance of our understanding of the importance of Ca^{2+}-permeable channels. The field was further advanced by the development of Ca^{2+} channel antagonists by Albrecht Fleckenstein in the 1960s, allowing the selective block of Ca^{2+} channels and hence, the ability to determine their roles in cell physiology. These agents, including dihydropyrindines (DHPs), were put to effective use in binding to proteins in skeletal muscle to allow biochemical cloning of the first voltage-gated Ca^{2+} channel subunits. Molecular cloning identified a 170-kDa pore-forming DHP-binding α1-subunit, now known to belong to the Ca^{2+} channel Ca_V1 family. Additional α1 subunits, belonging to the Ca_V2 family, were cloned from brain in the early 1990s. The Ca_V3 channels were the last to be cloned using in silico techniques in the late 1990s. It is now recognized that an excitable cell is defined by the presence of Ca^{2+} channels. Intracellular Ca^{2+} concentration controls numerous processes, including gene expression and transcription, cell signaling and cell death, neurotransmitter and secretion release, muscle contraction, and learning and memory. In cardiac and smooth muscle, contraction is initiated by direct influx of Ca^{2+} through Ca^{2+} channels. In neurons, Ca^{2+} channels initiate synaptic transmission. Such a variety of physiological effects requires a level of Ca^{2+} channel diversity to ensure adequate control of Ca^{2+} influx; there are a number of different subtypes, as discussed in the following sections.

Ca^{2+} Channel Families, Structure and Function

There are two main families of Ca^{2+} channels, high voltage–activated (HVA) and low voltage–activated (LVA) calcium channels (Fig. 2.4). Despite this variety, there is a common molecular structure to Ca^{2+} channels. In a similar manner to Na^+ channels, the Ca_V α-subunit represents the main pore-forming Ca^{2+} channel subunit of about 2000 amino acids and a molecular weight of around 190–250 kDa. The subunit is arranged in a 24 transmembrane protein and divided into four homologous domains (I–IV), each with six transmembrane segments (S1–S6) and intracellular N- and C-termini. The S4 segments contain positively charged lysine and arginine residues, which act as "voltage sensors" much like Na^+ channels, detecting potential changes across the membrane. Under the influence of an electric field, each homologous segment undergoes a rotating movement thereby initiating the overall conformational change allowing the pore channel to open. The four segment S5–S6 loops form a hairpin structure, which lines the inside of the pore and determines ion conductance and

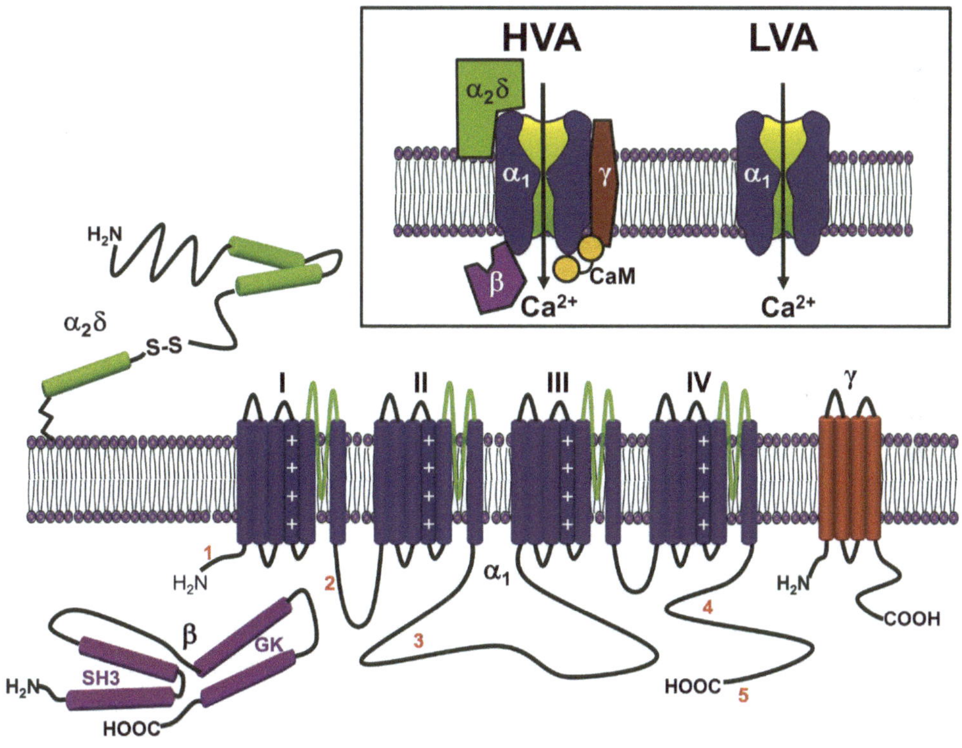

FIGURE 2.4 Topology of HVA and LVA voltage-gated Ca^{2+} channels. HVA channels are heteromultimers consisting of a pore-forming α1-subunit, extracellular α2δ- and intracellular Ca_Vβ-subunits, and sometimes a transmembrane Ca_Vγ-subunit. Modulation of the Ca^{2+} channel complex is largely dictated by α1-subunit molecular determinants: (1) the amino terminal Ca^{2+}-binding protein calmodulin association site in Ca_V1 channels; (2) the Ca_Vβ binding (α-interaction domain, AID); Ca_V2 family proteins also both G protein Gβγ-subunits at sites on the I–II linker region; (3) the so-called synaptic protein interaction site (synprint) in Ca_V2 channels; (4) calmodulin association site in Ca_V1 and Ca_V2 channels; and (5) scaffolding protein interaction sites in Ca_V2 channels. *Adapted from Simms BA, Zamponi GW. Neuronal voltage-gated calcium channels: structure, function, and dysfunction. Neuron 2014;82:24–45.*

selectivity and often contributes to major drug binding sites. Ca^{2+} ion selectivity is determined by a pair of glutamate residue on each segment, a feature unique to Ca^{2+} channels. HVA $Ca_V\alpha1$-subunits associate with auxiliary subunits, $Ca_V\beta$, $Ca_V\alpha2\delta$ (more commonly $\alpha2\delta$) and, potentially, $Cav\gamma$ subunits, which modulate $\alpha1$ properties. In contrast, LVA Ca^{2+} channels are thought to form in the absence of these auxiliary subunit. Both families of Ca^{2+} channels transduce electrical signals, HVA from action potentials and LVA from subthreshold depolarizing signals, respectively, into ionic exchange, thereby tightly regulating Ca^{2+} influx and their downstream effects.

Ca^{2+} Channel Isoforms

$Ca_V\alpha1$ Subunit

Ca^{2+} channels are defined fundamentally by their $\alpha1$-subunits, which are encoded by 10 separate genes, which divide into three functionally different families: Ca_V1 and Ca_V2 (HVA) and Ca_V3 (LVA) Ca^{2+} channels (Table 2.2). $Ca_V\alpha1$ isoform diversity is summarized in Table 2.2; together with defining features and selective pharmacological antagonists, channels can be broadly distinguished by electrophysiological properties. The Ca_V1 channel family encodes L-type Ca^{2+} current and consists of $Ca_V1.1$, expressed exclusively in skeletal muscles, and, respectively, the neuronal, cardiac, and endocrine, $Ca_V1.2$, $Ca_V1.3$, and $Ca_V1.4$ subtypes. L-type currents were originally designated "long" as they inactivate relatively slowly. It is now known that L-type current primarily shows Ca^{2+}-dependent inactivation (CDI), whereby the Ca^{2+}-binding protein, calmodulin, binds to an IQ motif in the carboxyl terminal, but it can also be affected by voltage-dependent inactivation (VDI), involving a conserved "inverted tepee" inactivation gate contributed by S6 helices extending into the I–II linker. There are three subtypes of Ca_V2 channels ($Ca_V2.1$, $Ca_V2.2$, and $Ca_V2.3$). Through alternative splicing and assembly with different combination of auxiliary subunits, the $Ca_V2.1$ gene encodes both P-type and Q-type Ca^{2+} currents. The P-type current is so-called because it was first identified in Purkinje neurons. The $Ca_V2.2$ encodes "neuronal" N-type Ca^{2+} current. Both $Ca_V2.1$ and $Ca_V2.2$ are expressed predominantly at the neuronal dendrites and presynaptic terminals while $Ca_V2.3$ (which encodes "residual" R-type Ca^{2+} current) is found at neuronal cell bodies and dendrites. Ca_V2 Ca^{2+} currents show VDI; R-type current is generally considered to exhibit faster inactivation than P/Q or N-type current. Ca_V3 Ca^{2+} channels all encode "transient" T-type Ca^{2+} current. As the name suggests, currents formed by the three isoforms, $Ca_V3.1$, $Ca_V3.2$, and $Ca_V3.3$, all show rapid VDI; in addition to the conserved inactivation gate, pore residues also affect Ca_V3 gating. These LVA Ca^{2+} channels are involved in pacemaking and repetitive firing. Ca_V3 Ca^{2+} channels are widely expressed in neurons of the PNS and CNS. Their roles are varied, ranging from control of vascular tone to pacemaking, but in the CNS, Ca_V3 act to regulate neuronal excitability.

Auxiliary Ca^{2+} Channel Subunits

$Ca_V\beta$-Subunit

The $Ca_V\beta$-subunit is a cytoplasmic protein (50 kDa), which consists of an Src homology (SH3) domain connected to a variable linker by a guanylate kinase domain. To date, four separate $Ca_V\beta$-subunit genes have been identified (Ca_V $\beta1$–4). The $Cav\beta$-subunit binds with nanomolar affinity via its variable β-interaction domain (BID) to the α-interaction domain (AID) located in the intracellular I–II linker of HVA Ca^{2+} channels. $Cav\beta$-subunits have major effects on Ca^{2+} channels; firstly, by acting as chaperones during Ca^{2+} channel assembly and trafficking to the plasma membrane and secondly as a modulator of channel gating both via direct channel interaction and via interacting with other regulatory proteins, such as the presynaptic protein RIM1.

TABLE 2.2 Nomenclature of Voltage-Gated Ca^{2+} Channels

Channel	Ca^{2+} Current Type	Localization	Specific Antagonists	Cellular Function	Inherited Disease
$Ca_V1.1$	L	Skeletal muscle, transverse tubules	Dihydropyridines, phenylalkylamines benzodiazepines	Excitation–contraction coupling	Hypokalemic periodic paralysis, malignant hyperthermia susceptibility
$Ca_V1.2$		Cardiac myocytes, smooth muscle myocytes, endocrine cells, neuronal cell bodies, proximal dendrites		Excitation–contraction coupling, hormone release, regulation of transcription, synaptic integration	Timothy syndrome: Cardiac arrhythmias, developmental abnormalities, and autism spectrum disorders. Brugada syndrome: abnormal electrocardiogram, sudden cardiac death
$Ca_V1.3$		Endocrine cells, neuronal cell bodies, cardiac atrial myocytes and pacemaker cells, cochlear hair cells		Hormone release, transcription regulation, synaptic regulation, cardiac pacemaking, hearing, neurotransmitter release from sensory cells	Sino-atrial node dysfunction and deafness
$Ca_V1.4$		Retinal rod and bipolar cells, spinal cord, adrenal gland, mast cells		Neurotransmitter release from photoreceptors	Stationary night blindness
$Ca_V2.1$	P/Q	Nerve terminals and dendrites, neuroendocrine cells	ω -agatoxin IVA	Neurotransmitter release, dendritic Ca^{2+} transients, hormone release	Familial hemiplegic migraine, spinocerebellar and episodic ataxias
$Ca_V2.2$	N	Nerve terminals and dendrites, neuroendocrine cells	ω -conotoxin	Neurotransmitter release, dendritic Ca^{2+} transients, hormone release	
$Ca_V2.3$	R	Neuronal cell bodies and dendrites	SNX-482	Repetitive firing, dendritic Ca^{2+} transients	
$Ca_V3.1$	T	Neuronal cell bodies and dendrites, cardiac and smooth muscle myocytes	TTA-A2, TTA-P2	Pacemaking, repetitive firing	Idiopathic generalized epilepsy, autism spectrum disorders
$Ca_V3.2$		Neuronal cell bodies and dendrites, cardiac and smooth muscle myocytes, other tissue	TTA-A2, TTA-P2	Pacemaking, repetitive firing	Absence epilepsy, autism spectrum disorders
$Ca_V3.3$		Neuronal cell bodies and dendrites	TTA-A2, TTA-P2	Pacemaking, repetitive firing	

α2δ-Subunit

α2- and δ-subunits are the product of a single gene transcription and translation, an α2δ preprotein is posttranslationally cleaved and reconnected by a disulfide bond. The α2δ-subunit is not only present extracellularly and is responsible for trafficking Ca^{2+} channels to the plasma membrane but also has roles in synaptic function. To date, four separate α2δ-subunit genes have been cloned (α2δ1–4). The topology of the different α2δ proteins stems from the information on α2δ1. Each subtype has a predicted signal sequence on the N-terminus, indicating that the protein is directed to the lumen of the endoplasmic reticulum and is expressed extracellularly. The α2δ protein is heavily glycosylated. The α2 protein is completely extracellular and the δ protein contains an 18- to 22-amino-acid-long hydrophobic region followed by a short sequence of hydrophilic amino acids on the C-terminus. Until recently, this hydrophobic region was believed to be a single-pass transmembrane domain, however, it has also been suggested that a glycosylphosphatidylinositol (GPI) anchor may be attached to δ, targeting α2δ to cholesterol-rich microdomains.

$Ca_V\gamma$-Subunit

The ~30 kDa $Ca_V\gamma$-subunit was co-purified with skeletal muscle Ca_V1 family Ca^{2+} channel. To date, seven Cavγ-like subunits have been cloned, but their specific role with Ca^{2+} channels has been controversial as they were also shown to associate with other membrane proteins such as glutamatergic AMPA receptors and may represent multifunctional, more promiscuous, membrane proteins.

Ca^{2+} Channels and Disease

The physiological significance of Ca^{2+} channels discussed earlier is reflected in the number of diverse neurological diseases that have been associated with abnormal activity or channel expression levels. A number of inherited CNS diseases have also been linked with Ca^{2+} channel mutations; the so-called Ca^{2+} channelopathies (Table 2.2).

Targeting Ca^{2+} Channels in Pain

The most studied Ca^{2+} channels involved in pain transmission are the $Ca_V2.2$ and Ca_V3 channels. $Ca_V2.2$ channels are densely expressed at the peptidergic terminal of C fibers on the dorsal horn of the spinal cord and their expression is increased following nerve injury and tissue inflammation. Due to the expression profiles of other HVA Ca_V2 channels, it is likely that they too may play a role in nociception. Thus, $Ca_V2.1$ Ca^{2+} channels are expressed predominantly at the synaptic terminals of laminae II–IV of the spinal cord dorsal horn and $Ca_V2.3$ Ca^{2+} channels are expressed on neuronal cell bodies and dendrites of sensory nociceptive neurons. The $Ca_V2.2$ blocker ziconotide, a small peptide inhibitor derived from cone snail conotoxin (see Table 2.2), has been successfully used by intrathecal route to treat complex neuropathies. Recently, the small molecule TROX-1 was developed as an oral state-dependent and use-dependent Ca_V2 inhibitor with a reduced peripheral side effect profile, which effectively alleviated pain in a number of animal models such as neuropathic pain, secondary allodynia, osteoarthritis, and acute nociception. The role of Ca_V3 Ca^{2+} channels in neuropathic pain has been quite difficult to unravel due to the complexity of the peripheral sensory system. In the brain, these channels regulate neuronal excitability and are highly expressed in the thalamus, the area responsible for relaying pain stimulus from the spinal cord to the cortex. Selective blockers of Ca_V3 channels demonstrated antinociceptive effects in a variety of animal pain models. Ca_V3 Ca^{2+} channels have also been implicated in peripheral and central sensitization resulting in neuropathic pain.

As an auxiliary subunit of HVA Ca^{2+} channels, it has been widely proposed that α2δ is involved in neuropathic pain. Both α2δ1 mRNA and protein levels in DRG neurons are highly upregulated in a peripheral nerve ligation animal model of neuropathic pain and correlated with the onset of allodynia. Gabapentin and the gabapentin analog pregabalin have been used to treat epilepsy, neuropathic pain, and restless leg syndrome. Gabapentin was licensed in the early 1990s as an antiepileptic drug. Originally designed as a GABA analog, it was shown subsequently that gabapentin does not bind to GABA receptors, or influence GABA reuptake. Rather, gabapentin has been shown to bind to the α2δ-1 and α2δ-2 subunits of Ca^{2+} channels and to cause a decrease in $Ca_V2.1$ and $Ca_V2.2$ Ca^{2+} channel currents. Gabapentinoids are proposed to prevent Ca^{2+} channels from trafficking to the membrane, thereby, reducing Ca^{2+} currents.

Familial Hemiplegic Migraine

Migraine is a common brain disorder of unknown etiology, which causes episodic headaches. Prevention and treatment of migraine using drug therapy is unsatisfactory despite the high incidence of this neurological disorder, which affects >10% of population. FHM is a monogenic type of inherited migraine with aura caused by mutations of the CACNA1A gene, which encode for the $Ca_V2.1$ channel (P/Q-type current) (Fig. 2.5, Table 2.2). These point mutations are varied, but they all result in a shift in activation to lower voltages and increase in Ca^{2+} influx. In mouse models, these point mutations resulted in a reduced threshold for an increased velocity of cortical spreading depression (CSD). This phenomenon underlies migraine aura and may activate migraine headache mechanisms.

Epilepsy and Ataxia

Mutations in the CACNA1A gene are also associated with episodic ataxia 2 (Table 2.2) and spinocerebellar ataxia 6, which typically cause a loss-of-function or an abnormal expansion of a polyglutamine repeats, respectively. $Ca_V2.1$ channels are heavily expressed in Purkinje neurons, the major output cells in the cerebellum,

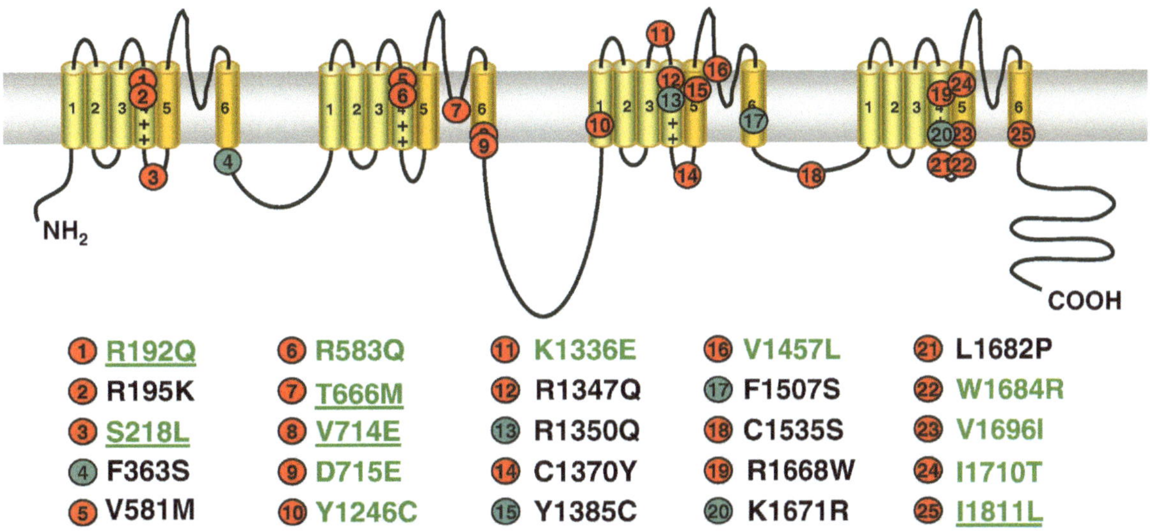

FIGURE 2.5 Schematic diagram showing identified CACNAIA gene mutations associated with disease. *Adapted from Pietrobon D. $Ca_v2.1$ channelopathies. Pflüg Arch 2010;71:841–7.*

this region of the brain being associated with maintenance of balance and motor coordination. Animal strains with mutant $Ca_V2.1$ channels, as well mice with genetic deletion of $Ca_V2.1$, suffer from movement disorders and ataxia. Moreover, mutations in the CACNA1G gene have been associated with idiopathic generalized epilepsy and mutations in the CACNA1H gene encoding $Ca_V3.2$ channels cause absence epilepsy in which the affected individuals transiently enter a sleep-like state that interrupts their normal activities.

K^+ Channels

The K^+ channel family comprises a large superfamily of transmembrane proteins, which conduct K^+ across the membrane of excitable tissues of the nervous and cardiac systems. In nonexcitable cells, they also play crucial roles in controlling the membrane potential, a key factor in cellular signaling. K^+ channels (usually, although not always) conduct K^+ ions out of the cell in response to a variety of stimuli such as voltage, intracellular calcium, pH, nucleotides, and Na^+ ions to name but a few. Here we describe the diversity in structure, biophysical properties, and tissue expression of the K^+ channel family. We highlight the pathological consequences of dysfunctional K^+ channels on human health and how drugs targeting K^+ channels can treat various diseases.

K^+ Channel Structure and Function

K^+ channels are formed by transmembrane helices spanning the cell's lipid bilayer and forming a pore through which K^+ ions flow. The most common K^+ channel structure is a tetramer (four subunits) with each subunit contributing a pore-forming domain (Fig. 2.6). The pore-forming domain of this family is extremely selective (at least 10,000 times more permeant for K^+ compared to Na^+) and can conduct K^+ ions rapidly (10^7/channel/s.). The selectivity filter of these channels bears a conserved motif (TVGYG), which is crucial for selectivity and is unique among ion channels.

K^+ Channel Isoforms

Tables 2.3–2.6 show the large diversity of K^+ channel subunits, which fall into four main families; voltage-gated, Ca^{2+}-activated, inward rectifying, and two-pore K^+ channels. Splice variation occurs in many K^+ channel subtypes. For example, multiple splice variants have been identified in the extended C-terminal domain of the large conductance Ca^{2+}-activated K^+ channel ($K_{Ca}1.1$) giving rise to different Ca^{2+} sensitivities and differential expression in the brain. Recent X-ray crystal structures of mammalian and bacterial K^+ channels have demonstrated that functional K^+ channels are formed from the interaction of four pore-forming domains, where each α-subunit consists of a single-pore domain giving rise to a tetrameric complex, or in the case of K_{2P} channels, each α-subunit consists of a two-pore domain where the functional channel is a dimeric complex. Native channels can be either homotetramers of a single class of α-subunit or heterotetramers of at least two distinct α-subunits, for example, the majority of the neuronal M currents are comprised of heteromultimers of the closely related $K_v7.2$ and $K_v7.3$ subunits. K_v5, 6, 8, and 9 subfamilies cannot form functional homomeric channels, but coassemble with K_v2 and K_v3 subunits to produce functional heteromeric channels. A number of membrane-spanning and peripheral proteins associate with the pore-forming α-subunits of K^+ channels. For example, with the voltage-gated $K^{\pm}$ family, cytoplasmic β-subunits are associated with K_v1 subtypes, MinK and the related MiRP subunits are associated with members of K_v3, K_v7, and K_v10 subtypes, while KChIP and KChAP subunits are associated with K_v4 subtypes (Table 2.3).

Voltage-Gated K^+ Channels

K_v currents have traditionally been classified electrophysiologically into two distinct groups, "delayed rectifiers" and "A" currents according to inactivation kinetics (the rate at which K^+ conductance is inhibited following channel

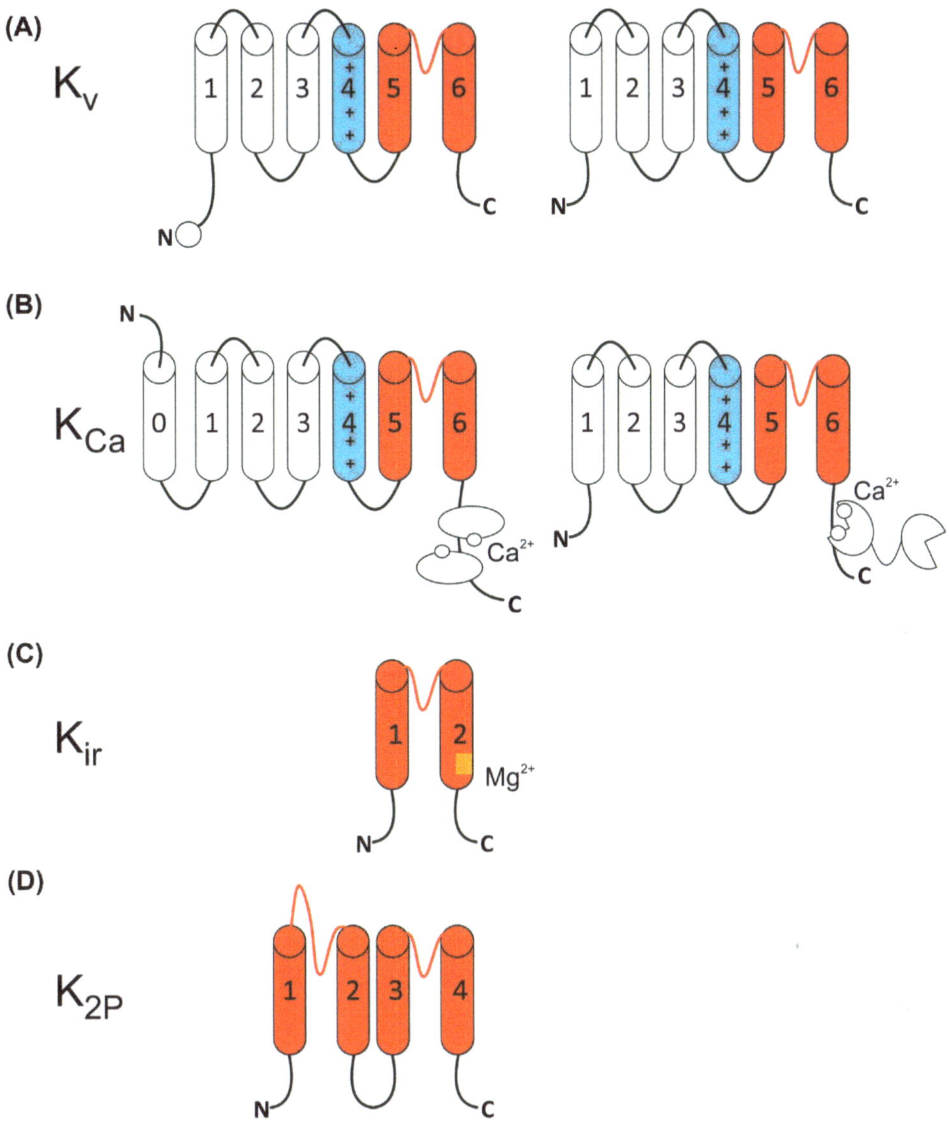

FIGURE 2.6 **K^+ channel subunit topology. K^+ channels can be subdivided into four principal families based on their subunit topology.** (A) The voltage-gated K^+ channel (K_v) family forms a large superfamily with 40 members. All K_v subunits contain six transmembrane domains with intracellular N- and C- termini, a voltage-sensing domain within TM4 (blue cylinders) and a pore-forming domain formed by transmembrane domains 5, 6, and the P-loop between these two domains (red cylinders and loop). The fully formed K_v ion channel is composed of four subunits, either homo- or heterotetramers with well-defined combinations (see text for details). K_v channels can conduct A-type currents in which the N-terminal blocking particle confers N-type inactivation (Fig. 2.1A, left panel), which delayed-rectifier channels lack (Fig. 2.1A, right panel). (B) Calcium-activated K^+ channels (K_{Ca}) are formed by tetramers of either seven transmembrane domains (1B, left panel) or six transmembrane domains (1B, right panel). The seven transmembrane domain $K_{Ca}1.1$ channels (formerly BK) are both calcium-activated (calcium binding to the C-terminus) and voltage-activated (by way of the voltage-sensing domain in TM4). The six transmembrane domain K_{Ca} channels are activated by calcium ($K_{Ca}2.x$, $K_{Ca}3.1$), and sodium/chloride ($K_{Na}1.1$ and $K_{Na}1.2$). KCa2.x and $K_{Ca}3.1$ are activated by calcium via a constitutively bound calmodulin molecule (bound to the c-terminus in Fig. 2.1B, right panel). (C) Inwardly rectifying K^+ channel subunits (K_{ir}) contain only two transmembrane domains comparable to the pore-forming domains of K_v and K_{Ca} subunits. Similarly, the mature K_{ir} channel is formed as a tetramer of α-subunits with heterotetramers being formed within subfamilies (eg, $K_{ir}3.1$ with $K_{ir}3.2$). They are strongly blocked by Mg^{2+} at depolarizing voltages (*orange block*), which underlies the inward rectifying mechanism (D) The K_{2P} subunit family is somewhat unique with the K^+ family in that its subunits are composed of two pore-forming domains (comparable to two covalently linked K_{ir} subunits). Unlike the other K^+ channel family members, the fully functional K_{2P} channel is composed of dimers of subunits.

TABLE 2.3 Voltage-Gated K^+ Channels

Family	Channel Name	Gene Name	Biophysical Properties	Accessory Subunit	Inhibitors	Activators	Pathology
Shaker-like	Kv1.1	KCNA1	Delayed rectifier	Kvβ1, Kvβ2	Tetraethylammonium		EA1
	Kv1.2	KCNA2	Delayed rectifier		Noxiustoxin (Kv1.2, Kv1.3)		EA1
	Kv1.3	KCNA3	Delayed rectifier		Margatoxin (Kv1.3)		
	Kv1.4	KCNA4	A current		Dendrotoxin-K (Kv1.1)		
	Kv1.5	KCNA5	Delayed rectifier		4-aminopyridine		
	Kv1.6	KCNA6	Delayed rectifier		Correolide (Kv1.3)		
	Kv1.7	KCNA7	Delayed rectifier				
	Kv1.8	KCNA8	Delayed rectifier				
Shab-like	Kv2.1	KCNB1	Delayed rectifier		Tetraethylammonium		
	Kv2.2	KCNB2	Delayed rectifier		4-aminopyridine		
Shal-like	Kv3.1	KCNC1	Delayed rectifier		Tetraethylammonium BDS-1 toxin 4-aminopyridine		SCA13
	Kv3.2	KCNC2	Delayed rectifier				
	Kv3.3	KCNC3	A current				
	Kv3.4	KCNC4	A current				
Shaw-like	Kv4.1	KCND1	A current	KChIP KChAP	4-aminopyridine		
	Kv4.2	KCND2	A current				
	Kv4.3	KCND3	A current				
	Kv5.1	KCNF1	Silent subunit				
	Kv6.1	KCNG1	Silent subunit				
	Kv6.2	KCNG2	Silent subunit				
	Kv6.3	KCNG3	Silent subunit				
	Kv6.4	KCNG4	Silent subunit				

KQT-like	Kv7.1	KCNQ1	M current (cardiac IKs)	MinK	Linopirdine	Flupirtine (Kv7.1 sparing)	LQT1/AF
	Kv7.2	KCNQ2	M current	MiRP2	XE991	Retigabine (Kv7.1 sparing)	BFNS
	Kv7.3	KCNQ3	M current		Tetraethylammonium		BFNS
	Kv7.4	KCNQ4	M current				DNFA
	Kv7.5	KCNQ5	M current				
	Kv8.1	KCNV1	Silent subunit				
	Kv8.2	KCNV2	Silent subunit				
	Kv9.1	KCNS1	Silent subunit				
	Kv9.2	KCNS2	Silent subunit				
	Kv9.3	KCNS3	Silent subunit				
Eag	Kv10.1	KCNH1	Delayed rectifier	MinK			
	Kv10.2	KCNH5	Delayed rectifier	MiRP1			
Erg	Kv11.1	KCNH2	Delayed rectifier (cardiac IKr)		E4031		
	Kv11.2	KCNH6	Delayed rectifier		Dofetilide		
	Kv11.3	KCNH7	Delayed rectifier		Ibutilide		
Elk	Kv12.1	KCNH8	Delayed rectifier				
	Kv12.2	KCNH3	Delayed rectifier				
	Kv12.3	KCNH4	Delayed rectifier				

TABLE 2.4 Ca^{2+}-Activated K^+ Channels

Family	Channel Name	Gene Name	Biophysical Properties	Accessory Subunit	Inhibitors	Activators
BK	KCa1.1	KCNMA1	Outwardly rectifying	KCNMB1-4	Charybdotoxin Iberiotoxin Tetraethylammonium	NS004 NS1619
SK	KCa2.1	KCNN1	Weak inward rectifier	Calmodulin	Apamin	EBIO
	KCa2.2	KCNN2	Weak inward rectifier			NS309
	KCa2.3	KCNN3	Weak inward rectifier			CyPPA (KCa2.1 sparing)
IK	KCa3.1	KCNN4	Weak inward rectifier	KCNMB1–4	ICA-17043	EBIO NS309
KNa	KCa4.1	KCNT1	Outwardly rectifying			
	KCa4.2	KCNT2	Outwardly rectifying			
	KCa5.1	KCNU1	Outwardly rectifying			

TABLE 2.5 Inward Rectifier K^+ Channels

Family	Channel Name	Gene Name	Biophysical Properties	Accessory Subunit	Inhibitors	Activators
IRK	Kir1.1	KCNJ1	Weak inward rectifier			
	Kir2.1	KCNJ2	Strong inward rectifier			
	Kir2.2	KCNJ12	Strong inward rectifier			
	Kir2.3	KCNJ4	Strong inward rectifier			
	Kir2.4	KCNJ14	Strong inward rectifier			
	Kir2.6	KCNJ18	Strong inward rectifier			
GIRK	Kir3.1	KCNJ3	Strong inward rectifier			
	Kir3.2	KCNJ6	Strong inward rectifier			
	Kir3.3	KCNJ9	Strong inward rectifier			
	Kir3.4	SKCNJ5	Strong inward rectifier			
	Kir4.1	KCNJ10	Strong inward rectifier			
	Kir4.2	KCNJ15	Strong inward rectifier			
	Kir5.1	KCNJ16	Silent channel			
KATP	Kir6.1	KCNJ8	Weak inward rectifier	SUR1 SUR2A SUR2B	Glibenclamide Tolbutamide	Minoxidil Nicorandil Cromakalim Diazoxide
	Kir6.2	KCNJ11	Weak inward rectifier			
	Kir7.1	KCNJ13	Weak inward rectifier			

TABLE 2.6 Two-Pore K^+ Channels

Family	Channel Name	Gene Name	Biophysical Properties	Inhibitors	Activators
TWIK	K2P1.1	KCNK1	Low expression		
	K2P6.1	KCNK6	Weak inward rectifier		
	K2P7.1	KCNK7	Nonfunctional		
TREK	K2P2.1	KCNK2	Outwardly rectifying	Fluoxetine	Riluzole
	K2P10.1	KCNK10	Outwardly rectifying		Halothane
	K2P4.1	KCNK4	Outwardly rectifying		
TASK	K2P3.1	KCNK3	Open rectifier		Halothane
	K2P9.1	KCNK9	Open rectifier		
	K2P15.1	KCNK15	Open rectifier		
TALK	K2P5.1	KCNK5	Nonfunctional		
	K2P16.1	KCNK16	Open rectifier		
	K2P17.1	KCNK17	Open rectifier		
THIK	K2P12.1	KCNK12	Nonfunctional		
	K2P13.1	KCNK13	Open rectifier	Halothane	
TRESK	K2P18.1	KCNK18	Strong outward rectifier		Halothane

opening; Fig. 2.6A; Table 2.3). Delayed rectifiers have absent or slow inactivation (over seconds) while A currents demonstrate rapid inactivation (over milliseconds). The classical A current has negative thresholds of activation, high sensitivity to 4-aminopyridine (4-AP), and insensitivity to TEA with highest similarity to K_v4 subunits. K_v4 subunits are unique among A currents in their Ca^{2+} sensitivity through assembling with the cytosolic accessory KChiP proteins. All K_v channels possess the same basic six transmembrane domain structure, where differences in rate of inactivation can be ascribed to differences in the N-terminal region of the channel protein. The N-terminus of channels underlying A currents have an N-terminal domain that forms a "ball-and-chain" and binds to the inner vestibule of the pore following channel activation preventing further K^+ flow and therefore undergoes a rapid N-type inactivation (see Fig. 2.2). K_v accessory subunits, $K_v\beta1$ and $K_v\beta3$, have inactivation domains within their N-terminus and can assemble with K_v1 delayed rectifiers giving rise to rapid inactivation. The M current is an outward current slowly activated at membrane potentials more negative than those that activate "A" currents and delayed rectifiers. The current is noninactivating and insensitive to TEA and apamin (the latter distinguishing it from SK channels) and is inhibited by interaction with Gq-coupled receptors. $K_v7.2$–$K_v7.5$ subunits underlie the neuronal M currents either as homomeric or heteromeric channels.

Ca^{2+}-Activated K^+ Channels

K_{Ca} channels share the transmembrane topology of voltage-gated K^+ channels, although one member of this family, the $K_{Ca}1.1$ channels have an additional transmembrane domain S0, which results in an extracellular N-terminus (Fig. 2.6B; Table 2.4). The family has been subclassified into three distinct groups according to single channel conductance, SK ($K_{Ca}2.1$), IK ($K_{Ca}3.1$), and BK ($K_{Ca}1.1$) (small, intermediate, and large conductance). $K_{Ca}1.1$ channels have high single channel conductance (in the range 150–300 pS) and are gated by both voltage and Ca^{2+}. The two gating mechanisms are interrelated as an increase in depolarization increases the Ca^{2+} sensitivity. Neuronal BK channels comprised of α-subunits co-assembled with accessory β2-subunits (KCNMB2) demonstrate inactivation while α-subunits alone or α-subunits co-assembled with accessory β4-subunits (KCNMB2) are noninactivating. $K_{Ca}1.1$ channels normally have high sensitivity to block by the scorpion toxin, iberiotoxin, and TEA, although β4-subunits confer iberiotoxin insensitivity (and martentoxin sensitivity). The Ca^{2+} sensitivity of $K_{Ca}1.1$ channels operates through direct binding to specific regions in the C-terminus (RCK domain and charged amino acid "calcium bowl"), which can be modulated by the presence of accessory subunits. $K_{Ca}2.x$ channels are gated by Ca^{2+}, but show no voltage-dependence. The channels have relatively small conductance (5–20 pS), and are usually insensitive to external TEA, but selectively blocked by the bee venom, apamin. Ca^{2+}-dependent K^+ channels with a conductance intermediate ($K_{Ca}3.1$) between BK and SK have been reported mainly in peripheral tissues. The Ca^{2+}sensitivity of $K_{Ca}2.x$ and $K_{Ca}3.1$ channels are conferred through binding of the Ca^{2+}-sensitive protein calmodulin to the C-terminus, rather than direct binding of Ca^{2+} ions to the channel. The Na^+-activated K^+ channels $K_{Ca}4.1/K_{Ca}4.2$ share a high homology with $K_{Ca}1.1$, but lack Ca^{2+} dependence (despite their channel name).

Inwardly Rectifying K⁺ Channels

K_{ir} channels allow significantly more influx of K^+ current, negative to the reversal potential, than efflux at depolarizing potentials. K_{ir} channels can be loosely classified into three groups: the highly voltage-dependent and constitutively active inward rectifiers; adenosine 5′-triphosphate (ATP)-sensitive inward rectifiers; and G protein-activated inward rectifiers (Fig. 2.6C; Table 2.5). Inward rectifiers can be subdivided according to the steepness of the voltage-dependence. The $K_{ir}1$ and $K_{ir}6$ subfamilies contain weakly rectifying channels, associated with ATP-sensitive K^+ currents, which display instantaneous gating. $K_{ir}2$, $K_{ir}3$, and $K_{ir}4$ subfamilies all display strong rectification and slow gating. Differences in rectification are due to different affinities of block by Mg^{2+} and polyamines (putrescine, spermine, and spermidine) while on- and off-kinetics of high-affinity polyamine block mainly account for slow gating kinetics.

$K_{ir}1$, $K_{ir}2$, and $K_{ir}4$ channels are constitutively active while $K_{ir}3$ channels (G protein-activated) are gated by binding of βγ-subunits through coupling to G protein–coupled receptors. ATP-sensitive K^+ channels, $K_{(ATP)}$, are sensitive to intracellular ATP, where direct binding of ATP to the Kir6 subunit causes the channels to close. The associated sulfonylurea receptor (a member of the ATP cassette–binding protein family with sensitivity to sulfonylurea compounds) is activated by binding of Mg^{2+}-nucleotides at two intracellular nucleotide-binding fold (NBF) domains which results in channel activation.

Two-Pore K⁺ Channels

K_{2P} channel α-subunits are comprised of two-pore domains, where a functional channel is a dimer of subunits (unlike all other families of K^+ channel that are tetrameric assembles of subunits; Fig. 2.6D; Table 2.6). The classic K_{2P} channel has open rectifier properties described by the Goldman–Hodgkin–Katz equation (giving rise to outward rectification); for example, $K_{2P}3.1$ and $K_{2P}9.1$. Such properties give rise to a constitutive leak current, which significantly contributes to cell resting membrane potential.

However, not all leak K^+ channels share these biophysical properties; for example, $K_{2P}18.1$ is a strong outward rectifier. Therefore K^+ leak channels may also contribute to the action potential waveform. K^+ leak channels are regulated by a diverse range of intracellular and extracellular factors, for example, the TREK subfamily ($K_{2P}2.1$, $K_{2P}4.1$, and $K_{2P}10.1$) are activated by lipids, heat and membrane stretch; the TALK subfamily ($K_{2P}5.1$, $K_{2P}16.1$, and $K_{2P}17.1$) are regulated by extracellular protons, while $K_{2P}18.1$ is activated by intracellular Ca^{2+} through direct binding of calcineurin to the channel (at the loop between M2 and M3). K_{2P} channels are insensitive to K^+ channel blockers such as TEA and 4-AP, however, $K_{2P}3.1$, $K_{2P}9.1$, $K_{2P}15.1$, and $K_{2P}18.1$ show high sensitivity to volatile anesthetics such as halothane.

Functional Roles of K^+ Channels in Action Potential Firing

Potassium channels underlie different phases of the action potential waveform as described in Fig. 2.1. K_{2P} channels and K_{ir} typically underlie resting membrane potential. K_v channels with negative activation thresholds contribute to the action potential threshold, for example, the rapidly activating K_v1 channels and slowly activating K_v7 subtypes. K_v subtypes with higher activation voltages underlie the repolarizing phase of the action potential, for example, K_v2 and K_v3 subtypes. $K_{Ca}1$ and $K_{Ca}2.x$ (following Ca^{2+} influx through Ca_V channels during the action potential) underlie the after hyperpolarization. A host of K^+ channels can control repetitive firing of action potentials, for example, the slow activation of K_v7 channels can terminate an action potential burst while KCa2.x channels determine the spacing between action potentials. K^+ channels can be regulated by G protein coupling, for example, the M current (carried by K_v7 channels) is inhibited by the action of acetylcholine or peptide hormones (through Gq coupling and regulation of PIP2 levels) giving rise to an increased excitability, while $K_{ir}3$ channels are directly activated by binding of G protein βγ-subunits leading to membrane hyperpolarization.

K^+ Channels and Disease

As K^+ channels have a major role in controlling excitable tissue, mutations in their genes often result in neurological and cardiac disorders. Here we describe some of the diseases that result from mutations in K^+ channel function.

Ataxias

KCNA1 encodes the α-subunit of $K_v1.1$; as discussed earlier these channels are tetramers and the $K_v1.1$ subunit can assemble as homo- or as heterotetramers with $K_v1.2$. These channels act as delayed-rectifier K^+ channels in neurons of the central and peripheral nervous system. In these tissues they contribute to membrane repolarization following the peak of an action potential, they are therefore essential for controlling excitability.

Multiple mutations in *KCNA1* have been identified in patients suffering with episodic ataxia type 1 (EA1). The majority of mutations result in either complete ablation of channel function or partial loss of function of $K_v1.1$ channels. Patients typically suffer from episodes of brief (seconds to minutes timescale) locomotor incoordination caused by dysfunctional cerebellar output. These episodes are triggered or aggravated by external stimuli such as stress, exertion, and alcohol. In addition to the cerebellar symptoms, patients suffer from uncontrolled muscle contractions (neuromyotonia or neurokymia).

KCNC3 encodes the α-subunit of $K_v3.3$, which forms a tetramer with characteristic A-type conductances. Due to its biophysical properties (high-voltage activation and fast inactivation), $K_v3.3$ channels are well tuned for sustaining high-frequency trains of action potentials in pacemaking cells. *KCNC3* is heavily expressed in cerebellar Purkinje cells, a cell type crucial for controlling locomotor activity. Spinocerebellar ataxia type 13 (SCA13) patients present with dysarthria, nystagmus, and ataxia of gait due to damage to the cerebellum. Unlike the majority of inherited ataxias, SCA13 is caused by point mutations in *KCNC3,* rather than expansion of polyglutamine tracts. To date, two mutations have been identified in *KCNC3,* which cause SCA13; a mutation of the voltage-sensor R420H resulting in a loss of function and F448L causing a change in activation and a slowing of channel inactivation. The mutations reduce the physiological role of Kv3.3 in controlling neuronal pacemaking resulting in hyperexcitability and neuronal cell death.

Epilepsy

KCNQ2 and *KCNQ3* encode the K^+ channel subunits $K_v7.2$ and $K_v7.3$, respectively which give rise to a conductance often referred to as the M current due do its sensitivity to muscarinic agonists. Expressed alone, these subunits can form homotetramers although current densities are greatly increased with co-expression. Similarly to the voltage-gated K^+ channels above, KCNQ2/3 ion channels are opened by depolarizing membrane voltages. Their activation is relatively slow (100ms range), do not inactivate, and open to small deviations from the resting neuronal membrane potential. These biophysical characteristics result in a channel well tuned for controlling the resting membrane potential and thus the threshold for action potential generation. Attacks are frequent and can involve blinking, staring, and tonic posturing. The disorder usually resolves after 6weeks and may increase the chance of developing adult onset epilepsy.

Benign familial neonatal convulsions (BFNC) is an autosomal dominant disorder characterized with short generalized seizures presenting and resolving perinatally (1–6 weeks). *KCNQ2* and *KCNQ3* were identified as causative in BFNC, and to date multiple loss of function mutations have been characterized electrophysiologically. The mutations in these genes may take the form of missense, deletion, or frame shift, but all result in either the partial or complete loss of functional $K_v7.2/3$ channels.

Dominant Nonsyndromic Autosomal Deafness

Similarly to $K_v7.2/3$ K^+ channels, $K_v7.4$ have biophysical properties well tuned to contribute to the resting membrane potential and so excitability of neurons. $K_v7.4$ is expressed in hair cells of the cochlear and appear to contribute to auditory physiology by controlling hair cell excitability. This is highlighted as loss of function mutations in *KCNQ4 and* lead to age-related DFNA, a progressive auditory dysfunction.

Pharmacological Modulation of K^+ Channels

K^+ channels comprise a large family of targets for drug development; with around 80 members and multiple genetic links to human disease, it is no surprise that they have received considerable attention. Openers of K^+ channels have been proposed in disorders where hyperexcitability is a key factor in pathology (epilepsy, stroke, ataxia, pain, amyotrophic lateral sclerosis) and conversely blockers are proposed where suppression of excitable tissue is disease-causing (multiple sclerosis (MS), spinal cord injury, cardiac arrhythmias). Here we highlight drugs that are clinically used, which modulate K^+ channel function.

Epilepsy

Retigabine is an opener of the K_v7 family of K^+ channels used for the treatment of epilepsy. As described earlier, Kv7 channels are associated genetically with rare-forms of epilepsy. Retigabine acts by suppressing hyperexcitable neurons in the CNS of epilepsy patients. Due to the clinical efficacy of Retigabine, there is now a great deal of effort to develop novel drugs targeting K_V7 for other indications such as stroke and pain.

Diabetes

Glibenclamide is an antidiabetic drug in a class of small molecules named the sulfonylureas; it is an oral treatment for type 2 diabetes. Glibenclamide binds to the sulfonylurea receptor 1 (SUR1), which in turn is the accessory subunit of $K_{ir}6.2$. These channels are expressed in pancreatic β-cells where they control the release of insulin into the blood system. Binding of glibenclamide to SUR1 inhibits $K_{ir}6.2$ conductance leading to β-cell membrane depolarization, activation of voltage-gated Ca^{2+} channels, and insulin secretion.

Multiple Sclerosis

4-AP is a blocker of a large number of voltage-gated K^+ channel family members (eg, $K_V1/2/3$). Such a broad pattern of inhibition would ordinarily be considered a safety risk; however, 4-AP is a slowly released drug, which is approved for use in MS—a disease in which neurons are progressively demyelinated resulting in conduction failure along axons. 4-AP is also under assessment in patients with spinal cord injury.

Other Voltage-Gated Channels

Chloride Channels

In mammals, there are nine members of the CIC family. Structurally, CIC channels are about 700–1000 Da and consist of between 10 and 12 transmembrane segments. Since they conduct Cl^- ions, which have a very negative equilibrium potential (about −70 mV), CIC channel opening causes hyperpolarization of the membrane potential, driving the neuron further away from threshold. Within the CNS, CIC-2 is expressed in neurons, and opening of these channels after a depolarizing event would allow repolarization of the neuron and potentially the termination of high firing rates. Mutations in *CLCN2*, the gene that encodes CIC-2, have been reported and may be associated with idiopathic epilepsy in humans.

Hyperpolarization-Activated Cyclic Nucleotide Channels

In mammals, there are four members of the HCN channel family (HCN1–4). Structurally, HCN channels resemble K^+ channels, consisting of four domains with each domain consisting of six transmembrane segments. In each domain, the S4 segment is positively charged and represents the voltage sensor. As the name suggests, these channels are activated by membrane hyperpolarization and once opened these channels pass Na^+ and K^+, which results in a depolarization of the membrane due to a reversal potential of around −30 mV. Their voltage sensitivity can be modulated by cyclic adenosine monophosphate (cAMP) binding to a site within the C-terminus, affecting the voltage dependence of activation. The channel was first discovered in cardiac sinoatrial node cells where its activation drives pacemaker cells to threshold voltages and activation of Ca^{2+} channels and action potential firing. In the brain, HCN channels are expressed in dendrites, somas, and axon terminals of neurons and they are likely to generate rhythmic activity, modulating network excitability. As such, a role for HCN in a number of neurological disorders, including epilepsy, is accumulating.

Transient Receptor Potential Channels

First discovered in *Drosophila*, transient receptor potential (TRP) channels represent the second largest family of voltage-gated ion channels. This large receptor family consists of 28 different channels divided into six distinct subfamilies. TRP channels consist of six transmembrane spanning proteins with a pore located between the 5th and 6th protein segment. Structurally, TRP channels resemble K_v channels but unlike K_v channels, the positive charges in the S4 transmembrane and the P-loop sequences of TRP channels display little conservation. TRP channel subunits assemble to form homo- and heterotetramers that create a receptor that is cation-selective. One of the unique features of TRP channels is the wide variety of exogenous and endogenous stimuli that activate these receptors. Excess heat, changes in pH, changes in pressure, and even molecules in commonly eaten foods (allicin in garlic, capsaicin in chili peppers) are just a few types of stimuli that can activate receptors in the TRP channel family. When TRP channels are activated there is a slow depolarization of a cell as well as a rise in intracellular Na^+ and Ca^{2+} (with the exception of two TRP channels that are strictly selective to monovalent cations). The exact mechanism of TRP channel activation still remains elusive and as a result less detail is known about this process.

Of the six distinct TRP subfamilies, the TRPC, or canonical, family of TRP receptors is most closely related to the family of TRP receptors characterized in *Drosophilia*. TRPC channels are nonselective and permeable to cations, with preferential selectivity for Ca^{2+} over Na^+. TRPC channels can be activated by phospholipase C stimulation, and it has long been hypothesized that these channels underlie the store-operated channels (SOC) located in the membrane of all nonexcitable cells. The TRPV channel family contains six members in mammals and is commonly referred to as the vanilloid family. TRPV1-4 are all heat-activated nonselective cation receptors that also function as chemosensors. TRPV5 and TRPV6 differ in that their heat sensitivity is low and they are permeable only to Ca^{2+}. The specific selectivity in these channels is crucial as TRPV5 and TRPV6 are responsible for Ca^{2+} absorbance in the kidney and intestine. The TRPM subfamily is composed of eight members and is sometimes referred to as the melastatin family. This group of TRP channels is unique in the fact that they possess a functional enzyme in their C-terminal domain. The TRPM channels can be activated by a variety of stimuli, and their main functional responsibilities include sensory transduction in taste cells and the regulation of Mg^{2+} in the kidneys and intestines. The TRPML, or mucolipin, family is composed of three members that are not well characterized. It has been proposed that TRPML receptors can act as proton leak channels as well as play a role in channel regulation. The TRPP, or polycystins, family of receptors is diverse, dissimilar collection of receptors that can be broken down into two subgroups: TRPP1-like and TRPP2-like. TRPP1-like receptors are composed of 11 transmembrane segments and contain multiple N-terminal adhesive domains that play a role in cell–cell contact. TRPP2-like receptors more closely resemble other TRP channels with six transmembrane domains and intracellular N- and C- termini. This subgroup of receptors is constitutively active Ca^{2+} permeable channels that are believed to play a role in sour taste perception. The last subfamily of TRP channels is the TRPA group, which is composed of a single member, the TRPA1. The TRPA1 receptor is composed of 14 N-terminal ankyrin repeats, which is believed to contribute to the receptors function as a mechanoreceptor that detects mechanical stress. TRPA1 is expressed in the DRG, trigeminal ganglion neurons, and in hair cells.

Ligand-Gated Ion Channels

Neurotransmission is traditionally mediated through the process of presynaptically released neurotransmitters relaying a signal to a postsynaptic counterpart. For this cell–cell signaling to occur, neurotransmitters must bind to specific cell-surface receptors. Cell-surface receptors can be divided into two structurally and functionally distinct groups: ligand-gated ion channels (LGICs) and G protein–coupled receptors. LGICs are a group of transmembrane ion channel proteins, which rapidly open in response to the binding of a chemical messenger, such as a neurotransmitter, allowing the passage of ions through the membrane. LGICs are critical for the control of the fastest synaptic events in the nervous system as the action of the transmitter reaches a peak in a fraction of a millisecond and usually decays within a few milliseconds.

Ligand-gated ion channels can be found on the cell membrane at postsynaptic, presynaptic, and extrasynaptic sites in a variety of cell types including nerve, muscle, and epithelial cells. LGICs interact with a range of chemical neurotransmitters, most notably acetylcholine (ACh), glutamate (Glu), γ-aminobutyric acid (GABA), 5-hydroxytryptamine (5-HT), and adenosine 5′-triphosphate (ATP). The need for LGICs to bind with a variety of neurotransmitters has resulted in the formation of a plethora of subtypes that collectively form a diverse superfamily of multisubunit transmembrane proteins. Despite LGIC diversity, all receptors in this family share some structural and functional properties. All LGICs are an assembly of three to five transmembrane subunits with varying topology. Subunit domains located extracellularly play a role in the binding of ligands whereas domains located intracellularly are important for interactions with intracellular modulatory proteins. All LGIC subunits contain between two to four hydrophobic transmembrane domains, which are important for ion channel properties such as ion selectivity (Fig. 2.7). Ion channel pores can be selective for either cations (Na^+, K^+, Ca^{2+}) or anions (Cl^-) and are often called excitatory or inhibitory based on their ability to promote or discourage the initiation of an action potential in a neuron. Although subunit composition varies between LGIC subtypes, it appears that all subunits undergo posttranslational modifications, such as glycosylation and the formation of intrachain disulfide bonds.

Despite some ubiquitous similarities in the LGIC superfamily, enough differences exist among receptors to group LGICs into three main subtypes: Cys-loop ligand–gated ion channels, glutamate-gated cation channels, and ATP-gated ion channels. Genome and cDNA sequencing analyses have revealed that each LGIC subtype is composed of subunits with a different number of transmembrane domains, and these domain differences contribute to subunits that are folded with a unique architecture. Extensive subunit diversity is a feature of LGIC subtypes. In vertebrate acetylcholine–gated ion channels alone, 17 different subunits have been identified. The extensive array of subunits in many LGIC subfamilies is responsible for considerable physiological and pharmacological diversity. This LGIC diversity can lead to multiple forms of ailment and disease, but it also provides plenty of promising drug targets.

Cys-Loop Ligand–Gated Ion Channels

The Cys-loop ligand–gated ion channel superfamily is composed of nicotinic acetylcholine, 5-HT_3, $GABA_A$, glycine, and zinc-activated ion channel (ZAC) receptors. Different Cys-loop receptors are the major targets for many active compounds including anesthetics, muscle relaxants, insecticides, and a range of drugs that treat neurological disorders. Cys-loop LGICs are composed of five protein subunits that form a pentameric conformation around a central ion pore. This structural template was first discovered when the first class of LGIC, the nicotinic ACh receptor, was purified and characterized by molecular cloning. Cys-loop receptors structurally have a characteristic loop formed by a disulfide bond between two cysteine residues separating 13 highly conserved amino acids located in the extracellular N-terminal domain of the α-subunit. The transmembrane topology

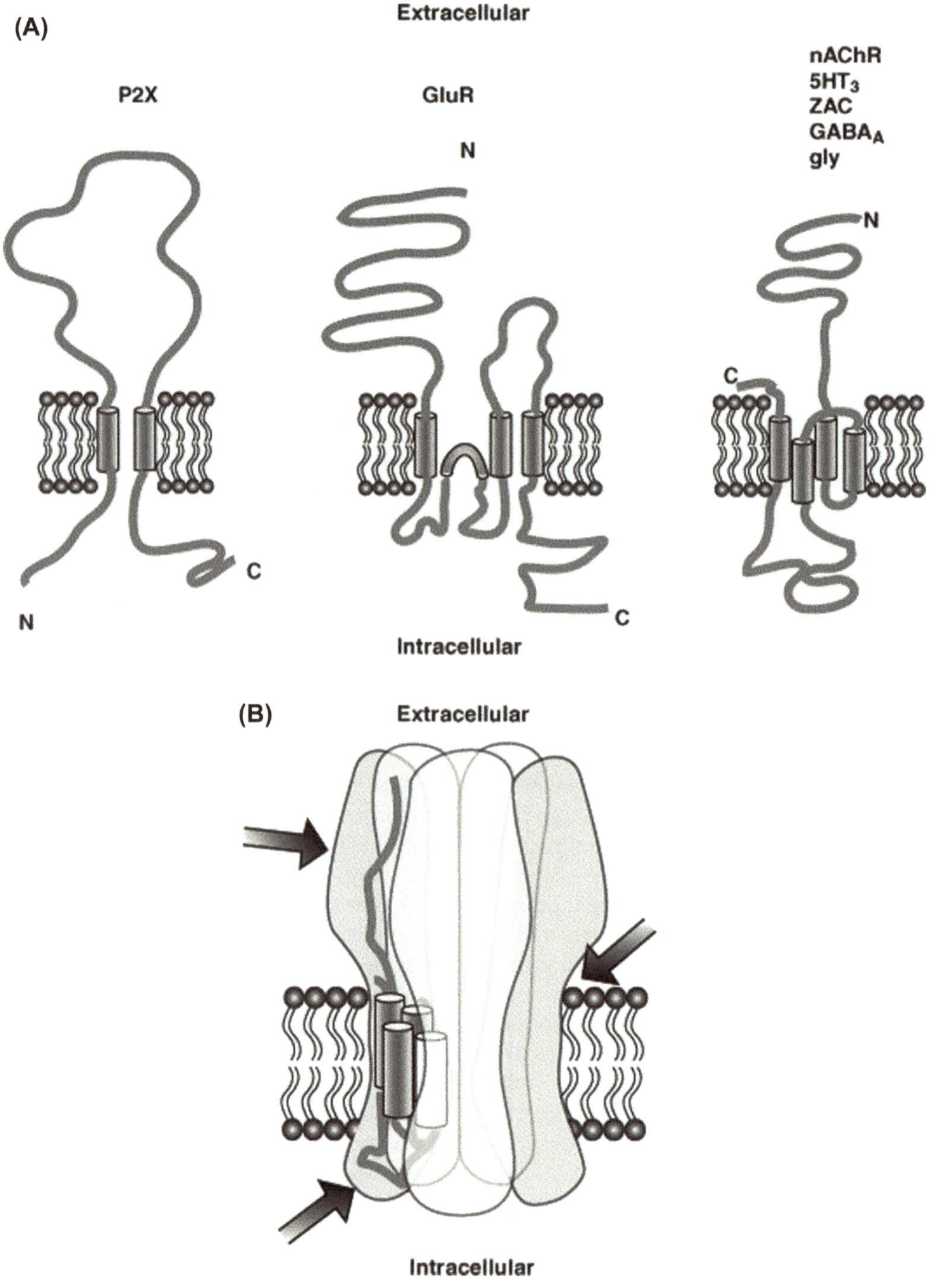

FIGURE 2.7 Structure of the different ligand-gated ion channels (LGICs). (A) Schematic representation of a single subunit from the three families of ligand-gated ion channels inserted in the cell membrane. Note the importance of both the extracellular and intracellular domains where different molecules can bind and interact. (B) Schematic representation of the typical structure of a four transmembrane LGIC such as the nAChR. Putative-binding sites for allosteric ligands are illustrated by the arrows. *Adapted from Hogga RC, Buisson B, Bertranda D. Allosteric modulation of ligand-gated ion channels. Biochem Pharmacol 2005;70:1267–76.*

of Cys-loop LGIC subunits has been widely debated, but there is now considerable experimental evidence that supports a structure in which each subunit contains four transmembrane domains (TM1–TM4) and extracellular N- and C- termini. With its four transmembrane domains, Cys-loop LGIC subunits are typically 430–600 amino acids in length with the greatest sequence diversity within the family occurring in the intracellular domain between TM3 and TM4.

Each region of the Cys-loop LGIC is structurally and functionally distinct. The four hydrophobic transmembrane domains are believed to form α-helices that make up the ion channel pore. The channel pore is lined with residues from the second transmembrane domain (TM2) from each of the five subunits in the assembled receptor. Ion channel selectivity is critical for the function of these channels and this selectivity is believed to arise primarily from residues within the second of the four hydrophobic transmembrane domains. The extracellular domain, which is largely composed of the N-terminus, contains residues that are the binding sites for neurotransmitters. Typically, when a ligand binds to the receptor, it causes a conformational change in the ion channel pore. The T2-helices twist moving the leucine residues, which usually block the pore, out of the channel pathway allowing the flow of ions to occur. Additionally, the binding of a ligand causes the constriction of the pore to widen to about 8 Å from its resting state of 3 Å. In this section, we highlight more closely the distinct properties of the major Cys-loop LGICs as well as disease implications and drugs that target each class.

Nicotinic Acetylcholine Receptors

Acetylcholine (ACh), like many other neurotransmitters, has the ability to bind to two structurally and functionally distinct types of receptors. The first is the metabotropic, G protein–coupled muscarinic ACh receptors. These receptors are classically defined as being activated by muscarine, a toxin from the mushroom *Amanita muscaria*. The second subtype of AChR is the fast ionotropic cationic nicotinic AChR (nAChRs), which is selectively activated by nicotine.

nAChRs have been well studied as they were the first neurotransmitter receptor to be purified biochemically and to be characterized by molecular cloning. Our early advanced knowledge of the nAChR actually originated from a natural oddity. The electric organ of some fish, such as the marine ray *Torpedo*, expresses nAChRs at such a high density that they approach a crystalline array. This extreme receptor density provided an unprecedented source of starting material for receptor purification that eventually enabled the nAChR to be well studied. The rapid advancement of cloning in the 1980s revealed that the *Torpedo* nAChR subunits are closely related to an extended family of mammalian cDNA that encode 16 homologous subunits. All of these subunits have the following in common: a strongly conserved, extracellular N-terminal domain of about 200 residues in length, a conserved three transmembrane domain structure, a cytoplasmic loop of differing amino acid sequence, and a fourth transmembrane domain with a short, nonconserved C-terminal domain. Subunits are classified into α- or non-α-subunits based on whether they have or lack a Cys–Cys pair near the entrance to TM1. This Cys–Cys pair is necessary for agonist binding, and subunits that have the pair are designated as α-subunits.

nAChRs can be subdivided into two groups based on the location of their expression: the skeletal neuromuscular junction or the central and peripheral nervous system. Neuromuscular nAChRs consist of five subunits: α1 and 4 non-α-subunits named β1, δ, γ, and ε. Although the combinations of potential receptors appear plentiful, only two naturally occur at the neuromuscular junction (α1–2β1γε or α1–2β1δε). This limited diversity is in contrast to nAChRs in the central and peripheral nervous systems. The nAChRs found in the nervous system are composed of subtypes of various homomeric and heteromeric combinations of 12 different nicotinic receptor subunits: α_2-α_{10} and β_2–β_4. At the neuromuscular junction, nAChRs have been shown to interact with the intracellular protein rapsyn, which plays an important role in synaptic clustering and tethering. Neuronal nAChRs have also been shown to interact with intracellular messenger proteins, but these interactions have been less characterized.

Several diseases have been linked to the disruption of nAChR function or expression. Mutations to the receptor are believed to cause autosomal dominant nocturnal frontal lobe epilepsy (ADNFLE), a rare form of genetic epilepsy, as well as slow channel myasthenic syndrome, a congenital neuromuscular disorder. The most well-known disorder that is associated with nAChRs, however, is myasthenia gravis, a disease that is caused by autoantibodies generated against the muscular nAChRs. nAChRs are also the target sites of multiple drugs. Targeting these receptors is an effective method of inducing skeletal muscle relaxation during anesthesia. Additionally, drugs aiming to combat schizophrenia, Alzheimer's disease, pain, Parkinson's disease, ADHD, and Crohn's disease all target nACh receptors.

5-Hydroxytryptamine–Gated Ion Channels

5-HT (serotonin) plays a variety of physiological roles in the central and peripheral nervous systems. The ability of serotonin to perform such an array of tasks stems from the neurotransmitter's ability to bind to a variety of receptor subtypes, most of which are G protein–coupled receptors (5-HT_1, 5-HT_2, 5-HT_4-5-HT_7). Serotonin does, however, bind to one subtype of 5-HT receptor that is part of the Cys-loop LGIC family, the 5-HT_3 receptor. 5-HT_3 is expressed throughout the central and peripheral nervous systems. Similar to other members of the Cys-loop family, the 5-HT_3 receptor is composed of five subunits that symmetrically enclose a central ion conduction pore that is cation-selective. Subunits of the 5-HT_3 receptor show sequence similarity with nAChR subunits, which may explain why the two receptors share similar characteristics. The 5-HT_{3A} receptor subunit is able to form functional homomeric receptors and may also combine with 5-HT_{3B} subunits to produce heteropentamers that display distinct electrical and pharmacological features. Genes for 5-HT_C, 5-HT_D, and 5-HT_E have been described, but less is known about the structure and properties of these subunits. Overall, 5-HT_3R subunits display less diversity than other channels in the Cys-loop family. In the peripheral nervous system, 5-HT_3 receptors are important in information transfer in the gastrointestinal tract; and in the enteric nervous system the receptors play a role in gut motility and peristalsis. In the central nervous system, 5-HT_3 receptors are involved in a variety of functions including emesis, cognition, and anxiety. Some early studies suggest that 5-HT_3 receptor dysfunction may play a role in various psychiatric and GI disorders. 5-HT_3 receptors are often the target of drugs used to control emesis and irritable bowel syndrome.

γ-Aminobutyric Acid–Gated Ion Channels

GABA is the predominant inhibitory neurotransmitter in the brain, and like many other neurotransmitters, it binds and activates both LGICs and G protein–coupled receptors. $GABA_A$ receptors are one of the most complex members of the Cys-loop receptor subfamily as they exhibit remarkable subunit diversity ($\alpha1$–$\alpha6$, $\beta1$–$\beta3$, $\gamma1$–$\gamma3$, δ, ε, θ, π, $\varrho1$–$\varrho3$). The large amount of identified subunits has the ability to assemble and form a significant amount of unique pentameric receptors. The exact subunit composition and distribution of innate $GABA_A$ receptors is yet to be clearly defined, but it is widely believed that $GABA_A$ receptors are heteromeric and consist of a subunit combination of $\alpha1$, $\beta2$, and $\gamma2$ or $\alpha1$, $\beta3$, and $\gamma2$ subunits. A small subset of $GABA_A$ receptors known as $GABA_{A\text{-}rho}$ (previously defined as $GABA_c$) receptors are homomeric. These receptors, which are located in the retina, consist only of rho (ϱ)-type subunits. $GABA_A$-gated ion channels are anion-selective channels that allow the permeation of chloride ions through the central pore. The binding of GABA and the subsequent influx of chloride ions into a neuron produces a net hyperpolarization that inhibits action potential firing (rare exceptions exist where chloride actually depolarizes a neuron). Dysfunction of the $GABA_A$ receptor has been linked to numerous diseases including anxiety, insomnia, and most notably epilepsy. Epilepsies induced by $GABA_A$ can vary in type, but the most commonly seen are febrile seizures and childhood absence epilepsy. Since $GABA_A$ dysfunction has been linked to multiple epilepsy disorders, the receptor has become a significant target for numerous antiepileptic drugs. In addition to epilepsy treatments, drugs that act on $GABA_A$ receptors are used to combat anxiety and insomnia.

Glycine-Gated Ion Channels

Glycine is the main inhibitory neurotransmitter of the spinal cord, brainstem, and retina. Like $GABA_A$ receptors, glycine receptors are believed to be pentameric complexes that are anion specific. Unlike $GABA_A$ receptors, however, glycine receptors lack subunit diversity as only four α-subunits (α1–α4) and one β-subunit are believed to exist. Glycine receptors can be homomeric complexes consisting of only α-subunits, or they can combine with a β-subunit to form a heteromeric channel. Glycine receptors are important for mediating synaptic inhibition in the brainstem and spinal cord. Within the large intracellular loop of the β-subunit, there exists a binding site for gephyrin, a glycine receptor–associated intracellular protein that plays a role in receptor clustering at postsynaptic density. Intracellular Ca^{2+} levels are also believed to act on glycine receptors in a way that increases receptor affinity and prolongs channel opening events. Mutations to glycine channel subunits have been linked to the rare neurological disorder hyperekplexia, or startle disease. Glycine receptors are target sites for general anesthetics as well as inhaled drugs of abuse.

Glutamate-Gated Cation Channels

Glutamate is abundant throughout the human body, but is most prominently found in the brain, where it acts as the main excitatory neurotransmitter as well as the precursor to GABA, the main inhibitory neurotransmitter. In the nervous system, glutamate has the ability to bind to both G protein–coupled receptors as well as ionotrophic glutamate receptors (iGluRs). The iGluRs can be pharmacologically separated into three categories based on their differential affinity to the exogenous ligands α-amino-3-hydroxy-5-methyl-4-isoxazole-propionic acid (AMPA receptors), *N*-methyl-D-aspartate (NMDA receptors), and kainic acid (Kainate receptors).

Initially believed to be pentameric complexes like the Cys-loop family of LGICs, there is now strong evidence that shows ionotropic glutamate receptors are composed of four subunits (>900 residues) that form a central ion channel pore. Glutamate receptors contain four discrete domains: the extracellular amino terminal domain, the extracellular ligand–binding domain, the transmembrane domain, and the intracellular C-terminal domain. Like Cys-loop receptor subunits, iGluR subunits contain four hydrophobic domains (M1–M4). Unlike their Cys-loop counterparts, however, there is evidence that suggests that only three of the four hydrophobic domains traverse the cellular membrane. The second of the four hydrophobic domains (M2) forms the receptor's reentrant loop, which contributes to the pore-forming region of iGluRs. The intracellular C-terminal plays a pivotal role in mediating protein–protein interactions, including interactions with proteins that regulate receptor trafficking and anchoring.

NMDA receptors are typically concentrated at postsynaptic sites where they assemble as heteromeric complexes constructed from GluN1, $GluN2_{A-D}$, or $GluN3_{A-B}$ subunits. The most commonly seen conformation usually involves the GluN1 subunit and at least one of the GluN2 subunits. NMDA receptors are unusual in the fact that they require two agonists to bind to efficiently activate the channel: glutamate and glycine. The glutamate binds to its selective binding site located on the GluN2 domain while the glycine binding site is located on the GluN1 subunit. NMDA receptors are considered mixed cation permeable channels, with a high relative permeability to Ca^{2+}. The pore of the NMDA receptor can be blocked by Mg^{2+} in a voltage-dependent manner. NMDA receptors play an important role in controlling synaptic plasticity and in memory function.

AMPA receptors and kainate receptors are often grouped into a subclass of receptors called the non-NMDA receptors. AMPA receptors are one of the most commonly found receptors in the brain, and most form heteromeric tetramers from a group of four subunits (GluR1–4). Each AMPA receptor has four sites, or one of each subunit, to which glutamate can bind. When glutamate binds, the AMPA receptor opens allowing the influx of mixed cations. The AMPA receptor's permeability to Ca^{2+} is determined by the GluR2 subunit. If an AMPA receptor lacks a GluR2 subunit, then it will be permeable to Ca^{2+}, whereas if it has a GluR2 subunit, it will be impermeable to the ion. AMPA receptors play an important role in mediating fast synaptic transmission in the CNS as well as contributing to synaptic plasticity. Kainate receptors were first identified as a distinct receptor type through their selective activation by the toxin kainate. Kainate receptors can be homomeric or heteromeric tetramer complexes composed from a combination of five different subunits (GluK1–5). The ion channel formed by kainate receptors is permeable to Na^+ and K^+ with only minimal Ca^{2+} permeability. The distribution of kainate receptors is relatively limited compared to AMPA and NMDA receptors and their exact function is not as well defined. It is believed, however, that kainate receptors have different pre- and postsynaptic functions, as presynaptic kainate receptors have been implicated in synaptic inhibition while postsynaptic receptors are involved in excitatory neurotransmission.

iGluR receptor dysfunction has been linked to a variety of CNS disorders. Defects in AMPA receptor trafficking has been linked to Fragile X syndrome while ectopic expression of kainate receptors is believed to contribute to temporal lobe epilepsy. Additionally, iGluR defects have been linked to neurodegenerative diseases, such as amyotrophic lateral sclerosis, and autoimmune disorders such as NMDA receptor encephalitis.

Adenosine 5′-Triphosphate–Gated Ion Channels

ATP is most widely known for its role in intracellular metabolic energy transfer, but it also can act as a neurotransmitter that binds to extracellular receptors. Just like other neurotransmitters discussed in this chapter, ATP binds to both ionotropic (P2X receptors) and G protein–coupled receptors (P2Y). The P2X receptor is a mixed cation–selective ion channel that can be found in glial cells as well as in pre- and postsynaptic nerve terminals throughout the central, peripheral, and autonomic nervous systems. Outside of the nervous system, P2X receptors are distributed throughout muscle and epithelia tissue. Just like its range of expression, the P2X receptors also have a range of functions, most notably modulating synaptic transmission, neuronal–glial interaction, and apoptosis. Like other LGICs, P2X receptors are oligomeric transmembrane proteins, but unlike other LGICs, there is evidence that P2X receptors only consist of three subunits. Additionally, P2X receptor subunits differ from other families of LGICs in the sense that they contain two hydrophobic membrane-spanning domains and intracellular N- and C-terminal domains. Both transmembrane domains have been shown to be able to modulate channel properties, but it is believed that the second transmembrane domain is primarily involved in forming the pore of the channel. Between the two transmembrane domains is a large extracellular loop that contains the binding sites for ATP. In mammals, seven P2X subunits have been identified (P2X1–7) and these subunits form both homomeric and heteromeric channels. Dysfunction of ATP-gated ion channels has been linked to chronic pain and has subsequently made the receptors a target for drugs aimed at combating this ailment.

Further Reading

[1] Hodgkin AL, Huxley AF. Currents carried by sodium and potassium ions through the membrane of the giant axon of *Loligo*. J Physiol Lond 1952b;116:449–72.

[2] Noda M, Shimuzu S, Tanabe T, Takai T, Kayano T, Takahashi H, et al. Primary structure of *Electrophorus electricus* sodium channel deduced from cDNA sequence. Nature 1984;312:121–7.

[3] Payandeh J, Gamal El-Din TM, Scheuer T, Zheng N, Catterall WA. Crystal structure of a voltage-gated sodium channel in two potentially inactivated states. Nature 2012;486:135–9.

[4] Catterall WA, Perez-Reyes E, Snutch TP, Striessnig J. International union of pharmacology. XLVIII. Nomenclature and structure-function relationships of voltage-gated calcium channels. Pharmacol Rev 2005;57:411–25.

[5] Dolphin AC. Calcium channel auxiliary $\alpha 2\delta$ and β subunits: trafficking and one step beyond. Nat Rev Neurosci 2012;13:542–55.

[6] Perez-Reyes E. Molecular physiology of low-voltage-activated T-type calcium channels. Physiol Rev 2003;83:117–61.

[7] Simms BA, Zamponi GW. Neuronal voltage-gated calcium channels: structure, function, and dysfunction. Neuron 2014;82:24–45.

[8] Hogga RC, Buisson B, Bertranda D. Allosteric modulation of ligand-gated ion channels. Biochem Pharmacol 2005;70:1267–76.

CHAPTER

2.1

Translational Correlation: Temporal Lobe Epilepsy and Hippocampal Sclerosis

J.C. DeToledo[1], *I. Kim*[2], *S. Tantikittichaikul*[1]

[1]Texas Tech University Health Sciences Center, Lubbock, TX, United States; [2]Vanderbilt University Medical Center, Nashville, TN, United States

Various types of cortex can generate seizures. The layered arrangement of the neocortex makes it a suitable substrate to initiate and propagate paroxysmal discharges. The six layers of cortex, however, are not a requirement and in fact, the phylogenetically older archicortex and paleocortex (hippocampus and parahippocampal gyrus) are a more common source of epileptic seizures in humans. In addition to having seizures with symptoms that correlate with temporal lobe involvement, many of these patients have severe neuronal cell loss and gliosis in the hippocampus that can be demonstrated in the MRI as mesial temporal sclerosis. The association of atrophic hippocampus and chronic seizures was proposed by Wilhelm Sommer in 1880 and subsequently confirmed by the work of Karl Stauder in 1936. He examined brains of 53 patients, 35 with history of temporal lobe epilepsy (TLE) and 18 without history of TLE. Stauder found that 94% of TLE patients had hippocampal sclerosis (HS), whereas only 16% of non-TLE group had HS on autopsy. These findings have been vastly expanded on since that time.

TEMPORAL LOBE EPILEPSY

Epileptic seizures are caused by abnormal brain electrical activities resulting in various symptoms including motor, sensory, autonomic, cognitive, or psychiatric manifestations. There are many reasons of seizure, and medical conditions that produce increased cortical irritability can cause seizures. Seizures induced by a systemic condition, such as sepsis, are called provoked (or symptomatic) seizure. Epilepsy is defined as recurrent unprovoked seizures.

TLE is one of the most common subtypes of epileptic seizures and the hippocampal discharges will often spread toward adjacent areas such as amygdala, hypothalamus, insular cortex, frontal operculum, cingulate cortex, or temporal association cortex. Clinically these patients present with episodes that initiate with an aura

Conn's Translational Neuroscience
http://dx.doi.org/10.1016/B978-0-12-802381-5.00003-8

(simple partial seizure) followed by behavioral arrest, automatisms, loss of consciousness, and postictal confusion. These seizures have a focal onset and may spread to other parts of the brain resulting in a secondarily generalized seizure. Auras are prodromal symptoms that patients experience at the beginning of seizures, and its content depends on the areas of the brain that are affected. For these reasons, the symptoms of the aura can help to localize the area of onset of seizures. Auras are technically simple partial seizures. For example, as epileptic discharges arise from the hippocampus and spread to the amygdala, patients can have olfactory auras, feelings of anxiety, déjà vu, and jamais vu. Fear is considered to be amygdaloid in origin. Autonomic auras such as palpitation, dyspnea, urinary urgency, and genital sensation usually come from spread into insular cortex and anterior cingulum. Auditory auras manifest as auditory hallucinations attributed to activation of the temporal association cortex. Vertiginous auras can occur with temporoparietal junction involvement. As the seizure spread after the aura, there is impaired consciousness; patients may have a fixed stare with widened palpebral fissures that is sometimes described as "blank staring." Automatisms manifesting as lip smacking, chewing, teeth grinding, or swallowing are common. Also patients can continue with ongoing motor activity or develop repetitive motor behavior. These symptoms define the complex partial stage of the seizure.

HIPPOCAMPAL SCLEROSIS

HS is neuronal loss and gliosis of hippocampus typically affecting CA1, CA4, and hilum, and is the most common histopathological finding in TLE. Although pathological change in hippocampus has been documented since early 1800s, the first microscopic description of HS was by Sommer in 1880. He had a young patient with frequent "petit mal" and "complete" seizures along with vivid hallucinations. One day he heard God telling him that he could fly. He jumped from a roof, survived, but died from infectious complications. At autopsy, Sommer recognized prominent loss of pyramidal neurons in CA1 and prosubiculum, which are sometimes referred as Sommer sector (SS). He also described other damaged areas, such as end folium (CA4 and hilum).

Many aspects of pathogenesis of HS have long been debated. One of the debates has been whether the HS is the cause or the consequence. While Sommer and Bratz suggested HS as a generator of seizures, other researchers, Epielmeyer, Scholz, Peiffer, Sutula, and Pitkanen argued that the pathological change in the hippocampus is only the damage from seizures.

According to the findings from University of California at Los Angeles group, HS arises from the early brain insult during the childhood, which is not necessarily related to seizures. However, they have also found that there is continued hippocampal cell loss from seizure activity, causing nonspecific change to the hippocampus different from HS. Difference is that in cases of HS, specific areas such as CA1, CA4, and hilum are typically affected while CA2, CA3, and subiculum are spared. However, in cases of damages from ongoing seizure activities, those areas will also be compromised. It is still not clear what kind of brain insult causes initial sclerotic damage to the hippocampus. The hypothesis febrile seizure in childhood initiating HS has been speculated but remains controversial.

Another interesting histological aspect of HS is mossy fiber sprouting. Mossy fibers are axons of dentate granule cells that synapse onto CA3 cells through hilum. As a result of neuronal loss in hilar region, mossy fibers sprout out and reinnervate. Abnormally reorganized mossy fibers synapse back to other dentate granule cells, or to theirs own dendrite, forming recurrent excitatory circuits. This excessive excitatory innervation is thought to be epileptogenic. It is still debated, however, whether it is the cause or consequence of TLE.

MANAGEMENT

TLE is treated with various antiepileptic medications. These medications can suppress the abnormal discharges or to suppress the spread of these discharges into other brain regions. It makes intuitive sense that a drug that can suppress the seizure focus would be better but unfortunately, most drugs used in clinical practice seem to prevent the spread of the seizures. In other words, these medications are not truly antiepileptic but rather, anticonvulsants. Another way to look at this is that medications treat the symptoms (the clinical seizure) but do not treat the cause (suppress or eliminate the seizure focus). Medications can be in isolation (monotherapy) or in combinations (polytherapy). While about 60% of patients with neocortical epilepsy respond well to medications, this success rate is substantially lower in patients with HS. These nonresponders are referred to as being drug-resistant epilepsy and are best treated with surgical resections. The resection typically includes affected hippocampus and amygdala. Even with the hippocampus has many important functions, resection of one hippocampus results in relatively minor deficits if the remaining hippocampus is healthy. Intelligence is not changed by surgeries, and actually successful epilepsy surgery can stop or improve the cognitive decline that patients with chronic epilepsy will often experience. On average, after a temporal lobectomy, 15% of patients achieve long-term antiepileptic drug discontinuation, 50% achieve seizure control with monotherapy, and 35% remained on polytherapy.

Further Reading

[1] Jerome Engle JR, Pedley TA. Epilepsy: A comprehensive textbook. 2nd ed. 1998.

[2] Thom M. Review: hippocampal sclerosis in epilepsy: a neuropathology review. Neuropathol Appl Neurobiol 2014;40(5):520–43.

CHAPTER

3

Neurotransmitters and Neurotransmission in the Developing and Adult Nervous System

S.L. Miller, H.H. Yeh

The Geisel School of Medicine at Dartmouth College, Lebanon, NH, United States

INTRODUCTION

Neurotransmitters shape the structure and function of the nervous system throughout life. Although not always the case, neurotransmitters typically elicit their response upon their effector cells via the process known as synaptic transmission. This fundamental concept of chemical transmission across the synapse as the principle mode of communication among neurons is relatively new. It was not until the Greek philosopher, Alcmaeon of Croton (Fig. 3.1), in 5th century BC proposed that the brain is the house of the mind and soul that scholars began to question the concept that the mind is contained within the heart. Alcmaeon was also the first to propose the concept of the animal spirit. Through dissection and the investigation of blood vessels, he posited that, like arteries and veins that contained oxygenated and deoxygenated blood, respectively, nerves must be hollow structures that contained a substance necessary to produce movement that he and others called the animal spirit. Although definitive evidence for the existence of the animal spirit was never provided for well over 2000 years, this remained the prevailing hypothesis for the mechanism by which nerves elicited motor activity.

Through an extensive series of elegant investigations, the chemical synapse (Fig. 3.2) at the neuromuscular junction (NMJ) has provided the foundation for understanding basic principles of rapid, chemical communication across the synapse. In this scheme, an action potential arrives at synaptic boutons of α-motor neurons, voltage-gated calcium channels open, resulting in rapid calcium influx, the synaptic vesicles containing acetylcholine (ACh) fuse with the plasma membrane, ACh is released into the synaptic cleft, and diffuses across the synaptic space to bind and activate nicotinic acetylcholine receptors (nAChRs) on the plasma membrane of skeletal muscle. The activation of nAChRs on the postsynaptic skeletal muscle cells results in depolarization and allows for contraction, a process referred to as "excitation–contraction coupling." This chapter introduces the basic principles of neurotransmission, neurotransmitters, their receptors,

Conn's Translational Neuroscience
http://dx.doi.org/10.1016/B978-0-12-802381-5.00004-X

FIGURE 3.1 Sculpture of Alcmaeon. *Reproduced with permission from Wolters Kluwer Health, Inc.*

and the role of glia in the regulation of synaptic transmission.

NEUROTRANSMISSION: A GENERAL OVERVIEW

To understand the effects of neurotransmitters, the steps of neurotransmitter release need to be introduced. Whether or not neurotransmitters are released is dependent on numerous factors, but one of the key processes is the generation of action potentials in the presynaptic neuron. The typical process begins with neurotransmitters binding to their ligand-gated receptor on the postsynaptic cell membrane. Depending on the transmitter–receptor interaction, this generates events known as excitatory postsynaptic potentials (EPSPs) or inhibitory postsynaptic potentials (IPSPs) (Fig. 3.3). For example, the neurotransmitter glutamate binding and activating the α-amino-3-hydroxy-5-methyl-4-isoxazolepropionic acid (AMPA) receptor results in EPSPs due to the influx of sodium, potassium, and sometimes calcium. This individual event

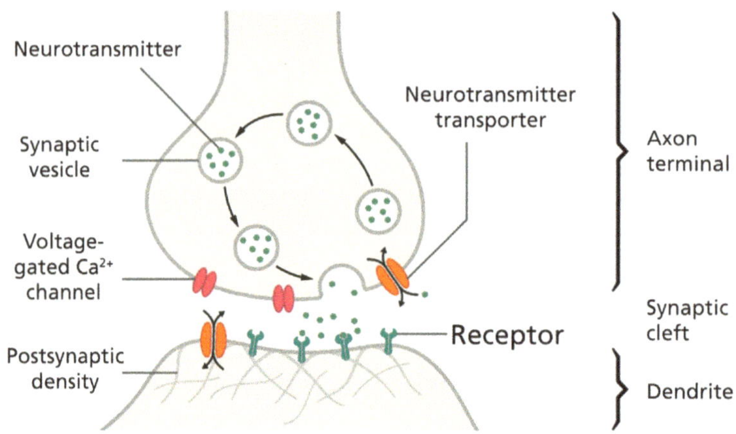

FIGURE 3.2 Diagram of the chemical synapse. Typical chemical synapses contain the presynaptic component (the axon terminal) and the postsynaptic component of the effector cell (eg, the dendrite). Synapses express voltage-dependent calcium channels that facilitate the exocytosis of neurotransmitter contained within synaptic vesicles. Following the release of neurotransmitter, activation of the postsynaptic neurotransmitter receptors is activated and mediates the action of the neurotransmitter. To terminate the action of the transmitter synapses contain transporters to reuptake the release neurotransmitter. *Adapted from T. Splettstoesser. Synapse schematic. Original image available at* https://commons.wikimedia.org/wiki/File:SynapseSchematic_lines.svg.

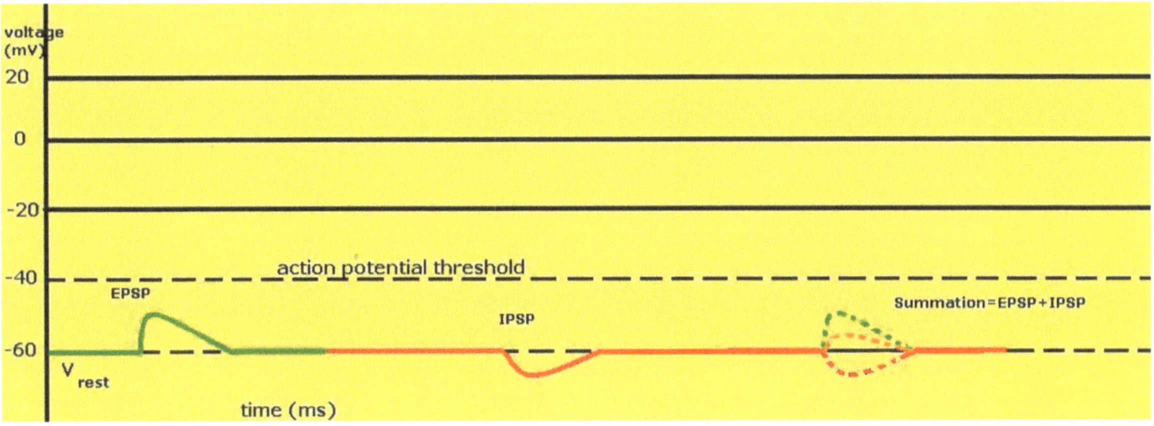

FIGURE 3.3 Example voltage changes following an EPSP, an IPSP, and summation when an EPSP and IPSP occur simultaneously. *Reproduced with permission from Wikimedia Commons. Available at:* https://commons.wikimedia.org/wiki/File:IPSPsummation.JPG.

produces an increase in membrane potential from rest (net positive change in voltage) but is typically insufficient to produce an action potential. To achieve this, multiple EPSPs must summate spatially and temporally in order to trigger an action potential. Spatial summation occurs when numerous presynaptic terminals synapse onto a single postsynaptic cell in close physical proximity and produce overriding EPSPs that add up to produce a sufficient voltage change required to induce an action potential. Temporal summation involves the same synapse onto a postsynaptic cell repeatedly producing EPSPs in close succession to bring the postsynaptic cell to the membrane voltage sufficient to produce an action potential (Fig. 3.4). Thus, spatial and temporal summation epitomize the concept of "action potential threshold."

The magnitude of the EPSP depends on the synapse, among other conditions, but since spike threshold can vary widely among neurons, the number of EPSPs required to elicit an action potential is also highly variable, depending on the type of neuron. For example, dorsal root ganglion (DRG) cells have a spike threshold of −11.6 mV, whereas that of a neocortical basket cell is −59.3 mV. The generation of action potentials is complicated by the fact that many neurons are simultaneously receiving inputs from inhibitory

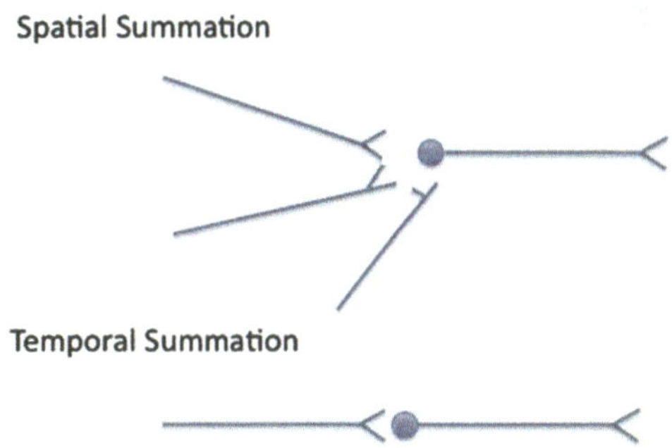

FIGURE 3.4 Schematic of the neuronal arrangement for spatial summation (above) and temporal summation (below). *Adapted with permission from Wikimedia Commons. Original image available at* https://commons.wikimedia.org/wiki/File:Interneuronalrelations.jpg.

neurons that are generating IPSPs onto the postsynaptic cell. However, once the voltage change is sufficient and threshold has been reached, voltage-gated sodium channels open at the axon initial segment and an action potential is initiated.

Propagation of the action potential along the axon occurs in two basic modes. In unmyelinated neurons, each segment of the adjacent axonal membrane must become sufficiently depolarized by nearby voltage-gated sodium channels in order to facilitate the opening of the sodium channels at that location. Since each segment of the membrane must be depolarized, this method of action potential propagation is relatively slow compared to that in myelinated axons (eg, 1–2 m/s vs. 120 m/s, respectively). In myelinated axons, a process known as "saltatory conduction" occurs whereby action potentials propagate by "jumping" through an internode (a small segment of unmyelinated axon) from one node of Ranvier to the adjacent one. Voltage-gated sodium channels are highly concentrated on the plasma membrane at each node, with limited distribution along the internode segments. This produces very large, rapid voltage alterations, specifically at the nodes that rapidly propagate to the subsequent node (Fig. 3.5).

Ultimately, in either myelinated or unmyelinated axons, the action potential arriving at the synaptic terminals and the ensuing voltage change result in the opening of voltage-gated calcium channels. Upon opening, calcium rapidly enters the terminals, diffusing down its electrochemical gradient to facilitate the fusion of synaptic vesicles to the plasma membrane and the release of neurotransmitter into the synaptic compartment. This sequence of events—the generation of EPSPs at the soma, the opening of voltage-gated sodium channels, the generation of the action potential axon initial segment and its propagation down the axon to the synaptic terminals, and neurotransmitter release at the synaptic terminals—is referred to as an orthodromic ("ortho" meaning straight, "dromic" meaning direction) action potential. In other words, neurotransmission occurs in the direction of soma to synaptic terminals (Fig. 3.5). This is likely to be the primary direction in which action potentials are directed during most physiological conditions; however, in experimental settings and rare physiological cases, antidromic action potentials can also occur ("anti" meaning opposed, "dromic" meaning direction).

Compared to orthodromic action potentials typically elicited at somatodendritic sites, antidromic potentials are elicited at the axonal membrane. These can occur through the stimulation of the axons to study various neuronal properties, but can also occur in limited situations in vivo. For example, within the hippocampus, sharp-wave ripple complexes (SPW-Rs) are high-frequency oscillations typically

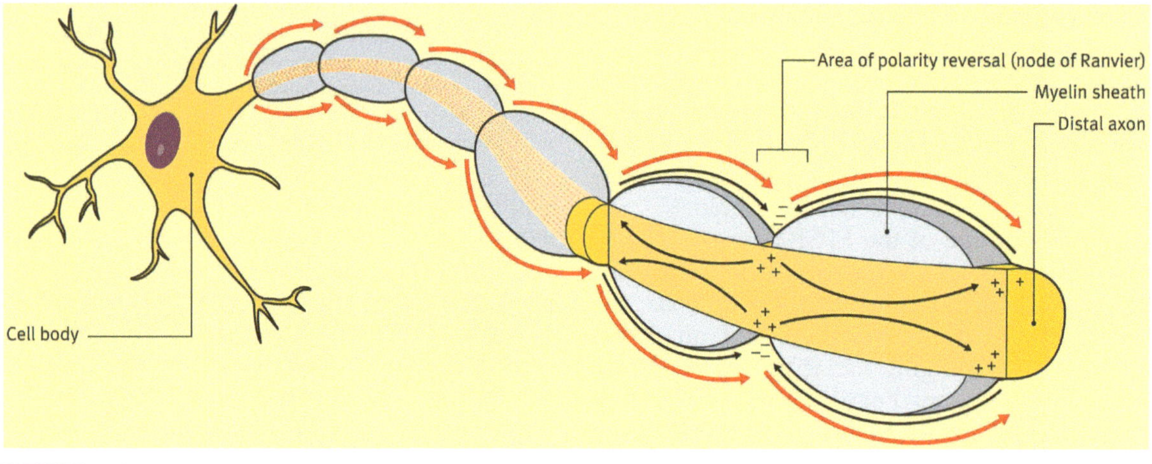

FIGURE 3.5 Saltatory conduction of action potentials. Action potentials are generated as the trigger zone at the cell body and depolarization spreads from each node of Ranvier to the next. This occurs quickly along the inner surface of the myelinated membrane and continues to the synaptic terminals. *Reproduced with permission from A. Fletcher. Action potential: generation and propagation. Anaesthesia & Intensive Care Medicine 2011;12(6):258-62.*

present during sleep, thought to be responsible for encoding sensory information into memories as their disruption prevents memory formation. Although the exact physiological role of these potentials is unclear, they can induce various forms of synaptic plasticity in a noncanonical fashion.

There are two episodes during an action potential waveform when another action potential cannot be elicited or is very difficult to elicit—the relative and absolute refractory periods, respectively (Fig. 3.6). Following triggering of the action potential at the axon initial segment, voltage-gated sodium channels open and sodium rushes into the cell, depolarizing it. Soon thereafter, voltage-gated sodium channels close and inactivate, voltage-gated potassium channels open and potassium leaves the cell along its electrochemical gradient, hyperpolarizing the membrane. Compared to sodium channels, potassium channels stay open longer and cause the cell to be hyperpolarized beyond its resting membrane potential. This period of time is called the after-hyperpolarization or AHP. The absolute refractory period spans the peak of the action potential until the resting membrane potential is reached. This is because during this period of time, the voltage-gated sodium channels are inactivated and cannot be activated regardless of the strength of the stimulus to initiate an action potential. As a result of this, the only voltage-gated sodium channels that can be depolarized are the ones present on adjacent axonal plasma membrane downstream from the soma. However, during the period of AHP, and before the cell returns to its resting membrane potential, a sufficiently strong stimulus can trigger an action potential. During this period, the sodium channels return to a closed confirmation from an inactivated state and can again participate in initiating another action potential. Since the potassium channels are still closing during this period, a stronger stimulus than normal is needed to reach action potential threshold. This period of in-the-action potential cycle is referred to as the relative refractory period because a new action potential can be initiated, albeit with more difficulty.

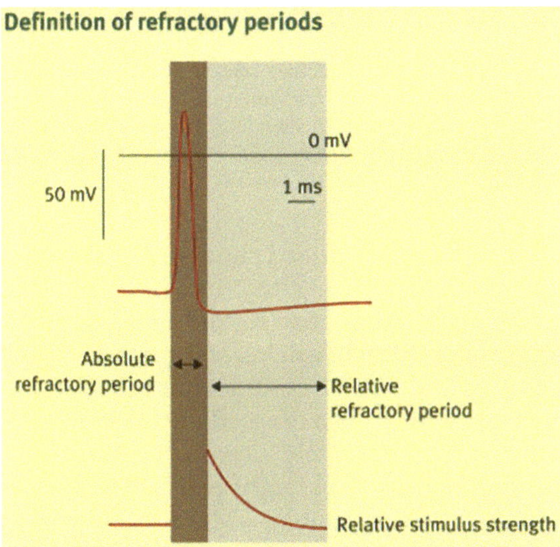

FIGURE 3.6 The darker and lighter shaded boxes superimposed on the action potentials show the absolute and relative refractory periods. The bottom trace represents the relative stimulus strength required to elicit a second action potential during the relative refractory period. *Reproduced with permission from C.H. Fry, R.I. Jabr. The action potential and nervous conduction. Surgery 2010;28(2):49-54.*

Neurotransmission in the Peripheral Versus Central Nervous System: Principles

Quantal Release

The release of neurotransmitter is thought to occur with individual units or "quanta." Each quanta represents a single synaptic vesicle packed with neurotransmitter and, upon its release, produces a response that can be statistically represented as a stochastic distribution at a given synapse (Fig. 3.7). The postsynaptic response resulting from a quantum can be displayed in a histogram by plotting the voltage of the response by the number of times this event

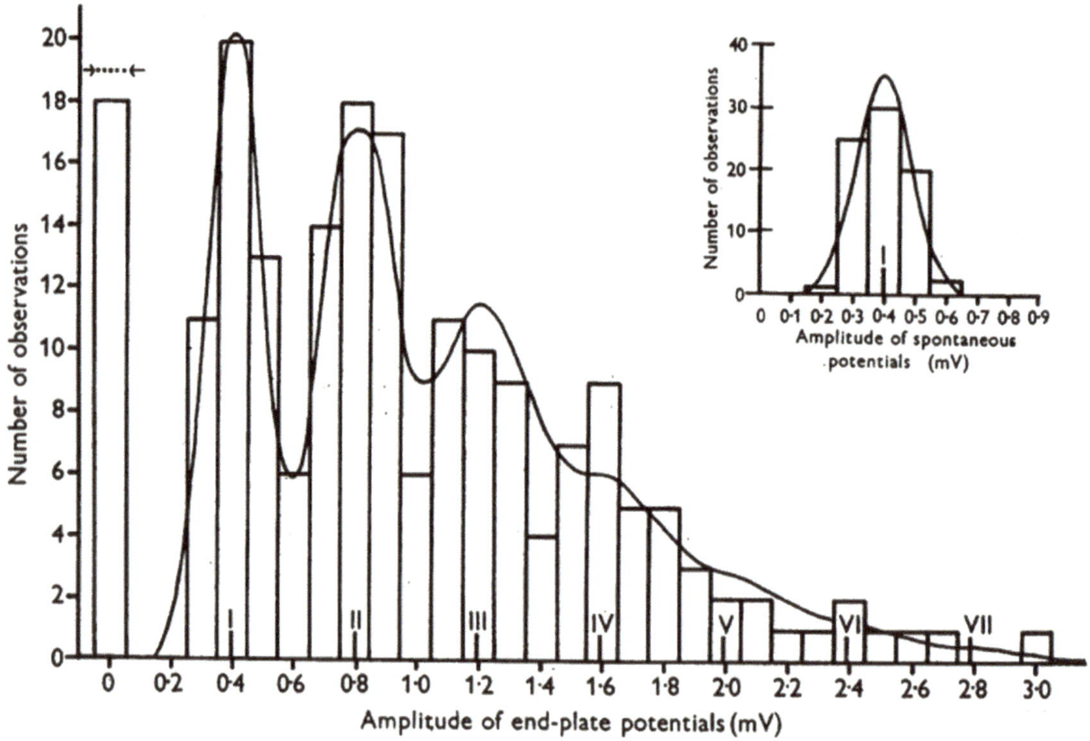

FIGURE 3.7 Histograms of EPPs and spontaneous potential amplitude distributions in a muscle fiber. Peaks in EPP amplitude distribution occur at 1, 2, 3, and 4 times the mean amplitude of the spontaneous miniature potentials. Gaussian curve is fitted to spontaneous potential distribution and used to calculate theoretical distribution of EPP amplitude (continuous curve). *Arrows* indicate expected number of failures of response to nerve stimuli. *Reproduced with permission from I.A. Boyd, A.R. Martin. The end-plate potential in mammalian muscle. J Physiol 1956;132:74-91.*

occurs. Typically, numerous peaks are observed in regularly graded intervals: one peak at 0.5 mV, another at 1 mV, another at 1.5 mV, and so on. This suggests that the average response produced by one quantum is 0.5 mV and the response is linear to the number of quanta released from the presynaptic site. In practice, however, the variance for each additional quantum may not be as linear due to differences at individual release sites, and distance of the release site from the electrode used to record the bioelectric potentials.

Neurotransmission Probability

At the NMJ, upon arrival of the action potential at the presynaptic terminals of an α-motor neuron, Ach release onto skeletal muscle cells is virtually guaranteed under physiological conditions. Calcium entry into the terminals can result in an almost 16-fold increase in the fusion of synaptic vesicles and release of Ach. This is evident as we contract our muscles, and the response is also immediate. Without the ability to immediately control muscles with immediate responses, fight-or-flight responses would be impossible. More importantly, inefficient release of Ach at the NMJ in the diaphragm would be incompatible with life. This dramatic increase in the synaptic release of Ach at the diaphragm has been termed the "safety factor" because the probability of release is so high that it guarantees the postsynaptic response necessary for respiration.

The situation in the central nervous system (CNS) is not as clear-cut. Action potential arrival at the terminals, with the subsequent increase in intracellular calcium, does *not* guarantee synaptic release of neurotransmitter. This concept is used routinely in experimentation investigating mechanisms of central neurotransmission. With minimal electrical stimulation, synaptic release of transmitter will occur at a particular rate, referred to as the failure rate—the frequency at which neurotransmission fails in the presence of minimal stimulation measured over a set number of trials. This is important because an altered failure rate may suggest whether the mechanism of action of certain drug or experimental condition is pre or postsynaptic. For example, a *decrease* in the failure rate indicates that the probability of neurotransmission has increased, and this may point to a compound promoting neurotransmission via a presynaptic mechanism of action.

NEUROTRANSMITTERS

Identification of Neurotransmitters

Certain criteria have been established in order to classify a substance as a neurotransmitter:

1. The putative transmitter must be synthesized in and localized to specific presynaptic neurons.
   a. Methods for directly determining the presence of transmitter within a presynaptic cell are lacking. Complicated by the fact that certain known transmitters are involved in cellular metabolism (eg, glutamate), many are small molecules that cannot be stained using traditional immunohistochemical techniques. However, one correlate that is often used in place of direct detection is immunohistochemical staining for the enzymes involved in the synthesis of a putative transmitter. For example, glutamic acid decarboxylase converts glutamic acid to γ-aminobutryic acid (GABA). Typically, the detection of this enzyme is used to determine whether or not a particular neuron is "GABAergic." Alternatively, immunostaining for proteins involved in packaging or storage of a putative transmitter into synaptic vesicles is routinely used to identify neurons containing specific transmitters. Staining for the vesicular glutamate transporter 1 (vGlut1) selectively identifies neurons that are packaging glutamate for vesicular exocytosis into the synaptic cleft.
2. Release of the putative transmitter from the presynaptic neurons.
   b. Satisfying this criterion is more straightforward in the peripheral nervous system (PNS), where large nerves are more accessible for stimulation and collection of compounds that may later be identified as a neurotransmitter. In the CNS, it is extremely difficult to collect substances released from specific presynaptic sites in a relatively noninvasive fashion. Despite this caveat, there is compelling evidence that certain neurotransmitters fulfill this criterion in the CNS.
3. Exogenous application of the putative transmitter on a receptive cell must elicit a response that mimics the synaptic response observed following presynaptic stimulation.
   c. Rapid delivery of ACh onto a skeletal muscle cell that has been denervated results in a motor end-plate potential (EPP) that resembles the response observed following the synaptic release of ACh from an α-motor neuron. Compared to the PNS, the intricate synaptic architecture of the CNS

makes it difficult to acquire evidence to fulfill this criterion. One strategy that is routinely employed is applying a minute amount of a substance onto a receptive cell by either regulated pressure or microiontophoresis. Pressure application, or "puffing," involves filling a micropipette with the putative transmitter and applying pressure pulses to rapidly release the substance onto a receptive cell. Microiontophoresis involves passing a brief pulse of current through a micropipette filled with an ionized solute in order to deliver a charged molecule to the immediate vicinity of a cell. Typically, these methods of delivery are accompanied by electrophysiological analysis of the receptive cell (eg, voltage clamp of the receptive cell to observe current profiles in response to puffing of the putative transmitter).

4. Pharmacological study related to the putative transmitter.
   d. Pharmacological manipulation of cognate receptors to a putative transmitter provides critical evidence in the identification of transmitters. Several pharmacological methodologies are routinely employed: (1) application of a nonendogenous, structural analogs to the putative transmitter, (2) application of a compound that is an agonist or antagonist to the cognate receptor to the putative transmitter, (3) application of a compound to facilitate or suppress activity of the putative transmitter's cognate receptor, and (4) pharmacological manipulation to enhance or suppress release of the putative transmitter.

It is often technically difficult or, in certain cases, impossible to fulfill all of the requirements mentioned in the preceding paragraphs due to the required experimental conditions. ACh has remained one of the classic examples that fulfill all of these conditions. In fact, some accepted neurotransmitters even violate these criteria for transmitter identification. As a specific example, endocannabinoids are released from a postsynaptic cell and exert their effects on a presynaptic cell. This so-called "retrograde transmission" was demonstrated in the early 1990s through a process known as depolarization-induced suppression of inhibition. Following the depolarization of a postsynaptic cell (the first example of which was the cerebellar Purkinje cell), a substance was released from the postsynaptic cell that resulted in a reduction of the release of GABA onto the Purkinje cell. Later, this was found to be due to cannabinoid receptor 1 (CB1) activation on presynaptic, GABAergic neurons. Although the mechanism of retrograde transmission has been well established for the endocannabinoid system, it is unclear whether other neurotransmitter systems utilize this mechanism of transmitter release.

The chemical structure of the different neurotransmitters has been used to classify them. Some of the examples of the different classes of neurotransmitter include:

- Ach.
- Amino acid neurotransmitters: γ-aminobutyric acid, aspartate, glutamate, and glycine.
- Monoamines: dopamine, epinephrine, norepinephrine, serotonin, and histamine.
- Peptides: oxytocin, dynorphin, and substance P.
- Other: nitrous oxide and anandamide.

Generally speaking, almost all neurotransmitters exert pre as well as postsynaptic actions through fast or slow mechanisms; by activating ionotropic receptors or metabotropic receptors, respectively. The specific details regarding how a neurotransmitter alters activity of the pre or postsynaptic neuron is dependent on the activity of the cell, the history of activity, the location of the synapse, the structure of the cognate receptor, and duration of its action at that receptor. We discuss some of the major neurotransmitters within the nervous system. There exist many other important neurotransmitters, and many of them have been covered in dedicated textbooks or monographs. The neurotransmitters highlighted in this chapter are selected to highlight some of the important, basic mechanisms of neurotransmitter action. It is important to note that, despite the multitude of references to the contrary, no neurotransmitter is inherently "excitatory" or "inhibitory." Indeed, it is the receptors for a particular neurotransmitter that dictate if neurotransmitter binding, and subsequent receptor activation would produce a depolarization or hyperpolarization.

ACETYLCHOLINE

ACh is one of the principal neurotransmitters of the nervous system. First discovered to be the transmitter responsible for producing cardiac muscle contraction, it is now widely accepted as one of the primary transmitters in the CNS. Beyond the original description, ACh actions were understood to occur though two modes of action, depending on the class of receptors activated: the muscarinic acetylcholine receptors (mAChRs) or the nicotinic acetylcholine receptors (nAChRs). ACh is synthesized from acetyl CoA and choline and its synthesis is catalyzed by choline acetyltransferase (ChAT). Due to its presence in neurons that release ACh, ChAT is often used for the immunohistochemical identification of cholinergic neurons.

mAChRs and Subtypes

Severe poisoning due to consumption of the fly agaric mushroom or *Amanita muscaria* results in excess saliva production, excessive tears, urination, defecation, gastrointestinal motility, vomiting, myoclonus, miosis, and seizures. These toxicological signs provided clues for localizing the mAChRs within the organism and provided insight into their function. For example, the sphincter pupillae, the circular smooth muscle fibers responsible for the constriction of the pupil, contain mAChRs and, following their hyperstimulation by ACh, produce excessively small pupils resulting in vision impairment. The alkaloid compound muscarine is now known to be responsible for this effect due to its action as a selective agonist of mAChRs.

mAChRs are class A or rhodopsin-like g protein-coupled receptors (GPCRs) and are monomeric proteins comprised of seven transmembrane domains (Fig. 3.8). Molecular cloning has identified five different mAChR subtypes. mAChRs contain a single orthosteric binding site for ACh and numerous allosteric sites on the transmembrane domains. Although ACh was the first transmitter to be discovered, and the existence of the mAChR subtypes has been known for decades, progress in understanding ACh neurotransmission has been severely hindered. In the case of mAChRs, the orthosteric binding site for ACh has a very high (>90%) sequence homology. As such, development of pharmacological tools specific to the different subtypes has been relatively slow.

M_1-, M_3-, and M_5-subtypes mediate the excitatory actions of mAChRs and are coupled to the $G_{q/11}$ heterotrimer. $G_{q/11}$ signaling activates phospholipase C (PLC), which liberates inositol phosphoinositides from the interior of the plasma membrane to induce the release of calcium from intracellular stores. The M_2- and M_4-subtypes are coupled to the $G_{i/o}$ heterotrimers and their activation results in the inhibition of adenylyl cyclase, decreasing the production of cyclic adenosine monophosphate (cAMP) (Fig. 3.9). mAChR activation

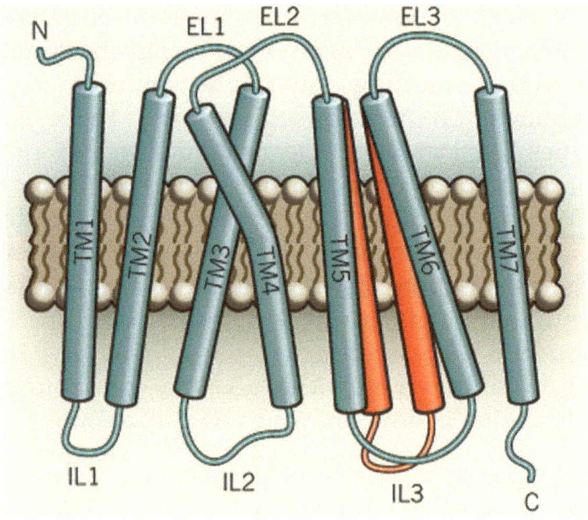

FIGURE 3.8 Schematic of the muscarinic acetylcholine receptor. G-protein-coupled receptors have seven transmembrane (TM) domains that span the cell membrane, giving rise to three intracellular loops (IL1–1L3) and three extracellular loops (EL1–EL3). The intracellular ends of TM5 and TM6 are farther apart in M_2 (blue) and other G_i-coupled receptors than in M_3 (red) and other G_q-coupled receptors. *Reproduced with permission from R.L. Kow, N.M. Nathanson. Structural biology: Muscarinic receptors become crystal clear. Nature 2012;482:480-81.*

can also indirectly modulate ion channel function through their second messengers. However, the specific nature of how and which ion channels and signaling pathways are altered in response to mAChR activation is tissue and cell-type specific.

M_1 mAChR

The M_1-subtype of mAChRs is the most abundant in the CNS. With its expression in cortex, hippocampus, and striatum, the M_1-subtype has been implicated in coordinating memory consolidation. Agonism of M_1 has been an attractive candidate for numerous clinical applications, including therapeutic management of Alzheimer's disease and schizophrenia. Present on the postsynaptic membrane of both excitatory and inhibitory neurons, stimulation of M_1 mAChRs produces a diverse array of responses. Hyperstimulation of the M_1-subtype is thought to be responsible for the generation of seizures induced by exposure to pilocarpine, a nonselective agonist of mAChRs. Homozygous mice lacking the M_1-subtype do not display seizures following pilocarpine administration, and $M_1{}^{+/-}$ are extremely resistant to seizures induced by pilocarpine. An outstanding question is how stimulation

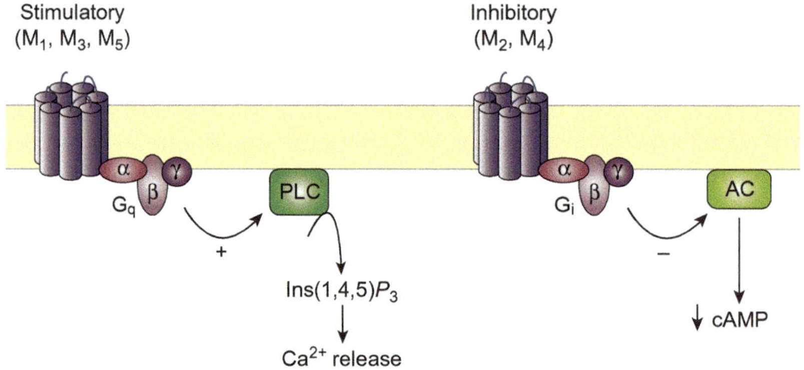

FIGURE 3.9 Five muscarinic acetylcholine receptors mediate the effects of acetylcholine. The M_1, M_3, and M_5 receptors are stimulatory and couple primarily to the mobilization of intracellular Ca^{2+}. An increase in cytoplasmic Ca^{2+} results from the coupling of these receptors with the heterotrimeric G protein G_q, which leads to stimulation of the effector enzyme PLC and the release of inositol-(1,4,5)-trisphosphate [Ins(1,4,5)P3]. The M_2 and M_4 receptors are inhibitory and negatively modulate adenylyl cyclase (AC) to reduce cytoplasmic concentrations of cAMP. *Reproduced with permission from R.M. Eglen, A. Choppin, N. Watson. Therapeutic opportunities from muscarinic receptor research. Trends Pharmacol Sci 2011;22(8):409–14.*

of the M_1-subtype and subsequent signaling via the $G_{q/11}$ heterotrimers leads to a hyperexcitable state.

The M_1 subtype–mediated $G_{q/11}$ signaling modulates ion channel conductance indirectly, to facilitate an excitatory state. In CA3 hippocampal pyramidal neurons, γ-oscillations can be generated with the application of muscarine, and are mediated by the modulation of the following Na^+/K^+ mixed currents: the hyperpolarization-activated current (I_h) and the Ca^{2+}-dependent, nonspecific cation current (I_{cat}). In $M_1^{-/-}$ mice, muscarine does not produce the change in I_h or I_{cat} observed in wild-type animals, and γ-oscillations cannot be elicited, indicating that I_h and I_{cat} are modulated by M_1 activation.

M_2 mAChR

M_2 mAChRs are located pre- and postsynaptically and are coupled to $G_{i/o}$ heterotrimers. They are expressed widely in the CNS and act as autoreceptors to decrease presynaptic ACh release, or as heteroreceptors to regulate the release of other neurotransmitters. Mice lacking the M_2-subtype display a 2.24 higher spontaneous and evoked release of ACh presynaptically that is independent of calcium. Pathophysiology of schizophrenia is thought to occur in part due to excessive dopamine neurotransmission. The M_2 mAChRs are an attractive target for the treatment of some of the symptoms of schizophrenia by virtue of their action as at dopaminergic terminals within the caudate putamen.

One mechanism by which the inhibitory presynaptic mAChRs reduce neurotransmitter release is through the inhibition of voltage-gated calcium channels. The calcium current ($I_{Ca^{2+}}$) in DRG neurons is dose dependently inhibited by muscarine. This effect is completely abolished by pertussis toxin (PTX). Since PTX blocks downstream signaling mediated by $G_{i/o}$ heterotrimers (associated with the M_2- and M_4-subtypes). The N-type calcium channels contribute to the $I_{Ca^{2+}}$ modulated in DRG neurons, since this effect is sensitive to ω-conotoxin, a selective blocker of these calcium channels. M_2 mAChRs in sympathetic neurons specifically modulate calcium conductance through N-type and P/Q-type calcium channels. This favors the M_2 mAChRs over the M_4-subtype as being responsible for suppressing $I_{Ca^{2+}}$ in DRG neurons.

There are also examples of the M_2 receptor subtype acting postsynaptically to mediate the inhibitory actions of ACh. On sympathetic and parasympathetic neurons, ACh-mediated hyperpolarization is due to an increase in potassium conductance that is mediated by the opening of G-protein inwardly rectifying potassium channels (GIRKs).

M_3 mAChR

M_3 mAChRs are located both pre- and postsynaptically, and are coupled to the $G_{q/11}$ heterotrimers. Mice lacking the M_3 mAChR consume fewer calories, have reduced adipose tissue, and weigh less at the start of their third postnatal week of life compared with wild-type littermates. M_3 mAChRs have also been implicated in mediating various forms of learning and memory. Although there is evidence for both the M_1- and M_3-subtype mediating different forms of learning and memory, it is unclear whether or not these subtypes act synergistically. In neurons from lamina II of the dorsal horn of the spinal cord, the application of oxotremorine, a nonselective mAChR agonist, reduced the frequency and amplitude of spontaneous inhibitory postsynaptic currents (sIPSCs) and the frequency of miniature inhibitory postsynaptic currents (mIPSCs) in wild-type and M_1/M_3 double-knockout mice. In M_2/M_4 knockout mice, however, oxotremorine increased the frequency of sIPSCs, mIPSCs and was blocked by the M_3 mAChR antagonist 4-DAMP. This indicates that this effect is specifically mediated by the M_3-subtype but is overwhelmed by the inhibitory actions of M_2 and M_4 mAChR in vivo.

M_4 mAChR

The M_4-subtype is inhibitory, located pre- and postsynaptically, and coupled to the $G_{i/o}$

heterotrimers. Their expression is wide in the CNS but is particularly high within the striatum, where they are involved in regulating the transmission of corticostriatal inputs. Stimulation of M_4 mAChRs by VU0152100, a selective positive allosteric modulator of the M4-subtype, reduces release of glutamate presynaptically and the amplitude of evoked excitatory postsynaptic currents (eEPSCs). The reduction in presynaptic release of glutamate is due to a suppression of $I_{Ca^{2+}}$ at the presynaptic corticostriatal terminals. Similarly, in hippocampus, activation of the M_4-subtype is responsible for a suppression of eEPSCs onto pyramidal cells in the CA1 subfield. M_4 activation can also act to liberate endocannabinoids from the postsynaptic membrane to inhibit the presynaptic cell in a retrograde fashion.

M_5 mAChR

The M_5-subtype has a relatively sparse pattern of expression in the CNS that is largely restricted to the hypothalamus, hippocampus, and midbrain. This mAChR subtype is postsynaptically located, coupled to the $G_{q/11}$ heterotrimers, and are excitatory. Information regarding the synaptic role of the M_5-subtype is extremely limited. In electrochemical recordings of the nucleus accumbens, stimuli applied to the laterodorsal tegmental nucleus, the primary cholinergic input, produces an initial rapid increase in dopamine release followed by a prolonged release of dopamine in wild-type mice. However, in M_5 knockout mice, the prolonged release of dopamine was lost, indicating that the postsynaptic actions of M_5 are required for the continued release of dopamine.

The nAChRs

The nAChRs are ion channels that mediate the rapid actions of ACh and are present both pre- and postsynaptically. So named due to their selective stimulation by nicotine, these ion channels conduct cations nonselectively, including sodium, potassium, and sometimes calcium. With a wide variety of subunit possibilities, the stoichiometry of most nAChR receptors are heteromeric structures with a few cases of homomeric pentamers. At the NMJ, subunit composition of the pentamer is age dependent. At the embryonic NMJ, nAChRs are composed of two α_1-, one β-, one δ-, and one γ-subunit. In the adult, the γ-subunit is replaced by the ε-subunit (Fig. 3.10).

In neuronal nAChRs, the subunit composition is more extensive. For the α-subunit, 10 different subunits have been identified (α_{1-10}) and four have been identified for the β-subunit (β_{1-4}). Pentamers can be formed from homomers of α_7-, α_8-, or α_9-subunits. Two ACh molecules bind the receptor and this typically occurs at the interface between α- and β-subunits. This can also occur between two α-subunits and explains the presence of functional homomeric nAChRs. Possible combinations for the heteromeric pentamers are potentially in the thousands, though based on available evidence this is likely not to be the case. In the mammalian nervous system, nine α-subunits (α_{2-10}) and three β-subunits (β_{2-4}) have been identified. Homomeric assemblies of neuronal nAChRs occur as pentamers of α-subunits (eg, five α_7-subunits). However, the heteromeric stoichiometry has been found to occur in two forms: (1) three α- and two β-subunits and (2) two α-, two β-, and an additional accessory subunit which could be either an α- or β-subunit (Fig. 3.11). nAChRs are capable of conducting sodium, potassium, and calcium, but the calcium conductance is dependent on subunit stoichiometry. The subunit composition present on skeletal muscle confers the lowest calcium to sodium conductance ratio, whereas the greatest is observed in the neuronal nAChR subunit arrangement. In particular, the α_7 homomeric nAChR has the highest calcium to sodium conductance ratio. Compared with the mAChRs, nAChRs can rapidly desensitize and be removed from the neuronal membrane.

Presynaptic Effects of nAChRs Activation

nAChRs present on the presynaptic compartment regulate the release of ACh or other

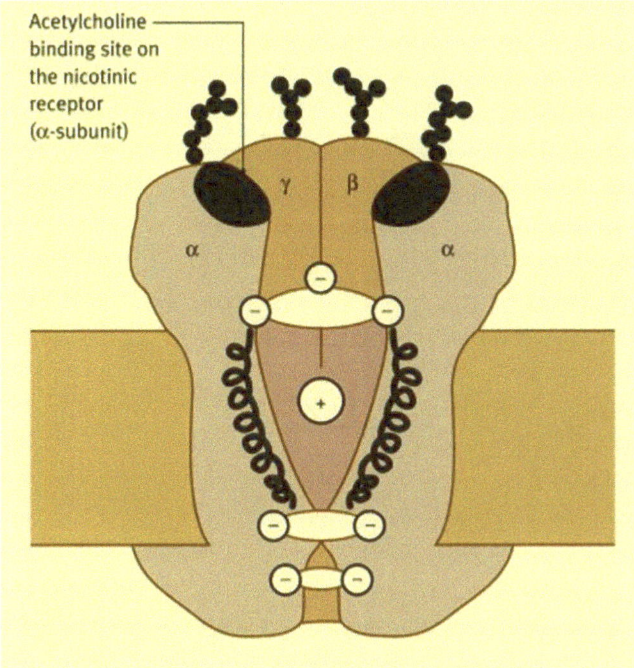

FIGURE 3.10 Section through a muscle nicotinic receptor. A muscle nicotinic receptor embedded in the cell membrane. Two α-subunits are shown, which carry the two binding sites for acetylcholine on the extracellular side of the receptor. Also shown are sugar residues linked to the extracellular terminals of the protein molecules, together with a β- and a γ-subunit (the latter is present in embryonic development but is replaced by an ε-subunit in the adult). The central pore of the receptor is lined by negatively charged amino acid residues to repel anions and allow cations (+) to pass through. The "gate," which opens when the receptor is activated by binding of acetylcholine, is situated deeper in the channel towards the intracellular opening of the receptor. *Reproduced with permission from A. Fletcher. Neuromuscular function and transmission. Anaesthesia & Intensive Care Medicine 2011;12(6):245-8.*

transmitters acting as auto- or heteroreceptors, respectively. Though limited evidence exists suggesting the presence of functional nAChRs at somatodendritic sites in the CNS, the prevailing view indicates the physiological role of nAChRs is to modulate transmitter release at presynaptic terminals. In the striatum, stimulation of presynaptic nAChRs with nicotine or other agonists facilitates the release of dopamine and can be partially blocked with α-conotoxin MII, a selective $\alpha_3\beta_2$ nAChR antagonist, indicating a role for subtype in facilitating dopamine release. Part of the release of dopamine that is facilitated by nicotine is not blocked by α-conotoxin MII. This release mechanism is insensitive to the $\alpha_3\beta_2$-subtype blocker and is likely mediated by the $\alpha_4\beta_2$-subtype, since it is one of the most common subtypes within striatum. Stimulation of the cortical-amygdalar afferents to the basolateral amygdala (BLA) produces large postsynaptic current that is mediated by glutamate receptors. Glutamate release at these synapses is modulated presynaptically by nAChRs, since application with nicotine facilitates an increase in the frequency of action potential-independent EPSCs and decreases the failure rate of glutamatergic transmission. This facilitation is not observed in mice lacking the α_7-subunit and demonstrates that the homomeric α_7 nAChRs are responsible for modulating glutamatergic transmission in the BLA.

Mechanistically, nAChRs modulate transmitter release through direct and indirect mechanisms. Many of the subtypes of nAChRs expressed presynaptically display high permeability to calcium (eg, the α_7-subtype) and enhance the calcium-dependent transmitter release upon their activation. Indirectly, nAChRs activation may produce a local depolarization that facilitates the opening of voltage-dependent calcium channels and produces an increase in calcium-dependent transmitter release.

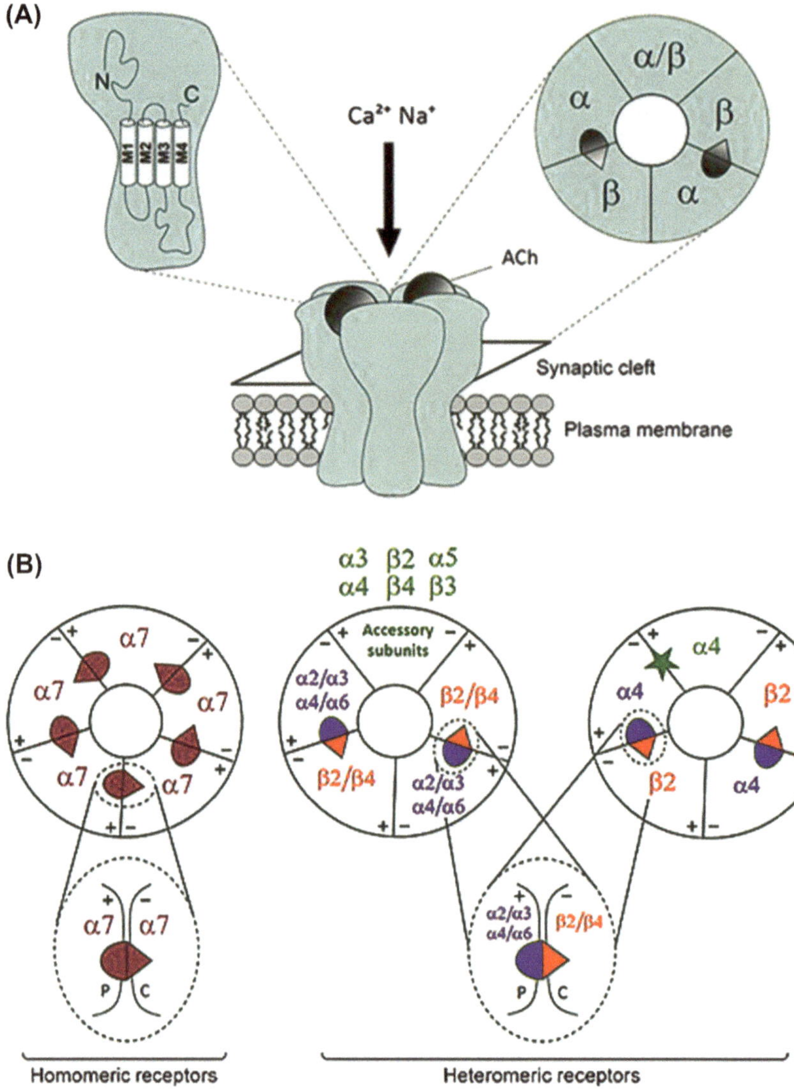

FIGURE 3.11 Basic structure of neuronal nicotinic receptors. (A) (Left) Diagram showing the topography of nAChR subunits. The extracellular amino terminal portion is followed by three hydrophobic transmembrane domains (M1–M3), a large intracellular loop, and then a fourth hydrophobic transmembrane domain (M4). (Middle) The arrangement of nAChR subunits in an assembled receptor. (Right) Localization and organization of the ACh binding sites in a heteromeric receptor. (B) Structure of homopentameric and heteropentameric neuronal subtypes. The pentameric arrangement of nAChR subunits in an α7 homopentameric subtype (left), heteromeric receptor subtype (middle), and the $(\alpha4)_3(\beta2)_2$-subtype (right). The localizations of the subunit interfaces of the orthosteric binding sites are indicated, together with the primary component P(+) carried by the α-subunits and the complementary component C(−)carried by an α- or nonα-subunit. In addition to the two orthosteric sites, the $(\alpha4)_3(\beta2)_2$-subtype has a binding site at the α4α4-interface (star). *Reproduced with permission from Zoli, M., Pistillo, F., Gotti, C. Diversity of native nicotinic receptor subtypes in mammalian brain. Neuropharm 2015;96(B):302-11.*

Postsynaptic Effects of nAChR Activation at the NMJ

At the NMJ, action potential arrival at the terminal results in the opening of voltage-gated calcium channels and a subsequent increase in intracellular calcium. The increase in intracellular calcium facilitates fusion of vesicles containing ACh to the plasma membrane resulting in the release of ACh into the synaptic cleft. Situated opposite the synaptic compartment to the sites of vesicular fusion on the plasma membrane of the presynaptic terminal, skeletal muscle cells contain specialized depressions on their plasma membrane to receive ACh upon release, named the junctional zone (Fig. 3.12). In the junctional zone, a high concentration of nAChRs is present. In the absence of an action potential in the presynaptic motor neuron, calcium-independent fusion of synaptic vesicles occurs infrequently. Each vesicle contains approximately the same amount of ACh and the resulting response is a single quantum. ACh diffuses across the synaptic compartment and binds the nAChRs within the junctional zone, producing a depolarization on the skeletal muscle cell referred to as the miniature end-plate potential (mEPP); that is, the minimal response elicited by one quantum (Fig. 3.13). The concentration of synaptic ACh is dramatically increased following an action potential which produces a much larger response, the EPP that is proportional to the quanta released. The EPP is mediated by the dramatic increase in sodium conductance through nAChRs into the skeletal muscle cell and indirectly facilitates the opening of nearby voltage-gated sodium channels due to local change in voltage at the motor-end plate. If the depolarization is of sufficient magnitude and duration, then the EPP spreads across the postsynaptic membrane of the muscle cell and results in the liberation of calcium ions from the sarcoplasmic reticulum and induces contraction.

GLUTAMATE

The amino acid neurotransmitter, glutamate, is the primary mediator of excitatory chemical transmission within the mammalian CNS and has been the subject of intense study. Due to its involvement in many metabolic processes, it has been historically challenging to identify glutamatergic neurons. Despite circulating glutamate in the blood, neurons obtain glutamate through two separate synthetic processes. Glial cells transport glutamate out of the synapse via glutamate transporters and convert it to glutamine by glutamine synthetase. Glutamine is subsequently released by glial cells and is taken up by local neurons. Glutamine is then catalyzed into glutamate via the action of glutaminase. The second synthetic pathway in neurons involves the transamination of α-ketoglutarate, an intermediate component of the Krebs cycle, into glutamate. α-ketoglutarate can also be obtained from local glial cells and actively transported in order to synthesize more glutamate.

Initially, the ionotropic glutamate receptors were subdivided into the *N*-methyl-D-aspartate (NMDA) and the nonNMDA receptors (Fig. 3.14). This classification was made based on the ability of synthetic NMDA to selectively bind a subclass of glutamate receptors with high affinity. Subsequently, the nonNMDA receptors were further divided into the α-amino-3-hydroxy-5-methyl-4-isoxazolepropionic acid (AMPA) and the kainate receptors, also named based on their selective stimulation by synthetic AMPA and kainate, respectively. The NMDA, AMPA, and kainate receptors mediate the fast responses of glutamate—they are ligand-gated ion channels permeable to sodium, potassium, and calcium to varying degrees, depending on the subunit composition. Similar to ACh, there are also a subset of glutamate receptors that mediate the slow actions of this transmitter, the metabotropic glutamate receptors (mGluRs). Eight subtypes have been identified (mGluR1–mGluR8), all of which are

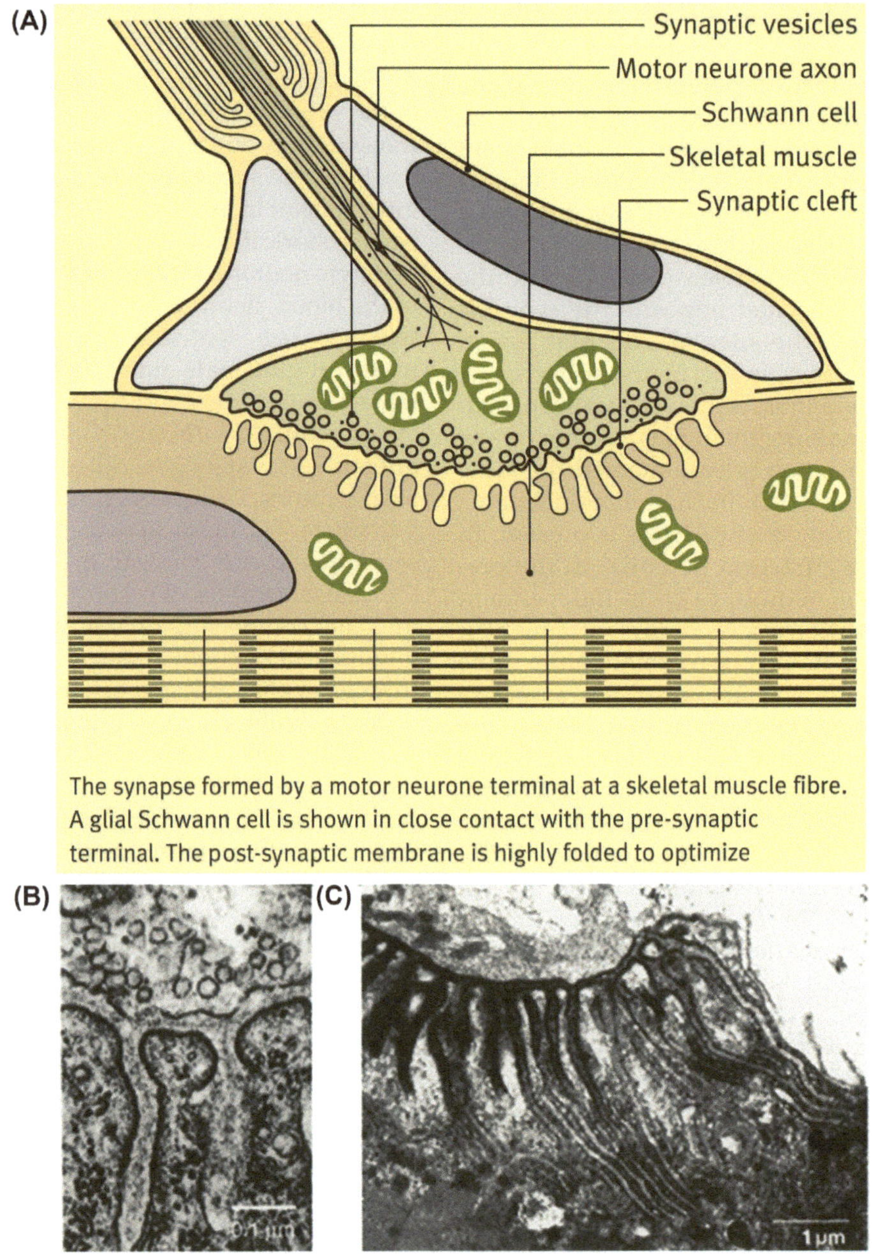

FIGURE 3.12 Schematic and electron micrographs of the NMJ. (A) Sagittal section through an NMJ. The synapse formed by a motor neuron terminal at a skeletal muscle fiber. A glial Schwann cell is shown in close contact with the presynaptic terminal. The postsynaptic membrane is highly folded to optimize nicotinic receptor activation, acetylcholine hydrolysis, and voltage-gated sodium channel activation. Electron micrographs of postsynaptic folds at NMJs in rat (B) and man (C). The membrane at the crests of the folds is thickened, probably reflecting the high density of AChRs there. In humans, the folding is often particularly extensive. In the example shown, periodic increases in density are seen in the interfold cytoplasm, suggesting attachment between the membranes of adjacent folds. A compact extracellular basal lamina extends into the folds. *Reproduced with permission (A) from A. Fletcher. Neuromuscular function and transmission. Anaesthesia & Intensive Care Medicine 2011;12(6):245-8; (B,C) from S.J. Wood, C.R. Slaterb. Safety factor at the neuromuscular junction. Progress in Neurobiology 2001;64(4):393-429.*

FIGURE 3.13 Miniature or spontaneous EPPs resulting from the release of a single quanta, in this case a synaptic vesicle containing acetylcholine.

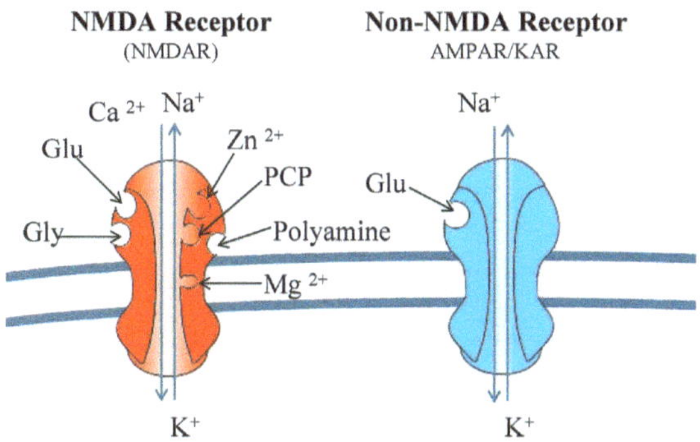

FIGURE 3.14 Glutamate receptors: structure and function. NMDARs bind glutamate, glycine, Mg^{2+}, Zn^{2+}, and polyamines. NMDARs form channels that are more permeable to Ca^{2+} than Na^+ and K^+. Kainate and AMPA receptors interact only with glutamate and their specific agonists, and their associated channels are more permeable to Na^+ and K^+ than Ca^{2+}. *Reproduced with permission from A.A. Kritis, E.G. Stamoula, K.A. Paniskaki, T.D. Vavilis. Researching glutamate – induced cytotoxicity in different cell lines: a comparative/collective analysis/study. Front Cell Neurosci 2015;* http://dx.doi.org/10.3389/fncel.2015.00091.

GPCRs and have been further classified into group I, group II, and group III mGluRs based on gene homology and second messenger signal transduction systems. The mGluRs have diverse synaptic effects and will not be covered here.

The AMPA Receptor

In the mammalian CNS, the AMPA receptor (AMPAR) is the primary mediator of excitatory synaptic transmission. Given its critical function, it is not surprising that mutations in genes that encode AMPAR subunit proteins result in profound deficits, including seizures and intellectual disability. The subunit make-up of the AMPAR include GluA1, GluA2, GluA3, and GluA4; in human, they are encoded by the genes *GRIA1*, *GRIA2*, *GRIA3*, and *GRIA4*, respectively. Alternative splicing produces two variants for each *GRIA* gene, referred to as the "flip" or "flop" forms, resulting in eight possible subunits for receptor assembly. Over the years, numerous studies have investigated the properties of recombinant AMPARs in in vitro expression systems, such as HEK cells or *Xenopus* oocytes. Wide subunit–dependent variations have been observed, underscored, and compounded by differences in auxiliary AMPAR proteins that are important for trafficking, membrane positioning, and anchoring of the AMPAR subunits.

Native AMPAR isoforms are thought to be primarily heteromeric tetramers, although pentameric AMPARs have been identified. Homomeric assemblies are thought to be far less common, as studies in vitro have reported that they display relatively low single-channel conductances compared to those of heteromeric assemblies. In addition, heteromeric AMPARs produce large inward currents mediated primarily by sodium with a smaller contribution from potassium. In some cases, certain AMPAR isoforms are permeable to calcium, determined entirely by the presence of an edited form of GluA2. Without GluA2, or with an unedited form of GluA2 [GluA2(Q)], AMPARs are incapable of conducting calcium. However, when the mRNA of the GluA2 sequence is edited by changing the region encoding a glutamine residue to one that encodes arginine, the resultant recombinant AMPAR assembly containing the edited GluA2(R) subunit diminishes calcium permeability. "Flip/flop" editing occurs in each gene in the area preceding the encoding of the M4 transmembrane domain of the AMPAR subunits. Although the differential roles of these isoforms are unclear, flip isoforms are capable of conducting more current upon channel opening compared to the flop isoform. These isoforms are dynamically regulated, and the ratio of their expression is dependent on a variety of factors including the history of activity, structure studied, disease state, and the age of the animal.

Presynaptic Effects of AMPARs

Reports on the presence of AMPARs are limited and have only been identified in two cases: the presynaptic terminals at the calyx of Held, and the GABAergic terminals innervating Purkinje cells in the cerebellum. In both cases, selective activation of AMPARs on presynaptic terminals reduces calcium entry and transmitter release. Remarkably, the reduction in voltage-gated calcium currents occurs indirectly through G proteins following AMPAR activation. In the cerebellum, presynaptic inhibition is to be due to the activation of $G_{i/o}$ signaling that suppresses P/Q-type voltage-gated calcium channels. The transmembrane AMPAR regulatory protein, stargazin, may be important in stimulating $G_{i/o}$ signaling following AMPAR activation, but the complete complement of proteins involved have yet to be identified.

Postsynaptic Effects of AMPARs

The presence of AMPARs at the postsynaptic site is typically required for fast excitatory transmission in the adult CNS. Subunit composition dictates the amplitude and kinetics of AMPAR-mediated currents. As stated previously, AMPARs containing the flip isoforms conduct more current. Selective activation of AMPARs with glutamate or AMPA produces evoked currents that are large amplitude, sharp, and rapidly desensitizing (Fig. 3.15). This class of glutamate receptors is also particularly important in a number of forms of synaptic plasticity. Depending on the subunit of the AMPAR, various kinases and phosphatases can phosphorylate or dephosphorylate the receptor and alter its conductance properties. The GluA1 subunit has been extensively studied due to its known role in increasing the postsynaptic response following the induction of long-term potentiation (LTP). Protein kinase C (PKC) and the calcium/calmodulin-dependent protein kinase II (CaMKII) both phosphorylate the GluR1 subunit at multiple sites to alter the conductance properties of GluR1 subunit–containing AMPARs.

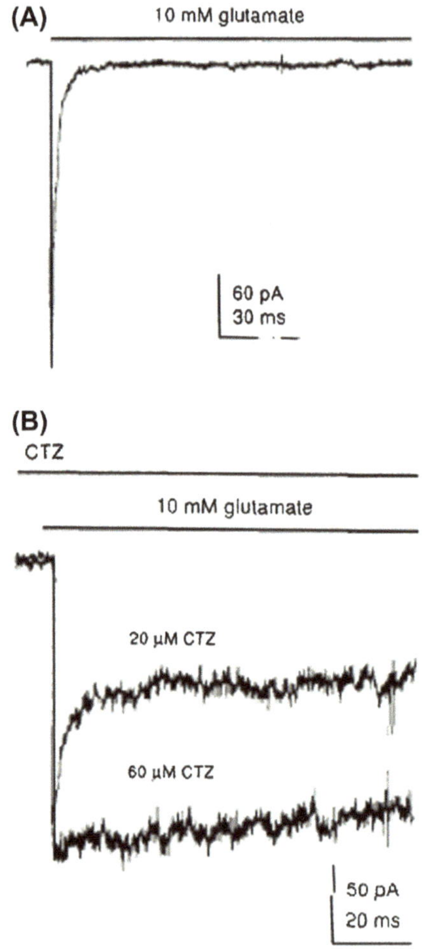

FIGURE 3.15 The effect of cyclothiazide on glutamate responses in outside-out membrane patches. Cyclothiazide (CTZ) blocks desensitization of glutamate receptors. (A) Top trace shows that 10mM glutamate activates an inward current that desensitizes almost completely. The duration of glutamate application is shown by the horizontal bar. (B) Lower traces show the glutamate-evoked responses from a different patch in 20- and 60-μM CTZ. *Reproduced with permission from L.O. Trussell, S. Zhang, I.M. Raman. Desensitization of AMPA receptors upon multiquantal neurotransmitter release. Neuron 1993;10(6):1185-96.*

AMPARs in Development

In embryonic and early postnatal rodent development, the flip isoforms of the AMPAR subunits predominate. The flop isoforms begin to be expressed during the early stages of postnatal development, and the ratios between the two isoforms change with age due to the relative increase in flop isoforms expression. In contrast to the adult CNS, AMPARs are not the primary receptors that mediate fast excitatory transmission in embryonic and early postnatal development. The presence of AMPARs early in development, especially when synaptogenesis is incomplete, suggests an alternative role. It is well known that glutamate plays key roles in differentiation, establishment of circuit architecture, and acts as a neurotrophic factor in development, most likely due to the expression of AMPAR isoforms that conduct calcium to mediate calcium-dependent signaling pathways in development. Consistent with this, GluA1 expression can be detected as early as embryonic day 15.5, GluA1 and GluA3 subunits are expressed in far greater quantities in early postnatal development compared to the GluA2 subtype, indicating that calcium conductance mediated by AMPAR activation is greater in early development.

The Kainate Receptor

Named for their ability to be stimulated selectively by kainic acid (kainate), the kainate receptor constitutes the other class of nonNMDA glutamate receptors. Overstimulation of kainate receptors produces strong, long-lasting seizures and are implicated in a wide variety of other psychiatric and neurological disorders. Seizures induced by organophosphate poisoning, for example, irreversibly inhibit acetylcholinesterase, leading to excessive stimulation of ACh receptors and inducing seizure activity. These long-lasting, drug-resistant seizures are eventually mediated by glutamate receptors including kainate receptors, specifically the GluK1 subunit–containing kainate receptors that potentiate and sustain ongoing seizure activity.

Kainate receptors can function as homomeric or heteromeric tetramers. In human, the subunits GluK1, GluK2, GluK3, GluK4, and GluK5 are encoded by *GRIK1, GRIK2, GRIK3, GRIK4*, and *GRIK5*, respectively. GluK1-3 can be assembled as homomers but bind kainate with low affinity. The GluK4 and GluK5 subunits must be assembled as heteromers with the GluK1–3 subunits, but have a much higher affinity for kainate. Similar to AMPARs, kainate receptor activation elicits inward, rapidly desensitizing currents primarily mediated by sodium and, depending on the subunit composition, may also conduct calcium. Similar to AMPARs, alternative splicing of the messenger RNA dictates the current magnitude and the current–voltage relationship depending on which final transcript is produced for a particular kainate receptor subunit. Amino acid substitutions within sequences encoding transmembrane domains lining the ion channel pore for both the GluK1 and GluK2 subunits alter calcium conductance, producing a wide range of calcium conductance values for the different homomeric or heteromeric GluK1 or GluK2 subunit–containing kainate receptors.

Presynaptic Effects of Kainate Receptors

Activation of kainate receptors on presynaptic terminals results in the facilitation or suppression of neurotransmitter release of a wide variety of transmitter systems. Infusion of 5- or 20-μM kainate into the prefrontal cortex increases dopamine concentrations to about 150% or 500% above baseline levels. However, infusion of either 20- or 100-μM AMPA produces an increase in dopamine up to about 250% or 1500% above baseline, respectively. This effect is blocked by CNQX, an antagonist to both AMPA and kainate receptors, demonstrating that the facilitation of dopamine release is mediated by both AMPA and kainate receptors.

Interneurons and pyramidal neurons within the BLA express kainate receptors that are localized to both somatodendritic and axonal sites. In particular, the GluK1 subunit expression is higher relative to other subtypes within this brain region. Application of ATPA, a selective agonist of GluK1 subunit–containing kainate receptors, to BLA pyramidal neurons differentially alters the failure rate of evoked inhibitory postsynaptic currents (eIPSCs) in a dose-dependent fashion. The 300-nM APTA decreases the eIPSC failure rate, whereas both 1-and 10-μM ATPA increases the eIPSC failure rate. This indicates that synaptic release of GABA is facilitated by the low-level activation of presynaptic kainate receptors on interneurons but is suppressed at higher levels of activation.

The presynaptic role of kainate receptors on glutamatergic neurons has also been studied. Evoked excitatory postsynaptic currents (eEPSCs) recorded from pyramidal cells are dose-dependently blocked by the application of UBP302, a selective antagonist to GluK1 subunit–containing kainate receptors. Application of ATPA decreases the failure rate of eEPSCs and increases the frequency of miniature excitatory postsynaptic currents (mEPSCs), demonstrating that kainate receptors are also acting presynaptically in the BLA to facilitate glutamate release. Taken together, these studies provide an explanation of how kainate receptor stimulation can produce seizures from the BLA, a known seizurogenic structure. As higher concentrations of agonist increase the failure rate of eIPSCs and decrease the failure rate of eEPSCs, this may suggest a shift in the BLA network's ratio of excitation to inhibition when kainate receptors are overstimulated.

Postsynaptic Effects of Kainate Receptors

Activation of somatodendritic kainate receptors mediates, in part, the action of the inward current resulting from glutamate stimulation of nonNMDA receptors at numerous synapses. Expression of kainate receptors is particularly high in the cornu ammonis 3 (CA3) subfield of the hippocampus, where they induce EPSCs that are resistant to the AMPAR antagonism. Though it appears to be less common than the fast excitation mediated by AMPAR activation, activation of kainate receptors can also lead to the triggering of action potentials.

Kainate Receptors in Development

With the exception of GluK4, all kainate receptor subunits are expressed as early as E12 in rodent. During development, GluK4 expression is generally restricted to the hippocampus and cortex but expression in cerebellar Purkinje cells is observed beginning at postnatal day 5 (P5). Widespread expression is observed for many of the other subtypes, and their relative expression levels stabilize beyond early postnatal development. Presynaptic kainate receptors mediate a form of short-term plasticity at CA3 to CA1 synapses acting to suppress glutamate release. However, this function of presynaptic kainate receptors in early postnatal development is lost either through the induction of LTP or by P14. This is likely due to the presence of the kainate receptor splice variant GluK1c that is present in early postnatal development. At pyramidal cells, the role of this GluK1c subtype is to tonically suppress glutamate release and at interneurons, GluK1c activation facilitates glutamate release by increasing network inhibition through dual roles: (1) suppressing excitation of glutamatergic neurons and (2) exciting GABAergic interneurons within CA3. This developmental switch has been studied extensively at the CA3–CA1 synapse, since this is one of the most important physiological roles of kainate receptors during development.

The NMDA Receptor

NMDA receptors possess unique properties and play a critical role in synaptic plasticity. In addition to binding glutamate, the NMDA receptor requires glycine as an endogenous co-agonist for its activation and maximizing its conductance. Their dysfunction has been associated with numerous pathologies including Alzheimer's and schizophrenia. Functional NMDA receptors are uniformly heteromeric, assembled from seven identified subunits: NR1, NR2A–D, and NR3A–B. Each subunit is alternatively spliced, making the number of potential NMDA receptor subunit combinations very large. As a result of this, the functional properties of the assembled NMDA receptors can vary widely. NMDA receptors typically form tetrameric structures with NR1 dimerizing with either another NR1 subunit or with a combination of NR2 and NR3 subunits.

It is important to highlight several features of the NMDA receptor in order to appreciate its physiological activation. NMDA receptors function as physiological coincidence detectors. Glutamate binding alone is insufficient to open the NMDA receptor on a particular neuron. The neuron must also be depolarized sufficiently to relieve a magnesium blockade of the ion channel pore. The presence of magnesium during physiological conditions confers a characteristic sharp outward rectification to the NMDA receptors, characterized by a reduced conductance at more negative membrane voltages. Interestingly, despite the fact that both are divalent cations, NMDA receptors conduct calcium but not magnesium. A combination of residues on the NR1 and NR2 subunits is responsible for conferring the calcium permeability and the magnesium blockade. Specifically, substituting the 598th arginine residue within the NR1 subunit with asparagine completely eliminates the conductance of calcium, while replacing the 595th asparagine residue of the NR2-A subunit with glutamine facilitates the conductance of magnesium ions and eliminates the blockade. The NMDA receptors can also conduct sodium and potassium upon activation.

Presynaptic Effects of NMDA Receptor Activation

The presence of NDMA receptors in the presynaptic terminal is controversial. Histological studies provided evidence for the presence of the NR1 and NR2 subunit on the presynaptic compartment of neurons within the spinal cord and at the CA3–CA1 synapse within the hippocampus. However, numerous contradictory physiological studies attempting to identify and categorize the presence and functional role of the presynaptic have not produced a consensus. The development of "caged" analogs to NMDA and MK-801, selective antagonists to NMDA receptors, have facilitated the identification of presynaptic NMDA receptors. Synapses between layer 4 and layer 2/3 within mouse barrel cortex display various forms of synaptic plasticity that are dependent on spike-timing; spike timing-dependent plasticity (STDP). LTP or LTD can be induced depending on when the presynaptic neuron and postsynaptic neuron fires. Induction of either STDP process in cortex is dependent on NMDA receptor activation but has been assumed to require postsynaptic NMDA receptors. However, photo-uncaging of MK-801 at axonal compartments blocks the LTD form of STDP indicating that this process requires presynaptic NMDA receptors. Overall, presynaptic NMDA receptors appear to play an important role acting as autoreceptors in regulating the release of glutamate.

Postsynaptic Effects of NMDA Receptor Activation

Compared to the nonNMDA glutamate receptors, NMDA receptors have a relatively high single-channel conductance but slower kinetics. NMDA receptors also have a higher affinity for glutamate. The increase in the conductance of calcium initiates numerous signaling cascades that, in turn, are critical for initiating various forms of synaptic plasticity. Due to its role in mediating LTP within the hippocampus, a critically important structure for learning and memory, hippocampal NMDA receptors have been intensively studied.

Upon NMDA receptor activation and concomitant increase in intracellular calcium, calcium binds the calmodulin-dependent kinases (CaM), which phosphorylates NMDA receptors, AMPARs and results in an increase in synaptic strength due to the recruitment of additional AMPARs (often the GluA1 subunit–containing AMPARs) in the postsynaptic membrane. This can be due to the targeting

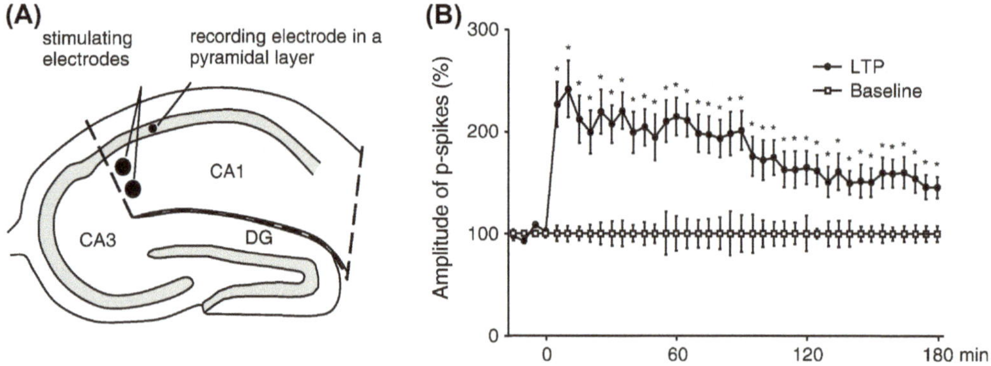

FIGURE 3.16 Long-term potentiation in rat hippocampal CA1 area. (A) Position of electrodes on a hippocampal slice. CA1, CA3, DG—hippocampus subfields. Hatched line indicates the direction of sections made at the end of incubation to separate CA1 for sample preparation. (B) The influence of tetanization on the amplitude of p-spikes in pyramidal layer of CA1. P-spikes were elicited by stimulation of Shaffer's collaterals. Abscissa, the time from the onset of tetanization (*LTP*, long term potentiation). Ordinate, the amplitude of p-spike, first, normalized to the average amplitude of p-spikes in four responses prior to tetanization ($n=5$), and then, to the mean amplitude of responses in the control group (baseline, $n=4$). $^*p<.05$ against baseline, Student's t-test. *Reproduced with permission from V.O. Pustylnyak, P.D. Lisachev, M.B. Shtark, O.I. Epstein. Regulation of S100B gene in rat hippocampal CA1 area during long term potentiation. Brain Research 2011;1394:33-9.*

and subsequent embedding of a pool of AMPARs that are readily available within the cytoplasm or due to the synthesis of new AMPAR molecules. In either scenario, EPSCs elicited from glutamate onto synapses that have undergone LTP are persistently increased in amplitude (Fig. 3.16). NMDA receptor activation can also elicit LTP by altering the single-channel conductance of the AMPARs. For example, NMDA receptor stimulation can induce the phosphorylation of serine residue 831 on the GluA1 subunit, thereby enhancing AMPAR conductance.

NMDA Receptors in Development

In the developing brain, there are "silent synapses." These synapses do not contain functional AMPARs but express NMDA receptors (Fig. 3.17). These synapses are considered functionally silent at rest because without an initial depolarization to relieve the magnesium block, their activation probability is extremely low. Driving the expression of AMPARs appears to depend on LTP induction which is dependent on NMDA receptors but it is unclear what drives the initial depolarization. A likely candidate for this is the presence of $GABA_A$ receptors at silent synapses. During early development, $GABA_A$ receptor activation paradoxically results in depolarization, rather than the hyperpolarization that its activation is often associated within the adult brain. Coordinated activity of the $GABA_A$ receptor and the NMDA receptor can produce calcium influx and trigger the activation of protein kinase A (PKA) resulting in the stabilization of AMPARs into the cell membrane.

Aside from the hippocampus, abundant functionally silent synapses have been identified during early development at thalamocortical synapses in sensory cortex. In barrel cortex, NMDA receptor-dependent LTP can be induced in thalamocortical synapses as late as P9. However, after P9, LTP can no longer be elicited at these synapses. Appropriate timing of unsilencing is dependent on the protein syngap1, which is involved in trafficking glutamate receptors to the synaptic compartment. The ratio of AMPA-to-NMDA-mediated current responses progressively increases during early postnatal development due to the recruitment of AMPARs to the cell membrane following NMDA receptor activation. However, mice that are heterozygotes for syngap1 display a

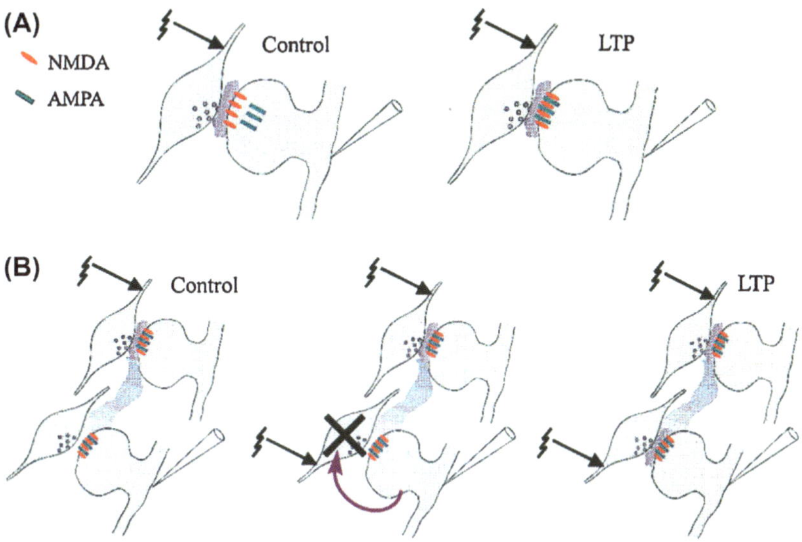

FIGURE 3.17 Models of LTP at silent synapses in the hippocampus. (A) Synapses express functional NMDA receptors but not AMPA receptors (left). Following LTP (right), functional AMPA receptors are expressed. (B) Glutamate can spill over from one synapse to another, resulting in the activation of NMDA receptors but not AMPA receptors, due to the higher affinity of NMDA receptors (left). Before LTP (middle), the synapse is silent because it cannot release neurotransmitter when activated. LTP results from the release of a retrograde messenger that turns on the presynaptic bouton, so that it can now release neurotransmitter (right). *Reproduced with permission from R.C. Malenka, R.A. Nicoll. Silent Synapses Speak Up. Neuron 1997;19(3):473-6.*

premature rise in the AMPA-to-NMDA receptor current ratio, demonstrating that syngap1 normally dictates a truncated development of excitatory synapses.

GABA

In many brain structures throughout the adult mammalian nervous system, GABA is the predominant neurotransmitter that mediates inhibitory synaptic transmission. Dysfunction of the neurotransmitter or its receptors is associated with a large spectrum of disease states, including autism, anxiety disorders, Alzheimer's disease, epilepsy, and schizophrenia. In any given brain region, the tendency is for neurons that release GABA often are present in fewer numbers compared to other cell types. Nonetheless, GABAergic neurons exert strong network influence.

Glutamic acid decarboxylase (GAD) catalyzes the conversion of glutamate into GABA and immunohistochemical staining for GAD is often used to identify GABAergic neurons. GABA binds to and activates three different receptors; the $GABA_A$, $GABA_B$, and $GABA_C$ subtypes, with the "A" and "C" subtypes being ionotropic and the "B" subtype being a metabotropic, G protein-coupled receptor. The $GABA_C$ subtype is primarily expressed within the visual system, notably the retina, and will not be discussed in this chapter.

The $GABA_A$ receptor is assembled into mostly heteromeric pentamers with two orthosteric binding sites, both of which must be occupied for receptor activation. These ionotropic receptors selectively conduct chloride and bicarbonate anions, and differ in their conductance and pharmacology depending on the subunit composition. An overwhelming array of possible subunits has been identified for the $GABA_A$ receptor, including

α_1–α_6, β_1–β_3, γ_1–γ_3, ε, θ, π, ϱ_1–ϱ_3. Assembly in vivo typically occurs through the combination of two α- and two β-subunits, and an additional, modulatory subunit, such as the γ-subunit (Fig. 3.18). Other combinations have been identified, but this composition appears to be the most prevalent. To add to the complexity, γ_2-subunits are produced in two different isoforms as a result of alternative splicing: the γ_2S (short) and γ_2L (long) isoforms with the L protein containing an additional eight amino acids within the cytoplasmic loop. Alternative splicing also regulates the presence of a short or long form of β_2-subunit resulting in the presence or absence of a 38 amino acid sequence. The junction of the α- and β-subunits contains the orthosteric binding sites for GABA, and is the reason for these subunits having been postulated to be the most common functional subunit assemblies found in vivo. Within in vitro expression systems, functional receptors have been produced from a combination of α- and β-subunits alone, but they lack sensitivity to benzodiazepines, a hallmark of most $GABA_A$ receptors.

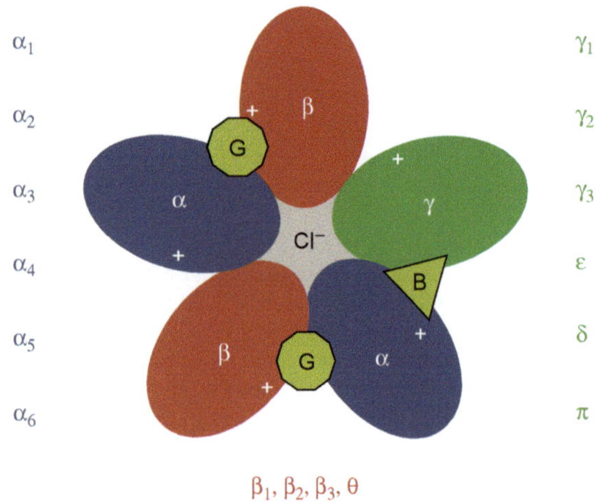

FIGURE 3.18 $GABA_A$ receptor structure. $GABA_A$ receptors belong to the Cys-loop superfamily of receptors which are known to be large protein complexes consisting of five subunits encircling a central ion conducting pore. Each subunit consists of two almost equally large domains: an N-terminal extracellular domain responsible for agonist binding and a C-terminal *trans*-membrane domain responsible for ion-channel pore formation. This figure shows the subunit arrangement of the most common $GABA_A$ receptor in the synaptic cleft. The "+" refers to the "plus" asymmetric side of each subunit, "G" indicates GABA binding sites and "B" indicates the benzodiazepine binding site. The ion-conducting pore of $GABA_A$ receptors selectively conducts chloride. Nineteen different $GABA_A$ receptor subunit isoforms have been cloned, of which 16 are known to assemble in a large number of permutations according to the designated color coding, and three subunits assemble to form a separate subgroup called the rho-receptors. Typically, the majority of $GABA_A$ receptors contain the α-subunits α_1, α_2, α_3, or α_5, together with a β-subunit and a γ_2-subunit in a 2:2:1 stoichiometry as shown. *Reproduced with permission from G. Munro, P.K. Ahring, N.R. Mirza. Developing analgesics by enhancing spinal inhibition after injury: GABAA receptor subtypes as novel targets. Trends Pharmacol Sci 2009;30(9):453-9.*

Postsynaptic Effects of $GABA_A$ Receptors

$GABA_A$ receptors are expressed at postsynaptic and extrasynaptic sites across the entire nervous system and are present on a majority of neurons. To date, no presynaptic $GABA_A$ receptors have been identified. The reversal potential for $GABA_A$ receptors are near rest and, in the adult, conduct outward current upon channel opening at depolarized membrane voltage values due to chloride influx into the cell or bicarbonate efflux out of the cell. This occurs with increasing magnitude at more depolarized membrane potentials and, therefore, the $GABA_A$ receptor displays outward rectification when chloride concentrations are near physiological levels. The current–voltage relationship for these channels is largely dependent on the chloride concentrations outside and inside the cell. For example, when the extracellular and intracellular chloride concentrations are equal, the current–voltage relationship is linear and does not display rectification. However, when extracellular chloride concentration is considerably higher than the intracellular concentration, outward rectification is observed with minimal current observed at membrane potentials near rest.

The reversal potential measured experimentally for $GABA_A$ receptors is typically very close to the reversal potential for chloride itself, indicating that these channels are primarily permeable to chloride with less contribution from bicarbonate. These ion channels are susceptible to desensitization, which can occur with prolonged exposure to an agonist, high concentrations of an agonist, or a combination of both. Upon the synaptic release of GABA, a rapid outward macroscopic current can be measured which has kinetics that are dependent on the synaptic architecture, cell type, receptor subunits, phosphorylation state of the receptor subunits, and the history of activity. In cerebellar granule cells, the predominant assembled $GABA_A$ receptor contain a γ_2-subunit, since their deletion dramatically reduces $GABA_A$ receptors in these cells. This subunit can be phosphorylated by either PKC or CAMIIK, but numerous other kinases modify single-channel $GABA_A$ channel function. Cyclic GMP-dependent protein kinase (PKG) and PKA are also involved in phosphorylating different $GABA_A$ subunits and can produce changes in: the number of receptors inserted into and stabilized in the postsynaptic or extrasynaptic membrane, desensitization, subunit stoichiometry, drug effects, drug sensitivity, and unitary conductance. Heterodimeric pentamers formed from $\alpha\beta$-subunits have the lowest conductance values compared to the heterotrimeric pentamers formed from $\alpha\beta\gamma$-, $\alpha\beta\delta$-, $\alpha\beta\varepsilon$-, and $\alpha\beta\pi$-subunit combinations, which have conductance values that are nearly 2.5-fold greater. Significant diversity also exists for the rates of desensitization and range from hundredths of milliseconds to seconds. Combinations of the δ-subunit with α_1- and β_3-subunits produce assemblies with the lowest desensitization, although the structural basis for $GABA_A$ receptor desensitization has not been determined definitively.

The pharmacology of $GABA_A$ receptors has a wide variety of applications in the research and clinical settings. Numerous drugs act on orthosteric and allosteric sites across the receptor. Analgesics, anesthetics, benzodiazepines, barbiturates, and neurosteroids are all known to modulate the $GABA_A$ receptor allosterically. Barbiturates are some of the first anticonvulsants and act as positive allosteric modulators. Barbiturate binding to the $GABA_A$ receptor facilitates an increase in the mean open time of the activated receptor (Fig. 3.19) resulting in a larger macroscopic current. At high concentrations, these compounds can directly facilitate the activation of the receptor and become increasingly nonselective.

Benzodiazepines are a widely used class of positive allosteric modulators to the $GABA_A$ receptor that do not have the deleterious side effects of barbiturates. Although typical positive allosteric modulators enhance the affinity for the agonist, benzodiazepines do not alter the affinity for GABA binding. Benzodiazepine binding increases the probability of the $GABA_A$ receptor being in the open confirmation (Fig. 3.19). Binding occurs at the junction of an α-subunit and a modulatory subunit, such as the γ-subunit, with the γ_2-subunit being associated with the highest affinity for benzodiazepines. Differential effects of mutations in benzodiazepine binding sites underscore the diversity of the $GABA_A$ receptor assemblies in vivo. Converting histidine residue 101 to arginine in the α_1-subunit prevents diazepam from producing motor impairment or sedation in mice. However, mutating the histidine residue at the same location in the α_2-subunit results in a loss of the anxiolytic effect of benzodiazepines.

Point mutation studies have also identified binding sites for anesthetics known to alter $GABA_A$ receptor function. Replacing asparagine residue 265 with serine in the β_2-subunit dramatically reduces the efficacy of the intravenous anesthetic etomidate. When $GABA_A$ receptor-mediated currents are recorded in the presence of etomidate, an enhancement in the decay time is observed, thereby increasing charge transfer and the overall inhibition. This effect is largely reduced in the mutant mice along with the sedative effects of etomidate and

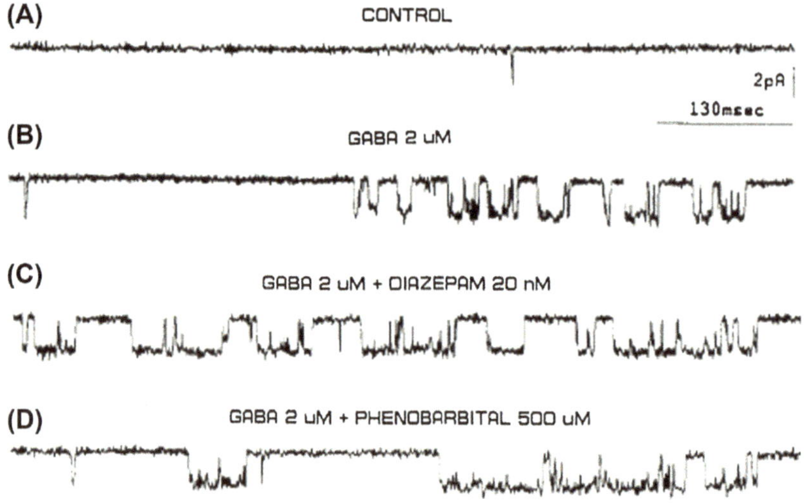

FIGURE 3.19 Single-channel chloride current evoked by GABA was enhanced by diazepam (DZ) and phenobarbital (PhB). (A) Prior to application of GABA, only rare brief spontaneous currents were present. (B) Following application of GABA, channel openings increased in frequency and occurred as individual openings or in groups of openings (bursts). (C) In the presence of DZ, GABA produced increased channel activity. (D) In the presence of PhB, GABA again increased GABA receptor channel activity. Membrane patches were voltage clamped at the potassium equilibrium potential (−75 mV). Downward deflections in all tracings represent channel openings and inward chloride currents. *Reproduced with permission from R.E. Twyman, C.J. Rogers, R.L. McDonald. Differential regulation of γ-aminobutyric acid receptor channels by diazepam and phenobarbital. Annals of Neurology 1989;25(3):213-20.*

highlights the role of this subunit in anesthesia. Many of these point mutations do not impair normal $GABA_A$ receptor activation, suggesting that there are a wide variety of allosteric binding sites. These binding sites potentially are of great interest for drug development.

Extrasynaptic Effects of $GABA_A$ Receptors

Within the nervous system, the primary site of expression for $GABA_A$ receptors is within the synaptic complex, specifically on the postsynaptic membrane. Synaptic $GABA_A$ receptors mediate phasic inhibition or the rapid inhibition resulting from the synaptic release of GABA. However, outside the synaptic compartment, $GABA_A$ receptors are present and are referred to as the "extrasynaptic" subtype. Extrasynaptic $GABA_A$ receptors mediate tonic inhibition due to GABA present extracellularly at ambient levels (Fig. 3.20). This class of $GABA_A$ receptors displays a much higher affinity for GABA, which facilitates their activation in the presence of low concentrations of ambient GABA. The δ-subunit is typically associated with the extrasynaptic $GABA_A$ receptors and associate with the α_{4-6}-subtypes. The δ-subunit confers higher affinity for GABA and reduces desensitization. These receptors display pharmacologically distinct properties—many of these receptors that are insensitive to benzodiazepines are activated by ethanol and volatile anesthetics, such as isoflurane. Sensitivity to ethanol is not restricted to δ subunit–containing $GABA_A$ receptor isoforms, but it displays high sensitivity for all of the $GABA_A$ receptor assemblies. In this light, ethanol targets both synaptic and extrasynaptic $GABA_A$ receptors. The discovery of extrasynaptic subtype of $GABA_A$ receptors is relatively recent and may provide pharmacological targets that can modulate inhibitory transmission and, thus, neuronal excitability.

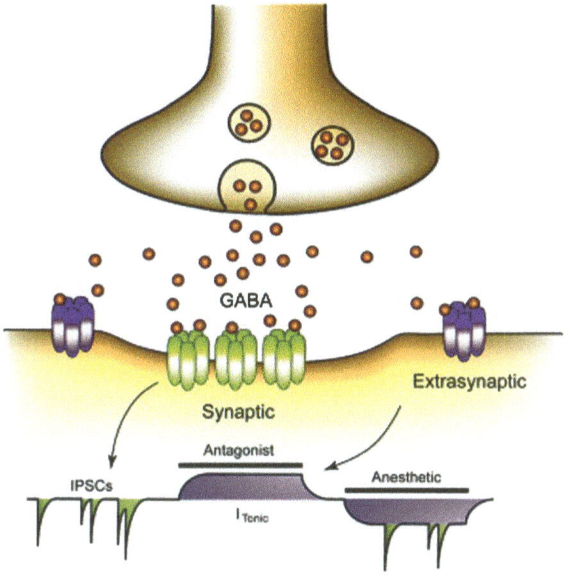

FIGURE 3.20 Synaptic and extrasynaptic activation of $GABA_A$ receptors. Action potential–dependent release of GABA into the synaptic cleft transiently activates $GABA_A$ receptors in the postsynaptic membrane. This generates inhibitory postsynaptic currents, which are typically of a brief duration due to GABA diffusion and uptake, and the desensitization of synaptic receptors. Extrasynaptic $GABA_A$ receptors are activated by low concentrations of GABA in the extracellular space arising from synaptic spillover or other nonsynaptic release mechanisms. These receptors have low desensitization rates and can produce a continuous or "tonic" current. The tonic current is revealed by application of a $GABA_A$ antagonist, which inhibits the current. Many anesthetics potently enhance the tonic current at clinically relevant concentrations. *Reproduced with permission from R.P. Bonin, B.A. Orsera. GABAA receptor subtypes underlying general anesthesia. Pharmacol Bioch Behavior 2008:90(1):105-12.*

The $GABA_A$ Receptor in Development

The prevailing thought is that GABA, via the activation of $GABA_A$ receptors, primarily mediates inhibitory synaptic transmission in the adult CNS. However, numerous studies have reported that GABA can be depolarizing during early development (Fig. 3.21). This appears to be conserved in vertebrates, as is the transition of GABA from being depolarizing to hyperpolarizing in the course of development. It should be noted that the paradoxical depolarizing action of GABA early in development has often been referred to as being "excitatory." While this may be true in many cases, a GABA-induced depolarization need not lead to neuronal excitation in terms of triggering action potentials.

In rodents, depending on the brain region, the transition begins during the second postnatal week and the canonical actions associated with GABA (mainly inhibitory) begin to be observed. In the adult, hyperpolarizing currents resulting from $GABA_A$ receptor activation are due a net chloride influx and a subsequent decrease (net negative shift) in the membrane potential. This is due to a lower intracellular chloride concentration intracellularly relative to the extracellular chloride concentration. The reversal potential for chloride in the adult is typically more negative than the resting membrane potential such that $GABA_A$ receptor activation does not impact significantly the membrane potential at rest. At more depolarized membrane potentials, IPSCs are typically of larger amplitude. This is not the case in the developing nervous system. The intracellular chloride concentration within developing neurons is greater compared to adult concentrations and sets the chloride reversal potential more depolarized. $GABA_A$ receptor activation thus produces a chloride efflux driven by the combination of the electrochemical gradient and the shift in membrane potential to approximate the chloride reversal potential (Fig. 3.22). This developmental shift in intracellular chloride concentrations is due to alterations in the expression of ion importers or exporters. Expression of the sodium–potassium–chloride co-transporters (Na^+–K^+–$2Cl^-$ co-transporters; NKCCs), NKCC1 in particular, drives the import of chloride and produces the high intracellular chloride concentrations observed in early development. In addition, potassium–chloride co-transporters (K^+–Cl^- co-transporters; KCCs) act to decrease intracellular chloride but has limited expression in the

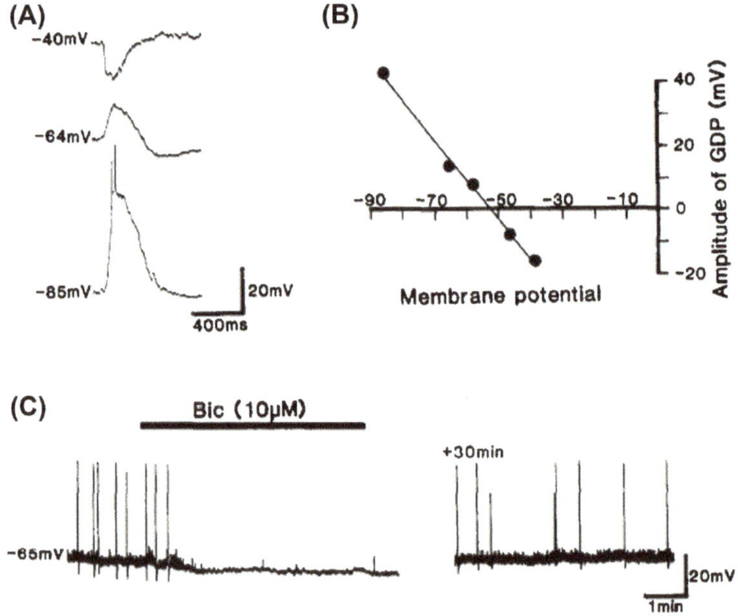

FIGURE 3.21 Giant depolarizing potentials (GDPs) are mediated by GABA acting on $GABA_A$ receptors. (A) GDPs recorded with K-methylsulfate electrodes at different potentials. (B) The amplitude of the potentials shown in (A) is plotted against the membrane potential. The response reversed at −52 mV slice was from a 4-day-old rat. (C) In another neuron, bath application of bicuculline (bar) reversibly blocked GDPs and induced a membrane hyperpolarization. *Reproduced with permission from E. Cherubini, C. Rovira, J.L. Gaiarsa, R. Corradetti, Y. Ben Ari. GABA mediated excitation in immature rat CA3 hippocampal neurons. Int J Develop Neurosci 1990;8(4):481-90.*

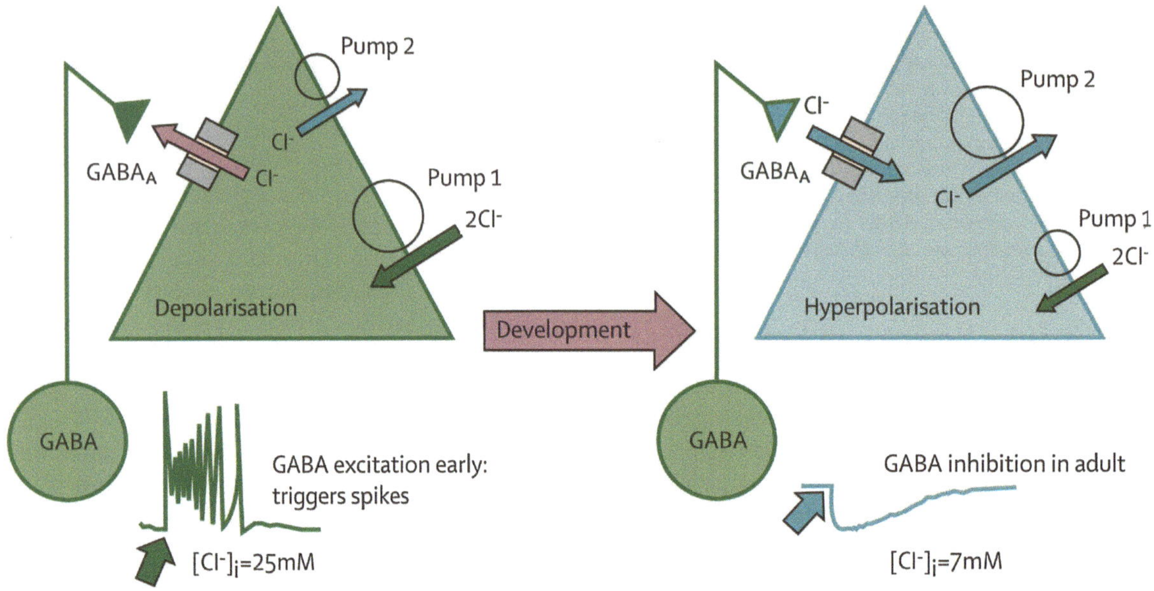

FIGURE 3.22 Excitatory actions of GABA and developmental alterations of the intracellular concentration of chloride. The diagram shows early operation of a co-transporter that imports and augments the chloride ion concentration in immature neurons, whereas the extruding co-transporter is not efficient. This leads to a higher concentration of chloride ions at an early stage and an efflux of chloride when GABA receptors are activated. Green represents the immature neuron and blue represents the mature neuron. $(Cl^-)_i$ = intracellular concentration of chloride. *Reproduced with permission from Y. Ben-Ari, G.L. Holmes. Effects of seizures on developmental processes in the immature brain. The Lancet Neurol 2006;5(12):1055-63.*

developing brain. Ectopic or early expression of the transporter KCC2 can switch the action of GABA from depolarizing to hyperpolarizing. Mice lacking this co-transporter maintain the excitatory actions of GABA throughout development and suggest that the enhanced expression of KCC2 in early development is critical for the "switch" of GABA function. Pharmacological manipulation of the NKCCs has also been reported to alter the function of GABA. Certain pathological conditions elevate the reversal potential for chloride including human epilepsies and a mouse model of Down syndrome. In either case, treatment with bumetanide, an NKCC1 blocker, reduces the reversal potential of chloride and restores the inhibitory effects of $GABA_A$ receptor activation. Numerous other neurodevelopmental disorders have been implicated to have alterations in the function of the different chloride co-transporters and are attractive targets for clinical treatment.

Although, as stated earlier, GABA is predominately inhibitory in the adult brain, some synapses have been found to be excited in response to GABA in the adult. The suprachiasmatic nucleus (SCN) is located within the hypothalamus and is responsible for encoding circadian rhythm in mammals. Neurons within the SCN generate action potentials at frequencies dependent on the time of day; high-frequency firing patterns during the day and low-frequency firing patterns at night. GABA being excitatory on SCN neurons during the day and inhibitory at night is probably due to a change in the intracellular chloride concentration of these neurons. In addition to few examples of excitatory synapses in the adult, a shift in the reversal potential towards more negative values has also been reported. With increasing age, a reduced reversal potential (more negative) for chloride has been identified in hippocampal neurons suggesting a greater inhibition effected by $GABA_A$ receptor activation. Activity can also alter the function of GABA, and specifically in hippocampus, normally inhibitory GABAergic inputs to mature pyramidal cells become temporarily excitatory following a high-frequency train of action potentials generated from the GABAergic neurons. The likely cause of this was the rapid, dramatic increase in somatic chloride concentration which could be blocked by pretreatment with $GABA_A$ receptor antagonists.

The $GABA_B$ Receptor

The metabotropic $GABA_B$ receptors are present pre- and postsynaptically and mediate the "slow" actions of GABA. These receptors are metabotropic g protein-coupled receptors but unique to this GPCR, functional $GABA_B$ receptors require dimerization of two different subunits: the $GABA_BR_1$- and $GABA_BR_2$-subunits—each of which contain seven transmembrane domains. Two isoforms exist for the R_1-subtype (R_{1a} and R_{1b}) but one R_1-subunit binds with an R_2-subunit by dimerizing at an intracellular domain on their C-termini. Numerous phosphorylation sites are present on the C-terminal domains of the $GABA_B$ subunits to regulate its stability in the cell membrane and function (Fig. 3.23). The role of the $GABA_B$ receptor in vivo is critically important for regulating neurotransmission. Deletion of either subunit can produce severe impairments in different forms of learning and memory, and produces seizures. Postsynaptically their activation is typically associated with the activation of GIRKs, thereby exerting an inhibitory effect on the postsynaptic cell (Fig. 3.24).

When the $GABA_BR_1$- and $GABA_BR_2$-subunits are co-expressed in *Xenopus* oocytes, GABA application produces large, inward currents that are blocked by CGP 55,845, a selective antagonist for $GABA_B$ receptors. These currents can be abolished in the presence of barium or pertussis toxin indicating that $GABA_B$ receptors modulate GIRK-mediated currents through a $G_{i/o}$ signaling pathway. On the presynaptic compartment, $GABA_B$ receptors act as autoreceptors or heteroreceptors to regulate the release GABA or other neurotransmitters. $GABA_B$ receptor activation presynaptically suppresses voltage-gated

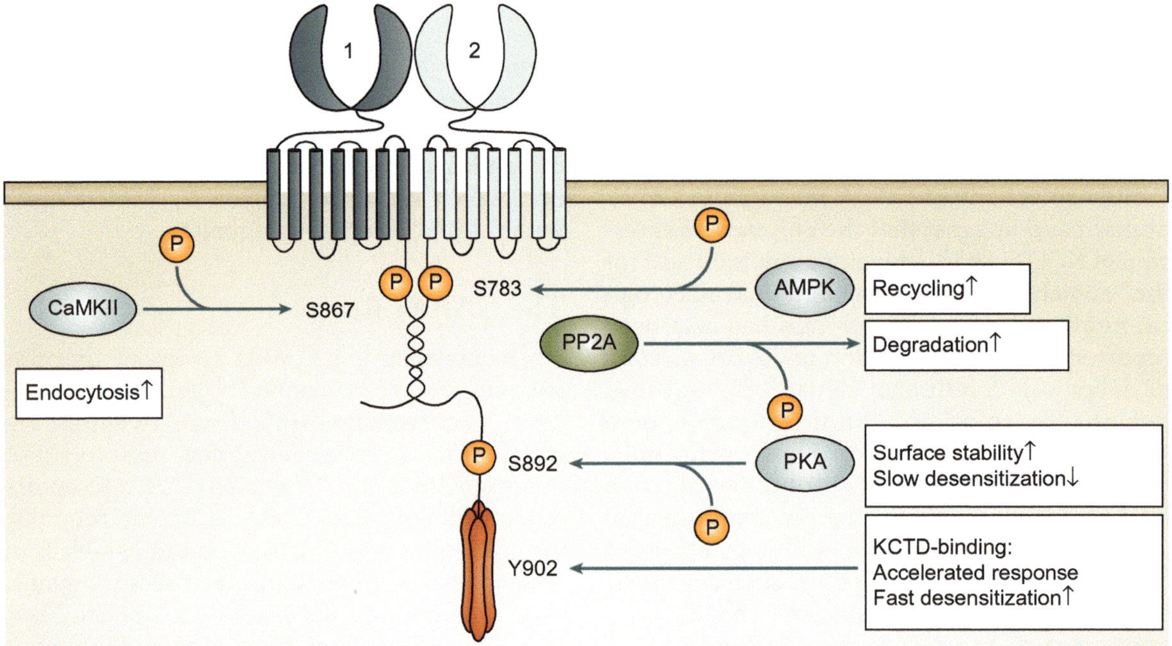

FIGURE 3.23 Ca^{2+}/calmodulin-dependent protein kinase II (CaMKII)-mediated phosphorylation of S867 in $GABA_{B1}$ triggers endocytosis and thereby impairs $GABA_B$ receptor ($GABA_BR$)-dependent activation of GIRKs. The number of $GABA_BRs$ at the plasma membrane is also determined by the phosphorylation status of S783 and S892 in $GABA_{B2}$. S783 acts as a phosphorylation switch at the level of postendocytic sorting. 5′-AMP-dependent protein kinase (AMPK)-dependent S783 phosphorylation under the condition of metabolic stress promotes $GABA_BR$ recycling, whereas dephosphorylation of this site by protein phosphatase 2A (PP2A) has been proposed to promote lysosomal degradation of $GABA_BRs$. Dephosphorylation of S783 has been implicated in the metamphetamine-evoked depression of $GABA_BR$-dependent GIRK signaling in GABAergic neurons of the ventral tegmental area. Surprisingly, however, despite the observed dephosphorylation of S783 in GABAergic neurons, no increase in the lysosomal degradation of $GABA_BRs$ has been observed. Protein kinase A (PKA)-mediated phosphorylation of S892 promotes the stability of $GABA_BRs$ in the plasma membrane, which in turn promotes $GABA_BR$-dependent activation of GIRKs. KCTDs constitutively bind to Y902 in $GABA_{B2}$ in a phosphorylation-independent manner. It is unknown whether the phosphorylation status of S892, which is in close proximity to Y902, influences KCTD-mediated effects on the kinetics of the $GABA_BR$ response. *Reproduced with permission from M. Gassmann, B. Bettler. Regulation of neuronal GABA(B) receptor functions by subunit composition. Nat Rev Neurosci 2012;13(6):380-94.*

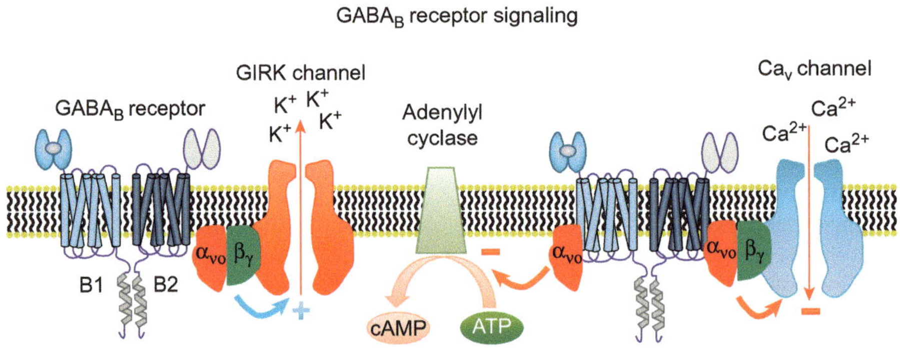

FIGURE 3.24 The primary $GABA_B$ receptor effectors. Activation of the G-protein-coupled $GABA_B$ receptor stimulates GTP-dependent G-protein ($G_{i/o}$) dissociation of the G_α and $G_{\beta\gamma}$ dimer. The $G_{\alpha i/o}$-subunit has been shown to inhibit adenylyl cyclase while the $G_{\beta\gamma}$ dimer is capable of modulating voltage-gated Ca^{2+} (Ca_V) or GIRK channels, resulting in potent neuronal inhibition. Effector specificity may be regulated by heterocomplex formation, guided by targeting protein partners and subcellular localization. *Reproduced with permission C.L. Padgett, P.A. Slesinger. GABAB Receptor Coupling to G-proteins and Ion Channels. Advances in Pharmacology 2010;58:123-47.*

calcium channels, thereby decreasing neurotransmitter release (Fig. 3.24).

Postsynaptic $GABA_B$ receptors have also been reported to modulate the function of extrasynaptic $GABA_A$ receptors. Baclofen, a selective $GABA_B$ receptor agonist, increases the tonic $GABA_A$ receptor-mediated current in neurons in rat hippocampal, cerebellar granule, and thalamocortical cells, the latter having been demonstrated to be transduced via $G_{i/o}$ signaling pathway. In addition, CGP 55,845 reduces $GABA_A$ receptor-mediated tonic current, indicating that $GABA_B$ receptors modulate extrasynaptic $GABA_A$ receptor function. On the other hand, baclofen cannot increase tonic inhibition if the δ-subunit is knocked out, strengthening the evidence that there is a role for the $GABA_B$ receptor in modulating the extrasynaptic effects of GABA. Baclofen remains the only clinically approved drug that acts upon the $GABA_B$ receptor, although further drug development is warranted due to the important physiological role of these receptors.

THE ENDOCANNABINOIDS

Endocannabinoids are important regulators of synaptic transmission within the CNS but are "atypical" neurotransmitters. Unlike canonical neurotransmitters, such as ACh, glutamate, and GABA, endocannabinoids differ in that they are: (1) not stored in and released into the synapse through synaptic vesicles, (2) not released at the presynaptic locus, and (3) act in an atypical "retrograde" fashion (Fig. 3.25). Δ^9-tetrahydrocannabinol (Δ^9-THC) is the active ingredient in marijuana and was discovered to act on two native GPCRs: cannabinoid receptor 1 (CB1) and cannabinoid receptor 2 (CB2). Not surprisingly, it was later discovered that these receptors have endogenous agonists named "endocannabinoids." Anandamide was the first endocannabinoid to be described and was found to recapitulate the effects observed with cannabinoids, such as Δ^9-THC. Numerous other cannabinoid compounds have been described, but anandamide and 2- arachidonoylglycerol (2-AG) are the predominant endocannabinoids in mammals.

Anandamide and 2-AG belong to the *N*-acyl ethanolamines (NAEs) and monoacylglycerols (MAGs) lipid families, respectively. The NAEs and MAGs are embedded in the plasma membrane and are liberated by different enzymes for the different lipid families. Synthesis of the NAEs, including anandamide, occurs through several different enzymatic pathways. The synthesis occurs in a two-step process with the first step being catalyzed by a calcium-dependent *N*-acyltransferase but it is unclear which enzymes are responsible for the second step in synthesis of anandamide. Synthesis of anandamide is likely due to collection of phospholipases including *N*-acylphosphatidylethanolamine-hydrolyzing phospholipase D (NAPE-PLD). Synthesis of 2-AG is also a two-step process: (1) PLC hydrolyzes phosphatidylinositol-4,5-bisphosphate into 1,4,5-trisphosphate and diacylglycerol; and (2) diacylglycerol lipase (DAGL) converts diacylglycerol to anandamide. 2-AG can be produced by a PLC-independent pathway but this will not be addressed here. Anandamide and

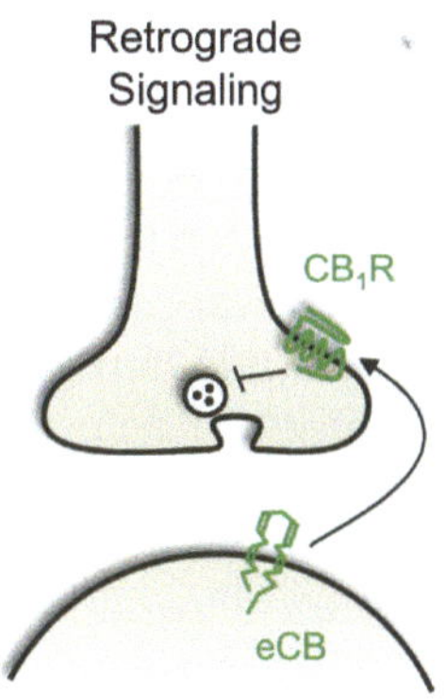

FIGURE 3.25 Endocannabinoid signaling at the synapse. Retrograde endocannabinoid (eCB) signaling. eCBs are mobilized from postsynaptic neurons and target presynaptic cannabinoid type 1 receptors (CB1Rs) to suppress neurotransmitter release. *Reproduced with permission from P.E. Castillo, T.J. Younts, A.E. Chávez, Y. Hashimotodani. Endocannabinoid Signaling and Synaptic Function. Neuron 2010;76(1):70-81.*

2-AG differ dramatically in their activation of either CB1 or CB2. The CB1 receptor subtype has the highest expression within the CNS and is coupled to the $G_{i/o}$ g-protein signaling pathway.

Presynaptic Effects of Endocannabinoids

The presynaptic action of endocannabinoids was the first function to be described for these transmitters and is now appreciated as one of their canonical actions. Endocannabinoids can mediate two separate but similar presynaptic phenomena: depolarization-induced suppression of inhibition (DSI) or depolarization-induced suppression of excitation (DSE). In CA1 of hippocampus, low-frequency train stimulation of GABAergic terminals synapsing onto pyramidal cells produces an eIPSC of similar magnitude until a depolarization is applied to the pyramidal cell. With the application of depolarization, the magnitude of the eIPSC decreases rapidly. This effect requires an intracellular calcium increase in the pyramidal cell and indicates that endocannabinoids synthesis is calcium dependent. Determining that this process does not require synaptic vesicular release requires the use of botulinum toxin, which acts by preventing vesicular fusion to the presynaptic membrane. In the presence of botulinum toxin, DSI is still observed when recordings are conducted on CA1 pyramidal cells. In addition to the use of botulinum toxin, application of the presynaptic neurotransmitter (eg, GABA) has been used to demonstrate that the observed DSI is not simply an alteration in the postsynaptic cell that changes the sensitivity to the presynaptic neurotransmitter. For example, application of GABA produces a postsynaptic response of equal magnitude in CA1 pyramidal cells before and after depolarization (before and after DSI is elicited). This was the first conclusive set of experiments to demonstrate that endocannabinoids act retrogradely through their postsynaptic release and subsequent modulation of presynaptic neurotransmitter release.

Since the initial description of DSI, DSE has also been identified. With DSE, the presynaptic cell releases a neurotransmitter that elicits an EPSC on the postsynaptic cell (eg, glutamate). Then, when the postsynaptic cell is depolarized, a reduction in the amplitude of the EPSC is observed that is mediated by the CB1 receptor (Fig. 3.26). CB1-dependent DSI and DSE has been described at a wide variety of synapses and modulate the release of numerous neurotransmitters, including GABA, glutamate, glycine, and ACh.

Activation of $G_{i/o}$ signaling at presynaptic terminals is often associated with a suppression of the conductance mediated by the voltage-gated calcium channels, thereby decreasing the synaptic release of neurotransmitter. CB1 receptors are coupled to the $G_{i/o}$ signaling pathways, since PTX abolishes CB1 effects. Mechanistically, CB1 activation has been found to suppress N- and P/Q-type calcium channel-mediated currents and reduce synaptic neurotransmitter release. Presynaptic CB1 activation has also been found to modulate different potassium channels, kinases, and other GPCRs.

GLIA AND SYNAPTIC TRANSMISSION

It is becoming increasingly clear that the traditional view of synaptic transmission between neurons and their effector cells requires the involvement of glial cells. Glial cells express many of the same neurotransmitter receptors as neurons but are inherently inexcitable. In other words, glia does not generate action potentials. However, they play a critical role in regulating neuronal activity and function. To highlight the novel understanding of the role these cells play in the nervous system, this chapter focuses on the role of a critically important glial cell type—the astrocytes. Astrocytes mediate and modulate transmitter release, and play a foundational role in development. These cells demonstrate a basic principle of the nervous system: glia play an

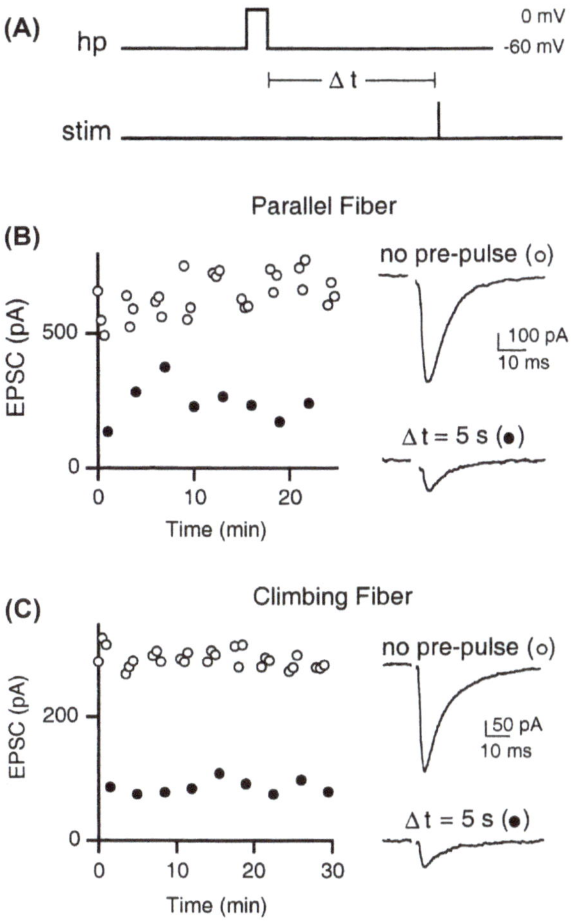

FIGURE 3.26 Postsynaptic depolarization inhibits excitatory cell afferents. (A) Stimulus protocol with the holding potential of the postsynaptic cell (hp; upper) and the stimulation timing (stim; below). Parallel fiber (B) and climbing fiber (C) EPSC amplitudes are plotted over time for control responses with no preceding prepulse to 0 mV (*open circles*) and test responses following Purkinje cell depolarization *closed circles*). Average parallel fiber (B) and climbing fiber (C) EPSCs are shown at the right. Stimulus artifacts are blanked for clarity. Parallel fiber and climbing fiber responses are from two representative experiments. The duration of the depolarization to 0 mV was 50 ms for parallel fiber experiments and 1 s for climbing fiber responses. The test stimulus followed the depolarization by $\Delta t = 5$ s. *Reproduced with permission from A.C. Kreitzer, W.G. Regehr. Retrograde Inhibition of Presynaptic Calcium Influx by Endogenous Cannabinoids at Excitatory Synapses onto Purkinje Cells. Neuron 2001;29(3):717-27.*

essential role in the function and development of the nervous system and are not simply the "brain glue" that they were once thought to be.

Astrocytes and Synaptic Transmission

Astrocytes are arguably the most important type of glia involved in modulating the activity of neurons. Astrocytes are highly arborized cells that extend across and encapsulate multiple dendrites from multiple neurons and can encompass over 10^5 synapses. In concert with the idea that glia can directly modulate the activity of neurons, astrocytes have been found to release many chemicals that were typically thought to be released exclusively from neurons, including glutamate. These cells regulate extracellular electrolyte and transmitter levels through the action of various transporters expressed on their cell membranes. For example, excitatory amino acid transporters (EAATs) are present on the membrane of astrocytes and are critical for moderating the concentration of glutamate in the synapse. Deletion of EAAT2 from astrocytes produces a high mortality rate accompanied by spontaneous seizures.

Action potentials are not observed in astrocytes but a rapid rise in their intracellular calcium can facilitate the release of chemicals into the synapse that directly modulate neuronal activity. These rapid, transient rises or oscillations in calcium levels can rapidly spread to adjacent astrocytes because many of them are directly connected through gap junctions. As such, rapid calcium oscillations initiated in a single astrocyte rapidly spread to nearby astrocytes to regulate and synchronize synaptic activity over a wide area. Astrocytes express a wide variety of functional neurotransmitter receptors and are directly activated by neurotransmitter release from presynaptic terminals. Glutamate released from presynaptic terminals can elicit a rapid rise of intracellular calcium in astrocytes, inducing them to release glutamate. Conversely, spontaneous or induced calcium transients in astrocytes can produce a rise of intracellular calcium within neurons, probably due to the transmitter release from astrocytes (Fig. 3.27). Release of glutamate from astrocytes is dependent on prostaglandin E2, which is notable since prostaglandins are not typically known to be involved in regulating neuronal excitability.

In the presence of TTX, neurons release neurotransmitters stochastically or in a random fashion that is independent of an action potential. The long-held view is that this is a random process that is inherent to the neuron. However, evidence now suggests that astrocytes can modify the frequency of action potential-independent release. With repeated action potentials, the postsynaptic response mediated by GABAergic neurons onto CA1 pyramidal cells can be enhanced and is dependent on the activation of the $GABA_B$ receptor. In this scenario, the $GABA_B$ receptors are recruited to the membrane of astrocytes and act to decrease the failure rate of the presynaptic release of GABA onto the CA1 pyramidal cells. The magnitude and frequency of the calcium transients in astrocytes dictate calcium-dependent transmitter release but

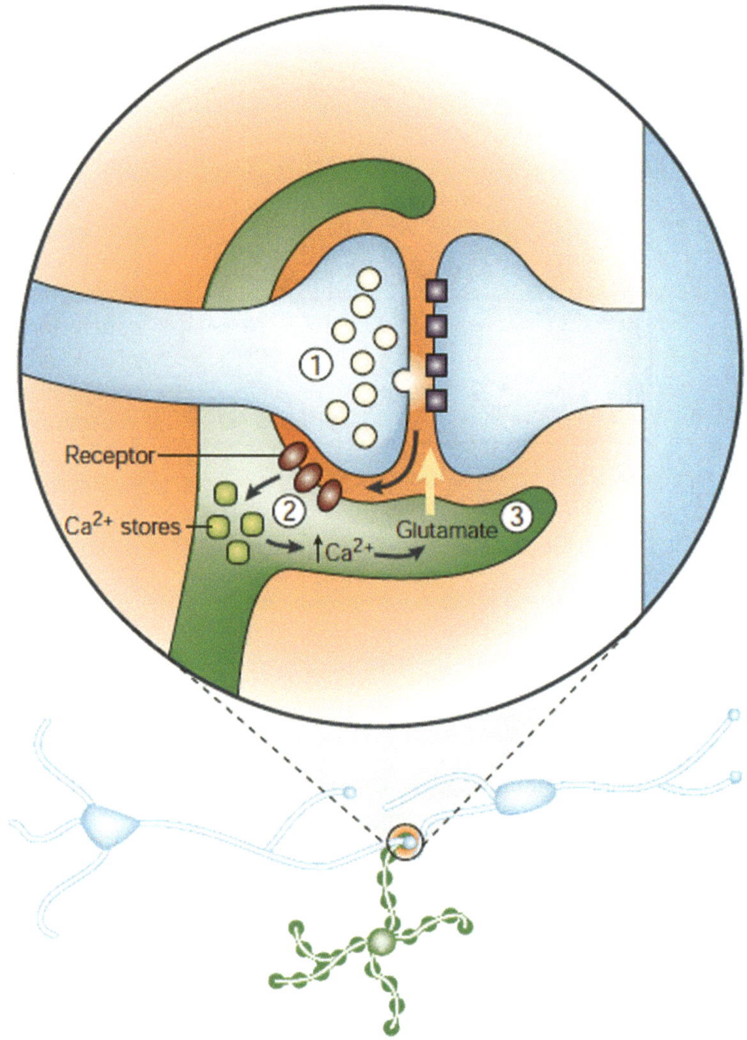

FIGURE 3.27 Intercellular signaling between neurons and astrocytes can have a local modulatory action, as well as participate in the communication between distant synapses. Representation of a tripartite synapse in which the process of an astrocytes wraps around a synaptic connection. Local signaling between neurons and astrocytes has the potential to modulate the synapse. For example, synaptic activity not only elicits postsynaptic potentials but GABA or glutamate can act on astrocytic receptors and trigger Ins(1,4,5)P_3-dependent Ca^{2+} release from internal stores to increase astrocytic Ca^{2+} levels. *Reproduced with permission from P.G. Haydon. Glia: listening and talking to the synapse. Nature Reviews Neuroscience 2001;2:185-93.*

it is not yet clear how the transmitters are released into synaptic compartments. Mechanisms mediating transmitter release by astrocytes may be similar to the molecular machinery responsible for neurotransmitter release in neurons. Selective stimulators or blockers of vesicular release have also been found to increase or decrease glutamate release from astrocytes, respectively. For example, α-latrotoxin promotes transmitter exocytosis in neurons but has the same action in astrocytes. Moreover, tetanus or botulinum toxins, which block the vesicular release of transmitter in neurons, also blocks glutamate release in astrocytes. Interestingly, histological experiments have identified some of the same vesicular proteins found in neurons on the membranes of vesicle-like organelles within astrocytes. Further work is required to understand the functional and bidirectional role of the synaptic communication between astrocytes and neurons.

Glia in Development

Glia are essential for the control of many developmental processes responsible for constructing the nervous system. Highlighted here are a few examples illustrating the importance of astrocytes during development.

Astrocytes produce a variety of factors implicated in synapse formation. Around the time of synaptogenesis, astrocytes appear near developing neurons and are crucial to their survival and construction of the synapse. Astrocytes are a heterogeneous population of cells that influence different aspects of synapse formation. For example, ablation of astrocytes in a small section of spinal cord during synaptogenesis selectively disrupts the development of glutamatergic synapses. Elegant culture experiments have demonstrated that astrocytes secrete factors into the culture medium that are responsible for increasing synapse formation between neurons. Thrombospondins were identified as one of the classes of factors secreted from astrocytes responsible for synapse formation in vitro. The contribution of these secreted factors has been studied in vivo and the genetic deletion of genes encoding thrombospondin 1 and 2 dramatically reduces synapse formation in early postnatal mice.

In addition to assisting in the physical assembly of synapses, astrocytes ensure their functional development. Tumor necrosis factor α, chondroitin sulfate proteoglycans, glypicans, and Wnts are produced by astrocytes and facilitate the recruitment of AMPA receptors to the cell membrane on developing neurons. Thrombospondins signal through neuroligins and the $\alpha_2\delta_1$-subunit of voltage-gated calcium channels. The factors astrocytes produce to regulate the development of GABAergic synapses have not been clearly defined.

During development, the number of neurons and synapses is typically larger than the numbers present in the adult. This reduction in neuronal number and synapses occurs through a process referred to as "pruning." Failure of synaptic pruning has been thought to be responsible for various disorders but most notably, the autism spectrum disorders. Astrocytes express the phagocytic receptors Megf10 and Mertk receptors that dictate the phagocytosis of the excessive synapses. In this light, Megf10 and Mertk receptors are important in synaptic remodeling. The process by which astrocytes regulate synaptic pruning is dependent on the particular developing synapse and the transmitter phenotype of the given synapse.

CONCLUSION

In this chapter, we have reviewed some of the basic concepts of neurotransmission at the chemical synapse. Neurotransmitter release into the synaptic space can produce depolarization or hyperpolarization of the postsynaptic cell. Simultaneously, neurotransmitters can modulate the presynaptic release of neurotransmitter from the same releasing cell or alter the release of neurotransmitter from neighboring cells. These effects can occur rapidly through ionotropic receptors or through a slower process mediated by metabotropic receptors. As we have discussed with GABA, no neurotransmitter is inherently excitatory or inhibitory–it is the receptor that determines the effect upon transmitter binding. In this regard, it is important to note that these effects are regulated throughout development. Finally, the emerging and important role of glia in regulating and perhaps even mediating neurotransmission is highlighted.

Further Reading

[1] Hammond C. Cellular and molecular neurophysiology. Boston (MA): Elsevier; 2015.

[2] Cooper JR, Bloom FE, Roth RH. In: The biochemical basis of neuropharmacology. 8th ed. Oxford (New York): Oxford University Press; 2003.

[3] Fryer AD, Christopoulos A, Nathanson NM. Muscarinic receptors. Heidelberg: Springer; 2012.

[4] Zoli M, Pistillo F, Gotti C. Diversity of native nicotinic receptor subtypes in mammalian brain. Neuropharmacol 2015;96(Pt B):302–11.

[5] Fletcher A. Neuromuscular function and transmission. Anaesth Intensive Care Med 2011;12(6):245–8.

[6] Kano M, Ohno-Shosaku T, Hashimotodani Y, Uchigashima M, Watanabe M. Endocannabinoid-mediated control of synaptic transmission. Physiol Rev 2009;89(1):309–80.

[7] Allen NJ. Astrocyte regulation of synaptic behavior. Annu Rev Cell Dev Biol 2014;30:439–63.

[8] Corty MM, Freeman MR. Cell biology in neuroscience: architects in neural circuit design: glia control neuron numbers and connectivity. J Cell Biol 2013; 203(3):395–405.

CHAPTER

3.1

Translational Correlation: Myasthenia Gravis and Other Neuromuscular Junction Disorders

A.M. El-Dokla, S.T. Ali

Texas Tech University Health Sciences Center, Lubbock, TX, United States

INTRODUCTION

Myasthenia is a Greek term for muscle weakness. Gravis is a Latin word meaning severe. Myasthenia is classified as either immune myasthenia gravis (MG) or congenital myasthenic syndrome. The focus in this chapter is on immune MG, an autoimmune disease that targets primarily the neuromuscular junction (NMJ).

The prevalence rate of MG is estimated at about 50–400 per million. It has a bimodal age of onset with early (at ages 20–24 years in female and 30–34 years in male) and late peaks (at ages 70–75 years). The disease occurs more frequently in women, especially in younger patients.

Clinical Presentation of MG

The hallmark of MG is fluctuating, fatigable, painless muscle weakness that becomes worse with exertion or exercise and toward the end of the day. The weakness is dynamic rather than fixed. The weakness is commonly exacerbated by warm weather, emotional stress, and systemic infections.

Extra ocular muscle (EOM) weakness manifested as diplopia and ptosis are commonly seen as the initial presentation. During the course of the disease, more than 90% of patients will have ocular manifestations. If the only manifestation of myasthenia is ocular, the disease is called ocular MG that usually has a more benign course.

Patients with generalized MG may also develop or present with bulbar symptoms, breathing difficulty, and limb muscle weakness. Bulbar muscle weakness will present as dysarthria with nasal speech due to weakness of the palatal muscles, tongue, or cheeks. Dysphagia with regurgitation of liquid through the nose can be another manifestation of palatal and pharyngeal weakness that is usually accompanied by difficulty chewing, jaw dropping, and bilateral facial muscle weakness. When facial weakness occurs, patients become unable to whistle or puff out the cheeks, and the smile becomes transverse. Involvement of respiratory muscles is the hallmark of myasthenic crisis and results

Conn's *Translational Neuroscience*
http://dx.doi.org/10.1016/B978-0-12-802381-5.00005-1

in respiratory distress and insufficiency. If not treated appropriately, respiratory failure can lead to death.

Fatigue is a common but nonspecific symptom reported by MG patients. Other causes, such as hypothyroidism, sleep apnea, and depression, should also be considered.

Diagnosis

On physical examination, eliciting fatigable weakness is usually the key feature of MG. The weakness usually becomes more evident when exercising the muscle. For example, asking the patient to sustain gaze usually provokes the ocular muscle weakness and leads to fatigable ptosis, diplopia, or pseudo-nystagmus.

Intravenous administration of edrophonium to patients with postsynaptic NMJ dysfunction usually results in improvement of muscle weakness. Edrophonium is a short-acting acetylcholinesterase inhibitor that inhibits the breakdown of acetylcholine and increases its availability at the NMJ. This test is less commonly used now because of the cholinergic side effects including tearing, sweating, cramps, and rarely bradycardia. False-positive test results can be seen in multiple other conditions, including motor neuron disease. A false-negative test is also common.

In patients with ptosis, placing an ice pack over the eye for 5 min (ice pack test) decreases the body temperature, which improves NMJ function and reduces ptosis.

The MG antibody (Ab) testing is commonly used in clinical practice for diagnosis. Acetylcholine receptor blocking, binding, and modulating Abs are commercially available. Acetylcholine binding Abs are likely the most sensitive and are found in about 80–85% of patients with generalized MG and in about 50% with ocular MG. There are other Abs found less commonly such as IgG versus muscle-specific tyrosine kinase (MuSK), low-density lipoprotein receptor–related protein 4 (LRP4), agrin, and antistriational muscle Abs. Antistriational Abs are frequently associated with thymoma and MG, especially in younger patients. MG with IgG Abs versus MuSK usually presents with bulbar and ocular symptoms and tends to have a more aggressive course than generalized AChR Ab–associated MG.

Electrodiagnostic testing is very useful in the diagnosis of patients with Ab-negative MG. Repetitive nerve stimulation (RNS) has been used for many years in neurophysiological laboratories to evaluate NMJ abnormalities. The slow (2–3 Hz) RNS with more than 10% reduction of compound muscle action–potential (CMAP) amplitude is present in about three-quarters of patients with generalized MG, and in only about 50% of patients with ocular MG. Single-fiber electromyography (SFEMG) increases the sensitivity of detecting NMJ abnormalities in generalized MG to about 95–99% and more that 95% in ocular MG. SFEMG is currently the most sensitive test for detecting NMJ abnormalities associated with MG.

Management

MG is caused by Abs directed against the postsynaptic NMJ membrane, especially the skeletal muscle acetylcholine receptors (AChRs). This phenomenon will result in reduction of available AChRs to bind with the acetylcholine and lead to reduction of the endplate potential, and therefore failure to generate an action potential and depolarize muscle.

The main goals of the treatment are to increase the availability of acetylcholine in the NMJ and to decrease the immune response against the postsynaptic NMJ membrane.

Pyridostigmine is an acetylcholinesterase inhibitor that inhibits the breakdown of acetylcholine in the NMJ and therefore increase its availability.

Different types of immunotherapy have been used successfully to treat MG. A detailed discussion of the immunotherapy for MG is beyond the scope of this textbook. Prednisone, azathioprine, and mycophenolate mofetil are the most commonly used medications in clinical practice. Azathioprine and mycophenolate mofetil are used as steroid-sparing agents. The choice of the proper medication for treatment will depend mainly on the side-effect profile and onset of action. Prednisone has a short onset of action (1–3 months), mycophenolate mofetil has a moderate onset of action (1–12 months), and azathioprine has a long onset of action (6–24 months). The common side effects of prednisone include weight gain, hypertension, diabetes, and osteoporosis. The common side effects of azathioprine are nausea, vomiting, fever, abdominal pain, pancreatitis, hepatic toxicity, bone-marrow suppression, and small increase of the risk oncogenicity, especially hematological malignancy. Mycophenolate mofetil is a relatively safe medicine with less reported hepatotoxicity or nephrotoxicity. However, the common side effects are diarrhea, abdominal cramps, fever, nausea, leukopenia, and peripheral edema. There is a rare risk of progressive multifocal leukoencephalopathy.

Plasma exchange (PE) and intravenous immunoglobulin (IVIG) are used mainly to treat MG crisis or generalized MG that is unresponsive to immune therapy or sometimes as a bridge to immunosuppression with slow to act agents. Plasma exchange removes Abs from patients' serum. IVIG has a neutralizing effect on IgG Abs against AChRs. It also decreases the production of those Abs by inhibiting globulin production.

Some studies have suggested that PE and IVIG are equally effective in the treatment of MG crisis while others suggested that the PE might be slightly more effective.

Thymus gland abnormalities are common in patients with MG. Thymectomy has been accepted as an optional treatment modality for autoimmune MG without thymoma. It is always recommended for the treatment of thymoma associated with MG. The evidence of thymectomy in treating autoimmune MG without thymoma is not based on double-blinded controlled trials. However, review of literature and some of the available evidence indicates that it may be associated with improvement in nonthymomatous autoimmune MG. An American Academy of Neurology statement indicates that it should be considered a treatment option. There is an active ongoing double-blinded controlled trial to test the role of thymectomy in nonthymomatous MG.

Presynaptic NMJ Disorders

Autoimmune Lambert–Eaton myasthenic syndrome (LEMS) and botulism are the most common presynaptic NMJ disorders. Autoimmune LEMS is caused by IgG Abs directed against voltage-gated calcium channels (VGCCs) at the presynaptic motor nerve terminal leading to decreased ACh fusion to synaptic vesicles and release from the presynaptic membrane. Patients present primarily with proximal limb muscle weakness more evident in the legs. The weakness can improve by a brief sustained exercise. The disease usually begins with ocular muscle weakness, but it is usually milder than in MG. Also in contrast to MG, autonomic symptoms are common and include blurred vision, dryness of the mouth and eyes, and impotence in males. LEMS can be associated with malignancy, especially small-cell carcinoma of the lung. On examination, deep-tendon reflexes are usually diminished or absent and a brief period of exercise will result in reappearance of the reflexes. Serum IgG Abs directed against VGCCs can be detected in about 85% of patients with LEMS. The slow (2–3 Hz) RNS shows reduction of CMAP amplitude; however, after fast (30–50 Hz) RNS or sustained muscle contraction, CMAP amplitude increment (60–100%) is appreciated. Treatments that have been used in clinical practice are 3,4-diaminopyridine, pyridostigmine, immunosuppression medications, IVIG, and PE.

Botulism is caused by the bacterium *Clostridium botulinum*, which secrets different types of botulinum neurotoxins (BTXs). BTXs inhibit the release of acetylcholine by destroying certain proteins (soluble NSF (*N*-ethylmaleimide-sensitive factor) attachment protein receptor, SNARE) essential for its release from the motor nerve terminal into the NMJ. Foodborne, infant, wound, iatrogenic, and inhalational are different types of botulism. The presentation of botulism usually starts with ocular and bulbar muscle weakness. The weakness is usually described as descending paralysis. In foodborne botulism, symptoms usually start after 12–36 h with significant diarrhea, nausea, and abdominal cramps. Early recognition of the disease is the key for good prognosis. BTX can be detected in samples from serum, gastric content, or stool. Fast RNS (30–50 Hz) usually results in a CMAP amplitude increment that may be relatively mild. Equine-derived botulinum antitoxin, when given early in the disease course especially in cases of food botulism, is usually associated with good prognosis.

Further Reading

[1] Pestronk A. Neuromuscular Disease Center. St. Louis (MO): Washington University School of Medicine; http://neuromuscular.wustl.edu.

[2] Kusner LL, Puwanant A, Kaminski HJ. Ocular myasthenia: diagnosis, treatment, and pathogenesis. Neurol 2006;12:231–9.

[3] Vincent A, Palace J, Hilton-Jones D. Myasthenia gravis. Lancet 2001;357:2122–212.

[4] Titulaer MJ, Lang B, Verschuuren JJ. Lambert-Eaton myasthenic syndrome: from clinical characteristics to therapeutic strategies. Lancet Neurol 2011;10:1098–107.

CHAPTER

4

Synaptic Development

A. Dabrowski, E. Johnson-Venkatesh, H. Umemori
Harvard Medical School, Boston, MA, United States

INTRODUCTION

The word "synapse" (συν + απτειν, or clasping together) was first coined in the late 19th century, amidst an explosion of anatomical studies into the structure of neurons in the brain. The etymology reflects its structure and function—the tight apposition of an axon onto its postsynaptic target, which is required for the synapse to become a portal of communication. A functional synapse requires correct alignment of presynaptic and postsynaptic specializations. Synapse development encompasses both a huge number of synapses and a great diversity of synaptic types (Fig. 4.1). During development, the appropriate number and type of synapses need to be formed in the correct place at the correct time in order for the central nervous system (CNS) to function properly. In this chapter, we overview the basic processes of synaptic development and mechanisms underlying them. We highlight the development of synapses in the human brain based on experimental knowledge from animal models and insights from human diseases. Specifically, this chapter gives a basic overview of synaptic structure, the stages of synaptic development, diversity of synapses, the roles of glia, and finally diseases associated with defects in synaptic development.

SYNAPSES AT A GLANCE

Zooming in on individual synapses reveals that they are highly specialized structures; as depicted in Fig. 4.2. Simplistically, the activated presynaptic neuron releases neurotransmitter into the synaptic cleft, a space 20–40-nm wide between the two cells; the neurotransmitter then crosses the cleft, and is received by the postsynaptic neuron. Hence, at the presynaptic side, there are neurotransmitter-laden synaptic vesicles and the machinery required for activity-dependent release of these vesicles at the "active zone" of the axon terminal; and on the postsynaptic side are the receptors that respond to the neurotransmitter and the associated machinery important for subsequent downstream signaling.

Neurons in the CNS form an extraordinary number of synapses with each other. For example, in the visual cortex of an adult human, there are about 300 million synapses in 1 mm^3, and an individual cortical neuron can receive on the order of 10,000 synapses. Overall, the number of synapses in the human brain is estimated to be on the order of trillions because each neuron can make hundreds to thousands of synapses, and there are billions of neurons

Conn's Translational Neuroscience
http://dx.doi.org/10.1016/B978-0-12-802381-5.00006-3

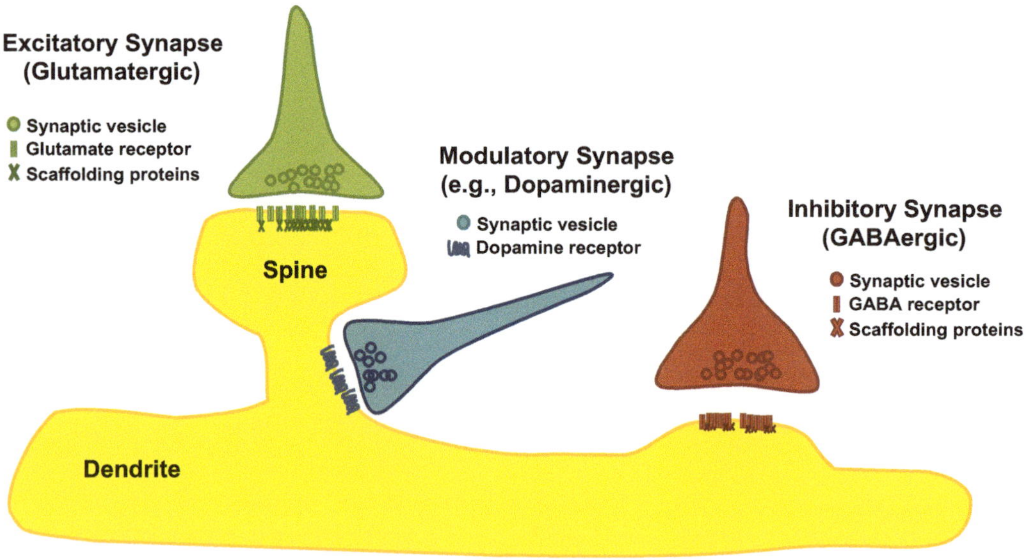

FIGURE 4.1 **Diversity of chemical synapses within the central nervous system.** The main types of synapses within the brain are excitatory (glutamatergic) and inhibitory (GABAergic). There are also many neuromodulatory synapses within the brain, represented in this figure by a dopaminergic synapse. Glutamatergic synapses usually form on specialized structures called spines, GABAergic synapses target the dendritic shaft (or soma), and neuromodulatory synapses form on either the dendritic shaft, or, as depicted here, the neck of a spine.

within the adult human brain. Thus, in terms of quantity, synaptic development is a significant undertaking.

Within this large number of connections, synapses are structurally, molecularly, and physiologically diverse. Presynaptically, neurons produce many separate types of neurotransmitters, which instruct their role in the neural circuit. In the CNS, excitatory neurons release glutamate, which activates the postsynaptic neuron; inhibitory neurons release γ-aminobutyric acid (GABA) or glycine, which suppresses the postsynaptic neuron; and neuromodulatory neurons release molecules, such as acetylcholine, dopamine, serotonin, epinephrine, or norepinephrine, which modulate transmission at excitatory and inhibitory synapses (Fig. 4.1). In addition, acetylcholine can act as an excitatory neurotransmitter when signaling through the ionotropic nicotinic receptors. Some synapses can even co-release multiple neurotransmitters, such as dopamine and GABA. Neurons may co-release additional neuromodulators, such as substance P, endorphins, and somatostatin to further regulate communication between neurons within the neural network. Furthermore, different types of presynaptic terminals (called boutons) throughout the nervous system contain a variable number of synaptic vesicles, from hundreds to tens of thousands of synaptic vesicles, creating synapses with vastly different physiological properties.

Synaptic structures (Fig. 4.2) are very complex at the molecular level, containing more than 1000 distinct proteins, many of which are distinctly expressed at subsets of synapses (eg, gephyrin at GABAergic synapses and PSD-95 at glutamatergic synapses). In the presynaptic terminal, the active zone contains protein machinery important for docking synaptic vesicles

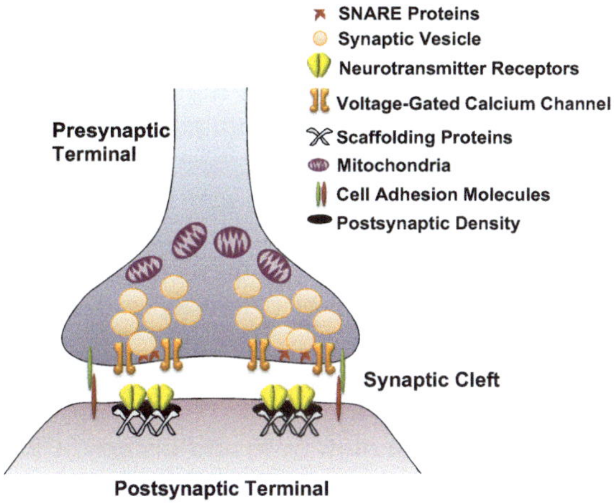

FIGURE 4.2 **Basic structure of a synapse.** The presynaptic terminal contains mitochondria, synaptic vesicles, and active zones where voltage-gated calcium channels and SNARE proteins localize. These specializations allow for the release of neurotransmitters into the synaptic cleft. At the active zone, voltage-gated calcium channels allow for calcium influx, which lets the SNARE proteins fuse the synaptic vesicle with the membrane and release neurotransmitter. Specializations on the postsynaptic side include the neurotransmitter receptors and postsynaptic density proteins including scaffolding proteins. Cell-adhesion molecules also continue to support and maintain the presynaptic and postsynaptic alignment.

and releasing them in response to an increase in calcium (usually elicited by an action potential). The major proteins of the presynaptic terminal include voltage-gated calcium channels, the major source of calcium in response to an action potential; the SNARE proteins, the major vesicle docking and release machinery; the synaptic vesicles themselves; and an assorted number of regulatory proteins. Postsynaptically, the appropriate neurotransmitter receptors cluster on dendrites opposed to the presynaptic active zone: glutamatergic (metabotropic, N-methyl-D-aspartate [NMDA], kainate, or α-amino-3-hydroxy-5-methyl-4-isoxazolepropionic acid [AMPA]) receptors at excitatory synapses, GABA or glycine receptors at inhibitory synapses. These receptors exist within a dense mesh of proteins called the postsynaptic density, a protein-rich domain attached to the postsynaptic membrane originally identified as an electron-dense region visualized during electron microscopy. Excitatory postsynaptic densities have numerous proteins, including scaffolding proteins, glutamate receptors, receptors for neuromodulatory molecules, adhesion molecules, signaling molecules, potassium channels, and even elements of local translational machinery. The morphology of excitatory postsynaptic sites is often further specialized into a dendritic spine—a protuberance from the shaft of the dendrite that serves to electrically isolate individual excitatory synapses. Although less dense and morphologically distinct than the excitatory postsynaptic density, the inhibitory postsynaptic density at inhibitory synapses also contains a large complex of receptors, scaffolding proteins, and signaling molecules.

Many synapses are also in close contact with glia, a group of "support" cells that are integral to the development and function of the brain. The types of glia within the CNS are astrocytes, microglia, and oligodendrocytes. Within the CNS, astrocytes are the most common glia and are a very diverse group of cells. The roles of astrocytes include the regulation of potassium homeostasis, modulation of glutamatergic signaling, serving as an interface between blood vessels and neurons, regulation of neural development, and supporting synaptic maintenance. Astrocytes are often interconnected by gap junctions and signal to each other via calcium waves. Microglia are the resident immune cells of the CNS, and can function similar to macrophages when activated. Oligodendrocytes, or Schwann cells in the peripheral nervous system, wrap neuronal axons with myelin, which facilitates action potential propagation down the axon. Research since the early 2000s demonstrates that even though glia are not directly within the synapse, they are critically important for the development, regulation, and maintenance of the synapse.

STAGES OF SYNAPTIC DEVELOPMENT (EXCITATORY SYNAPSES)

Given the exquisitely complex structure of the synapse, it is not surprising that the development of such an assembly is a multifaceted and highly regulated process. While there are certainly regional differences in the specifics of the progression, the overall process of synaptic development involves four major steps: (1) neurite outgrowth and targeting, (2) synaptic differentiation and specialization, (3) synaptic refinement, and (4) synaptic maturation and maintenance (Fig. 4.3).

Timing of Synaptogenesis Within Humans

In humans, synapse development occurs during late gestational and early postnatal development, and encompasses a phase of remarkable neural network emergence en route to a mature brain. Careful observation of postmortem human brain of various ages using electron microscopy by Patricia Goldman-Rakic and Pasko Rakic, among others, demonstrated that the first synapses appear at 5 weeks gestation and continue to increase exponentially in infancy until they peak in number at 2 years of age—in visual cortex at nearly 200% of mature brain count—and then steadily decrease until they finally achieve the "mature" number of synapses in the third decade of life.

In terms of absolute numbers, synapse development can be divided into stages. First, synapses appear at a slow rate and occupy a low density of neural tissue, at a time when neurogenesis (the birth of new neurons) is still ongoing. Next, a phenomenal acceleration of synapse formation occurs, coinciding with the end of neurogenesis and overlapping, but extending past axon and dendrite elaboration. In humans, this huge addition of synapses starts in the second trimester of pregnancy and continues until the second year of age, when synapse number peaks. These synapses are then slowly pruned away over the next decade of human life, through

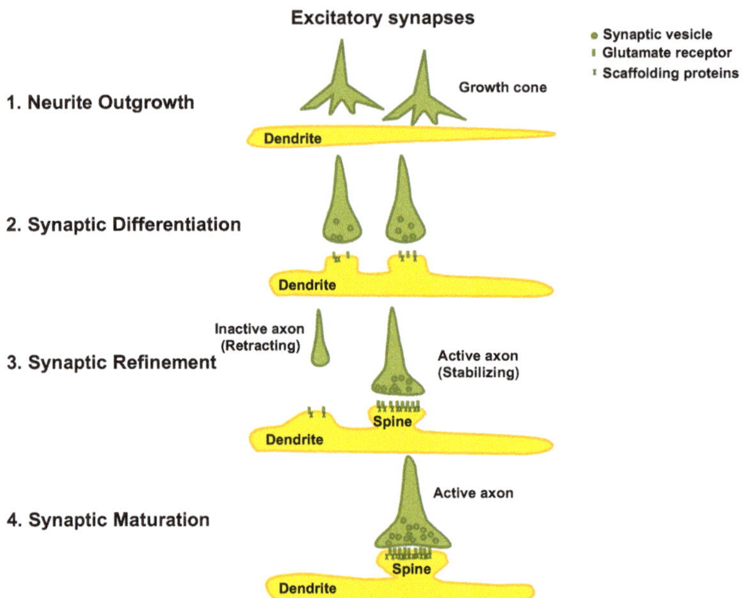

FIGURE 4.3 **Basic overview of the stages of excitatory synaptic development within the central nervous system.** The process of synaptogenesis begins with: (1) neurite outgrowth, whereby axon terminals approach the postsynaptic target, usually a neuronal dendrite. (2) Then a period of synaptic differentiation begins, and presynaptic and postsynaptic specializations (see Fig. 4.2) start to form. (3) Then synaptic refinement occurs and generally, inactive synapses are retracted and active synapses are stabilized. (4) Finally, the active synapses mature, further strengthen, and are maintained.

puberty, and then until the third decade of life, when the "mature" state is finally achieved. Synapse formation and elimination does not occur simultaneously in different brain regions. Even within the cerebral cortex of humans, the dynamics of synapse development vary, with areas of cortex for sensory input developing first and prefrontal areas last.

The general pattern of cortical synapse development is conserved among many species—for example, in rodents, the exponential phase of synapse addition extends for 2 weeks starting after birth, peaks at postnatal day 14–28 (P14–P28), and stabilizes at around P40. Furthermore, many of the molecules involved in synapse formation and function are highly conserved across species. Thus, rodents provide important experimental vertebrate models for understanding normal and pathological synapse development.

Neurite Outgrowth

Axon Differentiation, Outgrowth, Pathfinding, and Targeting

The cell bodies of synaptic partners often lie far apart, which means axons extend over long distances to form synapses onto their targets. To reach a target in the periphery, axons can extend as long as about 1 m in length. Frequently, the axons grow out to their postsynaptic target in a carefully coordinated process called axon guidance. In other cases, such as the cerebellar granule cells, the axon remains behind as the cell body migrates to a new location.

The process of axon guidance over great distances has been closely studied and involves both attractive and repulsive molecular cues. Axon guidance cues are detected at a specialized process at the axon tip, known as the growth cone—a highly active structure that forms the leading edge of the progressing axon. The growth cone responds to the environmental cues with considerable turnover of cytoskeletal components as it scans and moves through the environment. Chemoattractants can attract the axon growth cone along a gradient of molecules to grow toward the source; chemorepellants can prevent an axon from growing into an inappropriate location and effectively form corridors along which the axons grow. There are both short- and long-distance attractive and repulsive cues that serve as axon guidance molecules.

The environmental sampling by the growth cone as it navigates requires considerable turnover of cytoskeletal components and sensitivity to environmental cues. These environmental cues largely consist of the four families of axon guidance molecules: netrins, semaphorins, slits, and ephrins. Netrins have an important role in coordinating midline crossing in the nervous system, which is necessary for contralateral cross-wiring of the brain. Netrins can attract axons through the deleted in colorectal cancer (DCC) receptor or repel axons through the UNC5 receptor, and regulation of which receptor is expressed mediates the change in axon response as it crosses the midline. *De novo* mutations in netrins have been found in some patients with autism.

Semaphorins can either be membrane bound or secreted. They signal through multimeric signaling complexes that prominently feature plexins and neuropilins. Semaphorins primarily act as a local guidance cue, although they can play a wider variety of roles as well. In addition to their role in axon guidance, semaphorins can also regulate synapse formation. Changes in semaphorin, plexin, and/or neuropilin expression levels have been linked to schizophrenia, Alzheimer's disease, human T-lymphotropic virus-type I–associated myelopathy, autism, and glioblastoma. Gene mutation of Sema3E is associated with CHARGE (coloboma, heart defect, atresia choanae, retarded growth and development, genital abnormality, and ear abnormality) syndrome, and Cri du Chat syndrome is associated with the deletion of part of the Sema5A gene. In addition, genome-wide association studies have linked Sema4C to autosomal-dominant retinitis pigmentosa, Sema5A to Parkinson's disease, and PlexinA2 to schizophrenia.

Slits are secreted molecules that signal through robo receptors. Like netrins, slits were originally identified by defects in midline crossing; only they tend to be repulsive. As study of slits expanded, it has become clear that slits play a broader role than as simple axon guidance cues. They are also important for axon elongation, axon branching, and neuronal migration. Mutations in genes encoding robos have been linked to dyslexia and horizontal gaze palsy with progressive scoliosis.

Ephrins are membrane-bound ligands that signal through Eph receptors, which are receptor tyrosine kinases. There are two types of ephrins: ephrin-As are linked to the cell membrane through a glycophosphatidylinositol anchor and ephrin-B receptors have a transmembrane domain. Because of their transmembrane structure, ephrin-B/EphB signaling can be bidirectional. Ephrin-As were originally characterized as attractive molecules in the retinal–tectal system. They are expressed in a gradient that serves to steer axons. As with the other families of axon guidance molecules, the ephrins may play wider and more diverse roles than the originally identified role in axon guidance.

Guidance molecules use a variety of downstream signaling pathways to encode motility, and different patterns of intracellular signaling pathway activation can lead to growth cone attraction, repulsion, or collapse. The ratio of cyclic nucleotides (cyclic guanosine monophosphate and cyclic adenosine monophosphate [cGMP and cAMP]) and cytoplasmic gradients of calcium can both determine the decision between net attraction and repulsion. Ultimately, axon outgrowth is crucial to ensuring that synaptic partners are in contact with each other.

Dendrite Differentiation and Growth

In addition to axon outgrowth, dendritic processes must also elongate and elaborate for most of the synaptic connections in the CNS to form. An increase in the number of dendritic branches provides increased surface area for incoming synapses. Furthermore, excitatory synapses onto distal dendritic branches exert less excitatory drive than proximally located synapses, suggesting the topology of dendritic branching affects circuit function. Mutations in the gene disrupted in schizophrenia 1 (DISC1) lead to altered dendritic branching patterns in neurons. Dendritic fields often develop via tiling, which prevents overlapping of dendritic branches from the same cell (self-avoidance) or of a similar class of cells, and create distinct receptive areas for individual neurons. Contact-mediated regulators aid in tiling and dendritic development largely by marking "self": this process has been extensively studied in *Drosophila*, where Down syndrome cell-adhesion molecule (Dscam) can regulate dendrite self-avoidance via homophilic binding. *DSCAM* can generate 38,016 isoforms by alternative splicing in *Drosophila*, and this allows each cell to be "tagged" with a distinct isoform, facilitating "self"-recognition and avoidance. In mammals, a set of cell-adhesion molecules called protocadherins may play similar roles. Numerous genes, including numerous transcriptional regulators, which regulate dendritic development, have been identified. What determines the diversity of dendritic branching that exists between all of the different classes of neurons remains an active area of research.

Synaptic Differentiation and Specialization

Once axons have reached their synaptic target, synapse development ensues through reciprocal release of secreted molecules and through interaction of cell-adhesion molecules between the axon and its target. Synaptic differentiation is the process by which the specialized cellular machinery that makes up both the presynaptic and postsynaptic domains of the cell is synthesized and recruited to the presynaptic or postsynaptic compartments (Figs. 4.2 and 4.3). Differentiation in the presynaptic compartment is often assessed by examining the clustering of synaptic vesicles or accumulation of presynaptic active zone proteins at presynaptic terminals. On the postsynaptic side, scaffolding proteins, in particular Postsynaptic density protein 95 (PSD95) at excitatory synapses and gephyrin at inhibitory synapses, or neurotransmitter receptors, are used to examine postsynaptic differentiation. The majority of excitatory (glutamatergic) synapses in the CNS form onto dendritic spines—changes in morphology of dendritic spines are often used as a corollary to their maturation, as discussed in the following section.

Dynamics of Synapse Formation

Time-lapse studies of individual synapse assembly in vitro and in vivo have revealed that it is a multistep and highly dynamic process. In neuronal culture, synapses appear to assemble within hours of initial contact between axons and dendrites, with the presynaptic components typically preceding the postsynaptic components. Presynaptic components are trafficked along the axon as transport packets, containing a number of synaptic vesicles or active zone proteins forming a complex that accumulates at the nascent synapse. The trafficking of presynaptic components occurs even in the absence of postsynaptic contact, suggesting an intrinsic predisposition to form hemisynapses within the axon. However, in the presence of appropriate postsynaptic contacts, synaptic vesicles accumulate at nascent synapses and outcompete inappropriate sites lacking postsynaptic contact, preventing ectopic synapse formation.

Not all the components of the presynaptic terminal assemble simultaneously at the nascent synapse. First, synaptic vesicles assemble, then active zone proteins, followed by components of the synaptic vesicle release machinery, and finally mitochondria. Synaptic vesicles and active zone proteins are trafficked separately, suggesting that distinct stages of presynaptic assembly can be regulated by distinct signals. Furthermore, elements of the active zone can recruit other presynaptic proteins. For example, the active zone protein Bassoon directly recruits voltage-gated calcium channels.

The elements on the postsynaptic side also arrive at the nascent synapse independently from presynaptic elements. Clusters of NMDA receptors are transported in packets (akin to the synaptic vesicle and active zone packets presynaptically), bound to postsynaptic density components. The recruitment of NMDA receptor components can be independent of recruitment of PSD95, suggesting that there may be numerous and simultaneous processes of postsynaptic protein recruitment as well. PSD95 is not only transported along the dendrite as a packet, but also accumulates at the synapse from a cytosolic pool or from neighboring synapses. Thus, the postsynaptic density could form de novo, from the cytosolic pool, or by splitting from preexisting postsynaptic densities.

The fact that presynaptic components assemble before postsynaptic components (at least in vitro) and that they spontaneously exocytose neurotransmitter during synaptogenesis suggests that the released neurotransmitter could facilitate the accumulation of the appropriate receptors (eg, glutamatergic receptors to glutamate, GABAergic receptors to GABA) at the nascent synapse on the dendrite. Indeed, NMDA-type glutamate receptors that are trafficked along dendrites are routinely exposed to cell surface to sample the extracellular space. As further support, blocking neurotransmitter release in retinal ganglion cells during development leads to fewer synapses in the retina. This suggests that neurotransmitter release contributes to even the earliest stages of synapse development.

Molecular Mechanisms of Synaptic Differentiation

The process of synaptic differentiation involves the presentation of synaptic organizing molecules from the postsynaptic target that recruit synaptic vesicles and related machinery to the presynaptic terminal. Then, via reciprocal signaling via secreted molecules and cell-adhesion molecules, the postsynaptic structures form and aggregate. Molecules that are able to induce presynaptic or postsynaptic differentiation are called synaptic organizers, and simplistically can be categorized into secreted molecules and cell-adhesion molecules. Secreted molecules are thought to induce the initial synaptic differentiation because they can be released prior to contact between the presynaptic and postsynaptic neurons. They represent a diverse group of signaling molecules (Table 4.1). Secreted molecules function through activating specific receptors that then induce presynaptic and/or postsynaptic assembly to induce distinct signaling pathways that are still largely unknown. Secreted molecules can have both an autocrine (onto the same neuron it was secreted from) and a paracrine (onto the neuron across the synaptic

TABLE 4.1 Secreted Molecules That Regulate Synaptic Development Within the Central Nervous System. These Molecules Either Promote Presynaptic or Postsynaptic Differentiation

Class of Molecule	Molecule	Role in Synaptic Development
Fibroblast growth factors	FGF2	Presynaptic vesicle clustering
	FGF7	Presynaptic GABAergic vesicle clustering
	FGF22	Presynaptic glutamatergic vesicle clustering
	FGF10	Presynaptic vesicle clustering
Wnts	Wnt7a	Presynaptic differentiation and PSD95 clustering
	Wnt3	Presynaptic development
	Wnt5a	PSD95 clustering and GABA receptor clustering
	Wnt/*lin-44*	Inhibition of presynaptic development
Neurotrophic factors	GDNF	Presynaptic vesicle clustering
	BDNF	Presynaptic vesicle clustering and mobilization, postsynaptic differentiation
	NT-3	Postsynaptic receptor clustering
Pentraxins	Narp	AMPA receptor clustering
	NP1	AMPA receptor clustering
	NPR	AMPA receptor clustering
Semaphorins	Sema3F	Spine formation
Netrins	UNC-6	Presynaptic assembly
Sonic hedgehog	Sonic hedgehog	Induces presynaptic differentiation
Thrombospondins	Thromospondin-1	When released from astrocytes, promotes presynaptic vesicle clustering
Complements	C1q	When released from microglia, promotes pruning of excitatory synapses
Cerebellins	Cerebellin-1	Formation and stabilization of excitatory synapses, bidirectionally
LGI	LGI1	Synaptic maturation and refinement

cleft) effect. In addition to regulating synaptic differentiation, many synaptic organizers continue to influence and modulate synaptic maturation, maintenance, and plasticity.

Once contact has formed between two neurons, cell-adhesion molecules can signal for the differentiation and maturation of synaptogenesis (Table 4.2). The synaptogenic cell-adhesion molecules are transmembrane proteins, with relatively short intracellular domains compared to their long, extracellular domains that contain many "sticky" protein motifs. A general principle for their mechanism of action is transsynaptic binding—either homophilic (the same molecule is expressed both presynaptically and postsynaptically and binds to itself across the synapse, eg, SynCAM1 or cadherins) or heterophilic (different molecules are expressed presynaptically and postsynaptically, eg, neurexins, which are presynaptic, bind to neuroligins, which are postsynaptic). Transsynaptic binding also allows recognition of specific synaptic partners. For example, two cell-adhesion molecules, NGL-1 and NGL-2, localize to distinct regions of the CA1 pyramidal neuron dendrite, which form connections with axons from different regions of the brain.

TABLE 4.2 Cell-Adhesion Molecules, How They Promote Synaptic Differentiation and Development, and Their Link to Diseases

Presynaptic Molecule	Postsynaptic Receptor	Role	Disease Link
Neurexins	Neuroligins	Can trigger both presynaptic and postsynaptic differentiations. Different isoforms function at excitatory and inhibitory synaptic development.	NLGN3 and NLGN4 mutations have been associated with autism spectrum disorder. NLGN4 mutation has also been associated with intellectual disability.
SynCAMs	SynCAMs	Triggers presynaptic differentiation.	Not yet known.
Receptor protein tyrosine phosphatases, Slit- and Trk-like proteins, and various others	Leucine-rich repeat (LRR) transmembrane proteins	Regulates the development, function, and plasticity of both excitatory and inhibitory synapses. Thought to be particularly important for the maintenance of excitatory/inhibitory balance.	Linked to autism spectrum disorder, schizophrenia, and attention deficit hyperactivity disorder. In addition, LRRTM3 has been linked to Alzheimer's disease.
Cadherins	Cadherins	Important for cell–cell recognition and maintenance of the synaptic structure.	Have been associated with epilepsy, intellectual disability, autism, bipolar disease, autism spectrum disorder, and schizophrenia.
Ephrins	Ephs	Axon guidance, NMDA receptor clustering, dendritic morphology, and synaptic plasticity.	In animal models have been linked to spinal cord injury, demyelinating disease, and Parkinson's disease.
Netrin-G, LAR, PTP	Netrin-G ligands (NGLs)	Excitatory synapse formation via interaction with PSD-95.	Mutations have been correlated with schizophrenia, Rett syndrome, and bipolar disorder.
Basal lamina proteins, Fasciculin 2, SPARC	Integrins	Presynaptic and postsynaptic maturation and dendritic spine plasticity.	Associated with Alzheimer's disease, epilepsy, stroke, and Down's syndrome.

On the postsynaptic side, cell-adhesion molecules and synaptic density proteins reciprocally affect each other's synaptic localization—the postsynaptic density protein Dlg5 is required for dendritic spine formation by delivering N-cadherin to the surface. Cotransport of cell-adhesion molecules and glutamate receptors also occurs. Overall, synaptogenic cell-adhesion molecules are important for organizing presynaptic and postsynaptic densities and for recognition of synaptic partners via direct binding in the extracellular matrix; and disruption of these molecules has been implicated in many diseases (Table 4.3).

Synaptic Refinement

Initially, an overabundance of synapses is formed followed by a period of refinement, where more active synapses are stabilized and weaker synapses eliminated. The phenomenon of synapse elimination was initially characterized at the neuromuscular junction. Muscle fibers are originally innervated by several axons, but a period of activity-dependent refinement whittles this down to a single innervating axon for each muscle fiber. A similar phenomenon is observed at the climbing fiber to Purkinje cell synapse in the developing cerebellum: during development, a Purkinje cells receives synaptic input from several climbing fibers, but Purkinje cells in the mature brain receive input from only a single climbing fiber. If climbing fiber refinement onto Purkinje cells fails in mice, the mice display ataxia, demonstrating that loss of appropriate refinement has functional consequences. Thus, the process of synapse elimination is critical for the proper functioning of an organism.

The process of synaptic refinement is also critical for the development of sensory systems within the brain, by sharpening the connectivity of receptive fields. Similar to the innervation of muscle fibers, sensory inputs overgrow into the target brain area and initially intermingle with each other. Then through synaptic pruning the connectivity is whittled down to nonoverlapping targets. In rodents, each whisker has a clear corresponding map within the barrel fields of the mature brain. If refinement fails, the map does not clearly differentiate and sensory deficits occur. In the visual system of a wide variety of mammalian species, refinement is critical for eye-specific segregation of inputs into distinctly separate rather than overlapping regions. In the lateral geniculate nucleus, axon terminals of retinal ganglion cells separate into eye-specific

TABLE 4.3 Genes Associated With Schizophrenia. The Genes That Have Been Linked to Schizophrenia Represent Very Disparate Functions Both in Terms of Synaptic Function and Type of Synapse

Gene	Protein	Function
CHRNA7	Nicotinic acetylcholine receptor, alpha 7 subunit	Excitatory synaptic transmission
COMT	Catechol-*O*-methyltransferase	Degradation of catecholamines
DISC1	Disrupted in schizophrenia 1	Regulates neurite outgrowth and neural proliferation. By regulating the mTOR pathway, modulates synaptic and dendritic development
DTNBP1	Dysbindin-1	Regulates spine development
Npas3	Neuronal PAS domain protein 3	Transcription factor
SP4	Sp4 transcription factor	Transcription factor
NLGN1	Neuroligin 1	Cell-adhesion molecule that regulates excitatory synaptic development

regions after refinement. The neurons from the lateral geniculate nucleus then project to the visual cortex where synaptic pruning also occurs. This process leads to the formation of ocular dominance columns in the visual cortex, with each column favoring input from one eye.

As with the neuromuscular junction and cerebellum, the refinement of receptive fields requires neural activity. During early stages of development, prior to sensory input, spontaneous neural activity contributes to refinement, such as retinal waves. Later, sensory-driven refinement peaks during the critical period, a phase when synapses are especially plastic and sensitive to the environment. After the critical period, receptive fields are less malleable. The critical period was initially discovered by Hubel and Wiesel in the developing cat visual system. When a kitten is visually deprived by occluding one eye, the visual field of the open eye takes over that of the closed eye. This is not reversed if the eye is opened after the critical period. If visual occlusion happens later (eg, during adulthood), the visual fields are not rewired. Because many developmental events are time-sensitive, sensory deprivation of infants or children can also have long-term consequences, and extending the critical period would have clear clinical applications.

One theory for how activity is translated into a molecular readout for synaptic refinement is that there exist protective (trophic) and punishment (toxic) signals that are released in response to activity (Fig. 4.4). The trophic molecules are released at highly active synapses that aid in the strengthening and maturing of the synapse. Conversely, toxic molecules attack inactive synapses, and promote synaptic retraction. What these molecules are and how they signal remains an active area of study, although candidates are emerging. The neurotrophins are one family of molecules that have been implicated in promoting stabilization of active synapses. In addition, tagging and elimination of weaker (less active) synapses by microglia and astrocytes is an important regulator of synaptic pruning. In particular, signaling through the complement cascade can induce microglial engulfment of synaptic terminals.

In addition to neural activity from excitatory (glutamatergic) synapses, the activity by GABAergic signaling, especially from parvalbumin interneurons sculpt the network. Parvalbumin interneurons are a subclass of GABAergic

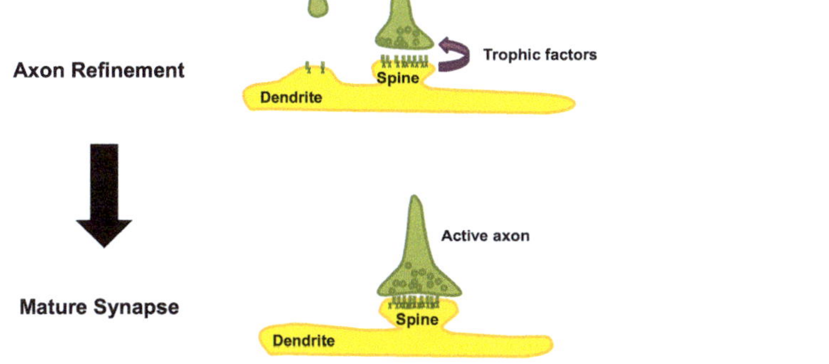

FIGURE 4.4 **A model of activity-dependent competition underlying synaptic refinement at glutamatergic synapses.** During development, an excessive number of synaptic connections is formed. However, these connections do not all have the same synaptic strength, some being weaker (less active) than the others. Stronger (more active) synapses drive synaptic refinement. Trophic factors send protective signals selectively to the stronger axon terminal that will stabilize and further mature that synaptic connection. Toxic factors preferentially target weaker synapses, where they promote retraction of the weaker synaptic terminal. The result is the proper number of mature connections between axon and dendrite.

interneurons that have a strong inhibitory effect on pyramidal neurons. When the development of parvalbumin neurons is impaired, synaptic refinement is perturbed and the critical period is deregulated. In the visual system, this results in impaired visual acuity because of incomplete separation of ocular dominance columns. Thus, patterns of network activity in the circuit have important effects on individual synapses.

While the role of activity driven by sensory input as a driver of synaptic refinement is well-supported, current research is finding that activity-dependent refinement is also clearly happening in other, nonsensory areas of the brain, such as the cerebellum, other cortical areas, and the hippocampus.

Synaptic Maturation and Maintenance

The final stage of synaptic development is the maturation and maintenance stage, marked by the strengthening and stabilization of synaptic connections followed by the continuous replenishment and upkeep of synaptic proteins. Maturation overlaps with synaptic refinement, as stronger connections are more likely to be maintained, and molecules that promote stabilization can also act as a maturation signal. Many molecules that guide early synaptogenesis continue to function during maturation and maintenance.

Synaptic Maturation

Maturing synapses take on distinct electrophysiological properties, compared to their younger selves. In excitatory synapses, there is a shift in the composition of glutamate receptors: the ratio of AMPA to NMDA glutamate receptors increases with maturation. In fact, many nascent excitatory synapses are silent synapses, which are synapses that are electrically silent because they express only NMDA receptors. NMDA receptors have magnesium blockade that is voltage dependent; in the presence of AMPA receptors, the magnesium blockade is alleviated, but without AMPA receptors present, glutamate signaling fails to depolarize the postsynaptic membrane.

Not only does the composition of receptors change at maturing synapses, but even the subunit composition of receptors changes. NMDA receptor subunits switch as part of neural maturation, with a gradual replacement of NMDA-type glutamate receptor 2B (GluN2B) subunits (which contain a C terminus CaMKII-binding region) by GluN2A subunits (which do not have this C terminus region). Functionally, mature receptors (containing GluN2A) have shorter ion channel open time, changing the electrophysiological dynamics at the postsynaptic membrane. The switch in NMDA receptor subunit composition is critical to synapse development: prolonged overexpression of GluN2B induces prolonged spine motility, while premature expression of GluN2A decreases spine density and volume, and decreases spine motility. Interestingly, GluN2B has opposite effects in synapse formation in excitatory and inhibitory neurons: GluN2B delays the formation of glutamatergic synapses onto excitatory neurons, but promotes excitatory synapse formation onto inhibitory synapses. The switch in NMDA-type glutamate receptor subunits may be an important mechanism contributing to balanced formation of excitatory and inhibitory tone in the neural network.

The shift in GluN2B and GluN2A subunits does not happen in all regions of the brain, and it is correlated with a shift in NMDA-type glutamate receptor co-agonist: in the hippocampus, the switch occurs in the synapses between CA3 and CA1 pyramidal neurons, but not at the synapses from entorhinal cortical neurons onto dentate granule cells. There is a concurrent shift in NMDA-type glutamate receptor co-agonists released in these regions—serine (preferentially used by GluN2A) starts to be expressed at the time of the subunit switch in the CA1, but glycine (preferentially used by GluN2B) remains released in the dentate gyrus. This suggests that there is coregulation of presynaptic and postsynaptic aspects of the physiological change at glutamatergic synapses.

Concomitant with maturation in receptor composition at synapses, mature forms of synaptic plasticity, such as long-term potentiation (LTP) also begin to appear. LTP is difficult to induce in immature circuits by conventional stimulation protocols, like high-frequency stimulation, and, when induced, is often mechanistically distinct from the plasticity present at mature synapses. In CA1 of the hippocampus, LTP in mature, but not juvenile, circuits is dependent on GABAergic signaling and somatic spiking. Synaptic plasticity involves an activity-dependent rearrangement in the composition of neurotransmitter receptors at synapses analogous to synaptic maturation, and thus, unsurprisingly, there is a large overlap in the molecular cues that guide synapse maturation and plasticity processes.

The morphology of synapses changes with maturation; at excitatory synapses, this includes maturation of dendritic spines. Spines have diverse shapes, often described as stubby, thin, or mushroom, and are highly dynamic structures. Spines initially form as filopodia, long thin protrusions without postsynaptic machinery, while mature spines typically have a mushroom appearance with a wide head and a thin neck; the wide head accommodating a larger number of receptors, and the thin neck facilitating the compartmentalization of the spine. Spine shape changes directly in response to exposure to glutamate. Spines are highly dynamic both in vitro and in vivo, and immature spines tend to be more dynamic than mature spines. Live imaging in the cortex of developing mice revealed that dendritic spines can form and retract within 2h. Younger animals have higher dynamics of spine formation and retraction than older animals, although turnover does persist into the adult brain, supporting the notion that elevated dynamics of synaptic structure turnover are an important part of development. In fact, filopodia-like spines have a higher motility and are more likely to form transient contacts with presynaptic boutons, compared to spines with more "mature" structure, which are less likely to be motile and more likely to form more stable contacts with presynaptic boutons. Proper spine maturation is necessary for healthy brain function as spine defects are present in many psychiatric and neurological diseases.

Synaptic Maintenance

The maintenance of synapses occurs beyond the developmental period, lasting throughout life. Even once mature, synapses are not stable entities. They can strengthen and weaken in response to LTP and long-term depression, which are thought to be the mechanisms underlying learning and memory processes. Structural and molecular changes lead to functional plasticity changes. Synapses have a high energy requirement, because synaptic vesicle turnover expends energy in addition to replenishing postsynaptic specific molecules. Because maintenance of functional synapses has such a high cost, it is often a point of vulnerability: synaptic dysfunction precedes cell death in many neurodegenerative diseases.

Molecules Involved in Synapse Maturation and Maintenance and Their Link to Neuropsychiatric Disorders

Many synaptic molecules regulate synapse maturation and maintenance and have been connected with a variety of diseases. For example, there are three *SHANK* genes that all encode for scaffolding proteins that are part of the postsynaptic density of excitatory synapses, making them important in maintenance of postsynaptic structure and molecular composition. Phelan–McDermid syndrome is caused by a 22q13 deletion that encompasses the *SHANK3* gene, and is characterized by severe intellectual disability, developmental delays, absent or delayed speech, neonatal hypotonia, and facial dysmorphism. Patients also frequently have autism and epilepsy. Mutations in all three *SHANK* genes have been associated with autism spectrum disorders.

SHANK3 mutations or deletions have also been associated with schizophrenia, bipolar disorder, and autism. Based upon the clinical observation that many people with *SHANK3* deletion have a loss of cognitive skills during adulthood, a link with Alzheimer's disease has also been postulated. Synaptic cell-adhesion molecules, such as N-cadherin, are also involved in the maturation and maintenance of the synaptic structure.

Neurotrophic factors have a wide variety of roles in modulating and maintaining the nervous system, although the exact role that they play in synaptic maturation is not fully understood. Brain-derived neurotrophic factor (BDNF) can regulate both excitatory and inhibitory maturation, depending on the brain region. BDNF is linked to a variety of disorders including bipolar disorder, major depressive disorder, obsessive compulsive disorder, eating disorders, intellectual disability, and schizophrenia. A specific V66M polymorphism in BDNF is associated with impaired BDNF trafficking, impaired learning and memory, and neuropsychiatric diseases. Other neurotrophins have also been linked with intellectual disability, sensory axonal neuropathy, and anxiety disorders.

Furthermore, local protein translation is of critical importance to synaptic maintenance and function. The presence of ribosomes at dendritic spines and the translation of mRNA locally, in or near spines, is critically important for spine morphogenesis and synaptic plasticity. Mammalian target of rapamycin (mTOR) is a protein kinase that is critically important for the regulation of translation at spines. In animal models of tuberous sclerosis and fragile X syndrome, mTOR appears to be hyperactive, while it is downregulated in models of trisomy 21 and Rett syndrome. Interestingly, rapamycin, an mTOR inhibitor, can rescue deficits in synaptic plasticity and behavior seen in tuberous sclerosis mouse models. Thus, regulation of local translation by mTOR appears to be critical for synaptic function, and its deregulation contributes to a wide variety of neurodevelopmental disorders.

In addition to the critical role of activity in synaptic refinement, the neural network requires constant neural activity for its survival. Thus, activity is an important factor driving the later stages of neural development. Deletion of a SNARE protein that is required for the exocytosis of synaptic vesicles, Mammalian uncoordinated-18 (Munc18-1), does not impair the initial formation of brain circuits, but without neurotransmitter release, the neurons then progressively degenerated through apoptosis, suggesting that the maintenance of a functional brain requires constant activity and/or that neurotransmitters themselves are trophic. Neuron depolarization controls gene expression: depolarization of dendrites induces an increase in intracellular calcium, which regulates the transcription factors, nuclear factor of activated T-cells (NFAT), cAMP response element-binding protein (CREB), and neuronal PAS domain protein 4 (NPAS4), and these in turn control both synapse maturation and excitatory versus inhibitory synapse balance. Thus, activity of the circuit and molecular control of synapse maturation form a reciprocal relationship in controlling synapse maintenance.

DIVERSITY OF SYNAPTIC DEVELOPMENT

The basic process of synaptic development and the majority of mechanistic insight into how synaptogenesis is regulated are derived from the study of glutamatergic synapses in the CNS as described previously. Yet, there are many different types of synapses to form within the nervous system (Fig. 4.1). The regulation of synaptic development at nonglutamatergic synapses is described in the following section.

Synapses of the Peripheral Nervous System

The specialized synapse whose development that has been best studied is the neuromuscular junction, and much of our understanding of how synapses form is based upon work initially conducted here. The basic process of synaptic development at the neuromuscular junction has been well characterized. The axon of the motor neuron approaches a muscle fiber, which has some prepatterning of acetylcholine receptors. Then as contact is made, the motor neuron release agrin, which leads to the further clustering of acetylcholine receptors by signaling through a receptor tyrosine kinase—muscle-specific kinase, MuSK—and an associated protein, rapsyn. Acetylcholine receptor clustering is further enhanced by increased transcription of the receptors locally by a secreted molecule, neuregulin, and its signaling through erythroblastic leukemia viral oncogene homolog B isoform (ErbB) receptor tyrosine kinases. The muscle fiber also releases signals, including extracellular matrix proteins that lead to the differentiation of the presynaptic terminal. The motor neuron can release neurotransmitter prior to reaching the muscle fiber, but as development progresses, the strength of this connection is enhanced. A period of maturation and refinement then begins, whereby an excess number of axons are eliminated from the muscle fiber via an activity-dependent competition. Maturation involves a morphological change of the nerve terminal, and expansion of its innervation field. Finally, neuromuscular junctions are maintained throughout life; the basal lamina is important for this maintenance. Failure of neuromuscular junction can lead to many diseases. Myasthenia gravis is an autoimmune disorder that attacks the neuromuscular junction, causing neurodegeneration. Specifically, autoantibodies against the postsynaptic proteins: acetylcholine receptors, MuSK, and low-density lipoprotein receptor-related protein 4 (Lrp4) cause the body to attack the neuromuscular synapse, which leads to decreased function. The hallmark of this disease is muscle weakness, which highlights the importance of neuromuscular function.

Study of synaptic development in other areas of the peripheral nervous system has been less well studied and is not well understood. An interesting phenomenon is neurotransmitter switch: a switch in neurotransmitter and neurotransmitter receptors occurring within the peripheral nervous system as the system matures. For example, mesenteric neurons switch from adrenergic receptors at the immature synapse to purinergic receptors at the mature synapses.

Inhibitory Synaptogenesis

The development of GABAergic synapses are less well understood than excitatory synaptic development, although there are many similarities between the two processes. Similarities include the stages of development, dynamics, and some of the mechanisms. The overall process of synaptogenesis follows the same general process outlined earlier in this chapter. Dynamics of inhibitory synapse formation in culture are also similar to excitatory synapse formation, although the dynamics may be slower. In culture, de novo synapses appear over 8h after contact, with presynaptic protein accumulation preceding postsynaptic protein accumulation by 2h, and synapses are often mobile as a presynaptic–postsynaptic unit.

As with excitatory synaptogenesis, neural activity is an important regulator of inhibitory synaptic development. However, unlike excitatory synapses where local activity, such as competition between two neighboring axons, is critical, inhibitory synapses are usually regulated by global activity patterns. The molecules that govern inhibitory synaptogenesis are not as well-known as those that govern glutamatergic synaptogenesis. Molecules important for inhibitory synaptogenesis include: BDNF, a neurotrophic factor, and NPAS4, an activity-dependent transcription factor, both of which are critical for the development of inhibitory synaptic transmission; and Slitrk, which signals through protein tyrosine phosphatase delta (PTPδ) to induce inhibitory synaptic differentiation. There are also groups of signaling molecules where different family members regulate excitatory versus inhibitory synaptic development. This may be one way to preserve mechanism while allowing for specificity. For example, fibroblast growth factor (FGF) 7 can induce inhibitory synaptic differentiation, while FGF22 induces excitatory synaptic differentiation, and neuroligin-2 preferentially promotes inhibitory synaptogenesis, while neuroligin-1 promotes excitatory synaptogenesis. Wnts can also regulate excitatory versus inhibitory synaptic development: Wnt5a induces PSD-95 clustering at excitatory synapses through a mechanism involving Jun amino-terminal kinase (JNK), but GABA receptor clustering at inhibitory synapses through CaMKII. Components of the inhibitory postsynaptic densities can recruit other postsynaptic proteins, like at excitatory synapses. For example, gephyrin, the postsynaptic density protein at inhibitory synapses, is required for clustering of GABA receptors at inhibitory postsynaptic densities.

Overall, inhibitory synaptogenesis is of particular importance to preserve the appropriate excitatory/inhibitory balance within the brain. This balance is necessary for the brain to function properly: too much excitation and the brain is epileptic, too little and no learning or plasticity may result. An imbalance between excitatory and inhibitory synapses is implicated in a diverse set of diseases including autism, schizophrenia, and epilepsy. A major remaining question in the field of neural development is how inhibitory and excitatory synaptic development interacts to sustain this balance.

Neuromodulatory Synapses

The overall development of the synapses that release transmitters other than glutamate or GABA in the CNS are thought to follow the same general process of guidance, differentiation, pruning, and maturation. Maintenance of these connections is then particularly important for proper brain function, as these neuromodulatory neurotransmitters are often lost in neurodegenerative disorders. However, to date, understanding the specific processes and mechanisms involved in the development of neuromodulatory synapses is not well understood. While these modulatory neurotransmitters represent a small percentage of synapses within the brain, their broad distribution and functions make understanding their development a critical question for future research.

What is clear is that the neuromodulatory neurotransmitters regulate neuronal development, GABAergic synaptogenesis, and glutamatergic synaptogenesis. For example, dopamine signaling is critical for neurogenesis and synaptogenesis throughout much of the rest of the CNS. Dopamine regulates cell proliferation, migration, differentiation, dendrite growth, and excitatory/inhibitory balance. Because dopamine can influence all of these processes, impaired dopamine development has been linked to many psychiatric disorders, such as attention deficit hyperactivity disorder, addiction disorders, schizophrenia, major depressive disorder, and bipolar disorder. In many cases, signaling through D1- and D2-type dopamine receptors has opposite effects on the neural development. For example, agonism of D1 receptor promotes tangential migration of GABAergic interneurons, whereas agonism of D2 receptors inhibits migration. Similarly, adrenergic signaling can regulate excitatory and inhibitory neurogenesis and synaptogenesis in the CNS. Serotonin and acetylcholine regulate neurite outgrowth. Cholinergic transmission is also thought to be important for cortical thickness and lamination, and noradrenergic neurons regulate lamination.

Electrical Synapses

The majority of this chapter focuses on chemical synapses, which are the most prevalent form of communication in the CNS; however, electrical synapses also exist in both neurons and glia.

Electrical synapses, also called gap junctions, are tightly apposed pores between two cells formed by connexins, which allow direct cytoplasmic exchange, including passage of electric current, as well as other small biomolecules. Electrical synapses synchronize the activity of connected neurons. Gap junctions are critical for the development of the nervous system, and, in neurons, are transiently expressed at higher levels during the first 2weeks of life in rodents. This expression appears to be controlled by the opposing action of GABA signaling and metabotropic glutamate signaling. Connexin36 is the most common neuronal connexin and is thought to be important for regulating synchrony in many regions throughout the brain. Mutations in connexin32 can cause a form of X-linked Charcot–Marie–Tooth disease (CMTX), an inherited demyelinating neuropathy marked by weakness and muscle atrophy in the legs and feet. Later in the disease, the hands and tongue may also be affected. Disease onset is usually during adolescence. In CMTX, communication between the myelinating Schwann cell and axon of the motor neuron is impaired. While critical for brain function, how electrical synapse formation is regulated remains an area of active research.

THE ROLES OF GLIA IN SYNAPTOGENESIS

Much of the research on synaptic development focuses on the direct interaction of the presynaptic and postsynaptic compartments, but it is becoming increasingly clear that glial cells, which surround and support synapses, are also critical for synaptic development. In addition to molecular cues exchanged between the presynaptic and postsynaptic neurons, the balance between specific synapse formation and broad coordination of synapse formation throughout the brain is also informed by cues from astrocytes and microglia. Because of their ability to sample broadly across populations of neurons by forming extensive contacts with thousands of neurons, astrocytes and microglia may provide an important mechanism of coordination of synapse development in parallel. The role of glia in synaptogenesis was first demonstrated by Barres and colleagues, when they demonstrated that neuronal media in which glia were grown had a very strong synaptogenic effect in vitro. Since then, the role of astrocytes and microglia has been increasingly appreciated.

Astrocytes closely contact synapses, poised for their roles in modulating synapse formation and function. Astrocytes can promote both excitatory and inhibitory synaptogenesis through contact via the γ-protocadherins—loss of astrocytic expression of γ-protocadherin leads to lower excitatory and inhibitory synapse number embryonically that recovers in early postnatal development of mice. Astrocytes also secrete numerous factors to induce synaptogenesis. Thrombospondins are secreted during early synapse development and induce excitatory synapse formation through binding to the calcium channel subunit Cacna2d1, although the synapses induced by thrombospondins are silent, and require further signaling for full maturation. Transforming growth factor beta (TGF-β) secreted by astrocytes onto nascent synapses induces increased excitatory synapse number through upregulating D-serine expression, which is a co-agonist of NMDA receptors. During later stages of development, hevin and secreted protein acidic and rich in cysteine (SPARC) are secreted from astrocytes to modulate excitatory synaptogenesis. Interestingly, hevin induces growth of excitatory presynaptic boutons, but SPARC directly antagonizes it. Furthermore, in cortical layer 2/3, hevin specifically promotes presynaptic maturation of inputs from the thalamus over inputs from cortex, suggesting a role of astrocytes in controlling synapse specificity amidst a population of neurons. Thus, astrocytes can use numerous mechanisms to coordinate synaptogenesis throughout development across a larger population of neurons that are in parallel neural circuits, but not necessarily in direct contact with each other.

Microglia, once thought of as merely the macrophages of the CNS, also have important roles in synapse formation. Microglia can directly contact synapses in a highly dynamic manner. Microglia express trophic signals, such as Insulin-like growth factor 1 (IGF-1) that may have important instructive roles in synapse maturation. Indeed, loss of the fractalkine receptor, CX3CR1, in microglia leads to delayed maturation of excitatory thalamocortical synapse maturation in barrel cortex—the synapses have a delay in the increase of AMPA-type to NMDA-type glutamate receptor ratio and GluN2B to GluN2A subunit switch. Similarly, loss of the transmembrane polypeptide DAP12 in microglia causes a failure in the functional glutamate receptor rearrangement in maturating synapses in hippocampus. Furthermore, microglia have an important role in later stages of excitatory synapse elimination: engulfment of synapses to be eliminated through a mechanism co-opting the immune system complement signaling. Thus, microglia, like astrocytes, have the ability to coordinate synapse formation across populations of neurons. Whereas astrocytes appear to control synapse formation throughout development, and especially in early stages of synaptogenesis, microglia appear to have a role predominantly in coordinating later steps of synapse development, such as functional glutamate receptor rearrangements and synapse elimination.

SYNAPTOGENESIS AND DISEASE

Synapses are essential to how information is processed in the brain; therefore, it is intuitively obvious that defective synapse formation leads to neuropsychiatric or neurological disease. There are many ways in which failure of synapse development can lead to disease. Failure to form connections between the correct synaptic partners (or formation of connections between inappropriate synaptic partners) can lead to an alteration in the areas of the brain that are connected, resulting in altered information processing (which is thought to underlie both schizophrenia and autism). Similarly, failure to form fully functional synapses can cause the connections with inappropriate synaptic weight, leading to failures in information propagation (which can underlie schizophrenia, autism, mental retardation, and epilepsy). Failure of synapses to acquire the ability to have changes in synaptic plasticity can lead to altered learning (which is thought to be a contributing factor to addiction). Finally, failure of synapses to maintain healthy structure throughout life can lead to neurodegeneration (eg, in Alzheimer's disease).

There are four lines of evidence for defects in synapses leading to neuropsychiatric and neurological disorders: (1) disease time course; (2) synaptic changes found in human patients; (3) genetic studies identifying synapse-related genes in human patients; and (4) animal models with mutated or deleted synaptic genes. For example, the time course of emergence of such diseases as autism or schizophrenia in humans coincides with important events in synapse development.

Disorders of Axon Guidance

Many mechanisms linked to axon pathfinding and outgrowth have been linked to human disorders including: agenesis of the corpus callosum, L1 syndrome, Joubert syndrome, Kallmann syndrome, and albinism, among others. In many cases, genetic mutations underlying these disorders are linked to molecules associated with axon guidance or growth cone regulation. For example, L1 syndrome arises from mutation to the *L1CAM* gene. L1 can act as a short-range guidance molecule through interactions with integrins, neural cell–adhesion molecule (NCAM), contactin, neuropilin, and transient axonal glycoprotein-1 (TAG-1). Through these binding partners, L1 enhances axon outgrowth and attraction.

In the specific case of deficiency in midline crossing, such as disorders with congenital mirror movements or horizontal gaze palsy with progressive scoliosis, underlying mutations have been linked to either the *DCC* or the *Robo3* genes. DCC is a receptor for the classic guidance molecule netrin-1. Robo3 functions to inhibit slit repulsion and facilitates netrin attraction. In particular, *Robo3* knockout mice recapitulate many of the phenotypes in midline crossing seen in patients with horizontal gaze palsy with progressive scoliosis.

Disorders of Differentiation, Refinement, and Maturation

Autism Spectrum Disorders

Autism is a group of developmental disorders characterized by impaired social skills and repetitive behaviors that typically manifest around two years of age after overtly normal infantile development. This onset corresponds to the peak of synaptogenesis. In fact, children with autistic brains appear to have initially smaller brains at birth that then grow to be larger than average. Synaptic spine development at this time appears to be particularly altered. In syndromes linked to autism, there are changes in dendritic spine morphology and density, but the changes differ between syndromes—in tuberous sclerosis there is an increased density in dendritic spines; in fragile X, spine density is not changed but the synapses stay thin and filopodial without maturing; and in Rett and Angelman syndromes, spine density is decreased and spines are less mature. The mechanism underpinning these spine changes is an area of intensive research.

Genome-wide association studies, which are large-scale studies of common genetic variants, of autism have found mutations that lead to impairments in steroidogenesis, neurite outgrowth and guidance, and glutamatergic synaptic development and transmission. Of particular importance to synaptic development and function, proteins implicated from these genetic studies include many different cell-adhesion molecules, such as neurexin and neuroligin, glutamate receptor subunits, and voltage-gated calcium channels. In addition, severe forms of autism are associated with missense and dosage sequence variants in the gene for the adhesive junction-associated δ-catenin protein, which binds to and regulates cadherins. All of these proteins either regulate synaptic development or are involved in synaptic function.

In addition to using large scale genomic studies to understand the molecular underpinnings of the autism spectrum disorders, syndromes associated with autism provide potential clues to the genetic causes of autism spectrum disorders in general. Syndromic forms of autism include fragile X syndrome, Rett syndrome, and Angelman syndrome. The clinical features of fragile X syndrome include intellectual disability, hyperactivity, hypersensitivity to sensory stimuli, anxiety, impaired visuospatial processing, and developmental delay. The disease is caused by a CGG expansion of more than 200 base pairs upstream of the *fmr1* gene that serves to silence transcription and, therefore, protein expression. In addition to the spine deficits seen in patients with fragile X syndrome, animal models of the disease suggest that there are a variety of other synaptic deficits as well. Loss of *fmr1* leads to deficits in axon branching, dendritic branching, synaptic differentiation, and synaptic pruning. Thus, almost every phase of synapse development appears to be impaired in fragile X syndrome. One of the long-term consequences of improper synaptic development within these models is impaired synaptic plasticity, which may cause some of the disease features.

Rett syndrome usually only affects females, and usually manifests after the first year of life. Features of Rett syndrome include loss of language skills, motor skills, and intellectual disabilities. Rett syndrome is almost always caused by a mutation in the gene encoding for methyl CpG-binding protein 2, *MECP2*, which like fragile X mental retardation protein (FMRP) regulates gene transcription. One of the genes best associated with *MECP2* is BDNF, which modulates many aspects of synaptic development and function. The *CDKL5* and *FOXG1* genes have also been associated with Rett syndrome. Both glutamatergic and GABAergic synaptic transmission is impaired in models of Rett syndrome.

The major features of Angelman syndrome include intellectual disability, lack of speech, and seizures. Angelman syndrome is usually caused by deletion or inactivation of *Ubiquitin-protein ligase E3A* (*UBE3A*). This gene encodes for an E3 ubiquitin protein ligase that is essential for protein degradation. Ubiquitin-mediated protein degradation is critical for the regulation of synapses. This system controls receptor turnover and is critical for the regulation of synaptic strength and plasticity. Decreased excitatory postsynaptic differentiation occurs in Angelman syndrome.

Intellectual Disabilities

Intellectual disabilities have many causes, including genetic causes and environmental toxin exposures (such as fetal alcohol syndrome). Intellectual disabilities are comorbid with many neurodevelopmental disorders, such as the autism spectrum disorders. Major genetic causes of intellectual disability include trisomy 21 (also known as Down's syndrome) and fragile X syndrome, described previously. In trisomy 21, a third copy of chromosome 21 is present. In addition to developmental delays, people with trisomy 21 often have variety of other health problems. The brain, particularly the cerebellum, is smaller in patients with trisomy 21. Synaptic structure is abnormal in animal models of trisomy 21. Dendritic spines do not develop normally, which leads to deficits in synaptic plasticity in animal models of Trisomy 21. A candidate molecule for causing many of these developmental defects is DSCAM (Down's syndrome cell-adhesion molecule). DSCAM is a member of the Immunoglobulin superfamily that can function homophilically or as a receptor for netrin-1. In animal models of Trisomy 21, DSCAM is dysregulated, leading to its overexpression. Further study in animal models suggests that DSCAM regulates dendritic tiling, dendritic branching, dendritic self-regulation, axon arborization, axon pathfinding, synaptic plasticity, and spine morphogenesis (see previous section).

Schizophrenia

Schizophrenia, a psychiatric disease that includes symptoms, such as delusions, hallucinations, and disorganized thinking, typically manifests in late adolescence and early adulthood. In a theory first put forth by Feinberg, the onset of schizophrenia is thought to be linked to aberrant pruning of synapses. In particular, glutamatergic innervation of parvalbumin interneurons appears to be impaired in animal models of schizophrenia. Anatomical findings in postmortem tissues from affected human patients also point to defective synapses. In schizophrenic brains, density of dendritic spines appears to decrease specifically on cortical layer 3 pyramidal neurons, suggesting defective connectivity within a subset of circuits. Problems with synaptic development also lead to impaired synaptic plasticity. There is evidence for altered glutamatergic, GABAergic, dopaminergic, and cholinergic synaptic transmission. Genome-wide association studies of schizophrenia have pulled out a variety of genes involved in neuronal development and maturation (Table 4.3). Thus, although the precise cause of schizophrenia is still unclear, synaptic dysfunction is undoubtedly at the heart of the disorder.

Epilepsy

Seizures during development are different than those later in life, because both glutamate and GABA are excitatory and can activate neurons early in development. Thus, the neonatal brain is highly excitable. Seizures in the immature brain often lead to synaptic reorganization rather than cell death. Seizure disorders can have a wide variety of underlying causes. Some genetic causes have been linked to channelopathy, but others to a more general problem with neural development. For example, mutations in the gene encoding for Leucine-rich, glioma inactivated 1 (*LGI1*) are associated with epilepsy characterized by focal seizures with acoustic auras. The exact role of LGI1 is unknown, but it is linked to the cytoskeleton as well as to interactions with cell-adhesion molecules, such as integrins and syndecans. LGI1 is also associated with neurite outgrowth and dendrite growth and branching. Thus, impaired synaptogenesis could contribute to the development of epilepsy. Disruption of the excitatory and inhibitory balance underlies the development of epilepsy.

Disorders of Synaptic Maturation and Maintenance

Depressive Disorders

Major depressive disorder is characterized by persistent feelings of sadness, and loss of interest or pleasure, and affects about 10–18% of the general population. Both patients and animal models have altered density, morphology, and organization of synaptic spines. Antidepressant drugs function to increase levels of neuromodulators, including serotonin, suggesting that misregulation of neuromodulatory synapses or abnormal neuromodulation or function of excitatory synapses underlies depression. Animal model studies suggest that antidepressant drugs alter spine and dendritic morphology. Thus, synaptic dysfunction may underlie depression.

Bipolar disorder is a psychiatric disorder characterized by intermittent periods of mania and depression with a prevalence of about 1–2% in the general population. The age of onset is usually during late childhood to early adulthood. One hypothesis for the cause of this disorder, the permissive hypothesis, postulates that a deficit in central serotonin neurotransmission is a necessary, but not sufficient, cause. Reduced central serotonergic neurotransmission permits the expression of the disorder, but the phase of the illness is then determined by catecholaminergic neurotransmitter expression. The manic phases are characterized by high levels of catecholaminergic neurotransmission, while the depressive phases are characterized by low levels of norepinephrine and dopamine, as well as by reduced serotonin neurotransmission.

Addiction

Drugs of abuse can also lead to changes in spines and synaptic proteins. Initially, drugs of abuse trigger dopamine release in the reward pathway. To compensate for the increase of dopamine, postsynaptic neurons eventually will either decrease their dopamine receptors or increase the number of dopamine transporters to try and decrease the amount of dopamine within the reward pathway. Animals injected with cocaine show an increase in the number of spines in the nucleus accumbens of the striatum, and synaptic strength is increased. All drugs of abuse eventually lead to some type of physical circuit rewiring and change in synaptic connectivity, and these hardwired changes are thought to cause the cravings, drug seeking behaviors, tolerance, withdrawal, and addiction. Because of these physical changes, drug users have extreme difficulty in ceasing drug use.

Alzheimer's Disease

The clinical features of Alzheimer's disease include memory deficits and cognitive decline. The onset is usually after the age of 65. Familial mutations in amyloid precursor protein, presenilin1, and presenilin2 can cause the disease. The neuropathology of the disease includes the loss of synapses, then the loss of neurons in the brain. Amyloid-β plaques and neurofibrillary lesions are associated with the disease. It is thought that inappropriate amyloid-β processing leads to excess amyloid-β in the synapse, leading to neuronal toxicity. The early stages may also include impaired plasticity. Thus, Alzheimer's can be thought of as a disorder of synaptic maintenance.

Huntington's Disease

Like Alzheimer's disease, Huntington's disease is a disorder of neurodegeneration. It is caused by a heritable mutation (trinucleotide CAG repeats) in the *huntingtin* gene that leads to polyglutamine expansion of the huntingtin protein. Disease onset is generally 30–40 years old and is characterized by motor, cognitive, and often psychiatric disorders. While later stages of the disease are associated by intraneuronal inclusions and neuronal cell death, early symptoms may originate from synaptic dysfunction. The huntingtin protein can interact with a variety of cytoskeletal and synaptic proteins. Mutations to the protein impair its ability to facilitate microtubule transport along the axon. Animal models of Huntington's disease reveal an impairment in excitatory synaptogenesis, specifically synapses mature more quickly in the striatum and in the sensory and motor cortices. As disease progresses, synapse loss becomes prevalent within the striatum and cortex. Exactly how huntingtin regulates excitatory synapses remains an active area of research. Nevertheless, synaptic loss and deregulation are early signs of Huntington's disease.

Parkinson's Disease

Parkinson's disease is a neurodegenerative disorder with a specific loss of dopaminergic neurons within the brain. The major clinical symptoms are tremor, bradykinesia, rigidity, and postural instability. Disease onset is usually after the age of 60. As dopaminergic neurons are the ones lost, treatment is aimed at increasing the available amount of dopamine in the brain of patients. What causes dopamine neurons to degenerate in Parkinson's disease remains unknown, but synapses are affected at both the pre- and postsynaptic sides. Presynaptically, there is experimental evidence that transport and integrity of dopamine-containing synaptic vesicles is impaired. In the postsynaptic cell, the striatal medium spiny neurons, NMDA receptor assembly is impaired at glutamatergic synapses. NMDA receptor assembly and function appears to be particularly sensitive to the loss of dopamine, and the resulting impaired plasticity may underlie the early symptoms of Parkinson's disease.

Overlapping Etiology Within Neurodevelopmental Disorders

Many of the neurodevelopmental disorders, especially mental retardation, autism, epilepsy, and schizophrenia, appear to have overlapping etiologies, and sometimes even cooccur in patients. Autism incidence overlaps with both mental retardation and epilepsy, especially in certain syndromic causes of autism, such as tuberous sclerosis, fragile X syndrome, or Rett syndrome. Schizophrenia can overlap with epilepsy, not only due to common circuits being engaged in epilepsy-associated psychosis, but also due to overlapping genetic causes. The broad disease phenotypes, the overlaps between the neuropsychiatric diseases, and the multiple genetic and developmental routes underlying neuropsychiatric diseases reflect the complexity and delicacy of neural circuit formation.

As described, there is strong evidence that aberrant synapse development can lead to neuropsychiatric disease. Targeting synaptic function and the overall neural circuitry has therapeutic benefits in neuropsychiatric disease. Indeed, many neurological and psychiatric drugs target the synapse. The future holds great promise of new therapies, such as genetically reprogramming stem cells derived from patients and reintroducing them into diseased synaptic circuitry in cases of otherwise unmanageable neurological disease, and there is still a great need for more precise and effective therapies based on the mechanisms underlying synaptic development and function.

Further Reading

[1] Nestler EJ, Hyman SE. Animal models of neuropsychiatric disorders. Nat Neurosci 2010;13(10):1161–9. http://dx.doi.org/10.1038/nn.2647.
[2] Shen K, Scheiffele P. Genetics and cell biology of building specific synaptic connectivity. Annu Rev Neurosci 2010;33:473–507. http://dx.doi.org/10.1146/annurev.neuro.051508.135302.
[3] Sheng M, Kim E. The postsynaptic organization of synapses. Cold Spring Harb Perspect Biol 2011;3(12). http://dx.doi.org/10.1101/cshperspect.a005678.
[4] Tau GZ, Peterson BS. Normal development of brain circuits. Neuropsychopharmacology 2010;35(1):147–68. http://dx.doi.org/10.1038/npp.2009.115.
[5] van Spronsen M, Hoogenraad CC. Synapse pathology in psychiatric and neurologic disease. Curr Neurol Neurosci Rep 2010;10(3):207–14. http://dx.doi.org/10.1007/s11910-010-0104-8.
[6] Walmsley B, Alvarez FJ, Fyffe RE. Diversity of structure and function at mammalian central synapses. Trends Neurosci 1998;21(2):81–8.
[7] Williams AJ, Umemori H. The best-laid plans go oft awry: synaptogenic growth factor signaling in neuropsychiatric disease. Front Synaptic Neurosci 2014;6:4. http://dx.doi.org/10.3389/fnsyn.2014.00004.

CHAPTER

5

The Peripheral and Central Nervous System

S. Franklin

University of Kentucky, Lexington, KY, United States

The human nervous system can be divided into the peripheral nervous system and the central nervous system. The central nervous system includes the brain and the spinal cord and the peripheral nervous system includes everything else including the peripheral nerves, cranial nerves, and autonomic structures.

In this chapter, we consider the surface structures, support structures, regional organization, and blood supply of the brain and spinal cord. The focus of this chapter provides a broad anatomical foundation on which the complex connectivity and function of the human nervous system can be built.

THE NEUROCRANIUM

The human brain is protected and supported by the cranium so that it is not damaged during daily activities. The cranium can be divided into a visceral cranium that is comprised of the bones of the face and a neurocranium that is comprised of the bones and meninges that support the brain. The bones of the neurocranium include the frontal bone, the ethmoid bone, the sphenoid bone, the occipital bone, the paired parietal bones, and paired temporal bones.

The domed dorsal portion of the neurocranium is called the calvaria (skull cap) and is formed by the flat portions of the bones of the neurocranium. The ventral portion of the neurocranium is called the cranial base and is much more irregularly shaped. The cranial base contains numerous foramina of various sizes for the passage of the cranial nerves, blood vessels, and the spinal cord.

Using the bony features of the bones of the neurocranium the cranial base can be generally divided into different regions. A basic understanding of the divisions of the cranial base can be used to appreciate the spatial relationships between the different areas, vessels, and structures of the brain.

The cranial base is divided into three fossae. The anterior cranial fossa begins at the frontal bone and ends at the lesser wing of the sphenoid bone. The shallowest of the cranial fossae, the anterior cranial fossa, mainly supports the frontal lobe and allows passage for the olfactory nerves as they join the olfactory bulb (CN I).

The anterior border of the middle cranial fossa is the lesser wing of the sphenoid and the posterior border is the superior border of the petrous ridge

Conn's Translational Neuroscience
http://dx.doi.org/10.1016/B978-0-12-802381-5.00007-5

of the temporal bone. The floor of the middle cranial fossa is formed mostly by the sphenoid bone and is riddled with foramina that allow for the passage of cranial nerves and blood vessels. The middle cranial fossa contains a midline structure called the sella turcica that translates into "turkish saddle." The sella turcica is formed by central part the sphenoid bone and has four raised clinoid processes, like bed knobs, and a central depression. The pituitary gland resides in its central depression. The cavernous sinuses are located on the lateral surface of the sella turcica.

The dorsum sellae of the sphenoid bone and the petrous ridge of the temporal bone define the anterior boundary of the posterior cranial fossa while the occipital bone forms its posterior and ventral wall. The posterior cranial fossa is the deepest of the three cranial fossae and contains the brainstem and cerebellum. A large opening in the ventral aspect of the occipital bone called foramen magnum allows for the passage of the spinal cord into the posterior cranial fossa.

MENINGES OF THE BRAIN

Like the spinal cord, the brain is a relatively soft structure and must be supported within the bony neurocranium and vertebral column by membranous layers called the meninges. The outer most layer often fused to the overlying bone is called the dura mater.

Deep to the dura mater is a very thin layer called the arachnoid membrane. Deep to the arachnoid membrane is the subarachnoid space that is filled with cerebrospinal fluid (CSF). The subarachnoid space of the neurocranium is continuous with the subarachnoid space of the spinal cord. The arachnoid membrane is not fused to the overlying dura mater but the pressure of the CSF in the subarachnoid space prevents this layer form collapsing on to the surface of the brain or spinal cord.

Arising from the arachnoid layer are thin filaments that traverse the subarachnoid space called arachnoid trabeculae. Arachnoid trabeculae fuse with the 3rd layer of meninges called pia mater that intimately covers the surface of the brain and spinal cord. The meninges generally function to support the brain and spinal cord, support vasculature structures, and contain the CSF as it surrounds that brain and spinal cord. A more detailed discussion of the function of the meninges is addressed later, but an anatomical understanding of the meninges clarifies the basic venous pattern of the brain, the return of CSF to the venous sinuses, and the position of dural folds.

Dura Sinuses

The outermost layer of meninges called the dura mater can be subdivided into the outer periosteal dura mater that is fused to the overlying bone and the inner meningeal dura that is opposed to the arachnid membrane. In most regions, these layers of dura mater are fused together but in certain areas the layers of dura mater separate and create dural folds and dural venous sinuses. Dural venous sinuses are endothelial-lined spaces that receive the venous drainage from the superficial cerebral veins.

The falx cerebri is the largest dural fold that attaches rostrally to the crista galli and passes along the inner surface of the calvaria to attach to the tentorium cerebelli caudally. Along its extent the falx cerebri extends vertically from the inner surface of the calvaria to the corpus callosum separating the two cerebral hemispheres. The large superior sagittal sinus is a dural venous sinus created within the superior edge of the falx cerebri between the periosteal dura and meningeal dura. The superior sagittal sinus sweeps caudally and drains into the confluence of sinuses located at the occipital pole. The inferior sagittal sinus is formed entirely of meningeal dura at the inferior free edge of the falx cerebri. It sweeps posteriorly and first drains into the straight sinus located at the junction between the falx cerebri and the tentorium cerebelli, before draining into the confluence of sinuses.

The tentorium cerebelli forms a horizontal dural tent over the cerebellum protecting it from the cerebral hemispheres above and separating the middle cranial fossa from the inferior cranial fossa. Most of the caudal and lateral attachments of the tentorium cerebelli are to the occipital bone and parietal bone. As the tentorium sweeps forward, it attaches rostrally to the posterior clinoid processes of the sella turcica leaving a large gap called the tentorial notch on the midline. The midbrain is continuous with the diencephalon at the tentorial notch. As described previously, the straight sinus is located at the junction between the falx cerebri and the tentorium cerebelli. Located below the tentorium cerebelli is a vertically oriented dural fold called the falx cerebelli that does not truly separate the hemispheres of the cerebellum. Along the posterior edge of the falx cerebelli is the occipital sinus that drains into the confluence of sinuses. The confluence of sinuses drains to the left and right transverse sinuses that run within the lateral edge of the tentorium cerebelli. The transverse sinuses drain to the sigmoid sinuses that wind ventrally along the lateral wall to ultimately drain into the internal jugular veins.

The last dural fold is called the diaphragm sella. This dural fold is attached to all for clinoid processes of the sella turcica and creates a horizontal sheet separating the central depression of the sella turcica from the remainder of the middle cranial fossa. The pituitary stalk or infundibulum is the only structure that connects the pituitary to the hypothalamus. The diaphragm sella is also the roof of an irregular-shaped dural venous sinus in the region of the sella turcica called the cavernous sinus. The cavernous sinus drains to the transverse sinus or the sigmoid sinus by the super and inferior petrosal sinuses.

Brain Herniation: Clinical Correlate

In the event of elevated intracranial pressure, areas of the brain may herniate into a different compartment. Brain herniation is often found associated with dural folds because of their ridged nature and their proximity to the spaces between compartments. A subfalcine herniation results occurs when a part of the medial aspect of a cerebral hemisphere is pushed under the falx cerebri. This type of herniation is often referred to as a cingulate herniation, because the cingulate gyrus is the most common structure herniated. The uncus is a part of the parahippocampal gyrus located on the ventral aspect of the temporal lobe close to the tentorial notch. An uncal herniation refers to the herniation of the uncus downward through the tentorial notch. The displaced uncus typically puts pressure on the midbrain and the oculomotor nerve (CN III). A central herniation involves the displacement of the diencephalon and cortical structures through the tentorial notch. Central herniations are typically fatal. While not associated with a dural fold, a tonsillar herniation results from the tonsils of the cerebellum being displaced through the foramen magnum. This herniation puts pressure on the caudal medulla, the cervical parts of the spinal cord, and vertebral arteries. Tonsillar herniations may also be observed in a developmental condition called Chiari malformation where the ventral part of the posterior cranial fossa is poorly developed.

DEVELOPMENT OF THE BRAIN VESICLES

The human nervous system develops through a complex process of induction, neurulation, and rapid growth in which a hollow neural tube is formed and reshaped from neuroectoderm cells. A basic understanding of the developmental structures that ultimately form the central nervous system (CNS) lead to an appreciation of the terminology used to describe the adult brain.

By the end of the 4th week of development the cranial end of the neural tube is re-modeled into three primary vesicles called, from cranial to caudal, the prosencephalon, the mesencephalon, and the telencephalon. The central core of these vesicles remains hollow and will eventually give rise to the ventricular system of the brain.

By the end of the 5th week additional growth has led to the primary vesicles differentiating into secondary vesicles. The prosencephalon gives rise to the telencephalon that will become the forebrain (cerebral hemispheres, subcortical white mater, and basal ganglia) and the diencephalon that will become the thalamus, hypothalamus, subthalamus, and epithalamus. The growth of the telencephalon is so prolific that it overgrows the diencephalon "burying" the diencephalon deep to the forebrain in the adult.

The other significant effect of overgrowth is that it helps create the cephalic flexure between the diencephalon and the mesencephalon. By convention, the directional nomenclature of the CNS changes at this flexure. Above the cephalic flexure, the top of the brain associated with the calvaria is dorsal and the cranial base is ventral. The frontal pole of the brain is rostral and the occipital pole of the brain is caudal. Below the cephalic flexure, the terminology used reflects the longitudinal axis of the body as a whole. For the brain and spinal cord, dorsal refers to the posterior aspect of these structures while ventral refers to the anterior aspect. The cerebellum, for example, arises from the posterior aspect of the brainstem.

The terms caudal and rostral are used below the cephalic flexure to describe the proximity of the structure to the head. The spinal cord, for example, is caudal to the brainstem.

The mesencephalon is not further divided and will develop into the midbrain of the brainstem. The rhombencephalon, however, will develop a posterior flexure called the pontine flexure that will divide it into the metencephalon that will become the pons of the brainstem and the myelencephalon that will become the medulla of the brainstem.

THE CEREBRAL HEMISPHERES

The cerebral hemispheres are part of the forebrain that developed from the massive growth of the telencephalon. The cerebral cortex refers to the laminated outer layers of the cerebral hemispheres that contain mostly neurons and glial cells. Deep to the cerebral cortex is a large network of axonal fibers generally referred to as the white matter. White matter of the brain includes projection fibers, association fibers, and commissural fibers.

The surface of the cerebral hemispheres is not smooth, but rather has numerous crevices, called sulci, and ridges, called gyri. Larger crevices may also be referred to as fissures. Remarkably, the general position of the gyri, sulci, and fissures are consistent between human brains. As a result, they have been used to name, organize, and divide the cerebral hemispheres into gross regions, lobes, and areas. Histological and electrophysiological examination of the gyri of the cerebral hemispheres supports the basic gross organization.

The cerebrum is divided into two cerebral hemispheres by the longitudinal fissure. There are very few gross anatomical differences between the left hemisphere and the right hemisphere but the two hemispheres are concerned with similar, but different, cognitive processes. Each cerebral hemisphere has a lateral, medial, and ventral surface and can be divided into lobes using major anatomical landmarks.

Lobes of the Cerebral Hemispheres

The frontal lobe is the largest and most rostral lobe of the cerebral hemispheres. The most rostral point can be referred to as the rostral pole. The ventral and lateral boundary of the frontal lobe is the lateral sulcus (Sylvian fissure). The caudal boundary of the frontal lobe is the central sulcus that separates it from the parietal lobe.

The precentral gyrus is a fairly straight gyrus located just rostral to the central sulcus. Most of the precentral gyrus is on the lateral surface of the hemisphere, but the superior part of the gyrus passes into the longitudinal fissure and is continuous with the paracentral lobule on the medial surface of the hemisphere. The precentral gyrus is considered the primary motor cortex of the brain. Rostral to the precentral gyrus, the gyri of the frontal lobe can be further divided into a superior, middle, and inferior gyri. The middle and superior gyri are referred to as the prefrontal area while the inferior gyri are referred to as Broca's area. The gyrus rectus and orbital gyrus are located on the ventral aspect of the frontal lobe that rests on orbital portion of the frontal bone.

The parietal lobe is bounded rostrally by the central sulcus that clearly separates it from the frontal lobe. When the lateral side of the cerebral hemisphere is viewed, a poorly defined line passes superiorly from the parieto-occipital notch to separate the parietal cortex from the occipital cortex. However, when the medial side of the cerebral hemisphere is viewed, the parieto-occipital sulcus clearly separates the parietal lobe from the occipital lobe.

The postcentral gyrus is a fairly straight gyrus located just caudal to the central sulcus. Most of the postcentral gyrus is on the lateral surface of the hemisphere but the superior part of the gyrus passes into the longitudinal fissure and is continuous with the paracentral lobule on the medial surface of the hemisphere. The postcentral gyrus is considered the primary sensory cortex of the brain. Caudal to the postcentral gyrus, the parietal lobe is divided into a superior and inferior parietal lobule. The inferior parietal lobule can be divided into the supramarginal gyrus and the angular gyrus. The supramarginal gyrus curves around the most caudal part of the lateral sulcus. The angular gyrus is caudal to the supramarginal gyrus and separates the parietal lobe from the temporal lobe.

The temporal lobe is separated from the frontal lobe by the lateral sulcus (Sylvian fissure). The superior, middle, and inferior gyri of the temporal lobe can be identified on the lateral surface of the temporal lobe passing from rostral to caudal. If the lateral sulcus is pulled open two gyri oriented perpendicular to the more lateral gyri can be identified on the superior aspect of the temporal lobe. These gyri are named the transverse gyri of Heschl that is considered to be primary auditory cortex.

The largest gyrus on the ventral surface of the temporal lobe passes from the temporal lobe to the occipital lobe and is called the occipitotemporal gyrus. Medial to the occipitotemporal gyrus is the parahippocampal gyrus. This gyrus has a characteristic medial bulge called the uncus. Collectively, the parahippocampal gyrus and uncus are considered the primary auditory cortex.

The occipital lobe is the most posterior lobe of the cerebral hemisphere. It rests on a top of the dural fold called the tentorium cerebelli separating it from the cerebellum. On the medial surface of the occipital lobe, the parieto-occipital sulcus clearly separates it from the parietal lobe. The calcarine fissure is also located on the medial surface of the occipital lobe and separates the more dorsal gyrus called the cuneus from the more ventral gyrus called the lingual gyrus. These gyri are considered the primary visual cortex. On the lateral surface of the occipital lobe no clear landmark separates it from the parietal lobe or the temporal lobe.

In addition to the four main lobes described before, the insular lobe is buried deep inside the lateral sulcus (Sylvian fissure) and is composed of long and short gyri.

Medial Surface

Important structures of each lobe can best be seen on the medial aspect of the cerebral hemisphere. However, the medial aspect of the cerebral hemisphere also provides an opportunity to view unique structures that cannot be seen on the lateral surface.

White mater tracts form a significant part of the cerebral hemispheres and can be categorized as: (1) projection fibers that project from the cortex to distant targets, (2) association fibers that interconnect regions within the same cerebral hemisphere, and (3) commissural fibers that interconnect similar areas between the cerebral hemispheres. On the medial aspect of the brain the corpus callosum is a thick and wide structure that is comprised of commissural fibers that interconnect the cerebral hemispheres. The corpus callosum can be divided from rostral to caudal into the rostrum, the genu, the body, and the splenium. The remaining white fiber tracks are not apparent on a dissected specimen and are discussed in more detail later.

The cingulate gyrus follows the corpus callosum but is separated from it by the cingulate gyrus. The cingulate gyrus has a wide variety of reciprocal connections to areas in the frontal lobe and temporal lobe that are associated with limbic function. Because of its position within limbic circuitry, the cingulate gyrus is considered a major part of the limbic lobe.

Basal Ganglia

The basal ganglia refer to a group of nuclei and white mater tracts derived from the telencephalon that developed deep to the overlying cerebral cortex and white matter. The structures comprising the basal ganglia are the caudate nucleus, the putamen, and the globus pallidus. The most characterized circuitry and function of the basal ganglia relates to their role in voluntary motor activity, but they also have connections to limbic areas suggesting that they have a role in emotion.

The putamen is the largest and most lateral component of the basal ganglia while the globus pallidus is the most medial component. Anteriorly, the putamen is continuous with the head of the caudate nucleus. In a coronal section of brain, the putamen and globus pallidus look like a lens just lateral the internal capsule with the narrow end oriented medially. As a result, the globus pallidus and putamen are collectively referred to as the lenticular nucleus.

The caudate nucleus is a C-shaped structure that follows the lateral ventricles. The caudate nucleus can be divided into three parts. The head of the caudate is the most rostral part and is associated with the anterior horn of the lateral ventricle. The body of the caudate is dorsal and lateral to the thalamus and is associated with the wall of the lateral ventricle in that region. The tail of the caudate nucleus follows the inferior horn into the temporal lobe.

DIENCEPHALON

The diencephalon is the caudal part of the forebrain that was derived from the prosencephalon. It is composed of the thalamus, the epithalamus, the hypothalamus, and the subthalamus. Each of these can be divided symmetrically into a right and left half. The gross anatomy of the diencephalon can most easily be seen on the medial surface of the brain below the cortex and the large white mater tract called the corpus callosum and above the midbrain of the brainstem.

Thalamus

The thalamus is the largest part of the diencephalon and functions as a processing center for projections entering and exiting the cerebral hemispheres. If the cerebral hemispheres are stripped away, the two halves of the thalamus look like egg-shaped masses perched on the brainstem. A thin vertical space filled with CSF called the 3rd ventricle is positioned midline between the two halves of the thalamus. The lateral and posterior parts of the thalamus are continuous with the underlying midbrain (mesencephalon), however the internal capsule serves as the lateral boundary of the thalamus and the posterior commissure serves as the posterior boundary. The thalamus extends to the rostral extent of the 3rd ventricle and is separated from the frontal cortex by the anterior commissure and the linea terminalis. A shallow horizontal impression in the thalamus called the hypothalamic sulcus is the ventral boundary of the thalamus separating it from the hypothalamus below. In some human brains the medial surface of the left and right thalamus is partially fused by a structure called the massa intermedia. While not evident on gross specimens, the thalamus can be divided into nuclear groups by thin layers of myelinated fibers called the intermedullary lamina.

Hypothalamus

The hypothalamus lies below the hypothalamic sulcus separating it from the thalamus above. Like the thalamus, a thin vertical space filled with CSF called the 3rd ventricle is positioned midline between the two halves of the hypothalamus thalamus. The hypothalamus extends anteriorly to the optic chiasm and posteriorly to include the mammillary bodies on the ventral surface of the brain. The hypothalamus is continuous with the pituitary gland via the pituitary stalk (infundibulum). While not apparent in gross specimens, the hypothalamus is divided into nuclear regions that are related to monitoring and controlling homeostatic function.

Subthalamus and Epithalamus

The subthalamus is the most ventral part of the diencephalon. It lies in between the thalamus and the midbrain. The largest division of the subthalamus is the subthalamic nucleus that is involved with integration of somatic motor function.

The epithalamus is the most dorsal and posterior part of the diencephalon. The epithalamus can be divided into the pineal gland and the habenular nuclei. The pineal gland is primarily a secretory gland whose secretions have been best characterized as playing a role in circadian rhythms. The best characterized circuit involving the habenular nuclei involves inputs from limbic regions of the forebrain and outputs to dopaminergic regions in the midbrain.

BRAINSTEM

The brainstem includes the medulla, the pons, and the midbrain. Gross anatomically, the spinal cord appears continuous with the medulla of the brainstem as it passes through the foramen magnum of the skull and the midbrain appears continuous with the diencephalon as it passes through the tentorial notch. The tentorial notch is a space created in the dural fold called the tentorium cerebelli that separates the posterior cranial cavity below from the middle cranial cavity above. Thus, the brainstem and associated structures are located in the posterior cranial fossa.

When examining the gross anatomy of the brainstem, it is very important to remember that the brainstem is comprised of distinct nuclear regions and longitudinal tracts that contain fibers passing upward toward the diencephalon or down toward the spinal cord. It is possible to divide all the regions of the brainstem into an anterior (ventral) basilar region that is largely composed of long descending tracts and a posterior (dorsal) tegmentum that is comprised of cranial nerve nuclei, relay nuclei, and axonal fibers. Gross anatomical features of the brainstem can be used to identify and describe the different regions of the brainstem, describe and localize where cranial nerves arise from, and provide some insight with regard to the underlying neuroanatomical organization within each region.

Medulla

On the anterior surface of the medulla two longitudinal ridges called the pyramids can be observed running along the midline. The pyramids of the medulla are created mostly by the descending corticomotor fibers as they pass through this region to the spinal cord. Just before the foramen magnum, as the medulla transitions into the spinal cord, the pyramids are obliterated as the large majority of corticospinal fibers decussate to the contralateral side. Just lateral to the pyramid of the medulla is a swelling called the olive. Functionally, the olive is involved with motor integration, but it is an important landmark that can be used to identify cranial nerves in this region. The glossopharyngeal nerve (CN IX) and vagus nerve (CN X) can be seen arising from the pre-olivary sulcus while the hypoglossal nerve (CN XII) can be seen arising from the post-olivary sulcus.

With the cerebellum removed the posterior aspect of the medulla can be observed. The most obvious feature is the termination of the diamond-shaped 4th ventricle at the obex. The inferior cerebellar peduncle can also be seen arising from the posterior lateral aspect of the cranial medulla.

Pons

The anterior aspect of the pons appears rounded and swollen compared to the medulla. The middle cerebellar peduncles can clearly be seen passing from the pons to the cerebellum. The trigeminal nerve (CN V) can be seen arising from the lateral aspect of the pons in the vicinity of the middle cerebellar peduncles. The abducens nerve (CN VI) can be seen arising from the anterior aspect of the pontomedullary junction while the facial nerve (CN VII) and vestibulocochlear (CN VIII) nerve arise from the lateral aspect of the pontomedullary junction. Like the medulla if the cerebellum is removed, the 4th ventricle can be appreciated in the region posterior to the pons.

CEREBELLUM

The cerebellum is not considered part of the brainstem but it is intimately connected to the posterior aspect of the medulla, pons, and midbrain via three distinct cerebellar peduncles: the inferior cerebellar peduncle, the middle cerebellar peduncle, and the superior cerebellar peduncle. The cerebellar peduncles contain the input and output fibers for the cerebellum.

Gross anatomically, the bumps of the cerebellum are much smaller than the gyri of the cerebrum and are remarkably thin layers called folium. The cerebellum can be divided into two cerebellar hemispheres separated by a vertical midline structure called the vermis. Each hemisphere can be further divided into an anterior lobe and a posterior lobe by the primary fissure. The final lobe is separated from the posterior lobe by the posterior fissure. Generally speaking, the cerebellum is involved with motor sensory integration and nonconscious motor control that leads to coordinated movements.

Midbrain

Due to the prolific growth of the cerebral hemispheres during development much of the midbrain is obscured from view as it passes through the tentorial notch. On the anterior surface, two large cerebral peduncles connect the diencephalon to the midbrain. The optic chiasm of optic nerve (CN II) is easily observed in this region but is a structure of the diencephalon rather than the midbrain. The oculomotor nerve (CN III) arises at the level of the midbrain and can typically be seen passing between the posterior cerebral artery and superior cerebellar artery. The proximity of these structures to the relatively ridged tentorial notch makes them vulnerable to damage the overlying brain herniates from the middle cranial fossa into the posterior cranial fossa.

The posterior aspect of the midbrain is dominated by square-shaped mass with four bumps, or hills, called the tectum. If the midbrain is sectioned along its midsagittal plane, the tectum is clearly separated from the anterior parts of the midbrain by the cerebral aqueduct. The cerebral aqueduct is a CSF-filled space that connects the 3rd ventricle of the diencephalon to the 4th ventricle of the brainstem. The paired colliculi "little hills" of the tectum can be subdivided into the superior colliculi and inferior colliculi that are involved with sensory integration. The trochlear nerve (CN IV) is unique among cranial nerves, in that its fibers cross to the contralateral side and that it arises from the posterior aspect of the midbrain in this region.

VENTRICLES

The nervous system developed from the neural tube that is composed of a layer of neuroectoderm that surrounds a neural canal. As the rostral end of the neural tube grows and folds into the primary and secondary vesicles, the neural canal within expands as well to form the ventricles of the brain. The ventricles of the brain remain connected to the neural canal of the developing spinal cord which is renamed the central canal in the adult.

Three openings develop in the roof of the 4th ventricle to connect the ventricular spaces to the subarachnoid space that surrounds the brain and the spinal cord. The median aperture (foramen of Magendie) opens on midline into a subarachnoid space inferior to the cerebellum called the cisterna magna and into the quadrigeminal cistern in the region of the tectum of the midbrain. Two lateral apertures (foramen of Luschka) also open into the cisterna magna.

A choroid plexus forms in each of the developing ventricles as local blood vessels interact with the ependymal cells that line the wall of the ventricles. In the choroid plexus the tight junctions between endothelial cells of blood vessels that help create the blood–brain barrier in other regions of the brain are not present. This allows the choroid plexus to secrete CSF directly into the ventricular system.

The ventricles are fluid-filled spaces, thus their shape is easily discerned using imaging modalities. Disruption of the flow of CSF through the ventricular system or overproduction of CSF results in dilation of the ventricles that impinges on the surrounding structures causing serious neurological pathologies. This condition is called hydrocephalus (water on the brain).

Each cerebral hemisphere contains a C-shaped lateral ventricle that can be divided into an anterior horn, a body, a posterior horn, and an inferior horn. The majority of the anterior horn resides in the frontal lobe. Its medial boundary is a thin membrane that separates it from the lateral ventricle on the other side called the septum pellucidum. The head of the caudate nucleus forms the lateral wall of the anterior horn while the corpus callosum forms the roof and rostral wall. The interventricular canal (foramen of Monroe) is located in the anterior horn of each lateral ventricle and allows CSF to drain to the 3rd ventricle.

The body of the lateral ventricle passes caudally from the interventricular foramen following the corpus callosum into the parietal lobe. In the region of the splenium, the inferior horn and posterior horn diverge. The inferior horn continues to curve into the temporal lobe and the hippocampal formation of the temporal lobe can be found in its medial wall. The posterior horn passes posteriorly into the occipital lobe.

The 3rd ventricle is vertically oriented and positioned between the thalamus and hypothalamus of the diencephalon. The 3rd ventricle is slit like when viewed in the frontal plane but is quite large when seen in the midsagittal plane. It extends rostrally to the lamina terminalis and anterior commissure and caudally to the posterior commissure. The floor is in part created by the optic chiasm. The 3rd ventricle is connected to each lateral ventricle by an interventricular foramen. The cerebral aqueduct arises from the posterior inferior aspect of the 3rd ventricle and passes through the midbrain to drain into the 4th ventricle in the region of the brainstem.

The 4th ventricle is diamond shaped beginning in the rostral pons and terminating at the obex on posterior medulla. The floor of the 4th ventricle is formed by the pons and medulla and is called the rhomboid fossa. The cerebellum forms the large majority of the roof of the 4th ventricle. About half way between the superior and inferior points of the 4th ventricle, in a region called the cerebellopontine angle, the lateral recesses will expand the 4th ventricle and the lateral apertures (foramen of Luschka) will connect it to the subarachnoid space. Just caudal to the cerebellopontine angle, the median aperture (foramen of Magendie) will also connect the 4th ventricle to the subarachnoid space. These apertures allow CSF to flow into the subarachnoid space and around the brain and spinal cord.

CSF is not static but rather flows from the deeper positioned ventricles to the more superficial subarachnoid space and ultimately into the venous system through the superior sagittal sinus. In fact, one of the major factors that drives the flow of CSF through this system is the pressure difference created as CSF drains from the subarachnoid space into the superior sagittal sinus. The drainage of CSF into the superior sagittal sinus is made possible by specialized structures called arachnoid villi. In general, a tough layer of dura mater creates the walls of the superior sagittal sinus. However, scattered along the length of the superior sagittal sinus are gaps in the dura mater that allow tufts of the underlying arachnoid membrane to push through. These arachnoid villi within the superior sagittal sinus contain one-way valves that allow the CSF to drain into the venous sinus but prevent blood from regurgitating back into the subarachnoid space.

BLOOD SUPPLY: THE BRAIN

The brain receives its blood supply from the left and right internal carotid arteries and the left and right vertebral arteries. The route these arteries take in order to reach the brain and the territories of the brain they supply are discussed in more detail but, in general, the branches of the internal carotid arteries supply the anterior structures of the brain while the branches of the vertebral arteries supply the posterior structures of the brain.

At the base of the brain a symmetrical anastomotic circle of arteries called the *circle of Willis* is formed that joins the internal carotid artery system to the vertebral artery system. The arteries involved in the *circle of Willis* are the internal carotid arteries, the anterior cerebral arteries, the anterior communicating artery, the posterior communicating artery, and the posterior cerebral arteries. The anterior part of the arterial loop resides just rostral to the optic chiasm while the posterior loop of arteries resides at the junction between the midbrain and the pons. In addition to the main arterial branches many smaller perforating arteries arise from the *circle of Willis* to supply deep brain regions of the brain.

Aneurysms are blood-filled bulges in the wall of a vessel. In general, the weakest point in the wall of a vessel is at a branch point. The arteries of the *circle of Willis* are prone to a type of aneurysm called saccular aneurysms (Berry aneurysms). As saccular aneurysms grow, they put pressure on the surrounding neurological structures and cause an array of neurological signs depending on their location. For example, saccular aneurysms associated with the anterior communicating artery can affect vision by compressing the optic chiasm. If a saccular aneurysm ruptures, blood will enter the subarachnoid space. This is called a subarachnoid hemorrhage. Subarachnoid hemorrhages can occur rapidly and without warning or previous neurological signs and symptoms.

Due to its remarkable route and clinical importance, the internal carotid artery has been divided into four parts; the cervical part in the neck, the petrous part as it passes through the base of the skull, the cavernous part as it traverses the cavernous sinus, and the cerebral part as it supplies the brain.

The cervical part of the internal carotid artery and external carotid artery branch from the common carotid artery in a region of the neck called the carotid triangle. At this location the walls of the internal carotid artery contain high concentrations of baroreceptors, which measure vascular blood pressure. Nearby a population of chemoreceptors called the carotid body can be found. The carotid body measures dissolved gasses within the blood as well as pH and temperature. Collectively, these structures provide crucial sensory feed back to the cardiorespiratory centers so that the brain receives enough blood during daily activities.

The internal carotid artery enters the base of the skull through the carotid canal, which is now referred to as the petrous portion of the internal carotid artery. The carotid canal does not pass directly through the skull, but rather bends the internal artery in two places as it enters the middle cranial fossa. The exit of the carotid canal opens into a dural venous sinus called the cavernous sinus. At this point, the internal carotid artery is within the cavernous sinus, completely surrounded by venous blood, and referred to as the cavernous portion of the internal carotid artery. The cavernous portion of the internal carotid artery gives rise to the hypophyseal artery and a meningeal artery.

The cerebral part of the internal carotid artery exits the cavernous sinus just rostral to the optic chiasm. The cerebral part of the internal carotid gives rise to the ophthalmic artery, the anterior choroidal artery, and the posterior communicating artery. The ophthalmic artery passes into the optic canal with the optic nerve (CN II) and provides blood supply to the retina. The anterior choroidal branch passes deep into the brain and supplies deep structures of the telencephalon, the diencephalon, and the mesencephalon. The posterior communicating branch joins with the posterior cerebral artery. The internal carotid artery terminates as the middle cerebral artery and the anterior cerebral artery.

Anterior Cerebral Artery

The anterior cerebral artery is one of the major arteries of the brain. It passes rostrally in the midsagittal plane from the internal carotid artery into the longitudinal fissure between the cerebral hemispheres. Soon after it has entered the longitudinal fissure, typically in the region of the optic chiasm, the anterior communicating artery branches from the anterior cerebral artery and connects it to the cerebral artery on the opposite side. From its rostral and ventral position between the hemispheres the anterior cerebral artery gives rise to the orbital branches followed by the frontopolar branches. The anterior cerebral artery terminates as the callosal marginal and pericallosal arteries that sweep caudally along the medial surface of the cerebral hemispheres reaching the parieto-occipital sulcus.

Middle Cerebral Artery

The middle cerebral artery is typically the largest branch of the internal carotid artery and it passes latterly toward the insular cortex that lies deep in the Sylvian fissure. Along its path, the middle cerebral artery gives rise to perforating branches called lenticulostriate arteries that supply subcortical structures of the brain like the basal ganglia and the internal capsule. As the middle cerebral artery enters the Sylvian fissure, it branches extensively. Branches of the middle cerebral artery emerge from the Sylvian fissure and supply the lateral surface of the frontal lobe, the parietal lobe, and the temporal lobe of the brain.

Vertebral Arteries

The vertebral arteries arise from the subclavian arteries at the root of the neck and ascend to the base of the skull through the transverse foramen of the cervical vertebra. The vertebral arteries enter the posterior cranial fossa through the foramen magnum of the occipital bone. Branches of the vertebral artery supply the medulla, the cerebellum, and the spinal cord before merging together to form the basilar artery at the pontomedullary junction.

As the vertebral arteries traverse the medulla, each side gives rise to a branch that merges to form the anterior spinal artery. The anterior spinal artery is a longitudinal artery that passes inferiorly along the anterior aspect of the medulla through the foramen magnum. It then continues along the anterior aspect of the spinal cord in the anterior median sulcus. The paired posterior inferior cerebellar arteries are the largest branches of the vertebral arteries and arise just before the vertebral artery merges to form the basilar artery. This artery wraps posteriorly toward the cerebellum and thus supplies the dorsolateral portion of the rostral medulla, the inferior cerebellar peduncle, and the inferior aspect of the cerebellum. In addition, another the longitudinal artery called the posterior spinal artery arise from the posterior inferior cerebellar artery (PICA) to supply the posterior aspect of the medulla and spinal cord.

As described earlier, the merger of the two vertebral arteries forms the basilar artery. It passes along the anterior surface of the pons and terminates by dividing into the left and right posterior cerebral arteries. Just rostral to the pontomedullary junction, the anterior inferior cerebellar artery arises from the basilar artery to supply the anterior aspect of the cerebellum including the flocculonodular lobe. Many small pontine arteries arise along the length of the basilar artery to supply the pons. Just before its termination into the posterior cerebral arteries the basilar artery gives rise to the paired superior cerebellar arteries that supply the superior aspect of the cerebellum and caudal portions of the midbrain.

The posterior cerebral arteries arise from the basilar artery just rostral to the oculomotor nerve (CN III) and the tentorial notch. The posterior cerebral artery supplies rostral portions of the midbrain and the subthalamus, the medial and lateral aspects of the occipital lobe, and the ventral aspect of the temporal lobe. Perforating branches from the posterior cerebral artery called thalamogeniculate arteries supply the majority of the thalamus and hypothalamus. The posterior cerebral arteries are connected to the internal carotid arteries by the posterior communicating arteries.

Venous Return

Blood is ultimately drained from the brain by the left and right internal jugular veins. As described previously, blood is drained to the internal jugular veins through a system of interconnected dural venous sinuses. Blood drains to the dural venous sinuses from a vast network of superficial, deep veins, and bridging veins. Damage bridging veins that results in blood filling the subdural space is called a subdural hematoma. Acute subdural hematomas can be fatal but smaller subdural hematomas can become chronic. In contrast, epidural hematomas result from blood pooling between the dura mater and the bone of the skull and are typically cased by the rupture of an artery.

The superficial surface of the cerebral hemispheres is covered with cerebral veins. Some of these veins drain directly into the closest dural venous sinus. Other cerebral veins drain into larger veins called and anastomotic veins that subsequently drain to the superior sagittal sinus, the transverse sinus, and the cavernous sinus depending on their proximity to each sinus. The three anastomotic cerebral veins are typically larger than a typical cerebral vein and reside in a characteristic location. The superior anastomotic vein of Trolard is generally vertically oriented in the region of the parietal lobe. It not only drains to the superior sagittal sinus but also connects inferiorly to the inferior anastomotic vein of Labbe and superficial middle cerebral vein. The inferior anastomotic vein of Labbe is horizontally oriented and positioned on the lateral surface of the temporal lobe. It drains to the transverse sinus. The superficial middle cerebral vein is also horizontally oriented and follows the Sylvian fissure on the lateral surface of the brain. At the anterior extent of the Sylvian fissure the superficial middle vein drains into the cavernous sinus.

Much like the cerebral veins, there is a tremendous network of deep veins that drain the brain. The internal cerebral veins, the basal veins (of Rosenthal), and the great cerebral vein of Galen are the most important deep veins of the cerebrum. In general, deep veins are concerned with draining the structures that create the wall of the ventricles, the choroid plexus, the basal ganglia, and the thalamus.

There are two internal cerebral veins that are located on the roof of the 3rd ventricle. The two internal cerebral veins arch caudally over the thalamus and join together to join the Great vein of Galen.

The basal veins of Rosenthal arise as the anterior and middle cerebral veins join together on the ventral aspect of the frontal lobe. The basal veins pass caudally just medial to the temporal lobe and lateral to the midbrain to join the Great vein of Galen. This system of veins drains deep structures that are positioned on the ventral surface of the brain and the midbrain.

In addition to draining the internal cerebral veins and the basal veins, the Great vein of Galen is joined by the inferior sagittal sinus and together they drain into the straight sinus.

SPINAL CORD

The spinal cord is located within the vertebral canal of the vertebral column for its protection. The vertebral column is created by a series of vertically stacked bones called the vertebra. The vertebral column begins at the base of the skull where it articulates with the occipital condyles. It then extends inferiorly along the posterior aspect of the body. Typically, there are 33 vertebra organized in groups from superior to inferior. Most superior, in the region of the neck there are 7 cervical vertebrae, in the region of the thorax there are 12 thoracic vertebrae, below the thoracic vertebra there are 5 lumbar vertebrae, below the lumbar vertebrae is one sacrum (which is 5-fused sacral vertebra). Below the sacrum is the coccyx. Posterior to the vertebral bodies, and lateral to the neural arch, the stacking of vertebrae creates intervertebral foramina. It is through these foramina that spinal nerves pass out from the spinal cord into the periphery.

The spinal cord is continuous with the caudal medulla at the foramen magnum. The spinal cord is divided into segments in much the same way as the vertebral column. From superior to inferior, the spinal cord has a total of 31 segments: 8 cervical segments (C1–C8), 12 thoracic segments (T1–T12), 5 lumbar segments, 5 sacral segments (S1–S5), and 1 coccygeal segment. Each spinal segment gives rise to spinal roots which merge to form a pair of spinal nerves that exit through the intervertebral foramen. Spinal nerves are mixed nerves of the peripheral nervous system that communicate with the CNS. They contain axonal fibers that carry sensory information to the spinal cord and they contain motor fibers that carry motor commands from the spinal cord to the periphery.

The surface of the spinal cord has a few interesting features. It is easily noticed that the spinal cord is not uniform in diameter along its length. Swellings called the cervical enlargement and the lumbosacral enlargement reflect the increased sensory and motor fibers required in the regions that innervate the upper and lower limb respectively. The anterior (ventral) surface of the spinal cord has a relatively deep longitudinal trench, positioned on the midline, called the anterior median fissure. The posterior (dorsal surface) of the spinal cord has a longitudinal trench, on midline, called the posterior median sulcus. The posterior surface of thoracic and cervical regions of spinal cord has a shallow addition trench called the posterior intermediate sulcus. This trench demarcates the separation between the two white matter tracts of the dorsal columns.

The spinal cord extends inferiorly within the vertebral canal until vertebral level L1/L2. Thus, the spinal cord only inhabits the upper 2/3 of the vertebral canal. The most inferior tip of the spinal cord is called the conus medullaris. Extending from the conus medullaris is the thin filum terminal that is formed by the pia mater of the spinal cord and attaches to the dural sac around vertebral level S2. At the cervical level, the spinal cord segments line up fairly well with the appropriate intervertebral foramen. However, at the caudal end of the spinal cord, the intervertebral foramen may be located well below the corresponding spinal cord segments. Many spinal nerve roots can be seen descending past the conus medullaris within the vertebral canal in order to exit through an intervertebral foramen at the appropriate level. Collectively these descending roots are called the cauda equina.

MENINGES OF THE SPINAL CORD

The spinal cord is a relatively soft structure and must be supported within vertebral column by membranous layers called the meninges. The outermost layer is called the dura mater. It is attached to the rim of the foramen magnum and extends down the length of the vertebral canal and ends around vertebral level S2 creating a closed dural sac. The epidural space of the spinal cord is positioned between the dura mater and the vertebra that is typically filled with fat and the internal vertebral venous plexus.

Deep to the dura mater is a very thin layer called the arachnoid membrane. Deep to the arachnoid membrane is the subarachnoid space that is filled with CSF. The subarachnoid space of the neurocranium is continuous with the subarachnoid space of the spinal cord. The arachnoid membrane is not fused to the overlying dura mater, but the pressure of the CSF in the subarachnoid space prevents this layer from collapsing on to the surface of the spinal cord.

Arising from the arachnoid layer are thin filaments that traverse the subarachnoid space called arachnoid trabeculae. Arachnoid trabeculae fuse with the 3rd layer of meninges called pia mater that intimately covers the surface of the spinal cord. The pia mater forms two special structures to support the spinal cord within the dural sac. Denticulate ligaments arise from the pia mater on the lateral edge of the spinal cord and fuse to the overlying dura mater and the filum terminale extends from the conus medullaris to the end of the dural sac in order to anchor the inferior tip of the spinal cord.

BLOOD SUPPLY OF THE SPINAL CORD

The spinal cord receives its blood supply from longitudinal arteries that arise from the vertebral arteries and from the aorta via segmental arteries. As the vertebral arteries traverse the medulla, each side gives rise to a branch that merges to form the anterior spinal artery. The anterior spinal artery is a longitudinal artery that passes inferiorly along the anterior aspect of the medulla through the foramen magnum. It then continues along the anterior aspect of the spinal cord in the anterior median sulcus. The central branches of the anterior spinal artery are responsible for supplying the anterior two-thirds of the spinal cord including the anterior funiculus, most of the lateral funiculus, and most of the gray matter. The posterior spinal arteries arise directly from the vertebral artery or from the PICA and pass inferiorly through the foramen magnum on the posterior aspect of the spinal cord. The central branches of the posterior spinal arteries supply the dorsal funiculus and a small part of the dorsal horn of the spinal cord.

Segmental branches arise from the entire length of the aorta. A spinal branch will arise from the segmental artery and pass through the intervertebral foramen following the spinal nerve. Anterior and posterior spinal medullary arteries and small radicular arteries will arise from the spinal branch of the segmental artery.

Spinal medullary arteries join the longitudinal arteries and supplement the blood supply to the spinal cord. Anterior and posterior radicular arteries will follow and supply the anterior and posterior roots of the spinal cord. The artery of Adamkiewicz is a large anterior spinal medullary artery located around T10. If this artery is obstructed, the anterior spinal cord inferior to this level may be severely compromised.

The venous return of the spinal cord consists of three veins on its ventral aspect and three veins on its posterior aspect that ultimately drain to the inferior vena cava. The anterior and posterior aspects of the spinal cord each have a median spinal vein and paired intermediate spinal veins. The veins on the surface of the spinal cord communicate with the internal vertebral venous plexus that resides outside the dural sac but within the vertebral canal. The internal vertebral venous plexus in turn communicates with the external vertebral venous plexus. None of the venous structures associated with the spinal cord have valves, thus providing a route for the spread of cancer and infection.

Overview of the Neuroanatomy of the Spinal Cord

A cross section provides the best view of the internal organization of the spinal cord. The spinal cord contains a butterfly-shaped core of gray matter surrounded by white matter. Gray matter of the spinal cord is comprised of neurons and glial cells while the white matter is comprised of myelinated axonal fiber tracts. The small central canal in the middle of the spinal cord is all that remains of the neural canal of the neural tube.

Gray Matter Versus White Matter

The gray matter of the spinal cord can be roughly divided into a dorsal horn and a ventral horn. The region between the dorsal horn and the ventral horn is often called the intermediate gray. In general, the neurons of the dorsal horn receive the majority of the incoming somatosensory sensory afferent fibers from the spinal nerve. The ventral horn contains lower motor neurons whose fibers join the spinal nerve in order to innervate skeletal muscle. It is important to recognize that the lower motor neurons of the ventral horn are the final common pathway for all skeletal muscle activity.

The gray matter can be divided into layers called Rexed laminae. The dorsal horn contains laminae I–VI. Lamina I is very thin when compared to Lamina II that sits immediately below it. Lamina II is often referred to as the substantia gelatinosa. Laminae III and IV are positioned below lamina II and can also be referred to as the nucleus proprius. Importantly, the dendrites of neurons in laminae III and IV project into the laminae above. In this way, the neurons of the nucleus proprius are sensitive to a wide variety of sensory inputs. Layers V and VI are positioned below Lamina IV and contain a large population of interneurons that connect the dorsal horn to the ipsilateral and contralateral ventral horn. The laminae of the dorsal horn will change in size in different regions of the spinal cord. For example, the substantia gelatinosa and nucleus proprius are much expanded in the region of the cervical enlargement and the lumbosacral enlargement as a result of the large amount of sensory fibers that are coming from the limbs.

The intermediate gray is all considered lamina VII. The intermediolateral nucleus refers to a column of neurons in lamina VII between spinal segments T1 and L2. These neurons are part of the autonomic outflow from the spinal cord. Another group of neurons called the dorsal nucleus of Clarke also resided with in lamina VII.

The ventral horn is comprised of laminae VIII and IX. Lamina VIII is comprised of interneurons while lamina IX is comprised of lower motor neurons. The lower motor neurons of lamina IX can be grouped into nuclei that share similar functions. There are two spatial maps that organize motor nuclei within the ventral horn. Motor nuclei that innervate proximal muscles are located in a medial position while nuclei that innervate more distal muscles are located in a more lateral position. An overlapping spatial map organizes the ventral horn such that motor nuclei that are involved with flexion reside most ventral within the ventral horn while motor nuclei involved with extension are positioned just above them. The cervical and lumbosacral enlargements contain large lateral nuclei that reflect the number of lower motor neurons needed to innervate the upper and lower limbs. In contrast, the lateral nuclei of the thoracic region are quite modest.

The white matter of the spinal cord is comprised of myelinated and unmyelinated fibers that are ascending or descending within the spinal cord. The white mater can be divided into four funiculi as they surround the inner gray matter. A single dorsal funiculus is located on the dorsal aspect of the spinal cord and a single ventral funiculus is located on the ventral aspect of the spinal cord. Each side of the spinal cord has a lateral funiculus. Each funiculus can be subdivided into smaller function groups called a fasciculi.

White matter tracts of the spinal cord convey sensory information from the periphery to nuclei in the spinal cord, the brainstem, the cerebellum, or the thalamus. The particular target depends on the submodality carried in the tract.

Fibers in the dorsolateral fasciculus and the fasciculus proprius ascend and descend one or two segments to synapse in nuclei of the spinal cord. The majority of fibers in the dorsolateral fasciculus are unmyelinated primary afferent fibers while the fibers within the fasciculus proprius arise from a variety spinal cord lamina. These fasciculi are responsible for creating local circuits that interconnect multiple levels of spinal cord. In addition to the distribution of important sensory information across spinal cord segments, these tracts play an important role in coordinating spinal cord reflexes.

The dorsal funiculus is comprised almost entirely from the ascending tracts of the dorsal column medial lemniscus (DCML) pathway called the fasciculus gracilis and fasciculus cuneatus. From the coccygeal segment to the thoracic segment T7, the dorsal funiculus is mostly comprised of the fasciculus gracilis. The ascending fibers of the fasciculus gracilis are actually primary sensory afferents whose cell bodies are located in the dorsal root ganglion of spinal nerves. Because these fibers have yet to synapse in the CNS, they are referred to as first-order fibers. At thoracic level T6, the ascending first-order fibers of the fasciculus cuneatus are added just lateral to the fasciculus gracilis. Ascending fibers in the fasciculus gracilis synapse in the nucleus gracilis of the caudal medulla while fibers within the fasciculus cuneatus synapse in the nucleus cuneatus of the medulla.

The nucleus gracilis and cuneatus send second-order fibers across the midline in the medulla. These fibers then ascend through the brainstem to synapse in the ventral mediolateral (VPL) thalamus. The thalamus sends third-order fibers through the internal capsule to cerebral cortex. This system is responsible for conveying conscious somatosensory information concerning discriminative touch, proprioception, and vibration.

The lateral funiculus both contains ascending sensory tracts and descending motor tracts. Let us begin by discussing the ascending sensory tracts. The spinothalamic tract is located deep to the rim of the lateral funiculus. It conveys pain, temperature, and crude touch. The spinothalamic tract is comprised of second-order fibers that originated from neurons located in the contralateral dorsal horn. In order to cross to the contralateral side of the spinal cord, these fibers used the anterior white commissure. The spinothalamic tract ascends through the brainstem to synapse in the VPL of the thalamus. The thalamus sends third-order fibers through the internal capsule to the cerebral cortex.

In order to best coordinate movements, the cerebellum requires sensory information from the periphery. Because the target of this system is the cerebellum rather than the cortex, it is convention to refer to it as nonconscious sensation. Nonconscious sensation from the lower limb is conveyed to the cerebellum by the anterior and posterior spinocerebellar tracts. These tracts are located on the rim of the lateral funiculus and are comprised of fibers from the dorsal nucleus of Clarke and the spinal border neuron respectively. The cuneo-spinocerebellar tract that conveys nonconscious information from the upper limb is best seen in the medulla.

The lateral corticospinal tract is a large tract that occupies the large majority of the dorsal part of the lateral funiculus. The lateral corticospinal tract has a spatial topography in which the fibers that innervate the lower limb are positioned lateral in the tract while the fibers that innervate the upper limb are positioned medial in the tract.

The lateral corticospinal tract contains upper motor neuron fibers that originate in the contralateral precentral gyrus (primary motor cortex). Upper motor neurons have the most control over lower motor neurons in the spinal cord. It has become common practice to refer to the corticospinal fibers as pyramidal fibers to emphasize the control they have on the lower motor neuron.

Ventral to the lateral corticospinal tract is a small extrapyramidal tract called the rubrospinal tract. Descending fibers in the rubrospinal tract originate in the contralateral red nucleus of the midbrain. The rubrospinal tract is more important in animals than in man.

The anterior funiculus contains the anterior corticospinal tracts, the medial and lateral vestibulospinal tracts, the tectospinal tracts, and the reticulospinal tracts. The anterior corticospinal tracts are pyramidal tracts and thus contain upper motor neuron fibers from the precentral gyrus. Anterior corticospinal fibers descend in the anterior funiculus and cross at their target level in the spinal cord to innervate lower motor neurons in the medial nucleus of the spinal cord.

The vestibulospinal tracts are extrapyramidal tracts that arise from the medial and lateral parts of the vestibular nucleus and target lower motor neurons in the ventral horn. The medial vestibulospinal tract descends in the medial longitudinal fasciculus (MLF) to the cervical segments of the spinal cord. The lateral vestibulospinal tract descends as its own fasciculus the entire length of the spinal cord. The vestibulospinal spinal tracts are concerned with coordinating movements of the head and body as the head moves in space.

The tectospinal tracts are extrapyramidal tracts that accompany the medial vestibulospinal tracts in the MLF of the anterior funiculus. The tectospinal tracts originate in the contralateral superior colliculus and descend to innervate lower motor neurons in cervical segments of the spinal cord that orient the head to a visual or auditory stimulus.

The reticulospinal tracts are extrapyramidal tracts that are diffusely organized in the anterior funiculus. The reticulospinal tracts originate in the reticular formation of the pons and medulla and typically stimulate a variety of targets in the spinal cord to coordinate posture, aid in locomotion, and modulate autonomic outflow. If the brain stem is damaged at or below the red nucleus destroying the pyramidal tracts and the rubrospinal tracts, the reticulospinal tracts are allowed to take control and the individual assumes a decerebrate posture in which the neck and both upper and lower limbs are locked in extension. If the lesion occurs above the red nucleus leaving the rubrospinal tract intact, the patient will assume a decorticate posture in which the head and lower limbs are locked in extension but the upper limbs are locked in flexion.

CHAPTER

5.1

Translational Correlation: Peripheral Neuropathy

A.M. El-Dokla, S.T. Ali

Texas Tech University Health Sciences Center, Lubbock, TX, United States

INTRODUCTION

Peripheral neuropathy (PN) is a common disease that results in dysfunction of the peripheral nerves. The etiologies of PN can be categorized as metabolic, toxic, infectious, nutritional, autoimmune, hereditary, paraneoplastic, endocrine, and idiopathic. In fact, the vast majority of patients with PN may present with similar symptoms despite different etiologies. PN involves different nerve fibers (motor, large sensory, small sensory, and autonomic), each of which has different presentations. In PN, there may be positive symptoms due to abnormal function or negative symptoms due to lack of the function. For example, pins and needle sensation, burning, and tingling are examples of positive symptoms of large and small fiber sensory neuropathy; whereas numbness, decreased pain sensation, and decreased position sensation are examples of negative symptoms. Weakness is an example of negative symptom of motor neuropathy, whereas muscle cramps is an example of a positive symptom (Table 5.1.1). The typical pattern of muscle weakness and sensory loss appreciated on the physical examination has a distal to proximal gradient. Ankle deep tendon reflexes are usually diminished or absent. The vast majority of PN caused by toxic, metabolic, endocrine, and hereditary etiologies present with gradually progressive sensory and motor symptoms in a length-dependent pattern. The symptoms usually start in the feet and later spread proximally to the legs. When the neuropathy reaches the knee, the symptoms usually occur in the fingertips and spread proximally. This stocking-glove distribution of symptoms is likely secondary to dying back axonal neuropathy as the longest nerves are affected first. Since the distance from the lumbosacral spinal cord to the tip of the toes is longer than the distance from the cervical spinal cord to the fingertips, the symptoms of dying back axonal neuropathy usually start in tip of the toes. The diagnosis of PN requires at least a detailed history, neurological examination, and electrophysiological testing. The answers from a few important questions will be very useful in narrowing the

Conn's *Translational Neuroscience*
http://dx.doi.org/10.1016/B978-0-12-802381-5.00008-7

TABLE 5.1.1 Symptoms of Peripheral Neuropathy

Fiber Type	Positive	Negative
Sensory	Pins and needles Tingling Burning Shooting pain	Numbness Decreased pain sensation, decreased position sensation Sensory ataxia
Autonomic	Hyperhidrosis Hypertension	Hypohidrosis Impotence Urinary retention Constipation Hypotension
Motor	Muscle cramps Fasciculation/muscle twitches	Weakness Fatigue

differential diagnosis and providing a clue to specific etiologies.

WHAT IS THE TEMPORAL PROFILE (ACUTE, SUBACUTE, OR CHRONIC)?

Most metabolic, nutritional, and hereditary neuropathies present with a chronic, gradually progressive course with sensory and motor symptoms that start in the feet and gradual spread proximally. The PN is usually symmetric. Acute onset of neuropathy is a red flag and usually narrows the differential diagnosis. Guillain–Barré syndrome (GBS) and vasculitis are among the common etiologies of acute neuropathy. Diphtheria is now rarely seen. Vasculitic neuropathy is caused by inflammation of the wall of the blood vessels that leads to ischemia and primarily axonal nerve injury. Vasculitic neuropathy presents with acute or less likely subacute onset of painful and asymmetric sensory motor PN. The distinctive clinical picture is multiple nerve involvement in an asymmetric pattern (mononeuritis multiplex or overlapping mononeuropathies). Systemic manifestations (namely, fever, skin rash, or organ involvement (eg, renal and liver)) are usually present in the course of the disease. However, vasculitic neuropathy can be present without associated systemic manifestation and is called isolated peripheral nervous system vasculitis. Muscle can also be affected in these cases. A stepwise progressive course can occur with involvement of multiple nerves and early diagnosis and treatment is the key to avoid long-term complication or even death. In about 20% of cases, vasculitic neuropathy can be present with a length-dependent distal and symmetric pattern, especially with extensive involvement of multiple nerves or late in the course of the disease.

Another rare example of acute onset of neuropathy is porphyria. Acute porphyria usually presents with a variety of symptoms like abdominal pain, PN, psychiatric symptoms, seizures, confusion, and coma. There is a high level of porphyrin precursor, delta-aminolevulinic acid (ALA), and porphobilinogen (PBG) in the urine.

SYMMETRIC OR ASYMMETRIC?

Most neuropathy starts distally in the feet and progress in a length-dependent pattern with a distal-to-proximal gradient. The presence of significant asymmetry is not a feature of most of the toxic, metabolic, and hereditary neuropathy, although diabetic neuropathy may have subtle asymmetry. As stated above, mononeuritis multiplex is a common presentation of vasculitis, which usually follows an asymmetric pattern. The presence of radiculopathy on top of PN commonly will give a false impression of asymmetric neuropathy. A thorough history, medical examination, and electromyography and nerve conduction studies (EMG/NCS) are usually needed to sort this out. L4–S1 and C5–C8 are the common radiculopathies that affect both lower and upper extremity respectively. L5 radiculopathy presents with numbness of the dorsum of the foot and lateral aspect of the calf, and weakness of the tibialis anterior muscle. Hence, for example, the patient with mild distal sensory motor neuropathy will have sensory symptoms affecting the toes and foot up to the ankle and mild weakness of toe flexion. In the presence of a superimposed right L5 radiculopathy, there would also be right foot drop and sensory symptoms on the lateral calf in addition to the earlier described neuropathic symptoms. Another important cause of asymmetric PN is multifocal CIDP (a variant of CIDP), which is discussed below. Superimposed mononeuropathies of the upper or lower extremity can also cause asymmetric symptoms and signs.

WHAT IS THE TYPE OF NERVE FIBER AFFECTED?

Most of the polyneuropathies affect both sensory and motor fibers clinically and by electrophysiological testing (ie, EMG/NCS). Predominant involvement of one nerve fiber type (sensory, motor, and autonomic) can be a clue to specific etiologies. Involvement of the dorsal root ganglia, which is the mother cell of the sensory nerves, will result in pure sensory neuropathy or more precisely sensory neuronopathy. Among the commonest causes of sensory neuronopathy are postviral infection, paraneoplastic with anti-Hu antibodies, and Sjogren syndrome. Small cell–lung cancer has a strong association with anti-Hu antibodies. Sjogren syndrome usually presents with dry eye and dry mouth (xerostomia) and may be associated with swollen salivary glands. IgG (IgG1) antibodies against SS-A or SS-B are usually present in the serum.

It is worth noting that patients with amyloid neuropathy have significant involvement of autonomic and small fibers, especially earlier in the course of the disease.

IS IT AXONAL VERSUS DEMYELINATING?

EMG/NCS is crucial in the initial evaluation of PN as it can differentiate between the axonal and demyelinating neuropathies. The human nerves are divided into myelinated and unmyelinated fibers. The myelinated fibers are composed of axons surrounded by myelinating Schwann cells, which secrete myelin. Myelin will act like an insulating layer around the nerve and therefore the nerve impulse will be conducted faster. The unmyelinated fiber is composed of nerve axons surrounded by non-myelinating Schwann cells. In primary axonal neuropathy, the pathogenesis starts with primary involvement of the nerve axon followed by axonal degeneration and secondary degeneration of its myelin sheath coverage. Electrophysiological testing and nerve conduction studies will show small compound muscle and sensory nerve action potential *amplitudes*. On the other hand, in demyelinating neuropathy, the pathology starts in the myelinating Schwann cell and/or the covering myelin sheath with macrophages being commonly invading the cell in autoimmune conditions, stripping the myelin off the axon. Electrophysiological testing and nerve conduction studies will show markedly *slowed conduction velocity* and markedly prolonged distal latency, which are markers of slow nerve conductivity. Conduction block may also occur.

The commonest examples of demyelinating neuropathy are chronic and acute inflammatory demyelinating polyradiculoneuropathy (CIDP and AIDP). CIDP presents with subacute to chronic onset of progressive sensory loss and motor weakness. The motor weakness usually predominates the picture, and proximal muscle weakness can be present. The multifocal variant of CIDP presents with asymmetric muscle weakness. The course can be relapsing and remitting. Autonomic fibers are involved to lesser degree in CIDP than in GBS. CIDP can be differentiated from AIDP based on the time from the onset till the nadir of weakness. In AIDP, the nadir of muscle weakness is usually reached within month. However, in CIDP, it is usually at least 2 months. Between 1 and 2 months is a gray zone. However, most of these cases turn out to be CIDP. It is important to differentiate between CIDP and AIDP for diagnostic and management purposes. AIDP is usually a monophasic disease treated with intravenous immunoglobulin (IVIG) or plasma exchange. Corticosteroids are not effective in the treatment of AIDP. However, CIDP can be treated with IVIG or steroids as first-line agents. Plasma exchange is used less frequently. CIDP requires long-term immunotherapy unlike AIDP.

IS THERE A FAMILY HISTORY OF NEUROPATHY?

Hereditary neuropathy is an important subtype of polyneuropathy. The clues to the diagnosis of hereditary neuropathy are the presence of family history of PN, slow course with few positive sensory symptoms, and presence of food deformity (ie, hammer toes and pes cavus). The most common type of hereditary neuropathy is Charcot–Marie–Tooth (CMT) disease. The disease usually starts in the first three decades of life with distal foot weakness associated with foot deformity. Patients usually have few positive sensory symptoms and do not complain of tingling sensation. The motor symptoms predominate. Patients are clumsy in childhood or toe walk. As the disease progresses, they develop foot drop and eventually become wheel-chair bound. CMT is classified into four major types according to the mode of inheritance and the pathophysiology (ie, axonal or demyelinating). CMT1 is primary autosomal dominant demyelinating; CMT2 is autosomal dominant axonal; CMTX is X-linked, and CMT4 is primarily autosomal recessive axonal neuropathy. There are more than 30 identified genes known to cause CMT; however, studies from France, UK, and USA found that in more than 90% of cases in which the genetic diagnosis was made, the genetic mutation was found in one of four genes. The four genes are *PMP22*, *MPZ*, *GJB1*, and *MFN2*. CMT1A is the most common type of CMT disease and accounts for about 50% of CMT cases. The disease is caused by duplication or point mutation of *PMP22* gene on chromosome 17. On the other hand, deletion in the *PMP22* gene will cause hereditary neuropathy with liability to pressure palsy.

IS THERE ASSOCIATED SYSTEMIC DISEASE?

The most common cause of PN in developed countries is diabetes, while leprosy is the most common cause of PN in developing countries. Diabetic polyneuropathy, diabetic amyotrophy or diabetic lumbosacral radiculoplexus neuropathy, CIDP, and mononeuropathy with focal nerve compression are different types of neuropathies associated with diabetes mellitus. Diabetic polyneuropathy is the most common diabetic neuropathy. It is a slowly progressive and length-dependent neuropathy. Patients may be asymptomatic for many years and the symptoms usually start with numbness and tingling distally in the toes and ball of the feet and then gradually progress proximally. Burning sensation, sensitivity to touch, stabbing pain, and nonhealing skin ulcer are among different clinical manifestations. Autonomic neuropathy including small fiber neuropathy, orthostatic hypotension, and cardiac autonomic neuropathy are common in diabetic patients. Cardiac autonomic neuropathy is associated with increased mortality from sudden death in diabetic patients.

RED FLAGS IN PERIPHERAL NEUROPATHY

Acute onset
Rapid progression of the disease
Asymmetrical presentation and findings on neurological examination

Presence of associated systemic manifestation (skin rash, fever, and liver or kidney disease).

Further Reading

[1] Bromberg MB. An approach to the evaluation of peripheral neuropathies. Semin Neurol 2010;30:350–5.

[2] Preston D, Shapiro B. Polyneuropathy. In: Electromyography and Neuromuscular Disorders: Clinical–Electrophysiologic Correlations. Elsevier Saunders; 2013. p. 384–416.

[3] Vavra MW, Rubin DI. The peripheral neuropathy evaluation in an office-based neurology setting. Semin Neurol 2011;31:102–14.

CHAPTER

5.2

Translational Correlation: Guillain–Barré Syndrome

A.M. El-Dokla, S.T. Ali

Texas Tech University Health Sciences Center, Lubbock, TX, United States

INTRODUCTION

Guillain–Barré syndrome (GBS) or acute inflammatory demyelinating polyneuropathy (AIDP) is an immune disease of the peripheral nervous system with incidence of 1–2/100,000/year. About two-third of patients have an antecedent upper respiratory tract infection. Others have an antecedent gastrointestinal tract infection. The antecedent infections, usually about 2–6 weeks prior to the onset of the GBS, may trigger the autoimmune response. *Campylobacter jejuni* (30%), cytomegalovirus, *Mycoplasma pneumoniae*, and Epstein–Barr virus are commonly reported infectious agents associated with the disease. Vaccination, especially the influenza vaccine, may also trigger the immune response in GBS; but the incidence of GBS after vaccination is low.

CLINICAL PICTURE

Patients present with acute or subacute symmetric weakness and paresthesia. The weakness typically starts distally in the feet and leg then gradually progresses to involve upper thigh muscle and upper extremities (ie, ascending paralysis). The progressive weakness can result in quadriparesis or quadriplegia, bulbar muscle weakness that leads to dysphagia, and diaphragmatic weakness that leads to respiratory failure. Early manifestations of respiratory failure include shortness of breath, mild tachycardia, and mild tachypnea. The classic ascending paralysis is only seen in about half of the patients. However, the clinical presentation may vary widely. The sensory symptoms are usually mild although paresthesia is common and pain is a prominent feature that is frequently present early and through the course of the disease. In a retrospective series of GBS patients, deep aching or throbbing low back pain was present in 61.8%, paresthesia in 49.1%, myalgia in 34.5%, and visceral pain in 20%. Deep tendon reflexes are lost during the first week; when the reflexes are spared after the first week, the diagnosis of GBS should be questioned.

The cranial nerves are affected in about 70%, lower motor neuron facial weakness occurs in about 50–70%, and ophthalmoplegia occurs in about 15% of patients.

Conn's Translational Neuroscience
http://dx.doi.org/10.1016/B978-0-12-802381-5.00009-9

Autonomic neuropathy occurs in about 60% of GBS patients. The common manifestations are sinus tachycardia and other cardiac arrhythmias, orthostatic hypotension, hypertension, and adynamic ileus.

GBS Variants

Miller–Fisher syndrome is a variant of GBS with the classic triad of ataxia, ophthalmoplegia, and absence of deep tendon reflexes. Serum anti-GQ1b IgG is associated with Miller–Fisher syndrome.

Acute motor axonal neuropathy (AMAN) is another variant of GBS that has poor prognosis and more aggressive course. The disease is common in Asia and is strongly associated with preceding diarrhea due to *C. jejuni* infection. As the name indicated AMAN is predominantly motor neuropathy with evolving paralytic muscle weakness. The sensory system is largely spared in contrast to typical GBS.

Acute motor and sensory axonal neuropathy and pharyngeal–cervical–brachial involvement are less common variants of GBS.

DIAGNOSIS

The gold standard diagnostic testing for GBS is nerve conduction studies (NCS), electromyography (EMG), and cerebrospinal fluid (CSF) analysis. The typical CSF findings are high protein content with normal white blood cell count. This is classically called albuminocytological dissociation that may be absent in the first few days after the onset of the symptoms. Early in the first week, the prolonged late reflexes (ie, H reflexes and F waves) may be the only abnormality seen in NCS. Other abnormalities include slowing of conduction velocities, prolongation of distal latencies, temporal dispersion, and conduction block that indicate demyelination become evident later in the course of the disease. In AMAN, the typical NCS finding is small compound muscle action potential (CMAP) *amplitude* that mainly indicates axonal damage.

Circulating antibodies in GBS are present in higher titer than in noninflammatory neuropathies. Several anti-ganglioside antibodies including GM1 and GM2 have been found in patients with AIDP. IgG against GM1 ganglioside is present in 40–60% of patients with AMAN and serum anti-GQ1b IgG is present in most patients with Miller–Fisher syndrome.

Most cases of GBS do not require a nerve biopsy for the diagnosis. However, nerve biopsy is occasionally ordered to rule out other diagnoses such as vasculitis.

PATHOLOGY AND PATHOGENESIS

Macrophage, lymphocyte, humeral factor, and alteration of the blood–nerve barrier are involved in the pathogenesis of AIDP leading to primary myelin damage. The typical pathological findings in GBS are macrophage-mediated demyelination, mononuclear inflammatory infiltrate with variable degree of axonal damage. When autopsy is performed, ongoing demyelination is also seen in dorsal nerve root.

In AMAN, the motor axon in the region of nodal and paranodal gangliosides is the target for the immune attack.

MANAGEMENT

Hospitalization for further evaluation and monitoring is indicated as the progression of the weakness is expected to occur for 1–3 weeks after the onset of symptoms and may lead to respiratory failure. ICU care is indicated for patients with rapid disease progression, facial and bulbar weakness, significant autonomic neuropathy, and breathing difficulties. Intubation should be considered early and electively for patients who have ineffective cough, significant decrease in vital capacity, or unable to handle their secretions.

Intravenous immunoglobulin (IVIG) and plasma exchange are equally proven to be effective treatment for GBS. Both are found to speed up the strength recovery and decrease the duration of mechanical intubation. The treatment is indicated early in the course of the disease, especially in the first 2 weeks for patients who cannot walk unassisted. Plasma exchange can be useful if given within 1 month of the onset of the disease. Corticosteroid is ineffective and has no role in the treatment of GBS.

Gabapentin, nonsteroidal anti-inflammatory medications, acetaminophen, and carbamazepine are commonly used for pain management. For refractory pain, tricyclic antidepressants, tramadol, or narcotic analgesics can be used.

OUTCOME AND PROGNOSIS

One year after the diagnosis, about 60% of patients have full recovery and about 85% walk unassisted. Old age, severe weakness with high grade on the GBS Disability Scale at neurological examination, and antecedent diarrhea are all associated with poor outcome. Mortality for GBS is estimated at 3–7%.

Further Reading

[1] Pestronk A. Neuromuscular Disease Center. St. Louis, MO: Washington University School of Medicine, http://neuromuscular.wustl.edu.

[2] van den Berg B, Walgaard C, Drenthen J, et al. Guillain-Barré syndrome: pathogenesis, diagnosis, treatment and prognosis. Nat Rev Neurol 2014;10(8):469–82.

[3] Yuki N, Hartung HP. Guillain–Barré syndrome. N Engl J Med 2012;366:2294–304.

[4] Rinaldi S. Update on Guillain–Barré syndrome. J Peripher Nerv Syst 2013;18:99–112.

CHAPTER

6

The Blood–Brain Barrier: A Restricted Gateway to the Brain

K. Shah, T. Abbruscato

Texas Tech University Health Sciences Center, Amarillo, TX, United States

INTRODUCTION

One of the vital functions of the central nervous system (CNS) is to integrate and coordinate information for the regulation and function of peripheral organ systems in the human body. The CNS compartment consists of the brain and spinal cord. The brain comprises only 2% of total body mass, yet has demanding metabolic requirements. About 20% of the oxygen and 25% of the glucose consumption in the human body is accounted for by the CNS [1]. For normal CNS development and function, a highly regulated and controlled extracellular and intracellular microenvironment is required to establish normal brain function. This microenvironment is regulated mainly by controlling solute flux into and out of the CNS at the interfaces between the CNS and the circulating blood. These barriers, essentially formed between the CNS and the systemic circulation, function as a dynamic regulator of nutrient/ion delivery, waste removal, while restricting potentially toxic xenobiotic or pathogen entry. Unfortunately, this selective barrier that protects and maintains proper neuronal function can create a major limiting factor for drug delivery into the CNS.

This concept of compartmentalization of the CNS microenvironment from the systemic circulation was mainly perceived to function solely as a physical barrier in the late 19th century. Max Lewandowsky was first to designate this barrier with the term "Bluthirnschnanke" meaning "blood–brain barrier," an anatomic barrier, now known as "blood–brain barrier (BBB)," while studying the limited permeation of potassium ferrocyanate into the brain [2]. Before this, early experimental evidence supporting this concept was first described by Paul Ehrlich, who noted upon systemic administration of water-soluble dyes, they stained all peripheral organs except the brain and spinal cord [3]. This observation was attributed to what was perceived as a low affinity of dye to nervous tissue. Further experimentation by Edwin Goldmann (1913) demonstrated that when trypan blue dye was injected directly into the cerebrospinal fluid (CSF), the dye stained the brain and spinal cord profusely but was absent in the peripheral organs [4]. These early experiments suggested the existence of a barrier between the CNS and peripheral circulatory system as well as between peripheral circulatory system and CSF.

Conn's Translational Neuroscience
http://dx.doi.org/10.1016/B978-0-12-802381-5.00010-5

The two main barriers of the CNS were identified: the brain microvascular endothelial cells forming the BBB and the epithelial layer of choroid plexus, the cerebral ventricles, and arachnoid mater covering the outer brain surface forming blood–CSF barrier (BCSFB). Over the next several years, the concept of the BBB gained insight into the cytochemical, anatomical, and functional characteristics through studies in both animals and humans. Landmark studies by Reese and Karnovsky (1967) demonstrated, using electron microscopy techniques, that the tight junction (TJ) between endothelial cells functions as a physical barrier, and this barrier was not provided directly by the astrocytic end feet that ensheathe cerebral endothelial cells. This was deduced by experiments that showed systemic horseradish peroxidase failed to enter CNS, whereas intra-cerebroventricular injection of horseradish peroxidase into the CSF stained only the brain extracellular fluid and diffused past the astrocytic end feet and basement membrane, and deposited on to the abluminal membrane of the brain microvascular endothelial cells [5]. With these breakthrough experiments, we now know that the BBB stringently restricts entry of substances while providing nutrient/ionic homeostasis necessary for the proper functioning of the CNS. Importantly, these dynamic characteristics of selective permeability are influenced by complex molecular and cellular interaction among brain endothelial cells and a variety of other brain cells such as astroglia, pericytes, and neurons. These required molecular and cellular interactions were demonstrated in another essential experiment by Steward and Wiley [6] where immature vascular brain tissue from embryonic quails was grafted into the coelom of chick embryos. What was observed is that the abdominal vascular endothelial cells developed nonleaky BBB characteristics such as TJs. Therefore, the former notion that the blood–brain functions solely as a physical interface is now considered with the understanding that the brain vascular barrier serves as a dynamic neurovascular unit (NVU), influenced by multiple companion cells (Fig. 6.1) The milieu of neurons, astrocytes, pericytes, microglia, and other components of the brain parenchyma induce and regulate the expression of a BBB phenotype

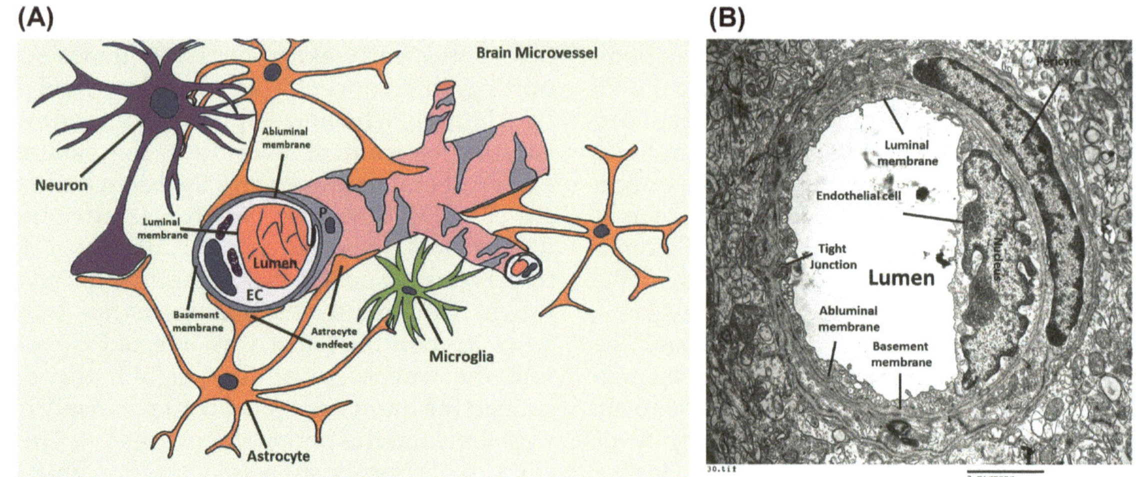

FIGURE 6.1 **The neurovascular unit (NVU).** (A) An illustration of the NVU. (B) An electron micrograph of normal mouse brain, representing the close spatial relationship and cellular physiologic interactions between the brain microvascular endothelial cells (EC) and pericytes (P), astrocytes, microglia, and neurons. *Electron micrograph generated in the Abbruscato Lab.*

within the brain endothelial cells. This complex structural arrangement creates a high electrical resistance interface of 2000–8000 Ω*Cm², which restricts the entry of solutes across the brain. Dysfunction of any one NVU component can profoundly influence normal BBB function and the overall pathology of a given disease state.

Most importantly, due to restrictive and protective properties imposed by the BBB for normal function of brain, it remains a challenging task to bypass the BBB for the delivery of therapeutic drugs to treat patients suffering from tumors, infections, and neurodegenerative diseases.

COMPONENTS OF THE NEUROVASCULAR UNIT

Brain Vascular Endothelial Cells

Brain microvascular endothelial cells that form the BBB possess distinguished characteristics in terms of structure, metabolic, transport, and adhesion properties compared to nonneural endothelial cells. The brain microvascular capillaries are located in close proximity (8–25 µm) to neurons to supply blood and form the largest diffusion barrier in the brain and body as whole. The circumference of the microvascular capillary lumen is enclosed by one or more endothelial cell, which are only 100–200 nm thick forming a very thin restrictive barrier. Therefore, activities of the BBB are central to brain homeostasis. A glycocalyx at the luminal endothelial surface is also present similar to the other peripheral microvascular endothelial cells, however, cerebral microvascular endothelial cells are distinct from the peripheral microvascular endothelial cells by increased mitochondrial content, a lack of fenestrations, minimal pinocytotic activity, and the presence of TJs. Complex combinations of transmembrane and cytoplasmic proteins linked to an actin-based cytoskeleton between two adjacent endothelial cells form tight junctions and a molecule seal between the outer leaflets of each endothelial cells. Claudin 1, 2, and 5, occludin, and junction adhesion molecule (JAM) are the integral proteins that form tight junctions. The tight junction between the adjacent cells in the barriers restricts diffusion of polar or small molecule solutes through the intercellular cleft (paracellular permeability) and maintains cellular polarity, a separation of luminal/apical (blood-facing) and abluminal/basolateral (brain-facing) membrane from the surrounding medium (Fig. 6.1B). The differential expression of solute transporters on the apical and basolateral membranes together with membrane-bound or intracellular enzymes regulate the transport of solute across the barrier (Fig. 6.2).

The two major transporter superfamilies: the ATP (adenosine triphosphate)-binding cassette

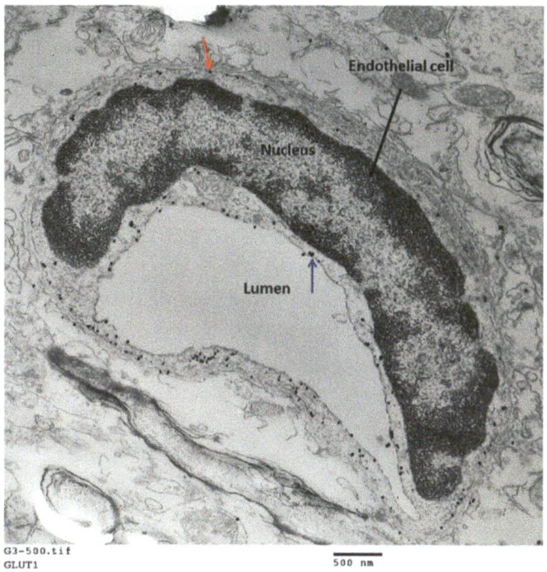

FIGURE 6.2 **The endothelial cell glucose transporter-1 (GLUT-1) expression.** A representative electron micrograph of normal mouse brain microvascular endothelial cell showing polarized expression of GLUT-1 transporter in endothelial cells. *Red arrow* shows abluminal membrane with 6-nm silver enhanced immunogold -abeled GLUT-1 protein; *Blue arrow* shows luminal membrane with immunogold labeled GLUT-1 protein. *Electron micrograph generated in the Abbruscato Lab.*

(ABC) and solute carrier (SLC) are known to be present in brain endothelial cells. The SLC transmembrane proteins help carry small molecules across the intracellular and/or extracellular membrane, for example, facilitative glucose transporters—GLUT1 (*SLC2a1*)—whereas ABC transporters use ATP to actively pump molecules including many drugs out from cells, for example, P-glycoprotein (Pgp). Large molecules, such as peptides or proteins with signaling roles within CNS, gain entry via absorptive and receptor-mediated transcytosis. Moreover, brain endothelial cells also function as a checkpoint and help in regulating the innate immune response, as well as the recruitment and entry of leukocytes in response to immune-reactive functions.

The function of transporters at the BBB in a variety of brain disease states has been explored. For example, the members of ABC transporters have been known to actively pump out drugs and are a major limiting factor for adequate CNS drug delivery of many anticancer agents. On the other hand, changes in function and expression of solute carriers at the BBB are also known to contribute to some disease pathogenesis. For example, increased function of facilitative glucose transporter type 1 (GLUT-1) at the BBB has been shown in cerebral ischemic stroke, whereas reduced glucose transport due to GLUT-1 deficiency has been shown to be associated with epilepsy. The BBB remains a major challenge in brain drug discovery by restricting many large molecules, such as monoclonal antibodies against Aβ peptide in treatment of Alzheimer's disease. Even small molecule neurotherapeutics and potential diagnostics for the treatment/detection of a variety of brain diseases can be limited in their ability to gain CNS access. To overcome these challenges, new alternatives for drug delivery have been in development such as utilizing BBB transporters for drug delivery. Recently, a bispecific antibody against transferrin receptors and Aβ-peptide has been tested to increase the uptake of large molecules via receptor-mediated transcytosis across the BBB. A major limitation to this approach remains the ability to identify a transport system at the BBB that has enough capacity to deliver an efficacious dose of active agent to a given receptor or target.

Astrocytes

The most abundant CNS macroglial cells, astrocytes (astroglia) are in close interaction with other brain cells and function to regulate and maintain nutrient/ionic/neurotransmitter homeostatic balance of the neural microenvironment. The adult human brain cerebral cortex is comprised of a total of about 17 billion neuronal and about 60 billion nonneuronal cells, which is the ratio of 3.72 nonneuronal/neuronal cells. Astrocytic end-feet projections cover about 80% of the abluminal side of the brain microvascular endothelial cells and assist in maintaining metabolite levels and water content in brain extracellular fluid. Close-cellular interaction of brain endothelial cells with astrocytes helps regulate and enhance the BBB phenotype and integrity in adult brain. Although astrocytes are not important in the development of TJs in immature brain endothelial cells, they are required for maintenance of TJs in adult endothelial cells.

The role of astrocytes in several disease conditions has been examined. During brain injury/inflammation, astrocytes release ATP which in turn activates local microglial cells. Moreover, besides their supportive and metabolic functions in the brain, astrocytes are involved in maintaining osmotic balance in the brain. Astrocytes' pericapillary end feet abundantly express specific channel proteins, the aquaporins, which have been shown to play a vital role in brain swelling and water imbalance following cerebral ischemic injury, further highlighting the interaction between

NVU components in physiological and pathophysiological states.

Basement Membrane

The basement membrane (BM) is made up of a 30–40 nm thick layers of extracellular matrix that is contiguous with the plasma membrane of the astrocyte end feet and entirely unsheathe the cerebral capillary endothelial cell. The BM is composed of structural proteins containing laminin, collagen IV, proteoglycans, heparan sulfate, fibronectins, and entactin, secreted by astrocytes, pericytes, and endothelial cells. The BM plays crucial roles in development and maintenance of the structural and functional elements of the brain microvasculature barrier function. The basal lamina of microvascular endothelial cells forms a negatively charged surface, which controls the movement of charged solutes between brain interstitial fluid and blood. This was evident from experiments testing large charged molecules, whereby positively charged ferritin movement was restricted across the basal lamina.

The BM provides an integral component of the neurovascular unit, any dysfunction of BMs are associated with developmental defects and even increased susceptibility to microvascular stroke. Recently, the mutations in type IV collagens (*COL4A1*) on chromosome 13q34, encoding the alpha1 chain of type IV collagen, have been linked to increased incidences of cerebral microvascular diseases in humans, including microbleeds, and lacunar strokes.

Pericytes

Pericytes are perivascular cells of mesodermal origin, present around the endothelium at irregular intervals on the abluminalmembrane of adult brain microvascular endothelial cells. The functional role of pericytes has emerged as a central NVU component not only in the development of barrier in embryonic brain but also in regulating capillary diameter, cerebral blood flow, maintain extracellular matrix protein levels. Importantly, pericytes are vital to BBB integrity and also function to guide astrocytic end-foot process to microvessel walls and facilitate polarization of astrocytic end feet, again emphasizing the intricate relationships among all components of the NVU.

Pericytes dynamically and actively respond to injury in various brain diseases such as ischemic stroke and Alzheimer's disease. As mentioned, pericyte interactions with brain endothelial cells are essential to retain the physical and functional characteristics of BBB. Accumulating evidences suggest early loss of pericytes at the NVU significantly contributes to both Alzheimer's disease and ischemic stroke pathogenesis.

Microglia

Microglial cells play an important role in innate immune responses and are the main line of active defense past the BBB in the brain. They are a type of glial cells that are believed to derive from monocytes that invade the CNS during embryonic development and persist as resident macrophages over the adult life. Microglia constitutes 10–15% of all the cells in the CNS. Microglial cells reside in close proximity to brain microvascular endothelial cells and their potential role in maintaining and regulating BBB function is not very well known. Mostly, microglia are maintained in the resting state but at times function to survey the surrounding microenvironment with motile processes and protrusions. Activated microglial cells, similar to macrophages, can take on an amoeboid shape and respond to tissue injury and invading pathogens by inducing their own phagocytic and cytotoxic mechanisms. When tissue injury or invading pathogens are detected, microglia exist in one or two active states and proliferation begins.

In the classically activated M1 pathway, microglia respond to inflammation primarily by releasing pro-inflammatory cytokines like interleukin-1β (IL-1β) and tumor necrosis factor-α (TNF-α). When tissue injury, invading pathogens, or damaged neurons are detected, microglial cells are activated by the M2 alternative pathway, thereby releasing chemokines that induce the neurotrophic pathway and vascular endothelial growth factor. Recent studies are increasingly identifying an active role of microglia in neurodegenerative disease like Alzheimer's disease and multiple sclerosis, both associated with BBB leakage and neuroinflammation.

Further Reading

[1] Belanger M, Allaman I, Magistretti PJ. Brain energy metabolism: focus on astrocyte-neuron metabolic cooperation. Cell Metab 2011;14(6):724–38.

[2] Lewandowsky M. Zur lehre von der cerebrospinalflussigkeit. Z Klin Med 1900;40:480–94.

[3] Ehrlich P. Das sauerstufbudurfnis des organismus. Eine Farbenanalytische Studie. Berlin: Hirschwald; 1885.

[4] Goldmann EE. Vitalfarbung am zentralnervensystem. Abhandl Konigl Preuss Akad Wiss 1913;1:1–60.

[5] Reese TS, Karnovsky MJ. Fine structural localization of a blood–brain barrier to exogenous peroxidase. J Cell Biol 1967;34(1):207–17.

[6] Stewart PA, Wiley MJ. Developing nervous tissue induces formation of blood–brain barrier characteristics in invading endothelial cells: a study using quail–chick transplantation chimeras. Dev Biol 1981;84(1):183–92.

CHAPTER

7

Autoregulation of Cerebral Blood Flow

D. Ruthirago, J.C. DeToledo

Texas Tech University Health Sciences Center, Lubbock, TX, United States

INTRODUCTION

Autoregulation of cerebral blood flow (CBF) is a process by which cerebral arteries maintain a constant CBF during variation of cerebral perfusion pressure (CPP). This process exists in several organs and is especially well developed in brain as it highly requires the stability of blood flow to bring nutrients and oxygen to its tissue. If autoregulation is impaired, the change in mean arterial pressure (MAP) can cause passive decrease and increase of CBF, which lead to cerebral ischemia and edema, respectively.

CPP is derived from MAP minus intracranial pressure (ICP). In physiological condition, cerebral resistance arteries (arterioles) dilate when CPP decreases, but constrict when CPP increases. Therefore, CBF is maintained relatively constant over a certain range of CPP. This range is called autoregulatory plateau and covers CPP between 50 and 150 mmHg in normotensive person. When CPP is beyond the limits of autoregulatory plateau, cerebral arterioles diameter changes passively to the CPP, causing passively increase or decrease of CBF.

MECHANISMS OF CEREBRAL AUTOREGULATION

Several factors were discovered to play role in the regulation of CBF.

Myogenic Factor

Smooth muscles in resistance arteries (arterioles) constrict during the increase in CPP and dilate during the decrease in CPP. This response occurs within a few seconds in an experiment which supports this hypothesis. Mechanoreceptor properties or stretch responses has been studied in smooth muscle cells, particularly the stretch-activated ion channels. In one study, membrane depolarization causes influx of Ca^{2+} through voltage-gated Ca^{2+} channels, results in vasoconstriction.

Metabolic Factor or Flow-Metabolism Coupling

Several previous studies found that CBF varies with cerebral metabolism. Several vasoactive

Conn's Translational Neuroscience
http://dx.doi.org/10.1016/B978-0-12-802381-5.00011-7

substances are released from brain tissue during the period of CBF reduction. These substances—such as potassium ion, hydrogen ion, carbon dioxide, adenosine, and nitric oxide (NO)—cause dilatation of cerebral arterioles.

Potassium ion (K^+) and hydrogen ion (H^+) are produced by synaptic transmission. ATP-sensitive K^+ channels were found in vascular smooth muscle, supporting the link between neuronal activation and CBF regulation. Vasodilatory response to hypercapnia and acidosis was believed to be mediated through action of H^+ on cerebral arteries, possibly via NO. Adenosine is a metabolite of neuronal activity. Extracellular level of adenosine increases with neuronal activity and with administration of glutamate, a main neurotransmitter in cerebral cortex. This increased level causes cerebral vasodilatation. It is believed that NO has a role in supplying other mediators, such as adenosine, as its second messenger for vasodilatation.

Endothelial Factors

Some experiments reported that intact endothelium is important in dilatation or constriction of cerebral resistance arteries in response to changes in CPP while other experiments reported different outcome.

The endothelium produces several vasomodulatory factors that have effects in cerebral autoregulation, such as NO, endothelium-derived hyperpolarization factor (EDHF), eicosanoids, and endothelins.

NO is a diffusible second messenger that operates via guanylate cyclase and causes smooth muscle relaxation. There are several isoforms of nitric oxide synthase (NOS). Endothelial NOS is found in endothelium of blood vessels, while neuronal NOS is found in neurons, and inducible NOS is found in the brain during hypertension or exposure to toxin. EDHF is believed to be another diffusible molecule that causes vasodilatation via hyperpolarization of vascular smooth muscles and activation of K^+ channel. The eicosanoids are vasoactive mediators created from arachidonic acid of cell membranes. Some mediators cause vasoconstriction and some cause vasodilatation. The cyclooxygenase (COX) enzyme system of eicosanoids is best studied in cerebral blood vessels. Prostacyclin (also called prostaglandin I2, PGI2) and prostaglandin E2 in intact endothelium are well-known vasodilators, especially in pathological condition such as subarachnoid hemorrhage. The endothelin system includes two receptors (ET_A and ET_B) and three ligands (ET-1, ET-2, and ET-3). ET_A receptors in the smooth muscles mediate vasoconstriction and are activated by ET-1 and ET-2. ET_B receptors within vascular endothelium mediate vasodilatation via NO and are activated by ET-1, ET-2, and ET-3. The endothelin system was found to function in pathological conditions such as cerebral ischemia and cerebral vasospasm.

Some studies reported mechanoreceptor property of endothelium that transmits transmural pressure and shear stress from blood vessel lumen into myogenic response, such as the endothelium decreases the release of endothelium-derived relaxing factor (EDRF) and increases the release of endothelium-derived constricting factor (EDCF), resulting in vasoconstriction.

Neurogenic Factor

The foot processes of astrocytes adjoin the cerebral blood vessels, forming a functional unit called "neurovascular unit" which can help regulate CBF. The nerves involved in this unit can be categorized into extrinsic and intrinsic perivascular innervation.

Extrinsic innervation are the vessels innervate outside the brain parenchyma. For example, the trigeminal ganglion, the superior cervical ganglion, and the sphenopalatine ganglion, which are believed to carry sensory, sympathetic, and parasympathetic nerves, respectively. Sympathetic nerves are believed to maintain increased tone to keep CPP below the upper limit of autoregulatory plateau when MAP is high. Parasympathetic nerves are hypothesized to function in pathological states. The trigeminal ganglion and its sensory nerves are believed to be involved in migraine and its vasodilatory mechanism. The intrinsic innervation starts when the cerebral blood vessels penetrate deep into the brain parenchyma. The majority of nerves in this area connect to astrocytic foot processes and do not adjoin directly on the blood vessels. The local interneurons are also believed to help regulate cerebral microvascular tone and transduce signal from subcortical pathways.

Astrocytes uptake excess extracellular potassium and shunt the potassium to their foot processes that abut cerebral vessels. The vasodilatory effect of potassium may explain neuronal effect on regulation of CBF. Other studies reported that electrical stimulation of astrocytes or administration of glutamate result in increased intracellular calcium, causing vasodilatation.

MODULATION OF AUTOREGULATION

Autoregulation of CBF can be changed by several factors.

Autonomic Nerves

Stimulation of sympathetic nerves causes shifting of upper and lower limits of autoregulation to higher CPP, while inhibition of sympathetic nerves cause shifting to lower CPP. This process helps prevent blood–brain barrier (BBB) disruption in acute MAP elevation by maintaining vasoconstriction ability. Function of parasympathetic nerves in autoregulation of CBF is still not clear, but its stimulation may have protective role in acute hypotensive period.

The Renin–Angiotensin System

In animal study, utilization of angiotensin-converting enzyme inhibitor shifted upper and lower limits of autoregulation to lower CPP.

EFFECTS OF CHRONIC HYPERTENSION ON AUTOREGULATION OF CBF

In hypertensive patients, autoregulatory plateau is shifted to the right, thereby CBF is maintained within normal range despite high MAP and CPP. The possible mechanisms are increased response to vasoconstrictor stimuli and decreased response to vasodilator stimuli. Structural alterations of cerebral arterioles in chronic hypertension result in alteration of vascular responsiveness. Hypertrophy of arterial wall with intrusion of smooth muscle on vascular lumen changes baseline resistance to higher level. Remodeling or reduction in external diameter of vessels is another major process that increases cerebral vascular resistance, resulting in rightward shift of autoregulatory plateau.

WHEN CPP IS TOO LOW: IMPAIRED CEREBRAL VASCULAR PERFUSION

CPP decreases in several conditions including ischemic stroke. When CPP falls below the lower limit of autoregulation, the brain is at risk of inadequate blood flow. Compensatory mechanism by increase extraction coefficient of oxygen from blood to cerebral tissue is important in order to sustain its metabolic demand. If CPP continues to fall beyond this point, clinical signs and symptoms of ischemia are expressed.

Lower limit of autoregulation is the point when CBF begins to decrease. It is not the point when cerebral resistance arteries (arterioles) are fully dilated. Some cerebral arterioles continue to dilate after CPP dropped beyond the lower limit of autoregulatory plateau.

WHEN CPP IS TOO HIGH: DISRUPTION OF BBB

When MAP is too high, CPP also increases. In some extreme conditions, such as hypertensive encephalopathy, CPP increases beyond the upper limit of autoregulation. The cerebral resistance arteries cannot sustain constriction and become dilated in some segments and remain constricted in some segments. This sausage-stringing pattern occurs because the dilated segments lose autoregulatory function and become passively dilated while other intervening segments still preserve their function. At later stage, the entire arterioles become dilated, and CBF increased. This causes damage of cerebral vascular endothelium and BBB disruption. Plasma proteins extravasate into the brain tissue, causing edema and neuronal dysfunction. In later stage, CBF is decreased when vasoconstrictor hormones reach vascular smooth muscles because of BBB disruption, and cerebral ischemia ensued.

CLINICAL CHALLENGES

Even though many of the mechanisms governing CPP are understood, the clinical management of increased or decreased CPP has proven to be complicated. This has been particularly true in the management of traumatic brain injuries where a number of factors contribute to alterations of intracranial dynamics, reduction of the CBF and oxygenation. The "Lund concept" introduced in 1990, for example, was a therapeutic approach to lower the CPP. It focused on the reduction of ICP by decreasing intracranial volumes, and this way reducing the risk of vasogenic edema and of elevating ICP. In practice, however, the lowered CPP caused reduction in CBF and increased ischemic complications in the injured brain. Subsequent trials that maintained CPP higher than 70 mmHg were associated with five times increase in and seemingly unrelated complication, acute respiratory distress syndrome.

CHAPTER

7.1

Translational Correlation: Stroke

J. Kim

Texas Tech University Health Sciences Center, Lubbock, TX, United States

NORMAL CEREBRAL BLOOD FLOW

The brain is one of the most highly perfused organs in the body. Neuronal cells normally utilize glucose as their primary energy source. The brain accounts for about 2% of body weight but about 18% of the total blood volume. About 15–20% (about 700 mL/min) of cardiac output circulates in the brain. Regional cerebral blood flow (CBF) normally averages 50–60 mL/100 g/min. Flow in gray matter is about 80 mL/100 g/min, whereas flow in white matter is estimated to be 20 mL/100 g/min. As energy reserves within the brain are negligible, adequate blood flow is essential for the provision of a continuous supply of energy-producing substrates and for the removal of the byproducts of cellular metabolism.

Blood flow studies confirm that regional CBF parallels metabolic activity and can vary from 10 to 300 mL/100 g/min. Motor activity of a limb is associated with a rapid increase in regional CBF of the corresponding motor cortex. Flow rates below 20–25 mL/100 g/min are usually associated with cerebral impairment, as evidenced by slowing on the electroencephalogram (EEG). CBF rates between 15 and 20 mL/100 g/min typically produce a flat (isoelectric) EEG, whereas rates below 10 mL/100 g/min are usually associated with irreversible brain damage.

CBF is determined by blood viscosity, cerebral perfusion pressure (CPP), and vessel radius. Decreases in CPP result in cerebral vasodilatation, whereas elevations induce vasoconstriction. This mechanism is known as autoregulation. In normal individuals, CBF remains nearly constant between mean arterial pressure of about 60 and 160 mm Hg.

PATHOLOGY AND ETIOLOGY OF STROKE

Stroke is the fourth leading cause of death and the leading cause of disability of adults in the United States. Strokes can be divided into hemorrhagic and ischemic strokes. Ischemic strokes account for 80–85% of all strokes. An ischemic stroke is the sudden onset of focal neurological dysfunction due to infarction in cerebrum, brain stem, or cerebellum from lack of CBF. A hemorrhagic stroke is caused by rupture of cerebral arteries and includes intraparenchymal hemorrhage and subarachnoid hemorrhage.

Conn's Translational Neuroscience
http://dx.doi.org/10.1016/B978-0-12-802381-5.00012-9

Ischemic stroke is a heterogeneous group of vascular diseases that embraces large-artery atherosclerosis (16.3%), cardiogenic embolism (29.1%), penetrating small-artery disease (lacunar infarcts, 15.9%), stroke of other determined etiology (2.6%), and stroke of unknown etiology (36.1%).

Large-artery strokes result from atherogenic embolus (artery-to-artery embolism) or hypoperfusion. They may manifest with large thrombus burden and more severe neurological deficit. Cardiogenic embolism results from an atrial or ventricular thrombus secondary to atrial fibrillation, mechanical or prosthetic heart valves, recent myocardial infarction, or cardiomyopathy. Lacunar infarcts are due to occlusion of small penetrating arteries, such as lenticulostriate artery and are the result of hypertension, diabetes, and dyslipidemia. Other determined etiology includes hypercoagulopathy, genetic disorders, such as sickle cell disease or fibromuscular dysplasia, dissections of intra- and extra-cranial arteries, illicit drug use, and autoimmune disorders.

The penumbra is poorly perfused, but viable tissue at risk of impending infarction surrounding a core of infarction; the goal of acute stroke care is the revival and rescue of the ischemic penumbra by rapid restoration of blood flow.

CLINICAL SYMPTOMS AND SIGNS

The neurological impairment in stroke reflects the area of the brain affected by the involved blood vessels.

Clinical symptoms and signs in anterior circulation stroke are different from that of posterior circulation stroke.

ANTERIOR CIRCULATION STROKE

The anterior and middle cerebral arteries (ACA and MCA, respectively) form the anterior circulation that supplies the forebrain. Large vessel strokes cause cortical strokes and small vessel strokes cause subcortical strokes.

MCA STROKE SYNDROME

Superior division stroke results in contralateral weakness and sensory loss involving mainly the face and arm but also can involve the leg. Both the right and left hemisphere are specialized and dominate the other in some specific functions. The left hemisphere controls language and speech in most people; the right hemisphere leads in interpreting three-dimensional images and space. Dominant hemisphere involvement will cause Broca aphasia, and nondominant hemisphere involvement causes contralateral hemi-neglect.

Inferior division stroke results in superior quadrantanopia or homonymous hemianopia, Wernicke aphasia, impaired cortical sensory functions or graphesthesia on the contralateral side of the body, and neglect of the contralateral limbs. Lesion to nondominant parietal lobe may cause an isolated agitated delirium.

PROXIMAL MCA OCCLUSION

Proximal MCA occlusion involves both cortical and penetrating branches. It results in contralateral leg, arm, and face hemiplegia or hemiparesis, eye deviation toward the lesion due to involvement of the frontal eye field, contralateral homonymous hemianopia due to involvement of the lateral geniculate body of the thalamus or optic radiation. Dominant hemisphere involvement causes global aphasia; nondominant hemisphere involvement causes hemi-neglect, anosognosia, and constructional apraxia.

ACA STROKE SYNDROME

Strokes involving ACA cause contralateral leg weakness more than in the arm and face.

If both ACA arise from a single ACA stem, the occlusion causes bilateral paraparesis or quadriparesis, profound abulia, akinetic mutism, disconnection syndrome, transcortical expressive aphasia, and urinary incontinence.

LACUNAR STROKE SYNDROME

Small penetrating branches of large arteries give different types of stroke syndrome known as lacunar stroke syndrome, namely pure motor, pure sensory, mixed sensory motor, dysarthria–clumsy hand, and ataxic hemiparesis (Table 7.1.1).

POSTERIOR CIRCULATION STROKE

Twenty percent of ischemic strokes are due to posterior circulation involvement.

BRAIN STEM STROKE

There are two key findings: one involves the cranial nerve and the other a long tract sign (Tables 7.1.2–7.1.4. Midbrain, Pons, Medulla).

TABLE 7.1.1 Lacunar Syndrome

Syndrome	Lesions
Pure motor hemiparesis	Corona radiata, genu or posterior limb of IC, anterior pons, medullar pyramid
Pure sensory	Thalamus
Sensorimotor	Thalamus, IC
Ataxic hemiparesis	Corona radiata, anterior limb of IC, posterior limb of IC, upper anterior pons
Dysarthria–clumsy hand syndrome	Upper paramedian base of pons, genu of IC, corona radiata

IC, internal capsule.

CEREBELLAR STROKE

The hallmark of cerebellar lesion is tremor with intended movement without paresis or paralysis. The vermis controls the axial and proximal musculature of the limbs and stroke involving the vermis results in difficulty maintaining posture, gait, or balance. The intermediate part of the hemisphere controls distal musculature; its lesion causes intention tremor, dysmetria (inability to stop a movement at the proper place), dysdiadochokinesia (reduced ability to perform alternating movement), and scanning dysarthria caused by asynergy of the muscles responsible for speech. The lateral part of the hemisphere involves motor planning. The flocculonodular lobe is involved in control of balance and eye movements.

INITIAL EVALUATION

After stabilization of airway, breathing, and circulation, an assessment of neurological deficits and possible comorbidities should be done on patients with suspected stroke. The goals are to identity patients with possible stroke, exclude stroke mimics, identify other conditions requiring immediate intervention, and determine potential causes of the stroke for early secondary prevention.

The focused history taking regarding the time of onset is crucial. The past medical history and clinical findings may lead to another cause for the patient's symptoms (stroke mimics). Information related to eligibility for therapeutic interventions in acute ischemic stroke are equally important (Table 7.1.5).

A neurological examination is important to confirm the patient's chief complaint and to further locate any additional deficit that is not noticed by the patient or family. The National Institute of Health Stroke Scale (NIHSS) is a scale designed as a quantitative and qualitative communication tool to monitor symptoms of stroke patients over time to help make medical decisions. This scale is composed of 11 items to evaluate level of

TABLE 7.1.2 Midbrain

Syndrome	Lesion	Cranial Nerve Sign	Long Tract Sign
Weber	Midbrain basis	Ipsilateral oculomotor palsy • Mydriasis • Ptosis • Eye pointing down and out	CST—Contralateral spastic hemiparesis CBT—Contralateral spastic hemiparesis of lower face
Claude	Midbrain tegmentum	Ipsilateral oculomotor palsy • Mydriasis • Ptosis • Eye pointing down and out	Red nucleus and superior cerebellar peduncle—contralateral ataxia
Benedikt	Midbrain basis and tegmentum	Ipsilateral oculomotor palsy • Mydriasis • Ptosis • Eye pointing down and out	CST—Contralateral spastic hemiparesis CBT—Contralateral spastic hemiparesis of lower face Red nucleus and superior cerebellar peduncle—contralateral ataxia
Parinaud	Dorsal midbrain	Vertical gaze centers • Fixed pupils, paralysis of upward gaze, and accommodation	
Nothangle	Midbrain tectum	Paralysis of gaze depends on cranial nerve III involvement	Cerebellar ataxia

CBT, corticobulbar tract; *CST*, corticospinal tract.

TABLE 7.1.3 Pons

Lesion	CN Sign	Long Tract Sign
Medial pons	Sixth CN—ipsilateral sixth nerve palsy	CST—Contralateral hemiparesis of arm and leg MLS—Decreased contralateral vibration and position sense
Lateral pons	Seventh CN—ipsilateral facial palsy	ICP—Ipsilateral limb ataxia
	Eighth CN—hearing loss	ST—Contralateral hemibody pain and temperature loss
	Vestibular CN—vomiting, vertigo, and contralateral nystagmus	Descending hypothalamic—ipsilateral Horner syndrome
	Spinal fifth CN—ipsilateral facial pain and temperature loss	

CN, cranial nerve; *CST*, corticospinal tract; *ICP*, inferior cerebellar peduncle; *MLS*, medial lemniscus; *ST*, spinothalamic tract.

TABLE 7.1.4 Medulla

Lesion	CN Sign	Long Tract Sign
Medial medulla	12th CN—ipsilateral tongue deviation	CST—Contralateral hemiparesis of the body ML—Contralateral loss of position and vibration sense on the body
Lateral medulla (Wallenberg)	Spinal fifth CN—ipsilateral facial pain and temperature loss	ICP—Ipsilateral limb ataxia ST—Contralateral hemibody pain and temperature loss
	Vestibular CN—vertigo, vomiting and contralateral nystagmus	Descending sympathetic tract—ipsilateral Horner syndrome
	Ninth and tenth CNs—ipsilateral vocal cord paralysis, dysphagia, and palate droop	

CN, cranial nerve; *CST*, corticospinal tract; *ICP*, inferior cerebellar peduncle; *MLS*, medial lemniscus; *ST*, spinothalamic tract.

TABLE 7.1.5 Stroke Mimics

Mimics	Clinical Features
Conversion disorder	Lack of cranial nerve findings, neurological findings in a neurovascular distribution, inconsistent examination
Hypertensive encephalopathy	Headache, delirium, significant hypertension, cerebral edema
Hypoglycemia	History of diabetes, serum glucose low, decreased level of consciousness
Migraine with aura, hemiplegic migraine, migrainewith brainstem aura	History of similar events, preceding aura, headache
Seizures	History of seizures, witnessed seizure activity, postictal period

consciousness, gaze, visual field, presence of facial palsy, motor strength of arms, motor strength of legs, ataxia, sensation, language, presence of dysarthria, and presence of neglect.

Several tests should be performed to identify systemic conditions that may mimic or cause stroke or that may influence therapeutic options (Table 7.1.6).

IMAGING AND DIAGNOSTIC TESTS

Noncontrast-enhanced computerized tomography (CT) of the brain accurately identifies the majority of cases of intracranial hemorrhage and helps discriminate stroke mimicries, such as brain tumors. Because of wide availability, easy access, and its ability to distinguish acute ischemic stroke from hemorrhagic stroke, CT is commonly used for acute stroke patients. MRI is better than CT in distinguishing acute from chronic ischemia and in identifying acute, small cortical, small deep and posterior fossa infarcts, and old silent infarcts. MRI with diffusion weighted imaging has a higher sensitivity for early ischemic changes.

Ultrasonography or vascular imaging of the cervical vessels should be performed in all patients with ischemic stroke in order to screen for stenotic extracranial and intracranial vessels that would warrant interventions, such as endarterectomy for internal carotid artery. An echocardiogram should also be done to screen for the presence of cardiac embolism.

TABLE 7.1.6 Tests for Systemic Conditions That May Mimic or Cause Stroke

All Patients	Selected Patients
Noncontrast computerized tomography (NCCT) of head or brain magnetic resonance imaging (MRI)	Hepatic function tests
Blood glucose	Toxicology screen
Serum electrolytes/renal function tests	Blood alcohol level
Lipid profile	Pregnancy test
Electrocardiography (ECG)	Arterial blood gas (for hypoxia)
Markers of cardiac ischemia	Chest radiology (for lung disease)
Complete blood count (CBC), including platelet count	Lumbar puncture (LP, if subarachnoid is suspected and CT is negative for blood)
Prothrombin time (PT)/international normalized ratio (INR)	Electroencephalography (EEG, if seizures are suspected)
Activated partial thromboplastin time (aPTT)	
Oxygen saturation	

TREATMENT OF ACUTE ISCHEMIC STROKE

After hemorrhagic stroke and stroke mimics are ruled out by neuroimaging and other diagnostic tests, the initial treatment of ischemic stroke depends on the type of stroke and eligibility for thrombolytic therapy. The administration of intravenous tissue plasminogen activator (IV tPA) is the standard of care when the onset neurological deficit is known to be less than 3h prior to presentation without any other contraindications. Other recently approved approaches for recanalization of acute ischemic stroke include mechanical thrombectomy (within 6h). If the patients are not eligible for thrombolytic therapy, antiplatelet therapy should be started unless there is a contraindication.

General treatment for patient stabilization and prevention and management of concomitant conditions are important. Additional studies to elucidate the etiology of stroke, such as echocardiogram, carotid ultrasound, lipid panel, and other tests, should be performed.

COMMON RISK FACTORS OF STROKES AND SECONDARY PREVENTION OF ISCHEMIC STROKES

Because another primary objective in the treatment of ischemic strokes is prevention, efforts to control the risk factors must continue after stroke. The most common nonmodifiable risk factors of strokes are age (>55 years), gender (men > women), and ethnicity (Black > Hispanic > White). The most common modifiable risk factors for strokes are hypertension with a relative risk (RR) of 2–5, atrial fibrillation (RR = 1.8–2.9), diabetes (RR = 1.8–6), smoking (RR = 1.8), hyperlipidemia (RR = 1.8–2.6), and asymptomatic carotid stenosis (RR = 2). Controlling blood pressure reduces the incidence of atheroembolic infarction and intracranial hemorrhage.

Diabetes and smoking were found to increase the incidence of strokes by contributing to the atherosclerotic process in large and small arteries. With regard to hyperlipidemia, LDL has the most impact on the incidence of stroke just as in the case of coronary artery disease. Atrial fibrillation is the most common risk factor for cardioembolism. In patients with atrial fibrillation or cardioembolism, anticoagulation should be considered.

Further Reading

[1] Steve Roach E, Bettermann K, Biller J. Toole's cerebrovascular disorders. 6th ed. Cambridge University Press; 2010.

[2] Mohr JP, et al. Stroke: pathophysiology, diagnosis, and management. 5th ed. Sanders; 2011.

[3] Caplan LR, van Gijn J. Stroke syndromes. 3rd ed. Cambridge University Press; 2012.

CHAPTER

7.2

Translational Correlation: Migraine

D. Ruthirago, P. Julayanont, J. Kim

Texas Tech University Health Sciences Center, Lubbock, TX, United States

DEFINITION AND CLASSIFICATION

Migraine is an episodic, unilateral or bilateral, moderate to severe throbbing headache, generally associated with nausea, vomiting, light sensitivity, or sound sensitivity. Migraine derives from the Greek word "hemikranios," meaning pain on one side of the head, which emphasizes the unilateral headache in many migraine patients. Without treatment, migraine can persist from 4 to 72h. There is variation in the clinical presentation, frequency, and intensity of migraine in each patient.

Migraine can be broadly divided into migraine with and without aura. In reality, some patients can have attacks of both types. According to the second Edition of The International Headache Classification (ICHD-2) by International Headache Society, migraine is classified into seven types. Some types of migraine can be further classified into subtypes.

1. Migraine with aura
   Typical aura with migraine headache
   Typical aura with nonmigraine headache
   Typical aura without headache
   Familial hemiplegic migraine
   Sporadic hemiplegic migraine
   Basilar-type migraine
2. Probable migraine with aura
3. Migraine without aura
4. Probable migraine without aura
5. Childhood periodic syndromes that are commonly precursors of migraine
   Cyclical vomiting
   Abdominal migraine
   Benign paroxysmal vertigo of childhood
6. Retinal migraine
7. Complications of migraine
   Chronic migraine
   Status migrainosus
   Persistent aura without infarction
   Migrainous infarction
   Migraine-triggered seizures

EPIDEMIOLOGY

Migraine is a common problem that affects about 12% of the general population. The prevalence of migraine in Europe and the United States is 6% in men and 18% in women. The initial attack of migraine usually occurs during

Conn's Translational Neuroscience
http://dx.doi.org/10.1016/B978-0-12-802381-5.00013-0

adolescence. Ninety percent of migraine patients have had their first attack before age of 40. However, migraine can occur at any age. Migraine is most common during age of 30–39 years. It occurs equally in male and female before puberty and becomes three times more common in women after menarche. Genetic factors may play role in the occurrence of migraine as some studies reported positive family history in 90% of patients. The most common type is migraine without aura, which accounts for 75% of cases.

PATHOPHYSIOLOGY

There is no unified mechanism that can explain pathogenesis of migraine at this time. The long-lasting hypothesis of cerebral vasoconstriction during aura and vasodilatation during migraine headache is no longer accepted since multiple subsequent studies found conflicting results. Several hypotheses are raised in literature which illustrate some aspects of pathophysiology of migraine. These include the concepts of cortical spreading depression, activation of trigeminovascular system and neurogenic inflammation, sensitization process, and the roles of serotonin and calcitonin gene-related peptide (CGRP).

Cortical Spreading Depression

Cortical spreading depression (CSD) is a propagating wave of depolarization of neurons and glial cells that spreads across cerebral cortex. The spreading of this wave is hypothesized to activate the trigeminal nerve afferents, causing inflammatory changes in pain-sensitive structures that create migraine headache. Several molecular cascades are activated by cortical spreading depression including the opening of pannexin-1 megachannels, the release of proinflammatory mediators. This phenomenon is believed to originate aura, alter permeability of blood–brain barrier, and finally prolong the activation of trigeminal nociceptor that causes migraine headache. Migraine without aura may be explained by occurrence of this phenomenon in some parts of the brain where depolarization wave cannot be perceived.

Studies using magnetic resonance spectroscopy support hypothesis that aura is a neuronal event that causes decreased cortical circulation in response to decreased metabolic activities due to defective mitochondrial oxidative phosphorylation. Magnesium also involves in mitochondrial metabolism and N-methyl-D-aspartate (NMDA) receptors function. Magnesium deficit can cause instability of neuronal polarization, leading to neuronal hyperexcitability and increased tendency for spontaneous depolarization.

Activation of the Trigeminovascular System and Neurogenic Inflammation

Trigeminovascular system consists of small pseudounipolar sensory neurons that originate from trigeminal ganglion and upper cervical dorsal nerve roots. These neurons innervate cerebral blood vessels, pial vessels, dura mater, and large venous sinuses. The projections from upper cervical nerve roots and trigeminal ganglion converge at trigeminal nucleus caudalis (TNC), which explains the distribution of migraine headache in anterior and posterior part of head and upper neck. Fibers from TNC travel to ventroposterior medial nucleus of thalamus and to sensory cortex. Other fibers also project to several subcortical areas such as reticular formation, cerebellum, and midbrain. Nociceptive information is also transmitted to limbic system, which explains emotional response in migraine headache.

Activation of trigeminal ganglion causes the release of several vasoactive neuropeptides that are associated with neurogenic inflammation such as substance P, CGRP, and neurokinin A. Vasodilatation and extravasation of protein are the two main mechanisms of neurogenic inflammation, which prolongs and worsens the migraine headache. Other mechanisms include increased platelet aggregation, activation of local immune response, and mast cell degranulation.

Sensitization

Sensitization is the process that occurs after neurogenic inflammation when neurons become more responsive to both nociceptive and non-nociceptive stimuli, namely decrease in thresholds of response, increase in magnitude of response, expansion of receptive field, and emergence of spontaneous activity. The sensitization processes occur in primary afferent neurons (peripheral sensitization) and in TNC and higher-order neurons (central sensitization). Sensitization is believed to play a role in transformation of episodic to chronic migraine and in each migraine attack. This process may explain the character of throbbing pain, worsening symptom with physical activities, hyperalgesia, and cutaneous allodynia in migraine headache.

Role of Serotonin in Migraine Headache

Serotonin is believed to play a role in pathogenesis of migraine headache because it has direct effect on cerebral vasculature, some action in central pain control, and its well-known projections from brain stem nuclei to cerebral cortex. The use of tricyclic antidepressants, which block serotonin reuptake, for prophylactic treatment of migraine also supports its role. However, its role is still inconclusive as selective serotonin reuptake inhibitors do not have the same prophylactic effect as tricyclic antidepressants.

Role of CGRP in Migraine Headache

CGRP is a neuropeptide that exists in trigeminal ganglia nerves and a potent vasodilator of cerebral blood vessels. CGRP is believed to mediate vasodilatory mechanism of neurogenic inflammation and the transmission of trigeminovascular nociceptive signal from intracranial vessels to central nervous system. Stimulation of trigeminal ganglion causes the release of CGRP. Infusion of CGRP caused migraine attack in patients in one study. However, other studies involving intracranial and extracranial arteries yielded conflicting results.

The fact that sumatriptan, the serotonin 1b/1d receptor agonist, normalizes the elevated level of CGRP in migraine patients supports its effect in the treatment of migraine by inhibiting CGRP release.

GENETICS AND MIGRAINE HEADACHE

The common types of migraine are complex genetic disorders, which means multiple genes and environmental factors have their roles in the development of migraine. Migraine patients may have genetic abnormalities of some membrane channels, receptors, and enzyme systems that make them more susceptible to migraine attack when they expose to environmental factors. There is no single Mendelian inheritance identified in common migraine at present. Some uncommon types of migraine are caused by mutations of certain genes: first three types of familial hemiplegic migraine are channelopathies, which are caused by mutations of CACNA1A gene, ATP1A2 gene, and SCN1A gene, respectively.

In conclusion, the brain of migraine patients may have some abnormalities such as in mitochondrial oxidative phosphorylation, neuronal ion channel dysfunction, and/or deficiency of intracellular magnesium that make them more susceptible to neuronal excitability and spontaneous depolarization. This process leads to activation of CSD and changes in regional cerebral blood flow. CSD activates the trigeminovascular system, which generates neurogenic inflammation by release of several neuropeptides including CGRP (a potent vasodilator). Migraine headache is triggered by both vasodilatory substances and activation of brain stem nociceptive centers by CSD.

CLINICAL SYMPTOMS AND SIGNS

The typical characteristic of migraine is moderate to severe, unilateral, pulsatile headache. The headache is usually aggravated by movement or physical activities and accompanied by nausea, vomiting, photophobia or phonophobia. The duration of headache ranges from 4 to 72 h if untreated. The patients may report some aggravating factors such as alcohol intake, sleep deprivation, and smell of perfume. Menstruation can trigger migraine in some patients.

Migraine without aura is the most common type of migraine. Some patients may develop neurological symptoms or aura just prior to or at the onset of migraine headache. Aura must completely resolve. There are various types of aura including visual aura, sensory aura, and reversible dysphasia. The visual aura can be positive symptoms (lights, spots, or flashing) and negative features (dimming or visual loss). Sensory aura includes tingling, pins and needles sensation, or numbness. Some patients experience reversible motor weakness in addition to the aforementioned aura which are diagnosed as hemiplegic migraine. Basilar-type migraine is a subtype of migraine with aura. The patients with basilar-type migraine may have brain stem–originated aura (dysarthria, vertigo, tinnitus, hypacusia, diplopia, ataxia, or decreased level of consciousness) or bilateral hemisphere–originated aura (visual symptoms simultaneously in both temporal and nasal fields of both eyes, decreased level of consciousness, or simultaneously bilateral numbness/tingling).

Retinal migraine is repeated attacks of monocular visual disturbance associated with migraine headache. Amaurosis fugax or transient monocular blindness must be differentiated.

Some features exhibited in patients with the history of migraine are not always fit to the typical migraine headache either in terms of headache characteristics or duration of headache and aura.

Chronic migraine is the migraine that occurs >15 times/month for at least 3 months. Patients with symptoms of typical migraine that lasts longer than 72 h are diagnosed as status migrainosus. Migrainous infarction is the migraine complicated by the cerebral infarction at the relevant territory.

EVALUATION AND DIAGNOSIS

Migraine is classified as the primary headache syndrome according to International Classification of Headache Disorders (ICHD). Thus prior to making diagnosis of migraine, the structural or organic causes should be excluded, in particular to the patients with atypical headache characteristics, prolonged duration of headache/aura, or headache associated with neurological symptoms.

Lumbar puncture may be required to exclude other secondary causes of headache if encephalitis, meningitis, subarachnoid hemorrhage, low-pressure headache, or idiopathic intracranial hypertension are suspicious.

The following algorithm summarizes the diagnostic approach for different subtypes of migraine according to ICHD. Childhood periodic syndromes, which are composed of cyclical vomiting, abdominal migraine, and benign paroxysmal vertigo of childhood, are not included in this algorithm. These conditions are considered as precursors of migraine (Fig. 7.2.1).

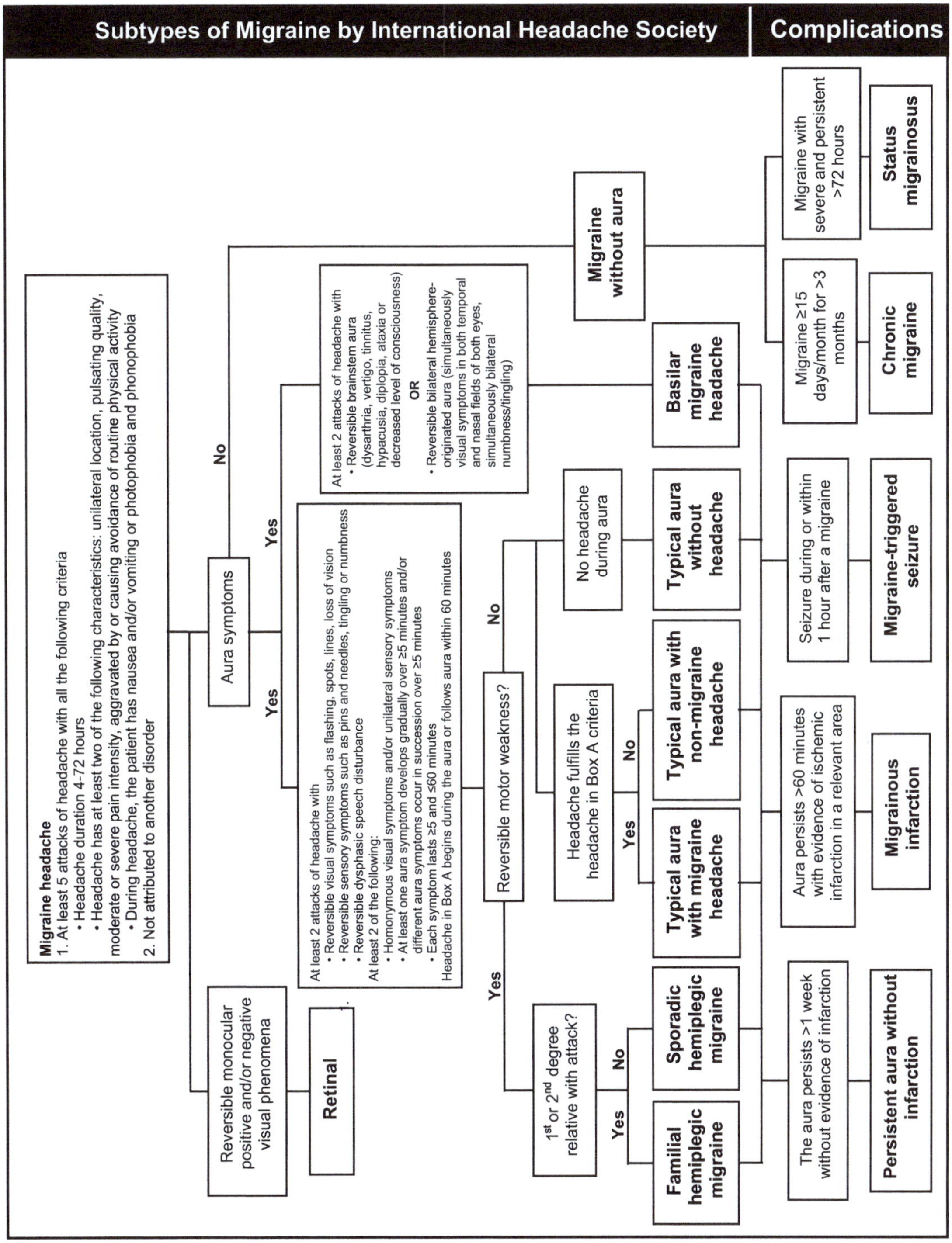

Adapted from the International Headache Society classification.

FIGURE 7.2.1 Subtypes of migraine by International Headache Society and complications.

TREATMENT

The treatment strategies for migraine consist of acute or abortive treatment and preventive treatment. The aims of acute or abortive treatment are to stop the progression of the attack and relieve to the current attack while the goal of preventive treatment is to reduce the frequency, duration, and severity of the future attack.

Abrtive Treatment

The abortive treatment includes both specific and nonspecific medications. The nonspecific medications are the medications that can treat any pain disorders which are not merely specific to the migrainous pain. The nonspecific treatment includes nonsteroidal anti-inflammatory drugs (NSAIDs), opioids, corticosteroids, and antiemetics. Medications specific to the migraines include dihydroergotamine (DHE) and triptans.

The acute treatment should be administered as early as possible to more effectively reduce the duration and the severity of the attacks. The optimal dose and appropriate route of administration are very important and affect the treatment outcomes.

Triptans

Triptans are working via the agonistic effect to the serotonin receptors (5-HT). They are selective 5-$HT_{1B/1D}$ receptor agonists. The vasoconstrictive effect of triptans via the 5-HT_{1B} receptor at the smooth muscle of intracranial arteries is one of its mechanisms to reduce the migraine headache. Triptans activate the 5-HT_{1D} in trigeminal nucleus caudalis, inhibiting the activation of nociceptive receptors and the release of vasoactive peptides including substance P, CGRP.

Triptans are the first-line therapies for patients whose migraine attack does not respond to analgesics. They will effectively work if administered with adequate dose early after the onset when the headaches are mild to moderate. Seven triptans commercially available in the United States includes almotriptan (Axert), eletriptan (Relpax), frovatriptan (Frova), naratriptan (Amerge), rizatriptan (Maxalt), sumatriptan (Imitrex), and zolmitriptan (Zomig). Sumatriptan is the first triptan used clinically and is available in tablet, nasal spray, and subcutaneous injection. The onset of action of triptans is within 20–60 min. In patients with severe headache or nausea/vomiting, subcutaneous sumatriptan should be considered as it has the fastest onset. From the Cochrane review, all triptans have similar effectiveness and tolerability for acute migraine treatment. As individual patients may have different responsiveness to different triptans, trial of different triptans may be effective in patients who failed to respond to the one triptans.

Due to the vasoconstrictive mechanisms, triptans are contraindicated in patients with poorly controlled hypertension, severe hepatic and renal impairment, hemiplegic migraine, basilar migraine, and coronary artery disease.

Nonsteroidal Anti-inflammatory Drugs

NSAIDs are also first-line medications for abortive treatment available over-the-counter in the United States. They are indicated in mild to moderate migraine but are less effective in severe migraine compared to triptans. NSIADs are nonspecific cyclooxygenase (COX) inhibitors. Blocking of the COX-2 isoenzyme is the desirable mechanism of NSAIDs. Inhibition of COX-2 reduces the production of prostaglandins which are the mediators of pain and inflammation. Ibuprofen and naproxen are the common oral NSIADs used in migraine. In the emergency department, intravenous ketorolac effectively reduces the severity and duration of migraine headache within 1 h after the injection.

Ergotamines and DHE

Ergot alkaloids are the first available migraine-specific medications that have been used in clinical practice for more than 60 years.

The precise mechanism of action is not clear. Its vasoconstrictive effects are believed to be the mechanism of action. Due to the lack of consistent evidence for their efficacy, the use of ergotamine and DHE has been largely replaced by triptans. However, it may remain beneficial in patients with prolonged attack.

The side effects of these medications are nausea/vomiting, bradycardia, and systemic vasoconstriction including cerebral, cardiac, and peripheral vessels. They are contraindicated in patients with peripheral vascular disease, coronary artery disease, uncontrolled hypertension, stroke, and severe renal and hepatic impairment. They should not be used in complicated migraine, hemiplegic migraine, and basilar migraine. Ergotamines and DHE should not be taken within 24 h of the triptans use.

Other Medications for Acute Treatment

Antiemetics is not a migraine-specific medication. Intravenous metoclopramide is commonly used in the emergency department to improve nausea and vomiting in migraine patients. The effectiveness of intravenous steroid is not consistent and is not recommended as a standard acute migraine treatment. In some clinical practice, intravenous steroid may be beneficial in status migrainosus.

Preventive Treatment

Strong Evidence of Efficacy

Many medications have been established as effective migraine preventive medications; the American Headache Society and the American Academy of Neurology recommend that these medications should be offered to patients with frequent migraine attacks.

For antiepileptic drugs, divalproex sodium/sodium valproate and topiramate are established as effective medications to prevent migraine. Propranolol, timolol, and metoprolol are beta-blockers with strong effectiveness in reducing attacks. An herbal drug named Butterbur, which is the *Petasites hybridus* extract, has been studied in many double-blind, randomized, controlled studies that have established its strong effectiveness in reducing frequency of migraine attacks.

Moderate Evidence of Efficacy

The medications in this category should be considered in patients who require migraine prophylaxis. Venlafaxine and amitriptyline are the antidepressants with moderate evidence in reduction of migraine attacks. For beta-blockers, atenolol and nadolol are probably effective to prevent further migraine attacks.

Further Reading

[1] Garza I, Swanson JW. Headache and Other Cranifacial Pain. Bradley's Neurology in Clinical Practice. 6th ed. Elsevier; 2012. p. 1715–30.

[2] The International Classification of Headache Disorders. 3rd ed. (beta version). Cephalalgia 2013;33:629–808.

[3] Silberstein SD, Holland S, Freitag F, Dodick DW, Argoff C, Ashman E. Evidence-based guideline update: pharmacologic treatment for episodic migraine prevention in adults: report of the Quality Standards Subcommittee of the American Academy of Neurology and the American Headache Society. Neurology 2012;78:1337–45.

CHAPTER

8

The Vestibular System

R.S. Swenson

Geisel School of Medicine, Hanover, NH, United States

Balance is a particularly important challenge for a bipedal species such as humans, and injuries due to falls represent a major health risk, especially in the elderly. There are three main systems that must collaborate in order to permit safe ambulation. Proprioceptive somatosensory nerves detect joint position and pressure on those body parts that are in contact with the environment, such as the feet with the floor. Visual cues are also particularly important in orientating one's position in space and in assessing movement both of the individual and of the environment. The vestibular system is the third main system that is critical for balance and for detection of movement. Phylogenetically, the vestibular system arose from a line of cells along the lateral side of fish that are specialized to detect movement and vibrations in the surrounding water. This has evolved to a system that is very sensitive to rapid changes in position and motion of the head. In addition to permitting perception of such motion, the vestibular system has important reflex connections to portions of the nervous system that adjust body position and muscle tone in response to movement and to the brain stem regions controlling eye movements. This is critical in order to stabilize the eyes during motion of the head. Because the vestibular system also connects to the cerebral cortex where the sensation of movement is processed, acute abnormalities of the vestibular system (especially, asymmetrical abnormalities) are characterized by an illusion of movement, termed vertigo. This not only leads to distress but also can be disabling and even precipitate falls.

This chapter discusses the several elements of the vestibular system, first focusing on the organization of the peripheral receptive organs before considering the transduction of the mechanical signals to electrical impulses, the transmission of these electrical signals to the central nervous system, and then the central connections of the vestibular system. These central connections are considered in light of their role in reflex stabilization of the eyes, in regulation of muscle tone and autonomic functions (including effects of motions sickness), and also in their contribution to the conscious perception of movement. Finally, the clinical manifestations of damage to the vestibular system including nystagmus and vertigo are explained, along with some of the methods for clinically evaluating the system.

THE INNER EAR

The peripheral vestibular organ is contained within a bony cavity (the bony labyrinth) of the inner ear (Fig. 8.1). Part of the bony labyrinth

Conn's Translational Neuroscience
http://dx.doi.org/10.1016/B978-0-12-802381-5.00014-2

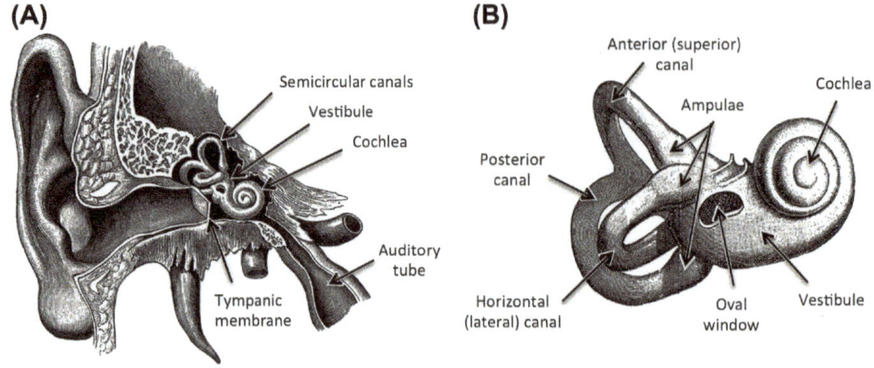

FIGURE 8.1 Bony labyrinth. (A) Cutaway image of the petrous temporal bone from the anterior. The hard bone surrounding the perilymph (the otic capsule) encases the structures of the inner ear. Note the relative positions of the three components of the bony labyrinth, with the cochlea anteromedial to the vestibule, which gives rise to the three semicircular canals. Also note the relative orientation of these canals. (B) The outlines of the bony labyrinth from the lateral aspect.

extends as a coiled tube into the cochlea, the part of the inner ear responsible for hearing (see Chapter 26). The vestibular part of the bony labyrinth consists of a large cavity, called the vestibule, and three semicircular canals that are located at right angles to one another. These canals are termed the horizontal (lateral), anterior (superior), and posterior canals and are located at an angle to the planes of the head (Fig. 8.2A and B). The cavity of the bony labyrinth contains a fluid called perilymph that is in communication with the cerebrospinal fluid through the perilymphatic duct. Therefore, the perilymph has composition very similar to that of cerebrospinal fluid, which is similar to the composition of extracellular fluid (Table 8.1). Because of the continuity between the cerebrospinal fluid and the perilymph, conditions that affect the cerebrospinal fluid, such as meningitis, can severely damage the inner ear.

The membranous labyrinth, which is contained within the bony labyrinth, consists of a watertight membrane-lined sac that is filled with endolymph. The membrane separates the endolymph that fills the inside of the membranous labyrinth from the perilymph that is within the bony labyrinth but outside of the membrane. The membranous labyrinth can be conceptualized as a complex fluid-filled balloon inflated in a container that is, itself, filled by a different fluid (Fig. 8.3). The ionic composition of endolymph is very different from the perilymph since the potassium concentration is much higher and sodium concentration is lower (Table 8.1). Endolymph is actively secreted by the stria vascularis, a specialized plexus of blood vessels located within the spiral ligament of the cochlear duct. Cells of the stria vascularis contain ion pumps that actively secrete the endolymph. These ion pumps are electrogenic—that is, they produce a voltage difference of +80 mV between the endolymph and perilymph. Ultimately, endolymph fills all portions of the membranous labyrinth, separated from perilymph only by the very thin lining of the membranous labyrinth.

There are two dilations or chambers of the membranous labyrinth that reside within the vestibule of the bony labyrinth. These are the saccule and utricle, interconnected by the utriculosaccular duct. Semicircular ducts arch from the utricle back to the utricle, with each semicircular duct having a dilation termed the ampulla. Thus, endolymph is free to circulate within any semicircular duct, passing the ampulla in its course.

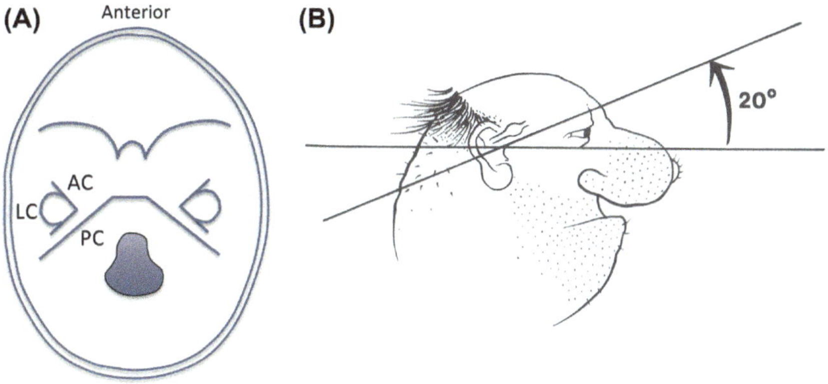

FIGURE 8.2 Planes of the semicircular canals. (A) Cartoon depicting the skull from the inside, showing the relative orientations of the anterior (AC), posterior (PC), and lateral (LC) semicircular canals. Note that the anterior canal of one ear is approximately parallel to the posterior canal of the other. (B) Depiction of the orientation of the lateral (horizontal) semicircular canal, the anterior side of which is angulated about 20 degrees upward from Reid's horizontal baseline. *Reproduced with permission from Reeves, A.G.m and Swenson, R.S. Disorders of the Nervous System: a Primer. Dartmouth Medical School.*

TABLE 8.1 Comparison Between the Approximate Ionic Composition of Cerebrospinal Fluid, Perilymph, and Endolymph

Approx. Ion Conc. (mM)	Cerebrospinal Fluid	Perilymph	Endolymph
Sodium	150	145	1.5
Potassium	3.0	4.8	160
Chloride	130	120	130
Bicarbonate	19	20	31
Calcium	0	0.8	0.02

After filling the membranous labyrinth, the endolymph is resorbed into venous blood at the endolymphatic sac, which is located adjacent to the sigmoid sinus in the posterior fossa of the skull. Endolymph reaches this sac via the endolymphatic duct that extends from the utriculosaccular duct through the petrous temporal bone to the sac. Restriction of flow of endolymph through the endolymphatic duct is the leading theory for the development of Meniere's syndrome, a condition that results in increased pressure in the membranous labyrinth due to diminished resorption of endolymph. This high pressure is not only damaging to receptive organs of the inner ear but also results in periodic disruptions of the integrity of the membranes ("blowouts"), with sudden changes in ion concentrations producing severe attacks of vertigo.

HAIR CELLS

Hair cells are the transducers that change mechanical energy of motion to electrical signals conveyed as vestibular nerve action potentials. These hair cells come in two shapes, a goblet-shaped type I cell and a thinner, more columnar type II cell (Fig. 8.4A). Both of these cells contact the peripheral terminals of vestibular

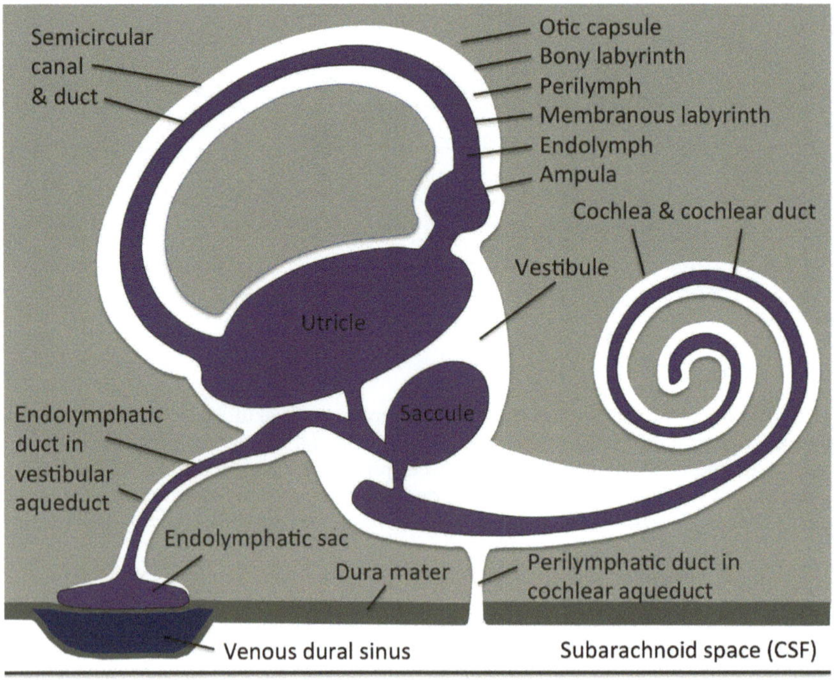

FIGURE 8.3 Membranous labyrinth. Cartoon depicting the inner ear in cross section showing the outlines of the bony labyrinth filled with perilymph and the membranous labyrinth (in blue) containing endolymph. Notice that the perilymph is continuous with the cerebrospinal fluid through the perilymphatic duct and that the membranous labyrinth continues through the endolymphatic duct to the endolymphatic sac where it can be resorbed into the venous blood at the dural sinus (endolymph is produced by the stria vascularis of the cochlear duct). *Adapted with permission from O'Rahilly, Müller, Carpenter and Swenson, Basic Human Anatomy.*

sensory axons. However, the base of the type I cell is essentially encased by the end of the nerve terminal (a calyx ending) while the type II cell is contacted by a much smaller, more normal appearing terminal bouton of the sensory nerve fiber. The type I cells respond to more intense vestibular stimulation while the type II cells respond to weaker stimuli. Nevertheless, the mechanisms of response are similar for both of these cell types.

Both type I and type II hair cells are part of the wall of the membranous labyrinth where they and their supporting cells replace the flattened cells that line the remainder of the membranous labyrinth. The apices of the hair cells face the inside of the membranous labyrinth and are therefore bathed by endolymph. The bases of hair cells are exposed to the perilymph that is outside of the membranous labyrinth. Therefore, the apex and the base of hair cells exist in very different electrical and chemical environments.

The apex of any one hair cell is highly specialized. The "hairs" for which these cells are named are critical to their mechanical sensitivity. The hairs consist of a series of stereocilia, which are basically giant microvilli, having a core of actin filaments. Each hair cell also has a single, eccentrically placed kinocilium that is positioned toward one edge of the apex of the hair cell. This kinocilium is longer and has the structure of a true cilium (ie, a core consisting of nine microtubular doublets surrounding two additional microtubules). The longest of the stereocilia (hairs) is situated adjacent to the kinocilium

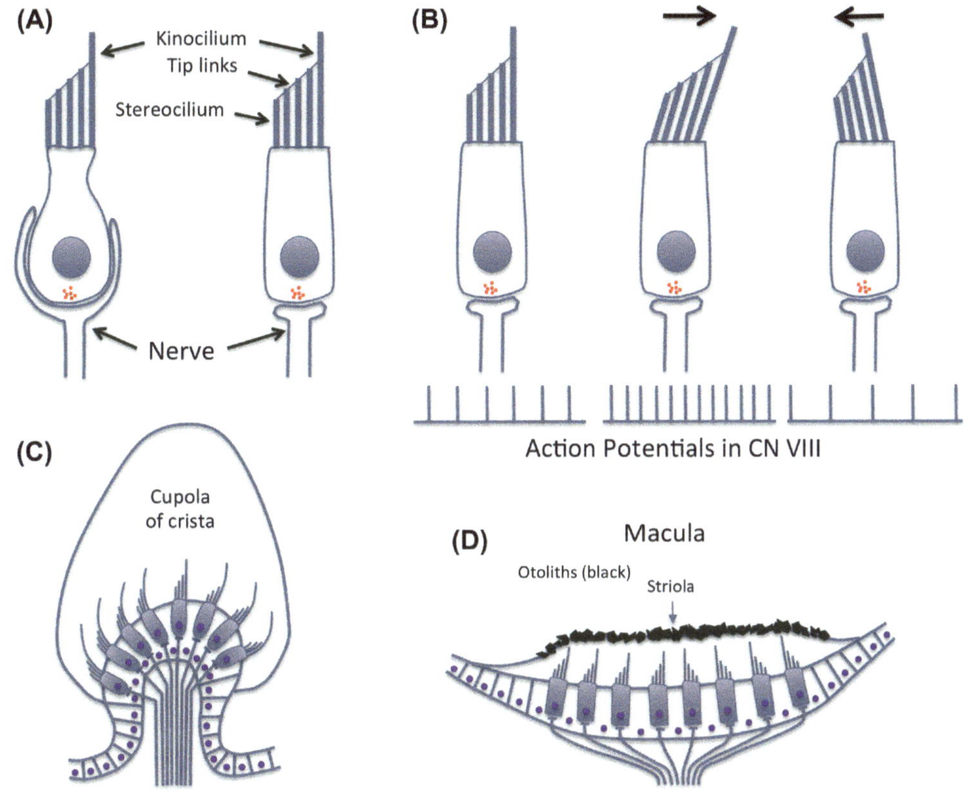

FIGURE 8.4 Cartoons depicting aspects of hair cell structure, function, and organization. (A) Type I (left) and type II (right) hair cells. Notice that all hair cells are polarized by the eccentric position of the kinocilium. (B) The effect of hair cell deflection on transmitter release and nerve activity. When hairs are deflected toward the kinocilium, tension is placed on the tip links. This tension will open potassium channels, resulting in an increase in transmitter release and in action potentials in the nerve (below). Also note that there is a baseline resting level of action potential generation. (C) Structure of a cristae, which is the receptive organ found in the ampulla of semicircular ducts. Note that all of the hair cells are polarized in the same direction. (D) Structure of a macula, a receptive organ in the wall of the utricle and saccule. Note that the hair cells have different polarization on either side of the striola (a *curved line* in the middle of the receptive organ).

while the stereocilia are progressively shorter at a greater distance from the kinocilium. This asymmetrical organization of the apex of hair cells, with the kinocilim off to one side, leads to a polarization of each hair cell. This is critical to the ability of the hair cell to signal the direction of mechanical forces applied to the hairs.

The process of transducing mechanical force on the hairs into electrical signals is dependent on some unique features of the tips of hair cells. Mechanically sensitive potassium channels located in the tips of the hair cells are attached to protein filaments ("tip links") that are, themselves, anchored to the side of the taller adjacent stereocilium (that is, the one located closer to the kinocilium). Forces that deflect the stereocilia toward the kinocilium result in tension on these tip links and an opening of the potassium channels (Fig. 8.4B).

The hairs are bathed in endolymph, which is very high in potassium. Also, the endolymph is very electrically positive relative to the negative inside of the hair cell (around 135 mV potential difference). This results in a very large electrochemical driving force that causes rapid entry of potassium into the hairs from the endolymph whenever the channels are open. Potassium entry results in depolarization of the hair cell which, in turn, opens voltage-gated calcium channels. The calcium entry through these channels results in the increased release of neurotransmitter (glutamate) from the base of the hair cells, exciting the peripheral terminals of afferent fibers of the vestibulocochlear nerve. Larger deflections of the hairs toward the kinocilium will result in more open potassium channels, greater depolarization, more calcium influx through the voltage-gated calcium channels, greater neurotransmitter release, and a higher frequency of action potentials in the vestibular nerve fiber connected to that hair cell (Fig. 8.4B).

Deflection of the hairs away from the kinocilium decreases tension on the tip links, closing the potassium channels to which they are attached. This hyperpolarizes the hair cell and decreases both calcium entry and neurotransmitter release. Furthermore, this results in fewer action potentials being generated in the associated axon of the vestibulocochlear nerve. Thus, each hair cell can signal deflection of hairs toward or away from the kinocilium by increasing or decreasing the frequency of action potential generation, respectively (Fig. 8.4B).

There is a delicately balanced system to establish a baseline level of transmitter release from hair cells. It is essential to have a steady baseline of activity if the system is going to be able to signal deflection of hairs away from the kinocilium since this signaling is accomplished by decreasing action potential generation in the nerve. Therefore, the process of establishing a baseline level of activity must be highly regulated and should also be capable of adapting to new baseline conditions. This baseline level of activity is a property of the hair cell, itself, requiring an intricate mechanism of adjustment of tension on the tip links that attach to the potassium channels in the tip of the hairs of the hair cells. The opposite ends of the tip link anchor to proteins in the membrane on the sides of the adjacent, longer stereocilium. These membrane proteins are connected to motor proteins on the inside of the stereocilium. These motor proteins can move toward or away from the tip of the hair based on intracellular calcium concentration. For example, when intracellular calcium is low, the motor protein will pull the anchor point of the tip link toward the tip of the hair. This, in turn, will place tension on the tip link and slightly open the potassium channels to which they are attached. This will produce enough depolarization to open some voltage-gated calcium channels, reestablishing a baseline level of intracellular calcium in the hair cell. This adjustment not only results in a baseline release of neurotransmitter but also stops the movement of the motor protein toward the tip of the stereocilium. By this mechanism, the hair cell will establish a new baseline level of activity in the associated nerve fiber whenever a new static condition is encountered. High intracellular calcium will do the opposite, resulting in decreased tension on the tip link and less opening of the potassium channels. There are two very important consequences of the fact that hair cells adapt fairly rapidly to static deflection of the hairs. Most significantly, this allows each hair cell to be responsive to movement in either direction from this new static position. However, it also makes the inner ear relatively insensitive to static positions or to very slow movements.

Finally, when considering the physiology of hair cells and the transduction of movement to neural impulses, it is important to recognize that there is an efferent control mechanism that innervates the ends of the vestibular sensory nerve fibers and/or the hair cells. These efferent nerve fibers arise from neurons located in the caudal pons and release acetylcholine and a number of other neurotransmitters, potentially including ATP, peptides [such as calcitonin gene-related peptide (CGRP) and opioids], nitric oxide, and GABA (gamma aminobutyric acid). The net result of activation of these efferent axons is to regulate sensitivity of hair cells by altering nerve activity in the vestibular nerve. However, their role in normal function and pathological states remains a topic of active investigation.

MACULAE AND CRISTAE

Hair cells are aggregated in two types of receptive organs in the inner ear, the maculae (sometimes referred to as the otolithic organs) and the cristae. These organs have some structural similarities and several important differences (Fig. 8.4C and D). Hair cells are the receptive element in both types of organs and the hairs are embedded in a gelatinous cap that, itself, protrudes into the endolymph. Movement of the gelatin results in deflection of the hairs. However, there are several important differences between maculae and cristae.

In the maculae (Fig. 8.4D), the gelatinous covering is studded with calcium carbonate crystals (otoliths or otoconia). These crystals are heavy and gravity will pull down on them and, therefore, on the gelatin. This deflects the embedded hairs. Additionally, linear acceleration, such as accelerating or decelerating at a stoplight, will cause the otoliths to lag behind, also deflecting the hairs. The result of this organization is that maculae are specialized to detect the orientation of the head in relation to gravity and the direction of linear acceleration. Since each hair cell is specialized to detect movement in one direction, the hair cells within the maculae are arranged such that there are individual hair cells oriented to detect virtually any head position or direction of acceleration. This is accomplished by a complex three-dimensional arrangement of each maculae and by the arrangement of hair cells within the macula. The two maculae (one in the saccule and the other in the utricle) are applied to the curved wall. Therefore, each of the maculae contains hair cells oriented in multiple planes. Additionally, there is a curved line in the middle of each of the maculae (the striola). Hair cells on one side of the striola are polarized in one direction and on the other side are polarized the other way. The net result of the curvature of the macula, coupled with the curvature of the striola is that there are hair cells polarized in all three dimensions and, regardless of the position of the head or the direction of linear acceleration, there will be some vestibular afferent nerve fibers from the maculae that will increase their frequency of action potential firing and others that will decrease firing rate. This pattern of activity is relayed to the central aspects of the vestibular system for interpretation.

Cristae are located in the ampullae of the semicircular ducts (three in each inner ear) and are different from maculae in several ways. First, the hair cells of cristae protrude into a flame-shaped gelatinous cap (termed the cupola) that lacks otoliths (Fig. 8.4C). Second, all of the hair cells in a crista are polarized in a single direction, which is along the plane of the semicircular duct. The cupola protrudes into the endolymph within the semicircular duct. Movement of the endolymph through the semicircular duct will deflect the cupola and, therefore, the hairs. Since all of the hair cells in the crista are polarized in the same direction, fluid movement in one direction through the semicircular duct will excite all of the hair cells in the crista while movement in the opposite direction will inhibit all of them.

A third way in which cristae are different from maculae is in the kind of stimuli that they are specialized to detect. The semicircular ducts are specialized for the detection of angular acceleration—that is, movements of yaw, pitch, or roll (eg, rotation, nodding, or lateral bending of the head). This results from the sensitivity of individual crista to fluid movement within their respective canal. Normally, this happens when fluid stays still (through inertia) and the head is rotated in the plane of the canal. Because each canal is roughly circular, it requires rotation in the plane of the canal in order to create this relative movement between the canal and the fluid. During angular acceleration of the head, the lag in fluid movement will bend the crista in the opposite direction to movement of the head and will either increase or decrease nerve impulses in the nerve associated with that canal based on whether the movement is in one direction or the other. This relative movement of fluid and the change in frequency of action potentials in the associated nerve will be greatest if the head movement is directly in the plane of the semicircular duct.

It is important to note that the three semicircular canals of each inner ear are roughly at right angles to one another (permitting detection of three dimensions of angular movement). However, they are not strictly in the horizontal, coronal, or axial planes. For example, the lateral (horizontal) canals are oriented at about 20 degrees to the horizontal plane, with the anterior side elevated (Fig. 8.2). The posterior and anterior (superior) canals are oriented in a manner such that the anterior canal of one ear is in a nearly parallel plane to that of the posterior canal of the other ear. This allows each ear to function in all three dimensions of angular movement. Therefore, one inner ear would suffice for the detection of movement in any angular direction. It is an interesting fact that, for any particular movement, one ear will signal the motion by increasing the firing rate of its nerve, while the complementary canal of the contralateral ear will signal the same movement by decreasing its firing rate. Since nerve activity has a greater range to increase than to decrease, there are subtle signs that can be detected when a patient is functioning with one inner ear, even though overall function ultimately returns to near normality after such damage.

Before concluding this discussion of the receptors of the inner ear, there is a common disorder that directly affects these receptors. In some patients, otoliths can become detached from the macula of the utricle. This may result from trauma, inflammation, or age. These become free to move about in the endolymph and have been termed canaliths. Occasionally, canaliths may enter the semicircular ducts (particularly, the posterior one) and can, during certain movements (particularly, rolling over in bed, looking up, or looking down), contact the cristae and deflect the hairs. This produces an intense sensation of spinning (vertigo) that lasts less than a minute and that has been termed "benign paroxysmal positional vertigo." This is accompanied by complex nystagmus (see later) that usually has a very evident rotatory component. The Nylan–Barany (Hall Pike) maneuver (Fig. 8.5) can usually reproduce the vertigo because it places the crista of the posterior canal in the most dependent position. It is important to recognize this cause of vertigo because it can usually be helped by the "canalith repositioning maneuver" which is designed to return the canaliths to the utricle.

CENTRAL VESTIBULAR CONNECTIONS

Vestibular receptors would not be of much use if they were not connected to the brain. As described earlier, the vestibular part of the vestibulocochlear nerve consists of nerve fibers that have contacted hair cells of the maculae and cristae. These sensory fibers have bipolar cell bodies located in the vestibular ganglion just medial to the inner ear. The centrally projecting axons from these bipolar sensory neurons traverse the internal acoustic meatus and the cerebellopontine angle to enter the brain stem at the medullary–pontine junction (Fig. 8.6). Most of these vestibular sensory axons synapse in subregions of the vestibular complex, a group of nuclei located immediately deep to the floor of the fourth ventricle Fig. 8.7. These nuclei straddle the rostral medulla and the caudal pons. Fig. 8.8 shows the relative position of the vestibular complex superimposed over a drawing of the dorsal aspect of the brain stem.

The vestibular complex of the brain stem consists of four main subnuclei: the medial (Schwalbe);

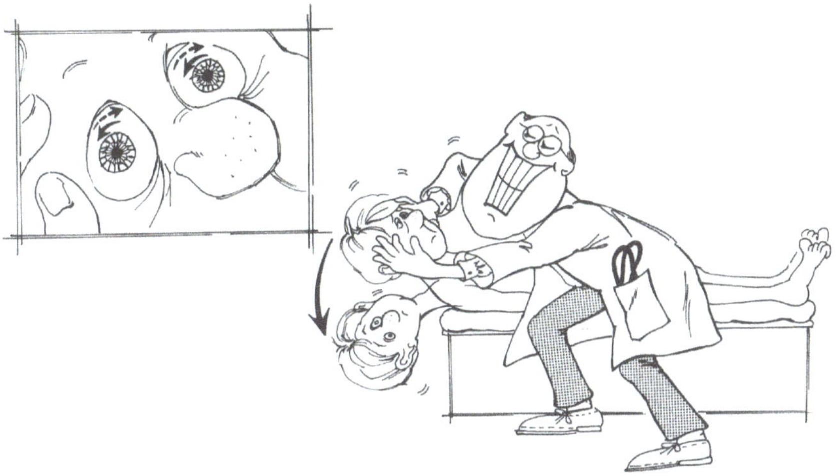

FIGURE 8.5 Depiction of the Nylan–Barany (Hall Pike) maneuver. Notice that the nystagmus is in a rotatory direction because of activation of the receptors in the ampulla of the posterior semicircular duct. *With permission of the authors: Reeves and Swenson. Disorders of the Nervous System.*

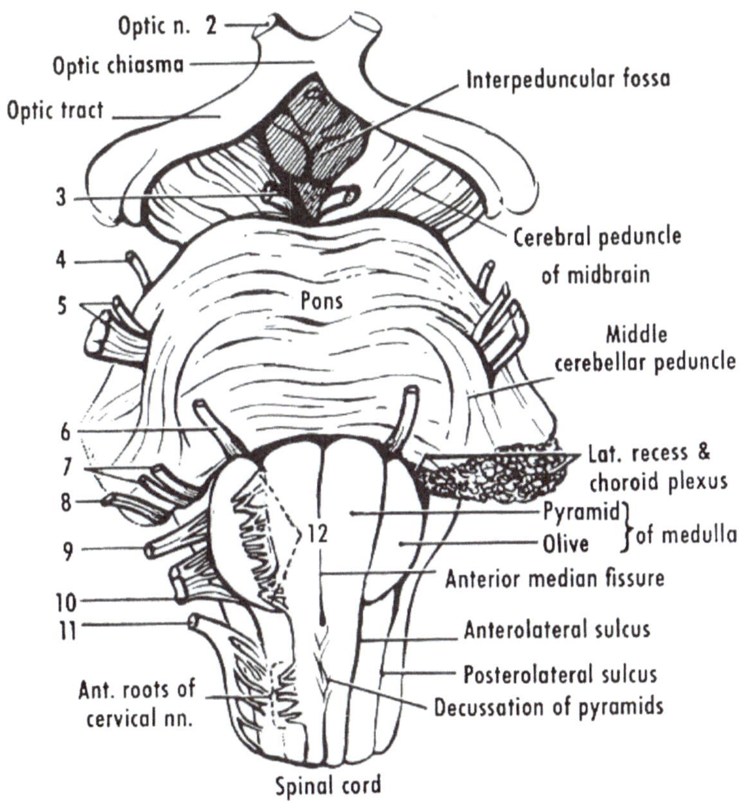

FIGURE 8.6 View of the brain stem from the ventral side. Note the location of the vestibulocochlear nerve (cranial nerve 8) adjacent to the facial nerve (cranial nerve 7) just lateral to the junction of the medulla and pons.

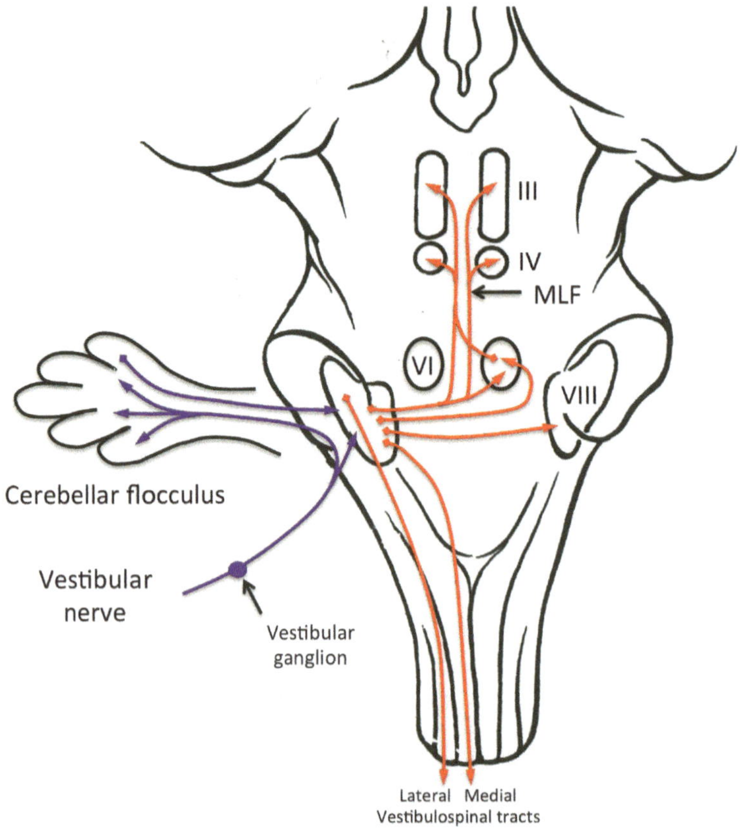

FIGURE 8.7 This is a view of the dorsal surface of the brain stem from the perspective of the cerebellum depicting the entry of the vestibular nerve (in blue) and selected central connections of the vestibular nuclear complex (in red). The vestibular nuclei are labeled VIII on the right of the brain stem. Note that the vestibular nuclei are located in the region of the rostral medulla and caudal pons. In addition to terminating in the vestibular nuclear complex, vestibular nerve fibers can be seen to enter the cerebellar flocculus. The flocculus is a major source of input to the vestibular complex. Nerve fibers arising from neurons in the vestibular nuclei (shown in red) connect through the media longitudinal fasciculus (MLF) with nuclei controlling eye movement: the abducens nucleus (VI); the trochlear nucleus (IV); and the oculomotor nucleus (III). Also, note that the vestibular nuclei have projections that descend through the medulla toward the spinal cord. These include the lateral and the medial vestibulospinal tracts. These tracts are important for regulating body muscle tone in response to vestibular stimuli. This dorsal view of the brain stem will be used in later figures to illustrate important pathways for vestibular reflex responses.

lateral (Deiters); superior (Bechterew); and inferior (descending) vestibular nuclei; as well as several accessory nuclei. There are differences between these subnuclei both in terms of connectivity and function. The most important functional division is between the nuclei receiving input predominantly from the cristae (especially, the superior and medial vestibular nuclei) and those receiving input primarily from the maculae and proprioceptors of the body (the lateral and inferior nuclei). This distinction functionally divides the vestibular complex into regions mainly associated with angular acceleration and stabilization of the head and eyes during rotational movement (the superior and medial vestibular nuclei) and regions more associated detecting static position and linear acceleration (the lateral and descending vestibular

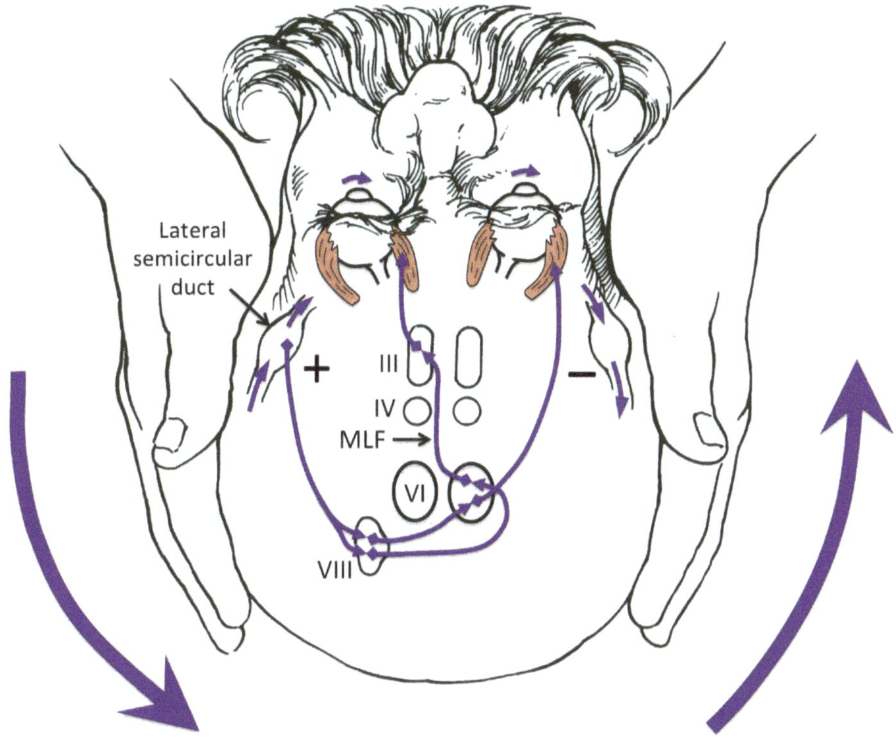

FIGURE 8.8 Cartoon depicting the pathways of the vestibulo-ocular reflex (VOR). Sudden rotation of the head to the left results in an equal and opposite movement of the eyes to the right. Note the direction of relative movement of the fluid in the lateral aspect of the lateral (horizontal) semicircular duct. Also note that this rotation will increase activity in the left vestibular nerve and will decrease activity in the right.

nuclei) and that adjust muscle tone to adapt to body position relative to gravity. The following discussion mainly focuses on these functionally distinct regions of the vestibular complex.

The vestibular nuclear complex receives a constant stream of input from vestibular receptors because of the baseline level of action potential generation in the axons associated with the hair cells of each inner ear (see section: Hair Cells). At rest, this input is balanced, resulting in a perception of stability. Anything that upsets this balance will result in the perception of movement as well as compensatory reflex responses. This may be normal, such as occurs with actual movement or may be pathological, as in the case where damage to one inner ear decreases input on that side. Thus, the inner ear may be thought of as a comparator of input from the two sides. Additionally, it is important to note that there are mechanisms mostly involving vestibular parts of the cerebellum that are capable of reestablishing balance in the nucleus when there is a constant pathological imbalance.

In addition to input from the inner ear via the vestibular nerve, the vestibular complex receives several other inputs. For example, input from the spinal cord allows proprioception (particularly from the neck) to influence vestibular functions. Early in life, there are reflex effects of head position (tonic neck reflexes) produced by activation of these spinal proprioceptors. This reflex results in increased extensor muscle tone on the side of the body toward which the head is rotated or tipped. The reflex is suppressed over the first several months of life and persistence beyond 6 months may be pathological.

The cerebral cortex, particularly those regions involved in control of eye movement also project to the vestibular nuclei, presumably providing a mechanism for regulation of the vestibular system based on conscious attention. Also, the flocculonodular lobe of the cerebellum makes an important input to the vestibular complex. This input arises from a region of the cerebellum that receives direct vestibular input (often termed the vestibulocerebellum). The vestibulocerebellum is unusual for two reasons. First, it is the only portion of the cerebellum that receives direct primary afferent input. Second, this is the only region of cerebellar cortex to contain Purkinje cells that project directly to the brain stem, since most of these axons terminate directly in the vestibular complex. The feedback from the cerebellum to the vestibular nuclei is essential for adaptation to, or compensation for changing vestibular environments (including to the effect of damage to the vestibular system).

EFFERENT CONNECTIONS OF THE VESTIBULAR NUCLEI

The efferent connections of the vestibular complex are critical to several reflexes and also to the conscious perception of motion. The connections of the vestibular complex that are involved in the vestibulo-ocular reflex (VOR; ie, reflex movements of the eyes in response to head movement) are discussed in the following section. There are additional projections from vestibular nuclei to the spinal cord (producing motor responses to vestibular stimulation), to the thalamus (for relay to the cerebral cortex, where movement can be consciously processed), and to the brain stem where a variety of autonomic responses can be generated, particularly to intense vestibular stimulation.

There are two main vestibulospinal tracts, the lateral and the medial vestibulospinal tracts (Fig. 8.7). The lateral vestibulospinal tract arises primarily from the lateral vestibular nucleus and terminates in the ventral horn of the spinal cord at all spinal levels after descending the spinal cord through the ventral part of the lateral funiculus. This tract powerfully facilitates extensor motor neurons and can be activated by tipping of the head toward one side where there is extension of the limbs on the side of the body to which the head is tipped. This reflex has been termed the vestibular righting reflex and it complements the tonic neck reflexes described above. The lateral vestibulospinal tract also appears to be responsible for the intense extensor posturing that occurs when the motor systems above the caudal brain stem are damaged. In this circumstance, the vestibulospinal tract becomes the dominant motor tract, producing decerebrate (extensor) posturing in response to noxious stimuli.

The medial vestibulospinal tract arises from the medial vestibular nucleus and descends through the medial longitudinal fasciculus into the ventral funiculus of the cervical spinal cord. It terminates bilaterally in the medial aspect of the ventral horn of the cervical and upper thoracic spinal cord. This tract is involved in adjusting neck muscle tone in response to vestibular stimulation, presumably to stabilize the head during angular movements.

The vestibular complex projects to brain stem regions involved in control of autonomic nervous system activity. These regions include portions of the brain stem reticular formation and the nucleus of the solitary tract. Intense vestibular stimuli can produce changes in blood pressure, heart rate, and sweating through the autonomic nervous system. Presumably, these reflexes exist to help the body adjust to new positions. Furthermore, there are vestibular connections with the area postrema (a region involved in nausea and vomiting) where stimulation can contribute to motion sickness.

Finally, there are projections from vestibular nuclei to the thalamus that are important in the perception of motion. These terminate in a small region of the ventral posterior nuclear region of the thalamus, sometimes called the ventral posterior intermediate nucleus. This region relays to poorly defined portions of the inferior parietal lobe, particularly the perisylvian region and the posterior insular cortex. Vestibular stimulation also results in activation of portions of the frontal lobe that are associated with eye movement. While the cortical input from the vestibular system is no doubt important in detecting motion, it is important to remember that the vestibular system is only one of the systems that the cerebral cortex uses in the conscious perception of position and motion. The cortex integrates visual and other proprioceptive information with the vestibular input in performing this function. Processes that result in conflicting sensory signals or conditions that enhance cortical excitability, such as migraine aura or seizures, can result in the illusion of movement (vertigo) when these cortical systems are affected.

VESTIBULO-OCULAR REFLEX

One of the most important roles of the vestibular complex is to stabilize the eyes during head movement. The reflex that is responsible for this is termed the VOR (Fig. 8.8). Any jerking movement of the head in one direction is compensated by an equal and opposite eye movement in order to stabilize the visual image. While this might seem to be an uncommon phenomenon, this is actually both a commonly elicited reflex and critically important whenever one's head is moving. Even walking or driving on a relatively smooth road results in some angular movement of the head and this must be compensated for if the visual image is to remain clear. Of course, this is more important when head movement increases in amplitude (eg, running or riding on a bumpy road). Patients who have bilateral vestibular damage (ie, who have lost normal vestibular signaling from the inner ear) complain of a potentially incapacitating inability to focus whenever they are moving.

The VOR is mediated via connections from the vestibular complex to the extraocular nuclei (ie, those nuclei that control eye muscles). Many of these connections traverse the medial longitudinal fasciculus (Fig. 8.8). It should not be surprising that portions of the vestibular nuclear complex that receive input from the horizontal (lateral) semicircular canal, which is most responsive to movements near the horizontal plane, connect mostly to the extraocular nuclei that move the eyes in the horizontal plane (ie, the abducens nucleus and the portion of the oculomotor nucleus that controls the medial rectus muscle). On the other hand, the portions of the vestibular nuclei that receive input from the posterior and anterior (superior) semicircular canals connect mostly to the trochlear nuclei and the other portions of the oculomotor nucleus. These are the nuclei that control the eye muscles that are involved in vertical eye movements, (ie, the superior and inferior recti and the superior and inferior oblique muscles). It should be noted that these vertically moving eye muscles are also involved in torsion of the eyes. This torsion is necessary to stabilize the eyes when the head is tipped to the side.

Because the VOR is so important to stabilize the eyes during movement, it must be highly accurate and also must be adaptable to situations such as damage to the inner ear hair cells with disease or age. Therefore, there are mechanisms to adjust the "gain" (ie, the relative strength) of the VOR in order to assure accurate eye movements. This "fine-tuning" of the reflex is critically dependent on the vestibulocerebellum, which is essential in order to assure that the reflex continues to produce accurate eye movements and to compensate for damage. A corollary of this is that compensation for damage to the vestibulocerebellum is slow and usually less complete than it is for damage to the vestibular organs of the inner ear.

NYSTAGMUS

No discussion of the vestibular system would be complete without some discussion of nystagmus, which is the most common sign of vestibular dysfunction. Nystagmus is a to-and-fro movement of the eyes and presents in two ways—pendular nystagmus and jerk nystagmus. Pendular nystagmus is an oscillation of the eyes that produces essentially equal movement (in speed and amount) to both sides. This is not due to damage to the vestibular system and it is not considered further in this discussion. Jerk nystagmus is produced by vestibular stimulation and consists of a relatively slow movement in one direction and then a very rapid correction (a "jerk"). The relatively slow movement is due to the vestibular stimulation and the rapid jerk is a correction generated by cortical activity when the image of objects of interest in the environment moves across the retina.

Nystagmus is entirely normal in certain situations (physiological nystagmus). The most common example is when someone is actually spinning around. When rotating to the left as in (Fig. 8.8) the eyes will move to the right at the same rate as the head is moving to the left. This is due to the normal VOR described in the preceding section. When the eyes approach the end of their normal range, they will quickly dart back to the right in order to fix on a new target. The quick corrective movement is generated by visual areas of the cerebral cortex to allow the eyes to focus on objects in the environment. After the quick correction, the eyes will again track to the left due to continued vestibular stimulation. As a person spins, therefore, there is a relatively slow movement of the eyes opposite to the head movement, followed by a rapid jerk of the eyes back toward the mid-position. It should be emphasized that this jerk nystagmus is entirely normal. In Fig. 8.8, you can see that the rotation to the left produces imbalance in vestibular nerve activity from the horizontal semicircular canals, with increased activity in the left ear and decreased in the right. In Fig. 8.9 on the lower right, you can see that destruction of the left inner ear or the vestibular nerve that conducts the signals can produce a similar imbalance. Therefore, such damage will produce not only the illusion of spinning to the right (vertigo) but also pathological jerk nystagmus with the fast phase to the right. This vertigo and nystagmus will not last forever, because the cerebellum will compensate over a period of hours to days and reestablish the balance of activity at the level of the vestibular nucleus. Hence, the presence of vertigo suggests an acute (sudden) and asymmetric change of vestibular function.

To complete the story of rotation-induced nystagmus, it is necessary to consider what happens if a person is rotated for several revolutions and then suddenly stopped. After several rotations, fluid in the semicircular canals begins to catch up with head movement due to drag on the fluid. If the individual were stopped suddenly, the inertia of the fluid would cause it to continue moving after the canal was stopped, producing relative fluid movement in the opposite direction. The person would instantly feel as if they were spinning in the opposite direction and nystagmus would also instantly change direction until the fluid stopped moving relative to the canal. This is also an example of normal, physiological nystagmus.

To summarize, jerk nystagmus occurs due to competing neural systems. The vestibular system attempts to drive the eyes opposite to head movement (or perceived head movement) via the mechanism of the VOR. The cerebral cortex attempts to fixate on objects and will rapidly jerk (saccade) the eyes back in order to fixate the vision on an object of interest. The slow phase of the movement is produced by the VOR and results from imbalance in vestibular activity. This can occur either due to movement of the

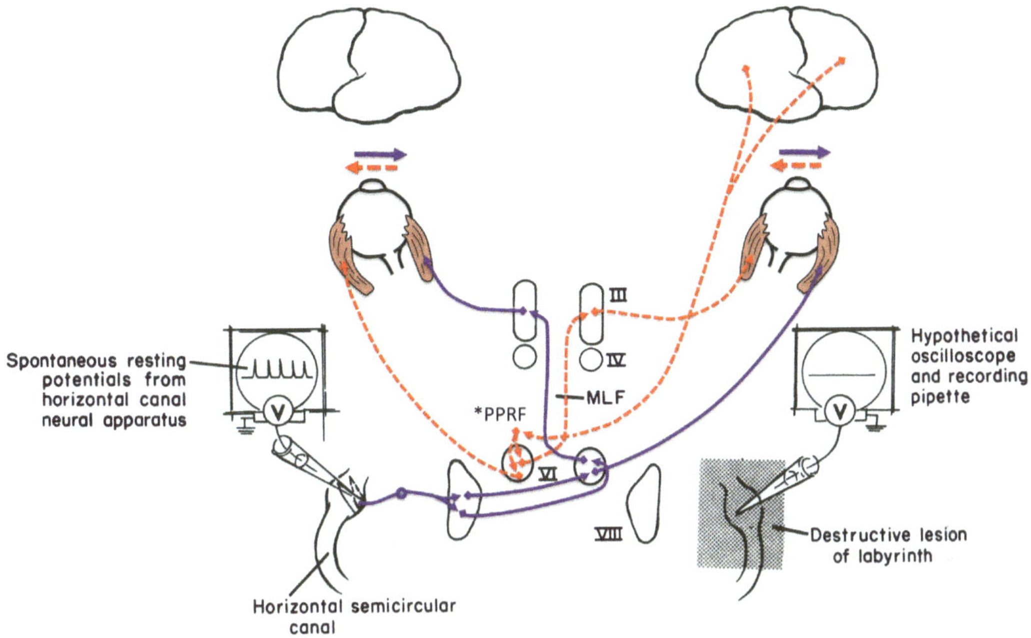

FIGURE 8.9 Cartoon depicting the results of a lesion of the right vestibular apparatus. Notice that the imbalance created (greater nerve activity in the right vestibular nerve that the left) is similar to the imbalance created by rotating the head to the left (Fig. 8.8). The *blue lines* indicate the pathways responsible for driving the eyes to the right. When the visual image begins to move, the cerebral cortex will drive the eyes suddenly to the left to recenter the image. This pathway is indicated by the *dashed red line*. The combination of slower movement in one direction, with sudden compensatory movement in the opposite direction, is "jerk nystagmus." *PPRF*, Paramedian pontine reticular formation/pontine lateral gaze center.

head (physiological) or to damage of one of the inner ears or vestibular pathways (pathological). The fast corrective phase that is generated by cerebral cortical activity will not occur if a patient's cerebral cortical activity is suppressed, such as by sedative overdose.

CALORIC TESTING

From the preceding discussion it should be apparent that it is possible to test the vestibular apparatus of the inner ear and its connections by observing eye motions in response to head movement. However, the fact that each inner ear can detect movement in any direction and can, through connection with the vestibular apparatus, drive eye movement in any direction complicates the ability to test each inner ear independently. Nevertheless, it is often necessary to test each ear in patients with vertigo because most pathology is unilateral. The most convenient and practical method to test one inner ear is by a procedure called caloric testing.

Caloric testing involves the creation of thermal convection currents in one inner ear by injecting cold or warm water or air into the external ear canal in order to stimulate the vestibular apparatus on that side in an isolated fashion. If a patient's head is placed in a semi-recumbent position, such that the horizontal (lateral) semicircular canal is vertically oriented (Fig. 8.10 inset), warm air or water introduced to the external ear canal will warm the fluid in the lateral aspect of this semicircular canal (Fig. 8.10A). Since warm fluid is less

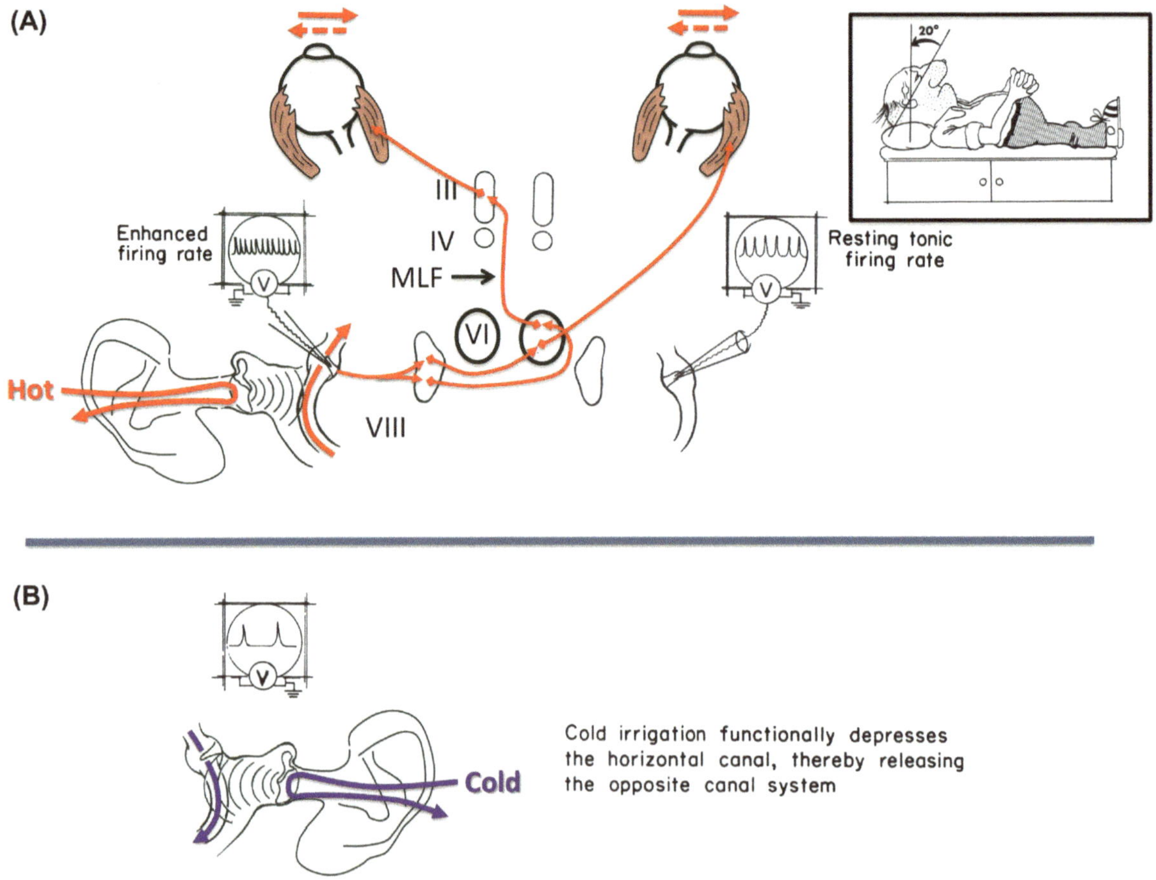

FIGURE 8.10 Cartoon showing caloric testing with warm (A) and cold (B) stimuli. This is done with the patient lying down (inset at the top) such that the horizontal canal is in a vertical position. In figure A, in which warm air or water is introduced to the ear canal, convection currents are produced similar to those provoked by rotation of the head toward that side. Nerve activity is increased on that side and the eyes will be driven slowly away from the side being tested (*solid arrow*), with a rapid correction (*dashed arrow*) if the patient is alert. Cold air or water (lower cartoon) will decrease nerve activity on the tested side and the patient will perceive rotation away from the stimulus.

dense than cooler fluid, this would produce a current in the lateral semicircular canal that mimics the current produced by rotation of the head toward the warm side. In fact, the person being tested will report the illusion of spinning toward the side being tested. This vestibular stimulation would drive the eyes opposite to the perceived direction of motion via the VOR (*solid arrow* at the top of Fig. 8.10A). If the patient is awake and if their visual areas are processing normally, the eye movement provoked by vestibular stimulation would cause movement of the visual image. This movement of the visual image will result in an extremely fast jerk (a saccade) to recenter the visual image (*dashed arrow* at the top of Fig. 8.10A). Therefore, in the case of caloric testing with warm air or water, the eyes would normally move relatively slowly away from the ear being tested followed by a jerk back toward that ear. The fast movement that is due to cortical activity is more easily seen than the slow movement produced directly by the vestibular stimulation and therefore the nystagmus is named by the fast component. The nystagmus would be toward the side of the ear being tested with warm water or air. Caloric testing with the use of cold air or water (Fig. 8.10B) would produce convection currents and eye movements in the opposite direction to that described for warm stimuli. The rotational velocity (degrees/second) and duration of nystagmus produced by caloric testing is a reflection of the integrity of the inner ear vestibular apparatus on the side tested. The caloric response can be suppressed consciously or by visual fixation, so the patient is cognitively challenged during the test and the exam is done without visual cues (using infrared videonystagmography). Since the absolute responsiveness of the inner ear is variable, the caloric test is interpreted according to the percent asymmetry between the two ears to bithermal stimulation with greater than 30% difference in response considered abnormal.

Finally, caloric testing has an additional use in clinical practice for the assessment of the comatose patient. There are two main categories of problems that produce coma and each requires a very different management approach. The first is damage to the brain stem, which can produce coma by damage to the ascending reticular activating system (ARAS). This is usually due to structural lesions that probably require urgent imaging and possible neurosurgical intervention. The other type of coma results from toxic, metabolic, or drug-related suppression, of cerebral cortical activity. This type of coma, often just called a metabolic coma, requires metabolic support while the specific cause is investigated. Caloric testing is useful in distinguishing these two basic causes of coma. Brain stem damage will result in abnormal caloric tests because the wiring of the VOR extends from the medulla through the midbrain. Structural damage to the ARAS is virtually incompatible with a normal VOR. On the other hand, the brain stem is relatively resistant to metabolic, toxic, or drug-related suppression, even when cerebral cortical activity has been severely depressed. Therefore, the eye movement that is produced during the normal slow component of the VOR is usually preserved until very late in metabolic coma while the fast component (the jerk) would be abolished early in this type of coma. Thus, the use of caloric testing can be very important in helping to distinguish the patient requiring an urgent scan of the head and neurosurgical consultation from the patient who requires metabolic support and an extensive workup for the many factors that can suppress cerebral cortical activity.

SUMMARY

Dizziness and instability are common complaints that range from distressing to disabling. They contribute to falls, particularly in the elderly and, therefore, are a major public health issue. The vestibular system, the focus of this chapter, is one of the important contributors to this problem. Since this system is involved in the detection of movement of the head, the main symptom is the illusion of movement: that is, vertigo. In patients with vertigo, the main focus should be on the evaluation of the vestibular system. However, it must be understood that balance is a complex interaction between the vestibular system and other sensory systems, including somatic proprioception and vision. Furthermore, the feeling of light-headedness (which is also often called "dizziness" by patients) suggests insufficiency of metabolic substrates or of blood flow to the brain. All of these must be considered in the systematic evaluation and management of the dizzy or unstable patient.

CHAPTER

8.1

Translational Correlation: Vertigo

B. Hwang, J. Kim

Texas Tech University Health Sciences Center, Lubbock, TX, United States

PERIPHERAL VERTIGO: BENIGN PAROXYSMAL POSITIONAL VERTIGO

Dizziness is one of the common reasons patients seek health care. Its prevalence is increasing with age. Dizziness can be classified into presyncope (near-faint dizziness due to reduced cerebral blood flow), psychophysiological dizziness (feeling of dissociation), disequilibrium (imbalance of standing or walking unrelated to an abnormal head sensation), and vertigo (illusion that the body or environment is spinning or tumbling). Vertigo is the most common type. It can be either due to loss of peripheral input caused by damage to the labyrinth or vestibular nerve or unilateral impairment of vestibular nuclear or vestibulocerebellar activity. It is essential to determine if the vertigo is central or peripheral because etiologies of central vertigo (ie, cerebellar/brain stem stroke) can be life threatening and require emergent intervention. Most patients with central vertigo have sudden-onset focal neurological symptoms and signs and are often unable to stand without support. Patients with peripheral lesion can typically stand but veer/tilt to the side of the lesion.

Epidemiology

Benign paroxysmal positional vertigo (BPPV) is the most common cause of peripheral vertigos. One-third of patients with vertigo were diagnosed as BPPV in multinational population study (the Registry to Evaluate the Burden of Disease in Vertigo, REVERT registry) and BPPV accounts for one-half of patients with peripheral vertigo. One German national study shows its prevalence and incidence were 1.6% and 0.6%, respectively. The prevalence of BPPV increases with age. BBPV is twice as common in women as men in all age group, and patients with family history are five times more likely to have BPPV than people without family history. Migraine, hypertension, hyperlipidaemia, and stroke were independently associated with BPPV.

Pathophysiology

The peripheral vestibular organs consist of three semicircular canals and two otolithic organs, the saccule and the utricle. The semicircular canals detect angular acceleration during rotational head movements, whereas the saccule and utricle are sensitive to detect linear acceleration

Conn's Translational Neuroscience
http://dx.doi.org/10.1016/B978-0-12-802381-5.00015-4

and head movements in the vertical and horizontal plane, respectively. Macula of the utricle and saccule houses the hair cell, the sensory receptors, which project into an otoconial membrane on which the calcite crystals, otoconia, rest. Otoconia get displaced during head movements. These are composed of organic and inorganic matrix. Otoconin 90 around otolin scaffolds is a key component of organic matrix and facilitates to form calcite crystals. These get anchored together with otolin-based fibrils on otolithic membrane.

Loosed fragments of otoconial lead to BPPV and those can freely float from utricular macule into the semicircular canals. These particles should be heavy enough to alter vestibular nerve activity. This leads to erroneous sensation of spinning with head movement with respect to gravity. The most common affected canal is the posterior semicircular canal.

Majority of BPPV are idiopathic. Vibration and percussive force (eg, head trauma and ear/maxillary operation) and residual effect of vestibulopathy (eg, Meniere's disease) can cause BPPV. Considering its association with giant cell arteritis, migraine, and stroke, BPPV is believed to have a vascular ischemic mechanism. Osteoporosis is one of the potential etiologies because bone loss can be the source of calcium carbonate to create debris.

Canalolithiasis is more common mechanism of BPPV than cupulolithiasis. In canalolithiasis, fragmented otoconia move freely in the canal, stimulating utriculopetal endolymph flow, whereas in cupulolithiasis otoconia are attached to the cupula or proximal canal near to the cupula.

Clinical Symptoms

Characterized symptom of BPPV is a sudden-onset, rotatory vertigo (86%), triggered by head movement, such as looking up, lying down, or getting up from bed or turning head. These episodes are mostly brief, lasting 1min or less, and self-limited with a median duration of 2weeks. The resorption rate of otoconia is dependent on calcium concentration in endolymph. More than half of the patients suffer from recurred periodical spells for weeks to months and the recurrence risk is about 15% per year. This is possibly because only 8% of all patients received effective therapy. The spells may be worse in the morning and the evening, and they tend to be wax and wane over time. Other accompanying symptoms are oscillopsia (31%). nausea (33%), vomiting (14%), imbalance (49%), fear of falling (36%), and falls (1%).

Evaluation and Diagnosis

Provocative maneuvers, such as Dix–Hallpike and head roll test, are standards to identify the involved canal by observing latency, duration, and direction of nystagmus. Majority of cases involve posterior canal.

In posterior canal BPPV, rotatory nystagmus toward the affected ear is commonly observed. This nystagmus usually appears with a latency of a few seconds and lasts less than 30s. After a patient sits up, the nystagmus will recur in the opposite direction. When this maneuver is repeated, the intensity and duration of nystagmus will diminish (fatigability).

Diagnosing the affected side in lateral canal BPPV can be challenged. In this type of BPPV, lateral head turn in the supine helps to identify the affected side. In geotropic lateral canalolithiasis, horizontal nystagmus beats toward the floor when turning the affected ear down. When turning the normal ear, nystagmus tends to be milder. In apogeotropic (away from the undermost ear) lateral canalolithiasis, the nystagmus beats toward the uppermost ear. In addition, for this type of BPPV, bow and lean test can be helpful. The patient sits with bending head forward, and then leans head backward. Right lateral canalolithiasis will show a right beating nystagmus during the bow and a left beating nystagmus during the lean.

Treatment

The purpose of treatment is to migrate debris and exit it into the utricular cavity. Drugs (vestibular suppressants) are usually used as only premedication before particle repositioning maneuvers, not used as curative treatment. Particle repositioning maneuvers (eg, Epley maneuver) are proven to be effective in most cases. Success rate is over 90%. After repositioning, postural restrictions (using a neck brace, avoiding bending head-up/down, and instructions to sleeping upright) are also useful. Home exercise (Brandt–Daroff exercise) was developed based on the same principle, though this is less effective than particle repositioning maneuvers. A very small number of intractable patients require operation. In these cases, surgical occlusion of the canal with bony plug or transection of the posterior ampullary nerve can be considered.

CENTRAL VERTIGO: POSTERIOR CIRCULATION INFARCTS AND WALLENBERG SYNDROME (LATERAL MEDULLARY SYNDROME)

Vertigo is one of the common clinical symptoms in patients with this syndrome. However, stroke presenting with vertigo may occur all along the vestibular pathways from the vestibular nuclei in the medulla oblongata to the rostral midbrain, the vestibulocerebellum, the thalamus, and multisensory vestibular cortex area in the temporoparietal cortex. These structures are mainly supplied by the vertebrobasilar circulation. Medullary infarction (lateral medullary infarction, medial medullary infarction, or vestibular nuclear infarction) and cerebellar infarction, involving posterior inferior cerebellar artery (PICA) or anterior inferior cerebellar artery (AICA), should be suspected when patients with vascular risk factors visit for dizziness.

Epidemiology

Stroke accounts for 3.2% of all dizzy patients who visit emergency room. Patients with dizziness have two times higher risk of stroke or cardiovascular events than those without dizziness. About 20% of ischemic strokes occur in the territory of vertebrobasilar circulation and patients with vertebrobasilar infarction present with vertigo.

Pathophysiology

PICA territory infarction is the most common cause of isolated acute prolonged vertigo. In general, proximal branches of PICA supply brain stem and longer circumferential branches supply the cerebellum. AICA territory infarction mostly causes vertigo with unilateral or bilateral hearing loss because the internal auditory artery, one branch of AICA, supplies the inner ear. The anterior spinal artery, vertebral artery, and PICA supply medulla oblongata. The anterior spinal artery supplies the medial region and the PICA supplies the posterolateral region of medulla. The vertebral artery supplies the area between the anterior spinal artery and PICA. Vertebral artery disease is the main cause of lateral medullary syndrome, and PICA disease can cause Wallenberg syndrome as well.

Atherosclerosis is the most common cause of posterior circulation infarcts. In young patients under 45 years of age, vertebral artery dissection can cause 20% of strokes. Head and neck trauma (whiplash injury), chiropractic manipulation, or even scuba diving can cause vertebral dissection as tortuous distal segment of vertebral artery is susceptible to mechanical distortion, such as cervical spine flexion/extension with rotational movement.

Clinical Symptoms

Patients with isolated nodulus or tonsilar infarction due to PICA lesion may have unidirectional spontaneous nystagmus and mild imbalance. Compared to patients with vestibulopathy, head impulse test on the patients is normal.

Lateral medullary infarction (Wallenberg syndrome) is usually accompanied with other neurological signs. Lateral medulla contains several nerve nuclei and tracts. Depending on involved lesion, clinical symptoms of this syndrome could be variable.

Lateral Medullary Infarct

Symptom or Sign		Structure Involved
Ipsilateral	Facial pain, dysesthesia, or anesthesia, reduced corneal reflex	Descending root of CN V
	Dysphagia and dysarthria	CN IX or X, nucleus ambiguous
	Paralysis of palatal elevation, pharyngeal constrictors, and vocal cord	
	Ataxia, dysmetria, intention tremor, hypotonia	Spinocerebellar tracts and cerebellar hemisphere
	Horner syndrome of miosis and anhidrosis of the face	Descending sympathetic tract in lateral reticular formation
	Nystagmus	Vestibular pathways
	Ipsilateral lateropulsion	Vestibulospinal or dorsal spinocerebellar tract
Contralateral	Loss of pain and temperature	Lateral spinothalamic tract
Nonlateralized	Nausea, vomiting, vertigo, and hiccoughing	Reticular formation and vestibular nuclei

Diagnosis

Vertigo with focal neurological sign suggests central-origin vertigo. Acute vertigo with ambiguous neurological deficit can be challenging. Neuroimaging, such as MRI or MRA, becomes essential to confirm stroke but diffusion-weighted MRI can miss about 20% of posterior circulation stroke within the first 24–48h. A refined bedside examination protocol, HINTS to INFARCT (Head Impulse test, Nystagmus and Test of Skew and Impulse Normal, Fast-phase Alternating or Refixation on Cover Test, respectively), can be helpful in ambiguous cases.

Treatment

Treatment for acute infarction with vertigo follows general acute stroke guideline and secondary prevention (antiplatelet therapy or anticoagulation). Antivertiginous medications are less effective for central vertigo than for peripheral vestibular lesions. Vestibular rehabilitation should be followed and have functional goals to reduce dizziness and improve gait balance.

Further Reading

[1] Kim JS, Zee DS. Benign paroxysmal positional vertigo. N Engl J Med 2014;370:1138–47.
[2] Von Brevern M, Radtke A, Lezius F, et al. Epidemiology of benign paroxysmal positional vertigo: a population based study. J Neurol Neurosurg Psychiatry 2007;78:710–5.
[3] Oghalai JS, Manolideis S, Barth JL, et al. Unrecognized benign paroxysmal positional vertigo in elderly patients. Otolaryngol Head Neck Surg 2000;122:630–4.
[4] Balof RW, Honrubia V. Clinical neurophysiology of the vestibular system. 3rd ed. New York: Oxford University Press; 2001.
[5] Baloh RW. Differentiating between peripheral and central causes of vertigo. Otolaryngol Head Neck Surg 1998;119:55–9.
[6] Epley JM. Human experience with canalith repositioning maneuvers. Ann N Y Acad Sci 2000;942:179–91.
[7] Li JC, Li CJ, Epley J, Weinberg L. Cost-effective management of benign positional vertigo using canalith repositioning. Otolaryngol Head Neck Surg 2000;122:334–9.
[8] Lee SH, Kim JS. Acute diagnosis and management of stroke presenting dizziness or vertigo. Neurol Clin 2015;33:687–98.
[9] Kim JS. Pure lateral medullary infarction: clinical radiological correlation of 130 acute, consecutive patients. Brain 2003;126:1864–72.

[10] Kameda W, Kawanami T, Kurita K, et al. Lateral and medial medullary infarction: a comparative analysis of 214 patients. Stroke 2004;35:694–9.

[11] Thomke F, Marx JJ, Iannetti GD, et al. A topodiagnostic investigation on body lateropulsion in medullary infarcts. Neurology 2005;64:716.

[12] Kettah JC, Talkad AV, Wang DZ, et al. Hints to diagnose stroke in the acute vestibular syndrome: three-step bedside oculomotor examination more sensitive than early MRI diffusion-weighted imaging. Stroke 2009;40:3504–10.

CHAPTER

9

The Cerebellum

C.R. Goodlett[1], G. Mittleman[2]

[1]Indiana University–Purdue University Indianapolis, IUPUI, Indianapolis, IN, United States;
[2]Ball State University, Muncie, IN, United States

INTRODUCTION

The cerebellum ("little brain") is present in all vertebrate species, and the basic organization—circuitry and distinct laminar patterning of the cerebellar cortex (molecular layer, Purkinje cell layer, and granule cell layer)—is remarkably similar across species. Across tetrapod vertebrates, the cerebellar cortex has a uniform, repetitive anatomical geometry, but multiple, distinct functional systems have been identified in roughly longitudinal planes across the cerebellum that are based on differences in extrinsic afferent and efferent connections. Cerebellar function is generally implicated in five main domains: (1) maintaining balance and equilibrium during standing and locomotion and control of gaze, including adjustments of gaze when the head is moving to keep the retinal image stable; (2) monitoring and correcting predicted and executed movements by comparing intended movement with actual performance; (3) coordinating timing and sequencing of complex voluntary movements; (4) motor learning and acquisition of skilled movements; and (5) from evidence generated over the last three decades, regulating nonmotor processes including linguistic, cognitive, and affective functions.

The uniformity of the cerebellar cortical anatomy, together with the fact that the Purkinje cells—the sole output of the cerebellar cortex—generate and express exquisitely precise, circuit-specific, activity-dependent functional plasticity, suggests that the cerebellum evolved by adding modules that perform similar kinds of computations across different functional domains. From that perspective, the evolutionary increase in size of the cerebellar hemispheres in mammals conserved the organization and computational properties of the cerebellar cortex circuitry, but expanded in concert with the expanding and diverging neocortical circuits. This evolution of cortico-cerebellar modules likely provided the parallel neural computational power necessary to support complex technical skills and tool use, the integration of visuomotor and sensorimotor functions for visual attention and guidance of skilled actions, the increased cognitive capacity to use skills appropriately, emergence of communication systems and language with complex recognition and expression, and the regulation of complex social interactions. Cerebellar damage typically can interfere with balance, equilibrium, or gait, with the timing, accuracy, or coordination of movements, with the acquisition of motor learning and motor skills, or certain

Conn's Translational Neuroscience
http://dx.doi.org/10.1016/B978-0-12-802381-5.00016-6

aspects of cognitive function, depending on the region or fiber tracts involved.

OVERVIEW OF EVOLUTION AND FUNCTIONAL ORGANIZATION

The cerebellum overlies the pons, the medulla, and the fourth ventricle and occupies the posterior cranial fossa, separated from the cerebral hemispheres by the tentorium cerebelli (an extension of the dura mater). As seen in Figs. 9.1 and 9.2, the human cerebellum has enlarged cerebellar *hemispheres* joined across the mid-sagittal plane by the narrow *vermis*. The transverse curved *primary fissure* divides the *anterior lobe* and the *posterior lobe*, and the prominent horizontal fissure separates the superior and inferior semilunar lobules of the posterior lobe. The region of the hemispheres immediately adjacent to the midline vermis is known as the *intermediate zone* or *paravermis*. The *posterolateral fissure* separates the phylogenetically older *flocculonodular lobe* composed of the midline *nodulus* and two lateral *flocculi*.

The size of the cerebellum increases across the phylogenetic tree, clearly evident after controlling for differences in body size and scaling relative to hindbrain size. Its evolution has been correlated both with the addition of subcortical motor systems (eg, the red nucleus and rubrospinal tract) and with the increasing size of the neocortex associated with expanding feedforward and feedback functional loops that interconnect the neocortex and the cerebellum. In lampreys, the cerebellum is little more than an indistinct swelling, whereas in cartilaginous and teleost fishes it is much expanded and foliated, yet without the deep cerebellar nuclei typical of tetrapods. Mormyrid fishes, which have a highly developed system of electroreception, are remarkable exceptions to the phylogenetic ordering of cerebellar size. The mormyrid cerebellum is are massively expanded and extensively folded to constitute about 80% of the mass of the mormyrid brain, reflecting its major role in electroreception. Among tetrapods, amphibians typically have a small cerebellum evident as a transverse medial lobe and a single deep nucleus, corresponding to the phylogenetically older vestibulocerebellum. In reptiles, the cerebellum is expanded as a single midline lobe with the appearance of two deep cerebellar nuclei (medial and lateral) correlated with the expansion of spinocerebellar control systems. Birds and mammals have enlarged and convoluted

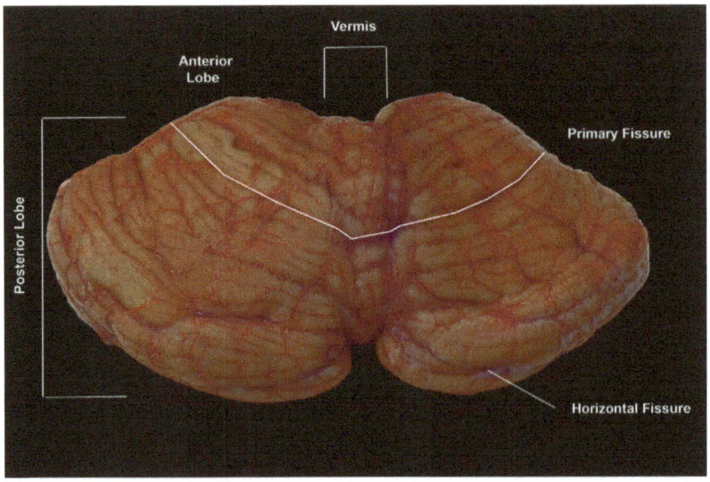

FIGURE 9.1 The superior surface of the cerebellum shown for a freshly removed cerebellum.

lobes and three deep nuclei, with expansion of the cerebellar hemispheres and the dentate nucleus (of the deep cerebellar nuclei) reflecting the evolution of the cortico-cerebellar systems of the neocerebellum.

The human cerebellum comprises about 10% of the mass of the brain, but contains upward of 75% the brain's neurons. Its extraordinary density of neurons in a comparatively small volume is accomplished by the extensive folding of the cerebellar cortex to form many transverse curved lobules subdivided into folia, such that most of its surface is hidden. The lobules have been identified by names and (in the vermis) by Roman numerals I–X (see Fig. 9.3), serving mainly as topographic landmarks for cerebellar locations; lobules as a whole do not correspond to precise functional divisions nor provide a simple basis for clinical correlates. If unfolded as a sheet, the human cerebellar cortex would stretch more than a meter in length with an area of nearly 50,000 mm^2. Recent analyses by Barton and colleagues of evolutionary rate changes along phylogenetic branches have shown that the cerebellum underwent dramatic size increases over the evolution of great apes and humans, faster even than changes in the size of the neocortex. When considering these new evolutionary insights in combination with burgeoning findings from neuroscience that implicate the cerebellum in learning, planning, and executive control over complex behavioral and cognitive sequences (eg, action sequences crucial to technical mastery of tools, social interactions and play, spatial cognition and navigation, language processes, verbal working memory, and even emotional regulation), it is evident that the traditional view of the cerebellum as a neuronal machine involved primarily in sensorimotor control and motor skill learning underestimates its significance in hominid evolution.

The three major functional subdivisions of the mammalian cerebellum (see Fig. 9.2), the vestibulocerebellum (archicerebellum), the spinocerebellum (paleocerebellum), and the cerebrocerebellum (neocerebellum), derive from its phylogenetic origins and correspond to differences in functional connectivity with other CNS

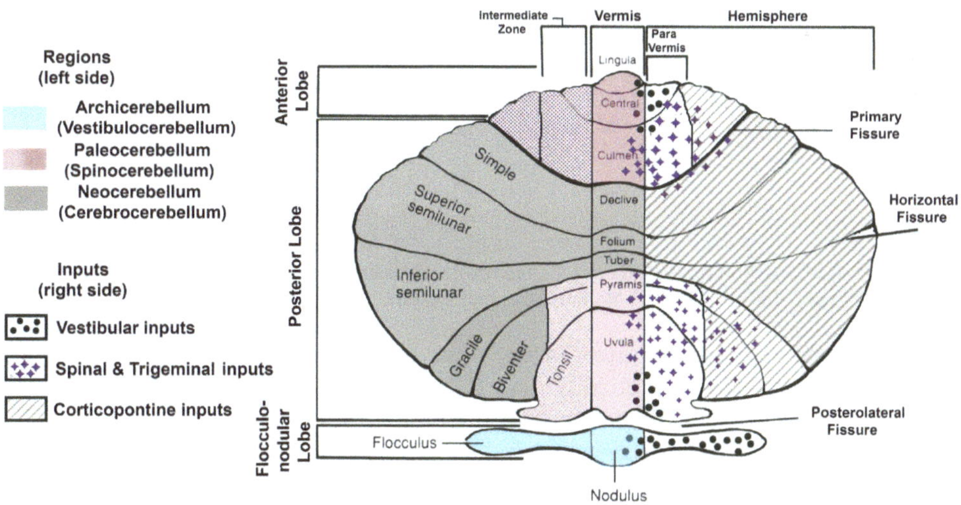

FIGURE 9.2 Schematic of the flattened cerebellum showing the fissures, lobes, and lobules of the cerebellum. The *shaded areas* on the left designate the three main evolutionary zones of the cerebellum, and the terms on the *left side* of the diagram refer to terminology used for the human cerebellum. The *right side* shows the general termination zones of the vestibular, spino- and trigeminocerebellar, and corticopontine inputs.

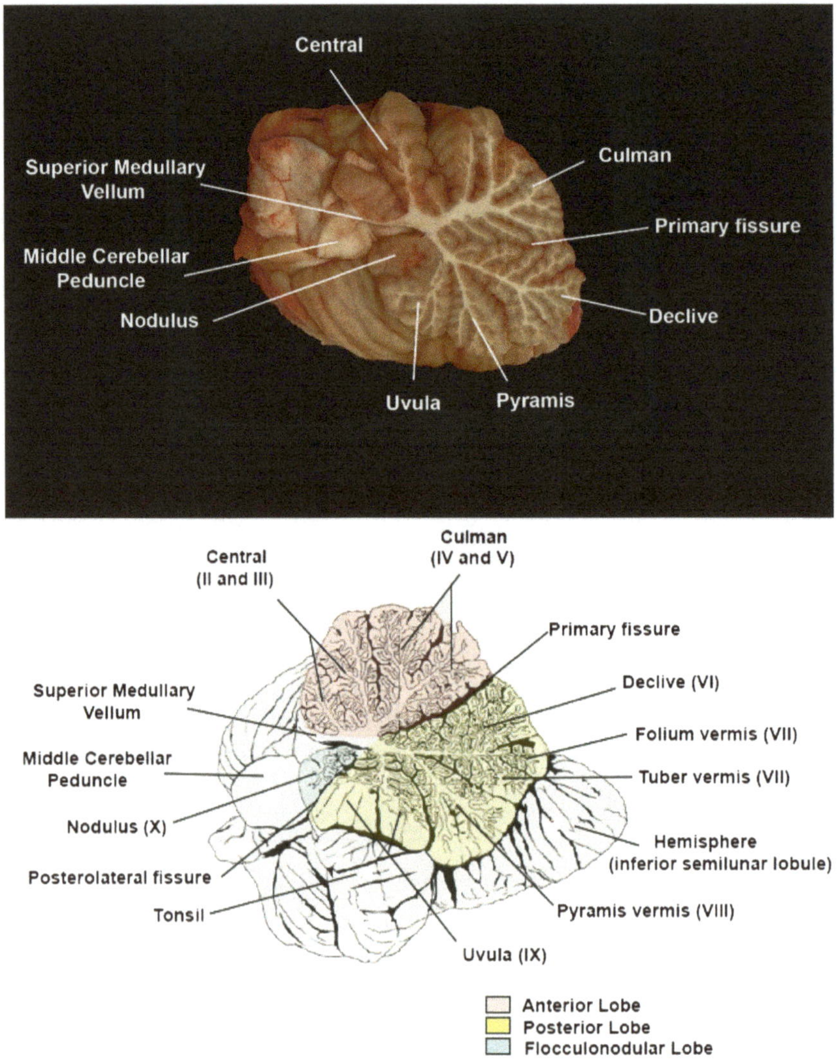

FIGURE 9.3 A midsagittal section through the cerebellar vermis of freshly dissected cerebellum, and a schematic showing the lobules by conventional name and Larsell's Roman Numeral system.

regions. The flocculonodular lobe comprises the *vestibulocerebellum* and is phylogenetically the oldest division and varies little across mammalian species. It receives direct connections from the vestibular nerve, is reciprocally connected with brainstem vestibular nuclei, and has substantial visual input. It integrates vestibular signals in control of standing and walking, and coordinates eye movements with movements of the head through its connections to portions of the vestibular nuclei that control the extraocular muscles.

The *spinocerebellum* is phylogenetically the next region of the cerebellum to appear and includes most of the vermis (excluding the portion just posterior to primary fissure), the intermediate zone, and the anterior lobe (Fig. 9.2). The vermis receives input from the spinal cord (trunk/limbs) and from the trigeminal system (head/neck). The intermediate zone receives converging input from motor cortex and proprioceptive/kinesthetic spinal systems, forming circuitry for its comparator function assessing the match between intended movement and actual movement. The spinal projections to the spinocerebellum come both from monosynaptic inputs from fast-conducting proprioceptive sensory afferents (muscles spindles, tendon organs, mechanoreceptors) and from spinal interneurons that themselves are targets of descending motor pathways. The spinocerebellum integrates this somatotopic input to control posture, muscle tone, and patterned movements (walking, reaching, and grasping) that are organized at the spinal level.

The *neocerebellum* is phylogenetically youngest and is expanded in humans and higher primates. Crossed projections come from neurons in the contralateral pontine nuclei and inferior olivary nuclei, which themselves are the targets of direct projections from ipsilateral cerebral cortex. Efferent projections from the cerebellum return to the contralateral motor regions of the cortex via the deep nuclei and the thalamus. These circuits form the "cortico–ponto–cerebellar loops" that provide the structural basis for feedforward/feedback regulation of cortical function, mediating comparisons of intended or anticipated actions (from cortex) to actual outcomes (from sensory inputs) either to properly anticipate and confirm the force, timing, and direction of intended movements or to generate a feedback error signal that results in subsequent correction. Such comparator functions of these loops likely mediate learning and regulation of skills and actions that reflect increasingly precise anticipatory control.

The striking expansion of the hemispheres of the neocerebellum in the hominoid lineage supports the view that cerebellar evolution is closely linked to emergence of neural substrates involved in complex cognitive and behavioral control of skills and cognitive regulation through expansion of these parallel loops with the neocortex. Such functions are now recognized to include regulation of timing and accuracy of complex planned movements, coordinated execution of skilled movements, regulation of speech and language, acquisition of motor skills, and learning and execution of accurate, nonstereotyped actions. Consistent with the neuroanatomical evidence of loops between the neocerebellum and nonmotor cortical regions such as the prefrontal cortex, functional imaging and neuropsychological assessment have demonstrated cerebellar regulation of various cognitive processes, including verbal working memory, executive functioning, and affective processes. From an evolutionary perspective, the direct connections between the cerebellum to brainstem motor nuclei that give rise to descending spinal pathways (red nucleus, vestibular nuclei) play a comparatively modest role in the regulation of movement that mainly comes from phylogenetically older regions. In contrast, the neocerebellar circuits involve heavy feedback projections to all major cortical regions involved in complex actions, as well as to prefrontal and limbic cortical regions (from the posterior vermis), via thalamic nuclei.

STRUCTURAL ORGANIZATION OF THE CEREBELLUM

The surface of the cerebellum contains a distinct, three-layered cortex (gray matter), and extensive core of internal white matter containing the afferent and efferent axons to and from the cerebellar cortex. Embedded within the deep white matter are four pairs of *deep cerebellar nuclei*, which provide the vast majority of the output axons that leave the cerebellum. The *cerebellar peduncles* are three bilateral pairs of white matter tracts (superior, middle, and inferior), which connect the cerebellum to the rest of the central nervous system. A thin, white lamina—the *superior medullary velum*—stretches between the superior cerebellar peduncles, forming the cranial portion of the roof of the fourth ventricle. The small *inferior medullary velum* forms the caudal portion of the roof of the fourth ventricle and contains the *foramen of Magendi*, a midline opening through which cerebrospinal fluid flows from the fourth ventricle into the external subarachnoid space.

The structural and functional organization of the cerebellum is now recognized to follow a general plan of longitudinal, medial-to-lateral sagittal zones that subdivide the general medial-to-lateral divisions (vermis, intermediate, and lateral zones) but which are much more precise than those macroscopic divisions. Seven morphologically distinct longitudinal (translobular) zones that run parallel in the parasagittal plane can be identified in the cerebellar cortex, each receiving climbing fiber input from a specific portion of the inferior olive. The Purkinje cells of these longitudinal zones (medial-to-lateral) project to specific portions of the deep cerebellar nuclei or vestibular nuclei in a roughly medial-to-lateral pattern. This spatial segregation of olivary inputs and the topographic specificity of olivary connections to Purkinje cells as well as the additional topographic specificity of Purkinje cell efferent projections to specific deep nuclei (or locations within the deep nuclei) is the neuroanatomical basis for the functional modular organization.

All of the inputs and outputs of the cerebellum are carried by the superior, middle, and inferior cerebellar peduncles. The superior cerebellar peduncle is the principal efferent pathway of axons leaving the globose, emboliform, and dentate nuclei (the intermediate and lateral nuclei), and also carries afferents into the cerebellum from the ventral spinocerebellar tract, the tectocerebellar tract, the rubrocerebellar tract, and the noradrenergic projections from the locus coeruleus. The middle cerebellar peduncle carries afferent fibers from the *contralateral* pontine nuclei, forming the pontine-cerebellar limb of the cortico-ponto-cerebellar loop, relaying information to the cerebellum that originates in the contralateral cerebral cortex. The inferior cerebellar peduncle carries both afferent and efferent fibers and connects the cerebellum with the brain stem through several pathways, including the *restiform body*, which conveys afferent fibers to the cerebellum from spinal cord and brainstem, and the *juxtarestiform body*, which carries vestibular fibers to and from the cerebellum.

The deep cerebellar nuclei receive inputs from two sources, the Purkinje cells of the cerebellar cortex (the sole output of the cerebellar cortex) and the extracerebellar sources including collaterals from climbing fibers and some mossy fibers that innervate the cerebellar cortex. In higher primates, there are four pairs of nuclei—from medial to lateral—the *fastigial*, *globose*, *emboliform*, and *dentate nuclei*. In many mammals, the emboliform and globose nuclei are combined into one grouping and called the *interpositus nucleus*. The deep cerebellar nuclei, together with the brain stem vestibular nuclei (that receive inputs directly from Purkinje cells of the flocculonodular lobe and the vermis), are responsible for the entire output from the cerebellum to other parts of the central nervous system. The neurons of the deep cerebellar nuclei are typically tonically active and provide excitatory drive to their targets. The *fastigial nucleus* (medial nucleus in lower mammals) receives input from Purkinje cells in the vermis and projects *bilaterally* through the juxtarestiform body of the inferior cerebellar peduncle to the vestibular nuclei and reticular formation. The small *globose nucleus* and more lateral *emboliform nucleus* receive inputs from Purkinje cells in the intermediate zone and project through the superior cerebellar peduncle to brainstem motor nuclei, primarily the *contralateral* red nucleus. The laterally positioned *dentate nucleus,* largest and most conspicuous of the deep nuclei, is greatly enlarged and dramatically convoluted in primates, reflecting the evolutionary expansion associated with the lateral cerebellar hemispheres. The dentate nuclei receive topographically organized input from Purkinje cell axons from the longitudinal zones of the lateral portions of the ipsilateral cerebellar hemispheres, and the efferent fibers make up the majority of efferent axons in superior cerebellar peduncle, with projections to the *contralateral* thalamus, including the ventral lateral (VL) nucleus, and the ventral anterior (VA) and mediodorsal (MD) nuclei. Smaller projections reach the red nucleus, the reticular nucleus, and the oculomotor nucleus.

The deep nuclei serve as the "effectors" of the computational processes of the cerebellum. They receive collateral inputs from climbing fibers and mossy fibers that project to the cerebellar cortex and send tonic excitatory output that is shaped by the activity-dependent plasticity of Purkinje cells that exert descending inhibitory control. In addition to the efferent feedback information provided by the deep nuclei via projections back to the extracerebellar target brain regions, collateral axons from all deep cerebellar nuclei also project back to the same areas of the cerebellar cortex from which they received Purkinje cell projections. Thus, the computational end points of output from the deep cerebellar nuclei also represent the intrinsic feedback loops within the cerebellum itself.

CYTOARCHITECTURE AND DEVELOPMENT OF THE CEREBELLAR CORTEX

The gray matter of the cerebellar cortex is a little more than 1 mm thick and has a uniform geometric cytoarchitecture that is remarkably similar in all mammalian species and across cerebellar zones. In the mature cerebellum, the *molecular layer* is the outermost layer, between the pial surface and the underlying Purkinje cell layer. It contains densely packed dendrites and axons but relatively few neuronal cell bodies (only *stellate cells* and *basket cells*). The *Purkinje cell layer* contains a single layer of *Purkinje cell* bodies; beneath them lies the *granular layer*, thick and densely packed with the enormous number of small *granule cells* along with inhibitory *Golgi cells* located deeper than the Purkinje cell layer, along with the *unipolar brush cells* and *Lugaro cells*.

Embryonically, the cerebellum develops from two germinal zones, the cerebellar anlage from which all cerebellar neurons except the granule cells are generated, and the later emerging rostral rhombic lip of the hindbrain zone that provides the proliferating precursors that form the *external germinal layer* that generate granule cells. During embryonic development, the neurons of the deep cerebellar nuclei are generated first, followed shortly thereafter by generation of Purkinje cells that migrate toward the pia dorsal to the roof of the fourth ventricle, past the neurons forming the deep cerebellar nuclei. Subsequent migrations yield unipolar brush cells, Golgi cells, basket cells, stellate cells, and glia. After the Purkinje cells arrive in the primordial cerebellum, they eventually settle into a single layer by the time period roughly comparable to the transition between the second trimester and third trimester in human fetal development, and then form short-lived perisomatic processes that make contact with the arriving climbing fibers. This provokes stimulation of the dramatic dendritic outgrowth of Purkinje cells into the molecular layer and their innervation by developing parallel fibers of the late-developing granule cells, marking the dynamic development of Purkinje cell synaptogenesis that proceeds over the human third trimester.

After the formation of the primordial cerebellum, the transient *external germinal layer* (EGL) (or *external granular layer* from Ramon y Cajal) is formed by migration of proliferating progenitor cells from the rostral rhombic lip, covering the cerebellar primordium. The EGL gives rise to granule cells. After a terminal division during granule cell generation, the undifferentiated postmitotic granule cells migrate from superficial zone of the EGL to the deeper zone of the EGL where they begin to differentiate. The differentiating granule cells migrate down through the molecular and Purkinje-cell layers, guided by Bergmann glia, to reach the granular layer, leaving trailing axonal processes that bifurcate to form the parallel fibers that extend along the longitudinal axis of the folia. The later-formed axons stack on top of and are superficial to the earlier formed axons, and the granule cells born earliest are found in the deeper portions of the granular layer with later-born cells in more superficial portions. As granule cell neurogenesis progresses toward completion, the EGL disappears.

The generation, migration, differentiation, and synaptogenesis of the neurons of the cerebellar cortex occur over a protracted period that, in humans, extends through the third trimester and into the first two years of life. This relatively late period of rapid cerebellar growth, involving acquisition of granule cells, elaboration of Purkinje cell dendrites, formation of the synaptic circuitry and synaptogenesis in the cerebellar cortex likely makes experience-dependent modifications in cerebellar circuitry in early infancy an important part of its functional maturation.

FIVE MAIN NEURONS OF THE CEREBELLAR CORTEX

Purkinje cells have very large cell bodies and enormous fan-shaped dendritic trees that fill the molecular layer, spreading broadly in the transverse plane of the folia but flattened in the horizontal plane. Purkinje cells are the efferent projection neurons of the cerebellar cortex because they provide its sole output; all other neurons in the cerebellar cortex are intrinsic neurons. Purkinje cell axons are myelinated, and the large majority synapse on neurons in the deep cerebellar nuclei, on which they provide the large majority of the total number of synapses, all of which are inhibitory with gamma-aminobutyric acid (GABA) as the neurotransmitter. The inhibitory control exerted by the Purkinje cells on deep cerebellar nuclei neurons can powerfully regulate and sculpt their efferent activity to other brain sites. Purkinje cell axons from the flocculonodular lobe, and some from the vermis, bypass the deep cerebellar nuclei and synapse directly onto neurons in the vestibular nuclei.

Cerebellar *granule cells* are the smallest, most densely packed, and by far the most numerous neuron type in the entire brain. The heavy *parallel fibers* are axons of the granule cells and project in the molecular layer in an orientation that runs medial-to-lateral along the long axis of the folia. The granule cells are the only intrinsic neurons of the cerebellum to contribute excitatory input to Purkinje cells, with glutamate as the primary neurotransmitter. The terminal branches of granule cell dendrites receive mossy fiber input; granule cell axons are small and unmyelinated. The axons penetrate the Purkinje cell layer, bifurcate into a T and the two branches travel several millimeters in opposite directions horizontally through the molecular layer in the direction of the long axis of the folium, side by side with many thousands of other similar fibers. A single parallel fiber makes synaptic contacts with the dendrites of numerous Purkinje cells that are aligned in a row along the long axis of the folium. This organization allows each granule cell to synapse with the dendritic spines of several hundred Purkinje cells. Parallel fibers also make synaptic contact with basket, stellate, and Golgi cells.

Basket cells are found in the deeper parts of the molecular layer near the Purkinje cell bodies. They have small cell bodies and their dendrites are broad in the transverse plane, similar to the orientation of the dendritic trees of Purkinje cells. Basket cells are so named for the characteristic pericellular basket that their axons form around Purkinje cell bodies on which they can exert powerful inhibition. The primary axonal projection is along the anterior–posterior direction in the folia (ie, orthogonal to the plane of parallel fiber projections), from which collateral branches descend to innervate the cell bodies of Purkinje cells, contacting about a dozen Purkinje cells in the transverse plane of the folia. Additional collateral branches travel perpendicularly in the longitudinal plane of the folia, contacting about three to six rows of adjacent Purkinje cells. This organization allows each basket cell to inhibit a local patch of about 100–200 Purkinje cells.

Stellate cells are usually located in the outer two-third of the molecular layer and synapse only on Purkinje cell dendritic shafts where they are inhibitory. The inhibitory influence of stellate cells is less than that of basket cells, in part because they synapse more distally on Purkinje cells than do the basket cells. *Golgi cells* are large neurons scattered in the superficial part of the granular layer just below the Purkinje cell bodies. Although Golgi cell dendrites are relatively simple, compared to Purkinje cells, their dendrites extend farther in the cerebellar cortex reaching both the molecular and granular layers. Golgi cells receive inputs from mossy fibers as part of a complex synaptic structure called a *glomerulus* that includes mossy fiber terminals and granule cell dendrites.

AFFERENTS TO THE CEREBELLUM

Inputs to the cerebellum outnumber its outputs (about 40:1 in humans), and inputs are directed mainly to the cerebellar cortex via two major types of afferents, *climbing fibers* (the olivocerebellar tract) and *mossy fibers* (all other inputs). A third, small group of *aminergic fibers* also reaches the cerebellar cortex, including dopaminergic, noradrenergic, and serotonergic projections. All of the afferents to the cerebellum arrive through one of the three cerebellar peduncles, but primarily through the inferior and middle peduncles.

The vast majority of climbing fibers come from neurons in the inferior olive and accessory olivary nuclei, which project to the contralateral cerebellar cortex. A given climbing fiber innervates only few (often only one) Purkinje cells, on which the climbing fiber entwines itself like a vine around the dendritic tree of the target Purkinje cell(s); collaterals also project to the deep cerebellar nuclei. Early in development, a Purkinje cell may be innervated by multiple climbing fibers, but in the mature cerebellum one Purkinje cell is innervated by only one climbing fiber and it makes hundreds of synaptic contacts with the dendrites of the Purkinje cell. The major transmitter of climbing fibers is glutamate, and the activation of the climbing fiber can generate very powerful excitatory postsynaptic potentials that stimulate complex-spike action potentials involving superimposed calcium spikes that invade the dendrites. This anatomical arrangement provides for powerful, precise activation of small sets of Purkinje cells, with no convergence and limited divergence of the climbing fiber input. Climbing fibers also provide a weaker excitatory input to the Golgi cells and basket and stellate cells, and to the deep cerebellar nuclei through collateral branches. They are critical for inducing synaptic long-term depression (LTD) in the cerebellar cortex, through concurrent synaptic activation of target Purkinje cells by powerful excitation from climbing fibers and weaker excitation from parallel fibers from the granule cells. LTD results from enduring modification of the efficacy with which the parallel fiber synapses suppress their subsequent excitatory control over the target Purkinje cells. LTD has been implicated as a mechanism of motor learning and adaptive changes in cerebellum.

The mossy fiber projections are the largest source of input to the cerebellum and they enter the cerebellum through all three peduncles. Myelinated mossy fiber axons branch many times in the white matter, and then lose their myelination as each branch gives off numerous collaterals in the granular layer. Large clump-like swellings, called *rosettes*, are specialized synaptic endings that occur repeatedly along the course of the axon branches. Rosettes are the core of dendritic contacts from many granule cells, Golgi cell dendrites, and terminal branches of Golgi cell axons. This synaptic organization results in considerable divergence and convergence of input, because each mossy fiber synapses with several hundred granule cells and each granule cell receives input from several different mossy fibers. Mossy fibers also send collaterals to the deep cerebellar nuclei. In contrast to climbing fibers, mossy fibers only activate Purkinje cells indirectly through the granule cells, which are excited by the mossy fibers.

Mossy fibers originating from the spinal cord and trigeminal system provide the information on the position and status of muscles, tendons, and joints. The ventral spinocerebellar tract transmits proprioceptive information from all parts of the trunk and limbs, ascending bilaterally and entering the cerebellum through the superior cerebellar peduncle. The dorsal spinocerebellar tract conveys proprioceptive information from the lower trunk and lower limbs, ascending on the ipsilateral side and entering the cerebellum through the inferior cerebellar peduncle. The cuneocerebellar tract transmits proprioceptive information from the upper trunk and upper limbs through the inferior cerebellar peduncle. All spinal cord afferent pathways to the cerebellum arise and terminate on the *ipsilateral* side, even those axons of the ventral spinocerebellar tract that initially crossover and ascend on the contralateral side of the spinal cord, only to recross and enter the ipsilateral cerebellum. There are two main regions where proprioceptive and tactile stimuli project to the cerebellar cortex, as two somatotopic maps. One is in the vermis of the anterior lobe and the other a bilateral, mirror-image representation in the posterior lobe.

The vestibular nuclei of the brain stem that receive major input from the vestibular labyrinth via the cranial nerve VIII send mossy fiber projections bilaterally to the flocculonodular lobe and most of the vermis. A smaller set of direct projections from the inner ear via nerve VIII also reaches the ipsilateral cerebellum. These vestibular sources of mossy fibers provide important information pertaining to equilibrium.

Information from multiple regions of the cerebral cortex reaches the cerebellum indirectly from three different pathways, the (cortico)pontocerebellar tract, the (cortico)olivocerebellar tract, and the (cortico)reticulocerebellar tract. The (cortico)pontocerebellar pathway that originates across many regions of the cerebral cortex projects to neurons in the pontine nuclei, and the axons of those neurons cross the midline and project to the contralateral cerebellum through the transversely oriented middle cerebellar peduncle. The middle cerebellar peduncle is a massive fiber tract evident on the base of the pons, and is the main pathway for the mossy fibers of the cortico-ponto-cerebellar loops. The olivocerebellar pathway relays information from the cerebral cortex that projects to the inferior olive. Inferior olivary fibers cross the midline to enter the cerebellum through the contralateral inferior cerebellar peduncle where they terminate directly on Purkinje cell dendrites as climbing fibers.

THE OUTPUTS FROM THE CEREBELLUM ARE LIMITED

The sole output from the entire cerebellar cortex is via axons of Purkinje cells, the majority of which project to neurons in the deep cerebellar nuclei and a minority to the neurons of the vestibular nuclei in the brain stem. The large majority of efferents that leave the cerebellum comes from the neurons of the deep cerebellar nuclei. All outputs from the cerebellum (either from the deep nuclei or from the Purkinje cells of the vestibulocerebellar zone) exit through either the superior or inferior cerebellar peduncles, with more via the superior cerebellar peduncle. The minority of Purkinje cell axons that do leave the cerebellum, directly travels through the juxtarestiform body of the inferior cerebellar peduncle, and synapses in the vestibular nuclei. The nonvestibular Purkinje cells—constituting the large majority—project to the deep cerebellar nuclei in a specific medial-to-lateral pattern. The fastigial nucleus receives Purkinje cell projections from the vermis, the globose and emboliform nuclei from the intermediate zone, and the large dentate nucleus from the lateral cerebellar hemispheres. Some Purkinje cells from the flocculus and nodulus project to both the fastigial and dentate nuclei as well as to the vestibular nuclei.

The outputs from the fastigial nucleus project mainly to the brainstem to the vestibular nuclei bilaterally and in the reticular formation. The globose, emboliform, and dentate nuclei project via the superior cerebellar peduncle. Some project axons to the inferior olive and reticular formation, but most project rostrally and cross to the contralateral side in the midbrain at the level of the inferior colliculus. Many of the axons that originate in the emboliform and globose nuclei, and a few fibers from the dentate nucleus, synapse in the contralateral red nucleus. Most of the axons from the dentate nucleus project beyond the red nucleus to terminate in the contralateral thalamus, primarily in the VL nucleus, with a smaller contingent terminating in the VA and MD nucleus. In summary, each fastigial nucleus receives input from the vermis and projects bilaterally to the vestibular nuclei. The globose and emboliform nuclei on each side receive input from the ipsilateral intermediate zone and project mainly to the contralateral red nucleus. Each dentate nucleus receives input from the ipsilateral hemisphere and projects to VL, VA, and MD nuclei of the contralateral thalamus.

CEREBELLUM AND COORDINATED MOVEMENT

Since early in the 19th century it has been well accepted that the cerebellum is importantly involved in movement. Using lesions of the cerebellum in a variety of animals, Flourens and other early experimental physiologists observed that following these ablations, animals developed a staggering gait and clumsiness and would fall from side to side. Signs of cerebellar dysfunction in humans was considered a clinical "cerebellar syndrome" that was historically identified as ataxia (lack of motor coordination) usually on the same side of the body as the lesion, disturbances of gait and equilibrium, and an array of impairments in motor function, again ipsilateral to the lesion, typically including deficits in timing and sequencing movements, and impaired eye movement and speech control. Subsequent to the early observations in humans and animals, the understanding of cerebellar roles in movement has expanded greatly. Given the clinical deficits exhibited by patients with cerebellar damage, it is now generally believed that the cerebellum contributes to the timing of movement, to acquisition and integration of sensory information to guide movement, and to the anticipation of the sensory consequences of action. Motor deficits associated with cerebellar damage can be broadly described in multiple categories that include oculomotor deficits, speech deficits, limb deficits, and ataxia of stance and gait. In addition, a growing consensus now extends the clinical features of the cerebellar syndrome to include linguistic functions and cognitive, executive, and affective functions.

The cerebellum is involved in most classes of eye movements and the fixation of gaze. An inability to maintain gaze is indicated by the presence of nystagmus while gaze is fixed. Patients with cerebellar lesions often also display errors in the adaptive control of eye movement as indicated by over- or undershooting of gaze (respectively, hypermetria or hypometria) directed at the intended position of a hand, arm, leg, or fixation point. Different subdivisions of the cerebellum appear to be involved in these abilities. The vestibulocerebellum mediates steady gaze holding, smooth pursuit, and the vestibulo-ocular reflex (VOR). Control over saccades is mainly under the control of the "oculomotor cerebellum," which includes the dorsal vermis (lobules VI/VI in Fig. 9.3), the ansiform lobule (superior and inferior semilunar lobules of Fig. 9.2), and the fastigial nucleus.

Much of the vermis is associated with posture as controlled by trunk (axial) musculature and proximal limb muscles, head position, extraocular movements, equilibrium, and locomotion. Damage to the vestibulocerebellum (flocculus, nodulus, and uvula) results in a significant loss of equilibrium, swaying while standing or staggering while walking with a tendency to fall backwards (truncal ataxia), and wide-based stance or gait, reflecting an underlying inability to incorporate vestibular information with body and eye movements. If damage is restricted to the vermal zone, truncal deficits will be present without disruption of function in individual limbs when tested separately. There are a number of common disturbances associated with damage to the lateral hemispheres. *Ataxia* designates clumsiness or disturbances in coordination of voluntary movement. *Asynergia*, or limb ataxia, is the impaired execution (eg, jerkiness) of coordinated sequential movements. *Hypotonia* is the reduced resistance to passive movement. *Intention tremor* is an oscillating movement of a limb, which is present only during limb movement such as reaching and is more evident at the end of the movement. *Dysdiadochokinesia* refers to uncoordinated, irregular movements that occur in which patients are incapable of performing rapid movements that require alternating contractions of agonist and antagonist musculature, such as pronating and supinating the forearm. The term *dysmetria* describes the inability of patients to execute complex movements in which the judgment of distance is impaired and expressed as inaccurate control of the range and direction of movement. These can produce a *decomposition of movement* in which the complex movement must be broken down into its various individual components, which appear jerky and uncoordinated rather than smooth and flowing, resulting in either undershoot or overshoot of the desired point ("past-pointing"). A related sign is known as "impaired check and rebound" elicited clinically under conditions when a limb (or axial support) is released from tension. For example, a clinician will have a patient lean their chest firmly against the examiner's hands; upon abrupt removal of the support from the examiner, the patient will be unable to check their forward movement and will lose their balance or rock back and forth. Similar rebound is evident when a limb ipsilateral to the side of cerebellar damage is released from tension, and it will flail uncontrollably. There also can be ocular signs, especially *nystagmus*, which is of greatest amplitude when looking to the same side as the lesion. *Dysarthria* refers to disorders in the mechanical component of the articulation of speech resulting from ataxia of the muscles controlling generation of speech and results in a typical pattern involving explosive, halting speech with slurred or garbled words and difficulty in modulating volume.

In his 1917 article describing patients with gunshot damage to the cerebellum, Holmes was the first to describe speech deficits in patients with cerebellar damage. Speech symptoms were described as slow, monotonous, staccato, scanned, indistinct, jerky, explosive, slurred, and labored. Subsequent research on the involvement of the cerebellum in speech production consistently shows that patients have a reduction in the maximum speaking rate coupled to slowed articulation, and deficits in the coordination of articulatory motor movements. Based on these deficits, it has been suggested that the main role of the cerebellum in the production of speech is to regulate the timing and sequencing of basic speech movement patterns and to synthesize them into linguistically meaningful words, phrases, and sentences. The precise cerebellar substrate mediating these abilities is yet to be identified.

The cerebellum plays a major role in the predictive control of grip forces when grasping and manipulating objects in the environment. Control of grasping requires two distinctive modes of control. For example, when a weight that is held in one hand is dropped into the other hand, we use predictive control mechanisms to anticipate the appropriate grip strength to catch the falling weight. In this case, grip forces increase simultaneously with load forces without a time delay. Mechanisms used to grasp an object of unpredictable weight are different (eg, a weight dropped by another person into the receiver's hand) and instead use sensory feedback from the weight to adjust grasp forces with some time delay. These two mechanisms are respectively referred to as predictive and reactive control of grasp or grip. Patients with cerebellar damage commonly show deficits in predictive control of grasp, typically exerting more force than necessary in object manipulation tasks. This deficit is generally associated with damage to the cerebellar Purkinje cells and/or dentate nucleus. This deficit has been characterized as a failure to predict the consequences of a voluntary motor act in situations where relying on direct sensory feedback would slow responding because sensory feedback is, of necessity, time delayed.

Both single- and multijoint limb movements are impaired in patients with focal lesions or degenerative disorders of the cerebellum. As noted previously, limb movements in these patients are characterized by undershooting or overshooting. The onset of movement toward a target is also slowed. Evidence from neuroimaging studies in these patients indicates that deficits in limb movements are associated with cerebellar cortical atrophy. Upper limb ataxia is associated with lesions in lobules IV–VI, whereas lower limb ataxia appears to be related to lobules III and IV (anterior lobe).

THE CEREBELLUM AND MOTOR LEARNING

The cerebellum plays an important role in the acquisition of motor skills marked by rapid, automatic, and accurate control of complex actions, such as playing Bach's Violin Partita #3 on a banjo. There are multiple forms of activity-dependent plasticity in the cerebellar circuitry, LTD of the efficacy of parallel fiber–Purkinje cell synapses has been most extensively studied in terms of cellular mechanisms. LTD can occur on a Purkinje cell that has converging activation from a set of parallel fibers in temporal association with activation from its climbing fiber. This activity-dependent LTD in the subsequent synaptic response of the Purkinje cell to the mossy fiber input can serve as a form of memory about coinciding signals and shape the pattern of subsequent cerebellar output to the same pattern of sensory input (via the mossy fibers), as might occur with trial-to-trial error correction.

Ito and others have extensively studied processes of neuroplasticity underlying the adaptation of eye movements of the VOR, which keeps the visual image on the retina stable during periods of head rotation by automatically moving the eyes in the opposite direction at the same speed, preventing retinal slip of the visual image. Vestibular information about head movement reaches the flocculus via mossy fibers and converges with information about intended eye position that is conveyed by climbing fibers to the same region, to control and correct eye position as the head undergoes rotational movement.

Certain conditions of mismatch between retinal image and direction of gaze, as when a subject puts on prism glasses that displace the retinal image, involve adaptive changes occurring over time that depend on LTD in Purkinje cells of the flocculus. Adaptation to prism glasses produces a striking pattern of behavioral performance in tasks such as throwing an object at a target. Initial throws when the prism glasses are first donned deviate in the direction opposite to the direction of the bent light path, but with practice, the normal subject regains accuracy. Then, upon the removal of the glasses, deviations in the opposite direction transiently occur until adaptation to the absence of prism glasses occurs. Patients with damage to the vestibulocerebellum show impaired or absent adaptation.

Classical conditioning of eyeblink (or nictitating membrane) responses is one of the most extensively studied and best understood forms of cerebellar-dependent learning, and is especially suited for translational studies between animal models and humans. Decades of research across multiple species (humans, rabbits, rats, mice, and sheep) have identified the behavioral properties and the neural circuitry essential for eyeblink conditioning, and detailed knowledge of the cellular and molecular processes of learning-related neuroplasticity has been developed. With eyeblink conditioning, a subject is given numerous paired presentations of a "neutral" conditioned stimulus (CS), for example, a tone that does not cause a reflexive blink, and an unconditioned stimulus (US), for example, a puff of air to the cornea of the eye that does cause a reliable blink, called an unconditioned response (UR). Over repeated trials of paired presentations, associative learning is expressed by the occurrence of an eyeblink response upon presentation of the CS in advance of (or in the absence of) the US, termed a conditioned response (CR), that typically becomes timed to occur just before the onset of the US. The rate of acquisition of this learning can be used as an index of the functional integrity of a defined neural circuitry in the cerebellum involved in eyeblink conditioning.

Experimental studies using lesions, reversible inactivation, genetically deficient mice, neural stimulation studies, neural recordings, tract-tracing, and functional neuroimaging have provided converging evidence that a relatively specific circuitry in the cerebellum ipsilateral to the side undergoing eyeblink conditioning, together with its functional brainstem connections, is essential for learning the CS–US association. Currently, a two-site model of learning-related neuroplasticity in the cerebellum is thought to account for normal acquisition of the conditioning, with long-term potentiation (LTP) occurring in the lateral anterior interpositus nucleus and LTD occurring in the portions of the cerebellar cortex intermediate zone (anterior lobe and hemispheric lobule VI) that project to the interpositus.

Neural signals encoding the CS are projected to discrete portions of the pontine nuclei, and the mossy fiber axons from these pontine neurons project to discrete regions of cerebellar cortex with collaterals also going to the interpositus nucleus. Neural signals encoding the US are projected from the trigeminal system to neurons of the dorsal accessory inferior olive that give rise to the climbing fibers that project to discrete regions of cerebellar cortex with collateral branches to the interpositus nucleus. Note that though the cerebellum is not required for unconditioned eyeblinks, neuroplasticity arising from converging CS and US input projections to the lateral portion of the anterior interpositus and to the cerebellar cortex (anterior and hemispheric simplex lobules) are essential for normal rates of acquisition of learning and performance of conditioned eyeblinks. The efferent projections of the interpositus neurons (to the magnocellular division of the red nucleus) provide the signals that reach the cranial nerve motor nuclei to generate the conditioned eyeblink. Neuroplasticity in the form of long-term potentiation (LTP) at the interpositus nucleus appears to be necessary for acquisition and performance of the eyeblink CR; interpositus lesions or reversible inactivation prevent acquisition or expression of conditioning. Neuroplasticity at synapses converging on Purkinje neurons, including LTD in Purkinje cells engaged by the conditioning stimuli, appears to regulate temporal specificity and amplitude of the CRs. Notably, *pcd* mutant mice that lose all of their Purkinje neurons by the fourth week of life, but retain deep cerebellar neurons, can acquire eyeblink CRs, but acquisition proceeds slowly and the CRs are neither strong nor well timed. Functional neuroimaging studies have confirmed that similar zones in the cerebellum are functionally activated in humans during classical conditioning, and patients with cerebellar damage ipsilateral to the eye undergoing classical conditioning show deficits in eyeblink conditioning. Because eyeblink conditioning can be used in parallel studies in animal models and humans, it provides one of the most direct experimental models to study diseases or disorders involving cerebellar function. For example, identification of cerebellar damage and associated eyeblink conditioning deficits in a rat model of binge-like alcohol exposure led to examination and confirmation of similar deficits in children with fetal alcohol spectrum disorder.

THE CEREBELLUM AND COGNITIVE ABILITIES: CLINICAL STUDIES

Although initially regarded mainly as involved in motor activity and regulation of balance, equilibrium and eye movement control, there is increasing data indicating the cerebellum's role in cognition and higher function. As discussed earlier, the cerebellum is essential for classical conditioning of eyeblink responses, and is involved in learning complex motor patterns involving multiple muscle groups and joints. Studies in humans with focal cerebellar lesions confirm that this structure is not simply involved in stimulus-response (or procedural) learning, but in cognitive behaviors as well. Thus, patients with focal cerebellar damage do poorly in neuropsychological tasks measuring visuospatial abilities and planning, such as block design and copying the Rey figure. They also show impaired performance on tests measuring spatial abilities such as the Raven Standard Progressive Matrices. This is noteworthy because the Raven test requires no movement from the patient, which eliminates the possibility that motor deficits can account for the impaired performance. More recently, deficits in executive function including planning, set-shifting, and working memory have been confirmed in patients with diseases confined to the cerebellum, and have also indicated marked differences in the severity of symptoms that are related to the anterior–posterior locus of damage.

Functional neuroimaging studies have also indicated that the cerebellum is importantly involved in cognitive function. Activation of the cerebellum has been found during tests of linguistic function including verb for noun substitution and verbal fluency. Cerebellar activation has also been reported during tests of working memory, verbal memory, mental imagery, shifting attention, cognitive planning/problem solving, and sensory discrimination. Neuroimaging data also indicates that there is, along with some overlap, anatomical specificity in the patterns of cerebellar activation for different cognitive tasks, suggesting that, within the limits imposed by neuroimaging methodology, different areas of the cerebellum have unique contributions to different cognitive functions.

One limitation of imaging studies is that many tested tasks also require either movement or motor planning; conclusions about the precise role of the cerebellum in cognitive function have been controversial. This issue was addressed in an experiment by Courchesne and colleagues in 1997 in which functional neuroimaging was carried out on subjects while they performed an attention task that did not require motor learning, motor response selection, error detection, error correction, imagined motor action, or motor system guidance. Results obtained during this task were compared to those found during a motor task. The results demonstrated a double dissociation between areas of the cerebellum activated by the attention or motor tasks indicating that different regions of the cerebellum were involved in distinctly different neurobehavioral functions.

An association between developmental cerebellar pathology and alterations in cognition can also be gleaned from the symptomatology of numerous disorders associated with developmental cerebellar damage. Included in this list are tuberous sclerosis, Fragile X, and Joubert syndromes and autism (although links between cerebellar neuropathology and autistic symptoms can also be found for Angelman, Rett, Williams, and Moebius syndromes). Tuberous sclerosis (TS) is an autosomal dominant disorder caused by the mutation of either the TSC1 or TSC2 gene. TS is associated with hamartomatous lesions (tubers), which are often present in the brain as well as numerous other organs. In 17–68% of cases, TS is associated with autism. In patients with TS, an association between cerebellar tubers or dysfunction and autistic behaviors has been demonstrated.

Fragile X syndrome (FXS) is an X-linked genetic disorder caused by a mutation in the *FMR1* gene. The behavioral phenotype of *FXS* is highly variable, but includes mild to severe mental retardation, and autistic features such as hyperactivity, attention deficits, speech and language difficulties, poor eye contact, hand biting, and tactile defensiveness. The expression of the *FMR1* gene and its protein product, FMRP, is ubiquitous although the highest levels of expression are found in the hippocampus and the cerebellum. MRI studies have consistently revealed a significant reduction in the size of the posterior cerebellar vermis in the brains of FXS patients. In addition, microscopic analysis has revealed focal Purkinje cell loss and Bergmann gliosis in the cerebella of FXS brains. FXS patients often display great somatic variability in the methylation of the *FMR1* gene. As hypermethylation of the *FMR1* gene causes the failure of its transcription, FXS patients are often mosaic in their expression of FMRP. This mosaicism is likely the main reason for the large variability of FXS symptoms.

Joubert syndrome (JS) is an autosomal recessive disorder that results in the agenesis or dysgenesis of the cerebellar vermis and malformation of multiple brainstem structures. Although early reports lacked thorough behavioral investigations, perseveration and self-mutilation were mentioned along with hyperpnea, abnormal eye movements, tongue protrusion, hypotonia, ataxia, and mental retardation as characteristics found in cases of JS. More recently, several cases of JS have been dually diagnosed with autism, and JS has been characterized by autistic behaviors such as sensory integration deficits, tactile defensiveness, perseveration, difficulty shifting attention between tasks, and stereotypy.

In particular, mounting data from Courchesne and from others demonstrate the cerebellum's involvement in autism and point clearly to the importance of the cerebellum in the integration of complex behaviors. Autism spectrum disorders (ASD) are a broad range of generally idiopathic neurodevelopmental disorders that affect one in 68, with males more frequently affected than females. Virtually unchanged from when this disease was first described, the diagnosis of autism is based on the triad of social, communication, and imagination impairments, along with restricted repetitive activities. Due, in part, to failures to control the degree of mental retardation that co-occurs with autism, these impairments were initially viewed as representing the "core" deficits of this disease. As summarized by Happe and Frith over 15 years ago, modern cognitive theories of ASD and neuropsychological research has raised the alternative possibility that a number of basic cognitive and attentional deficits underlie these diagnostic symptoms. Thus, for example, the idea that language deficits drive the impairments in social behavior and communication is not supported. Rather, lack of communication is probably the important feature; autistic individuals who have language skills fail to use those skills for intentional communication, which appears more consistent with an underlying motivational deficit. Along with this putative deficit in the motivation, impairments in attention, memory, and a marked tendency to perseverate in the face of changing contingencies have been documented in autistic patients.

Autism is now recognized as having an organic basis. Several brain regions have been causally implicated in autism, especially the hippocampus, and the cerebral and cerebellar cortices. The most extensive and reproducible neuropathology has been found in the cerebellum, and includes cerebellar hypoplasia and reduced neuron numbers. In MRI analyses, specific lobules of the cerebellum have been found to be hypoplastic in high-functioning autistic patients. In a study from Carper and Courschene reported in 2000, of a total of 20 postmortem autism cases that were examined, 19 showed evidence consistent with developmental abnormalities that consisted of substantial reductions in the number of Purkinje cells. Other groups have also reported very similar postmortem pathology in people with autism that consistently involves the loss of cerebellar Purkinje cells. In addition to the cerebellar neuropathology, related structural abnormalities in the frontal lobe have recently been reported. Quantitative MRI in the Carper and Courschene study also found that the degree of frontal lobe abnormality correlated with the degree of cerebellar abnormality in a large sample ($n=42$) of autistic children. Specifically, in many of these children, it was found that the volume of the frontal cortex was increased in proportion to tissue loss in the cerebellum. Although the mechanisms accounting for the changes in frontal cortex are unknown, such a reciprocal relationship between the frontal cortex and cerebellum may not be surprising based on the multiple, parallel neural cortico–ponto–cerebellar loops between the cerebellum and frontal cortex, which have recently been identified. Strick and colleagues have directly addressed the issue of cerebellar–prefrontal cortex (PFC) interactions using herpes simplex virus to map pathways from primate cerebellum to frontal cortex. Importantly, he found that the cerebellum projects via the thalamus to areas 9 and 46d in PFC, and that these projections originate in areas of the dentate nucleus that are spatially segregated from those that project to motor areas of the cortex. This result is suggestive that there are separate cerebellar output channels that influence motor and cognitive processes.

Based primarily on similarities between the cognitive neuropsychological deficits observed following cerebellar or dorsal prefrontal cortical lesions in humans, Strick suggests that damage to areas of the (a) dentate nucleus, (b) areas of the cerebellar cortex that innervate it, or (c) dorsal PFC yield similar deficits because they all interrupt input to the PFC from the cerebellum and thereby disconnect the prefrontal cortex and the relevant components of the cerebellum (referred herein as the "disconnection hypothesis"). It should be noted in this regard that dopaminergic activity in the PFC plays a critical role in normal cognitive processes and neuropsychiatric pathologies. Although speculative, it is possible that disruption of this cerebellar–PFC projection could lead to reductions in dopaminergic activity in medial PFC as well as account for the associated deficits in anticipatory behavior and shifting attention found in autistic children. The effects of dopamine on neuronal activity in the PFC are rather complex. Both increases and decreases in baseline and task-related neuronal activity in the presence of elevated dopamine levels have been observed.

CEREBELLUM AND COGNITIVE ABILITIES: ANIMAL MODELS

One avenue of exploration in determining the involvement of the cerebellum in cognitive function is the use of neurogenetic mutations in the mouse that affect cerebellar development and function. The mutant model system recommends itself at several levels: (1) The perturbation in structure is highly discrete and reproducible (a so-called genetic lesion), (2) The primary lesion and secondary cascade of events can be thoroughly described, (3) A whole host of such genetic lesions are available in the mouse, and many of these affect the cerebellum, and (4) Each mutation has different effects upon the cerebellum, giving rise to a rather complete series of cerebellar pathologies in these mice. To this partial listing, one must add the exquisite ability to manipulate the murine organism through experimental transgenic embryological techniques. One of these techniques, the production of experimental mouse chimeras, offers a unique facet to the murine model of cerebellar structure and function. In this technique, eight-cell embryos from two distinct genotypes of mice are combined to form a single embryo that is composed of cells that come from either of the parental genotypes. This technique allows one to overcome a big drawback of using neurological mutant mice in the study of the cerebellum's role in higher learning, which is: These mutations typically knock-out much of the cerebellar circuitry, creating an ataxic mouse whose behavior is difficult to analyze. First, ataxia is a confounding element in behavioral work. Second, the severe cell loss experienced in neurological mutants is not typical of alterations in human conditions such as those seen in autism. However, by making chimeras between a neurological mutant mouse and a congenic wildtype mouse, one can obtain a whole range of chimeras that have a variety of cerebellar phenotypes intermediate between the mutant and wildtype. These chimeras are typically not ataxic, and can thus be tested behaviorally. Furthermore, the wildtype cells are restored to the mutant cerebellum in varying numbers and locations. Therefore, these mouse chimeras (1) possess cerebella that are better approximations of the developmental cellular changes seen in people with behavioral disorders such as autism and (2) are additionally associated with a heterogeneous phenotype that could be viewed as similar to the heterogeneity seen in diseases such as autism. Two such models are highlighted here.

Lurcher Mutant Mice

One cerebellar mutant mouse, Lurcher (gene symbol: *Lc*), has been widely used to explore the role of the cerebellum in simple behaviors. In the Lurcher mouse, due to a mutation in δ2 glutamate receptor gene, there is a specific and complete loss of cerebellar Purkinje cells that occurs during the 2nd to 4th weeks of postnatal life. The only other detectable cell losses are the cartwheel neurons in the dorsal cochlear nucleus and those that occur secondary to Purkinje cell death. For example, cerebellar nuclear cells are only mildly affected. Lurchers have been reported to show deficits in a variety of simple behavioral tasks including rope climbing and runway tasks. Since these mice are ataxic and display characteristic lurching movements, chimeric mice, which have a variable loss of Purkinje cells dependent upon the incorporation of the wildtype lineage, have been used to examine the behavioral impact of Purkinje cell loss. Lurcher mice and chimeras display impaired memory in a radial maze task, impaired executive function as measured by a serial reversal learning task (with errors negatively correlated with Purkinje cell number), and a negative correlation between repetitive behaviors and cerebellar Purkinje cell number in chimeric mice. Thus, these mice exhibit executive dysfunction and repetitive behaviors directly related to cerebellar Purkinje cell numbers and thus approximate some of the symptoms commonly observed in autism.

Fmr1 Knockout Mice

The *Fmr1* gene codes for the production of the fragile X mental retardation protein known to be involved in cognitive development, and mutations of this gene result in fragile X syndrome. Patients with fragile X syndrome exhibit autism-like cognitive symptoms and have an increased probability of an autism diagnosis. Because of this, *Fmr1* knockout (KO) mice have been used to model autism. These mice display several behavioral deficits similar to those seen in autism such as hyperactivity, perseverative behavior, reduced spatial learning abilities, memory deficits, and reduced fear conditioning. *Fmr1* KO mice have cerebellar abnormalities such as elongated spines on cerebellar Purkinje cells and decreased volume of cerebellar nuclei. Also, eyeblink conditioning, known to be dependent on synaptic plasticity of cerebellar neurons, is attenuated in *Fmr1* KO mice suggesting that these cerebellar abnormalities have functional relevance. Abnormalities also exist in the PFC of *Fmr1* KO mice. Layer 5 pyramidal neurons display hyperconnectivity in the mPFC, and synapses between these neurons are not able to recover from LTD as rapidly as the same neurons in control mice.

CEREBELLUM AND COGNITIVE ABILITIES: A PUTATIVE NEUROPHYSIOLOGICAL SUBSTRATE FROM ANIMAL MODELS

Many of the behavioral deficits observed in ASD have been found to be related to frontal cortical function, and dopamine receptor antagonists have shown clinical efficacy in controlled trials in ASD. Based on that, Mittleman and colleagues have explored the hypothesis that disruption of cerebellar–PFC projections could lead to a dysregulation of dopaminergic-mediated activity in PFC and result in some of the behavioral deficits typically observed in ASD. The hypothesized affected circuitry is shown in Fig. 9.4. Dopamine neurons in the midbrain ventral tegmental area (VTA) comprise the mesocortical dopaminergic pathway. One possible neuronal circuit for cerebellar modulation of this pathway may involve cerebellar projections to the reticulotegmental nuclei of the pons (RTN), which receive glutamatergic input from deep cerebellar nuclei (CN), and the pedunculopontine tegmental nuclei (PPT), which receive input from and project back to the RTN and also projects to dopaminergic neurons in the VTA. A second pathway involves glutamatergic CN projections to the mediodorsal (md) and ventrolateral (vl) thalamic nuclei (ThN) that in turn send glutamatergic inputs to the PFC, which make close appositional synapses with VTA dopaminergic inputs onto PFC pyramidal cells. The PFC then projects back to the cerebellum via pontine nuclei.

Initially, lurcher (*Lc*) mutant (*Lc*/+), wildtype (+/+) mice were used as the experimental system

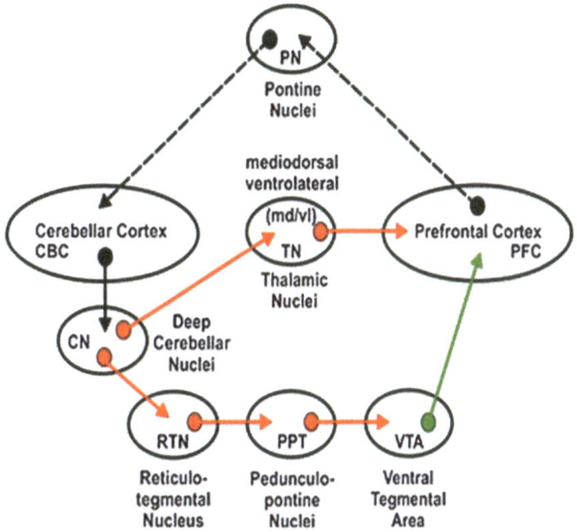

FIGURE 9.4 Neuronal circuitry underlying cerebellar modulation of medial prefrontal cortex (PFC) dopamine neurotransmission. Cerebellar modulation of PFC dopamine release may occur via polysynaptic inputs from cerebellar nuclei to dopamine-containing cells in the ventral tegmental area or via a monosynaptic input to thalamic projections making close appositions with dopamine terminals synapsing onto PFC pyramidal cells. Glutamatergic pathways are shown as *red lines* and the dopaminergic pathway as a *green line*. *Dashed lines* represent PFC projections to cerebellum via the pontine nuclei. Nuclei abbreviations are shown in the *ovals*.

to explore the disconnection hypothesis. To test the functional importance of the circuitry linking the cerebellum to the PFC, Mittleman and colleagues used an electrochemical recording procedure (fixed potential amperometry), which permits measurement of dopamine release and reuptake in dopaminergic terminal sites in the brain. In *Lc* wildtype mice, electrical stimulation of the cerebellar cortex evoked a significant increase in PFC dopamine release, an effect that persisted well after the stimulus had ended. It was also found that when stimulating the dentate nucleus of these mice, similar electrical stimulation could evoke PFC dopamine release of a comparable magnitude, but briefer duration. In marked contrast, stimulation of the cerebellar cortex in *Lc* mutant mice *failed* to increase PFC dopamine release; stimulation of the dentate nucleus in *Lc* mutant mice increased PFC dopamine release, but compared to wildtype mice, this response was significantly attenuated.

Additional studies determined which cerebellum-to-PFC pathway depicted in Fig. 9.4 was more important in mediating PFC dopamine release. Baseline levels of dentate nucleus electrical stimulation–evoked PFC dopamine release were recorded using fixed potential amperometry in combination with carbon fiber dopamine recording electrodes in *Lc* wildtype mice prior to and after inactivation by microinfusion of lidocaine initially into the ventrolateral (vl) or mediodorsal (md) thalamus (ThN), or the ventral tegmental area (VTA) of individual mice. Stimulation of other deep cerebellar nuclei (eg, interposed and fastigial) failed to evoke significant PFC dopamine release. Infusions of the local anesthetic lidocaine is an effective means of blocking all axonal transmission to or from specific areas and pathways and thus provided a unique means to determine the functional anatomy of cerebellar modulation of PFC dopamine release. Blocking transmission through the VTA or combined ThN permitted unequivocal determination that the RTN–PPT–VTA and thalamic routes contribute approximately 50% each to cerebellar modulation of dopamine release in the PFC. To confirm that these effects were mediated predominantly by glutamatergic inputs to these nuclei, we replicated these studies with infusions of the broad-spectrum ionotropic glutamatergic receptor antagonist kynurenate, which resulted in a comparable 50% influence each of the two cerebellar neuronal pathways mediating PFC dopamine release.

A separate study determined the relative contributions of DN–RTN–PPT–VTA–PFC and DN–ThN md/vl–PFC circuits to cerebellar modulation of PFC dopamine release using kynurenate microinfusions in *Fmr1* and Lurcher mutants, and wildtype mice. Both circuits in *Fmr1* and Lucher wildtype mice exhibited an equivalent contribution to cerebellar PFC dopamine modulation (~50% via the VTA–PFC circuit vs. ~50% via the thalamic (vl plus md)–PFC circuit). Most significantly, there was a functional reorganization of these pathways in lurcher and *Fmr1* mutant mice that appeared remarkably similar. Specifically, both mutations exhibited a comparable shift in pathway contribution to dentate-evoked PFC dopamine release away from the VTA–PFC circuit to the ThN vl–PFC circuit. These findings suggest that, regardless of the nature of the mutation, diverse types of cerebellar neuropathology yield similar dysregulation of cerebellar–PFC pathways.

Considered together, it seems reasonable to conclude that the disconnection hypothesis of Strick and colleagues is supported. Developmental damage to the cerebellum is consistently associated with a disruption of dopaminergic input to the PFC. This disruption is characterized by a shift in the strength of projections through the thalamus and a reduction in strength of pathways projecting through the VTA. It is certainly possible that this disorganization of cerebellar–PFC projections plays an important role in the magnitude and variety of cognitive deficits observed in patients with cerebellar damage.

SUMMARY

Because the cerebellum has a geometric simplicity of its neuronal architecture and patterns of inputs and outputs, follows distinct developmental patterning, and has a remarkable capacity for physiological and synaptic neuroplasticity to alter the function of specific neuronal circuits, it has been a popular target of study. Because the pattern of local circuitry throughout the cerebellar cortex is organized as essentially identical modules, a strict localization of function based on differences in intrinsic circuitry or even the types of computational functions performed likely is untenable. Instead, understanding the diverse functions and the clinical correlates of local cerebellar damage likely will require understanding the differences in information mediated by the relatively segregated and parallel afferent and efferent connections that constitute the cortico–ponto–cerebellar loops (or those that involve oculomotor or vestibulomotor loops).

In light of the importance of the anatomical and functional organization into generally parallel modules of the cerebellar circuitry, the evolutionary trends of expanding neocerebellar modules appear to conserve the organizational and computational properties of the cerebellar cortex circuitry in general while expanding the cortical–cerebellar modules. From this perspective, the evolutionary evidence suggests that the conjoint expansion of the neocortical–cerebellar modules in the hominoid lineage may have provided the essential computational power needed to accommodate and support the emergence of complex technical skills, advanced visuomotor attention and skill, increased cognitive capacity (needed to use the skills appropriately), communication and language systems involving complex recognition and expression, and the regulation of complex social interactions. It is now clear that the cerebellum has a role in cognitive, executive, linguistic, and affective regulation, beyond the well-established role in motor coordination and motor control, fitting with new perspectives on its evolutionary role. The precise operations and the essential computational functions carried out by the cerebellum are still not certain, but regulation of wide-ranging cortical functions through feedforward/feedback loops is now recognized, including prefrontal cortical functions.

Cerebellar circuitry has an exquisitely precise capability for experience-dependent plasticity, and cerebellar lesions can cause an array of deficits in motor learning and cognitive effects. Neuroplasticity in defined cerebellar circuitry in the vestibulocerebellum is essential for adaptation of the VOR; learning-related plasticity in defined circuits of the spinocerebellum is necessary for classical conditioning of discrete motor responses. Likewise, the lateral cerebellar hemispheres are important for learning complex motor and cognitive skills acquired. The function of the cerebellum is now recognized to extend beyond motor coordination and motor control and includes domains such as verbal working memory, language and linguistic function, regulation of planning of goal-directed behavior, and correction of that behavior when conditions change, and even to regulation of emotion and mood states. Disordered development of the cerebellum may be a key contributor to neurodevelopmental disorders such as autism or fetal alcohol spectrum disorders. Advances in understanding the role of the cerebellum in the motor and nonmotor domains will require more precise knowledge of the functional mapping of the specific modular topography, the target actions and experience-dependent changes that underlie the cortical-cerebellar loops, and the computational processes occurring within and between modules, and combining those with advances in understanding molecular mechanisms of cerebellar plasticity in relation to behavioral change.

Further Reading

[1] Allen G, Buxton RB, Wong EC, Courchesne E. Attentional activation of the cerebellum independent of motor involvement. Science March 28, 1997;275(5308):1940–3.

[2] Barton Robert A, Venditti C. Rapid evolution of the cerebellum in humans and other great apes. Curr Biol 10/20/2014;24(20):2440–4.

[3] Carper RA, Courchesne E. Inverse correlation between frontal lobe and cerebellum sizes in children with autism. Brain April 2000;123(Pt 4):836–44.

[4] Desmond JE, Fiez JA. Neuroimaging studies of the cerebellum: language, learning and memory. Trends Cogn Sci September 1, 1998;2(9):355–62.

[5] Fine EJ, Ionita CC, Lohr L. The history of the development of the cerebellar examination. Semin Neurol 2002;22(04):375–84.

[6] Freeman JH, Steinmetz AB. Neural circuitry and plasticity mechanisms underlying delay eyeblink conditioning. Learn Mem 2011;18(10):666–77.

[7] Happe F, Frith U. The neuropsychology of autism. Brain August 1996;119(Pt 4):1377–400.

[8] Holmes G. The cerebellum of man. Brain 1939;62:1–30.

[9] Manto M, Bower JM, Conforto AB, et al. Consensus paper: roles of the cerebellum in motor control–the diversity of ideas on cerebellar involvement in movement. Cerebellum June 2012;11(2):457–87.

[10] Mosconi MW, Wang Z, Schmitt LM, Tsai P, Sweeney JA. The role of cerebellar circuitry alterations in the pathophysiology of autism spectrum disorders. Front Neurosci September 1, 2015;9.

[11] Rogers TD, Dickson PE, Heck DH, Goldowitz D, Mittleman G, Blaha CD. Connecting the dots of the cerebro-cerebellar role in cognitive function: neuronal pathways for cerebellar modulation of dopamine release in the prefrontal cortex. Synapse November 2011;65(11):1204–12.

[12] Schmahmann JD. Disorders of the cerebellum: ataxia, dysmetria of thought, and the cerebellar cognitive affective syndrome. J Neuropsychiatry Clin Neurosci August 1, 2004;16(3):367–78.

CHAPTER

9.1

Translational Correlation: Disorders of the Cerebellum

H. Wilms, P. Laengvejkal

Texas Tech University Health Sciences Center, Lubbock, TX, United States

INTRODUCTION

While the main function of the basal ganglia is the selection of motor programs, the cerebellum is required specifically for planning and implementation of motion sequences. For this, movement characteristics and coordinates are predetermined. During the execution of a specific movement simultaneous integration of multisensory input for these movements occurs, and the movement characteristics are reported back to the cerebral cortex. Furthermore, the cerebellum plays an important role in motor learning. Cerebellar lesions are characterized by the following basic disturbance patterns: Muscle power cannot be precisely portioned and maintained, resulting in fragmentation and decomposition of movement into component parts. The recruitment of antagonist muscles required for a precise movement execution fails so that the resulting movement effect is uncoordinated. (Therefore, the clinical cerebellar symptom *intention tremor* is based on a disturbed coordination of proximal muscle agonists/antagonists of the shoulder–arm girdle when carrying out the finger-to-nose test.)

MOTOR SYMPTOMS OF CEREBELLAR DYSFUNCTION

Ataxia (from the Greek "a taxis" meaning "without order") is an impairment of coordination in the absence of significant weakness. With *truncal ataxia* the patient is not able to maintain an upright sitting posture of the trunk but falls backward or to the side. Severe *stance ataxia* with unsystematic or systematic swaying is often already evident while the patient stands with normal foot positioning. Mild *stance ataxia* only becomes evident when the patient is asked to stand with feet together, eyes open and hands by his side. *Gait ataxia* is characterized by a widened base, as well as considerable strike stride friability. Targeted movements are called *dysmetric* when arms/legs/eye movements over- or undershoot a target (*hypometric* or *hypermetric*) due to lack of coordination. Characteristically, the patient's finger or toe/heel oscillates with a low frequency of 3–4 Hz with increasing side-to-side movements as it approaches the target (*intention tremor*). Extremity ataxia is usually more pronounced in the arms than the legs due to the fact that in humans the major part of

Conn's *Translational Neuroscience*
http://dx.doi.org/10.1016/B978-0-12-802381-5.00017-8

spinocerebellar tracts originates from the lumbar and lower thoracic spine, since with upright gait legs more than arms are responsible for balance control. With a unilateral lesion of the cerebellum, this movement disorder is ipsilateral only. Moreover, with acute unilateral, cerebellar lesions *reduced muscle tone* is observed: This is best tested by gripping the standing patient by his shoulders and turning his trunk side to side. With the arms hanging down loosely, passive excursions of the arm will be wider on the ipsilateral side of the lesion. *Cerebellar dysarthria* is characterized by either slurring articulation or *scanning speech.* Due to the impairment of cerebellar coordination speech loses its natural fluency and modulation, but sounds slow, labored, hesitant, and staccato-like. Vocalization and sometimes respiration in speech can be affected as well since the cerebellar deficit embraces the respiratory and laryngeal muscles as well. *Charcot triad,* named after the famous French neurologist, is a characteristic neurological syndrome consisting of intention tremor, nystagmus, and scanning speech. It is often, but not exclusively, seen in multiple sclerosis.

CONTRIBUTION OF THE CEREBELLUM TO COGNITION

The cerebellum plays a key role in implicit learning and therefore contributes to procedural memory. This means that well-trained, automated motion sequences can be retrieved without conscious control. Examples for this are the coordination of muscles while skiing, dancing, or riding a bike. Consequently, with cerebellar lesions acquisition of new motor skills becomes more difficult. A neuropsychological syndrome consisting of deficits in attention, abstract reasoning, verbal fluency, and affect is sometimes seen after cerebellar lesions and has been summarized as *cerebellar cognitive affective syndrome* (CCAS).

For the topographical mapping of individual cerebellar symptoms, remembering the anatomy of the cerebellum is helpful. There are three phylogenetic different parts, the archicerebellum ("vestibular afferents"), the paleocerebellum ("spinal and trigeminal afferents"), and the neocerebellum ("corticopontine afferents").

The archicerebellum includes the flocculonodular lobe, which is located in the medial zone. As it is closely interconnected with the vestibular nuclei, it is responsible for the maintenance of equilibrium and coordination of eye movements. Dysfunction of the vestibulocerebellum and its efferents leads to a combination of truncal, stance, and gait ataxia but not extremity ataxia, the body position cannot be stabilized by visual control. Oculomotor deficits include saccadic pursuit, downbeat nystagmus, and impaired suppression of the vestibulo-ocular reflex. Speech is dysarthric.

The paleocerebellum integrates information from spinocerebellar afferents, the visual system, and the vestibular system. If one of the three aforementioned systems fails, the remaining two systems are sufficient to enable normal stance. (A healthy person is able to stand with his eyes closed!). With additional failure of a second system, for example, due to a lesion of the *anterior lobe* stance ataxia occurs. While affected patients are able to stabilize their trunk with open eyes when standing, they are not able to do so with their eyes closed. They start swaying but in contrast to patients with lesions of the archicerebellum do not fall, their speech is scanning.

The neocerebellum (sometimes called cerebrocerebellum) receives afferents mainly from premotor centers in the frontal lobe (premotor cortex and supplementary motor area). In these areas movements are planned, the "rough movement draft" is then projected into the lateral cerebellar hemispheres, where it is further processed, modulated, and the planned activity of involved muscle groups is coordinated. Feedback loops involving the red nucleus and inferior olive contribute to this as well. The efferents project via thalamic pathways back to the precentral cortex. In patients with lesions of the neocerebellar cortex, dentate nucleus or superior cerebellar peduncle extremity movements are dysmetric, most often hypermetric. With targeted purposeful movements an intention tremor is noted.

CLINICALLY IMPORTANT DISORDERS OF THE CEREBELLUM

Pragmatically, with regard to differences in clinical management, ataxia of acute onset (minutes/hours) has to be distinguished from ataxia with subacute onset (days/months/years). With regard to etiology congenital, hereditary, and acquired (vascular, infectious, paraneoplastic, and metabolic) forms of ataxia can be distinguished.

ATAXIA WITH ACUTE ONSET

Ataxia of acute onset is worrisome for a cerebellar stroke (infarction or hemorrhage), particularly if hemiataxia or other brain stem signs are present on clinical examination. The causative vascular lesion interrupts the cerebello-thalamic pathway. This constitutes a neurological emergency and warrants immediate neuroimaging by brain CT. Management includes control/normalization of hypertension, hyperglycemia, and in the case of space-occupying lesions due to brain edema neurosurgical decompression. Intoxication, particularly not only from alcohol, but also from prescription drugs (lithium or phenytoin), can cause acute ataxia as well. Ataxia is usually bilateral and mental status is altered (somnolence, coma) in cerebellar disorder due to intoxication.

ATAXIA WITH SUBACUTE ONSET

Dandy–Walker syndrome is a congenital CNS malformation characterized by complete absence of the cerebellar vermis resulting in cystic expansion of the fourth ventricle. Affected children suffer from developmental delay and ataxia.

Infections triggering cerebellar deficits are most commonly seen in children: *Acute disseminated encephalomyelitis (ADEM)/cerebellar ataxia of childhood* usually affects children before the age of 10 years. It is triggered by viral infections or vaccinations (varicella). Typically fever, limb/gait ataxia, and dysarthria develop within hours to days. In about two-third of the cases, full spontaneous recovery occurs within months while a minority of patients continues to suffer from ataxia or learning difficulties even after years. Similar forms of *parainfectious cerebellitis* have been described as case reports in adults triggered by either viral (eg, Epstein–Barr and rubella) or bacterial (eg, *Borrelia*, *Listeria*, and *Mycoplasma*) infections. In these cases CSF studies show lymphocytic pleocytosis, while brain MRI is noted for swelling of the cerebellum, sometimes with contrast enhancement. The clinical course in these cases is usually benign.

Paraneoplastic cerebellar degeneration (PCD) is an autoimmune process directed against the Purkinje cells in the cerebellum. In females, in whom the incidence is slightly higher, it is most commonly associated with gynecological tumors. In male patients it is associated with pulmonary and colorectal cancers as well as lymphomas. Autopsy shows almost complete loss of Purkinje cells, atrophy of the dentate nucleus, and glial proliferation. Typically, middle-aged patients develop a neocerebellar syndrome of subacute onset with cerebellar dysarthria and extremity/gait ataxia. During the diagnostic workup Purkinje cell (Anti-Yo), antibodies can be detected. Oligoclonal bands and signs of autochthonous IgG production are often found in CSF studies since the antibody is often produced intrathecally. Treatment is aimed at the underlying tumor disease since immunosuppressive therapies (corticosteroids, IVIG and plasmapheresis) usually do not better the cerebellar syndrome. Due to the nature of the underlying disease prognosis is bad in most cases, in PCD associated with lymphomas it is a little bit better.

Numerous endocrine or metabolic disorders such as hypothyroidism, hypoparathyroidism, vitamin deficiency (B1, B12, and E) as well as heavy metal intoxication (ie, lead, mercury, and thallium) can manifest with subacute ataxia. These have to be distinguished from inherited cerebellar ataxias.

Examples of autosomal-recessive ataxias are *Friedreich's ataxia (FRDA)*, *ataxia telangiectasia (AT)*, and *cerebrotendinous xanthomatosis*. Autosomal-recessive *FRDA* is the most common hereditary ataxia. Prevalence is 1–2/100,000 newborn. Diagnosis of FRDA is based on the following cardinal symptoms: Slowly progressive ataxia usually starting around puberty (age range 2–25 years), loss of vibrational/positional senses and deep tendon reflexes, followed by paresis particularly of the lower extremities. Common non-neurological manifestations include progressive cardiomyopathy, kyphoscoliosis, and diabetes mellitus. Most cases of *FRDA* are caused by loss-of-function mutations in the frataxin (FXN) gene located on chromosome 9. The vast majority of affected individuals have an expanded trinucleotide (GAA) repeat in intron 1 of the frataxin gene. The repeat expansion results in reduced transcription of the gene (ie, silencing) and decreased expression of the gene product frataxin. Frataxin plays a role in mitochondrial iron metabolism. By their late 20s, patients may be confined to a wheelchair. Death, often due to arrhythmia or heart failure, usually occurs by middle age. *AT* is the most common form of infantile-onset cerebellar ataxia. The defective gene in AT has been mapped to chromosome 11 and designated the ATM gene (for AT mutated). The gene mutation leads to defective DNA repair. Gait ataxia begins in the second year of life, followed by poor hand coordination and dysarthria. Associated symptoms include oculocutaneous telangiectasia, immune deficiency (leading to recurrent pulmonary infections), and an increased incidence of (lymphoreticular) malignancies. Due to this multisystem involvement AT is difficult to treat and has a poor prognosis. As with all hereditary ataxias, no specific treatment exists for the ataxic syndrome or the progressive cerebellar neurodegeneration. Patients die from recurrent pneumonia or malignancy at a median age of 25 years. Clinical symptoms in *cerebrotendinous xanthomatosis*, also called Refsum's disease, include cerebellar ataxia, severe polyneuropathy, and hearing loss. It is caused by mutations in the *PHYH* gene, leading to impaired α-oxidation activity and accumulation of phytanic acid, a branched chain fatty acid, in nervous tissues. Workup for this condition in selected individuals with sporadic adult-onset ataxia is important since a phytanic acid–restricted diet and regular plasmapheresis can lead to stabilization of the disease and sometimes even reverse the symptoms.

As of August 2015 autosomal-dominant *spinocerebellar ataxias (SCA)* can be subdivided into SCA 1–40 based on the responsible underlying Gen mutation. They constitute a genetically heterogeneous group of neurodegenerative disorders, all characterized by slowly progressive gait more than stance ataxia usually after age 25. Several types of SCAs (1, 2, 3, 6, 7, and 17) are associated with expansion of CAG repeats in the region that encodes for polyglutamine tracts in the protein products. Cerebellar atrophy with variable degrees of brain stem atrophy is the most common autopsy finding. Based on the particular subtype other distinguishing clinical features may include dysarthria, oculomotor impairment, peripheral neuropathy, seizures, retinal degeneration, tremor, and altered muscle tone. SCA 3 also known as Machado–Joseph disease is the most prevalent form of SCA in the USA. Autosomal-dominant *episodic ataxias (EA)* are characterized by short episodes of intermittent ataxia triggered by emotional or physical stress. Two subtypes with different mutations can be distinguished. *Episodic ataxia type 1 (EA-1)* is based on a mutation of the potassium channel gene KCNA1 on chromosome 12. It is characterized by episodic ataxic episodes lasting for seconds to minutes triggered by physical exertion or startle response during early childhood. These spells are often accompanied by myokymia (rippling of facial muscles). Mutations of the calcium channel gene CACNA1a on chromosome 19 manifest with *episodic ataxia type 2 (EA-2)*. In affected patients ataxic episodes last longer for hours to days, other symptoms include nystagmus and vertigo. Treatment attempts with acetazolamide or 4-aminopyridine have been successful in controlled trials in EA-2.

In contrast to this, in most cerebellar disorders therapy is only available if a causative noxious stimulus or treatable underlying metabolic condition can be identified. Supportive physical/speech/occupational therapy is indicated for all patients suffering from cerebellar ataxia to train and preserve remaining functional abilities.

Further Reading

[1] Klockgether T. Sporadic ataxia with adult onset: classification and diagnostic criteria. Lancet Neurol 2010;9(1):94–104.

CHAPTER

10

The Basal Ganglia

I. Noorani

University of Cambridge, Cambridge, United Kingdom

INTRODUCTION

How we generate our movements is a fundamental question in biology. Almost all of our responses to stimuli require some form of movement. A group of structures essential to producing this is the basal ganglia. The basal ganglia constitute a large subcortical motor system. Although the cerebellum and basal ganglia both influence movement by the same cortical areas, they do so via separate connections to the thalamus, and they tend to have opposing effects, for example the basal ganglia output is inhibitory whereas that of the cerebellum is excitatory.

The idea that the basal ganglia contribute to motor control dates back to the early 20th century, when the British neurologist S.A. Wilson described a condition with damage to the liver and putamen that causes rigidity, tremor, and weakness. Of course, we now know this as "Wilson's disease." Given that there were differences in this condition compared with damage to the corticospinal tracts, Wilson postulated the basal ganglia to be an "extrapyramidal system" in contrast to the "pyramidal" (corticospinal) system. Later, when modern anatomical techniques were developed, it was found that most basal ganglia outputs were to motor cortical areas. Therefore the basal ganglia can be seen as "prepyramidal" rather than extrapyramidal, and it is now believed that these nuclei are important for producing movements. Although it was initially believed that they were the initiators of motor programs, we now know this is unlikely to be the case and that the basal ganglia function at a later stage than initiation itself in the neural production of movement.

This chapter describes the anatomy of the basal ganglia and discusses ideas on how they influence movement and how lesions of specific areas lead to distinct clinical phenotypes.

ANATOMY

The basal ganglia receive numerous cortical inputs and they are processed to produce an output to the frontal lobes and brainstem, needed for planning and generation of movement. The terminology of basal ganglia regions has changed over the years and some structures are still referred to by more than one name, which is often a source of confusion. In this section, we introduce consistency in the nomenclature of the basal ganglia structures.

The structures of the basal ganglia include the striatum (caudate nucleus, putamen, and nucleus accumbens), the subthalamic nucleus (STN),

Conn's Translational Neuroscience
http://dx.doi.org/10.1016/B978-0-12-802381-5.00018-X

the globus pallidus (GPi = internal segment; GPe = external segment; and ventral pallidum), and the substantia nigra composed of the pars compacta (SNpc) and pars reticulata (SNpr). The striatum and STN receive most inputs from the cerebral cortex, although the striatum also gets inputs from thalamic nuclei. The majority of basal ganglia outputs originate from GPi and SNpr. These are inhibitory outputs to the thalamus and brainstem.

The striatum is found in the forebrain and is so-called because it appears striped owing to the axonal fibers passing through it. In primates, the caudate and putamen of the basal ganglia are separated by the internal capsule, although in rodents this is not so. There are four main types of neuron in the striatum: the medium spiny neurons (constitutes 95% of neurons in this region), large aspiny neurons, the medium aspiny cell, and the small aspiny cell.

Most areas of cortex have inputs to the basal ganglia via the striatum; more specifically these inputs are on to the dendritic spines of the medium spiny neurons with glutamate as the neurotransmitter. Somatotopical organization of motor and sensory cortex is preserved in the basal ganglia, which may allow for independence of functional circuits in the basal ganglia, for instance, oculomotor, cognitive, limbic, and somatosensory circuits. However, there is also a substantial amount of convergence and divergence of information, such as by diverse input to the large dendritic fields of medium spiny neurons from multiple cortical areas, which allows for integration of information from several different regions. The striatum gives a convergent inhibitory output to the GPi and SNpr.

The STN is at the junction of the diencephalon and midbrain and receives excitatory glutamatergic input from the frontal cortex. Such input is largely from motor regions, such as the frontal eye fields, premotor and supplementary motor areas, and primary motor cortex. The GPe also provides an inhibitory GABA input to the STN. From this description it is evident that there are two routes from the cortex to the GPi, GPe, and SNpr: one via the striatum and the other via the STN. The former is inhibitory and slow (15–20 ms), whereas the latter is excitatory and comparatively fast (5–8 ms).

The globus pallidus is medial to the putamen and rostral to the hypothalamus, and in primates is separated into the internal and external segments by fibers of the internal medullary lamina. The GPi output is an inhibitory GABAergic one, and 70% of this output is to the thalamus and brainstem. In contrast, the GPe mainly projects its output to the STN. This sets up an "indirect" striatopallidal pathway from the striatum to the GPe, then to the STN, and from there to the GPi. The contrasting "direct" striatopallidal pathway is from the striatum directly to the GPi.

In much the same way that the globus pallidus is in two parts, the substantia nigra is also divided into two. The pars compacta (SNpc) is densely populated with cells whereas the pars reticulata (SNpr) is not so well populated. The SNpr has inhibitory inputs from the striatum and excitatory inputs from the STN; it outputs to the medial part of the ventrolateral thalamus (VLm) and the magnocellular part of the ventral anterior thalamus (VAmc), which send projections to the prefrontal and premotor cortex. The lateral SNpr also has inhibitory outputs to the superior colliculus and medial dorsal thalamus (this projects to the frontal eye fields), and these connections are important for eye movement control to which SNpr contributes.

The SNpc is well known because it is the area whose dopamine-neurons degenerate in Parkinson's disease. "Substantia nigra" means "black substance" this region was named so because it appears black due to neuromelanin contained in dopamine-neurons here. The SNpc has inhibitory input from the striatum, and its output is a topographically arranged set of projections to the caudate and putamen. The dopamine neurons can alter activity in multiple circuits because they have inputs from one striatal circuit and output to the same as well as neighboring circuits.

MOTOR CONTROL THEORIES

It was widely believed that the basal ganglia initiate movement by sending motor commands to motor cortex for execution. This theory stemmed largely from clinical observations of patients with movement disorders: the lack of movement in Parkinson's disease was attributed to an inability to initiate movements because of substantia nigra degeneration, and conversely the excess and uncontrollable movements of ballism and chorea were thought to be caused by a release of motor systems from basal ganglia control leading to initiation of unwanted movements. However, some key observations led to this theory falling out of favor. In particular, the observations that basal ganglia damage does not delay movement initiation and that neural recordings in the basal ganglia indicate rather later activation of this region compared with other known areas important for triggering movements. There are several more modern theories regarding how the basal ganglia operate in relation to motor control, and although the field is still a controversial one, there are three main theories, which are currently prevalent and being intensely investigated. These are discussed further.

THE DIRECT AND INDIRECT BASAL GANGLIA PATHWAYS

The notion that the basal ganglia is composed of functionally opposing direct and indirect pathways has stimulated much research over the past few decades. This concept, although still regarded as rather controversial, has been helpful in interpreting deficits seen in clinical neurology disorders by providing a conceptual framework for understanding basal ganglia functioning. The direct pathway is an inhibitory one from the striatum to the GPi/SNpr, and so increased activity in this pathway leads to decreased GPi/SNpr output. The indirect pathway has more components than the direct one, traveling from the striatum to the GPe, then to the STN, and from there to the GPi/SNpr. This is a net excitatory pathway such that increased striatal activity enhances GPi/SPpr output via this pathway. It is proposed that these pathways are normally in balance, and by adjusting the activity in one pathway compared to the other, this balance can be shifted to trigger or inhibit the cortical targets of the basal ganglia. In this way, movement itself can either be facilitated or reduced, and it is argued that aberrations of the activity in these pathways can explain excess movement in chorea and ballism, for example, and reduced movement, such as in Parkinson's disease.

A useful example of how these distinct pathways may operate in parallel for functional responses is provided by a recent and elegant experiment in rodents. In the study, a task called "countermanding" was employed, in which the rat subject was required to make a response to a stimulus except in cases where a signal was also presented instructing the rat not to make an impending movement. Such a task is a model of an important aspect of our everyday behavior: stopping a response we are about to make because it has become out of date. One only has to think about a scenario where one is going to step on the gas pedal of the car as the green light is about to appear when suddenly a pedestrian enters the street: in this instance, one must immediately cancel their upcoming intention of accelerating the car if they are not to harm the pedestrian! This experiment is therefore a succinct way of modeling such cancellation responses. Using neurophysiological recordings, it was demonstrated that in trials where the stop signal was presented, the STN was active shortly after presentation of the stop signal, irrespective of whether the movement was subsequently canceled or not. In contrast, neurons in the downstream SNpr were active only in trials where the movement was successfully canceled, suggesting a specific pathway from STN to SNpr for canceling responses when needed. It was similarly demonstrated that on trials where the rat was supposed to move, there was higher striatal input to the SNpr, implying that this pathway from striatum to SNpr generates the "go" movements. One can therefore imagine a competitive "race" between these two pathways: the winner of the race, that is whichever provides the dominant input to the SNpr, determines whether the upcoming movement is canceled or not. It is certainly conceivable that this competition between basal ganglia pathways may underlie similar behaviors outside of countermanding. For example, antisaccades are saccadic eye movements in the opposite direction to a stimulus, in which one must stop their natural tendency to make an eye movement toward the stimulus in order to successfully make the antisaccade itself. It has thus been suggested that competition between these two basal ganglia pathways in a race manner may similarly be responsible for whether a subject makes antisaccades correctly. This is important clinically because it has been noted that patients with Parkinson's disease display increased error rates in the antisaccade task, which is likely related to the basal ganglia dysfunction impairing the operation of these pathways.

THE PRODUCTION OF LEARNED MOVEMENT SEQUENCES

Marsden proposed that the basal ganglia are important for producing learned sequences of movements. There are many examples of where movement sequences as opposed to individual movements are prominent, such as using a knife and fork to cut a piece of food and place it in your mouth, or pressing on the clutch of a car with your foot before shifting gear with your hand and then stepping on the accelerator. Such learned movement sequences are omnipresent. Notably, patients with Parkinson's disease, who have an overall loss of GPi output due to reduced striatal input, do indeed have more difficulty with producing movement sequences than they do with individual movements. This can be tested experimentally by asking patients to generate sequences or individual movements, and they are slower in the former than the latter. This has also been used to explain the micrographia, which these patients display: their writing gets smaller and slower the more they do it, and given writing is indeed a fine and lengthy sequence of dexterous movements, having this problem would be compatible with the hypothesis that the basal ganglia are required for sequential movements.

INHIBITING COMPETING MOTOR PROGRAMS

Another popular hypothesis on basal ganglia functioning is that they tonically inhibit their targets in the cortex and brainstem, and they facilitate movement by releasing this inhibition for the desired movement set and correspondingly increase their inhibition of potentially competing movement patterns. An example would be for turning one's head toward a pedestrian while driving on the road: one needs to turn one's head, and at the same time stop certain posture-holding mechanisms, such as the tonic neck reflexes. The proposed mechanism in which this happens is that the direct basal ganglia pathway enables the desired motor program to be activated, while the indirect pathway is activated in parallel to disinhibit the GPi, thus inhibiting the potentially competing motor paradigms.

EFFECTS OF BASAL GANGLIA LESIONS

Although clinical disorders of the basal ganglia are very valuable in establishing the functions of these nuclei, it can be difficult to tease apart the roles of specific nuclei given that such disorders often affect several of these. In this regard, inducing focal lesions in the basal ganglia in animal models are extremely useful for learning about their functionality.

Unilateral lesioning of the striatum, in particular the putamen, produced slowed movement of the contralateral limb, but with intact reaction time indicating that the initiation of movement is unaffected. Bilateral and large striatal lesions result in the paucity of movement, bilateral reduction in the speed of movement, and abnormalities of posture, although these results are not always replicated.

Pharmacological and electrolytic lesions of the STN in monkeys or humans result in hemiballism: large-scale, random, and sudden involuntary jerking movements of the contralateral limb, with often normal voluntary movements. A possible explanation for this behavior is that lesioning the STN causes reduced excitatory input to the GPi, resulting in disinhibition of cortical and brainstem motor control regions. In support of this, lesioning the GPi in hemiballism or chorea patients helps reduce these involuntary movements.

As the GPi and GPe are immediately next to each other, it is difficult to specifically lesion one of these nuclei without damaging the other and this is a potential confounding factor with many studies attempting to study the effect of one of these nuclei. However, in most cases, damaging both of these nuclei gives similar effects to lesioning the GPi alone, which is the slowing of contralateral movement but normal reaction time (and therefore normal movement initiation).

The SNpr can be inactivated without affecting other basal ganglia nuclei using the GABA agonist muscimol. This causes involuntary saccades (rapid eye movements for gaze shifting) and hence the inability to maintain gaze fixation. The SNpr normally inhibits oculomotor control programs in the superior colliculus, and inactivating the SNpr causes disinhibition of the superior colliculus and therefore excess saccades.

Damage to the SNpc causes the symptoms and signs of Parkinson's disease, which is covered in the next section.

DISORDERS OF THE BASAL GANGLIA

The list of known disorders of the basal ganglia is expanding as our knowledge of the functions of this region increases. Originally, such diseases were fundamentally motor disorders, but it is now known that more cognitive disorders and also emotional and cognitive deficits in primarily motor disorders, such as Parkinson's disease are due to abnormalities in the basal ganglia. In this section, the main disorders of the basal ganglia are discussed.

Parkinson's disease (PD) is one of the commonest progressive neurodegenerative disorders, and is characterized by slowness and paucity of movement, rigidity, and a low-frequency 4–6 Hz typically unilateral resting tremor. The origin of this disease has been linked to degeneration of the dopamine neurons of the SNpc, and damaging these neurons in animal models produces a Parkinsonian phenotype. When the striatal dopamine deficit reaches around 70%, there is a shift in the striatopallidal activity toward the indirect circuit, causing excessive STN activity, and thus increased stimulation of the GPi and SNpr. The higher output from these areas excessively inhibits the thalamocortical projection, with resultant cortical activation needed for the execution of automatic and voluntary movements. The degenerating neurons commonly contain aggregating proteins called Lewy bodies, and the death of these neurons may be linked to oxidative stress and mitochondrial dysfunction. Although the majority of PD cases are sporadic in nature, genetic abnormalities have been linked to the disease; in particular there are clear associations with α-synuclein and parkin genes. Despite being a classical motor disorder, PD has other symptoms including a dysexecutive syndrome, wherein patients have difficulties in planning, problem solving, shifting attention, and suppression of automatic responses; and this syndrome has been associated with dopaminergic loss and the striatofrontal loop disturbance. The disease can be treated by dopaminergic agonists or supplementing dopamine as L-dopa, although these do not delay treatment progression. Deep brain stimulation, usually of the subthalamic nucleus, has proven to be an effective treatment for the movement disorder in selected and otherwise medically refractory cases of Parkinson's disease.

Huntington's disease is another basal ganglia disorder, however, patients with this have a very different syndrome compared with Parkinsonian patients. In Huntington's disease, patients have the dyskinetic movements of chorea, as well as behavioral disorders and cognitive problems. The primary pathology is degeneration of the medium spiny GABA neurons in the striatum, in addition to loss of neurons in cortical layers V and VI. This results in the disruption of network activity in basal ganglia–cortical pathways. The underlying cause is an expanded polyglutamine-coding CAG repeat in the mRNA from the huntingtin gene. The disease tends to affect those of middle-age and is unfortunately fatal with a prognosis of up to 20 years.

Another hyperkinetic disorder linked to basal ganglia dysfunction is myoclonic dystonia, 30–50% of which cases have mutations in the epsilon-sarcoglycan gene (SGCE). There is a high incidence of neuropsychiatric features in this disease if such mutations in SGCE are present, including anxiety and obsessive-compulsive disorder. These features may well be due to dysregulation of pathways from the striatum to the cortex as part of the limbic system, which disrupts the functioning of structures associated with emotion, including the orbitofrontal cortex and anterior cingulate. The movement disorder seen in this patient group involves proximal and axial muscle dystonic spasms and distal myoclonic jerks, which are ameliorated by drinking alcohol.

Tourette's syndrome involves impulsive tics and commonly neuropsychiatric disturbances, such as obsessive-compulsive disorder and attention deficit hyperactivity disorders. It is thought to be a disorder in which there is abnormal release of inhibition of inappropriate movements or behaviors. The tics are often preceded by feeling uncomfortable, which is relieved by carrying out the tic or repeating it many times. Research demonstrates that inhibitory GABAergic pathways in striatum and cortex are deficient in these patients, generating these unwanted movements and behaviors, and there is ongoing interest in using deep brain stimulation of the globus pallidus internus as a potential treatment.

In this chapter, we have seen how the basal ganglia are organized anatomically and the functions they perform in generating voluntary, controlled movements. Current understanding is that the basal ganglia are arranged as functional circuits in which there is a major direct and an indirect pathway to the thalamus, which then outputs to cerebral cortex for motor control. Such knowledge is useful in clinical practice for relating the underlying pathology of movement disorders, such as Parkinson's disease and Huntington's disease to the clinical phenotype, using our understanding of these basal ganglia–cortical pathways. However, it is increasingly believed that the basal ganglia contribute to more than motor control; in particular, they may regulate more cognitive functions and executive behaviors, giving possible explanations for the mechanism behind certain neuropsychiatric disorders, for example obsessive-compulsive disorder. Much still remains to be explored in this field, with major outstanding questions being whether the basal ganglia inhibit competing motor programs and if so, how they may do this in relation to their known circuitry. Nevertheless, there are concerted ongoing efforts to explore such questions further, and research promises to generate more knowledge on this subject, which will undoubtedly be of benefit for treating a vast array of movement disorders in neurology.

Further Reading

[1] Wilson S. Progressive lenticular degeneration: a familial nervous disease associated with cirrhosis of the liver. Brain 1912;34:295–509.

[2] Marsden CD. The mysterious motor function of the basal ganglia: the Robert Wartenberg Lecture. Neurology 1982;32:514–39.

[3] Albin RL, Young AB, Penney JB. The functional anatomy of basal ganglia disorders. Trends Neurosci 1989;12:366–75.

[4] Hikosaka O, Isoda M. Switching from automatic to controlled behaviour: cortico-basal ganglia mechanisms. Trends Cogn Neurosci 2010;14:154–61.

[5] Nelson AB, Kreitzer AC. Reassessing models of basal ganglia function and dysfunction. Annu Rev Neurosci 2014;37:117–35.

[6] Obeso JA, Rodriguez-Oroz MC, Stamelou M, Bhatia KP, Burn DJ. The expanding universe of disorders of the basal ganglia. Lancet August 9, 2014;384(9942):523–31.

[7] Noorani I, Carpenter RH. Basal ganglia: racing to say no. Trends Neurosci September 2014;37(9):467–9.

CHAPTER

10.1

Translational Correlation: Disorders of the Basal Ganglia

H. Wilms, A.O. Claudio

Texas Tech University Health Sciences Center, Lubbock, TX, United States

INTRODUCTION

Movement disorders have been linked to the intricate connection between the basal ganglia and other areas of the CNS, specifically thalamus and frontal cortex. In summary, the direct striatopallidal projection system constitutes a positive while the indirect system constitutes a negative feedback system to precentral motor areas. Therefore, disturbances in basal ganglia function can manifest as either hyperkinetic (excessive involuntary movements) or hypokinetic (decreased motion) movement disorders. *Hypokinetic movement disorders* are thought to arise from the degeneration of dopaminergic neurons resulting in the loss of striatal dopaminergic stimulation as seen in Parkinson's disease (PD). This leads to a preponderance of the indirect striatopallidal system with increased inhibition of the globus pallidus internus and disinhibition of the subthalamic nucleus. As a consequence, increased stimulation of excitatory efferent projections from the basal ganglia (pars regulata of the substantia nigra, globus pallidus internus) occurs leading, via increased inhibition of thalamic nuclei, to decreased excitatory efferent projections to the frontal cortex. Chorea, dystonia, myoclonus, and tics are examples of *hyperkinetic movement disorders*. It is postulated that in these, the direct striatopallidal system predominates, which results in the disinhibition of thalamic efferents to frontal cortical areas.

HYPOKINETIC MOVEMENT DISORDERS

Parkinson's Disease

With an incidence of 100–200/100,000, PD is one of the most common neurological disorders. Prevalence is 0.2% in the seventh decade; it rises to 2.5% in the ninth decade of life. The etiology of PD is not yet fully understood, genetic and environmental/toxic factors play a role: families in whom Parkinsonism is inherited in either an autosomal dominant, autosomal recessive, or X-linked mode have been described. In these families, 15 Gen loci (PARK 1–13) have been identified. They contribute to an earlier age of onset in familial Parkinsonism and are often associated with higher grades of

Conn's Translational Neuroscience
http://dx.doi.org/10.1016/B978-0-12-802381-5.00019-1

cognitive impairment as compared to sporadic cases. Regarding environmental risk factors, Parkinsonian syndromes can be induced in animal experiments in primates by exposure to pesticides/herbicides such as rotenone and paraquat.

Motor symptoms in PD are thought to arise from degeneration of dopaminergic neurons in the substantia nigra, leading to low levels of dopamine at striatal receptors. However, other neurotransmitters are involved in PD as well, since low levels of acetylcholine have been linked to the development of dementia, while deficiency of serotonin contributes to depressive symptoms in PD.

Cardinal motor symptoms of PD are bradykinesia, rigidity, tremor, and postural instability. *Bradykinesia*, defined as slowness of movement execution, manifests in activities of daily living such as dressing and eating, which take PD patients longer to perform. Bradykinesia can be evaluated clinically by asking the patient to perform standardized movements such as tapping his thumb and index finger in rapid succession: The observed movement is noted for reduced amplitude, slowing, and early fatigability. Moreover, bradykinesia can manifest as a soft, low-volume voice (*hypophonia*), loss of facial expression with development of a poker face (*hypomimia*), and small handwriting (*micrographia*). PD patients adopt a stooped posture and short-stepped, narrow-based, shuffling gait with reduced arm swing (*hypokinetic gait*). With passive movements of the arm, a "waxy, lead pipe-like" increase in muscle tone or *rigidity* is observed. In contrast to spasticity, an increase in muscle tone seen after upper motor neuron lesions, rigidity can be detected through the whole range of motion (*cogwheel phenomenon*). *Tremor* in PD is characteristically a 4–5 Hz *rest tremor*, which can be distinguished from the lower-frequent 3 Hz intention tremor in cerebellar lesions and the higher-frequent 10–12 Hz postural/action tremor in hereditary, essential tremor. *Postural instability* in PD manifests as a feeling of unsteadiness when walking. Clinically, it can be tested for using the "pull test" in which the standing patient is briskly pulled backward by his shoulders and the degree of compensation for the sudden imbalance is noted. Optional nonmotor symptoms include sensory deficits (dysesthesia/pain), dysfunction of the autonomous nervous system (orthostatic hypotension, urinary incontinence), and cognitive deficits (frontal-executive dysfunction early on, dementia in advanced stages).

Drug treatment in PD is aimed at restoring striatal dopamine levels through different mechanisms: since dopamine does not pass the blood–brain barrier, its precursor *L-dopa*, which is metabolized in the central nervous system (CNS) into dopamine, is prescribed. It is combined with *carbidopa*, a decarboxylase inhibitor, which blocks L-dopa conversion in the periphery to reduce systemic side effects (*carbidopa/levodopa*). *Dopamine agonists*, acting as direct stimulators at postsynaptic receptors, are the drugs of choice in young-onset PD. Alternative treatment strategies are aimed at blocking the breakdown of dopamine (monoamine oxidase B [MAO-B] inhibitors; catechol-O-methyl transferase [COMT] inhibitors). While initially all hallmark symptoms of PD respond favorably to aforementioned dopaminergic therapy (honeymoon phase), after 4–8 years of treatment, motor complications characterized by paradoxical, hyperkinetic movements (ON-dyskinesia) occur. These are thought to arise from the nonphysiological stimulation of dopamine receptors by the oral dopamine replacement therapy. At this disease stage, neurosurgical intervention should be considered: Historically, ablative procedures such as pallidotomy and thalamotomy have been used to alleviate PD symptoms. These have largely been replaced by chronic electrical deep brain stimulation targeting either the globus pallidus internus or subthalamic nucleus.

HYPERKINETIC MOVEMENT DISORDERS

Chorea

Choreatic movement disorders (from Greek "choreia," dance) are characterized by rapid, involuntary, purposeless movements that flow from one part of the body to another. Typically, they are amplified by intentional movements or emotional stress and subside during sleep or general anesthesia. Primary chorea is most commonly seen in *Huntington's disease* (*HD*), which with a prevalence of 2–10 patients/100,000 inhabitants is not a rare disease, but the most common inherited movement disorder. It is inherited in an autosomal dominant mode with full penetrance: HD is caused by an unstable trinucleotide (CAG) repeat expansion (>38 repeats) in the gene encoding for Huntingtin on the short arm of chromosome 4. While Huntingtin is a ubiquitous protein throughout the body, pathology appears to be limited to the CNS, with atrophy of the caudate and putamen (the neostriatum) being the most prominent. As a consequence, GABAergic projections to the globus pallidus and substantia nigra are reduced. This leads indirectly to decreased inhibition of thalamic nuclei and via increased cortical activation to chorea. The pathophysiology of neuron loss in HD is incompletely understood. Huntingtin protein is thought to become toxic with the CAG expansion ("gain of function"). At the cellular level, protein aggregates are seen both in the cytoplasm and nucleus. Age of disease onset is usually between 40 and 50 years although HD can manifest during childhood (Westphal variant). Anticipation, that is earlier disease manifestation in descendants, is particularly seen with inheritance of the mutation through the father. During the early disease stage, features of chorea include mild grimacing, involuntary contractions of eyebrows, and shoulder shrugging. At this point HD patients typically try to integrate the developing jerky movements of arms and legs into pseudo-purposeful movements (*parakinesia*) to conceal the movement disorder. With progression of the disease, nonchoreatic motor symptoms such as eye movement abnormalities, dystonia, dysarthria, and dysphagia develop. Psychiatric symptoms are another clinical hallmark of the disease: Increased irritability and social disinhibition manifest early, followed by depression, anxiety, apathy, and cognitive decline. HD is a relentlessly progressing, chronic neurodegenerative disease with an average survival time of 12–15 years. While no cure exists, in 2008 the FDA approved tetrabenazine for the treatment of Chorea in HD. Its mechanism of action involves blocking the transport of dopamine into vesicles in the presynaptic terminal. This depletes dopamine from the vesicles and reduces dopamine transmission. Historically, dopamine receptor antagonists (neuroleptics) have been used for the treatment of chorea, as well. *Sydenham chorea*, an immune-mediated complication of rheumatic fever, is the most common cause of acute secondary chorea in children worldwide. Typically, children develop generalized chorea and behavioral changes several weeks after streptococcal pharyngitis due to structural similarities between the basal ganglia antigen and streptococcal antigens. In contrast to HD, Sydenham chorea is a self-limited illness with good recovery. Other infectious causes of chorea include bacterial/viral encephalitis and HIV infection.

Dystonia

Dystonia is a hyperkinetic movement disorder characterized by intermittent, tonic or phasic, muscle contractions leading to abnormal postures or repetitive movements. While hyperkinetic movements in chorea are unpredictable and chaotic, dystonic movements are patterned, repeatedly involving the same muscle groups. Dystonia is classified based on the number of body parts affected. Focal dystonia involves a single part of the body: *cervical dystonia* (*spasmodic torticollis*) is characterized by painful muscle contractions of neck muscles leading to either intermittent neck flexion (*antecollis*), neck extension (*retrocollis*), or sideways rotation of the head (*laterocollis*). Intermittent involuntary eyelid closure (*blepharospasm*) and grimacing of the lower parts of the face, usually involving mouth, jaw, and platysmal muscles (*oromandibular dystonia*), are other examples of focal dystonia. Multifocal dystonia involves two or more noncontiguous body regions while involvement of two or more contiguous body regions is referred to as segmental dystonia. In regard to etiology, dystonia can be divided into primary (idiopathic, hereditary) and secondary (symptomatic) dystonia. Autosomal dominant early onset dystonia (<26 years old) is the most common inherited form of primary dystonia. It has been linked to a GAG deletion within the DYT1 gene on chromosome 9 causing abnormal function in the ATP-binding protein of torsin A. The function of this protein is still not fully understood, but it is believed to play a role in the secretory pathway of synaptic vesicles within the basal ganglia. Secondary dystonia can result from structural basal ganglia lesions (stroke, tumor, and infection), which usually involve the putamen.

Chemodenervation of overactive muscles by repeated injections of botulinum toxin A is the treatment of choice in focal dystonia. Botulinum toxin A is a potent neurotoxin produced by *Clostridium botulinum*. It acts as a zinc endopeptidase, cleaving specific proteins involved in vesicular fusion, thereby interfering with acetylcholine release at the neuromuscular junction, leading to weakness 1–2 weeks after intramuscular injection. While the nerve terminal is irreparably damaged, reinnervation of the paralyzed muscle occurs and reinjection has to be repeated every 8–12 weeks. In generalized dystonia, oral drug treatment or neurosurgical intervention are the treatments of choice. Oral medications used in management of dystonia include benzodiazepines (clonazepam), dopamine depleting agents (tetrabenazine), and GABA-agonists (baclofen). However, the anticholinergic trihexyphenidyl is the best studied and most widely used oral drug in generalized dystonia. Patients with severe generalized dystonia, who failed or cannot tolerate oral drug treatment, might be suitable candidates for deep brain stimulation targeting the globus pallidus internus.

Myoclonus

Myoclonus is a sudden, brief, lightning-fast involuntary movement caused by either contraction (*positive myoclonus*) or brief inhibition (*negative myoclonus*) of facial, truncal, or extremity muscles. Hiccups and sleep jerks are examples for *physiological myoclonus*. Myoclonus is categorized based on the side of occurrence as *cortical* (involving a focal muscle group), *subcortical* (generalized), *spinal* (confined to one dermatome), or *peripheral* (within distribution of an irritated nerve) myoclonus. *Essential myoclonus*, an autosomal-dominant inherited disease, is characterized by generalized myoclonus involving arms and axial muscles. The onset is usually during the youth of the patient; the clinical course is benign. Different transmitters are involved in the emergence of myoclonus. Lack of inhibitory glycine is thought to be involved in some forms of spinal myoclonus. GABA deficiency most likely contributes to the development of cortical and subcortical myoclonus, explaining why treatment attempts with GABAergic substances such as benzodiazepines or valproic acid are often helpful.

Tics

Tics are sudden, purposeless brief stereotypical motor movements and/or vocalizations. *Simple tics* such as wrinkling one's nose, blinking, or clearing one's throat are distinguished from *complex tics* with involvement of multiple muscle groups (trunk flexion, gestures). If pharyngeal, laryngeal, or oral muscles are affected, vocalization can occur (*vocal tic*).

In contrast to other hyperkinetic movement disorders, tics can be temporarily suppressed by the patient. Characteristically, this is often associated with an urge to make the movement (*sensory tic*). In most cases of tic disorder, the etiology remains unclear (*primary, idiopathic tics*). However, tics can be induced by street drugs or can occur after CNS trauma/encephalitis. An autoimmune origin is suggested by detection of antineuronal serum antibodies directed against basal ganglia tissue in single cases. *Transient tic disorder*, lasting for more than one but less than 12 months, affects up to 15% of children, boys more often than girls. Per definition it involves motor or vocal tics but not both. In contrast, in the more severe, chronic *Gilles de la Tourette syndrome*, multiple motor tics and at least one vocal tic are associated with an obsessive-compulsive and attention deficit hyperactivity disorder. Tourette syndrome is thought to be a hereditary condition (autosomal dominant with incomplete penetrance). When considering drug therapy for tics, the often benign, transient character of the movement disorder should be considered and weighed against possible side effects. Drug treatment with neuroleptics (fluphenazine, pimozide) is only indicated when chronic tics interfere with activities of daily living and school/job performance.

CHAPTER

11

Choroid Plexus–Cerebrospinal Fluid Transport Dynamics: Support of Brain Health and a Role in Neurotherapeutics

C.E. Johanson

Alpert Medical School at Brown University, Providence, RI, United States

INTRODUCTION

Even though having a mass about 500 times smaller than brain, the adult choroid plexus (CP) performs an astonishing number of functions that support the health of neurons and glia and their recovery following injury. Long recognized has been the hydrodynamic role of CP to form the cerebrospinal fluid (CSF) that percolates throughout the entire central nervous system (CNS). Much less appreciated have been the multiple ways that the CP–CSF furnishes neurotrophins and growth factors for brain growth and restoration as well as hormones and proteins that mediate a wide spectrum of actions on neuronal networks. Knocking out certain genes expressed by CP epithelial cells, with their many homeostatic functions, can devastate cerebral metabolism and behavior.

This chapter addresses the micro- and macrostructures of CP and the ventricles, the array of transporters that affect movement of water, solutes, and cells between plasma and CSF, the circulation of CSF and other aspects of hydrodynamics, multifunctional aspects of the CP–CSF nexus, injuries to the blood–CSF barrier (BCSFB) and ventricular wall, and the promising role of translational therapies to promote brain recovery by manipulating the CP–CSF dynamics in many disorders. Several comparative points are made of BCSFB versus blood–brain barrier (BBB), but the former is emphasized in this chapter.

NEUROANATOMIC AND HISTOLOGICAL ASPECTS OF CHOROID PLEXUS–CEREBROSPINAL FLUID

When viewed by scanning electron microscopy, the CP organ three dimensionally takes on the appearance of a cauliflower. By this analogy, the individual "flowers" correspond to the separate fronds of CP that are grouped together in clusters. Each leaf-shaped frond consists of an outer single layer of choroid epithelial cells surrounding an inner core of interstitium

Conn's Translational Neuroscience
http://dx.doi.org/10.1016/B978-0-12-802381-5.00020-8

vasculature. A stalk-like structure attaches the CP base, where cells proliferate, to the ventricular wall ependyma.

Normally dangling in the ventricles from a foundation affixed to a brain–ependymal transition zone, the CP spreads out widely like a fishnet in the sea of CSF. Pulsations of CSF cause slight to-and-fro movement of the CP, thus maximizing the tissue exposure to potentially exchangeable materials in the ventricles. This continual motion of CSF is dynamically important for mixing substances both secreted and reabsorbed at the blood–CSF interface. Lateral ventricle CP lies on the floor of the cavities (curving with the caudate), while third and fourth choroidal tissues are attached to the roofs of their respective ventricles.

In humans, the total CP system weighs about 3 g. In rats, the CP collectively has a mass of about 0.004 g, and in mice an order-of-magnitude less. The diminutive size of CP in rodents and its relative inaccessibility situated deeply within the ventricular system have discouraged direct physiological experimentation. Consequently, CP transport modeling has lagged that of other epithelia, for example, kidney and intestine. Moreover, research of the BCSFB has been less extensive than that of the BBB. However, these gaps in research knowledge on the CP are rapidly being filled by current investigators who increasingly recognize the importance of integrating BCSFB and BBB data when modeling CNS homeostasis and disease.

Blood flows to the choroid plexuses at a brisk rate. Vascular perfusion of the plexuses is about 5–10 times the mean cerebral blood flow. High-level transport and synthetic activities in CP epithelial cells require great vascular perfusion to supply essential substances for the abundant metabolism and manufacturing. Disrupted CP blood flow seriously interferes with a plethora of CSF homeostatic processes. From fetal life to advanced aging, the brain counts upon a robust CP–CSF system to provide life-sustaining growth factors, hormones, enzymes, and micronutrients.

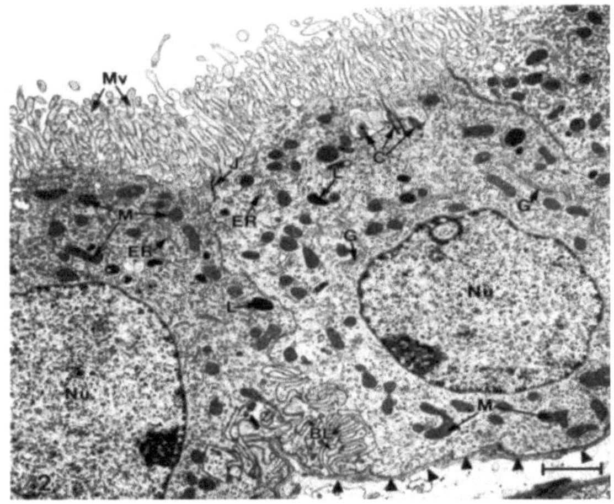

FIGURE 11.1 Ultrastructure of typical mammalian choroid plexus epithelial cells: the fine structure of the epithelium from an untreated adult Sprague–Dawley rat lateral ventricle CP (image presented) has similar size and configuration as human counterpart. Portions of three cells are depicted. Tissue was fixed for electron microscopy with OsO_4. The hallmark feature is an abundant microvillous (Mv) apical membrane, immersed in CSF, providing an enormous surface area for molecular exchange between epithelial cells and ventricles. Numerous mitochondria (M) provide energy-rich ATP for transport and secretory activity. ER and G, respectively, refer to endoplasmic reticulum and Golgi synthetic apparatus. Basal labyrinth (BL) intertwines the membranes of adjacent cells. The oval nucleus (Nu) contains a nucleolus. Basal lamina (see *arrowheads*) divides the cell above from the ISF below. *C*, centriole scale: bar is 2 μm, *J*, tight junction. *From Johanson C. Cerebrospinal Fluid Res 2008;5:10.*

Structure intimately relates to function. Therefore, analyzing CP epithelial ultrastructure (Fig. 11.1) reveals a lot about BCSFB work. First and foremost, the CP specializes in solute and water transport. Transporters need considerable membrane surface area to move solutes in a high-capacity fashion. In the choroidal epithelium, a substantial surface area is provided by lush microvilli facing the CSF, and by an extensively intertwining basolateral membrane facing the interstitial fluid (ISF) between and under the epithelial cells.

Mitochondria provide energy to translocate molecules among BCSFB compartments (Fig. 11.1). Choroid cell cytoplasm is studded with mitochondrial organelles. A high density is observed at the apical side to provide energy substrate, ATP, to a myriad of CSF-facing transporters and to the nearby endoplasmic reticulum/Golgi complexes with their synthetic machinery. The general appearance of CP epithelial ultrastructure resembles the proximal tubule. Similar to the kidney, the CP transfers copious water and substrates to maintain a stable extracellular (CSF) environment.

FIGURE 11.2 Cross-sectional schema for a choroid plexus villus: diagram is reconstructed electron micrograph of an adult CP villus, comprised of an epithelial cell ring encompassing the inner interstitial–vascular core. Innermost capillaries lack tight junctions, allowing penetration of water-soluble agents including proteins. Endothelial cells (*unfilled arrowheads*) surround capillaries that contain red blood cells (RBC). Intermediate interstitial fluid (ISF) zone contains extracellular matrix components, including collagen and fibroblasts. ISF accommodates dendritic cells and macrophages–monocytes that conduct immunological surveillance and effector actions impacting CSF–brain immune status. Epithelial cells are fastened together by tight junction structures (see *arrows*), the true anatomic substrate of the BCSFB. Each cilium has motor (whip-like flagellar) or sensory function. Microvilli (*arrowheads*, black fill) stud the apical CSF-facing membrane, providing substantial surface area to mediate the transport/secretory phenomena undergirding CSF formation and biochemical homeostasis.

BLOOD–CSF BARRIER VERSUS BLOOD–BRAIN BARRIER: FUNDAMENTAL DIFFERENCES

Some authors consider the BCSFB as part of the generic BBB. As a first approximation, it is proved that both barriers have tight junctions (zonulae occludentes) that restrict the diffusion of most water-soluble solutes into the CNS. But the similarity stops there, because CP and cerebral/spinal microvessels have widely diverging functions. Choroidal capillaries are highly permeable; cerebral capillaries are markedly impermeable. The parenchymal cells of the BCSFB are the choroid epithelium; on the other hand, BBB is comprised primarily of endothelium. Astrocyte feet covers and modulates most of the BBB, but not the BCSFB.

Physically, the configuration of CP with parenchymal epithelial cells (Fig. 11.2) markedly differs from capillary wall histology, with the neurovascular unit of the BBB-CP, unlike the BBB, does not contain enveloping pericytes and astrocyte foot processes. Rather, the choroidal cells stand "independently" of ensheathments, and arrange as a single epithelial layer circumferentially surrounding the inner interstitial-vascular core (Fig. 11.2). This villous-like arrangement of CP, quite unlike the BBB/capillary endothelial wall, is anatomically suited for a great turnover of fluid.

CP specializes in forming a large volume of CSF. Contrastingly, the brain capillaries secrete ISF slowly by a process insensitive to carbonic anhydrase inhibition by acetazolamide. This great difference points CP as the main driver of fluid movements in the CNS. Moreover, the cortical subarachnoid CSF flow along penetrating arteries into Virchow–Robin spaces not only promotes mixing with brain ISF but also the eventual clearance distally along glia-enveloping veins that exit the brain. Such fluid clearance is designated as the glymphatics. In this fashion, CSF originating in CP flows, in part, to superficial cortical sites apparently to promote extracellular fluid clearance via the glymphatic system.

Perhaps the greatest differences between BCSFB and BBB exist in vast arrays of transporters that exchange water, solutes, and peptides throughout the CNS. Endothelial transporters mainly carry glucose, amino acids, and free fatty acids across BBB to nearby neurons and glia. On the other hand, BCSFB transporters deliver plasma glucose and organic acids for the local CP epithelial metabolism and also provide growth factors, hormones, and peptides to CSF for global brain distribution.

Intriguingly, water traffic into the CNS is differentially regulated via distinctive isoforms of water channels in the two major barrier systems. Aquaporin 4 (AQP4) facilitates water diffusion bidirectionally across the BBB astrocyte feet. On the other hand, AQP1 controls water fluxes via CSF-facing membrane channels in choroid epithelial cells. Isoforms one and four allow differential expression and thus control of water fluxes that is region-specific. Water exchange at CNS transport interfaces is dynamic. It is important to distinguish rapid exchange of water (BBB) from the net formation of the salt-rich fluid at the BCSFB.

Permeability differences exist in barrier tight junctions. The zonulae occludentes of BBB are extremely tight. They are a crucially important line of defense for the immediate neuronal environment. Tight junctions in the BCSFB are the "leaky type." Therefore, the electrical impedance of CP is 10 times less than the impervious BBB. The more permeable BCSFB allows slow leakage of plasma proteins into CSF. Due to dissimilar molecular constituents of tight junction complexes, the BCSFB and BBB are differentially vulnerable to pathophysiological breakdown caused by physicochemical insults to the CNS.

BASIC TRANSPORT AND SECRETORY MECHANISMS IN CHOROID PLEXUS EPITHELIUM

Na^+ pumping on the CSF side of CP impacts fluid dynamics within the CNS. In addition, Na^+ pumping indirectly enables carrier-mediated transport of plasmaborne vitamins and other substances relying on the steep transmembrane Na^+ gradient for active accumulation in CSF. Hydrolysis of membrane-bound Na–K-ATPase generates high-energy P_i for producing ATP to transport Na^+ electrochemically uphill from choroid cell to CSF. Na^+ extrusion across the apical membrane lowers the epithelial concentration to about 20–30 mM. At the steady state then, there is set up a downhill Na^+ chemical gradient (directed inward across the basolateral membrane) of >100 mM; moreover, the electrochemical gradient downhill for cationic Na^+ is even steeper since the choroid cell interior is about 45 mV negative to ISF. This enables basolateral cotransporters (eg, Na-glucose, Na-ascorbate, and Na-HCO_3) to use the Na^+ gradient to carry organic substrates and anions into the cell. Thereafter, other carriers in the CP apical membrane complete the delivery of intracellular substrates, for example, ascorbate and HCO_3^- into CSF.

Na^+ pump activity is also integral to CSF formation. The primary step in generating CSF is active Na^+ extrusion into the ventricles. Agents such as digoxin directly inhibit the Na^+ pump and thereby lower CSF formation rate. Other drugs, like amiloride, exert an indirect effect on Na^+ pumping by blocking the basolateral Na^+ channel. Such blockage allows less Na^+ into the cell (from the interstitium) to feed the Na^+ pump on the other side of the cell facing the CSF. Other diuretics (eg, acetazolamide and benzolamide) alkalinize choroid cell pH (by elevating HCO_3^-), thereby slowing down basolateral Na–HCO_3 transport into the cell. Accordingly, this diminishes availability of interstitial Na^+ for key Na-K-ATPase activity.

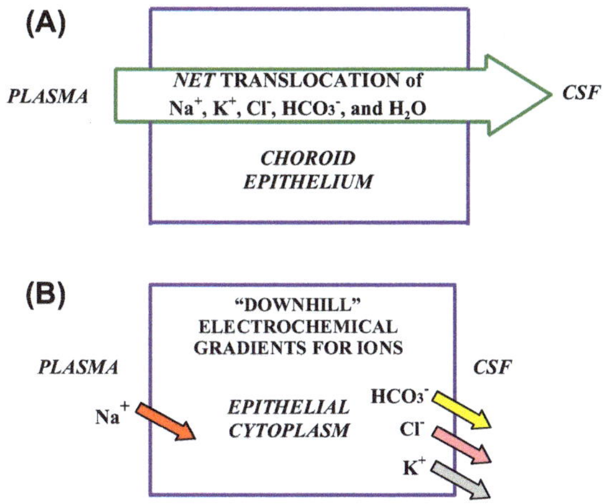

FIGURE 11.3 Ion gradient driving forces for fluxes across external limiting membranes of choroid plexus epithelium: CSF formation is essentially a net flux of Na^+, K^+, Cl^-, HCO_3^- and water from choroidal plasma into ventricular CSF (A). Transmembrane gradients are depicted for each major ion species in CSF production (B). An energetically downhill ion electrochemical gradient obtains for Na^+ across the plasma-facing membrane. The basolateral gradient for Na^+ is steep, directed inward, thereby allowing Na^+ cotransporters and the amiloride-sensitive Na^+ channel to move Na^+ (and cotransported ions or molecules) into choroidal cytoplasm. At the other pole, that is, the apical membrane, there are electrochemical gradients (directed outward to CSF) for Cl^-, HCO_3^-, and K^+. Accordingly, cellular Cl^- and K^+ diffuse through apical membrane via ion-specific channels (Fig. 11.3) into the ventricles. HCO_3^- is actively transported into CSF and may also diffuse outward via an anion channel. Ion electrochemical gradients (*arrows*) are set up by active ion transport into the cell (Cl^-, HCO_3^-, and K^+) or out of the epithelium (Na^+). *From Johanson C. Cerebrospinal Fluid Res 2008;5:10.*

Many other choroid cell membrane transporters and ion channels elaborate the CSF secretion. Carriers move plasma (interstitial) Cl^- into the cell against an electrochemical gradient for Cl^-. Such actively accumulated Cl^- by the epithelium sets up a favorable downhill electrochemical gradient for Cl^- to diffuse via an anion channel into CSF. CSF formation is essentially the transport of Na^+, Cl^-, and HCO_3^- from plasma to choroid cell to CSF; the gradients are depicted in Fig. 11.3. The net transport of ions into CSF sets up an osmotic gradient (ie, increased osmotic pressure relative to plasma and cell cytosol) that attracts water into the ventricles via AQP1 channels in the apical membrane. Conversely water moves down its chemical potential gradient from epithelial cell to hypertonic nascent CSF. Fig. 11.4 summarizes CP transporters and channels relevant to CSF formation.

UNIQUE FEATURES OF CEREBROSPINAL FLUID CIRCULATION

CSF is originally extracted from vascular perfusion by way of the anterior (lateral CP) and posterior (4VCP) choroidal arteries (Fig. 11.5). Choroidal tissues specialize in forming fluid then circulates by bulk flow within the cerebral interstitial compartment and the ventriculo-subarachnoid system in brain/spinal cord. Eventually the CSF flows into cervical lymph glands and the craniospinal venous sinuses in the dural coverings. This completes the circulation back to the systemic vasculature.

CSF is 99% water. This highly watery medium enables volume transmission of hydrophilic substances transported by carrier systems across BCSFB into the ventricles. Such bulk flow of CSF is driven by several propelling "macro" forces: arterial pulsations in the plexuses, positive hydrostatic pressure gradients between CSF and venous blood, and the force generated by respiratory inspiration. Body posture, by affecting hydrostatic pressure gradients, also contributes to flow dynamics. On a "micro" scale, CSF propulsion occurs in an anterio-caudal direction along the ventricular borders (CP and ependyma) by rhythmic beating of cilia that thrust fluid forward. CSF moves to and fro with the systolic and diastolic phases of cardiac output, but the net motion of fluid proceeds inexorably toward drainage sites in the meninges.

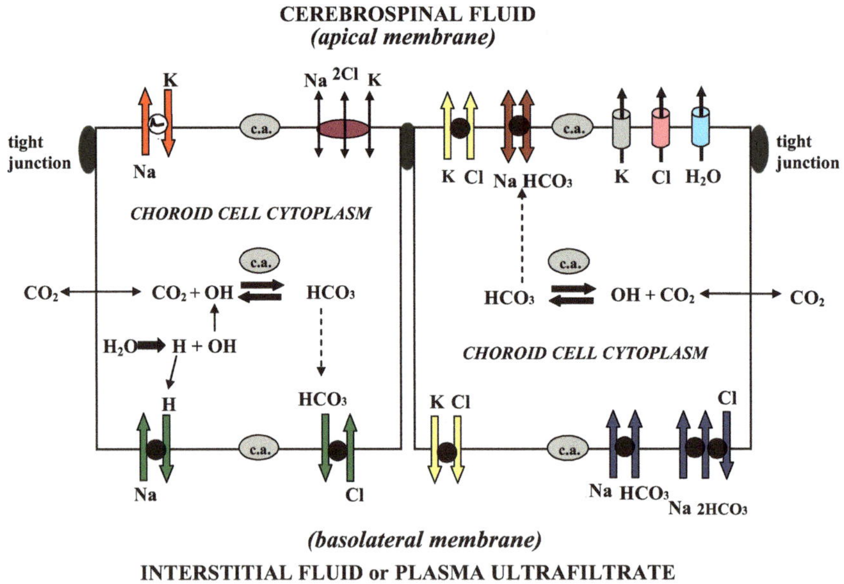

FIGURE 11.4 Ion transporters and channels that mediate CSF formation: an asymmetric localization (basolateral vs. apical) in membrane-bound transport proteins regulates ion and water fluxes. CSF formation results from the net transport of Na^+, Cl^-, HCO_3^-, and water from blood to CSF. On the interstitial side, the Na^+ channel, ENaC (not depicted) or Na–H exchange allows regulated diffusion of Na^+ into cell. Cellular HCO_3^-, generated by carbonic anhydrase (c.a.) activity, feeds membrane transporters to get Cl^- into (basolateral membrane), and Na^+ out of (apical membrane), the choroid cell. Cotransporters (Na-linked Cl–HCO_3 and Na–HCO_3) and ion exchangers (Cl–HCO_3) move Na^+, Cl^-, and HCO_3^- from plasma into the epithelium. Following cytoplasmic swirling of ions to the apical membrane, Na^+ is transported by carriers (Na–K–2Cl and $NaHCO_3$ cotransporters) and the Na^+ pump, from the choroid cell into CSF. In addition, ion-specific channels promote diffusion of K^+, Cl^-, and HCO_3^- into CSF. Water diffuses into CSF via AQP1 channels by "chasing" the osmotically active ions into the ventricles, to complete CSF formation. *From Johanson C. Cerebrospinal Fluid Res 2008;5:10.*

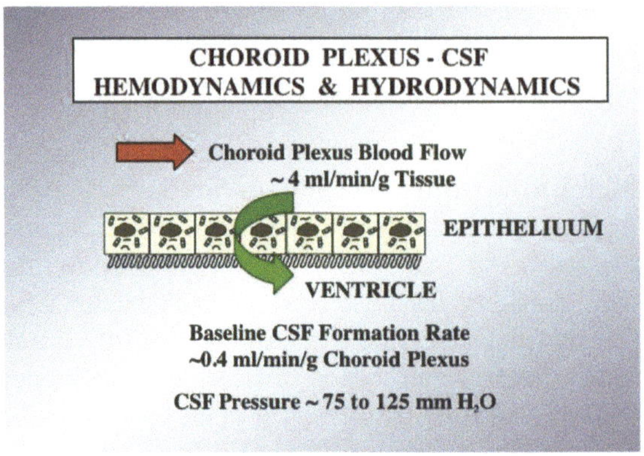

FIGURE 11.5 Coupling of cerebrospinal fluid turnover to vascular perfusion of choroid plexus: blood flow in adult mammals is autoregulated at about 4 ml/min/g of lateral ventricle choroid plexuses, considerably greater than the mean CBF of about 0.5 ml/min/g. Brisk blood flow to the plexuses sustains a relatively high rate of epithelial transport of solutes and water into the ventricles. About 10% of the vascular input is extracted as CSF produced by the CP epithelium. CSF formation, normalized for the mass of CP tissue, is fairly constant at about 0.4 ml/min/g for several laboratory mammals analyzed, but this ratio is less for adult humans.

When CSF formation rate is normal and the circulatory pathways not occluded, the intracranial pressure (ICP) in adults is about 100–120 mm H_2O. ICP can be elevated by any combination of the following: augmented CSF production by CP, blockage of flow pathways (especially Sylvian aqueduct narrowing), and increased resistance to outflow by inflamed meninges. Rising central venous pressure, due to a sinus thrombus, exerts "back pressure" that thwarts efficient CSF drainage. Intracranial hypertension consequently ensues. As ICP increases, the functions of CP, ependyma, and periventricular regions become impaired. Pathophysiology associated with several factors that impede CSF dynamics, and their disease states, are discussed later. Disabling disturbances in CSF circulation highlight the fact that the homeostatic roles of the CP–CSF for brain, for example, anabolite provision and catabolite clearance, depend upon smoothly flowing CSF.

CHOROID PLEXUS AS A MULTIFUNCTIONAL ORGAN SERVING THE BRAIN

CP is well known for producing CSF. Another dominant role, however, is to stabilize CNS extracellular fluid composition. To accomplish homeostasis for the brain, CP carries out a variety of transport and stabilizing functions (Table 11.1). In mediating homeostatic, nutritive, and neurotrophic roles, the choroid epithelial cells engage in both transfer and synthetic functions. In transferring vitamins, hormones, and growth factors from plasma to CSF, the epithelium uses a variety of specialized transporters (and endocytosis/exocytosis mechanisms) to move polar solutes across the basolateral and apical membranes. Synthesizing/secreting peptides, proteins, and small organic molecules from CP to CSF is another mechanism for furnishing neurotrophic and metabolism-supporting agents to neurons and glia. Transcriptomes for protein synthesis/secretion vary between lateral and fourth CP epithelium, presumably due to forebrain versus hindbrain differences in modulatory effects of CSF proteins, respectively, on target cells in cerebrum versus cerebellum. Table 11.2 categorizes polypeptide/protein factors by mechanisms of distribution, as well as trophic versus hormonal and other modulatory effects.

Neurotrophic Stimulation of Fetal Brain Growth

Certain growth factors and neurotrophins are integral to normal brain growth and to restoration following CNS disruption. CP–CSF has a major role in biochemically driving the fetal brain development of glial and neuronal networks. A vast array of factors (Table 11.2), secreted by choroid epithelial cells into CSF, modulates CNS cell survival, growth, mitosis, neurogenic lineages (differentiation), and migration. Subgranular and subventricular regions are primary targets for polypeptides conveyed by CSF flow. Choroidal secretions are especially pivotal for brain development in early embryonic life when the cerebrovascular blood supply is primitive. Members of the fibroblast- and epidermal-growth factor families regulate the neurogenesis-driven mitotic expansion of brain in utero. The critical importance of these CSF neurotrophic systems in early life is evident when gross deterioration of brain follows experimental removal of the life-supporting choroid plexuses.

Upon injury to adult brain, members of the transforming growth factor-β and insulin-like growth factor families, respectively, promote central inflammation control and wound healing. TGFβ2, an immunosuppressant, is upregulated by CP after acute ischemia. Following traumatic brain injury, a biosensing function of CP detects the distant disturbance and responds by upregulating IGF-2 for CSF distribution of the peptide to repair the wound. Another choroidal expression altered by cerebral trauma is

TABLE 11.1 Choroid Plexus, a Multifunctional Organ Supporting the CSF–Brain Compartments

Function Type	Receptors, Transporters and Enzymes Involved	Functional Impact on CNS
Kidney like	CSF K and Cl homeostasis, mediated by Na/K pumping activity and Na-Cl-K cotransporter in apical membrane. Acid–base homeostasis, CP and CSF, regulated by Na-H exchanger, Na-HCO_3 cotransport and anion channel activities in CSF-facing membrane.	CSF K concentration regulation is important for stabilizing neuroexcitability. CSF pH and HCO_3 levels have a modulatory effect in the brainstem on respiratory control.
Liver like	NADPH-cytochrome P-450 and reductase 7-benzoxyresorufin-O-dealkylase activities, similar to liver, found by Ghersi-Egea[a] to express in CP. Membrane-bound epoxide hydrolase and UDP–glucuronosyltransferase activities, like the hepatic levels, have also been measured in CP.	The BCSFB epithelium has a metabolic and enzymatic component. CSF entry of drugs and xenobiotics, as well as endogenous bilirubin, are reduced or regulated to minimize harm to neurons.
Neuroendocrine	CP basolateral receptors for leptin, prolactin, and IGF-1 mediate uptake of these hormone "signals" from the blood. Efflux mechanisms then extrude these peptides into ventricular CSF for hormonal distribution to target cells around the CNS.	CP acts like a "relay station," conveying plasma hormone signals from blood, across the BCSFB, into the ventricles for flow distribution to hypothalamic nuclei for integrating "feedback signals." Consequently, pituitary hormone release adjusts for reproductive and metabolic needs.
Immunological	Ten toll-like receptors in CP, for pattern recognition, recognize certain pathogens, then trigger an immune response. Immune cells in CP participate in immuno-detection and immuno-effector actions. Dendritic cells and monocyte-like macrophages in the interstitium relay immune signals across CP into CSF–brain. Antigen-presenting Kolmer cells embedded in apical membrane detect CSF changes that reflect central immune status.	In adjusting to infections and injuries, involving introduction of "foreign" antigens to blood–CNS barriers, the BCSFB is an important immune–neural interface to minimize neural damage and neuroinflammatory complications. Disease relapse associates with modified leukocyte and cytokine traffic across CP.
Excretory	Reabsorption transporters in both CP membranes facilitate the removal of catabolite waste products from CSF. Organic anion transporter-1 and transporter-3, and the multidrug resistance-associated protein-1, -4, and -5 are inserted, respectively, on the apical and basolateral membranes.	Buildup of organic anions and other brain metabolites in CSF harms neuronal function. Active removal of excess metabolites and uremic toxins, by OAT and MRP transport at BCSFB, improves brain milieu. Avid removal of antibiotics in CSF, to subtherapeutic levels, is a big pharmacological challenge.

[a] *Ghersi-Egea JF, Leninger-Muller B, Suleman G, Siest G, Minn A. Localization of drug-metabolizing enzyme activities to blood–brain interfaces and circumventricular organs. J Neurochem 1994;62:1089–96.*

the newly discovered augurin polypeptide. An ecrg4 gene product, augurin, displays unusually high expression in CP. After stab wounding, the choroidal expression of augurin (a mitosis inhibitor) is downregulated, presumably to allow the production of new cells in targeted neurogenic regions. This likely fosters recovery in brain regions afflicted by traumatic injury.

Micronutritional and Trace Element Support

Brain requires an exceptionally high level of ascorbate or vitamin C. Neurons, having a concentration of about 10-mM ascorbate, use vitamin C as an antioxidant. Ascorbate is also a key modulator of DNA metabolism, serving as a cofactor for

TABLE 11.2 Choroid Plexus Growth Factors, Hormones, Neurotrophins, and Proteins, Transported or Secreted Into CSF, That Support Choroidal and Cerebral Metabolism and Growth

GROWTH FACTORS SYNTHESIZED/SECRETED BY CP
Basic fibroblast growth factor-1 (FGF-1)
Acidic fibroblast growth factor-1 (FGF-2)
Transforming growth factor β (TGFβ1; β2; β3) and TGF latent binding proteins (LTBP-1 and LTBP-3)
Hepatocyte growth factor (HGF)
Insulin-like growth factor-2 (IGF-2) and binding protein
Vascular endothelial growth factor (VEGF) and VEGFC ligand
Granulocyte colony-stimulating factor (G-CSF), a hematopoietic factor
Amphiregulin, a member of epidermal growth factor family
GROWTH FACTORS TRANSPORTED/DISTRIBUTED BY CP FROM BLOOD TO CSF
Insulin-like growth factor-1 (IGF-1)
HORMONES SYNTHESIZED/SECRETED BY CP INTO CSF
Arginine vasopressin (AVP) fluid regulator
Augurin, an ecrg4 product that regulates mitosis
Hepicidin, an iron regulator
Orexin A or hypocretin-1, an arousal and appetite hormone
HORMONES TRANSPORTED/DISTRIBUTED BY CP FROM BLOOD TO CSF
Leptin, an appetite and metabolism regulator
Prolactin, a reproductive hormone and lactotroph
Adipsin and adiponectin, fat hormones that regulate neuroinflammation
NEUROTROPHINS/NEUROTROPHIC FACTORS SYNTHESIZED/SECRETED BY CP INTO CSF
Brain-derived neuronal factor (BDNF)
Glial-derived neuronal factor (GDNF)
Neurotrophin-3 (NT-3)
Neurotrophin-4 (NT-4)
Nerve growth factor (NGF)
Ciliary neurotrophic factor (CNTF)
PROTEINS SYNTHESIZED/SECRETED BY CP INTO CSF
Transthyretin, thyroid hormone carrier and β-amyloid conformation modulator
Cystatin C, a protease inhibitor upregulated in neural injuries
Klotho, a calcium metabolism and aging regulator
CYTOKINES SYNTHESIZED/SECRETED BY CP INTO CSF
Cardiotrophin-1, a member of interleukin-6 (IL-6) superfamily

the 10–11 translocation enzyme reaction (Fig. 11.6). Vitamin C, not synthesized by humans, is water soluble and therefore does not penetrate the BBB. Rather, sodium ascorbate transporters (SVCT-2) at the BCSFB actively transport blood vitamin C into CP; then, another carrier moves ascorbate from choroid epithelium to CSF for an accumulation level 4–5 times above plasma. Such thorough regulatory transport assures stability of ascorbate in CSF–brain in the face of systemic deficiency.

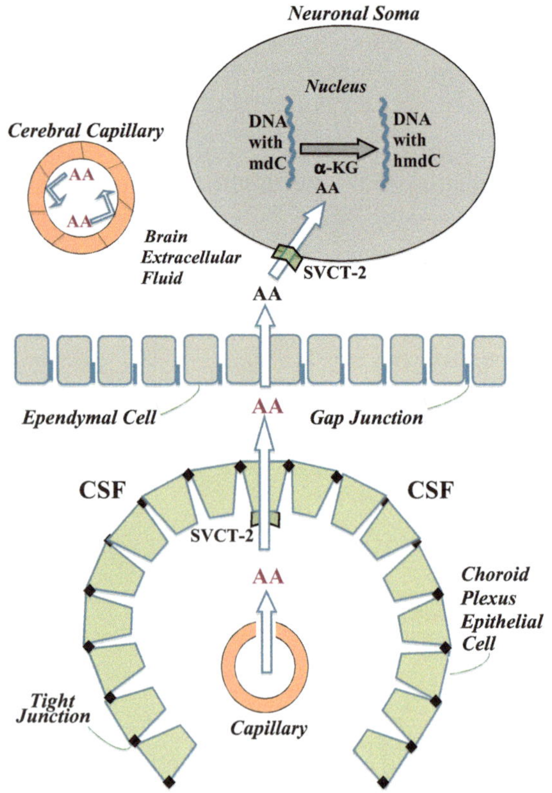

FIGURE 11.6 Distribution route from blood into CNS of ascorbic acid (AA) that serves as a cofactor for TET enzymes in neuronal nuclei: AA in plasma does not permeate BBB but reflects off the endothelial luminal membrane and tight junction (upper left). Rather, AA penetrates the BCSFB reaching brain by a circuitous route involving CP, CSF, and ependyma. Thus AA is transferred (*unfilled arrow*) from leaky blood vessels in CP through epithelial cells into ventricular CSF by the Na^+-dependent vitamin C transporter 2 (SVCT-2) in the basolateral membrane. From CSF, AA diffuses freely through permeable gap junctions in ependyma into brain extracellular fluid. There, AA reaches parenchyma and is greatly concentrated by SVCT-2 in neuronal membranes. Upon accessing neuronal cytoplasm, AA penetrates the nuclear membrane. Within the nucleoplasm, AA acts as an essential cofactor for TET (with Fe^{2+} and α-KG) to oxidize certain methyldeoxycytidine (mdC) molecules in the same DNA strand to hydroxymethyldeoxycytidine (hmdC). hmdC is further oxidized and removed from DNA by thymine DNA glycosylase with subsequent DNA repair. SVCT-2 transporters for AA in CP and brain are concentrative homeostatic systems for AA in CSF and neurons. *From Spector R, Johanson C. Mol Brain 2014;7:3.*

Folic acid, or vitamin B9, is also actively transported from plasma into CSF–brain. CP, not the BBB, mainly provides folate for the CNS parenchyma. Three distinct, sequential steps translocate folate across the BCSFB. They are, in sequence, starting from the basolateral membrane: the folate receptor-α (FRα) binding the methyltetrahydrofolate (MTHF) into a pinched-off vesicle, the proton-coupled folate transporter (PCFT) releasing the vesicle-enclosed folate into cytoplasm, and finally the extrusion of MTHF into CSF by the reduced folate carrier in the apical membrane. CSF volume transmission then carries folate throughout the entire ventriculo-subarachnoid space, from which it readily penetrates the permeable ependyma or pia-glia to access neurons for uptake.

Folate is vitally important for neural tube development, acting as a 1-C group for methylation steps in DNA and RNA synthetic reactions. Cerebral folate deficiency associates with spina bifida–hydrocephalus and other CNS disorders. When certain genes are mutated (eg, PCFT or FRα), the CP cannot concentrate vitamin B9 in CSF at a level 3–4 times that of plasma. Patients with neurological disease- or death-threatening cerebral folate deficiency syndromes benefit from therapeutic mega doses of folate. With such supplementation, the increased flux by diffusion (not the usual active transport) attains a CSF folate level meeting the needs of cerebral metabolism.

Trace elements have essential roles in neural metabolism. CP and brain microvessels display markedly-different capabilities for transporting elements into CNS. Iron (Fe^{2+}) in plasma is transported across CP much more slowly than the BBB. On the other hand, manganese (Mn^{2+}) and copper (Cu^+) are carried through the BCSFB considerably more rapidly than across brain microvessels. Mn^{2+} is a trace element that penetrates both barriers in a regulated fashion, albeit at substantially different rates. Too much Mn^{2+} in brain, or manganism, is toxic to neurons. Permeation of Mn^{2+} across BCSFB and

BBB depends upon molecular species, that is, whether free (Mn^{2+}) or bound. Mn^{2+} binding to transferrin, citrate or albumin alters permittivity into brain. In rats, the K_{in} for $^{54}Mn^{2+}$ uptake by CP was about 100 times greater than at BBB, as assessed by in situ perfusion. In a steady-state intravenous Mn^{2+} loading/brain uptake investigation, Mn^{2+} uptake by carriers in CP was about 700 times greater than BBB. Accordingly, Mn^{2+} penetration rate into CP–CSF is >2 orders of magnitude more than cerebral capillaries. Rapid permeation of Mn^{2+} across the BCSFB enables MRI analysis that tracks the Mn^{2+} distribution pathway from plasma into CP–CSF, then reaching periventricular areas, and finally distributing to all brain regions. This uptake kinetics profile is reminiscent of the model that CSF serves as a source for distributing CP-derived materials (eg, Na^+, Cl^-, ascorbate, and folate) throughout the brain.

Another trace element, Cu^+, also rapidly accesses the CSF from blood. Copper has an essential cofactoring role for many enzyme reactions in brain. In two studies, Cu^+ had a K_{in} at the rat CP considerably greater than BBB. Substantial uptake of plasma Cu^+ by the CP epithelial copper transporter-1, however, does not necessarily lead to subsequently extensive throughput into CSF (especially if the apical ATP7A efflux transporter is not expressed/inserted sufficiently, as in Menkes disease). Mice with a Menkes-like phenotype (Cu^+ deficiency in CSF–brain) can be rescued by transgenetic insertion, via a CSF vector, of a healthy *ATP7A* gene. This restores ability of CP to secrete Cu^+ across apical membrane into CSF to attain Cu^+ homeostasis in brain. *ATP7A*-mutated infants have a low CSF [Cu^+]. By analyzing tissues from multisources, investigators concluded that the CP *ATP7A* transporter in CP provides essential Cu^+ for developing brain viability. The regulated carrier system of Cu^+ delivery across barriers assures that sufficient, but not excessive (toxic oxidative stress) Cu^{2+} is available to neurons and glia.

Neuroendocrine Regulation

Hormonal regulation of CSF dynamics essentially stabilizes ICP and ventricular volume. Several neurohormones impact the rate of fluid formation by CP, the pivotal parameter in maintaining ICP within a normal range. Na^+ flux at the BCSFB is primary in generating water movement into the ventricles. Accordingly, drugs and hormones that diminish Na^+ transport in CP also reduce CSF formation rate. Amiloride, a weak diuretic, blocks the Na^+ channel in CP basolateral membrane. Consequently, radiolabeled tracer ^{22}Na turnover from plasma to ventricles is curtailed; CSF production is thereby reduced. Aldosterone, a supporter of Na^+ permeation via ENaC channels, is antagonized by the CSF-formation inhibitor, spironolactone.

Atrial natriuretic peptide (ANP) is the prime candidate as the main endogenous inhibitor of CSF flow. In the kidney, ANP is a natriuretic/diuretic agent that augments urine flow by blocking Na^+ reabsorptive transport in the tubule. Diuretic agents usually reduce CSF formation rate because CP acts like a reverse kidney (since the choroidal Na^+ pump is reversed in polarity being on the CSF-facing, not plasma-facing membrane).

Substantial evidence supports a major ANP role in modulating Na^+ transport at the BCSFB. Over 70 studies have characterized binding sites/receptors in CP for the natriuretic hormones—ANP, BNP, and CNP. ANP exerts many inhibitory actions in adult CP: decreases ^{22}Na uptake, blocks the basolateral ENaC and apical AQP1 channels, and converts a proportion of epithelial cells to the dark, neuroendocrine-like genre (lower CSF-producing capacity). In both in vivo and in vitro (culture) preparations, ANP slows down CSF formation and/or lowers ICP. Four clinical studies revealed that the ANP titer in CSF rises in proportion to ICP augmentation. A servomechanistic model proposed by the author describes the increased ANP in CSF (derived from a homeostatic action by hypothalamus to secrete ANP into the ventricles in response to incremented CSF pressure) feeding back to the BCSFB to curtail CSF production. Consequently, ICP reflexly drops.

Arginine vasopressin (AVP) and basic fibroblast growth factor (FGF2) are two other hormones, potentially interactive, in CSF-regulatory functions. AVP and FGF2 colocalize in CP epithelium and in certain hypothalamic nuclei; in both cases, modulations of fluid dynamics are implicated. Both AVP and FGF2 induce dark, neuroendocrine-like cells in CP; and each one separately curtails CSF production by CP in animals and cell cultures. Their hormonal inhibitory effects on CP–CSF are receptor mediated. Endocrine interactions between CP and hypothalamus are known for many hormones but deserve more attention for the AVP-FGF2 functional tandem. Ang II is another hormone that interacts with AVP. Ang II, like AVP, reduces CP blood flow; this partially explains Ang II inhibition of CSF production. ANP, AVP, FGF2, and Ang II are on a lengthy list of hormones that curtail CSF dynamics. Due to many adverse effects of elevated ICP on brain function, nature has provided multiple mechanisms to regulate fluid movement into the ventricles. This affords pharmacological opportunities to reduce ICP by intervening therapeutically at many control points in these CSF homeostatic systems.

Systemic Immunological Interfacing/ Interaction with Neural Systems

CP–CSF is at the heart of immunosurveillance in the semi-immune-privileged CNS. Choroid epithelial cells receive inflammatory signals from both peripheral immune cells and systemic (plasma) cytokines. BCSFB permeability is altered by tight junction protein rearrangements, induced by disease and infection, that allow diapedesis of "activated" leukocytes through modified paracellular tight junctions. Neuropathologists regard the BCSFB as a primary gateway for pathogens to enter the CNS. Chasing viruses and bacteria into CSF are chemical/cellular components of the immune system that attempt to counteract the perturbing effects of microbes on brain homeostasis. Altogether, the CP portal is a physiologically strategic point for peripheral immune system communication with the CNS.

Even though barrier tight junctions protect the delicate neuronal networks against gross systemic inflammatory reactions, there are mechanisms that orderly convey immune system information across the BCSFB. Cytokines in plasma bind to choroidal membrane toll-like receptors (TLR), thereby generating "immune signals" for conveyance by CSF volume transmission to multiple brain targets. TLRs are pattern recognition receptors that crucially identify pathogens and then trigger the innate immune response. As many as 10 members of the TLR family have been identified and cloned. TLRs in CP sense the presence of plasma bacterial and viral components and then for brain protection effect responses to eliminate the invading microorganisms.

Considerable leukocyte trafficking transpires in the BCSFB, subsequent to spinal cord and traumatic brain injury (TBI). Expression of CP adhesion molecules and chemokines controls the immune gateway for leukocyte transepithelial migration into CSF. After spinal cord injury, M2 monocyte-derived macrophages from CP home, via CSF flow, to the spinal trauma site to promote healing. Thus, by a BCSFB epithelial transmigration involving VCAM-1-VLA-4 and CD73 enzyme as mediators, Schecter et al. demonstrated that a favorable anti-inflammatory milieu for spinal repair was precluded by impeded CSF flow delivery of M2 macrophages from CP. Following TBI, it was observed by Szmydynger–Chodobska et al. that chemoattractant CCL2 released into CSF attracts monocytes, occasionally along with neutrophils, for recruitment/movement to brain injury sites. Thus the homing of monocytes and macrophages to injured spinal cord/brain via CP–CSF critically affects the immunological status related to repair.

Dendritic cells, in addition to macrophages, reside in the interstitial–vascular core of CP. They have an "interfacing role" in protecting the CNS against neurodegeneration, infections, and brain tumors. Dendritic cells present antigen on their surface for processing by T cells; as such, they are "messengers" between the innate and adaptive immune systems. Dendritic cells in CP also penetrate CSF to serve a variety of immune functions for the CNS. One function proposed for dendritic cells is to restimulate T cells that invade CSF–brain. In experimental autoimmune encephalitis (EAE) and multiple sclerosis (MS), T cells penetrating the BCSFB are accompanied by dendritic cells coinvading the CSF. The coinfiltration of myelin-specific T cells and dendritic cells targets periventricular sites. Quantification of dendritic and T-cell copenetration into regions around the ventricles affected by EAE/MS, as carried out by Clarkson et al. has provided valuable insight on immunological mechanisms and treatment of myelin-associated diseases.

Cytokines and peptides secreted by CP into CSF diffuse readily across the ependyma and gain access to multiple periventricular regions (Fig. 11.7). Cardiotrophin-1, a member of the IL-6 superfamily, exemplifies CP-derived peptides carried by CSF to sites in the ventricular wall and subjacent brain for modulatory functions. Cardiotrophin-1, originally discovered in cardiac hypertrophy studies, is pleiotropic with neuroprotectant properties to counter glucose-uptake defects and cognitive decline. When injected intracerebroventricularly, cardiotrophin attenuates memory/learning losses induced by either β-amyloid or high-fat diet. It reduces neuroinflammation by decreasing levels of pro-inflammatory TNFα and IL-1β, and increasing anti-inflammatory IL-10. Cardiotrophin-1 mRNA and protein are constitutively high in rat from birth through adulthood in CP, but undetectable in brain parenchyma. Also present in human CP and CSF, the cardiotrophin cytokine shows promise as a neuroprotective agent in adjusting the inflammatory balance in brain.

Cross-Talking (Signaling) Among Blood, CP–CSF and Brain

Interposed between blood and the CSF–brain nexus, the CP is ideally situated (and histologically structured) to receive immune signaling from both the periphery and the CNS; and then to integrate this information for executing the appropriate

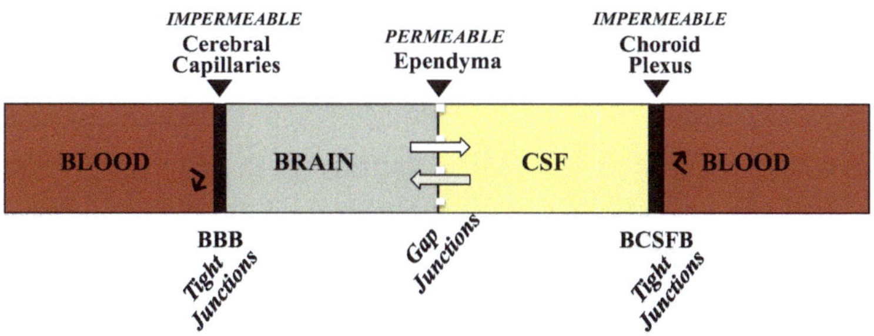

FIGURE 11.7 Relative permeabilities of the blood–brain–CSF cellular interfaces in adult mammals. Two impervious barriers (ie, low permeability to water-soluble materials) are the BBB and the BCSFB, respectively, comprised of endothelial and epithelial cells joined by tight junctions. Hydrophilic solute diffusion from plasma across the cerebral microvessels and CP is impeded by tight junctions (depicted by angular arrows to portray deflection of diffusing solutes). In contrast, readily free diffusion of even large water-soluble molecules (including proteins) occurs between ventricular CSF and brain interstitium, as demonstrated by bidirectional arrows (unfilled) through the discontinuous ependymal border. The great permeability of the ependyma interfacing the ventricles and brain allows CSF proteins, hormones, antibodies, cytokines, and so on to permeate rapidly and extensively the periventricular regions of hippocampus and hypothalamus, in order to modulate neurogenic, neuroendocrine, and autonomic functions.

action on the BCSFB interface to regulate CSF immunity. Due to extensive exchange of materials between the leaky choroidal capillaries and interstitium, there is facilitated transfer of immune signals (via leukocytes and cytokines) directly to receptors in the parenchymal epithelium of CP. Thus the ISF milieu in CP (unlike the brain interstitial milieu) reflects peripheral immune status.

In analogical contrast, the BBB with tight capillary junctions does not have the same histological and functional relationship to ISF and targeted neurons/glia. Moreover, the CSF, flowing more freely than brain ISF, is better suited than the later to convey immune signaling by convection to target cells (eg, to the spinal cord, following spinal injury). Accordingly, the CP–CSF is particularly well suited as an immunosurveilance system for the brain, and as a regulator of leukocyte traffic across a selective, immunologically educative barrier to remote injury sites via bulk flow.

The choroidal epithelium, unlike the cerebral endothelium, has antigen-presenting Kolmer cells at the apical (CNS-facing) surface. Kolmer cells take self-antigens, brought by CSF to CP from injured brain and present the "self" material to T cells in, or trafficking through, the BCSFB. This is important CSF immunosurveillance of the brain. Homeostatic responses by CP to present the appropriate inflammation-resolving monocyte-derived macrophages to CSF–brain curtail neuroinflammatory events that exacerbate neurodegeneration.

At another level of cross-talk, the CSF-mediated exchange of immune information [via neural interferon (IFN) forms] between CP and brain triggers adaptive gene expression in choroid epithelial cytokines. This feedback signaling protects neurons from adverse biochemical changes linked to cognitive loss. Brain has the proclivity to convert metabolic signaling from a healthy IFN-II in young adults to a less viable IFN form (IFN-I) in old age. Choroidal secretion of factors, such as BDNF and IGF-1, which favorably affect neuronal growth, differentiation, and survival, is reduced by IFN-I modulation at the BCSFB. Baruch et al. report that as a result of the rising level of IFN-I in aging brain–CSF, the neurotrophic effect of CP secretions on the CNS is curtailed. This manifests in experimental animals as declining neurogenesis in hippocampus and reduced cognition. Such a cognitive/neurogenic decline in aging mice reverses when IFN-II is reestablished. Lower levels of IFN-II are found in aging human CP. This raises the intriguing possibility that therapeutic manipulation of IFN-II/IFN-I ratios in human aging might produce better cognitive outcomes. Cerebral IFN status is thus communicated/adjusted by bidirectional cross-signaling between brain and CP via CSF.

Excretory Mechanisms for Catabolite Disposal

Optimal neurotransmission requires healthy, clean synapses. Therefore, to prevent synaptic clogging, it is essential for CNS catabolite and protein disposal systems to remain fully operative in aging. Efficient clearance systems prevent neurotoxic catabolites from accumulating in the neuronal networks, cerebrovasculature, and CP/meninges. CP epithelial transporters remove unwanted molecular debris from the ventricles. In addition, the CSF bulk flow route clears unneeded proteins and other macromolecules into lymph and venous blood. Similarly, there are analogous BBB transporters to remove organic anions and peptides from ISF, and a glymphatic outflow pathway to convect large molecules into cervical lymph glands. Later, we exemplify how the potentially toxic product of amyloid precursor protein, β-amyloid (Aβ), is cleared from the CNS into blood at four distinct loci.

Choroid plexus Epithelial Transporters

CP apical membrane is studded with transporters not only to move beneficial substrates into CSF, but also to reabsorb unneeded substances from ventricular fluid. Several studies have characterized Aβ removal from CSF by the lipoprotein receptor-related protein, LRP-1, at the apical side. Such homeostatic reabsorptive transport by LRP-1 is relevant in neurodegeneration with excessive Aβ in brain. Consequently, the lowered concentration of Aβ in CSF, in Alzheimer's disease (AD) for example, helps to set up CSF sink action on free Aβ (unbound, nonsequestered by interstitial plaque) in ISF. Vigorous removal of Aβ by CP in aging and AD compensates for the simultaneous deficient LRP-1 activity in cerebral capillaries.

CSF Bulk Flow

Normally circulating CSF exerts a sweeping effect, or sink action, on brain substances. Sink action importantly eliminates unwanted products of cerebral catabolism and harmful plasma proteins that incrementally leak across more permeable interfaces in advanced aging. As CSF formation rate declines with aging, the sink action of the ventricles becomes progressively less effective in clearing solutes. In senescence, NPH, and AD, there are the common denominators of expanding ventricular volume and decreasing CSF turnover rate. The reduced fluid dynamics promotes Aβ retention by brain and the encompassing meninges. Severely diminished CSF bulk flow impairs cerebral metabolism if misfolded proteins accumulate in the CSF–brain compartments.

Brain Microvessel Endothelial Transporters

The LRP-1 and P-gp transporters, respectively, in the abluminal and luminal membranes of the cerebral endothelium remove Aβ from ISF. Downregulation of these Aβ-removing transporters occurs in aging, NPH, and AD. A biochemically vicious cycle ensues such that as greater amounts of Aβ are retained, there is consequently more downregulation of the Aβ-removing transporters LRP-1 and P-gp. The outcome is greater damage to vessels walls and brain parenchyma. As a result of injured tissue, the Aβ burden in CNS continually climbs upward in late-life stages. Accordingly, Aβ progressively builds up in cells and ISF with a devastating pathology that includes burgeoning neuroinflammation. As the BBB fails, more pressure is placed on the BCSFB to maintain cerebral homeostasis. Significantly, the CP retains or even upregulates LRP-1 expression in NPH and AD. Sound expression of LRP-1 at the BCSFB helps to buffer the Aβ dys-homeostasis in the environment of glia and neurons.

ISF Flow Along Glymphatic Pathways

Bulk flow clearance of brain ISF is an emerging model. Although certain aspects of glymphatic drainage need bolstering, there is sufficient evidence to warrant attempts to consolidate this model of brain fluid dynamics. Subarachnoid CSF flows along pial/cortical arteries into Virchow–Robin spaces. There CSF mixes with ISF and then drains outward along glial sleeves around veins that exit back into subarachnoid space. Glymphatic fluid follows vascular pathways into cervical lymphatic drainage depots. When glymphatic clearance is disrupted, Aβ builds up in cortex. The energy for driving the glymphatic circulation starts with the pressure head of CSF flow into the Virchow–Robin spaces. Therefore, the diminishing CSF dynamics with aging/senescence likely provides less "drive" on glymphatic flow/drainage. In addition, the pulsations of arteries and veins, with each cardiac cycle of systole and diastole, may contribute to glymphatic fluid movement toward lymphatic glands. Healthy glymphatics is evidently important in the physiological integration that fosters normal cerebral Aβ homeostasis.

Stabilization of the Declining Brain in Aging

With aging, the brain loses the full capacity of homeostatic mechanisms operating via CSF distributive functions that target many CNS regions. Part of this curtailment is due to reduced transporter/enzyme expression in CP slowing the local epithelial metabolism, another part to the reduced dynamics of CSF streaming across the BCSFB into the ventricles, and a third part to the damaged ependyma–subependyma that normally maintains balance between CSF and periventricular ISF. Still, despite these functional attenuations, the CP–CSF–ependyma nexus remains surprisingly able to aid brain in warding off infections, diseases, and trauma.

One system, however, that merits greater attention is the downturn in subventricular neurogenic regions to spawn the birth and migration of new neurons/glia. Neurogenesis in the dentate gyrus is modulated by factors derived from the cerebral capillaries on one side and the CP-CSF on the other (Fig. 11.8). Because CSF content of growth factors and neurotrophins critically modulates neurogenesis (Fig. 11.9), it follows that a promising aspect of translational research is finding ways to sustain CP peptide secretions in aging.

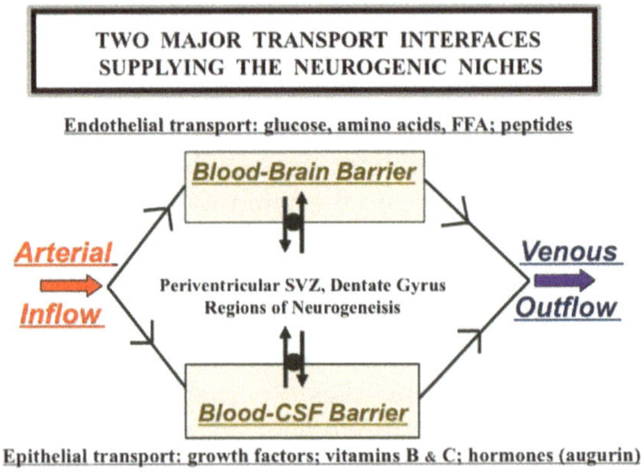

FIGURE 11.8 Idealized schema to portray the functional parallelism in the two major vascular systems serving brain tissue (BBB) and the choroid plexuses (BCSFB): capillaries of brain and spinal cord contain arrays of transporters to carry amino acids, glucose, and free fatty acids into the CNS to supply glia and neurons. Plasma peptides and micronutrients also penetrate the BBB, by vesicular or facilitated mechanisms. Choroid epithelial cells in BCSFB have transporters to convey (actively or by facilitation) plasma-derived growth factors and hormones into ventricular CSF. Plasma amino and fatty acids, as well as glucose, are carried into choroid cells mainly for the local epithelial metabolism rather than for cells in the more distant CSF–brain. CP also synthesizes/secretes proteins and peptides into CSF, for transependymal distribution into brain (including neurogenic regions proximate to the ventricles). Thus neurogenic niches receive peptides from both vascular and CSF circulations. *Adapted from Johanson C. Cerebrospinal Fluid Res 2008;5:10.*

PRINCIPLES AND STRATEGIES TO GUIDE CSF TRANSLATIONAL RESEARCH

During early brain development, there is profuse expression of many growth factors and neurotrophins by CP that are secreted into CSF to help build the CNS. These trophic factors are distributed to fetal periventricular neurogenic regions for the regulation of mitosis, differentiation, and migration of nascent cells. Much of this peptide/protein secretory activity by CP, prevalent before birth, becomes relatively quiescent in young adulthood when the brain is generally healthy. With onset of middle age, and especially later stages of life, there is

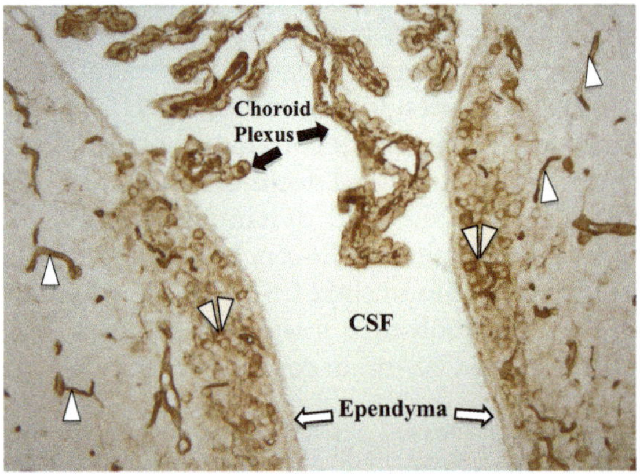

FIGURE 11.9 The interior CSF ventricular system interfacing the ependymal border: CSF secreted by lateral ventricle CP (*dark-filled arrows*) contains growth factors, peptides, and hormones (Table 11.2) delivered by bulk flow to neuronal, neuroendocrine, and progenitor cell targets. These are predominantly in regions bordering the lateral, third and fourth ventricles. The ependymal lining (*unfilled arrows*), separating CSF from underlying brain, can be disrupted early in development (congenital hydrocephalus) or late in life (NPH and AD). Single wedges point to brain microvessels. Brown reaction product, nestin, marks for the type VI intermediate filament protein essential for stem cell self-renewal (immature neural cells). Although less efficient in aging, neurogenesis/renewal persists in adult rat periventricular or subependymal niches (*double wedges*). CP–CSF proximity to subependymal niches facilitates trophic and hormonal regulation of neurogenesis.

considerable reactivation or "reexpression" of growth factors and other neuron-supporting peptides as the aging brain faces disorders at an ever-increasing frequency. These debilitations require the birthing of new neurons to replace dying ones. Accordingly, basic knowledge gained from ontogenetic peptide–expression/regulatory patterns can apply in therapeutic strategies to promote restitution of the weakening CNS in aging.

CP is often compared to kidney, and for good reasons. Choroidal fluid secretory activities are comparable to the renal fluid reabsorptive functions. Drugs that enhance urinary fluid outflow, that is, diuretics, also curtail CSF formation rate. Therefore, the pharmacology of controlling CSF volume and ICP relates closely to the kidney control of plasma volume and arterial blood pressure. Consequently, certain drugs approved by FDA for modifying renal and cardiovascular functions may be useable in CNS fluid–imbalance disorders once dosing and administration routes are refined.

Moreover, the ciliary body-aqueous secretory system closely mimics epithelial transport phenomena in the CP–CSF. Similarly, certain drugs used to control intraocular pressure, for example, antiglaucoma agents, find usage for controlling CSF production and ICP. At the very least, basic physiological mechanisms delineated for ciliary, renal, and other epithelia may well abet construction of CP physiological paradigms for transport and secretion.

Finally, multiple factors need evaluation for the best interpretation of CSF phenomena. Diurnal and even seasonal (light-sensitive) variations in transport can be challenging when comparing data from other species to human findings. Titers of many fluid-regulating hormones in CSF oscillate over a 24-h cycle and may explain how the CP formation of fluid varies considerably, day versus night, in humans. Gene expression, or the transcriptome, varies with CP region: lateral versus third and fourth ventricle. Not surprisingly then, the transport activity differs from one choroidal tissue to another.

A protein expression variable should be considered when deciding to pool CPs from lateral and fourth ventricles; and upon sampling fluid at specific regions within the ventricular and spinal CSF systems. Diverse inflammatory responses at the BCSFB introduce the regionality variable in choroid epithelial and immune–cell interactions. For reliable translational research, careful attention should be paid to many complexly related variables (structure, region, biochemistry, species, gender, experimental time, and life stage) encountered in the dynamic, heterogeneous CP–CSF nexus of mammals.

CSF INVOLVEMENT IN BRAIN PATHOPHYSIOLOGY AND NEUROTHERAPEUTICS

CSF malfunction initiates or exacerbates many brain disorders/diseases. Structural and functional damage to CP, ependyma, arachnoid, and pia-mater—all CSF interfaces—can profoundly affect the course and severity of neural diseases. It stands to reason, then, that translational therapeutic strategies to alleviate or cure brain disorders should include improving the distorted functions of these diverse CSF-bordering regions.

Hydrocephalus

Hydrocephalus is CSF retention, usually with ventriculomegaly and elevated ICP. Complications relate more to faulty functions of CP, ependymal, and arachnoid, than to BBB disruption. In congenital hydrocephalus, there is commonly an ependymal histopathology (dyscrasia) of the third ventricle–aqueductal region impeding CSF flow through the midbrain region. In chronic adult, metabolic hydrocephalus or NPH, there are expanded ventricles along with increased resistance to CSF outflow at arachnoidal drainage sites. ICP is raised in NPH, but not to the extent occurring in congenital hydrocephalus.

For more than a half century, CSF shunting via catheters to the peritoneal cavity has been frequently done to unload excess CSF and ICP. Over the past decade, however, cauterized ablation of the lateral ventricle CPs, sometimes with endoscopic third ventriculostomy, has successfully reduced CSF formation and relieved ICP. By the same principle, extirpation of CP papillomas attenuates both CSF volume and pressure. The mainstay pharmacological agent for hydrocephalus is acetazolamide, an inhibitor of carbonic anhydrase that curtails CSF production by CP. However, systemic acidosis and other side effects of acetazolamide have limited use mainly to acute situations in which this CSF inhibitor brings down high levels of ICP.

New therapeutic strategies to minimize surgeries associated with shunt complications are needed to manage hydrocephalus. Acetazolamide would be more valuable for ICP control upon minimizing side effects caused by systemic administration-related multiorgan actions. Recently intranasal instillation of peptide drugs, which rapidly access the CSF–brain, has ameliorated neural disease. Intranasal administration of "diuretic" agents like acetazolamide would produce a more local (topical) effect on CP compared with the widespread inhibition of systemic renal function. Intracerebroventricular ANP, an inhibitor of CP-CSF function, reduces ICP in experimental animals. Although there is a BBB and BCSFB to ANP, intranasal instillation in humans circumvents these barriers to plasma peptide penetration into the CNS. It is worth testing intranasal ANP, with or without acetazolamide, for the palliation of elevated ICP without major systemic side effects.

Intracranial Hypertension

Otherwise known as pseudotumor cerebri, the rare disorder (1:100,000) of intracranial hypertension (IH) is characterized by elevated ICP >250 mm H_2O without ventricular expansion. Overweight females of child-bearing age comprise most of the IH patient cohort. Although IH etiology is not crisply defined, several factors are implicated: overproduced CSF, under-reabsorbed CSF at arachnoidal sites, cerebral interstitial edema, and impaired cerebral venous drainage.

Pharmacological remediation of IH is not well developed. Neuroendocrine imbalance has been described for many IH patients. Therefore, therapeutic approaches would benefit from prospective systematic studies of how levels of certain hormones (estrogen, progesterone, testosterone, and corticosteroids) modulate CSF dynamics and brain fluid balance. Acetazolamide and shunting benefit some IH patients but have limitations due to the same undesirable effects encountered with usage in hydrocephalus. Many IH patients suffering from poor control of ICP await new pharmacological breakthroughs.

The low incidence of IH poses challenges in new drug development for this "orphan disease." However, some pharmacological opportunities with already approved drugs present themselves. Intranasally instilled acetazolamide is expected to lower CSF formation and ICP without causing the difficult side effects of systemic usage. ANP not only reduces CSF formation and ICP, but also has the attractive property of being antifibrotic. Fibrosis in the arachnoid membrane reduces the facility of CSF outflow into the venous sinuses; this occurs, for example, after subarachnoid hemorrhage. Thus, using intranasal ANP for posthemorrhagic hydrocephalus might confer the double benefit of reducing CSF formation/ICP and maintaining normal outflow across arachnoidal interfaces with lymph and venous circulations.

Choroid Plexitis and Meningitis

Microbes have a great proclivity for binding and/or penetrating the CP. The neuropathologist Seymour Levine alerted his colleagues that the BCSFB is a preferential site of pathogen passage into CNS via CSF. CP is the initial dissemination point for many infections, such as intracranial tuberculosis, cryptococcosis, cytomegalovirus, bacteria, and parasites. Cryptococcal fungal infections cause choroid plexitis. Imaging findings (enlarged Virchow–Robin spaces) differ between immunocompetent and immunodeficient individuals; inflammatory responses are greater in the former cohort. Inflammation, however, is generally subdued since the polysaccharide capsule of the fungus hinders phagocytosis and leukocyte migration. Anti-cryptococcal factors are in serum, but not CSF, hence the unrestrained spreading of cryptococcus, once in CSF, throughout the ventricles, ependyma, and meninges. Drugs to mimic antifungal factors deserve exploration for usage in the deficient ventricular CSF. Cryptococci and their associated gelatinous pseudocysts can impede CSF flow, isolating the lateral ventricle temporal horn. Cryptococcus-induced choroid plexitis is currently treated with antifungal azole compounds, for example fluconazole.

Viruses and bacteria also permeate the CP epithelial barrier, transcellularly, and eventually gain access by CSF bulk flow to the meninges. Viral meningitis is much more common than bacterial inflammation of the meninges. Bacterial meningitis is treated with certain antibiotics, for example, ceftriaxone or imipenem, because the later drugs are not extensively reabsorbed from CSF (by CP uptake transporters) after having gained access to the ventricles. Inflamed meninges, by particularly virulent strains of bacteria, can be life threatening. Because CSF circulation, starting in CP, has a widespread reach to the corners of the CNS, it is imperative that research be directed toward preventing the initial binding of microbes and parasites to the CP epithelial membranes. Design/development of antipathogenic agents to minimize movement of viruses and bacteria into CSF is a paradigm shift to identify new approaches to block initiation of CNS infections at the BCSFB. A pharmacological blocking strategy directed at the "upstream" lateral CPs would likely help to prevent "downstream" CSF-facilitated infections in periventricular and meningeal regions.

Stroke/Ischemia

Acute forebrain ischemia, with reperfusion, devastates not only the CA1 region of hippocampus but also the nearby lateral ventricle CP. Programmed cell death, or apoptosis, occurs in both regions. Many CA1 neurons are lost after severe ischemia, with the destructive biochemical cascades and structural damage culminating 2–3 day after the insult. Concurrently in CP following an acute ischemic episode, there is widespread epithelial damage and breaching of the BCSFB; however, the severely injured epithelial cells are restored so that by 24-h postischemia, the CP regains ultrastructural and homeostatic integrity. Still, it is therapeutically desirable to minimize CP damage in ischemia so that homeostasis-mediated phenomena can provide trophic support quickly to acutely stressed brain areas.

The proximity of the lateral CPs to hippocampal regions enables many functional interactions between these two areas. Many growth factors and neurotrophins secreted by lateral CP into CSF (Table 11.2) modulate the development of the forebrain, including the hippocampus, and repair after injury. Pharmacotherapeutic strengthening of the BCSFB just after stroke, possibly with supplemental growth factors, has been proposed. The notion is that a more rapid recovery of the disrupted BCSFB would accomplish at least two things: (1) attenuate the leakage of potentially deleterious plasma proteins and cytokines, immediately following ischemic damage to the CP epithelial/tight junction barrier, into the CSF and adjacent hippocampal regions and (2) provide a quicker, more robust secretory response to provide healing agents to damaged CA1 neurons. Pharmacologically boosted trophic support of CP, then, would have the goal of quickly restoring CSF-mediated homeostasis to minimize hippocampal damage.

Traumatic Brain Injury

Acute physical trauma to brain, as with concussive cortical injury, leads to breakdown of both the BBB and BCSFB, even at a considerable distance from the traumatized CNS region. The totality of the damage manifests as structural harm to the barriers, leakage of plasma proteins into CSF, ventriculomegaly, tissue edema, and behavioral/cognitive deficits (Fig. 11.10). Damage to the barrier and epithelial functions of CP, with choroidal edema in TBI patients, was initially recognized by Hoff, and by Rand and Courville, in the early 1930s. Modern investigations have confirmed BCSFB breakdown in TBI, but with the CP in many instances still retaining ability to participate in the recovery of the physically traumatized brain. The ability of CP to biosense injury to neurons/glia (over a considerable distance away) prompts the question: what are the injury signals (cytokines?) conveyed to the BCSFB by way of ISF, CSF, blood, or innervation? Even more intriguing is the ability of CP to respond to TBI by upregulating/secreting growth factors into CSF for flow delivery back to the targeted neural tissue injury for repair.

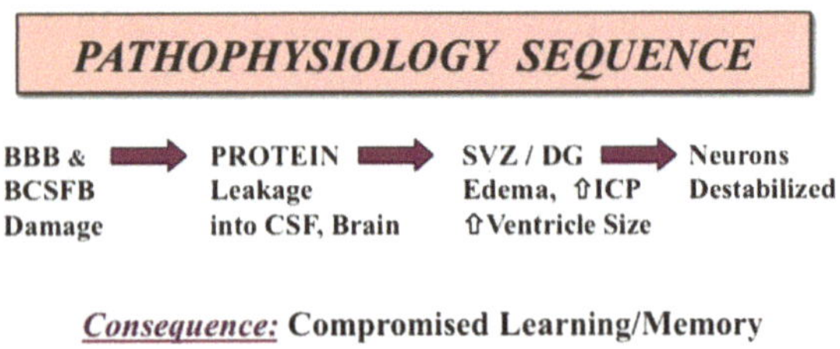

FIGURE 11.10 Injury-model schema for the flow (sequence) of pathophysiological events in the anatomically interconnected compartments of the CP–ventricular CSF–ependyma–periventricular regions: in the experimental insult models of forebrain ischemia, TBI and hyperthermia, there is a sequence of "common denominator" events in the CSF–brain interior. Soon after the experimental stressor is imposed on the CNS, the lateral ventricle CP tight junctions break down, interstitial edema forms, and parenchymal epithelium destabilizes (vacuole formation, microvilli stripping, cell swelling, gain of Na^+, loss of K^+). BCSFB permeability increases, allowing plasma proteins into the ventricles; this attracts water, leading to ventriculomegaly and elevated ICP that harm periventricular zones. Cerebral microvessels (BBB) also break down, allowing protein leakage, with consequent exacerbation of subependymal and brain injuries. The collapse in barrier integrity perturbs cerebral interstitial homeostasis, synaptic phenomena, and neurotransmission. Cognitive/behavioral deficits ensue but can be minimized by early-experimental intervention with exogenous growth factors and neurotrophins.

Choroidal secretions have been shown time and again, in various models of healthy and diseased CNS, to provide CSF trophic and maintenance support for distant brain parenchyma. The critical importance of choroidally derived peptide and protein factors for brain development was demonstrated by Russian physiologists 6–7 decades ago in infant dogs that suffered extensive neural pathophysiology upon surgical removal of CP tissues. The same growth factor systems in CP highly expressed in fetal/neonatal development are reactivated in later adult life in reaction to harm, for example, TBI.

Accordingly, physiologists and neurosurgeons alike have suggested administering a "cocktail" of regenerative growth factors and glio-/neurotrophins to patients severely compromised by TBI or other injuries. Many growth factors and peptide hormones in blood do not significantly penetrate the BCSFB and BBB. To circumvent the barriers, though, a "cocktail" mixture of healing factors can be given by indwelling ventricular catheters or by intranasal instillation following human brain injury. The growth factor "cocktail" approach for patients has been encouraged especially by the success of many investigators injecting "protective peptides" into the CSF of animals with injured brains, and observing favorable experimental outcomes (Fig. 11.11) to minimize central damage. Such rapid CSF-access regimens would expedite delivery of "repair peptides" to the CSF–brain system not only for TBI and stroke but also for certain psychiatric disorders.

Autism/Social Disorders

The prevalence of autism and related spectral disorders (ASD) in the United States, as of

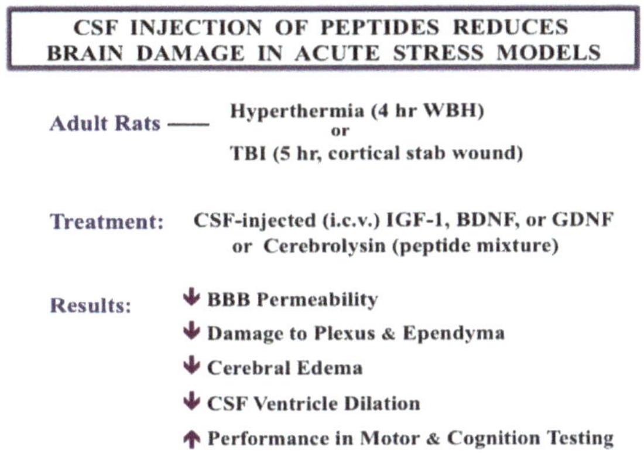

FIGURE 11.11 Acute experimental protocol for adult rat barrier disruption by whole-body hyperthermia or traumatic brain injury (TBI): experiments evaluated extent of injuries to CP, CSF and brain, and sought minimization of neural damage by pretreatment with, or early postinsult administration of, healing factors. Agents tested were a growth factor (IGF-1), neurotrophins [brain-derived neuronal factor (BDNF) or glia-derived neuronal factor (GDNF)], and Cerebrolysin (a mixture of neurotrophic peptides). Test factors injected into a lateral ventricle reduced damage, that is, by minimizing barrier breakdown, edema, tissue pathology, ventriculomegaly, and learning/motor deficits. Pharmacology effects are explained in the text. Intracerebroventricular injection of neuroprotective factors experimentally mimics the "boosted secretion" of endogenous peptides by CP into CSF.

2015, is one in 68 children. Key social functions integrated by brain involve oxytocin modulation of specific neuronal networks. Like the nonapeptide AVP secreted by hypothalamic nuclei into CSF, the nonapeptide oxytocin undergoes neuroendocrine diurnal rhythms and targets receptors in CNS. Oxytocin is considered a prosocial or "social-bonding" hormone. Oxytocin deficiency in CSF exacerbates certain social and psychological disorders. As a social-affiliation hormone, oxytocin has been analyzed before and after birth, often in the context of mother–child bonding. Just before birthing, CSF oxytocin tends to increase in mothers to be. Mean CSF oxytocin is about half the plasma-level prepartum, but twofold greater postpartum (conscious sheep). Mother-reared rhesus monkeys have more positive emotional and social behaviors than nursery-reared counterparts; this is reflected by persistently greater levels of CSF oxytocin titers in the mother-reared animals compared to nursery-raised infants.

Low levels of oxytocin in CSF have been consistently related to problematic socialization. A CSF deficiency in the neuropeptide oxytocin in children predicts anxious behavior syndromes; moreover, blood levels of oxytocin predict CSF levels, pointing to the utility of analyzing blood specimens in anxiety-ridden children (when CSF samples are unavailable). Some studies report an inverse relationship between childhood adversity (including suicide attempts) and CSF–plasma oxytocin levels. A recent study of childhood trauma and revictimization, however, has stressed complex rather than straightforward negative (inverse) associations between CSF oxytocin levels and untoward behaviors. In a long-term follow-up study (>20 years) of adults, lifetime violent behavior negatively correlated with CSF oxytocin.

ASD manifests with many aspects of social dysfunction and therefore has been the subject of extensive approaches to oxytocinergic evaluation and treatment. Oxytocin concentrations in plasma of children with autism is considerably less than in controls. One study of autism found an increased level of the inactive form of oxytocin, indicating a defect in peptide processing. An association has been found between autism and two SNPs of the oxytocin receptor (OXTR). Hypermethylated suppression of OXTR gene expression has also been observed in autism, indicating possible epigenetic modulation. Gene defects (deletions) are also found in the metabolic generation of oxytocin peptide. To date, the numerous deficits in oxytocin neuroendocrinology in patients with ASD point to a great need for rectifying specific malfunctions of the oxytocinergic system.

Pharmacological supplementation of neuropeptide-deficient patients is an area of wide clinical interest. CSF is commonly suggested as pharmacological access point. Oxytocin, an enhancer of emotional perception in autistic children, is under scrutiny as a potential pharmacotherapy to improve social functioning. Since plasma oxytocin, like vasopressin, does not readily penetrate CNS barrier systems, various routes of peptide administration have been evaluated. A human assessment revealed that oxytocin reached a maximal level in CSF about 75 min after intranasal administration, at a time when plasma oxytocin had begun to fall; the half-life of oxytocin in CSF was a favorable 2–4 h (versus 3 min in plasma), reflecting a slower rate of decomposition in enzyme-deficient CSF. In macaque monkeys, intranasal spray and intranasal nebulizer were equally effective in raising CSF concentrations of oxytocin. In another study of monkeys, aerosolized oxytocin (delivered by oro-nasal mask) was more effective than intravenous and even intranasal administration to elevate the CSF concentration of oxytocin. It is important to emphasize the caution that oxytocin intranasal supplementation to CSF has been effective in some but not all clinical trials. Additional formation is needed on oxytocin doses, timing (in development), and neuropsychological response profiles in the complex constellation of ASD.

Multiple Sclerosis

A prominent feature of MS is the altered trafficking of self-antigenic (myelin) T cells across the BCSFB, sometimes even preceding the BBB crossing. Is this aberrant immune cell traffic through CP the main cause of MS, as suspected for decades? Autoreactive, myelin-specific T helper cells reside interstitially in peripheral lymphoid organs such as CP. Penetration of T "helpers" into CSF–brain likely initiates an immune response that destroys myelin sheaths. However, T immune cells also have a normal CNS-patrolling function, via the CSF, to detect brain disturbances and then facilitate any repair process needed. Therefore, it is a challenge to distinguish the dual roles of the CSF T-cell population in resolving versus fulminating specific CNS immunological problems.

The variable movement of T cells across CP in disease stages of relapse versus remission is well known. The rodent EAE model of MS informs on the nature of leukocyte movements through the BCSFB into CSF. Starting the immunopathological process, myelin components (antigens) stripped from nerve fibers drain by bulk flow from ISF into CSF, and thence into the superior cervical lymph glands. There, antibodies formed against the self-antigenic "foreign" myelin, gain entrance to the venous blood for circulatory transit back to the CNS (efferent arm of the immunological response). Upon gaining entrance to CSF by permeating the CP, the myelin antibodies (T cells) reach/destabilize periventricular regions, cortex and cord where the ongoing myelinopathy is destructive.

Myeloid cells, including antigen presenters residing in CP, perform immunosurveillance and immunomodulatory functions. Dendritic cells in the choroidal interstitium (on the blood side of the epithelial BCSFB) present their antigens to specific T cells. In addition, CP-surrounding antigen-presenting elements, for example, Kolmer cells fastened to the apical membrane of the BCSFB, also participate in the stages of initiation, progression and resolution of MS. Current CP/MS research emphasis places priority on elucidating the phenotypes/functions of various myeloid cells that stream through CP.

Immunopathological mechanisms in CP are being sought. Animal and human data intimate how the BCSFB is breached by immune system components in EAE and MS. The C—C chemokine receptor 6, regulating entry of TH-17 cells across CP, is essential for EAE initiation in mice; accordingly, abolishing receptor six thwarts EAE development. Disassembly of a tight junction claudin three component in human CP and BBB (in gliomas), resulting in enhanced barrier permeability, may be directly induced by immune cells. While these promising leads await further pursuit, it is becoming clearer that certain T cells have both deleterious and potentially helpful functions. An immunological research challenge introduced by Schwartz and Baruch, and reinforced by Kooij and de Vries, is to fully characterize various immune cell phenotypes in CP; and how each participates temporally, functionally and spatially in their distribution across CP in normal immunophysiology versus disease-associated immunopathologies.

Amyotrophic Lateral Sclerosis

A long-observed phenomenon has been CNS barrier breakdown relatively early in the course of amyotrophic lateral sclerosis (ALS). Increasing credence in the significance of barrier breaching in ALS progression is upheld by the continually reported barrier disruptions that exacerbate MS and AD. Nearly half of ALS patients show evidence of BCSFB breakdown, as manifested by rising levels of plasma-derived albumin or IgG in CSF. However, allusion has also been made to disrupted CSF flow in the subarachnoid space, another factor potentially contributing to elevated concentration of plasma proteins and immunoglobulins in CSF. Thus both altered CSF dynamics and barrier permeability need differential analysis for specific roles in modifying the CSF neurochemistry in ALS.

Up to 90% of ALS is sporadic rather than familial. Given that the CP controls CSF formation and regulates fluxes of plasma solutes and leukocytes into the ventricles, it is especially pertinent then to understand how epigenetic factors cause a malfunctioning CP–CSF system capable of exacerbating ALS. This could happen, for example, by inappropriate T-cell penetration into CSF, or the excessive movement of cytokines, albumin, globulins, and oxidative products into the ventricles. In addition, the role of CP-secreted exosomes, via apical membrane pinched-off vesicles, deserve examination as "vectors" of CSF-transferred proteins (mutated or misfolded, such as superoxide dismutase 1, SOD1?) to spinal motor neurons. Toxic CSF procured from ALS patients experimentally degenerates cultured neurons. By CSF bulk flow to upper and lower spinal cord, there could be seeding and spreading of harmful materials (including inappropriate T cells) to the extracellular microenvironment of vulnerable motor neurons. Consequent neuroinflammatory events would destabilize vulnerable motor neurons.

Deficient CSF neurochemistry also harms ALS patients. Transthryetin and cystatin C in CSF originate mainly in CP. Both of these neuroprotectant CSF proteins are lowered in ALS. Decreasing levels in CSF beg the question whether the CP synthesis of these protective proteins is curtailed in ALS by a malfunctioning BCSFB. Transthyretin stabilizes protein and β-amyloid conformation and therefore is likely protective against AD neurodegeneration. Unfortunately, post-translational transthyretin is altered in ALS patients. Cystatin C, a cysteine protease inhibitor, undergoes longitudinal changes in CSF concentration in ALS that reflect fast-slow stages of the disease, and predict survival. These clinical and research observations prompt the consideration of beneficial "booster" therapy to shore up the impoverished cystatin C and transthyretin levels in ALS-patient CSF.

Parkinson's Disease

Aggregated α-synuclein (α-Syn), including the toxic oligomer of this presynaptic protein, is the characteristic focal point in Parkinson's disease (PD) pathophysiology. Normally α-Syn coordinates with the SNARE proteins in the basal ganglia neurons to manage the vesicular packing, fusion and release of dopamine from axon terminals into adjacent synapses. Epigenetic factors, coupled with genetic predispositions, can lead to conditions that alter the three-dimensional structure of α-Syn. Neurotransmission is adversely affected when α-Syn undergoes aberrant protein conformation; such misfolding predisposes to α-Syn aggregation (oligomers and fibrils) and consequently the onset of PD and Lewy body disease.

Upon export (leakage) from neurons, the extracellular misfolded α-Syn can spread widely by flow into CSF and thereafter to many brain regions. Ependymal and supependymal α-Syn immunoreactivity (5G4 antibody to β-sheet component of α-Syn) correlates with Braak stages of Lewy-related pathology. Moreover, human astrocytes and neurons internalize disease-associated extracellular α-Syn. This prompts consideration of cell-to-cell spread of the misfolded α-Syn among CNS regions, and evaluation of the possible destabilizing role of ISF–CSF bulk flow in seeding/spreading the PD or Lewy Body neurodegeneration phenomena.

Epigenetic and familial factors in trace element handling by the BBB and BCSFB are subjects of wide interest in regard to impact on PD, ALS, and AD. Dyshomeostasis of Fe, Cu, and Mn distribution between blood and CNS regions affects the onset and exacerbation of several types of neurodegeneration, including PD. Excessive trace element concentrations in the CSF–brain result from barrier breakdown or inappropriate expression of transporters that move Fe, Cu, and Mn into and out of CNS. Ceruloplasmin and transferrin, respectively, in CP facilitate Cu and Fe movement across BCSFB into CSF. Choroidal apical membrane transporters, both secretory (ATP7a) and reabsorptive, control net flux of Cu into the ventricles; their expression levels require precision and coordination to stabilize the CSF level of Cu. Excesses or deficiencies of trace elements in CNS distort neuronal functions and provoke neuroinflammation. Copper-dependent enzymes require precise amounts of the element. Too much Cu binding to α-Syn and Aβ predisposes the formation of heterodimers that promote peptide (eg, α-Syn) aggregation and thereby increment harmful reactive oxygen species. Manganism, or excessive Mn ions in the CNS, has also been strongly implicated in PD. In terms of pathophysiological cascades that worsen PD (ie, the "upstream" vs. "downstream" metabolic defects), the failure of the CNS barrier–regulatory systems for Cu, Fe, and Mn translocation likely ranks as one of the initial ("upstream") factors that impact early stages of neurodegeneration.

Alzheimer's Disease

Several facets of CP–CSF functioning have a bearing on the cascading neurodegeneration of AD. Foremost is the progressive decrease in CP viability and reduced CSF formation with advancing age and senescence (Fig. 11.12). Coupled with curtailed production of CSF by CP is the enlarging ventricular system as the brain

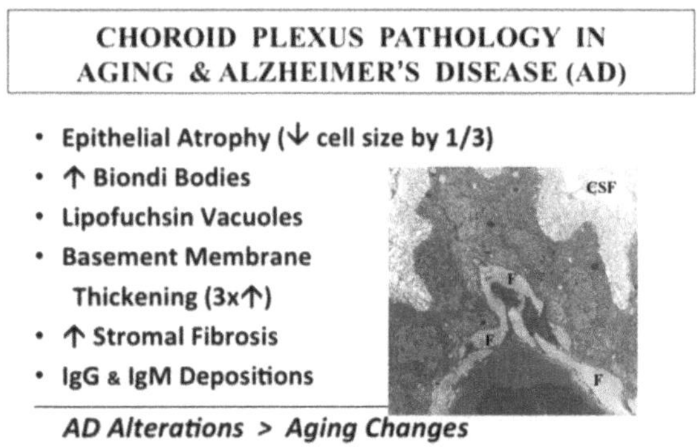

FIGURE 11.12 Histological and biochemical aspects of choroid plexus senescence: aging exacts a toll on human CP, leaving interstitial immune deposits, collagenous fibrosis, calcification, and pathological cellular inclusions (ring-like lipoid Biondi bodies and lysosomal lipofuscin granules) in the epithelium. Deposits of immunoglobulins and collagen in extracellular space impair diffusion of substrates exchanged between plasma and CSF. In the electron microscopy image inserted in this diagram of aging (36 months) rat CP, there is substantial interstitial fibrosis (F) between the outer (top of figure) epithelial cells and the inner (bottom) choroidal vasculature. Choroid cells in some humans undergo atrophy in advanced aging, bringing about epithelial shrinkage, and vacuole formation. These structural changes accompany curtailed expression of membrane-associated enzymes (Na–K-ATPase), the water channel AQP1, and functional proteins, such as transthyretin. Such structural and biochemical alterations lead to reduced CSF turnover and lower secretion of certain proteins. Alzheimer's disease-altered CP is usually more damaged than in aging. Significantly, though, the ability of CP to retain stability of CSF Na^+, K^+, pH, HCO_3^-, folate, and ascorbate remains remarkably intact in AD. *Adapted from Johanson C. Cerebrospinal Fluid Res 2008;5:10.*

TABLE 11.3 Cerebrospinal Fluid Dynamics and Volumetrics in Aging and Neurodegeneration

	Rat Aging[a]			Human Disease[b]		
	3 Month	19 Month	30 Month	Normal	NPH	AD
CSF formation rate (ml/min)	0.00121	0.00148	0.00065	0.40	0.25	0.20
CSF volume (space) (ml)	0.156	0.196	0.308	150	300	250
CSF turnover rate (volumes/day)	11	10.8	3.0	4.0	1.2	1.2

[a] *CSF formation rate and volume in rats were determined by indicator dilution during ventriculo-cisternal perfusion.*
[b] *CSF formation rate and volume in patients were quantified by a modified Massermann technique.*
AD, Alzheimer's; *NPH*, normal pressure hydrocephalus.
Volumes/day = number of times each day that the entire CSF volume is renewed.
CSF turnover rate = formation rate/volume.
Johanson CE, Duncan JA 3rd, Klinge PM, Brinker T, Stopa EG, Silverberg GD. Multiplicity of cerebrospinal fluid functions: new challenges in health and disease. Cerebrospinal Fluid Res 2008;5:10.

undergoes atrophy; together these factors set a lower turnover rate of CSF (Table 11.3) with a concomitant reduction in sink clearance action on brain catabolites. The consequent buildup in "neural debris" in the CNS extracellular fluid (ISF + CSF), resulting from less flushing action by CSF, adversely affects cerebral metabolism and neurotransmission.

Permeabilities of both the BCSFB and BBB increase with aging and degeneration of the parenchymal and vascular structural elements. Aβ oligomers, for example, augment BCSFB disassembly/permeability by stimulating production of matrix metalloproteinases and upregulating inflammatory genes/cytokines in CP–CSF. Moreover, the expression of six human CP-regulatory proteins implicated in apoptosis and mitochondrial dysfunction has been identified in AD. Thus barrier breakdown is accelerated by the aging-associated greater damage from cumulative oxidative and nitrogen-reactive products; and by interactions of pathogens, immune cells and cytokines with barrier cells/tight junctions. Barrier disruption in advanced aging and AD enables leakage of plasma proteins and other large molecules (normally excluded) into CSF and ISF. Such protein leakage increases osmotic pressure and stresses the central microenvironment in AD, already compromised by neuronal dyshomeostasis (eg, misfolded proteins) and slower excretion of catabolites secondary to compromised CSF sink action.

How effectively can CP mediate CSF homeostasis even in senescence and AD? Neuropathologists have commented on the gross structural changes in CP histopathology in autopsied tissue from very old and AD patients (Fig. 11.12). The presence of interstitial fibrosis, epithelial cell flattening, psammoma bodies, immune molecule accumulation (IgG, IgM and C1q) and amyloid deposition give the impression of a tissue incapable of effectively carrying out a wide array of homeostatic roles. Yet, CSF compositional studies of AD patients reveal concentration stability of ascorbate, folate, Na^+, K^+, pH, and HCO_3^-. CSF transthyretin tends to be lower in AD, but nevertheless still at a level likely to help counter further instabilities in brain Aβ conformation. Consistent changes in CSF Aβ, tau, and F2-isoprostane concentrations in AD almost certainly reflect brain more than CP metabolic distortions. Interestingly, the ability of CP in senescence and dementia states continually to remove Aβ efficiently from CSF (in contrast to the dwindling number of LRP-1 and P-gp transporters in BBB) speaks to persisting choroidal transporters for preventing an even greater Aβ burden in CNS.

NEW OPPORTUNITIES FOR MODIFYING THE CHOROID PLEXUS–CSF NEXUS IN BRAIN DISORDERS

With an ever-increasing appreciation of the diverse ways that CP serves the brain, to confer stability of fluid dynamics and a healthy environment for neurons, there is the growing realization that a sound BCSFB is fundamentally important for brain viability. Working hand-in-hand, the CP and BBB microvessels are complementary partners to keep the neural parenchyma supplied with essential substrates and protected from perturbations in extracellular solutes and pathogens. Breakdown of one or both barriers places the neuronal networks at risk.

Although homeostatic and compensatory systems in the CNS protect and heal the CSF–brain, it is sometimes necessary to have additional medicinal support to rescue the failed systems "inside" the barriers. Pharmacological repair of the injured/infected brain has been hampered by the protective barriers that minimize drug penetration to needy targets. However, new approaches in CSF pharmacology offer considerable hope for more efficacious therapies of neural disorders.

Hydrocephalus and intracranial hypertension (pseudotumor cerebri) both wreak havoc on the CNS due to elevated ICP. Patients of neurosurgeons and neurologists would benefit from new agents to reduce ICP. Since diuretics have limited efficacy in controlling ICP, the search is on for agents from other pharmacological classes. A neuroendocrine approach, using recombinant peptides (or analogs) of endogenous fluid-regulating hormones, may provide the answer to the decades-long question of what agents can best attenuate ICP. ANP, or similar agents that bind to and activate natriuretic peptide receptors, show promise in relieving ICP. Especially attractive is the potential use of intranasal instillation to enable peptides to circumvent the BBB and gain access to CP–CSF by directly penetrating the interior of the brain. This approach bypasses the BBB and BCSFB that typically preclude permeation of plasma-borne peptides across the tight junctions into brain and CSF.

Meningitis usually involves an inflamed CP and meninges. Antibiotic drugs to treat bacterial meningitis have been designed to "escape" reabsorption from CSF by CP transporters, once the drug has penetrated the ventricles from blood. Standard doses of drugs, such as penicillin and cephalothin, however, have limited usefulness in meningitis due to their rapid regulatory removal from CSF; however, newer antibiotics, such as imipenem and ceftriaxone stay in CSF, once transported there, and attain therapeutic concentrations against bacteria in the CNS. A novel approach to minimize or prevent entry of pathogens across the BCSFB into CSF, and then their ventriculo-subarachnoid transit "downstream" to the arachnoid membrane, would be to thwart binding (the initial step in barrier translocation) of the virus or bacteria to the epithelial cells at the interface. Now recognized by neuropathologists as the main port of entry of certain pathogens into the CNS, the CP is a logical target for (1) assessing ways to preclude the penetration of viruses/bacteria across the BCSFB and (2) attenuating neuroinflammatory responses in CP to reduce cytokine access to CSF.

Stroke/ischemia and TBI have been the subjects of scores of investigations to characterize the postinsult homeostatic upregulation of numerous growth factors and neurotrophins in CP, ependyma, CSF, and brain parenchyma. With acquired information about endogenous responses, investigators have successfully used boosting doses of the candidate factors (introduced into CSF by injection, perfusion, or osmotic pump infusion) to ascertain the nature of specific repair responses. Dozens of peptides, alone or in combination and delivered before or after injury, have proven effective in minimizing damage or restoring normal functions. This encourages "peptide cocktails," instilled directly into the CNS (by intraventricular, intraparenchymal, or intrathecal routes), for the purpose of expediting healing responses. As more of these "cocktails" are tested, the benefits of such exogenous peptide supplementations to CSF will become more evident as to usage by neurosurgeons and other clinicians. A methodological corollary of this approach is to enhance peptide presentation via the BCSFB either by the expression of transgenes delivered to CP/ependyma via vectors, or by identifying new therapeutic peptides through bacteriophage display.

Peptide therapy for autism and other social disorders, to shore up levels of oxytocin deficiency in the CNS, will also likely benefit from regimens to adjust CSF–brain levels of peptides. Intranasal instillation of oxytocin has the considerable advantage of bypassing the BBB, but additional investigations are necessary to ascertain why not all ASD patients benefit from the supplemented peptide. Is this due to limited access of exogenous oxytocin to receptors in some but not all children, or to an unappreciated pharmacodynamic variable? As recently treated by Young and Barrett, context is critical in hormonal therapy for ASD; and so oxytocin treatment should be combined with behavioral therapy. Moreover, an important pharmacokinetic/pharmacodynamic consideration is neuronal network maturation, which may be less harmed by intermittent than with continuously administered oxytocin (in this regard, early-developmental, sustained exogenous oxytocin reduces cognate receptor expression). Combining oxytocin with another peptide, the melanocortin-4 receptor (MC4R) agonist, deserves consideration because the later enhances several positive responses associated with oxytocin. Furthermore, the reliability/efficiency of intranasal delivery of peptides to CSF in young children versus adults deserves attention in rigorously-controlled studies of age and developmental effects.

MS and the experimental counterpart, EAE, have been extensively analyzed for the BCSFB role in regulating immune cell traffic and cytokines into ventricular CSF. The IFN system, α- and β-families, has a critical anti-inflammatory role in the CNS. IFN therapy shows promise for controlling adverse inflammatory phenomena in human MS and animal EAE. Central versus peripheral aspects of IFN modulation need clarification. Mice injected intracisternally with polyinosinic-polycytidylic acid (poly I:C) demonstrated stimulation of dendritic cells/macrophages (CD45/CD11b positive) in CP and meninges to upregulate IFN-β. The CP–CSF–brain IFN-β surge was associated with neuroprotection as the anti-inflammatory IL-10 and CXCL10 were also increased enabling microglia and astroglia to reclaim axonal function. Early EAE in this mouse model was stabilized as long as IFN-α/β was expressed. This Danish study by Korooshi et al. exemplifies how the ability to access CNS-innate IFN-driven regulatory processes in CP is a promising clinical immunological approach to target MS and other neural diseases.

ALS benefits from MS research that has elucidated how CP controls leukocyte movement into the ventricles and effects cytokine neuroinflammatory responses impacting the brain. Neuroinflammation in ALS hastens the development of spinal cord pathology. Immunization with a myelin-derived antigen activates CP, resulting in recruitment of immunoregulatory cells into the CNS that attenuates disease progression in an ALS mouse model. Mutations in SOD1 have been implicated as a risk factor for ALS development. However, the intracerebroventricular infusion of monoclonal antibody or the derived Fab fragment, against misfolded forms of mutant SOD1, delays mortality in a mouse model of ALS. Viral vectors have been used to carry growth factor transgenes to cells surrounding the CSF system: CP and the ependyma surrounding the ventricles and spinal cord central canal. In this regard, the adeno-associated virus serotype-4 (AAV4)-mediated expression of VEGF and IGF-1 by CSF-bordering cells helped to improve survival outcome in familial ALS mice of the SOD1(G93A)-mutated strain. In several experimental studies, then, the introduction of antibodies or viral vectors to CSF has helped to remediate ALS-like symptoms and associated neuropathology.

PD and AD share pathological features, such as misfolded protein accumulation and consequent interference with synaptic neurotransmission. The subsequent aggregation and deposition, of both the Syn and Aβ proteins, remain a primary problem awaiting better therapeutic resolution. Any successful attempt to increase CSF turnover rate, perhaps by intervening at the regional level of the CP or arachnoid in the aging brain, will likely lessen the burden of misfolded protein conglomerates in the brain extracellular space. Another pharmacological approach is augmenting the expression of BCSFB and BBB transporters that remove protein fragments associated with Syn and Aβ. For example, intravenous immunoglobulin contains autoantibodies against Aβ, removable from the brain of AD and MS patients; the removal mechanism, demonstrated in transgenic AβPP (precursor protein) mice, is actively reabsorbed Aβ in CSF by the LRP-1 transporter in CP apical membrane. An alternative but perhaps supplemental approach is to regulate immunologically the entrance of macrophages and leukocytes across CP:epithelial and CSF:ependymal membranes to accelerate removal of misfolded protein debris, by phagocytosis or antigen presentation to the disposing cells. Clearly, pharmacological manipulation of barrier transporters, leukocyte-transfer mechanisms and CSF dynamics shows promise for cleaning and repairing the neuronal interstitial microenvironment.

SUMMARY AND OVERVIEW

CP-CSF has been an under-utilized system for treating a variety of neurological disorders. The "cross-talk" or signaling between CP and brain, involving hormones, antibodies, vesicles, trophic proteins, and leukocytes, points to numerous targets or CSF-mediating "vehicles" for remediating brain diseases. The experimental effectiveness of intranasally instilled compounds, including macromolecules, to access rapidly the CSF and brain, opens many doors of opportunity for "topically" treating CNS pathology. The intranasal route thus overcomes a plethora of difficulties, historically encountered, to access neurons and glia by designing agents to "outwit" the BBB impenetrability.

Pharmacological control over neurogenic regions, to modify the nascence, subtype molding and migration of cells, is most promising to promote healing of neural/glial injuries or deficits. Ependymal cells that are neurogenically dormant are identifiable by single-cell transcriptome analysis in regard to the specific molecules (such as VEGF) that activate them following injury. Endogenous CSF modulation of fetal periventricular neurogenesis predicts that exogenous hormone/drug presentation to the ventricular system will be effective in fostering neuron birthing in disease-compromised adults. Here, indwelling ventricular catheters can aid in delivering pharmacological agents into CSF of neurosurgical patients (eg, parenchymal stroke, subarachnoid hemorrhage, and TBI). Introduction of therapeutic neurospheres, and quite possibly CP explants, to the ventricles represents an emerging new clinical paradigm in strategies to restore brain viability.

Finally, the current explosive growth in identifying extensive and highly specific panels of human CSF chemical biomarkers is a key translational area of modern neuromedicine. The wider availability of newer-generation technology facilitates the acquisition of finer clinical data for differential diagnosis. A panel of seven tau peptides in human CSF, for example, is now available by antibody-free, mass spectrometry. With the advent of more sophisticated methodologies to delineate CSF peptides, proteins and glycoproteins, and resolve a wider array of lipids, greater progress will be made on distinguishing metabolic groups of CSF neurochemical biomarkers and subsets of neurodegenerative disorders. CSF biomarking needs to separate the origins of biochemical molecules, that is, those derived from CP and ependyma versus those emanating from neurons and glia. Significant advances in CSF biomarker identification, classification, and computational informatics are on the horizon. Thus the future looks sound for advancing CSF tools to aid neural disease diagnosis, prognosis and therapy evaluation.

Acknowledgments

Appreciation is expressed to J. Johanson for help in constructing illustrations and figures. The National Institutes of Health is thanked for sponsoring NIH RO1grants, C.E.J., P.I. (NINDS NS27601 and AG 027901) that have furnished choroid plexus and blood–brain barrier data, some of which is recapitulated in this chapter.

Further Reading

[1] Baruch K, Deczkowska A, David E, Castellano JM, Miller O, Kertser A, et al. Aging. Aging-induced type I interferon response at the choroid plexus negatively affects brain function. Science 2014;346:89–93.

[2] Clarkson BD, Walker A, Harris M, Rayasam A, Sandor M, Fabry Z. Mapping the accumulation of co-infiltrating CNS dendritic cells and encephalitogenic T cells during EAE. J Neuroimmunol 2014;277:39–49.

[3] Gonzalez AM, Podvin S, Lin SY, Miller MC, Botfield H, Leadbeater WE, et al. Ecrg4 expression and its product augurin in the choroid plexus: impact on fetal brain development, cerebrospinal fluid homeostasis and neuroprogenitor cell response to CNS injury. Fluids Barriers CNS 2011;8:6.

[4] Johanson C, Stopa E, Baird A, Sharma H. Traumatic brain injury and recovery mechanisms: peptide modulation of periventricular neurogenic regions by the choroid plexus-CSF nexus. J Neural Trans 2010;118:115–33.

[5] Lun MP, Johnson MB, Broadbelt KG, Watanabe M, Kang YJ, Chau KF, et al. Spatially heterogeneous choroid plexus transcriptomes encode positional identity and contribute to regional CSF production. J Neurosci 2015;35:4903–16.

[6] Spector R, Snodgrass SR, Johanson CE. A balanced view of the cerebrospinal fluid composition and functions: focus on adult humans. Exp Neurol 2015;273:57–68.

[7] Stolp HB, Liddelow SA, Sá-Pereira I, Dziegielewska KM, Saunders NR. Immune responses at brain barriers and implications for brain development and neurological function in later life. Front Integr Neurosci 2013;7:61.

CHAPTER

12

The Cerebral Cortex

R.S. Swenson, A.T. Gulledge
Geisel School of Medicine, Hanover, NH, United States

INTRODUCTION

The human cerebral cortex consists of a 2–5-mm-thick layer of gray matter containing 10–14 billion neurons that, for most of its surface, is stratified in six layers. The six-layer cortex is described as "neocortex" to distinguish it from a much less extensive region of cortex having fewer layers (allocortex). The mammalian neocortex has shown a gradual evolution in relative size, such that in humans only a very small region, mostly in the medial temporal region, remains as allocortex. These areas include much of the parahippocampal gyrus (particularly the piriform and periamygdalar cortex), as well as the hippocampus. Befitting the relative proportions of neocortex and allocortex, this chapter focuses primarily on the neocortex, with brief reference to the allocortex in the discussion of the temporal lobe and other areas, including the anterior insula and cingulate gyrus. The olfactory cortex is described in greater detail in Chapter 17, which deals with olfaction, and the subiculum and hippocampus are discussed in the context of learning and memory in Chapter 32.

We first consider the basic anatomical organization of the neocortex, including its laminar and tangential structures. Subsequently, the basic circuitry represented within the cortex is discussed along with the importance of neuromodulators in the regulation of cerebral cortical activity. Lastly, the regional topography of the cortex is discussed, correlating this with both classical and more recent functional understandings.

While only few millimeters thick, the human neocortex would, if flattened out, cover about 2500 cm^2; about the size of a coffee table. In order to fit into the confines of the cranium, the thin cortex has multiple folds with convexities (gyri) and concavities (sulci) in the surface. Although there is sufficient reproducibility of gyri and sulci to define landmarks of the cortex via these folds (Fig. 12.1), there is also substantial individual variability in the pattern. Importantly, some of the larger anatomical subdivisions based on the folding pattern correlate roughly with cytoarchitectonic divisions of the cortex, as defined by Brodmann in the early part of the past century (Fig. 12.2). Furthermore, these cytoarchitectonic divisions correlate reasonably well with major functional subdivisions of the cortex, and are listed with the prefix "BA" in the following discussion, in reference to Brodmann's areas presented in Fig. 12.2.

Through the late 18th and 19th centuries, it became apparent to clinicians that various brain areas, including subregions of the cerebral cortex, were specialized in terms of their function.

Conn's Translational Neuroscience
http://dx.doi.org/10.1016/B978-0-12-802381-5.00021-X

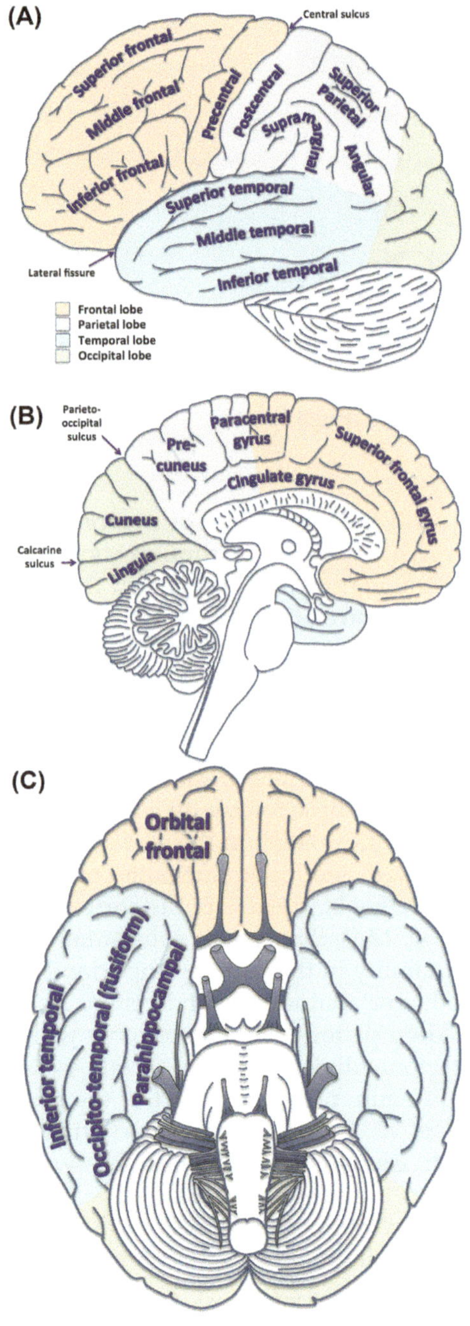

FIGURE 12.1 Simplified schematic of the cerebral cortex with major gyri labeled. (A) Lateral hemisphere. (B) Medial hemisphere. (C) View from the ventral aspect of the brain.

This recognition evolved into a modular understanding of cortical functions where functions were defined by the clinical deficits resulting from damage. This understanding was further informed by observations following electrical cortical stimulation and, occasionally by the clinical manifestations of focal seizures. The modular organization of the cortex was also found to correlate well with cytoarchitectonic structure (eg, Brodmann's areas), which still provides a useful framework for understanding cortical function, although newer techniques for functional assessment of the brain have significantly refined our understanding of cortical organization. One interesting and important observation is that many of the functionally defined cortical modules exhibit selective expression patterns of clusters of genes. This brings the crude, clinically defined macroscopic organization of the cortex down to the cellular and molecular level.

An additional organizational principle that has become increasingly apparent segregates the cortex into a posteriorly located perceptual component that is significantly larger than the anteriorly positioned executive portion. The dividing line separating the perceptual from the executive follows the central sulcus and then the lateral fissure, thereby defining the anteriorly positioned frontal lobe to be the great majority of the executive portion of the cortex. The perceptual portion of the cortex is responsible for the processing of sensory input and then the recognition of the sensation, assigning significance to it, integrating it with other sensory inputs, and directing appropriate amounts of attention toward the sensation. The executive portion of the cortex is mostly responsible for how (and whether) to respond to cues perceived in the environment, and then planning and initiating appropriate responses, while at the same time suppressing inappropriate responses. It should be apparent that both the perceptual and executive components of the cortex need to interact closely together

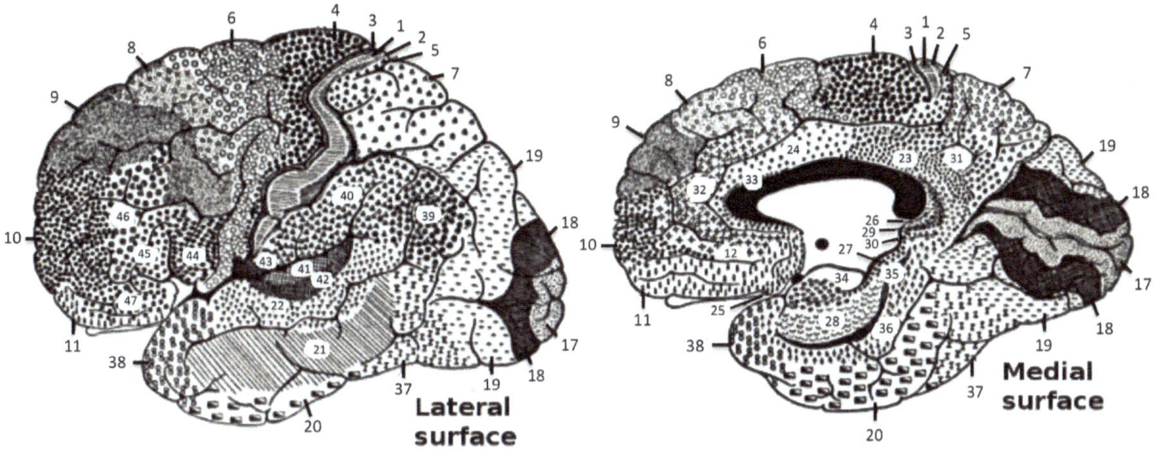

FIGURE 12.2 Cortical areas defined on histologic basis by Korbinian Brodmann [Brodmann K (1909). Vergleichende Lokalisationslehre der Großhirnrinde: in ihren Prinzipien dargestellt auf Grund des Zellenbaues]. Note that the text refers to these regions by their numbers prefixed by "BA" for Brodmann's areas.

for optimal behavioral responses to be produced. This functional interaction explains why even small lesions of the cortex can produce complex behavioral deficits that involve both perceptual and executive function. Indeed, functional imaging of the cortex during higher-order cognition typically reveals an intricate interplay between regions of the anterior and posterior cortex; that is, the perceptual and executive regions.

Despite the enhanced insight afforded by tools such as functional neuroimaging, classically defined regional topography of function remains useful in explaining neurological and/or neuropsychiatric effects of cortical damage. The following discussion is largely organized according to these structural and functional cortical subdivisions. Please be aware that this represents but the most superficial level of understanding of a highly complex whole. One must also be cognizant of the interrelationships of cortical regions when considering cerebral cortical involvement in vision (Chapter 19), hearing (Chapter 26), olfaction (Chapter 17), language function (Chapter 31), motor function (Chapter 5), and behavior (Chapter 29).

CORTICAL ARCHITECTURE

Before considering the gross topography, circuitry, and functional anatomy of the cortex, it is important to understand the general organization of the neocortex. Histological techniques of the 19th century revealed a laminar cytoarchitectural structure of the cortex that can be divided into six layers, numbered (classically with Roman numerals) sequentially from superficial to deep (Fig. 12.3). Although all areas of the neocortex have layers, the thickness and complexity of these layers vary widely from region to region, with some cortical areas lacking a given layer, while other areas may have individual layers that can be further subdivided into multiple sublayers having different physiological and morphological properties. The laminar structure of the cortex not only results directly from layer-specific differences in neuron density, but also reflects differences in neuron subtype and afferent inputs.

The most superficial cortical layer, residing at the border of the brain and pia mater, is the molecular layer, layer I, which contains relatively few neuronal perikarya. Instead, layer I

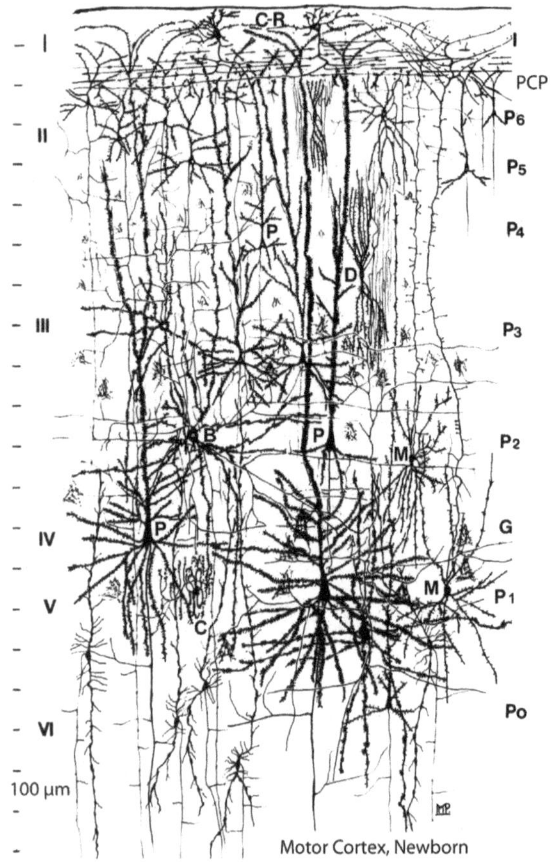

FIGURE 12.3 Montage of camera lucida drawings from a Golgi preparation illustrating the location (distance from pial surface) and the size of the main neuronal types of the newborn motor cortex. The current laminar nomenclature from layer I to layer VI (superficial to deep) is included on the left side of the figure. Note the very large pyramidal neurons in layers V and VI and the smaller size of pyramidal neurons in layers II and III. Also note that the pyramidal neurons have a long process (apical dendrite) that extends toward and, usually, into layer I. On the right side of the figure is a numbering system for cortical layers based on the developmental sequence of pyramidal neuron migration to the cortex (P1 being the earliest arriving pyramidal neurons and P6 being the latest). Note also that there are scattered smaller neurons that are not pyramidal in shape (inhibitory). Scale (left) = 100 μm increments. *B*, Basket cells; *C*, chandelier cells; *C–R*, Cajal–Retzius cell; *D*, double tufted neurons; *M*, Martinotti cells (a variety of somatostatin-positive neurons); *PCP*, remnants of the pyramidal cell plate; *Po*, polymorphous neuron layer. *Figure courtesy of Professor Miguel Marin-Padilla.*

is comprised of a rich plexus of dendrites from neurons in deeper cortical layers, and axons carrying afferent input from both local and remote locations. Beneath layer I is the external granular layer, layer II, which is densely packed with neurons, followed by the thicker, but less dense, layer III, the external pyramidal layer. Beneath layer III is the internal granular layer, layer IV, which varies in thickness and complexity depending on cortical region, and the infragranular layers, V and VI, which exist between layer IV and the white matter beneath the cortex. The internal pyramidal cell layer, layer V, is a wide swath of large neurons of medium density, while layer VI, the multiform or fusiform layer, is less organized and has many neuron morphologies. Each of these cortical layers contain a variety of different neuronal shapes, sizes, and densities, as well as different organizations of local and afferent inputs. The varying thickness and appearance of cortical layers, in turn, provide the foundation for identification of Brodmann's areas (Fig. 12.2). The following discussion considers the gross topography and then the cellular and synaptic organization within and between the layers of the cortex.

CORTICAL LOBES

Broadly speaking, the cerebral cortex is divided into four main lobes easily visible on the surface of the brain (Fig. 12.1). There is a much deeper region, the insula or Isle of Rile (sometimes considered to be a fifth lobe of the brain) that is not visible from the surface. Although the majority of the cortex is visible from the lateral aspect (Fig. 12.1A), extensive portions of the temporal, frontal, and occipital lobes also appear on the ventral surface of the brain (Fig. 12.1C). Also, the frontal, parietal, and occipital lobes continue onto the medial surface, abutting the falx cerebri in the midline (Fig. 12.1B).

The lobes of the brain are defined by important surface landmarks that tend to be conserved across individuals. The central sulcus is visible on the lateral aspect of the hemispheres, extending ventrally from the dorsal aspect of the hemisphere and nearly reaching the lateral fissure. This is the dividing line between the frontal lobe anteriorly and the parietal lobe immediately posterior to the sulcus. The very deep lateral fissure separates the frontal and parietal lobes from the temporal lobe, which is immediately ventral to this fissure. The occipital lobe is the posterior-most part of the cortex. While there is no gross anatomical boundary that clearly separates the occipital lobe from the posterior parietal lobe on the lateral aspect of the hemispheres, a prominent parieto-occipital sulcus is visible on the medial aspect of the brain. The insula is often described as the fifth lobe of the brain. This consists of small gyri located in the depths of the lateral fissure that can only be visualized by separating the inferior frontal lobe from the superior temporal (ie, by opening the lateral fissure).

As described previously, the perceptual regions of the cortex include the parietal, occipital, and temporal lobes and the posterior insula, and are predominately involved in perceiving, recognizing, and interpreting the environment. The executive cortex mostly consists of the frontal lobe and anterior part of the insula, and is involved in evaluating possible courses of action, determining their suitability for the situation, and then planning, initiating, and generating appropriate motor responses.

CEREBRAL CORTICAL ORGANIZATION

A general principle of cortical organization, both in the perceptual and executive portions of the brain, is that cortical regions that have more easily recognized functions (primary cortices)

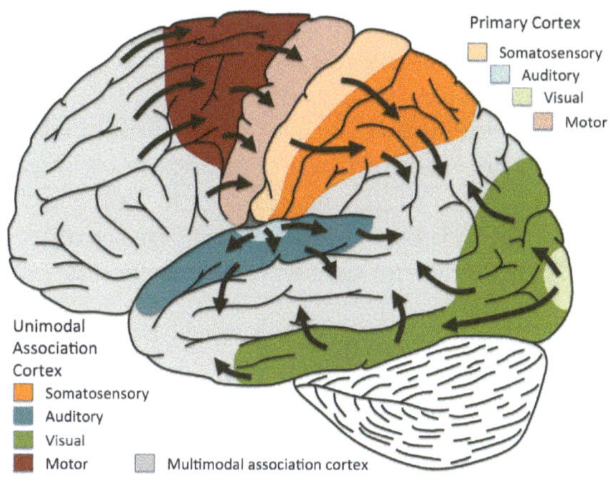

FIGURE 12.4 General location of the primary sensory (somatosensory, auditory, and visual) and motor cortex, and of the unimodal and multimodal association cortex. *Arrows* indicate the primary direction of information transmission, however, this information flow is bidirectional. Note that the largest proportion of the cortex is classified as multimodal association cortex.

are bordered by regions having more complex organizational functions, termed "association cortex" (Fig. 12.4). In this understanding, the primary cortices are often referred to as "lower-order" while the association cortices are termed "higher-order" regions of the cortex. The primary and association cortices are interconnected by short corticocortical connections consisting of axons extending from the cortex of one gyrus to adjacent regions of cortex through the white matter at the core of the gyrus. As one progresses further away from a primary cortex, the degree of complexity of function within the association cortex increases. For example, a primary sensory cortex (ie, the region of termination of axons conveying information relating to one sensory modality) is bordered by the cortex that is involved in recognizing the nature of the sensation and assigning importance to the sensation. This latter region is termed unimodal associational cortex. Adjacent to regions of unimodal association cortex are areas of multimodal association cortex that integrate information from two or more sensory modalities, allowing ever more complex representations of the environment. Multimodal sensory association cortex is responsible for prioritizing the significance of environmental cues and in determining which sensory inputs should be attended to. Thus, cortical areas of lower functional complexity are bordered by regions exhibiting progressively more complex sensory processing and integration.

A similar hierarchy of organization exists in the executive (ie, motor) regions of the brain, which are mostly located in the frontal lobe (Fig. 12.4). The primary motor cortex, at the most caudal edge of the frontal cortex, is responsible for generating relatively simple motor responses. Adjacent cortical regions rostral to the primary motor areas are involved in generating patterned responses and initiating action. Cortical regions that are even further removed from the primary motor cortex are involved in analyzing the appropriateness of responses and preventing inappropriate responses. Thus, on both the perceptual and the executive side, the cerebral cortex has a fundamentally hierarchical organization, reflected in primary ("simple" or "lower-order") cortex surrounded by association areas with progressively more complex "higher-order" function. It should not be surprising that damage to the higher order–associational areas are likely to result in more complex and multimodal deficits.

THE CORTICAL CIRCUIT

Although the horizontal laminar structure of the cortex has long been known (Fig. 12.3), subsequent investigations of the cellular structure of the cortex have revealed an even finer translaminar (ie, vertical) organization. This is most apparent in primary sensory areas where stacks of neurons orthogonally aligned to the pial surface have similar receptive field properties. This vertical organization also exists in motor cortex where neurons at different depths along the orthogonal axis exhibit correlated activity during specific limb movements. Based on these observations, it is now widely accepted that vertically oriented assemblies of cortical neurons sharing sensory and/or motor responsiveness represent a fundamental computational building block for the cerebral cortex. These modules are commonly referred to as "columns" of cortex although there is large area-to-area variability in their size (from ~10 μm to ~1 mm diameters), the sharpness of their boundaries, and the degree of uniformity in their cellular responsivity. Therefore no consensus yet exists as to exactly how columns should be defined. Nevertheless, numerous studies have revealed an exquisite vertical organization within the neocortex that is largely conserved across species and between neocortical areas. This vertical architecture, coupled with lamina-specific distributions of afferent inputs from distinct sources, form a generalized "cortical circuit" that includes a diverse array of excitatory and inhibitory neuron subtypes.

Functionally, columnar cortical circuits appear to act primarily as integrators of two broad classes of afferent input conveying "bottom-up" and "top-down" information. Bottom-up input arrives from primary nuclei of the thalamus, which relays input from most sensory systems and the spinal cord, cerebellum, basal ganglia and brain stem, and corticocortical input from lower-order–cortical areas. Top-down input arises primarily from higher-order areas of the cerebral cortex and secondary thalamic areas conveying information from the basal ganglia. Because it is not practical to probe cortical circuit function at the cellular level in humans, most of our knowledge in this area comes from animal studies. However, the tremendous homology of architecture across mammalian species suggests that neuronal connectivity patterns and functions within the cortical circuit that are studied in lower mammals are likely conserved in humans. In the later sections, we discuss the major neuron subtypes in the cerebral cortex, their synaptic connectivity into a cortical circuit, and mechanisms by which neuromodulators may influence circuit function.

Excitatory Neurons

A fundamental feature of cortical structures, including the neocortex and hippocampus, is that they are dominated by networks of excitatory neurons that make up about 80% of the total neuron population. While these neurons share many morphological and physiological features, including glutamatergic neurotransmission and spiny dendrites, subpopulations of these neurons are adapted for specific roles within cortical circuits. Broadly speaking, excitatory neurons in the cerebral cortex can be divided into four functional groups, each specialized in their synaptic input and projections to occupy a unique niche in the overall circuit. Later we discuss how each of these excitatory neuron subtypes contributes to cortical circuit function.

The first group of excitatory neurons includes spiny stellate neurons that are unique to layer IV, and which are generally considered relay cells for thalamocortical input. As their name implies, spiny stellate neurons radiate numerous spiny dendrites that are largely confined to layer IV where they are a preferred target for thalamocortical afferents from primary sensory nuclei. However, spiny stellate neurons receive the bulk of their excitatory input from the axon collaterals of neighboring spiny stellate neurons, with limited input from other cortical layers. In addition to their intralaminar collaterals, the axons of spiny stellate neurons project superficially to innervate the basal dendrites of pyramidal neurons in layers II and III. Thus, networks of spiny stellate neurons are positioned to integrate and amplify input from "core" (primary sensory) thalamic nuclei, and to relay this "bottom-up" information to pyramidal neurons in superficial layers (Fig. 12.5).

The second class of excitatory cortical neuron includes pyramidal neurons that have axonal projections that are limited to telencephalic target areas, including the neocortex, striatum, and/or amygdala. Many of these cells also project through the corpus callosum or anterior commissure to the contralateral cerebral cortex. The majority of intratelencephalic (IT) neurons reside in layers II and III. Like all pyramidal neurons, IT neurons have basal and apical dendrites receptive to different types of afferent input. The basal dendrites of IT neurons receive feed-forward, bottom-up input from spiny stellate neurons of layer IV, recurrent excitatory input from neighboring IT neurons, and long-distance corticocortical afferents from IT neurons in lower order–cortical areas. Regions of higher functional complexity ("higher-order" cortex) tend to have more IT cells and, therefore, thicker superficial cortical layers (I–III). The apical dendrites of IT neurons end in tufts of dendritic branches in layer I (see Box 12.1). These tufts receive excitatory, top-down feedback input from higher order–cortical areas and from so-called "matrix" (second-order) thalamic nuclei (Chapter 13). Due to their strong local recurrent connectivity, networks of IT neurons form a rich substrate for associative plasticity, and may participate in the reverberatory networks thought to underlie working memory. In addition to recurrent connectivity, local axon collaterals of IT neurons project to infragranular cortical layers to innervate pyramidal neurons in layer V while their long distance efferent axons target higher order–cortical regions in both the ipsilateral cortex and homologous cortical areas in the contralateral cerebral hemisphere. In vivo, IT neurons tend to exhibit infrequent stochastic firing in response to sensory input, with information thought to be "sparsely" encoded across a large population of IT cells.

The third class of excitatory neuron in the cortex are corticofugal (CF) pyramidal neurons in layer V that project to deep subcortical targets, including the spinal cord, brainstem, and tectum. Ipsilateral collateral branches innervate the striatum, matrix nuclei of the thalamus, or ipsilateral lower order–cortical areas. CF neurons

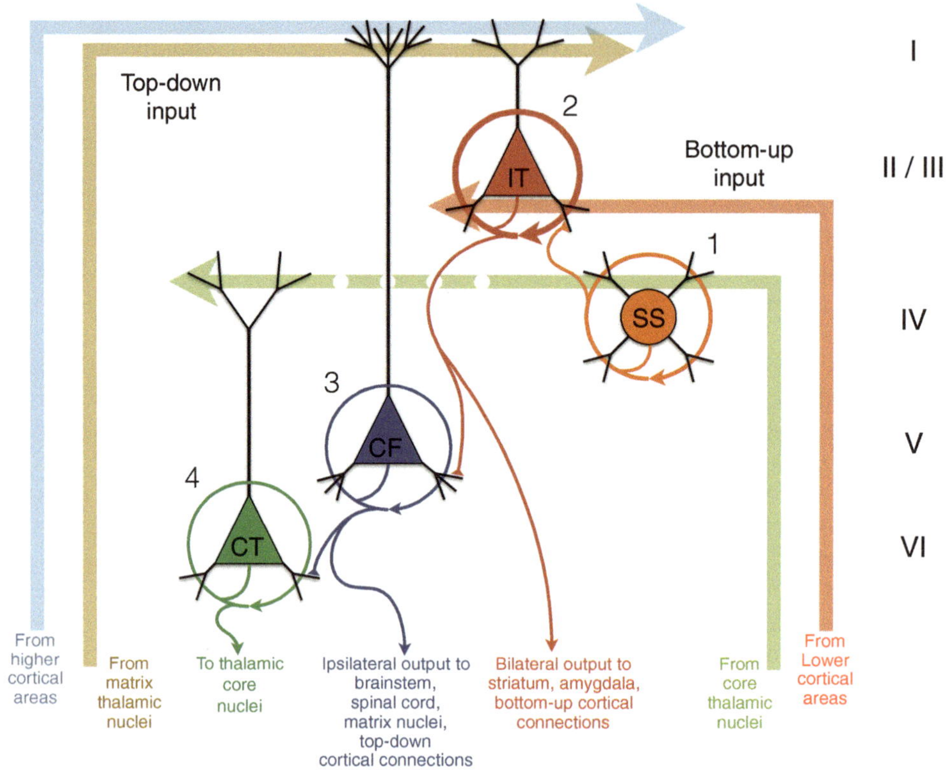

FIGURE 12.5 Simplified schematic of excitatory cortical circuitry in primary sensory cortex (lamina indicated by Roman numerals at right) shown as a series of unidirectionally connected recurrent excitatory networks. (1) A recurrent network of spiny stellate (SS) neurons in layer IV receives "bottom-up" sensory input from sensory-specific ("core") thalamic nuclei. (2) SS neurons project upward to innervate recurrent networks of IT pyramidal neurons in layers II/III. Descending local collaterals from IT neurons innervate corticofugal (CF) networks in layer V, while long-range IT projections provide input bilaterally to the striatum and amygdala, and also form bottom-up projections within the cortex. (3) CF neurons are responsible for subcortical projections to the brainstem, spinal cord, tectum, and "matrix" nuclei of the thalamus and also provide "top-down" corticocortical projections to the ipsilateral cortex. In addition to their recurrent connectivity, axon collaterals of CF neurons innervate corticothalamic (CT) neurons in layer VI. (4) CT neurons project back to core thalamic nuclei. Pyramidal neurons at all stages integrate top-down (corticocortical and thalamic matrix) input arriving at their apical tuft with progressively more processed bottom-up information arriving at their basal dendrites (see Box 12.2). In motor and associative cortical areas lacking a defined layer IV, core thalamic input typically arrives at the basal dendrites of IT neurons in layer III.

tend to have more elaborate dendritic trees than IT neurons, but exhibit a similar division in the types of bottom-up and top-down afferent drive arriving at their basal and apical dendrites, respectively. Unlike IT neurons, CF neurons are not primarily recipients of layer IV output, but instead receive a unidirectional projection from IT neurons at their basal dendrites. In addition, they are reciprocally connected to each other to form recurrent networks of CF neurons. Finally, relative to IT neurons, CF neurons tend to be more excitable and fire at higher frequencies in response to sensory stimuli in vivo.

The final major group of excitatory neurons includes corticothalamic (CT) pyramidal neurons primarily located in layer VI. Many of these neurons have a pyramidal morphology, with the exception that their apical tufts may

BOX 12.1

Pyramidal neurons are positioned to compare top-down expectations with bottom-up sensory input. Pyramidal neurons have two spatially remote dendritic fields receptive to different types of afferent input. "Top-down" input from higher cortical areas and matrix thalamic nuclei arriving at apical tuft dendrites are thought to convey prospective information that predicts current states. On the other hand, "bottom-up" input from lower cortical areas and core thalamic nuclei arriving at basal dendritic locations conveys updated interpretations of sensory input. The apical dendrite, which electrically couples the basal and tuft dendritic fields, together with two specialized spike-initiation regions in the cell, allows pyramidal neurons to detect and report coincident top-down and bottom-up input. The first specialized region, common to most mammalian neurons, is the axon initial segment (AIS), which is enriched in voltage-gated sodium channels and is the site of action potential initiation. In pyramidal neurons, the AIS is located at or near the base of the soma and therefore close to the basal dendritic field. The AIS effectively transduces strong bottom-up excitatory input into action potential output. Because the AIS is electrically distant (up to 1 mm) from the apical tuft dendrites, top-down input on its own typically fails to generate action potentials at the AIS. However, a second specialized region, enriched in voltage-gated calcium channels, exists in the distal apical dendrite near the base of the apical tuft. This "apical initiation zone" is a coincidence detector, generating broad calcium spikes lasting tens of milliseconds when top-down input at the tuft occurs at the same time as bottom-up-driven action potential initiation at the AIS. Calcium spikes occurring in the apical initiation zone, in turn, drive the generation of bursts of action potentials at the AIS. These bursts are thought to functionally convey meaning to target neurons downstream. This happens throughout the cortex, with each pyramidal neuron acting as an in-parallel coincidence detector for top-down and bottom-up inputs.

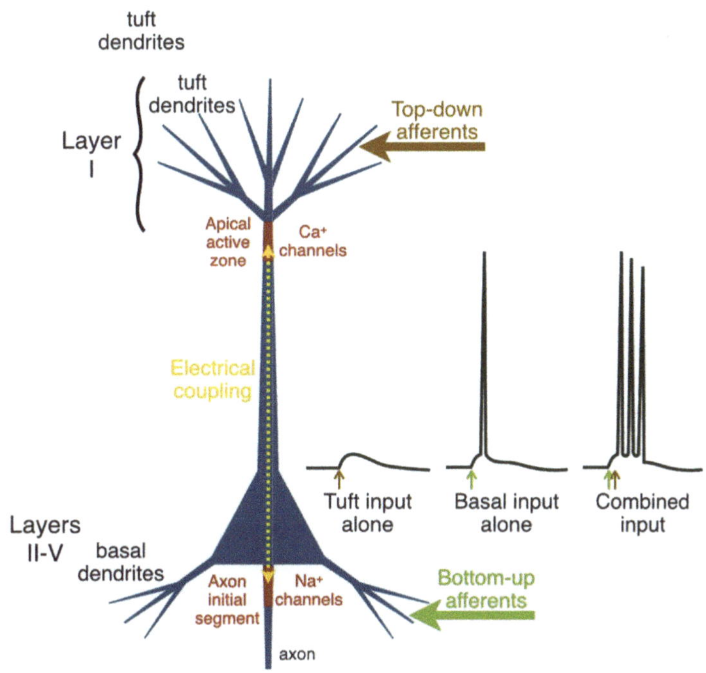

branch within layer IV, rather than in layer I. Thus, the tuft dendrites of many CT neurons are positioned to receive the same thalamocortical drive as received by spiny stellate cells while their basal dendrites receive input primarily from the CF neurons of layer V and reciprocal connections from neighboring CT neurons in layer VI. Thus, CT neurons complete a CT loop, by integrating the raw bottom-up information from the thalamus arriving at their tuft dendrites with the processed output of the cortical circuit arriving at their basal dendrites, and then directing their output back to the same core thalamic nuclei that target their cortical area.

Therefore, as represented in Fig. 12.5, the cortical circuit can be viewed as a hierarchy of unidirectionally connected recurrent excitatory circuits that integrate bottom-up sensory information with top-down intracortical activity. In primary sensory cortices, the circuit begins with thalamic input from sensory-specific thalamic nuclei arriving at spiny stellate neurons in layer IV, providing them with about 20% of their overall excitatory drive. Millisecond-by-millisecond, this stream of new sensory information from the thalamus is integrated with ongoing recurrent activity within the spiny stellate network and is projected upward to superficial IT pyramidal neurons. Recurrent networks of IT neurons, in turn, integrate this bottom-up information from layer IV with top-down input from higher order–cortical areas arriving at their apical tuft dendrites, and project their output to networks of CF neurons in layer V. CF neurons in layer V provide a final round of top-down/bottom-up integration before directing their output to subcortical targets in the midbrain and brainstem, and spinal cord. At the same time, this cortical output drives CT neurons in layer VI that provide feed-back to the thalamus. Similar circuitry exists in motor areas, with the exception that core thalamic drive arrives directly into the basal dendrites of IT neurons, as layer IV is largely absent.

Pyramidal Neurons as Coincidence Detectors

To understand information processing in cortical circuits, it is important to know how afferent inputs to specific cortical layers influence the output of the circuit elements described earlier. Pyramidal neurons have two prominent dendritic receptive zones: a basal dendritic tree close to the cell body and an apical tuft of dendritic branches in layer I. These two regions receive distinct types of information. Afferents arriving layer I convey "top-down" corticocortical input from higher (eg, more multimodal and associative) cortical areas, or from matrix thalamic nuclei, while afferents to layers III and IV convey "bottom-up" input from lower (eg, more unimodal) cortical areas and from core thalamic nuclei. Similarly, output from any given cortical area is relayed both "upward", in the form of feed-forward projections to increasingly associative higher order–cortical areas and/or association ("matrix") thalamic nuclei, and "downward", in the form of feed-back projections to more unimodal, lower order–cortical areas and/or to specific ("core") thalamic nuclei.

Top-down afferent input arriving at tuft dendrites is thought to convey "prospective" predictions for the current state of affairs while bottom-up afferents innervating basal dendrites convey newly acquired "perceptive" information distilled from sensory systems. By receiving both kinds of input, pyramidal neurons are positioned to compare top-down internally derived expectations with bottom-up real-world data. Pyramidal neurons are able to accomplish this comparison due to the specialized electrical properties of their apical dendritic trees that generate synergistic bursts of activity in response to coincident top-down and bottom-up input (see Box 12.1). In this way, millisecond-by-millisecond, pyramidal neurons are able to compare expectations to current events, and send that output to both higher-order (as updated bottom-up information) and lower-order (as updated top-down predictions) cortical areas.

Inhibition in the Cortex

Cortical networks are notable for the robust recurrent connectivity within and between the populations of excitatory neurons described previously (see also Fig. 12.5). These excitatory networks, combined with activity-dependent synaptic plasticity, provide a rich substrate for associative learning and the generation of complex behavior. However, recurrent synaptic connectivity between cortical neurons presents its own challenges, as excessive excitation can lead to epileptic activity and excitotoxicity (cell death). To keep activity under control, the cerebral cortex contains diverse specialized inhibitory GABAergic "interneurons" that make up about 20% of the total neuron population. Subtypes of cortical interneurons can be differentiated based on their morphological and physiological properties, as well as by their differential expression of calcium-binding proteins and peptidergic cotransmitters. While there may be more than 20 identifiable interneuron subtypes in the neocortex, they can be segregated into the three broad functional groups described in the following sections.

The largest group of inhibitory interneurons, found throughout layers II through VI, are known as fast-spiking (FS) interneurons due to their ability to generate action potentials at very high (>>50 Hz) frequencies (Fig. 12.5). FS interneurons are identifiable by their selective expression of the calcium-binding protein parvalbumin, and can be divided into two subtypes based on their distinctive clustering of axon terminals at pyramidal cell bodies (characteristic of "basket" FS neurons) or at the axon initial segments of pyramidal neurons (characteristic of "chandelier" FS neurons). By forming inhibitory synapses at or near axon initial segments, FS neurons are positioned to effectively gate excitatory output from pyramidal neurons. Further, FS neurons can fire action potentials quickly (~1-ms delays) in response to excitatory afferent drive, and are innervated by both long-distance afferents (eg, thalamocortical input) and local collaterals from pyramidal and spiny stellate neurons. Thus, FS neurons are positioned to rapidly provide both "feed-forward" and "feed-back" inhibition (see Box 12.2) to limit the temporal window over which pyramidal neurons respond to afferent excitatory input. In this manner, FS neurons are thought to "pace" cortical output and the network oscillations reflected in electroencephalograph (EEG) recordings.

The second group of interneurons, also found in layers II through VI, are defined by their selective expression of the peptide cotransmitter, somatostatin. Rather than targeting perisomatic regions, the axons of somatostatin-positive (SOM) neurons tend to innervate dendrites, including the apical dendritic tufts of pyramidal neurons. It is thought that SOM neurons are activated by the excitatory output of local neurons to provide "feed-back" inhibition preferentially to dendritic portions of pyramidal neurons. By targeting dendrites, particularly those in the apical dendritic tuft, SOM neurons can suppress the synergistic integration of top-down and bottom-up input in populations of pyramidal neurons (see previous sections).

The third group of interneurons is found primarily in superficial lamina (ie, cortical layers I through III). While these neurons exhibit tremendous heterogeneity in their morphological, physiological, and synaptic properties, including diverse expression of peptides such as vasoactive intestinal polypeptide, cholecystokinin, and calretinin, they are unique from FS and SOM neurons in their selective expression of ionotropic receptors for serotonin (5-HT, via the 5-HT_{3a} receptor) and acetylcholine (ACh, via expression of nicotinic receptors). As 5-HT and ACh release are associated with mechanisms regulating attention in the cortex (see later sections), it is likely that this group of neurons,

BOX 12.2

Two forms of inhibition within cortical networks. (A) Many PV^+ and 5-HT_3R^+ GABAergic interneurons receive excitatory input from thalamic or long-distance corticocortical afferents. These interneurons form "feed-forward" inhibitory synapses on neighboring neurons that are also receptive to the excitatory afferent drive. This generates rapid excitatory–inhibitory responses in cortical neurons receiving both excitatory afferents and feed-forward inhibition, with the inhibitory component narrowing the temporal window of opportunity for action potential genesis. (B) Many SOM^+ GABAergic interneurons have reciprocal connectivity with populations of local excitatory neurons. Thus, these neurons are situated to produce "feed-back" inhibition to the local network by releasing GABA in proportion to the network's excitatory output. (C) Some interneurons, including PV^+ neurons, may be positioned such that they are receptive to both local and afferent excitatory drive, allowing their output to deliver both feed-back and feed-forward inhibition to the network. Together, feed-forward and feed-back inhibitions sculpt network output in response to both extrinsic and intrinsic drives.

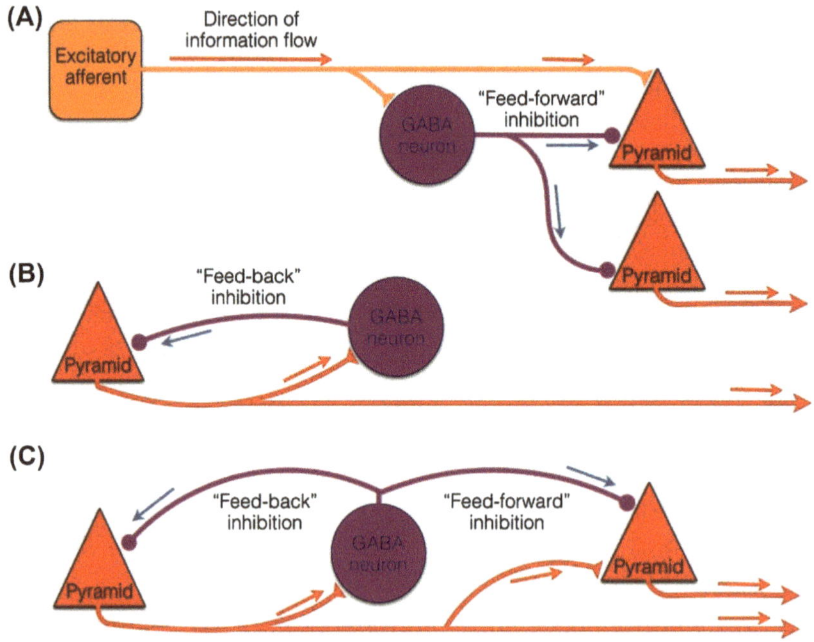

often referred to simply as 5-HT_3-expressing interneurons, shapes information flow in superficial corticocortical circuits during conscious attention. However, the diversity of 5-HT_3-expressing interneurons suggests that different subpopulations are specialized for distinct roles, with some subpopulations positioned to provide feed-forward inhibition to tuft dendrites in response to top-down afferent input, while others may disinhibit the cortex by preferentially targeting, and therefore inhibiting, other local GABAergic interneurons. Nonetheless, together with PV and SOM neurons, 5-HT_3-expressing neurons contribute to an overall inhibitory balance that maintains stability in cortical circuits.

NEUROMODULATION OF CORTICAL CIRCUITS

Information flow through cortical circuits relies primarily on glutamatergic and GABAergic synaptic transmission among and between the cortical neurons subtypes described earlier. However, there are many additional intrinsic and extrinsic transmitters, including monoamines, ACh, peptides, and endocannabinoids that modulate circuit function by altering cortical neuron excitability and/or synaptic transmission within the cortex (Table 12.1). Several of these neuromodulatory transmitters, including ACh, 5-HT, and the catecholamines, norepinephrine (NE) and dopamine (DA), are clinically important, as they are directly implicated in a variety of debilitating neurological and psychiatric conditions. These transmitters act primarily through "metabotropic" G-protein-coupled receptors, which, by triggering multiple intracellular biochemical cascades, alter the way cortical neurons release and/or respond to glutamate and GABA. Some neuromodulators, such as 5-HT and ACh, can also act through "ionotropic" ligand-gated cation channels that directly excite postsynaptic neurons. The diversity of receptor subtypes, and their selective expression among cortical neuron subpopulations, allows for complex and precise regulation of cortical circuits by promoting activity in some circuit elements while inhibiting output from others. In the following sections, we describe broadly what is known about the best described neuromodulatory systems.

Acetylcholine

ACh acts as a neuromodulator in the cortex to facilitate arousal, attention, and memory formation. Drugs that block ACh receptors promote delirium and amnesia while drugs that enhance cholinergic signaling are used as cognitive enhancers. The majority of cholinergic input to the cortex arises from neurons in the basal forebrain, including the nucleus basalis of Meynert and the diagonal band of Broca. Cortical neurons respond to ACh via metabotropic and/or ionotropic ACh receptors that are expressed in a cell type–specific manner (Table 12.1). Muscarinic ACh receptors (mAChRs) are metabotropic receptors that, depending on their

TABLE 12.1 Direct Effects of Neuromodulators on Intrinsic Excitability of Cortical Neuron Populations

Transmitter	SS	IT	CF	CT	PV^+	SOM^+	5-HT_{3A}^+
ACh	M2-like ↓	M1 ↑	M1 ↑↑ Area-dependent nAChR ↑	M1 ↑ nACh ↑↑	??	??	nAchR ↑
5-HT	??	5-HT_{2A} ↑ 5-HT_{1A} ↓	5-HT_{1A} ↓	??	??	??	5-HT_{3A} ↑
DA	??	D1 ↑	D2 ↑	??	D1 ↑ D2 ↓	??	??
NE	??	β ↑	α2 ↑ β ↑	??	α ↑	α ↑↑	α ↑

Indicated are the receptor subtypes (if known) and direction (↑ and ↓) of neuromodulator-induced changes in intrinsic neuronal excitability, as determined in animal models. *5-HT*, serotonin; *ACh*, acetylcholine; *CF*, corticofugal; *CT*, corticothalamic; *DA*, dopamine; *IT*, intratelencephalic; *NE*, norepinephrine; *PV^+*, parvalbumin-positive; *SOM^+*, somatostatin-positive; *SS*, spiny stellate; *??*, data insufficient or unavailable.

coupling to G-protein α-subunits, can inhibit or excite postsynaptic neurons, while nicotinic ACh receptors (nAChRs) are ligand-gated cation channels that directly excite postsynaptic neurons. Pyramidal neurons in the cerebral cortex express G_q-coupled "M1" mAChRs in somatodendritic compartments. These receptors enhance neuron excitability by reducing potassium conductances and opening nonspecific cation channels. Muscarinic excitation appears to be common to all pyramidal neuron subpopulations (ie, IT, CF, and CT neurons), but is most prominent in CF neurons, suggesting that ACh may preferentially enhance cortical output to subcortical structures. In addition, depending on cortical area, some pyramidal neurons, including CT neurons, express nAChRs that provide a second source of cholinergic excitation. Thus, ACh release in the cortex potently enhances the excitability of pyramidal neurons. However, ACh also influences the activity of cortical interneuron subpopulations via activation of muscarinic receptors (both excitatory M1 receptors and inhibitory, $G_{i/o}$-coupled "M2-like" receptors) and/or nAChRs. Nicotinic responses appear selective to populations of superficial interneurons, primarily in layers I–III, that may inhibit pyramidal neuron apical dendrites while at the same time disinhibit pyramidal neuron somata via vertically oriented axon collaterals that preferentially target interneurons in deeper layers. Finally, M2-like receptors on some, but not all, glutamatergic and GABAergic release sites in the cortex act to reduce transmitter release.

Together, these diverse cellular actions of ACh sculpt information flow in cortical circuits in ways that are not yet fully understood. However, behaviorally, the net result of cortical ACh release is enhanced attentiveness to environmental stimuli and memory formation. Drugs that block mAChRs, such as scopolamine, induce amnesia while nicotine (an agonist at nAChRs) enhances memory performance. On the other hand, reduced cholinergic input to the cortex contributes to cognitive deficits associated with Alzheimer's disease while acetylcholinesterases that enhance synaptic concentrations of ACh are used therapeutically in the early stages of Alzheimer's disease.

Serotonin

5-Hydroxytryptamine, or 5-HT, is a monoamine transmitter in the cortex that contributes to emotional regulation, mood, and inhibitory control. 5-HT neurons in the dorsal and median raphe nuclei of the brainstem project serotonergic axons throughout the cortex, where postsynaptic signaling is achieved via several classes of metabotropic receptors, as well as the 5-HT_3 ligand-gated cation channel. The most prominent metabotropic 5-HT receptors in the cortex are the inhibitory ($G_{i/o}$-coupled) 5-HT_{1A} (1A) and excitatory (G_q-coupled) 5-HT_{2A} (2A) receptors although other metabotropic 5-HT receptors are expressed at lower levels at both pre- and postsynaptic sites. In animal models, pyramidal neurons tend to primarily express 1A and/or 2A receptors that are located in separate subcellular domains. Inhibitory 1A receptors located near the axon initial segment directly inhibit action potential generation. On the other hand, 2A receptors in the apical dendrites of pyramidal neurons promote action potential output by enhancing postsynaptic responsiveness to excitatory synaptic input. In some animal models, 2A receptors are preferentially expressed by IT neurons while CF neurons tend to exhibit 1A-dependent inhibition; however, it is not known whether similar response segregation exists in the human cortex.

Cortical interneurons also express combinations of 1A and 2A-like receptors (Table 12.1), with one subclass of interneuron being defined by its selective expression of the 5-HT_{3A} (3A) ligand-gated cation channels. 5-HT_3-expressing interneurons, which coexpress cholinergic nAChRs, are most numerous in superficial cortical layers, including layer I, and are thought to regulate the excitability of pyramidal neuron apical dendrites while also disinhibiting deeper cortical layers. Finally, as with the cholinergic system, presynaptic metabotropic 5-HT receptors on some glutamatergic and GABAergic terminals suppress transmitter release to directly influence synaptic transmission in cortical circuits.

While the complex interactions of these serotonergic mechanisms remain to be explored, cortical 5-HT release is thought to have a generalized inhibitory effect on cortical output (mediated directly or indirectly by 1A and 3A receptors), with limited 2A-dependent amplification of behaviorally relevant subcircuits. Alterations in the normal balance of 1A and 2A signaling are thought to contribute to a number of neuropsychiatric conditions, such as anxiety, depression, and schizophrenia. Drugs that enhance extracellular 5-HT levels in the cortex (eg, selective serotonergic reuptake inhibitors) act as anxiolytics and antidepressants, and reduce impulsivity, while 5-HT-depletion promotes anxiety and impulsive behavior. These effects are thought to reflect a normal dominance of inhibitory 1A receptor signaling in the cortex that acts to reduce anxiety and impulsivity. Enhanced 2A receptor activation in the cortex is anxiogenic, and in excesses promotes impulsivity and contributes to the etiology of schizophrenia. Indeed, 2A antagonists are used as antipsychotics and have anxiolytic properties while hallucinogenic 2A agonists, such as LSD and psilocybin, are used to model the positive symptoms of psychosis.

Catecholamines

Two catecholamines, DA and NE, work in parallel to enhance arousal and attention in the cortex. Dopaminergic afferents to the cortex arise from the ventral tegmental area and the substantia nigra while the majority of cortical NE afferents stem from the locus coeruleus in the pons. DA and NE neurons are tonically active during wakefulness, but increase their firing in response to both rewarding and adverse stimuli. DA and NE are released throughout the cortex, but appear to be especially important in regulating executive function in the frontal cortex.

A wide variety of receptors for DA and NE are expressed in the cerebral cortex (Table 12.1). DA receptors include G_s-coupled D1-like (D1 and D5 receptors) and $G_{i/o}$-coupled D2-like (D2, D3, and D4) receptors, while NE receptors include G_q-coupled $\alpha 1$ and $G_{i/o}$-coupled $\alpha 2$, and G_s-coupled β-adrenergic receptors. As with ACh and 5-HT, there is evidence from rodent studies for segregated expression of DA and NE receptors among the pyramidal neuron subtypes in the cortex. IT neurons selectively express D1-like receptors while CF neurons preferentially express D2-like and $\alpha 2$-receptors. Cortical pyramidal neurons also express β-adrenergic receptors although their relative distribution among pyramidal neuron subpopulations is unknown. Functionally, postsynaptic DA and NE receptors appear to universally promote excitation in pyramidal neurons, albeit via engaging different signaling mechanisms and ionic effectors. Consequently, psychostimulants, such as methylphenidate, that increase extracellular levels of DA and NE tend to increase action potential generation in cortical neurons. However, catecholamines also regulate cortical interneurons, making the net effect of DA and NE on cortical circuit performance difficult to predict. PV^+ interneurons tend to be excited by D1-like receptors while all interneuron populations appear to be excited via α adrenergic receptors, with excitation being particularly robust in SOM^+ interneurons. Further, D2-like and $\alpha 2$ receptors present on presynaptic terminals may regulate glutamate and GABA release in the cortex to further sculpt information flow through cortical circuits.

Dysfunction of DA and NE input to the cortex contributes to a number of neurological and psychiatric disease states. Reduced catecholaminergic input to the prefrontal cortex, as occurs in Parkinson disease, can lead to dysexecutive syndrome, which manifests as distractibility, perseverative behavior, and impaired response inhibition. Deficits in cortical catecholamine transmission also contribute to schizophrenia and attention-deficit hyperactivity (ADHD) disorder. All three of these disorders involve deficits in a form of immediate recall memory, known as "working memory," that depends critically on DA and NE signaling in the cortex. Unlike other forms of memory, working memory stores task-relevant information in the form of reverberatory network activity for only brief periods of time (on the order of seconds to minutes). Cortical depletion of DA or NE, or blockade of DA or NE receptors, impairs working memory. On the other hand, excessive DA and NE signaling is also detrimental to working memory performance. A number of studies have demonstrated that working memory is facilitated by an intermediate level of catecholaminergic stimulation in the cortex. Consistent with this conclusion, manipulations that moderate DA and NE signaling (eg, low, but not high, concentrations of agonists and antagonists for DA and NE receptors) tend to enhance both working memory performance and task-associated reverberatory activity of cortical pyramidal neurons.

TOPOGRAPHIC ORGANIZATION OF THE CORTEX

From the preceding discussions, it should be apparent that the types of neurons and the connections between the cells of a cortical column are similar across cortical areas. What, then, is the basis for functional differentiation between cortical regions? The most important feature responsible for functional specialization of cortical regions is the specific inputs and outputs of each area. This is reflected in the relative sizes (thickness and density of cells) of the cortical layers in the region, which provides clues as to the function of the region. For example, regions of primary sensory cortex have large internal granular layers (layer IV) for termination of large amounts of thalamocortical input, while the primary motor cortex has large infragranular layers (particularly layer V), which is the origin of output to the basal ganglia, brain stem, and spinal cord, and association cortex (ie, regions involved in synthesizing sensory experiences and motor functions) have large supragranular layers because these layers are associated with the interconnections between cortical regions. The structure of cortical layers is the foundation of Brodmann's map of the cortex (Fig. 12.2), and therefore this map defines regions of functional significance. The following discussion refers to both gross topography (Fig. 12.1) and Brodmann's areas (Fig. 12.2) in describing the functionally significant regions of the cortex.

The following discussion considers primary and association areas of each lobe of the brain as well as the general functions and deficits produced by damage. Please bear in mind that cortical areas do not function in isolation; multiple cortical areas cooperate as a larger, multimodal network in the performance of all but the simplest of functions. Particularly, all but the simplest actions involve a perception-action loop that involves both posterior (perceptive) and anterior (executive) parts of the cortex. This can complicate the task of clinically defining the location of a lesion producing a complex behavioral abnormality.

FRONTAL LOBE

The frontal lobe is immediately anterior to the central sulcus and has several anatomical regions, as depicted in Fig. 12.1. The gyrus immediately anterior to the central sulcus, the precentral gyrus, extends a very short distance onto the medial aspect of the hemisphere, ending as part of the paracentral lobule. On the lateral hemisphere, the frontal lobe is divided into superior, middle, and inferior frontal gyri. The superior frontal gyrus overlaps onto the medial aspect of the frontal lobe, abutting the cingulate gyrus, which itself is adjacent to the corpus callosum. The inferior frontal gyrus abuts the lateral fissure, and is divided (from posterior to the anterior) into an opercular, triangular, and orbital portion. The middle frontal gyrus is located between the superior and the inferior frontal gyri. Collectively, these three frontal gyri (superior, middle, and inferior) extend to the frontal pole. The orbital frontal gyri occupy the inferior (ventral) portion of the frontal lobe, resting over the orbits. The medial frontal cortex is quite heterogeneous. The cingulate gyrus parallels the corpus callosum, originally being dorsal, and then anterior and, finally, inferior to the corpus callosum where it has been termed the subcallosal gyrus. For functional reasons, the remainder of the medial prefrontal cortex is often subdivided into the infralimbic cortex posteriorly and the prelimbic cortex anteriorly.

From a functional perspective, the precentral gyrus and anterior paracentral lobule contain the primary motor cortex (BA4), also known as M1 (Motor 1). This is the region in which relatively simple body and head movements can be evoked by electrical stimulation at very low stimulus intensities when mapping the cortex. Although individual muscles are not represented in the cortex, cortical stimulation provokes particular movements, such as flexion or extension of a joint, producing coordinated contraction of some muscles and relaxation of others in order to generate the movement. There is a somatotopic map of the body represented in M1, with muscles of the head being closest to the lateral fissure and those of the lower extremities overlapping the superior part of the cortex onto the medial aspect of the hemisphere. The hand, then, would be represented in the precentral gyrus about halfway from the lateral fissure to the dorsal aspect of the lateral hemisphere. It is important to note that those body parts having finer motor control (eg, the hand) have a geographically greater area of cortical representation. The dimensions of these representations are somewhat "plastic" and can be increased by use, such as by learning a new motor skill. Damage to the precentral gyrus results in weakness that is most severe in muscles of fine control, especially those of the distal limbs. Very proximal muscles may be spared due to alternative motor control pathways for these muscles. If lesions are restricted to the primary motor cortex, the pattern of weakness is generally associated with decreased muscle tone and diminished deep tendon reflexes although larger lesions that include adjacent premotor areas usually increase muscle tone and reflexes. This apparent paradox relates to the connectivity of the primary motor cortex, which is largely excitatory to motor neurons, versus that of the premotor cortex, which connects extensively to regions of the brain stem that are inhibitory to muscle tone.

The "premotor cortex" (generally corresponding to BA6) includes the posterior portions of the superior and middle frontal gyri. The premotor cortex extends onto the superior part of the medial hemisphere where it has been termed the supplementary motor area (SMA). Movements can be produced from stimulation of the premotor cortex, although it requires higher current intensities, and typically it produces more complex and somewhat delayed movements relative to those produced by stimulation of the primary motor cortex. During voluntary movements, the premotor cortex becomes active significantly before movement takes place, and this region is active in the planning stage of movements regardless of whether the movement actually takes place. For example, this is the region that becomes most active when imagining the performance of a movement. Lesions in the premotor cortex can affect the sequencing of movement, resulting in a loss of fluidity of normal movements, a deficit termed apraxia. Damage to this region also results in increased muscle tone and exaggerated deep tendon reflexes. Lesions in the superior and medial parts of the premotor cortex, especially the SMA, often severely impair the initiation of voluntary movements (a deficit called abulia, when it is severe).

The "frontal eye fields" comprise the cortex immediately anterior to the premotor cortex (mostly coextensive with BA8). This region becomes active shortly before voluntary horizontal gaze to the contralateral side. Lesions of this area diminish the patient's ability to voluntarily look toward the contralateral side and, when the lesion is acute and severe, their eyes are likely to be directed toward the side of damage. This eye deviation is caused by the unopposed action of the intact hemisphere. Seizures affecting the frontal eye field of one hemisphere commonly drive the eyes contralaterally, that is, away from the side of epileptic activity. It is now recognized that this region contributes to the ability to attend to the visual environment, and to some other aspects of prefrontal cortical function (see the following sections).

Broca's area is an important region for language generation within the dominant hemisphere (the hemisphere with greatest language function). This area is located in the posterior aspect of the inferior frontal gyrus, immediately anterior to the precentral gyrus, including the opercular and triangular gyri (areas 44 and 45). This region is devoted to motor aspects of language, that is, the expression of language, including speech, writing, and sign language. Isolated damage to Broca's area produces motor aphasia (also known as expressive, or Broca's, aphasia) even though comprehension of language would be normal. Such patients have difficulty generating meaningful language including writing, as well as difficulty repeating phrases. Few words may be successfully generated with great effort and hesitation ("telegraphic speech"), and some emotional words and phrases may be patterned from limbic areas of the brain. These patients are very frustrated because they retain the capacity to recognize that they are unable to communicate effectively.

The remainder of the frontal lobe anterior to the premotor cortex has been classically called the prefrontal cortex. This term inappropriately implies that it is in front of the frontal lobe. However, the prefrontal cortex is actually just the anterior (rostral) part of the frontal lobe. Ever since the time of the publication of the clinical observations on Phineas Gage, a railroad worker who damaged his frontal lobe in an injury, the prefrontal cortex has been recognized as important in regulating behavior by selecting appropriate responses and inhibiting inappropriate responses. This region is involved in organizing and planning activities, and contributes to what is perceived by others as "personality." The prefrontal cortex also reveals the most impressive structural difference between the human brain and that of other mammals. Because of this complexity of human behaviors, it should not be surprising that this is the region of the frontal lobe with the largest number of competing theories as to specific localization of function. Nevertheless, the prefrontal cortex can broadly be divided into the dorsolateral prefrontal cortex (DLPC), the orbital frontal cortex, and the medial prefrontal cortex.

The DLPC is visible on the lateral aspect of the hemisphere and mostly consists of the anterior portions of the superior and middle frontal gyri, centered on BA9, but also including adjacent regions of BA10 and BA46. This region is most important for executive functions of the brain. These functions include the ability to plan and sequence activities, the ability to focus on tasks (as well as to return to them after brief periods of distraction), the ability to shift and adapt patterns of performance to changing circumstances, and working memory. Working memory is the capacity to maintain things in memory for brief periods by continuing to mentally rehearse them. For example, individuals can typically store at least seven items (such as the digits of a phone number) for brief periods via working memory, and can remember at least five such items in reverse order. Forcing someone to focus on another cognitive task, even if this interruption is quite brief, effectively disrupts working memory. It is now known that reverberatory circuits interconnecting the DLPC with the association areas of the parietal, temporal, and/or occipital lobes (depending on what is being remembered) maintain information in working memory although damage to the DLPC produces the most severe and persistent working memory deficits.

The orbital frontal cortex (BA11) is the ventral part of the frontal lobes, located directly superior to the eye sockets (orbits). This region has extensive interconnections with so-called limbic regions of the brain, involved in emotional responses. These regions appear to have a major role in comparing the likely results of various behavioral responses, facilitating responses that have the most favorable outcomes. Damage to this region often results in socially inappropriate or maladaptive behaviors that may include outbursts (including profanity), inappropriate sexual activity, compulsive addictive behaviors, diminished ability to respond to social cues, and lack of empathy.

The medial prefrontal cortex is a complex region anterior to the SMA on the medial hemisphere. This includes anterior portions of the cingulate gyrus (see the following sections) as well as those parts of cortical areas BA8–12 that overlap onto the medial hemisphere. The posterior part of the medial frontal cortex includes the anterior portions of BA32.

Given the cytoarchitectonic heterogeneity of the medial PFC, it is not surprising that there are several competing theories as to functions of the region. One function appears to be conflict monitoring, in which intended outcomes are matched with reality. This region then interacts with effector areas of the brain to implement necessary adjustments to behavior. Another function appears to include the valuation of outcomes particularly in terms of negative consequences of behaviors. Yet another function is risk prediction in order to minimize the risk of adverse outcomes of behaviors. This is similar but not identical to cost-benefit analysis where both positive and negative outcomes must be weighed. Finally, this region also appears to be important in regulating autonomic and behavioral features appropriate to the situation. Lesions of this region often result in inappropriate or reckless activities and a lack of consideration for others (deficit of empathy). Given this wide range of functions, it is not surprising that this region is connected extensively with other cortical areas as well as subcortical regions involved in emotional responses, including the amygdala.

The cingulate gyrus is an anatomically continuous gyrus on the medial aspect of each hemisphere, adjacent to the corpus callosum and extending from the frontal lobe into the parietal lobe. It is cytoarchitectonically heterogeneous. The subgenual region (BA25) is the most ventral portion of the cingulate gyrus, located in a subcallosal position, while BA33 and BA24 portions of the cingulate gyrus wrap around the anterior and then the superior aspects of the corpus callosum. These areas are continuous with the BA23 and BA31 portions of the cingulate gyrus located superior to the splenium, which is the posterior part of the corpus callosum. The retrosplenial part of the corpus callosum includes BA26, BA29, and BA30.

The cingulate gyrus remains an area of intense interest and study, particularly related to clinical disorders of mood as well as chronic pain. One example of this relates to the subgenual region (BA25), which has extensive connections with subcortical regions, including the hypothalamus and amygdala that are involved in endocrine, autonomic, and behavioral responses to emotion, and also to cortical regions including the insula and surrounding medial frontal cortex. This region also has a dense innervation by 5-HT -releasing axons and a high concentration of 5-HT receptors, and is overactive in certain depressive states. As a result, it has become a target for brain stimulation that appears to regulate this excessive activity.

Another example of evolving research interest in the cingulate gyrus relates to the increased activity of anterior cingulate regions (mostly BA24 and BA33) in chronic pain states. This heightened activity appears to correlate with the affective components of pain, that is the experience of suffering. It has been proposed that this region is responsible for assessing the importance of the unpleasant experience, and assigning a relative (negative) value to it. It is noteworthy that these same regions increase their activity when observing other individuals experiencing pain and, therefore, may be part of a network important in empathy.

Furthermore, the anterior cingulate gyrus and adjacent portions of the medical frontal lobes (the paracingulate cortex) are important in "theory of mind." This is the ability to recognize perspectives in oneself and others and to distinguish between the two. Deficits in theory of mind appear as an inability to comprehend and respond appropriately to the desires, beliefs, and intentions of others. This is of particular interest in our attempts to understand the neurological underpinnings of autistic spectrum disorders. Functional imaging studies have identified several regions that are particularly important for this function. In addition to the anterior cingulate and paracingulate cortex, the region around the superior temporal sulcus, the anterior pole of the temporal lobe, and the region of the precuneus of the parietal lobe are now recognized as particularly important for these functions.

The posterior and retrosplenial portions of the cingulate cortex have not been as well studied as the anterior. The posterior cingulate region is active when individuals retrieve autobiographical information and upon exposure to familiar words, places, or objects. Animal studies show that cells in this region increase activity when making choices that entail some element of risk. Damage to the more posteriorly situated retrosplenial cortex has a much more profound effect on memory and can also affect spatial navigation. Some (but certainly not all) of these symptoms might reflect damage to the connections between this cortical region and the hippocampus, which pass through the white matter of the medial temporal lobe in a bundle of axons called the cingulum. The cingulum follows the course of the cingulate gyrus and interconnects the several portions of the gyrus. The tight connection between these regions suggests integration of functions across the cingulate gyrus, and may partially explain difficulties in assigning specific functions to individual subregions.

PARIETAL LOBE

The bulk of the parietal lobe is viewed on the lateral hemisphere situated immediately posterior to the central sulcus and continuing to a somewhat indistinct group of parieto-occipital sulci on the lateral hemisphere that separate the parietal from the occipital lobe. Ventrally, the parietal lobe is limited by the lateral fissure. The postcentral gyrus is immediately posterior to the central sulcus, with BA3, BA1, and BA2 reflecting separate areas as one progresses from the depths of the central sulcus posteriorly across the top of the gyrus. The posterior portion of the parietal lobe can be divided into the superior and inferior parietal lobules by an interparietal sulcus. The superior parietal lobule (BA5 and BA7) extends from the postcentral sulcus to the occipital lobe. The Inferior parietal lobule includes two regions: the supramarginal gyrus, which is an inverted "U"-shaped gyrus that extends dorsally from the posterior aspect of the lateral fissure (mainly consisting of BA40), and the superior part of the angular gyrus (BA39). The postcentral gyrus continues onto the medial aspect of the hemispheres as the posterior part of the paracentral lobule. Similarly, the superior parietal lobule also extends onto the medial aspect of the hemisphere where it is called the precuneus. The precuneus abuts the more posterior portions of the cingulate cortex (BA31 and BA23). There is a very distinct parieto-occipital sulcus on the medial aspect of the hemisphere.

The primary somatosensory cortex encompasses most of the postcentral gyrus, receiving the greatest concentration of thalamocortical fibers conveying somatic sensation. Similar to the motor homunculus described earlier in the precentral gyrus, somatosensory areas in the postcentral gyrus are organized topographically according to relative sensitivity, with head structures represented closest to the lateral fissure, the legs represented at the most superior (dorsal) aspect of the gyrus, and the feet represented on the medial aspect of the lobe in the paracentral lobule. It should be noted that there are essentially three overlapping sensory representations, one for each of the histologically defined cortical regions (BA1, BA2, and BA3) that comprise the postcentral gyrus. These areas have slightly different response patterns to sensation. There is a ventral secondary somatosensory cortex located in the parietal operculum (the part of the inferior parietal lobe that dips into the lateral fissure; ie, the lowest part of BA40 and BA43). This has a cruder representation of the body and interconnects with the primary somatosensory cortex and the sensory association areas.

The remainder of the parietal lobe is comprised of association cortex. The anterior portion of the superior parietal lobule, particularly BA5 and the anterior part of BA7, is the somatosensory association cortex that interprets what is being felt (stereognosis). Also, this region appears to be important in maintaining a map of the location of the body and, because of commissural connections with the contralateral hemisphere, where the hands are in relation to one another. As you might imagine, lesions of this region not only can prevent one from recognizing objects by touch (astereognosia), but also can result in uncoordinated movement. The most posterior portion of the superior parietal lobule is multimodal association cortex, which also receives input from visual association regions. These higher-order areas of association cortex, located between sensory modalities, are important in determining the overall significance of the sensory input and selecting particular sensory inputs to attend to. Therefore, posterior parietal lobe damage can result in inattention. What might be surprising is that this is a lateralized function where damage to the nondominant hemisphere produces contralateral inattention. This may be so severe as to render the patient completely inattentive to sensory input from the contralateral side of the world to the point that they may even deny ownership of that side of their own body despite incontrovertible proof to the contrary! This occurs because, although the nondominant posterior parietal lobe attends to sensory input from both sides of the body and world, the posterior parietal lobe of the dominant hemisphere is focused almost exclusively on attending to sensory input from the contralateral side of the body and of the world. Damage to the dominant posterior parietal lobe does not produce such severe neglect because the nondominant hemisphere is able to distribute attention bilaterally.

The inferior parietal lobule (IPL) is even more complex multimodal association cortex, residing between visual, auditory, and tactile association areas and consisting of the supramarginal gurus and the superior part of the angular gyrus. In the dominant hemisphere, much of the function of the IPL revolves around the intersection of written and spoken language. The supramarginal gyrus is particularly important for processing the sounds that comprise language (phonology). In the nondominant hemisphere, the supramarginal gyrus has several functions including the multimodal appreciation of position in space and the understanding of the postures and gestures of others (part of the mirror neuron network that is active both when performing an action and when observing another individual perform that action).

The angular gyrus is very complex and has long been known to be important in recognizing the meaning of written and spoken language (semantics) and functioning in concert with the supramarginal gyrus in recognition of the phonemes that comprise language. Indeed, the angular gyrus is active bilaterally during language tasks, and is essential for the ability to read. However, this region also participates in many other tasks, including number processing and performance of arithmetic. It is also important in attention and spatial orientation. This is particularly true for the nondominant hemisphere, with lesions to the nondominant angular gyrus contributing to neglect of the contralateral world (which can also occur due to lesions more superiorly in the parietal lobe). This gyrus is also important in certain memory functions, including working memory for semantic information, which involves collaborative activation of the DLPC and the angular gyrus. Furthermore, the angular gyrus is active during retrieval of episodic memory (memory of events).

Certain more complicated tasks can be disrupted by lesions of the angular gyrus. Go/no-go tasks (ie, where one stimulus requires a response and another requires suppressing the same response) are affected. Also, this region is active during tasks that require the ability to infer the emotional state of another person. Finally, the angular gyrus is very active during passive reflection (ie, it is part of the "default mode network"). Along with the medial frontal cortex and the posterior cingulate cortex, these regions are deactivated by behaviors that are directed toward a goal. Given these multiple functions, it should not be surprising that lesions to these areas can produce complex symptoms. Dominant angular gyrus lesions can produce Gerstmann syndrome, which includes difficulty writing, problems performing arithmetic calculations, confusion in distinguishing left from right, and difficulty distinguishing the fingers of a hand. Usually, there are significant problems with language comprehension, as well.

The precuneus of the medial parietal lobe is a region that is active during "theory of mind" tasks, that is, tasks that require the individual to take a perspective different than their own by being aware of what another person is thinking or intending to do. In that regard, this region is very active in situations requiring empathy. Additionally, the precuneus is highly active during resting states of the brain (as part of the so-called "default mode network"), showing decrease in activity when shifting from this "resting state" to more goal- and task-oriented states.

TEMPORAL LOBE

The temporal lobe is located inferior to the lateral fissure and anterior to the occipital lobe. There is a superior part that extends into the lateral fissure as the temporal operculum. This includes several gyri that extend transversely into the depths of the fissure called the transverse gyri (of Heschl). Within the transverse gyri, BA41 and most of BA42 are known as the primary auditory cortex and have a tonotopic representation of sound. The posterior portion of the temporal operculum has been termed the "planum temporale." This is larger in the dominant hemisphere and is important in the interpretation of speech. The lateral aspect of the temporal lobe is roughly subdivided into superior, middle, and inferior temporal gyri that roughly correspond with BA22, BA21, and BA20, respectively. These are demarcated by the superior and inferior temporal sulci. The posterior aspect of the superior temporal gyrus has additional areas that are adjacent to the transverse gyri (including the posterior part of the planum temporale) that is termed the secondary auditory cortex (mostly BA22). Most posteriorly, the angular gyrus (BA39) surrounds an upward extension of the superior temporal sulcus. At the anterior pole of the temporal lobe, the superior temporal gyrus ends in BA38.

The inferior aspect of the temporal lobe is dominated by the temporo-occipital (fusiform) gyrus (most of which includes BA37 and some

BA36), which is separated from the medially located parahippocampal gyrus (mostly BA27, BA28 and BA35) by the collateral sulcus. The anterior aspect of the parahippocampal gyrus protrudes medially as the uncus (BA34) and has been termed the piriform cortex (a four- to five-layer region of allocortex). The entorhinal portion of the parahippocampal gyrus extends a short distance onto the medial aspect of the temporal lobe, ending by dipping into the hippocampal sulcus and transitioning to the three-layer cortex (ie, archicortex) of the subiculum and hippocampus (see Chapter 32).

Functionally, the temporal lobe is complex. The posterior part of the lateral hemisphere of the temporal lobe is mostly involved in auditory processing and, near the border with the occipital lobe, visual processing. In the dominant hemisphere, the posterior part of the superior temporal gyrus (BA22) comprises the majority of Wernicke's area, a region critical to language comprehension. The more posterior portions of this area (ie, closer to the occipital lobes) are critical to reading, particularly the angular gyrus (see the earlier sections). Also, there is growing recognition that there are subregions within Wernicke's area that are involved in recognition of different categories of words and verbal constructs. Higher-level synthesis is then necessary in order to extract meaning from language.

Patients with damage to Wernicke's area typically have difficulty with language comprehension even when they, themselves, are speaking. They cannot repeat phrases accurately although they often create words that have some similarities in rhythm or sound to those which they are attempting to use (paraphasic errors). Spontaneous speech is usually copious and contains similar errors. There are usually deficits in recognition of written language that is similar to the deficit in understanding speech.

Language function is often not completely lateralized, and there is often some level of preservation of function even after large lesions. Furthermore, the portions of the nondominant hemisphere that are located in the same general cortical regions as Wernicke's area in the dominant hemisphere are involved in evaluating the significance of tone of voice and inflection in verbal communication.

As mentioned earlier, portions of the temporal lobe are activated during "theory of mind" tasks. These tasks require subjects to predict actions, perspectives, and responses of others. The region around the superior temporal sulcus (particularly in the nondominant hemisphere) and the anterior poles of both temporal lobes are activated during such tasks and, therefore, likely play a role in these functions.

The temporo-occipital (fusiform) gyrus has several functions. The more posterior portions, particularly within the occipital lobe, are part of the color-processing system. Neurons that are important in facial recognition are located somewhat more anteriorly, especially in the nondominant hemisphere. Lesions to this region can produce inability to recognize faces (prosopagnosia). In the dominant hemisphere, the fusiform gyrus is extensively interconnected with the angular gyrus and is important in word recognition. It is interesting that the fusiform gyrus and angular gyrus are often highly active in a form of synesthesia where certain combinations of letters are perceived to have color.

The parahippocampal gyrus (mostly containing BA27, BA28, and BA35) is a heterogeneous region located on the inferomedial aspect of the temporal lobe, immediately medial to the fusiform gyrus. The anterior part includes the prepyriform and periamygdalar cortex, often referred to collectively as the piriform lobe. This region represents one termination of the olfactory tract and is the primary olfactory cortex. The piriform cortex can be easily identified by inspection of the inferior aspect of the temporal lobe since it covers the uncus, a prominent bump over the amygdala that is located within the anteromedial temporal lobe. It should not be surprising that the primary olfactory cortex is interconnected with the amygdala, a region controlling autonomic, endocrine, and behavioral responses to certain salient stimuli, including olfactory.

The entorhinal cortex is more posterior in the parahippocampal gyrus adjacent to the subiculum, which forms an anatomical and functional bridge between the entorhinal cortex and the hippocampus. Superficial layers of the entorhinal cortex project to the dentate gyrus of the hippocampus while many axons of hippocampal pyramidal neurons terminate in the deeper layers of the entorhinal cortex. There are several hypotheses regarding the specific importance of what appears to be a tight feed-back connection. Since the entorhinal cortex has connections with broad regions of association cortex, some consider the entorhinal cortex to be an interface between the hippocampus and these broad areas of association cortex. Probably the best studied aspect of this interrelationship between the entorhinal cortex and the hippocampus is their role in spatial navigation. The entorhinal cortex contains a representational spatial grid organization while hippocampal CA1 pyramidal neurons known as "place cells" respond only when an animal is in a particular location in the environment, with different CA1 neurons showing different place preferences such that the entire environment is "mapped" by activity across a population of CA1 neurons. These observations have led to several theories and ongoing research into the precise nature and significance of interconnections between hippocampal place cells and the spatial grid cells of the entorhinal cortex. Nevertheless, it is clear that damage to either the entorhinal cortex or the hippocampus can produce severe deficits in the ability to orient to the environment and to navigate within it. The lateral portions of the posterior parahippocampal gyrus along with medial aspects of the fusiform gyrus have been termed the "parahippocampal place area." This region is highly associated with the recognition of scenes (objects in context). More medial portions of parahippocampal cortex are important in recognition of contextual associations including short-term episodic declarative memory (ie, recall of facts, events, and directions). Short-term memory is dependent on functional synaptic plasticity within the circuitry between the entorhinal cortex and the hippocampus. Short-term memory and orientation to the environment are common in Alzheimer's disease; it should not be surprising that the pathology of the disease usually involves these regions of the medial temporal lobe.

OCCIPITAL LOBE

The medial aspect of the occipital lobe is separated from the more anteriorly situated parietal and temporal lobes by a distinct parieto-occipital sulcus. The calcarine sulcus is a very important landmark on the medial aspect of the occipital lobe, extending posteriorly from the parieto-occipital sulcus to the occipital pole. This sulcus provides a landmark for defining the location of the primary visual cortex. The lingual gyrus and the cuneus are the gyri immediately inferior and superior, respectively, to the calcarine sulcus. These gyri encompass the primary visual cortex (BA17), the region receiving thalamocortical connections from the lateral geniculate nucleus of the thalamus. These terminate in a topographically accurate pattern relative to the contralateral visual world. The upper visual world is represented immediately inferior to the calcarine sulcus (in the lingual gyrus) and the lower visual world immediately superior (in the cuneus). Central vision is represented near the occipital pole, overlapping slightly onto the lateral hemisphere. It is important to note that, despite a well-documented topography of visual input, the response of neurons to visual input is not simply a point-to point transmission. Instead, visual information is highly synthesized, with cortical neurons responding to individual features of the visual scene, such as lines, edges, colors, and directions of movement. These have to be synthesized into a useful pattern by surrounding areas of visual association cortex (mostly BA18 and BA19) that extend onto the lateral aspect of hemisphere. Furthermore, multimodal association areas in the parietal and temporal lobes are important in aspects of visual processing and integration with input from other sensory modalities.

The occipital lobe continues onto the lateral and inferior aspects of the hemisphere where the border between the occipital lobe and the posterior aspect of the parietal and temporal lobes is rather indistinct. The cortex in this region is part of BA18 and BA19 (the visual association areas), which is described in greater detail in Chapter 19.

INSULAR CORTEX/INSULAR LOBE

The gyri of the insular cortex (BA13 and BA14) are found at the depths of the lateral fissure. This area is hidden from view and represents only a tiny fraction of the cortical surface. Therefore, its functions have been less well studied than most other cortical areas. Nevertheless, as more has been learned of its functions, a very complex picture is emerging. The connections of the insula tell a story of engagement with the internal processes of the body. The posterior portions of granular insular cortex receive input from thalamic nuclei relaying sensations that are mostly, but not exclusively, focused on monitoring the internal environment, that is, interoception. This sensory input also includes an enhanced representation of sensations of taste and smell. The insular cortex has extensive reciprocal connections with the amygdala and with the anterior cingulate cortex, suggesting an important role in emotional responses. The insula is active in functions as diverse as the analysis of painful stimuli, generating adverse responses to unpleasant tastes and smells, the ability to recognize and respond to social emotions, the generation of cravings for addictive substances, and in production of certain aspects of speech. To date there is no satisfactory singular functional role for this comparatively tiny cortical region, although Damasio has suggested that the insula may be an important node in connections between feelings and emotions and the internal responses of the body to those emotional states, forming a loop with other cortical and subcortical brain regions that are involved in visceral sensations and responses.

SUMMARY

The cerebral cortex represents the highest level of complexity of the human brain. Nevertheless, the great majority of the cortex consists of the same six layers, with a predominance of excitatory cells along with scattered smaller inhibitory neurons. The wiring of these six layers leads to a vertical processing organization (often termed columnar), which permits integration of novel inputs from the thalamus (bottom-up input) with activity arriving from other regions, especially other parts of the cortex (top-down input). Conceptually, this system integrates predictions with ongoing perception at the level of pyramidal neurons to determine the output of the cortical column. The thickness of specific layers of the cortex within any cortical column varies in different regions of the cortex because the cortical layers have specific roles to play in processing and because different cortical regions have different functions. The functional organization of the cortex has both modular and integrated aspects where modules of relatively simple function (primary cortex) connect with regions of progressively higher-level function (unimodal and then multimodal association cortex). These regions are further organized into perceptual cortex (parietal, occipital, and temporal cortices) and executive cortex (mostly frontal cortex). Normal function requires integration of the perceptual and executive regions of cortex (a perception-action loop) in order to generate a behaviorally appropriate response. Thus, the clinical ramifications of disease of the cortex must be understood both in terms of damage to functional modules (where specific functions may be lost) and in terms of the capacity to disrupt larger networks that contribute to normal behavior.

Further Reading

[1] Berridge CW, Arnsten AF. Psychostimulants and motivated behavior: arousal and cognition. Neurosci Biobehav Rev 2013;37:1976–84.
[2] Dembrow N, Johnston D. Subcircuit-specific neuromodulation in the prefrontal cortex. Front Neural Circuits 2014;8:54.
[3] Douglas RJ, Martin KA. Neuronal circuits of the neocortex. Annu Rev Neurosci 2004;27:419–51.
[4] Feldmeyer D, Brecht M, Helmchen F, Petersen CC, Poulet JF, Staiger JF, et al. Barrel cortex function. Prog Neurobiol 2013;103:3–27.
[5] Harris KD, Shepherd GM. The neocortical circuit: themes and variations. Nat Neurosci 2015;18:170–81.
[6] Horton JC, Adams DL. The cortical column: a structure without a function. Philos Trans R Soc Lond B Biol Sci 2005;360:837–62.
[7] Larkum M. A cellular mechanism for cortical associations: an organizing principle for the cerebral cortex. Trends Neurosci 2013;36:141–51.
[8] Puig MV, Antzoulatos EG, Miller EK. Prefrontal dopamine in associative learning and memory. Neuroscience 2014;282:217–29.
[9] Rudy B, Fishell G, Lee S, Hjerling-Leffler J. Three groups of interneurons account for nearly 100% of neocortical GABAergic neurons. Dev Neurobiol 2011;71:45–61.
[10] Teffer K, Semendeferi K. Human prefrontal cortex: evolution, development, and pathology. Prog Brain Res 2012;195:191–218.
[11] Gallagher HL, Firth CD. Functional imaging of 'theory of mind'. Trends Cogn Sci 2003;7:77–83.
[12] Fuster JM, Bressler SL. Past makes future: role of pFC in prediction. J Cogn Neurosci 2015;27:639–54.

CHAPTER

13

The Thalamus

J.H. Chien, J.J Cheng, F.A. Lenz

Johns Hopkins Hospital, Baltimore, MD, United States

The thalamus is a structure at the center of each cerebral hemisphere and is a relay for sensory pathways, and for brain stem, cerebellar, and subcortical pathways to cortex. The thalamus also serves as a relay between cortical structures. There have been substantial translational advances in surgery related to the anatomy, physiology, electrophysiology, and imaging of the thalamus. The basic approaches to surgery of the thalamus are to lesion or stimulate specific nuclei for the treatment of movement disorders such as tremor. The basic problem with surgery of the thalamus is that functionally distinct and therapeutically important structures are located in a small area. Therefore, small differences in the location of a lesion or stimulating electrode may lead to irreversible and significant limitation of therapy, or to complications. Surgeries to lesion or disrupt parts of the thalamus for Parkinson's disease, tremor, and pain were carried out frequently in the 1950s and 1960s.

In the early 1970s, the introduction of levodopa and the complications of surgical procedures led to disappearance of these procedures except at a handful of centers worldwide. The limitations of medical approaches to the treatment of Parkinson's disease, such as drug-related dyskinesias and fluctuations in therapeutic effect, have led to the resurgence of surgical therapies (pallidotomy and thalamotomy) in the 1980s. In addition, complications were decreased by the introduction of MRI and of electrophysiological monitoring technologies such as microelectrode recordings. In the late 1980s, the introduction of deep-brain stimulation (DBS) addressed the concern long associated with these lesioning procedures since side effects can be decreased and therapeutic effects improved by changing the intensity and pattern of stimulation. Finally, combinations of modeling of activations produced by the applied electric fields, of imaging, and electrode or stimulator technologies has again transformed thalamic surgery.

The basis of thalamic surgery is the functional distinctions between individual nuclei or nuclear groups. The thalamus is composed of a number of groups of nuclei, each of which is often specifically related to different connections; these groups of nuclei are segmented into their component nuclei. These structures have often been renamed by anatomists publishing articles or atlases, and by other neuroscientists studying the thalamus; this is a source of confusion, particularly among persons who are reading the thalamic literature for the first time. In this chapter, we limit ourselves to human and monkey thalami in which the anatomy is similar

Conn's Translational Neuroscience
http://dx.doi.org/10.1016/B978-0-12-802381-5.00022-1

although the size of some structures are different such as the pulvinar which is larger in humans. The main nomenclature systems for monkeys are those of Olszewski, Carpenter, and Jones, while those for humans of Hassler, Talairach, and Jones are most often used by surgeons and imagers.

In general, thalamic nuclei can be divided into specific, multimodal, and nonspecific. The specific and multimodal nuclei are identified with different functional systems by their inputs and by the projections of their output neurons from the thalamus to the cortex through the axons of thalamic relay cells. The specific nuclei have focused projections upon well-defined motor or sensory cortical areas while multimodal areas have widespread connections with associational cortical areas, which are themselves defined by input from pathways related to multiple sensory modalities. Thalamic multimodal nuclei are related to higher cognitive functions such as language, attention, or working memory through connections including parts of the temporal and parietal lobes or prefrontal cortex. Finally, nonspecific nuclei include the intralaminar nuclei, which have stronger projections with the caudate putamen more than with the cortex.

The thalamus is divided in the sagittal plane by the internal medullary lamina, which is the white linear structure on the dorsal (top) surface of the thalamus in Fig. 13.1; it divides anteriorly to define the borders of the anterior nucleus. The nuclei which are located within this lamina(intralaminar nuclei) include the globoid central median posteriorly, and then the paracentral, and central lateral nucleus which is a thin sheet of cells within the internal medullary lamina. These nuclei are nonspecific and receive diffuse inputs from a broad range of sensory pathways and reciprocal connections with the caudate putamen more than with

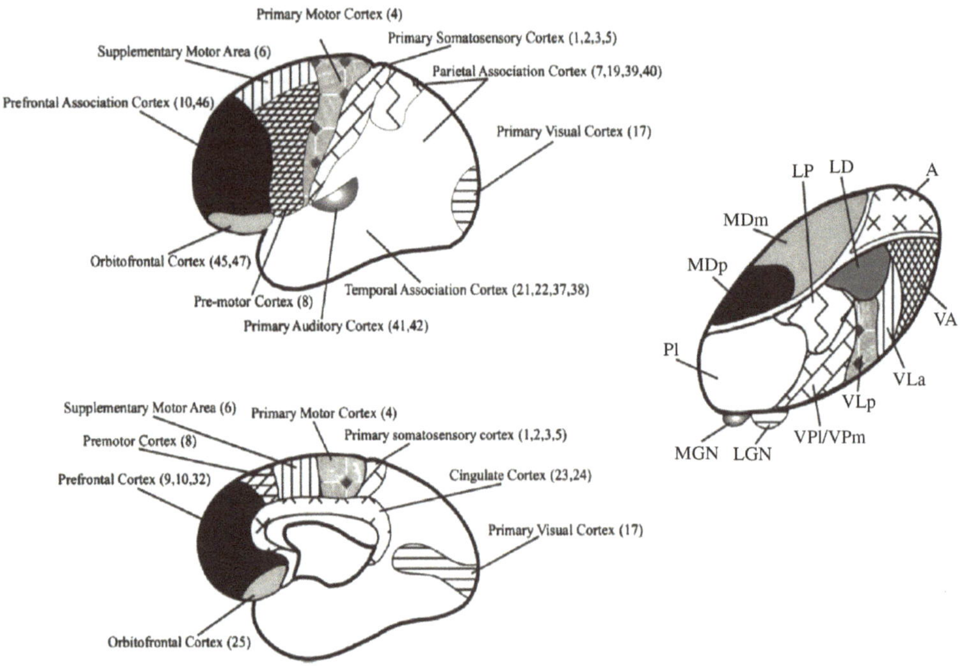

FIGURE 13.1 **Thalamic nuclear nomenclature.**

the cortex. Lateral to the internal medullary lamina are the dorsal lateral group and the ventral nuclear group. The ventral nuclear group of the thalamus is largely composed of specific nuclei and includes, from anteriorly, the ventral anterior which relays inputs from the basal ganglia to premotor cortex, the ventral lateral anterior and posterior which relay inputs from the pallidum and the deep cerebellar nuclei to the supplementary motor cortex and motor cortex, respectively.

Posteriorly, the ventral posterior lateral receives inputs from the dorsal column nuclei and projects on primary somatic sensory cortex. The spinothalamic tract terminates in the ventral posterior lateral, particularly the posterior part and posteriorly adjacent nuclei; the lateral and medial geniculate nuclei project upon primary visual and auditory cortex. The medial dorsal nucleus and the dorsal tier of thalamic nuclei is composed of multimodal nuclei including, from anterior, the lateral dorsal nucleus, the lateral posterior nucleus, and the pulvinar.

Lateral to the dorsal and ventral nuclear groups, and separated by a white matter layer, is the reticular nucleus which is a jacket to the thalamus covering the lateral aspect and extending partly inferiorly. The reticular nucleus is comprised of a net of (inhibitory) cells which receive inputs from traversing fibers from relay neurons to cortex, and from cortex to the thalamus, and which then project back upon intrinsic interneurons and thalamic relay cells. In general, the thalamic relay cells receive excitatory inputs from structures in the spinal cord, brain stem or cerebellum, subcortical structures, and cortex and then have an excitatory projection upon cortex. The exception to these excitatory inputs to the relay cells is the inhibitory input to the ventral lateral anterior nucleus from the internal segment of the globus pallidus.

Medial to medial dorsal nucleus are the midline thalamic nuclei which are larger in rodents than in primates in which they consist of a number of ill-defined clusters named median central, paraventricular, reuniens, and rhomboid nuclei.

THERAPIES TARGETING THE THALAMUS

The translational approaches to the thalamus are related to surgery of the thalamus, which are particularly effective and durable in the case of tremor. The effectiveness of thalamic stimulation for the treatment of tremor has been demonstrated in a randomized controlled trial of DBS versus thalamotomy and multiple large prospective uncontrolled studies. In addition, the durability of this therapy in both parkinsonian tremor and essential tremor has been shown by prospective nonrandomized studies with a 6-year follow-up. The safety of this technique in the long term has been established in studies of DBS for the treatment of chronic pain with follow-up lasting over decades.

Advances in the neurosciences have partially explained the effectiveness in the tremor of Parkinson's disease and essential tremor, the two most common conditions treated by thalamic surgery. In the case of parkinsonian tremor, as with most tremor mechanisms, there are possibilities of an intrinsic pacemaker or the oscillation of an unstable feedback mechanism.

In the case of Parkinson's disease, the proposed thalamic oscillator is activated by the recovery from hyperpolarization of thalamic relay cells in ventral lateral anterior caused by increased inhibitory input from the internal pallidal segment. When they are hyperpolarized, these cells produce a calcium spike which leads to a burst of action potentials itself followed by a potassium-mediated inhibition. This hyperpolarized state again leads to recovery through passive depolarizing currents and

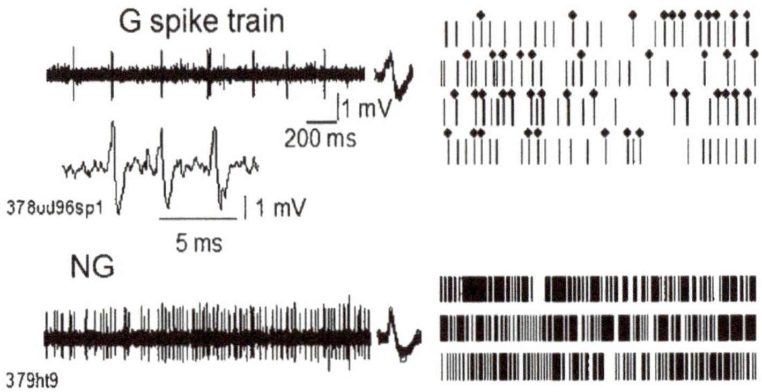

FIGURE 13.2 **Thalamic spike trains** illustrating the single spike (NG, lower panel) and inhibition-related bursting (G, upper panel), mixed burst and single spike patterns of firing. Within each panel, the top left shows the single unit recording while the right shows the times of occurrence of action potentials by vertical bars and of bursts by diamonds above the vertical stroke(s) indicating the action potentials. The inset in the top panel shows a thalamic burst on a small scale.

so to recurrent oscillations. These inhibition-related bursts in the spike train (Fig. 13.2 top row left) are composed of action potentials (top row middle), as indicated by diamonds above the times of occurrence of discriminated action potentials (top row right). These bursts of action potentials have a characteristic pattern such that the intervals between spikes become longer between successive intervals in the burst (Fig. 13.2, inset left). However, such bursts are rarely found in recordings from Vim and Vop of patients with parkinsonian tremor undergoing thalamotomy or DBS implants. These neurons more commonly fire as a train of single action potentials at a relatively constant firing rate (Fig. 13.2 bottom row).

Another possible mechanism of parkinsonian tremor is feedback hypothesis through a peripheral long loop reflex pathway. This mechanism proposes that the oscillation of unstable stretch reflex arcs, termed transcortical reflex arcs, are transmitted through motor cortex in somewhat the same way that tendon tap reflexes are transmitted from muscles to the spinal cord and back to the muscle. In this model, the increased transmission of activity through these reflexes might lead to parkinsonian rigidity and tremor just as exaggerated spinal reflexes may cause spasticity and tremor. This hypothesis is consistent with the results of recordings during surgery for tremor which demonstrate that the activity of thalamic neurons leads to tremors among thalamic neurons with sensory inputs (Fig. 13.4), and not neurons without those inputs. Finally, an engineering analysis has demonstrated that the majority of cells show activity consistent with a peripheral feedback loop.

Essential tremor has often been compared with one animal model of tremor, which is produced by injection of a nonspecific drug (harmaline). Harmaline tremor replicates many components of human essential tremor. In harmaline-treated animals, a pathway from the cerebellum to the spinal cord through synapses in the deep nuclei of the cerebellum and the brain stem plays a role in maintaining of this tremor. In humans, the deep nuclei of the cerebellum project primarily to the thalamus, and human thalamic activity in essential tremor is often directly correlated with EMG at tremor frequency. The highest proportion of neurons showing this correlation is found in ventral lateral posterior which receives inputs from the deep cerebellar nuclei and projects to the motor cortex (Fig. 13.1). These neurons show higher firing rates and tremor-related activity (Fig. 13.3A) during posture or active movements than at rest (order of cells in rows of Fig. 13.3B correspond to those in Fig. 13.3A). Fig. 13.3C shows activity in the thalamus of patients with chronic pain and no abnormality of movement. There is no evidence of tremor frequency activity which was only found during pointing in patients with essential tremor. This association of

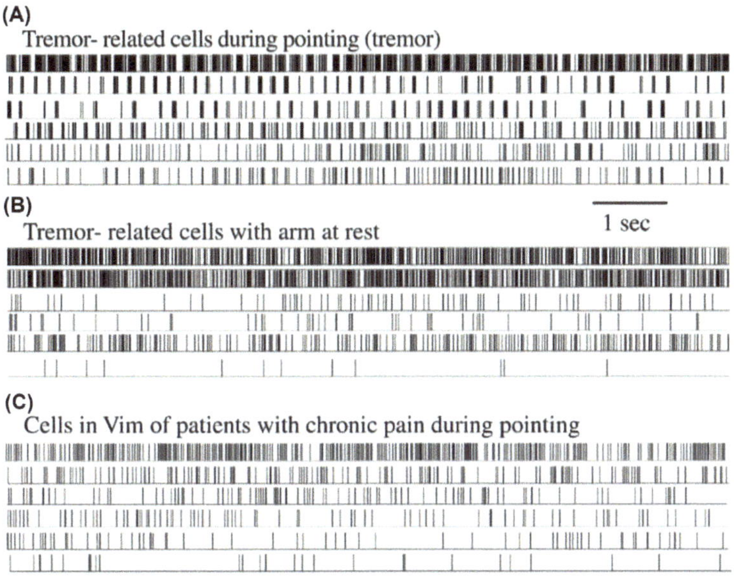

FIGURE 13.3 **Thalamic spike trains in patients with essential tremor.** Firing of thalamic single neurons during an upper extremity pointing movement with tremor (A) and with the arm at rest for the spike train of the same neuron at the corresponding line in (B) and (C). Thalamic activity during pointing in patients with chronic pain.

movement with tremor-related activity suggests that essential tremor may result from tremor-related thalamic neuronal activity when it is enabled by circuits involved in the generation of active movements.

Thalamic activity during the essential tremor, which occurs particularly at the target of pointing movements to a visual target, is similar to the thalamic activity which occurs in tremor secondary to cerebellar disruption. This is a nice parallel to the primary clinical characteristic of essential tremor that it is related to either postural movements or active movement. Finally, many of the neurons that are activated during active movement are also activated during sensory stimuli so that the feedback from the periphery might also influence essential tremor-related neuronal activity (Fig. 13.4).

Microelectrode recording is often used to validate the target nuclei predicted by the MRI scan and so improves the accuracy of the surgical procedure. Microelectrode recording is particularly important in thalamic procedures since the targets measure 2–4 mm along the anterior posterior axis at about 6 cm deep to the surface of the brain. In addition, the MRI is unable to resolve these nuclei so that direct MRI targeting is not possible, although nuclear locations are sometimes inferred from radiological landmarks. Microelectrode recordings can distinguish between cellular and axonal structures, and can segment nuclei based upon neuronal firing patterns and stimulation through the microelectrode, and compensate for shifts in the location of nuclei occurring during surgery due to loss of CSF. These microelectrodes are usually modifications of the types used over many years in animal studies and consist of platinum–iridium wire, which is sharpened electrolytically and glass-coated with an exposed tip of 2–4 μm. This leads to an impedance of approximately 1 MΩ although electrodes of lower impedance, semi-microelectrodes, are often employed.

Recordings from the ventral lateral posterior nucleus or ventral lateral anterior nuclei neurons show neuronal activation during active movement, and in response to joint movement, or deep manual stimulation of muscles in the absence of a response to stimulation of overlying cutaneous

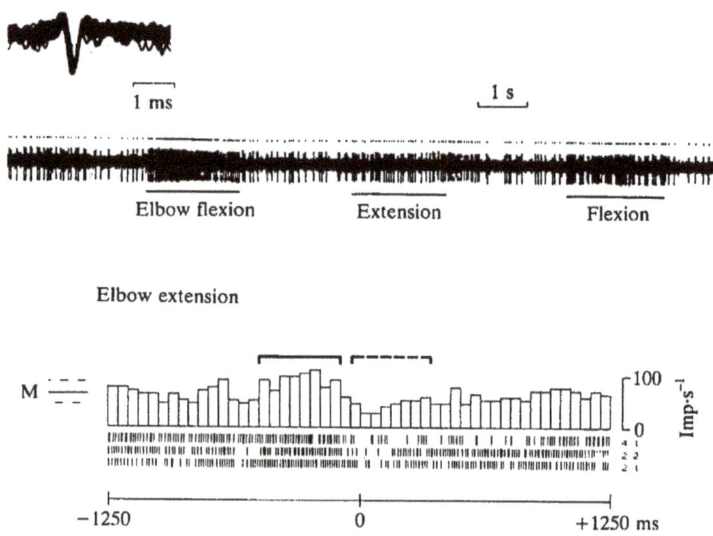

FIGURE 13.4 **Thalamic sensory and motor-related neuronal activity.** The Top panel shows neuronal responses to passive flexion and extension of the elbow, as indicated under the spike train. Each vertical line is a neuronal action potential or spike (inset action potential shape). The bottom line shows a raster histogram (mean ± 2SD) of the spike train as the subject extends the elbow in response to the verbal command to move.

structures. Stimulation in ventral lateral nuclei may produce muscle contraction, joint movement, or a change in the patient's movement disorder, such as tremor arrest, which is a dramatic confirmation of the surgical target.

Stimulation is also used to localize side effects such as tingling sensations related to stimulation in the ventral posterior, muscle contractions lateral to the thalamus in the internal capsule, and perhaps dyskinesias below the lower border of ventral lateral. The responses to stimulation through a microelectrode estimate but do not exactly replicate the response to the larger clinical DBS electrode (1.5 mm contact) so that stimulation must also be carried out through the DBS electrode prior to implantation. Overall, this microelectrode-based localization compels clinicians to apply the basic neuroscience as a confirmation of the target predicted by imaging, and as an exercise in translational neuroscience.

DEEP-BRAIN STIMULATION

The introduction of DBS has transformed movement disorder surgery of the thalamus and is based upon the anatomy and physiology reviewed earlier. In fact, it represents a reintroduction of a refinement of a technique first applied in the 1950s. Perhaps the first use of DBS targeted the hypothalamus for the treatment of chronic pain by Pool at Columbia. Over the intervening years, DBS was used most frequently to stimulate the periventricular gray or the somatic sensory thalamus—ventral posterior as a treatment for chronic pain.

During the course of these procedures, surgeons observed stimulation-evoked improvements in movement disorders, which are associated with chronic pain syndromes such as the dyskinesias associated with chronic pain following thalamic infarcts (Dejerine–Roussy syndrome). Thereafter, the use of lesions experienced a resurgence in the treatment of tremor and Parkinson's disease, followed, the 1980s, by the return of stimulation techniques.

Many translational approaches to the thalamus are technical innovations seeking to optimize the placement of electrodes and stimulation parameters in DBS therapy. At present, approaches to predict the best parameters for stimulation in each patient are limited by the neural elements at the location where the

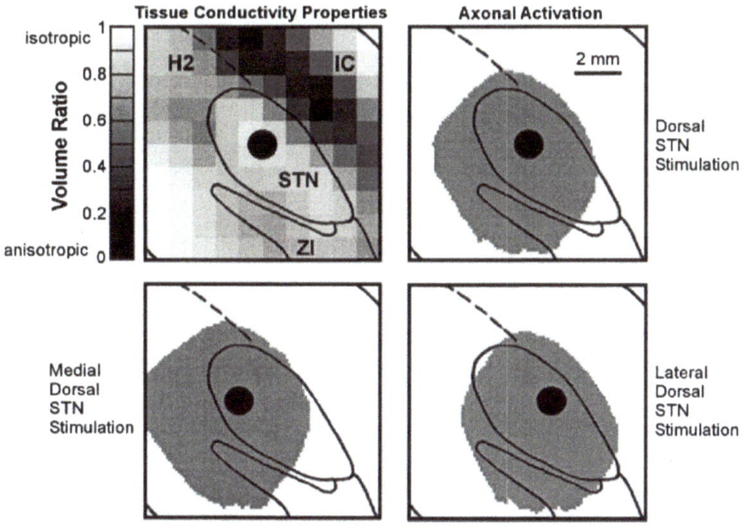

FIGURE 13.5 Model of the extent of excitation of neural elements surrounding the monkey subthalamic nucleus (STN), fields of Forel (H2), internal capsule, zona inserta (ZI), as determined anatomically. Left upper panel indicates the extent to the tissue is composed of axons with a defined directionality, which influences the extent to which fibers having one direction (ZI) are excited versus those with another direction (IC). The DBS stimulating electrode (*black dot*) at three separate locations was projected to produce excitation as a function of location and as indicated in gray shading in the upper right and lower panels.

electrode is implanted. This includes the neurons and their component somata, axons, and synapses, any or all of which may be excited by stimulation at that location; these complexities are further complicated by differences between individuals. These approaches incorporate the charge density limit, which is a safety factor for stimulation parameters which can safely be applied in DBS protocols, and which is measured by histological evidence damage to brain tissue at autopsy. The charge density limit is $30\,\mu C/cm^2$, which is the product of the applied voltage and the pulse width over the impedance times the surface area of the electrode. Of course, the nature of the neural elements that will be activated at any location is an important factor in modeling the effect of DBS.

Among these elements, the myelinated axons are the most easily excited by applied electrical fields. Currents crossing the excitable membrane of a neural element result in changes of the transmembrane voltage as the second derivative of the applied potential. The orientation of the DBS electrode relative to axons in the neighborhood, as determined by diffusion tensor imaging, has been included in models which predict that the applied current at recommended therapeutic parameters will activate elements over a large volume, as reported by Mcintyre. As an example, these models predict that at 3 V excitation of neural elements will extend well beyond the borders of structures like the subthalamic nucleus, which is a dense focus of neurons surrounded by white matter tracts (Fig. 13.5).

Human brain tissue has diverse electrical characteristics which may produce side effects when the effect of stimulation extends outside the subthalamic nucleus. The same effect may occur with stimulation within the subthalamic nucleus since it is surrounded by large dense tracts of axons, which may respect or traverse the subthalamic nucleus. Therefore, the stimulation does not excite a well-defined region but must be described by gradients of excitation which result from the location and the type of neural elements within the electrical field. To a degree, measurements of time and applied current jointly (chronaxie) can also be used in predicting the excitation produced by stimulation in the CNS since it is different between cell bodies and axons.

The elements excited by DBS in a particular patient are predicted by a complicated calculation since cell bodies and axons excited primarily by the electrical field may excite or inhibit other cells as a result of trans-synaptic transmission. These trans-synaptic mechanisms may involve other thalamic relay cells, thalamic cortical thalamic circuits, and interneurons both intrinsic to the specific nuclei and in the reticular nucleus. Furthermore, these effects may be specific to the nucleus being stimulated and the effect of stimulation may be greater upon the surrounding white matter than upon nucleus itself. The complexity has been nicely pointed out by Kiss who demonstrated two separate populations of neurons inactivated in a rodent ventral thalamic DBS model. In one, there was rapid spike firing during stimulation which soon returned to resting membrane potential. In the other, a long-lasting depolarization without action potentials was described. Despite the differences, both responses interrupted ongoing patterns of activity and information transmission in these cells.

Alternately, patterns with both stimulation evoked hyperpolarization and depolarization of cell bodies may result in decreases in neuronal firing from the cell somata resulting in part from inhibitory trans-synaptic mechanisms in the internal pallidum (GPi). Nevertheless, the firing zone and axon may be excited at the same time as stimulation effects in the neuronal body; this argues that stimulation effects might be independent of trans-synaptic inhibition and that intranuclear stimulation is an effective modulator of "pathological network activity." Since beta-frequency oscillations (13–30 Hz) are prominent in the basal ganglia of patients with Parkinson's disease, stimulation at higher frequencies may disrupt these "pathological oscillations."

These results have been incorporated in a model of capacitative and resistive (together impedance) properties of the interface between the electrode, scar, and the adjacent tissue after the scar tissue capsule has formed around the DBS electrode. This model also includes the orientation and directional biases of the fibers in the electrical field of stimulation based upon three dimensional diffusion tensor imaging. The results of this model have been contrasted with the activations in upper and lower extremity muscles produced by stimulation of internal capsule of a patient with Parkinson's disease. This study is a reliable test of such models because the internal capsule is composed of myelinated axons with coherent orientations, which are directly connected to the spinal motor apparatus. Therefore, stimulation in the internal capsule may be a first approximation for the effect of stimulation in structures which are less well defined and homogeneous. This model is currently being upgraded both to optimize the placement of electrodes and to predict the best parameters of stimulation in combination with types and doses of drug therapy.

Acknowledgments

Some of the studies included in this chapter were supported by the National Institutes of Health–National Institute of Neurological Disorders and Stroke (RO1s NS28598, NS38493 and NS40059 to FAL). None of the authors has conflicts of interest related to this work. The figures are adapted with permission as follows: Fig. 13.1 from Smith (2008) in Conn's Translational Neuroscience, and Fig. 13.5 from McIntyre with Vitek (2004) in Journal of Clinical Neurophysiology, and from the Journal of Neurophysiology: Fig. 13.2 from Kim with Lenz (2009), Fig. 13.3 from Hua and Lenz (2005), and Fig. 13.4 from Lenz with Tasker (1988).

Further Reading

[1] Anderson T, Hu B, Pittman Q, Kiss ZH. Mechanisms of deep brain stimulation: an intracellular study in rat thalamus. J Physiol 2004;559:301–13.

[2] Bendok B, Levy RM. Brain stimulation for persistent pain management. In: Gildenberg PL, Tasker RR, editors. Textbook of stereotactic and functional neurosurgery. New York, St. Louis, San Francisco, Auckland, Bogotá, Caracas, Lisbon, London, Madrid, Mexico City, Milan, Montreal, New Delhi, San Juan, Singapore, Sydney, Tokyo, Toronto: McGraw-Hill; 1998. p. 1539–46.

[3] Hirai T, Jones EG. A new parcellation of the human thalamus on the basis of histochemical staining. Brain Res Rev 1989;14:1–34.

[4] Lenz FA, Casey KL, Jones EG, Willis Jr WD. The human pain system: experimental and clinical perspectives. New York (NY): Cambridge University Press; 2010.

[5] Lenz FA, Dostrovsky JO, Kwan HC, Tasker RR, Yamashiro K, Murphy JT. Methods for microstimulation and recording of single neurons and evoked potentials in the human central nervous system. J Neurosurg 1988;68:630–4.

[6] McIntyre CC, Foutz TJ. Computational modeling of deep brain stimulation. Handb Clin Neurol 2013;116:55–61.

[7] Schuurman PR, Bosch DA, Bossuyt PM, Bonsel GJ, van Someren EJ, de Bie RM, et al. A comparison of continuous thalamic stimulation and thalamotomy for suppression of severe tremor. N Engl J Med 2000;342:461–8.

[8] Zakaria R, Lenz FA, Hua S, Avin BH, Liu CC, Mari Z. Thalamic physiology of intentional essential tremor is more like cerebellar tremor than postural essential tremor. Brain Res 2013;1529:188–99.

CHAPTER

14

Cranial Nerves and Brainstem

R.A. Feroze, A. Ozpinar, N. Alan, R.F. Sekula Jr.
University of Pittsburgh School of Medicine, Pittsburgh, PA, United States

INTRODUCTION

There are 12 cranial nerves that control motor and sensory function of the head and neck, innervating voluntary and involuntary muscles and receiving general and special sensory information. Cranial nerves carry six different forms of fibers: general somatic efferent, general somatic afferent, general visceral efferent, special visceral efferent, visceral afferent, and special afferent. It is essential to understand the function of each cranial nerve and their entire course throughout the skull base to understand their respective pathologies. The advent of advanced imaging, specifically magnetic resonance imaging (MRI), has allowed demonstration of detailed anatomy of the basal cisterns, brain stem, and skull base enhancing evaluation of cranial neuropathies.

All cranial nerves, except cranial nerves I and II, have their respective nuclei located in the brain stem. The brain stem consists of three sections: midbrain, pons, and medulla. Along with cranial nerve nuclei, it contains other intrinsic nuclei, a ventricular system that holds cerebrospinal fluid, and ascending/descending white matter tracts which convey neuronal signals between the brain and spinal cord.

The midbrain also known as mesencephalon extends from the superior to inferior colliculi and consists of tectum, tegmentum, crus cerebri, and cerebral aqueduct. The cranial nerve nuclei that reside in the midbrain are: oculomotor nuclear complex, trochlear nucleus, and mesencephalic nucleus of trigeminal nerve. The pons extends from superior colliculi rostrally to the pontomedullary sulcus caudally. In addition to multiple fiber tracts, the pons houses the motor and principal sensory nucleus of the trigeminal nerve, spinal trigeminal nucleus, abducens nucleus, facial nucleus, superior and inferior vestibular nuclei, and auditory nucleus. The medulla, also known as medulla oblongata, extends from pontomedullary sulcus to the spinomedullary junction and contains medial and lateral vestibular nuclei, dorsal motor nucleus of vagus nerve, hypoglossal nucleus, and accessory nucleus. The relationship of the cranial nerve nuclei and white matter fiber tracts to the brain stem remains a topic of active investigation. Advances in imaging techniques have allowed for a better understanding of the structures that are affected by complex pathologies of the brain stem and cranial nerves.

In this chapter, our goal is to provide an up-to-date overview of common cranial nerve and brain stem pathologies and their respective clinical presentations, pathophysiologies, surgical/pharmacological treatments, and various

Conn's Translational Neuroscience
http://dx.doi.org/10.1016/B978-0-12-802381-5.00023-3

imaging tools used to elucidate disease severity and effect of treatment.

COMMON CRANIAL NERVE PATHOLOGIES

Trigeminal Neuralgia

The trigeminal nerve is the largest cranial nerve entering the brain stem. Consisting of sensory and motor fascicles, it enters the brain stem at the mid-lateral surface of the pons. The sensory root synapses in the trigeminal ganglion located in the Meckel's cave, which is posterior and lateral to the either cavernous sinus. The motor fascicles pass through the trigeminal ganglion (ie, Gasserian ganglion) and join part of the sensory nerve to make up the mandibular division of the trigeminal nerve. Ophthalmic and maxillary divisions of the trigeminal nerve synapse in the trigeminal ganglion.

The ophthalmic branch, purely sensory, courses through the superior orbital fissure providing sensory innervation to the forehead, nose, eyelid, conjunctiva, lacrimal glands, conjunctiva, and a portion of the nasal mucosa. The maxillary nerve enters the base of the skull through the foramen rotundum and branches into various nerves that reach the pterygopalatine fossa through the inferior orbital fissure and pterygopalatine canal providing sensory innervation to the middle third of the face. The mandibular division travels into the cranium through the foramen ovale and provides sensory innervation to the lower third of the face, the tongue, jaw, and floor of the mouth. It also contains the only motor component of the trigeminal nerve, providing innervation to the muscles of mastication.

One of the most debilitating condition involving the trigeminal nerve is trigeminal neuralgia (TN). The TN is a facial pain syndrome associated with brief, paroxysmal attacks of pain often described as "electrical" and "shock like."

The etiopathogenesis of TN is not well understood. Physiological abnormalities including demyelination and axonal loss along the centrally myelinated portion of the trigeminal nerve have been noted in patients with and without vascular compression. This injury to the trigeminal nerve may lead to the paroxysmal transmission of sensory information and production of ectopic action potentials resulting in sensations of paroxysmal pain.

Treatment for TN includes pharmacological therapy, microvascular decompression (MVD), and all of the ablative treatments. Pharmacological therapy for TN includes carbamazepine, a voltage-gated sodium channel blocker, which often provides relief, for a period of time. Response to carbamazepine may also serve as predictor of successfulness of MVD. The MVD is considered the "gold standard" treatment for TN by relieving compression of the nerve caused by an offending vessel. The blood vessel is separated from the nerve and held in place by soft implants. The superior cerebellar artery is the most common compressive vessel. Although MVD provides the most durable relief for TN (ie, about 70% of patients remain free of pain without medication), the procedure requires a craniotomy and the presence of offending vessel. Ablative procedures include those percutaneous procedures directed at the trigeminal ganglion (ie, balloon rhizolysis, glycerol rhizolysis, and stereotactic radiosurgical rhizolysis utilizing the Cyberknife or Gamma Knife.)

Balloon compression involves applying pressure to one or more of the divisions of the trigeminal nerve in Meckel's cave, which is expected to injure the medium and large myelinated fibers leading to reduced transmission of pain. Glycerol rhizotomy involves the injection of glycerol in Meckel's cave and was incidentally discovered to cause relief of pain, most likely due to demyelination and axonal damage of the diseased trigeminal ganglion leading to reduced ectopic action potential generation and propagation of pain signals. Finally, radiofrequency thermo-coagulation involves the use of electrodes to identify areas generating pain signals followed by thermocouples to induce lesions in these areas blunting impulse propagation. All three treatments have been shown to provide significant pain resolution but carry increased side effects compared to MVD, given their ablative nature.

Choice of treatment often depends on etiology of the neuralgia. While neurovascular compression is the most common cause of TN, other etiologies include tumors compressing the cisternal portion of the nerve/ganglion or demyelinating lesions precipitated by multiple sclerosis. For such conditions, MVD is not an appropriate treatment. Surgical resection of compressive tumors is expected to improve symptoms. TN due to multiple sclerosis have been shown to require significantly more treatments than non-MS cases.

TN is a clinical diagnosis; however, there are many etiologies that can cause this syndrome. MRI has been traditionally used in order rule out confounding diagnosis, such as vascular malformation, demyelinating diseases, and tumor around the nerve. The development of steady-state free precession (SSFP) sequence has improved the spatial resolution providing better illustration of the cisternal intracranial structures (Fig. 14.1). These technical advances in MRI of the cranial nerves has led to increased interest in used imaging as an adjuvant modality for identifying appropriate surgical candidates. Recent studies utilizing this concept have demonstrated that in patients with a clinical diagnosis of TN, and imaging findings consistent with neurovascular compression at the root entry zone of the trigeminal nerve, MVD provides excellent and durable outcomes.

Although experimental white matter tractography has gained popularity in being able to show three-dimensional (3D) representation of fiber tracts as well as to identify changes in neural microstructure. Processes relevant to TN, such as demyelination and axonal injury result in microstructural changes that can be quantitatively analyzed using tractography. Recent advances have allowed the use of tractography to detect focal changes in the trigeminal nerve following treatment. A significant decrease in fractional anisotropy, a measurement used to reflect fiber density, axonal diameter, and myelination in white matter was found following Gamma radiation at the root exit zone, while contrast-enhanced MR images were not able to detect changes in the trigeminal nerve following treatment. Tractography has been applied to cranial nerves in very few cases and remains a promising area of investigation for cranial nerve disorders.

Hemifacial Spasm

The facial nerve exits the brain stem at the pontomedullary sulcus and traverses superiorly across the pons. Eventually the facial nerve detaches from the pons traveling across the cerebellopontine angle cistern and finally enters the internal acoustic meatus. Hemifacial

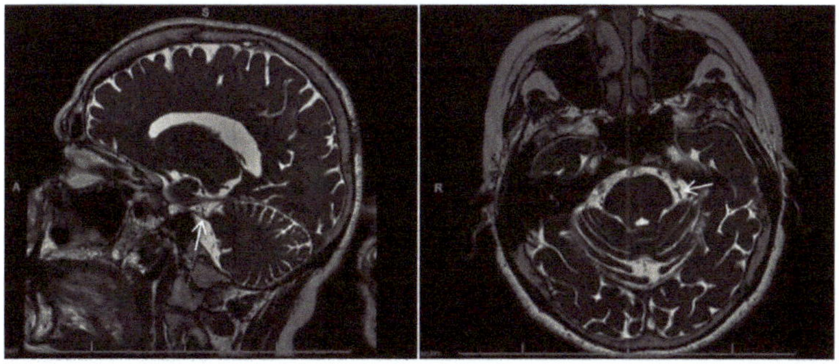

FIGURE 14.1 Trigeminal neuralgia. Sagittal (left) and axial (right) SSFP images demonstrating compression of the trigeminal nerve by superior cerebellar artery resulting in symptomatic trigeminal neuralgia.

spasm (HFS) is characterized by hyperactivity of the facial nerve resulting in involuntary contractions of one side of the face. Unlike TN, the etiology of HFS seems to be more consistently related to vascular compression of and its effects on the centrally myelinated portion of the facial nerve.

Various theories exist regarding the cause of ectopic excitation produced in HFS. One hypothesis postulates that injury to the facial nerve causes retrograde charges to the facial nucleus that lead to hyperexcitability and production of ectopic potentials. A more peripheral cause has also been proposed which implicates cross-excitation between fibers at the site of lesions as the cause of ectopic impulse generation. A third hypothesis suggests a leakage of neurotransmitter from sympathetic terminals on arteries which are damaged in HFS as the culprit. Further studies are required to elucidate the mechanism of ectopic signal generation in HFS.

Partial and temporary relief of HFS has been observed through injection of botulinum toxin. Following injection, the botulinum toxin is cleaved into a heavy and light chains which prevent release of acetylcholine from the presynaptic terminal by preventing the opening of presynaptic calcium channels and inhibiting the function of SNARE [acronym derived from "SNAP (soluble NSF attachment protein) receptor"] proteins, respectively. While botulinum toxin is one of the oldest and most common remedies for HFS, it is an imperfect treatment due to the need for continuing injections and eventual development of tolerance. Therefore, it is now reserved for patients who do not qualify for surgical intervention.

The surgical treatment for HFS is MVD and is used for cases where the facial nerve is being compressed by an offending blood vessel. Complete symptom resolution following MVD has been observed in 91.1% of cases. Symptom resolution can be immediate or delayed. Remyelination at the site of injury is thought to be the cause for delayed symptom resolution.

High-resolution MRI images have been used successfully to identify neurovascular compression in HFS cases (Fig. 14.2). Any mechanical perturbation of the facial nerve between the pontomedullary sulcus to within 4 mm of the detachment point from the pons is considered clinically relevant. Hemodynamic changes in the posterior circulation accompany neurovascular compression in HFS. Color-coded duplex ultrasound imaging has revealed hemodynamic differences between the affected and unaffected sides. These measurements were made transcranially by aligning the ultrasound transducer against the foramen magnum acoustic bone window and imaging using a transnuchal approach. Blood-flow velocities in the anterior inferior cerebellar artery (AICA), posterior inferior cerebellar artery (PICA), vertebral and basilar arteries were measured. The vertebral and basilar arteries give rise to PICA and AICA which are more commonly involved in HFS. Higher velocities were detected across AICA and PICA on the side with HFS with no significant differences in the basilar and vertebral arteries, indicating the presence of local

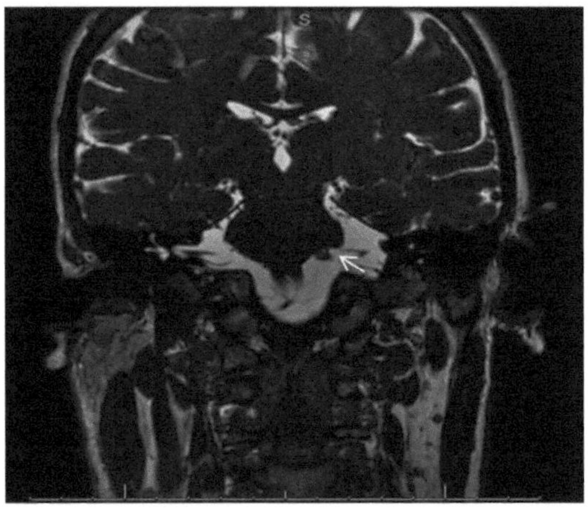

FIGURE 14.2 Hemifacial spasm. Coronal SSFP image demonstrating the anterior inferior cerebellar artery contacting and deforming the pons at the lateral aspect of the facial nerve (*white arrow*) along the root detachment point resulting in hemifacial spasm.

disease in the offending blood vessels instead of broad hemodynamic changes originating from more proximal locations.

An association has been found between Chiari I patients and HFS suggesting that an overcrowded posterior fossa may lead to impingement of the facial nerve. Furthermore, 3D MRI volumetric analysis of the posterior fossa has been used to detect reduced average posterior fossa volumes in HFS subjects compared to matched controls. A smaller posterior fossa increases the likelihood of impingement of the facial nerve and is a possible explanation for increased incidence of HFS in the female gender and children.

Glossopharyngeal Neuralgia

The glossopharyngeal nerve emerges from the lateral medulla, travels anterolaterally, and leaves the cranium through the jugular foramen. The jugular foramen contains the superior and inferior ganglia of the glossopharyngeal nerve. These ganglia contain pseudo-unipolar cell bodies that transmit afferent signals. The glossopharyngeal nerve then passes between the superior and middle pharyngeal constrictors and enters the posterior third of the tongue.

Containing both sensory and motor components, the glossopharyngeal nerve provides somatic motor innervation to the stylopharyngeus muscle, visceral motor innervation to the parotid gland, and carries afferent sensory fibers from the posterior third of the tongue, pharynx, and tympanic cavity. Finally, it is involved in the regulation of the baroreceptor and chemoreceptor reflexes by carrying sensory information from the carotid body and the carotid sinus.

Glossopharyngeal neuralgia (GPN) is much less common than TN, but it is a similar disease in many ways. Patients also experience brief, paroxysmal "electrical and shock-like" pain originating from the pharynx and posterior tongue radiating up to the Eustachian tube and inner ear. Episodes of pain last for several seconds followed by pain-free intervals. Chewing, talking, or sudden movements of the jaw can trigger pain. Patients may also experience vertigo, vomiting, tinnitus, and involuntary movements. The pain may be more oropharyngeal or tympanic, providing clues to the location of nerve damage. As in TN, pain in GPN may be episodic or continuous depending on the type of insult to the glossopharyngeal nerve.

Due to the anatomical relationships between the vagus and glossopharyngeal nerves, asystolic episodes, bradycardia, hypotension, and syncope have also been observed in GPN patients. Abnormal hyperstimulation of the glossopharyngeal nerve due to injury leads to increased feedback to the nucleus of the solitary tract, a major sensory nucleus located in the medulla, which interacts with multiple cranial nerves and receives cardiovascular, visceral, respiratory, and gustatory input. The increased feedback to the nucleus of the solitary tract is transmitted via collaterals to the dorsal motor nucleus of the vagus nerve. Abnormal activation of this feedback loop, normally functioning to regulate heart rate in response to changes in blood pressure or oxygen content, leads to a heightened vagal response causing bradycardia and subsequent hypotension. Such cardiovascular changes may occur with a pain episode or manifest without typical neuralgia symptoms. In some cases, these can lead to life-threatening complications and should be ruled out using electrocardiogram studies. Resolution of cardiovascular symptoms has been shown following pharmacological and surgical intervention.

GPN may be idiopathic, caused by mass effect of a tumor, neurovascular compression, or demyelinating diseases, such as multiple sclerosis. GPN has also been associated with TN and Chiari I malformation. As with TN, MRI with SSFP imaging has allowed for better visualization of injured glossopharyngeal nerves and demonstration of offending structures. At the origin of the glossopharyngeal

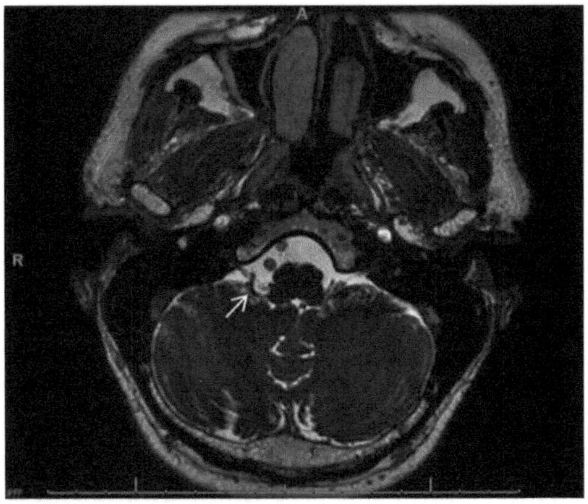

FIGURE 14.3 Glossopharyngeal neuralgia. Axial SSFP image demonstrating right glossopharyngeal nerve being compressed by the posterior inferior cerebellar artery (*white arrow*) causing glossopharyngeal neuralgia.

nerve, vascular compression may occur due to the posterior inferior cerebellar artery or anterior inferior cerebellar artery (Fig. 14.3).

Secondary GPN can occur due to Eagle's syndrome where an elongated styloid process and/or calcification of the stylohyoid ligament impinges the glossopharyngeal nerve causing symptoms. 3D reconstructions from computational tomography images may be used to characterize the abnormal styloid process in Eagle syndrome and provide clues to interference with the glossopharyngeal nerve.

Pharmacological treatment for GPN includes antiseizure medications, such as carbamazepine, gabapentin, or phenytoin and selective serotonin reuptake inhibitors. In addition, localized nerve blocks using anesthetic or neurolytic agents can also provide pain relief and are administered intraorally or below the styloid process. Nerve blocks can have side effects, such as difficulty swallowing or vocal cord paralysis due to impairment of the baseline function of the glossopharyngeal nerve. Numerous patients report only temporary improvement following pharmacological treatment requiring the need for surgical intervention. Depending on the primary etiology of GPN, surgical treatment may involve MVD, tumor resection, posterior fossa decompression in the case of Chiari malformation, or stylectomy for Eagle's syndrome. All such procedures carry the risk of neurovascular injury and can cause long-lasting dysphonia and dysphagia.

Vestibular Schwannomas

The vestibulocochlear nerve contains fibers, which carry information regarding hearing, balance, equilibrium, and motion. It emerges from the pontomedullary sulcus, crosses the cerebellopontine angle and enters the internal acoustic meatus traveling adjacent to the facial nerve, internal auditory artery, and the superior vestibular nerve. In the internal acoustic meatus, the vestibulocochlear nerve separates into the vestibular and cochlear nerves. The cochlear nerve enters the cochlea where its axons branch to make contact with ciliated cells on the spiral lamina and provide sensory innervation.

The vestibular nerve consists of the central processes of bipolar neurons of the vestibular ganglion. The peripheral processes of these neurons synapse with the ciliated sensory cells of the inner ear. Specifically, these cells are found in the ampulla of the three semicircular canals and in the utricle and saccule. The ciliated cells are stimulated by the movement of endolymph during rotational and linear head movements, generating impulses that are carried by the vestibular nerve to the brain stem. Sensory information from various structures innervated by the vestibulocochlear nerve is delivered to specific nuclei in the brain stem for further processing.

Vestibular schwannomas, formerly known as acoustic neuromas, are the most common tumors of the cerebellopontine angle and arise from the Schwann cell sheath surrounding the vestibular nerve. These tumors are benign, grow slowly, and may compromise function of the trigeminal and facial nerves. They occur with an incidence of 1.1/100,000 people with a documented increase in the last few decades. Vestibular schwannomas may occur unilaterally or bilaterally. They cause vertigo, hearing loss, facial paralysis, tinnitus, TN, and can also lead to hydrocephalus, and brain stem compression. Rare cases have undergone malignant transformation as well.

Bilateral vestibular schwannomas are found in neurofibromatosis 2 (NF2), an autosomal-dominant disorder associated with the *NF2* gene on chromosome 22. *NF2* acts as a tumor-suppressor gene and both alleles must be mutated to result in disease. *NF2* normally encodes for a protein called Merlin which is believed to maintain membrane stability, cell-to-cell attachment, and cell motility through cytoskeletal and integrin interactions. Among its many functions, Merlin localizes to the cell membrane and stabilizes cadherin-dependent cell–cell junctions. It is inhibited from phosphorylation by cadherin binding. Nonphosphorylated Merlin inhibits activation of Rac1, a Rho family GTPase mediating loss of contact inhibition between cells and promoting cell proliferation. Inhibition of Rac1 results in prevention of mitogenic signaling and tumorigenesis. A protein called PAK causes phosphorylation of Merlin when it is not bound to cadherins, resulting in reduced inhibition of Rac1 and cell proliferation. Cell studies show that at high cell densities, Merlin is hypophosphorylated, causing inhibition of the Rac1 pathway and cell proliferation. In contrast, PAK phosphorylates Merlin at low cell densities, preventing it from stabilizing cell junctions and inhibiting Rac1 signaling, leading to loss of contact inhibition and increased cell division, respectively. In addition, Merlin has also been shown to work by interacting with the CD44 cell receptor, preventing it from binding to hyaluronan, and promoting cell-dependent contact inhibition. Vestibular schwannomas have also been shown to express neuroregulin-1 and its receptor ERBB2, which promote cell proliferation. Merlin inhibits ERBB2 activation and downstream expression of paxillin, a mediator for Rac1 binding. Loss of Merlin leads to increased expression of paxillin and Rac1 signaling. Proteomic studies reveal that missense mutations, which may be found in Neurofibromatosis 2, disrupt proper folding of Merlin leading to loss of tumor-suppressive function. This would result in uninhibited Rac1 signaling leading to uncontrolled cell proliferation and schwannoma formation.

Histological examination of schwannomas from both sporadic and NF2-related cases reveal the presence of spindle-like cells with elongated nuclei and a transformation in Schwann cell function from myelinating to nonmyelinating. In addition, islands of neoplastic cells are surrounded by areas of eosinophilic fibrosis with tumor cells in the margins. NF2-related schwannomas have been reported to encapsulate nerve fibers, while sporadic tumors around are associated with nerve fibers outside the tumor capsule.

Contrast-enhanced MR imaging is the most commonly used way of visualizing vestibular schwannomas, measuring tumor size, and identifying impingement or potential scarring of surrounding structures, such as the facial nerve (Fig. 14.4). Enlargement of the internal auditory canal may also be noted in CT scans. Given the route

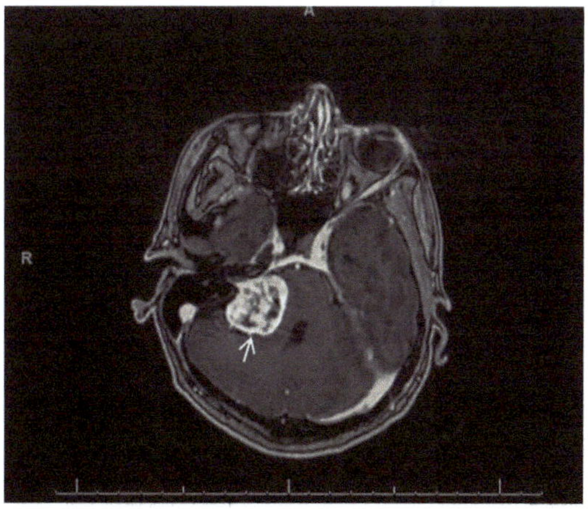

FIGURE 14.4 Vestibular schwannoma. Contrast-enhanced T1-weighted axial MRI demonstrating contrast-enhancing lesion (*white arrow*) at the right cerebellar–pontine angle consistent with a diagnosis of vestibular schwannoma.

of the facial nerve through the internal auditory canal, a schwannoma extending into this opening would compress the facial nerve leading to symptoms similar to HFS. Angiography can be used to visualize vascularization of tumors and presence of nearby vessels subject to mass effect.

Various molecular-targeted therapeutics are under investigation as treatment options for vestibular schwannoma. Bevacizumab, a humanized monoclonal antibody against the angiogenesis-promoting factor VEGF, is widely used for many types of tumors and has been shown to cause regression of progressive vestibular schwannomas and improve hearing in isolated cases. Vestibular schwannomas have been found to express the VEGF receptor at high levels and therefore Bevacizumab is a promising treatment option. However, larger studies are required to establish its efficacy. Inhibitors of EGFR and HER2 receptors including trastuzumab, erlotinib, and lapatinib have been well established in lung/breast cancer and are also under consideration for large trials on vestibular schwannoma patients. Such therapeutics may serve as an important adjunct to surgical treatment.

Microsurgical resection and stereotactic radiosurgical ablation (eg, Gamma Knife) are the most common forms of treatment for vestibular schwannomas. While microsurgical resection offers a chance for complete removal of the tumor, risks of facial nerve injury have been reported between 7% and 58%. Radiosurgical ablation is sometimes preferred over microsurgery for smaller tumors or to preserve hearing and facial nerve function with schwannomas that are present in the auditory canal. Size, stage of tumor, extra-meatal tumor extension, and angle of internal acoustic meatus play a role in damage to the facial nerve during surgical excision of vestibular schwannoma. All of these factors can be evaluated from contrast-enhanced MR images. Larger tumors have been associated with poor facial nerve outcome. Anterior-growing schwannomas cause displacement of the facial nerve and increase risk of injury. A narrow angle of the internal acoustic meatus with respect to the vestibular schwannoma increases chances of compression of the facial nerve between the wall of the meatus and the tumor leading to facial nerve-related symptoms. The consistency of tumor has also been correlated with surgical outcome. Cystic tumors have been found to be more aggressive and associated with more surgical complications compared to solid tumors due to rapid, unpredictable expansion of the cystic spaces within the tumor. With improvement in imaging techniques and microsurgery, a decrease in facial nerve complications has been associated with schwannoma excision. Completely removing this complication remains a major goal of improving vestibular schwannoma treatment.

BRAIN STEM PATHOLOGIES

Brain Stem Cavernous Malformation

Cavernous malformations (CMs) are clusters of dilated sinusoidal low-flow blood vessels that have a high risk of hemorrhage. These vessels are characterized by inadequate smooth muscle and extracellular matrix components leading to increased risk of leakage. Bleeding from CMs may lead to an array of symptoms including headaches, neurological deficits, and hemorrhagic strokes. Cerebral CMs occur in the brain cortex as well as the brain stem but, unlike arteriovenous malformations, do not encompass brain tissue. Blood flow through CMs is slow, making them difficult to identify on angiograms and CT scans unless bleed has occurred. Therefore, MRIs are most commonly used for making diagnoses (Fig. 14.5). Both sporadic and genetic cases of CMs exist with a higher risk of symptoms associated with accompanying genetic predisposition.

While most CMs are supratentorial, 9–35% are located in the brain stem and have higher bleeding rates than their superior counterparts.

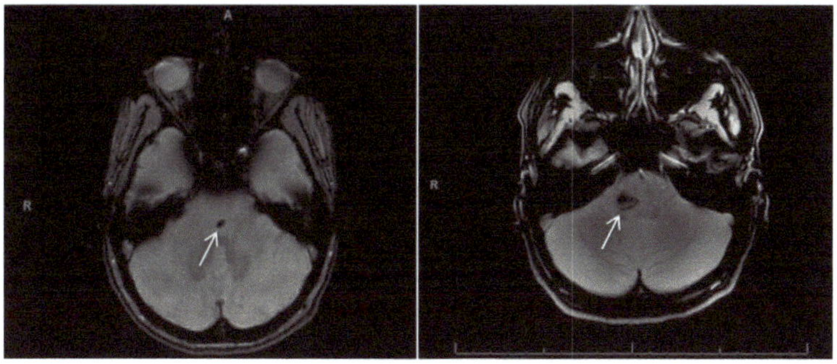

FIGURE 14.5 Brain stem cavernous malformation. Gradient-recalled echo (GRE) T2-weighted axial MRI images of brain stem cavernous malformation (*white arrows*) at time of initial diagnosis (left) and 12-month surveillance (right).

Management is also more challenging due to the risk of permanent neurological damage associated with surgical resection in the brain stem. Surgical resection has been the gold standard of treatment when neurological deficits resulting from mass effect are present.

Most common presenting symptoms of brain stem CMs include headaches, cranial nerve deficits, and hemiparesis followed by ataxia and dizziness. Such neurological deficits occur due the mass effect of CMs or repeated hemorrhages, resulting in injury of cranial nerve nuclei and white matter tracts in the brain stem. The annual risk for hemorrhage of brain stem CMs is 2–21%. After a primary neurological event, the frequency of subsequent events increases, epitomizing the importance of early treatment. For asymptomatic patients, conservative treatment in the form of surveillance MRIs has been considered due to the lack of parenchymal damage associated with every bleeding episode, however this approach remains a subject of debate. Mortality rate following conservative treatment is much higher than surgical resection. For patients in whom CMs cannot be easily resected, stereotactic radiosurgery has been successful in reducing CM size.

Various factors of CMs have been used to predict risk of hemorrhage and success of treatment. Favorable long-term outcomes have been associated with small-size brain stem CMs (<10 mm). The timing of surgical resection has also been found to be relevant. While no definitive consensus exists regarding the ideal time to perform surgery, it is believed that over time the hematoma undergoes fibrosis and is surrounded by glial scarring and calcifications which obscure a previously well-demarcated area for dissection. However, acute surgical resection is also challenging due to the presence of edematous tissue which comprises the dissection plane between the CM and the brain stem. Therefore, subacute resection appears to be associated with the least amount of surgical complications. However, this may not be possible in certain cases where a high risk of hemorrhage is present at acute stages. Local administration of steroids prior to surgery has been used successfully to reduce edema making the plane between the CM and brain stem easier to dissect.

For surgically treated brain stem lesions, a 42% rate of early neurological morbidity and a 5–32% risk of long-lasting morbidity have been reported. While many factors may contribute to these values, a definite cause is believed to be damage to crucial white matter tracts due to the CM or surgery. Identification of such tracts prior to and during surgery is expected to reduce risk of long-term damage. Diffusion-tensor imaging has become an adjunct to surgical treatment but is unable to represent the crossing of multiple fibers and their terminations. Recent studies have reported that the use of high-definition fiber tractography (HDFT) has overcome some limitations of diffusion-tensor imaging. HDFT allows for accurate tracking of fiber pathways through complex crossings and allows for better characterization of changes in perilesional white matter tracts. It has been used to guide surgical planning by providing a quantitative and qualitative assessment of corticospinal tracts around brain stem CMs. Disruption and displacement using tractography reconstructions have been used to qualitatively assess damage to corticospinal tracts. A measure called quantitative anisotropy (QA), which is a measurement used to reflect fiber density, axonal diameter, and myelination in white matter, has also been used to evaluate corticospinal tract fiber integrity before and after surgical treatments.

Longitudinal HDFT reconstructions in patients with brain stem CMs have shown displaced and partially disrupted corticospinal tracts. Improvement in the displacement relative to the unaffected side was observed following surgical resection. It is believed that displaced white matter tracts still have the potential to regain function if the offending stimulus is removed. Improvement of displacement on HDFT reconstructions was correlated with improvement in motor function, unilateral hemiparesis, speech apraxia, paresthesias. Mean QA values were found to be lower on the affected side compared to the contralateral side in patients with brain stem CMs. Following surgical resection, differences in QA values between the affected and unaffected sides stabilized at subsequent 6 months of follow-up indicating the efficacy of surgical CM removal. Application of HDFT to other brain stem and cranial nerve conditions is a promising area of investigation and is expected to improve diagnoses and serve as an important tool in the guidance of surgical treatments.

Chiari Malformation

Chiari malformation is a deformity of the posterior fossa originally reported in the 1890s from autopsy cases of infants with myelodysplasia and caudal dislocation of the medulla and cerebellum. Over the next century, Chiari malformation referred to any pathological herniation of brain stem tissue through the foramen magnum and into the cervical canal. In recent years, classification of such disorders has evolved.

Chiari malformation type I is now defined as cerebellar tonsil herniation of 5 mm or greater below the foramen magnum (Fig. 14.6). Chiari II is considered to be a pathologically distinct disorder with herniation of the medulla and vermis in addition to the cerebellar tonsils. This section explores the pathophysiology of Chiari I and how new imaging techniques have provided insight into disease burden.

There are many predisposing conditions to Chiari I that may result in cranial constriction, spinal cord tethering, intracranial hypertension, or intraspinal hypotension. All these mechanisms lead to a change in the volume of the posterior fossa and reduction in the foramen magnum outlet area. The hallmark symptom of Chiari I is suboccipital headache that is exacerbated by cough, Valsalva maneuver, or exercise.

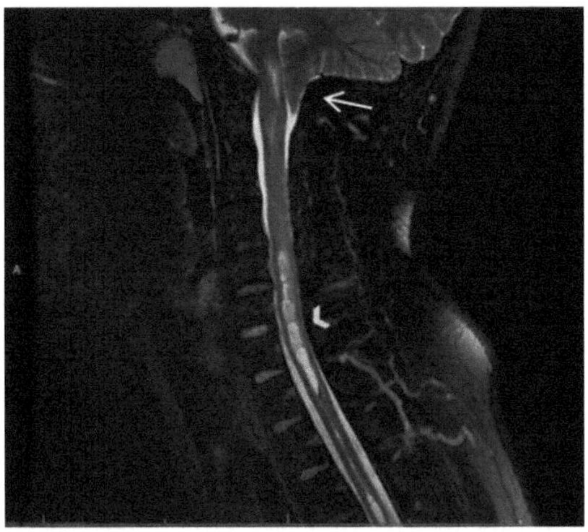

FIGURE 14.6 Chiari I malformation. Sagittal noncontrast T1-weighted MRI demonstrating downward herniation of the cerebellar tonsils (*white arrow*) below the foramen magnum consistent with a diagnosis of Chiari I malformation. Syrinx (*white arrowhead*) present in cervical portion of spinal cord indicative of accompanying syringomyelia.

Any increase in intrathoracic pressure leads to an increase in intracranial pressure (ICP) which causes pain due to reduced posterior fossa volume. Symptoms in Chiari I patients vary greatly and may include various types of headaches in addition to motor and sensory loss.

Multiple theories regarding the cellular origins of Chiari I have been proposed. Initially, this condition was considered to be a primary neural malformation. However, more recent evidence indicates that Chiari I arises from an imbalance in the formation of connective tissue. One theory suggests that inadequate mesodermal development is responsible for a hypoplastic posterior fossa. Other studies have implicated an undeveloped endochondrium of the posterior fossa as causing Chiari I. In contrast, another hypothesis suggests that a disorder in neural crest cells migration to the neck musculature may result in Chiari I. Premature fusion of ossification centers has also been considered as a cause of underdeveloped posterior cranial fossa. While these proposed mechanisms are all very different, there is general agreement that Chiari I is due to abnormal morphometric changes of the posterior fossa and not due to any defect in neural/brain tissue. Chiari I is also believed to have a genetic component due to the abundance of monozygotic and dizygotic twin studies in the literature. Finally, Chiari I has been shown to be co-segregated with various genetic disorders, some of which are characterized by failure of differentiation of neural crest cells.

Using MRI images, a reduction in posterior fossa and 4th ventricle volumes has been observed in Chiari I patients compared to healthy subjects. The abnormal architecture of the posterior fossa due to Chiari I is associated with altered cerebrospinal fluid (CSF) dynamics. Abnormal CSF flow is thought to lead to headaches and specialized MR imaging has been used to study hydrodynamics in cases of Chiari I to better elucidate underlying pathology and make correlations to symptoms. Such studies have used measures, such as intracranial volume change during the cardiac cycle, intracranial compliance, CSF velocity, CSF stroke volume, cerebral blood flow/venous drainage, and cerebral pressure gradient. Post processing of MRIs allows for such measurements to be made. Pulsatility-based segmentation methods can be used to differentiate the lumens of blood vessels and CSF reservoirs in the brain from the surrounding environment. Flow rate and velocity can then be determined by integrating all the pixels in the lumen. Time integration of CSF flow rate results in calculation of the CSF stroke volume during one cardiac cycle.

Due to the reduced volume of the 4th ventricle in Chiari I, an abnormally high ICP is expected which contributes to symptoms. MRI measurements have also shown an increase in spinal cord motion in Chiari I patients. A region of interest of the spinal cord is chosen and the velocity and displacement is measured over a cardiac cycle. The mechanical perturbation on the spinal cord by an overcrowded posterior fossa may impinge on the dorsal columns, corticospinal tracts, and spinothalamic pathways, contributing to the sensory and motor deficits observed in a subset of Chiari I patients. More studies are needed to confirm this mechanism.

Reduced intracranial volume change and intracranial compliance have been reported in Chiari I patients. Intracranial compliance can be determined by difference between the net arterial minus venous blood flow and the craniospinal CSF flow waveforms. These waveforms are the inflow and outflow in the skull and would be equal in a noncompliant system. The greater difference between these waveforms indicates a higher compliance. The smaller difference between the waveforms indicates a less-compliant CSF circulation. The CSF drains into the jugular veins and to the rest of the venous circulation. The reduced intracranial compliance is expected to manifest increased drainage through the jugular veins. MR venography images of the neck veins showed this phenomenon in patients with Valsalva-induced headaches.

A 3.4-mm Hg higher ICP as measured by MRI was found in Chiari I patients with Valsalva-induced headaches compared to those who had non-Valsalva-related headaches. The higher ICP is another indication of low intracranial compliance. With reduced compliance, the intracranial space loses its ability to accommodate CSF without a large increase in pressure. The pressure increase results in headaches. Similar fluid dynamics have been observed with cough-induced headaches, which are also characterized by higher ICP produced following a transient increase in intrathoracic pressure. The interplay between intracranial compliance and pressure explains how decompression of the posterior fossa through surgery improves headaches in Chiari I patients. After surgical expansion of the posterior fossa, the intracranial compartment is able to accommodate a greater volume of fluid without the same magnitude of pressure increase causing a reduction in high ICP-induced headaches.

While MRI-based flow and pressure measures improve our understanding of how Chiari malformation changes CSF dynamics resulting in symptoms, they are also beginning to be used clinically as marker of severity of disease and can be extended to other anatomical defects of the brain.

CONCLUSION

Pathologies of the brain stem and cranial nerves are common and potentially debilitating, including pain syndromes, spasmodic, vascular, and congenital conditions. Advances in imaging results in more accurate diagnosis, allowing for timely interventions. Neurosurgical procedures remain the mainstay of treatment. However, pathophysiology of these conditions is yet to be fully elucidated which could potentially result in innovations in more accurate, early diagnostic tests and more effective treatment modalities.

Funding Statement

The authors did not receive any grant funding from any agencies in the public, commercial or not-for-profit sectors.

Competing Interests Statement

The authors have no personal or institutional interest with regards to the authorship and/or publication of this manuscript.

Further Reading

[1] Alperin N, Loftus JR, Oliu CJ, Bagci AM, Lee SH, Ertl-Wagner B, Green BA, Sekula R. Magnetic resonance imaging measures of posterior cranial fossa morphology and cerebrospinal fluid physiology in Chiari malformation type I. Neurosurgery 2014;75(5):515–22.

[2] Faraji AH, Abhinav K, Jarbo K, Yeh FC, Shin SS, Pathak S, Hirsch BE, Schneider W, Fernandez-Miranda JC, Friedlander RM. Longitudinal evaluation of corticospinal tract in patients with resected brainstem cavernous malformations using high-definition fiber tractography and diffusion connectometry analysis: preliminary experience. J Neurosurg 2015:1–12.

[3] Fong B, Barkhoudarian G, Pezeshkian P, Parsa AT, Gopen Q, Yang I. The molecular biology and novel treatments of vestibular schwannomas. J Neurosurg 2011;115(5):906–14.

[4] Frederickson A, Gold MS, Sekula RF. Pathophysiology of trigeminal neuralgia. In: Li S, Zhong J, Sekula RF, editors. Microvascular Decompression Surgery. Netherlands: Springer; 2016.

[5] Hughes MA, Branstetter BF, Frederickson AM, Oskin JE, Yankevich U, Sekula RF. Imaging hemifacial spasm. Neurographics 2014;5(1):2–8.

[6] Hughes MA, Branstetter BF, Taylor CT, Fakhran S, Delfyett WT, Frederickson AM, Sekula RF. MRI findings in patients with a history of failed prior microvascular decompression for hemifacial spasm: how to image and where to look. Am J Neuroradiol 2014;36(4):768–73.

[7] Sekula Jr RF, Arnone GD, Crocker C, Aziz KM, Alperin N. The pathogenesis of Chiari I malformation and syringomyelia. Neurol Res 2011;33(3):232–9.

[8] Sekula Jr RF, Frederickson AM, Jannetta PJ, Quigley MR, Aziz KM, Arnone GD. Microvascular decompression for elderly patients with trigeminal neuralgia: a prospective study and systematic review with meta-analysis. J Neurosurg 2010;114(1):172–9.

CHAPTER

14.1

Translational Correlation: Bell's Palsy

J. Kim

Texas Tech University Health Sciences Center, Lubbock, TX, United States

Bell's palsy is the most common disease of the facial nerve. It is a clinical syndrome with acute unilateral paralysis of muscles innervated by the facial nerve. About half of all cases develop maximum paralysis within 48h. The diagnosis is mainly based on clinical symptoms. Oral steroids are used for patients with acute Bell's palsy. The majority of patients will have full recovery.

Bell's palsy is the most common acute mononeuropathy. It was named after Sir Charles Bell who described the anatomy of the facial nerve and its association with unilateral facial paralysis.

EPIDEMIOLOGY

In the general population, the incidence of Bell's palsy ranges from 11.5 to 40.2 per 100,000. Bell's palsy affects both genders equally. In general, 15- to 45-year-old age group is affected most, but there tends to be a slightly greater occurrence in men older than 40 years and in women younger than 20 years. It affects about 1 in 65 people in a lifetime. Bell's palsy involves both sides of the face equally but bilateral involvement is rare. The recurrence is not common. Bell's palsy is more common in patients with diabetes, hypertension, immunocompromised status, after upper respiratory viral infection, and in pregnancy. There is a possible association of chilling or cold temperature with its development.

The majority of patients with Bell's palsy have spontaneous resolution; 84% regain near normal facial function and 71% resolve completely. About 94% of patients with incomplete paralysis recover completely within 4 months of onset. In patients with complete paralysis, 61% have complete recovery.

ETIOLOGY

The cause of idiopathic Bell's palsy remains unclear. One of the suggested causes is reactivated herpes simplex virus (HSV-1) infection centered around the geniculate ganglion. HSV-1 induces edema of the facial nerve that leads to the facial dysfunction. HSV-1 enters the body through mucocutaneous contact and has an affinity for peripheral nerves. The virus can remain dormant in the ganglia of the peripheral nerves until reactivated. The presence of HSV-1 in intratemporal facial nerve endoneural fluid harvested during nerve decompression, and the ability to incite facial

Conn's Translational Neuroscience
http://dx.doi.org/10.1016/B978-0-12-802381-5.00024-5

palsy in animal model through primary infection and reactivation induced by immune modulation support the theory. The other etiologies for facial nerve palsy include infections, ischemic insults, autoimmune disorders, and trauma:

- Infection: Human immunodeficiency virus, Lyme disease, poliomyelitis, meningitis/encephalitis, herpes, varicella zoster (Ramsay Hunt syndrome), Epstein–Barr virus, mumps, rubella, otitis media.
- Trauma: Soft tissue trauma, temporal bone fracture, surgery, birth trauma, barotrauma.
- Autoimmune disease: Guillain–Barre syndrome, Melkersson–Rosenthal syndrome, multiple sclerosis, amyloidosis, sarcoidosis.
- Neoplasm: Acoustic neuroma, facial nerve tumor, neurofibroma, hemangioma, glomus tumor, parotid tumor, central nervous system tumor—meningioma, head and neck cancer, squamous cell carcinoma, rhabdomyosarcoma, metastatic cancer.
- Congenital: Mobius syndrome, congenital lower lip paresis, hemifacial macrosomia, CHARGE (coloboma, heart defects, atresia of the choanae, retardation of growth and development, genital and urinary abnormalities, ear abnormalities and/or hearing loss), VACTERL (vertebral, anal atresia, cardiac, trachea, esophageal, renal and limb defects), Chapple syndrome, brachiootorenal, mononeural agenesis, congenital absence facial musculature, Poland syndrome, thalidomide.
- Otologic: Acute otitis media/mastoiditis, chronic otitis media, cholesteatoma.
- Vascular: Stroke.

SYMPTOMS AND SIGNS

The diagnosis of Bell's palsy should be clinical. The onset of Bell's palsy is acute or subacute. The typical presentation is an acute onset of facial weakness that is unilateral and fairly progressive. About half of all cases develop maximum paralysis within 48h and the facial weakness tends to reach its peak within 72h. Pain behind the ipsilateral ear or in the mastoid area can precede the paralysis by a day or two. If the pain is significant and accompanies Bell's palsy, the diagnosis of Ramsey–Hunt syndrome, which is thought to be caused by varicella zoster infection, should be considered. Some patients have decreased sensation in one or more branches of the trigeminal nerve. Taste impairment is present to some degree. If the stapedius muscle is paralyzed, the patient will report hyperacusis or distortion of sound in the ipsilateral ear. The incomplete eye closure due to loss of facial muscle weakness can result in corneal exposure and desiccation. The associated loss of the orbicularis oculi muscle function impairs the adequate handling of tears. The weakness of orbicularis oris deprives the patient of the ability to control their lips and mouth, so they have difficulty eating and drinking. They drool or bite their own buccal mucosa. Patients may have labial dysarthria—trouble pronouncing certain words that have the letter "b" or "p" in them.

When the patient develops an abrupt onset of unilateral facial palsy, the diagnosis of stroke is important. The patients with the preservation of upper facial movement (frowning and frontalis contraction) more likely have central-type facial weakness due to cerebrovascular disease.

DIAGNOSIS AND DIFFERENTIAL DIAGNOSIS

When evaluating a patient with facial weakness/paralysis for Bell's palsy, the following should be considered:

1. Bell's palsy is rapid in onset (<72h).
2. Bell's palsy is diagnosed when no other medical etiology is identified as a cause of the facial weakness.
3. Bilateral Bell's palsy is rare.
4. Currently, no cause for Bell's palsy has been identified.

5. Other conditions may cause facial paralysis, including stroke, brain tumors, tumors of the parotid gland or infratemporal fossa, cancer involving the facial nerve, and systemic and infectious diseases, including zoster, sarcoidosis, and Lyme disease.
6. Bell's palsy is typically self-limited.
7. Bell's palsy may occur in men, women, and children, but is more common in those of the 15- to 45-year-old age group, those with diabetes, upper respiratory ailments, or compromised immune systems, or during pregnancy.

An alternative diagnosis should be carefully considered if:

1. Stepwise, fluctuating, or gradually progressive (>72h) facial palsy: infections.
2. Bilateral facial palsy: GBS.
3. Recurrent facial palsy: facial nerve neuroma.
4. Prolonged complete palsy without any hint of improvement (>4months).
5. Sudden complete facial palsy (Table 14.1.1).

TABLE 14.1.1 Differential Diagnosis of Facial Palsy

Features	Classic Bell's Palsy	Alternate Diagnosis
Onset	Acute and rapid progress (<72h)	Stepwise, fluctuating, or gradually progressive (>72h): infections Sudden complete facial palsy
Laterality	Unilateral; rarely bilateral	Bilateral: GBS
Recovery	Typically self-limited	Prolonged complete palsy without any hint of improvement (>4months)
Etiology	No other causes	Other conditions: stroke, brain tumors, tumors of the parotid gland or infratemporal fossa, cancer involving the facial nerve, and systemic and infectious diseases, including zoster, sarcoidosis, and Lyme disease.
Recurrence	Rare	Recurrent facial palsy: facial nerve neuroma

ASSESSMENT AND GRADING

There are many facial paralysis grading scales, including Facial Nerve Grading Scale or the House–Brackmann (HB) scale. Each scale is subject to misinterpretation and interobserver variability if not well understood or used correctly. The mostly widely accepted and used scale is the HB scale. The clinicians use this scale to describe patient's facial dysfunction and to objectively monitor patients during the course of the treatment. The HB scale has been widely used in research for facial palsy.

The ubiquitous availability of smartphones with camera enable patients and clinicians to document via photo and video and these are essential in documenting and monitoring the extent of facial paralysis. Photograph and video views should include frontal, lateral, oblique, frontal view with brow elevation, eye closure, smiling, nasal snarl, puckering the lips, and showing one's teeth.

DIAGNOSTIC TESTS

Routine laboratory or diagnostic imaging for patients with new-onset Bell's palsy are not recommended. History and physical examination are the essential aspects of making the diagnosis of Bell's palsy. Acute facial paresis/paralysis with no abnormal history or physical findings is idiopathic in the vast majority of cases. Clinicians should not routinely perform diagnostic imaging for patients with typical clinical presentations. Laboratory testing may be indicated in selected patients with identifiable risk factors or atypical presentation of sudden onset unilateral facial paresis/paralysis: if Lyme disease is suspected, blood tests and analysis for antibodies to *Borrelia burgdorferi* should be performed.

The routine use of diagnostic imaging is not recommended at the time of initial presentation. MRI studies of Bell's palsy may commonly show enhancement along the involved (ipsilateral) facial nerve, especially around the area of the geniculate ganglion, but this finding does not influence the course of therapy. MRI of the entire course of the facial nerve, with or without contrast, is warranted, (1) if the patient has suggestive history or physical findings (ie, history of tumor or persistent localized pain in the temporal area); (2) if the paralysis fails to recover in the expected time frame or worsens; or (3) if the patient has any presentation of facial weakness inconsistent with Bell's palsy such as a second paralysis on the same side, paralysis of isolated branches of the facial nerve, paralysis associated with other cranial nerve involvement, or no sign of recovery after 3 months. If an MRI is contraindicated, a contrast-enhanced CT can be used.

Electroneuronography, or nerve conduction study of the facial nerve, is not routinely recommended, but it can be useful with patients with complete paralysis. Greater than 90% reduction of compound muscle action potential amplitude of the affected side compared to the normal side suggests a 50% chance of incomplete recovery. While the presence of a residual response on neurophysiology implicates neuropraxic injury with higher chance of good recovery, the lack of neurophysiological response suggests complete degeneration or axonotmesis with a prolonged paralysis and incomplete recovery.

TREATMENT

In the early phase of Bell's palsy during the period of facial dysfunction, treatment is targeted toward improving facial recovery with interventions, such as steroids, antivirals, and protection of the eye. The late phase is directed toward treating any residual facial movement deficit, and addressing synkinesis, facial contractures, autonomic dysfunction such as crocodile tears or hemifacial spasm.

As the majority of patients with Bell's palsy will experience improvement of facial movement about 3 weeks after onset of the paralysis, ocular protection is essential to protecting vision in patients with facial paralysis. Measures to prevent corneal exposure and desiccation include frequent application of artificial tears and lubrication combined with taping the eyelid shut at night. Measures such as application of artificial tears and lubrication combined with taping the eyelid shut at night will prevent corneal exposure and desiccation. Consultation with an ophthalmologist at the early stage is important to characterize any predisposing ocular diseases and assist with ocular protection if the patients have severe or complete paralysis. The early use of an eyelid weight should be encouraged in those with diabetes, patients with preexisting eye disease, complete facial palsy with no response on neurophysiology, and the presence of ongoing irritation despite the use of the eye protective therapies.

Because inflammation and edema causing compression of the facial nerve as it travels through the facial canal is the leading posited mechanism of Bell's palsy, therapy targeting the inflammatory process remains the main aim for the treatment. Oral corticosteroids are presumably to decrease nerve edema and thereby facilitate the return of facial nerve function. American Academy of Neurology concluded that steroids are highly likely to be effective and should be offered to increase the probability of recovery of facial nerve function. American Academy of Otolaryngology–Head and Neck Surgery Foundation recommend that oral steroids be prescribed within 72 h of symptom onset for Bell's palsy patients 16 years and older. They recommend a 10-day course of oral steroids with at least 5 days at a high dose (either prednisolone 50 mg for 10 days or prednisone 60 mg for 5 days with a 5-day taper) initiated within 72 h of symptom onset. The benefit of treatment after 72 h is less clear.

There seems no benefit from adding antivirals to corticosteroids in comparison with corticosteroids alone for patients with Bell's palsy. Combination of antiviral agents and corticosteroids may possibly have a beneficial role in patients of severe Bell's palsy.

Physiotherapy includes heat therapy, electrostimulation, massage, and biofeedback. These treatment options are not recommended for all patients with Bell's palsy, but they can be used for patients who have incomplete recovery and who have developed hypertonia, hyperkinesis, or synkinesis.

Surgical decompression is not recommended as a standard approach to patients with Bell's palsy.

PROGNOSIS

About 85% of patients experience some recovery in the first 3 weeks and the remaining 15% after 3–5 months. Seventy percent of all patients experience full recovery. Of the patients who do not recover, sequela was slight in 12% of patients, mild in 13%, and severe in 4%. Recovery of motor function tends to be preceded by recovery of taste. Recovery of some motor function in the first week is the most favorable prognostic sign. The patients with severe facial palsy, protracted length of time prior to onset of recovery, and persistent pain tend to have incomplete recovery. The patients with incomplete recovery may have persistent facial weakness, synkinesis—unintentional movement of a segment of the face during volitional movement of another segment—or facial contracture.

Further Reading

[1] Peripheral nerve disorders. Facial nerve palsy and hemifacial spasm. Handb Clin Neurol 2013;115:367–80.

[2] Baugh RF, et al. Clinical practice guideline: Bell's palsy. Otolaryngol Head Neck Surg 2013;149:S1–27.

[3] Gronseth GS, Paduga R. Evidence-based guideline update: steroids and antivirals for Bell palsy: report of the guideline development Subcommittee of the American Academy of Neurology. Neurology 2012;79:2209–13.

CHAPTER

15

Hypothalamic–Pituitary Regulation

S. Melmed

Cedars-Sinai Medical Center, Los Angeles, CA, United States

The pituitary gland, weighing about 600 mg and situated at the base of the brain, is located within the sella turcica and abuts the cavernous sinuses, cranial nerves, and optic chiasm. These anatomic constraints confine an expanding pituitary mass, and result in compression of pituitary tissue leading to compromised pituitary hormone reserve function. As the gland is encased by bony structures, the path of least resistance for intrasellar mass extension leads to invasion of the cavernous sinuses laterally, and lifting of the optic chiasm dorsally.

Hypothalamic-releasing and -inhibiting hormones are synthesized centrally and secreted into pituitary stalk portal vessels to impinge on specific pituitary cells to release or inhibit pituitary trophic hormone release into the systemic circulation. The anterior pituitary gland comprises at least six functional and terminally differentiated cell compartments, each expressing a unique gene product encoding a trophic hormone (Fig. 15.1). These hormones are synthesized and secreted into the systemic circulation to regulate the repertoire of endocrine gland function, including thyroid, adrenal, gonadal, breast, metabolic, growth, and body composition functions. This complex central endocrine control system is mediated by six hormones, each produced by a highly differentiated cell type: somatotrophs produce growth hormone (GH); lactotrophs produce prolactin (PRL); corticotrophs express proopiomelanocortin (POMC), the precursor of adrenocorticotropin (ACTH), melanocortin hormone (MSH), lipotropic hormone (LPH), and endorphins; gonadotrophs produce luteinizing hormone (LH) and follicle stimulating hormone (FSH); and thyrotrophs produce thyroid stimulating hormone (TSH). The heterodimeric glycoprotein hormones TSH, LH, and FSH comprise a common α-subunit and a specific β-subunit, stabilized by disulfide bonding. Specificity of hormone action is determined by the β-subunit which also is the interactive moiety binding a cognate receptor for each of the three hormones. GH, PRL, and ACTH are single-chain polypeptides. Each pituitary hormone elicits distinct trophic responses in target glands, which in turn exert negative and positive feedback control at the hypothalamus and pituitary to regulate pituitary synthesis and secretion (Fig. 15.2).

Pituitary hormones are synthesized and then secreted into the systemic circulation in a controlled fashion to regulate the repertoire of endocrine gland cell survival and hormone synthesis and secretion (Fig. 15.3). Overall, thyroid, adrenal, gonadal, breast, metabolic, growth, and body composition functions are maintained in homeostasis by finely tuned central pituitary hormone actions (Table 15.1).

Conn's Translational Neuroscience
http://dx.doi.org/10.1016/B978-0-12-802381-5.00025-7

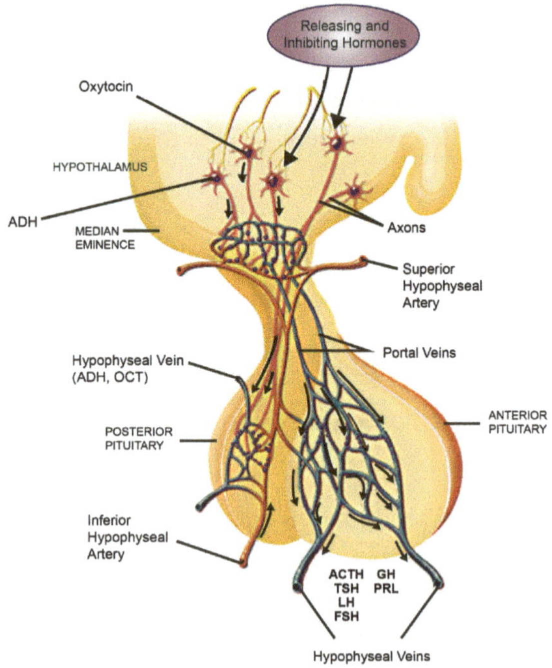

FIGURE 15.1 The hypothalamic–pituitary unit and relationships of neural structures and hormone secretion. Hypothalamic-oxytocin and -vasopressin neurons terminate directly in the posterior pituitary. Hypothalamic neurons produce growth hormone-releasing hormone (GHRH), corticotropin releasing hormone (CRH), thyrotropin-releasing hormone (TRH), and gonadotropin releasing hormone (GnRH) which traverse the hypophyseal–portal circulation surrounding the pituitary stalk and penetrate the anterior pituitary. Hypothalamic-neurohormones signal through cognate GPCRs to differentiated anterior pituitary cells to secrete growth hormone (GH), adrenocorticotropin hormone (ACTH), thyroid stimulating hormone (TSH), luteinizing hormone (LH), and follicle stimulating hormone (FSH), respectively. Dopamine neurons signal tonically to inhibit PRL secretion, while hypothalamic somatostatin inhibits both GH and TSH release. *Adapted from Melmed S, Pituitary. In: Dale DC, Federman DD, editors. ACP Medicine, vol. 1; 2006. pp. 571–86.*

REGULATION OF PITUITARY HORMONE SECRETION

The six functional and highly differentiated cell compartments each expressing a unique gene product are all regulated by specific signals. The pituitary response, acting as the master conductor, transduces central neuronal signals that effectively impact peripheral endocrine target glands by biochemical signaling. Integrated control of pituitary function is subserved by three tiers of regulation (Fig. 15.4). The first tier comprises control by hypothalamic hormones, the second tier incorporates paracrine and autocrine intrapituitary hormones, cytokine and growth factor signals, and the third tier comprises feedback regulation enabled by peripheral hormones which regulate their respective pituitary- or hypothalamic-control systems. The integrated net result of these three signaling levels is to enable synthesis of appropriate hormone abundance, and generation of uniquely timed and sized pituitary hormone secretory pulses into the systemic circulation.

PITUITARY TRANSCRIPTION FACTORS

Highly differentiated nuclear transcription factors determine pituitary trophic cell-specific development, differentiation, and specific hormone production. POUIF-1 determines lactotroph and somatotroph specificity, while T-Pit directs cell-specific POMC transcription in corticotroph cells. Gonadotroph cell differentiation is mediated by DAX-1 and STAR proteins, and thyrotrophs by several factors including GATA-1. Disruptions of these factors by hereditary mutations leads to failed pituitary function. For example, inactivating POU1F-1 mutations lead to GH deficiency and short stature. Some of these inherited mutations may involve more proximal transcription factors (eg, PROP-1) leading to multiple pituitary hormone deficiencies (Fig. 15.2).

Pituitary hormone synthesis and secretion may proceed in an uncontrolled fashion leading to hypersecretory syndromes, especially caused by pituitary tumors. In these patients, there is loss of physiological control

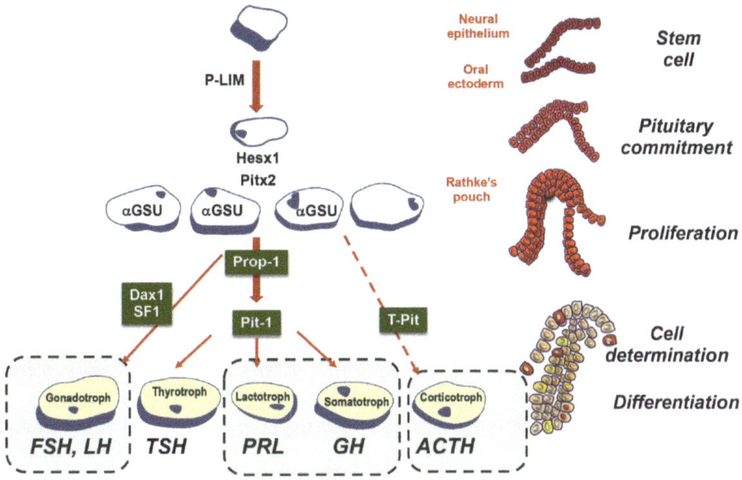

FIGURE 15.2 Human anterior pituitary gland development and cell lineage determination by cell-specific transcription factors. Pituitary trophic cells are depicted with transcription factors which determine cell-specific hormone gene expression. *From Melmed S, Kleinberg DL, Ho K. Pituitary physiology and diagnostic evaluation. In: Melmed S, Polonsky KS, Larsen PR, Kronenberg HM, editors. Williams Textbook of Endocrinology, 12th ed. Philadelphia: Elsevier-Saunders Publication; 2011. pp 175–228.*

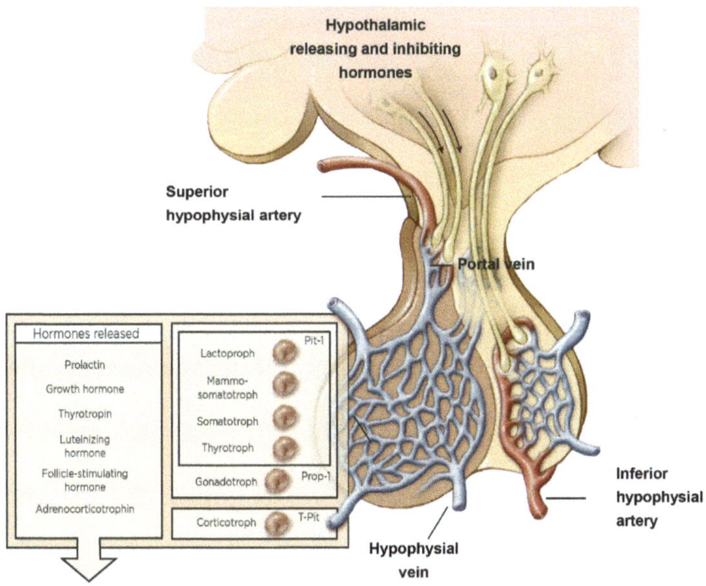

FIGURE 15.3 Control of anterior pituitary hormone secretion is achieved by hypothalamic hormone, which traverse the portal vein, cell-specific transcription factors, and negative feedback control POU1F1 (pituitary-specific transcription factor) denotes POU domain, class 1, transcription factor 1; Prop-1 Prophet of Pit-1. *From Melmed S. Acromegaly. N Engl J Med 2006;355:2558–73. Available from: http://www.nejm.org/doi/full/10.1056/NEJMra062453.*

TABLE 15.1 Anterior Pituitary Hormone Expression and Regulation

Cell	Corticotroph	Somatotroph	Lactotroph	Thyrotroph	Gonadotroph
Tissue-specific transcription factor	T-Pit	Prop-1, Pit-1	Prop-1, Pit-1	Prop-1, Pit-1, TEF	SF-1, DAX-1
Fetal appearance	6 weeks	8 weeks	12 weeks	12 weeks	12 weeks
Hormone	POMC	GH	PRL	TSH	FSH LH
Protein	Polypeptide	Polypeptide	Polypeptide	Glycoprotein α, β subunits	Glycoprotein α, β subunits
Amino acids	266 (ACTH 1–39)	191	199	211	210, 204
Stimulators	CRH, AVP, gp-130 cytokines	GHRH, ghrelin	Estrogen, TRH, VIP	TRH	GnRH, activins, estrogen
Inhibitors	Glucocorticoids	Somatostatin, IGF-I	Dopamine	T_3, T_4, dopamine somatostatin glucocorticoids	Sex steroids, inhibin
Target gland	Adrenal	Liver, bone, other tissues	Breast, other tissues	Thyroid	Ovary, testis
Trophic effect	Steroid production	IGF-I production, growth induction insulin antagonism	Milk production	T_4 synthesis and secretion	Sex steroid production, follicle growth, germ cell maturation
Normal circulating range in humans	ACTH, 4–22 pg/L	<0.5 μg/L[a]	M < 15; *F* < 20 μg/L	0.1–5 mU/L	M, 5–20 IU/L, F (basal), 5–20 IU/L

[a] *Hormone levels integrated over 24 h.*

M, male; F, female. For other abbreviations, see text.

Modified from Melmed S, Polonsky KS, Larsen PR, Kronenberg HM, (eds.). Williams Textbook of Endocrinology, 12th ed. Philadelphia: Elsevier-Saunders Publication; 2011.

and erratic adenomatous hormone secretion ensues with resultant clinical disturbances. Pituitary tumors may arise from any of the hormone-secreting cell types. These are invariably benign adenomas which may be small (<10 mm) micro-adenomas or larger macro-adenomas. Some of these tumors are particularly aggressive and lead to compressed normal pituitary function, with invasion of vital vascular structures (eg, cavernous sinus) or cranial nerve compression leading to visual impairment. Clinical phenotypes associated with pituitary adenomas are determined by the cell of origin and the specific overproduced hormone. Thus, somatotroph tumors lead to acromegaly/gigantism, lactotroph tumors to infertility and lactation, corticotroph tumors to hypercortisolism with Cushing disease, and thyrotroph tumors to hyperthyroidism and goiter. Tumors arising from gonadotroph cells are usually nonsecreting, but may present with hypogonadism. This may be due to several factors. Sustained nonpulsatile secretion of gonadotrophin hormones may lead to downregulation of their peripheral receptors. Furthermore, the adenoma may also not efficiently synthesize gonadotrophins. Finally, an expanding pituitary mass, regardless of etiology, may compress normal pituitary tissue and compromise gonadotropin secretion.

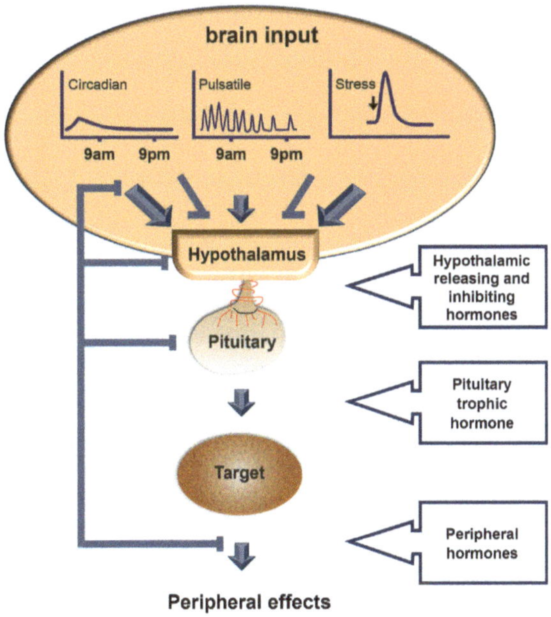

FIGURE 15.4 Hypothalamic–pituitary–target axes are regulated by three tiers of control. Hypothalamic neurons synthesize hormones which are released into the hypophyseal–portal circulation to impinge on anterior pituitary cells to mediate brain-pituitary signal transduction resulting in release to the circulation, ultimately reaching target glands and organs. Within the pituitary, local cytokines, and growth factors also regulate anterior pituitary hormone synthesis and release. Distal target glands secrete hormones that elicit the required effects on peripheral tissues, and also feedback to regulate respective hypothalamic–pituitary control hormones. *From Ben-Shlomo A, Melmed S. Hypothalamic regulation of anterior pituitary function. In: Melmed S, editor. The Pituitary, 3rd ed.; 2010. pp 21–45.*

HYPOTHALAMIC FACTORS REGULATING PITUITARY FUNCTION

Cajal proposed the functional concept of hypothalamic regulation of the anterior pituitary, which was subsequently proven by Harris, who demonstrated that direct hypothalamic electrical stimulation was sufficient to induce ovulation. This finding supported the hypothesis for existence of a humoral mechanism whereby chemical messengers traverse the hypothalamic–pituitary portal venous system to impinge upon and signal to the anterior pituitary. The role of the hypothalamic–pituitary unit has now been well defined, and mechanisms underlying pituitary control well established.

The hypothalamus, located below the third ventricle, comprises distinct yet heterogeneous neurosecretory nuclei, whose axons extend to the median eminence which extends to the infundibular stalk. Complex electrical signals emanating from brain centers terminate at hypothalamic nuclei to regulate synthesis and release of hypothalamic-peptide hormones. The derivation of these brain signals is diverse and intricate, and may include stress, temperature sensation, fluid and electrolyte balance, as well as higher functions.

Hypothalamic-neuronal bodies synthesize and package pituitary control factors: thyrotropin-releasing hormone (TRH), gonadotrophin releasing hormone (GnRH), corticotropin releasing hormone (CRH), growth hormone-releasing hormone (GHRH), and somatostatin (somatotropin releasing–inhibiting factor, SRIF), mostly located in the medial hypothalamus. Each of these hormones binds to a cognate receptor expressed on a respective differentiated pituitary target cell membrane. The median eminence functionally links the hypothalamus and the pituitary. Superior hypophyseal arteries which derive from the internal carotid arteries, supply the hypothalamus, and form median eminence capillaries outside the blood–brain barrier. Hypophyseal portal vessels form the hypothalamic–portal circulation, the predominant anterior pituitary blood source. Hypothalamic-axonal termini are functionally linked to fenestrated pituitary portal venous capillary endothelium, enabling release of hypothalamic peptides into the hypophyseal–portal circulation to reach anterior pituitary trophic hormone-secreting cells (Fig. 15.3). Accordingly, concentrated levels of hypothalamic-releasing and -inhibiting hormones traverse these vessels to reach the pituitary trophic hormone–producing cells without significant systemic dilution. Further control systems for hypothalamic hormone vascular transit include regulated portal capillary contractility, as well as bidirectional hypothalamic–pituitary control enabled by retrograde flow of portal blood from the pituitary dorsally towards the median eminence. This complex vascular mechanism allows exquisite regulation of anterior pituitary hormone secretion by highly specific hypothalamic-releasing and -inhibitory factors.

The overarching transduction of brain signals to affect peripheral endocrine chemical messages is characterized by both qualitative, temporal, and quantitative features. The abundance of hypothalamic, pituitary, or peripheral messenger molecules at each level, as determined by both half-life as well as concentration synthesized, increases incrementally. At each level of pituitary control, an increasing magnitude of amplified hormonal release occurs. Thus, nanomolar concentrations of hypothalamic factors are transduced to elicit micromolar concentrations of secreted pituitary hormones, which in turn elicit millimolar concentrations of peripheral hormones. Furthermore, the circulating half-life of hypothalamic peptides is very short (minutes), pituitary hormone half-life is usually minutes (eg, 12 min for GH) to hours, while peripheral hormone circulating half-life may be days (eg, thyroid hormones; IGF-I). Pituitary cells are therefore targets for sensitive and highly leveraged hypothalamic regulation by timed hypothalamic-hormone secretory patterns.

Brain signaling to peripheral hormone-secreting target glands transduced by the hypothalamic–pituitary unit is characterized by several physiological features. For example, target gland actions are controlled by unique patterns of secretion at each level. Thus, circadian, pulsatile, or acute central stimuli signal to secrete specific hypothalamic hormones into the hypophyseal–portal circulation with specific secretory profiles. Subsequent pituitary hormone secretion to the systemic circulation is controlled in a diurnal and/or pulsatile fashion, and then bind distal receptors expressed on target organs which regulate peripheral endocrine hormone production. Disrupted secretory patterns may lead to endocrine dysfunction or failure of peripheral target glands to appropriately respond to pituitary signals. Thus, continued erratic secretion of gonadotrophin hormones by a pituitary adenoma may lead to downregulation of gonadal function. Negative feedback regulation of hormone secretion exists at all levels of control, but mostly by target gland hormones which usually suppresses both pituitary and hypothalamic-hormone synthesis and secretion. Negative feedback suppression of pituitary hormone production usually overrides positive central signals. Accordingly, hypersecretion of peripheral hormones by secreting tumors (eg, thyroid, adrenal) leads to suppressed pituitary hormone production as well as suppressed trophic cell proliferation. Conversely, ablation or failure of peripheral gland function results in exuberant hypersecretion of pituitary trophic hormones.

HYPOTHALAMIC–PITUITARY–ADRENAL AXIS

This vital pituitary axis is essential for cell survival and response of the organism to stress (Fig. 15.5). Adrenal cortisol regulates multiple cellular processes essential for maintaining fuel and energy metabolism, immunity and cardiovascular homeostasis, as well as central nervous system functions.

Corticotroph cells comprise about 20% of the functional hormone-secreting anterior pituitary cells, and appear early by the 8th week of gestation. These are the first human pituitary cells to develop, differentiate, and mature. Prominent neurosecretory granules (150–400 nm) within the corticotroph contain POMC gene products, including 1–39 ACTH, β-lipotropin, and endorphins. Each of these hormones contains

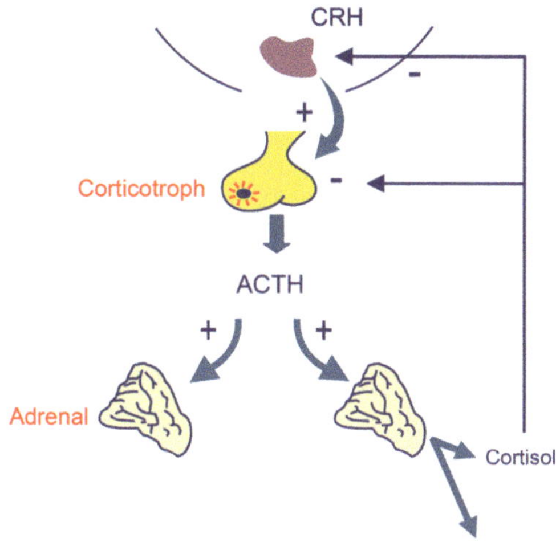

FIGURE 15.5 Hypothalamic corticotropin releasing hormone (CRH) stimulates the corticotroph to synthesize and release adrenocorticotropin hormone (ACTH), which in turn stimulates the adrenal glands to produce and release cortisol and androgens. Cortisol acting in a negative-feedback loop inhibits further release of both CRH and ACTH. *Melmed S. Mechanisms for pituitary tumorigenesis: theplastic pituitary. J Clin Invest 2003;112(11):1603–18.*

a prominent carbohydrate moiety. Within the human pituitary, POMC is exclusively expressed in corticotroph cells. TPit/Tx19 is the critical cell-specific transcriptional factor for corticotroph cell differentiation and POMC transcription.

CRH, a 41-aa peptide synthesized in the paraventricular nucleus as well as in higher brain centers, is the central regulator of the hypothalamic–pituitary–adrenal (HPA) axis. Neuronal CRH production is controlled by multiple afferent inputs from hypothalamic, limbic forebrain and brain stem inputs to enable fine-tuned control of basal and stress-induced CRH action to induce POMC-derived peptides including ACTH and β-endorphin.

Pituitary CRH action is mediated by the transmembrane corticotroph receptors, CRHR1, which signals to induce corticotroph POMC transcription and ACTH production. CRHR1, a G-protein coupled receptor (GPCR), is also widely distributed in the brain and is coupled to the Gαs unit, induces cAMP and activates phosphokinase-A (PKA). CRHR1 is desensitized and internalized upon prolonged CRH stimulation. CRH-binding protein (CRH-BP) transiently binds with high affinity to CRH and also urocortins, and may competitively inhibit CRH action. Intravenous CRH administration elicits rapid ACTH release peaking in the circulation at 30 min, and followed by adrenal cortisol secretion which usually peaks at 60 min after injection. The adrenal GPCR for ACTH (melanocortin-2 receptor) signals to induce a cascade of steroidogenic enzymes leading to cortisol production. Both urocortin 1 and arginine vasopressin (AVP) also act to induce ACTH and cortisol albeit less robustly than CRH.

In the corticotroph cell, the POMC prohormone is processed by convertase enzymes to yield ACTH and β-lipotropin, as well as inactive peptide fragments. β-lipotropin is further cleaved to yield β-endorphin and other peptides, especially in the intermediate pituitary region. ACTH is secreted to the systemic circulation to stimulate adrenal gland steroid hormone synthesis and secretion by maintaining adrenal cortical cell integrity and inducing steroidogenesis pathways. POMC synthesis is suppressed by glucocorticoids acting as negative feedback regulators, and stimulated mainly by CRH, as well as by AVP, and pro-inflammatory cytokines.

CRH release controls a 24-h diurnal rhythm for both ACTH and cortisol secretion required. This secretory pattern is an important determinant for maintaining adrenal function ACTH and cortisol levels start rising between 1 and 4 am and hormone levels peak at about 6 am. A lower cortisol peak is evident in the early afternoon, continues to fall during the afternoon and evening, then reaches a midnight nadir level. Circadian rhythmicity is likely more characterized by changes in hormone pulse amplitude rather than pulse frequency. Interestingly, CRH infusion to CRH KO mice restores diurnal corticosterone patterns pointing to the CRH requirement for maintaining normal circadian ACTH rhythm.

The HPA axis functions to sustain the neuroendocrine stress response and maintain overall metabolic homeostasis. The HPA stress response is a summated outcome of hypothalamic, intrapituitary, and peripheral hormone and cytokine signals acting to regulate adrenal cortisol production with subsequent appropriately integrated immune, cardiovascular, and hormonal functions. Inflammatory cytokines activate hypothalamic-CRH and -AVP secretion, pituitary POMC gene expression, and ACTH release. Inflammatory-associated central glucocorticoid receptor resistance also attenuates the ability of glucocorticoids to suppress the HPA, further enabling ACTH secretion. Inflammatory cytokines and neurotransmitters, thus, increase CRH release to stimulate ACTH and adrenal cortisol secretion. Induced cortisol levels, in turn, limit inflammation and confer host protection. Several other peptides also regulate ACTH by acting centrally. For example, leptin stimulates CRH release, while somatostatin inhibits hypothalamic-CRH release.

CRH release and ACTH levels are enhanced by both pathologic and physiological stressors including cytokines, hemorrhage, adrenalectomy, and hypoglycemia. In contrast, glucocorticoid-mediated negative feedback results in dual suppression of both hypothalamic-CRH, pituitary POMC synthesis, and ACTH release. As a clinical sequel, loss of cortisol feedback inhibition relieves the corticotroph cell of negative restraint, and results in high ACTH levels. This is exemplified clinically as encountered after surgical adrenalectomy or primary adrenal failure. Although severe stress and subsequent increased CRH levels can override glucocorticoid negative feedback inhibition, steroids act to physiologically suppress CRH gene and protein expression, CRH secretion to the portal system and ultimately adrenal function.

An important clinical consequence of this potent central glucocorticoid action is that patients receiving chronic glucocorticoid therapy (eg, for asthma and arthritis) are unable to mount an effective stress response due to suppressed ACTH.

HYPOTHALAMIC–PITUITARY–THYROID AXIS

The thyroid axis regulates development, energy metabolism, and growth and the axis is controlled by complex central and peripheral signals mediated mainly by hypothalamic TRH and pituitary TSH (Fig. 15.6). Thyrotroph cells express TSH and comprise about 5% of the functional hormone-secreting anterior pituitary cells with relatively small secretory granules ranging from 120 to 150 μm. TSH comprises a specific β-subunit and common α-subunit shared with LH and FSH. TSH binds to a GPCR

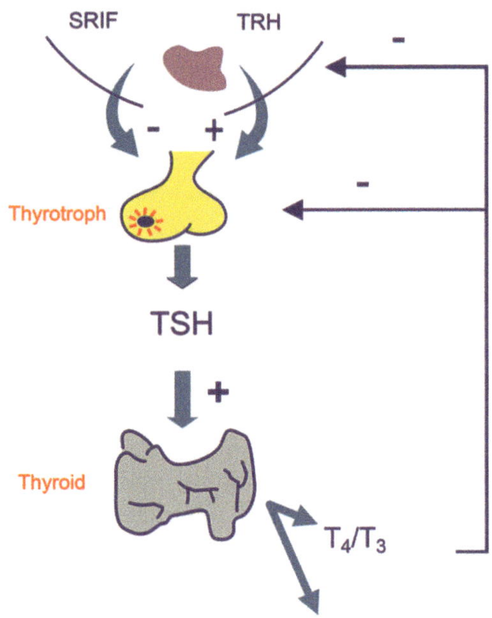

FIGURE 15.6 Hypothalamic TRH stimulates pituitary thyrotrophs to synthesize and release TSH which induces the thyroid gland to produce thyroid hormones which negatively regulate further release of both TSH and TRH. SRIF also inhibits both TRH and TSH. *Melmed S. Mechanisms for pituitary tumorigenesis: theplastic pituitary. J Clin Invest 2003;112(11):1603–18.*

on thyroid follicular cells to stimulate thyroid hormone (T4) synthesis and release. T4 is converted in the periphery to the more active T3 which elicits most thyroid cellular actions.

TRH is a hypothalamic tripeptide (pyroglutamyl histidyl-prolinamide) that acts through a transmembrane rhodopsin family GPCR expressed on the thyrotroph surface to stimulate TSH synthesis and secretion; it also stimulates the lactotroph to secrete PRL. TRH is derived from hypothalamic preproTRH (242 aa) which is cleaved to generate the tripeptide. TRH synthesis is induced by cold, thereby activating the thyroid hormone response to hypothermia. Hypothalamic-TRH signaling is mediated primarily by $G_{q/11}$ proteins, hydrolysis of phosphatidylinositol 4,5-P_2 (PIP_2), and intracellular Ca^{2+} and phosphokinase C (PKC) activation. TRH also regulates TSH glycosylation. TSH synthesis and production are markedly suppressed by several factors including thyroid hormones, dopamine, somatostatin, and glucocorticoids. These molecules act to inhibit TSH production overriding central TRH induction. Intravenous TRH injection in humans rapidly elicits up to 22-fold increase in TSH synthesis and secretion, reflected as a biphasic release pattern measured as circulating TSH levels.

Thyroid hormones feedback to potently suppress the thyroid axis by inhibiting TRH gene transcription, and TSH synthesis and release. During states of hypothyroidism, SRIF also inhibits TRH-induced TSH. As thyroid hormone negative feedback inhibition on the axis is so powerful, thyrotroph proliferation and TSH production are both released from restraint when thyroid hormones are removed. Thus, untreated thyroid gland failure relieves the thyrotroph of negative feedback inhibition and results in exuberant TSH synthesis with resultant elevated TSH levels. Furthermore, as thyroid hormones suppress thyrotroph cell proliferation, patients with low thyroid hormone levels may also exhibit thyrotroph hyperplasia. Increased levels of TSH are seen with thyroid damage, postoperatively, after radioactive ablation, thyroiditis, and thyroiditis.

HYPOTHALAMIC–PITUITARY–GH AXIS

GH acts to regulate linear growth in children, and to maintain homeostatic body composition and muscle mass in adults (Fig. 15.7). Although GH acts directly on peripheral tissues (especially liver, muscle, cartilage, adipocytes, heart), most growth-promoting functions of GH are mediated by GH-induced IGF-1.

Cells expressing both PRL and GH (lactosomatotrophs) arise from acidophilic stem cells, and mature somatotrophs comprise 35–45% of

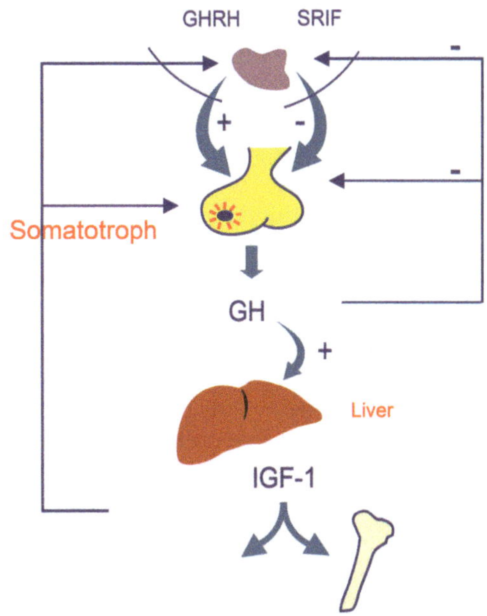

FIGURE 15.7 Hypothalamic GHRH stimulates the somatotroph to synthesize and release growth hormone (GH), which in turn stimulates the liver to produce IGF-1. Peripheral GH action may be either IGF-I dependent or independent. Both GH and IGF-1 act in a negative feedback loop to suppress both GHRH and GH, while SRIF also inhibits both GHRH and GH release. *Melmed S. Mechanisms for pituitary tumorigenesis: theplastic pituitary. J Clin Invest 2003;112(11):1603–18.*

pituitary cells, that is, these are the most preponderant pituitary cell type. Somatotrophs contain abundant GH secretory granules up to 700 μm in diameter.

Hypothalamic-GHRH neurons in the arcuate and ventromedial nuclei project into the median eminence, and terminate on hypophyseal–portal capillaries to release GHRH into the portal venous system. The precursor GHRH pre-prohormone comprises 108 amino acids, and is subsequently processed to GHRH (1–40)-OH and GHRH (1–44)-NH_2, with fragments as short as GHRH (1–29)-NH_2 exhibiting potent GH-releasing activity. GHRH is structurally similar to the secretin family, including glucagon and vasoactive intestinal polypeptide (VIP). GHRH receptor (GHRHR), a transmembrane GPCR mostly expressed on pituitary somatotroph cells, signals to induce both cell proliferation as well as GH synthesis and secretion via intracellular calcium fluxes, increased cAMP, and activated phosphatidyl inositol and MAPK pathways. The pancreatic GHRH signals to regulate insulin and glucagon release.

The GH gene encodes a 191 aa polypeptide in the pituitary, and about 15% of circulating GH comprises a shorter 176 aa peptide derived by alternative splicing. Basal GH secretion as well as periodic pulse patterns are engendered by coordinated hypothalamic GHRH– and SRIF–release patterns. GH synthesis, secretion, and secretory patterns are controlled by a myriad of hypothalamic and peripheral factors. Inducing factors include hypothalamic *GHRH*, which primarily stimulates somatotroph GH synthesis and release, and Ghrelin, an octanoylated gastric-derived peptide, which also induces GH. Inhibitory signals derive mainly from hypothalamic *somatostatin* (SRIF)] which inhibits GH secretion. Somatostatin binds five distinct receptor subtypes (SST1 to SST5) and somatotrophs expresses predominantly SST2 and SST5 subtypes which signal to suppress GH secretion. GH synthesis is also suppressed by negative feedback exerted by IGF-1, the peripheral target hormone for GH which directly suppresses GH, while indirectly inducing hypothalamic SRIF to suppress GH.

Peripheral GH tissue responses and especially growth rates appear to be gender specific. For example, male somatostatin and GHRH expression are higher than in females, leading to higher pulsatile GH peaks and lower troughs. Consequently, GH levels are higher in men, compared with continuous basal GH secretion observed in women. These observations may at least in part explain gender differences in linear growth pattern differences and liver enzyme induction. These differences may also be accounted for by sex steroids which regulate both somatostatin and GHRH neurons, and which may also contribute to sexually dimorphic GH secretion patterns.

Both GH and IGF-1 cause negative feedback suppression of hypothalamic-GHRH release. Hypothalamic SRIF is also released in response to GH, thus indirectly contributing to the negative feedback loop suppressing GHRH and GH itself. Interestingly, mutant mice which are null for SSTR2 are refractory to negative GH feedback implying that GH-stimulated SRIF directly inhibits GHRH release, in addition to releasing SRIF into the hypophyseal–portal circulation to further inhibit pituitary GH.

Abundant IGF-1 receptors are expressed in both the pituitary and hypothalamus, and IGF-1, the GH target hormone directly inhibits pituitary GH synthesis at the level of the GH gene, as well as acting indirectly via the hypothalamus by inducing SRIF.

GH is also induced by several neuropeptides and amino acids including ghrelin, L-arginine, glucagon, dopamine, galanin, and apomorphine, as well as α-adrenergic agonists induce GH. β-Adrenergic blockade also induces basal GH and enhances GHRH- and insulin-evoked GH release. Ghrelin, a gut-derived 28-aa GH secretagogue signals via the ghrelin receptor to increase intracellular pituitary calcium via inositol phosphate to dose-dependently induce GH release. Ghrelin also induces hypothalamic GHRH to stimulate GH, and Ghrelin and GHRH cotreatment is synergistic for GH induction likely acting to allosterically regulate pituitary GHRH receptors.

Most GH secretory peaks occur at night, and GH is elevated with deep sleep, exercise, physical stress, and sepsis. Integrated 24-h GH secretion is higher in women and is also enhanced by estrogens which likely increase peripheral GH resistance.

Randomly measured GH values are very low or even undetectable in about 50% of daytime samples, especially in obese and elderly patients. Nutritional status is a strong determinant of GH secretion. Thus, a glucose load will normally suppress serum GH to <0.7 μg/L in women and even lower to <0.07 μg/L in men. In contrast, chronic malnutrition, or undernutrition, or hypoglycemia, are associated with increased GH levels.

GHRH dose-dependently elicits secretory serum GH peaks at 15–45 min after injection, returning to baseline levels within 90–120 min. Unlike CRH, repeated or constant GHRH administration does not down-regulate the GH response. Administration of estrogens, glucocorticoids, or the presence of malnutrition, all lead to enhanced GH responses to GHRH. In contrast, GH response to GHRH is blunted by somatostatin administration, obesity, and aging. GHRH also acts as a somatotroph trophic factor and leads to induced proliferation and if persistently oversecreted (eg, by a hypothalamic or neuroendocrine tumor) results in pituitary hyperplasia. Transgenic mice overexpressing GHRH exhibit somatotroph hyperplasia in enlarged pituitary glands. Clinically, neuroendocrine tumor ectopic GHRH production leads to somatotroph hyperplasia and GH hypersecretion with acromegaly. GHRH action may also be constitutively enhanced as a result of $G_{\alpha s}$ subunit mutations observed in up to 40% of GH-secreting pituitary adenomas.

The GH receptor dimer exhibits structural homology with the cytokine/hematopoietic superfamily and is abundantly expressed on liver, cartilage, and adipocytes. GH signals through JAK/STAT pathway to regulate expression of target genes mediated by both STAT and also non-STAT signaling pathways. Importantly, IGF-I, the potent growth factor, is generated in response to GH, mainly in the liver. Chondrocyte cell proliferation and linear growth are determined by GH and IGF-I as well as other hormones including thyroid hormone and sex steroids.

HYPOTHALAMIC–PITUITARY–GONADAL AXIS

The gonadal axis regulates sex steroid synthesis and secretion, as well as germ cell development and function. Thus, axis controls sexual development and function, as well as fertility (Fig. 15.8).

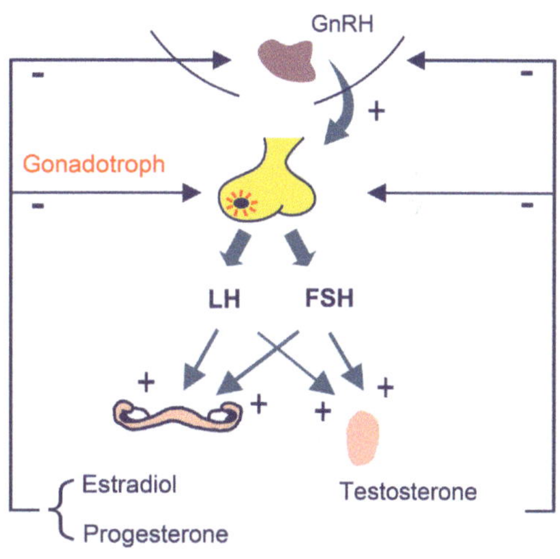

FIGURE 15.8 Hypothalamic GnRH stimulates the gonadotroph to synthesize and release both FSH and LH which stimulate the gonads to produce sex steroid hormones and regulate germ cell function. Estradiol and testosterone regulate release of central GnRH, FSH, and LH in both females and males. *Melmed S. Mechanisms for pituitary tumorigenesis: theplastic pituitary. J Clin Invest 2003;112(11):1603–18.*

GnRH neurons arise from the olfactory placode and migrate to the hypothalamus upon a scaffolding of neurons and glial cells, as well as in response to chemotactic and adhesion signals. GnRH-I neurons have been proposed to also arise from the anterior pituitary placode, as exemplified in zebrafish mutants lacking a pituitary but with intact olfactory organs, yet devoid of hypothalamic GnRH-I neurons. Hypothalamic GnRH-I neuronal networks coordinate a pulse generator for GnRH release and project to the median eminence and stalk to terminate on the hypophyseal–portal capillaries.

Hypothalamic GnRH is essential for synthesis and release of the gonadotrophin hormones FSH and LH. GnRH-I is the predominant hypothalamic peptide and while GnRH-II is expressed mostly outside the brain. The *GnRH-I* gene encodes a 92 aa protein that is processed to the 10 aa GnRH molecule, retaining a pyroGlu at the N-terminus and a Gly-amide at the C-terminus, both required for bioactivity.

The GnRH receptor, a transmembrane GPCR, lacks a typical intracellular C-terminal cytoplasmic domain and mostly binds $G_{\alpha q/11}$ to activate phospholipase C, diacylglycerol–PKC, and inositol-3-phosphate pathways to increase intracellular calcium and induce glucoprotein synthesis. Continuous GnRH exposure down-regulates GnRH receptors and desensitizes gonadotrophs mainly by receptor trafficking including internalization and degradation.

GnRH neurons exhibit a highly coordinated pulsatile pattern of GnRH release into the hypophyseal–portal circulation discreet pulsing every 60–120 min elicits rhythmic LH and FSH secretion patterns. In fact, discreet pituitary-derived LH pulses are observed within minutes after a hypothalamic-GnRH spike. GnRH pulse patterns and frequency determine rates of FSH/LH release, ratios of FSH and LH, and LH and FSH glycosylation, which all are required to determine appropriate gonadal actions of the gonadotrophins. LH rather than FSH is a more reliable indicator of GnRH pulsatility. Gonadotroph GnRH has also been proposed to communicate with hypothalamic GnRH by retrograde neurohypophyseal flow exemplifying a short-loop regulatory feedback that controls both LH and GnRH secretion. Discreet GnRH pulses prime basal gonadotroph secretory responsiveness, in contrast to continuous GnRH exposure which may down-regulate the axis by inducing GnRH receptor desensitization. This may explain why patients with androgen-dependent prostate cancer experience suppressed gonadal function induced by parenteral long-acting GnRH agonist treatments.

Kisspeptin enhances GnRH pulse amplitude and duration and kisspeptin neurons also express leptin receptors. This central pathway links the reproductive system to nutrition and energy metabolism at the level of the hypothalamus.

In general, estradiol, progesterone, and testosterone attenuate GnRH pulse frequency and GnRH release to the hypophyseal–portal circulation. In humans, pituitary LH/FSH are suppressed by chronic estradiol exposure while acute estrogen peaks, as characterized by the preovulatory mid-estrous surge, result in positive feedback to enhance both GnRH release and subsequent FSH/LH pulse frequency and amplitude. Progesterone acts mainly at the hypothalamus to attenuate GnRH pulse frequency, while testosterone negative feedback regulation occurring mainly at the hypothalamic and pituitary levels is mediated in part by conversion to estrogens.

Numerous other factors are known to regulate the hypothalamic–pituitary gonadal axis. Galanin-like peptide induce GnRH in the presence of estradiol, excitatory amino acids, catecholamines, NPY, and neurotensin may also all act to enhance estradiol stimulation of gonadotrophins. Noradrenergic and adrenergic neurons stimulate GnRH release, while neural inhibitors of estradiol effects on GnRH include GABA and opioids, which also suppress the LH surge.

Gonadotroph cells express both the gonadotroph hormones, FSH and LH, and comprise about 10–15% of functional anterior pituitary cells. They contain two groups of secretory granules; large 350–450nm and smaller 150–250nm granules packaged in vesicles. Both bihormonal and monohormonal groups of gonadotrophs are evident. Gonadotroph-specific differentiation is characterized by GnRH receptor expression and by steroidogenic factor 1 (SF-1) and DAX-1 nuclear receptors. Gonadotroph cells synthesize and secrete both LH and FSH, glycoprotein hormones that comprise α and β subunits. Hormone specificity of action at the level of the gonads is conferred by the unique β subunits, while the α subunit is common to FSH, LH, and TSH.

Activins secreted by the gonads also induce, while inhibin suppresses FSH. Gonadotropins bind to respective GPCRs expressed in the ovary and testis signaling to regulate germ-cell development and function and estrogen or testosterone synthesis. Ovarian follicle development and ovarian estrogen production are regulated by FSH, and ovulation and maintenance of the corpus luteum are controlled by LH. Testicular Leydig cell testosterone synthesis and secretion is induced by LH, and seminiferous tubule development and spermatogenesis are under FSH control.

SOMATOSTATIN

Hypothalamic somatostatin is produced predominantly in the anterior periventricular nucleus, as well as in paraventricular, arcuate, and ventromedial hypothalamic nuclei. These neurons project into the median eminence, and their fibers terminate on the hypophyseal–portal circulation to release the peptide into the portal circulation reaching anterior pituitary cells. Endocrine transport of the peptide to the anterior pituitary is also enabled by CSF leakage from the third ventricle to the portal system, peripheral somatostatin, and short portal blood vessels crossing from the posterior to the anterior pituitary.

Somatostatins are cyclic peptides with a single covalent bond between two Cys residues and a bioactive core composed of a Phe–Trp–Lys–Thr amino acid sequence. Both SRIF14 and SRIF28 are derived from posttranslational cleavage of pro-somatostatin by PC1 and PC2 and carboxypeptidase E. Somatostatin signals through five GPCR somatostatin receptor subtypes, SSTR1, 2, 3, 4, and 5. Somatostatin 14 exhibits high binding affinity to all the receptor subtypes, with highest affinity to SSTR2, the subtype responsible for inhibiting GH. SRIF receptors bind the $G_{\alpha i/o}$ subunit of the $G_{\alpha\beta\gamma}$ tetramer, reduce cAMP levels, opening potassium channels and closure of calcium channels. Pituitary SSTR2 is suppressed by chronic glucocorticoids. SSTR2 also signals constitutively in the absence of ligand to suppress GH and ACTH secretion. The primary function of somatostatin is to inhibit pituitary hormone secretion. There is also evidence that SRIF may restrain pituitary tumor cell growth, especially somatotrophs. SRIF inhibits secretion of GH, TSH, as well as ACTH in the presence of low glucocorticoid levels. SRIF also exerts ubiquitous inhibitory effects on multiple gastrointestinal, pancreatic, and vascular functions. In patients with acromegaly, somatostatin analogs are administered to control excess GH secretion, normalize IGF-I levels, and stabilize or shrink the pituitary adenoma mass.

PROLACTIN

Lactotroph cells comprise about 15–25% of functioning anterior pituitary cells and likely arise from a common GH-producing precursor cell. PRL secretory granules (250–800nm) are evenly distributed, while smaller (200–350nm) granules are more scattered. Mammosomatotroph cells co-secrete both PRL and GH within the same granule. Lactotroph hyperplasia occurs during pregnancy and lactation, and resolves within several months of delivery.

Hypothalamic dopamine is released from the arcuate and anterior periventricular nuclei. The anterior pituitary also receives dopaminergic signals derived from the posterior pituitary through interconnecting short portal veins. Dopamine binds to lactotroph dopamine receptor subtype 2 (D2R) to inhibit prolactin synthesis and release, and lactotroph proliferation. This potent inhibition is further reinforced by observations that D2R null mice develop lactotroph hyperplasia, hyperprolactinemia, and pituitary lactotroph adenomas. D2R, a GPCR, couples the $G_{\alpha i/o}$ subunit to decrease cAMP, increase potassium influx, inhibit inositol phosphate production, and decrease intracellular calcium to acutely inhibit prolactin synthesis and secretion, usually within minutes.

As the predominant hypothalamic control of PRL secretion is inhibitory, mediated by hypothalamic dopamine, PRL levels are high with pituitary stalk section compromising portal dopamine passage, as seen with compressive mass sellar lesions. Lactotroph dopamine type 2 (D_2) receptors signal to inhibit PRL synthesis and secretion. Targeted knockout of the D_2 receptor results in murine hyperprolactinemia and associated with lactotroph proliferation. Lactotroph hyperplasia induced by estrogen occurs during pregnancy and lactation. Serum PRL levels are about 10–25 μg/L in women and 10–20 μg/L in men.

Pituitary prolactin secretion is under tonic inhibition by hypothalamic dopamine (Fig. 15.9). As a consequence, compromised dopamine portal transit, or antagonized dopamine receptor signaling, as seen with pituitary stalk damage or dopamine antagonist drugs, leads to marked increase in lactotroph production and PRL secretion.

As dopaminergic neurons express PRL receptors, a short negative feedback loop of prolactin on dopamine synthesis is operative. Estrogen increases prolactin mainly by reducing expression of lactotroph dopamine receptor expression. Most neuroleptic drugs that inhibit dopamine secretion or receptor signaling also increase prolactin levels. Thus, patients receiving chronic anti-psychotic medications often exhibit hyperprolactinemia and galactorrhea. Other central peptides inconsistently stimulate prolactin secretion including TRH, vasopressin, VIP, angiotensin II, galanin, substance P, bombesin-like peptides, and neurotensin.

PRL is a 199 aa polypeptide with about 40% homology to GH. PRL signals through a dimeric transmembrane receptor (PRLR), a member of the cytokine/emythropoeitin family. Subsequent signaling is mediated by JAK2 and StatS which translocates to the nucleus to elicit target gene transcription. In the primed breast, PRL is required to initiate and maintain lactation. PRL induces growth of and maintains the developed

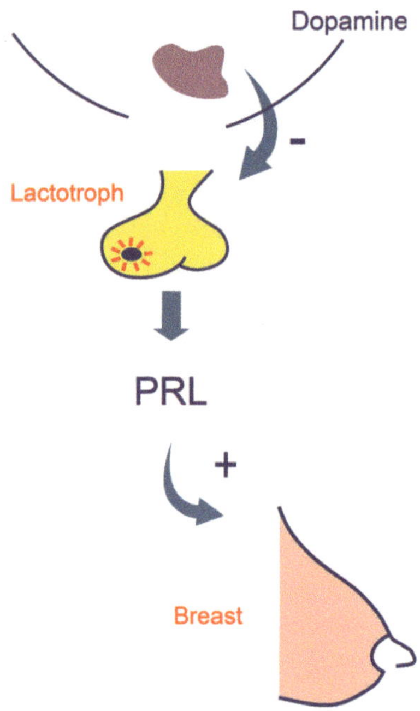

FIGURE 15.9 Hypothalamic dopamine traverses the portal vasculature to exert tonic suppression of lactotroph PRL synthesis and release. *Melmed S. Mechanisms for pituitary tumorigenesis: theplastic pituitary. J Clin Invest 2003;112(11):1603–18.*

mammary gland to enable breast milk production and suckling. In lower vertebrates, PRL also regulates fluid and water balance.

High PRL acts to suppress female and male gonadal functions. This role for PRL is thought to preserve species survival by preventing pregnancy in a lactating mother. Thus, females exhibit ovulation inhibition and amenorrhea and males develop azoospermia and impotence. Accordingly, hyperprolactinemia, especially encountered in patients with PRL-secreting pituitary adenomas, leads to galactorrhea and infertility.

Further Reading

[1] Melmed S. Mechanisms for pituitary tumorigenesis: the plastic pituitary. J Clin Invest 2003;112(11):1603–18.
[2] Millar RP, Lu ZL, Pawson AJ, Flanagan CA, Morgan K, Maudsley SR. Gonadotropin-releasing hormone receptors. Endocr Rev 2004;25(2):235–75.
[3] Clemmons DR. The diagnosis and treatment of growth hormone deficiency in adults. Curr Opin Endocrinol Diabetes Obes 2010;17:377–83.
[4] Brooks AJ, Waters MJ. The growth hormone receptor: mechanism of activation and clinical implications. Nat Rev Endocrinol 2010;6:515–25.
[5] Melmed S. Pathogenesis of pituitary tumors. Nat Rev Endocrinol 2011;7(5):257–66.
[6] Langlais D, Couture C, Kmita M, Drouin J. Adult pituitary cell maintenance: lineage-specific contribution of self-duplication. Mol Endocrinol 2013;27(7):1103–12.
[7] Gittleman H, Ostrom QT, Farah PD, et al. Descriptive epidemiology of pituitary tumors in the United States, 2004–2009. J Neurosurg 2014:1–9.
[8] Ho KK. The year in pituitary 2014. J Clin Endocrinol Metab 2014;99(12):4449–54.
[9] Ntali G, Capatina C, Grossman A, Karavitaki N. Clinical review: functioning gonadotroph adenomas. J Clin Endocrinol Metab 2014;99(12):4423–33.
[10] Bernard V, Young J, Chanson P, Binart N. New insights in prolactin: pathological implications. Nat Rev Endocrinol 2015;11(5):265–75.

CHAPTER

16

Neuroendocrinimmunology: A Primer

M.D. Lumpkin

Georgetown University School of Medicine, Washington, DC, United States

The field of *neuroendocrinimmunology* (NEI) encompasses a number of related and complementary disciplines that seek to understand the bidirectional regulatory relationships that exist between the nervous system, the endocrine system, and the immune system. Each component of this wide-ranging field involves a focus on the molecules of communication operating between these three systems that will facilitate the maintenance of whole body homeostasis. Some examples of these molecules of communication are cytokines, lymphokines, and chemokines of the immune system; hormones of the hypothalamic–pituitary–peripheral gland axes; and neurotransmitters of the central and peripheral nervous systems. Some investigators and physicians are primarily concerned with the presence and actions of immune cell cytokines and chemokines in altering the activity of nerve cells and how this might affect both physiology and behavior in the organism. Another subarea of examination is one that probes the nature of the neuronal innervation and regulation of immune organs such as the bone marrow, thymus gland, and spleen. NEI also has a specific realm of focus on the ability of hormones of the hypothalamic–pituitary unit to govern the proliferation and activities peripheral macrophages, monocytes, lymphocytes, and microglial cells of the brain. Other researchers and clinicians direct their attention to the ability of immune cells such as lymphocytes to actually manufacture their own neuroendocrine hormones (and their receptors) such as adrenocorticotropic hormone (ACTH), thyroid-stimulating hormone (TSH), growth hormone (GH), prolactin, and corticotropin-releasing hormone (CRH). An additional subdiscipline is one that specifically explores how immune system components such as antibodies can destroy acetylcholinergic receptors at the neuromuscular junction and attack the myelin sheaths of neurons that compose the white matter of nervous tissue. The key to understanding the role of NEI in maintaining healthy bodily function is to appreciate the simultaneous bidirectionality of communication that exists between the cells of the nervous system, the endocrine system, and the immune system. Imagine information, carried by all the categories of messenger molecules referenced earlier, flowing both clockwise and counterclockwise in a circle between these three great systems of the body.

What type of event might be well served by this mutual system of regulation? One answer is when the body needs to respond to stressful episodes in an efficient and coordinated

Conn's Translational Neuroscience
http://dx.doi.org/10.1016/B978-0-12-802381-5.00026-9

fashion that will ideally maintain healthy functioning of the organism. Stressors of all types will stimulate peripheral and brain-derived immune cells to produce and secrete a cluster of pro-inflammatory cytokines such as interleukin-1 (IL-1), interleukin-2 (IL-2), tumor necrosis factor (TNF), and numerous others. These pro-inflammatory cytokines, whether manufactured by peripheral immune cells or by neurons and glial cells, will communicate with the hypothalamus to stimulate the neurosecretion of the hypothalamic hormone CRH, considered by some to be the "master stress hormone" of the body. The CRH then stimulates the release of the anterior pituitary hormone ACTH. The ACTH will next increase the synthesis and output of the hormone cortisol by the adrenal gland. This glucocorticoid will then act in a negative feedback fashion to not only reduce the further secretion of CRH and ACTH, thereby restoring cortisol to its baseline level, but the previously elevated cortisol will also act to inhibit further cytokine secretion by monocytes, lymphocytes, and glial cells, as well as reduce their increased proliferation rates. Thus, the "bidirectional communication" between the nervous system, the endocrine system, and the immune system is demonstrated (Fig. 16.1).

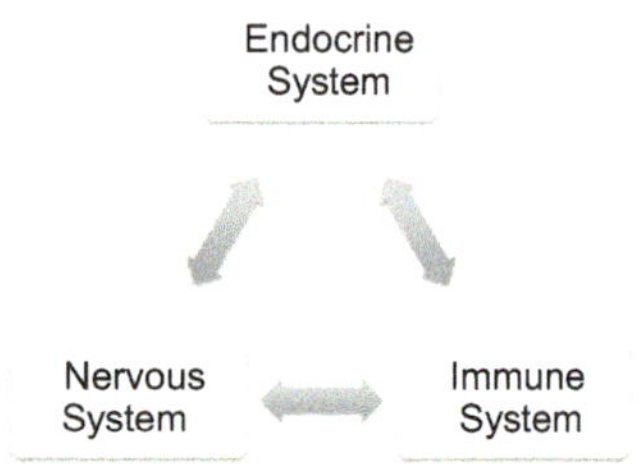

FIGURE 16.1 Molecules of communication, especially hormones, neuropeptides/neurotransmitters, and cytokines, flow bidirectionally between the endocrine, nervous, and immune systems in order to maintain whole-body homeostasis. Perturbations in any one of these systems will provoke reactions by the others in attempts to provide appropriate regulatory responses.

NEUROENDOCRINIMMUNOLOGY I: INNERVATION OF LYMPHOID ORGANS

Some of the earliest findings showing a relationship between the nervous system and the immune system came from the laboratory of Dr. David Felten. He and others first described *noradrenergic* (*NA*) nerve fibers originating from postganglionic sympathetic nerves that were providing direct innervation to immune tissues of the spleen, lymph nodes, thymus gland, and intestinal Peyer's patches. These norepinephrine-containing nerve endings were seen making contact with B lymphocytes, macrophages, and thymocytes. That both α- and β-adrenergic receptors were found on the surfaces of these immune cells further supported the notion that they were subject to regulation by the autonomic nervous system. It was also found that stimulation of the sympathetic nervous system and administration of epinephrine would produce leukocytosis, lymphopenia, and suppression of natural killer (NK) cell activity. This indicated that binding and activation of catecholaminergic β-receptors on lymphocytes suppress their proliferation, thereby contributing to an impaired immune status. Because catecholamines can suppress lymphocyte activity, this has prompted some to hypothesize that this is why chronic stress, which prolongs activation of the sympathetic nervous system, makes one more susceptible to infections, cancer, and other chronic diseases. Consistent with this notion are the observations that the specific stressors of maternal deprivation, cold exposure and foot shock, will cause thymus gland involution, diminished antibody production, and decreased lymphocyte proliferation. In contrast, however, sympathetic denervation of immune organs results in the proliferation of bone narrow, spleen, and lymph node cells, and increased immunoglobulin production upon antigen-stimulation. A question that is often asked is whether stress causes cancer? As of mid-2016, there is no firm evidence that stress initiates cancer cell proliferation or tumor origination. However, Sloan et al. have shown that restraint stress in mice will increase the body-wide metastasis of breast cancer tumor cells implanted into mammary tissue. This group further determined that such cancer cell spread could be inhibited by the β-adrenergic antagonist propranolol or promoted by administration of the β-adrenergic agonist isoproterenol. Thus, stress-induced elevations of catecholaminergic stress hormones may encourage the dissemination of cancerous cells throughout the body.

It has also been noted that the use of stress reduction techniques typical of mind–body medicine practices such as biofeedback, breath control, imagery, and meditation can decrease sympathetic outflow from the central nervous system (CNS) to the immune organs. These stress management methods have been seen to increase the number of CD4 T lymphocytes and NK cell activity, thus improving the overall immune protection of stressed individuals such as medical students and hospital residents. The regular use of such mind–body stress reduction techniques has also been shown to prevent the rise of cortisol experienced by medical students as they advance in their training, possibly explaining in part why their immune cell parameters improve with stress management training.

Additional studies from Dr. Felten's research have identified a number of neuropeptide hormones in autonomic nerves innervating the lymph nodes, spleen, and thymus gland that also appear to participate in the regulation of immunity. Specifically, *neuropeptide Y* (NPY), *substance P* (SP), and *vasoactive intestinal peptide* (VIP) have been found in neurons that are in direct contact with lymphocytes, macrophages, and thymocytes in the thymus gland and spleen. Receptors for these neuropeptides have been located on lymphoid cells. When VIP is released from these nerve terminals onto its lymphoid receptors, lymphocyte proliferation is decreased, antibody production is reduced, and NK cell activity is inhibited.

Another immune modulator found in nerve fibers that innervate lymphoid tissues is the pro-inflammatory neuropeptide *substance P*. Being an inflammatory agent and a mediator of pain responses transmitted through peripheral nerves, it follows that substance P stimulates lymphocyte proliferation and mast cell activity resulting in the production of the additional pain mediators, histamine and leukotrienes. As a countervailing influence to the actions of substance P, the peptide *somatostatin*, found in many tissues including central and peripheral nerves, and due to its ability to inhibit many different cell types, can suppress the release of substance P from nerve endings.

Sympathetic nerves containing norepinephrine and that innervate the lymph nodes, spleen, and thymus gland have also been found to contain the peptide NPY. NPY and norepinephrine are secreted together from sympathetic nerve endings and act synergistically to inhibit lymphocyte proliferation and cytokine release, again showing the role of stress hormones in decreasing immune function.

A further indication of the cross-talk between the immune system and the CNS is the presence of the same hormones and their receptors in both locations. This situation has been described for CRH, ACTH, vasopressin, oxytocin, enkephalins , β-endorphin, and neurotensin, all of which are elevated after stressful challenges to the organism and, for the most part, act to suppress immune cell activity. Their role in immune inhibition no doubt contributes to the increased vulnerability to infection and disease seen after an individual's exposure to chronic stressors.

NEUROENDOCRINIMMUNOLOGY II: CYTOKINES IN THE NERVOUS SYSTEM

While cytokines, especially those of the pro-inflammatory variety, were originally believed to be produced only by immune cells in the periphery of the body for the purpose of regulating other peripheral immune cells, it is now recognized that as activated macrophages, monocytes, lymphocytes, and NK cells circulate through the vasculature of the brain, they may release their cytokine contents into CSF and brain tissue. Peripherally manufactured cytokines, though large in molecular size, can also gain access to neurons and glial cells in the hypothalamus because of specific neurovascular transport mechanisms and the absence of the blood–brain barrier at circumventricular organs like the organum vasculosum of the lamina terminalis (OVLT) and the median eminence (ME) that are associated with the hypothalamus. Astroglial and microglial cells of the brain, and especially those of the hypothalamus, produce interleukins-1, 2, 4, and 6 and tumor necrosis factor alpha (TNF-α). Receptors for these cytokines are also present in the brain. IL-1 is also synthesized by neurons in the hypothalamus and hippocampus and these neurons appear to communicate with various hypothalamic-releasing factor neurons via IL-1 receptors.

Emotional stress and physical stressors such as infection, inflammation, and trauma stimulate the production of pro-inflammatory cytokines such as IL-1, IL-6, and TNF by glial cells and neurons. If the stressful insults last for extended periods, as one sees in a variety of chronic pathological conditions, the ongoing elevations in these cytokines can adversely impact normal neural functions resulting in such problems as fever, somnolence, anorexia, dementia, and even neuronal apoptosis. Two examples of this are Alzheimer's disease and human immunodeficiency virus (HIV) infection of the brain parenchyma (NeuroAIDS) where an increase in the CNS levels of IL-1 have been implicated in causing some of the behavioral disturbances observed in these conditions. Fortunately, not all the inflammatory events surrounding increased neural IL-1 and the cytokine cascades that it provokes are negative in nature. Following stress-related elevations of IL-1 in the brain, certain neural cells can seek to protect themselves from excessive IL-1-induced pathology by producing the cytokine known as the endogenous IL-1 receptor antagonist (IL-1Ra). Because IL-Ra is homologous in structure to the IL-1 ligand and has similar affinity for the IL-1 receptor, this antagonist effectively competes for the same receptor. In doing so, it blocks the receptor and dampens any further damaging IL-1 activity.

Normal short-term physiological fluctuations of cytokine levels in response to intermittent acute stressors typical of a "fight or flight" situation also appear to be beneficial to the organism. Physiological concentrations of IL-1, IL-2, and IL-6 in the hypothalamus will stimulate CRH and ACTH by the hypothalamic–pituitary unit. This, of course, increases synthesis of cortisol by the adrenal gland cortex. The cortisol then acts in negative feedback fashion to signal neurons, glial cells, monocytes, and macrophages to reduce their output of CRH-stimulating levels of cytokines. Cortisol's inhibitory effects will also reduce the proliferation of immune cells and immunoglobulin production. The purpose of cortisol negative feedback on these immune cell components is to ensure that stress-induced increases in immune system activity do not proceed unchecked. With cortisol keeping a tight rein on immune cells, two goals are achieved. First, the possibility of a normal immune response turning into an autoimmune response is diminished. Second, an overly robust inflammatory state is prevented during a "fight or flight" episode that could impede defensive movement.

Another protective mechanism that neurally active IL-1 puts into play against its own potentially tissue-damaging effects is that it stimulates the production of nerve growth factor (NGF). This is a neurotrophic substance that promotes the repair, proliferation, and survival of damaged neurons.

Pro-inflammatory cytokines IL-1, IL-2, IL-6, TNF, and interferon have all been shown to influence anterior pituitary hormone secretion primarily by altering the hypothalamic-releasing and -inhibiting factors that regulate them. In particular, these same cytokines have been shown to increase ACTH and cortisol production by directly stimulating CRH neurons. CRH, being the "master stress hormone," is in a pivotal position for directing stress responses because its neurons project to and synapse on growth hormone-releasing hormone (GHRH), luteinizing hormone-releasing hormone (LHRH), and thyrotropin-releasing hormone (TRH) nerve cells in the hypothalamus. When stress-induced IL-1 and IL-6 stimulate CRH, not only are ACTH and cortisol concentrations increased, but also the other hypothalamic releasing factors and their target pituitary hormones are inhibited by such increases in CRH activity. In an acute stress response, this scenario is important for the immediate survival and well-being of the individual. The higher cortisol levels mobilize metabolic substrates such as glucose to fuel the stress-induced acceleration of cardiovascular, neural, and skeletal muscle functioning that includes raising the blood pressure and blood flow to critical organs and tissues. At the same time the stress hormone axis is being activated, the IL-1/CRH system will suppress the neuroendocrine axes governed by GHRH, LHRH, and TRH. In the short term, this accomplishes the desired goal of shifting metabolic fuel to supporting the stress response (ie, "fight or flight"), while temporarily reducing the use of metabolic substrate for supporting somatic growth, reproduction, and nonspecific whole-body metabolism.

Stress-induced activation of hypothalamic CRH also triggers sympathetic nervous system outflow from the brain through the spinal cord and coursing to visceral organs and immune tissues. This further aggravates immune system inhibition as catecholamines are released onto immune cell adrenergic receptors, thereby making an individual even more susceptible to infection and disease. IL-1 and CRH, acting alone or together, can also enhance somatostatin neurosecretion, while at the same time they both inhibit the secretion of GHRH. The net effect is to suppress GH release from the adenohypophysis. Recalling that GH is anabolic for the maintenance and growth of both boney and soft tissues (somatic growth), and that chronic stress will inhibit GH secretion by the aforementioned mechanisms, it is not surprising then to see protein wasting of muscle in adult patients and somatic growth failure in children afflicted with a chronic disease stressor like HIV/AIDS. This is a condition where pro-inflammatory cytokines and CRH are elevated in the brain. Specifically, increases in IL-1 and CRH in the hypothalamus also reduce the output of LHRH and TRH. A chronic viral stress like HIV, with attendant cytokine and CRH induction, can manifest as anovulation and amenorrhea in females and impaired spermatogenesis in males with low secretion levels of pituitary-luteinizing hormone (LH) and follicle-stimulating hormone (FSH) and gonadal sex steroids, the end result being a loss of fertility and libido. In similar fashion, IL-1 and TNF-α have been observed to inhibit hypothalamic TRH and pituitary TSH release. The ensuing reduction in thyroid hormone concentrations contributes to decreased overall metabolic activity that worsens the lethargy, somnolence, and fatigue already being driven by IL-1 and TNF acting directly on neural elements in the brains of individuals like AIDS patients.

Another problem resulting from continuous stimulation of the cytokine/CRH/ACTH/cortisol axis due to chronic stress-related maladies is the overexposure of hippocampal neurons to high levels of cortisol. These hippocampal neurons are involved both in mediating cortisol negative feedback on CRH and in memory function. Excessive exposure to and binding of cortisol by the glucocorticoid receptors there will result in hippocampal neuronal damage and death. This can impair cortisol negative feedback, resulting in even higher cortisol levels, memory disorders, and further immune system suppression from the ongoing cortisol elevations.

In theory, the use of receptor antagonists aimed at blocking IL-1 and CRH receptors during chronic stress (and the diseases resulting therefrom) could be used to interrupt the cytokine/CRH cascade of deleterious events characterized earlier. In fact, such compounds are currently under therapeutic development and testing.

NEUROENDOCRINE PEPTIDES IN IMMUNE CELLS

Another intriguing part of the NEI story was the finding by Dr. Edwin Blalock and colleagues that some immune cells may manufacture their own neuroendocrine hormones. It was found that CRH could stimulate the production of ACTH and enkephalins from B lymphocytes while cortisol would inhibit their secretion from these cell types. T cells have been observed to produce LH, FSH, TSH, and growth hormone. Somatostatin, prolactin, and VIP secretion has been detected from activated monocytes. Interestingly, in an analogous fashion to the hypothalamic–pituitary relationship, there are also releasing factor hormone facsimiles found in immune cells that appear capable of stimulating their corresponding pituitary-like hormones located in the same immune cells. For instance, GHRH identified in lymphocytes can be secreted *locally* among other lymphocytes, bind to GHRH receptors present on lymphocyte plasma membranes, and stimulate the *local* release of GH from the lymphocyte population. This local regulatory mechanism also seems to exist for TRH and TSH, as well as LHRH and LH, in certain immune cells. This local production of hypothalamic and pituitary-like hormones among monocytes and lymphocytes, while providing for paracrine and autocrine regulation, once diluted in the systemic circulation, typically do not reach concentrations capable of exerting effects on the body's endocrine glands.

IMMUNOMODULATION BY NEUROENDOCRINE PEPTIDES

Some of the local regulatory actions of immune cell-produced neuroendocrine peptides have been determined. For example, CRH can suppress the endotoxin-stimulated release of IL-1 and IL-6 by human monocytes. Also in the inhibitory category, ACTH can suppress the activity of monocytes and inhibit the B lymphocyte production of antibodies. LH and FSH have been observed to reduce the activity of T cells and NK cells. T-cell proliferation can be suppressed by somatostatin and VIP. On the stimulatory side, GH and prolactin can both stimulate antibody synthesis, lymphocyte proliferation, and antagonize the inhibitory effects of glucocorticoids on cytokine production and lymphocyte proliferation. Prolactin and growth hormone can also stimulate the growth and activity of the thymus gland and lymphoid cells. Further, GH stimulates macrophage activity and primes monocytes to enhance their production hydrogen peroxide, while prolactin increases macrophage tumoricidal activity, the synthesis of interferon-γ, and stimulates T-cell proliferation. TSH enhances immunoglobulin synthesis while β-endorphin stimulates the activity of T, B, and NK cells and increases interferon-γ production.

NEUROIMMUNE AUTOIMMUNITY

Another subdiscipline of NEI focuses on the immune system phenomenon of pathologies caused by autoimmune production of autoantibodies that attack various components of the nervous system. This area of study is typically referred to simply as neuroimmunology. A vivid example of a neuroimmune event is in the disease of myasthenia gravis. In this neuromuscular disorder, autoantibodies attack and destroy acetylcholine (ACH) receptors at the neuromuscular junction and this prevents the secreted acetylcholine (ACH) from stimulating muscle contraction. It has been suggested that possible mechanisms leading to ACH receptor dysfunction include the blocking, destruction, or cross-linking of receptors and complement-mediated lysis of the receptors. While the triggering event that results in myasthenia gravis is not evident at this time, it does appear that T cells of the thymus gland are involved in generating much of the pathology of this condition. A significant majority of myasthenia gravis patients have an enlarged thymus gland and 10% actually have noncancerous thymus gland tumors. The following chapter discusses this autoimmune disorder in greater detail and the reader is referred there for more information.

A second neuroimmune disorder that is characterized by attacks on neuronal components is multiple sclerosis (MS). Patients with this disorder experience the gradual but progressive loss of myelin on CNS neurons with additional neuronal destruction that will ultimately impair mobility and other nerve-controlled functions. This autoimmune disease results from autoantibodies generated against the antigenic components of myelin. This process may involve the generation of T cell–mediated immune responses during infections with bacteria or viruses, followed by antibody generation against the pathogen's antigenic proteins. These antigens appear to mimic those of the myelin and therefore the myelin components come under immune cell and antibody attack with ultimate neuronal destruction occurring. Antibodies to myelin basic protein and other myelin fragments have been measured in CSF and other body fluids. Immune cell activation in MS increases the levels of pro-inflammatory substances like interferon-γ and prostaglandin E_2. As might be anticipated from earlier information in this chapter, immune cell stimulation in MS patients drives up their production of ACTH, prolactin, β-endorphin, and substance P. The roles of these hormones in the clinical profile of MS patients are not currently understood. A lack of adequate Vitamin D is often seen in MS patients and its administration seems to have a positive effect in MS patients. Since stress can trigger symptomatic episodes in MS patients, and because stress increases the levels of pro-inflammatory immune cell products, this may explain why the use of stress-reducing mind–body medicine skills such as meditation can prevent or lessen stress-related MS episodes, perhaps by lowering pro-inflammatory agents.

Further Reading

[1] Taub DD. Neuroendocrine interactions in the immune system. Cell Immunol 2008;252(1–2):1–6. http://dx.doi.org/10.1016/j.cellimm.2008.05.006.

[2] Petrovsky N. Towards a unified model of neuro-endocrine-immune interaction. Immunol Cell Biol 2001;79:350–7.

[3] MacLaughlin BW, Wang D, Noone AM, Liu N, Harazduk N, Lumpkin M, et al. Stress biomarkers in medical students participating in a mind body medicine skills program. Evidence-Based Complementary Altern Med 2011. http://dx.doi.org/10.1093/ecam/neq039.

[4] Blalock JE. A molecular basis for bidirectional communication between the immune and neuroendocrine systems. Physiol Rev 1989;69:1–32.

[5] Sloan EK, Priceman SJ, Cox BF, Yu S, Pimentel MA, Tangkanangnukul V, et al. The sympathetic nervous system induces a metastatic switch in primary breast cancer. Cancer Res 2010;70:7042–52.

[6] Lumpkin MD. Cytokine regulation of hypothalamic and pituitary hormone secretion. In: Foa PP, editor. Humoral factors in the regulation of tissue growth: blood, blood vessels, skeletal system, and teeth. Endocrinology and metabolism; 5. New York: Springer-Verlag; 1993. p. 139–59.

[7] Lumpkin MD. Clinical approaches to hormonal and neuroendocrine imbalances. The hypothalamus-pituitary-adrenal axis. In: Jones DS, Quinn S, editors. Textbook of functional medicine. Gig Harbor, WA: The Institute for Functional Medicine; 2005. p. 610–8.

[8] Scarborough DE. Cytokine modulation of pituitary hormone secretion. In: O'Dorisio MS, Panesai A, editors. Annals of the New York Academy of Sciences, vol. 594. 1990. p. 169–87.

[9] Raine CS, editor. Advances in neuroimmunology. In: Annals of the New York Academy of Sciences, vol. 540. 1988.

[10] Hubbard JR, Workman EA, editors. Handbook of stress medicine: an organ system approach. Boca Raton, Florida: CRC Press; 1998.

[11] Reichlin S. Neuroendocrinology. In: Wilson J, Foster D, Kronenberg H, Larsen P, editors. Williams Textbook of Endocrinology. Philadelphia: W.B. Saunders Co.; 1998. p. 229–32.

[12] Peisen JN, McDonnell KJ, Mulroney SE, Lumpkin MD. Endotoxin-induced suppression of the somatotropic axis is mediated by interleukin-1 beta and corticotropin releasing factor in the juvenile rat. Endocrinology 1995;136:3378–90.

[13] Mulroney SE, McDonnell KJ, Pert CB, Ruff MR, Resch Z, Samson WK, et al. HIV gp 120 inhibits the somatotropic axis: a possible GH-releasing hormone receptor mechanism for the pathogenesis of AIDS wasting. Proc Natl Acad Sci USA 1998;95:1927–32.

CHAPTER

16.1

Translational Correlation: Clinical Neuroimmunology

J.C. Hoyle, D.M. Welch, M.K. Racke

The Ohio State University, Columbus, OH, United States

NEUROMYELITIS OPTICA

NMO or Devic's syndrome is a neuroinflammatory demyelinating disease that selectively targets the optic nerves and spinal cord with clinical presentations classically of bilateral optic neuritis and/or longitudinally extensive myelitis.

NMO used to be diagnosed as a subtype of multiple sclerosis (MS) due to the similar clinical presentations of the two diseases and the lack of disease-specific biomarkers. In 2004, serum NMO-Ig was identified as a highly specific biomarker for NMO patients, but was absent in patients with MS. NMO Ig is an autoantibody that targets the aquaporin-4 (AQP-4) water channel expressed on astrocytic end feet at the blood–brain barrier (BBB). NMO-Ig is thought to contribute to the pathogenesis of NMO in several aspects. By targeting AQP-4, NMO-Ig increases BBB permeability, subsequently inducing immune cell (neutrophil and eosinophil) infiltration, which leads to damage of the central nervous system (CNS). In addition, lack of AQP-4 alters glutamate homeostasis. Excitatory amino acid transporter 2 (EAAT2), the major transporter for extracellular glutamate uptake, is essential for maintenance of low levels of extracellular glutamate. Evidence showed both the expression and function of EAAT2 are impaired in AQP-4-deficient mice. As a result, extracellular glutamate levels are enhanced, which leads to necrosis of CNS cells, such as neurons and oligodendrocytes.

NMO differs from MS in several aspects. NMO tends to be more common in patients of African and Asian descent, while MS is much more biased to patients of northern European descent. Autoimmune disorders, such as systemic lupus erythematosis (SLE), Sjogren's syndrome, and thyroid autoimmune disorders, can coexist with NMO, and are subsequently referred to as an NMO-spectrum disorder. Of concern with NMO is that it is an inflammatory disorder that has the potential to be mistaken for MS. This is important, because treatments of the two diseases are quite different. MS is typically treated with immunomodulatory agents, whereas NMO treatment is primarily by immunosuppression. Agents used to treat MS, such as interferon-β have actually been shown to make NMO worse, which is another reason that one needs to make an accurate diagnosis.

NMO can be distinguished from MS by noting that NMO lesions in the spinal cord tend to involve multiple vertebral levels, whereas MS tends to only involve approximately one or two vertebral levels. Furthermore, as noted above, NMO-IgG testing can help differentiate

Conn's Translational Neuroscience
http://dx.doi.org/10.1016/B978-0-12-802381-5.00027-0

NMO from MS, although it can be negative in up to a third of patients with NMO. Typically, a first attack of NMO will demonstrate no brain lesions or a lesion pattern inconsistent with MS. It should also be noted that when NMO patients have brain lesions on MRI, they do not have the typical appearance of MS brain imaging. The likely location of the brain lesions in NMO are in the AQP4-rich regions around the hypothalamus or around the third or fourth ventricle. The cerebrospinal fluid findings in NMO also differ from those observed in MS. NMO may have over 50 WBCs/μL and will often demonstrate neutrophils or eosinophils, which are never observed in MS CSF. While MS patient will typically demonstrate oligoclonal bands in the CSF, they are only observed in 15–30% of patients with NMO.

NMO lacks the secondary progressive course observed in MS patients, however, the prognosis of NMO patients is worse than that of MS patients. Therefore, relapse prevention is vital in treating NMO patients. While administration of steroids is a common strategy for acute relapses in both NMO and MS, immunosuppressant treatment is the major therapeutic intervention for relapse prevention in NMO patients. One medication used in these patients is Azathioprine (2.5–3 mg/kg/day), which suppresses the actively dividing B and T cells. Mycophenolate (1000 mg twice a day) is also often used as a baseline agent to treat NMO. In addition, the B cell depleting monoclonal antibody rituximab has shown promising results in open label studies and is becoming more accepted as treatment for NMO. Pulse cyclophosphamide or mitoxantrone are also used in NMO patients with active disease.

IDIOPATHIC ACUTE TRANSVERSE MYELITIS

Common clinical presenting symptoms of transverse myelitis (TM) would be paraparesis that may be asymmetric, along with progressive weakness, sensory changes, and urinary dysfunction. These symptoms typically occur below a specific spinal level, although they do not have to be symmetric. A recent onset of a viral-like infection, which has resolved is a common precipitating factor in TM, and vaccine-related triggers have been described. Evidence of inflammation is necessary for the diagnosis. CSF and MRI are first-line diagnostic tests. Imaging should exclude extra-axial compression as the cause of the symptoms.

One should also consider several other secondary causes of myelitis when making the diagnosis of TM, depending on the clinical presentation of the case. These include infectious etiologies (syphilis, Lyme, HIV, HTLV, HSV, *Mycoplasma*, hepatitis, tuberculosis, varicella, CMV, fungal infection, and EBV), mineral and vitamin deficiencies (copper, B12, and E), vascular disease (infarct of the cord, aortic pathology, and dural arteriovenous fistula), MS (differentiated by imaging of the brain and radiological features of the myelitis, see later), paraneoplastic causes, and other inflammatory diseases. If the clinician has suspicion for a systemic inflammatory disorder, then appropriate tests may also be ordered to confirm such a diagnosis (ie, antinuclear antibody, anti-DNA, anti-Ro, anti-La, anti-RNP, anti-Jo antibodies). Chest CT and ACE levels can be assessed to evaluate for sarcoidosis.

One should make sure that the distribution of the clinical deficit is not in the distribution of the anterior spinal artery. In addition, another vascular cause that should be excluded is arteriovenous malformation, and imaging should evaluate for the presence of abnormal flow voids on the surface of the spinal cord or a spinal angiogram if highly suspicious for a dural arteriovenous fistula (ie, a stuttering course of myelopathic symptoms in an older age male).

A common clinical problem with the initial presentation may be ambiguity between signs and symptoms of TM and that of acute inflammatory demyelinating polyradiculoneuropathy (AIDP), the most common form of Guillain–Barré syndrome (GBS). The initial clinical presentation of TM can mimic AIDP due to the lack of reflexes, since the patient may be in spinal "shock."

Distribution/Presentation

A pure paraplegia or paraparesis with a corresponding distribution of sensory loss favors TM, while AIDP may present with a gradient of motor and sensory loss involving the lower extremities greater than the upper extremities. In AIDP, dysesthetic pain, with involvement of the upper extremities and cranial nerves, and absent deep tendon reflexes involving the upper extremities are more common findings, although reflexes may be preserved in early or variant forms of GBS. Sensory loss may occur on the trunk in GBS, but that tends to be in a ventral, "shield"-like pattern rather than a well-defined band-like sensory level around the entire trunk.

Urinary Effects

Urgency or retention is a common early finding in TM and is less common in AIDP.

MRI Findings

TM lesions are most commonly seen in the cervical and thoracic cord with increased signal on T2 and gadolinium enhancement on T1 imaging at the site of acute inflammation. The signal in cross section typically occupies more than 50% of the area of the cord in idiopathic TM. An eccentric location of lesions with partial cord involvement is more suggestive of a concern for MS. Furthermore, MS tends to be localized to one or two vertebral levels. NMO or other secondary disease associations of longitudinally extensive myelitis (such as lupus, Sjogren's, sardoidosis, dural arteriovenous fistula) generally occupy three to four or more segments in the spinal cord.

Lumbar Puncture

CSF pleocytosis and/or elevated IgG index indicate the inflammatory nature of TM. If the CSF is not indicative of inflammation, it can be repeated in 2–7 days to determine whether inflammatory changes are present.

Treatment

The use of intravenous methylprednisolone (1000 mg) or dexamethasone (200 mg) for 3–5 days is typically used for acute TM. The decision to offer continued steroids or add a new treatment is often based on the clinical course and MRI appearance at the end of 5 days of steroids. If the patient does not improve and meets the criteria of severe TM, then additional treatment with plasmapheresis is employed.

IMMUNE-MEDIATED NEUROMUSCULAR DISORDERS

Neuromuscular disorders have a wide variety of etiologies, including genetic, degenerative, toxic-metabolic, and immune mediated. It is critical to be able to diagnose immune-mediated neuromuscular disorders correctly, as these conditions may have very successful treatment options that can have a profound impact on a patient's functional level and their overall quality of life. In addition, some immune-mediated neuromuscular disorders may be life-threatening in the acute phase, such as GBS and myasthenia gravis exacerbations. We will outline the clinical background of the various neuromuscular immune-mediated disorders and then discuss the management strategy for these conditions as an aggregate.

CLINICAL CONSIDERATIONS

Peripheral Nerve

Guillain–Barré Syndrome or Acute Inflammatory Polyradiculoneuropathy

GBS or AIDP is a vital disorder to diagnose accurately as it represents one of the neurological emergencies. It has an incidence of about one–two per 100,000 with a slight male predominance. It has an average age of presentation of 40 years old, though may present essentially at any age. In most cases there is a history of an antecedent infection, most commonly an upper respiratory infection 2–4 weeks prior to the onset of symptoms. This is a disorder that is presumed to be related to a T cell-driven process targeting myelin with molecular mimicry to an antigen causing an autoimmune cascade. This is best demonstrated with *Campylobacter jejuni* as an example of an inciting antigen that provokes this autoimmune process, seen as an association with the axonal variants of GBS. Immunoglobulin and complement deposition is also seen in GBS.

Clinically, GBS most typically presents as an acute demyelinating inflammatory polyradiculoneuropathy (AIDP) with the onset of paresthesias/sensory loss distally and the evolution of ascending weakness affecting the distal and proximal lower extremities first. Demyelination may occur anywhere along the course of the peripheral nerves from distal zones up to the nerve roots and even motor nerves that may affect the face or respiratory muscles. Reflecting that, weakness may be both proximal and distal of the arms and legs, 50% of patients have facial weakness, a sizable number of patients get oropharyngeal weakness, and 30% develop respiratory failure. Reflex loss is typically seen on examination, although preserved reflexes may be seen early in the clinical course or with some variants of GBS. Radicular back pain is commonly reported. By definition, symptoms in GBS progress over a period of time of less than 4 weeks with about half of patients reaching their nadir after 2 weeks. Two-thirds of patients get autonomic dysfunction. The most common manifestation of this dysautonomia is sinus tachycardia, which does not require treatment. But, labile blood pressure, cardiac arrhythmias, and bradycardia are more malignant manifestations of this that may require supportive care. Careful attention is given to the respiratory status of the patient, which may require either noninvasive ventilation like BiPAP or invasive ventilation with intubation (see section: Management).

The diagnosis of GBS is clinical. This is an important consideration as diagnostic test results may miss early, acute phase GBS. For instance, cerebrospinal fluid results that show an elevated protein without pleocytosis (cytoalbuminological dissociation) is supportive of a diagnosis of GBS, though up to half of patients may have a normal cerebrospinal fluid protein in the 1st week of symptoms. An unexpected pleocytosis, such as cases that have a white blood cell count in the cerebrospinal fluid greater than 10–20 cells/μL, should trigger consideration of alternate or coexistent diagnoses, such as HIV, Lyme, or sarcoidosis. Electromyography (EMG) classically demonstrates acquired demyelinating electrophysiological changes, but again it may take time for those findings to mature and in the acute phase, the yield might be low for these characteristic findings. Early electrophysiological signs of GBS prior to the development of frank demyelinating changes could include one or more of the following: a prolonged or absent H reflex, prolongation of F waves, drop of motor amplitudes, or prolongation of distal motor latencies. In addition, a pattern of sural sensory sparing in the distal legs with relatively lower sensory responses in the arms, reflecting the nonlength-dependent pathophysiology of GBS, may be seen. Changes of classic demyelination with slowing of conduction velocities and conduction blocks may not be apparent initially, and may evolve over time. If the patient is significantly weak, then one of the most reliable acute findings on an EMG is identifying neurogenic recruitment on the needle electrode examination of a weak muscle.

Given this is a clinical diagnosis, it is important to differentiate it from other causes of acute weakness, such as spinal cord injury. Spinal cord dysfunction may also cause weakness of the limbs, as well, and hyperreflexia might not occur initially, as the patient may be in spinal shock. A sensory level in a band-like pattern around the entire trunk or urinary incontinence favor evaluating for a spinal cord etiology (see section: Acute Transverse Myelitis). GBS may cause numbness on the trunk, but this is typically in more of a ventral/shield pattern and not a clear sensory level around the entire torso. Porphyria can cause acute motor predominant neuropathy, but this is axonal with other clinical features that may include an upper extremity predominance of symptoms, abdominal pain, autonomic symptoms, recurrent spells of symptoms with resolution in between, or mental status changes. Exposure to toxins, such as arsenic (white lines on the fingernails—Mees' lines, confusion, abdominal symptoms, and hair thinning), tick paralysis (acute motor paralysis typically in a younger patient- screen for ticks in all acute weakness patients), acute neuromuscular junction disorders (see below), and vasculitis (see later) also need to be in the differential.

There are less common variants of GBS in addition to AIDP. Miller Fisher syndrome presents with ophthalmoparesis, ataxia, and areflexia. It is associated with the GQ 1b antibody in 90–95% of patients. GQ1b antibodies may also be positive in 66% of patients with Bickerstaff syndrome, which presents with encephalopathy, brainstem dysfunction, and hyperreflexia. Other axonal variants of GBS include acute motor axonal neuropathy (AMAN) and acute motor and sensory axonal neuropathy (AMSAN), which again are more common with *C. jejuni* exposure. Another variant, pharyngeal–cervical–brachial GBS, may mimic botulism. Ganglioside antibody testing (for antibodies other than for GQ1B) can be positive in some GBS patients, but is lower yield.

Chronic Inflammatory Demyelinating Polyradiculoneuropathy

Chronic inflammatory demyelinating polyradiculoneuropathy (CIDP) presents classically with sensory disturbance, proximal and distal weakness, and areflexia, but by definition has progression or relapses of clinical symptoms over greater than an 8-week period. Four to eight weeks of progression or relapse reflects an indeterminate condition known as subacute inflammatory demyelinating polyradiculoneuropathy. CIDP has an incidence of 1–8 per 100,000. The precise mechanism of autoimmune dysfunction is unknown, but appears to involve T-cell activation with other important contributions from macrophages and humoral factors. The diagnosis is supported by features of acquired demyelination on EMG, including conduction velocity slowing and conduction blocks. Lumbar puncture also typically reveals an elevated protein with cytoalbuminological dissociation, and again if there is an unexpectedly high amount of pleocytosis then other conditions should be investigated, including HIV, Lyme, and sarcoidosis.

Variants of CIDP include a multifocal asymmetrical presentation, known as the Lewis Sumner variant (multifocal acquired demyelinating sensory and motor neuropathy—MADSAM) and also a distal predominant variant that has a chief presentation of sensory ataxia and is associated with an IgM monoclonal antibody and MAG antibody. POEMS syndrome (polyneuropathy, organomegaly, endocrinopathy, monoclonal protein, skin changes) should be considered in patients with a demyelinating polyneuropathy, a monoclonal gammopathy, and other systemic manifestations, including hyperpigmented skin, white nails, organomegaly, or endocrine disease. Monoclonal gammopathy is almost always IgG or IgA with a lambda pattern. Demyelinating polyneuropathy that is resistant to immunotherapy should also trigger suspicion of this diagnosis. A skeletal survey needs to be performed to look for an underlying osteosclerotic myeloma. In addition, many of these patients will have an elevated serum vascular endothelial growth factor level (VEGF).

Vasculitis

Peripheral nerve vasculitis presents with a multifocal, mononeuritis multiplex pattern. The pathophysiology involves vascular inflammation causing occlusion and infarction of peripheral nerve axons. This is very painful and presents with a stepwise, asymmetric accumulating nerve disability, such as a unilateral foot drop followed by unilateral wrist drop, etc., until the patient develops more diffuse symmetric weakness and sensory loss. This condition may be secondary to systemic vasculitic conditions, such as polyarteritis nodosa, Wegener's, Churg–Strauss, or other connective tissue diseases, such as rheumatoid arthritis or lupus. Involvement of the kidneys, the gastrointestinal tract, the lungs, or the sinuses may be clues to a systemic process. Sinus and respiratory disease could implicate Wegener's disease. Eosinophilia and asthma-like symptoms would favor Churg–Strauss. An association with hepatitis B may be seen with polyarteritis nodosa. Cryoglobulinemia may present with skin findings, including purpura, and may have an association with hepatitis C. A large laboratory panel including general chemistries, a sedimentation rate, hepatitis panel, complement, cryoglobulins, ANA, ANCA, serum immunofixation, and other infectious disease labs in some cases, such as HIV, West Nile, HTLV, Lyme, CMV, and tuberculosis may be tested to cast a large net to pick up secondary causes of vasculitis. Peripheral nervous system vasculitis may also be a primary condition, not associated with any systemic associations. A sedimentation rate is not as high with primary nervous system vasculitis.

Sensory Ganglionopathy

Severe sensory loss due to disease of the dorsal root ganglia has a rather restricted differential, including paraneoplastic disease, chemotherapy toxicity (such as cisplatin), excess vitamin B6, but also may be seen as an acute autoimmune manifestation of a GBS-like illness or in association with Sjogren's disease. Clinically, patients present with severe sensory ataxia, manifested with a positive Romberg when standing or sensory ataxia of the arms, such as not being able to touch one's nose with eyes closed appropriately. An EMG can help differentiate sensory ganglionopathy from other causes of sensory ataxia involving the dorsal column of the spinal cord.

Multifocal Motor Neuropathy

Multifocal motor neuropathy (MMN) is a rare disorder (less than 1:100,000 incidence) that presents with subacute asymmetric motor limb weakness. Patients may also experience cramping of muscles. This follows a peripheral motor nerve pattern and tends to involve the upper extremities more than the lower extremities. For instance, a wrist drop (radial nerve prominent) pattern is commonly seen. Bulbar and respiratory involvement is rare (this can help differentiate this disorder from amyotrophic lateral sclerosis (ALS), which also tends to present with a myotomal pattern of weakness rather than the multifocal motor nerve pattern seen in MMN). EMG demonstrates motor nerve demyelination and conduction blocks, and there also is an association with GM1 antibodies in the serum.

Autoimmune Autonomic Impairment

Patients with acute to subacute, dramatic presentations of autonomic dysfunction may have an autoimmune etiology to their symptoms. Acute to subacute development of orthostatic hypotension, dry eyes/dry mouth, GI dysautonomia, large pupils, and/or a neurogenic bladder should warrant testing for the nicotinic acetylcholine receptor antibody as part of a paraneoplastic evaluation to look for further confirmation of an autoimmune process underlying the patient's symptoms.

Neuromuscular Junction (Excluding Myasthenia Gravis)

Lambert–Eaton myasthenia gravis has an incidence of one to two per million. Fifty percent of cases are paraneoplastic, most typically associated with small cell lung cancer. Clinically, patients with Lambert–Eaton present with less oculobulbar symptoms with a proximal weakness pattern that affects the lower extremities more than the arms. Unlike myasthenia gravis, which presents with normal reflexes, hyporeflexia is seen with Lambert–Eaton syndrome. As this is a presynaptic neuromuscular junction disorder, weakness and reflexes may actually improve with facilitated exercise temporarily. In addition, autonomic dysfunction, such as dry mouth and constipation, are frequent complaints. Serum voltage-gated calcium channel antibody testing is highly sensitive and an incrementing pattern on high-frequency repetitive nerve stimulation is seen on EMG, as well as motor amplitude increases on nerve conduction studies immediately after 10 s of exercise.

Botulism is also a very rare consideration in the differential diagnosis of neuromuscular junction dysfunction. This is typically associated with foodborne illness, wound infections, and enteric or infantile presentations. Like Lambert–Eaton, this is a presynaptic neuromuscular junction defect with the botulinum toxin inhibiting the release of acetylcholine. It presents with oculobulbar symptomatology and descending paralysis. Like Lambert–Eaton, there is an incrementing response on high-frequency repetitive nerve stimulation, but this is more modest in the 20–60% increment range, rather than the 100% plus increment range typically seen with Lambert–Eaton. If this condition is suspected, please contact the CDC for guidance on antitoxin therapy and confirmatory testing of the toxin. Of note, with infantile cases less than 1-year old, botulism immune globulin therapy is utilized.

Muscle

The four main inflammatory myopathies are dermatomyositis, polymyositis, inclusion body myositis (IBM), and necrotizing myopathy. The basic pathophysiology of these disorders will be discussed first, followed by clinical and diagnostic consideration.

With dermatomyositis, a type 1 interferon dysregulation is favored over an antibody/complement-mediated microangiopathy in its pathogenesis. Polymyositis and IBM both involve T lymphocyte CD8+ cytotoxic damage, though IBM also has degenerative findings with rimmed vacuoles and amyloid deposits out of proportion to the inflammation, leading some to hypothesize that this might be more of a degenerative muscle condition, particularly since it does not tend to respond to immunomodulatory therapy. Necrotizing myopathy demonstrates sparse inflammation with macrophages in the region of necrotic fibers, but no perivascular or endomysial inflammation. Regardless, this category of myopathy may still respond to immunomodulatory treatment.

Clinically, dermatomyositis is associated with rash in most cases, such as a heliotrope rash of the eyelids, V sign rash on the chest, a shawl rash on the upper back, or Gottron's papules on the knuckles. It has a nonspecific proximal weakness pattern on presentation, and like other inflammatory myopathies, generally presents subacutely rather than indolently. Muscle biopsy may have a classic association of perifascicular atrophy or other features, such as decreased capillary density or membrane attack complex (MAC) deposition. Some patients with dermatomyositis do not present with rash and thus the clinical neuromuscular presentation with biopsy findings suggest the diagnosis. Subcutaneous calcifications may be seen in dermatomyositis, but it is more common with children.

Polymyositis also nonspecifically presents with subacute proximal muscle weakness. A hallmark biopsy finding is inflammatory invasion of nonnecrotic fibers. IBM tends to occur more commonly after the age of 50, and presents more asymmetrically and with a slower onset of symptoms. Neuromuscular examination demonstrates out of proportion weakness of the quadriceps and finger flexor muscles compared to the other myopathies. A creatine kinase (CK) level is only modestly elevated in IBM, whereas in an active myositis from dermatomyositis or polymyositis, the CK level may be several thousand plus prior to treatment. With that being said, some cases of dermatomyositis, for instance, might present with CK levels that are only mildly elevated and in some cases even normal at diagnosis. Of note, high CK levels may falsely elevate liver enzyme levels, such as AST and ALT. Obtaining a GGT is helpful to suggest that this is a false elevation from the muscle disease rather than liver pathology. Malignancy may be associated with dermatomyositis or polymyositis in a minority of patients, particularly with older patients. A malignancy screen should be performed on presentation and annually for a total of 2–3 years in these patients. Necrotizing myopathy may also be paraneoplastic, and malignancy screening should be performed in these patients as well.

Necrotizing myopathy may also be triggered by statin medication. Statins may nonspecifically cause myalgias/myopathy that resolve with discontinuation of the offending toxic statin, but in some patients a significant proximal weakness continues despite discontinuation of the statin and then a subsequent muscle biopsy reveals a necrotizing myopathy that responds to immunomodulatory treatment.

Myositis may also be seen in overlap syndromes, along with established connective tissue disease. Testing for rheumatological disease with clinical clues is important, though false positive ANA testing may be seen in isolated dermatomyositis or polymyositis, so this is ordered in the context of the appropriate clinical context. Dermatomyositis and polymyositis may have cardiac or pulmonary complications with 10–20% of patients having interstitial lung disease. Anti Jo-1 antibody testing is positive in 50% of cases with interstitial lung disease. A myositis antibody panel can be helpful diagnostically. Salient antibodies include Jo-1 as mentioned; signal recognition particle (SRP), which is associated with necrotizing myopathy and is less responsive to treatment, requiring more aggressive immunotherapy; and Mi-2, which is associated with dermatomyositis that has a more favorable response to treatment.

MANAGEMENT

Nonpharmacological Issues

Assessment and management of the respiratory status of a patient with immunological neuromuscular disease is critical in patients with acute neuromuscular dysfunction, such as GBS and myasthenia gravis exacerbations. Following oxygen saturations and even arterial blood gas labs is inadequate and not sensitive enough for picking up early respiratory weakness concerns. At the bedside, assessing neck flexion strength and having the patient count as high as they can in one breath are good methods for evaluating the status of the patient. To assess neck flexion strength, the patient must be lying flat. If the patient is able to give good resistance with full neck flexion, that is reassuring, whereas if the patient has trouble lifting his or her head up off of the bed, there is concern for impending neuromuscular respiratory compromise. In addition, if the patient can count to greater than 20–30 in one breath at a rate of two counts per second, that is reassuring as well. The best determination, though, is to obtain pulmonary function testing with negative inspiratory flow and vital capacity measurements assessing for significant restrictive lung disease. This should be monitored closely, such as every 4h to see the trend of the respiratory parameters. BiPAP can be used as a noninvasive strategy, though invasive evaluation may be necessary in patients progressing to respiratory failure. It is important to monitor whether the patient is in any distress, for example, using accessory breathing muscles, or unable to lie flat due to exacerbation of the patient's shortness of breath from diaphragm weakness. Finally, it is also important to be cognizant of the rate of symptom progression. Deteriorating respiratory parameters in the setting of rapidly progressive weakness is much more concerning than would be the case in a patient that is more stable in their symptom progression.

Patients with neuromuscular weakness with bulbar involvement need to have a swallowing study to determine oral intake capabilities. Patients with dysphagia are at an increased risk for aspiration and development of pneumonia. Myasthenic patients that are presenting with an exacerbation should be screened for an underlying infection that triggered the exacerbation and their history should be reviewed for any new medications that may have unmasked their myasthenia further. Botulinum toxin, penicillamine, high-dose magnesium, nondepolarizing blockers during anesthesia, antiarrhythmia medications, β-blockers, and certain antibiotics may worsen myasthenia gravis, classically including aminoglycosides and fluoroquinolones. In addition, corticosteroids can paradoxially worsen myasthenia gravis if they are started at a high dose too rapidly.

Pharmacological Approach

Immunomodulatory treatment for neuromuscular disorders includes intravenous immunoglobulin (IVIG), plasmapheresis, corticosteroids, or cytotoxic medications. Much of the time, a wide variety of these treatments may be effective and their use may translate rather nonspecifically across many of the inflammatory neuromuscular conditions. As general principles, IVIG or plasmapheresis may be indicated for severe acute or relapsing time frames. Steroids are commonly employed, but if they are ineffective or if there is concern of long-term corticosteroid side effects, then cytotoxic medication for "steroid sparing" is indicated. In addition to these general thoughts, some of the inflammatory neuromuscular conditions respond very specifically to some agents and not others, and each of the conditions have their own treatment nuances, which will be outlined next.

GBS only responds to IVIG or plasmapheresis. Corticosteroids are not helpful. IVIG and plasmapheresis have similar efficacy for GBS, and so IVIG might be a first-line preference given that this is less invasive (not requiring a central line) and is a faster treatment course than plasmapheresis. However, other comorbid situations, such as kidney disease or thrombotic risk, would make plasmapheresis preferable. MMN also responds much more specifically to IVIG and does not respond well to steroids or plasmapheresis. It may respond to cyclophosphamide or rituximab. CIDP responds very well to IVIG, plasmapheresis, or corticosteroids. There is about a 90% efficacy rate to respond to at least one of these agents. Steroid sparing, cytotoxic medications (ie, azathioprine, mycophenolate, cyclosporine), have less evidence behind them for treatment of CIDP, but are still utilized for refractory disease or difficulty in managing the condition long term with conventional options. The DADS variant of CIDP does not respond well to typical immunomodulation, but rituximab has shown some promise. Vasculitis necessitates the use of high-dose steroids and in cases with rapid progression of weakness or systemic involvement, cyclophosphamide for 6 months should be added in combination with high-dose steroids.

Lambert–Eaton myasthenia gravis should first revolve treatment around searching for an underlying malignancy (small cell lung cancer) and addressing treatment of the malignancy if found. Symptomatically, 3,4 diaminopyridine is effective and acts as a potassium channel inhibitor to increase acetylcholine release and adding pyridostigmine can also be tried, though the efficacy of pyridostigmine appears marginal. Further immunomodulatory therapy with IVIG or oral immunosuppressants can then be considered as needed.

Dermatomyositis and polymyositis can respond to prednisone and with azathioprine, methotrexate, or IVIG (rituximab or other cytotoxic medications are used in refractory cases). IBM does not tend to respond to immunomodulatory treatment. Necrotizing myopathy often requires aggressive immunomodulation with a combination therapy of prednisone, cytotoxic meds, plus or minus IVIG.

General Principles of Immunosuppressive Therapy

All patients initiating immunosuppression should be warned of the risks for infection. For patients on 20mg daily of prednisone or higher combined with another immunosuppressant, sulfamethoxazole/trimethoprim (Bactrim DS) should be given on Mondays, Wednesdays, and Fridays for *Pneumocystis carinii* (PCP) prophylaxis (atovaquone for sulfa allergy). We recommend yearly inactivated flu vaccines and the pneumovax vaccination every 5years. Avoid live vaccines, such as nasal flu mist, MMR, or varicella while on immunosuppression. The varicella vaccine can be considered (principally for patients 50years and older) if it can be administered greater than 4weeks prior to the start of immunosuppression or if the immunosuppression is mild.

Corticosteroids

Corticosteroids are inexpensive and successful in many cases, but do have many possible side effects, particularly with chronic use. In addition to infection, the main concerns with chronic use include the following:

Diabetes, hypertension, insomnia, gastrointestinal abnormalities, such as ulcers, cataracts, weight gain, Cushingoid appearance, psychiatric problems (ie, mania, depression, and even psychosis), osteoporosis, avascular necrosis of the hip, and corticosteroid myopathy. With inflammatory myopathies, if a patient has worsening of weakness on moderate to high-dose corticosteroids, then an EMG and CK level can help differentiate a flare of myositis versus corticosteroid myopathy. If the CK is normal and the EMG shows a nonirritable myopathy, then corticosteroid myopathy may be causative.

Further preventive care while on steroids is listed as follows:

For prophylaxis, a stomach acid medication (a "proton pump inhibitor" such as omeprazole) should be given with any gastrointestinal symptoms or risk factors (acid reflux, stomach ulcer risk). DEXA bone scans should be obtained at baseline optimally and at least every 1–2years (depending on individual osteoporosis risk factors) and the dosing of the prednisone. All patients on steroids chronically should be on vitamin D 800 units a day and calcium 1g a day to help prevent osteoporosis. If a patient has osteoporosis, we generally give alendronate 70mg once weekly. For osteopenia, we give alendronate 35mg once weekly. This applies particularly to postmenopausal females and males above the age of 50. Younger patients may be followed more conservatively if they have not had a fracture and they do not have accelerating bone loss on serial DEXA scans.

Intravenous Immunoglobulin

For patients needing IVIG, it is dosed initially at 2g/kg total given in a divided regimen over 4–5days. Patients may also then be on "maintenance" IVIG, receiving 0.2–0.5 g/kg every 1–4 weeks, titrating based on response and chronic needs. Improvement typically occurs by day five and is maximal at day 8–9. Treatment effects last 8–12 weeks (half-life is 3 weeks). IgA-deficient patients theoretically could have an anaphylactic reaction to the medication. Other side effect concerns include the following: headache—this ranges from a mild headache with an infusion to severe headache with "aseptic meningitis"; renal failure; significant rash; thrombotic tendency (myocardial infarction, stroke, deep venous thrombosis, cerebral venous thrombosis, or pulmonary embolism); and anemia, thrombocytopenia, or pancytopenia. IVIG may spuriously elevate ESR and also may give a false Coombs test. Finally, viral transmission through the infusion is theoretically possible, but highly unlikely.

Plasmapheresis

Plasmapheresis is performed every other day for about five exchanges for exacerbation/acute management of autoimmune neuromuscular disorders, though occasionally is utilized once every 1–4 weeks as a single exchange for maintenance therapy. Plasmapheresis is generally well tolerated. IV complications, like infection (including sepsis), line perforation, or thrombosis (clotting) may occur. Fluid shift changes are part of the procedure and patients with heart failure or labile blood pressure are at more risk because of these vascular issues. Transient electrolyte abnormalities may occur (particularly low calcium or low potassium). Allergic reactions and those including lung injury reactions to blood plasma products may occur, which may require medical intervention and though rare, could be a serious side effect. Removal of circulating clotting factors may be relevant and clotting factors may be monitored with frequent or extensive plasmapheresis. This also needs to be taken into account if plasmapheresis is performed within proximity to a surgical procedure. ACE inhibitor medication (such as lisinopril) needs to be held for 24 h before treatment to avoid a reaction resembling anaphylaxis (flushing, drop in blood pressure, abdominal cramping).

Cytotoxic Medications

Immune-suppressing cytotoxic medications not only increase the risk of infection but they also have a small increased risk of cancer with long-term usage, typically many years of use (this is seen especially with skin cancer and lymphoma; one should recommend avoidance of excessive sun exposure and encourage application of sunscreen). Cyclophosphamide has a higher risk of cancer in general (especially bladder cancer) with chronic use. For complete details on cytotoxic medications, please see Table 16.1.1.

Pregnancy Considerations

Corticosteroids are generally safe in pregnancy, though cleft palate in the first trimester is a concern and premature rupture of membrane can occur with high doses. We recommend utilizing as low of a dose as possible in pregnancy and avoiding during the first trimester if possible. IVIG and plasmapheresis could increase the risk of spontaneous abortion due to a 24-h period of associated coagulopathy with treatment, and increased clotting risk goes up further in pregnant patients who receive IVIG. Fluid shifts and hypotension need to be monitored closely with plasmapheresis. Nevertheless, these all typically are well tolerated in pregnancy. We recommend reserving these treatments for severe or life-threatening symptoms.

It is best to generally avoid any cytotoxic medications during pregnancy. Mycophenolate and methotrexate are particular absolute contraindication concerns as compared to azathioprine and cyclosporine.

TABLE 16.1.1 Immunosuppressive Agents for Neuromuscular Disorders

Medication	Dosing Strategy	Side Effects	Monitoring	Special Considerations
Mycophenolate mofetil	1000–1500 mg bid May take 6–12 months for effectiveness	Diarrhea/GI irritation, tremor, infection, category D in pregnancy- women should use 2 forms of birth control as mycophenolate may alter oral contraceptive levels	Periodic CBC (monitor for leukopenia, anemia, and thrombocytopenia), and LFTs	Cholestyramine, antacids, sevelamer, calcium, iron, magnesium, and rifampin decrease mycophenolate levels, while valtrex, acyclovir, and probenecid increase levels. Doses greater than 1 g bid should be avoided if creatinine clearance is less than 25
Azathioprine	2–3 mg/kg/day- start at 50 qd then go up in 50 mg increments every 1–2 weeks in bid dosing to goal dose • slow onset of therapeutic action (maybe over one year)	• GI complaints • An idiosyncratic flu-like reaction in a minority of patients requiring discontinuing the drug • Pancreatitis	• Thiopurine methyltransferase (TPMT) level prior to initiation of therapy- if low use other therapy due to increased risk of life threatening leukopenia • Regular CBC and LFTs	• Adjust dose with renal failure • Avoid allopurinol as it increases the risk of bone marrow or liver side effects. Coumadin levels may be decreased by imuran. Use with caution with ACE inhibitors and aminosalicylates as they also may increase leukopenia
Metho-trexate	7.5–25 mg once a week (given in three divided doses 12 h apart); dosage increases are made in 2.5 mg increments every week or two	Liver disease, stomatitis, GI complaints, pulmonary fibrosis, alopecia, rarely lymphoma. Caution with use of NSAIDs, which may potentiate toxicity. Severe diarrhea and severe ulcerative stomatitis require discontinuation of therapy	• Liver disease, alcohol history, and hepatitis B and C screening • LFTs should be monitored closely and dose adjusted accordingly. If persistently elevated, consider liver biopsy • Periodically, pulmonary function testing should be obtained. If patient has anti-jo-1 antibody syndrome (interstitial pulmonary disease), then would avoid use of methotrexate	Give folate 1–5-mg qd to reduce side effect toxicity. Side effects, such as elevated LFTs, stomatitis (mouth sores), nausea, vomiting, and abdominal pain may all respond. Would give prophylaxis with 1 mg daily of folate and increase as needed to help control side effects

Cyclosporine	4–6 mg/kg/day- typically 150–200 mg bid (comes in 25 and 100 tabs). Started at 3–4 mg/kg/day and increased towards 6 mg/kg/day to achieve a trough serum level of 50–200 ng/ml. It generally is used in refractory inflammatory syndromes. Benefits start within 2 weeks and are maximal by 3 months	GI side effects, hypertension, renal disease, tremor, excessive hair growth, glucose intolerance, bone loss, gum hypertrophy, hyperkalemia, hypomagnesemia. Of note, rarely posterior reversible leukoencephalopathy syndrome can be seen with cyclosporine	Monitor bun/cr (electrolytes) and blood pressure readings bun/cr to monitor for side effects. LFTs are followed, as well. If blood pressure or creatinine are elevated then decrease or discontinue medication. Nephrology consult for renal disease	In addition, NSAIDs/ACE inhibitors may potentiate renal disease in cyclosporine and cyclosporine may enhance and cause digoxin, statin, or colchicine toxicity
Rituximab	Dosing is 750 mg/m^2 up to 1 g IV and then repeated after 2 weeks; may consider repeat dosing after 6 months	An infusion reaction mimicking an allergic reaction all the way up to an anaphylactic reaction is a concern for rituximab infusions. Pretreatment with 650 mg tylenol, 50 mg of benadryl, and 100 mg of methylprednisolone (high dose steroid is avoided in myasthenia patients) is recommended along with a slow infusion rate that is slowly titrated up. Treat any subsequent allergic infusion reaction or concern for anaphylaxis	Consider monitoring IgG levels to assess infection risk	Progressive multifocal leukoencephalopathy (PML) is a risk in about less than 1:20,000 patients. Could consider JC virus screening Hepatitis B should also be ruled out and addressed as needed prior to treatment as reactivation can occur Inactivated vaccines typically utilized in immunosuppressed patients (such as influenza or pneumovax) or vaccines such as the hepatitis B vaccine all have less response while on rituximab

Continued

TABLE 16.1.1 Immunosuppressive Agents for Neuromuscular Disorders—cont'd

Medication	Dosing Strategy	Side Effects	Monitoring	Special Considerations
Cyclophos-phamide	PO dose of 1.5–2 mg/kg/day IV- start dosing at 750 mg/m^2 in general (or 500 mg/m^2 in patients greater than 70 years old, obese patients- you should base more on ideal body weight, or patients with renal dysfunction particularly if creatinine clearance is less than 40) If nadir wbc at 2 weeks is below 3.5 (or absolute neutrophil count below 1500) then decrease next infusion by 250 mg/m^2 or if above 4 and disease not fully controlled, then increase by 250 mg/m^2 to a maximum of 1 g/m^2 WBC below 3.5 while on the oral regimen should also lead to dose reductions	Nausea should be treated with zofran 4–8 mg 30 min prior to infusion and again 4–8 h afterwards. Decadron 10 mg 4 h after infusion can be used, as well Other possible side effects include neoplasia (especially bladder cancer), sterility, alopecia, pulmonary fibrosis, cardiac myocarditis	Periodic UAs to monitor for hemorrhagic cystitis and later after therapy complete, still periodically long term to screen for bladder cancer Monitor periodic CBC, LFT, and chemistries	The IV infusion route has less side effect potential. Hemorrhagic cystitis is a risk on cytoxan. Hydrate thoroughly when giving the cytoxan. 1 L of fluid IV and then patient to drink 1 L of fluid every 6–8 h for 24 h. Patient recommended to void as frequently as possible. Check UA for blood for hemorrhagic cystitis concern prior to each infusion. May consider giving MESNA with each infusion if there is significant concern for hemorrhagic cystitis risk. Check CBC at each 2-week nadir and at time of each infusion Caution with medications that induce hepatic enzymes which can alter cyclophosphamide levels and with succinylcholine which may lead to prolonged neuromuscular junction blockade in patients on cyclophosphamide

OTHER NEUROIMMUNOLOGICAL CONDITIONS

Neurosarcoidosis

Sarcoidosis results from antigenic stimulation that drives a TH1 dominant immunological response, facilitated by interferon-gamma, interleukin 2, and tumor necrosis factor alpha, to result in noncaseating granuloma formation in tissues in the body.

The most common sites of involvement are the lungs, skin (ie, erythema nodosum), eyes (ie, uveitis), lymph nodes, and in 5% of cases the nervous system. There is a predilection for African-Americans. Neurological involvement may occur in the context of other systemic clues or in isolation. Neurological clues to the diagnosis include cranial nerve involvement (particularly Bell's palsy that may be bilateral or recurrent) or optic neuritis, particularly if bilateral; meningeal disease/inflammation; longitudinally extensive myelitis; subacute peripheral neuropathic weakness; and hypothalamic–pituitary axis dysfunction. Parenchymal brain masses, seizures, hydrocephalus, and encephalopathy may also be seen.

Diagnosis may be suggested by an elevated ACE (angiotensin-converting enzyme), but that is nonspecific and lacks sensitivity. Cerebrospinal fluid is typically inflammatory with pleocytosis and an elevated protein (in some significant cases a low glucose), and testing of this fluid also helps to evaluate for other conditions in the differential, such as fungal, tuberculous, and carcinomatous meningitis. Magnetic resonance imaging (MRI) studies may show white matter changes (and cerebrospinal fluid oligoclonal bands may be seen) providing diagnostic confusion with MS, but contrasted MRI studies may differentiate if they show classic findings of neurosarcoidosis including enhancement of the meninges, cranial nerves, or hypothalamic–pituitary axis. If neurosarcoidosis is considered likely, then a biopsy is needed to confirm presence of noncaseating granulomas consistent with the disorder. This may include a systemic transbronchial or lymph node biopsy. If nervous system tissue is easily accessible, that may be considered, but in many cases that is high risk. Blind biopsies of muscle, bone marrow, or conjunctiva may have 30–50% yields as well. Corticosteroids are first line for neurosarcoidosis. Cytotoxic medications may be required and there is increasing literature supporting the use of tumor necrosis factor inhibitors targeting the specific pathophysiology of sarcoidosis. IVIG may be considered for certain neuropathy presentations of sarcoid like small fiber neuropathy and GBS-like presentations.

Central Nervous System Vasculitis

Like peripheral nervous system vasculitis, CNS vasculitis may be secondary to a connective tissue disease or systemic vasculitis, or it may be isolated to the nervous system (primary CNS angiitis). Patients with CNS vasculitis may present with headache, encephalopathy, and a wide range of focal neurological deficits. This correlates with multifocal infarcts in the brain bilaterally with multifocal bilateral diffusion lesions on MRI. This may be seen with embolic disease, but may occur with vasculitis as well. Hemorrhage may be seen, but is less common. Another MRI clue may be contrast enhancement, such as leptomeningeal or other patterns of enhancement, but overall MRI findings are nonspecific and imaging is most useful to help exclude other diagnoses. Cerebrospinal fluid generally shows inflammation and is important to help rule out infectious or malignancy mimics. Angiography may reveal "beading" of the blood vessels, but this is not sensitive or specific enough for final diagnosis. The gold standard is biopsy of the meninges or brain. Once a diagnosis is secure, the patient is treated with aggressive immunosuppression with high-dose steroids and cytoxan.

Other Autoimmune Encephalopathies

Antibody-mediated encephalopathies represent a relatively new category of described immune-mediated disorders. They are often associated with cancer, may respond favorably to immunomodulatory treatment, and can be diagnosed serologically.

A classic version of a paraneoplastic or autoimmune encephalitis would be anti-Hu limbic encephalitis. This is commonly associated with small cell lung cancer. It may present with limbic encephalitis (mesial temporal lobe signal may be seen on MRI), causing symptoms of behavioral changes, cognitive dysfunction, or seizures. This can occur as part of a larger range neurological involvement, such as having comorbid sensory ganglionopathy, myelitis, or brainstem involvement, but it may also occur in isolation. Treatment of the underlying tumor is the paramount issue in these cases, though immunotherapy may be tried as well.

A hallmark autoimmune encephalopathy that responds well to immunotherapy is anti-NMDAR encephalitis. It affects mainly young patients with the overwhelming majority being under the age of 45. In females, there is an association with ovarian teratoma, especially above the age of 18. In males, it is much rarer to detect a tumor.

Patients will present subacutely with seizures, confusion, amnesia, behavioral changes, or psychiatric features, and many progress to later develop movement disorders. A large number will also present with prodromal symptoms, such as a fever or headache, and others will require intensive care at initial presentation for seizure management or reduction in consciousness. MRI results are often normal or only have nonspecific transient signal changes, but EEG results most commonly demonstrate a slow, disorganized background or less frequently, epileptiform discharges. CSF analysis ranges from normal to increased lymphocyte levels, and may have oligoclonal bands suggestive of autoimmunity. Diagnosis is confirmed by the detection of anti-NMDAR antibodies (or to other cell surface antigens) in the serum or most reliably followed in the CSF.

It has been suggested that autoimmune antibody synthesis, particularly those targeting the NMDAR, is triggered by some precipitating event, particularly neoplasm or viral infection. Patients' anti-NMDAR antibodies then cause a selective and reversible decrease in receptor density and synaptic localization, particularly in the hippocampus. The loss of these receptors impairs synaptic transmission, resulting in the characteristic clinical presentation.

The detection of an underlying neoplasm is actually a better prognostic factor than that of the nonparaneoplastic disease. Resection of the tumor along with first-line immunotherapy (corticosteroids, IVIG, plasma exchange) often results in very favorable outcomes. When no underlying tumor is detected, most patients respond favorably to first-line immunotherapy with the majority demonstrating a complete recovery or only minor impairments after 24 months. To those unresponsive to initial first-line therapy, rituximab and cyclophosphamide have been effective.

Hashimoto encephalopathy (HE), also known as steroid-responsive encephalopathy associated with autoimmune thyroiditis (SREAT), is a rare immunotherapy-responsive encephalopathy associated with high-serum antithyroid antibody (ie, antimicrosomal and antithyroglobulin) concentrations. The mechanism of HE is unknown, but it appears to be unrelated to altered thyroid gland function, as most patients are euthyroid at presentation. An autoimmune etiology is suspected due to the presence of antithyroid antibodies and a positive response to immunotherapy. Nearly all cases of HE have a positive outcome except those in which diagnosis and treatment are delayed for many years.

Thus, early recognition is critical to preserve cognitive function.

Patients typically present with an acute or subacute onset of confusion or reduced consciousness, seizures, myoclonus, stroke-like signs, or incoordination. Diagnosis depends on the presence of elevated antithyroid antibody titers while other causes of encephalopathy must be excluded. Results from MRI, EEG, and CSF evaluation are variable and nonspecific, yet helpful in excluding other potential causes. Treatment with corticosteroids is effective in most cases (daily oral prednisone and/or high-dose IV methylprednisolone), however, due to the limited number of cases, standard dosages have not been established. In patients intolerant of corticosteroids or with ineffective treatment, other immunomodulatory treatments have been utilized, such as azathioprine, IVIG, plasmapheresis, cyclophosphamide, and rituximab.

Finally, of note, a paraneoplastic panel of the serum and CSF provides a broad stroke to evaluate for antibody associations in suspected clinical cases of autoimmune encephalopathy. In cases without clear antibody associations, classic clinical features with a subacute course of symptoms may still justify an empiric treatment trial of immunomodulatory therapy to assess for diagnostic and therapeutic response.

AUTOIMMUNE EPILEPSY

New Onset Refractory Status Epilepticus

The term new onset refractory status epilepticus (NORSE) describes a very small percentage of patients with status epilepticus (SE) who present de novo, are refractory to standard treatment, and for which no clear explanation is found. The typical working diagnosis in these cases is viral encephalitis, even though diagnostic testing is often unrevealing. This diagnosis is made based on the findings of CSF pleocytosis, variable MRI abnormalities, and the absence of an alternative explanation. Recent studies have sought to investigate if other noninfectious processes, such as an immune-mediated mechanism, may also cause this syndrome.

While much is still unknown about NORSE, there have been several antibodies implicated in the pathogenesis of the disease: in particular, antibodies against voltage-gated potassium channels, NMDA receptors, and glutamic acid decarboxylase (GAD). Due to the diverse collection of purported causes and a lack of definitive diagnostic criteria, responses to treatment, including immunotherapy, have been variable and inconclusive. However, after standard antiepileptic drug (AED) therapy has proven ineffective, early and aggressive immunotherapy may offer an option that could be effective in some patients.

Rasmussen Encephalitis

Rasmussen encephalitis (RE) is a rare, debilitating disease in which dysfunction from one brain hemisphere results in focal seizures and progressive neurological deterioration. It is most often a disease of childhood. Patients will present with intractable seizures (simple partial motor), hemiparesis, and unilateral cerebral atrophy on MRI. There is an association with anti-GluR3 autoantibodies in many patients offering a possible clue into the pathogenesis of the disease. This hypothesis has prompted the use of immunotherapy after initial AED therapy has proven ineffective. While the only definitive cure is hemispherectomy, options, such as early immunotherapy (ie, corticosteroids, IVIG and plasmapheresis) may offer a potential treatment to slow disease progression as an adjunct.

Further Reading

[1] Amato AA, Russell JA. Neuromuscular Disorders. 2008.

[2] Daroff RB, Fenichel GM, Jankovic J, Mazziotta JC. In: Bradley's Neurology in Clinical Practice. 6th ed. 2012.
[3] Hoyle JC, Jablonski C, Newton HB. Neurosarcoidosis: clinical review of a disorder with challenging inpatient presentations and diagnostic considerations. Neurohospitalist April 2014;4(2):94–101.
[4] Khawaja AM, DeWolfe JL, Miller DW, Szaflarski JP. New-onset refractory status epilepticus (NORSE) — the potential role for immunotherapy. Epilepsy Behav June 2015;47:17–23.
[5] McKeon A, Lennon VA, Pittock SJ. Immunotherapy responsive dementias and encephalopathies. In: Lewis SL, editor. Continuum: Lifelong Learning in Neurology. American Academy of Neurology, vol. 16 (2). April 2010. p. 80–101. Dementia.
[6] Pachner AR. A Primer of Neuroimmunological Disease. Springer; 2012.
[7] Tawil RN, Venance S. Neurology in practice. In: Gross RA, Jonathan WM, editors. Neuromuscular Disorders. 2011.

CHAPTER

16.2

Translational Correlation: Multiple Sclerosis

M. Avila, A. Okai, J.C. DeToledo

Texas Tech University Health Sciences Center, Lubbock, TX, United States

Multiple sclerosis (MS) is an inflammatory, immune-mediated disease which involves the central nervous system (CNS), causing demyelination, inflammation, axonal loss, and neurodegeneration. It is characterized by multiple focal areas of myelin loss in the CNS, which are referred to as plaques or lesions.

Demyelination is accompanied by variable gliosis and inflammation and by relative axonal preservation. Most common areas involved in MS include optic nerve, periventricular and juxtacortical, subpial spinal cord, brain stem and cerebellum white matter regions. MS has historically been considered a primary white matter disease, but there is increasing evidence that cerebral gray matter is also involved. The cortical involvement probably accounts for the cognitive impairment and progression of motor disability that is seen in many cases.

THE ROLE OF IMMUNE SYSTEM IN MS

Experimental autoimmune encephalomyelitis (EAE) is the animal model used to study MS. This model has identified the role of T cells, in particular CD4+ T cells polarized to a T helper (Th) 1 in driving the autoimmune process. However, there is evidence that other cell types such as Th17B cells also contribute to the pathogenesis of MS.

Th1 or Th2 cells are CD4+ effector cells that are differentiated in response to exposure to specific interleukins (ILs), after which they continue to express and secrete specific ILs. Th1 cells have been involved in MS pathogenesis and in other autoimmune conditions like Crohn's disease.

In MS, Th2 cells exhibit regulatory function as Th2 differentiated cells. In addition to T cells, there are a number of other effector cell populations that appear to be involved in the pathogenesis of MS, including CD56+, natural killer (NK) cells, invariant NK T cells that express both NK and T-cell markers, and stem cells.

Intrathecal immunoglobulin G (IgG) synthesis, persistence of oligoclonal bands, and clonal expansion of memory B cells are well-characterized clinical features of the humoral response in the cerebrospinal fluid (CSF) of MS patients.

Conn's Translational Neuroscience
http://dx.doi.org/10.1016/B978-0-12-802381-5.00028-2

Many of the new treatments for MS target the various molecules involved in this co-stimulatory or signal transduction cascade.

DIAGNOSING MULTIPLE SCLEROSIS

About 80–85% of the with MS follow an initial relapsing remitting course (RRMS), which is characterized by episodes with fairly rapid onset of new or recurrent neurological followed by partial or complete recovery. Onset of disease is typically between 20 and 40 years old, with clear female predominance.

Diagnosis is based on clinical, laboratory, and radiographic information. The revised McDonald criteria defines an attack as a patient-reported symptom or objectively observed signs typical of an acute inflammatory demyelinating event in the CNS with duration of at least 24 h in the absence of fever or infection. Most common presentation includes optic neuritis, brain stem (internuclear ophthalmoplegia), and spinal cord (partial transverse myelitis).

About 10–15% of the patients experience a progressive course of MS from the onset. This is defined as primary progressive MS, which is characterized by insidious onset of symptoms followed by gradual deterioration over time. Spinal cord is frequently involved in these cases, and these patients commonly have gait difficulties, balance, spasticity, weakness, and bladder and/or bowel involvement. Patients tend to be older when compared with RRMS, with no clear gender predominance. MRI lesions are similar in appearance, but there is a tendency for fewer gadolinium-enhancing lesions compared to RRMS.

About 25–40% of patients with RRMS develop a secondary progressive course after an average of 20 years. Secondary progressive MS is diagnosed when after an initial RRMS, a patient demonstrates disease progression independent of relapses for at least 6 months.

OTHER IMPORTANT CONSIDERATIONS

Clinical Isolated Syndrome

Clinical isolated syndrome (CIS) is defined as the first clinical occurrence of neurologic symptoms which last for at least 24 h and is caused by CNS inflammation and demyelination. Patients with CIS may have risk of developing MS. Determining which CIS patient has high risk can be difficult.

Optic neuritis is one of the most frequent presentations of CIS. MRI of the brain is very helpful as the presence or absence of multifocal brain lesions helps to predict the likelihood of developing MS.

The Optic Neuritis Treatment Trial (ONTT) showed that the 10-year risk of developing MS after initial episode of optic neuritis was 38% for the entire study group.

There are currently no biomarkers that are pathognomonic for a diagnosis of CIS or MS. Diagnostic workup of a CIS classically begins with a brain MRI. Additional MRI studies of the spine can be helpful in case-by-case bases. MRI are also useful to document dissemination in space.

CSF evaluation for oligoclonal bands (OCBs) and elevated IgG index as well as evoked potentials may further assist with the diagnosis, but are not always performed since diagnosis can be usually made based on the clinical presentation, MRI findings, and ruling out potential potentially confounding differential diagnosis.

The MRI scan can show clinically silent T2-bright areas more frequently than abnormalities can be detected by clinical history and examination alone; 60–70% of brain lesions and 30% of spinal cord lesions develop without corresponding clinical attacks. New silent lesions appear 5–10 times more frequently than do lesions associated with clinical relapses. Between 50% and 80% of patients with CIS have one or more clinically silent T2-bright abnormalities on their baseline brain MRI. Both the presence and the number of clinically silent lesions reflect older subclinical disease activity and are predictive of development of MS in the next 5–14 years.

BALANCING RISK AND BENEFITS OF THERAPY

There are several factors associated with aggressive MS, some of which include male gender, African-American ethnicity, frequent relapses with poor recovery, high burden of MRI lesion at presentation, older age on onset, spinal cord lesions, presence of two or more gadolinium enhancing, brain atrophy, and low vitamin D.

The treatment of MS has evolved considerably in the past decades. As of mid-2016, there are injectable, infusions, and oral options for patients, all of which have proven to decrease the number or relapses when compared to placebo. Treatment choices now depend on a full understanding of a patient's clinical condition and perceived understanding of each patient in order to give an individual or "personalized" approach.

CONCLUSION

MS is a leading cause of disability in young patients. Treatment from early stages, including high risk of CIS, is advisable. There is a wide range of agents to choose from for the treatment of relapsing forms of MS. Treatment should be individualized and patient should be followed for evaluation of disease duration and underlying subclinical activity since response to treatment is variable.

Further Reading

[1] Popescu BF, Lucchinetti CF. Pathology of demyelinating disease. Ann Rev Pathol 2012;7:185–217.

[2] Sobel RA, Moore GRW. Demyelinating diseases. In: Love S, Louis DN, Ellison DW, editors. Greenfield's neuropathology. 8th ed. London, England: Oxford University Press; 2008. 1513Y1608.

[3] Kasper LH, Shoemaker J. Multiple sclerosis immunology, the healthy immune system vs the MS immune system. Neurology 2010;74(Suppl. 1):S2–8.

[4] Polman CH, Reingold SC, Banwell B. Diagnostic criteria for multiple sclerosis: 2010 revisions to the McDonald criteria. Ann Neurol 2011;69(2):292Y302.

[5] Fisniku LK, Brex PA, Altmann DR. Disability and T2 MRI lesions: a 20-year follow-up of patients with relapse onset of multiple sclerosis. Brain 2008;131(Pt 3):808Y817.

CHAPTER

17

The Olfactory System

C. *Trimmer*[1], *J.D. Mainland*[1,2]

[1]Monell Chemical Senses Center, Philadelphia, PA, United States; [2]University of Pennsylvania School of Medicine, Philadelphia, PA, United States

INTRODUCTION

Defining the olfactory system is not a straightforward task. Although humans tend to think of the olfactory system as designed to detect volatile molecules and the gustatory system to detect nonvolatiles, the distinction between the two systems is based on anatomy rather than stimulus properties. In fish the olfactory system detects molecules in solution, not in air, but fish maintain anatomically distinct olfactory and gustatory systems. The olfactory system is not, however, a unified anatomical structure, but rather a collection of subsystems with distinct anatomy, receptor repertoires, transduction mechanisms, and central projections. All of these subsystems are united by their function: detecting chemicals in the environment and transducing them into neural signals.

Chemicals provide a rich source of information about the environment, and many animals trust their sense of smell with their most important tasks: finding prey, avoiding predators, identifying mates, marking territory, navigating their environment, recognizing kin, and establishing social hierarchies. Although humans tend to rely more heavily on other senses, such as vision and audition to gather information about their surroundings, we retain the capacity to detect and discriminate a large number of odorous molecules (odorants) and complex mixtures of molecules (odors). This discriminatory ability adds complexity to our perception of both odors and flavor, which is largely a function of our detection of the volatile molecules released from our food and beverages.

One of the major challenges in the field has been the seeming lack of structure to the stimulus. While the primary colors and basic tastes were largely worked out decades ago, early efforts to organize the olfactory perceptual space were largely failures. The discovery of the olfactory receptor family in 1991 has given the field new experimental control over the system and order has begun to emerge from the chaos. The receptors serve as molecular "tags" that allowed researchers to divide the system into smaller parts, isolating the anatomy and function of one out of the hundreds of input channels. That said, there are still a number of fundamental problems that remain unanswered. We don't know the essential features of a molecule that the olfactory system encodes and there is not a perfumer or scientist who can predict the smell of a novel chemical structure or the chemical structure of a novel smell.

Conn's *Translational Neuroscience*
http://dx.doi.org/10.1016/B978-0-12-802381-5.00029-4

In this chapter, we examine how the various subsystems of the olfactory system detect chemicals in the environment and subsequently encode them using neural signals. In addition, we discuss disorders that disrupt this encoding, resulting in olfactory dysfunction. We focus on the main olfactory system of mammals.

OLFACTORY STIMULUS

In order for an odor to stimulate the main olfactory system, it must activate receptors in the dorsal nasal cavity that are surrounded by a thin aqueous layer of mucus. Thus, an odor must be volatile enough to evaporate, yet not so volatile that it never sorbs across the mucous layer. In addition, mammalian olfactory receptors (ORs) have a hydrophobic-binding pocket, and therefore prefer hydrophobic compounds. The field is currently poor at predicting the boundaries of these characteristics that define odorous molecules. In addition, determining how the olfactory system translates the molecular structure of an odorant molecule into perceptual characteristics has proved difficult—a number of structurally similar molecules smell quite different and a number of similar smelling molecules are structurally unrelated. Thus, the field is also very poor at predicting the perceptual characteristics of a given odor, such as its character or intensity.

Scientists and philosophers have long attempted to correlate the physical properties of odors to their perceptual characteristics. In vision, wavelength translates into color; in audition, frequency translates into pitch; in olfaction, the mapping from chemical structure to olfactory percept is unknown. That said, there are tantalizing clues and relationships. The presence of a certain functional group in a molecule sometimes correlates with a given odor character: thiols impart the odor of rotten eggs, amines impart a fishy odor, and esters impart a fruity odor. Still, while the functional group is highly relevant to the perceived odor in some cases (Fig. 17.1A–C), in other cases functional group seems irrelevant (Fig. 17.1D–H).

In addition, monomolecular odors are difficult to place on a continuous axis, as we can with color (wavelength) or pitch (frequency). This predictive framework allows researchers to systematically vary wavelength, for example, and probe the receptive range of a primary sensory neuron. In addition, researchers can then interpolate and extrapolate to untested wavelengths by fitting a function that varies with wavelength. In olfaction, initial attempts to organize odors attempted to define primary odors—a small set of odors that could be combined in different proportions to yield all possible olfactory perceptions. This approach failed to reliably organize perceptual space, leading researchers to describe the space with a large number of features, and then use dimension-reduction techniques to identify perceptual axes. In broad strokes, this work discovered that if humans are asked to sort compounds by similarity, they typically resort to using pleasantness. In other words, if I give you a 100 vials and ask you to place them in a line with the most similar compounds close to each other and the most different compounds far apart, at one end of the line the vials will all smell unpleasant, and at the other end of the line the vials will all smell pleasant. In addition, this change in pleasantness correlates with molecular size—pleasant molecules tend to be larger than unpleasant molecules.

Most stimuli encountered in the natural world are not monomolecular, but rather are complex mixtures. One might envision the olfactory system as analyzing a complex mixture and breaking it down into its component odors, much as an expert chef tastes an unknown dish and reverse engineers the recipe. This is not, however, how the olfactory system deals with complex mixtures. Although a typical food odor may contain hundreds of volatile compounds, many of those volatiles are present at levels that are too low to be perceptually relevant. Typically, recreating

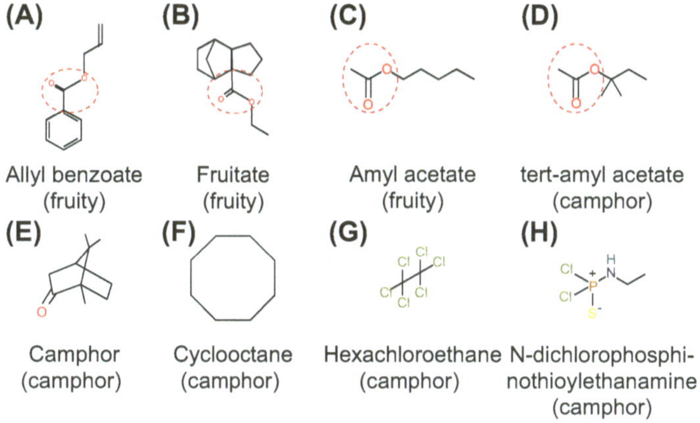

FIGURE 17.1 Structure–odor relationships. Molecules (A–C) contain an ester functional group (*circled*), which correlates with a fruity percept despite structural variation elsewhere in the molecules. (D) is an isomer of amyl acetate (C), but has an odor that is more similar to (E–H).

a food odor only requires between 3 and 40 of the volatiles in the original food. The field has begun to develop models that can describe the perceptual similarity of complex mixtures, but they remain noisy and are not yet universal to all types of mixtures.

In addition, not only are many natural stimuli complex odors, but they frequently occur in the context of a background odor. This is a major failing point of the electronic noses currently available. While it is fairly straightforward to design a broad molecular detector that can differentiate between target odors, it is much more difficult to maintain discrimination between target odors in the presence of background odors.

Finally, although we have mostly focused on describing volatile stimuli that activate the main olfactory system, the accessory olfactory system also detects nonvolatile compounds including peptides and small proteins.

PERIPHERAL OLFACTORY SYSTEM

The mammalian nose can contain multiple distinct subsystems, including the main olfactory system, the accessory olfactory system, the trigeminal system, the septal organ of Masera, and the Gruenberg Ganglion (GG) (Fig. 17.2). The function of many of these systems has been only poorly characterized, and even for the well-characterized systems new receptor families mediating chemosensation were discovered in 2006 and 2016. This suggests that we may not yet have a complete picture of all of the peripheral chemosensory mechanisms in the nose.

The Main Olfactory System

Main Olfactory Epithelium

In the main olfactory system, odorants must sorb from the air into the mucous layer of the nose. The odorants are then detected by the olfactory sensory neurons (OSNs) in the main olfactory epithelium (MOE), a specialized neuroepithelial tissue located at the back of the nasal cavity below the cribiform plate (Fig. 17.3). The MOE is home to millions of OSNs, in addition to microvillar and supporting sustentacular cells, all of which overlie a basal layer of stem cells. The OSNs are bipolar neurons, with their cell body found in the basal portion of the MOE. A dendrite extends from the apical pole of the neuron to the epithelial surface. At the surface, the dendrite forms the olfactory knob, from which 10–12 cilia extend into the mucous layer that lines the MOE. A single axon extends from the basal pole of the OSN, where it merges with the axons of other OSNs, passing through the cribiform plate before joining other bundles (termed

FIGURE 17.2 Lateral view of the human nasal cavity. The dorsal area of the nasal cavity is lined with olfactory epithelium. Axons from olfactory sensory neurons in the epithelium coalesce to form the olfactory nerve, which projects through the bony cribiform plate to the olfactory bulb.

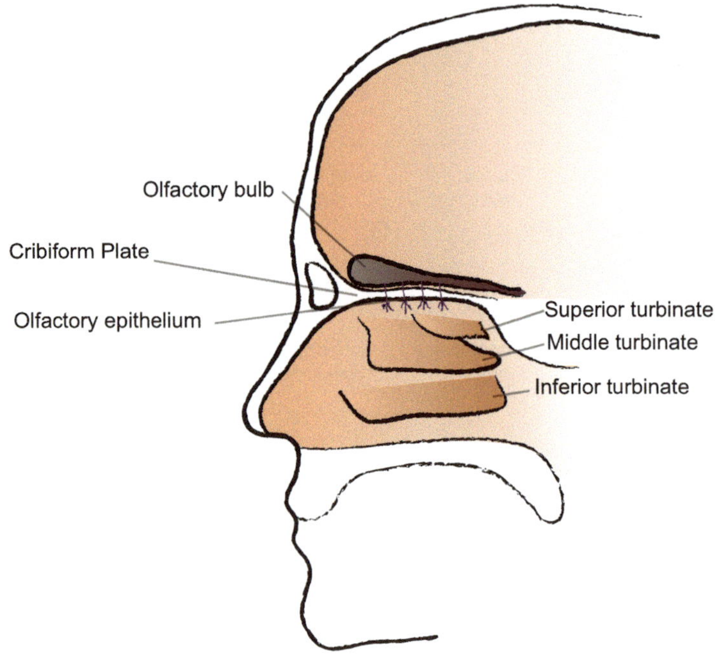

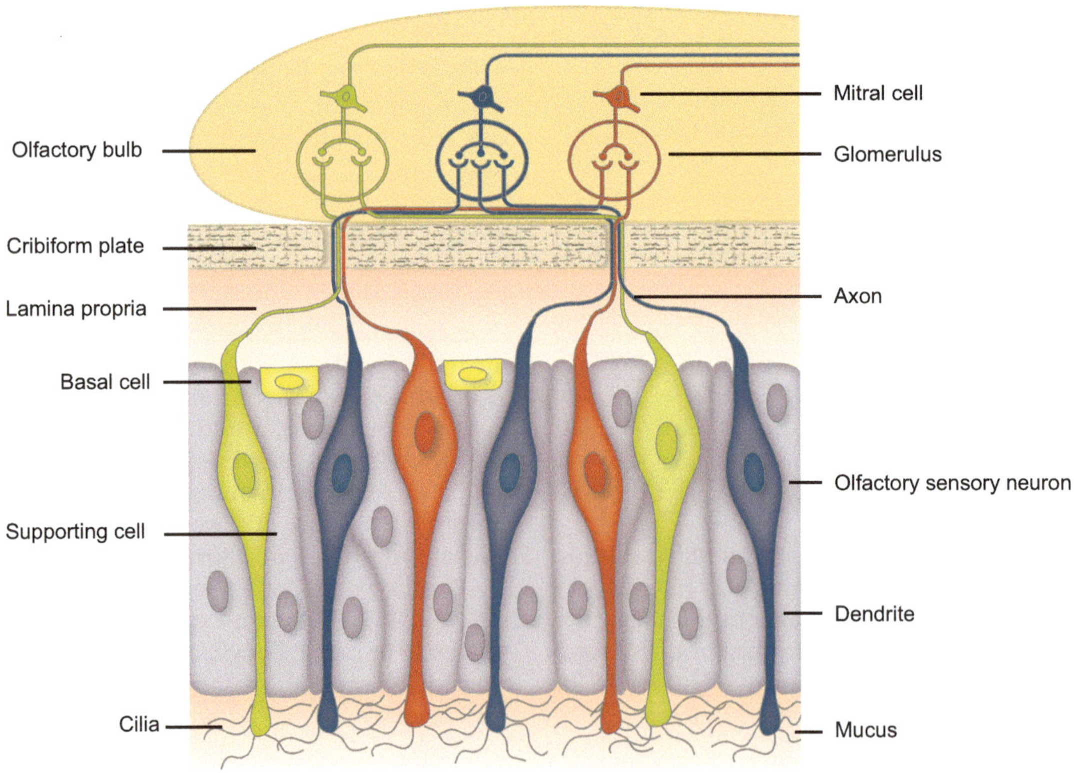

FIGURE 17.3 Schematic of the main olfactory epithelium showing the major cell types.

olfactory fila). These fila comprise the olfactory nerve, which synapses on mitral/tufted (M/T) cells in the main olfactory bulb (MOB). The OSNs are unique, in that they are the only neurons that directly contact both the external environment and the central nervous system; as such, they may constitute a route for the entry of foreign substances into the brain.

The cilia in the mucous layer of the MOE serve as the primary site for odor detection; these projections are covered in receptors specialized for odor recognition and contain the downstream machinery needed to activate the neuron following odor detection. The Bowman's glands, which lie in the lamina propria underneath the epithelial layer, and the supporting cells of the MOE secrete the mucous covering the MOE. The mucous layer is primarily aqueous, but the binding cavities of ORs are somewhat hydrophobic. Odorant-binding proteins, secreted in the mucous layer, increase the solubility of hydrophobic molecules, allowing them to reach the receptors buried in the mucous layer. The mucous layer also contains enzymes that act on compounds in the mucous. These enzymes break down odorant molecules, and inhibition of these enzymes has been shown to alter perceived odor character in humans.

The OSNs are surrounded and separated by the supporting sustentacular cells in the MOE. In addition, microvillar cells, which project microvilli into the mucous layer in lieu of cilia, are interspersed in the human MOE and constitute roughly 10% of the OSN population. Although they are assumed to be neurons, whether they project axons to the MOB is unknown.

Underlying all of the cell populations of the MOE is a niche of basal stem cells, which function to renew the OSN population that is undergoing constant turnover. OSNs are short-lived, having a lifespan of about 40 days in mice, and therefore must be regularly replaced by the basal stem cell population. The regular development and maturation of new OSNs is unique to the mammalian olfactory system, and is thought to reflect the damage it regularly sustains due to its exposure to the external environment. OSNs are frequently exposed to airborne toxins, viruses, harsh chemicals, and other sources of injury that induce their degeneration and replacement from the stem cell population. Current research is exploring the possibility of exploiting the regenerative capacity of the OSN population for therapies in other neuronal cell types.

Odor perception begins with the activation of the OSNs. An odor percept comprises many types of information, including identity, quality, intensity, and pleasantness. For odors to be discriminated, this information must be odor specific. The olfactory system accomplishes this task by the differential activation of the OSNs. Volatile molecules activate specific populations of OSNs, which vary in relation to both odor identity and odor concentration, and this pattern of activation determines the information sent to the brain for further processing.

The Odorant Receptors

Odorant receptors are expressed in the cilia of the OSNs and function to detect odors dissolved in the mucous overlying the MOE. Two different families of odorant receptors are expressed by the OSNs: the ORs and the trace amine-associated receptors (TAARs). The ORs are a large, multigene family of G-protein-coupled receptors (GPCRs) predicted to have a seven-transmembrane domain topology. In humans, the visual system uses three receptor types to represent color, while the gustatory system uses an order of magnitude more (~40), and the olfactory system uses two orders of magnitude more. A central question in the field is why so many receptors are required to represent the stimulus space. The size of the OR gene family is not only large, representing 2–3% of all genes in the genome, but also highly variable. Across mammals, OR gene families range in size from fewer than 30 in the dolphin to nearly 4000 in the elephant. Humans have roughly 800 ORs. This gene family is characterized by a high degree of pseudogenization, in which a gene is rendered nonfunctional due to genetic changes, the extent of which also varies across species. Almost half of the ORs in the human genome are pseudogenes, leaving about 400 ORs that code for functional receptors. Mammalian ORs are classified based on sequence homology into two broad classes: class I and II. Class I receptors are the only class of OR present in fish and are therefore thought to have evolved earlier and be more responsive to water-soluble ligands.

In the mammalian olfactory system, ORs are expressed monoallelically, leading to the view that each receptor type represents one input channel for the olfactory system. OSNs expressing the same type of OR are broadly dispersed throughout the MOE, although the distribution is not uniform. The MOE is divided into a number of broad zones, with any given receptor being randomly expressed within a given zone, but not necessarily expressed in all zones. The clearest distinction is zone I—class I receptors are essentially not expressed outside of this dorsal zone. The number of additional zones and the sharpness of the border between zones is less clear. While each population expressing a given receptor has a boundary, it is unclear if the boundaries of multiple populations overlap. Some receptors are expressed in all zones, and some only in one. No receptor is expressed in noncontiguous zones. It is unclear if this zonal structure is also present in humans. One theory behind this zonal expression pattern is that the MOE can perform chromatographic separation of an odor. A receptor expressed early in the airstream would be exposed to higher concentrations of hydrophilic odorants than a receptor expressed late in the airstream, which would allow the olfactory system to extract additional information about the composition of a mixture. Note that odors also enter the nasal cavity from the oral cavity, termed retronasal olfaction. This occurs primarily with food odors from the mouth, and the chromatographic separation is altered as a result of the change in the direction of airflow.

The OR gene family is characterized by tremendous sequence diversity in the transmembrane domains predicted to mediate ligand binding. This allows the receptor family the ability to respond to the incredibly diverse range of odorants encountered in the environment. In addition, this diversity mediates the tuning properties of these ORs, that is, their receptive range. ORs show a variety of tuning breadth—in one heterologous screen of 1200 structurally diverse compounds, about 8% of ORs were relatively broadly tuned, activated by more than 50 compounds, while 8% were narrowly tuned, responding to only one to three ligands. Though ORs may be optimally tuned to certain structural features, they are also able to respond suboptimally to related structural features. This observation has led to the hypothesis that the olfactory system uses a combinatorial mechanism to encode odor information (Fig. 17.4). Each OR has a distinct receptive range in its response to odorants, and each odorant activates specific, but overlapping, patterns of ORs. A given OR also has different response profiles to different odorant concentrations, responding to many more odors at high concentrations than low. Thus, odor information, including both identity and concentration, is encoded by the distinct pattern of OSN (and therefore OR) activation. The field has had limited success in matching mammalian ORs to odor ligands—ligands have been published for fewer than 15% of the approximately 400 intact human ORs. In other words, for over 85% of human ORs, there are currently no published ligands that activate that receptor. Given this lack of data, it is unclear what molecular features are detected by the OR repertoire.

A subset of OSNs express members of a second odorant receptor family, the TAARs, that respond preferentially to amines. Similar to the ORs, TAARs are seven-transmembrane GPCRs, but are more closely related to serotonin and dopamine receptors than to ORs. The TAAR family has only six intact members in humans and 15 in mice.

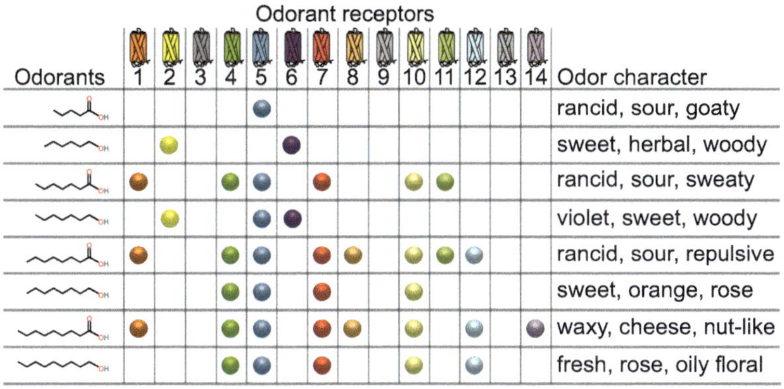

FIGURE 17.4 Different odorants are recognized by unique, but overlapping ensembles of odorant receptors. These codes are translated into diverse odor percepts, but the rules underlying this transformation are not known. *Adapted from Malnic B, Hirono J, Sato T, Buck LB. Combinatorial receptor codes for odors. Cell 1999;96(5):713–23.*

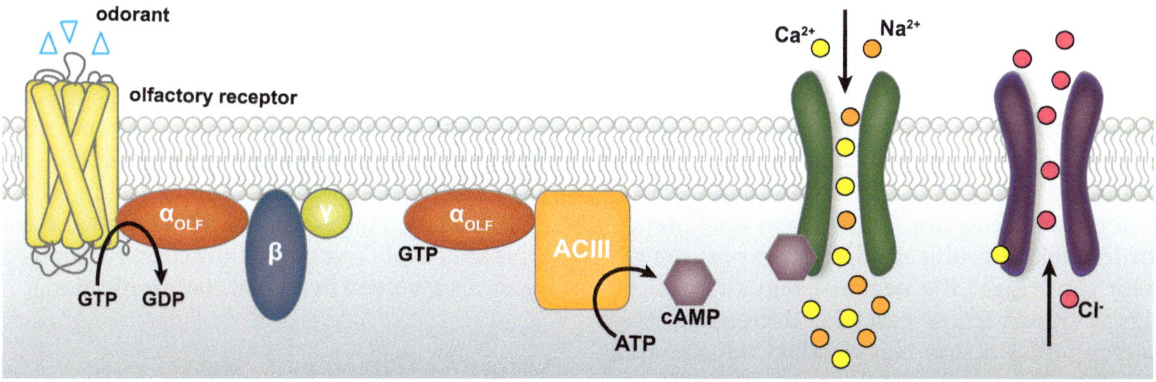

FIGURE 17.5 Schematic of the transduction cascade in the main olfactory system.

OSNs expressing TAARs show zonal expression patterns but are intermingled with each other in the MOE. The OSNs expressing TAARs share many similarities with OSNs expressing ORs: OSNs express only one type of TAAR, TAARs are expressed monoallelically, TAARs are localized to cilia, and TAARs project to glomeruli in the MOB. While OSNs expressing a given OR project to two glomeruli per bulb, OSNs expressing a given TAAR project to four to six glomeruli per bulb, in a specific region of the dorsal MOB. TAARs appear to respond primarily to primary and tertiary amines. In mice, loss of a Taar gene can have significant behavioral consequences. Mice innately avoid phenylethyl amine, a component of urine in carnivorous animals. With the genetic deletion of Taar4, the receptor for this odorant, mice no longer find phenylethyl amine aversive, demonstrating that this group of receptors may influence innate behavior.

Signal Transduction

Odorants binding to an OR or TAAR in the cilia of the OSN initiate a cyclic adenosine triphosphate (cAMP) signal transduction cascade, propagating depolarization along the neuron and culminating in the release of neurotransmitters at its synapse in the MOB (Fig. 17.5). The ORs and TAARs are GPCRs, a class of receptors which function through their association with a group of G-proteins (α,β,γ). Upon odorant binding, the OR undergoes a conformational change, allowing it to function as a guanine nucleotide exchange factor. The OR activates its associated G-protein—$G_{\alpha OLF}$—by exchanging its bound guanine diphosphate for guanine triphosphate. Activated $G_{\alpha OLF}$ can then dissociate from its β- and γ-subunits and activate adenylyl cyclase III (ACIII). ACIII stimulates cAMP production, which binds to and opens a cyclic-nucleotide-gated channel (CNGA2). CNGA2 preferentially allows the influx of positive calcium (Ca^{2+}) and sodium (Na^{2+}) ions from the mucous layer, initiating depolarization of the neuron. Depolarization is further increased when Ca^{2+} binds and activates Ca^{2+}-activated chloride (Cl^-) channels, causing the efflux of negative Cl^- ions from the cell. The influx of Ca^{2+} and Na^{2+} ions and efflux of Cl^- ions creates an action potential that is propagated down the length of the neuron, terminating in the release of neurotransmitter at the synaptic terminal in the MOB.

The olfactory system must rapidly turn off responses to a stimulus in order for the neuron to represent temporal information and prepare for the next stimulus. Prolonged exposure to an odor can trigger the negative regulation of the signal transduction cascade, decreasing odor perception, a phenomenon called odor adaptation. The increase in intracellular Ca^{2+} initiated by odor/OR binding is essential to the adaptation process. Ca^{2+} binds to the calcium-binding protein calmodulin, which subsequently binds CNGA2, decreasing its affinity for cAMP and decreasing Ca^{2+} infux in the cell. In addition, calmodulin activates calmodulin kinase II, resulting in the phosphorylation of ACIII, which decreases cAMP production, and a cyclic nucleotide phosphodiesterase, which degrades cAMP. In addition, odorants are also removed by mucus flow, absorption and uptake in the circulatory system, and degradation by enzymes in the mucus.

There are four additional subpopulations of OSNs in the MOE, each defined by a distinct transduction mechanism. The first is defined by the expression of the receptor guanylyl cyclase GC-D. These neurons have typical OSN morphology and are found in the MOE as well as the septal organ of Masera. They do not express several of the key transduction proteins found in the canonical OSNs, including ACIII, CNGA2, PDE1C2, and PDE4A. Instead, they express the cGMP-activated channel subunit CNGA3 and PDE2. These neurons project to a subset of atypical glomeruli that form a ring in the caudal MOB, known as the necklace glomeruli. These cells have been shown to respond to two peptides and CO_2, but the nature of the receptor and the function of this system are largely unknown.

The remaining three subpopulations each express a particular transient receptor potential (TRP) channel, namely TRPC2, TRPC6, and TRPM5. TRPC2-expressing cells appear to be restricted to the basal layer of the MOE, and may be immature neurons. TRPC6-expressing cells are bipolar microvillar cells in the apical MOE. TRPM5-expressing OSNs are enriched in the ventrolateral MOE and project to M/T cells in the MOB that target the vomeronasal amygdala. These OSNs selectively respond to low concentrations of chemicals involved in animal communication.

Accessory Olfactory System

Many species, excluding humans, have a secondary system, termed the accessory olfactory system, or vomeronasal system. The functional distinction between the main olfactory systems and the accessory olfactory system is not entirely clear—although some researchers suggested that the main olfactory system detects environmental odors and the accessory olfactory system detects pheromones, there is a large degree of overlap in terms of the stimuli they detect and the behaviors they mediate. The accessory olfactory system appears to be more narrowly specialized for social communication, as compromising the vomeronasal organ (VNO) leads to deficits in territorial and reproductive behaviors. The VNO is located at the base of the nasal septum and contains vomeronasal sensory neurons (VSNs). The VNO detects nonvolatile stimuli via a pumping mechanism that draws chemicals that contact the nose into the VNO by means of vascularized erectile tissue in the lateral side of the VNO lumen. In mouse, the neuroepithelium of the VNO is divided into two nonoverlapping layers. VSNs express three different receptor families: vomeronasal receptors 1 (V1Rs), vomeronasal receptors 2 (V2Rs), and formyl-peptide receptors (FPRs). The apical layer expresses V1Rs and four of the FPRs, while the basal layer expresses V2Rs and one of the FPRs. Although distinct from the OR and TAAR gene families, the V1Rs, V2Rs, and FPRs are GPCRs with a 7-transmembrane domain topology and exhibit sequence diversity, conferring the ability to respond to diverse molecules. The organization of the VNO is similar to the MOE, as a VNO neuron expresses only one type of V1R, V2R, or FPR. The axons of the VNO neurons form the vomeronasal nerve and project to the accessory olfactory bulb (AOB). The distinct layers are maintained in the AOB, with apical VSNs projecting to the rostral AOB, and basal VSNs projecting to the caudal AOB. Notably, certain species lack this anatomical segregation, and have a uniform vomeronasal epithelium. V1Rs appear to respond to small organic molecules, V2Rs respond to peptides and proteins, and FPRs respond to formylated peptides, which are produced by bacteria and organelles, such as mitochondria.

Trigeminal System

The oral and nasal cavities are innervated by neurons of the trigeminal system. Receptors are located on unmyelinated free nerve endings of the maxillary and ethmoid branches of the trigeminal nerve as well as solitary chemosensory cells that reach the surface of the nasal epithelium and synapse on trigeminal afferent nerve fibers. This subsystem responds to chemical and mechanical sensations of warmth, cooling, and irritation, and the primary function is thought to be avoidance of chemicals that can damage the tissues of the airway. In line with this, stimulation of the trigeminal receptors in the nasal cavity can elicit protective reflexes, such as a decrease in respiratory rate, and increase in nasal secretion, vasoconstriction, and sneezing. However, trigeminal stimulation contributes to our appreciation of some foods and beverages, such as the burn of peppers, coolness of menthol, pungency of onions, and stinging of carbonation. These sensations are largely mediated by the TRP channel family of receptors. Upon activation, the TRP channels allow the influx of Ca^{2+} and Na^{2+} ions, depolarizing the neuron and sending an electric signal along the TG nerve. The combination of pain, touch, and thermal sensations in the mucous membranes of the nose, mouth, and eyes is collectively referred to as chemesthesis.

Septal Organ of Masera

The septal organ of Masera (SO) is a small, isolated patch of sensory epithelium at the base of the nasal septum. The epithelium expresses a subset of ORs that are also expressed in the MOE. The epithelium of the SO is largely similar to the MOE, with ciliated OSNs, microvillar cells, supporting cells, and Bowman's glands. The majority of the OSNs express the transduction machinery found in OSNs in the MOE as well as a subset of ORs that are also expressed in the MOE. The OSNs project to a small subset of glomeruli in the posterior ventromedial region of the MOB, and some of these glomeruli also receive input from the MOE. The SO has not been found in primates, but is present in many mammals. The functional role of the SO is unknown.

The Gruenberg Ganglion

The GG is a collection of cells located at the entrance of the nasal cavity that express olfactory marker protein, a marker for mature OSNs. The cells are embedded in connective tissue and appear to have no access to the nasal lumen, but are activated by volatile alarm pheromones and cool temperatures. The axons of GG neurons project to necklace glomeruli in the caudal MOB, but to a population that does not receive innervation from GC-D neurons. The GG has been studied almost exclusively in mice, and it is unknown if it is present in other mammals.

OLFACTORY BULB

Odor information sent from the nose is first processed in the MOB, a bilateral structure located on the ventral surface of each frontal lobe above the cribiform plate (Fig. 17.6). The MOB is composed of six concentric layers featuring different neuronal populations. OSNs project from the MOE, pass through the cribiform plate, and synapse on the M/T neurons, the primary output pathway of the MOB. The most distinctive anatomical features of the MOB are the glomeruli, located in the glomerular layer, where the OSN axons contact dendrites of the M/T neurons and periglomerular cells. Each OSN sends one axon to one glomerulus. Although in the periphery, OSNs expressing a given type of OR are scattered in the MOE, at the level of the MOB these scattered neurons all converge to a few glomeruli. Thus, glomeruli are genetically specified channels that encode molecular features. Because each OSN expresses one type of OR and projects to only a few glomeruli, presynaptic glomerular-activation patterns are a direct reflection of the tuning properties of the OR. Thus, odorants tend to activate distinct but overlapping patterns of glomeruli. In addition, glomeruli receiving input from OSNs expressing a specific receptor are located symmetrically in the lateral and medial halves of the bulb within the same animal and found in roughly the same topographical location across different animals. Glomerular-activation patterns, therefore, serve to encode odorant information spatially across the MOB in a pattern that is stable across animals.

Within the glomeruli, the axons of the OSNs synapse with the dendrites of different neuronal populations, including M/T cells and the periglomerular interneurons (Fig. 17.6). M/T cells have different odor tuning and response properties, with tufted cells responding more quickly and with broader tuning than mitral cells. Each glomerulus has 20–50 M/T cells, which serve to relay information from the OSNs along the lateral olfactory tract (LOT) to the olfactory cortex. M/T cells interact with three main types of interneurons—periglomerular cells in the glomerular layer, short axon cells, and granule cells in the deeper external plexiform layer. The dendrites of the periglomerular cells, located around the glomeruli, form dendrodendritic synapses with the M/T cells, inhibiting their synaptic relay of OSN information. The granule cells, located in a deeper layer of the MOB, form dendrodendritic synapses with the secondary dendrites of M/T cells and function to negatively modulate their activity. In addition, the MOB receives input that can modulate activity based on the behavioral state of the animal using cholinergic, noradrenergic, and serotonergic systems.

The topography of OSN expression in the MOE is largely preserved in the MOB. The rodent MOB has roughly two glomeruli for every type of OR. Thus, there are two complete mirror-image glomerular maps in the bulb—one medial and one lateral. The medial glomeruli receive input from OSNs in the medial MOE, and lateral glomeruli receive input from OSNs in the lateral MOE. In addition, topographical information from zonal expression in the MOE is preserved in the bulb, with dorsal glomeruli receiving input from OSNs that project from more dorsal portions of the MOE. While the general organization of the human MOB is similar to that of mice, current evidence shows that humans have more than 5000 glomeruli on average, rather than the expected 800 (2 glomeruli for each of our approximately 400 intact ORs).

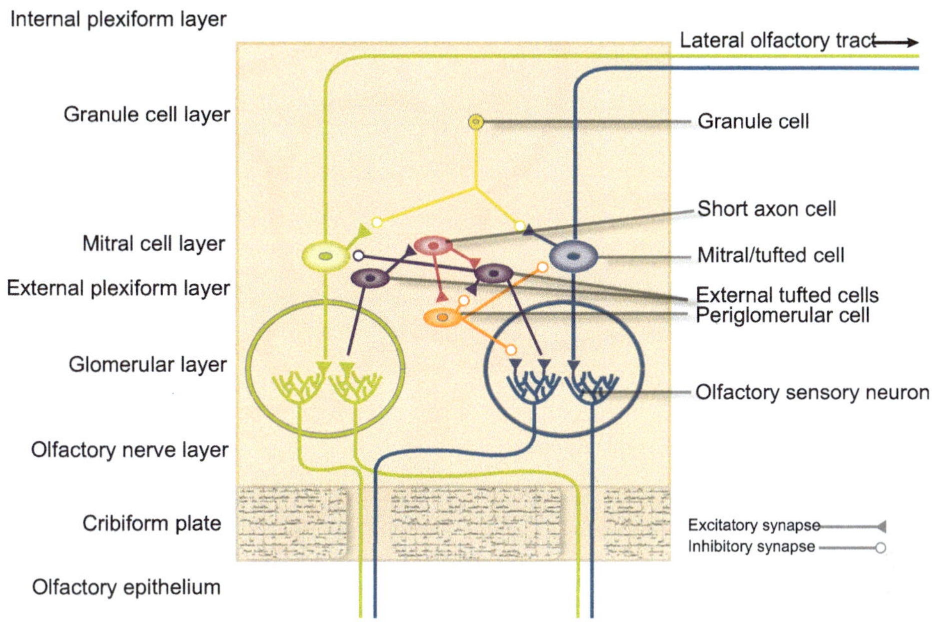

FIGURE 17.6 Circuit diagram of the main olfactory bulb.

CENTRAL OLFACTORY SYSTEM

In rodents, the M/T cells project directly, and mostly ipsilaterally, to the cortex. This is unusual as in all other mammalian sensory systems, information passes through the thalamus before reaching cortex. Downstream targets receiving input from the MOB include the piriform cortex, the anterior olfactory nucleus (AON), the nucleus of the LOT, periamygdaloid cortex, the olfactory tubercle, the anterior and posterior cortical amygdala, and the entorhinal cortex (Fig. 17.7). All of these structures, with the exception of the olfactory tubercle, also send descending projections back to the MOB. The MOB does not send centrifugal projections back to receptor neurons.

The target of the largest MOB projections is the piriform cortex, a three-layered paleocortex on the ventrolateral surface of the brain. The MOB projections to piriform cortex appear to lack topographical organization. This is striking as the olfactory system starts with the most structured wiring diagram in the nervous system, guiding millions of OSN axons to thousands of glomeruli with a convergence ratio of about 22,000 to one in rodents. The output of the MOB, however, then projects to random locations in the piriform cortex, scrambling this precise ordering. The piriform is roughly divided into anterior and posterior divisions. Anterior piriform cortex receives more afferent inputs from the MOB and fewer associational inputs than the posterior division, suggesting that anterior piriform encodes odor identity, while posterior piriform compares the odor to stored templates and detects and learns correlations between olfactory objects. The diffuse projections from the MOB to piriform cortex presumably allows the piriform cortex to create unified odor objects distinct from the component odors. Functionally, while glomeruli in the rat MOB show cross-habituation between structurally similar odorants, neurons in piriform cortex

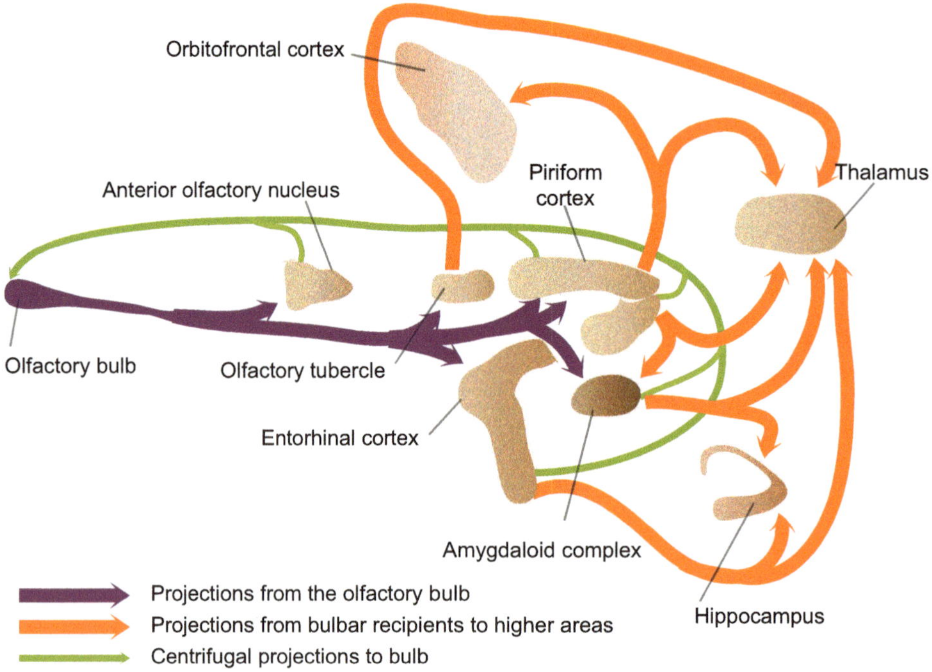

FIGURE 17.7 Projections between the olfactory bulb and higher areas.

show better discrimination. Similarly, when a single component of a complex mixture is removed, responses are more highly correlated with the original mixture in the piriform cortex than in the MOB. In humans, anterior piriform habituates to odors with the same structure, while posterior piriform habituates to odors with the same odor quality. Piriform cortex also responds to nonolfactory information, such as trial initiation, reward, and respiration, suggesting that it incorporates top–down information and does more than simple feature extraction. Given that the piriform cortex is involved in associative memory, the MOB may function as the feature extractor, while the anterior piriform creates representations of complex odors, and the posterior piriform functions as association cortex, matching odor representations from the anterior piriform with behavioral information from other brain areas.

The second largest MOB projection targets the AON. Although the majority of the AON projects back to both olfactory bulbs, one region has a very structured projection pattern. The pars externa region of the AON (AONpe) receives projections from the ipsilateral MOB and projects to the contralateral MOB. More precisely, two glomeruli receiving input from the same receptor type, but in different nostrils, are connected to each other via the AONpe. This provides a substrate for interhemispheric comparisons between isofunctional regions of the two MOBs. The two nostrils are connected through the back of the mouth, but are exposed to distinct samples during orthonasal (in through the nostrils) breathing. The two nostrils typically have distinct flow rates due to a swelling in the ethmoid sinus that changes from one side to the other roughly every 4h in humans. The AONpe is poised to compare samples measured at the two different flow rates. The AON is also notable in that it receives relatively little descending input.

Both the cortical amygdala and olfactory tubercle preserve crude topography from the MOB; with some exceptions, dorsal glomeruli project to ventral regions, and ventral glomeruli project to dorsal regions. The olfactory tubercle also receives input from the piriform cortex and the cortical amygdala. The olfactory tubercle has projection patterns that resemble the nucleus accumbens; it receives input from the prefrontal cortex, hippocampus, and amygdala and projects to the ventral striatum, pallidum, thalamus, hypothalamus, and brain stem nuclei that control feeding, drinking, and locomotion. The olfactory tubercle does not send associational axons to other olfactory areas and the circuits within the tubercle are largely uncharacterized.

The orbitofrontal cortex is often referred to as the secondary olfactory cortex, and receives both direct connections from the piriform cortex and indirect connections through the dorsomedial nucleus of the thalamus. The direct projections from piriform cortex are combined with input from other sensory modalities as well as information regarding the animal's behavioral state. For example, responses to food odors in this region decrease when the animal is fed to satiety. Lesions in this region, but not other prefrontal areas, result in impairment in odorant discrimination and identification. The orbitofrontal cortex encodes value as well as odor identity.

OLFACTORY CODING

In the current model of the olfactory system, odorant features are represented by a combinatorial code (Fig. 17.4). Each odorant has a combination of features that are extracted by the receptors and represented as distinct glomeruli in the MOB. Although the molecular features of an odorant are thought to be a key determinant of olfactory perception, the rules governing this relationship remain unknown. The correlation between perceptual properties, molecular structure, and neural response (or any combination of these) is not well understood.

A given receptor also has different response profiles to different odorant concentrations, responding to many more odors at high concentrations than at low concentrations. Thus, the activation pattern across receptors encodes both the odor identity and also the concentration. How the olfactory system reconstructs these two features from the activation pattern is unclear. Indeed, there are few, if any, explicit predictions relating OR activity patterns to olfactory perception. Although a number of researchers have attempted to develop quantitative structure–activity relationships for odorants, these models typically focus on a particular quality (what structural features lead to a "woody" percept), rather than mapping odor space globally. One challenge is that even when receptor responses are known for hundreds of odors, it is still difficult to predict how it will respond to a new odor. It is important to note that the features that the receptors extract may not be straightforward: there may be no receptor that detects if an odor has an ester group, or a "musk" receptor that represents a salient perceptual category. Instead, the receptor may require a complex configuration of these descriptors that seem to be elemental from the perspective of organic chemistry or perfumers and flavorists.

Olfactory information is high dimensional—in humans, about 400 input channels transmit features of the olfactory stimulus. How is this information organized in the brain? The olfactory system may use unique coding principles to make sense of the stimulus. First, while most organisms are excellent at learning a complex mixture and discriminating it from other complex mixtures, the olfactory system is poor at breaking down a mixture into its components. Once a mixture has four or more components, humans are typically quite poor at determining the composition of the mixture. Second, olfactory transduction is much slower than in vision or audition due to the timing of the sniff cycle. This allows the system to use temporal encoding, in combination with spatial encoding, to increase the capacity of the system.

OLFACTORY DYSFUNCTION

Humans display tremendous variability in olfactory perception including differences in odor detection threshold, quality, intensity, and pleasantness. Olfactory ability can be affected by many factors, including genetics, gender, age, environment, and health, and can vary for specific odors or for overall olfactory acuity. Specific anosmias are the inability to perceive a certain odor, and are relatively common in the population. For example, up to 30% of the population is unable to smell the odor androstenone, the primary component of boar taint, and about 6% of the population has a specific anosmia to isovaleric acid, a component of body odor.

Detection thresholds, the lowest concentration of an odor that can be detected, can vary by several orders of magnitude in the population. The perceived quality and pleasantness of some odors also differs greatly between individuals. Reduced sensitivity to an odor is called hyposmia and increased sensitivity is termed hyperosmia. This perceptual variation occurs, in part, because individuals have unique repertoires of ORs, which mediate odor perception. ORs are characterized by a tremendous amount of genetic and functional variation. The roughly 400 functional genes in the human genome each have on average five segregating variants, including pseudogenes, copy number variations, and single nucleotide polymorphisms (SNPs) that are predicted to alter OR function. In addition, any two humans differ functionally at roughly 30% of their OR genes. Genetic changes have been directly linked to differences in olfactory acuity. For example, two amino acid changes in the olfactory receptor OR7D4 are associated with the specific anosmia for androstenone.

In addition to changes that affect the perception of a single odor, a proportion of the population experiences a complete loss of smell, termed general anosmia. This condition affects over 6 million Americans, of which about 3% are congenital (termed congenital anosmia, CA). The primary genetic cause of CA is Kallmann syndrome, a condition characterized by the failure to start or complete puberty, hypogonadism, and infertility in addition to smell loss. In families without Kallmann syndrome, genetic studies of one family with inherited CA have demonstrated mutations in CNGA2, a member of the olfactory signal transduction pathway, but the etiology in other CA families is unknown. Anosmia can also be acquired following a severe upper respiratory infection or head trauma damaging the frontal lobe or severing the olfactory nerve.

Partial or total olfactory dysfunction can result from a variety of other demographic, environmental, and health factors. Olfactory ability declines with age, although the degree of decline varies among individuals. Over 70% of individuals over the age of 80 are reported to have smell dysfunction, attributed to changes in the anatomy of the olfactory system over time and the long-term exposure to airborne viruses and toxins. Damage from viral infection and chronic exposure to chemicals, such as heavy metals, can irreversibly decrease olfactory ability by injuring the olfactory epithelium. In addition, the OSNs may serve as an entry point for these agents into the brain. Finally, significant decreases in olfactory acuity are reported in people with neurological disorders, such as Parkinson's disease (PD) and Alzheimer's disease (AD). Because these smell deficits appear in the preclinical stage of PD and AD development before the onset of other neurological symptoms, they may serve as an early warning sign for these disorders. In summary, while variation in odor perception is characteristic of the human population, a decline in general olfactory ability can be symptomatic of aging, chronic damage to the OSNs, or neurological disease onset.

Further Reading

[1] Buck L, Axel R. A novel multigene family may encode odorant receptors: a molecular basis for odor recognition. Cell 1991;65(1):175–87.

[2] Mombaerts P, Wang F, Dulac C, Chao SK, Nemes A, Mendelsohn M, et al. Visualizing an olfactory sensory map. Cell 1996;87(4):675–86.

[3] Malnic B, Hirono J, Sato T, Buck LB. Combinatorial receptor codes for odors. Cell 1999;96(5):713–23.

[4] Haddad R, Khan R, Takahashi YK, Mori K, Harel D, Sobel N. A metric for odorant comparison. Nat Methods 2008. http://doi.org/10.1038/nmeth.1197.

CHAPTER

18

Intranasal Trigeminal Chemoreception

S. Al Aïn[1], *J.A. Frasnelli*[1,2]

[1]Université du Québec à Trois-Rivières, Trois-Rivières, QC, Canada; [2]CÉAMS, Research Center of the Hôpital du Sacré Cœur de Montréal, Montréal, QC, Canada

INTRODUCTION

Throughout life, the ability to perceive and respond to harmful or noxious thermal, mechanical, and chemical cues in the environment is critical for survival. Like the mouth, the nose is the source of stimulation for latter multimodal sensory integration, such as pressure, temperature, irritation, and humidity. Chemosensory processing is mainly mediated by three neural systems, namely the gustatory, the olfactory, and the trigeminal system. Gustatory sensations stem from taste cells within the taste buds in the oral cavity, whereas olfactory sensations derive from olfactory nerves in the nasal cavity. The trigeminal system is a third chemosensory system next to these two, which conveys sensations such as stinging, burning, or pungent evoked by chemical substances. These sensations arise from interactions of the chemical substances with specific chemoreceptors, in analogy to smell and taste. While gustatory and olfactory receptors are limited to the oral and nasal cavity, respectively, trigeminal receptors can be found in both anatomical areas. The two chemosensory systems arising from the nasal cavity—olfactory and trigeminal system—are strongly interconnected at different levels, and these interactions contribute substantially to the global perception of chemosensory sensations. In fact, the majority of odorous substances stimulate both the olfactory and the trigeminal system. Accordingly, disentangling the implication of the olfactory and trigeminal processing is a relatively delicate enterprise.

Contrary to the olfactory system, which has received much more attention, our knowledge of the intranasal trigeminal chemosensory system and its relationship with olfaction is far from complete. This chapter reviews recent findings on intranasal chemosensory function of the trigeminal nerve in the human adult and its relation to olfaction. It is divided into seven sections covering (1) definitions, (2) neuroanatomy and physiology, (3) trigeminal pathways, (4) interactions between olfactory and trigeminal systems, (5) assessment, (6) trigeminal responses, and (7) factors influencing trigeminal perception. Finally, possible applications of trigeminal stimulation are proposed.

Conn's Translational Neuroscience
http://dx.doi.org/10.1016/B978-0-12-802381-5.00030-0

DEFINITIONS

Some key terms need to be defined to clarify their meanings to the reader. First, an *odorant* refers to any chemical substance that stimulates the olfactory system. As a consequence, an *odor* designs the sensation resulting from the stimulation of the olfactory system by an odorant. Second, a *trigeminal stimulus* refers to any chemical substance that stimulates the trigeminal system, sometimes referred to as *irritants*. Accordingly, a *trigeminal sensation* corresponds to the sensation induced by the stimulation of the trigeminal system via a trigeminal stimulus. Third, a *mixed olfactory/trigeminal stimulus* or *bimodal stimulus* refers to any chemical substance that stimulates both the olfactory and trigeminal systems.

A normal sense of smell is described by the term *normosmia*, whereas *hyposmia means* reduced olfactory function. The term *anosmia*, finally, describes a complete loss of olfactory function.

NEUROANATOMY AND PHYSIOLOGY

In different mammal species, up to four different chemosensory systems can be found within the nasal cavity. Next to the main olfactory system, they comprise the trigeminal system, the vomeronasal system (VNO), and the terminal system. The olfactory system detects chemical volatiles (odorants), generally at very low concentration, and provides several functions for humans, including food detection, danger avoidance, or social communication. The VNO is the main organ for the reception and processing of pheromones in many species including some mammals. In humans, this system seems not to be functional and only remnants of the VNO can be observed in humans. They appear to have no function in chemosensory perception and processing. Finally, the sensory function of the terminal system may be tuned to detect pheromone in human thought to play a role in reproductive behavior although this remains highly controversial. The present chapter focuses on the trigeminal system, and we begin by describing the trigeminal nerve's basic anatomy.

Neuroanatomy

The trigeminal nerve is the fifth cranial nerve (CNV) and the largest nerve among the 12 cranial nerves. Its free trigeminal nerve endings innervate the facial mucosa and the skin and convey sensory information arising from contact of a stimulus (physical, thermal, or chemical) with the facial skin or the face's mucosa. The trigeminal nerve's main function is the sensory innervation of the face, such as the perception of touch or pain; it also detects and conveys chemosensory signal generated from ingested or inhaled substances. A very minor portion of the trigeminal nerves' fibers contribute to the motor functioning, such as the control for the muscles responsible for masticating. The trigeminal nerve is therefore identified as a mixed sensory/motor nerve.

Sensory and Somatosensory Branches

The trigeminal nerve has three major branches, which is the actual origin of the name trigeminal. The ophthalmic (CNV1) and maxillary (CNV2) nerves are purely sensory while the mandibular nerve (CNV3) has sensory and motor functions. The nasal mucosa is innervated by branches of the ophthalmic and the maxillary nerves, whereas branches of the maxillary and the mandibular nerves innervate the oral mucosa.

Regarding the nasal mucosa, the nerves stemming from the ophthalmic branch such as the anterior ethmoidal nerve and the infraorbital nerve innervate the anterior part of the mucus membrane from the nasal cavities and to the nasal pyramid while the nerves coming from the maxillary branch such as the posterior superior medial nasal nerve and the nasopalatine nerve supply the posterior part of the nasal cavity, comprising the septum and the superior, middle, and inferior turbinates. Interestingly, the anterior part of the nasal cavity is more sensible to chemosensory than mechanical stimuli, whereas posterior part of the nasal cavity reacts equally to both kind of stimulation. Therefore, the sensitivity of the nasal mucosa varies as a function of the nature of the stimuli and the location, suggesting the existence of different types of trigeminal nerve fibers specifically distributed within the nasal epithelium.

Trigeminal Nerve Fibers

As aforementioned, the trigeminal nerve exhibits mechanosensory and chemosensory fibers. Mechanosensory fibers are represented by large fast-conducting Aβ-fibers. Chemosensory fibers refer to thin and slow-conducting unmyelinated C-fibers and thin and fast-conducting myelinated Aδ-fibers that are involved in thermoreception (temperature perception) and nociception. Different types of chemosensory trigeminal fibers mediate different trigeminal sensations; for instance, C-fibers seem to preferentially mediate dull burning sensations, whereas Aδ-fibers are involved in sharp stinging sensations. The free endings of these C- and Aδ-fibers carry receptors, which can be activated by different types of stimuli, such as temperature, pressure, humidity, and certain chemical substances, in a specific manner.

Receptors

The intranasal trigeminal system is activated via polymodal nociceptors located on free nerve endings or solitary chemoreceptor cells located in the oral and nasal mucosa. Some of them are known; it is, however, possible that additional trigeminal receptors exist although they have not been identified so far.

Polymodal Nociceptors

The activation of specific polymodal nociceptors, via a ligand–receptor interaction, induces distinct sensations in term of touch, temperature, and nociception. With regard to nasal chemoreception, four trigeminal receptors belonging to the subfamily of transient receptor potential (TRP) receptors have been identified so far. TRP receptors have six transmembrane regions and are relatively nonselectively permeable to cations (see Table 18.1 for an overview):

1. **TRPV1**: At its discovery, this receptor was called VR1 (vanilloid receptor 1). It responds to capsaicin, camphor, eugenol, piperine, acids, and other substances as well as heat above 43°C and evokes burning, stinging, and tickling sensations.
2. **TRPV3**: Eugenol, thymol, and similar substances as well as temperatures above

TABLE 18.1 Typical Stimuli for Four TRP Receptors

	TRPA1	TRPM8	TRPV1	TRPV3
Temperature range	<17°C	<28°C	>43°C	>39°C
Agonist (found in)	Allicin (garlic)	Menthol (peppermint)	Capsaicin (hot peppers)	Thymol (thyme)
	Allyl-thio-isocyanate (horseradish)	Eucalyptol (eucalyptus)	Piperine (pepper)	Borneol (camphor)
Evoked sensation	Dull burning	Cooling, refreshing	Hot, burning, stinging	Warming

Temperature range: needed to activate the receptor. Chemical stimuli, agonist: at the respective receptor; (found in): origin of the substance in herbs and spices. Evoked sensation: sensation that is evoked if humans are exposed to the agonists.

39°C activate this ion channel and produce a sensation of warmth.

3. **TRPM8**: This ion channel responds to exposition to eucalyptol, menthol, and other cooling substances, as well as cool temperatures below 28°C and is formerly known as cold and menthol receptor (CMR1). Its activation elicits a sensation of freshness and cooling.
4. **TRPA1**: It is activated cinnamaldehyde, allyl thioisocyanate (mustard oil), or various irritant gas and cold temperature below 17°C; its activation induces a stinging and burning sensation.

Interestingly, TRPV1 and TRPA1 receptors are often coexpressed in the same sensory neurons, influencing each other's activation patterns; in sharp contrast to this, TRPA1 and TRPM8 receptors are exclusively expressed in different sensory neurons. This picture is further complicated by complex agonist–receptor interactions as some molecules can bind with different chemoreceptors; for example, carvacrol is an agonist for both TRPA1 and TRPV3. It is therefore difficult to predict the exact perceptual property of a given stimulus.

In addition to the TRP channels, non-TRP receptors have been described on trigeminal nerve endings, such as the ion channel sensible to acids (**ASIC/DRASIC**) that play an important role in the sensation of cutaneous acid-induced pain, the purinergic receptor (**P2X**), and the nicotinic family receptors, especially nicotinic acetylcholine. Finally, the nasal trigeminal nerves carry histamine H1 receptors. The exact sensation associated with the activation of these receptors has not yet been described. Nicotine typically is described to evoke a dull burning sensation; this may be conveyed via the nicotinic receptor family.

Solitary Chemoreceptor Cells

The existence of solitary chemoreceptor cells located in the anterior respiratory epithelium of the nasal cavity and their function has been well documented in nonhuman species, such as rodents. In turn, little is known in humans. In rodents, these cells establish synaptic connection with trigeminal afferent nerve fibers on the surface of the nasal epithelium. They express T2R bitter-taste receptors, gustducin—a G-protein subunit expressed in taste receptor cells—and TRPM5, a calcium-activated nonselective cation channel belonging to the family of TRP channels, which is involved in taste transduction and in the sweet, bitter, and umami perception. These cells may contribute to perceptions mediated by the trigeminal system although it is not known which exact sensations may arise from the activation of solitary chemosensory cells in the nose.

Function

The intranasal trigeminal system's primary function is to protect the individual from potential dangerous/noxious chemical substances via several physiological and neural mechanisms, such as (1) protective mucosal responses, (2) brain stem reflex circuits, (3) breathing control, and (4) olfactory–trigeminal interactions. It also plays an important role in (5) perception of food quality and pleasure associated with nutrition.

Protective Mucosal Responses

Neurons of the trigeminal system mediate protective responses on the nasal mucosa. Indeed, nociceptive, parasympathetic, and sympathetic nerves innervate nasal mucosa and regulate glandular and vascular responses such as increasing secretions from tears, saliva, and nasal mucosa, which are relevant for cleaning and humidifying ambient air, by the efferent axon response mechanism. For example, people typically report increased production of nasal mucus when they are eating very spicy substances such as Mexican food.

Brain Stem Reflex Circuits

Neurons of the intranasal trigeminal system project to the brain stem and are the afferent branch of reflex circuits. Via these circuits, strong trigeminal stimulation leads, next to painful sensations, to the elicitation of protective reflexes, such as reflex apnea, coughing, sneezing, gagging, or vomiting. This can be observed during campfire smoke, which induces coughing and, in higher concentrations, reflex apnea.

Breathing Control

The intranasal trigeminal system plays an important role in breathing control. In fact, breathing patterns are modulated/affected as a function of air quality such as irritant content concentration: the more the irritant, the slower the respiratory rhythm. While we are probably not aware of it, we could observe this when inhaling campfire smoke in smaller concentrations.

Olfactory–Trigeminal Interactions

The intranasal trigeminal system further interacts with the olfactory system in order to provide sensory information about inhaled air in terms of quality and composition. Again, this can be observed when we are exposed to strong odorants, which are considered irritant if they stimulate the trigeminal system in addition to the olfactory system. Such stimulation may be caused by the exhaust of factories or pig farms.

Perception of Food Quality and Pleasure Associated With Nutrition

Finally, trigeminal perceptions play a role in quality of life, especially regarding the appreciation of food and drinks. In fact, some foods are known to be particularly strong trigeminal stimuli, such as wasabi, chili peppers, or curry and play a dominant role in Ethnic cuisines, such as Japanese, Mexican, or Indian food.

CHEMOSENSORY PATHWAYS

Stimuli

As we say later, the olfactory and trigeminal systems are intimately connected. It is therefore important to distinguish among (1) *selective/pure trigeminal stimuli*, (2) *selective/pure odorants*, and (3) mixed olfactory–trigeminal or *bimodal stimuli*, which stimulate both the olfactory and the trigeminal nerve. In fact, only few substances belong to category 1 and are completely odorless trigeminal stimuli. The substance having the most widespread use is carbon dioxide (CO_2), an odorless irritant gas that exclusively stimulates the chemosensory trigeminal nerve; it evokes a stinging sensation. In some animals (mosquitoes, frogs), however, CO_2 receptors are found in the olfactory system; these receptors have not been observed in humans. Another pure trigeminal stimulus is capsaicin, the pungent ingredient of hot chili peppers. This substance, however, is not volatile and therefore mainly stimulates the oral mucosa in the context of food consumption. Similarly, only few substances belong to category 2 and are truly stimulating the olfactory system without any trigeminal costimulation. Among these, phenyl ethanol (which has a smell of rose), hydrogen sulfide (H_2S, which smells of rotten eggs), or vanillin (vanilla odor) are commonly used as pure olfactory stimuli. However, when delivered at high concentrations, phenyl ethanol also stimulates the trigeminal system. Although this has not yet been observed for the other two substances, it cannot be excluded that they also would activate the trigeminal system in higher concentrations.

Therefore, the vast majority of volatile substances belongs to the third category, at least at high concentration levels, as they stimulate both the olfactory (cranial nerve I) and the trigeminal (cranial nerve V) nerves. In other words, by increasing concentration, virtually all odorants will become trigeminal stimuli (or better, mixed olfactory–trigeminal stimuli), and in turn, by reducing their concentration, most mixed olfactory–trigeminal stimuli will become pure olfactory stimuli. For some substances, the difference in concentration between trigeminal and olfactory threshold is smaller (eg, ammonia) than for others (eg, phenyl ethanol).

Peripheral and Central Processing

We have already learned trigeminal sensations arise from the interactions of volatile substances with specific chemoreceptors located in endings of the trigeminal nerve. The same nerves convey not only chemosensory but also somatosensory information to central structures. The incoming fibers are pseudo-unipolar sensory fibers whose cell bodies are located in the Gasserian ganglion (trigeminal ganglion and semilunar ganglion) in Meckel's cave. The fibers' central portion enters the central nervous system at the brain stem between pons and the middle cerebellar peduncle. They project directly to the subnuclei of the trigeminal sensory nucleus that extends from the rostral spinal cord to the midbrain. Here they synapse; the information is then relayed to the contralateral lateral and the medial thalamic nuclei. From lateral thalamic structures, neurons project to the primary (SI) and secondary (SII) somatosensory cortex, the brain regions involved in sensations such as pain, touch, and heat. Somatotopy is preserved throughout the trigeminal pathway, as sensory information transmitted by the ophthalmic nerve is processed in the inferior portion of the postcentral gyrus, while that of the mandibular nerve is conveyed in the central gyrus. The medial thalamic nuclei neurons, however, project to the prefrontal cortex, insula, cingulate gyrus, brain stem, and to the limbic system. Together with SI and SII, these areas belong to the pain matrix. The activity in this network represents the intensity and unpleasantness of painful perceptions and is responsible amongst others for the emotional aspects of pain perception.

Since mid-2000s, a growing body of research has investigated neural circuit activated by intranasal trigeminal stimulation using noninvasive methods. Brain-imaging studies have reported that trigeminal stimulation typically leads to brain activation which partially overlaps neural pathways involved in the processing of (1) *somatosensory stimuli*, such as brain stem, thalamus, SI/SII, anterior cingulate, and (2) *pure olfactory stimuli*, with activation of the piriform cortex, the medial orbitofrontal cortex, periinsular regions. Much as olfactory processing, pure trigeminal stimuli also produced stronger right-sided activation following by bilateral presentation, indicating that the right hemisphere is involved to the processing of chemosensory information. This predominance of the right hemisphere therefore represents a common characteristic of both trigeminal chemosensory and olfactory systems and is not found for somatosensory trigeminal stimuli.

Explanations for these observations have been put forward; for example, some suggest the existence of connections between the somatosensory pathway and the olfactory processing brain regions. Such a connection was indeed found in rats: here some trigeminal ganglion cells with sensory nerve endings in the nasal epithelium possess branches reaching the olfactory bulb, thus reentering the central nervous system with their peripheral ending. Therefore, irritant substances may act on the processing of olfactory stimuli through this collateral innervation of the epithelium and the bulb. Such collaterals have not yet been found in humans. However, it is far from understood via which channels trigeminal stimuli lead to activation of olfactory brain areas.

INTERACTIONS BETWEEN OLFACTORY AND TRIGEMINAL SYSTEM

Associations Between Olfactory and Trigeminal System

As previously addressed, interactions between the trigeminal and the olfactory system have been evidenced insofar as almost all stimuli activate both systems. In addition to this, interactions between both chemosensory systems take place in the mucosa as well as in the central nervous system. As a consequence of these interactions, information from one system can suppress or enhance the activity in the other, depending on the quality and the nature of the stimuli.

Periphery

Only few studies investigated the interaction in the periphery between both sensory systems. When measuring electrophysiological responses in the nasal mucosa, patients with congenital or acquired anosmia showed larger responses to the pure trigeminal stimulus CO_2 compared to individuals with a normal sense of smell. These observations suggest a reactive upregulation of trigeminal sensitivity when olfactory function is lost. However, the exact mechanisms remain to be elucidated. For example, an increase of trigeminal receptor density, or an increase of trigeminal receptor responsiveness has been put forward as possible causes of increased peripheral trigeminal susceptibility in olfactory loss.

In summary, individuals with loss of olfactory function exhibit increased peripheral trigeminal responsiveness.

Central Nervous System

Much more data is available on the interactions between both sensory systems in central nervous processing areas. One of the methods used to measure neuronal activity in the central nervous system is the assessment of electrical activity of the cortex by means of EEG-derived event-related potentials (ERPs). Patients with congenital anosmia show similar electrical responses to trigeminal stimuli compared to participants with a normal sense of smell, whereas individuals with acquired anosmia exhibit smaller responses. This suggests that the trigeminal system depends on a functioning olfactory system, but not so in patients with congenital anosmia. This is corroborated by magnetic resonance imaging of the brain: pure trigeminal chemosensory stimuli activate brain regions which are usually involved in the processing of olfactory stimuli, including orbitofrontal, piriform, and insular cortex, and they do so mainly in the right hemisphere. In patients with acquired anosmia, these activations are weaker. In summary, individuals with acquired anosmia have weaker responses, but individuals with congenital anosmia have normal responses, similar to individuals with normal olfactory function. These findings seemed paradox until one study showed that the effect in acquired anosmia wears out over time: In fact, the trigeminal system of individuals with acquired olfactory dysfunction recovered sensitivity over time but without reaching normal levels; this suggests the existence of adaptive mechanisms.

In summary, loss of olfactory function affects trigeminal sensitivity negatively. Over time, this effects wears out and is weaker in congenital anosmia.

Behavior

The effects of interactions between olfactory and trigeminal systems on behavioral measures can be studied in (1) individuals with *normosmia* and in (2) patients with *hyposmia or anosmia.*

NORMOSMIA

Several lines of evidence support the existence of intimate olfactory–trigeminal interactions. Studies on individuals with normal olfactory function showed that sensation stemming from stimulation of one of the chemosensory channels can be affected if another chemical sense is stimulated simultaneously. Generally speaking, trigeminal co-stimulation reduces or inhibits olfactory sensations; in turn, olfactory stimulation enhances sensitivity to trigeminally mediated sensations. For example, in a binary mixture of the pure trigeminal stimulus CO_2 and the pure olfactory stimulus H_2S, the stinging evoked by CO_2 suppresses the smell of H_2S, which is perceived weaker in the mixture than when presented alone. In turn, the presence of H_2S increases the trigeminal sensation produced by CO_2, compared to the sensation when CO_2 is presented alone. These interactions are not as straightforward when using mixed olfactory trigeminal stimuli. When either CO_2 or H_2S were combined with the mixed olfactory–trigeminal stimulus carvone, the intensity of both CO_2 and H_2S was suppressed, whereas that of carvone was increased.

These findings suggest the existence of a relative hierarchy between systems with a dominance of (1) trigeminal sensation over olfactory sensation, and (2) mixed stimulation over either system alone.

HYPOSMIA AND ANOSMIA

Several studies on individuals with olfactory dysfunction have demonstrated that loss of one sense is usually accompanied by a reduced sensitivity of the other chemical senses, similar to what is known about gustatory sensitivity in patients with anosmia. Therefore, patients with both acquired and congenital olfactory dysfunction exhibit lower trigeminal sensitivity; typically, patients with acquired anosmia have weaker results than individuals with congenital anosmia. This is consistently found with psychophysical methods, and corroborates the electrophysiological measures and brain-imaging techniques mentioned before. This underlines again the interaction between both the chemosensory systems.

In addition to this, odorants rated as more intense by individuals with normal olfactory function are also the ones which are best detected by individuals with olfactory dysfunction. This suggests that perceived odor intensity is, at least partially, caused by trigeminal stimulation. In other words, olfactory intensity ratings are strongly influenced by the trigeminal component of the stimulant.

Model of Olfactory Trigeminal Interaction

Based on these findings, we proposed a model with mixed sensory adaptation/compensation, to explain the paradox finding of increased peripheral trigeminal susceptibility paralleled by reduced central responsiveness of the trigeminal system in acquired anosmia. The model stipulates that, in healthy subjects, (1) peripheral trigeminal susceptibility is inhibited by olfactory input due to adaptive processes; further, (2) the olfactory system is involved in an amplification of trigeminal signal at the central nervous level, even when there is no concomitant olfactory stimulation. This amplification does not work in the case of acquired olfactory dysfunction, reducing the signal strength in the trigeminal system. As a result, compensatory processes increase trigeminal susceptibility on a mucosal level. In other words, according to this model trigeminal susceptibility is continuously downregulated in the case of an intact sense of smell. In patients suffering from acquired anosmia, this excitatory circuit of intrabulbar trigeminal collaterals is not functioning, resulting in compensatory upregulation and thus in increased peripheral responsiveness of the trigeminal system.

While this model remains hypothetical, it puts forward several testable hypotheses. For example, it requires a direct interaction in the periphery of the olfactory and the trigeminal system. In fact, it may occur via the constant excitation of trigeminal fibers that reenter the central nervous system and innervate the olfactory bulb, so-called intrabulbar trigeminal collaterals.

In summary, a model of mixed habituation/adaptation effects has been put forward to explain the prima facie paradox findings of effects of olfactory loss on trigeminal sensitivity.

Dissociation Between the Chemosensory and the Somatosensory Portion of the Trigeminal Nerve

Interestingly, the interaction between the olfactory system and the trigeminal nerve seems to be limited to the chemosensory portion of the latter. For example, the ability to detect somatosensory (air puffs) and either olfactory or trigeminal chemosensory stimuli is not correlated; this was observed in both patients with olfactory loss and individuals with a normal sense of smell. Similarly, both groups do not differ with regard to thresholds to somatosensory electrical cutaneous stimulation, which is in striking contrast to findings on the chemosensory portion of the trigeminal system, which is strongly affected by olfactory status. Finally, chemosensory trigeminal and somatosensory trigeminal stimulations evoke distinct brain-activation patterns when measured with functional MRI: intranasal trigeminal chemosensory stimulation leads to activations of brain regions which are usually activated after olfactory stimulation, such as the orbitofrontal cortex, piriform cortex, and insula; somatosensory stimulation, on the other hand, does not activate these brain areas, but rather the thalamus and the primary and secondary somatosensory cortices after cutaneous stimulation. When both stimuli are directly compared, chemical trigeminal stimulation leads to higher activation in the left insula and the right frontal lobe compared to mechanical trigeminal stimulation.

Taken together, these findings indicate that the close connection between olfactory system and trigeminal system is restricted to the chemosensory portion of the trigeminal system, and that the chemosensory and the somatosensory trigeminal system are relatively independent from each other. Both the chemosensory trigeminal and the olfactory systems share neuroanatomical and functional characteristics, whereas the somatosensory and the chemosensory portions of the trigeminal nerve have been shown to exhibit different ones.

ASSESSMENT

Neuroscientists and clinicians dispose of a large number of techniques to assess trigeminal sensitivity. They include psychophysical tests and objectives techniques. The main challenge when assessing the trigeminal system is the interference from olfactory costimulation, which is nearly invariably observed when the trigeminal system is stimulated by volatile substances.

Psychophysical Tests

Psychophysical tests generally refer to the assessment of trigeminal sensitivity by (1) ratings of chemosensory stimuli in individuals with normal olfactory function; (2) ratings of chemosensory stimuli in individuals with anosmia; (3) odor localization; (4) respiratory reflexes; (5) reflexive changes in nasal blood flow; (6) eye irritation.

Ratings of Chemosensory Stimuli in Individuals With Normal Olfactory Function

The most straightforward procedure to evaluate the trigeminal system is to instruct individuals to focus on the trigeminal properties of a stimulus and to evaluate them. For example, one could instruct participants to assign numerical value proportional to the stimulus magnitude they perceive, not only in terms of overall sensation, but also in terms of distinct trigeminal sensations such as stinging or burning. This technique requires training of the participants, because they have to disregard the concomitant olfactory stimulation which cannot be avoided with most of the stimuli. However, a possible confusion of trigeminal and olfactory sensations can never completely be ruled out.

Ratings of Chemosensory Stimuli in Individuals With Anosmia

One method to overcome this is the testing of individuals with anosmia, as this condition, by definition, does not lead to concomitant olfactory stimulation. However, this approach is problematic due to the fact that sensitivity to trigeminal stimuli is reduced in participants with anosmia, as several studies showed. Further, especially in acquired anosmia, residual olfactory sensations are possible.

Odor Localization

One particularly elegant method is based on observations made nearly a century ago by German physiologist von Skramlik in 1924. He observed that, if only one nostril was stimulated, the correct identification of which nostril had been stimulated depends on the capacity of the stimulus to activate the trigeminal system. This technique was later used by Nobel laureate von Bekesy when he asked whether or not humans are able to localize odors. While he determined that we can do so, more recent work showed that we are only able to do so with stimuli which activate the trigeminal system—unlike other animals such as rats and sharks. In fact, the degree to which a stimulus is localizable is directly correlated to the degree to which it stimulates the trigeminal system. Therefore, this method allows to assess and to compare "trigeminality" of different stimuli as well as to determine and compare trigeminal responsiveness between different individuals or groups of individuals. Today, the localization task is used in a semi-quantitative fashion. Typically, subjects are asked to localize a stimulus which is presented 40 times in a row. Testers have to make sure to avoid habituation between stimulations, therefore an interstimulus interval of 40 s should be maintained.

The other methods described herein rely on measuring responses based on trigeminal reflexes.

Respiratory Reflexes

Respiratory reflexes can be assessed by the measurement of nasal and oral respiration using separate pneumotachometers. These devices enable to record breath mean airflow velocity, breath maximum airflow velocity, breath volume, and breath duration, during stimulus exposure.

Reflexive Changes in Nasal Blood Flow

More indirectly, trigeminal responsiveness can be measured by assessing reflexive changes in nasal blood flow, by using laser Doppler techniques. Here, trigeminal stimulation occurs as soon as nasal irritation is indicated by a reduction of tidal volume (volume per breath).

Eye Irritation

Another method consists in measuring trigeminal sensitivity of the eyes to exposure to irritant. Here, the cornea is exposed to increasing concentrations of an irritant until the individual claims a distinct irritation or blinks. This method has the drawback to test another tissue, the cornea, as a proxy for nasal sensitivity.

Objective Techniques

In addition to psychophysical tests, growing body of research relies on objective techniques to assess the trigeminal chemoreception. Such methods rely less on the active participation of the study subjects and enable the recording of responses even in children or aphasic patients. These techniques include (1) electrophysiological measures, such as negative mucosa potential (NMP) and chemosensory ERPs, and (2) neuroimaging techniques such as functional MRI and PET. In addition to sophisticated machinery for the evaluation of the nervous system's responses, these methods also rely on automated delivery of the stimuli, which is feasible by the use of an olfactometer.

Odorant Delivery Method

Olfactometers are devices built to present odor stimuli in a standardized computer-controlled manner with determined air flow, odor concentration, odor duration, onset, and offset. Sometimes even humidity, temperature, and other features can be controlled. Stimuli are typically embedded in a constant flow of odorless, humidified air of controlled temperature in order to avoid concomitant mechanical or thermal stimulation. Further, it is also possible to present odor at a specific moment depending on physiological measures, such as respiratory rhythm; for example, odor releasing may occur just during inhalation. Next to custom-built prototypes, several companies sell commercial olfactometers; these machines, however, are typically beyond the financial possibilities of average clinicians and are nearly exclusively used in specialized labs. Therefore, the techniques mentioned in the next paragraphs are usually nearly exclusively used in such specialized laboratories.

Negative Mucosa Potentials

NMP is a noninvasive method that records changes in the periphery of the trigeminal system using an electrode inserted on the nasal mucosa. NMP recordings reflect summated receptor potentials of chemoreceptors of the trigeminal nerve within the respiratory nasal mucosa in reaction to chemosensory stimuli eliciting trigeminal-meditated sensations. NMP are the result of activation of both C-fibers and Aδ-fibers. NMP highlight peripheral activation of the trigeminal system exclusively. The recording sites should be chosen in function to accessibility, low signal-to-noise ratio, and/or absence of stimulus-induced artifacts. Measures of interest are latencies and amplitudes of the slow negative wave that occurs after 1 s. The amplitude of the NMP is sensitive to stimulus quality (eg, CO_2, menthol, and ethanol), stimulus concentration, and perceived intensity. Given that responses to the NMP decrease after multiple presentation of the stimulus, interstimulus interval should be at least 30–40 s.

Trigeminal Event-Related Potentials

Trigeminal ERPs reflect an electrical change recorded from the brain in response to trigeminal-mediated sensations, by means of one to five or more electrodes placed on the human scalp. Specifically, trigeminal ERPs are EEG-derived poly-phasic signals induced by the activation of cortical neurons that generate electromagnetic fields. The trigeminal ERP are thought to reflect the central nervous processing of trigeminal stimulations. The trigeminal ERP response refer to a minor positive peak (P1) with a latency superior to 300 ms, followed 400 ms later by a high negative peak (N1), and then, 650 ms later by a late positive complex (P2 or P2/P3). The majority of the studies have mainly examined amplitudes and latencies of the N1 peaks. In individuals with normal olfactory function, largest trigeminal ERP in response to trigeminal stimuli are recorded in central and parietal brain areas. Similar to the NMP, stimulus presentation has to be separated by a relatively long interstimulus interval of at least 30–40 s, otherwise habituation effects are likely to happen. The main advantage of ERP is the very high temporal resolution in the range of milliseconds. By contrast, its spatial resolution is poor, compared to brain-imaging techniques such as functional MRI or PET. The localization of the specific neural areas appears to be difficult by using ERP since number of possible generation sites for any one voltage distribution are almost infinite. So far, it's not possible to determine the location and the orientation of a set of neurons by only knowing the scalp distribution of the electrophysiological responses (inverse problem).

Neuroimaging

The use of functional magnetic resonance imaging (MRI) since 1990s has provided new insight into the processing of chemosensory information. Functional MRI is a noninvasive technique measuring to map brain activity. The underlying mechanism—which is not yet understood in full detail—is that neural activity elicits large change in blood flow and leads to a more oxygenated blood. Accordingly, the ratio of oxygenated hemoglobin/deoxygenated hemoglobin in specific brain areas reflects neural activity. Functional MRI is measuring this blood oxygenation level–dependent (BOLD) effect. Without going into much detail, the minimal volume unit of analysis is represented by a voxel—the 3D analogy of a 2D pixel. It typically has the size of a cube with 3 mm edge length.

Although this is fairly small, a voxel still contains tens of thousands of neurons; the activation of a voxel thus reflects the average activation of tens of thousands of voxels. Still, compared to other techniques, functional MRI has a high spatial resolution, but its temporal resolution is relatively weak, about 3–5 s.

Other neuroimaging methods include positron emission tomography (PET) and magnetoencephalography (MEG); for the research of the trigeminal system these techniques have, however, been used just in a handful of studies. PET measures metabolic changes that reflect neuronal activity, by tagging fluorodeoxyglucose (FDG) with a positron-emitting isotope (^{18}F) or by tagging water with a positron-emitting oxygen isotope ($H_2{}^{15}O$). This technique consists in measuring the concentration of radioactive markers using PET cameras, which exhibit a good mapping resolution of about 5 mm. However, the signal-to-noise ratio needs to be improved by smoothing applied to the data, resulting in limited spatial resolution ranging from 10 to 20 mm. In addition, an adequate signal to noise ratio can be detected from 30 s to 1 min.

MEG measures neuronal activity by measuring the changes in magnetic fields related to electrical activity changes over the surface of the scalp. The advantages and disadvantages of MEG therefore resemble to those of ERP measurements. Unlike PET and functional MRI, which are the result of changes in blood flow or blood oxygenation parameters, MEG measurements depend on neuronal electrical activity; as such MEG provides fast responses in the millisecond range. However, one limitation is the difficulty in measuring changes in electrical activity/magnetic fields in deep brain structures.

In summary, each method (psychophysical, electrophysiological, and neuroimaging) has advantages and disadvantages. Psychophysical and electrophysiological techniques have clear limitations insofar as their relationship with functional processing is not clearly validated and the spatial resolution is poor. Conversely, functional brain imaging provides a higher spatial specificity, but its weakness is the temporal resolution. Researchers have therefore to choose the technique with regard to the scientific question they seek to answer.

TRIGEMINAL RESPONSES

In this following section, we expose a nonexhaustive overview of trigeminal responses in terms of detection, localization, intensity rating, and hedonics judgment of mixed olfactory/trigeminal or pure trigeminal stimuli administered nasally to individuals.

Detection

The olfactory system is more sensitive than the trigeminal system as it reacts at relatively low concentrations of chemical substances while the trigeminal system is tuned to react at higher concentrations; in turn the physiological and behavioral responses are substantially stronger as, for example, one can see with the coughing reflex. Interestingly, chemosensitivity for different trigeminal stimuli shows a strong correlation between each other, whereas such an association was not proved in olfaction. As a result, higher sensitivity to an agonist of one trigeminal receptor will predict higher sensitivity to agonists of the same receptor and of another receptor.

Mixtures

When subjects are exposed to a bimodal mixture and its individual components (trigeminal stimulus, olfactory stimulus), the detection thresholds are equivalent for each stimulus. This indicates that in the bimodal mixtures; the pure trigeminal stimulus does not inhibit detection of the olfactory components, which is in contrast to suprathreshold intensity ratings. In contrast to this, trigeminal ERP responses to bimodal mixtures appear earlier than responses to either component alone. Brain imaging clarifies the effects: the anterior portion of the orbitofrontal cortex is activated for binary odor mixtures and deactivated for single components while its lateral portion is more activated with increased impurity. Therefore, mixture impurity is coded in this lateral portion, whereas the anterior portion detects odor mixtures.

In summary, binary odor mixtures and their individual components are processed differently by the brain.

Intensity

As mentioned earlier, the nasal trigeminal system participates in the overall intensity evoked by volatiles. The trigeminal system modulates the perception of odor intensity in a near-additive manner with time-intensity trading. In fact, there is a close association between stimulus concentration and duration and perceived intensity of a trigeminal stimulus. A more concentrated stimulus is perceived as stronger than one with a lower concentration; similarly, a longer-lasting stimulus is perceived as stronger than a shorter one. Therefore, the trigeminal system is a mass detector rather a concentration detector. As it turns out, intensity ratings and stimulus duration/concentration are linearly correlated. In other words, the total amount of molecules—which is the result of the multiplication of stimulus concentration, stimulus duration, and stimulus volume—will determine perceived intensity rather than concentration, duration, or volume alone.

This can be observed on a behavioral level, but even with electrophysiological techniques assessing peripheral susceptibility. The NMP reflects summated receptor potentials of chemoreceptors of the trigeminal nerve, and thus peripheral susceptibility. Accordingly, stimuli with higher concentration evoke larger NMP amplitudes as do stimuli which are perceived as more intense. Pain killers decrease perceived intensity and reduce NMP amplitudes, even when stimuli are held constant. On central levels, the activity of several trigeminal processing brain regions is correlated with perceived intensity, including amygdala and different areas of the cingulate cortex. As trigeminal stimuli with high intensity are typically perceived as unpleasant, it remains unclear whether this result is due to either intensity or unpleasantness as these subregions of the cingulate cortex may be involved in the processing of emotions and pain.

Mixtures

One of the key questions when looking at mixture of trigeminal stimuli is whether there is any enhancement or suppression of intensity when stimuli are mixed with the same or different qualities together. In order to understand trigeminal processing it is helpful to look at the other chemosensory systems, which are better known. When two equally strong olfactory stimuli of similar chemical composition are mixed together, the mixture is perceived as more intense than the single components, but less intense than the sum of the intensities of the single components. The same phenomenon occurs in gustation when two taste stimuli were mixed together. With regard to mixtures involving the trigeminal system, different scenarios are possible: (1) *mixtures of trigeminal stimuli*; (2) *mixtures of trigeminal and olfactory stimuli*; (3) *mixtures of stimuli with different intensities*. Real-world perception would typically be a combination of these three cases.

MIXTURES OF TRIGEMINAL STIMULI

When looking at mixtures of trigeminal stimuli, suppression and enhancement effects depend on the involved trigeminal receptors. A mixture of stimuli activating the same receptor (eg, TRPV1) is perceived as less intense as a mixture activating different trigeminal receptors (eg, TRPV1 and TRPM8).

MIXTURES OF TRIGEMINAL AND OLFACTORY STIMULI

The intensity of binary mixtures of pure odorants, pure trigeminal stimuli, and olfactory trigeminal stimuli is not well predictable. It seems as if the intensity of pure odorants is suppressed when mixed with a pure trigeminal stimulus or with a mixed olfactory–trigeminal stimulus. By contrast, the intensity of a pure trigeminal stimulus is suppressed when mixed with a mixed olfactory–trigeminal stimulus but enhanced when mixed with a pure odorant. A mixed olfactory–trigeminal stimulus, finally, is enhanced when mixed with a pure trigeminal stimulus.

A mixture of a pure trigeminal stimuli and a pure odorant activates a different neural pathway than the two stimuli presented separately. Compared to its individual components a bimodal mixture leads to larger activations in chemosensory areas such as the orbitofrontal cortex as well as multisensory integration centers such as the intraparietal sulcus. In other words, a bimodal mixture induces higher cortical activations than the sum of its single components. This supremacy of the bimodal stimulation might be due to the fact that information is encoded into both systems providing a superior memory trace and reducing the possibility of interference.

MIXTURES OF STIMULI WITH DIFFERENT INTENSITIES

The stimulus with the strongest intensity before mixing is dominant in the mixture and suppresses the other substance. At low concentrations however, trigeminal stimuli increase pungency ratings of the odors. Possibly, this is the result of integration processes in the piriform cortex, which receives input from both the trigeminal system and from olfaction. Nonetheless, first integration may take place as early as in the olfactory bulb since this structure possibly receives input from trigeminal sensory neurons in addition to olfactory ones.

In summary, there appears to be a relative hierarchy between the different systems. Trigeminal stimulation dominates olfactory stimulation while bimodal stimulation dominates either system separately. Within the trigeminal system, activation of several receptor systems leads to larger activations.

Hedonic Valence

One of the key features when perceiving a chemical substance is its hedonic valence. Hedonic valence of a variety of stimuli is processed in the amygdala and the primary sensory cortex. Both the olfactory and the trigeminal system have intimate connections to the amygdala leading to a high degree of functional connectivity between both the trigeminal/olfactory and limbic systems. Both trigeminal and olfactory stimuli produce pleasant or unpleasant sensations, but unpleasant stimuli appear to be clearly linked to high concentrations. During the past decade, a growing number of studies have shed light on sensory and cognitive mechanisms underlying hedonic valence perception of chemosensory stimuli. Intersubject variations in hedonics judgment appear to be extremely high due to personal experience (eg, familiarity and culture). Moreover, hedonic judgments of mixed olfactory–trigeminal stimuli are associated to odor identification.

At the central level, unpleasant bimodal odorants activated a different cerebral pathway compared to pleasant bimodal odorants, including the piriform cortex, the amygdala, the orbitofrontal cortex, the thalamus, and the cingulate gyrus. Importantly, brain activity is lateralized with regard to reported pleasantness, especially in the piriform, orbitofrontal, and insular cortex. Specifically, unpleasant odors are associated with greater responses of regions in the left hemisphere. These findings are in line with some theories on emotions, which hypothesize a left hemisphere dominance for unpleasant stimuli.

Mixtures

So far, little is known about the sensory mechanisms and its neural correlates that underlie the pleasantness perception of bimodal or trigeminal mixture. Again there appears to be a left to right difference with regard to hedonics, but this difference is most prominent in the insular cortex.

Localization

In other sensory systems, such as vision and audition, object localization takes place in higher-order parietal brain regions, whereas object identification is processed in temporal brain areas (dual stream theory). Consistent with this dual stream theory, the intraparietal sulcus and the superior frontal sulcus are involved in localization of odorous substances, but are not related with their identification. It is important to remind the reader that the localization of an odorant is possible only when the odorous stimulus activates the trigeminal system. In other words, it is impossible to localize pure odorants, at least for humans, which activate the olfactory system exclusively.

Stimulus localization performance is determined by stimulus concentration, stimulus duration and stimulus volume. In other words, longer or more concentrated stimuli, or stimuli with a high volume or flow rate, are more easily localized, in line with the notion of the trigeminal system being a mass detector. When any of these parameters is large enough, even stimuli which are usually considered to be pure olfactory stimuli can be localized. This is the case for phenyl ethanol.

Active sniffing improves olfactory perception, probably due to intake of a higher number of molecules as a result of increased volume and duration. However, this does not hold true with regard to the localization task: here mixed olfactory/trigeminal stimulus are more easily localizable when the stimulus containing air is blown into the nose (passive condition) rather than when the subject is sniffing the air (active condition). Interestingly, some individuals seem to be more sensitive than others and are able to localize stimuli which not everybody else is able to localize. Furthermore, odor localization can be trained as a 10–20-day exercising paradigm associated with improved localization abilities even for pure odorants.

Mixtures

Olfactory localization ability depends also on the complexity of the chemosensory stimulation. As a consequence, mixtures containing substances activating two different types of trigeminal receptors are more easily localized. For example, localization of a mixture of menthol and mustard oil mixture, activating TRPM8 and TRPA1, respectively, is easier than localization of a mixture of menthol and eucalyptol, which both stimulate TRPM8. In fact, when stimuli in a mixture activate the same receptor suppression occurs, whereas a mixture additive effect takes place when they stimulate different receptors.

FACTORS INFLUENCING TRIGEMINAL PERCEPTION/ SYSTEM

Internal Factors

Intranasal Anatomy

Intranasal anatomy plays a role in intranasal trigeminal sensitivity. For example, the size of the nasal cavity is positively correlated with amplitudes of trigeminal ERPs in response to suprathreshold CO_2 and menthol stimuli. Nevertheless, a larger nasal cavity does not predict lower CO_2 thresholds. Additionally, perception of trigeminal stimuli in such CO_2 appears to be most accurate in the anterior portion of the nasal cavity, whereas sensitivity to mechanical stimuli appears to be highest in the posterior portion. This seems to have a potential implication on differences in the perception of orthonasal and retronasal olfactory stimulation; the latter being particularly important for food and aroma perception.

Sex

It is generally established that women have a better sense of smell than men, as women tend to outperform men in most odor tasks (sensitivity, identification, and memory retrieval). In the same way, clear sex-dependent differences in trigeminal sensitivity are found for behavioral, physiological, and brain-activation responses. In general, psychophysical studies showed women have better trigeminal ability than men in terms of sensory trigeminal sensitivity, perceptual acuity, intensity, discrimination, intensity (CO_2), apnea reflex, and lateralization. In addition, several ERP studies showed sex differences to trigeminal stimulation.

However, sex-difference effects remain controversial. Indeed, numerous studies failed to reveal sex differences on sensitivity or localization performances. Importantly, women exhibit lower intranasal trigeminal thresholds than men in the subjective method of trigeminally mediated detection exclusively (ie, responses of subjects who focused on sensations stemming from the trigeminal system) but no sex difference is found using the more objective lateralization method (ie, the ability of subjects to identify the side of the nose receiving a unilaterally presented stimulus). A potential explanation is that women pay more attention to trigeminally mediated sensations and, as a consequence, detect the intended sensation sooner than men, even though women's and men's sensitivity may not be different from each other. In addition to psychophysical investigations, NMPs (NMP), the peripheral correlate of trigeminal activation, indicates similar patterns of pain-related electrical potentials in the respiratory nasal mucosa for women and men.

At a central processing level, women exhibit larger amplitudes of the early positive component (EPC) and of the late positive component (LPC) of the ERPs in response to intranasal trigeminal stimuli. Given that the LPC amplitude mirrors the cognitive processing of a stimulus, such as involuntary (re)allocation of attention, context updating in memory, and event categorization, the enhanced LPC amplitude in women may suggest differential subjective assessment and/or emotional classification of an object. In addition, studies on neuroimaging highlight clear differences in brain-activation patterns between women and men in different tasks. For example, women and men exhibited specific bilateral activation of the insula and in the left piriform–amygdala region, women exhibited also activation in the left OFC when passively smelling emotional odors (120 odorants; almost all of them were bi-modal olfactory–trigeminal odorant) or performing odor hedonic judgments.

Taken together, these findings may reveal that sex differences would not be mainly mediated by differential peripheral processing. As proposed by several authors, it may be worth considering that women allocate more attentional resources to irritant stimuli than men do, resulting in differences in cognitive processing and subjective perception. In this line, a study noticed a sex difference effect for the LPC when the stimuli (trigeminal and bimodal) were expected, with no such difference when stimuli were unpredicted.

Aging

Elder persons have reduced olfactory abilities such as odor detection, discrimination, identification, and memory perception. Similar findings are known from taste. In addition to smell and taste, the nasal trigeminal system declines with age. When trigeminal and/or bimodal odorants are presented, elderly subjects show reduced performance in thresholds using subjective and objective lateralization methods, in perceived intensity and higher threshold of reflex apnea. More precisely, sensitivity—as expressed as olfactory detection and trigeminal lateralization thresholds—in both chemosensory systems is affected with aging, especially in subjects older than 60 years. Nevertheless, the age-related impairment is more pronounced for the olfactory system compared with trigeminal system.

It is not clear on which level this age-related impairment of trigeminal sensitivity actually takes place. It can be observed even on the level of the nasal mucosa, with smaller NMP responses in elder subjects. So far, the mechanisms underlying the age-dependent decline of trigeminal sensitivity are still unclear. The decrease of intranasal trigeminal sensitivity with age may be due to (1) a decrease of nociceptive function that affects mainly the responsiveness of Aδ-fibers, but not C-fibers, and/or (2) age-related changes of the mucosa of the nasal cavity. Further, age-related differences may also be the result of alterations in central processing, as suggested by chemosensory ERP studies, but this implication is still unclear.

Handedness

Handedness is considered to reflect cerebral lateralization/functional hemispheric asymmetries, although this relation is indirect and complex. Several studies have assessed lateralization in olfactory/trigeminal sensitivity as a function of the subjects' handedness. The findings concerning detection thresholds are controversial and the interpretations remain unclear. Some studies report higher sensitivity for bimodal stimuli on the side contralateral to the preferred hand; other studies found an advantage on the ipsilateral side. Some studies found no difference. Similarly mixed findings can be found regarding hedonic judgments.

With regard to the localization task, there appears to be a tendency towards dominant side: right-handers have a tendency to localize stimuli to the right side, and left-handers to the left. It is, however, unclear whether this is truly superior sensitivity on the dominant side or rather an attentional effect as we know from other sensory systems.

Taken together, there is some indication that olfactory/trigeminal functions may be differentially processed by the two hemispheres, as it is suggested by the differences in lateralization. As a consequence, handedness might be thought as one of the determining factors in some lateralized olfactory/trigeminal function such as hedonic judgment, but its implication in detection threshold need further investigations.

External Factors

Olfactory Status

Olfactory status is one of the factors that determinate intranasal trigeminal sensitivity, for both perithreshold and suprathreshold stimuli. Hyposmia and anosmia are common phenomena in the general population. People suffering from hyposmia, exhibiting lower ability in olfactory perception, are estimated at 15% of the population, whereas anosmia, the complete loss of olfactory function can be observed in approximately 5% of the population. The most common causes of partial or total loss of smell can be acquired at some point during lifetime following sinonasal disease, head trauma, upper respiratory tract infections, or neurodegenerative diseases. Anosmia can also be congenital, meaning subjects are born without a sense of smell. Congenital anosmia appears in rare cases, as it occurs in around 1% of all cases of anosmia. It is well established that patients with olfactory dysfunction exhibit lower trigeminal sensitivity compared with individuals with normal olfactory function, in terms of trigeminal detection thresholds, intensity ratings, breathing pattern/respiratory reflexes, forced choice stimulus localization, NMP amplitudes, and ERPs. It is important to note that only the chemosensory portion of the trigeminal system seems to be affected, as somatosensory sensitivity remains intact in anosmia and hyposmia.

The etiology of olfactory loss also influences trigeminal sensitivity. Interestingly, the least affected seem to be individuals with Isolated Congenital Anosmia, whose trigeminal sensitivity is virtually unaffected. The groups of patients with olfactory dysfunction following a viral infection of the upper respiratory tract and sinunasal disease have significantly reduced trigeminal sensitivity compared to individuals with normosmia, but less than the most heavily affected group, which are the patients with olfactory dysfunction following a traumatic brain injury. It is conceivable that head trauma probably causes more pronounced damage including those brain regions directly involved in the processing of intranasal trigeminal stimuli.

The effects of etiology are further overlapped by the well-known age effect: In elder individuals—even those with normosmia—reduced trigeminal sensitivity is more common and therefore group differences are much less pronounced.

It is noteworthy to point out that the effect of olfactory loss on trigeminal sensitivity wears out over time. This may be due to at least two factors: first, the reference group, individuals with normosmia, also have the tendency to lose trigeminal sensitivity over the years, due to the aging effect. Second, the olfactory system has the ability for recovery, even after years, and improved olfactory function precedes the improvement of the trigeminal deficit.

Trigeminal Dysfunction

Unlike the very common olfactory dysfunctions, a loss of trigeminal chemoreception is a very rare event. It is typically observed in the context of loss of the trigeminal nerve. For example, one study reported that subjects with total unilateral destruction of the trigeminal nerve after the removal of an acoustic neuroma assessed odors less intense compared to heathy controls. Another study identified a general olfactory-related impairment in one subject suffering from a total destruction of the trigeminal nerve after meningioma on her left side: peripheral and central electrophysiological responses to pure odor or pure trigeminal stimuli were lower, and accordingly, the odor sensitivity (thresholds) was better on the right side than the left side. The patient's gustatory sensitivity was otherwise similar on both sides of the tongue. Although admittedly not a lot of data is available, these findings lend further support to the intimate connection between the trigeminal and the olfactory systems.

Neurodegenerative Diseases

It is well documented that hyposmia is one of the earliest manifestations of certain neurodegenerative diseases of the central nervous system, including Alzheimer's disease, Parkinson's disease, or Huntington's chorea. Olfactory deficits have a considerable prevalence in neurodegenerative diseases, with over 90% for patients with both Alzheimer's and Parkinson's disease. Reduced olfactory function typically precedes other clinical symptoms by up to 10 years and is present for years at the time of diagnosis.

As opposed to the well-investigated olfactory dysfunction in neurodegenerative diseases, almost no data is available on trigeminal perception in these conditions. Interestingly, ERPs reveal the trigeminal system to be relatively unaffected by Parkinson's disease, as opposed to the olfactory system. This is relatively surprising since, as aforementioned, the trigeminal and olfactory systems are closely interconnected; thus one could have been easily expected that olfactory dysfunction would be mutually associated with trigeminal deficit as well. Further investigations need to be conducted to better comprehend trigeminal perception in neurodegenerative diseases, and the mechanisms underlying a possible dissociation between olfactory and trigeminal sensitivity. This dissociation may provide an interesting tool in the future development of early screening tools for neurodegenerative diseases.

CONCLUSIONS AND OUTLOOK

The trigeminal system plays an important role in humans. Its primary function is to detect inhaled harmful substance and to help clearing the airways from them. As a second function, it adds to the global perception of most odorants. The trigeminal and olfactory systems are closely interconnected and influence each other. In this sense, olfactory dysfunction affects trigeminal perception and vice versa. Psychophysical, electrophysiological, and neuroimaging techniques have been used to attain knowledge of the peripheral and central processing. Importantly, olfactory/trigeminal deficits are also closely associated with aging, viral infections, neurodegenerative diseases, head injuries, chronic sinusitis infections, and nasal obstructions.

The loss of sense of smell is not harmless as subjects with olfactory dysfunction report lower quality of life, especially regarding food and drink appetence, socialization, and intimate relationships. Therefore, olfactory dysfunction can lead to changes of dietary habits that may in turn exacerbate diseases or contribute to nutritional deficiencies. Taking into account that dietary preference can be related to chemosensory perception, olfactory/trigeminal impairments may contribute to lower food appetence/consumption and so malnutrition in elderly. In parallel, a preference for higher concentration of sugar and salt was noticed in elderly compared to younger subjects. Consequently, adding some chemosensory food enhancers with trigeminal components would enable individuals to detect sensory information and thus enhance appetite and food acceptance/consumption, on the one hand, and decrease the salt/sugar overconsumption, on the other hand. Further, since up to 20% of the population suffer from hyposmia or anosmia, one may consider adding trigeminal stimuli as warning agents to, for example, household gas in order to increase detection.

One currently hot topic in olfactory research is plasticity in the sense of smell. Olfactory training improves olfactory function, such as olfactory thresholds, odor localization, odor intensity, olfactory ERPs, scent tracking, in healthy subjects and in patients suffering from neurodegenerative diseases or trauma injuries. Most of these studies applied bimodal odorants. It will be extremely relevant to investigate the impact of repeated trigeminal stimulations on the activation of both the chemosensory trigeminal and olfactory systems. Pure trigeminal stimuli, or bimodal stimuli with a strong trigeminal component, may produce remarkable beneficial effects on both olfactory and trigeminal perception and processing and thus may enhance both sensitivity and higher-order cognitive functions.

Finally, the dissociation for olfactory and trigeminal impairment in neurodegenerative disease may provide an extremely interesting avenue for the possible development of early screening tools in order to distinguish neurodegeneration from normal aging.

Further Reading

[1] Kobal G, Van Toller S, Hummel T. Is there directional smelling? Experientia 1989;45:130–2.

[2] Caterina MJ, Schumacher MA, Tominaga M, Rosen TA, Levine JD, Julius D. The capsaicin receptor: a heat-activated ion channel in the pain pathway. Nature 1997;389(6653):816–24.

[3] Frasnelli J, Schuster B, Hummel T. Interactions between olfaction and the trigeminal system: what can be learned from olfactory loss. Cereb Cortex 2007;17(10):2268–75.

[4] Frasnelli J, Hummel T. Age-related decline of intranasal trigeminal sensitivity: is it a peripheral event? Brain Res 2003;987(2):201–6.

CHAPTER

19

Vision

J.F. Hejtmancik[1], P. Cabrera[1], Y. Chen[1], O. M'Hamdi[1], J.M. Nickerson[2]

[1]National Institutes of Health, Bethesda, MD, United States; [2]Emory Eye Center, Atlanta, GA, United States

INTRODUCTION

The visual system consists of optical components including the anterior segment—cornea, aqueous humor, and lens—and the posterior segment—the vitreous body and the retina. Posterior to the globe, the neural components include the optic nerves, optic tracts, optic radiations, the visual cortex, and a variety of nuclei within the brain. Receiving and interpreting visual signals requires each of these structures to function appropriately.

Light passes through the cornea, aqueous, lens, and vitreous and is focused on the retina, where the process of converting light into neural signals, called transduction, occurs. The neural signals pass through the optic nerves and tracts to central structures where more elaborate processing occurs and where information from other senses is integrated with those from the visual pathways. Finally, the oculomotor system—the efferent limb of the visual system—is responsible for maintaining eye position and executing eye movements. This chapter discusses visual processes and the structures that carry them out, as well as the critical periods for development of intercellular communication throughout the visual pathway from the retina to the visual cortex. The emphasis in this chapter will be more on general principles and less on experimental results. Detailed accounts about the experimental bases of principles to be discussed here may be found in the Selected Readings section at the end of this chapter. From Neuroscience in Medicine, 2nd ed. (P. Michael Conn, ed.), © 2003 Humana Press Inc., Totowa, NJ.

REFRACTION AND THE ANTERIOR SEGMENT

The eye needs to be able to focus light on the retina in order to have clear vision, much like a camera needs to be in focus for a photograph to be clear. Much of the refractive or focusing power of the eye can be attributed to anterior segment components, specifically the cornea and the lens.

Light is *refracted* or bent as it passes through the cornea, aqueous humor, lens, and vitreous body (Fig. 19.1) because of the differences in refractive indices of these structures. The refractive index is a physical property of a substance and is defined

Conn's Translational Neuroscience
http://dx.doi.org/10.1016/B978-0-12-802381-5.00031-2

as the ratio of the velocity of light in a vacuum to the velocity of light in that material. When a light wave travels from one medium to another medium that has a different index of refraction, the velocity of light changes and the waveform is bent at the interface of the two media. The angle at which light is bent depends on the angle at which it hits the second medium and on the ratio of the refractive indices of the two media. The curvatures of the cornea and lens greatly affect the path that light takes as it travels through the eye. It is important that these and other structures in the anterior and posterior segments be intact so that light rays that emanate from a single point in the distance are focused onto a single point on the retina.

The image of an object is projected onto the retina in an inverted fashion (Fig. 19.1). Thus the inferior part of the visual field is projected onto the superior part of the retina, the nasal visual fields are projected onto the temporal part of the retina, and the temporal visual fields are projected onto the nasal retina. Human eyes are fixed in the front, the major part of each visual hemifield is perceived by both retinas and is defined as the binocular visual domain.

Four surfaces in the eye strengthen the refractive potential, including the anterior and posterior surfaces of the cornea and the lens. Because the extent of refraction is determined by the difference of the refractive index between two substances, most refraction happens on the anterior surface of the cornea, when light travels from air into the denser cornea. One-third of the refractive potential occurs in the lens where vision is focused on near or distant objects. The degree of lens curvature depends on the contraction of the circular ciliary muscle. This movement decreases the diameter of the lens, allowing it to assume a more spherical shape. This phenomenon is

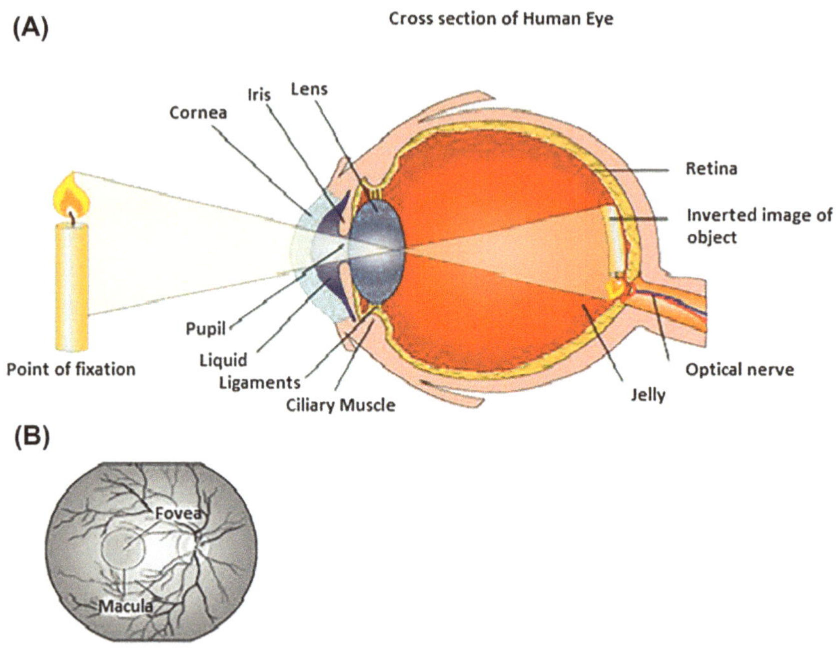

FIGURE 19.1 (A) Summary of eye anatomy and the refraction process. The light refraction trace from the fixation point to the target point on the retina represents all the effects including the surfaces of the cornea and lens (posterior and anterior). In the right, the optic nerve represented by the optic disc emerging vessels. (B) Aspect of the central retina, showing the macular region and fovea on the left.

called accommodation, and it allows the eye to focus from a far point to a near point.

Many refractive errors can be distinguished: myopia (ie, nearsightedness), is the most common refractive error of the eye and it appears when the anteroposterior diameter of the eye is elongated. Hyperopia (ie, farsightedness) is a condition in which distant objects are usually seen clearly but near objects are out of focus. It is the result of the visual image being focused behind the retina rather than on it. Astigmatism is usually caused either by an irregular shape of the cornea or by a curvature of the lens, in which the vision is blurred because a point source of light is focused into a line instead of a point. The options for correcting these refractive errors include appropriate lenses or surgery.

STRUCTURE OF THE RETINA

Vertebrate Retina Consists of Distinct Layers

Both the vertebrate and invertebrate retina and its component cells have been described in great detail. The following discussion mainly focuses on vertebrate retina, although studies of invertebrate models have shed light in understanding many aspects of the visual process.

The vertebrate retina structure is shown schematically in Fig. 19.2. The adult structure of the retina consists of many different cell types. The processes of retinal development are being elucidated more and more but are outside the range of this discussion. For our purposes, it is sufficient to know that the retina is divided into two structural and functional components: the retinal pigment epithelium (RPE, the non-neural part) and the neural retina (ie, sensory component).

The RPE cells contain melanin granules, which prevent light passing through the retina from being reflected off the sclera and degrading vision—a problem in pigment disorders such as albinism. RPE cells also perform housekeeping for the photoreceptors by resynthesizing visual pigments and phagocytosing shed outer-segment tips. This requires the photoreceptor outer segments to be closely approximated to the RPE layer. Hence, the neural processing networks of the retina are located anteriorly, and light must pass through

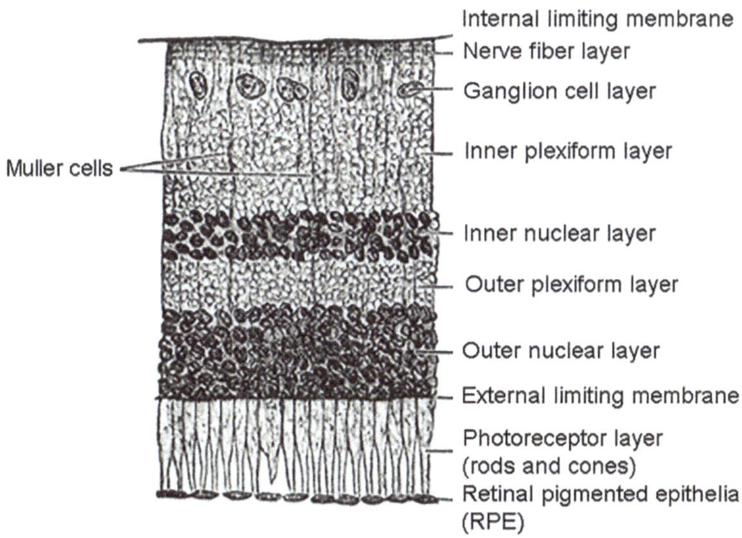

FIGURE 19.2 Structure of the vertebrate retina. *Modified from H. Gray (1918). Anatomy of the Human Body. Figure available at* https://commons.wikimedia.org/wiki/File:Gray881.png.

them prior to stimulating the photoreceptor cells. The neural retina is composed of six neuronal cell types and glial (Müller) cells arranged into three nuclear layers with interspersed plexiform layers. Cell-to-cell interactions between the layers will be discussed later on in this chapter.

The photoreceptor cells carry out the biochemical process of transforming light to electrical energy, which is known as *phototransduction*. The highly specialized photoreceptor cells fall into one of two types—rods or cones—depending on their morphology and physiological function. Most notably, these cells contain outer segments consisting of a series of stacked membranous discs (Fig. 19.3A). A pigment called rhodopsin, found in the disc membrane of the rod photoreceptors, absorbs light energy and initiates the transduction cascade. Rods are able to detect light in low levels and are therefore critical for night vision. They are mostly concentrated in the periphery of the retina, while cone cells densely populate the central retina, known as the macula. Cone cells mediate fine vision under strong illumination (daylight).

In humans, each cone contains one of three different opsins that vary in the wavelength at which they maximally absorb light. Different wavelengths of light stimulate these receptors to varying degrees, making it possible to distinguish between different colored stimuli. Cones are concentrated in the fovea, the central 1.5 mm of the macula. The fovea is devoid of retinal elements other than photoreceptors—a feature that greatly enhances central fine vision. Axons from the peripheral retina arc around the macula to minimize light absorption and scattering in this critical area.

Neural signals originate from the outer retina and are transmitted to the inner retina. As mentioned earlier, phototransduction occurs in the photoreceptor cells, whose cell bodies make up the outer nuclear layer. Axons from these cells synapse in the outer plexiform layer of the retina with the dendrites of the horizontal cells and bipolar cells. The horizontal and bipolar cell bodies form the inner nuclear layer. They synapse in the inner plexiform layer (IPL) with the dendrites of the ganglion cells, whose cell bodies form the innermost nuclear layer—the ganglion cell layer. The ganglion cell axons make up the nerve fiber layer of the retina, which form bundles that arc around the macula and eventually congregate

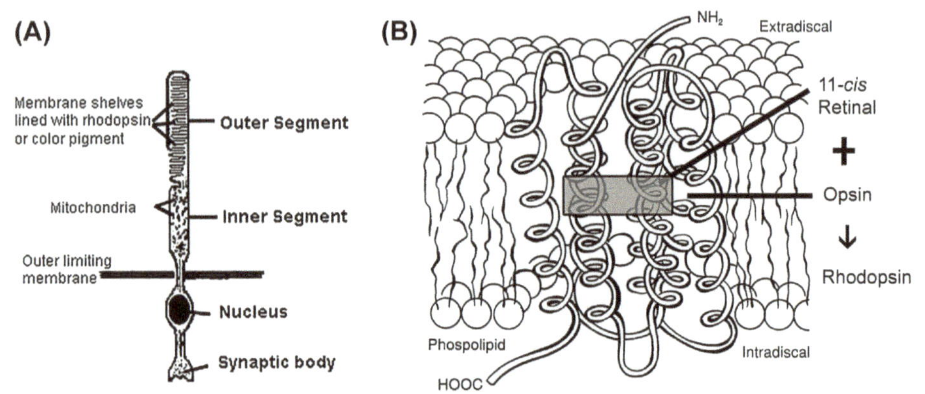

FIGURE 19.3 (A) Schematic of a rod photoreceptor cell. Photoreceptor cells are highly specialized structures that contain an inner and an outer segment. Outer segments are closely associated with the retinal pigment epithelium and are made up of membranous discs that contain the retinal pigments essential for vision. The synaptic body is located anteriorly and synapses with the bipolar cells and horizontal cells in the neurosensory retina. *Image entitled "Functional parts of the rods and cones" courtesy of "Kosigrim" from the Wikipedia project.* (B) The rhodopsin molecule is made up of Opsin, a transmembrane G protein–coupled receptor, and 11-*cis* retinal, a chromophore.

at the nasal retina to form the second cranial nerve—the optic nerve—which has connections to the brain. These neural interactions will be discussed in more detail in the following sections.

SIGNAL TRANSDUCTION

Phototransduction Is Initiated by Activation of Opsin–Chromophore Complexes in the Photoreceptors

In the photoreceptors, the information contained in light absorbed is converted into neural signals in a process called *phototransduction*. Phototransduction is a good model for understanding other signal-transduction processes. For example, opsins are membrane-bound G protein–coupled receptors (GPCRs), showing homology to a family of hormone receptors that are also GPCRs, including adrenergic receptors that activate adenylyl cyclase. The structure of rhodopsin was shown in Fig. 19.3B. Fig. 19.4 is an overview of the process of phototransduction. The opsin–chromophore complex (eg, rhodopsin in rods, cone opsins in cones) in the disc membrane absorbed a photon of light in the outer segment of the cone cell or rod cell, which causes the chromophore retinal to change from the 11-*cis* to the all-trans conformation. This conformational change activates transducin (eg, photoreceptor-specific G protein) by the exchange of a bound GDP for GTP. A cGMP phosphodiesterase (PDE) is then activated by transducin, which cleaves cyclic guanosine monophosphate (cGMP). Decreased cGMP levels cause closure of the cGMP gated–ion channels, leading to intracellular hyperpolarization. This in turn closes calcium channels and subsequently decreases glutamate release in the synaptic terminal of the photoreceptor (ie, signal generation).

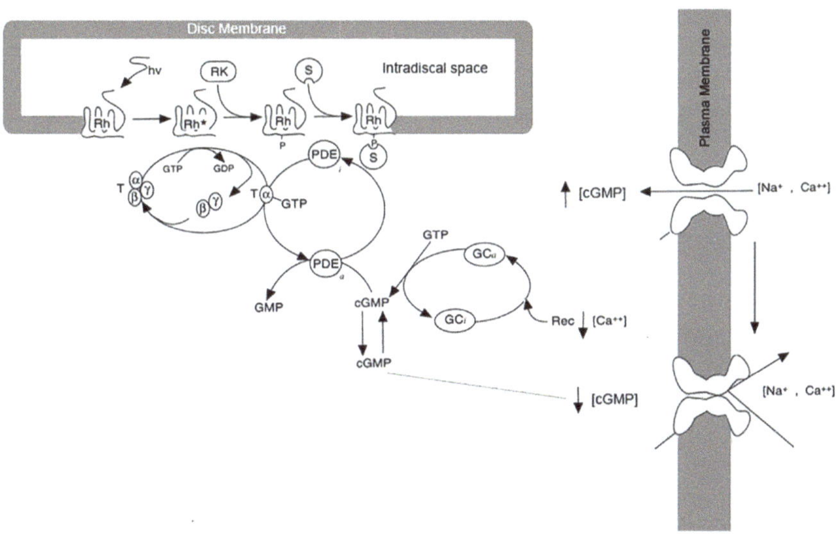

FIGURE 19.4 Phototransduction. Light (*hv*) activates rhodopsin (Rh→Rh*). Rh* activates transducin (T), the photoreceptor-specific G protein, converting GTP to GDP in the process. The α subunit separates from the β and γ subunits, then activates cGMP PDE, cleaving cGMP. Decreased cGMP closes the gated-ion channels, which results in hyperpolarization of the plasma membrane and subsequently decreases neurotransmitter release. The following pathways modulate the photoresponse and are also shown in the diagram: (1) inhibition by phosphorylation of rhodopsin by rhodopsin kinase (RK), whereby the resulting Rh-P molecule is bound by S-antigen (S) a.k.a. Arrestin or 48-kDa protein, and (2) the regeneration of cGMP through stimulation of guanylate cyclase (GC) by recoverin (Rec) a.k.a. 26-kDa protein, in low [Ca^{2+}] conditions.

The rods react readily to low intensities of light because they have a low threshold of excitation, playing an important role in vision at twilight and at nighttime. In contrast, the cones, which require a much higher intensity of light, are critical for fine vision and color discrimination. The posterior pole of eye, including the macula and fovea centralis with high densities of cones, is responsible for central vision.

The basic transduction process is modulated by several mechanisms to provide the eye with sensitivity over a range of light intensities and wavelengths. Phosphorylation of rhodopsin by rhodopsin kinase upon light stimulation is essential for effective quenching of the signal through the interaction of phosphorhodopsin and arrestin (S antigen). The light-induced response is also modulated by the phosphorylation or dephosphorylation of a PDE subunit. There are several other mechanisms involved in the transduction cascade, such as: Ca^{2+} regulation for dark and light adaptation, cGMP concentration regulation, and opsin association and disassociation with its chromophore after photon activation. The phototransduction cascade is also regulated by phosphorylation and dephosphorylation.

The photoactivation of rhodopsin apparently triggers other biochemical cascades, such as the activation of retinal phospholipase C that eventually leads to the release of arachidonic acid from phosphatidylinositol bisphosphate. The role of these pathways and by-products in normal retinal physiology is unclear. They are likely involved in some type of second-order modulation of the photo response or in housekeeping-type functions, such as signaling the turnover of outer segment discs. Understanding the molecular basis of vision, and consequently identifying the possible causes of retinal lesions, is contingent on deciphering these processes and identifying the components.

Opsin combines with 11-*cis* retinal via a Schiff's base linkage to reform a functional rhodopsin pigment. Rhodopsin is photoisomerized by a photon of light, which initiates the phototransduction cascade. The visual cycle (Fig. 19.5) is composed of several processing steps that ultimately regenerate the rhodopsin precursors, 11-*cis* retinal and opsin. Retinol isomerase, an important enzyme in this cycle, is only produced by RPE cells, thus all-trans-retinol needs to be transported from the photoreceptor to the RPE cell, and 11-*cis* retinal from the RPE back to the photoreceptor.

The visual cycle depicted in Fig. 19.5 generally applies to rod photoreceptors. The mechanism in cones is analogous but not identical, and it has much faster kinetics. The isomerase activity in cones appears to be localized to the Muller cell rather than the RPE cell.

Cell Interaction in Neural Retina

Visual stimuli are transduced as continuously graded changes in membrane potential from photoreceptors to other cells of the neural retina. The ganglion cells (efferent retinal neurons) send a series of all-or-none signals (eg, action potentials) with enhanced color, motion, and contrast detection through the optic pathways. The interaction of different neural cells in the retina is depicted in Fig. 19.2. The receptive field of a ganglion cell consists of all the photoreceptors that synapse with it, in which the light stimulation causes an increase (for "on" ganglion cells) or decrease (for "off" ganglion cells) in its firing rate. It is a roughly circular area of retina less than 1 degree at the fovea to 3–5 degrees at the retinal periphery. The "on" ganglion cells are on center/off surround cells, which means light that strikes the center of the receptive field increases the cell's response, while light striking the periphery inhibits the cell's response. An "off" ganglion cell has the opposite arrangement. Both types of ganglion cells are better stimulated by images with large contrasts striking the center and periphery of their receptive fields, compared with an evenly spread illumination. It is a general principle that the visual

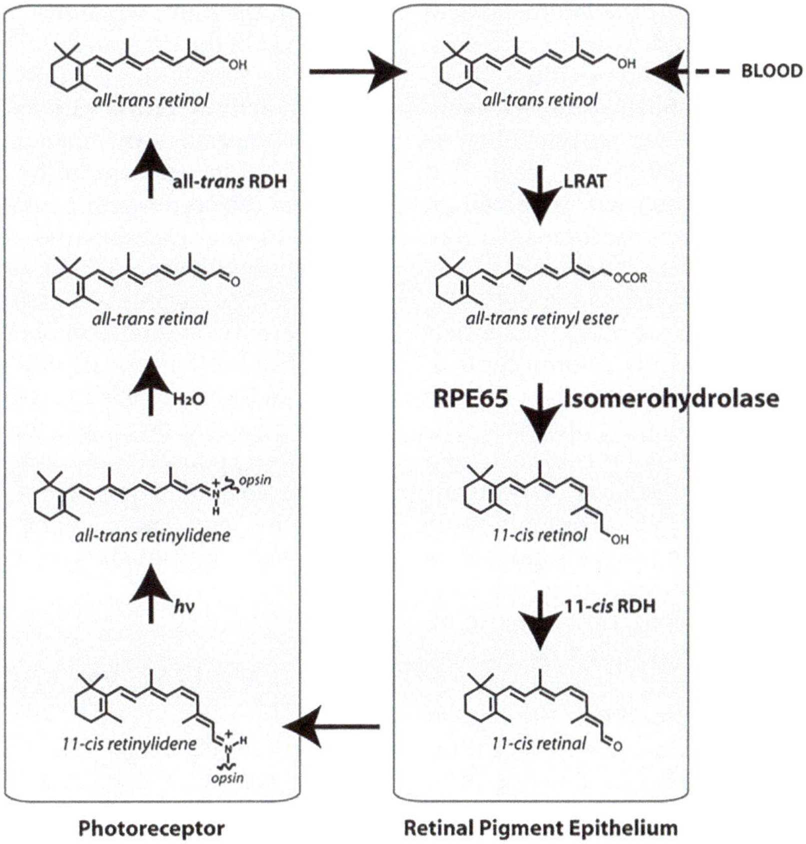

FIGURE 19.5 The visual cycle, also known as the vitamin A cycle. RPE65 is the retinol isomerase found only in RPE cells. *Image modified with permission from Moiseyev G, Chen Y, Takahashi Y, Wu BX, Ma J-x. RPE65 is the isomerohydrolase in the retinoid visual cycle. Proc Natl Acad Sci USA 2005;102(35): 12413–418.*

system responds primarily to four aspects of a stimulus: type, duration, location, and intensity.

The concept of a receptive field is important when investigating the retinal circuitry in detail. The output of photoreceptors is transmitted to bipolar and horizontal cells. The later cells exhibit inhibitory functions, using gamma-aminobutyric acid (GABA) to perform negative-feedback control in the distal retina. They receive input from photoreceptors in the periphery of a receptive field and then send an inhibitory signal back to the photoreceptors and bipolar cells. This enhances contrast, making the edges sharper and compensating for the blurring of the image due to light scattering by the various optical media of the eye.

For bipolar cells, some are stimulatory (eg, on), synapsing to ganglion cells with dendrites in the inner portion of the IPL; some are inhibitory (eg, off), synapsing to ganglion cells with dendrites in the outer portion of the IPL. Then the ganglion cells convert the graded light signals into all-or-none signals (action potentials), which is a typical way of handling information in the brain. Some ganglion cells synapse with both types of bipolar cells, just like some amacrine cells. Even though the interactions of amacrine cells are restricted to within the IPL, they synapse with all the cells types found there.

Amacrine cells also have different subsets: excitatory cells that use acetylcholine as a neurotransmitter and inhibitory cells that use glycine. Some of these are involved in light adaptation in different mechanisms from photoreceptors. The time course of visual responses are also controlled by amacrine cells. Some amacrine cells generate action potentials just like ganglion cells, while photoreceptors, bipolar cells, and horizontal cells generate sustained and graded voltage potentials. The ganglion cells receive input from bipolar and amacrine cells and send the final relay to the brain for further processing. Interestingly, more than 50% of ganglion cells receive input solely from the foveal and parafoveal area, which is only about 5% of the retina. This leads to an extreme bias toward central vision rather than peripheral visual perception.

Compared with photoreceptors, the bipolar cells are much fewer, and there is an even further reduction in the number of ganglion cells. As many as 140 million rods and cones exist in human retina, but only 1 million ganglion cells. Each photoreceptor interacts with a number of different neurons, which also have several different synapses. These integrative features of the retinal neurons are probably essential for sorting out the signals.

Because light is usually discussed as the "activator" in phototransduction, the following statement may seem conflicting, but it is merely paradoxical: photoreceptor cells are depolarized in the absence of light and hyperpolarized in the presence of light. In the "*on*" or dark state, the photoreceptors release glutamate at the bipolar- and horizontal-cell synapses. The horizontal cells are also "*on*" in the dark as a result.

Upon the photoreceptor cells' release of glutamate, the "on" bipolar cells are inhibited and the "off" bipolar cells are excited. Therefore, the "on" bipolar cells are inhibited in the dark and excited in the light, while the "off" bipolar cells are excited in the dark and inhibited in the light.

The electroretinogram (ERG) measures the electrical activity of the various retinal neurons. The ERG measures the changing potential difference between two electrodes (specifically one placed on the cornea and a reference electrode), similar to the electrocardiogram (ECG) and electroencephalogram (EEG). An example of an idealized ERG tracing is shown in Fig. 19.6. The downward a-wave reflects photoreceptor activity, the b-wave reflects activity primarily of the bipolar and Muller cells as well as other cells in the inner nuclear layer, and the d-wave reflects the "off" response of the inner nuclear layer. With the ERG, it is possible to identify retinal dysfunction and, more importantly, to narrow the focus to specific cell types when diagnosing visual problems.

VISUAL PATHWAYS

Ganglion cell axons collect in fine bundles that arc around the macula and converge at the optic disc to form the optic nerve. The optic disc is insensitive to light because it does not contain any photoreceptors; this causes a blind spot 15 degrees temporal to the midline noticeable only on monocular viewing. The nerve fiber bundles maintain their spatial relations with the visual field as they traverse the optic nerve and the optic radiations to the visual cortex: fibers from the right hemifield of each retina are located on the right half of the optic nerves, and the fibers from the left hemifield pass along the left half of the optic nerves. The left and right optic nerves converge at the optic chiasm after passing through the optic foramina (Fig. 19.7). At the chiasm, nerve fibers corresponding to the nasal halves of each retina cross to the opposite side, while the temporal nerve fibers remain uncrossed. This arrangement results in the entire right visual field (projected onto the temporal hemifield of the left retina and the

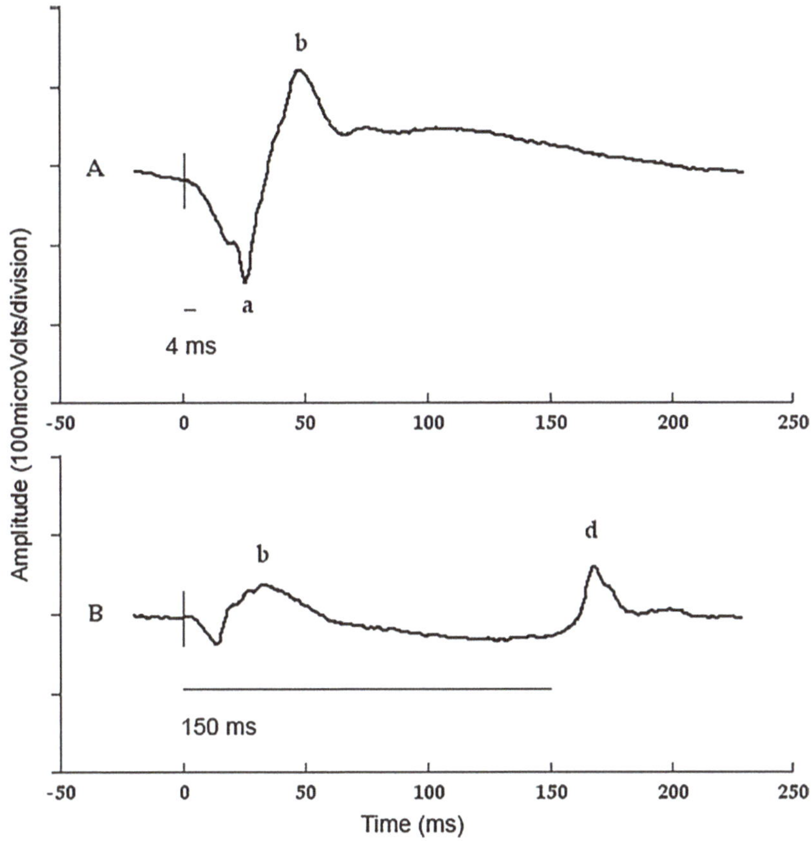

FIGURE 19.6 Normal electroretinogram (ERG) tracing in a human. (A) Dark-adapted ERG elicited by a brief (4 ms) flash on a dark-adapted retina. The downward a-wave (a) is generated mainly by photoreceptor hyperpolarization; the b-wave (b) is mainly generated by depolarization of cone on-bipolar cells and rod bipolar cells. (B) Light-adapted ERG elicited by a longer (150 ms) flash on a light-adapted retina. The b-wave (b) is generated by the light-stimulated depolarization of cone on-bipolar cells; the d-wave (d) reflects the depolarization of cone off-bipolar cells because of stimulus offset.

nasal hemifield of the right retina) being relayed to the left hemisphere of the brain and the left visual field (projected onto the temporal hemifield of the right retina and the nasal hemifield of the left retina) being relayed to the right hemisphere. Posterior to the optic chiasm, the optic fibers continue as the optic tract to the thalamus.

Some fibers separate from the optic tract and continue on to the pretectal area of the midbrain and the superior colliculi (tectum). These pathways activate the pupillary light reflex response and mediate the direction of eye movements in response to visual, acoustic, and somatosensory stimuli, respectively. Most of the optic tract fibers form synapses within the lateral geniculate nucleus (LGN), which is the most important subcortical region for visual stimuli processing. The spatial correspondence between the retina and the visual fields are maintained in the LGN, and thus it is said to be retinally mapped. LGN neurons are also segregated according to eye laterality, with separate layers of neurons receiving inputs from each eye. Ganglion cell subtypes in each retina send axons to three distinct cell

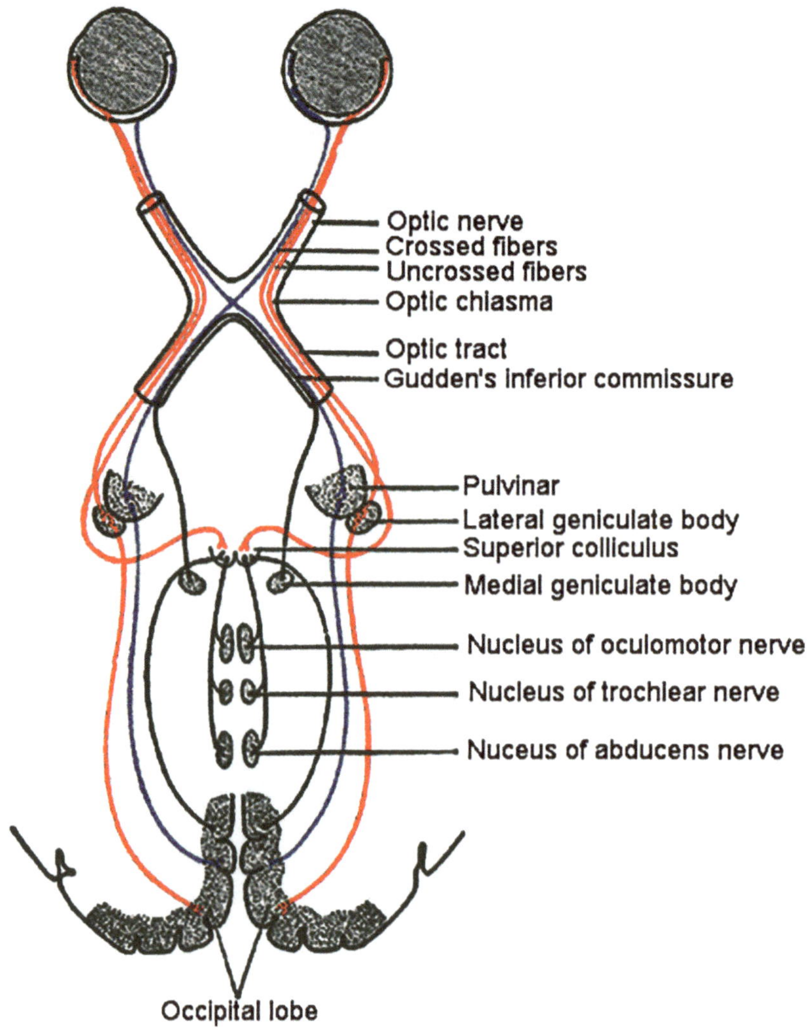

FIGURE 19.7 The mammalian optic visual system. *Modified from H. Gray (1918). Anatomy of the Human Body. Figure available at* https://commons.wikimedia.org/wiki/File:Gray722.png.

layers in the LGN, which has a total of six layers all in all. The retinotopic maps of the binocular fields from the two eyes are congruent except for a small region—the monocular segment—that represents the far periphery of the visual field of the contralateral eye (Fig. 19.8).

Axons of the upper quadrants of the optic radiation, corresponding to the lower visual field quadrants, project either to the medial laminae as the geniculostriate pathway or to the cerebral cortex at the superior edge of the calcarine sulcus. Axons of the lower quadrants, corresponding to the upper visual field quadrants, synapse at the lateral half of the lateral geniculate body, whose cells project to the inferior lip of the calcarine sulcus. The primary visual cortex (V1), also known as the striate cortex and Brodmann's area 17, receives the input from the LGN.

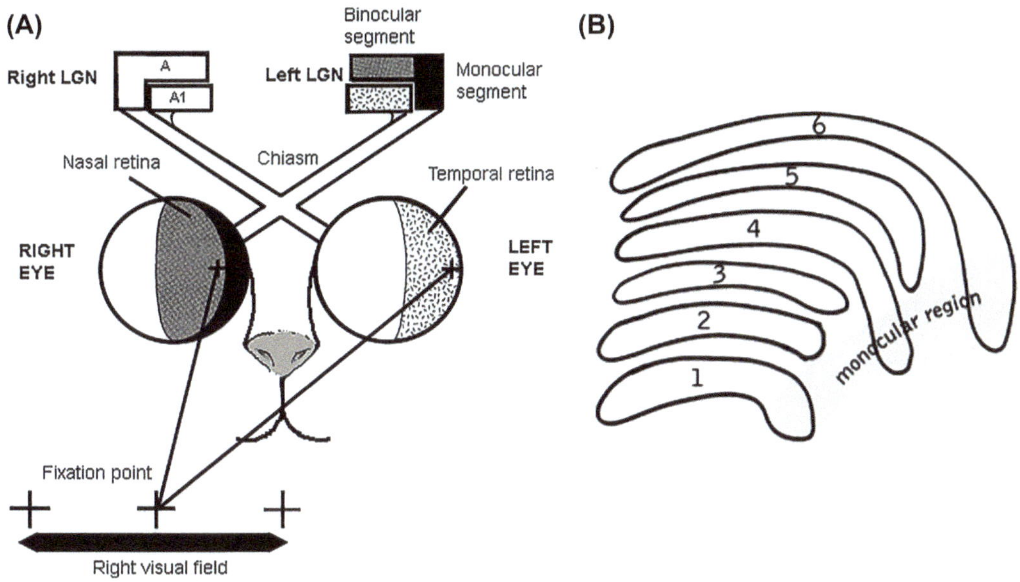

FIGURE 19.8 LGN organization. Each retinal hemifield projects to a separate cell layer in the LGN. (A) In the cat, the principle layers are called A and A1. The contralateral nasal retina projects to layer A, whereas the ipsilateral temporal retina projects to layer A1. The LGN is, hence, retinotopically mapped. The maps in both LGN are in perfect register, except for the monocular segment in the lateral region of layer A. This region is activated by stimuli from the far temporal visual field of the contralateral eye which falls outside the binocular visual field. (B) In humans, LGN layers 1, 4, and 6 receive information from the contralateral nasal retina (temporal visual field). Layers 2, 3, and 5 receive fibers from the ipsilateral temporal retina (nasal visual field). The monocular region (highlighted) only receives information from the far temporal visual field of the contralateral eye that falls outside the binocular visual field. *Modified from J. Hutchins. Lateral geniculate nucleus. Figure available at* https://commons.wikimedia.org/wiki/File:Lateral_geniculate_nucleus.png.

LGN neurons target a band of cells in the midcortex known as layer IV. Input to the different LGN layers do not intermingle in layer IV.

This was demonstrated by Hubel and Weisel by the injection of radioactive proline into one eye which was finally transmitted to eye specific–dominance columns in layer IV of the striate cortex. An illustration of the ocular-dominance columns is found in Fig. 19.9. There are about 1 million ganglion cells in each eye and a similar number of lateral geniculate neurons. The synaptic relay causes little transformation in the receptive fields. Moreover, the synaptic input from the LGN is sent to small, circular, monocular receptive fields in the layer IV neurons, which transmit the information to more superficial cortical layers, for example layer III, in which the same cells connect with axons from different ocular-dominance columns.

Most neurons in layer III receive stimuli from corresponding retinal areas in both eyes, and the receptive fields from the two eyes are mapped to the same location in space. Binocular vision is initiated, wherein information from both eyes are merged forming a single perception of visual space. Layer III neurons project to other areas in the cortex. The projection from layer IV to layer III is the first important elaboration of receptive fields. Neurons in layer III have elongated receptive fields and are sensitive to long, high-contrast bars or edges oriented along the receptive field axis, known as orientation selectivity. Hence, most cortical neurons have poor or no

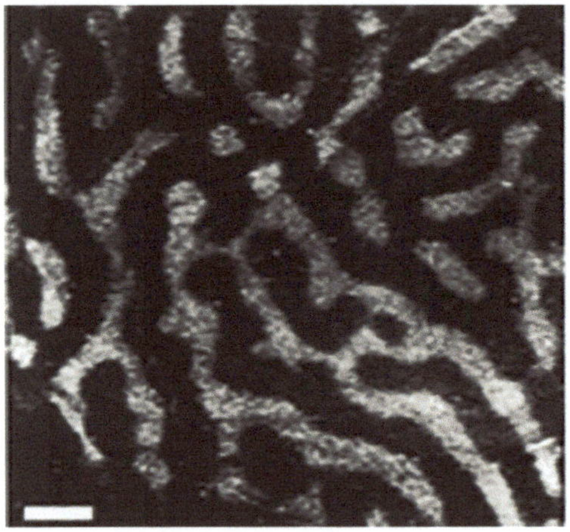

FIGURE 19.9 Magnified section across layer IV of the LGN showing ocular-dominance columns. Light colored bands represent nerve endings from one eye and the dark-colored bands represent nerve endings from the other eye. Arrows show examples of blood vessels. *Reproduced with permission from Stiles J, Jernigan TL. The basics of brain development. Neuropsychol Rev 2010;20(4):327–48.*

response at all to variations in diffuse illumination of the retina. Cortical neurons respond to contrast borders well because of excitatory and inhibitory intracortical connections.

There are additional projections from the striate cortex to a large number of extra striate visual areas of the cortex. For example, areas V1 and V2 integrate most of the information regarding color, form, and movement, and other areas fine-tune that information. The most important pathway for the conscious experience of vision in mammals is the geniculo-cortical pathway. People with striate cortex lesions perceive total or almost total blindness. However, they are still partially able to locate or identify objects—a sign that an intact retinotectal pathway provides some type of subconscious vision experience. These pathways are not independent of each other and are connected by a corticotectal pathway from the cortex to the superior colliculi. Several minor pathways, in addition to these two, exist but are not yet well understood. They lead to the ventral lateral geniculate (or pregeniculate) nuclei, hypothalamus, and tegmentum. These hypothalamic projections may possibly be involved in the synchronization of the circadian rhythm.

The cortex receives and relays information to the same cortical areas, including the opposite cortical hemisphere through the corpus callosum. Locus ceruleus neurons use epinephrine as a neurotransmitter and the Raphe nuclei use serotonin. The axons from the brain stem spread out widely and connect with a variety of cortical layers. Using acetylcholine as a neurotransmitter, the projection from the nucleus basalis of Meynert is organized in a similar pattern. Due to their anatomy and the effects induced by applying the corresponding neurotransmitters to cortical neurons, these inputs seem to play an important role in modulating visual processing in different behavioral states.

VISUAL SYSTEM DEVELOPMENT

A critical period in development may be considered a period of time wherein intercellular communication alters a cell's fate. This concept is attributed to Hans Spemann, an experimental embryologist during the turn of the 20th century, who showed that if a piece of early embryonic tissue was transplanted from a donor to a host, it would take on the characteristics of the host tissue but only if transplantation took place within a well-defined period. He further demonstrated that the transformation is not reversible. Cell-to-cell contact and chemical signaling were shown to mediate the transformation of the transplanted cells.

The work of Konrad Lorenz in the mid-1930s gave new significance to the concept of critical period with respect to brain development. Lorenz was interested in how graylag goslings form a social attachment to their mother. He observed that, in the mother's absence, the goslings formed a social attachment to any moving object, including Lorenz himself, and followed it and behaved toward it as they did to their mother. Lorenz used the term "imprinting" to stress the permanence by which this first visual image was etched in the young bird's nervous system. Lorenz found that imprinting is limited to a finite time (eg, within the first 2-day post hatching), which he called the critical period for formation of social attachment. It was Lorenz himself who drew the analogy between the induction of tissue to change its developmental fate at critical periods and the process of imprinting the external environment on the nervous system.

The impact of this work in the field of developmental psychology was tremendous. "Imprinting" and "critical period" imply that "behavioral phenotype" changes caused by early sensory experience were irreversible and permanent, much like the irreversibility of tissue phenotype determination during embryogenesis. The concept of critical period was extended by numerous studies to characteristics of mammalian psychosocial development. The prevailing theory was that the fates of CNS neurons and neural circuits depended on the animal's experiences during its early postnatal life. Research in this area was not only scientifically significant; it had a political impact as well.

The determination of neuronal fate requires the stimulation of peripheral sensory receptors and communication by chemical synaptic transmission. That synaptic activity can cause alterations in neuronal connectivity during CNS development was a concept that eventually received solid support from neurobiologists, beginning with Hubel and Wiesel, whose experiments partly earned them the 1981 Nobel Prize in medicine. They discovered that visual experience, or lack thereof, during early postnatal life, is one of the important determinants of neural pathways in the central visual system. Much work has been dedicated to the examination of experience-dependent plasticity of neural connections in the visual system, making it an excellent model to demonstrate the principle of critical period in the development of the nervous system.

ORGANIZATION OF THE VISUAL PATHWAY FROM RETINA TO CORTEX

Visual Pathway's Anatomical and Physiological Organization

In the first part of this chapter, the structure and function of the visual system were presented in detail. Some developmental aspects are presented in this section to give an overview on what should be accomplished in each developmental period to establish correct wiring and function of the sensory system. The visual pathway starts with the projection of the ganglion cell axons from the neurosensory retina to the optic nerve. The ganglion cells that receive stimulation from the right hemifield project to the left (contralateral) hemisphere. Similarly, the ganglion cells receiving stimulation from the left hemifield project to the right hemisphere. Because the human eyes are localized in the front of the head; the visual fields of the two retinas overlap in the central part of each visual hemifield, called the binocular visual field. Each retina projects to the two hemispheres; the ganglion cells localized in the nasal retina send their projection through the midline to the contralateral hemisphere, while the ganglion cells of the temporal retina send axons to the hemisphere of the ipsilateral side.

ACTIVITY-INDEPENDENT DEVELOPMENT OF ORDER IN THE VISUAL SYSTEM: PATHWAY, TARGET, AND ADDRESS SELECTION

The preceding discussion highlights the great precision in the neural connections making up the mature visual pathway. Here are several examples:

1. It is only the ganglion cell axons from nasal retinas that cross at the chiasm.
2. The diverse population of axons in the optic tract is segregated at the LGN by eye laterality and by retinotopic position.

3. Within the cortex, the LGN axons synapse at specific layers and within each layer, they are further separated according to retinotopic position and by eye laterality.
4. Layer IV cells connect with cells in other layers to facilitate binocular vision and are specialized to detect contrast borders.

Does the establishment of these highly specific connections depend on synaptic activity during critical periods of development? In the development of the CNS, activity often precedes experience. Chemical synaptic transmission in the visual pathway occurs *in utero*, even prior to photoreceptor cell development. Therefore, the fact that a developmental process occurs before visual input does not denote that it is independent of activity. Consider how the drug tetrodotoxin (TTX), which binds to voltage-sensitive sodium channels and blocks action potentials, differentiates developmental processes. Processes that are TTX-insensitive are considered to occur independently of intercellular communication, whereas those that are TTX-sensitive are dependent on synaptic activity.

There are three phases in the development of long-range connections in the CNS: pathway selection, target selection, and address selection. Some situations that exemplify pathway selection are: growing axons that originate in the nasal retina cross to the other side at the optic chiasm, and LGN axons project to the cortex via the optic radiations instead of to the spinal cord through the cerebral peduncle. Target selection examples include the decision of LGN axons to innervate cortical layer IV instead of V, and the decision of optic tract axons to form connections with the LGN instead of part of the thalamus. Address selection is exemplified by the segregation of inputs according to retinotopic location in the cortex and LGN, the separation of axons from the two eyes in the LGN and layer IV of the cortex, and the convergence of retinally mapped input from both eyes onto layer III neurons.

There is now evidence that pathway and target selection take place entirely in the absence of neural activity. Several aspects of address selection (eg, the formation of crude retinotopy) are also independent of synaptic activity. However, these do not indicate that these phases in development are not dependent on intercellular communication; cell–cell contact and gradients of diffusible chemicals are critical in virtually all phases of visual-pathway development. The phenomenon that neurons make connections by differential chemical attraction or repulsion is called "chemoaffinity."

The extracellular matrix (ECM) of the ventral wall of the optic stalk forms a molecular highway for retinal axons to grow on. The growth cones, which are the growing tips of the axons, use surface molecules such as integrins to bind laminin in the ECM. This interaction contributes to axonal elongation (Fig. 19.10). The next mechanism that makes axons that are growing together to stick together is termed "fasciculation," caused by the expression of specific cell-adhesion molecules on the membrane surface of axons.

Depending on the different cell-surface receptors the axons express, it may manifest different

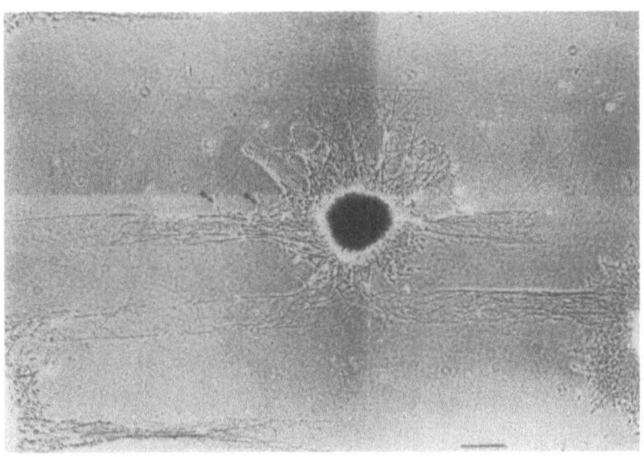

FIGURE 19.10 Neurites that grow in a patterned substrate with parallel strips of laminin applied to a type IV collagen-coated culture dish. The neurites branch out but restricted to the strips of laminin. Small fascicles extend onto the collagen substrate (arrows), indicating a slight compromise of the ability of the neurites to duplicate the laminin pattern. *From Gunderson RW. Response of sensory neurites and growth cones to patterned substrata and fibronectin in vitro. Dev Biol 1987;121:423.*

chemoaffinity to the ECM. Temporal axons from the retina travel toward the chiasm, but they receive a signal there that shifts them away. On the other hand, nasal axons converse across the midline and continue into the contralateral optic tract. Such differences are contributed by differential expression of cell-surface molecules based on cell position in the retina. It is believed that the nasal–temporal gradient in surface markers of axons is matched to the complementary gradients on the surfaces of cells in target structures, which generates retinotopy.

Once axons reach their target destination, they receive signals from their new extracellular environment that tells them to stop further growth. These signals may include the absence of specific glycoproteins in the ECM (eg, laminin), or diffusible signals from nearby cells. Neurotransmitters can also inhibit axonal growth, probably by increasing calcium concentrations. Evidence shows that diffusible factors can promote axon growth and guide axons to distant targets. It was previously shown that Zic2, a transcription factor, together with the guidance receptor EphB1 and its ligand ephrin B2, are critical components of a scheme that prevents ventral temporal RGC fibers from crossing at the optic chiasm. There are several other pairs of factors that regulate growth of axons outside the eye.

These molecular mechanisms give rise to a very organized visual system. Synaptic activity appears to refine the patterns of connectivity according to functional criteria.

ACTIVITY-DEPENDENT DEVELOPMENT OF ORDER IN THE VISUAL SYSTEM

An important part of the organization of the visual pathway is identified without any participation of from neural function and synaptic communication. The lateral geniculate axons and the retinal ganglion cell axons project through the proper pathways to end in retinotopic order in the correct destination structures. The important characteristics of the visual system architecture that depend on retinal function are the projection of axons in the LGN and the cortex in concordance with the eye that transmits different signals to them and the installation and preservation of connections in the visual cortex that create binocular and stimulus-selective receptive fields. Before birth, eye segregation starts in the LGN.

Activity Influences Axon Segregation in the Lateral Geniculate Nucleus

The axons from the contralateral retina are usually the first to reach the LGN. At the LGN, they disperse and occupy the entire nucleus. The ipsilateral projection arrives at a later time and intermingles with the contralateral eye axons. Over a period of several weeks, axons from the left and right eyes segregate into specific domains. It has been shown that intraocular injection of TTX prevents segregation because it is dependent on synaptic activity in the retina. What is then the source of the activity, and how does it orchestrate segregation?

It is certain that, because segregation occurs before birth, and before photoreceptor development, it is not driven by light stimulation. Ganglion cells already show nonrandom activity during fetal development; they fire in quasi-synchronous waves that propagate across the entire retina. The wave origin and direction may be random, but during each surge, ganglion cell activity is highly associated with the activity of cells in its vicinity. These waveforms are generated independently in the two retinas, thus there is no correlation between the two different activity patterns.

Segregation is believed to depend on a model of synaptic stabilization first proposed by Donald Hebb in the 1940s, wherein only retinal terminals that are active simultaneously with their postsynaptic partner LGN neurons are retained. This is underscored in the mnemonic: "Neurons that fire together wire together," coined by Sigrid Lowel and Wolf Singer. Neural connections are said to employ Hebb synapses when they are modified according to this rule. The underlying hypothesis is that when a postsynaptic LGN neuron is activated by a wave of retinal activity, the active retinal inputs are consolidated in the neuron. Since the activity from the two eyes occurs independently, the inputs operate on a winner-takes-all pattern so that only one input is retained while the other is quenched, resulting in the complete segregation of the two inputs.

Retinotopic Features in Cortical Layer IV

Segregation of axons not only occurs in the LGN, but also in cortical layer IV (Fig. 19.11). The development of ocular-dominance columns can be affected by vision deprivation, especially monocular deprivation (eg, such as that caused by a unilateral congenital cataract), in which image formation is greatly impaired.

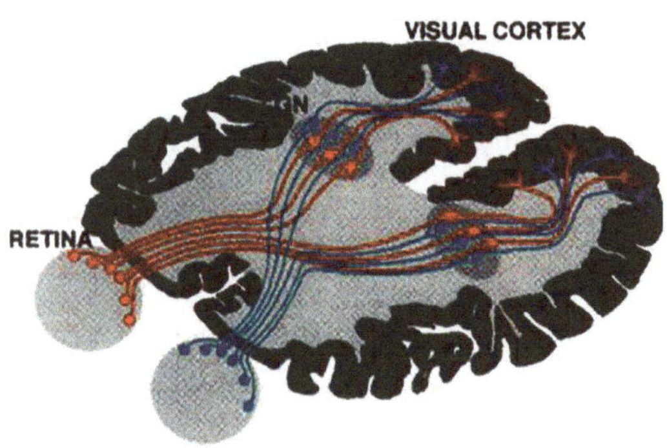

FIGURE 19.11 Retinotopic representation in the visual pathway. The nasal axons cross over in the midline of the optic chiasm and travel to the contralateral LGN, whereas the temporal axons project to the ipsilateral LGN. Then the LGN axons project to the layer IV of the primary visual cortex where they are organized in alternating eye-specific regions. *From Anna AP, Carla JS. Brain waves and Brain wiring: the role of endogenous and sensory-driven neural activity in development. Pediatr Res 1999;45:447–58.*

If experimental animals have one eye sutured shut after birth and it remains sutured in the period of axonal segregation, the ocular-dominance columns for the open eye will develop to be much wider than columns corresponding to the sutured eye (Fig. 19.12). However, if deprivation occurs later than the period of axonal segregation, the anatomical change in ocular-dominance columns are not observed, implying a critical period for the development of retinotopic features in cortical layer IV.

Within this critical period, it is possible to reverse the effects of monocular deprivation by opening the previously closed eye and closing the previously open eye. This can cause the shrunken ocular-dominance columns in the stimulus-deprived side to expand, and those in the expanded columns to shrink. This shows that even after ocular-dominance column segregation appears to be complete anatomically, the dynamic equilibrium established in the afferents serving the two eyes can still be disrupted by visual deprivation. Beyond the critical period, the afferents appear to lose their capacity for retraction and growth.

The sizes of the LGN neurons that relay information to the visual cortex also change with alterations in the cortical ocular-dominance columns. LGN cells shrink when deprived of normal visual input. This reflects the decreased axonal connections of cells in layer IV. Shrinkage, oddly enough, is observed only in the LGN fraction where both retinas are represented. Cells that receive monocular input are not substantially affected by sensory deprivation. In addition to this, when the central retina in the nondeprived eye is destroyed, a region of the LGN becomes relatively free of the effects of binocular competition, resulting in less shrinkage of LGN neurons. These observations imply that the loss of territory in layer IV of sensory-deprived afferents is not merely caused by disuse, but rather by binocular competition. The synaptic disconnection of afferents that serve the closed eye is seemingly fostered by activity in the open eye.

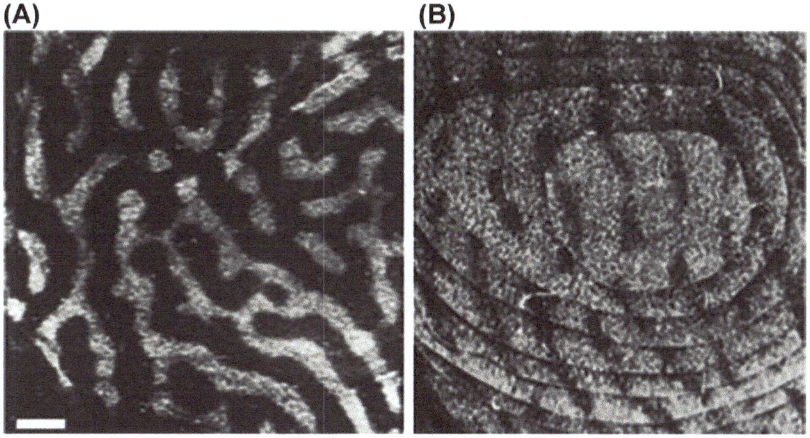

FIGURE 19.12 Dark field autoradiographs of ocular-dominance columns (ODC) in two monkeys after injection of one eye with ^{3}H-proline. (A) Normal pattern in a 6-week old monkey. (B) ODC pattern from a monkey that was monocularly deprived for 18 months starting at 2 weeks of age. The tracer was injected into the nondeprived eye at 18 months. The ODC for the nondeprived eye expanded (*light bands*) whereas those of the deprived eye shrunk (*dark bands*). *Reproduced with permission from Stiles J, Jernigan TL. The basics of brain development. Neuropsychol Rev 2010;20(4):327–48.*

Influence of Visual Experience on the Establishment and Maintenance of Binocular Connections

The connections that promote binocular vision are the last to be molded in the retinal–geniculate–striate pathway. Sensory experience in early postnatal life influences the formation and modification of these connections. The establishment of binocular receptive fields depends on generating analogous activity patterns between the two eyes, unlike the segregation of eye-specific domains which depends on asynchronous patterns of activity in the two retinas. For example, monocular deprivation by replacing pattern vision with "white noise" in one eye greatly disrupts the binocular connections in the striate cortex. Even a brief time of monocular deprivation will cause neurons outside of layer IV to respond only to stimulation from the normal eye. This change of the binocular organization of the cortex is termed "ocular-dominance shift" (Fig. 19.13).

A critical period of postnatal life exists for a shift in ocular dominance to occur. In cats, it peaks at 1 month of age and reduces to very low levels by 3–4 months of life. But in humans, it is estimated that this critical period extends to about 10 years of age, which coincide with the period of fastest growth of the head and optical axes. It is believed that the plasticity of binocular connections is necessary to maintain proper binocular vision during this period. The activity-dependent fine-tuning also causes these connections to be highly susceptible to deprivation.

Strabismus, or misalignment of one or both eyes, is a condition that disrupts cortical binocularity.

In chronic, uncorrected cases, there is a total loss of binocular receptive fields even though the two eyes retain equal cortical representation (Fig. 19.14). This is because the visually evoked patterns arrive at the cortex out of sync. This shows that competition, rather than disuse is responsible for the disconnection of inputs from one eye. Strabismus, if early in onset, can also make the segregation of ocular-dominance columns in layer IV more distinct.

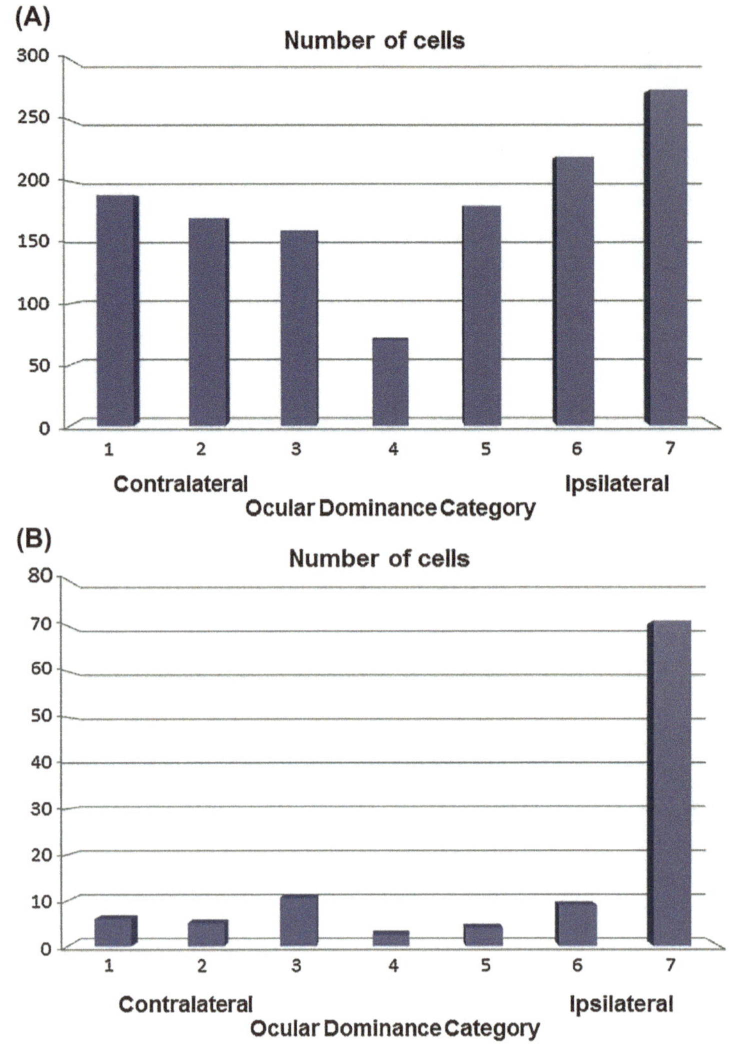

FIGURE 19.13 Ocular-dominance shift after monocular deprivation. Illustrated are histograms of ocular-dominance data obtained from the striate cortex of (A) normal monkeys and (B) monkeys that had been monocularly deprived early in life. The bars show the number of neurons outside of layer IVc in each of the seven ocular-dominant categories. Cells in groups one and seven are activated by stimulation of the left or right eye, respectively, but not both. Cells in group four are activated equally well by either eye. Cells in groups two and three and in groups five and six are binocularly activated, but they show a preference for the left or right eye, respectively. The histogram in A reveals that most neurons in the visual cortex of a normal animal are driven binocularly. The histogram in B shows that a period of monocular deprivation leaves few neurons responsive to the deprived eye. *Data from Wiesel TN. Postnatal development of the visual cortex and the influence of the environment. Nature 1982;299:583.*

Changes in binocularity and ocular dominance after sensory deprivation are associated with physiological consequences. The loss of binocularity in conjunction with strabismus completely eliminates stereoscopic depth perception, whereas monocular deprivation stimulates an ocular-dominance shift that favors the healthy eye and leaves the affected eye visually impaired. This effect can be reversible if it is corrected during the critical

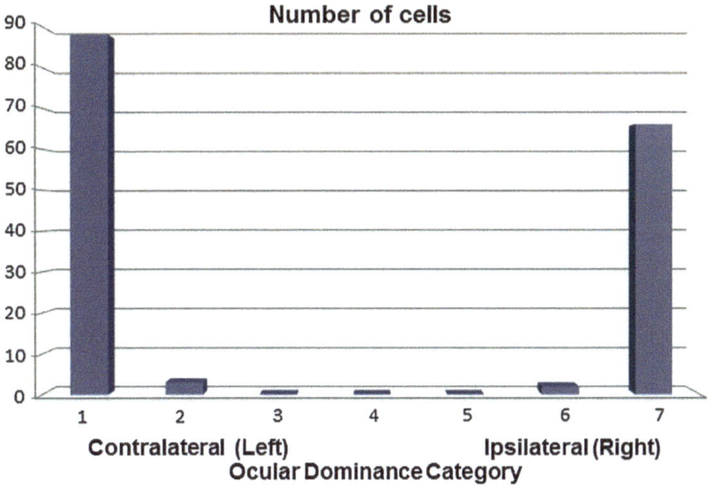

FIGURE 19.14 Ocular-dominance histogram of cells in the striate cortex of a 3-year old monkey in which the lateral rectus of the right eye was transected at 3 weeks old. The bars represent the number of neuronal cells outside of layer IVc in each of the seven ocular-dominance categories. Cells in groups one and seven are activated by stimulation of the left or right eye, respectively, but not both. Cells in 2, 3, 4, 5, and 6 are binocularly activated. This diagram shows that binocular cells are almost completely absent. *Data from Wiesel TN. Postnatal development of the visual cortex and the influence of the environment. Nature 1982;299:583.*

period. The take-home message is that conditions that cause monocular sensory deprivation, such as unilateral congenital cataract or disorders of ocular misalignment, must be corrected during early childhood to avoid permanent vision loss.

Extraretinal "Enabling Factors" Influence the Development of Binocularity

With increasing age, additional restraints on the forms of activity seem to modify cortical circuits. In the prenatal period, spontaneous bursts of retinal activity are sufficient to direct aspects of address selection in the LGN and cortex. In the neonatal period, an interaction with the visual environment, such as attending to visual stimuli or using vision to guide behavior, begins to play an important role. During this period, retinal activity derived from visual stimulation may be insufficient for modifications of binocularity. In anesthetized animals, monocular stimulation could induce a brisk response in cortical neurons, but did not affect modifications of binocularity. These findings have led to the hypothesis that behavior-related release of extraretinal "enabling factors" is of critical importance for the synaptic plasticity in the cortex. Some evidence suggested that eye movements may induce this release.

Evidence points to the probability that these factors might be related to numerous modulatory systems congregating at the striate cortex, including cholinergic input from the basal forebrain and noradrenergic input from the locus ceruleus. The trajectory of these axonal projections is distinct from that of the optic radiations. Upon surgical transection of these modulatory inputs in animal models, ocular-dominance plasticity outside of cortical layer IV is substantially impaired, while transmission in the retinal–geniculate–cortical pathway remains normal. Acetylcholine and norepinephrine increase the excitability of cortical neurons and may lower the firing threshold of these neurons in response to visual stimuli.

Elementary Mechanisms of Synaptic Plasticity During the Critical Period May Involve NMDA Receptors

Hebb synapses, which have been explained earlier in this text, account for much of the activity-dependent development of the visual system, particularly the growth of binocular connections. To illustrate this, afferents synapsing on the same cell are joined together only if they convey synchronous activity patterns. The neurotransmitter employed at all of the modifiable synapses (eg, geniculocortical, retinocortical, and corticocortical) is likely glutamate, aspartate, or another amino acid, and recruits excitatory amino acid (EAA) receptors. This receptor family is categorized into metabotropic and ionotropic receptors. Metabotropic receptors are linked to intracellular second-messenger systems by G proteins. Ionotropic receptors are ion channels that conduct positively charged ions into the postsynaptic cell. Ionotropic receptors may be subclassified into AMPA (α-amino-3-hydroxy-5-methyl-4-isoxazole propionic acid) receptors and NMDA (N-methyl-D-aspartate) receptors, named after the compounds that act as agonists at their respective receptor sites. AMPA and NMDA receptors colocalize at numerous synapses. Activation of the AMPA receptor at the postsynaptic junction triggers a cascade of monovalent cation conductance with a linear current–voltage association and a reversal potential of about 0 mV. Activation of this same receptor by glutamate stimulates a depolarizing current whose amplitude diminishes over time.

There are two unusual features that distinguish the NMDA receptor from the AMPA receptor: its voltage dependence and its Ca^{2+} conduction. It is voltage-dependent because at the resting membrane potential, Mg^{2+} enters the cell, lodges in the channel and physically obstructs it, thereby blocking conduction, whereas during depolarization, ion conduction occurs because the Mg^{2+} block is relieved. The conduction of current through the NMDA receptor channel is contingent on the release of glutamate at the presynaptic terminal and the depolarization of the postsynaptic membrane. The second unusual feature is that the NMDA receptor is a Ca^{2+}-ion channel. The level of Ca^{2+} passing through the NMDA receptor channel determines the degree of presynaptic and postsynaptic coactivation. NMDA receptors in the cortex and LGN serve as sensors of presynaptic and postsynaptic activity and Ca^{2+} entry through the NMDA receptor channel generates the biochemical mechanisms that modify synaptic effectiveness. The connections from layer IV to layer III neurons in the visual cortex have been demonstrated to show Hebbian enhancement.

Conditioning synapses between layer IV and III neurons results in long-term potentiation (LTP) (Fig. 19.15). This is prevented by the application of 2-amino-5-phosphonovaleric acid (APV), an NMDA receptor antagonist. That the application of APV in vivo can also disrupt the following mechanisms suggests that they contribute to innate synaptic remodeling: binocular competition in layer IV, segregation of eye-specific inputs in the LGN, ocular-dominance domains after monocular deprivation, and the modification of binocular connections in the superficial layers.

Simultaneous firing of presynaptic and postsynaptic neurons strongly activates NMDA receptors, and this contributes to the formation of a synapse between them during development. Low levels of NMDA receptor activation, on the other hand, lead to less Ca^{2+} influx, which in turn triggers long-term depression (LTD), by which the effectiveness of the active synapses is decreased. The preservation of a synaptic connection formed during development may therefore be dependent on its ability to surpass the threshold for an NMDA-receptor-mediated response. Failure to achieve this threshold leads to disconnection. Both processes depend on activity in the retinofugal pathway and on postsynaptic Ca^{2+} entry. After a connection was formed during development, it may need to evoke an NMDA receptor-mediated response beyond

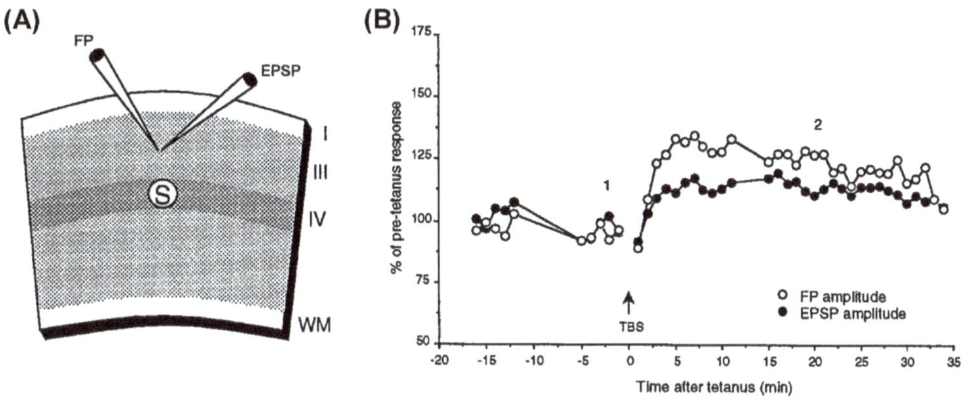

FIGURE 19.15 Stimulation of layer IV results in a long-term potentiation (LTP) of synaptic responses in layer III. (A) Electrical stimulation (S) was applied to layer IV, and intra- and extracellular synaptic responses were monitored in layer III. (B) Theta burst stimulation (TBS) evoked a sustaining increase in the intracellular (*solid circles*) and the extracellular (*open circles*) responses. *EPSP*, excitatory postsynaptic potential; *FP*, field potential. *From Kirkwood A, Bear MF. Hebbian synapses in visual cortex. J Neurosci March 1994;14:1634–45.*

some threshold level to maintain the connection, otherwise disconnection will happen. Activity in the retinofugal projection and postsynaptic Ca^{2+} entry will influence both processes.

Plasticity of Visual Cortex After Critical Periods

Multiple critical periods exist for visual system development. The critical period for activity-dependent anatomical modification of geniculate axonal arbors in layer IV is followed by the critical period for the experience-dependent change of binocular connections. It is important to mention that the primary visual cortex of the adult brain is still plastic, although it has passed beyond the critical period for a specific developmental process. The concept of critical periods must be discussed based on what intercellular interaction being considered.

The adaptation of striate cortical circuits following a circumscribed lesion in the retina gives an illustration of plasticity in the adult brain. Activity of the cortical neurons at the receiving point in the retinotopic region initially falls and become unresponsive to any form of visual simulation. These neurons regain responsiveness in a few weeks, although to visual stimulation from areas of the retina surrounding the lesion.

The presence of a retinal lesion causes the brain to "fill in" the defect in the cortical retinotopic map (Fig. 19.16). This map plasticity is believed to be stem from the modification of horizontal projections that interconnect the layer III neurons throughout the retinotopic map. This demonstrates that, even beyond the critical period, cortical circuits can be modified by peripheral lesions.

It is assumed that, throughout life, the cortex employs synaptic adjustments to reorganize itself as an adaptation to changes in its sensory environment. Synaptic plasticity mechanisms—LTP and LTD—persist in the adult striate cortex's superficial layers. There is a possibility that the experience-dependent synaptic remodeling mechanism in the cortex contributes to environmental adaptation, as well as to memory and learning in the adult brain.

Why Critical Periods End

The plasticity of visual connections is maintained in the adult brain, but the range over

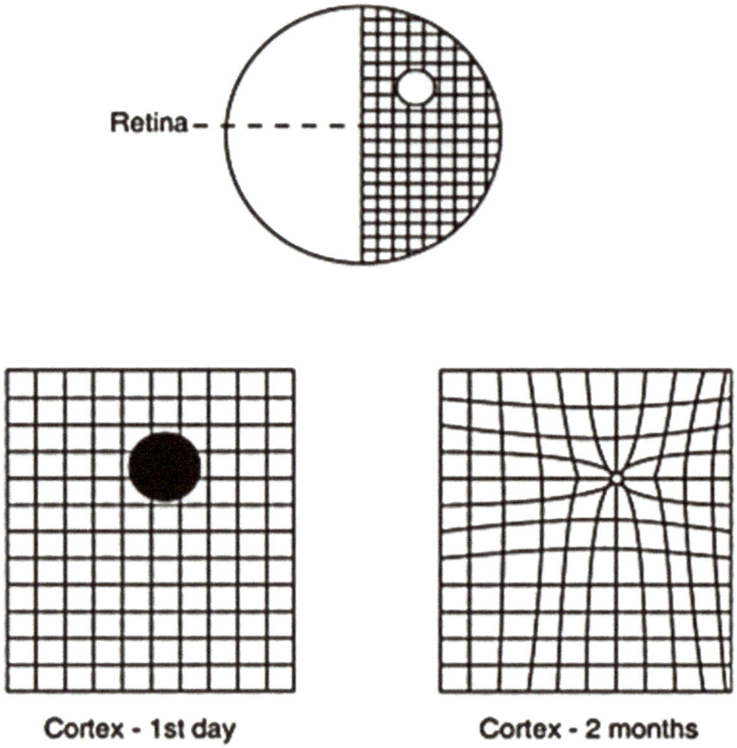

FIGURE 19.16 Visual cortex plasticity in the adult. A retinal lesion causes the corresponding region in the cortex to be initially silent. Eventually, these silent neurons learn to respond to signals from neighboring retinal areas, and "fill in" the defect. *Modified with permission from Charles Gilbert, Rockefeller University. Image originally printed in Barinaga M. The brain remaps its own contours. Science 1992;258:216.*

which this plasticity occurs decreases with advancing age. In early development, anatomical rearrangements of axonal branches are possible, but in adulthood, the plasticity seems to be restricted to changes in local synaptic efficacy. Furthermore, as the brain matures, the adequate stimulus for evoking a change appears to be increasingly limited. As an example, whereas patching one eye causes a profound alteration in cortical circuitry during infancy, patching during adolescence fails to elicit the same effect.

It is still unclear why critical periods end. With increasing knowledge about the basic mechanisms of axonal path finding and synaptic plasticity, more will be learned about how these processes are regulated. Although some of the rate-limiting factors are already identified, modulating the activity-dependent plasticity in the developing visual system. During a critical period, the initial activity-dependent development of vast connections in the LGN and cortex achieve the final form. To answer the question why critical periods end, for example, the formation of ocular-dominance column in layer IV, one hypothesis is that once segregation finishes and the afferents disconnect from the same postsynaptic cells, the substrate for the "winner-takes-all" mechanism disappears. If that is true, then monocular deprivation, which suppresses a competing input, only "saves" the initially open-eye axonal arbors from being retracted. According to the assumption that axons fail to grow after they have reached their target structure and achieve their initial forms of connection, when segregation is complete, the final pattern of connectivity is "imprinted" permanently due to loss of synapses and axon retraction.

However, reverse suture experiments found that re-expansion of ocular-dominance columns could occur after segregation is complete, which indicates that there are more explanations for the critical period ending. Another hypothesis appears to be changes in the ECM or surface glycoproteins expressed by the geniculate axons, which may explain the loss of the capability for axonal elongation in layer IV. A third possible explanation for the ending of critical period is based on the changes in the elementary mechanisms of synaptic plasticity. It has been shown that some EAA receptors change during postnatal development. For example, there is evidence that when binocular connections are most vulnerable to monocular deprivation, EAA-stimulated phosphoinositide turnover reaches a peak in the visual cortex; but at the end of the critical period, it is almost absent. At the same time, the effectiveness of NMDA receptors in layer IV decreases, which can be detected at the single-channel level. It may be caused by a developmental change in the NMDA-receptor subunit composition.

With increasing age, certain types of activity can no longer activate NMDA receptors or other elementary mechanisms to trigger plasticity, possibly due to filtration by successive synaptic relays. Some neurotransmitters, such as acetylcholine and norepinephrine, may facilitate synaptic plasticity in the superficial cortical layers, possibly through enhancing polysynaptic intracortical transmission. The decline in plasticity may be explained by a decline in the effectiveness of these neurotransmitters, or a change in the background condition under which they are secreted.

Some evidence suggests that the injection of the adult cortex with norepinephrine can restore some degree of plasticity in the cortex. However, there is evidence showing that intrinsic inhibitory circuitry matures late in the visual cortex. Thus activity patterns with modifiable synapses in superficial layers in early postnatal development may be inhibited in the adult. Whatever the reason for the decrease in synaptic plasticity during the end of the critical periods, it seems that the duration of the critical step is not determined by the precise postnatal age. For example, dark rearing of kittens appears to slow the critical step for modification of binocular connections. It also inhibits the segregation of ocular-dominance columns and some of the modifications of properties in EAA receptors. The duration of this critical step is more closely related to the history of stimulation more than temporal age.

OCULOMOTOR SYSTEM

The oculomotor system is the part of the CNS, which functions mainly in maintaining visual stability and controlling eye movements. It is a very delicate system. The fovea, which has the highest visual acuity on the retina, can only discern an area with the size of a quarter held at arm length. To gain precise and clear vision, the eyes must be directed precisely at an object of interest and held in a stable state. Otherwise it will cause blurred vision or diplopia if the eyes do not properly align with each other.

The oculomotor system is composed of the efferent limb of the visual system and the vestibular system. It is important to remember that the senses of hearing or touch also have effects on eye movement. Five distinct eye movement systems are responsible for voluntary and reflex selection of a visual target: vestibular, optokinetic, smooth pursuit, saccadic, and vergence. The extraocular muscles control all movements of the eyes. Several brain stem structures in the vertical and horizontal planes precisely modulate the actions of these muscles. No skeletal muscles that are innervated by spinal cord nerves can compare to the fine structure and types of extraocular muscle fiber. The premotor structures also connect with motor neurons in the cervical spinal cord which controls the muscles of the neck, enabling the motor behavior of gaze. Gaze involves both eye and head movement, which is produced by a complex interaction between the visual and vestibular systems.

Understanding of the brain stem, cerebellar and cortical structures of the oculomotor system can help clarify the mechanisms underlying eye movements. Sometimes it is clinically possible to distinguish between the various components of the oculomotor system as lesions in different part of the brain stem, cerebellar and cerebral cortex produce specific gaze impairment. This chapter will illustrate the consequences of these lesions and link the neuroanatomical pathways to the five different eye movements.

FIVE EYE MOVEMENTS

There are five different types of eye movements: vestibular, optokinetic, smooth pursuit, saccadic, and vergence. The first four are conjugate movements, which mean both eyes move in the same direction simultaneously. These eye movement types share some parts of the brain stem neural circuitry. Vergence, on the other hand, is disjunctive movement, which is mediated by different neural control mechanisms from those used by conjugate eye movement. These two systems only share the same motor neurons and extraocular muscles.

Vestibular eye movements are elicited by head movements that activate the semicircular canals. The *vestibulo-ocular reflex* (VOR) leads to eye movements in the direction opposite from head movement, thus stabilizing images on the retina. The interaction of the VOR and vision can be tested by fixating on the thumb with the arm outstretched while rotating the head. As the VOR compensates for head movement, note that the thumb stays in focus but the peripheral images are blurred. If the VOR is impaired, any movement of the head causes blurred vision, just like shaking a camera blurs the photos. Because the semicircular canals are most sensitive to brief or rapid movements, the VOR also functions best in the same movements.

Optokinetic eye movements are elicited when the entire visual field is moving, such as when looking out the window of a moving car. The optokinetic reflex includes a steady eye movement in the direction at which the scene is moving that serves to stabilize the image on the retina. Optokinetic nystagmus is defined by a slow eye movement in the direction of a moving object and a rapid return of eye position in the opposite direction. The vestibular system works with the optokinetic system to maintain the focus of the image on the retina during head movements. The optokinetic reflex is able to compensate for sustained or slow head movements which are poorly detected by the vestibular apparatus.

Smooth–pursuit eye movement follows images as they move across the field of vision. These movements keep the target images moving in the visual field centered on the fovea. When smooth-pursuit eye movements occur with head movement in the same direction the VOR must be stopped or suppressed.

The *saccadic* eye movement is characterized by their rapidity in scanning that modify foveal fixation from one point in the middle of a visual field to another target in the periphery. Saccades are used when you scan a picture or when you skip from word to word in reading a book, but also replace or increase smooth pursuit when the picture moves quickly in the visual field. Intentional saccades are those made to a recalled target, whereas reflexive saccades are carried out in response to a new object in the peripheral visual field.

The *vergence* eye movement is elicited when shifting focus from a distant to a near object or vice versa. They are disconjugate eye movements because they do not cause the eyes to move in the same direction, but in opposite directions. There are two types of vergence movements: *convergence* is when both eyes move toward the middle to fixate on a near object, whereas *divergence* is when both eyes move away from the midline to focus on an object in the distance. Vergence movements are accompanied by changes in the shape of the lens and the diameter of the pupil. The near response or the near triad is composed of convergence, an increase in the lens thickness (also known as accommodation), and constriction of the pupil (miosis).

EXTRAOCULAR MUSCLES

All changes in eye position result from the contraction and the relaxation of extraocular muscles. There are six extraocular muscles responsible for eye movements: the superior rectus, lateral rectus, medial rectus, inferior rectus, superior oblique, and inferior oblique muscles. The two muscles which play the role of the primary elevators and depressors of the eyes are the superior and inferior recti, respectively. The medial rectus and lateral rectus adduct the eye. The activity of the oblique muscles is more complex. We note that the superior oblique is considered to be the primary intorter (inward rotation of the eye) and the inferior oblique is the primary extorter (outward rotation). We note that some defects in the extraocular muscles, or to one of the three cranial nerves involved in eye movement can be detected in routine ophthalmic examination. Downward deviations of the eye are called hypotropias, and upward deviations of the eyes are called hypertropias. Lateral deviations are called exotropias, while medial deviations are known as esotropias. Another extraocular muscle named the levator palpebrae superioris, elevates the upper eyelid. Dysfunction of the levator muscle can cause eyelid droop, or ptosis.

MOTOR NUCLEI OF THE EXTRAOCULAR MUSCLES

Inputs from premotor structures that are related to the control of different types of eye movements converge in a final common pathway on the motor neurons of the CN III, IV, and VI nuclei. The three cranial nerves mentioned provide general somatic efferent innervation to the extraocular muscles.

Oculomotor Complex

The motor neurons of the oculomotor complex (eg, cranial nerve III) innervate the ipsilateral medial rectus, inferior rectus, and inferior oblique, as well as bilateral superior recti and levator palpebrae superioris muscles with contralateral predominance. The axons destined for the superior recti and levator palpebrae superioris muscles decussate near the oculomotor nucleus before traveling ventrally and exiting the brain stem with the rest of the third cranial nerve branches. The oculomotor nerve exits the brain stem from the anterior surface of the midbrain, through the interpeduncular fossa. The nerve passes between the posterior cerebral and the superior cerebellar and arteries as it exits the brain stem. Note that oculomotor nucleus lesions would affect eye movements in both eyes, whereas oculomotor nerve lesions would affect only the ipsilateral eye.

The organization of neurons within the oculomotor nucleus mirrors the topography of muscles they innervate; the motor neurons of the medial rectus and inferior rectus are adjacent to each other, as well as that of the superior rectus and levator palpebrae superioris. Inferior oblique motor neurons are also situated close to the superior rectus motor neurons, which is largely a reflection of the synergistic actions of the two muscles. The motor neurons supplying the levator palpebrae superioris occupy the *central caudal subnucleus*.

The somatic oculomotor nucleus and the supraoculomotor region contain *internuclear neurons* that have connections with the abducens nucleus and facial nucleus in the brain stem. A fraction of the internuclear neurons are joined by collaterals from motor neuron axons, and may be involved in carrying out the synchronized activities of the facial and extraocular muscles (ie, during blinking).

Overlying the rostral portion of the somatic oculomotor nucleus is the *Edinger–Westphal* nucleus, which contains preganglionic parasympathetic neurons. An additional population of autonomic neurons is found in between the somatic portions of the oculomotor nucleus along the midline. The general visceral efferent axons of these neurons travel via the third cranial nerve to the ciliary ganglion, where they synapse with postganglionic fibers that innervate the sphincter pupillae of the iris and ciliary body. The parasympathetic fibers to the sphincter pupillae of the iris carry out pupillary constriction and those fibers to the ciliary body carry out lens accommodation. The autonomic and somatic divisions of the oculomotor nucleus carry out the near response—a triad consisting of convergence, pupillary constriction, and accommodation.

A third-nerve palsy involves somatic deficits and loss of parasympathetic functions, such as pupillary constriction. Typically, when parasympathetic fibers are compromised, a patient would manifest with pupillary dilatation (mydriasis), the absence of direct and consensual light reflexes for the affected eye, and inequality of pupil size or anisocoria. The somatic portion of the third cranial nerve innervates majority of the extraocular muscles and the levator palpebrae superioris, thus a lesion in this portion would give rise to strabismus and a droopy eyelid (ptosis). At rest, the motility deficits result in an exotropia (lateral outward eye deviation caused by an unopposed lateral rectus). Vertical muscle involvement only becomes apparent when the patient is asked to look up or down; lateral downward gaze is still intact because of the uninvolved superior oblique but upward gaze and downward gaze in other directions are compromised. Extraocular muscle testing, which tests eye movements in six cardinal directions, is part of routine ophthalmic examination because it can aid in diagnosing the site of nerve or brain stem lesions.

Trochlear Nucleus

The trochlear nucleus (cranial nerve IV) supplies most of its motor neurons to the contralateral superior oblique muscle, and supplies only 10% of its motor neurons to the ipsilateral muscle. The axons originate from the dorsomedial portion of the nucleus, travel around the medial longitudinal fasciculus (MLF) and the cerebral aqueduct, cross over in the anterior medullary velum, and exit the dorsal surface of the brain stem. It is the only cranial nerve that exits from the dorsal surface of the brain stem. The trochlear nerve is especially vulnerable in traumatic injury because of its small size and its anatomical relationship to the tentorium cerebelli. In cases of fourth nerve palsy, the affected eye is higher and extorted because of an ipsilateral overacting inferior oblique muscle. A V-shaped pattern will be observed if the lesion is bilateral, the eyes drift apart if looking up and converse if looking down.

Abducens Nucleus

The abducens nucleus is the center for conjugate horizontal movements of the eye. About 70% of the abducens nucleus are motor neurons that control the ipsilateral lateral rectus muscle. The remaining 30% are abducens internuclear neurons, which innervate the medial rectus motor neurons in the contralateral side (Fig. 19.17). Thus, the abducens nucleus can elicit contraction of the ipsilateral lateral rectus muscle and the contralateral medial rectus muscle at the same time, achieving cooperative movement of both eyes in the same direction. The axons of abducens motor neurons exit the ventral part of the brain stem at the level of pontomedullary junction. It is vulnerable to injury not just because of its size; it also has the longest intracranial course among all the cranial nerves. The most commonly injured region is where it courses ventral to the basilar pons and where it

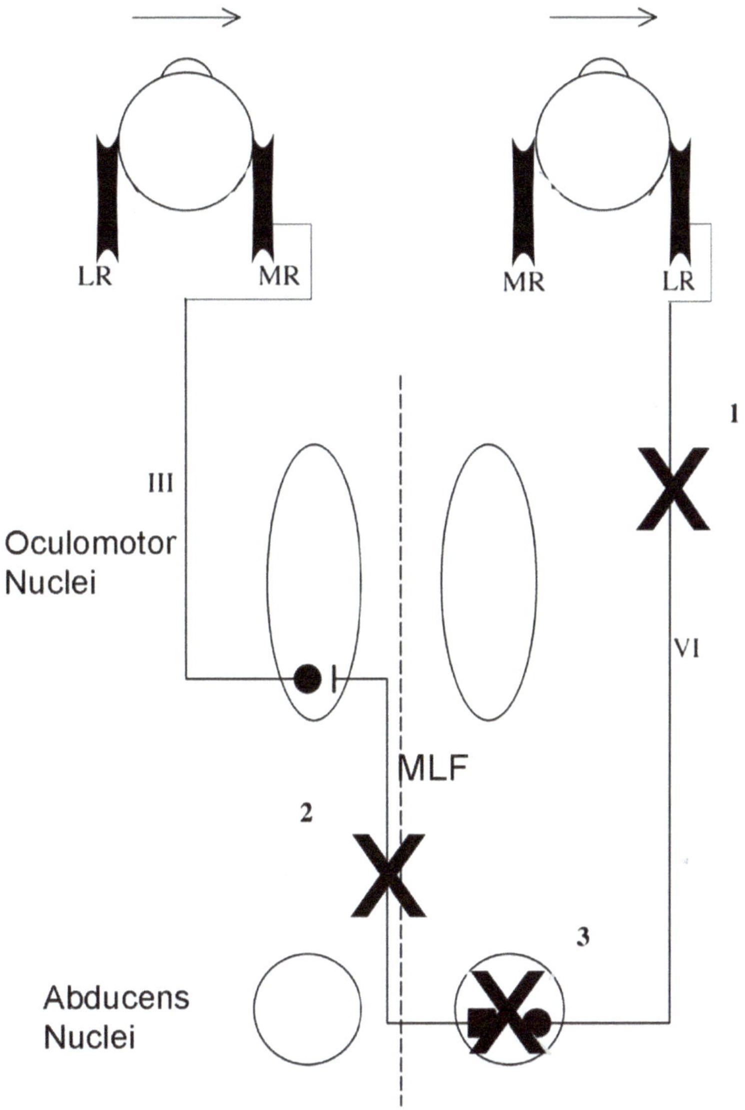

FIGURE 19.17 Control of conjugate horizontal eye movement. Filled circles represent motor neurons in the oculomotor and abducens nuclei that send projections through CN III and CN VI, respectively. CN III innervates the medial rectus and CN VI innervates the lateral rectus. Axons from the abducens internuclear neurons (*filled square*) cross the midline and traverse the MLF to synapse with the contralateral medial rectus motor neurons. Conjugate horizontal signals are equally received by the abducens motor and internuclear neurons, which relay the signals to the contralateral medial rectus motor neurons. Activation of the abducens nucleus causes abduction of the ipsilateral eye and adduction of the contralateral eye. There are different consequences of abducens nerve (1), MLF (2), and abducens nucleus (3) lesions (see section: Abducens Nucleus).

crosses the petrous portion of the temporal bone. The axons of the abducens internuclear neurons ascend in the contralateral MLF and form extensive synaptic connectivity with the contralateral medial rectus motor neurons.

Dysfunction of the abducens nerve (cranial nerve VI) can result from lesions occurring anywhere in the abducens nucleus or the abducens nerve (Fig. 19.17). If it occurs in the nucleus, the conjugate horizontal eye movements toward the side of lesion are paralyzed (3). If a lesion occurs in the abducens nerve, it will cause esotropia at rest and failure of ipsilateral attempted abduction perhaps accompanied by diplopia due to unopposed action of the medial rectus muscle (1). If it occurs in the MLF, it causes internuclear ophthalmoplegia, manifesting as loss of adduction while conjugate movements of the eye (2). In young people, internuclear ophthalmoplegia is commonly caused by multiple sclerosis; while in old people, it is frequently seen in strokes.

PREOCULOMOTOR NUCLEI

Four brain stem premotor areas are associated with the control of the eye movement and gaze in the horizontal and vertical planes. We note that these different structures present direct monosynaptic connections with motor neurons of the extraocular motor nuclei.

Control of the Vestibulo-Ocular Reflex by the Vestibular Nuclei

The VOR originates from the semicircular canals in the inner ear (see Chapter 26). This reflex is supervised by a group of three neurons. The circuit starts from the primary vestibular neuron to the secondary vestibular neuron to the eye-movement motor neuron. Aside from the afferents from the primary vestibular neurons, the second-order vestibular neurons also receive central input from the cerebellar flocculus and fastigial nuclei. The efferent arm of the VOR starts from the canal-specific efferent projections of the second-order vestibular neurons, which target motor neurons in the extraocular motor nuclei. Each semicircular canal is associated with two pairs of extraocular muscles. This relationship is based on the spatial orientation of the semicircular canals relative to the direction of movement of the muscles.

Axons from second-order neurons in the vestibular nuclei may be either inhibitory or excitatory. This is important in coordinating eye movement since in order for both eyes to be able to look in one direction, the stimulation of contraction of an eye muscle must be accompanied by the inhibition of its antagonist (eg, when the lateral rectus contracts, the medial rectus in the same eye must be inhibited). Fig. 19.18 demonstrates how the vestibulo-ocular pathway controls compensatory horizontal movement of the eyes.

The normal functional operation of the VOR requires not only the basic three-neuron reflex arc (eg, first-order vestibular ganglion neuron→second-order vestibular nucleus neuron→extraocular motor neuron) that effects a compensatory eye movement after a head-turn, but also a mechanism to maintain gaze in the new position. Input provided by the prepositus hypoglossi nucleus is also necessary for the normal functional operation of the VOR. When head rotation prompts a compensatory eye movement, the head velocity signals of vestibular neurons are insufficient to maintain gaze in the new position. Located in the caudal pons, the nucleus prepositus hypoglossi (NPH) plays a fundamentally important role in the system to hold horizontal gaze (Fig. 19.19). The NPH forms widespread reciprocal connections with the vestibular nuclei. It functions as a neural integrator that transforms the head velocity signals from vestibular neurons to eye movement signals that the oculomotor neurons can interpret and carry out. NPH damage is associated with difficulty in maintaining gaze positions. As the NPH is involved primarily in horizontal gaze holding, the abducens nucleus is the main target of its

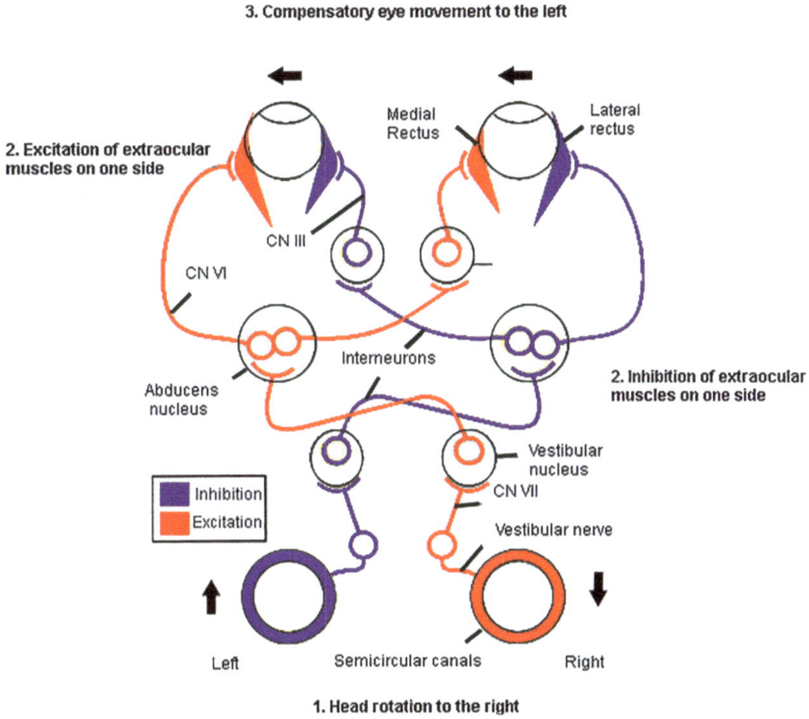

FIGURE 19.18 The vestibulo-ocular pathway relating the different semicircular canals to the horizontal gaze centers and the horizontal rectus muscles. Head rotation (1) triggers an inhibitory signal to the extraocular muscles on one side and an excitatory signal to the muscles on the other side (2). The result is a compensatory movement of the eyes in the direction opposite to that of the head rotation (3). *From "ThreeNeuronArc" by Tvil at the English language Wikipedia. Licensed under CC BY-SA 3.0 via Commons* - https://commons.wikimedia.org/wiki/File:ThreeNeuronArc.png#/media/File:ThreeNeuronArc.png.

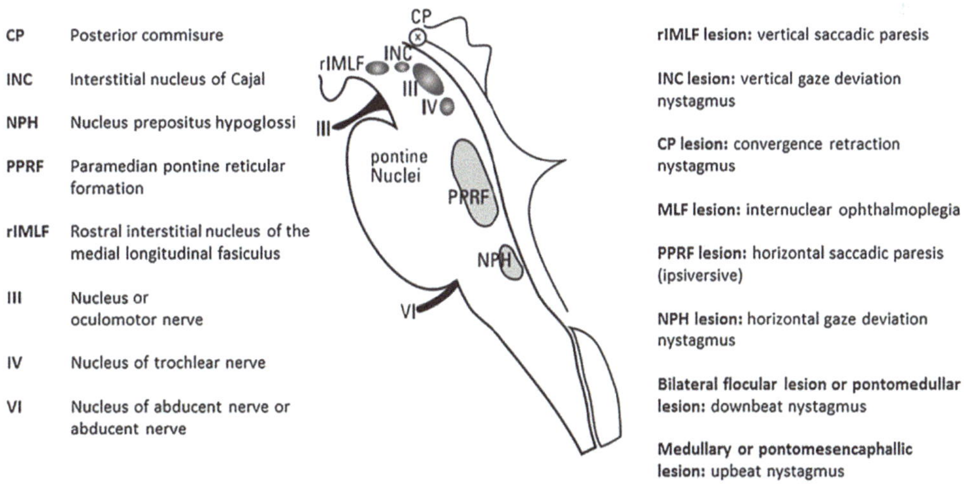

FIGURE 19.19 A midsagittal view of the brain stem showing structures critical for holding the gaze. *From Strupp M, Kremmyda O, et al. Central ocular motor disorders, including gaze palsy and nystagmus. J Neurol September 2014;261(2):S542–58.*

excitatory and inhibitory connections. The interstitial nucleus of Cajal (INC), which controls vertical gaze, may function similarly as neural integrator.

Lesions involving the vestibular nerves, nuclei, or MLF lead to remarkably different types of dysfunction (Fig. 19.19). Damage to the vestibular nerve or nucleus may cause spontaneous nystagmus, positional nystagmus, and an enhanced vertical nystagmus in the caloric reflex test. The balance of inputs from the bilateral semicircular canals and otolith organs are disturbed in these deficits. But it is important to know that the peripheral lesions only cause transit deficits because of compensation achieved by the commissural connections between vestibular nuclei.

As mentioned earlier, damage to the MLF causes the clinical syndrome of internuclear ophthalmoplegia (Fig. 19.17). The affected eye shows minimal adduction when attempting to gaze contralaterally, the opposite eye abducts with nystagmus. But vergence eye movement is not impaired because only the axons of the abducens internuclear neurons are damaged, while the motor nerve innervation of the medial rectus and lateral rectus muscles is preserved. The premotor neurons and motor neurons that are responsible for vergence eye movements are located rostral to the lesion.

Control of Vertical Gaze by the Mesencephalic Reticular Formation

The part of the mesencephalic reticular genesis in the vicinity of the oculomotor complex includes two structures that are linked to the innervation of vertical upward and downward gaze (see Fig. 19.19). The rostral interstitial nucleus of the MLF is situated at the intersection of the mesencephalon and diencephalon oblique to the field H of Forel and the periventricular gray matter in the region of the subthalamus. From this location in the rostral midbrain, the INC is located in a position lateral to the MLF level with the rostral region of the somatic and visceral oculomotor nuclei. In the two structures, neurons have different afferent synaptic links with the superior colliculus and the vestibular nuclei. The different excitatory and inhibitory efferent links are generated bilaterally with the different vertical motor neurons in the oculomotor nucleus and trochlear nuclei. The contralateral connections specifically linked to vertical upward eye movements cross the midline through the posterior commissure. The rostral interstitial nucleus of the MLF includes burst neurons that discharge before the beginning of the vertical saccadic eye movements. Physiologically these are distinguished from the tonic neurons in the INC, which has activity depending on the eye position. Neurons located in these structures establish many connections with the spinal cord, providing the basis for their function in the innervation of vertical gaze.

We note that lesions of the rostral interstitial MLF at the mesodiencephalic junction can cause paralysis of vertical upward or downward gaze (eg, Parinaud's syndrome). In addition, the direction is linked to the size of the lesion. Some lesions of the posterior commissure can have a selective result on vertical upward saccadic eye dynamic and can cause some deficiency in the pupillary light reflex as a result of being positioned close to the pretectal area (eg, pretectal syndrome).

Conjugate Horizontal Gaze Is Mediated by the Pontine Reticular Formation

Conjugate horizontal gaze is mediated by the paramedian zone of the pontine reticular formation. Excitatory burst neurons are found in the nucleus reticularis pontis caudalis rostral to the abducens nucleus and project to ipsilateral abducens neurons and the cervical spinal cord. Inhibitory burst neurons project to contralateral abducens neurons and are found caudal to the abducens nucleus. Both sets of pontine neurons receive inputs from the vestibular nuclei and the superior colliculus. Omnipause neurons, found in the vicinity of the burst neurons, inhibit the burst neurons except during saccadic eye movements. Omnipause neurons are therefore regulators that prevent false saccades. Lesions of the omnipause neurons produce constant saccadic oscillations of the eyes known as oscillopsia. The paramedian pontine reticular formation is known as the center for conjugate horizontal gaze. Lesions in this area lead to paralysis of ipsilateral horizontal gaze and head movements, whereas lesions of the abducens nucleus only affect conjugate horizontal eye movements and not head movement.

CEREBRAL CORTEX AND CEREBELLUM EXERT CONTROL ON EYE MOVEMENT

While brain stem structures are of primary importance for functions of the oculomotor system, their proper operation also requires input from the cerebral cortex. In addition to direct projections to the superior colliculus and pretectum from the retina, binocularity and directional selectivity of neuronal receptive fields in the pretectum and superior colliculus require the descending connections from the primary and secondary visual areas of the cortex. Voluntary motor behaviors that are mediated by subcortical structures also are influenced by association areas of the cortex, for example, the frontal eye fields and occipitotemporal and posterior parietal cortices.

Control of over spinal and other brain stem motor nuclei, for example, the trigeminal, facial, and hypoglossal, by Brodmann's area four through the descending corticospinal and corticobulbar tracts is definitive. In contrast, cortical motor control of eye movement by Brodmann's area eight of the prefrontal cortex, for example, the "frontal eye fields," is less direct. This is a major source of cortical input to the intermediate and deep layers of the superior colliculus and includes afferent cortico-cortical connections of area 8 with the visual cortical areas. Lesions within this region only produce transient defects in eye movements, mostly in eye movements requiring attention to a particular stimulus within the visual field. Combined lesions of the frontal eye field and superior colliculus are required to produce a significant effect on voluntary eye movements. These are manifested by deficits in visually guided horizontal saccadic eye movements.

In the *superior colliculus* inputs from any of the various sensory systems can produce a corresponding eye movement directed toward the original stimulus. Thus, it can be considered a site of sensorimotor transformation especially related to orientation behavior. Visual inputs from the retina and visual cortex are primarily associated with superficial layers of the superior colliculus, while the deeper layers receive auditory, somatosensory, and visual inputs from a variety of different cortical and subcortical areas (Fig. 19.20). The maps of the visual field, auditory space, and the body are superimposed on the motor map in a fashion so that they are in register with one another. Thus stimulation of different points in the superior colliculus can produce saccades differing in both amplitude and direction. This allows saccades to be directed precisely to a target of a visual, auditory, or other sensory modality.

The efferent neurons from the deep layers of superior colliculus also modify the direction, velocity and amplitude of saccades. They project through the tectospinal tract to the brain stem premotor areas, which are located in the mesencephalic and pontine reticular formation, to mediate vertical and horizontal saccadic eye movements. In addition, the output neurons in the deep layers also project to the cervical spinal cord motor neurons that control the neck muscles (Fig. 19.20).

The *optokinetic reflex* relies on cortical projections to the pretectum (eg, nucleus of the optic tract) and accessory optic nuclei. Visual projections through these structures are transmitted to the cerebellar flocculus by the nucleus reticularis tegmenti pontis, the inferior olive, and the

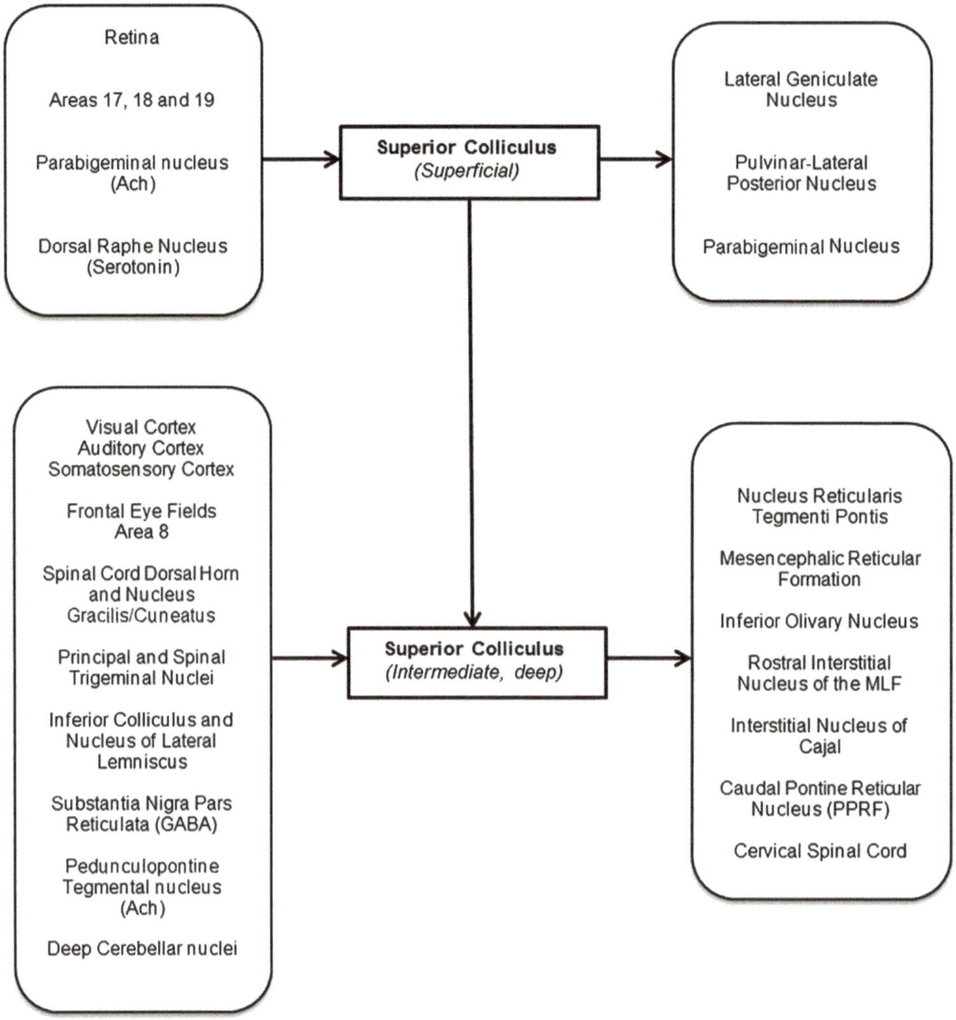

FIGURE 19.20 Major efferent and afferent connections of the two layers of the superior colliculus. *Ach*, acetylcholine; *GABA*, gamma-aminobutyric acid; *PPRF*, paramedian pontine reticular formation.

prepositus hypoglossi nucleus. Then all these neural signals generated by the optokinetic pathway converge upon the vestibular nucleus and are sent to motor neurons for the proper eye movements. The occipitotemporal and parieto-occipital cortex may also be involved in the optokinetic reflex, because lesions in this area can lead to loss of the bilaterality and specificity of direction normally shown by this reflex.

The vestibular neurons also act as second-order neurons for smooth pursuit eye movements. Visual inputs to cortical area 19 induce pursuit signals, which are transmitted to dorsolateral pontine nuclei, relay in the cerebellar flocculus, and then to the vestibular nuclei. The same descending pathways are used for sending out signals to motor neurons as the optokinetic reflex, although the pontine nuclei used are distinct (Fig. 19.21).

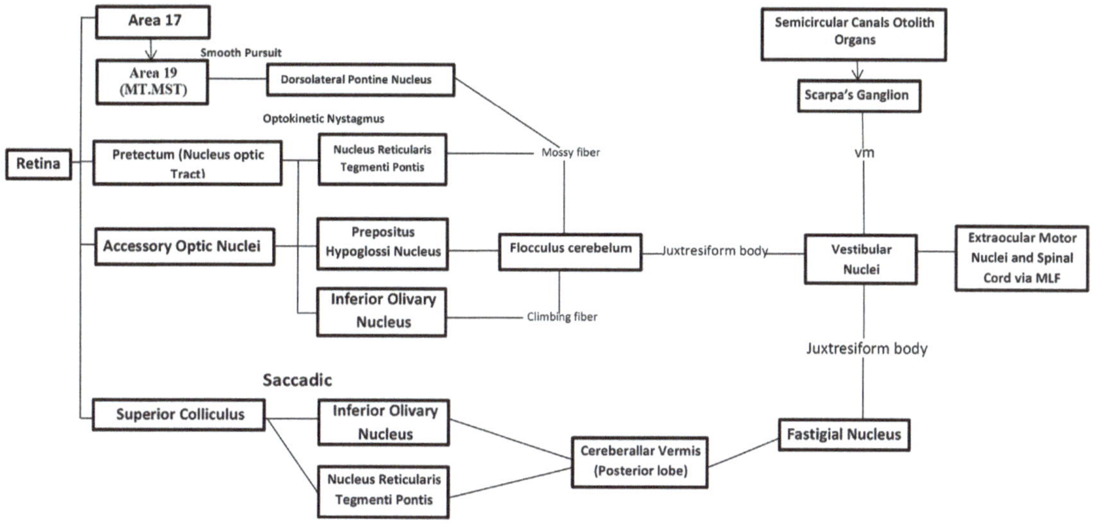

FIGURE 19.21 Structures and pathways mediating the convergence of vestibular and visual information in the cerebellum and their roles in optokinetic, smooth pursuit, and saccadic eye movements. The pretectum, accessory optic nuclei, and superior colliculus all project to precerebellar relay nuclei in the basilar pons (including the nucleus reticularis tegmenti pontis and the dorsolateral pontine nucleus) and medulla (inferior olivary nucleus). These regions transmit visual information to areas of the cerebellum that also receive input from neurons in the vestibular nuclei and vestibular ganglion. Output from the cerebellum is directed to vestibular nuclei neurons that in turn project to the extraocular motor nuclei and spinal cord through the MLF.

The cerebellum plays a crucial role in combining sensory inputs from many different sources. These inputs are used to recalibrate eyes movements. Cerebellar lesions can cause optokinetic nystagmus and deficiency in smooth pursuit and holding eccentric positions of gaze. In addition, the total loss cerebral function impedes the capacity to use the visual information to recalibrate eye movements in order to compensate for the paralysis of one or more extraocular muscles. The vestibulocerebellum, including the nodulus, floculus, and uvual, directs eye movements that maintain images in a stable position on the retina when the head is moving, for example, cancellation of the VOR, or still, in smooth pursuit. In the cerebellar flocculus, the primary vestibular fibers and axons coming from brain stem precerebellar nuclei (eg, nucleus reticularis tegmenti pontis through mossy fibers), inferior olivary nucleus (through climbing fibers), and dorsolateral pontine nucleus (through mossy fibers) converge. The optokinetic reflex and smooth pursuit are both smooth eye movements which resulting from visual stimulation, their projection at the flocculus provide for the combination of neurons into a single common pathway with similar purposes (Fig. 19.21). The cerebellar fastigial nucleus relates primarily to the visual portion of the vermis in the posterior lobe. The vermis also is characterized by the presence of afferents connections from the same brain stem nuclei, but primarily to those linked synaptically to the superior colliculus. Both the dorsal cerebellar vermis and fastigial nuclei play a crucial role in controlling saccadic amplitude and flexibility. In conclusion, the cerebellum serves primarily as the location of convergence of visual and vestibular stimuli (information), which are transmitted to the oculomotor system through a variety of cerebello-vestibular projections from the fastigial nuclei and flocculus to the vestibular nuclei.

BASAL GANGLIA AND EYE MOVEMENT

Manifestations of diseases of the basal ganglia such as Huntington's chorea and Parkinson's disease, typically include deficits in saccades. One pathway through which the basal ganglia influence oculomotor control is the GABA-mediated inhibitory pathway from the substantia nigra's pars reticulata to the intermediate layer of the superior colliculus. The caudoputamen (eg, striatum) and the subthalamus relay input to the pars reticulata region of the substantia nigra, forming part of an indirect corticotectal circuit that brings sensory information from throughout the cerebral cortex to the superior colliculus. Caudate nucleus and substantia nigra neurons discharge before intentional saccades.

Another important pathway between the oculomotor system and the basal ganglia arises from neurons in the caudal mesencephalic and rostral pontine tegmentum. This pathway relays an excitatory cholinergic transmission to the superior colliculus, corresponding in part to the pedunculopontine tegmentum nucleus. This region of the tegmentum also receives afferent connections from the pars reticulata of the substantia nigra and provides efferents to the subthalamus and the pars compacta of the substantia nigra. Lesions in this region, and only in this region so far, have been found to cause the resting tremor typically seen in Parkinson's disease.

VERGENCE EYE MOVEMENTS

Vergence eye movements consist of convergence, wherein the coordinated contraction of both medial rectus muscles brings both eyes toward the midline (ie, the nose), and divergence, wherein the coordinated contraction of both lateral rectus muscles brings both eyes toward the periphery. These are disconjugate eye movements and thus do not use the MLF pathway. While much remains to be understood about the control of vergence eye movements, it is known that the oculomotor nucleus is capped by a loosely organized group of neurons that fires tonically depending on the vergence angle (the angle formed between the two eyes as gaze converges or diverges). In addition, the control of vergence eye movement, especially combined conjugate-vergence movements, appears to be influenced by the cerebellum. Here, although the precise pathways have not been defined, neurons in the parietal and visual association cortex discharge in relation to vergence, and vergence is compromised when these areas are damaged.

BASIS OF DISEASE STATES

Visual Pathway Lesions Produce Characteristic Visual Field Loss Patterns

Lesions at specific points along the optic pathway give rise to distinctive visual field defects (Fig. 19.22). A retinal or optic nerve lesion anterior to the optic chiasm will result in visual loss in the eye on the same side. Bilateral prechiasmal lesions affect the visual fields in both eyes. Complete transection of the optic nerve results in total blindness of the ipsilateral eye. In this case, the consensual pupillary reflex in the affected eye would still be intact because the pretectum receives and sends fibers bilaterally. Direct pupillary reflex, however, would not be intact, hence the blind eye would have an afferent pupillary defect (Marcus–Gunn pupil). The pupillary light reflex is not affected by lesions beyond the pretectum and superior colliculus.

Lesions at or posterior to the optic chiasm result in specific patterns of bilateral visual field loss. Space-occupying lesions that compress the chiasm (eg, pituitary tumors) damage the crossing fibers and produce bitemporal hemianopia. A binasal hemianopia would result if there were lesions on both outer sides of the chiasm, but this is very rare.

Retrochiasmal lesions cause bilateral, homonymous defects on the contralateral hemifield. For example, a left optic tract transection would

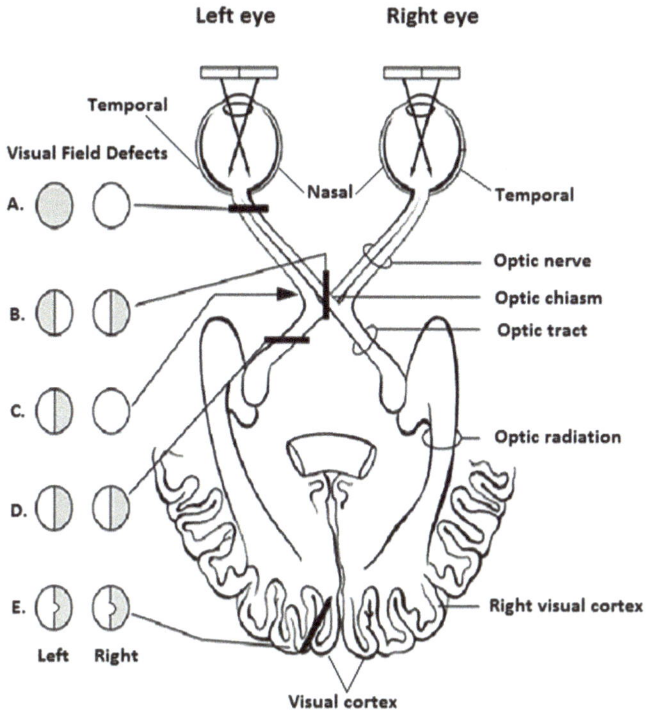

FIGURE 19.22 The optic pathways in the brain. Lesions at specific points along the pathway produce characteristic patterns of visual fields loss: (A) Total visual loss of left eye. (B) Bitemporal hemianopsia. (C) Left nasal hemianopsia. (D) Right homonymous hemianopsia. (E) Right homonymous hemianopsia with macular sparing.

produce a right hemifield defect in both eyes. The spatial relationships of the ganglion cells in the retina correspond to the locations of the fibers in the optic radiations. As the optic radiations form the Meyer's loop in the temporal lobe, inferior retinal fibers occupy the more anterior part of the path. Damage in this part can result in a homonymous field defect in the superior quadrants (superior quadrantanopia). Injuries to the optic radiations in the parietal lobes give rise to a homonymous field defect in the inferior quadrants (inferior quadrantanopia).

Lesions that affect the occipital lobe are often caused by cerebrovascular accidents involving the posterior cerebral artery and its branches. But because the tip of the occipital lobe, which interprets central vision, receives a collateral blood supply from the middle cerebral artery, central vision is spared while the peripheral hemifields are lost. Occipital lobe lesions, in other words, produce a contralateral homonymous hemianopia with macular sparing.

Visual system disorders are not solely caused by gross structural damage. Genetic visual disorders that affect the photoreceptors have been described. In the following text, we discuss two such disorders: retinitis pigmentosa (RP), and a form of color blindness known as blue cone monochromacy (BCM).

Genetic Lesions Resulting in Defects in the Color Pigments Can Cause Loss of Color Vision

BCM is an X-linked form of color blindness that is understood at a molecular level. Only the short-wavelength-sensitive ("blue-sensitive") cones are present in individuals with BCM.

Signs and symptoms are mostly manifested by males and include low visual acuity, glare-induced visual loss, nystagmus, and severely compromised color vision. Progressive macular atrophy was also found in affected individuals of certain families.

The most common mechanism that results in BCM is unequal homologous recombination between the chromosome segments containing the very similar red and green color pigment genes, that results in the deletion of all save for a single gene (in some cases, a red–green hybrid). Alternatively, two distinct genes may be present, but one is inactivated by a point mutation. Finally, gene inactivation can result from deletions in a locus control region situated about 4–18 kilobases upstream of the red pigment gene. Patients with these genetic defects can discriminate brightness since rod cells are preserved. However, color vision is severely impaired because only one type of the cone cells is present, the blue sensitive cones. In contrast, blue cone pigment gene mutations can cause tritanopia, in which discrimination between red and green is preserved, while blue and yellow discrimination is compromised. The dominant inheritance pattern implies that either the viability or fidelity of blue sensitive cones is impaired by these mutations.

The full explanation of the pathological mechanisms of BCM and tritanopia is a typical example of the advantages of combing the methods of molecular genetics with clinical insights, and provides solid evidence supporting the trichromatic theory of human color vision which was first stated by Thomas Young in 1802.

Retinitis Pigmentosa

RP is a large group of visual disorders that are heterogeneous clinically and genetically. Partial or total blindness arises, with primary loss in rod photoreceptors and a progressive retina degeneration that starts at the periphery. It affects about one in 2000–4000 persons worldwide. The age of onset and progression rate differs in individuals and families.

The typical symptoms include night blindness (usually the presenting symptom), abnormal ERG results with decreased amplitude or possibly extinguished (Fig. 19.23), retinal vessel attenuation, and the characteristic pattern of "bone spicule" pigment deposit on the retinal surface.

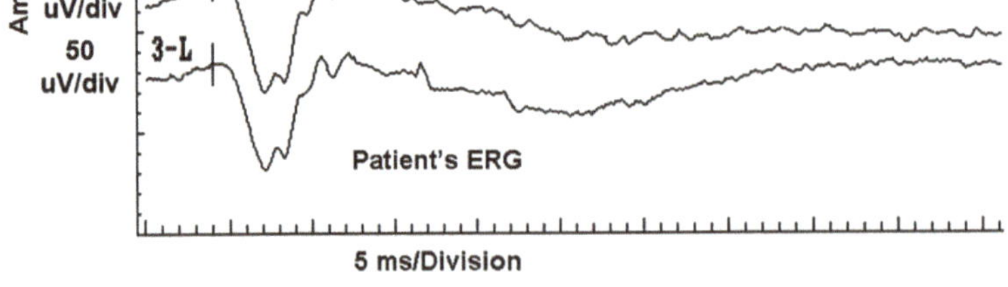

FIGURE 19.23 Electroretinogram (ERG). The normal dark-adapted ERG is at the top, and the bilateral ERG tracings (R and L eyes) of a patient with retinitis pigmentosa are at the bottom. The ERG of the patient is markedly reduced compared to the normal subject.

These findings are in accordance with initial deterioration of the rods, as they are responsible for night vision and located primarily in the periphery of the retina while the cones are sensitive to color and fine vision and are densely packed in the macula. Genetic heterogeneity is another feature of RP. It can be transmitted in autosomal dominant (ad), autosomal recessive (ar), and X-linked forms of inheritance pattern.

Many genes have been identified to cause RP. Peter Humphries et al. firstly identified a linked region on the long arm of chromosome three in a large Irish family with adRP. Since the gene for human rod opsin was previously mapped to this linked region, Dryjia et al. targeted this gene as the candidate gene and quickly found a mutation in the rhodopsin gene, which was identified as the causative mutation in 17 out of 148 unrelated adRP families. The C to A transversion causes a highly conserved proline at amino acid 23 substituted by histidine. In a transgenic mouse with the P23H rhodopsin mutation, it was found that although the photoreceptors developed normally, the length of the light-sensitive outer segments was short. With increasing age, the mouse had a slowly progressive retinal degeneration with abnormal light-evoked responses on ERG. The rhodopsin mutations accounted for about 25% of RP cases. Using linkage analysis and candidate genes sequencing, more and more genes associated with RP have been identified.

To date, more than 200 genes responsible for inherited retinal diseases have been reported with many more yet to be discovered, resulting in a constantly growing number of publications. An updated summary is posted on the RetNet Website (https://sph.uth.edu/Retnet/disease.htm). More recent studies are geared toward elucidating the molecular mechanisms that give rise to the RP phenotype. Some studies have shown that defective opsins are poorly transported to and incorporated into the membrane. Others result in a phototransduction cascade that is constitutively activated. Mutations in different photoreceptor-specific genes can directly affect vision through functional changes and may cause RP by indirectly inducing apoptosis in RPE cells. Conversely, mutations in genes primarily expressed in the RPE can cause damage to photoreceptors, which are dependent on the RPE for metabolic support.

Candidate RP genes often code for proteins involved in phototransduction or proteins with high expression levels in the retina. Aside from rhodopsin, mutations in peripherin, a protein that stabilizes the rod outer-segment discs, also cause RP. The following proteins have been implicated in the pathogenesis of RP: arrestin, rod cGMP-gated channel alpha subunit, rod outer segment membrane protein 1, the alpha and beta subunits of cGMP PDE, cellular retinaldehyde-binding protein, RPE-retinal GPCR, and cyclic nucleotide–gated channel alpha-1 (CNGA1). Genes encoding proteins ranging from transcription factors, such as NRL or neural retina leucine zipper, to pre-mRNA processing factors, such as three homolog PRPF3, have also been implicated. Interestingly, the inheritance pattern or RP caused by the same gene may vary with the amino acid mutation or sometimes even despite the same mutation. In addition, RP can be clinically heterogeneous, even among family members with the same RP-causing mutation. This phenomenon is still a mystery. This could be explained by the existence of factors such as modifying genes or environmental differences. Eventually, studies on these factors may lead to cures for certain cases of RP. Identical mutations can cause more than one type of retinal degeneration such as macular degeneration (in which the macula is primarily affected, eg, peripherin mutations) congenital stationary night blindness (rhodopsin mutations), etc.

Causes of Ocular Motility Disorders

Coordinated eye movements include convergence, saccades, smooth pursuit, and the VOR, and require proper functioning of supranuclear control mechanisms, brain stem gaze centers, cranial nerves, extraocular muscles, and the neuromuscular junctions. Malfunction of any of these structures and mechanisms not only impairs ocular motility with distinct visual and neurological signs and symptoms.

Supranuclear Systems: Saccadic, Pursuit, and Vestibular Systems

Saccadic eye movements are voluntary or reflexive rapid eye movements from the lateral direction of gaze to the center and back again. The saccadic system allows an object of interest seen in the periphery of the visual field to be focused on the fovea and in central vision. It involves interplay among the pursuit, vergence, and vestibular systems. The pursuit system controls the speed of eye movement so that the image of slowly moving object is maintained on the fovea. The vergence system governs convergence and divergence so that images of objects can be simultaneously maintained on the foveas of both eyes. The vestibular systems manage eye movement in response to changes in head movement so that fixation is maintained.

The structures primarily responsible for generating saccades include the frontal eye fields, supplemental eye fields, parietal eye fields, pontine, and mesencephalic reticular nuclei, the cerebellum, and the basal ganglia, so that lesions in any part of them may cause abnormal saccadic eye movements. Frontal eye field lesions are associated with the failure to initiate a voluntary saccade in the direction opposite the lesion. Hence, in a patient who suffered a stroke involving the right frontal lobe, the eyes cannot be voluntarily moved to the left past the midline and are deviated to the right or toward the side of the lesion. However, as long as the cranial nerve nuclei, cranial nerves, and the extraocular muscles involved in horizontal gaze are intact, reflex movements moving the eyes to the left may be evoked. Eye movement disturbances in the presence of intact cranial nerves and nuclei are known as supranuclear gaze palsies. Slowed saccadic movements are found in some degenerative cerebellar or basal ganglia disorders, such as Huntington's disease, in which they may be one of the early signs of neurological dysfunction.

Similarly, the pursuit system involves several CNS structures, such as the cerebellum, vestibular nuclei, pontine nuclei, and the striate and extrastriate cortex. When there are lesions in any of these structures, eye movement speed during pursuit is inadequate and cannot keep up with a moving target. Corrective saccades are generated to compensate for the deficiency, giving rise to jerky or ratchet-like eye movements called saccadic pursuit. Unilateral parietotemporal cortex lesions can impair ipsilateral pursuit, resulting in the saccadic pursuit response to objects moving toward the side on which the lesion lies. The localization of the pursuit deficit to the same side as the parietal lobe lesion can be explained by the double decussation of the brain stem pathways: in both the cortical projections between the pontine nuclei and cerebellum, and the vestibular and the abducens nuclei.

Conjugate Gaze Abnormalities Implicate the Brain Stem

Conjugate gaze occurs when both eyes move in the same direction and speed, maintaining a constant alignment. The horizontal conjugate gaze-controlling center, the paramedian zone of the pontine reticular formation, lies in the pons near the abducens nucleus. A unilateral lesion causes inability for either eye to move toward the affected side beyond the midline. As distinguished from supranuclear gaze paralysis, which is caused by frontal lobe lesions, involuntary reflex eye movements cannot circumvent this form of gaze paresis. The center controlling vertical conjugate gaze center lies in the midbrain. Lesions in this region, seen frequently with a pineal gland tumor, will compress the dorsal midbrain and induce conjugate vertical gaze paralysis, causing an inability to move the gaze up.

The MLF connects cranial-nerve nuclei on both sides of the brain involved in conjugate eye movement. Lesions in the MLF cause deficient inward eye movements, known as *internuclear ophthalmoplegia*. The most common cause of lesion in the MLF is multiple sclerosis, especially in patients with bilateral lesions. If the lesion is isolated the eyes may retain the ability to move in all directions, but conjugate movements in the horizontal plane are impaired because the nuclei are no longer connected. When a patient with a right MLF lesion attempts left lateral gaze the left eye moves normally, but the right eye cannot move inward across the midline. However, convergence on near objects is generally preserved, because CN III and the medial rectus muscles remain intact.

Defects in Individual Cranial Nerves Cause Distinct Patterns of Ocular Motility Impairment

Defects of the CN III (oculomotor), CN IV (trochlear), or CN IV (abducens) all have distinct patterns of impaired ocular motility. Palsy of CN III causes paralysis of the superior, inferior, and medial recti and inferior oblique muscles, resulting in a "down and out" position in the affected eye. This position results from the unopposed actions of the lateral rectus and the inferior oblique, which are controlled by the abducens and trochlear nerve, respectively. In addition, a dilated pupil and ptosis will be observed in complete oculomotor nerve palsy. Diplopia is a common complaint, which may manifest in several directions of gaze based on the number of muscles involved. The most common causes of oculomotor nerve damage are ischemic as seen in diabetes, or compressive in cases of arterial aneurysm or brain herniation. The latter is considered as a medical emergency because of imminent death arising from compression of other vital structures.

An abducens nerve palsy causes the affected eye to turn inward toward the nose with inability to move laterally beyond the midline, as the lateral rectus is impaired. The resulting diplopia is worse on lateral gaze toward the side of the lesion. The abducens nerve is easily injured due to its long intracranial course and also because it is angled over bony structures at the skull base.

An early sign of blunt trauma to the skull or increased intracranial pressure with downward herniation of the brain stem could be an abducens-nerve palsy.

A trochlear nerve palsy weakens the superior oblique muscle and causes an inability to depress the globe when it is in the medial (adducted) position. This may cause a hypertropia or upward deviation of the eye on the affected side. Diplopia may occur, and children may acquire a compensatory head tilt towards the direction of the normal eye in order to ameliorate the diplopia.

Disorders of Muscle or Neuromuscular Transmission

Muscle disorders often impair ocular motility, especially those muscle disorders arising from thyroid or mitochondrial disease. In these conditions, ocular muscle weakness may appear earlier and be more severe than truncal or appendicular muscle weakness.

In myasthenia gravis, a disease of neuromuscular transmission, ocular muscle weakness is often a presenting sign, and can involve a single muscle or all of the extraocular muscles. Drooping of the eyelid (ptosis) is also a common finding. There is a characteristic weakness associated with myasthenia gravis—it varies from day to day and hour to hour and is usually worse after prolonged use of the eyes, such as after watching television or during prolonged gaze in one direction. Patients may experience acute diplopia when ocular muscle weakness sets in. Ocular myasthenia gravis may be misdiagnosed as oculomotor palsy, but the absence of an afferent pupillary defect and the positive response to anticholinesterases such as pyridostigmine distinguish it from the later.

Nystagmus

Nystagmus is an involuntary and rapid to and fro oscillation of the eyes. It can occur linearly along either the horizontal or vertical axis, rotationally, or in multiple directions. There is a quick phase (saccadic movement) in one direction and a slow phase (smooth pursuit) in the other direction. The direction of the nystagmus is dictated by its quick phase. *Physiological nystagmus* appears in normal people usually when the eyes look to the extreme lateral or vertical direction for an extended period, and occasionally when the eye muscles are fatigued. Voluntary nystagmus, consisting of very rapid 10- to 25-s bursts of back-and-forth horizontal movements, is rare and found in only about 5% of the population. Pathological examples of nystagmus include *gaze-evoked nystagmus*, which is seen in certain directions of gaze, usually away from the midline along the horizontal or vertical plane. It may be found in sedative drug use, and in cerebellar diseases. *Downbeat nystagmus* is characterized by an upward drift of the eye and a faster downward corrective saccade. It is basically a defect in vertical gaze holding and is evident in the primary position. It occurs mainly caused with lesions in and around the craniocervical junction. *Pendular nystagmus* is characterized by the same speed of eye movement in both directions. It is most often congenital and associated with conditions that cause poor vision. Acquired pendular nystagmus usually indicates a pathology in the brain stem, such as stroke or multiple sclerosis.

Vestibular nerve dysfunction causes vestibular nystagmus. In unilateral lesions, the fast phase is directed toward the opposite side of the lesion. A rotatory component and vertigo, often a subjective sense of spinning, are also commonly associated. A variety of lesions can cause vestibular nystagmus, including lesions of cranial nerve VIII (eg, from compression by a tumor), and viral labyrinthitis. Caloric stimulation can induce vestibular nystagmus in normal individuals, usually by instilling ice water into the ear canal, which triggers convection currents in the endolymph, and results in nystagmus toward the ear stimulated. Severe vertigo may also be experienced. This procedure is a simple, although uncomfortable, test of the integrity of the vestibular system.

Further Reading

[1] Albert DM, Jakobiec FA, editors. Principles and proctice of ophthalmology. Philadelphia: W.B. Saunders Co.; 2000.
[2] Bornstein MH, editor. Sensitive periods in development: interdisciplinary perspectives. Hillsdale (NJ): Lawrence Erlhaum Associates; 1987.
[3] Glaser JS, editor. Neuro-opthalmology. 2nd ed. Philadelphia: JB Lippincott; 1990. p. 299, 361.
[4] Hubel DH. Eye, brain, and vision. New York: WH Freeman; 1988.
[5] Leigh RJ, Zee DS. The neurology of eye movements. 4th ed. Oxford; 2006.
[6] Nathans J. Rhodopsin: structure, function, and genetics. Biochemistry 1992;31:4923.
[7] Palczewski K, Benovic JL. G-protein-coupled receptor kinases. Trends Biochem Sci 1991;16:387.
[8] Pierrot-Deseilligny C, Ploner CJ, Muri RM, Gaymard B, Rivaud-Pechoux S. Effects of cortical lesions on saccadic: eye movements in humans. Ann N Y Acad Sci April 2002;956:216–29.
[9] Tessier-Lavigne M. In: Kandel ER, Schwartz JH, Jessell TM, editors. Phototransduction and information processing in the retina. 1993. p. 401.

CHAPTER

20

Oculomotor Systems and Control

D.M. Waitzman

University of Connecticut Health Center, Farmington, CT, United States

INTRODUCTION

Millions of evolutionary years before the arrival of a "voluntary extraocular motor control system," a much more fundamental system developed to stabilize vision and provide orientation during movement. Movements of the body were continually monitored, and subsequently the motion detection mechanism centered in the head. Phylogenetically, the vestibular system was one of the earliest sensory systems to develop. It is found in the earliest predecessors to the lamprey (earliest fish-like organisms) and has been conserved among all of the species that evolved from our deep sea ancestors. However, trouble arose with the development of visual perception. Initially the light-sensitive regions of the body were fixed to the head and thus signals from those sensors could be interpreted relative to head or body motion. However, as the visual apparatus itself became mobile, the need for a separate control system for the "eyes" became necessary. In terrestrial animals, a binocular oculomotor control system for moving the eyes developed and has been "grafted" onto the vestibular system, which provides balance and orientation relative to gravity. Refinement of a *voluntary* oculomotor system paralleled the shift of the two eyes from a lateral to a more frontal position on the animal.

The purpose of eye movements is to move the fovea (the region of highest acuity in the retina) in order to see. Eye movements were of intense interest to the earliest scientific observers. Indeed, Aristotle was one of the first to describe the binocular movement of the eyes. The ancients were fascinated by the fact that we had two eyes but a visual perception of single vision, except when things went wrong. Understanding exactly how a construct of single vision was formed in the mind has been a difficult problem for both researchers and clinicians to understand until the landmark work of Hubel and Wiesel (see also Chapter 19) demonstrating the intrinsic organization of the visual cortex and how signals from two eyes are fused into one. The high precision and complexity of the primate visual system has demanded the evolution of a complementary precise, rapid motor control system to voluntarily redirect the fovea from one region of interest to another. At the same time, once a single "visual construct" is formulated in the cerebral cortex, an ongoing argument has arisen surrounding the question of whether and how the brain stem develops the control

Conn's Translational Neuroscience
http://dx.doi.org/10.1016/B978-0-12-802381-5.00032-4

signals for eye movement. Namely, is this a conjugate signal that drives both eyes simultaneously, or does the brain stem develop signals for individual eye movement control?

Deficits in oculomotor control generate dramatic changes in visual perception, orientation, and our sense of stability. When eye alignment is incorrect, two primary clinical scenarios ensue. Children younger than five often develop amblyopia (central suppression and reduced acuity) in one of the eyes, and adults often experience diplopia, or double vision. Therefore, maintaining eye alignment throughout life is an essential, *neural* task requiring close coordination of the visual and oculomotor systems. Since eye movements are very easily recorded in patients, they provide an opportunity for diagnosis and treatment of a wide variety of neurological disorders. This chapter concentrates on the anatomy, neurophysiology, and behavioral control of this voluntary oculomotor system. Clinical conditions are described to demonstrate how understanding the organization of oculomotor system can provide insight into a variety of neurological diseases.

ANATOMY OF THE PRIMATE OCULOMOTOR SYSTEM

The execution of normal voluntary visually guided rapid eye movements (REMs; saccades) requires an intact visual system including the peripheral and foveal portions of the retina, the optic nerves, optic tracts, optic radiations, lateral geniculate nucleus (LGN), and the primary visual cortex (areas 17, 18, and 19). Thus, while the majority of the structures carrying the signals for visual cognition and voluntary movement decisions are located supratentorially, the majority of the machinery for carrying out the commands of the cerebral cortex is located in the brain stem (infratentorial) and are described first.

ORBITAL ANATOMY AND MOTOR NUCLEI

Six extraocular muscles control the movements of each eye: four rectus and two oblique muscles. Each of the rectus muscles inserts in front of the coronal plane dividing the eye in half. The oblique muscles, on the other hand, insert behind the vertical equator and thus their action is inverted. The medial and lateral rectii move the eye in the horizontal direction (adduction and abduction, respectively). Movements of the eye in the vertical direction are more complex and require the concerted action of four extraocular muscles (a rectus and an oblique muscle for each of up and down gaze). Moreover, the action of the two complementary vertical eye muscles changes with variation in the horizontal position of the eye. Upward eye movements activate the superior rectus and inferior oblique muscles. With the eye positioned laterally, the primary action of the superior rectus muscle is to move the eye up and out (elevation and abduction) while the inferior oblique muscle in lateral gaze will *extort* the eye (rotates the 12 o'clock position of the eye *around the axis of gaze* toward the temporal side of the head). On the other hand, if the eye is adducted (in the medial position toward the nose), the primary action of the inferior oblique muscle is to move the eye up, while in this position the superior rectus muscle intorts the eye. In a similar fashion for downward movements, if the eye is abducted (turned laterally), the inferior rectus muscle moves the eye down while the superior oblique muscle will intort the eye. If the eye is adducted (positioned toward the nose) then the superior oblique muscle will depress the eye while the inferior rectus muscle extorts the eye.

Four of the extraocular muscles, levator palpebrae, and constriction of the pupil are controlled by the oculomotor nerve (3rd cranial nerve). The lateral rectus muscle is innervated by the abducens nerve (6th cranial nerve), while the superior oblique muscle is innervated by the trochlear nerve (4th cranial nerve). The concerted action of *pairs* of extraocular muscles is required to produce the perception of single vision. For example, movements of the eyes to the right would require the action of the right lateral rectus and the left medial rectus muscle. For vertical movements, the pairing requires the *primary* action of a rectus muscle in one eye and contraction of an oblique muscle in the fellow eye. For example, to look up and to the left would require the primary action of the left superior rectus muscle and the right inferior oblique muscle. Such coordinated movements of the two eyes are called *versions*. However, action of single muscles in individual eyes are described as the *ductions* of the eye.

ANATOMY AND PHYSIOLOGY OF THE EXTRAOCULAR MUSCLES

Anatomic examination of the extraocular muscles has demonstrated that they are not homogenous. Morphologically, there are at least six different types of striated muscle fibers in each muscle. However, these fibers can be divided into two major groups: those that are innervated singly by a centrally located individual motor endplate (SIFs) and multiply innervated fibers (MIFs) in which there are multiple motor endplates distributed in grape-like clusters called "en grappe" along the length of the fiber. The MIFs are unique to extraocular muscles and are not present in other striated skeletal muscles of the body. In response to stimulation of one of the oculomotor nerves, MIFs respond with a prolonged graded contraction that is local to the motor endplates (nontwitch), while the SIFs respond with a single "on–off" contraction (twitch) which *propagates* throughout the muscle fiber. The muscle is also divided longitudinally into an outer "orbital" layer and an inner "global" layer. The orbital layer contains many SIFs and fewer MIFs. The global layer contains "global red SIFs," "global white SIFs," "global intermediate SIFs," and "global MIFs." The orbital layer inserts into Tenon's capsule which is a fibro-elastic connective tissue ring that forms "sleeves" or "pulleys" around each of the extraocular muscles. The larger muscle fibers of the global layer insert into the myotendinous junction that joins the eye muscle to the sclera. These junctions are innervated by nerves that provide numerous palisade-like endings and are thought to serve a sensory function in humans.

Recent work has demonstrated that the SIFs and MIFs of the extraocular muscle originate from different portions of the oculomotor nuclei. SIFs of the orbital layer originate within the classically defined boundaries of the oculomotor nucleus in the midbrain. However, the MIFs of the global layer that insert into the myotendinous junction originate from cell groups located at the periphery of the oculomotor nuclei (III, IV, and VI nuclei). For example, the motoneurons innervating the MIFs associated with the medial and inferior rectus muscles form a group of smaller neurons located dorsal-medial to the primary oculomotor nucleus, called the "C group." It has been hypothesized that the nontwitch fibers are more closely associated with slow movements and ocular fixation, while the twitch fibers are primarily activated during REMs. Not surprisingly, almost all of the single neuron recordings in the oculomotor nuclei have occurred from the large α-motoneurons innervating the SIFs, and the exact discharge pattern of the motoneurons innervating the MIFs remains unknown.

ANATOMY AND PHYSIOLOGY OF CONJUGATE SACCADIC EYE MOVEMENTS

Saccadic eye movements to targets located at a distance (typically greater than 1 m) require a sudden coordinated shift of both foveae onto a peripheral target. Saccadic eye movements can achieve eye rotational speeds of up to 700°/s and accelerations exceeding 20,000°/s^2. Eye movements of 10° amplitude last no more than 40–50 ms. To achieve this degree of speed, primates have developed a series of tightly coupled neurons in the brain stem whose sole function is ocular control. Orchestration of eye movements in the horizontal and vertical directions are relegated to separate regions of the brain stem. Coordination of horizontal saccades originates in the paramedian portion of the pontine reticular formation (PPRF) while the rostral interstitial nucleus of the MLF (riMLF) in the rostral midbrain provides the control signals for the generation of vertical saccades (Fig. 20.1).

The bulk of the information known about the primate control system both at the brain stem and cortical levels has been obtained via recordings and anatomic tracing in subhuman rhesus monkeys and confirmed in human patients. We acknowledge the major contribution that the sacrifice of these animals has made to medical science.

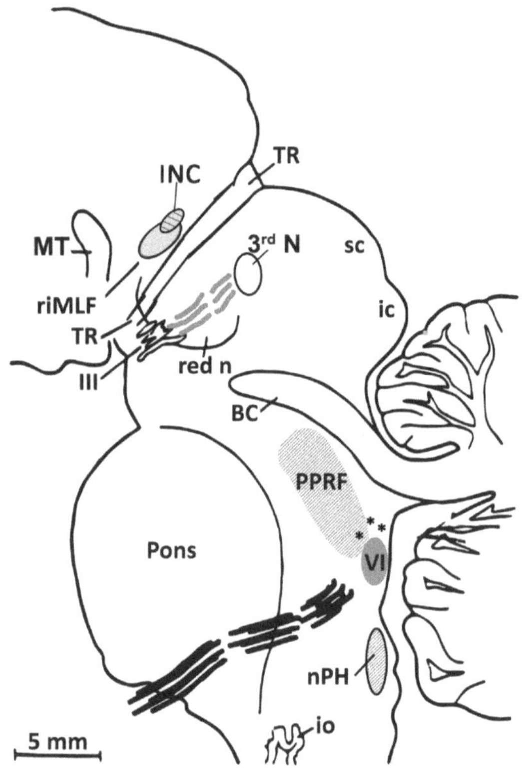

FIGURE 20.1 Anatomy of the brain stem regions mediating oculomotor control. Note that the tractus retroflexus serves as a major identifying landmark for the vertical saccadic eye movement system. Both the riMLF and the INC are located just rostral to this structure. * mark region just rostral to the abducens nucleus associated with paramedian tract (PMT) that projects to the flocculus of the cerebellum; *3rd N*, oculomotor nucleus; *III*, oculomotor nerve; *VI*, abducens nucleus; *BC*, brachium conjunctivum, alternatively called the superior cerebellar peduncle (outflow to the midbrain, especially the red nucleus); *ic*, inferior colliculus; *INC*, interstitial nucleus of Cajal; *io*, inferior olive; *MT*, mammillo-thalamic tract; *PPRF*, paramedian portion of the pontine reticular formation; *red n*, red nucleus; *riMLF*, rostral interstitial nucleus of the MLF; *sc*, superior colliculus; *TR*, tractus retroflexus.

HORIZONTAL SACCADES: BRAIN STEM CONTROL

Before an eye movement begins, the eyes are typically fixed on a visual target. To achieve fixation, inhibitory neurons (transmitter is glycine) located along the midline of the pons in the raphe interpositus (RIP) nucleus have a continuous discharge to suppress the depolarization of "rotation activator cells" (commonly referred to as the excitatory burst neurons, EBNs) for both the horizontal and vertical saccadic eye movement systems. For horizontal gaze, these neurons are located in the PPRF and for the vertical/rotatory movement system, they are located in the rostral interstitial nucleus of the median longitudinal fasciculus (riMLF) in the reticular formation of the rostral midbrain. Neurons in the RIP stop firing just prior to saccades in *all* directions and thus have been dubbed "omnipause cells." We provide an example of the signals required to generate a rightward horizontal saccade in response to a stimulus appearing peripherally in the right visual field with the subject starting in primary position (straight ahead gaze) (Fig. 20.2).

First, the omnipause neurons (OMNs) cease activity (ie, pause) so that a group or "pulse" of high frequency (>600 spikes/s) action potentials

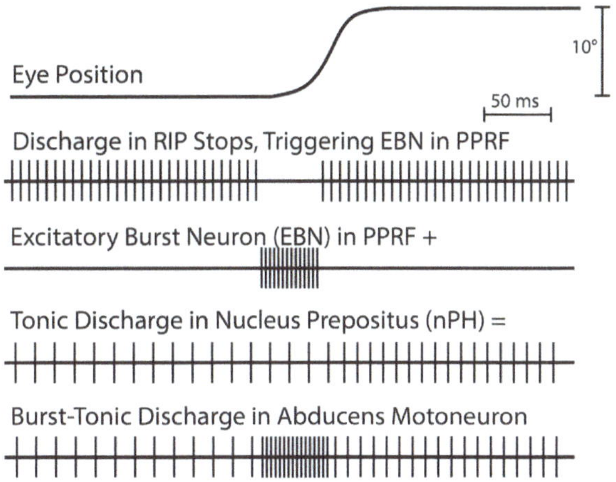

FIGURE 20.2 Timing of the neural activity associated with saccade execution. A saccade executed to a target to the right is shown in the Eye Position trace (Note upward deflections on the Eye Position trace are by definition rightward saccades). The 10° saccade shown in this illustration is completed within about 50 ms. The continual, tonic activity of an omnipause neurons is shown on the second trace. Note that activity stops about 15 ms before saccade onset, triggering the excitatory burst neurons in the PPRF (horizontal burst generator) and the riMLF (vertical burst generator) to discharge at high frequency. The burst is completed about 15 ms prior to saccade end. Activity from the EBNs of the PPRF is transmitted to both the motoneurons in the abducens nucleus and the nPH located at the ponto-medullary junction, and positioned just below the obex (triangular-shaped ependymal tissue at the caudal end of the 4th ventricle). The neurons in the nPH change the PPRF burst into a tonic discharge that is directed to the motoneurons of the abducens nucleus (4th trace down) where it is summed with the PPRF burst itself. This tonic activity counters the normal restoring forces of the eye muscles and fat to hold the eye stable at a new eccentric position.

can be generated from the EBNs lying in the right horizontal gaze center, the PPRF. According to the amplitude of the planned saccade, the OMNs return to their tonic level of activity about 15 ms prior to the end of the saccade. The horizontal EBNs project to two groups of cells in the ipsilateral abducens nucleus: small internuclear neurons and large α–motoneurons (Fig. 20.3).

The axons of the abducens α-motoneurons project laterally, and caudally through the pontine reticular formation to exit from the ventrolateral part of the pons at the ponto–medullary junction. Here, they form the 6th cranial nerve that travels rostrally along the clivus, through Dorello's canal (a specialization of the dura), into the cavernous sinus, and then through the superior orbital fissure to reach the right lateral rectus muscle. Large motoneurons distributed throughout the classically defined abducens nucleus impinge on singly innervated "twitch" muscle fibers (SIFs) while smaller motoneurons innervating multiply innervated "nontwitch" muscle fibers (MIFs) reside in a dorsal/medial shell over the traditionally defined abducens nucleus. The axons of abducens internuclear neurons (AINs) (located within the central portion of the abducens nucleus) cross the midline and almost immediately ascend within the highly myelinated, contralateral median longitudinal fasciculus (MLF) and target medial rectus α-motoneurons of the oculomotor nucleus. Interneurons of the paramedian tract (PMT) located at the rostral pole of the abducens nucleus (separate from the nucleus containing the motoneurons innervating the MIFs) project to the flocculus.

The high frequency (>600 spikes/s) pulse of excitatory activity from the EBNs initiates and accelerates eye rotation toward the new target. This is a rapid ballistic event, with no opportunity for correction during the movement. Inaccurate targeting produces successive corrective saccades. The pulse of excitatory activity is rapidly suppressed ("cut-off") by the resumption of the OMN discharge. To hold the eye at the new eccentric target position, the EBNs in the PPRF also project to a region of the medulla called the nucleus prepositus hypoglossi (nPH). This group of neurons lies just under the caudal portion of the 4th ventricle (the obex) and changes the burst of activity into a calibrated prolonged signal that holds the eye in its new eccentric position. These "hold" neurons (which actually perform the mathematical equivalent of an integration of the pulse into a "step" of

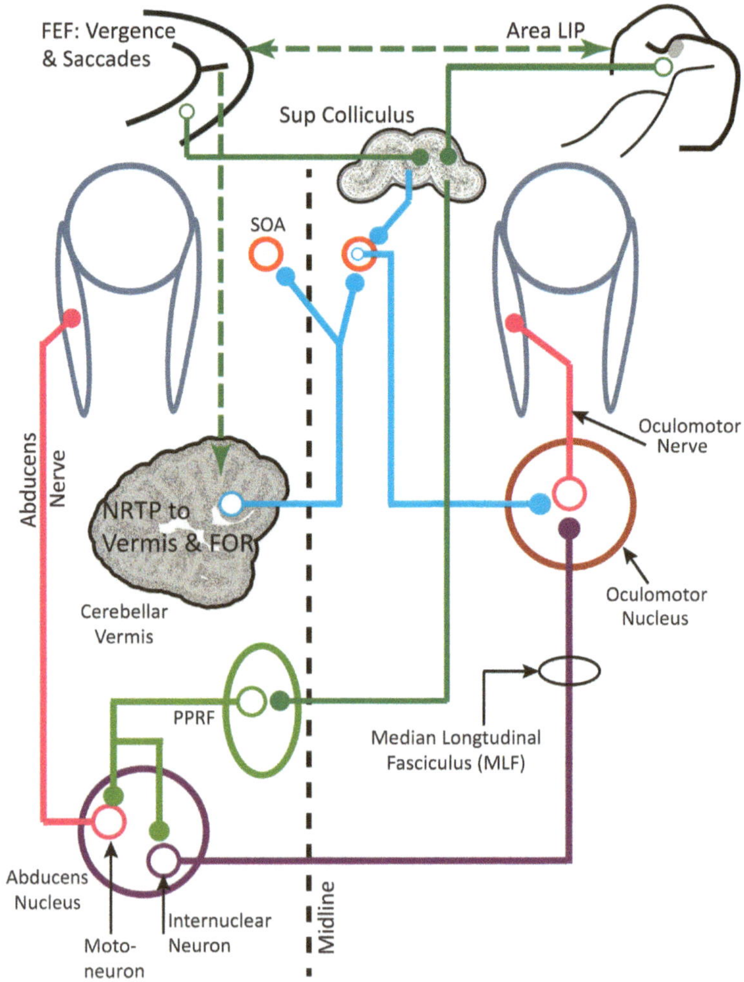

FIGURE 20.3 The anatomical interconnections of the neural structures participating in the control of horizontal gaze. The FEFs and area LIP (lateral interparietal region of the inferior parietal lobule marked by *small gray patch* on the diagram) are the primary cortical regions controlling voluntary eye movements. *Green lines* interconnect supranuclear structures of the oculomotor system related to saccades. Pathways thought to mediate the control of vergence (see text) are shown in *blue*. Motoneuron pathways to the extraocular muscles are shown in *deep pink*. Interneurons connecting the abducens with the medial rectus subdivision of the oculomotor complex are shown in *purple*. Note that the superior colliculus receives saccade-associated signals from both the FEF and Area LIP, but only LIP provides (absolute) vergence activity to the SC. Vergence activity originating from the anterior bank of the arcuate sulcus (just rostral to the saccade-associated regions) in the FEF make use of projections to the NRTP (nucleus reticularis tegmenti pontis) which then provides ascending mossy fiber connections to the deep cerebellar nuclei (primarily the fastigial oculomotor region, FOR) and the Purkinje cells of the oculomotor vermis (Lobules V–VII). *LIP*, lateral interparietal region of the inferior parietal lobule; *MLF*, median longitudinal fasciculus; *SOA*, supraoculomotor area associated with vergence and accommodation.

activity) in the nPH provide *tonic* excitation to the abducens α-motor and internuclear neurons, which is summed with the pulse of activity to hold the eye in the new eccentric position following the rapid saccadic shift. Should the hold neurons fail, a saccade to a peripheral target would be followed by a drift back toward primary (straight-ahead) position (secondary to the viscoelastic properties of the orbital fat and eye muscles themselves) producing what has been called "gaze-evoked nystagmus." Note that the pulse of activity from EBNs must be exactly matched to the tonic level of activity from the nPH for the saccade to end properly (ie, the concept of integration).

VERTICAL SACCADES: BRAIN STEM PATHWAYS

Vertical and torsional eye movements require the coordinated discharge and inhibition of four extraocular muscles on each eye, therefore the organization of the vertical control system is more complex, but does have similarities to the horizontal eye movement control system.

The first point to be established for the vertical system is that there are no separate regions of the riMLF devoted to upward or downward saccades. EBNs for up and down gaze are *intermingled* in the riMLF. On the other hand, intermingled with the up and down burst neurons are a smaller proportion of cells that discharge in association with rotatory movements. Neurons associated with torsion located in the right riMLF discharge in association with clockwise or *ipsi*-torsional (12 o'clock position of the eyes toward the right from the subject's point of view) saccades. Neurons in the left riMLF discharge in association with counter-clockwise or again *ipsi*-torsional movements (now the 12 o'clock position of the eyes rotate toward the left side). Realize that the torsional burst neurons have to drive two torsional muscles, one mediating extortion in the ipsilateral eye (either an inferior oblique or inferior rectus muscle) and another controlling intorsion (either a superior oblique or superior rectus muscle) for the contralateral eye.

A vertical/rotatory saccade is initiated by the supranuclear suppression [originating in the frontal eye fields (FEFs) or superior colliculus (SC), see the following sections] of activity in the OMNs located in the RIP that project to the EBNs in the riMLF. Therefore, in addition to its projections to the EBNs in the "vertical gaze center" and the PPRF in the "horizontal gaze center," the RIP serves as the physiological focal point for triggering saccades in all directions. During upward saccades, the axons from the neurons in the riMLF project caudally via the MLF to provide a burst of action potentials to the *ipsilateral* α-motoneurons in the superior rectus and inferior oblique subdivisions of the oculomotor nucleus. In addition, axon collaterals from the descending projections of the riMLF EBNs cross the midline *at the level of the oculomotor nucleus* to reach the *contralateral* α-motoneurons of the SR and IO subdivisions of the oculomotor complex. To depress the eyes, the descending projections of the riMLF EBNs target the *ipsilateral* IR subdivision of the oculomotor complex and the *ipsilateral* α-motoneurons of the trochlear nucleus (recall that the trochlear nerve crosses dorsally in the brain stem reaching the contralateral eye after exiting the brain stem). Thus EBNs of the riMLF that discharge for upward saccades directly activate four extraocular muscles across the two eyes that participate in eye elevation. In distinction, riMLF neurons discharging for downward saccades activate only two of the depressor muscles (an inferior rectus muscle on the ipsilateral eye and the superior oblique muscle on the fellow eye).

Similar to the horizontal control system, vertical/torsional saccadic eye movements are ballistic events, with no opportunity for correction during the movement. Inaccurate targeting produces successive corrective saccades. After the pulse of activity has been generated, the RIP neurons resume their activity to suppress the activity of the EBNs. Analogous to the horizontal system, the eye is held at the new target position by "hold" or tonically discharging neurons, which are located in the interstitial nucleus of Cajal (INC). These neurons receive axon collaterals from the descending projections of the riMLF EBNs. Then, neurons in INC project caudally to the α-motor neurons in the oculomotor and trochlear nuclei providing them with tonic activity. Also, similar to the horizontal system, the pulse or "burst" of the EBNs in the riMLF has to be precisely converted (ie, mathematically integrated) into the tonic or "step" discharge rate of INC neurons, which is maintained while the eyes are held in eccentric vertical positions. Vertical gaze evoked nystagmus results when the "pulse" and "step" activity are improperly matched.

Because of these innervation patterns, an isolated unilateral lesion of the riMLF will produce loss of downward and ipsitorsional saccades, most often leaving upward eye movements intact. On the other hand, *bilateral* lesions near the junction of the midbrain and diencephalon, which include both riMLFs, produce deficits in vertical gaze affecting upward, downward, and torsional movements in all directions leaving horizontal REMs intact. Lesions involving the INC will produce vertical gaze evoked nystagmus. These experimentally induced lesions are in contradistinction to many of the clinical observations that have more often reported illnesses causing impairment of upward more than downward gaze (see section: Clinical Applications).

SUPRANUCLEAR CONTROL OF SACCADIC EYE MOVEMENTS

A number of structures in the cerebral cortex and the brain stem participate in the supranuclear control of REM, but we focus primarily on three functional regions: SC, frontal, and parietal cortex. One of the key concepts to understanding supranuclear eye movement control is that signals of neurons in the PPRF and abducens nuclei, which cause the eyes to move, are functions of time (ie, a temporal or time-varying signal). In this circumstance, the rate and duration of the discharge of EBNs determine the speed and amplitude of the saccade. However, signals controlling eye movements in the cerebral cortex and SC ("two supranuclear" regions) are coded spatially. This means that the *locus* of cerebral or collicular activation determines the size (ie, amplitude) and direction of the subsequent saccade, *not* the *rate of discharge*. Experiments in both monkeys and man have demonstrated a specific topographic arrangement of neurons in the SC and the FEFs (see the following separate sections). For example, an image appearing in the right visual hemifield would activate a precise region of the left occipital cortex. Recall that there is a retinotopic map of the contralateral hemifield of vision across the primary visual cortex (areas 17, 18, and 19). While cortical signals themselves are modulated as they pass from one visual region of the cortex to another, the retinotopic map is roughly preserved in each subsequent visual region. Eventually, activation in the left occipital lobe will produce a topographically organized activation of at least one of the cortical eye movement regions (ie, the FEFs or the inferior parietal lobule). In other words, the signal relayed from the visual (and other sensory and cognitive regions of the cerebral cortex) will activate *specific portions* of the eye movement regions of the parietal and frontal cortex and the intermediate and deep layers of the SC. This spatially or topographically organized signal must then be changed into the appropriate temporal firing patterns required to activate the extraocular muscles. This is known as a "spatial–temporal" transformation. Exactly how this transformation occurs is a topic of active investigation in the oculomotor field.

A second major organizational feature of the supranuclear pathways is that the FEFs have both SC ***dependent and independent*** paths via which to generate saccades (Fig. 20.4).

Clinically, this is reflected in the finding that patients must have defects in both the FEF and the SC to produce a severe, long-lasting saccade deficit. Also note that the collicular-independent pathways must cross before reaching the omnipause region of the pons. However, the projections from the FEF to the SC (collicular-dependent pathways) are uncrossed. Descending projections to the PPRF and the RIP from the intermediate and deep layers of the SC cross in the MRF as they exit the ventral aspect of the SC. Different from the FEFs, the projections from the parietal cortex to the brain stem are collicular-*dependent* (there is *no* collicular-independent pathway from the parietal cortex). The parietal axons target primarily the ipsilateral intermediate and deep layers of the SC. Thus activation of the left hemisphere or left SC produces saccades to the right. But loss of activity in the left hemisphere (eg, a large stroke in the left frontal lobe) produces a reduction in the activity driving the

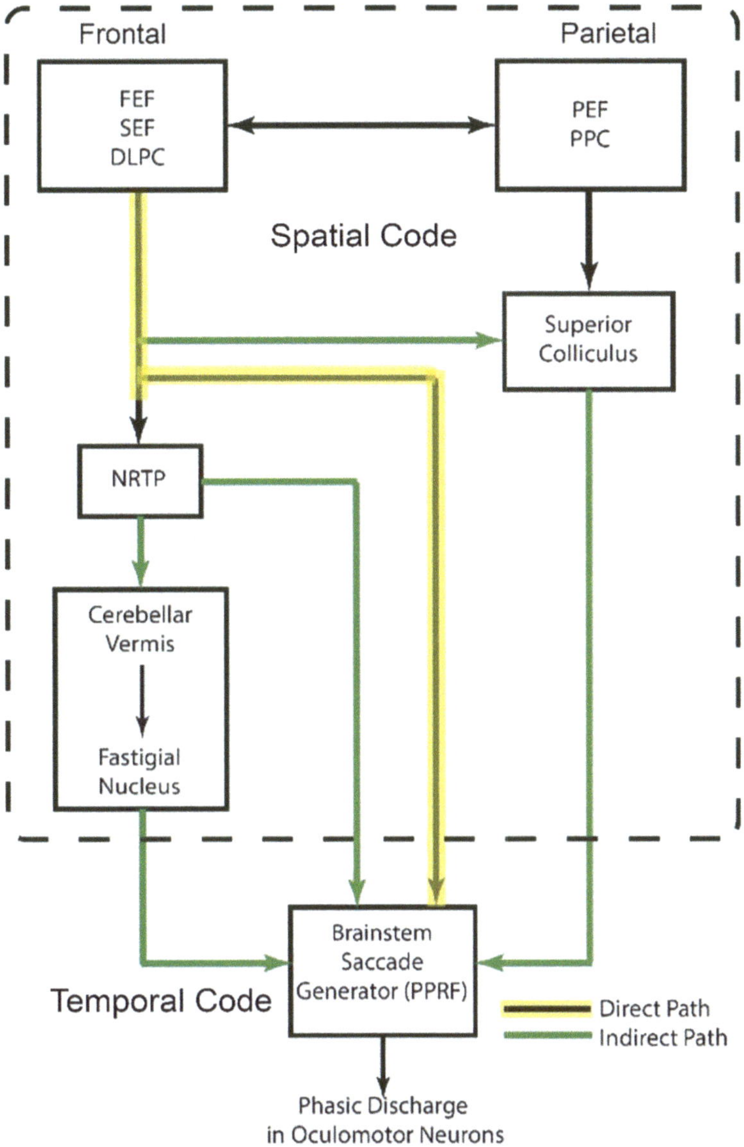

FIGURE 20.4 Diagram outlining how the topographically arranged activity of supranuclear regions such as the FEF and area LIP are transformed in the appropriate temporal activity required to drive the motoneurons and thus move the eyes. Direct pathways for cortical activity to reach the extraocular muscle nuclei are shown in *black and yellow highlighting*. Indirect pathways for the control of rapid eye movement are shown in *green*. A special weighting function has been hypothesized for how the output of the activity from the superior colliculus is changed into the temporal signals required to move the eyes. How the direct activity from the FEF to the pons is transformed remains an area of active investigation. *DLPC*, dorsolateral prefrontal cortex associated with memory-guided and antisaccades; *FEFs*, frontal eye fields; *PEF*, parietal eye fields; *PPC*, posterior parietal cortex; *SEF*, supplementary eye field.

eyes to the opposite side (to the right), and patients with an *acute* event will display a "gaze preference" toward the side of the damage (in the current example, the eyes and head would be deviated to the left). As opposed to the combined collicular/FEF lesions, this "gaze preference" usually disappears over the course of 6–8 weeks following the cortical infarction or damage.

POSTERIOR PARIETAL CORTEX

Topographic excitation representing an image from the right visual hemifield is sent from the left occipital visual association cortex to the left posterior parietal (inferior parietal lobule and the lateral intraparietal (LIP)) cortex. Activity in the posterior parietal area provides spatial targeting coordinates and allows the focus of the subject's attention on the image. The left posterior parietal cortex provides ipsilateral, excitatory projections to the intermediate layers of the left SC, which, in turn, provides bilateral excitatory projections to the vertical gaze centers (riMLF) and predominantly crossed (X) excitatory projections to the right horizontal gaze center (PPRF). Therefore, by activating SC neurons, the posterior parietal cortex can drive gaze in response to a visual stimulus with particular spatial coordinates.

Small lesions in the posterior parietal cortex result in an attention deficit, inaccurate targeting, and increased saccade latency. However, such deficits are typically transient not lasting longer than 1–2 weeks. Large lesions result in frank neglect of the contralateral visual field with a paucity of saccades directed to the contralateral hemifield of vision. Such deficits are often permanent because both the FEF and the intermediate/deep layers of the SC have been deprived of this activity (ie, interruption of both parietal–frontal and parietal–collicular projections).

FRONTAL EYE FIELDS

Excitation in the left occipital visual association area representing an image from the right visual hemifield is sent not only to the inferior parietal lobe, but also to the left FEF (Brodmann Area 8), located within the left premotor cortex. The FEF also receives spatial and attentional signals from the left posterior parietal cortex. The left FEF provides uncrossed excitatory corticobulbar projections directly to the left vertical gaze center (riMLF) and predominantly crossed excitatory projections to the right horizontal gaze center (PPRF). This pathway can initiate gaze in response to valid visual stimuli. The left FEF also provides uncrossed excitatory projections to the intermediate layers of the left SC ("cortico-tectal fibers"). Here, FEF efferents can augment or diminish the effects of an excitatory signal arising from the posterior parietal cortex, depending on whether or not the saccade was judged to be desirable.

The FEFs, in conjunction with neighboring areas of the prefrontal cortex, provide spatial memory, interpret the significance of the visual stimulus and provide voluntary control of movements. Should the visual image be judged to have significance, then the FEF will request that the saccade be executed. If not, it will suppress the saccade. This is often tested with a technique called antisaccades. The subject is requested to fixate. After a variable interval, a target is flashed (say in the left visual hemifield). The instructional set is to look away from the side that the visual stimulus appeared by an equivalent distance. Patients with frontal lesions find it very difficult to suppress unwanted saccades and will often look at the "real" target, rather than the equivalent distance to the opposite side. Persons suffering lesions of the FEFs also have great difficulty with "internally generated" eye movements. For example, they would have impairment of the initiation and execution of voluntary saccades to the remembered location of a visual stimulus (in total darkness).

SUPERIOR COLLICULUS

The SC is comprised of seven layers, but based on its inputs and function, the SC can be more simply understood as two primary regions. The superficial region which is on the dorsal surface of the (sSC) which receives primarily retinal input has three layers. The intermediate and deep regions of the SC (dSC) which receive a variety of other inputs participate in the control of eye and head movement and have four layers.

Electrical microstimulation in the dSC of monkeys whose head is restrained elicits contralateral saccades at short latency (<40 ms) whose maximal amplitude is primarily dependent upon the site and not the parameters of stimulation generating a *topographic map* of eye movement amplitude and direction. Stimulation closer to the rostral pole of the SC produced small amplitude saccades, while stimulation in the more caudal portions of the SC produced large amplitude–combined head and eye (ie, gaze movements) when the head was unrestrained. The dSC topographic map generated via electrical stimulation corresponds closely with the amplitude and direction sensitivity of the cells recorded at each collicular site. On the basis of the collicular map, a *restricted group* of dSC neurons discharge in association with a conjugate movement of particular amplitude and direction. Moreover, if the eye movement sensitivity of a particular dSC neuron is mapped, it forms a restricted region of the contralateral "eye movement" space for which the cell is active. In analogy to the visual fields of the sensory cells in the visual system, this region has been termed a "movement field." In other words, eye movements with amplitudes or directions larger or smaller than the boundaries of the "movement field" will *not* activate the dSC cell. The discharge of the dSC neurons is much higher when the saccade is made in the light than in the dark. In contrast to the EBNs located in the PPRF, the rate of discharge of dSC neuron is only *broadly* tuned to the velocity of the upcoming saccade. As a result, when saccades of similar amplitude but different velocities (eg, saccades to remembered targets) are generated, there is only a very moderate reduction in the peak discharge of dSC neurons in association with the slower movements.

Permanent lesions of the SC produced transient deficits in saccades that recovered within 2 weeks. However, the latency of contraversive saccades was persistently increased following these lesions. On the other hand, combined lesions of the FEFs and the SC permanently eliminated the ability to generate and execute contraversive saccades. The exact mechanism(s) via which the "topographic map" of the intermediate and deep layers of the SC is changed into the temporal pattern of activity necessary to generate REMs remains unknown. One suggestion is that a physiological and anatomic weighting of the activated groups of neurons in the SC are responsible for the generation of the activity needed to drive saccadic movements.

OCULOMOTOR VERGENCE

Significance of the Control of Binocular Movements

The presence of binocular retinal disparity (the difference in the projection of an object on each retina relative to the fovea) provides the perception of visual objects in three dimensions (stereopsis). The development of stereopsis is a critical step in early primate brain development that permits fine manipulations of tools or other objects with the upper extremities in the near field. It remains unclear to which degree stereopsis (the sensory perception of depth) or vergence eye movements (the motor activity required to align a visual object on the fovea of each eye) are necessary for the estimation of distances in three-dimensional space. The loss of depth perception at an early age can have serious impact on overall long-term vision outcome, overall school performance, and self-esteem. The primary *ocular* conditions that impair or abolish stereopsis include anisometropia (ie, a difference in the refraction between the two eyes) or other severe anterior eye disease (eg, congenital cataract or retinopathy of prematurity). In the absence of these conditions, the development of childhood strabismus (misalignment of the visual axes) and its associated cerebral accompaniment, amblyopia are typically the result. This may occur as a result of global damage to the central nervous system (eg, trisomy 21, hydrocephalus, cerebral palsy, intrauterine TORCH infections), changes in the innervation of the extraocular muscles (eg, Duane's syndrome), or some other genetic predisposition. Therefore, loss of stereopsis, development of amblyopia, and strabismus can have long-term consequences with a major impact on public health. This section provides insight into the neural mechanisms underlying the coordination of the eyes to view objects in the near field. In so doing, we will need to revise the organization of the oculomotor system outlined earlier.

Neural Mechanism for Maintenance of Retinal Correspondence

Up to this point we have discussed the anatomy and physiology of the control of REMs (saccades) during which the eyes move in the same direction and each eye is in "lock step" with the other. This is the definition of "conjugate eye movements" and applies primarily when the subject is viewing objects at a distance (typically optical infinity, but for most practical purposes, the eyes are conjugate when targets are positioned more than 10 m from the nose). However, when tracking at a distance of 10 m or closer one eye moves more than the fellow eye. Therefore, the brain cannot send precisely the same neural signal to each individual eye, otherwise the fovea of one or both of the eyes will not point to the target of interest. Instead, the underlying "conjugate eye movement" signal must be modified. Two conditions are recognized in which each eye moves a different amount (therefore called "disconjugate" or "disjunctive" eye movements) while viewing objects in the near field. First, objects can move along the same plane, along a special "circle" called the horopter, where the visual stimulus is projected onto points of each retina positioned precisely the same distance from each fovea (called "corresponding points"), and therefore while both eyes move in the same direction, one eye moves less than the other. In the second circumstance, objects can move in depth from one horopter to another (either closer or farther from the subject). Shifting from one horopter to another induces a disparity, such that the images are projected at noncorresponding points on the retina (ie, each eye perceives the same image at different distances from the fovea). This shift in disparity may occur slowly or rapidly. In either circumstance, the eyes move in opposite directions, called vergence. Tracking a target that moves away from the subject induces divergence, while a target moving closer induces convergence. Targets moving in depth along a plane positioned equally between the two eyes induces "symmetrical" vergence. However, in almost all other situations, the vergence demand will be different for each eye requiring different amplitudes of movement for each eye. In sum, the change in disparity when objects move from a distant to a near position (or vice versa) elicits *fusional* vergence eye movements. Binocular disparity is only one of the two primary sensory stimuli for eliciting vergence eye movements. The second sensory stimulus is visual blur that results from a lack of focus of objects on a single retina (ie, rays of light from a point source do not converge at the same point on an *individual* retina). In addition to the adjustments in accommodation (ie, changes in the thickness of the lens), blur also leads to the generation of *accommodative* vergence eye movements and constriction of the pupils (miosis). Together, vergence, accommodation, and miosis form the "near triad reflex." However, note that shifting the line of fixation from one distance to another induces *both* binocular disparity and visual blur. Therefore, the vergence movements elicited by this change in distance are both fusional and accommodative in origin.

Neural Mechanisms Underlying Fusional Vergence

Early observers of the oculomotor system described the vergence system as generating smooth eye movements that were executed over the course of 300–500 ms (velocities <60°/s) as opposed to the saccadic eye movement system in which the time course ranged from 30 to

90 ms. Moreover, it was hypothesized that the vergence and saccadic eye movement systems worked independently. In other words, if a target approached a subject rapidly, the required saccade and vergence movements would be summed linearly.

Experiments performed in the 1970s changed that view and suggested that vergence occurring during the saccade was accelerated. Therefore, these eye movements were comprised of a combination of a fast (saccade) and slow eye movement (vergence). Moreover, when the dynamics of an *individual* eye was analyzed, the direction of the *rapid* component of each eye was the same, but the amplitude of motion for each eye could be quite different and could account for a 40–90% reduction of the disparity. The remaining disparity was addressed by the "vergence component" of the movement during which the eyes moved in opposite directions and was dramatically speeded up (>200°/s) leading to its description as "fast vergence." Thus, a change in disparity when objects move from a distant to a near position would elicit *fusional* vergence eye movements which were comprised of a disconjugate saccade and "fast vergence" movement. These findings (change in time course, and different amplitudes for an individual eye) strongly argued against the "linear combination" of vergence and saccadic eye movement signals.

How the precise correspondence between optically equivalent points of the two retinas occurs remains uncertain, but it is essential for the development of stereopsis. Two competing theories for the development of retinal correspondence have been debated since the late 19th century. On the one hand, Ewald Hering insisted that the oculomotor system was hard wired at birth and in fact the eyes should be viewed as a "single organ" "the binoculum". On the other hand, Hermann von Helmholtz argued that regulation of each eye occurred separately, and that retinal correspondence was imposed on the oculomotor system as a result of a set of learning rules derived from the visual system. Hering's idea was recast as "Hering's law of equal innervation," namely that for the eyes to move together (ie, a conjugate shift of each eye in the same direction), the brain had to generate the same amount of activity for each agonist muscle.

Processing of Disparity Information Throughout the Visual System

Disparity signals are critical to visual function and have been documented at multiple sites within the primate visual system including the neurons of V1, V2, V3, V3A, V4, and middle temporal (MT) (V5), and middle superior temporal (MST) areas. MST has reciprocal connections with the lateral intraparietal area (area LIP) whose neurons are activated before and during saccadic eye movements. Neurons in area LIP of the parietal cortex are tuned for different absolute disparities and therefore provide the neural signal necessary to initiate saccades to specific points in 3D space. Area LIP is part of the dorsal stream of visual information and as such provides disparity information via dense projections to the FEFs and to the deep layers of the SC.

The Müller Experiment Used to Test "Hering's Law of Equal Innervation"

Consider the visual situation shown in Fig. 20.5 that was originally envisioned by Müller [1].

A target is rapidly moved from a distant to a closer location along the axis of vision of one eye (left). In this case, the right eye would make a large movement (a combination of a saccade and vergence), but the left eye would *not move at all*. In Hering's view, two signals were required to generate the appropriate movement. A conjugate gaze signal, ΔE_{Cong}, would be distributed to both agonist muscles equaling half of the angle required to point the fovea toward the target. The conjugate signal would be summed with a convergence signal (ΔE_{Verg}) calibrated to shift

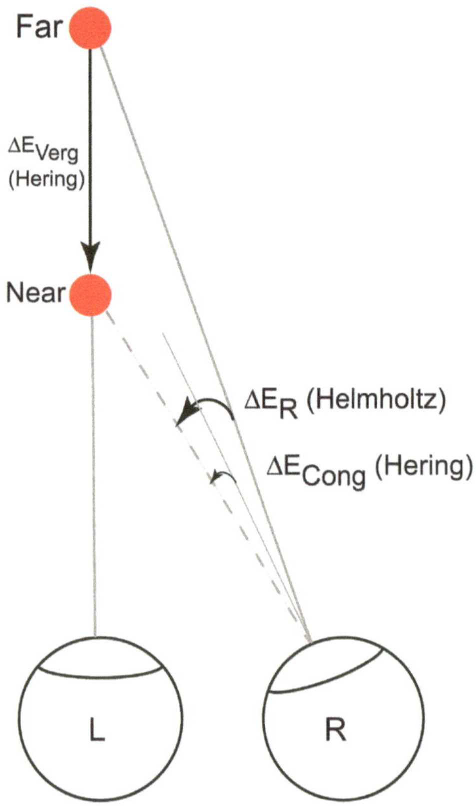

FIGURE 20.5 An extreme paradigm in which a target moves in depth along the line of site of one eye. Therefore, one eye makes a large eye movement while the fellow eye is required to make no eye movement. Helmholtz and Hering hypothesized major differences in the types of signals that the brain used to execute disconjugate or disjunctive saccades in this circumstance (see main text).

the fovea of the right eye; the remainder of the distance to be directed toward the near target. At the same time, the oppositely directed convergence signal to the left eye would cancel the input from the conjugate gaze system. This difference would result in little or no movement of the left eye. Helmholtz, on the other hand, suggested that the brain programmed separate signals for the right and left eyes. The signal ΔE_R rotated the right eye the full amount necessary to look at the near target. The left eye received no signal. Only in the last 3 decades using single neuron recording techniques in awake, behaving macaque monkeys have we begun to understand which of these mechanisms (separate conjugate and vergence = Hering; separate right and left eye movement signals = Helmholtz) is implemented in the brain circuitry and physiology.

Neural Substrate Supporting a Hering-Like "Horizontal Vergence System"

The wide distribution of disparity information throughout the cerebral cortex leaves open the question of how disparity information is translated into the motor signal required to direct the eyes at the same point in space. Neurons located at the caudal most part of the arcuate sulcus of the FEFs discharge in association with vergence and provide descending efferents to the nucleus reticularis tegmenti pontis (NRTP) where vergence responses have been documented. Projections from the NRTP are directed to the fastigial oculomotor region (FOR) and the interposed (IP) nuclei (the emboliform and globose nuclei comprise the "IP nuclei") regions of the deep cerebellar nuclei that are clearly associated with changes in vergence. The IP and the fastigial nuclei both project to the supraoculomotor area (SOA), a region of the MRF dorsal and lateral to the oculomotor nucleus. A subset of neurons in the SOA have been described as "near response" cells because of the close relationship of their discharge with pure changes in accommodation. Other neurons in the SOA are sensitive to vergence position as well as vergence velocity (VV) during symmetrical (both eyes moving equally but in opposite directions) vergence shifts.

There are projections from the near response neurons (both vergence velocity and position) in the SOA to the medial and inferior rectus subdivisions of the oculomotor nuclei. As noted earlier, two separate portions of the oculomotor nuclei associated with the medial and inferior rectus muscles have been identified, the large motoneurons (A group) located within the classically defined borders of the oculomotor nuclei (most of these neurons innervate the SIFs in the orbital layer of the muscle) and a group of smaller neurons located dorsal–medial to the primary oculomotor nucleus, called the "C group," which innervates the MIFs of the global layer. The *dendritic* fields of medial rectus and inferior rectus motoneurons in the "C group" of the oculomotor complex extend dorsally and medially to include much of the SOA neurons. The dendritic fields of motoneurons innervating the medial rectus muscle are distributed only to the ipsilateral SOA, while the dendritic fields of the inferior rectus motoneurons are distributed to the SOA on *both* sides. In sum, projections from the most caudal portion of the arcuate sulcus in the FEF can reach the SOA "vergence" region via a cortico-pontine–cerebellar–midbrain loop.

Parallel Pathway for Control of Horizontal Vergence

An alternative, more direct pathway for the control of vergence could make use of cortico-collicular projections. Neurons carrying absolute disparity information from area LIP directly target the intermediate and deep layers of the superior colliculus (dSC). In addition, the SC also receives disparity information from the FEF, which has strong projections to the intermediate and deep layers of the SC. Yet another source of disparity information to the dSC may arise from the superficial layers of the SC (sSC) where binocular visual neurons have been identified.

Evidence Supporting a Helmholtz Organization Into Right and Left Eye Signals

Early data suggested that medial rectus motoneurons receive two separate sources of input: one from the AINs providing conjugate signals and a second from the supra-oculomotor nucleus mediating vergence. Thus, a simplistic approach of adding vergence onto the underlying conjugate signal would be sufficient to explain the results of the Müller experiment. Surprisingly, the activity of the motoneurons *and* the internuclear neurons of the abducens was *suppressed* during vergence eye movements. In other words, the motoneurons located in the medial rectus subdivision of the oculomotor complex receive an inappropriate (ie, markedly reduced) signal during vergence saccades. The origin of this suppression signal targeting the abducens motoneurons *and* AINs remains unclear. There is no anatomic or physiological evidence for oculomotor internuclear neurons (IONs) (neurons in the oculomotor nuclei descending to reach the abducens nucleus) that could mediate such inhibition. Alternatively, this inhibitory signal might arise from GABAergic inhibitory neurons located in the SOA or the adjacent MRF which project to the abducens nucleus.

Despite the evidence for a "pure vergence" signal in the SOA, a number of pieces of data suggest that the interaction between the versional and vergence subsystems must occur before the level of the oculomotor neurons. Evidence for fast vergence was difficult to explain using separate vergence and version systems. The time course of vergence eye movements is much longer than for conjugate saccades. Saccadic eye movements even of the largest amplitude are completed within 70–80 ms while pure vergence movements can last hundreds of milliseconds. At the same time, the duration of the fast component of a disconjugate "saccade" from a far to near target is longer than expected for a conjugate saccade of similar amplitude, but hundreds of milliseconds shorter than a typical "slow vergence" movement. Furthermore, vergence velocity is enhanced and peak saccadic velocity is reduced during such movements. Therefore, disconjugate eye movements cannot be generated by a simple linear combination of conjugate and vergence neural signals.

Because the amplitude of the rapid horizontal component of movement for each eye during disconjugate saccades was quite different, Enright [2] suggested that this could be produced by providing a different burst ("pulse") and tonic discharge ("step") to each eye. Indeed, recent physiological evidence supports this idea. The discharge of PPRF burst neurons is most closely associated with an *individual eye* and not with a conjugate gaze signal as would have been predicted by Hering. Investigation of abducens motoneurons suggests that they receive an admixture of innervation from each of the "right" and "left" eye–driven PPRF neurons, thereby imposing ocular coordination at the level of the motoneurons. Additional data has demonstrated that neurons in the vestibular nuclei and the nucleus prepositus carry signals for movement of individual right and left eyes.

Role of the MRF in the Control of Saccades and Vergence

A region of the midbrain lateral to the oculomotor nuclei and ventral to the SC called the central MRF has been demonstrated to participate in eye movement. As noted previously, the SOA neurons are located in the medial aspect of the MRF just dorsal and lateral to the oculomotor nucleus, and their discharge is most closely associated with vergence and accommodative movements. However, microstimulation 1–3mm lateral to the oculomotor nucleus in the central portion of the MRF (cMRF) elicits "conjugate movements." Activation of the dorsal portion of the cMRF produced fixed amplitude contraversive saccades at short latency, while stimulation in the more ventral cMRF generated contraversive saccades whose amplitude was dependent upon the initial position of the eyes, so-called "goal-directed" and eye-centering saccades. These results have suggested that the MRF has a number of characteristically different regions: the SOA, the dorsal cMRF, and the ventral cMRF. Neurons in the cMRF have a clear burst of discharge associated with the contralateral direction, but have much larger movement fields than the comparable neurons in the SC. In distinction from the SC, the movement fields of MRF neurons have been demonstrated to be not only open (ie, discharge for contralateral eye movements of all sizes beyond a certain threshold), but also display a monotonic relationship between the spike number and amplitude. However, the initial stimulation and single neuron recording experiments in the cMRF did not examine the movements of individual eyes. More recently, single unit recording from neurons in the cMRF in monkeys prepared for binocular eye movement recording and capable of performing disconjugate saccades in depth has demonstrated a range of responses from neurons best driven by a single eye to clearly conjugate eye movement–related cells. Taken together, these findings strongly suggest that not only do the premotor generators of the burst and tonic level of activity harbor separate right and left eye activity, but supranuclear regions such as the cMRF are also encoded for individual eye activity.

An alternative possibility is that neurons in the cMRF could provide a parallel pathway for descending information to reach the omnipause region via direct, inhibitory projections to the omnipause region. Such inhibition of the OMNs would indirectly activate PPRF burst neurons. Furthermore, neuroanatomical data has demonstrated strong projections from the cMRF to the C group of the oculomotor nuclei where the motoneurons controlling the MIFs are located. Thus, discharge of cMRF neurons could enhance the velocity of the eyes as well as provide a regulatory role and/or muscle tension during disjunctive saccades made in depth.

Role for the SC in the Control of Vergence

Few models have suggested that the saccade–vergence interaction could occur without the need to resort to nonlinear interactions at the oculomotor plant or independent eye control.

One such model hypothesized that the saccadic system (particularly the fixation neurons at the rostral pole of the SC) utilized the OMNs to gate the vergence system. However, a careful analysis of OMN activity suggests that the tonic level of these neurons is *modulated* during vergence, and not gated (ie, not turned on or off). Alternatively, removing the inhibition generated by the OMNs would markedly increase the gain of the vergence system and this could be viewed as a multiplicative effect on vergence velocity. Based on an analysis of the dynamic interaction of saccades with vergence, it has been proposed that the SC might be the source of a cyclopean saccadic burst. In this schema, the topographic collicular map would be utilized as an appropriate weighting function to provide increased enhancement of vergence for larger amplitude saccades. The target locus of this multiplication of the collicular output with vergence has not been identified, however, an interaction at the level of the OMNs has been proposed.

In contra-distinction to this hypothesis, recent single neuron recording and electrical stimulation experiments have suggested that not only does the cMRF, but also the rostral pole of the dSC participates in the generation of vergence and/or disjunctive saccades and harbor signals that are tuned more for one eye than the other. Stimulation of the rostral pole of the dSC *during* eye movements in depth can partially suppress (not enhance the velocity of) vergence movements. The downstream projections from the dSC have recurrent collaterals that synapse on neurons in both the cMRF and in the near-response region (ie, the SOA), suggesting that the signals carried by the neurons in the dSC are actually tuned in three-dimensional space. Indeed, single neuron recording in the rostral portion of the SC has shown that these cells *are* active during vergence (both convergence and divergence) movements, but not disconjugate saccades (fast vergence). At the same time, neurons located in the more caudal portions of the SC have not been extensively examined for evidence of a discharge associated with disconjugate eye movements.

CEREBELLAR PATHWAYS FOR OCULOMOTOR CONTROL

The cerebellum participates in many aspects of eye movement including those induced by vestibular stimulation, pursuit of slowly moving objects, as well as saccades. Two main divisions of the cerebellum relevant to eye movement control are currently recognized, but have overlapping roles: (1) vestibulo-cerebellum (archi-cerebellum) and (2) the dorsal vermis of the posterior lobe and fastigial nuclei. The vestibulo-cerebellum is comprised of the flocculus and paraflocculus (tonsils) laterally and the uvula and nodulus medially. The flocculus receives its primary inputs from the vestibular nuclei, nPH (the horizontal position integrator), the inferior olive, and the cell groups of the PMT. The cells of the PMT are located at the rostral end of the abducens nucleus (see the *s in Fig. 20.1). These cells probably contribute an efference copy signal of the current position of the eyes to the flocculus. The effects of removal of the flocculus and paraflocculus are described later under the heading of Smooth Pursuit. The nodulus lies immediately caudal to the inferior medullary velum and the adjacent ventral uvula. The nodulus receives input from the vestibular nuclei, nPH, inferior olivary nucleus, and direct primary afferents from the vestibular nerve. The primary efferents of the nodulus and uvula are to the vestibular nuclei. Physiologically, the nodulus and the uvula participate in the generation of "velocity storage" in the brain stem. This means that they change short duration (about 6–10 s) signals from the semicircular canals into a signal which lasts 20–30 s or longer. This is reflected by the prolonged nystagmus observed while rotating someone in the dark. Typically, this nystagmus is stopped or reduced by quickly pitching the head forward and back. This effect of pitch on horizontal nystagmus (called "dumping") is lost after the removal of the nodulus and uvula. Another rare form of nystagmus called periodic alternating nystagmus in which nystagmus occurs first to one direction and after 2–3 min reverses direction to beat in the opposite direction develops in patients (eg, the Arnold–Chiari malformation, see section: Clinical Applications) in whom there has been damage to the nodulus and uvula.

The dorsal vermis is the second portion of the cerebellum that is devoted to oculomotor control and it primarily participates in the execution of accurate saccades and accelerating the eyes during smooth pursuit (see section: Smooth Pursuit). Lobules VI and VII of the vermis receive mossy fiber inputs from the PPRF, NRTP, the dorsolateral and dorsomedial pontine nuclei (DMPN), vestibular nuclei, and the nPH. The inputs from the PPRF, NRTP, and nPH are most likely important in the planning and execution of saccades when the eyes are stable, while those from the dorsolateral pontine nuclei (DLPN) and vestibular nuclei are important in the generation of saccades *during* smooth pursuit (see section: Smooth Pursuit).

Copies of the signals from the FEF and SC (recall that these are arranged topographically according to saccade amplitude and direction) requesting the desired amplitude saccade and the current motor command signals from the phasic (burst) neurons in the PPRF and riMLF are directed to the dorsal oculomotor vermis (lobules V–VII). Signals from the SC arrive via ascending mossy fiber inputs from the NRTP and probably also from the DMPN, whereas the signals from the FEF utilize the DMPN and the DLPN. The primary output of the oculomotor vermis is directed through the caudal fastigial nuclei (the so-called FOR) which distributes it to the appropriate inhibitory burst neurons (not the OMNs) located in the more caudal PPRF on both sides of the brain stem.

This configuration permits a direct comparison between the real-time, on-going saccade (from the PPRF) to the requested amplitude of eye movement (from the FEF and/or SC). The difference between these signals is called "motor error." The FOR receives additional climbing fiber inputs from the inferior olivary nucleus and axon collaterals from mossy fibers projecting to the dorsal vermis, namely from the DLPN and the NRTP. Current thinking suggests that there is a topographical arrangement of "motor error" arrayed across the fastigial nuclei. As the fastigial activity shifts from lateral (large motor error) toward the midline (small motor error), the "motor error" from the vermis reaches zero and the saccade ends. It is at this point that the fastigial neurons deliver a "stop" or "choke" signal that halts further movement of the eyes, via activation of the inhibitory burst neurons in the PPRF *bilaterally*. In other words, the oculomotor cerebellum participates in a feedback pathway which ensures that the eyes reach the desired goal relative to the head and then stops the eye movement. Moreover by virtue of the appropriate timing of the fastigial output, the oculomotor vermis enforces the speed, accuracy, and stability of saccades. As such, it is no surprise that cerebellar disease primarily affects these dynamic variables.

Clinical Applications

Disorders in saccadic accuracy, either hypo- or hypermetria are usually the hallmark of cerebellar disease (ie, it is a form of dysmetria). Hypometric saccades (the eye lands short of the intended target) are generated because the number of spikes in the burst ("pulse") of discharge from the EBNs in the PPRF or riMLF are reduced. Since the tonic discharge of the neurons in the nPH for the horizontal system and the INC for the vertical system is precisely matched to the burst the eye lands short of the target, but eye position remains stable. Slow saccades may occur in a variety of cerebellar, cerebral, or extraocular muscle diseases and result from a decrease in burst frequency even though spike number remains normal. Again, since the burst and tonic discharge are matched, the final position of the eyes remains stable (ie, no nystagmus). If the extraocular muscle range is normal, the disorder is central and might occur in a patient who has had a stroke in the pons affecting some of the EBNs or their inputs.

The oculomotor effects of loss of the dorsal vermis most likely explain the lateropulsion (hypermetria) encountered in *Wallenberg's* syndrome in which there has been an infarction of the lateral medulla. In that disorder, interruption of olivocerebellar climbing fibers in the inferior cerebellar peduncle ("restiform body") is thought to increase the activity of the Purkinje cells in the ipsilateral vermis. The dorsal vermis in turn inhibits the underlying fastigial nuclei producing the observed lateropulsion.

SMOOTH PURSUIT

The primary visual stimulus for inducing smooth pursuit movements is slip of an image across the retina. This can occur in two ways. First, there may be a slowly moving object in one portion of the visual field while the body and head remain still. A second way to induce retinal image motion is for the subject to be moving producing an optic flow field. In this situation, the purpose of the smooth pursuit system is to track (ie, stabilize) stationary objects in a constantly moving field. If the object of regard is in-line with the heading of the subject, then there is only a small degree of slip across the fovea. On the other hand, if the object is below or to the side of the direction of the heading, for instance an obstruction on the ground ahead, then considerable retinal slip would occur unless a specific system were working to stabilize the image on the retina and produce clear vision.

As was discussed in the chapters on the visual system, two primary divisions of the visual system have been described. One is used for analysis of features (ie, high acuity) and utilizes projections through the parvocellular layers of the lateral geniculate. Secondary neurons in the parvocellular layers of the LGN send their axons to layer 4Cβ of the primary visual cortex. The second division makes uses of the magnocellular layers of the LGN and projects to visual cortex layer 4Cα. One of the primary purposes of this pathway is to convey visual motion information to the brain. This is evidenced by a shorter latency for visual responses to be registered in layer 4Cα than in 4Cβ. Inactivation of the magnocellular layers of the LGN significantly impairs, but does not abolish smooth pursuit. Once the visual motion signal reaches the visual cortex, it is distributed to layer 4B and hence to V2. Visual area V2 (area 18) in turn projects to extrastriate visual area V5, also known as area MT. Damage to the primary visual cortex will abolish smooth pursuit in any direction in the blind hemifield.

Located in the superior temporal sulcus, the primary purpose of visual area MT is to develop an estimate of target speed. Monkeys with discrete lesions of area MT demonstrated an inability to estimate the speed of moving targets. Saccades to moving targets in the hemifield opposite the side of the lesion were inaccurate, and pursuit responses were delayed. Saccades to stationary stimuli were normal. The next layer of cortical motion processing occurs in an adjacent region of the superior temporal sulcus, area MST (medial superior temporal region), to which area MT projects. The major difference are the inputs to area MST, which receives signals of both eye and head movements (and position). As a result, these cells can determine the true heading from a moving visual scene and permit tracking small objects while a subject is in motion. Monkeys with discrete lesions in area MST portrayed two deficits for motion. The first was a retinotopic deficit similar to that, which occurred following MT lesions. The monkey was unable to estimate target speed in a visual region contralateral to the lesion. The second deficit was a unidirectional loss of horizontal smooth pursuit ***toward*** the side of the lesion regardless of the hemifield that the visual stimulus was projected upon. These results suggested that area MST provides an estimate of target speed independent of other movements to subsequent processing regions. This estimate of target speed is sent to both the FEFs and the DPLN in the brain stem to generate a matching eye speed command. The FEF probably provides a predictive component of the smooth movement to the DLPN.

The DLPN, which receives inputs from multiple regions of the cerebral cortex (FEF, MT, MST, and parietal cortex) and brain stem (nucleus of the optic tract) related to smooth pursuit, provides major inputs to the cerebellar flocculus and paraflocculus (tonsils). The vestibulo-cerebellum (flocculus and paraflocculus) sends efferents to the vestibular nuclei (ie, the medial vestibular nucleus) and a small group of cells (the y-group) that cap the inferior cerebellar peduncle and which in turn convey an eye position signal to the oculomotor and trochlear nuclei. The vestibulo-cerebellum is important in the transformation of visual and vestibular signals into a common frame of reference. Damage to the DLPN causes loss of ipsi-directional pursuit, and different from area MT and MST lesions, the saccade deficit is directional and not retinotopic. Similarly, removal of the paraflocculus produces a marked reduction in the gain of ipsi-directional pursuit. Entire removal of the cerebellum is necessary to completely abolish pursuit. The latter effects are probably mediated via additional projections from DLPN and the PPRF to the dorsal oculomotor vermis and subsequently to the caudal fastigial nucleus (see section: Cerebellar Pathways for Oculomotor Control). The cells in the fastigial nuclei discharge most vigorously during the initial acceleration of pursuit. On the other hand, the cells in the paraflocculus discharge most vigorously during steady state pursuit. This suggests that the fastigial nuclei are important for the initiation (acceleration) of the eye movement signal, while the paraflocculus is critical in the maintenance of the pursuit movement. The projections from the paraflocculus for controlling smooth movements are probably mediated via the medial vestibular nucleus and y-cell group and subsequently to the oculomotor neurons. At the same time, projections mediating smooth pursuit movements from the fastigial nuclei to the ocular motoneurons are not clear.

Clinical Applications

The vestibule-cerebellum is primarily targeted in patients who have the *Arnold–Chiari* malformation. In this condition, the tonsils (paraflocculus) and also the flocculus itself are drawn down into the foramen magnum and become compressed. The patients often complain of severe headaches, diplopia, and imbalance ("dizziness"). Examination reveals that they have gaze-evoked nystagmus as a result of a deficient neural integrator, probably the result of reduced output to the medial vestibular nuclei and the nPH (the horizontal system integrator). Gaze-evoked nystagmus is often observed in this situation. The burst of discharge is normal, but the precise match to the tonic discharge from the nPH or vestibular nuclei is either not sustained or is too low to keep the eye stable in eccentric position. Patients with the Chiari malformation may also have reduced smooth pursuit and deficient tracking using the eye and head together. Finally they often display downbeat and torsional nystagmus, especially on looking laterally and down.

EYE MOVEMENTS IN HUMAN DISEASE

As noted earlier, lesions of the structures that comprise the oculomotor system are often diagnostic of particular illnesses affecting the CNS. Stroke was mentioned previously in reference to loss of the frontal and the parietal cortices. While MRI is often used in clinical settings to establish a diagnosis, many times the MRI is "normal" but the patient has significant neurological signs and symptoms. The beauty of the oculomotor system is that since it is a very sensitive indicator of a patient's neurological status or function, it can often be utilized in making clinical diagnosis in the absence of structural abnormalities on MRI. The following reviews a select number of illnesses in which eye movements are critical to establishing the diagnosis. Moreover, measurement of eye movements may provide insight into the rate of development or existence of a variety of neurological diseases and could permit the introduction of disease-modifying therapies at a much earlier stage of the disease. Also, once therapies have been introduced, monitoring eye movements could prove to be a sensitive indicator of efficacy since the actual function of the CNS is not demonstrated on an MRI. This is one of the core concepts underlying the translation of laboratory findings to therapies to combat neurological diseases.

MOVEMENT DISORDERS: DEGENERATION OF NEURONS IN MOTOR CONTROL SYSTEMS (PARKINSON'S AND PROGRESSIVE SUPRANUCLEAR PALSY)

Parkinson's disease (PD) is the second most common neurodegenerative disease in the US after Alzheimer's disease. Nearly one million people are currently diagnosed with PD in the United States and about 15,000–60,000 people newly detected each year. Clinically, these patients demonstrate "Parkinsonism": tremor at rest, rigidity, bradykinesia, and impairment of postural reflexes. However, the pathological hallmark of PD is the accumulation of filamentous, cytoplasmic inclusions consisting mainly of α-synuclein aggregations in the form of Lewy bodies or Lewy neurites. This accumulation is associated with the loss of dopamine neurons in the substantia nigra pars compacta. The causes of sporadic PD, which accounts for about 90% of cases of Parkinsonism is unknown. From the standpoint of eye movements, patients with idiopathic PD often display problems in suppressing REM execution in an antisaccade task and "ratchet" smooth pursuit (intermittent saccadic interruptions in smoothly executed voluntary movements). Patients with PD may also have difficulty with internally generated saccades (so-called "memory-guided" movements) displaying longer latencies and hypometria. Furthermore, they may have significant problems with the direction of visual attention. Patients with PD have marked difficulty in planning a sequence of eye movements to remembered targets, possibly associated with an inability to rapidly choose between competing "plans." This could be the result of a loss of top-down inhibition of selective attention as a result of reduction in prefrontal–basal ganglia activity. Thus, the eye movement abnormalities are not the result of impaired saccade dynamics, but rather the inability to channel attention in the motor system to select a specific task.

Progressive supranuclear palsy (PSP) is often mistaken for idiopathic PD in its early stages. While there may be stiffness in the limbs and neck, bradykinesia, and postural instability (Parkinsonism), in distinction to people with idiopathic PD, patients with PSP most often do not have resting tremor or "cog-wheeling." In fact, one of the early features of PSP is the distinct loss of balance characterized by repeated falls *backward*. The gait abnormalities of PD typically do not appear until much later in the illness and most often the festinating gait has the patient bent *forward* (ie, the patient falls forward). In addition, the eye movements of patients with PSP are quite distinct. They often have dramatic loss of downward greater than upward gaze, but eventually all vertical eye movements are lost, with a rather curious preservation of horizontal saccades. However, fixation in patients with PSP is often interrupted by nearly continuous side to side small amplitude (<2°) "square wave jerks." Patients with PSP also display severe bulbar symptoms with loss of swallowing (dysphagia), reduction in the articulation of speech (dysarthria), and subcortical dementia. The prevalence of PSP is about 5 per 100,000, and men and women are equally affected. Average age at onset is 63 years, and mean time from symptom onset to death is 7 years. No pathological-proven cases have begun before the age of 40 years.

Another major distinction between PSP and PD is the underlying pathology and its distribution. Whereas neurons in PD display α-synuclein aggregations in the form of Lewy bodies (ie, synucleinopathy), neurons targeted in PSP harbor abnormal aggregates of τ protein (4-repeat τ, 4R tauopathy). Moreover, the degeneration found in PSP appears in several subcortical structures, including the substantia nigra, the subthalamic nucleus, and the rostral midbrain. Neurofibrillary tangles are present in these areas. Tufted astrocytes (Gallyas-positive) are the hallmark feature of PSP that differentiates it in pathology from other 4R tauopathies such as corticobasal degeneration (CBG) (astrocytic plaques). It is thought that loss of the neurons in the riMLF accounts for the hallmark loss of rapid, voluntary vertical saccades. Despite the significant loss of voluntary gaze and balance, the vestibular ocular responses both in the horizontal and vertical directions are completely preserved, thus justifying the use of the term "supranuclear" palsy. In sum, understanding the oculomotor and vestibular systems is critical to distinguish between a number of the primary degenerative diseases affecting motor control.

Oculomotor behaviors may also prove useful in monitoring and testing new therapeutic advances in these and other related diseases. One area of intense interest has been the severe disturbances in REM sleep found in patients with narcolepsy and cataplexy. Patients with reduced levels of REM sleep will display significant daytime sleepiness entering almost instantaneously into REM (narcolepsy) and may also lose postural tone falling to the floor (cataplexy). Normally during REM sleep, the motor system is paralyzed with the exception of the eye movements. However, patients with "REM behavioral disorder" (RBD) display severe, violent movements during sleep associated with the onset of REM, suggesting that the normal inhibition of the corticospinal tracts during REM has been removed. Interestingly, a significant number of patients with PD display one or both of these sleep disturbances. In fact, patients who develop RBD without Parkinson's features are at high risk (greater than 66%) for the development of PD or other synucleinopathy (eg, dementia with Lewy bodies) in the succeeding 8 years. This finding identifies specific subgroups of patients at high risk for the PD and therefore could form the basis for organizing trials of neuroprotective medications in PD. Recording eye movements over the course of treatment could be a significant adjunct to the efficacy of any neuroprotective agent. Proof of concept has already been demonstrated in macaques with MPTP-induced PD and patients in whom REM sleep has been restored following treatment with levodopa.

HEREDITARY, PARANEOPLASTIC, AND CEREBELLAR DEGENERATION: PARANEOPLASTIC DESTRUCTION OF PURKINJE CELLS

Cerebellar disease can arise from intrinsic degenerative processes which are often hereditary in nature. At the same time, patients may display similar oculomotor deficits from extrinsic factors often called "acquired cerebellar ataxias." Many of these outside modulators of cerebellar function have been well known for decades, for example ethyl alcohol. Others have come to light only more recently. Many of the cerebellar toxins such as alcohol have both acute and chronic effects. Thus, excessive ingestion of alcohol can cause acute ataxia, but prolonged use of alcohol is often associated with poor nutrition that can lead to Wernicke's disease which can have profound effects on the oculomotor system (eg, complete ophthalmoplegia) typically the result of involvement of the periaqueductal gray of the brain stem (PPRF and riMLF), but also atypically in the cerebellum most often in the vermis, and dentate nuclei. It is critical to understand that bilateral involvement of the thalamus, mammillary bodies, or the periaqueductal brain stem alone cannot explain the oculomotor signs noted in these patients. Indeed, in an autopsy study with Wernicke's disease, over 50% of the cases displayed damage in the dentate and vermal regions of the cerebellum. More recently, MRI imaging (FLAIR and DWI) has demonstrated similar damage in the cerebellum and deep cerebellar nuclei of living patients. Thus, "atypical" locations may in fact be critical to the development of many of the "cerebellar signs" described in patients with Wernicke's disease. Moreover, chronic alcohol intake can have a direct toxic effect on susceptible neurons, particularly those in the superior portion of the cerebellar vermis causing atrophy out of proportion to the rest of the cerebellum. It is thought that this may explain the more prolonged effects of alcohol on gait and balance.

While some of the lesions demonstrated using DWI MRI in patients with acute Wernicke's disease are reversible with prompt administration of large doses (>500 μg/day) of IV thiamine, continued B-1 deficiency will cause transient lesions to become permanent. It is unclear if it is this nutritional mechanism via which atrophy develops or some other direct toxic effect of alcohol.

A more recently recognized extrinsic factor with a profound effect on cerebellar function is immune-mediated antibody formation directed against either intracellular organelles or surface proteins of Purkinje cells. These antibodies are typically generated by small cell tumors arising elsewhere in the body. Some examples are shown in Table 20.1.

TABLE 20.1 Antibodies in Paraneoplastic Cerebellar Degeneration

Antibody	Associated Malignancy	Symptoms Other Than Ataxia
Anti-Hu	Small cell lung cancer	Encephalomyelitis
Anti-Yo (PCA1)	Breast and ovarian cancer	
Anti-Ri (ANNA-2)	Breast and ovarian cancer	Brain stem encephalitis
	Small cell lung cancer	Opsoclonus–myoclonus syndrome
Anti-Tr	Hodgkin's lymphoma	Limbic encephalitis
Anti-Ma2	Germ cell tumor	Limbic and brain stem encephalitis
Anti-VGCC	Small cell lung cancer	Lambert–Eaton myasthenic syndrome
	Breast and ovarian cancer	
Anti-CV2 (CRMP5)	Small cell lung cancer	Limbic encephalitis
	Breast cancer	
	Non-Hodgkin's lymphoma	Chorea, dystonia, uveitis, retinitis
Anti-GAD	Thymoma	Limbic encephalitis
		Stiff person syndrome

ANNA-2, antineuronal antibodies; *CRMP5*, collapsing response mediator protein 5; *GAD*, glutamic acid decarboxylase; *PCA*, Purkinje cell antibodies; *VGCC*, voltage-gated calcium channel.

For example, anti-Hu antibodies arise from small cell carcinoma of the lungs and are directed toward multiple targets in the peripheral and central nervous system including the peripheral sensory nerves, the dorsal root ganglia, temporal lobe of the cerebral cortex, and the Purkinje cells of the cerebellar hemispheres. As such, anti-Hu antibodies are frequently detected in a variety of neurological syndromes including: cerebellar ataxia, limbic encephalitis, brain stem encephalitis, myelitis, or intestinal pseudo-obstruction. Numerous other antibodies have now been identified and include a variety of CNS targets such as the voltage-gated calcium channels (VGCC) and glutamate decarboxylase (GAD). Therefore, they can interfere with synaptic conduction and in the case of antibodies directed against the VGCCs, they can produce a neuromuscular-type illness called the Lambert–Eaton syndrome. While it is uncommon for these patients to demonstrate ocular motor disorders, fatigable ptosis and hypometric saccades have been described. Interestingly, high titers of anti-Hu antibodies are absent in SCLC patients who *do not exhibit* neurological disorders. In other words, if these paraneoplastic processes can be recognized at an early stage, then treatment of the underlying cancer can not only ameliorate the neurological symptoms, but have a major impact on life expectancy (ie, allows the cancer to be recognized and treated at a much earlier stage).

A vast number of hereditary disorders have now been recognized that affect the cerebellum, currently the count exceeds 40 (see Table 20.2 for oculomotor-associated ataxias). Many of these hereditary disorders can affect the oculomotor system, retina, or both.

The spinal cerebellar ataxias (SCA) can be roughly divided by modes of inheritance into autosomal dominant, autosomal recessive, X-linked, and mitochondrial. Specific gene defects have also been identified for many of these disorders and are commercially available for testing. These disorders are clinically characterized by progressive incoordination of gait as well as problems with coordination of the hands, speech, and eye movements. Other features of these disorders have helped to distinguish one type of ataxia from another and include evidence of frank ophthalmoplegia, spasticity, neuropathy, and cognitive/behavioral difficulties. However, these features are not definitive and there is considerable clinical overlap making the use of genetic testing a cornerstone of their diagnosis.

Among the autosomal dominant SCAs (ADSCA), SCA2 and SCA6 are most closely associated with oculomotor deficits. Both disorders arise from CAG trinucleotide expansions in their respective gene targets: ATXN2 and CACNA1A, respectively. Both are most common in adulthood, with SCA2 occurring most often in the 3rd to 4th decade while SCA6 most often does not appear until the 5th to 6th decade. However, cases of SCA2 have been reported in children as young as 10 and SCA6 in teens as young as 19. Patients diagnosed with SCA2 often have a lifespan of 10–15 years following diagnosis, while SCA6 is associated with an almost normal lifespan (>25 years). SCA2 is one of the more common forms of spinocerebellar ataxia. SCA2 is characterized by progressive cerebellar ataxia, including nystagmus, and, in some individuals, ophthalmoparesis or Parkinsonism. Pyramidal findings are present; deep tendon reflexes are brisk early on and are absent later in the course. Saccadic eye movements are markedly slow, but accuracy is maintained. This suggests that it is the actual burst neurons in the brain stem reticular formation that are affected by this disease and not a specific portion of the cerebellum. In contrast, SCA6 patients display normal velocity saccades, but the accuracy (either hyper- or hypometria) is often reduced. This suggests that damage to the FOR and/or the oculomotor vermis (Lobules V to VII) has impaired the ability of the cerebellum to monitor the current progress of a saccade in progress. The ataxia found in patients with SCA6 can often be displayed intermittently and not continually.

TABLE 20.2 Hereditary Cerebellar Ataxias Associated With Oculomotor or Retinal Signs

Disease Name (Population Frequency)	Gene	Average Onset (Range in Years)	Average Duration (Range in Years)	Distinguishing Features
AUTOSOMAL DOMINANT ATAXIAS				
SCA1	ATXN1	3rd to 4th decade (<10 to >60)	15 years (10–28)	Pyramidal signs, peripheral neuropathy, slow saccades, possible Parkinson-like features
SCA2	ATXN2	3rd to 4th decade (<10 to >60)	10 years (1–30)	Slow saccadic eye movements, peripheral neuropathy, decreased DTRs, dementia
SCA3 (Machado–Joseph disease)	ATXN3	4th decade (10–70)	10 years (1–20)	Pyramidal and extrapyramidal signs; lid retraction, nystagmus, decreased saccade velocity; amyotrophy fasciculations, sensory loss
SCA6	CACNA1A	5th to 6th decade (19–71)	>25 years	Sometimes episodic ataxia, very slow progression, gaze-evoked and downbeat nystagmus. Possible symptomatic benefit with acetazolamide
SCA7	ATXN7	3rd–4th decade (0.5–60)	20 years (1–45; early onset correlates with shorter duration)	Visual loss with retinopathy, slow saccades, may have Parkinson-like features
SCA18	7q22–q32	Adolescence (12–25)	Decades	Ataxia with early sensory/motor neuropathy, nystagmus, dysarthria, decreased tendon reflexes, muscle weakness, atrophy, fasciculations, Babinski responses
SCA23	PDYN	5th to 6th decade	>10 years	Dysarthria, abnormal eye movements, reduced vibration and position sense; one Dutch family; neuropathologic
SCA26	EEF2	(26–60)	Unknown	Dysarthria, irregular visual pursuits; one Norwegian-American family; MRI: Cerebellar atrophy
SCA28	AFG3L2	Age 19.5 years (12–36)	Decades	Nystagmus, ophthalmoparesis, ptosis, increased tendon reflexes; two Italian families
EA1	KCNA1	1st–2nd decade (2–15)	Attenuates after 20 years	Myokymia; attacks lasting seconds to minutes; startle- or exercise-induced; no vertigo
EA2	CACNA1A; CA CB4	2–32	Lifelong	Nystagmus; attacks lasting minutes to hours; posture change–induced; vertigo; later, permanent ataxia
EA5	CACNB4	20–30	Lifelong	Nystagmus; acetazolamide helpful

Continued

TABLE 20.2 Hereditary Cerebellar Ataxias Associated with Oculomotor or Retinal Signs—cont'd

Disease Name (Population Frequency)	Gene	Average Onset (Range in Years)	Average Duration (Range in Years)	Distinguishing Features
AUTOSOMAL RECESSIVE HEREDITARY ATAXIAS				
Ataxia-telangiectasia (1:40,000 to 1:100,000)	ATM	1st decade	10–20	Telangiectasia, immune deficiency, cancer, chromosomal instability, increased α-fetoprotein, oculomotor apraxia, nystagmus, and peripheral neuropathy
Ataxia with oculomotor apraxia type 1 (unknown)	APTX/ apratаxin	Childhood	Decades	Oculomotor apraxia, choreoathetosis, mild intellectual disability, hypoalbuminemia
Ataxia with oculomotor apraxia type 2 (unknown)	SETX/ senataxin	Age 10–22 years	Decades	Cerebellar atrophy, axonal sensorimotor neuropathy, oculomotor apraxia, increased α-fetoprotein (distinguish from ataxia-telangiectasia)
IOSCA (rare, Finland)	C10orf2/ twinkle	Infancy	Decades or less	Peripheral neuropathy, athetosis, optic atrophy, deafness, ophthalmoplegia, seizures
Autosomal recessive spastic ataxia of Charlevoix–Saguenay (rare)	SACS/sacsin	Childhood	Decades	Spasticity, peripheral neuropathy, retinal striation
Refsum disease (rare)	PHYH; PEX7	1st–6th decade	Decades	Neuropathy, deafness, ichthyosis, retinopathy
MIRAS (rare Finland or Norway)	POLG1	Childhood to young adulthood	Decades	Nystagmus, dysarthria, epilepsy, cerebellar atrophy on MRI
SANDO (rare)	POLG1	Childhood to young adulthood	Decades	Abnormal eye movements, ragged red fibers, myopathy, dysphagia, neuropathy, myopathy

Of the autosomal recessive forms of SCA, two are most closely associated with oculomotor abnormalities: ataxia with oculomotor apraxia Type 1 (AOA1) and Type 2 (AOA2). Both of these disorders are typically diagnosed in childhood: AOA1 around the age of 7 and AOA2 between the ages of 10–22 years. The initiation of saccadic eye movements in Type 1 patients is often markedly prolonged allowing the head to account for more of the gaze shift. Eventually, these patients develop complete ophthalmoplegia. AOA1 is an autosomal recessive disorder that affects the aprataxin (APTX) gene. In addition to the ataxia, these patients often have evidence of hypoalbuminemia. Early in the course, about 80% of the patients display choreoathetotic movements which can often lessen as the disease progresses. Intellect remains unaffected in patients of

Portuguese ancestry while patients of Japanese origin often display progressive loss of mental capacity.

Patients with AOA2 display dysmetric saccades (typically hypometria with slow saccade initiation). Oculomotor apraxia is present in about 50% of individuals. There is typically a dissociation of eye–head movements in the "head-free" condition, in which the head reaches the lateral target before the eyes. Smooth pursuit may be "jerky," (ie, inability to maintain the target on the fovea requiring a "catch-up" saccade), gaze-evoked nystagmus, poor horizontal optokinetic nystagmus (OKN), and square-wave jerks saccadic eye movements made away and then back to a "fixation" target within a very short (<than 50 ms) time interval. Recall that saccade initiation time is at least 200 ms. Areflexia and later a peripheral axonal sensorimotor neuropathy (>90% of individuals) are found. There is marked atrophy of the cerebellum on MRI. Metabolically, more than 95% of individuals with AOA2 display a serum alpha-fetoprotein (AFP) concentration greater than 20 ng/mL. Serum total cholesterol concentration greater than 5.6 mmol/L is found in ~50% of affected individuals (normal value: 3.5–5.8 mmol/L). Serum creatine kinase (CK) concentration is increased in some affected individuals, and elevated immunoglobulin levels (IgG and IgA) have been found in several families. Mentation is typically preserved. If an elevated AFP level is found, it is critical to distinguish this form of ataxia from ataxia telangiectasia (A-T) which also displays oculomotor apraxia, but patients with A-T often have telangiectasias of the conjunctivae, immunodeficiency, choreoathetosis, and are at increased risk of leukemia and lymphoma. Karyotyping (if specific genetic testing is not available) will demonstrate a 7:14 chromosomal translocation. AOA2 is associated with CAG repeats in SETX, the gene that encodes the protein senataxin. An autosomal *dominant* form of ataxia has also been described associated with two SETX missense variants (p.Asn603Asp–p.Gln653Lys) shared in cis configuration by a mother and daughter. Both had cerebellar ataxia with atrophy of the cerebellum, dysarthria, oculomotor defects (saccadic pursuits and gaze-evoked nystagmus), and tremor. There was no evidence of elevation in AFP in the patients with this very unusual variant. Furthermore, variation in the SETX gene has been associated with amyotrophic lateral sclerosis 4 (ALS4), an autosomal dominant form of juvenile onset ALS. Therefore, caution has to be exercised in the interpretation of the results of genetic testing.

Further Reading

[1] Müller J. Elements of physiology. Philadelphia: Lea and Blanchard; 1843.

[2] Enright JT. The remarkable saccades of asymmetrical vergence. Vis Res 1992;32(12):2261–76.

[3] Michaelides M, Moore AT. The genetics of strabismus. J Med Genet 2004;41(9):641–6.

[4] Leigh RJ, Zee DS. The neurology of eye movements. Oxford University Press; 2015.

[5] Tang X, Buttner-Ennever JA, Mustari MJ, Horn AK. Internal organization of medial rectus and inferior rectus muscle neurons in the C group of the oculomotor nucleus in monkey. J Comp Neurol 2015;523(12):1809–23.

[6] Van Horn MR, Waitzman DM, Cullen KE. Vergence neurons identified in the rostral superior colliculus code smooth eye movements in 3D space. J Neurosci 2013;33(17):7274–84.

[7] Nachbauer W, Eigentler A, Boesch S. Acquired ataxias: the clinical spectrum, diagnosis and management. J Neurol 2015;262(5):1385–93.

[8] Jayadev S, Bird TD. Hereditary ataxias: overview. Genet Med 2013;15(9):673–83.

CHAPTER

21

A Modern Clinicopathological Approach to Traumatic Brain Injury

V.E. Koliatsos, L. Xu, J. Ryu, N. Ziogas

Johns Hopkins University School of Medicine, Baltimore, MD, United States

INTRODUCTION: TBI AS HETEROGENEOUS MORBIDITY

Traumatic brain injury (TBI) is a very common problem. In the United States alone, it is estimated that there are 2–3 million new cases of TBI requiring medical attention annually. Contrary to common perception, TBI is not a disease per se but rather an adverse event, a calamity, which causes a number of primary and secondary lesions and is responsible for transient, static, or, rarely, progressive clinical syndromes and illnesses. Organizing such complexity is difficult, but there are some meaningful ways to classify problems by cause, mechanism or pathology, and clinical severity or outcome. The clinicopathological cycle leading from patient observations to pathological hypotheses to modeling and then back to bedside for additional observations or new diagnostics and therapeutics has not been established for many TBI-related clinical problems (Fig. 21.1). The relationship among causes, mechanisms, and clinical outcomes is complex and the classification of TBI is usually based on variables from two or more categories, for example, cause and mechanism or mechanism and outcome. Acute problems associated with moderate or severe TBI involve elevations in intracranial pressure and other brain and nonbrain pathophysiologies and cause serious, life-threatening illness managed in critical care settings. This is a specialized topic that is beyond the scope of this chapter.

TBI is often classified as mild or moderate–severe based on several criteria, most commonly the presence and duration of post-TBI changes in consciousness. Mild TBI is practically identical to the construct of *concussion* that is featured by brief or no changes in mental status and symptoms that typically improve quickly and disappear in a few weeks. This is an extremely common problem in the community and most likely underrepresented in formal surveys because it is often not managed in hospital settings. In addition, in view of its overwhelmingly good prognosis, little effort has been placed into a more accurate diagnosis and research interest has been minimal. However, there is a measurable minority of patients that do not do well; many of the symptoms of this so-called *post-concussive syndrome* (PCS) resemble idiopathic mental illness, but it is far from clear that post-concussive illness is identical to mood or somatoform disorder. Repeat

Conn's *Translational Neuroscience*
http://dx.doi.org/10.1016/B978-0-12-802381-5.00033-6

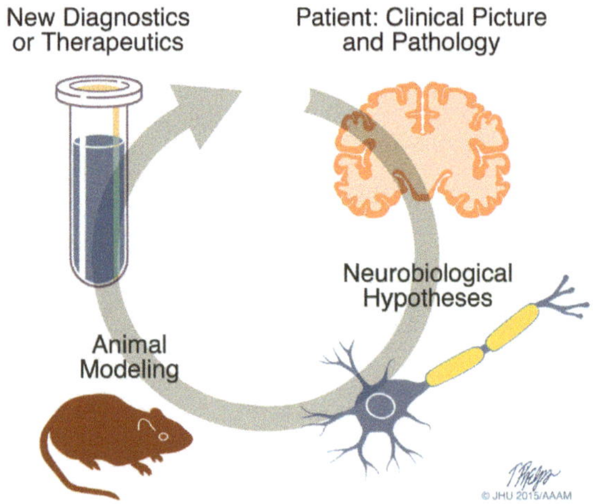

FIGURE 21.1 The clinicopathological model was pioneered as the pillar of scientific medicine in the first half of 19th century by Bichat, Dupuytren, and other great French physicians of the time, and was greatly popularized through the work of Rudolf Virchow in Germany and William Osler in the United States. In the case of the nervous system, the clinicopathological pioneer was probably Charcot. The main tenet of the model is that symptoms and signs in the patient correlate with or can be explained by organ pathology. A modern version of the clinicopathological model for neuropsychiatric disease may be illustrated as a series of repetitive cycles beginning with the patient and his brain (*top right*), from the study of which one can derive neurobiological hypotheses (*bottom right*) that can then be tested in models (*bottom left*) from which new diagnostic and therapeutic ideas can emerge (*top left*). The science jargon of the last part is "from bench to bedside." While some neurological disorders like multiple sclerosis and AD are through one or more iterations of the cycle, morbidities associated with TBI are very early in the process.

concussion, associated with particular lifestyles or medical problems, may have a cumulative effect on the brain that is very different from that of a single concussion.

Some types of TBI lead to chronic encephalopathies. Such conditions cause significant disability because of permanent damage to the brain or progressive disease. These problems are the focus of the present brief review and include focal contusions often caused by falls, *diffuse axonal injury* (DAI) due to rotational acceleration as it occurs in motor vehicle crashes, TBI due to boxing or other sports-related or unrelated repeat concussions, and blast injury to brain, an old problem that has resurfaced in the wake of recent conflicts in Iraq and Afghanistan. These forms of TBI are evolving conditions that may get worse over time, are associated with numerous secondary morbidities, and are best viewed as chronic disease. Hence, in a large number of cases, TBI is a calamity that sets forth processes leading to chronic illness.

KEY CLINICAL AND PATHOLOGICAL FEATURES OF MAIN TYPES OF TBI

Focal TBI: Contusions

Blunt TBI causes focal damage to the brain and its coverings (meninges), including hematomas and parenchymal contusions. Penetrating TBI may also be viewed as an extensive focal TBI defined by the ballistics of the projectile and other complex interface phenomena including the skull. The brain *contusion* (Fig. 21.2) is a very common TBI-associated lesion caused by direct impact to skull (in engineering terms, *contact loading*). Falls, strike by or against events and motor vehicle crashes are common causes. Contusions can take place at both side of impact (coup) and a remote brain site that is often the diametric opposite of coup (contrecoup). Striking the immobile head with blunt force, for example, a weapon, causes skull fractures and coup lesions, whereas if the moving head strikes a firm fixed surface there is predominance of contrecoup lesions irrespective of the site of coup. Contrecoup injuries are by far the commonest lesions and tend to occur at ventral frontal and temporopolar locations, perhaps because of the tight encasement of these brain regions in the anterior and middle brain fossae and friction with the irregular surfaces of sphenoid and ethmoid bones.

The brain contusion is hemorrhagic necrosis of brain tissue that ranges from multiple triangular hemorrhagic lesions near the cortical

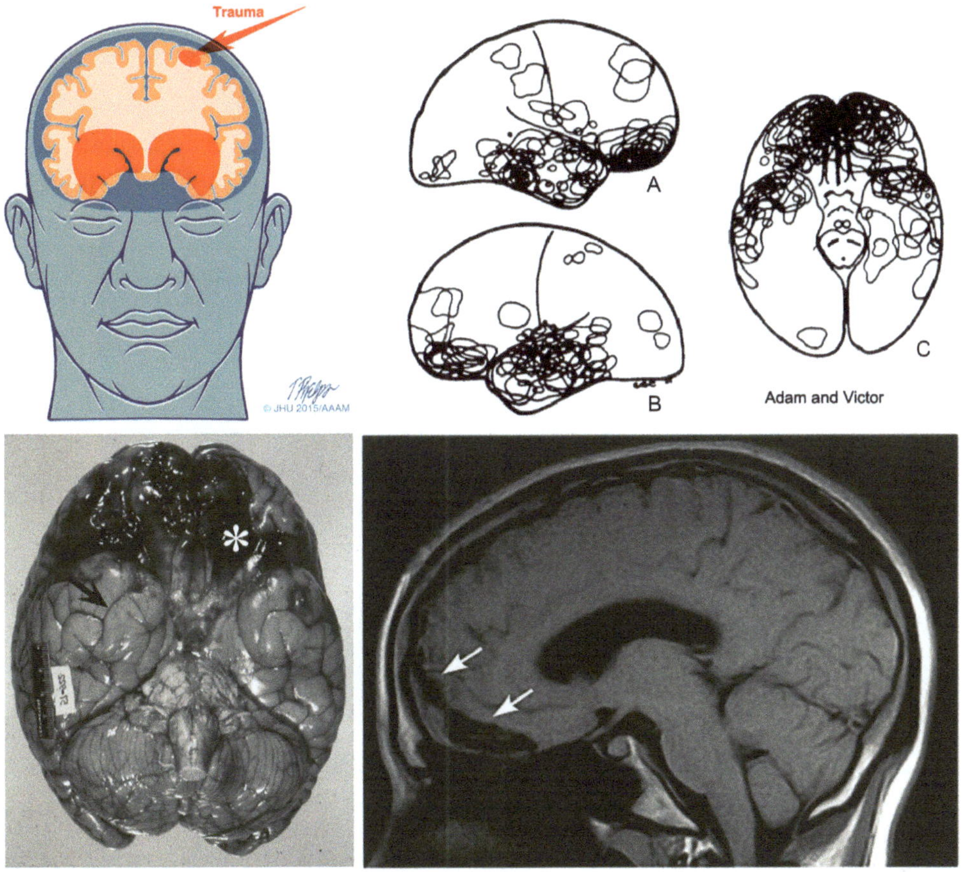

FIGURE 21.2 In the case of contusion, the impact that can be anywhere in the skull usually causes a small coup lesion and a larger contrecoup lesion in the diametrical position, very commonly affecting orbitofrontal and temporopolar regions (*top left*). Courville's classical lesion delineations (*top right*) are still valid. Brain contusions have high mortality rates due to local edema and rapid hemorrhagic expansion of brain tissue and, in some cases, uncal herniation (*bottom left*; asterisk shows the contusion and arrow the herniation). Many people with smaller contusions survive but have deficits in social comportment (*bottom right*; arrows point to orbitofrontal contusion in this 20-year-old patient with intact intelligence but impulsive inappropriate behaviors). *Pathology specimen courtesy of Dr. Juan Troncoso.*

surface to frank parenchymal hemorrhage and is commonly associated with subarachnoid and intraventricular bleeding. At the center of the contusion neurons undergo necrosis, whereas apoptosis may prevail at the penumbra. Efferent axons submit to secondary (Wallerian) degeneration. By virtue of their anatomical distribution, contrecoup lesions cause frontal presentations with cognitive, affective, and behavioral problems. Focal neurological signs are often absent. In contrast to cerebral infarcts (stroke), cerebral contusions are featured by a preferential involvement of superficial cortical layers.

Diffuse TBI: Primary Traumatic Axonal Injury

In contrast to brain contusions, *primary traumatic axonal injury (DAI)* is the result of inertial loading of axons (*impulse, not contact, loading*)

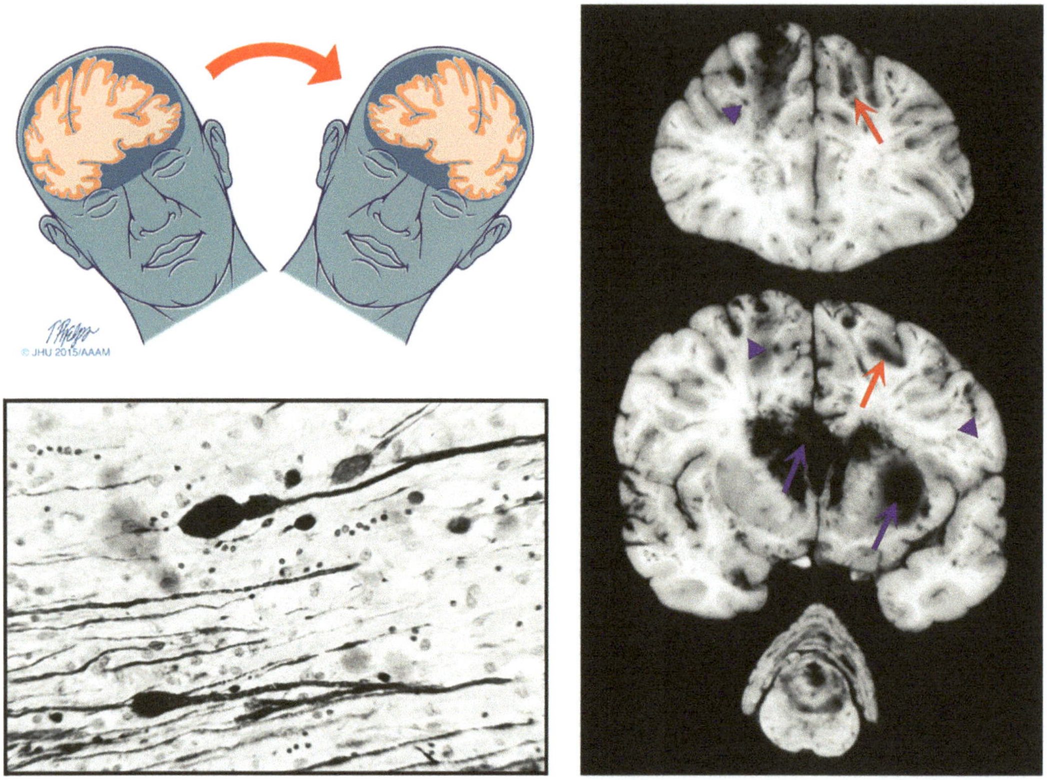

FIGURE 21.3 In the case of diffuse axonal injury, the impulse of rotational acceleration causes ultra-rapid deformation of brain tissue that tends to preferentially damage the white matter with forceful shearing and stretching (*left top*). Damage may be more serious in the primate brain because of the sheer bulk but also complexity of the white matter in which pathways cross each other in different directions. Extensive injury of this type leads to coma and death. Macroscopic pathology (*right*) features numerous hemorrhagic lesions from the shearing and breakage of small blood vessel (*blue arrowheads*) that tend to coalesce (*blue arrows*) in certain brain sites, for example, the genu and splenium of corpus callosum, as had been formulated by Strich in the late 1950s; in severe cases, the white and gray matter become separated along the ridge of gliding contusions (*red arrows*). The convention is that regions of shearing of blood vessels are also sites of axonal lesions independent of hemorrhage. Microscopic pathology features axonal breakage and the formation of terminal bulbs (*left bottom*, APP immunostaining). *Pathology specimens courtesy of Dr. Juan Troncoso.*

that occurs very rapidly, usually as a result of rotational acceleration of the head in the course of motor vehicle crashes and high-impact falls. DAI leads to multifocal axonal disruption and disconnection often associated with microhemorrhages or a separation between gray–white matter in the form of gliding contusions (Fig. 21.3).

DAI was initially identified with silver degeneration methods. However, because neuronal proteins undergoing fast anterograde transport, including neurofilament proteins and amyloid precursor protein (APP), accumulate in damaged axons, immunohistochemistry for these proteins has largely replaced silver methods in recent years. There is increasing evidence that DAI is a common denominator among various types of TBI, from concussions to blast and chronic traumatic encephalopathy (CTE, see later). The gyrencephalic human brain may be especially prone to DAI, in part because of the sheer bulk of white matter but perhaps

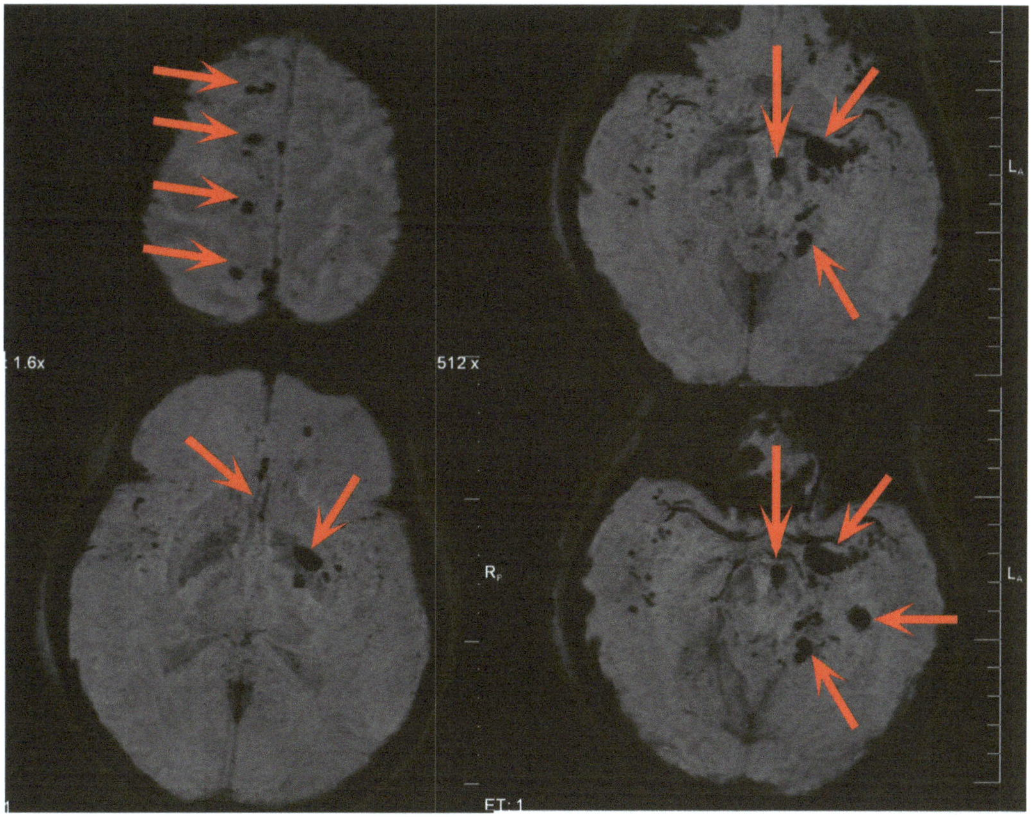

FIGURE 21.4 This is the brain MRI (susceptibility weighted sequence) of a 24-year-old female with a recent history of severe TBI due to a motor vehicle accident, in which she was ejected from her car on crash. *Red arrows* indicate some of the numerous microbleeds that are heme-positive and magnetic-field distorting lesions. Such lesions are caused by small-vessel breakage and they mark sites where axonal injury is also thought to have occurred. Confluence in the left internal capsule–peduncle–pons has caused right leg weakness. Patient had significant cognitive and behavioral symptoms.

also the three-dimensional (3D) complexity of intertwined white matter tracts. Therefore, DAI is one of the commonest TBI-associated pathologies and deserves a closer look. Because of the preferential involvement of associative and commissural tracts, extensive DAI causes cognitive, behavioral, and emotional disorders or complex motor and sensory deficits (Fig. 21.4).

The mechanisms of DAI have received considerable attention by biologists and bioengineers, with emphasis on the axons themselves and much less on myelin sheaths. Engineers tend to associate DAI with the viscoelastic properties of axons, that is, the fact that axons behave as viscous materials under slow tensile strength and as elastic entities when exposed to rapid tensile elongation (time-dependent strain). The former condition may either leave the axonal structure intact or in fact lead to microtubule polymerization and fast axonal growth during development, whereas the latter condition causes multiple-site breakage of microtubules leading to undulations, swellings, and axon bulbs (Figs. 21.5 and 21.6). In the case of myelinated axons, this process seems to start from the node of Ranvier. In some cases, the axon initial segment may also be implicated. Although emphasis has been placed historically

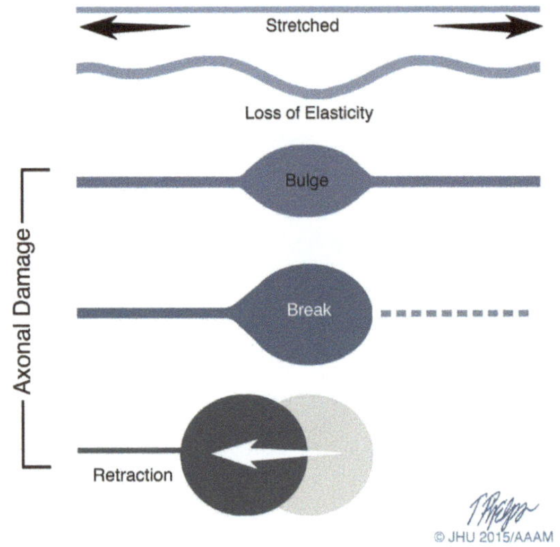

FIGURE 21.5 A sketch illustrating the main steps of traumatic (diffuse) axonal injury, from top to bottom. Ultrarapid stretch causes failure of the viscoelastic properties of axons. Note the progression from a more or less straight to an undulated "fractured" axon, then the formation of swelling (blue) and axon bulb (gray blue) and then the retraction ball (gray); in some models, ball-like boluses of axolemma appear to be pulled back toward the cell body. The distal portion of the axon undergoes Wallerian degeneration. "Retraction ball" is a historical term initially used to describe silver-positive globose microscopic structures that were isolated from or tenuously linked to axon segments. Retraction balls may be just axon bulbs.

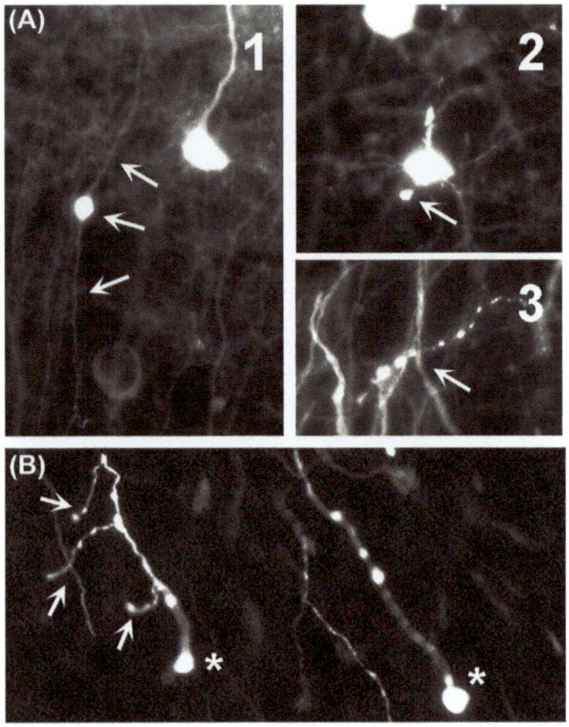

FIGURE 21.6 This preparation of a YFP transgenic mouse exposed to impact acceleration injury depicts injured axons in the corticospinal tract and exemplifies the range of axonal pathologies seen in diffuse axonal injury: in (A), panels 1 and 2 illustrate retraction balls, one of them (1) in the course of what appears to be an intact proximal axon segment and the other (2) right next to the cell body. These two structures probably represent different biological processes. Panel (3) illustrates a beaded (varicose) axon. (B) shows branches resembling regenerating sprouts (arrows). Asterisks depict axon bulbs or cross-sections of varicosities.

on myelinated axons, unmyelinated fibers are certainly also involved in DAI.

The Complex Problem of Blast TBI

Blast TBI is a 100-year-old problem, first surfacing during trench warfare in the western front in World War I. Trinitrotoluene had already been synthesized in 1863, but its use as a weapon was realized after it was packaged in artillery shells. Such shells were excessively used during the battles of Somme, Ypres, and Verdun and caused a tremendous number of casualties with minor military advantage. What made the medical news, however, was the appearance of a novel neurological illness that was initially dubbed "Not Yet Diagnosed Nervous," "Neurasthenia," or "Shell shock" and led to tens of thousands of soldiers being discharged home for recuperation. Many were British. The way the problem was handled in Great Britain is a landmark in the history of medicine and exemplifies the tension between Neurology and Psychiatry that is active to this day: as eloquently explained by Edgar Jones, veterans sent to Maudsley Hospital in London, under the influence of the neuropathologist

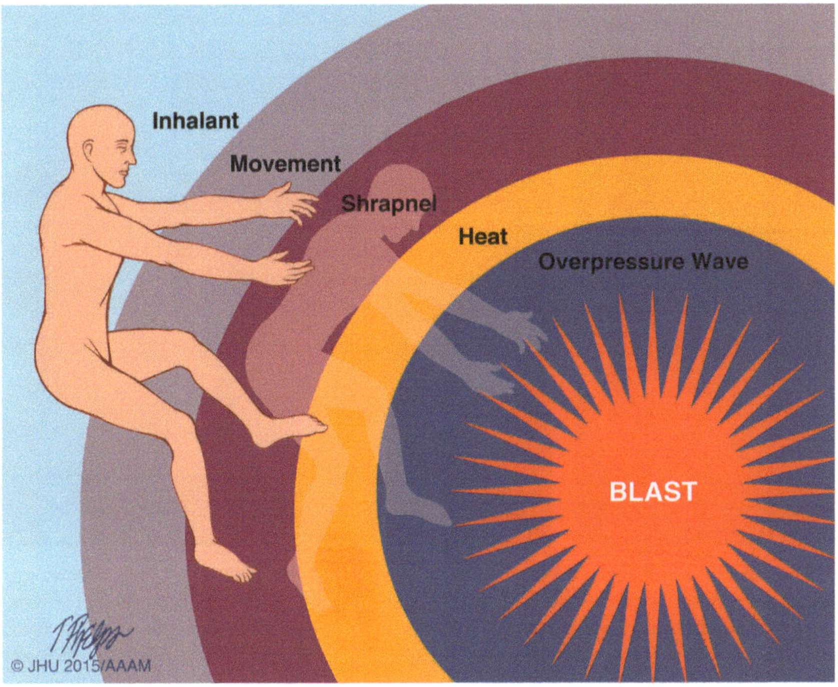

FIGURE 21.7 Blast TBI is a very complex injury that starts with the overpressure wave caused by the explosion (blue) that mobilizes the body of the individual first by positive (wave) and then negative (wind) pressure (gray) and also showers the body and head with shrapnel (purple). The excess heat from the explosion (orange) and asphyxia from toxic inhalants and lung injury (light blue) complete this grave picture.

Frederick Mott, were treated as neurological cases; veterans sent to Maghull Hospital outside Liverpool were managed as hysteria based on the tenets of the—then novel—technique of psychoanalysis. The grounds for this dilemma should become apparent to any clinician who watches video clips from these patients.[1] Ultimately, all these cases were administratively thrown out by the British government as incidents of malingering. The problem resurfaced in the Yugoslav war and was painfully experienced by the United States and its allies in the post-9/11 deployments. Although the bulk of TBI suffered by over 300,000 of US warfighters is due to blunt forces, blast TBI has been the defining or "signature" medical problem of the Iraq and Afghanistan wars.

[1] https://www.youtube.com/watch?v=S7Jll9_EiyA.

Blast TBI is a complex injury (Figs. 21.7 and 21.8). Besides the presumed primary effects of overpressure caused by the shock wave, there may be secondary injuries from the violent projection of mobile elements and debris (shrapnel) against the body, and tertiary injuries from the displacement of the body causing contrecoup contusions and DAI from rotational acceleration. There may also be quaternary injuries, including flash burns from the intense heat of the explosion, that may enhance the injurious impact of other forces, as well as asphyxiation and respiratory damage from the inhalation of toxic substances. Additional factors that play important roles in the clinical picture are *post-traumatic stress disorder* (PTSD), chronic pain, and mood and adjustment disorders. Poignantly, the admixture of so many psychiatric problems keeps the tension between neurological and

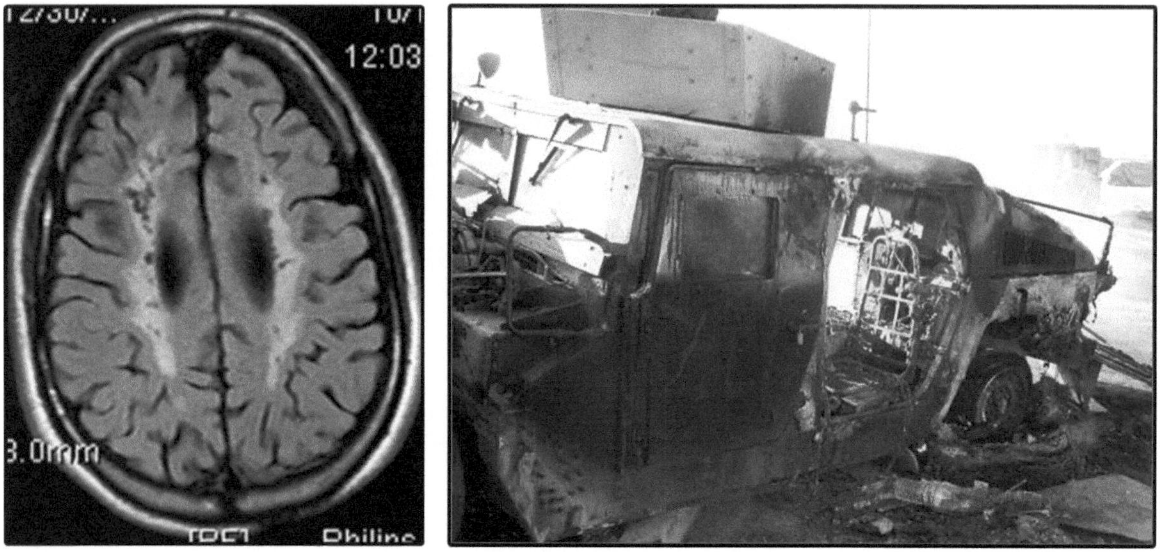

FIGURE 21.8 This is an example of mixed blast injury from the Iraq war. This 21-year-old Army gunner in a Humvee convoy encountered an improvised explosive device (IED) attack and was ejected from vehicle. He was incinerated in the explosive fire for at least half an hour and sustained 60% total body surface area burns requiring intensive care. His brain MRI (*left*; fluid attenuated inversion recovery sequence) shows diffuse axonal injury, a diffuse white matter signal that may signify white matter hypoxia, and brain atrophy. *Right* panel is a photograph of what was left of the Humvee.

psychiatric formulations alive a hundred years after the discrepant treatment of Somme veterans in Great Britain.

Although the secondary and tertiary components of blast TBI are identical to other types of TBI reviewed here, the neuropathology related to the primary effect of blast has not been characterized. Part of the problem is lack of sufficient high-quality autopsy material, especially from long-term survivors of blast injuries. When we examined brains of veterans with history of blast exposure we found APP (+) fiber abnormalities in the frontal lobe in a characteristic arrangement of clusters of axonal varicosities forming 3D honeycombs around brain arterioles (Fig. 21.9). Such lesions colocalize with reactive microglia, a pattern indicating active neuroinflammation. These configurations appear to be distinct from the large fronts of axonal undulations and bulbs seen in classical DAI and from axonal abnormalities seen on other types of TBI or in the case of opiate overdose (Fig. 21.10). This work that is suggestive of a vascular pattern of DAI that may be different from TBI due to classical impact or impulse forces needs to be confirmed in larger series of brains and interpreted based on findings from animal models or engineering simulations (*see later*).

The Problem of TBI-Associated Neurodegeneration

A number of morbid conditions and lifestyles are associated with repeat concussion (repeat mild TBI), for example, autism (head banging), epilepsy (falls), imprisonment and domestic and child abuse (exposure to repeat violence), and collision and contact sports like boxing. It is thought that repeat concussions have a cumulative impact, that is, the outcome is more than the

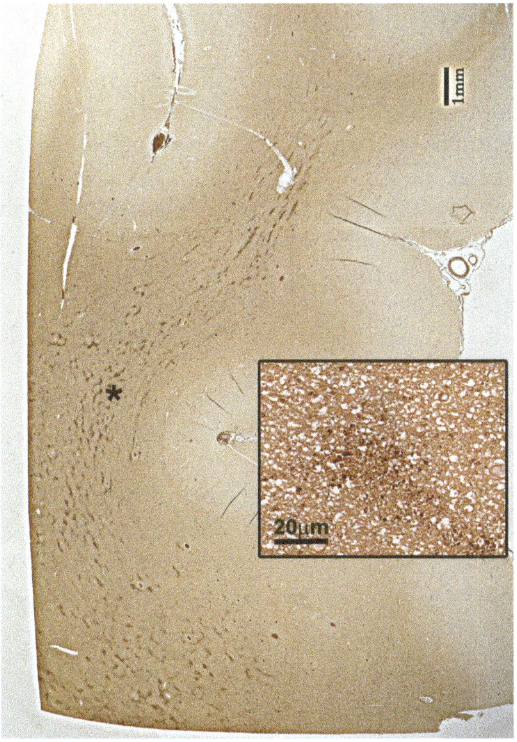

FIGURE 21.9 The typical honeycomb pattern of APP (+) staining in the superior frontal gyrus of an index case with a history of blast injury. Inset is enlargement of the area indicated with an asterisk. Note the groups of tens-hundreds of axonal varicosities in the deep white matter of frontal lobe.

sum of individual concussive effects and, in fact, subconcussive burden may be also important; in other words, the key factor may be the total burden of exposure or lifestyle, rather than individual "index" concussions. This is an important issue that deserves clearer hypotheses and more research. Repeat concussions are associated with two main clinical presentations: one is PCS, possibly a more prolonged or severe version of it compared to PCS after single concussion; the other is CTE, a progressive neurodegenerative disease thought to be caused by repetitive mild TBI. It is unclear whether PCS is associated with neuropathology, although DTI abnormalities

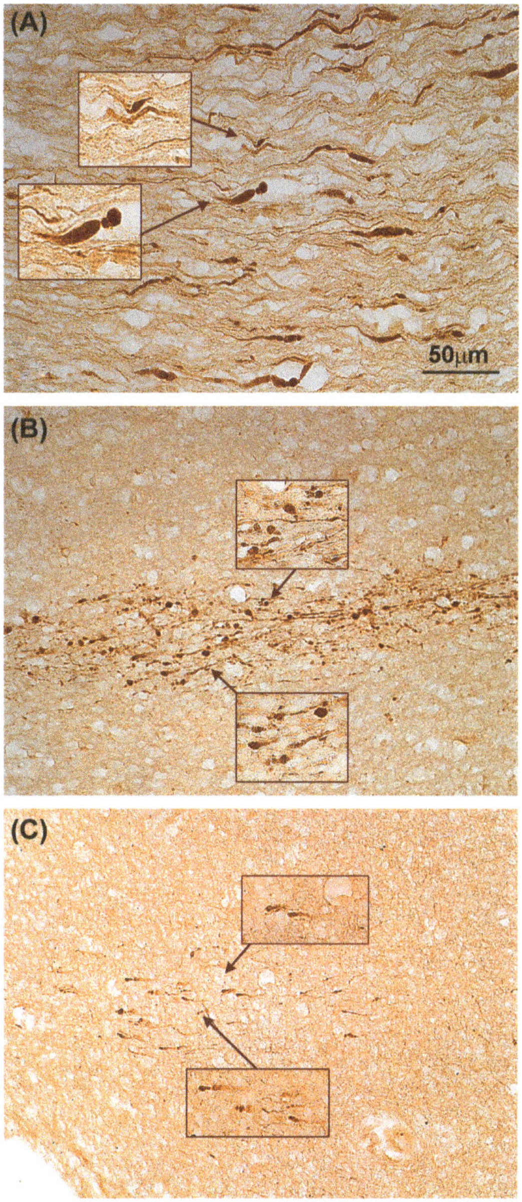

FIGURE 21.10 Microscopic features of APP (+) axon abnormalities in motor vehicle crash (A), blast (B), and opiate overdose (C) in typical cases. Insets represent enlargements of indicated areas. Note differences in size of axons among the three types of injury and also the prominence of undulations and axon bulbs in cases of motor vehicle crash (A).

have been reported in some white matter tracts. CTE is featured by tau and 43-kDa transactive response (TAR) DNA-binding protein (TDP-43) inclusions and neuronal cell loss in neocortex and the limbic system.

CTE has been primarily associated with boxing and has been given a variety of names, such as "punch drunk" by Martland in 1928 and "dementia pugilistica" by Millspaugh in 1937 before called CTE by Critchley in late 1950s and Corsellis in the 1970s. In the era of immunohistochemistry, Omalu and then McKee redefined the problem as a degenerative tauopathy, expanded its occurrence to several collision and contact sports beyond boxing, and launched greater efforts at a more thorough characterization. At the present time, tau inclusions in cell bodies and processes of both neurons and astrocytes are the hallmark of the condition and occur in characteristic anatomical locations (Fig. 21.11). Early changes seem to have a preferential perivascular distribution. Superficial cortical layers and regions at the bottom of cortical sulci tend to be affected more and this may be a distinguishing feature from classical degenerative "tauopathies" such as Alzheimer's disease (AD), frontotemporal degeneration (FTD), and progressive supranuclear palsy (PSP). It is also thought that the disease begins in cortical and corticoid regions in the frontal lobe and then progresses caudally.

Despite considerable efforts by McKee and colleagues to define clinicopathological stages of progression based on the gold standard of Braak staging in AD, the natural course of CTE remains elusive. This task requires prospective studies with minimal sampling bias and equal chance of including cases with relatively good

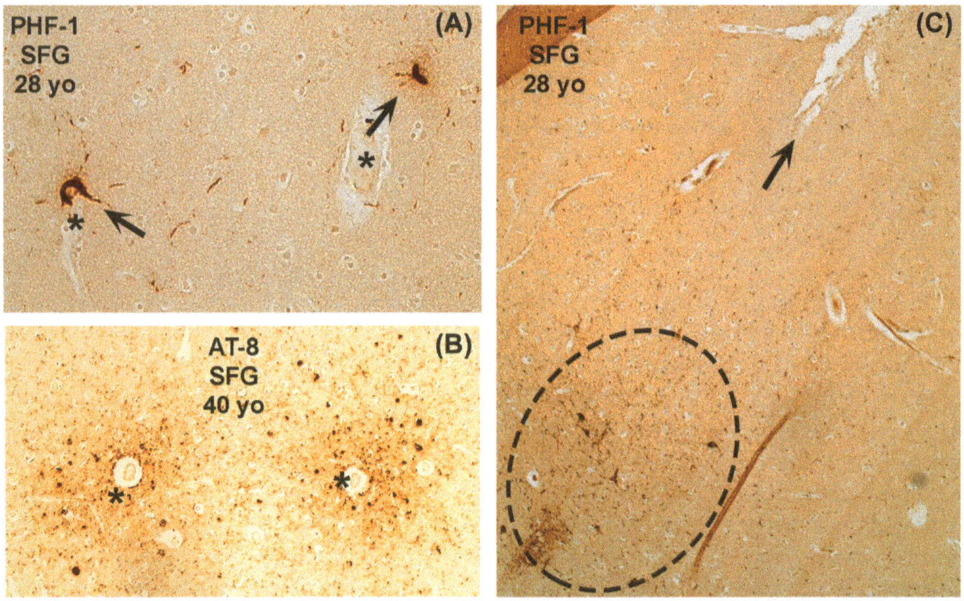

FIGURE 21.11 CTE-associated tauopathy may have unique anatomical features. All sections are taken through the superior frontal gyrus and stained with antibody PHF-1 against phosphorylated tau epitopes. (A) and (C) are from a 28-year-old man with a history of 14 amateur and 14 professional bouts without known history of knockouts or neuropsychiatric history, who died of cardiac arrhythmia. (B) is from a 40-year-old professional NFL player without neuropsychiatric history who died of metabolic coma. Note the very early perivascular tangles and neuropil whiffs in A and B and the deep sulcal location of tau accumulation in C. *Pathology specimens courtesy of Dr. Troncoso.*

outcomes (Fig. 21.12). Our present understanding of the problem is exclusively based on retrospective series, predominantly from brains of boxers who have come to autopsy and increasingly also brains of professional football players who have died prematurely due to suicide or accidents associated with misjudgment of circumstances. Examples of the latter are accidental gunshot while cleaning gun (John Grimsley) or jumping on and falling from the back of a moving track in the course of fiancé chase (Chris Henry). Omalu and others have written on the early presence of psychiatric symptoms, primarily depression and personality changes, but such problems are overrepresented in early adulthood, primarily due to the high prevalence of mood disorders. These conditions are commonly provoked by TBI and respond to routine treatment. If early CTE is featured by mood and behavioral symptoms, is the clinical presentation the harbinger of CTE or merely that of the common idiopathic psychiatric illness? On the other end of the age spectrum, there is no clear-cut demarcation of CTE from early-onset neurodegenerative disease, especially FTD. More work is needed to separate between CTE and other, more common, conditions, and this work requires cohort studies aided by careful clinical characterization and, ideally, input by biomarkers. One of the key questions in CTE is why only some athletes develop it and others not. For now, conventional wisdom is that the risk rises with the amount of concussive burden: in the case of boxing, "bad" boxers who cannot protect themselves from the punches of their opponents; in the case of football, either players who suffer the greatest number of hits like linebackers and linemen, or players who endure the most severe blows like running backs and quarterbacks. Genetic and other predispositions have not been adequately explored.

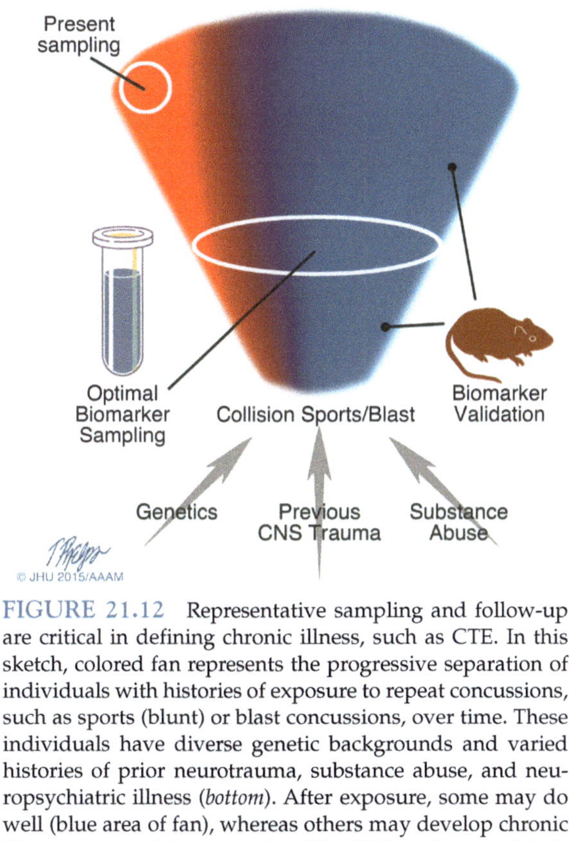

FIGURE 21.12 Representative sampling and follow-up are critical in defining chronic illness, such as CTE. In this sketch, colored fan represents the progressive separation of individuals with histories of exposure to repeat concussions, such as sports (blunt) or blast concussions, over time. These individuals have diverse genetic backgrounds and varied histories of prior neurotrauma, substance abuse, and neuropsychiatric illness (*bottom*). After exposure, some may do well (blue area of fan), whereas others may develop chronic illness, some of it progressive like CTE (*red* area of fan). At present, outcomes after repeat concussions are defined based on the examination of brains of individuals who have done poorly (*top left circle*). Instead, sampling ought to be performed earlier, such as to include cases with both good and poor prognosis and to be repeated over time in order to define course and outcome. This sampling in live patients would involve biomarkers that should be validated in animal models.

NEUROBIOLOGICAL APPROACHES AND HYPOTHESES REGARDING TBI

Generalities

It is difficult to synthesize the very large number of papers and opinions on biological mechanisms underlying common forms of TBI. As stated in the beginning, impact or impulse forces operating from outside the nervous system in most cases, or perhaps from inside in the case of blast, initiate dynamic processes that evolve over time, often long after the initial injurious event. Such processes were initially captured under the terms "primary" and "secondary" TBI that are certainly valid as long as we understand that such terms are simplifications that illustrate a process. Some forms of TBI, for example, contusions, may conclude faster than others. Here we discuss the most relevant mechanisms and hypotheses related to chronic TBI with emphasis on DAI, *neuroinflammation* and *proteinopathy* (Fig. 21.13).

Contusion

The pathobiology of contusion overlaps extensively with the molecular and cellular biology of cell death, as characterized in pioneering work in the 1990s. There are close similarities with stroke lesions. Both prototypical forms of cell death, that is, necrosis and apoptosis, are represented in contusions, with the former predominating. As noted previously, apoptosis may be more prevalent at the penumbra and progressively increases hours or days after the primary insult. Both mechanical and anoxic mechanisms have been implicated and they all involve massive influx of Ca^{2+} via the perikaryal and axonal membranes. Influx may proceed through physical breaks due to the forceful stretching and disruption of membranes, but also the opening of voltage and ligand-activated Ca^{2+} channels. Calcium entry activates different classes of hydrolytic enzymes, such as proteases, especially calpain, endonucleases, and lipases, such as phospholipase A2. The activation of nitric oxide synthase generates superoxide free radical. Intra-mitochondrial accumulation of Ca^{+2} leads to a collapse of ATP production.

Some of the previous effects initiate a progressive destruction of membranes and the nucleosomal fragmentation of DNA. These events lead to rapid necrosis or a more protracted apoptotic cell death, in a stochastic paradigm that we described in our previous work on apoptosis in Huntington's disease. Contusions may evolve in the first few hours or days after injury and some, for example, the ones associated with expanding hemorrhage or herniation of compressed brain structures may cause significant secondary morbidity and death. Although much emphasis has been historically placed on the contusions themselves, the location of traumatic contusions in highly connected frontal and temporal paralimbic regions invites the consideration of transsynaptic mechanisms, at least for the purpose of understanding progressive changes and clinical consequences (see section: DAI: Transsynaptic Degeneration and Disconnection).

DAI: Cellular and Molecular Approaches

The fundamental mechanisms of DAI have attracted a lot of attention since mid-1990s, but many issues remain unclear. Both structural and functional alterations have been proposed and the interactions of the two sets of mechanisms are complex. Based on in vitro evidence from unmyelinated axons, Smith and colleagues have emphasized the early role of fracture of microtubules, that is, the stiffest part of axon structure, caused by ultra-rapid stretching. Microtubules play a major role in axon transport and their breakdown causes an arrest in anterograde transport of organelles and molecules followed by their local accumulation and axonal swelling, then separation of the mechanically unsustainable volume, and secondary axotomy with the formation of axon bulbs or retraction balls (Fig. 21.6). Agents that facilitate microtubule polymerization ameliorate DAI, whereas compounds that inhibit microtubule polymerization exacerbate axonal damage. Additional primary structural effects were proposed by Povlishock and colleagues and include mechanoporation of the axolemma and neurofilament compaction and breakdown.

There are closely associated biochemical events, such as the influx of Ca^{2+} followed by activation of calcium-dependent cysteine proteases (calpain, caspases) and phosphatases, among else. Some of these enzymes actively participate in further digestion of cytoskeletal proteins and breakdown of microtubules or loss of neurofilament sidearms. Excess Ca^{2+} influx is caused both via mechanoporation of the axolemma and also opening of deformation-sensitive Na^+ channels followed by reversal of the Na^+/Ca^{2+} exchanger, membrane depolarization, and then opening of voltage-gated Ca^{2+} channels. Intraaxonal sources

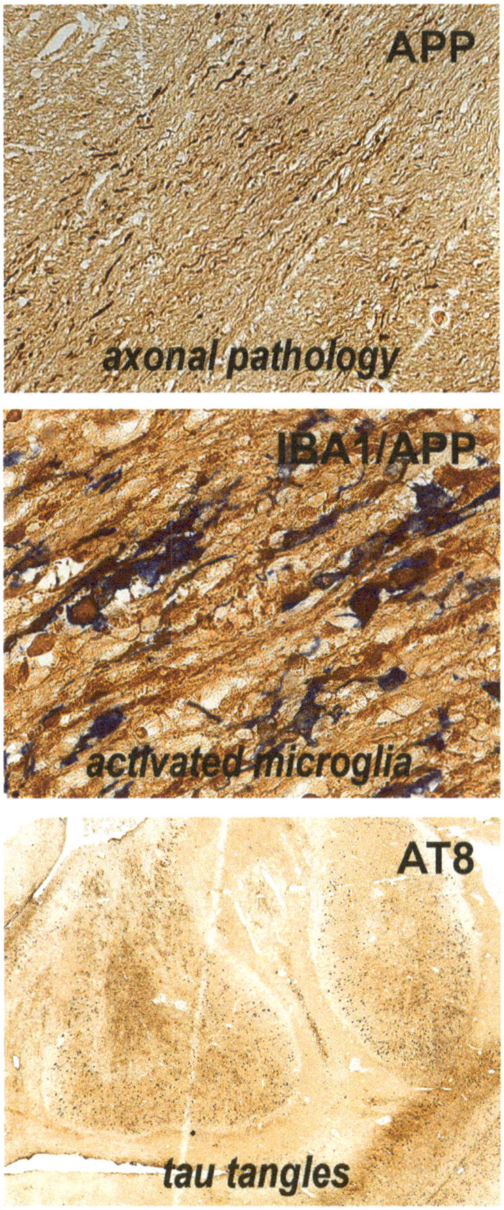

FIGURE 21.13 Three key pathologies associated with TBI chronicity: diffuse axonal injury with disconnection (*top panel*), neuroinflammation often associated with diffuse axonal injury (*middle panel*), and tau proteinopathy (*bottom*). Top depicts a front of undulated, bulbous APP (+) axons in the cingulum of a patient who died a few days after a motor vehicle crash. Middle panel illustrates middle frontal white matter with APP (+) diffuse axonal injury (brown) associated with IBA1 (+) activated/transformed microglia (blue). Bottom panel is taken through the hippocampus of a middle-aged man with history of dementia pugilistica and shows widespread tau tangles immunostained with antibody AT8 against phosphorylated epitopes of tau protein.

of Ca^{2+}, such as of mitochondria may also play a role. These complicated processes have been packaged into distinct scenarios of DAI severity: a milder one is featured by ionic channel dysfunction and cytoskeletal failure leading to anterograde transport arrest with local swelling; a more severe one is featured by mechanoporation of the axolemma and severe cytoskeletal breakdown with conversion of anterograde to retrograde axonal transport, in which case swelling is absent. The previous patterns are "work in progress" but serve to illustrate options that may differentiate between mild and severe TBI; there are many processes going on at the same time in all scenarios, including regenerative attempts of the proximal axon that may have limited effectiveness in the central nervous system (CNS). Retrograde degenerative effects on neurons (atrophy or cell death) often complete this picture of progressive axonopathy and are reminiscent of changes encountered in classical models of axotomy inflicted by nerve crush or transection.

DAI: Transsynaptic Degeneration and Disconnection

An important aspect of DAI is its secondary effects on populations of neurons with injured axons or damaged inputs. Such effects can lead to retrograde and orthograde (transsynaptic) degeneration; transsynaptic effects may continue upward or downward in the neural circuit. For example, in the first weeks/months after a severe acceleration–deceleration injury, there is evidence for progressive reduction in brain size with selectivity for particular brain regions. Advances in cellular and molecular neuropathology, primarily based on axotomy models that we have reviewed elsewhere, may shed light on these phenomena; for a number of reasons, such lessons have not been fully absorbed into the field of TBI. The characterization and timing of these secondary changes and the establishment of therapeutic time windows will not only allow for a better diagnostic assessment of TBI on an individual basis, but also help define opportunities for the prevention of secondary injury based on the temporal evolution of neurodegenerative processes.

As mentioned, transsynaptic effects can spread from primary retrograde or orthograde events to additional interconnected regions and whole circuits. Such changes occur at both the functional and structural levels. Transsynaptic effects are not limited to lesion and degeneration: the pathophysiological implications of axonal damage can also impact the function of circuits. The increasing appreciation of the importance of circuits began with the advent of parallel/distributive theories of brain function introduced by McClelland and Rumelhart in the 1980s, coinciding with powerful anatomical and metabolic methods to trace neural connections. These developments have challenged previous localizing or serial-connection notions and may serve as a new foundation to understand pathophysiology, to accurately formulate diagnosis, and to assign prognosis after moderate–severe TBI.

Focal lesions, that is, contusions, are not an exception to the parallel distributive processing paradigm. For example, the idea that inferior frontal/orbitofrontal contusions cause their notorious adverse effects on executive and social functions of the brain solely on the basis of the ablation of these regions, without taking into account their major neocortical and limbic connections, is problematic. Why one patient with a frontal contusion becomes impaired for life because of disinhibited or antisocial behaviors and someone else with very similar size or location of lesion can achieve good psychosocial and occupational adjustment? The explanation of such differences may require an understanding of the identity of the network that breaks down after lesion in a nodal region of frontal circuitries that process reward and related signals. Such circuit-based understanding of TBI may require nonhuman primate models or novel approaches to studying neuropathology and pathophysiology in the human brain.

Neuroinflammation and TBI

Neuroinflammation involves cellular innate immune responses to injury, changes in the blood–brain barrier, and systemic immune responses, lasting for various lengths of time and having features reminiscent of inflammation in other organ systems. Innate immune responses involve microglia and astrocytes. The microglial cell is the principal CNS macrophage and the central player in this response. Derived from myeloid precursors that invade the brain during embryogenesis, microglia and its processes form a continuous lattice that is ideal for probing the 3D environment. Upon sensing, via pattern recognition receptors, a "pathogen-associated" or "damage-associated" molecular pattern (PAMP or DAMP) linked to a bacterial product or injury, quiescent microglia becomes activated. Pattern recognition receptors include signaling as well as endocytic receptors; Toll-like receptors (TLRs) are one of the best studied receptor family in the former class. Microglial activation involves major changes in appearance and the manufacturing and release of pro-inflammatory cytokines and chemokines, the production of intracellular and extracellular reactive oxygen species, and the neutralization or phagocytosis of noxious stimuli.

The physiology of microglial activation is not well understood, but is thought to have both detrimental and protective sides. This idea is based on the pattern of responses of peripheral macrophages, including classical activation into M1 "killer" cells that secrete high levels of IL-12 and low levels of IL-10, and M2 cells with an inverse IL-12/IL-10 secretion profile that are involved in tissue repair and wound healing or suppress inflammation. A large burden of PAMPs and DAMPs or excessive response to such signals or their misinterpretation is thought to cause neurotoxicity. In the CNS, neurotoxicity related to microglial behavior has been best studied in neurodegenerative disease and multiple sclerosis. In the case of neurotrauma, neuroinflammation is perhaps better understood in spinal cord injury than TBI. It is possible that neuroinflammation contributes to TBI chronicity.

The complexity of the neuroinflammatory response, especially the balance between protective and detrimental effects but also the coexistence of intraneural and systemic effects, has been blamed by Popovich and others for recent failures of immunosuppressive or other intraspinally directed therapies to treat spinal cord injury. Based on the concepts of systemic inflammatory and compensatory anti-inflammatory response syndromes imported from sepsis, these researchers have emphasized the role of host defense in the pathogenesis and outcome of neuroinflammation. Such mechanisms are relevant, especially if we consider that patients with moderate–severe TBI also have significant nonneural injuries and other systemic complications. Therefore, processes like the ones triggered with sepsis should be taken into account and better formulated with future research. The role of systemic mechanisms is evident in the case of spinal cord injury-induced "immune depression syndrome" that predisposes spinally injured patients to infections and perhaps other complications related to immunodeficiency. Although there may be differences between TBI and spinal cord injury, such as the higher intensity of microglia activation, neutrophil recruitment, and accumulation of T and B lymphocytes in the latter, there are also key similarities, including the involvement of both detrimental and protective immune cell species. At the very least, spinal cord injury may serve as a model for TBI. In addition, concepts linking brain with the periphery in common pathophysiological cascades are very helpful, not the least because the disruption of the blood–brain barrier after TBI allows the infiltration of cellular and noncellular elements from peripheral blood, including blood-borne macrophages and neutrophils. In such a fashion, the "privileged" brain is directly subjected to systemic immune regulation.

We and others have found microglial activation in multiple types of clinical and experimental TBI, including blast TBI and repetitive mild TBI. A single mild blunt or blast TBI may be associated with transient neuroinflammatory response, whereas moderate–severe TBI can cause sustained neuroinflammation. Other investigators have studied astrocyte responses and still others have focused on the profile of cytokine and neurokine production and release, especially IL-1β, IL-6, IL-8, IL-10, IL-18, tumor necrosis factor, and vascular endothelial growth factor. Some of these cytokines, for example, IL-6, is thought to be neuroprotective. In many cases, activated microglia colocalizes with DAI, primarily the distal stump of a damaged axon that undergoes Wallerian degeneration (Fig. 21.14). Estes and McAllister have argued that the zeal of discovering signs and elements of an immune response in all kinds of neurological diseases and animal models and calling it "neuroinflammation" is an oversimplification that ignores the cellular complexity and the homeostatic or adaptive sides of immune response. At present, we have not adequately defined a pathogenic signature of neuroinflammatory response in the CNS that can be targeted for therapeutic intervention with a favorable side effect profile.

Mechanisms of Blast TBI

As explained in section "Key Clinical and Pathological Features of Main Types of TBI," blast TBI is a complex situation with several component parts that vary with respect to the nature of their harmful effects on the individual. Some of these, for example, secondary and tertiary injuries, are identical with other classical types of TBI, such as gunshot wounds, contusions, and DAI. Others, for example, the ones related to asphyxia, can cause anoxic encephalopathy. The intense heat can exacerbate the noxious effects of all other mechanisms. The effects of the overpressure wave, one of the primary phenomena associated with an explosion and an event unique to blast, are still unclear.

An animal model that we developed with Ibolja Cernak using a shock tube shows a sharp rise in central venous pressure leading to hemorrhagic infarcts in multiple parenchymatous organs including liver, spleen and kidneys, and acute right heart failure. Lesions are more severe when animals were positioned with their abdomen facing the blast wave. The main neuropathologies are axonal degeneration (traumatic axonal injury) in discrete white matter tracts and associated microglial activation. Although there is classical barotrauma in the lungs with more severe injuries, the brain can show axonal pathology even in cases with minimal, if any, lung barotrauma. Our examination of a handful of human brains with history of blast exposure has revealed the presence of APP (+) axonal abnormalities in sub-millimeter distance from arterioles diffusely in the white matter of hemispheres. These lesions are associated with activated microglia.

The reason for the periarteriolar distribution of axonal lesions in the human brain is unclear. We can hypothesize that the central venous pressure elevation gives rise to subsequent elevation in pressure in the brain sinuses and, perhaps, to venules, with back-loading of the capillaries and at least transient brain edema. The pulse of overpressure and brain edema may compress the parenchyma onto the stiffest part of the organ that is the arterioles, endowed as they are with muscular walls and containing blood under pressure. Pulsatile deformation of tissue against the arterioles may explain the periarteriolar distribution of DAI.

Another important aspect of blast neurotrauma is the very significant protection of DAI in experimental animals with stiff Plexiglas shields encasing thorax and abdomen. It is possible that such protection prevents the squeeze of the torso and, via this mechanism, it also prevents a sharp rise in venous pressure.

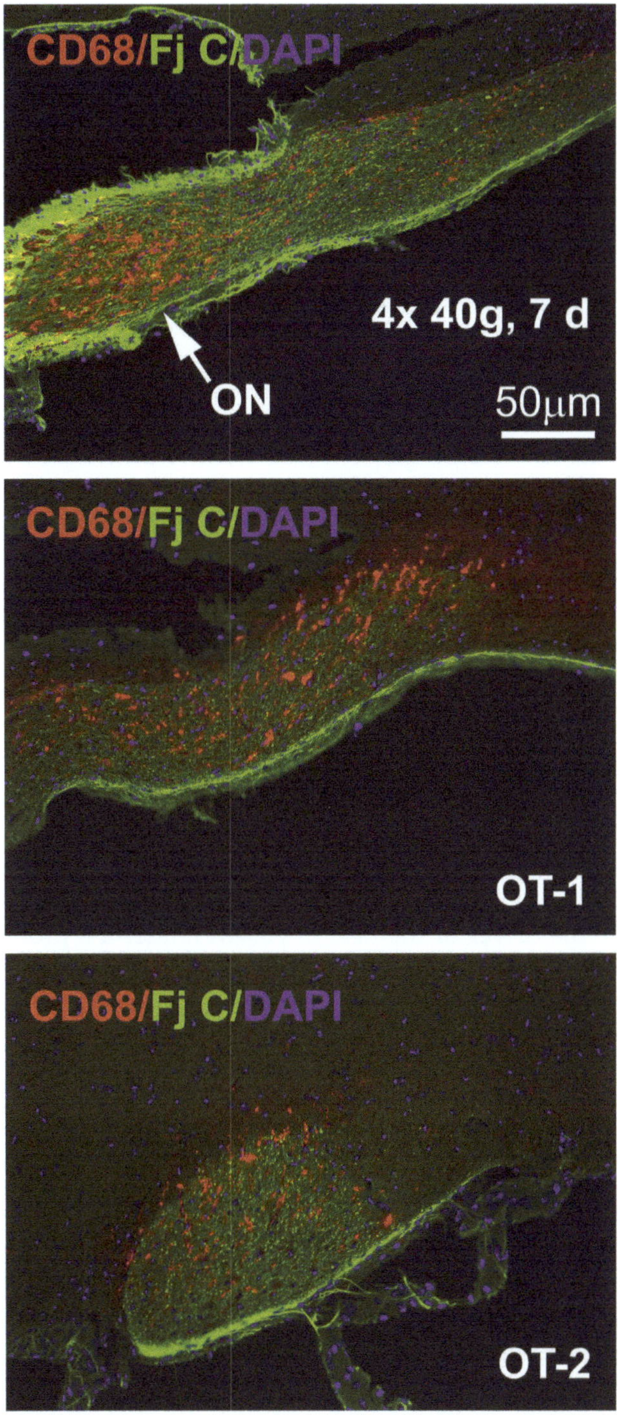

FIGURE 21.14 Neuroinflammation in white matter tracts undergoing traumatic (diffuse) axonal injury with the impact acceleration model. The example here is from the optic nerve (*ON*) and optic tract (*OT*). CD68 (+) reactive, phagocytic microglia colocalizes with Fluoro-Jade C (+) injured axons in both the ON and OT of experimental subjects exposed to repetitive mild injury at 7 days post final injury. Reactive microglia is labeled in red and injured axons in green. *Xu L, Nguyen JV, Lehar M, Menon A, Rha E, Arena J, Ryu J, Marsh-Armstrong N, Marmarou CR, Koliatsos VE. Repetitive mild traumatic brain injury with impact acceleration in the mouse: multifocal axonopathy, neuroinflammation, and neurodegeneration in the visual system. Exp Neurol 2016;275:436–49.*

Another possibility is the prevention of systemic inflammatory mechanisms arising in the abdominal cavity and transmitted into the brain in the same fashion as explained for the systemic inflammatory response syndrome in the previous subsection.

Proteinopathy in TBI (CTE)

Proteinopathy, that is, the acquisition of abnormal conformations by brain proteins and their tendency to aggregate, is increasingly recognized as a common problem across neurodegenerative disease and, for some investigators, the cause of such disorders. In fact, neurodegenerative diseases are often classified on the basis of corresponding aggregated proteins, for example, Aβ in AD ("amyloidosis"); synuclein in Parkinson's disease, Lewy body dementia, and multiple systems atrophy ("synucleopathies"); TDP-43, tau, and FUS in FTD and amyotrophic lateral sclerosis (ALS); TDP-43, SOD1, and ubiquitin in ALS; and tau in PSP. In the case of CTE, it is known for a decade that neuronal and astrocytic tangles and extracellular neurites contain abnormally aggregated tau. To what extent such protein abnormalities are the driver or the passenger of neuropathology is not exactly clear, but the primacy of proteinopathy is a popular view at this time and this is how we explain this hypothesis here. If proteinopathies are the cause of the corresponding neurodegenerative disease, the question arises why such protein configurations lead to widespread neuronal death. An attractive current theory is that of prion formation: that pathogenic misfolded proteins can first seed in stable conformational states and then spread, over and over again, from one part of the brain to another. While misfolded proteins self-template into such seeds, they can also hetero-template by corrupting normal proteins. In such a fashion, seeds increase in number and size and abnormal proteins take on fibrillar forms, disrupt normal neuronal function, and cause neuronal death. Such fibrils can be transmitted transynaptically or by exocytosis or after being extruded from dead nerve cells and picked up by neighboring neurons.

The microtubule-associated protein tau is a sine qua non of normal neuronal function and is involved, among else, in stabilizing microtubules and in the trafficking of organelles and molecules down microtubule trails. In the adult human brain, tau exists in six alternatively spliced tau isoforms differing by inserts at the N terminus and the microtubule-binding domain. Normal native ("unfolded") tau is soluble and its normal function in pacing axonal transport is thought to depend on cyclical dephosphorylation–phosphorylation associated with unbinding and binding to microtubules, respectively. The significance of tau in human neurodegenerative disease is readily appreciated by the fact that tau pathology is encountered in over 20 neurodegenerative diseases including AD and FTD. In the course of these disorders tau displays misfolding to β forms, hyperphosphorylation, and aggregation (fibrillization) in patterns that are characteristic for each one of these disorders and differentially affect distinct neuronal populations. Tau mutations that seem to cause familial FTD, such as P301L and V337M render the protein more prone to aggregation. The microtubule-binding repeat domain aggregates faster than full-length tau and that removal of the flanking regions increases aggregation. Repeat-domain tau with FTD-associated mutations is especially prone to fragmentation and β structure. Fragmentation may be initiated by a thrombin-like protease and then N-terminal and then C-terminal cleavage of repeat-domain tau. In addition, cell death–associated activation of caspase 3 cleaves tau at D421. Repeat domain–rich protein fragments nucleate ("corrupt") full-length tau and recruit it into the aggregate. Aggregation is modulated by phosphorylation but not dependent on it. Once formed, such "prion"-like tau continues to proliferate by seeding normal tau long after the initial event, in our case repetitive TBI, has ceased to exist. Frost and Diamond had initially shown that tau aggregates are taken up from the outside of cells and turn normal tau into misfolded, fibrillar forms; this work has been replicated by Virginia Lee and her colleagues at Penn. Newly aggregated tau gets transferred between cells. In a landmark study, Goedert and Tolnay showed that aggregated tau can be transmitted from tauopathic mice expressing FTD-P301S tau to normal mice expressing wild-type human tau by injecting brain extracts from transgenic mice into the brain of wild-type mice; pathology spread from injection site to other brain regions. The trans-synaptic mode of transmission has been further supported by work in the laboratories of Duff, Hyman, and others.

Proteinopathy may coexist with axonal injury and neuroinflammation and be caused by them, for example, via excessive phosphorylation of tau within neuronal perikarya. Lamb and colleagues have demonstrated that activated microglia release IL-1 that binds to IL-1 receptors in adjacent neurons where it induces p38 MAP kinase and tau hyperphosphorylation. Fractalkine receptors (CX3CR1) present in microglia that respond to the chemokine fractalkine (CX3CL1) released by neurons may be a significant regulator of this process. Importantly, lipopolysaccharide alone that causes sometimes long-lasting neuroinflammation can also cause tau hyperphosphorylation, a fact suggesting that systemic inflammation may predispose to proteinopathy. Prion- and neuroinflammation-based theories of proteinopathy in TBI, especially repeat DAI, have been illustrated in Fig. 21.15.

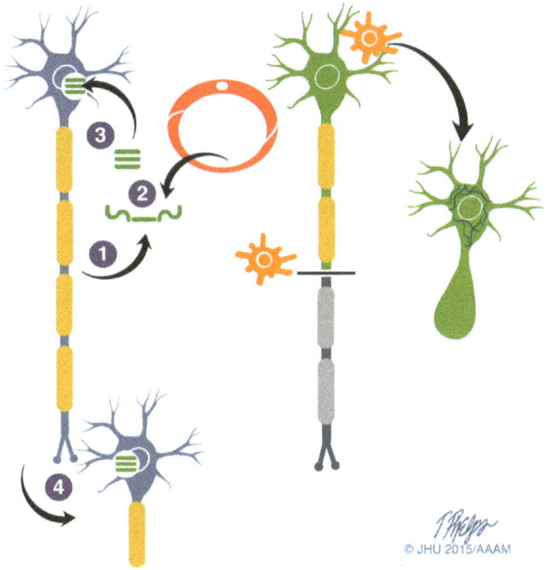

FIGURE 21.15 Prion (*left*) and inflammatory (*right*) approaches to understand tau aggregation and tauopathy. Left: Repetitive concussive injury can cause axonal disruption or damage that then allows the escape of tau protein in the extracellular space (1) and, in conjunction with microvascular damage or mere disruption of the blood–brain barrier (leaky capillary in sketch), may expose entire tau molecules to endogenous protease (thrombin) cleavage and truncation to repeat-domain, pro-aggregation tau (2), as described by Mandelkow and colleagues. This is a stable state of protein that leads to self-assembly in fibrils and subsequent recruitment and corruption of normal tau. Tau aggregates can be internalized from the extracellular to intracellular compartment (3) and then become propagated from one cell to another locally or along synaptic circuits long after the injurious insults have ceased (4). As entire neural systems become involved, individuals develop symptoms. The lengthy interval from aggregation to toxicity is consistent with the long incubation time required for the clinical manifestation of CTE. Right: Traumatic axonal injury leads to retrograde tau accumulation within the neuronal cell body. Neuronal injury activates adjacent microglia via fractalkine signaling and, in turn, activated microglia contributes to the phosphorylation of accumulated tau via IL1-p38 MAP kinase signaling. Hyperphosphorylation increases the probability of fibrillation and, once this happens, steps 3 and 4 of the left-side sketch may be initiated.

EXPERIMENTAL MODELS OF TBI

As stated in the beginning, the clinicopathological cycle concludes with experimental science. All science involves models, which are relatively simple (reductionist) systems serving to test scientific hypotheses. For many diseases, especially complex diseases that evolve over time, such as TBI, these models involve work in experimental animals (animal models). However, TBI modeling is broader and also involves in vitro systems and engineering simulations. Animal models of TBI have been superbly reviewed in several places, including Kobeissy's book on brain neurotrauma, and we only briefly address here some general points that serve to support the rest of this review. These models are conventionally divided into focal and diffuse, but the distinction is only partly useful. The initial weight drop and controlled cortical impact (CCI) models are *focal models* that reproduce cortical contusions. CCI is widely used across species, from mice and rats to nonhuman primates. Fluid percussion (FPI) models deliver fluid pulses onto the dura either at midline (midline FPI) or laterally (lateral FPI) and they generate a mixture of focal and diffuse effects. *Diffuse models* include the inertial acceleration model used in nonhuman primates and pigs and the impact acceleration model initially developed in rats and more recently adapted for use in mice. Acceleration models generate primary DAI. Many permutations of these models have been developed, especially for the purpose of replicating mild TBI conditions. Gennarelli's optic nerve stretch injury in the guinea pig is an interesting model that, although strictly fitting the description of a focal model, has been artfully used by Maxwell and others to shed light into the cellular and molecular mechanisms of DAI.

A never-ending debate is the appropriateness of the animal model for the question asked. TBI researchers have spent a considerable amount of time controlling the physics of injury, such as force, velocity, angular acceleration, time and frequency of exposure, etc. Without a question, having injury parameters that can be quantified allows valuable injury dose-response studies and improves experimental consistency. However, there is also need for an import of basic neuroscience in understanding downstream processes and mechanisms, for example, secondary axonal injury, transsynaptic degeneration, the role of neuroinflammation, etc. Models have had limited success in advancing our clinicopathological understanding of TBI and have contributed minimally to the development of new diagnostics (biomarkers) or therapeutics. Specific concerns have been voiced and range from the appropriateness of lissencephalic rodent species as models of DAI or as behavioral models in general, to the need of using better standardized models and exploring consensus processes and outcomes (common data elements). The use of large gyrencephalic animals such as nonhuman primates or pigs is not easy; besides animal welfare and similar concerns, a problem is the cost and limited availability of equipment capable of delivering high-magnitude torque like the pneumatic actuators that have been used at the University of Pennsylvania since the early 1980s. Other problems are lack of sharing of resources, lack of collaborative multi-institutional studies, and increasingly evident limitations in funding that have discouraged the entry of younger investigators or researchers from other neuroscience disciplines.

In vitro models have also been explored in the field of TBI and have been designed to transect, compress, stretch, and accelerate pieces of brain tissue or cell culture systems. This is a specialized topic that cannot be adequately reviewed here, except insofar as some of these models and paradigms relate to the clinicopathological analysis of this paper. For example, a couple of microfluidic systems using neuronal cultures and dynamic stretch or axonal transection have shed light into key structural properties of axons under normal or strained conditions or after transection, and have also begun to clarify the interactive role of nonneuronal cells, such as microglia in these processes. Tang-Schomer and Smith from the University of Pennsylvania have elegantly shown the essential role of microtubules for axonal integrity and ways in which microtubule fracture or dissolution contributes to axonal injury and degeneration under conditions similar to the ones encountered in DAI. With the caveat that such experiments involve nonmyelinated axons, this work has enlightened our understanding of late axotomy in DAI and has influenced the discussion of the topic earlier in this paper, as well as the conceptualization of Fig. 21.5 and the interpretation of profiles illustrated in Fig. 21.6. Hosmane and his colleagues at Hopkins developed microfluidic chamber platforms that enable a targeted access to perikarya or axons of cultured neurons for delivering media, pharmacological agents, or other cells such as microglia for cell-interactive experiments. Using these platforms, axons can be visualized at high resolution during injury and forces exerted upon the axon can be better understood and measured. Co-culturing microglia in the axon compartment has generated spectacular displays of interactions between microglial cells and injured axons and allowed the development of a reproducible quantitative assay to assess microglial clearance of injured axons and their debris. Such platforms hold great promise for understanding mechanisms of chronic TBI by combining more than one variable in the same model, for example, traumatic axonal injury and microglial response.

Bioengineering inputs have been of tremendous value in modeling TBI and some of the language we used here, for example, in describing different types of TBI in sections "Key Clinical and Pathological Features of Main Types of TBI" and "DAI: Cellular and Molecular Approaches," we directly imported from engineering papers. Modeling based on engineering is contributing to TBI prevention, for example, in ameliorating head impacts via the appropriate helmet design or, hopefully, ameliorating or preventing blast TBI with the design of optimal body gear. Engineering input and design have greatly helped the development of animal and in vitro models reviewed in the preceding paragraphs. Of great significance is the use of finite element analysis simulation models. Taking into account different strain properties of key tissue elements and their anatomical distribution, the finite element method divides the brain into a mesh of smaller domains (finite elements) and assigns to them mathematical signatures as smaller equations that can then be solved over larger domains. This methodology may be used to predict which parts of the whole would stretch, shear, expand, etc. under particular physical conditions representing, for example, impact, impulse, or blast injury but also the magnitude of distortion. Such analyses may add significantly in cases where tissues are not available or simply to aid interpretations and shape conclusions from biological experiments.

Modeling is the final step in the clinicopathological cycle before a next iteration begins, hopefully with more refined hypotheses and a better understanding of the TBI patient and his problems: the outcome of modeling may be a diagnostic or therapeutic product or simply a more finessed concept that can serve as a launcher of more precisely targeted work. It is clear from this chapter that, the greater clarity we have about a problem or hypothesis, and also the power, limitations, or advantages of an experimental model, the more likely it is that going from bench to bedside and again back to bench will be a forward process. The complexity of clinical problems associated with TBI requires greater clarity in our ideas and approaches and it is not compatible with a generic talk about TBI and its experimental models.

Further Reading

[1] Buki A, Povlishock JT. All roads lead to disconnection? Traumatic axonal injury revisited. Acta Neurochir 2006;148(2):181–93.

[2] Koliatsos VE, Ratan RR. Cell death and diseases of the nervous system. Totowa (New Jersey): Humana Press; 1999.

[3] Koliatsos VE, Xu L. The problem of neurodegeneration in cumulative sports concussions: emphasis on neurofibrillary tangle formation. In: Kobeissy FH, editor. Brain neurotrauma: Molecular, neuropsychological, and rehabilitation aspects. Boca Raton (Florida): CRC Press; 2015. p. 659–66.

[4] Ryu J, Horkayne-Szakaly I, Xu L, Pletnikova O, Leri F, Eberhart C, et al. The problem of axonal injury in the brains of veterans with histories of blast exposure. Acta Neuropathol Commun 2014;2:153.

[5] Smith DH, Hicks R, Povlishock JT. Therapy development for diffuse axonal injury. J Neurotrauma 2013;30(5):307–23.

[6] Walker LC, Diamond MI, Duff KE, Hyman BT. Mechanisms of protein seeding in neurodegenerative diseases. JAMA Neurol 2013;70(3):304–10.

CHAPTER

21.1

Translational Correlation: Disorders of the Spinal Cord

S. Živković[1], S.T. Ali[2], A.M. El-Dokla[2]

[1]University of Pittsburgh Medical Center, Pittsburgh, PA, United States; [2]Texas Tech University Health Sciences Center, Lubbock, TX, United States

Spinal cord is a continuous segmented structure covered by three protective membranes of the central nervous system (CNS), surrounded by cerebrospinal fluid, and encased within the spinal canal. Spinal roots originate from the spinal cord and continue as spinal nerves, branching out to plexuses and peripheral nerves. Diseases of the nervous system can involve spinal cord and spinal roots as isolated neurologic syndromes, or in the setting of systemic or multifocal neurologic disorders. Lesions affecting the spinal cord, spinal nerves, and roots result in motor and sensory segmental defects which allow anatomic localization, although multiple segments are often affected. Clinical presentation of disorders depends on topography of the injury and affected structures and pathways. Typically, the examination will show a loss of sensation below the level of spinal injury (sensory level) and various sensory modalities (pain/temperature vs. vibratory/proprioception) may get affected differently depending on the type and location of spinal cord injury (anterior vs. posterior). Additional symptoms include spastic or flaccid paresis (corticospinal tracts vs. lower motor neurons), hyperreflexia, and loss of sphincter control.

Anatomically, spinal cord is organized with centrally placed gray matter surrounded by white matter tracts (pathways). Major white matter tracts include sensory tracts which enter the spinal cord through dorsal roots and continue as *spinothalamic* and *spinoreticulothalamic* tracts, and *spinocerebellar* tracts, *fasciculus gracilis*, and *fasciculus cuneatus*. Additionally, descending major motor tracts control the tonic activation and volitional activity of motor neurons including *corticospinal*, *corticorubrospinal*, *reticulospinal*, and *vestibulospinal* tracts, exiting the cord through ventral roots.

PHENOTYPES OF SPINAL CORD INJURY

Clinical phenotypes of spinal cord injuries include (1) transverse myelopathies, (2) the hemicord syndrome, (3) anterior cord syndrome, (4) central cord syndrome, (5) syndrome

Conn's Translational Neuroscience
http://dx.doi.org/10.1016/B978-0-12-802381-5.00034-8

of the conus medullaris, (6) syndrome of the cauda equine, and (7) radiculopathies.

Transverse Myelopathy

Transverse myelopathy is associated with impairment of multiple ascending and descending tracts passing through the level of the lesion, and the complete injury will affect all of them. In most cases, transverse myelopathy is incomplete reflecting the nature of underlying injury. All sensory modalities are impaired below the level of the lesion, and spastic paresis is associated with hyperreflexia. Transverse myelopathy may be related to traumatic spine injuries and herniated discs, neoplasms, autoimmune, vascular disorders, and infections.

Hemicord Syndrome (Brown–Sequard Syndrome)

Hemicord syndrome (*Brown–Sequard* syndrome) is typically caused by extramedullary lesions or penetrating trauma, and less commonly by ischemia or inflammatory lesions (eg, multiple sclerosis). Brown–Sequard syndrome manifests with contralateral loss of pain and temperature sensation and ipsilateral spastic weakness and proprioceptive loss. Additional symptoms may include ipsilateral loss of sweating, distal to the injury; segmental lower motor neuron dysfunction, and sensory loss due to damage of the anterior horn cell and dorsal spinal roots at the level of injury.

Anterior Cord Syndrome

Anterior cord syndrome may be caused by direct (anterior) compression of ventral spinal cord or anterior spinal artery ischemia (*rare*). Clinical presentation consists of weakness, hyperreflexia, and loss of pain and temperature sensation with preserved vibratory sense and proprioception.

Selective involvement of anterior horn cells in patients with spinal muscular atrophy or polio manifests with exclusive motor symptoms including weakness, muscle atrophy, and flaccidity. Concurrent involvement of corticospinal tract with amyotrophic lateral sclerosis (ALS; Lou Gehrig disease) results also in hyperreflexia and spasticity. Hereditary spastic paraplegias are a group of genetic neurodegenerative disorders which present with progressive lower extremity weakness and spasticity as a result of distal axonopathy affecting corticospinal tracts.

Posterior and Posterolateral Column Syndrome

Posterior and posterolateral column spinal injury is characteristic for subacute combined degeneration of the spinal cord with vitamin B12, vitamin E, or copper deficiency, and for spinocerebellar ataxias and neurosyphilis (*tabes dorsalis*). Clinically, these patients present with paresthesias, gait ataxia, and loss of vibratory and position sense, while the sensitivity to pain and temperature remains preserved. Corticospinal tract involvement manifests with hyperreflexia and spasticity. Gait ataxia may be the primary clinical manifestation of epidural spinal cord compression due to spondylosis, protrusion of intervertebral discs, an epidural abscess/hematoma, or even tumor.

Central Cord Syndrome

Central cord syndrome may be caused by syringomyelia or intramedullary cord tumors. When cervical spinal cord is involved, the patients may present with upper motor neuron weakness with preferential involvement of upper extremities ("man-in-the-barrel syndrome"), bladder dysfunction, and variable sensory loss. Most often, syringomyelia follows a slowly progressive course with gradual progression of symptoms. Acute cervical central cord syndrome is uncommon and may result from neck hypertension leading to central hemorrhage and acute quadriplegia followed by leg strength improvement, sensory loss below the lesion level, and urinary retention (neurogenic bladder). Syringomyelia in the lower cervical or upper thoracic spinal cord may manifest with symmetric sensory loss in the upper extremities and upper thorax, sparing the lower extremities and abdominal and lumbosacral regions ("cape-like" sensory loss). Only rarely does syringomyelia extend into lumbosacral cord segments. Sacral sensation is often spared with intramedullary lesions.

Conus Medullaris and Cauda Equine Syndromes

In adults, spinal cord does not extend beyond L2 level, and lesions involving the most caudal part of the spinal cord may manifest as conus medullaris or more distally can cause cauda equina syndromes. Pure conus medullaris syndrome is typically due to an intramedullary lesion (eg, neoplasm and infarct). Conus medullaris lesions are associated with early sphincter dysfunction and weakness of pelvic floor muscles, with pain occurring relatively late. Typically, cauda equina syndrome typically causes early radicular pain in the distribution of the lumbosacral roots with asymmetric sensory loss and late loss of sphincter control. However, it is often difficult to distinguish conus medullaris manifestations from cauda equina based only on symptomatic presentation. Most common causes of conus medullaris and cauda equina syndromes are spondylosis and vertebral disc protrusions, and other causes include ankylosing spondylitis, tethered cord, and local neoplasms.

Radiculopathy

Radiculopathy is most commonly caused by foraminal encroachment of spinal nerves. Clinical symptoms include radicular pain and weakness presenting in a myotomal distribution depending on affected spinal root(s). Spinal nerves are formed by dorsal (sensory) and ventral (motor) spinal roots. Common etiologies of radiculopathy include spondylosis, intervertebral disc displacement, or trauma which may result in selective myotomal and dermatomal clinical presentation. Additionally, inflammatory, neoplastic, and infectious processes can also cause radiculopathies (radiculitis), which can result in spinal root compression due to inflammation and swelling. Neuroborreliosis and leptomeningeal spreading of carcinoma or lymphoma (leptomeningeal carcinomatosis and lymphomatosis) may also present with radicular symptoms. HIV patients may also develop CMV polyradiculopathy. Clinical symptoms include radicular pain and weakness presenting in a myotomal distribution depending on affected spinal root(s).

CAUSES OF SPINAL CORD INJURY

Spinal cord disorders are related to various types of structural injuries or inflammatory, toxic, or metabolic derangements. The most common etiologies of spinal injuries include spondylosis, traumatic spinal injury, inflammatory, and infectious disorders. Less frequent causes of spinal disorders include spinal vascular disorders, neoplasms, toxic-metabolic, and neurodegenerative disorders.

Spondylotic Myelopathy and Traumatic Spinal Cord Injury

Degenerative changes of vertebral column and intervertebral discs, also known as spondylosis, may manifest with compressive myelopathy, radiculopathy, or back (and neck) discomfort. Imaging studies show spondylosis of cervical spine in more than 90% of individuals over the age of 60 years, and most have minimal symptoms. Cervical spondylotic myelopathy is associated with cord compression as a result of degenerative changes of the vertebral column which become more common with aging and other factors, including trauma and repetitive motion. Clinical manifestations vary depending on the topology of cord compression, including weakness, pain, spasticity, gait ataxia, sensory loss, and loss of sphincter control. Surgical treatment is considered in patients with moderate to severe and progressive neurologic deficits.

Traumatic spinal injury is a source of major morbidity in affected patients with a reported annual incidence of 54 per 100,000. Highest risk was reported in elderly, attributable to traumatic falls. Clinical features depend on the type of trauma and topology of affected spinal regions and pathways. Surgical treatment may be needed for cord decompression and stabilization of the vertebral column.

Vascular Disorders of the Spinal Cord

Vascular supply of spinal cord is organized around anterior and paired posterior spinal arteries, with anterior spinal artery supplying the anterior two-thirds of the cord. Mid-thoracic spinal cord has the least extensive vascular network and is more susceptible to cord ischemia, especially T1–T4 and L1 segments. When compared to cerebral infarction, spinal cord ischemia is very rare and may account for up to 1–2% of all strokes. Spinal cord infarction can be precipitated by different traumatic events, medical procedures, systemic inflammatory disorders, or infections. Clinical presentation will depend on the level of infarction and signs of ischemia of anterior or posterior spinal cord may predominate (*anterior spinal artery*—sudden flaccid paresis with low pain and temperature sensation and preservation of vibration and proprioception; *posterior spinal artery*—sudden loss of proprioception and vibratory sense with hyporeflexia and preserved strength and sensation to pain and temperature). Spinal vascular malformations can rarely mimic motor neuron disease with stepwise gradual progression. Epidural hematoma may cause a compressive myelopathy.

Autoimmune Disorders of the Spinal Cord

Transverse myelitis is an inflammatory myelopathy which may present as an isolated monophasic event, or in the context of a systemic inflammatory disorder (eg, systemic lupus erythematosus (SLE) and Sjogren's syndrome).

Multiple sclerosis is an inflammatory disorder of the central nervous system which can also manifest with an inflammatory myelopathy that may be difficult to distinguish from other causes in the absence of typical brain lesions on imaging. Neuromyelitis optica (previously also known as Devic's syndrome) is an autoimmune disorder with a combination of optic neuritis and inflammatory myelopathy which may precede or follow each other by several years.

Infections of the Spinal Cord

Infectious disorders can also rarely affect the spinal cord and present with various phenotypes. Clinical phenotypes include encephalomyelitis (eg, EBV), transverse myelopathy, posterolateral cord syndrome (eg, neurosyphilis), motor neuronopathy (eg, polio), or polyradiculopathy (eg, cytomegalovirus). Spinal cord infections should be distinguished from postviral inflammatory syndromes, including postviral transverse myelitis. Additionally, epidural abscesses may cause spinal cord compression.

Neurodegenerative Disorders of the Spinal Cord

At this time, mutations of more than 50 genes have been described as cause distal axonopathy affecting corticospinal tracts and manifesting as hereditary spastic paraplegias with minimal or no involvement of other spinal cord structures. Spinocerebellar ataxias manifest with ataxia and posterior/posterolateral cord syndrome in addition to different combinations of extrapyramidal and corticospinal tract involvement, cognitive decline, peripheral nerve, or ocular involvement. To date, mutations of more than 40 genes have been described as causes of spinocerebellar ataxias.

Neoplasms of the Spinal Cord

Neoplasms of the spinal cord may be divided depending on their location into intramedullary and extramedullary tumors. Overall, metastases are more common than primary spinal tumors. Intramedullary tumors are primary spinal neoplasms arising from neuroectodermal tissues (eg, ependymoma and astrocytoma), and extramedullary intradural tumors are often benign meningiomas and nerve sheath tumors. Additionally, systemic neoplasms may manifest with leptomeningeal metastases and present with radicular symptoms. Epidural tumors are usually vertebral metastatic lesions which may result in a spinal cord compression.

DIAGNOSIS AND TREATMENT OF SPINAL CORD DISORDERS

Diagnostic evaluation of spinal cord disorders depends heavily on neuroimaging, especially magnetic resonance imaging (MRI) and computed tomography (CT). However, imaging results may be nonspecific and may not reveal the underlying cause of injury. Additional testing may include clinical neurophysiology testing (EMG/NCS, SSEP) and various laboratory tests depending on features of individual clinical cases (Table 21.1.1).

Treatment of spinal cord disorders depends on the underlying etiology and localization of injury. Treatment of traumatic spinal cord injury has evolved over the years, and the widespread use of high-dose corticosteroids for the treatment of acute spinal injury remains controversial. Additionally, there is no consensus on timing

TABLE 21.1.1 Diagnostic Evaluation of Spinal Cord Disorders

	Individual Tests	Disorders
Neuroimaging	MRI of the spine • nonenhanced imaging • contrast-enhanced imaging • diffusion-weighted imaging CT myelogram Spinal angiogram	Spondylosis Tumors, infections, inflammatory disorders Spinal cord infarct Spondylosis Vascular malformations
CT myelogram	EMG/NCS SSEP	Radiculopathy, motor neuron disorders
Laboratory testing—cerebrospinal fluid	Cell count, protein, oligoclonal banding VDRL Viral PCRs and serology (CMV, VZV, EBV, HTLV-1) Lyme's PCR and serology Viral cultures	Multiple sclerosis Neurosyphilis Myelitis Lyme's disease
Laboratory testing—blood	ANA, dsDNA, SSA/B, ACE, copper, vitamin E, NMO antibodies, Lyme's genetic testing	Systemic metabolic and inflammatory disorders Hereditary spastic paraplegia, spinocerebellar ataxias

and type of surgical treatment to maintain anatomic stability and integrity of the vertebral column, and decisions are made depending on clinical features of individual cases. The potential therapeutic role of stem cells, gangliosides, cytokines, and nerve growth factors in the spinal cord regeneration after injury remains a target of intense basic and clinical research, but it is still not ready for clinical use. Surgical and endovascular interventions are effective in treatment of spinal cord vascular malformations. Inflammatory and autoimmune disorders of the spinal cord are usually treated with immunomodulatory and antiinflammatory therapies including high-dose corticosteroids and plasmapheresis, while other agents including intravenous immunoglobulin, rituximab, and cyclophosphamide are used for treatment of refractory cases. Treatment of spinal tumors depends on underlying pathology and often includes surgical intervention and irradiation.

Further Reading

[1] Byrne TN, Benzel EC, Waxman SG. Disease of the spine and spinal cord. Oxford University Press; 2000.
[2] Cho TA. Spinal cord functional anatomy. Continuum 2015;21:13–35.
[3] Hurlbert RJ, Hadley MN, et al. Pharmacological therapy for acute spinal cord injury. Neurosurgery 2013;72:93–105.
[4] Silva NA, Sousa N, Reis RL, Salgado AJ. From basics to clinical: a comprehensive review on spinal cord injury. Prog Neurobiol 2014;11:25–57.
[5] Toledano M, Bartleson JD. Cervical spondylotic myelopathy. Neurol Clin 2013;31:287–305.

CHAPTER

21.2

Translational Correlation: Abnormalities of the CSF System

S. Tantikittichaikul, J.C. DeToledo

Texas Tech University Health Sciences Center, Lubbock, TX, United States

INTRODUCTION

The cerebrospinal fluid (CSF) system is the space comprised by the cerebral ventricles and the subarachnoid space that is delimitated by the meninges. The meninges are the protective tissue layers that cover the brain and extend caudally covering the entire spinal cord. The three layers of meninges from outermost to innermost layers are the dura mater, arachnoid, and pia mater, respectively. Besides protecting the central nervous system (CNS), the meninges also play an important role in the growth and development of the underlying brain tissue. The CSF suspends and cushions the brain and assists in the homeostasis of the interstitial fluid of the brain parenchyma.

DISORDERS OF CSF CIRCULATION

The total CSF volume in adults ranges from 90 up to 150mL and it is produced at a rate of 400–600mL/day. The CSF is produced by the choroid plexus inside the ventricles and absorbed by the arachnoid villi in the convexity of the hemispheres. Abnormal production or absorption of CSF can cause changes in intracranial pressure resulting in a variety of neurological symptoms.

Increased intracranial pressure (IICP): Because the CSF is contained in an unyielding space (skull and spinal canal), CSF absorption has to match production, otherwise CSF will accumulate and cause IICP. There are three main mechanisms causing IICP: increased CSF production, decreased CSF absorption, and obstruction to pathways of CSF circulation. Other less common etiologies include increased venous pressure (venous sinus thrombosis, heart failure), brain edema (cytotoxic edema), increased vascular permeability (vasogenic edema), or interstitial edema (Table 21.2.1). The symptoms of IICP in adults (closed skull sutures) depend on the acuteness of the increased ICP and the rate of progression (Table 21.2.2). Acute increases in ICP tend to cause more symptoms, whereas chronic IICP may go asymptomatic for months or years. Headache is the most common symptom reported by patients with IICP (stages 1–2). Headache is typically diffuse and characteristically worse in the morning and also aggravated

Conn's Translational Neuroscience
http://dx.doi.org/10.1016/B978-0-12-802381-5.00035-X

by coughing, straining to move bowels, and other Valsalva-type maneuvers. Nausea, vomiting, dizziness, and pulsatile tinnitus are not uncommon but are not very specific. More severe cases can have bilateral papilledema and progress to permanent vision loss if IICP is left untreated.

TABLE 21.2.1 Mechanism of Increased Intracranial Pressure

Mechanism	Example
Mass effect	Tumors, bleeding, infarction, abscess
Cerebral edema	Vasogenic edema, cytotoxic edema, or interstitial edema
Increased venous pressure	Venous sinus thrombosis, congestive heart failure
Obstructed CSF circulation	Intraventricular cyst, ependymoma, intraventricular hemorrhage
Increased CFS production	Meningitis, choroid plexus tumor
Decreased CSF absorption	Arachnoid cyst, arachnoiditis
Idiopathic	Pseudotumor cerebri, normal pressure hydrocephalus

TABLE 21.2.2 Stages of Increased Intracranial Pressure

	Stage	Clinical Symptoms/ Signs
1	Mild increased intracranial pressure due to compensatory mechanism	Asymptomatic
2	Significant IICP and overcome compensatory mechanism, CSF volume may increase greater than 100–120mL, rising of blood pressure to maintain cerebral perfusion pressure	Headache, nausea, vomiting, dizziness, tinnitus, blurry vision, and diplopia Hypertension, papilledema, and abducen nerve palsy
3	Reduced cerebral perfusion pressure cause brain damage	Altered mental status, coma, death Brain stem dysfunction, Cushing reflex

As the increase in the ICP persist (stage 3), patients may become confused and have altered mental status. The elevated intracranial pressure starts to compromise the cerebral perfusion pressure resulting in brain ischemia. Cushing reflex may be detected at this stage and present with elevated blood pressure, tachycardia that with prolonged high blood pressure activates baroreceptor reflex causing bradycardia in the late stages. At this point, abnormal respiratory pattern, especially Cheyne–Stokes respiration can occur due to dysfunction of voluntary respiration center.

Decreased intracranial pressure (intracranial hypotension): This is a common complication of lumbar puncture but may also occur with postsurgical CSF leaks and spontaneously as well. The patient presents with postural headache that is worsened by changing to the upright position. Magnetic resonance imaging (MRI) shows thickening or enhancement of the meninges in many of these cases.

Hydrocephalus: The increase in intracranial pressure causes dilatation of the cerebral ventricles. Hydrocephalus is often classified as communicating or noncommunicating based on the site of obstruction of CSF pathways. Communicating hydrocephalus presents as an enlargement of all ventricles without any defect in the intracerebral pathway of the CSF circulation. Noncommunicating hydrocephalus will show unequal dilatation of the ventricles and is often secondary to narrowing of the aqueduct of Sylvius. In these cases, the CSF cannot reach the subarachnoid space. In elderly adults, there is a clinical presentation of communicating hydrocephalus with a slow and insidious onset called normal pressure hydrocephalus (NPH). These patients present with gait instability, cognitive disturbances, and urinary incontinence. An important feature of this syndrome is that the symptoms are potentially reversible by surgical shunting of the CSF.

In children, congenital anomalies and perinatal hemorrhages are common causes of IICP. The clinical symptoms in these cases is different from adults, particularly if the skull is still pliable and the fontanels still open. These children may have no clinical symptoms other than bulging of the fontanels and enlargement of the head circumference.

ABNORMAL CSF PROFILES

The CSF is a clear and colorless fluid. In the normal CSF, this sterile fluid should not contain lymphocytes more than five per microliter. The number of protein is not greater than 45 mg per deciliter and glucose is about 60–70% of plasma concentration. Increased cellularity, altered protein, or glucose level in the CSF may indicate CNS abnormalities.

Meningitis: The inflammation of the leptomeninges from infection may present with various symptoms. Acute onset of fever, confusion, and meningeal irritation are the triad of CNS infection. Detection of at least two of this triad can be >90% accuracy for diagnosis. Increased intracranial pressure may present later during the course, especially in the chronic type of infection. Bacteria and virus are the most common pathogens. For bacteria, pneumococcus is the most frequent etiology in adult. The source of infection can be three main routes including hematogenous spreading, spreading from paracranial infection (eg, sinusitis), or direct invasion (eg, trauma). In viral infection, the inflammation tends to extend to the brain parenchyma and causing meningo-encephalitis. The patient increases risk to develop seizure due to involvement of cortex. The most common etiologies include enteroviruses and herpes simplex viruses. The other infectious etiologies of infection include fungus, tuberculosis, and parasitic. These tend to be more chronic presentation. Immunocompromised patient, such as HIV infected or with diabetes may have increased susceptibility to atypical infection. Imaging studies of the brain with contrast may show meningeal enhancement due to disruption of blood–brain barrier from infectious process. Abnormal brain parenchyma can be detected especially from viral meningo-encephalitis. CSF studies are the gold standard for diagnosis of meningitis. Different types of the parameter of the CSF can differentiate the etiology of infection (Table 21.2.3).

Ventriculitis: The inflammation of ventricles can lead to various neurological symptoms related to increased intracranial pressure. The symptoms can be in the range from mild (headache, dizziness) to severe (coma) or death. About

TABLE 21.2.3 CSF Parameter of Meningitis

Parameter	Bacteria	Virus	Fungus	Tuberculosis
Opening pressure	Elevated	Usually normal	Variable	Variable
WBC count	≥100 per mm^2	<100 per mm^3	Variable	Variable
Cell differential	Predominance of PMNs	Predominance of lymphocytes	Predominance of lymphocytes	Predominance of lymphocytes
Protein	Mild to marked elevation	Normal to elevated	Elevated	Elevated
Glucose level	Low (<40% of serum glucose)	Usually normal	Low	Low

WBC, white blood cell; *PMNs*, polymorphonucleocytes.

90% of this cases come from Staphylococci. The most associated risk factor is brain surgery, especially extraventricular drainage procedure. MRI or CT brain with contrast can detect the abnormal enhancement of ventricular lining, intraventricular pus, or debris. CSF analysis may have low yield in this situation because the profiles may show subtle changes and difficult to differentiate from postoperative inflammation. In this circumstance, a white count differentiation with more than 10% of PMNs may be 90% sensitive for predicting infection. CSF cultures can also help to identify organism.

Leptomeningeal carcinomatosis (LC): LC, sometimes also referred to as meningeal carcinomatosis, is a relatively uncommon complication of some systemic solid tumors and lymphoproliferative malignancies. The malignant cells reach the leptomeninges via hematogenous spread or by direct extension of a preexisting lesion. The frequency of LC complication has increased as more effective therapies have improved the survival of oncological patients and for this reason, it tends to occur late in the course of the disease, when disseminated and progressive disease are already present. Because LC can affect any part of the CNS that is covered by leptomeninges, including the cerebral hemispheres, brain stem, cranial nerves, spinal cord, and nerve roots, it often presents with multifocal neurological signs and symptoms. Lung, breast, melanoma, and hematologic malignancies are common causes of LC. The neurological symptoms vary depending on where the leptomeninges are maximally involved. Headache and altered mental status are common presentations with involvement of the cerebral hemispheres. Nausea, vomiting, ataxia, and cranial neuropathies are more common with involvement of brain stem and posterior fossa Back pain and radicular symptoms suggest involvement of the spinal cord. The diagnosis of LC can be challenging. Enhancement of the meninges on the MRI is a characteristic finding but can be absent in some cases. The classic CSF profile in LC is the combination of a high opening pressure, low concentration of CSF glucose, high protein with lymphocytic pleocytosis. The CSF cytology for malignancy cells is usually positive in these cases. The sensitivity of a single CSF cytology is about 70%, but it can be up to 90% with subsequent sampling (three spinal taps). The prognosis in these cases is guarded and radiation therapy to symptomatic sites with intrathecal chemotherapy increases the median survival to 3–6 months.

CHAPTER

22

Posttraumatic Stress Disorder

G.H. Wynn[1], D.M. Benedek[1], L. Johnson[2], K. Choi[1], L. Zhang[1], X.-Z. Hu[1], H. Pollard[1], R. Ursano[1]

[1]Uniformed Services University of the Health Sciences, Bethesda, MD, United States; [2]Queensland University of Technology, Brisbane, Australia

INTRODUCTION

Posttraumatic stress disorder (PTSD) is an event-related disorder occasioned by exposure to a significant traumatic event, and followed by the development of characteristic symptoms in the aftermath of the event. Examples of such traumatic events include disasters, criminal violence, combat, motor vehicle accidents, and sexual assault. As defined by DSM-V, PTSD symptoms include intrusive memories, avoidance of reminders, negative alterations in cognitions and mood, and marked alterations in arousal and reactivity. These symptoms generally occur shortly after the traumatic experience, but sometimes develop much later. The majority of individuals exposed to a given traumatic event are resilient and will recover from transient psychological symptoms without intervention. When considering the possibility of PTSD after a given trauma, important risk factors include the intensity of the threat, the duration of the threat, personal history of trauma, and the nature of the traumatic experience.

Our knowledge of PTSD continues to evolve as evidenced by the 2013 change to the diagnostic criteria with DSM-V. Although changing, the fundamental characteristics of PTSD have been recognized for centuries. In many ways, PTSD is a "forgetting disorder" in which "forgetting" is impaired. Or alternatively PTSD can be seen as a "cuing disorder" in which reminders of traumatic events have become overgeneralized. Historically, terms such as shell shock and battle fatigue are a few of the many constructs that have been offered by those attempting to describe the experience of PTSD. Despite the long history of knowing that trauma negatively impacts the human psyche, it is only in recent times that we have begun to understand how trauma and its repercussions impact the brain and human biology. Given the vast array of advances in our understanding of PTSD, this chapter covers the neurobiology of stress, animal models of PTSD, biomarkers of PTSD, the relationship between PTSD and cardiovascular disease (CVD), neuroimaging to discover the structural and functional basis of PTSD, and possible therapies for PTSD based on insights from the field of neuroscience.

Conn's Translational Neuroscience
http://dx.doi.org/10.1016/B978-0-12-802381-5.00036-1

THE NEUROBIOLOGY OF TRAUMATIC STRESS

From Selye to Stressed Synapses

Traumatic stress affects people by modifying their memories, by changing their self-perceptions, and by affecting how they respond to future events. Studies of stress, at both the human and model organism levels, show that stress can alter brain structure. Change occurs not only at gross and microscopic levels, both also at the genomic level. The consequence is that stress also changes brain function, resulting in modulation of synaptic transmission. Importantly, stress changes memory, which, if the stress is not catastrophic ("eustress") can have important positive impacts on survival. This impact can be measured at both the whole-organism level and in cellular models of memory.

Hans Selye famously posited that modest stress is advantageous to function. By contrast, sustained, chronic stress can be deleterious. The dose effect profile of stress is generally thought to follow an inverted U-shaped dose–response curve. Selye described stress as having three stages: alarm, resistance, and exhaustion. Selye observed that exhaustive phases of stress are associated with enlarged adrenal glands. Yerkes and Dodson, whose studies focused on arousal, showed that moderate stress facilitates, and high stress inhibits cognitive performance. Empirical studies have established that the alarm, resistance, and exhaustion phases of stress, represented by the inverted U-shaped dose–response of cognitive performance function, are mediated by glucocorticoids (GCs).

During stress—including traumatic stress—cortisol is released from the adrenal glands in concentrations greater than those released during nonstress circadian and ultradian cycles. Cortisol binds to both mineralocorticoid and glucocorticoid receptors (GRs). Both mineralocorticoid receptors and GRs are found in the brain where they are distributed within neurons and glia. The mineralocorticoid receptor has a higher affinity for cortisol than GRs. Therefore, at sub-stress or moderate stress levels, mineralocorticoid receptors are preferentially bound and activated by cortisol. However, during periods of high and traumatic stress, cortisol is released in higher concentrations and GRs are bound and activated. Intracellular mineralocorticoid receptors and GRs, when bound to cortisol, both form homodimers and translocate from the cytosol through nuclear pores to DNA where they act as transcription factors at the promoter level. GCs have been documented to both transactivate and transrepress a variety of genes, and to have consequences for the biology of the immediate gene products, cognate proteins, and ultimately behavior.

In neurons, mineralocorticoid receptors and GRs are found both within the cytosol and nucleus. They are also found in neuronal membranes such as the postsynaptic density of excitatory synapses. Like stress at the organism level, stress at the synapse is also likely to be controlled by an inverted U-shaped dose–response curve. Furthermore, increasing evidence suggest that activation of mineralocorticoid receptors *increases* synaptic excitability, whereas activation of GRs acts to *reduce* synaptic excitability. These synaptic-mediated mechanisms of cortisol action appear to act both prior to and in parallel with the genomic-mediated mechanisms that regulate cell structure. Structure and function of dendrites, dendritic spines, and synapses are very sensitive to stress and GCs. Thus, the high stress of a traumatic event can affect and regulate neuronal structure and function at multiple levels.

Effects of Acute Stress Are Mediated by Catecholamines

The stress response also involves activation of the sympathetic nervous system, which directly controls the release of epinephrine and norepinephrine. These transmitters and hormones also affect neuronal function and behavior. Memory systems are facilitated by activation of beta-adrenergic systems. Both memory consolidation and reconsolidation have been shown to be inhibited by propranolol in experimental models. The integration of the cortisol and catecholamine response is nowhere better manifest than at the level of the adrenal gland. While cortisol is secreted from the adrenal cortex following corticotropin-releasing hormone (CRH)-dependent release of ACTH from the pituitary gland, epinephrine is secreted from chromaffin cells in the adrenal medulla. Chromaffin cells have cholinergic receptors, located at terminals of the splanchnic nerve. The splanchnic nerve is part of the sympathetic nervous system, and is connected directly to neurons in the spinal cord. Being connected directly to the central nervous system, chromaffin cells can instantaneously and acutely drive the "flight or fight" mechanisms through action of secreted catecholamines on their respective receptors in the blood vessels, heart, sweat glands, smooth muscles in the gastrointestinal (GI) and respiratory tract, and elsewhere, both peripherally and centrally. However, sustained activation of these "fight or flight" systems results in trajectory from physiological "alarm" to the pathological processes of "resistance" and finally "exhaustion."

Sustained Stress Drives Structural Changes in the Brain

Experimental studies of intense stress in animals have demonstrated the susceptibility and plasticity of the brain to stress. Interestingly, animal models of stress are often used to model aspects of physiological stress since they are not physically harmed. A typical model of stress that can be used to induce plastic changes in the brain is restraint or confinement stress. The restraint stress model uses a mesh-enclosed structure that allows physiological thermoregulation, and also confines the animal for periods of minutes to hours. Animals are located in light vulnerable and threatening locations from where their instinct is to escape. The stress method is considered psychological because it is a perceived threat, rather than real physical impact, that challenges and stresses the animal. Variable stress is an alternative model utilizing restraint stress in combination with the stress of varying housing conditions for the animals. In these models, corticosterone levels (the rat cortisol analog) increase acutely to maximal levels. Under chronic conditions of 10 or more days of stress, brain micro-anatomy changes in interesting ways.

The hippocampus, for example, is significantly affected by stress. Chronic restraint stress in animal models leads to a significant reduction in dendrite branches. Thus, stress causes an apparent pruning back of neuronal dendrites. The CA3 region of the hippocampus is significantly affected and other subregions are also likely impacted by stress. Large pyramidal neurons of the CA3 region are strongly linked to spatial memory and spatial navigation. They also play important roles in other memories including spatial, temporal, and associative memories of traumatic events. Therefore, stress induces structural changes to hippocampal neurons that affect their ability to store and retrieve important behavioral information.

In addition, the prefrontal cortex, like the hippocampus, is also affected by stress, and neuronal hypotrophy is a frequent consequence. For example, large pyramidal neurons of the prelimbic region of the prefrontal cortex have been shown in experimental models to undergo significant dendrite retraction and loss. While the prefrontal cortex in humans is associated with executive function, the prelimbic region in animal studies has been shown to regulate the acquisition of amygdala-based fear behavior and regulate hypothalamic circuits controlling the hypothalamic–pituitary–adrenal (HPA) axis. Thus, stress induces changes to prefrontal cortex that are likely to impact traumatic memory acquisition and extinction as well as HPA-mediated stress responses.

Neurons of both hippocampus and prefrontal cortex have been shown to lose dendritic spines as a result of stress. Dendritic spines receive the majority of excitatory glutamatergic input to neurons and are critically involved in plasticity and memory. Studies of memory have shown that spines are formed and undergo changes in both volume and synaptic strength when memories are acquired. The loss of dendritic spines in both hippocampal and prefrontal cortex neurons during stress has the consequence of reducing neuron circuit interconnectivity, and also likely impacts the ability of these circuits to acquire and process new memories.

In the amygdala, some studies have shown that the stress effect is in the opposite direction to its effects on hippocampus and prefrontal cortex. Chronic variable stress results in hypertrophy of pyramidal neurons of the basolateral nucleus of the amygdala. By contrast, neurons of basolateral nucleus of the amygdala show more dendrites with greater length and branching. Moreover, dendritic spines are increased following stress, in contrast to the case of the hippocampus and prefrontal cortex, where they are lost.

The effects of stress on hippocampus, prefrontal cortex, and amygdala can be mimicked by GCs suggesting that the stress effect is mediated by this stress-induced hormone. While studies have focused on repeated exposure to higher levels of GCs, it is important to note that a single acute dose of corticosterone in a rodent model can also induce hypertrophy in amygdala dendrites.

By contrast, GCs can induce different effects on neuron structure dependent on their location in the brain. For example, the hippocampus is structurally altered in other ways by high levels of chronic stress. Prolonged stress and elevated GCs have been shown to cause neuronal death and inhibit adult neurogenesis in the dentate gyrus of the hippocampus. Antagonism of NMDA-type glutamate receptors can block both the stress-induced and the GC-induced effect on dendrite structure and neurogenesis. Thus, excitatory synaptic circuits respond to stress-induced increases in cortisol to structurally change and regulate neurons.

Fear Circuits and Traumatic Stress

Stress-modulated neuronal circuits are those circuits especially involved in the acquisition, expression, and extinction of fear memories. Traumatic memories frequently involve the association of unconditioned and conditioned stimuli. Experimental animal models of conditioned fear have therefore been informative for understanding the interaction between the brain's conditioned fear systems and its stress systems, particularly in response to circulating GCs. These studies have helped describe and inform the effects of stress on memory systems.

For example, stress and GCs change the response levels of fear memory acquisition and expression. Moreover, generalization to stimuli is also changed by stress. Increasing the level of GCs in animals followed by conditioned fear training leads to higher fear levels when tested. When comparing fear to the context or stimulus cues, fear to the context is more affected by GCs. When animals are exposed to chronic stress, they show increased levels of fear to new and known contexts. Importantly, stress reduces extinction to fear and reduces memory of fear extinction. Thus, after extinction, fear levels return more predominately in stressed individuals.

ANIMAL MODELS OF PTSD

There are a number of animal models of PTSD that model key aspects of the disorder, but there is no animal model of the entire disorder. These models have and continue to provide essential information on the neurobiology of fear and stress, and especially their interaction. Animal models are generally assessed by their ability to provide validity for simulation of pathological disorders. Many aspects of validity have been proposed in animal models. Face validity and predictive validity are two important components of animal model validity for PTSD. Face validity relates to observable behavioral and biological properties and includes aspects of ethological and biological measures. Predictive validity relates the outcomes and how closely the model corresponds to outcomes in the pathological state. In addition, dependent or independent variables in animal model experiments focus on aspects of the disorder. This is especially true in PTSD where the disorder is acquired pathology, thereby allowing aspects of acquisition and subsequent responses to be modeled.

Pavlovian Fear Conditioning

Pavlovian fear conditioning is an important animal model for the study of PTSD because it models core aspects of the disorder. When persons are exposed to a traumatic event, they acquire very strong memories of aspects of the event. These memories include specific cues associated with the event such as specific sights, sounds, smells, and other stimuli associated with the trauma. The fundamental properties of being exposed to an aversive event, together with other stimuli that were previously not important, lead to a long-term associative memory of the stimuli with the trauma. Pavlovian fear conditioning is therefore a model of associative memory. In animal models of Pavlovian conditioning, an aversive unconditioned stimulus (US), usually an electric foot shock, is paired with a neutral conditioned stimulus (CS), usually a sound or light. These two stimuli become associated such that the CS alone is able to elicit the same reaction as the US on its own. This means that a previously neutral cue (the sound or light) can elicit the same response as the electric shock. This model is also known as Pavlovian threat conditioning as it models the initial threat response common across all animals. Many aspects of the neuroanatomy and physiology of the neurobiology underlying Pavlovian fear/threat conditioning have been identified using animal models. The shared neural circuitry between Pavlovian fear conditioning and PTSD indicates good etiological validity for this model. Pavlovian fear/threat conditioning can also be performed in human subjects. A small shock is usually applied to the subject's wrist or fingers and it is simultaneously paired with neutral stimulus on a computer screen. Thus, aspects of fear/threat memory acquisition, expression, and extinction can be studied in both healthy subjects and persons with PTSD.

Fear-Potentiated Startle

The fear-potentiated startle model is a variation on Pavlovian fear conditioning, which has good predictive validity. In fear-potentiated startle, an animal is exposed to a loud aversive startling sound. In addition, the animal is also presented with a CS that is predictive of the loud startling sound. When presented with a loud startling sound, an animal will startle with increased movement. Similar to Pavlovian fear conditioning, there are also increases in other defensive systems. These include activation of the HPA axis and the sympathetic nervous system. The neutral stimulus can also elicit the same reaction after conditioning. From a therapeutic perspective, drugs that are effective for anxiety have also been shown to be effective in fear-potentiated startle. Relevantly, fear-potentiated startle has also been shown to be predictive in PTSD, whereby persons with PTSD have an increased startle response. The model can also be used in humans, both healthy subjects and those with PTSD, where the startle stimuli can be either a loud noise or alternatively a strong air blast to the throat.

Models of Multiple Unpredictable Stress

A variety of animal models of PTSD have been developed to provoke a wide range of behavioral responses following repeated stress exposure. For example, stress responses in rodents can be produced by electric shock, restraint, learned helplessness, social aggression and defeat, or exposure to a predator. The unpredictable stress models include a combination of those physical and/or psychological stressors. One model utilizes restraint stress and tail shock to produce physical and painful stress, which increases the HPA axis function in animals. Another model includes unpredictable environmental stimuli such as odors of the predator, noise, flashing light, or cage shaking.

Restraint With Tail Shock Paradigm

The repeated exposure to inescapable tail shock model in rodents appears to mimic core symptoms of PTSD observed in humans by producing long-lasting behavioral and physiological changes. For instance, exaggerated acoustic startle responses can be observed in stressed animals as well as individuals with PTSD. In this model, stress exposure consists of a daily 2-h session of immobilization and repeated tail shocks for 3 consecutive days. The animals are restrained in a Plexiglas tube, and shocked repeatedly on the tail at varying intervals over the course of 2h. The concept of this paradigm was derived from the learned helplessness paradigm in which animals experience stress under conditions where they cannot escape. Previous studies suggested that repeated chronic, rather than acute, stress is more effective in producing physiological and behavioral abnormalities. These include elevated plasma corticosterone levels, exaggerated acoustic startle, and reduced body weight. Thus, repeated stress exposure may be critical to produce greater physiological and behavioral responses to stress. It is important to note that 3days of stress exposure in rats may be equivalent to 3months of stress in humans, if one considers the life span differences between the species.

Warrior Stress Paradigm

This stress paradigm uses a combination of a predator stressor and unpredictable environmental stimuli. Although the predator stress does not cause physical pain, this is an effective stressor used in rodents that can be manipulated by presenting the odors of the predator such as introducing cotton balls with synthetic fox urine into a test cage with the rat. It has been shown that exposure to predator stress can produce behavioral and biochemical responses to stress such as reduced food consumption, enhanced acoustic startle response, freezing behavior, and elevated stress hormone levels. To model unpredictable environmental stimuli, and to avoid habituation to the repeated presentation of fox urine, different orders of environmental stimuli are introduced such as sudden noise, flashing light, and cage shaking each day. Repeated and unpredictable stressors lead to increased levels of corticosterone stress hormone in rats, and have face validity for human stress and behavior in rodent studies.

The caveat for such studies is that the animal models of stress do not and cannot fully and faithfully represent PTSD in humans. However, they can be used to gain insight into candidate biological mechanisms for PTSD. Therefore, utilizing various animal models of stress may enhance our understanding of the molecular mechanisms of PTSD and contribute to an improved treatment strategy for this debilitating stress-related disorder.

BIOMARKERS OF PTSD

PTSD can be a chronic and disabling anxiety disorder. However, the diagnosis of PTSD is currently based on clinical observation, the occurrence of specific symptoms, and the duration of these symptoms. Because there is presently no laboratory "test" for PTSD, potential biomarkers have been investigated. Knowing biomarkers for PTSD *susceptibility* may be useful for prevention by identifying subjects at high risk for PTSD after trauma exposure, such as combat and nature disasters. Also, biomarkers for PTSD diagnosis may benefit early identification for treatment. Biomarkers related to course and recovery are also important. In addition, biomarker studies can help us to understand the pathology of the disease from risk to onset, chronicity, and recovery and thus facilitate the discovery of new therapeutic targets for PTSD.

The Biomarker Concept

The term "biomarker" is a modern term for an old concept in which a biochemical entity or biophysical feature can be used for identifying a biological or pathological process. Just as for the results of conventional clinical chemistry or radiology measurement, results can be used individually or in combination to arrive at a diagnosis, determine prognosis, and provide clinically relevant endpoints for therapy. In the absence of mechanism, as is the case for PTSD, biomarker discovery is carried out in the context of many possible candidate biomarkers. This discovery-based approach, rather than a conventional hypothesis-based approach, is utilized in hopes that one or more of these random features will specifically stratify with disease presence, with disease severity, and/or with disease outcome.

From such a systems biology perspective, the clinical chemistry approach can include microRNAs (c. 2000 in the human genome), genes (c. 23,000 in the human genome, measurable as either individual gene sequences, individual mRNAs, individual noncoding RNAs), and a fraction of the proteins expressed by these genes (perhaps 100,000 total protein species, including proteins with posttranslational modifications). Biophysical features can include conventional X-rays, CAT scans, and functional MRIs. Bioinformatics strategies are then used to elicit individual biomarkers, or combinations of such biomarkers, that can yield statistically significant differences between those with PTSD and "healthy controls." The parametric statistics are also unique in that by measuring many samples from a few patients, a Bonferroni correction is necessary to determine significance. For example, if significance in a conventional *t*-test is $p = .05$, the Bonferroni correction depends on dividing 0.05 by the total number of features measured. Thus, for a 23,000 gene analysis, the Bonferroni-defined significance would be [$p = .05/23{,}000 = .00000217 = 2.17 \times 10^{-6}$]. For a result to be significant, the *p* value would therefore have to be *less* than this amount. One popular alternative to the Bonferroni correction includes calculation of a nonparametric false discovery rate (FDR). This is accomplished by comparing the data in known samples and controls with systematically random distributions of all the data. Significant differences depend on an FDR of less than 10%, and a q statistic, related to the FDR, of less than 0.05.

Importantly, however, psychiatric geneticists are far away from obtaining a comprehensive list of PTSD risk-susceptibility genes. The entire human genome is 3,000,000,000 (3billion) bases long, meaning that genome-wide association studies (GWAS) only address about 1.7% of the genome. An alternative approach, whole-exome analysis, has been to analyze sequences in all exons (coding sequences in the genome). However, exons only comprise about 2% of the entire genome, and overlap with the GWAS targets. Thus, about 98% of the genome remains to be to be studied. Fortunately, the cost of measuring whole-genome sequences (WGSs) has plummeted in recent times and with the precision medicine initiative (PMI) now being widely pursued, the standard of analysis will now include conducting WGSs on all patients.

Most of the known genetic biomarkers are imputed from GWAS. The way this approach has worked in the past is that polymorphisms in about 500,000 sites in the human genome (0.16% of the genome) are analyzed and queried as to an association with a function or a disease. Using WGSs from the 1000 genome database (namely, http:/www/1000genomes.org), possible genes that might be "near" these sites are queried as to relevance. Genes identified by this approach are said to be "imputed" or "inferred." Developing capacity using WGS rather than GWAS suggests that rare disease-associated alleles are missed about 50% of the time. Fortunately, using WGSs, the field is well on its way to filling in the missing information.

Potential PTSD Molecular Biomarkers

In the field of PTSD research, researchers have been trying to develop a single test, such as a blood or genetic test for PTSD. In other medical fields, certain biomarkers have already been developed for diagnosis. For example, cholesterol measurements and genetic tests have been used for predicting the risks for heart attacks and Huntington's disease, respectively. Researchers in psychiatry hope to use the same strategies to identify biomarkers for PTSD risk(s), and implement the information clinically. The search for PTSD biomarker is ongoing, though scientists are closer than ever to achieving this goal. Currently, diagnosis occurs through clinical-interview-based DSM diagnostic criteria. Studying biomarkers, at least in part, will help us to better understand PTSD, its etiology, and its pathologies, which may lead to development of an objective diagnostic tool for more effective treatments based on the individual's pathology. It is known that not all individuals will develop PTSD or any other trauma spectrum disorder after trauma exposure. The probability to develop PTSD depends on the balance between risk and resilience factors in an individual. Biomarkers might identify high-risk populations that had not manifested any symptoms or met PTSD diagnostic criteria. Those PTSD susceptibility biomarkers would be especially useful for identifying individuals at high risk for PTSD to prevent them from being exposed (primary prevention) to traumatic events or, if subjects are exposed to traumatic events, to start a timely preventive therapy before manifestation of PTSD symptoms (secondary prevention). Since PTSD biomarkers may be associated with severity of PTSD symptom(s), the biomarker could be used to monitor therapeutic responses. Therefore, the desirable properties of biomarkers for PTSD vary with their different usages. For example, a screening test requires having high sensitivity, specificity, predictive values, large likelihood ratios, and low cost to test a large population including normal and high-risk subjects. A diagnostic biomarker may be chosen for testing stages of the disease (acute and chronic). Sensitivity or specificity of monitoring biomarkers may be less important in tests where the marker is used to test the same individual serving as one's own control. Currently, several candidate biomarkers for PTSD have been reported. Here, rather than be comprehensive in a rapidly changing field, we provide a summary of just a few of these candidate biomarkers (Table 22.1). These candidate biomarkers conceptually fall into the categories of either conventional or modern biomarker definitions.

Glucocorticoids and the Glucocorticoid Receptor

Two of potential biomarkers for PTSD include concentrations of GCs in the blood, a classical biomarker, and the GR, which might be a genetic biomarker in the modern sense. For example, it has been shown that the propensity to develop PTSD can be associated with altered sensitivity of leukocytes to regulation by GCs. The relevant finding is that high levels of PTSD symptoms after deployment can be shown to be independently associated with a high dexamethasone (DEX) sensitivity of T-cell proliferation in these same soldiers when cells were analyzed before deployment. The caution here is that such alterations in T-cell proliferation were only observed in individuals who reported PTSD symptoms *without* depressive symptoms. In this case the difference in activity can be traced most likely to modifications in the GR gene itself, or to ancillary factors that affect GR function. Relevantly, before military deployment, the GR number in peripheral blood mononuclear cell (PBMC)

TABLE 22.1 Selected Potential Molecular Biomarkers of PTSD

Molecular Biomarkers of PTSD	Methods
High sensitivity of leukocytes for glucocorticoids (GCs) and a high number of glucocorticoid receptors (GR) in PBMC prior to deployment predicted PTSD development	Longitudinal study searching for PTSD risk and resilience markers by comparing trauma-exposed soldiers with PTSD to those without PTSD and to healthy individuals without prior trauma exposure
SNPs in FKBP5 and PTSD; mRNA levels of FKBP5 and the GR-inhibitor STAT5B reduced in PTSD	Childhood trauma and DNA demethylation in functional glucocorticoid response elements (GREs) of FKBP5; mRNA levels
LINE-1 and Alu differentially methylated in pre-deployment samples of individuals with and without PTSD	LINE-1 and Alu methylated in US military service members
Molecular risk of PTSD at the SLC6A3 locus	Influenced the risks of developing PTSD
Gender-specific association between neuropeptide PAC1 (a receptor of the neuropeptide pituitary adenylate cyclase-activating peptide (PACAP)) and PTSD susceptibility	Association study
High levels of corticotropin-releasing hormone (CRH) in cerebrospinal fluid (CSF) of PTSD patients	CRH levels in CSF
5-HTT and PTSD	Association study
CPT dysregulation in PTSD of US military population	Gene micro-array
BDNF associated with PTSD in US military population	Association study
P11 mRNA overexpression in the brain of PTSD	Quantitative PCR

was significantly higher in service members who developed high levels of PTSD symptoms after deployment relative to matched comparison subjects. The risk for inclusion in the PTSD group after deployment increased 7.5-fold with each GR increase of 1000. Furthermore, for those post-deployment soldiers whose PTSD symptoms were sustained in a 1–6 month–period post-deployment, the number of GR molecules was proportionately higher than their resilient colleagues. These results indicate that the sensitivity of leukocytes to GCs is a predictive factor for the development of PTSD and that a high GR number in PBMCs is a vulnerability factor for PTSD.

FKBP5

FK506 is an immune-suppressant drug used in organ transplants to suppress calcium/calmodulin/nuclear factor of activated T-cells (NFAT) driven gene expression. Of the numerous proteins found to bind this drug, functional polymorphisms in the gene for FKBP5 have been associated with an increased risk of developing stress-related psychiatric disorders in adulthood. Only risk allele A carriers of rs9296158 showed GR supersensitivity with PTSD in patients with the [GG]FKBP5 genotype. These data suggest that functional variants in FKBP5 are associated with biologically distinct subtypes of PTSD. This allele alters chromatin interaction between the transcription start site and long-range enhancers in the FK506-binding protein 5 (FKBP5) gene. The FKBP5 gene promoter contains a functional GC response element for which cytosine methylation at a CpG site in this sequence is associated with childhood trauma and is linked to increased stress-dependent gene transcription. The demethylation in FKBP5 gene is followed by a long-term dysregulation of the stress hormone system and a global effect on the function of immune cells.

LINE-1 and Alu

Serum contains measurable amount of DNA that has been released from dead or dying cells into the circulation, and sequencing this DNA can provide biologically important information. As mentioned earlier for the FKBP5 gene, DNA methylation can have critically important consequences, including a report of historical experience based on the epigenetic effects of acquired DNA methylation patterns. LINE-1 (Long INterspersed Element) is a non-LTR retrotransposon that makes up 17% of the human genome, and is also widely found in eukaryotic genomes. Importantly, RNAs transcribed from LINE-1 are essential functional components of nucleosomes. In recent studies of serum DNA methylation in PTSD cases and controls from US military service members recently deployed to Afghanistan or Iraq, it was found that LINE-1 was hypermethylated in controls post- versus pre-deployment, and hypomethylated in cases versus controls post-deployment.

Alu sequences are the most abundant transposable elements in the human genome, and are exclusively found in primates. Alu insertions into DNA have been associated with a variety of diseases, including inherited predispositions to various cancers metabolic diseases and Alzheimer's disease. Fortunately, however, most Alu alleles are operationally silent. As for LINE-1, Alu was also found to be hypermethylated for cases versus controls pre-deployment, but hypomethylated post-deployment. Thus, the methylation status of LINE-1 and Alu DNA can be a reporter of experience for the service member in which PTSD is a consequence is intriguing and may suggest resilience or vulnerability factors. These data thus warrant a larger sample size study to determine whether these patterns of methylation of LINE-1 and of Alu can be a biomarker for PTSD.

SLC6A3

The SLC6A3 (solute carrier family 6 (neurotransmitter transporter, dopamine), member 3) locus increases risk of PTSD. 9R allele carriers had almost double the risk of lifetime PTSD compared to 10R/10R genotype carriers, controlling for age, sex, race, socioeconomic status, number of traumas, smoking, and lifetime depression. In the subsample of participants with available methylation data, a significant interaction was observed whereby 9R allele carriers showed an increased risk of lifetime PTSD only in conjunction with high methylation in the SLC6A3 promoter locus, controlling for the same covariates. These data indicate a role for the 9R allele in increasing susceptibility to PTSD and provide preliminary evidence that a "double hit" model, including both a putatively reduced function allele and high methylation in the promoter region, may more accurately capture molecular risk of PTSD at the SLC6A3 locus.

PAC1

Pituitary adenylate cyclase-activating peptide (PACAP) is known to be associated with PTSD. Data suggested that the PACAP/PAC1 receptor pathway plays an important role in PTSD. It is found in heavily traumatized subjects, that there is a sex-specific association of PACAP blood levels with fear, PTSD diagnosis and symptoms in females. rs2267735, a single-nucleotide polymorphism (SNP) in a putative estrogen response element within ADCYAP1R1, predicts PTSD diagnosis and symptoms in females only. This study suggests that alteration of signaling in the PACAP/PAC1 pathway is associated with stress responses and PTSD. PACAP levels and ADCYAP1R1 SNPs may serve as useful biomarkers for PTSD.

Corticotropin Releasing Factor

There are also data showing that basal elevations of CRH concentrations in cerebrospinal fluid (CSF) are associated PTSD. In a study in which CSF was collected hourly by means of a flexible, indwelling subarachnoid catheter over a 6-h period in 11 well-characterized combat veterans with PTSD and 12 case-matched normal volunteers, mean CSF CRH levels were significantly higher in the subjects with PTSD than in normal subjects. Therefore, it is possible that high basal CSF CRH concentrations might be a potential conventional biomarker for PTSD.

SLC6A4

Multiple lines of evidence suggest that polymorphisms in the serotonin transporter gene (SLC6A4) promoter play a significant role in the susceptibility to developing not only PTSD but possibly certain other psychiatric disorders. The 5-HTTLPR region of SLC6A4 gene is a degenerate repeat polymorphic region of the promoter that occurs in both short and long versions. The long version (5-HTTLPR) results in higher levels of expression of SLC6A4 mRNA in human cells. The 5-HTTLPR polymorphism may play a role in moderating several categories of emotional response after stressful life events. This presumably may relate to susceptibility to PTSD. In a cross-sectional study, the triallelic system of 5-HTTLPR was genotyped in 582 European-American and 670 African-American individuals who reported experiences of childhood adversity, adult traumatic events, or both. The PTSD diagnoses were made by DSM-IV diagnostic criteria. Although the 5-HTTLPR genotype alone did not predict the onset of PTSD, it correlated significantly with adult traumatic events, and with a history of childhood adversity, to increase the risk for PTSD, particularly in those with high rates of both types of trauma exposure. Subjects experiencing both childhood adversity and adult traumatic events were more likely to develop lifetime PTSD compared with those who experienced either type of adverse event. Homozygous individuals with one or two copies of the S′ (S) (short) allele compared with the L′ (L) (long) polymorphism had higher risk for PTSD. These studies show different and possibly relevant effects of the short versus long allele on propensity or susceptibility to PTSD, and suggest that PTSD is influenced by the interactive effect of environmental and SLC6A4 gene.

BDNF

Brain-derived neurotrophic factor (BDNF), a gene whose gene product regulates neuronal survival, growth, differentiation, and synapse formation, has been frequently associated with PTSD. A common SNP in the BDNF gene (Val66Met) influences hippocampal volume, memory, and appears related to susceptibility to a variety of neuropsychiatric disorders, including PTSD. In a previous study ($n = 461$) looking at BDNF in PTSD, 9% participants met criteria for probable PTSD, while the remaining 91% subjects did not. In those with probable PTSD, the frequency of Met/Met was nearly threefold higher than in the controls. The allelic frequency of Met was twofold higher in individuals with probable PTSD than in controls. This candidate genetic biomarker was complemented by conventional biomarker BDNF levels in plasma of the subjects with PTSD. The PTSD patients had significantly higher levels of circulating BDNF protein than the non-PTSD controls. These results agree with evidence suggesting that BDNF is involved in the development of PTSD. Met carriers have a smaller hippocampus volume, relative to controls and perform more poorly than controls (Val/Val) on memory tasks. The frequency of Met alleles is higher in veterans with psychotic PTSD than nonpsychotic veterans with PTSD, or veterans without PTSD. However, a case–control genetic association study has shown no relationship between BDNF gene Val66Met and a PTSD diagnosis.

In another examination of the genotype/phenotype relationships between PTSD and BDNF, it was determined that the frequency of the BDNF val66met genotype was significantly different in those with PTSD who also had an exaggerated startle reflex. Relevantly, an exaggerated startle reflex is a core symptom of PTSD. The Met/Met genotype was also significantly more frequent among those with exaggerated startle compared to those without exaggerated startle, indicating that the Met/Met genotype may be associated with hyperarousal vulnerability. As a genetic control to this conclusion, the frequencies of the Val/Val genotype were significantly higher in the non-PTSD, and non-startle groups than in PTSD, and startle groups, respectively. Those data indicate that the Val/Val genotype plays may play protective role in PTSD and/or exaggerated startle reactions.

The relationship of BDNF Val66Met to exaggerated startle and reported severe stressful life events were also analyzed. The Met/Met genotype carries a higher startle score among those who experienced less than four stressful life events. However, no effects of G × E interaction on startle score were observed among those who experienced higher (four or more) stressful life events. These data therefore suggest that exposure to stressful life events is associated with a higher risk of hyperarousal in Met/Met, and lower risk of hyperarousal in Val carriers.

Finally, among those who reported fewer stressful life events, the Met/Met genotype was associated with significantly higher PCL total scores than Val carriers. It is therefore possible that at a lower stressful event exposure, Met/Met had a higher PTSD symptom severity, while Val carriers had lower PTSD symptom severity. Thus, the protective effects of the Val genotype are overwhelmed at higher levels of stress exposure. It is also possible that neuronal mechanisms may be different in minimally exposed and highly exposed individuals who develop PTSD.

S100A10

The p11 (S100A10) calcium-binding protein has been shown to be overexpressed in the postmortem brain of patients with PTSD. Such overexpression is regulated by GCs or stress, which results in significant increase of GR activation. Activated GR is transferred into the nucleus. The GR interacts with GC response elements (GREs) in p11. The results showed that a GR antagonist or a GRE mutation attenuated GC-induced p11 overexpression. As such p11 may play an important role in pathology of traumatic stress and PTSD, and may be a candidate biomarker for PTSD.

PTSD AND SOMATIC HEALTH: CARDIOVASCULAR DISEASE

PTSD has been related to various aspects of somatic health and aging, morbidity, and mortality. In particular, one focus has been CVD. Various aspects of CVD have been operationally associated in the medical literature since at least the time of the American Civil War. Da Costa studied over 300 soldiers, all housed at the Turner Lane Dispensary in Philadelphia, and most of whom had survived lengthy action during the Peninsula Campaign. Using the diagnostic methods then available, Da Costa developed one of the first ever "big data" analyses, showing not only that CVD frequently accompanied PTSD, but that it was also independent of other medical variables. Da Costa termed this condition "soldier's heart" or "irritable heart," and the term has made its way into psychiatric textbooks in chapters describing anxiety. With PTSD becoming increasingly diagnosed in both civilian and military contexts, the American Veterans Administration (VA) has popularly referred to the relationship as the "Brain-Heart Mystery" (http://www.research.va.gov/currents/spring2014/Spring2014-1.cfm). It is a mystery because the mechanisms linking PTSD and CDV are unknown.

However, based on epidemiological approaches to analysis, the PTSD/CVD relationship is firmly established. In longitudinal studies of large numbers of combat veterans with PTSD in a VA setting, the increased risk for CVD is about 50%. In addition, the associations between

PTSD and different aspects of CVD appear to be significantly independent of traditional risk factors such as smoking, hyperlipidemia, obesity, or exercise (SHOEs), and of other potential confounders including alcohol use, sleep quality, social support, depression, age, gender, diabetes, hypertension, and combat service.

An alternative approach to understanding why some individuals exposed to traumatic events are resilient, while others are susceptible, has been to consider the possibility that there might be a genetic basis for the propensity to develop PTSD. Some of the strongest evidence for this genetic hypothesis has been a study of 562 twins from the Viet Nam Twins Repository. In parallel with psychiatric assessment, myocardial perfusion was also assessed by positron emission tomography. These investigators showed that the incidence of objectively validated CVD in those with PTSD was more than double that of twins without PTSD. The PTSD/CVD linkage in affected twins was also found to be significantly independent of SHOE and other potentially confounding variables.

The problem has now turned to identifying genes that might be responsible for PTSD, and possibly also CVD. It is yet to be determined whether any of these PTSD-risk genes, or others, are actually associated with the PTSD/CVD linkage.

Based on the Viet Nam twins study, it is very likely that the mechanism will be found to be genetic. However, conclusive information on this genetic mechanism will await WGS studies on large numbers of well-phenotyped PTSD patients, and on objective measures of different aspects of the CVD that may be found to be typical of different ethnicities and genders.

NEUROIMAGING AND PTSD

Neuroimaging data from the past decade has shown a number of brain regions involved in the pathological process of PTSD. These anatomic regions, identified primarily through fMRI and PET, include the ventromedial prefrontal cortex (vmPFC) and anterior cingulate cortex (ACC), amygdala, hippocampus, and insula. While considerable research has shown these individual areas are altered in patients with PTSD, there is solid evidence that the interplay between these regions is more critical to understanding and treating PTSD than any specific regional alteration.

vmPFC and ACC

Researchers have used a variety of techniques to study neural activity of the vmPFC (namely, Broca Area 25) and the ACC in individuals with PTSD. Negative or aversive stimuli including script-driven imagery of trauma, trauma-related images, emotional faces, aversive descriptions, and fear-conditioning approaches result in decreased activity in the vmPFC and rostral ACC in fMRI and PET. Neutral stimulus also results in decreased vmPFC activation in individuals suffering from PTSD. Interestingly and most notably, there is an inverse correlation between vmPFC activation and PTSD symptom severity.

While there is clear evidence regarding the decreased activity of the vmPFC and the associated rostral ACC (rACC), there is some evidence that the dorsal ACC (dACC) has increased activity in response to negative stimuli. These findings have not been confirmed and speak to the vast amount the field has yet to learn about the brain in the context of various mental pathologies.

Amygdala

The amygdala is one of the most studied areas of the brain in the context of PTSD. In general, neuroimaging has found the amygdala to be overactive in PTSD. More specifically, there is increased amygdala activation in response to trauma-related stimuli. These trauma-related stimuli have included script-driven imagery, aversive sounds, and backward-masked faces. Additionally, trauma-unrelated emotional material has also resulted in increased amygdala activation in patients with PTSD compared to those without PTSD. Individuals with PTSD also have increased amygdala activation at rest and in response to neutral stimuli. Additionally, individuals with PTSD have relatively greater amygdala activation in response to fearful faces than to neutral faces compared to healthy control individuals without PTSD. During an extinction learning fear-conditioning paradigm, patients with PTSD showed greater amygdala activation. Conversely, decreased amygdala activation during the first few weeks after a traumatic was associated with a lower likelihood of developing PTSD and may be an indicator of neurobiological resilience to trauma.

Hippocampus

While the research involving the hippocampus in PTSD has been mixed, studies showing an association have more often found decreased activation and decreased hippocampal volume in those suffering from PTSD. In particular, several studies have shown decreased left hippocampal activation in response to script-driven imagery, trauma-related images, and emotional word retrieval. Individuals with PTSD also show increased amygdala activation in response to neutral stimuli, though as with many of the hippocampal findings, there are studies with opposing findings. More notably, some work has shown an increase in left hippocampal activity response to trauma-related stimuli after psychotherapy. These findings, along with the finding that hippocampal activation appears to be negatively correlated with PTSD symptom severity, reinforce the role of the hippocampus in the pathological process of PTSD.

Insula

Several studies have shown increased activation in the insula, particularly the right insula, in response to negative stimuli. Increased insula activation was found in response to script-driven imagery, trauma-related imagery, and emotional faces. A positive correlation was found between insula activity and PTSD symptom severity. Altered GABAergic transmission here and in the anterior cingulate has been proposed as potential contributors to states of persistent fear and/or anxiety. Interestingly, painful stimuli are also associated with increased insula activation in individuals with PTSD.

Connectivity

The interaction of the various regions involved in PTSD based on neuroimaging studies is both fascinating as well as vital to understanding this disabling disorder. The roles of the various participants can be oversimplified into the categories of executive (PFC), emotional (amygdala/insula), and memory (hippocampus). These categories can be instructive in understanding the interplay. The PFC, as part of the executive system, performs the vital role of inhibiting the amygdala. The amygdala and insula and their emotionally associated responses to traumatic experiences rely on PFC inhibition to tamp down escalating neuronal activity in the face of traumatic stimuli. In the midst of this functioning system, the hippocampus is activated in order to provide comparisons that provide perspective and further regulate the emotionally based activation of the overall system. In PTSD, the amygdala and insula are more active at baseline, while the PFC and hippocampus are less active at rest. In the context of a traumatic stimulus, the ability of the PFC to provide inhibitory control is insufficient to tamp down the already ramped up amygdala/insula activity. The hippocampus is inhibited in the context of this fight or flight charged system; such a process may provide an evolutionary advantage by focusing on the present during periods of heightened stress. However, for an individual suffering from PTSD, the heightened stress may be a misperception of the stimulus (eg, car backfiring perceived as a bomb exploding), and the resulting impact is lowered hippocampal involvement and in the long term, potentially reduced volume. The origin of this dysregulation of inhibitory controls is not yet well understood and further research will hopefully elucidate the causal relationship between traumatic experience and dysregulated feedback mechanisms within these brain regions. Such research may result in neuroimaging being used as a marker or measure of disease progression or response to treatment.

NEUROSCIENCE AND TREATMENT

PTSD treatments can be divided into three main categories. These categories include psychotherapy, pharmacotherapy, and complementary and alternative medicine (CAM). While a full discussion of each of these categories is not possible in this chapter, this section provides a brief review. In general, the mechanisms of how these therapies work are not known. However, it is reasonable to expect that insights into these mechanisms will be available through PTSD-specific biomarkers, and that some of the biomarkers will provide a way to measure therapeutic progress.

Psychotherapy

Psychotherapy, or talk therapy, is a cornerstone of the treatment of PTSD and has some of the strongest efficacy evidence among current treatments. Psychotherapeutic treatments for PTSD primarily consist of cognitive therapy and trauma-focused therapies. Cognitive therapies focus on changing cognition by actively confronting cognitive distortions through a series of homework and practical exercises. These exercises reinforce behavioral and cognitive changes aimed at decreasing misperceptions and cognitive distortions that can result in negative emotions. The cognitive therapy most widely used to treat PTSD is cognitive-processing therapy. Cognitive-behavioral therapies, including cognitive-processing therapy, have shown benefit in PTSD, although the exact neurological mechanism remains unclear. Known effects of cognitive therapies include improved neuronal activity in the prefrontal cortex and hippocampus, as well as increased BDNF, and improved hippocampal volume. Some of these findings have been sustained for as much as months after completion of treatment. Trauma-focused therapy, on the other hand, is aimed at confronting the traumatic experience and desensitization to the mental and physiological impact of the traumatic memory. The most common trauma-focused therapy is prolonged exposure (PE), which has shown benefit in multiple rigorous studies. Similar to cognitive therapies, trauma-focused therapies are known to have significant impact on the brain as evidenced by various neuroimaging studies, but the specific means by which PE causes decreased amygdala activity or increased hippocampal activity remains unknown.

Pharmacotherapy

Currently there are two medications, FDA approved, for the treatment of PTSD. These medications are paroxetine and sertraline, both in the class of drugs known as serotonin-specific reuptake inhibitors (SSRIs). Beyond these two are a large number of drugs being used off-label by clinicians seeking alternatives for a variety of reasons. These off-label compounds include serotonin norepinephrine reuptake inhibitors (SNRIs), tricyclic antidepressants (TCAs), antihypertensives, antiepileptics, antipsychotics, sedative hypnotics, and other SSRIs. With varying levels of evidence and understanding of the mechanism of action, this wide assortment of compounds represents the vast possibilities for accessing the brain systems involved in PTSD while speaking to the complexities facing clinicians treating those suffering from PTSD. These medications work through receptors including serotonin, dopamine, and norepinephrine. Many reregulate these receptors while activating second-messenger systems with follow-on stimulation of mRNA protein synthesis. Some of these proteins, such as BDNF, are involved in neuronal growth and regulation, and are known to improve overall arborization.

Despite an ever-increasing knowledge base, the mechanism of action by which psychopharmaceuticals reduce PTSD symptom levels or alleviate suffering remains elusive. From the monoamine hypothesis through to mRNA protein synthesis of neuronal regulators, the field has seen a steady progression towards the specific pathology underlying much of mental health including the vital issue of sleep disruption. Hopefully, with full understanding of the process by which these disabling conditions arise, effective treatments can be developed to counter the impact of PTSD.

Complementary and Alternative Medicine

There are a myriad of nontraditional treatment modalities for PTSD. These treatments are most often grouped together as CAM. The CAM includes more interventional treatments such as acupuncture and neurostimulation as well as more activity-based methods including meditation and recreational therapy. While the neuroscience of interventional treatments such as acupuncture and neurostimulation are somewhat understood, the mechanisms of other modalities such as meditation remains poorly understood. Neurostimulation, which includes transcranial magnetic stimulation (TMS), induces neuronal firing through electrical or magnetic stimulation. TMS uses a rapidly pulsing magnetic field to induce depolarization of superficial neurons in the brain. Activating these superficial neurons, such as the vmPFC, can result in increased inhibition of the amygdala and insula and subsequent increased activity in the hippocampus. These changes in regional functioning, induced overtime through repeated application of TMS, can result in decreased PTSD symptoms.

CONCLUSION

As the field of PTSD research continues to move forward, it is compelling to think about the considerable progress already made and the possible future avenues to pursue. As has been seen throughout this chapter, the depth of understanding has grown by leaps and bounds over just the past decade. When considering how our knowledge of PTSD has advanced, we can see the active advancements in the basic mechanisms of the neurobiology of stress. Researchers studying the foundational aspects of the various brain systems in the pathology of PTSD have shown us new ways to conceive of the disorder and are opening up new avenues for potential treatments. Similarly, we see various biomarkers, such as P11 and FKBP5, which may lead to better diagnostics and disease tracking. Work on PTSD and cardiovascular health may help uncover the root of the long known but poorly understood relationship between mental health disorders and a shortened life expectancy. Tracking and addressing these long-term impacts is a vital necessity to avoid the long-term costs of this disorder.

In regard to future research, there is much work remaining. More specifically, the next stages will include a move to more specific understanding of the brain in PTSD. This move will push our understanding beyond brain regions into the specific actions of cell groups and even specific cells. More will be understood about localized networks and the involvement of specialized pathways in the development of PTSD. The advent of newer technologies, such as magnetoencephalography, and technologies yet to be invented, will advance our understanding in unique and as yet inconceivable ways. This array of advances will all serve to help refine our understanding of how the brain works and provide a precise architecture of the brain's functioning in order to facilitate new and novel way of intervening for those suffering from PTSD.

This very specific knowledge about the interactions of neurons should be augmented by further exploration of the role of sleep and sleep-related deficits, in the development and sustainment of PTSD. These studies, too, should consider a genetic component in order to address the long-standing questions regarding the genetic underpinnings of PTSD and potentially answer at least a part of the age-old question of the relative contributions of nature versus nurture.

Further Reading

[1] Bremner JD, Licinio J, Darnell A, Krystal JH, Owens MJ, Southwick SM. Elevated CSF corticotropin-releasing factor concentrations in posttraumatic stress disorder. Am J Psychiatry 1997;154:624–9.

[2] Da Costa JM. On irritable heart; a clinical study of a form of functional cardiac disorder and its consequences. Am J Med Sci 1871:2–52. CXXI.

[3] Dias BG, Ressler KJ. PACAP and the PAC1 receptor in post-traumatic stress disorder. Neuropsychopharmacology 2013;38:245–6.

[4] Johnson LR, McGuire J, Lazarus R, Palmer AA. Pavlovian fear memory circuits and phenotype models of PTSD. Neuropharmacology 2012;62(2):638–46.

[5] Mehta D, Gonik M, Klengel T, Rex-Haffner M, Menke A, Rubel J. Using polymorphisms in FKBP5 to define biologically distinct subtypes of posttraumatic stress disorder: evidence from endocrine and gene expression studies. Arch Gen Psychiatry 2011;68:901–10.

[6] Prager EM, Johnson LR. Stress at the synapse: signal transduction mechanisms of adrenal steroids at neuronal membranes. Sci Signal 2009. http://dx.doi.org/10.1126/scisignal.286re5. PMID: 19724063.

[7] Vaccarino V, Goldberg J, Rooks C, Shah AJ, Veledar E. Post-traumatic stress disorder and incidence of coronary heart disease: a twin study. J Am Coll Cardiol 2013;62:970–8.

[8] van Zuiden M, Heijnen CJ, Maas M, Amarouchi K, Vermetten E, Geuze E. Glucocorticoid sensitivity of leukocytes predicts PTSD, depressive and fatigue symptoms after military deployment: a prospective study. Psychoneuroendocrinology 2012;37:1822–36.

[9] Zhang L, Zhou R, Xing G, Hough CJ, Li X, Li H. Identification of gene markers based on well validated and subcategorized stressed animals for potential clinical applications in PTSD. Med Hypotheses 2006;66:309–14.

CHAPTER

23

Somatosensation and Pain

S.A. Prescott, S. Ratté

The Hospital for Sick Children and University of Toronto, Toronto, ON, Canada

Our body continuously surveys its external environment and its internal state. This is referred to as *exteroception* and *interoception*, respectively. Interoception is crucial for homeostatic regulation of physiological processes like breathing, but this information is most often used without us being aware of it. Similarly, we are typically unaware of our joint position or movement yet continuous proprioceptive feedback is critical for balance and coordinated locomotion. We are more aware of touch, temperature, and pain. Pain is especially powerful in drawing our attention, although protective reflexes are often initiated before we perceive the painful stimulus. Somatosensation is not unique in this regard; phenomena like blindsight—the ability of cortically blind individuals to respond to visual stimuli without conscious awareness of the stimulus—emphasize how much sensory processing occurs subconsciously even in the visual system.

These issues highlight the distinction between sensing and perceiving. *Sensation* refers to the detection of stimuli of different modalities. *Perception* refers to how our brain interprets that sensory input, which can be shaped by situational context, expectations, and a multitude of other factors. The two processes—sensing and perceiving—are obviously linked, so much so that we take the link for granted. Yet we do not always perceive a stimulus in the way one would expect. This is exemplified by illusions like the thermal grill in which interleaved warm and cool bars are misperceived as burning hot, and by clinical conditions such as ciguatera in which innocuous thermal stimuli are mistakenly perceived as burning hot. Pain is often described in terms of the stimuli that would normally evoke that percept regardless of what the stimulus really is. For instance, burning pain is often described as if *one's skin is on fire*. Perceived as pressure, cardiac ischemia is often likened to *an elephant sitting on one's chest*. Of course, the proper medical responses do not involve fire hoses or elephant removal.

This chapter focuses on somatosensory modalities including proprioception, thermoception, touch, and nociception. Given the clinical importance of chronic pain, later parts of the chapter are dedicated to discussion of how changes in somatosensory processing lead to chronic pain and how understanding the mechanisms underlying those changes can be used to help alleviate such pain. We begin with an introduction of concepts important for sensory physiology.

Conn's Translational Neuroscience
http://dx.doi.org/10.1016/B978-0-12-802381-5.00037-3

BASIC CONCEPTS IN SENSORY PHYSIOLOGY

First-order sensory neurons—also known as primary afferents—are responsible for sensing stimuli on the skin or internal tissues and relaying that information to the central nervous system (CNS). Action potentials, or spikes, are used to convey that information over long distances, more than 1 m in some human axons and much farther in other species. Second-order sensory neurons in the spinal cord and brainstem receive that information via synaptic connections and relay signals to higher order sensory neurons deeper in the CNS. Fig. 23.1 illustrates the two basic pathways. Although Fig. 23.1 depicts strict feedforward pathways, local circuits and long-distance feedback loops play an important role in processing the information and modulating its transmission to the cortex. Central processing is critically important for how a given stimulus will ultimately be perceived but what

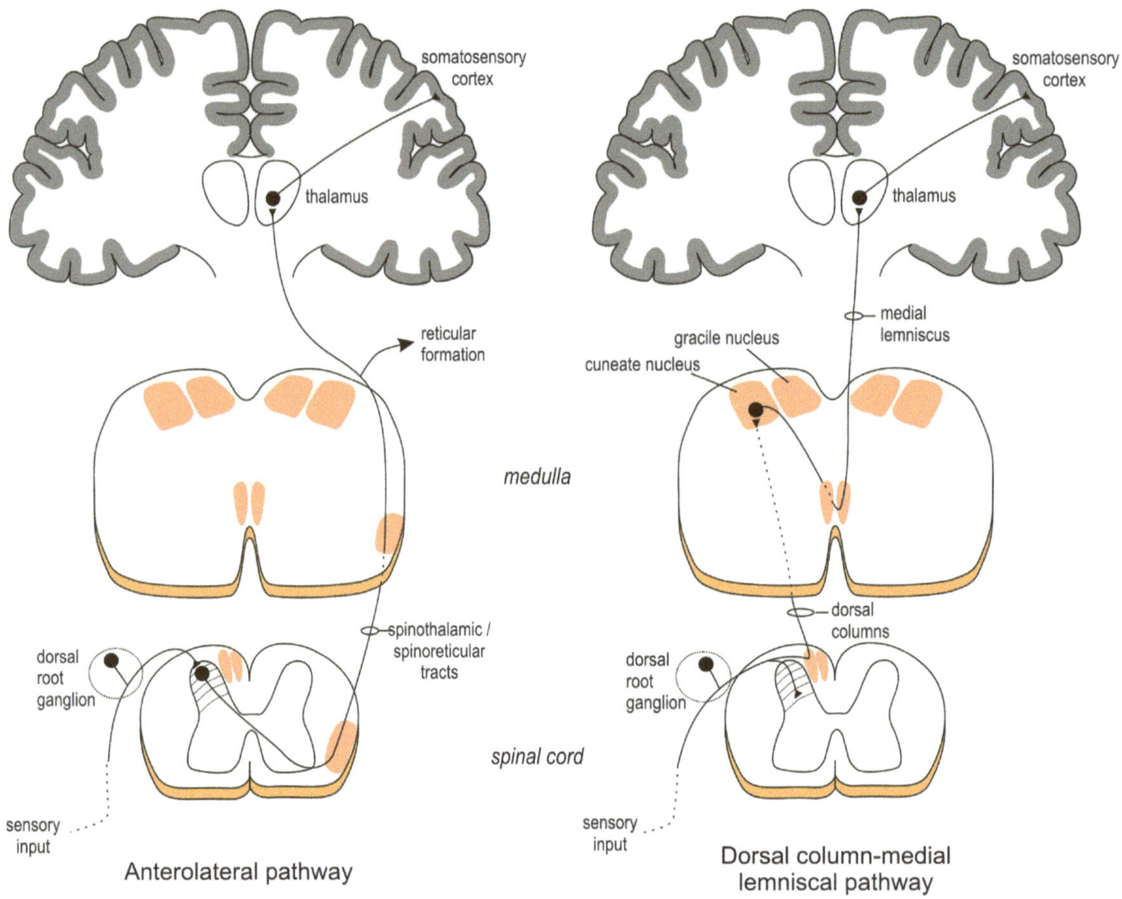

FIGURE 23.1 Somatosensory information from the periphery is routed through the central nervous system via two main pathways: the anterolateral pathway (left) and the dorsal column-medial lemniscal pathway (right). Fibers ascending in the dorsal columns carry information about proprioception and fine touch, while those ascending in the anterolateral tract carry information mostly about pain, temperature, and crude touch.

distinguishes sensory modalities (and submodalities) from one another is the type of physical stimuli to which primary sensory neurons respond. Accordingly, we dedicate a significant portion of this chapter to describing first-order sensory neurons.

Transduction is the process by which a physical stimulus is converted into a receptor potential in the peripheral endings of primary afferent neurons. This graded depolarization, also known as a generator potential, must be converted to a spike train in order for the sensory signal to reach the CNS. This second step depends on the local excitability, which reflects the complement of ion channels present in the axon terminal. The initial transduction step is mediated by the specific receptor molecules present in the fiber terminals and the association of those terminals with specialized end organs, that is, non-neural cells or capsules, such as Pacinian corpuscles (see section: Somatic Submodalities and Receptor Mechanisms). The *adequate stimulus* refers to the type of stimulus to which a fiber is most sensitive. This term has been largely replaced by the concept of tuning. A neuronal *tuning curve* shows the threshold stimulus intensity plotted against the stimulus parameter of interest, as illustrated for two different types of mechanoreceptors tested across a range of vibration frequencies in Fig. 23.2. The *x*-value at the minimum (trough) of the curve represents the stimulus to which the neuron is most sensitive. A narrow or broad tuning curve means that the neuron is more or less selective, respectively, for that stimulus parameter.

The *receptive field* of a neuron describes the portion of sensory space in which stimulation will activate that neuron (Fig. 23.3). In the somatosensory system, this typically refers to an area on the body surface, whereas in the visual system, for example, it refers to an area in visual space. The receptive field of a primary somatosensory afferent is determined by the area innervated by that particular fiber. Downstream neurons in the CNS can also be said to have a receptive field but, in their case, the receptive field reflects the sum of the receptive fields of upstream neurons converging onto the neuron of interest. Receptive fields can also be shaped by other aspects of the microcircuitry, most notably synaptic inhibition (see section: Central Pathways). The size of receptive fields affects tactile acuity: two points spaced closely together will be perceived as a single point if both points fall within the same receptive field. Therefore, two-point discrimination will be less precise where receptive fields are large, like on the torso, and more precise where receptive fields are small, like on the hands or lips. The smaller the receptive fields, the denser the innervation, implying that relatively more sensory neurons are dedicated to parts of the body with high acuity. The amount of somatosensory cortex dedicated to various body parts reflects the acuity of those body parts (see section: Thalamic and Cortical Regions Involved in Somatosensation).

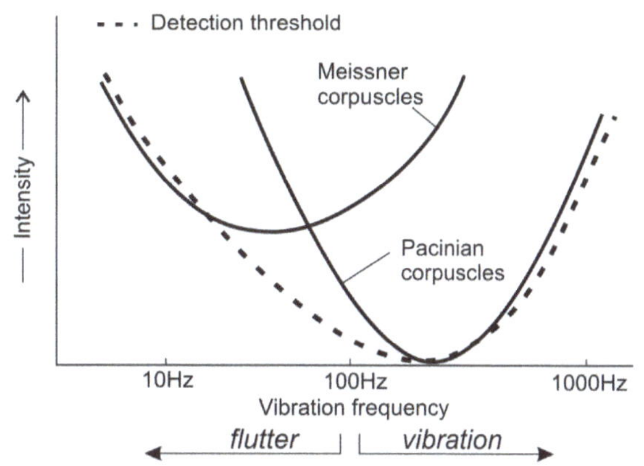

FIGURE 23.2 Tuning curves show the minimum stimulus intensity required to activate a given receptor type (*solid curves*) or to be detected by a person (*dashed curve*) at a given stimulus frequency. Meissner corpuscle-RA1 afferents are tuned to vibrations near 50 Hz, whereas Pacinian corpuscle-RA2 afferents are tuned to vibrations near 250 Hz. Vibrations <100 Hz are encoded by the former and are sensed as flutter. Vibrations >100 Hz are encoded by the latter and are sensed as vibration.

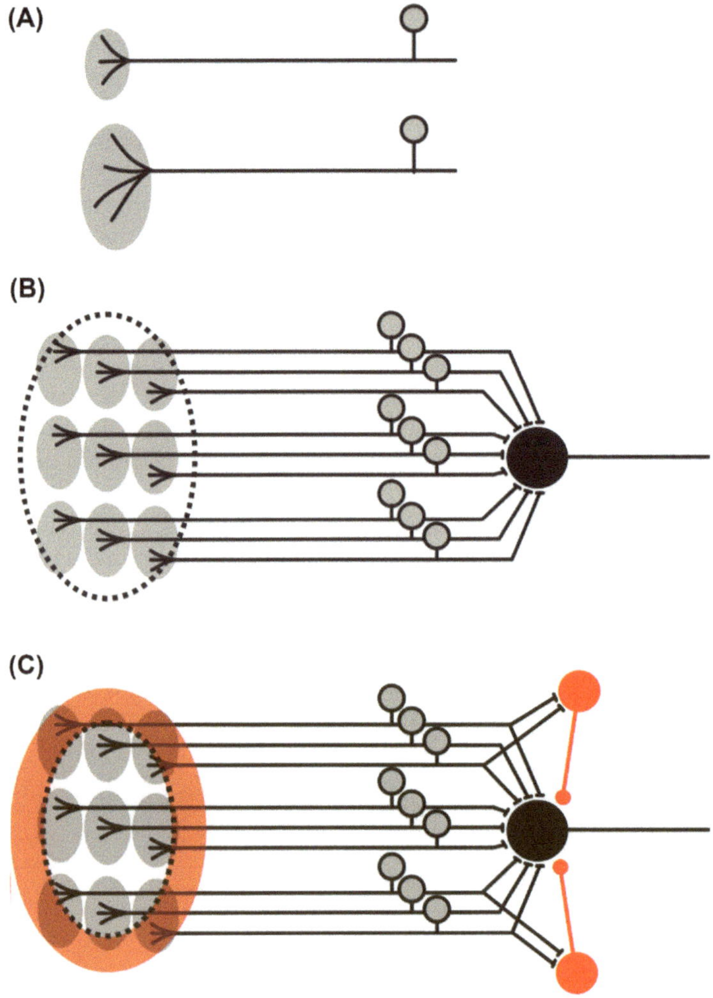

FIGURE 23.3 Receptive fields. (A) The receptive fields of first-order sensory neurons reflect the area innervated by their terminal branches. (B) For a second-order sensory neuron, its receptive field (*black dashed oval*) is the sum of the receptive fields of upstream neurons. Convergence of more first-order neurons onto second-order neurons translates into a larger receptive field. (C) Lateral inhibition evoked by first-order sensory neurons on the edge of the receptive field creates an inhibitory surround (*red shading*) that sharpens the second-order sensory neuron's receptive field.

FIRST-ORDER SOMATOSENSORY NEURONS

Unlike the other senses in which sensors are concentrated in dedicated sensory organs—the eyes, ears, nose, or tongue—primary somatosensory neurons innervate the entire surface of the body plus internal tissues. The cell body (soma) of each primary somatosensory neuron is located in a dorsal root ganglion (DRG) or trigeminal ganglion. Afferents from the trigeminal ganglia innervate the face and other cranial tissues like dura mater, and relay that information to trigeminal nucleus in the brainstem, while afferents from the DRG innervate the rest of the body and relay information into the dorsal horn of the spinal cord. Located close to but officially outside the CNS, both sets of ganglia contain only sensory neurons. These neurons are pseudounipolar, which means that a single fiber exits the cell body and splits at a T-junction, with one end innervating peripheral tissue and the other end connecting with second-order sensory neurons in the CNS (see Fig. 23.1). The peripherally projecting fibers innervate skin, muscle, joints, and internal organs. The types of sensory neurons innervating a given tissue dictate the sensibility of that tissue: sensory neurons innervating the colon respond to stretch but not to temperature change, which is why burning the colon is not painful whereas stretching it is; in contrast, the skin is innervated by diverse sensory neurons, hence the rich sensory experience associated with cutaneous stimuli.

The peripheral terminals of primary afferent fibers are specialized to detect certain stimuli. This "tuning" is a direct reflection of the types of receptors they express and the specialized end organs with which they associate. But different fibers are *not* specialized to produce certain percepts. For example, fibers known as nociceptors are specialized to detect noxious stimuli, and while their activation will typically evoke pain, this is not always the case: in the heat of battle, injury may go completely unnoticed despite nociceptor activation until the stress of the situation recedes; on the other hand, the thermal grill illusion demonstrates burning pain in the absence of any nociceptor-activating stimulus. These two examples illustrate, respectively, that nociceptor activation is neither sufficient nor necessary to produce pain. "Pain" fibers simply do not exist. The neural code for pain and its relation to other submodalities has been the focus of significant debate (see Box 23.1).

Primary afferent fibers are classified according to several interrelated features (Fig. 23.4). Aα and Aβ fibers have the fastest conduction velocities due to their large diameter and thick myelination. These fibers convey information important for proprioception and light touch. Aδ fibers are intermediate in their conduction velocity, consistent with their medium diameter and thin myelination. Different subsets of these fibers are involved in light touch, thermosensation, and nociception. C fibers have the slowest conduction velocity due to their small diameter and lack of myelination. C fibers are involved in nociception, itch, thermosensation, and affective touch (which is distinct from discriminative touch). Most nociceptors are C fibers but not all C fibers are nociceptors, contrary to what is commonly assumed. C fibers significantly outnumber A fibers and are often subdivided based on whether they express peptide neurotransmitters. Fiber size has implications beyond conduction velocity (in part because of the different voltage-gated ion channels expressed in different fiber types). For instance, large fibers have the lowest threshold and are therefore the first to be recruited by electrical stimulation of a nerve. The largest fibers are also the first blocked by ischemia (eg, when a blood pressure cuff is applied) but the last to be blocked by local anesthetics. The capacity to differentially activate or block certain types of afferents has been used to help work out their relative contributions.

Before proceeding, it is important to reiterate that sensory input is transduced at the peripheral end of each primary afferent neuron. Mechanical, thermal, or chemical activation of receptors generates a receptor potential that is converted locally to a spike train, which is then relayed past the cell body and into the CNS. This distribution of responsibilities—spike initiation in the distal end of the axon and spike propagation toward the cell body—is unlike what occurs in most central neurons, where inputs are received on the dendrites and integrated in the cell body, with spikes being generated in the axon initial segment (near the cell body) before being relayed down the axon. Spikes can originate in the cell body, axon, or central terminals of primary somatosensory neurons, but this is atypical. Spikes originating from the "wrong" location are termed *ectopic*.

BOX 23.1

THEORIES OF SOMATOSENSORY CODING

In 1840, Müller outlined the concept of specific nerve energies, according to which each sense organ evokes a specific sensation regardless of how it is activated. Extending this, von Frey proposed in the 1890s that there are four cutaneous modalities—touch, warmth, cold, and pain—each subserved by distinct skin receptors. This was the origin of *specificity theory*, which posits that each sensation arises from the activation of distinct sensory neurons. According to this theory, primary sensory neurons must be specialized to detect (ie, are tuned to) different types of stimuli. Those primary sensory neurons should connect to equally specialized second-order neurons, and so on, forming *labeled lines*. According to the theory, the brain knows which type of sensory neuron got activated based on which brain area receives the resulting signal via its dedicated input line.

Notably, if differently tuned primary afferent neurons converge onto a second-order neuron, the second-order neuron will have broader tuning, which is to say that "labeling" is lost (or at least reconfigured). Although there is no denying that primary afferents are highly specialized, most higher-order neurons ultimately receive input from more than one type of primary afferent. This is not to say that all projections become equivalently tuned, but they are less specialized than the primary afferents. Observations that touch modulates pain, temperature modulates touch, and so on, contradict labeled lines and Specificity Theory, and are the motivation behind pattern theories. The best known pattern theory is the *gate control theory* of Melzack and Wall, which argues that low-threshold inputs normally inhibit high-threshold inputs and that the ratio of the two inputs, reflected in the output of the projection neuron, will dictate whether a stimulus is perceived as painful. Many of the original details of the gate control theory have been disproven (after bitter debate) but the core ideas survive and have been instrumental in driving pain research since mid-1960s.

Primary sensory neurons are highly *specialized*, but this means that they respond preferentially to certain stimuli, not that they respond exclusively (ie, *specifically*) to those stimuli. Realistic stimuli co-activate differently tuned neurons; in fact, meaningful co-activation patterns require heterogeneously tuned neurons (otherwise all neurons would be equivalently activated). This is to say that specialization is necessary for, not mutually exclusive of, most pattern-based codes. According to *combinatorial coding theory*, the richest and most robust coding strategies would exploit afferent co-activation patterns, consistent with the evidence available from other sensory systems, like color vision, which boils down to the relative activation of three differently tuned—red, green, and blue—photoreceptors. This is still very much an active area of research.

SOMATIC SUBMODALITIES AND RECEPTOR MECHANISMS

Proprioception

Proprioception refers to the sense of limb position and movement, where the latter is specifically referred to as kinesthesia. Proprioceptive feedback is critical for proper balance and motor control. Innervated by fast-conducting Aα fibers, muscles have receptors involved in proprioception. *Muscle spindles* comprise a bundle of thin muscle fibers that are enclosed within a capsule. Mechanosensitive afferents wrap around the muscle fibers and are activated by stretching such that their firing rate is proportional to muscle length. Contraction of the thin (intrafusal) muscle

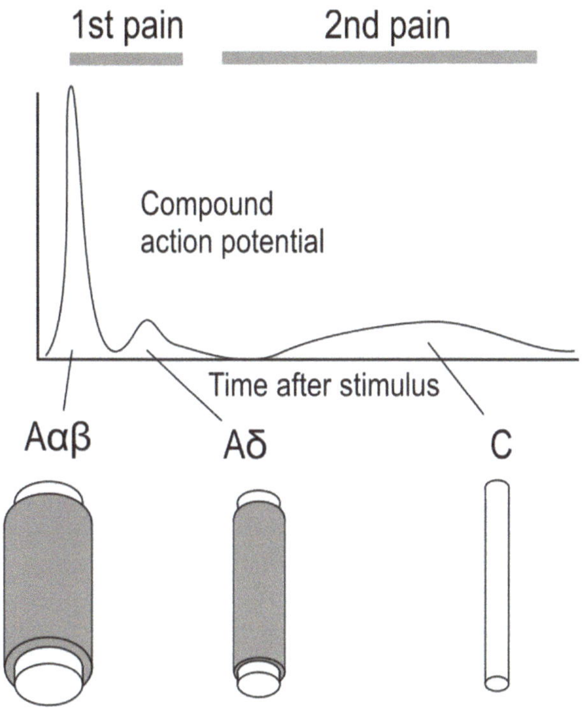

Myelinated	Thinly myelinated	Non-myelinated
Large diameter	Medium diameter	Small diameter
Light touch Proprioception	Light touch Thermosensation Nociception	Thermosensation Nociception Itch Affective touch

FIGURE 23.4 Primary afferents are categorized by their size, myelination, and conduction velocity. Each fiber type is associated with different somatic submodalities. After a noxious stimulus, the immediate (first) and delayed (second) pain are linked with myelinated (A) and unmyelinated (C) fiber activation, respectively.

fibers, which are innervated by motoneurons, can effectively modulate the sensitivity of the muscle spindles. *Golgi tendon organs*, which are located between muscle and tendons, sense muscle force rather than length, and are an important component of reflex circuits. Joints are endowed with *joint-capsule receptors* that sense tension but, interestingly, individuals with artificial joints can still sense the angle and movement of their artificial joint reasonably well on the basis of input from muscle spindles.

Thermoception

Thermoception refers to temperature sensation. The temperature of the air or contacting object is sensed relative to skin temperature, which is typically maintained around 32°C. Rapid temperature changes are more readily sensed than slower changes; in fact, slow changes between 31°C and 36°C tend not to be detected because they are compensated for by dilation/constriction of cutaneous blood vessels. Slowly reducing the temperature will give a sensation of coolness near 31°C, with the sense of cold increasing until pain is felt near 12°C. Moving temperature slowly in the other direction, warmth is first felt near 36°C and the feeling of hot will increase until it is painful near 45°C. These sensations are subserved by a number of *transient receptor potential (TRP) channels* that are located in the membrane of primary afferent terminals and are active across different temperature ranges (Fig. 23.5). Notably, many TRP channels are also activated by specific chemicals, most notably food products like chili peppers and mint that give characteristic burning and cooling sensations, respectively. Importantly, the temperature-sensitivity of TRP channels can be significantly modulated by numerous factors including inflammation.

Hot receptors and high-threshold cold receptors correspond to different sets of C fibers, whereas low-threshold cold receptors (ie, cool receptors) correspond to Aδ fibers. Contrary to simplistic notions of temperature coding, these fiber populations do not operate independently. This is nicely illustrated by the fact that input from C-cold fibers is inhibited by input from Aδ-cool fibers; this "masking" can be revealed by selectively blocking the Aδ-cool fibers (which can be done with a blood pressure cuff), which results in a given temperature feeling colder than it really is, often as burning cold. Anyone who has felt burning pain after washing numb hands in lukewarm water has experienced this unmasking phenomenon. This argues that temperature perception does not depend only on which type of TRP channel is activated at a given temperature. Further to this point, capsaicin (the active ingredient of chilli peppers and an agonist of TRPV1 channels) can cause burning pain or itch depending on whether it is applied to the skin in a diffuse or punctate

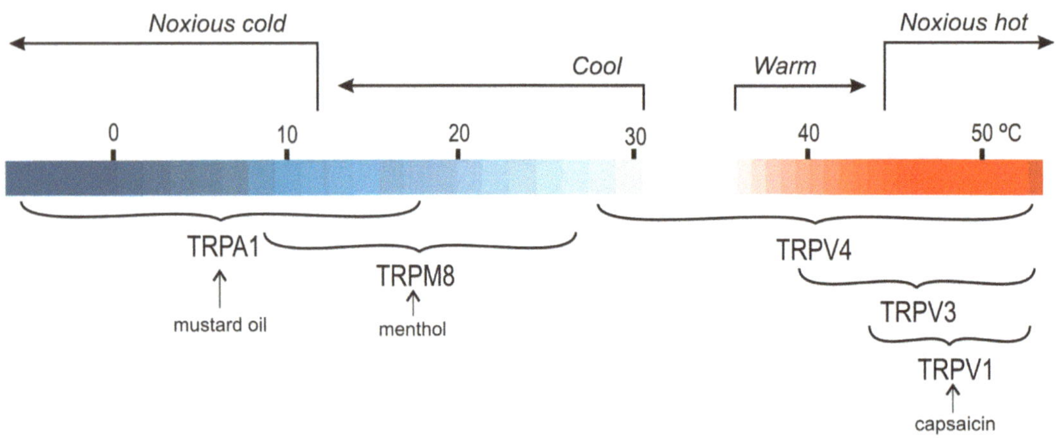

FIGURE 23.5 Thermal stimuli are sensed by transient receptor potential (TRP) channels tuned to different temperature ranges. Many of the channels also respond to chemicals. Temperature ranges associated with the sensation of noxious cold, cool, warm, and noxious hot are indicated.

manner, respectively. Unlike most other primary afferents that are silent unless stimulated, many thermosensitive afferents spike spontaneously and can modulate their firing rate up *or down* in response to temperature changes.

Touch

Our sense of touch relies on low-threshold mechanoreceptors (LTMRs). These LTMRs are subdivided into slow adapting (SA), which spike repetitively during a sustained stimulus, and rapid adapting (RA), which spike only transiently during stimulus onset and offset; each of these classes is subdivided based on subtler features like the regularity of spiking. Fig. 23.6 summarizes the distinguishing physiological and anatomical characteristics of the different fiber types and their associated end organs. Glabrous skin, the non-hairy skin covering the palms and soles, is endowed with a high density of end organs that underlie the discriminative capacity of this sort of skin. Hairy skin has the same end organs with the exception of Meissner corpuscles but the hair follicles themselves are also innervated by sensory fibers.

Rapid adapting type 1 (RA1) afferents are Aβ fibers associated with Meissner corpuscles. *Meissner corpuscles* are specialized end organs formed from Schwann cells arranged as horizontal layers embedded into the connective tissue, perpendicular to the innervating fiber. Indentation of the skin deforms the Meissner corpuscle, causing action potentials in its associated RA1 fiber. RA1 afferents are most sensitive to light touch repeating at frequencies around 50 Hz, which is perceived as flutter (see later). Whereas RA2 afferents innervate a single Pacinian corpuscle, a single RA1 afferent can innervate 30–80 Meissner corpuscles. RA1 afferents are very sensitive to movement across their receptive field and are thus thought to play an important role in sensing texture and detecting slip (which is an important feedback signal when grasping an object).

Rapid adapting type 2 (RA2) afferents are Aβ fibers associated with Pacinian corpuscles. *Pacinian corpuscles* have an onion-like structure, with layered Schwann cells encasing the terminal of the sensory fiber. This layering is parallel to the axis of the fiber, unlike in Meissner corpuscles, and Pacinian corpuscles are also located much deeper in the skin. Pacinian corpuscles are most sensitive (ie, have the lowest threshold) to vibration at 250 Hz. *Vibration* is perceived as a continuous sort of buzzing in response to mechanical stimuli repeating at rates >100 Hz, consistent with the activation of RA2 afferents via Pacinian corpuscles, whereas *flutter* is perceived as discontinuous (albeit rapid) tapping in response to mechanical stimuli repeating at rates <100 Hz, consistent with the activation of RA1 afferents via Meissner corpuscles (see Fig. 23.2). Impressively, RA2 afferents can fire at rates exceeding 500 Hz, thus allowing one spike per stimulus cycle such that stimulus frequency can be encoded by firing rate. That said, downstream neurons cannot fire at such high rates, which implies that the neural code for vibration must change as the signal passes from primary afferents to higher-order, central neurons. The receptive field of RA2 afferents is large due to the deep placement of Pacinian corpuscles within the dermis of the skin, which is consistent with our rather poor ability to localize vibratory stimuli.

Slow adapting type 1 (SA1) afferents are Aβ fibers associated with Merkel cells. *Merkel cells* are located near the basal layer of the epidermis. In glabrous skin, Merkel cells are located on rete ridges (epidermal thickenings that extend downward between the dermal papillae), while in the hairy skin, Merkel cells form touch domes. Merkel cells communicate with SA1 afferents through synapse-like contacts. Both Merkel cells and their innervating fibers are mechanosensitive but Merkel cells are now known to be primarily responsible for transducing the sustained component of the signal (ie, pressure), whereas SA1 afferents respond directly to the dynamic component of the signal. In other

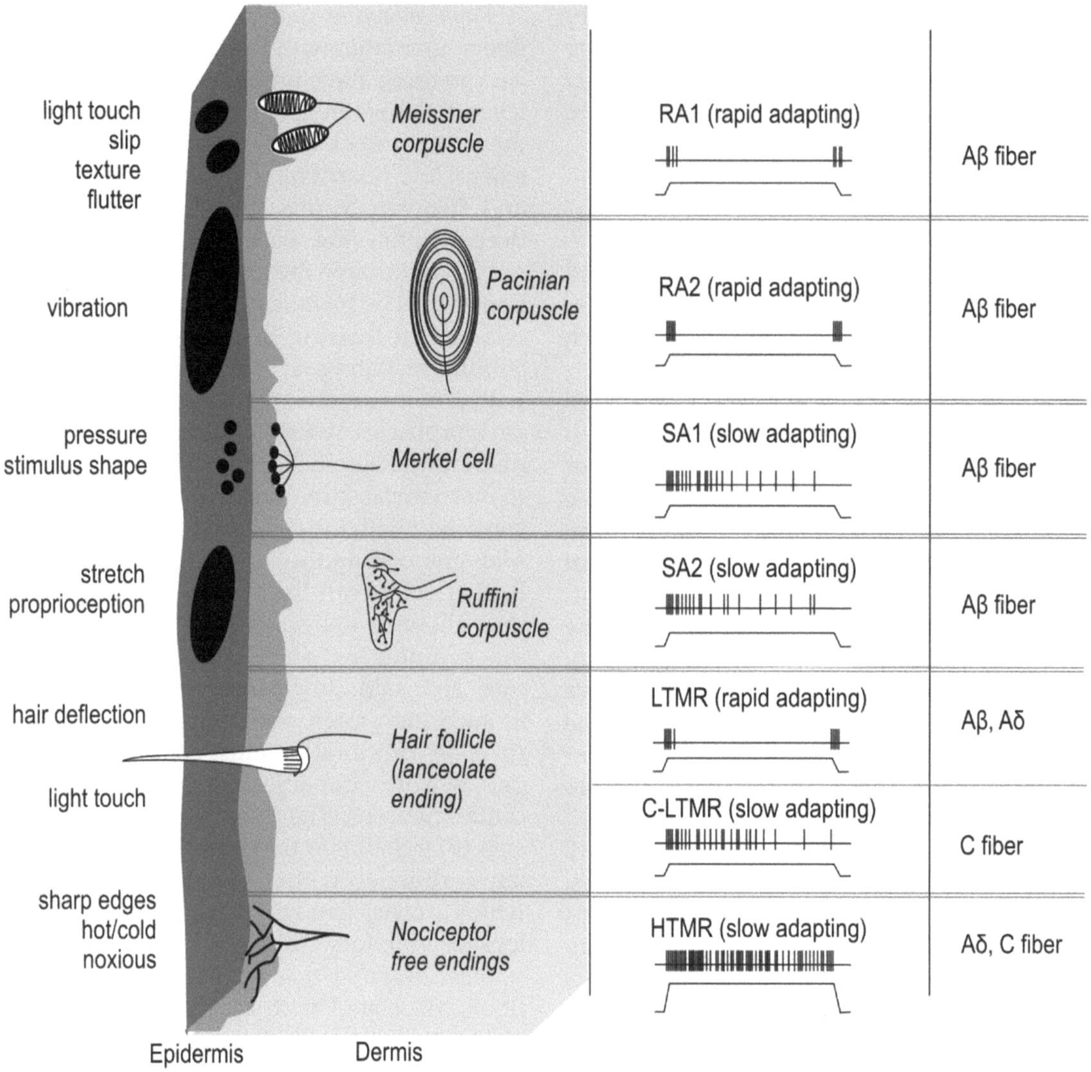

FIGURE 23.6 Skin is innervated by primary sensory fibers capable of detecting different aspects of mechanical stimulation based on their association with distinct end organs. Spiking patterns fall in two broad categories: slow adapting in which spikes occur throughout the stimulus and rapid adapting in which spikes occur at the stimulus onset and offset. Stimulus waveform is shown below each spiking response. *HTMR*, high-threshold mechanoreceptor; *LTMR*, low-threshold mechanoreceptor.

words, chemical transmission from the Merkel cell to the sensory fiber encourages the fiber to spike repetitively during sustained indentation, thus allowing that indentation to be encoded by the firing rate of SA1 afferents. Among the touch receptors, SA1 afferents have the highest spatial acuity and tend to respond strongly to edges and corners, which makes them a prime candidate to encode stimulus shape.

Slow adapting type 2 (SA2) afferents are Aβ fibers associated with Ruffini endings located deep in the skin. *Ruffini endings* are spindle-shaped cylinders that resemble Golgi tendon organs. Compared with SA1 afferents, SA2 afferents are less sensitive to skin indentation but more sensitive to skin stretch, and their receptive field is also several times larger. SA2 afferents are also sensitive to hand shape, which suggests a role in proprioception. Skin stretching normally results from shear force (ie, acting parallel to the skin surface), whereas indentation results from compressive force (ie, acting perpendicular to the skin surface). Encoding shear force is distinct from detecting movement across the skin (ie, slip), which is ascribed to RA1 afferents, but both senses provide important sensory feedback for modulating the grip when handling an object.

This discussion of fiber types and their association with certain sensations must be placed in context. For experimental purposes, stimuli can be designed to preferentially activate a certain type of end organ and the associated sensation can be ascertained. However, under natural conditions, complex stimuli co-activate sets of differently tuned afferents and the resulting sensation depends not on which one type of afferent is activated but, instead, on which combination is activated. Textures—the smoothness of silk or the roughness of wool—are good examples of complex stimuli whose encoding depends of multiple types of afferents. The fact that stimuli co-activate different afferents becomes even more obvious when considering the hairy skin, where each hair follicle is itself a specialized mechanosensory organ innervated by >1 type of afferent.

In mammals, there are three types of hairs: guard (or tylotrich), awl/auchene, and zigzag. The distinction is important insofar as the follicle of each hair type is innervated by a different complement of sensory fibers, which can include Aβ, Aδ, and C fibers. Guard hair follicles are innervated by Aβ fibers; awl/auchene hair follicles are innervated by Aβ, Aδ, and C fibers; and zigzag hairs are innervated by Aδ and C fibers. In each of the aforementioned cases, the fibers form *longitudinal lanceolate endings*, which are fence-like structures involving Schwann cells, that transduce hair deflection into an RA spiking response. In the case of Aδ fibers, their response to hair deflection is direction sensitive because their longitudinal lanceolate endings are not evenly distributed around the follicle. Each type of hair follicle is also innervated by fibers that form circumferential endings around the hair follicle, but the response properties of such fibers remain unknown. It should be emphasized that the C fibers innervating hair follicles are indeed LTMRs and are thought to underlie tickle and affective or emotional touch (as opposed to discriminative touch). C-LTMRs are absent from glabrous skin. In hairy skin, Merkel cells are concentrated in touch domes located near the top of guard hair follicles, and serve to translate guard hair deflection into a sustained (SA1-like) spiking response in the associated Aβ fibers. An important point alluded to at the end of the preceding paragraph is that stimulation of a single hair will necessarily co-activate the multiple fiber types innervating that hair. Moreover, a realistic stimulus that affects different types of hairs will activate an even larger array of fiber types, arguing that one's perception of a somatosensory stimulus depends not on any one fiber type, but, rather, on an orchestra of different fiber types.

The search for the ion channel(s) responsible for transducing mechanical forces into a receptor potential has been a difficult one. Certain channels like TRPA1 have been implicated in mechanotransduction but turned out to be only indirectly involved. There are several mechanically sensitive ion channels that could, in theory, act as the transducer but recent evidence points most convincingly to Piezo channels in the case of mammalian touch. Piezo2 is expressed in mechanosensitive afferents and in Merkel cells, which, as explained, help shape the responses of SA1 afferents. Piezo proteins oligomerize, forming a huge, mechanically activated channel complex estimated to have between 120 and 160 transmembrane segments. Mechanotransduction in other tissues is critical for proprioception, visceral sensations, such as bloating and pain, and blood pressure regulation (by baroreceptors), not to mention transduction in the auditory and vestibular systems. The molecular identity of the mechanotransducer in these other cases remains uncertain.

Nociception

Nociceptive pain is pain originating from the activation of high-threshold afferents known as *nociceptors*. Pathological forms of pain can occur without nociceptor activation, but we will defer discussion of those pathological conditions until later and focus here on high-threshold fibers activated by mechanical, thermal, and/or chemical stimuli. By definition, nociceptors have a high activation threshold, which is to say that they are activated only by intense (ie, noxious) stimuli that risk causing tissue damage. Nociceptors often correspond to C fibers, but can also be Aδ or even Aβ fibers; it should be clear from previous discussion that not all C fibers are nociceptors. The sharp pain felt almost instantaneously upon stubbing your toe is mediated by A fibers, whereas the slightly delayed but longer lasting pain is mediated by C fibers (see Fig. 23.4). Unlike LTMRs, which are associated with specialized end organs, nociceptors have free nerve endings (see Fig. 23.6). Mechano-nociceptors are activated uniquely by strong mechanical stimulation, especially sharp edges or points. Thermo-nociceptors are activated uniquely by noxious hot or cold (see section: Thermoception). Polymodal nociceptors respond to noxious mechanical, thermal, and chemical stimulation. Skin, muscles, joints, and many other tissues are innervated by nociceptors.

Itch

Also known as pruritus, itch is best described as an irritating cutaneous sensation that elicits the desire to scratch. This is distinct from pain, which tends to elicit withdrawal, yet both are protective. Itch originating in the skin is known as pruritoceptive itch and can be evoked by chemical, mechanical, thermal, or electrical stimulation. The sensation of itch relies on C fibers terminating superficially within the skin. Histamine, whether injected intradermally or released from mast cells (as occurs during an allergic reaction), activates a subset of C fibers that express histamine receptors. These particular C fibers are mechanically insensitive but respond to chemicals like capsaicin and mustard oil that normally elicit pain. Interestingly, capsaicin elicits different sensations depending on how it is applied to the skin, with punctate application evoking itch and diffuse application evoking burning pain. This contradicts the simplistic notion that specific C fibers underlie the sensation of each, and argues in favor of a more sophisticated neural coding scheme that remains as yet unclear. Like with pain (see section: Neuropathic Pain), abnormal neural processing can result in neuropathic itch, which occurs in the absence of a pruritogenic stimulus. The responsible changes can occur in the peripheral nervous system (PNS) or CNS. Unresponsive to interventions that would normally soothe pruritoceptive itch, neuropathic itch can be very difficult to treat and, undiminished by scratching, can lead to significant skin lesions and discomfort.

Visceral Sensations

Sensations originating from the viscera (ie, internal organs) differ from the skin, muscle, and joints. For instance, we can identify the location of cutaneous stimuli very precisely, and although we are less precise in localizing muscle or joint stimuli, visceral inputs are especially diffuse and poorly localizable. This results from a relatively small number of sensory neurons (mostly C fibers with free nerve endings) innervating large areas of a given organ, and branching extensively within the spinal dorsal horn to connect with many second-order neurons. Convergence of visceral and somatic inputs onto common second-order neurons can lead to referred pain (see section: Nociceptive Pain). Beyond pain, visceral sensations include dyspnea (shortness of breath), nausea, and bloating/distension, all of which function as warning signals yet are distinct from somatic sensations; the viscera do not give rise to sensations of touch, vibration, hot, or cold. Most mechanosensitive visceral afferents have low activation threshold but can encode (ie, continue to increase their firing rate) into the noxious range. These and other afferents that are normally unresponsive—referred to as "silent" nociceptors—can be sensitized by inflammation and presumably by other factors. This sensitization plays a significant role in dysfunctional pain syndromes, such as irritable bowel syndrome and painful bladder syndrome (interstitial cystitis). Parenchymous viscera like the liver and pancreas to not give rise to pain unless there is inflammation or involvement of the organ capsule. Likewise, the brain is insensate; electrically stimulating sensory areas can evoke sensations felt in the body part represented by stimulated neurons, but this is not equivalent to sensing the electrical stimulus as an electric shock to the brain. On the other hand, stimulation of the dura mater is painful and the sensitization of dural afferents contributes to headache and migraine.

CENTRAL PATHWAYS

As already explained, the cell body of each first-order sensory neuron is located in a DRG or trigeminal ganglion. Those ganglia contain only sensory neurons. One branch of each sensory neuron projects peripherally via a peripheral nerve, often joined by motor and autonomic fibers. The other branch projects centrally via a dorsal root, entering the brainstem (for afferents from the trigeminal ganglia) or spinal cord (for afferents from the DRG). Upon entering the CNS, different fiber types follow different routes. Aα fibers from golgi tendon organs connect directly to motoneurons in the ventral horn of the spinal cord, creating a monosynaptic circuit responsible for stretch reflexes. A variety of fibers enter the spinal dorsal horn, forming synapses on second-order, spinal neurons. Certain fibers ascend ipsilaterally, forming synapses in the dorsal column nuclei located in the medulla. Fig. 23.1 illustrates the major ascending pathways. The schematic in Fig. 23.7 highlights key cortical areas and subcortical structures, the organization and function of which are discussed further.

Anterolateral Pathway

Fibers branch upon entering into the spinal cord, sending collaterals a few spinal segments rostral and caudal to their entry level via Lissauer's tract. The dorsal (or posterior) horn is divided in six layers (laminae) of grey matter, with numbering starting at the dorsal-most edge. Different fiber types terminate in different lamina: fibers carrying noxious and thermal information tend to terminate superficially, in laminae 1 and 2, whereas fibers carrying low-threshold mechanosensory information terminate more deeply. All layers except lamina 2 (also known as the substantia gelatinosa) contain projection neurons that take input from primary sensory neurons, directly and via excitatory interneurons, and relay that information to the thalamus and other supraspinal targets. Those project neurons cross to the other side of the spinal cord and then ascend in the anterolateral section of the spinal cord white matter. Notably, projection neurons constitute a relatively small proportion of all neurons in the dorsal horn, the majority being local (segmental) excitatory or inhibitory interneurons. The variety of interneuron types and how they are synaptically interconnected are still an active area of research. That said, it is important to appreciate

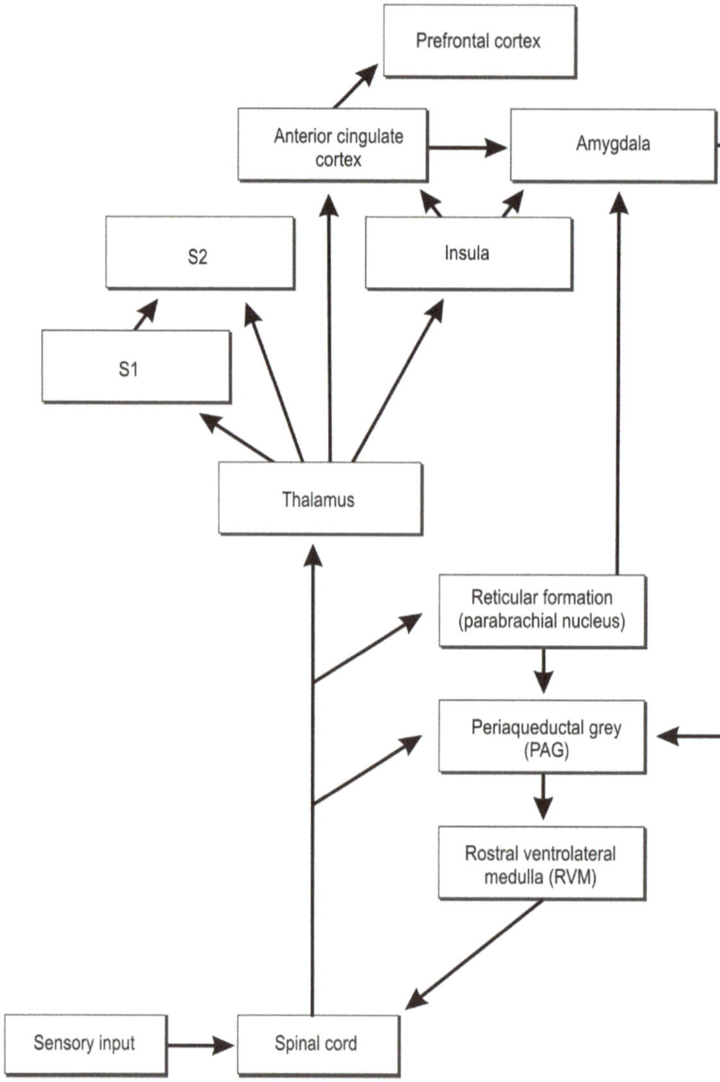

FIGURE 23.7 Schematic illustrating the key subcortical and cortical areas involved in processing ascending somatosensory information and for mediating descending modulation. *S1*, primary somatosensory cortex; *S2*, secondary somatosensory cortex.

the crucial role of synaptic inhibition in modulating the transmission of information through dorsal horn circuits.

Both gamma-aminobutyric acid (GABA) and glycine are used as inhibitory neurotransmitters in the spinal dorsal horn. In addition to being expressed on spinal neurons (see later), $GABA_A$ receptors are present "presynaptically," on the central terminals of primary sensory neurons. Because primary sensory neurons maintain a high intracellular chloride concentration, activation of presynaptic $GABA_A$ receptors causes depolarization; in contrast, $GABA_A$ receptor activation causes hyperpolarization in most central neurons because those neurons maintain a lower intracellular chloride concentration. However, primary afferent depolarization (PAD) is nonetheless inhibitory because it tends to reduce the excitability of the central terminals (by inactivating sodium channels and otherwise impeding spike propagation) so that less transmitter is ultimately released. Activation of the $GABA_A$ and glycine receptors on spinal neurons hyperpolarizes those neurons, or at least it reduces the depolarization caused by excitatory (ie, glutamatergic) input, which is referred to as shunting. Pharmacological blockade of synaptic inhibition in the spinal dorsal horn causes severe pain and hypersensitivity, consistent with the Gate Control Theory of pain (see Box 23.1), and as one might expect, inhibition is pathologically reduced in neuropathic pain conditions. Inhibition should not, however, be thought of as a simply gating the transmission of information through the spinal cord; instead, synaptic inhibition contributes to subtler yet important computations, such as shaping receptive fields by creating inhibitory surrounds (see Fig. 23.3).

Projection neuron axons ascend contralaterally in the anterolateral quadrant of the spinal cord. Some axons project to the thalamus, whereas others terminate in the brainstem. These tracts are named according to their origin and their target. The spinothalamic tract connects to third-order neurons in thalamus that project to the cerebral cortex, where continued processing of the signal ultimately leads to perception. The spinoreticular tract ends in the reticular formation of the medulla and pons, from which information is relayed to thalamus and elsewhere. Located in the reticular formation, the parabrachial nucleus routes information to the periaqueductal gray (PAG), amygdala, and hypothalamus. The spinomesencephalic tract ends near the midbrain tectum and PAG, where descending pathways that modulate function in the spinal cord originate (see section: Descending Modulation).

Dorsal Column–Medial Lemniscal Pathway

Many Aβ fibers ascend ipsilaterally in the dorsal columns, although they still give off collaterals that enter the spinal cord grey matter near the spinal segment where the fiber first enters. The dorsal column on each side includes the gracile fascicle, which is located medially and carries fibers from the lower limb and trunk, and the cuneate fascicle, which is located laterally and carries fibers from the upper limb and neck. The fibers terminate in the gracile and cuneate nuclei, respectively, which together constitute the dorsal column nuclei. Second-order neurons whose cell bodies are in the dorsal column nuclei send axons across the midline and up the medial lemniscus pathway to the thalamus. Like for the first synaptic relay point in the spinal dorsal horn (for the anterolateral pathway), there are interneurons in the dorsal column nuclei that process the incoming information before it is conveyed to the thalamus.

The Aβ fibers ascending in the dorsal columns carry information about proprioception and touch. By comparison, the fibers ascending in the anterolateral tract carry information mostly about pain, temperature, and crude touch. The separation is not complete but it is significant enough to be clinically recognizable in patients who experience hemisection of their spinal cord. The outcome, which is termed Brown–Séquard syndrome, is characterized by loss of fine touch and vibration sensation as well as proprioception ipsilateral to the injury, plus loss of pain, temperature, and crude touch sensation contralaterally. This results from interrupting the ascending pathways before or after the cross-over point (ie, decussation) of the ascending fibers—the dorsal column-medial lemniscal pathway decussates at the level of the medulla, whereas the anterolateral pathway decussates near the spinal levels where the afferent fibers arrive. Notably, pain and temperature sensation is often spared for the two or three segments immediately below (ie, caudal to) the lesion because primary afferents send collaterals this number of segments rostral, thus bypassing damage to the anterolateral tract on the contralateral side. Ipsilateral paralysis also occurs due to interruption of descending motor fibers.

Thalamic and Cortical Regions Involved in Somatosensation

The ascending fibers in the medial lemniscus terminate mostly in the ventral posterior nuclei of the thalamus, with inputs from the face in the medial division (ie, ventral posterior medial) and inputs from the rest of the body arriving in the lateral division (ie, ventral posterior lateral). Ascending fibers in the spinothalamic tract of the anterolateral system terminate mostly in ventral posterior nuclei, intralaminar nuclei, and posterior nuclei. Fibers from ventral posterior nuclei project to primary somatosensory cortex (S1), fibers from the posterior nuclei project to secondary somatosensory cortex (S2), while those from the intralaminar nuclei project elsewhere in cortex and to the basal ganglia. Somatosensory information can arrive in the cortex after crossing as few as two synapses but, as already explained, significant processing occurs prior to the signal arriving in cortex and the information is, evidently, broadcast to multiple cortical (and subcortical) areas. Moreover, despite ascending signals being partitioned between the dorsal column-medial lemniscus and anteroloateral pathways, to some extent, these signals re-converge in thalamus and cortex. The processing that occurs within the cortex is remarkably complex and not yet fully understood. Neocortex is organized into six layers, with thalamic input arriving primarily in layer 4 before being sent up to layer 3 and down to layer 5. In addition to processing the information with local microcircuits, neurons in superficial layers communicate directly with other areas of cortex while neurons in deeper layers send information back to thalamus and other areas outside cortex. Neurons with similar response properties (eg, rapid-adapting vs. slow-adapting) tend to be organized as columns that span across the layers.

Primary somatosensory cortex (S1) is located in the postcentral gyrus. It comprises four areas known as Brodmann's areas 1, 2, 3a, and 3b. Each area is somatotopically organized, meaning that the (contralateral) body surface maps in a spatially organized way to the cortex. This is often represented as a homunculus (little man) reclined across the cortex, with his foot positioned medially and his face more lateral. Notably, the amount of cortex dedicated to a certain body region is not proportional to the surface area of that region but, instead, is proportional to its innervation density and tactile acuity. Hence, the hands and face, which are densely innervated, are allotted more cortical real estate than the torso. This somatotopic organization is hinted at by the sequence of body regions affected during a Jacksonian seizure, with numbness progressing across one half of the body, from the fingers to the toes, as epileptiform activity shifts medially across the contralateral somatosensory cortex. Yet another complete representation of the body surface is present in the secondary somatosensory cortex (S2), which is located just posterior and lateral to S1. It should be noted that other areas of cortex, such as the insula and anterior cingulate are particularly important for pain, especially the affective/motivational aspects of pain. Lesioning connections to the frontal lobes (ie, lobotomy) was very effective in relieving the suffering caused by chronic pain (but had other significant negative effects). To be clear, lobotomized patients could still sense pain but did not perceive it as something unpleasant. The affective component of the pain experience—its unpleasantness—is ultimately what makes pain "pain." These sorts of clinical observations point to the importance of cortical processing for ultimately generating what we think of as true pain.

Descending Modulation

The PAG receives bottom-up signals from the spinal cord and top-down signals from the cortex. The PAG in turn connects to the rostroventral medulla (RVM), which sends fibers down the spinal dorsal horn, thus forming a feedback loop. Releasing serotonin, noradrenaline, and GABA, these descending fibers modulate the activity of spinal neurons, thereby modulating the ascending signals carried by spinal projection neurons. Neurons of the PAG and RVM both express opioid receptors, the activation of which by opiates like morphine produces analgesia by engaging the descending modulatory pathways. Opioid receptors are also present on certain spinal neurons. The endogenous opioids enkephalin and dynorphin are expressed by neurons in the RVM, PAG, and spinal cord. This provides for powerful endogenous modulation that can be engaged during fight-or-flight situations, such as in the heat of battle. This modulatory system is also important for the placebo effect, and for the emotional and attentional modulation of pain based on top-down signals from cortex. The direct engagement of descending modulation by bottom-up signals is illustrated by the phenomenon known as *diffuse noxious inhibitory control (DNIC)*: if a painful test stimulus is applied in the presence of a second painful (conditioning) stimulus applied elsewhere on the body, the test stimulus will feel less painful than if the test stimulus were applied alone. This spatially diffuse form of counter-irritation, whereby pain inhibits pain, involves the descending modulatory pathways described before. Although the activation of opioid receptors (by endogenous or exogenous ligands) causes powerful analgesia under normal conditions, this modulation becomes less effective under neuropathic conditions, for example, DNIC is reduced in patients with neuropathic pain. Opioids tend to be less effective against neuropathic pain and, paradoxically, can exacerbate pain under certain conditions. These changes seriously compromise our ability to treat neuropathic pain.

TYPES OF PAIN

Pain is defined as an unpleasant sensory and emotional experience associated with actual or potential tissue damage, or described in terms of such damage. Pain is a subjective experience that depends on sensory input and how that input is processed by the nervous system—from the transduction step in primary afferent neurons to complex distributed processing within neocortical networks. Pain can serve as a diagnostic tool, providing critical information about underlying diseases or injury, but pain can also be a difficult clinical problem in its own right. We distinguish between three types of pain on the basis of mechanistic differences—nociceptive pain, inflammatory pain, and neuropathic pain. Pain can be described in many other ways, including its duration—acute vs. chronic—and anatomical origin—somatic vs. visceral. Many of these descriptors are relatively self-explanatory but a number of noteworthy concepts will be introduced in the following text.

Nociceptive Pain

Nociceptive pain is the normal response to noxious (intense) stimulation. This type of pain can range from sharp, pricking, or shock-like to dull, aching, or burning depending on the responsible stimulus. Nociceptive pain is predictably initiated by the activation of nociceptors, which as explained in section "Nociception," are primary afferent neurons with a high activation threshold. Sharp pain is due to the activation of myelinated (mostly Aδ) nociceptors, whereas dull, aching pain is due to activation of unmyelinated (C) nociceptors (see Fig. 23.4). Nociceptor activation initiates protective withdrawal reflexes and draws our immediate attention, which helps us learn which stimuli are dangerous and should, in future, be avoided—the motivation to avoid pain is a strong one. In other words, nociceptive pain acts as an important alarm system, alerting us to danger and helping teach us what to avoid in the future. Such pain also warns us about internal problems, from cardiac ischemia to a burst appendix. The importance of this alarm system is illustrated by individuals with congenital insensitivity to pain who, in the absence of pain, sustain repeated injury and fail to seek medical attention, oblivious to infection and other conditions that would normally manifest pain.

Nociceptive pain continues only as long as the noxious stimulus is maintained. That said, sustained or recurrent noxious stimulation can occur in certain disease states like osteoarthritis, where changes in the joint can allow normal weight bearing to produce forces sufficient to activate nociceptors. While nociceptive pain originating from the skin can usually be well localized, enabling you to escape from or remove the noxious stimulus, nociceptive pain originating from deeper structures, including muscle and viscera, tends to be more diffuse and may in fact be very difficult to localize. *Referred pain* is pain attributed to somewhere on the body surface despite the noxious stimulus originating in a deep structure. In cardiac ischemia, for example, the noxious input (ie, ischemia) is detected by fibers innervating the heart but the tightness and crushing pain is felt in the chest and down the left arm. This pattern is no coincidence, as the sensory input from these regions of the body surface and sensory input from the heart both arriving in the lower cervical and upper thoracic levels of the spinal cord (Fig. 23.8). Certain projection neurons receive convergent input from skin and deep structures, but their activation is perceived as if stimulation originated from the skin (which is where most stimulation occurs). In the case of appendicitis, pain is initially felt near the umbilicus because both the appendix and umbilicus are innervated by sensory afferents entering the spinal cord as the T10 level; it is only when the adjacent peritoneum becomes inflamed that pain becomes more clearly localized to the right lower quadrant of the abdomen. It goes without saying that efficiently diagnosing a variety of conditions relies on learning the patterns of referral, which is facilitated by understanding the underlying anatomy.

Inflammatory Pain

Inflammatory pain refers to increased sensitivity due to the inflammatory response associated with tissue damage. Under these sensitized conditions, an innocuous stimulus can be perceived as painful—this is known as *allodynia*—and the pain evoked by a noxious stimulus is exaggerated in both amplitude and duration—this is known as *hyperalgesia*. These two forms of hypersensitivity do not necessarily reflect changes in the same neurons and, moreover, can apply to mechanical or thermal stimulation. After a bad sunburn, for example, clothes brushing lightly against your skin or a lukewarm shower can be painful, while a slap or hot shower can be downright excruciating. Inflammatory pain serves not so much as an alarm, but more as a reminder of recent injury, discouraging activities that risk reinjury so that recovery can proceed quickly.

Primary hyperalgesia refers to hypersensitivity at the site of injury, within the area of inflammation, whereas secondary hyperalgesia refers to hypersensitivity extending outside the area of injury. Primary hyperalgesia is a direct consequence of *peripheral sensitization*, the process whereby inflammatory mediators and other local factors—prostaglandin, bradykinin, cytokines (eg, interleukin), histamine, serotonin, protons, potassium, ATP, glutamate, and so on—trigger intracellular signaling pathways in the peripheral terminals of primary afferents, leading to changes in the function and/or expression of receptor molecules and voltage-gated ion channels, which ultimately causes those afferents to

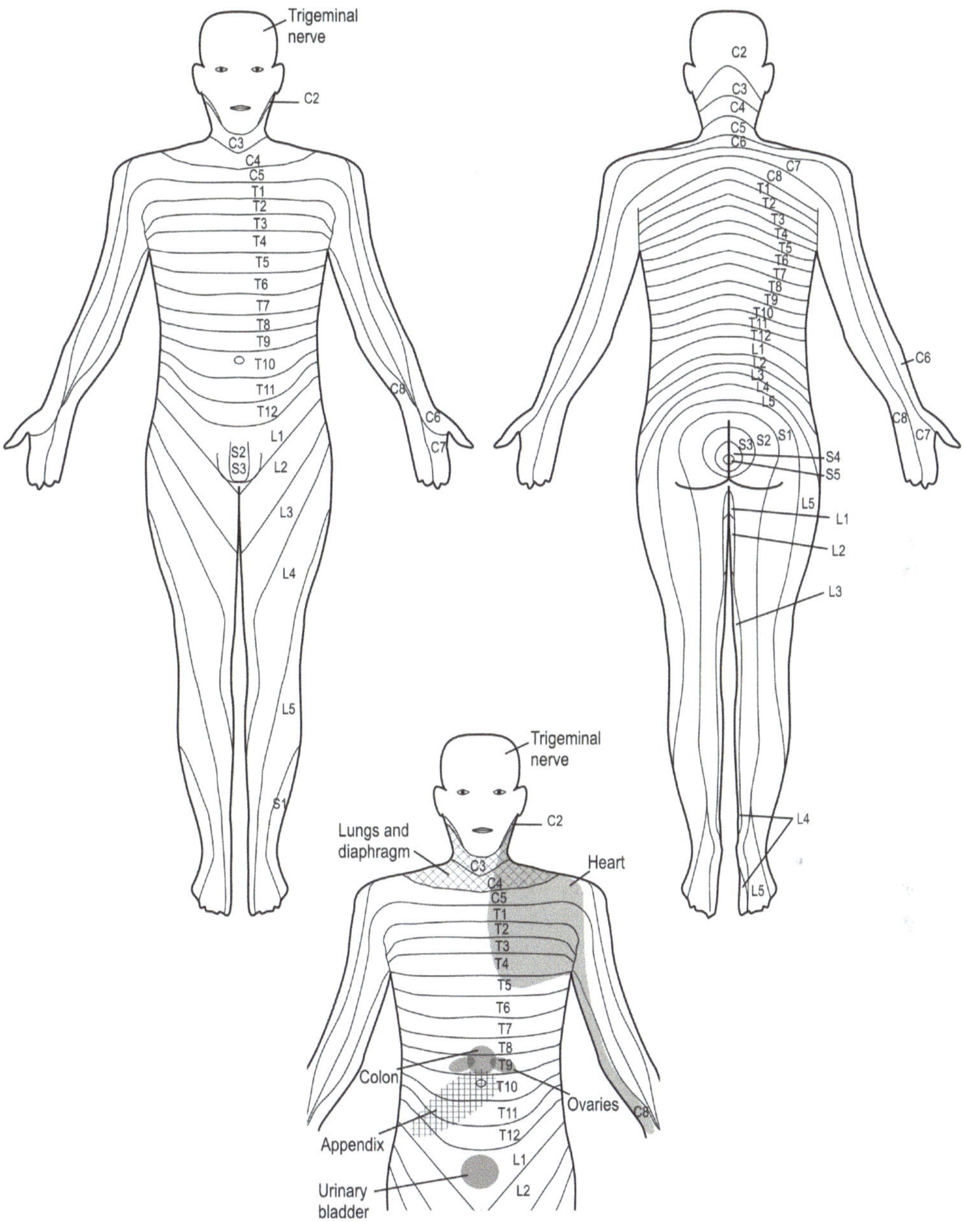

FIGURE 23.8 Each dermatome represents the body surface area innervated by primary afferents originating from a given pair of dorsal root ganglia (DRG) and projecting to the corresponding spinal segment. The face and other cranial tissues are innervated by afferents from the trigeminal ganglia. Primary afferents innervating internal organs often converge onto spinal neurons that also receive input from the body surface. This can result in pain that is referred (sensed in) the dermatomes corresponding to the spinal segments in which convergence occurs. *C*, cervical; *L*, lumbar; *S*, sacral; *T*, thoracic.

become hyperresponsive to stimulation. Notably, peptidergic C fibers can release substance P and calcitonin gene–related peptide (CGRP) from their peripheral terminals, causing *neurogenic inflammation*, thus forming a nasty positive feedback loop. Through a phenomenon known as an *axon reflex*, in which spikes initiated in one fiber branch propagate antidromically down neighboring fiber branches, neurogenic inflammation can extend throughout the receptive field of a peptidergic C fiber. The hypersensitivity can spread even further through the effects of *central sensitization*, which involves synaptic plasticity and other changes in the downstream central circuits. Synaptic plasticity and numerous other changes contribute to making the pain persistent but, generally speaking, inflammatory pain persists only as long as the inflammation. Reducing the inflammation is therefore a logical way to reduce the pain. Nonsteroidal anti-inflammatory drugs (NSAIDs), such as aspirin block the synthesis of prostaglandins and thromboxanes via the enzyme cyclooxygenase.

Neuropathic Pain

Neuropathic pain refers to pain arising from damage to or dysfunction of the nervous system. The pathological changes responsible for this type of pain can occur in the PNS or in the CNS, originating from any one of a number of etiologies including trauma (eg, nerve compression and spinal cord injury), disease (eg, diabetes and multiple sclerosis), infection (eg, HIV, varicella zoster/shingles), and neurotoxic chemicals (eg, chemotherapeutic and antiretroviral agents). The pain occurs spontaneously, either as a persistent burning sensation or as a paroxysm (sudden burst); the intense electric shooting pains associated with trigeminal neuralgia and sciatica are two examples of paroxysmal pain. Neuropathic pain is also associated with hypersensitivity—allodynia and hyperalgesia—to mechanical and/or thermal stimuli. Ironically, the most obvious consequence of nerve damage is loss of function (negative symptoms), not gain of function (positive symptoms). Spontaneous pain, allodynia, and hyperalgesia are all positive symptoms. But in fact, quantitative sensory testing has clearly established that many patients with neuropathic pain also experience negative symptoms evidenced by sensory deficits. In other words, those with neuropathic pain are often less able to detect a light touch or subtle temperature change, but when the stimulus is detected, it is more likely to be mistakenly perceived as painful. Another notable feature of neuropathic pain is that it often takes days, or even weeks or months, to manifest following the injury or exposure responsible for the neural damage, although this long latency is less obvious when the underlying pathology is subtle and more protracted. The timescale suggests that neuropathic pain really represents the maladaptive response of the nervous system to damage, as opposed to the immediate consequence of that damage. The result is a difficult-to-treat mix of varied symptoms. It should also be recognized that many chronic pain conditions can involve more than one type of pain (eg, a neuropathic component and an inflammatory component); each component is liable to respond to different therapeutic interventions.

Dysfunctional pain refers to chronic pain conditions in which there is no noxious stimulus, no inflammation, and no damage to the nervous system; in other words, nociceptive pain, inflammatory pain, and neuropathic pain are all excluded. The caveat here is that ruling out inflammation and neural damage is difficult, especially since neural "damage" might occur in different ways and at many different spots throughout the nervous system. Indeed, dysfunctional pain syndromes, such as irritable bowel syndrome, painful bladder syndrome (interstitial cystitis), and fibromyalgia share certain features with neuropathic pain. Primary erythromelalgia and paroxysmal extreme pain disorder might be considered dysfunctional pain syndromes but, in both these cases, the cause of the pain has been traced back to mutations in NaV1.7 sodium channels, which are expressed in primary sensory neurons. Although there is no damage per se, the channel mutation evidently alters how sensory information is processed and how it is ultimately perceived, which is the same problem as in neuropathic pain. Intriguingly, other mutations in the NaV1.7 channel are responsible for congenital insensitivity to pain.

NEUROPATHIC PAIN: AN UNMET CLINICAL CHALLENGE

For a given injury or disease, only a subset of unfortunate individuals will go on to develop chronic neuropathic pain. A person's genetic background is evidently very important but, with the exception of a few monogenic disorders, such as the channelopathies mentioned before, one's susceptibility to develop chronic neuropathic pain depends on numerous genes and environmental factors that interact in complex ways. Notably, sex is an important factor, with many chronic pain conditions being more prevalent in women than in men. The hope is to develop personalized, precise interventions that can alleviate neuropathic pain while leaving other sensations, including (beneficial) nociceptive pain, intact. We have a long way to go. Many patients, despite receiving the best available treatment, continue to experience unremitting pain and have significantly reduced quality of life because of it. According to a 2011 report from the Institute of Medicine, approximately 100 million adults in the United States suffer from chronic pain (much of which has a neuropathic component). Between the direct costs of treatment and lost productivity, chronic pain is estimated to cost $560–635 billion annually. This is a huge burden on the economy. The personal toll on those with chronic pain is incalculable. The reality is that despite intense research and the identification of many promising drug targets, development of new, more efficacious analgesics has been disappointingly slow.

Neuropathic pain is difficult to treat in part because it involves diverse pathophysiological changes occurring in both the PNS and CNS. In many cases, damage to primary afferents causes the surviving afferents to become hyperexcitable, spiking spontaneously and producing exaggerated responses to stimulation. Interestingly, Aβ fibers have been shown to underlie injury-induced mechanical allodynia. Whereas reducing the activation threshold of nociceptors would be a straightforward way to cause allodynia (by allowing innocuous input to produce a signal normally interpreted as pain), one would predict that enhancing the activation of Aβ fibers should evoke a sense of more intense touch, not burning pain. Furthermore, according to the gate control theory (see Box 23.1), touch is supposed to inhibit pain, not evoke it. This confusing situation points to changes in central processing. It is now firmly established that low-threshold afferents connect via polysynaptic pathways (ie, excitatory interneurons) to projection neurons that are normally activated only by high-threshold input. Those polysynaptic pathways are typically ineffective because of synaptic inhibition but, if that inhibition is compromised, the polysynaptic pathways can be opened and the resulting activation of ascending pathways causes the low-threshold input to be misperceived as painful. Notably, inhibition is reduced in many neuropathic pain conditions; however, inhibition can be reduced through different mechanisms, including reduced transmitter release and altered chloride regulation in the postsynaptic neuron. Correcting different disinhibitory mechanisms requires different therapeutic interventions, but it is not trivial to identify which disinhibitory mechanism is at play and which therapy to apply. Moreover, many other forms of plasticity occur at synapses and within individual neurons. That plasticity is driven (or at least accompanied) by a pronounced neuroimmune response. Reversing these changes has proven extremely difficult.

There is good evidence that abnormal input from hyperexcitable afferents initiates and helps maintain the downstream changes occurring in the spinal cord. Indeed, blocking this afferent input can relieve or significantly attenuate many instances of neuropathic pain. This provides significant motivation to try to therapeutically modulate afferent excitability. Nerve blocks using local anesthetics can provide relief but ideally one would reverse hyperexcitability rather than block all activity. It should be mentioned that the antiepileptic drug Carbamazepine, which acts by stabilizing the inactivated state of voltage-gated sodium channels, is effective against trigeminal neuralgia and is also used for other forms of neuropathic pain. There has been a major push to identify more selective blockers of ion channels, such as NaV1.7, but there is also reason to suspect that the therapeutic effects of very selective drugs will be circumvented by pathophysiological changes that involve up- or downregulation of several different ion channels. Some of the more recently developed drugs that are effective in reducing neuropathic pain—namely gabapentin, pregabalin, and ω-conotoxin—act via calcium channels involved in synaptic transmission.

Other commonly used pharmacological interventions include tricyclic antidepressants, SNRIs (serotonin–norepinephrine reuptake inhibitors), and opioids. As mentioned earlier, opioids tend to be less effective against neuropathic pain than they are against nociceptive or inflammatory pain; moreover, opioids have significant adverse effects and abuse potential that complicate their long-term use. There are also neuromodulation techniques, such as transcutaneous electrical nerve stimulation (TENS) and spinal cord stimulation (SCS). Both approaches involve stimulating primary afferents in order to engage inhibitory mechanisms in the spinal dorsal horn. In SCS, electrodes are placed over the dorsal columns in order to selectively activate myelinated sensory fibers (since those are the only fibers present in the dorsal columns). The spikes are conducted antidromically reaching caudal segments of the spinal cord via collaterals. The SCS-evoked paresthesia (sensation of "pins and needles") is used to help position the stimulating electrodes so that inhibition is engaged in the correct spinal segments, but that paresthesia can itself be unpleasant and the analgesic effects of SCS often decrease over time. New variations of SCS avoid causing paresthesia by applying short electrical pulses at very high frequency, but the mechanism by which this reduces pain remains unclear. Other therapies, from acupuncture to relaxation, are also beneficial. A comprehensive treatment plan will consider all of these options, combining interventions in ways that maximize benefit (ideally through synergistic interactions) while minimizing adverse effects. The problem is that without more effective treatment options, even the best treatment plans tend to fall short of providing adequate pain relief.

Despite the emphasis on pathological changes occurring in the peripheral and spinal cord, one must bear in mind that chronic pain is associated with changes through the somatosensory system. In fact, structural changes in the neocortex of chronic pain sufferers are well documented and are correlated with psychological changes. Descending modulatory systems are also compromised. With so many different changes occurring, it is difficult to disentangle which contribute to the abnormal perception of the pain and which are simply correlated with that abnormal perception.

SUMMARY

Humans rely heavily on their visual and auditory systems to communicate and to interact with the world. Although less appreciated, somatosensation is also critical in many ways. Proprioception and touch are used subconsciously to coordinate movement and to manipulate objects. Our very survival relies on nociceptive pain to alert us to danger. But the perception of somatosensory input is not as simple as receiving a signal generated through activation of a certain receptor; instead, incoming signals are processed at multiple points within the circuit, which includes modulation by top-down signals (reflecting our expectations, mood, and so on) before it gives rise to perception. If that processing goes awry, as in neuropathic pain, we can misperceive the lightest touch as excruciating pain. Our inability to alleviate such pain reveals the gaps in our current understanding of how somatosensory information is normally processed and how that processing becomes pathologically altered. These are intriguing basic scientific questions whose answers will have huge clinical implications for countless people suffering from chronic pain.

Further Reading

[1] Abraira VE, Ginty DD. The sensory neurons of touch. Neuron 2013;79:618–39.
[2] Bushnell MC, Ceko M, Low LA. Cognitive and emotional control of pain and its disruption in chronic pain. Nat Rev Neurosci 2013;14:502–11.
[3] Costigan M, Scholz J, Woolf CJ. Neuropathic pain: a maladaptive response of the nervous system to damage. Annu Rev Neurosci 2009;32:1–32.
[4] Melzack P, Wall PD. Pain mechanisms: a new theory. Science 1965;150:971–9.
[5] Prescott SA, Ma Q, De Koninck Y. Normal and abnormal coding of somatosensory stimuli causing pain. Nat Neurosci 2014;17:183–91.
[6] Relieving pain in America: a blueprint for transforming prevention, care, education, and research. The National Academies Press; 2011.
[7] Saal HP, Bensmaia SJ. Touch is a team effort: interplay of submodalities in cutaneous sensibility. Trends Neurosci 2014;37:689–97.

CHAPTER

23.1

Translational Correlation: Trigeminal Neuralgia

J. Kim

Texas Tech University Health Sciences Center, Lubbock, TX, United States

INTRODUCTION

Trigeminal neuralgia is a severe, paroxysmal, short-lasting, electric shock-like pain in the distribution of one or more branches of the trigeminal nerve. The name *"Tic douloureux"* was given because of the facial spasms that often develop with the attacks.

Trigeminal neuralgia can be divided into classical and symptomatic trigeminal neuralgia as per International Headache Society. In patients with classical trigeminal neuralgia, no other causes of symptoms other than microvascular compression are identified. In those with symptomatic trigeminal neuralgia, the clinical symptoms and signs are quite similar to that of classic trigeminal neuralgia but another underlying etiology or lesion is responsible (Table 23.1.1).

EPIDEMIOLOGY

The annual incidence of trigeminal neuralgia is 4–12.6 per 100,000. It is a common disorder of middle age and later life; it is rare in subjects younger than 18 years. Most cases develop symptoms after age 50 years, and the peak incidence is at about 60–70 years of age. Female is inflicted twice as common as male but there is no predilection for race. It is common in patients with multiple sclerosis. Patients with hypertension and migraines have higher incidence of trigeminal neuralgia than the general population. Patients with migraines with aura were at greater risk of trigeminal neuralgia than those without aura. The association with other cranial nerve neuralgias, notably glossopharyngeal neuralgia, is noticeable. Familial cases of trigeminal neuralgia are rare. There is possible association with chilling or cold temperature.

PATHOPHYSIOLOGY

The pathogenesis of trigeminal neuralgia is not yet clarified but the one of the proposed mechanisms of trigeminal neuralgia is demyelination of the nerve secondary to compression by aberrant vessels leading to ephaptic transmission of impulses. Pathological and radiological studies have demonstrated conflict to the nerve root by such vessels, most commonly the superior cerebellar artery. Alleviation of symptoms

Conn's Translational Neuroscience
http://dx.doi.org/10.1016/B978-0-12-802381-5.00038-5

TABLE 23.1.1 Classical and Symptomatic Trigeminal Neuralgia

Diagnostic Criteria	Classical Trigeminal Neuralgia	Symptomatic Trigeminal Neuralgia
Chronicity	A. Paroxysmal attacks of pain lasting from a fraction of 1 s to 2 min, with or without persistence of aching between paroxysms, affecting one or more divisions of the trigeminal nerve and fulfilling criteria B and C	A. Paroxysmal attacks of pain lasting from a fraction of 1 s to 2 min, with or without persistence of aching between paroxysms, affecting one or more divisions of the trigeminal nerve and fulfilling criteria B and C
Pains	B. Pain has at least one of the following characteristics: **1.** intense, sharp, superficial, or stabbing **2.** precipitated from trigger areas or by trigger factors	B. Pain has at least one of the following characteristics: **1.** intense, sharp, superficial, or stabbing **2.** precipitated from trigger areas or by trigger factors
Stereotypy	C. Attacks are stereotyped in the individual patient	C. Attacks are stereotyped in the individual patient
Neurological deficit	D. There is no clinically evident neurological deficit	D. There may be sensory impairment in the distribution of the appropriate trigeminal division
Causes	E. Not attributed to another disorder except compression of the trigeminal root by tortuous or aberrant vessels	E. A causative lesion, other than vascular compression, has been demonstrated by special investigations and/or posterior fossa exploration

by surgical techniques that separate the offending vessels from the nerve further supports this hypothesis. The demyelination was demonstrated in surgical specimens and demyelinated axons are prone to ectopic impulses. Ephaptic cross talk between fibers mediating light touch and pain fibers could be responsible for the precipitation of painful attacks by light tactile stimulation of facial trigger zones.

SYMPTOMS AND SIGNS

The symptom consists of paroxysms of intense, stabbing pain in the distribution of the second and third divisions (rarely the first division) of the fifth cranial nerve. The pain is nearly always unilateral. The pain seldom lasts more than a few seconds. The paroxysms recur frequently, hundreds of times each day, and the attacks rarely occur during sleep. Between episodes, patients are free of symptoms but some may complain of a more or less continuous discomfort, itching, and sensitivity to the face.

Trigger zones are pathognomonic for trigeminal neuralgia but not all patients have trigger zones. Stimulation of certain areas of the face, lips, or gums such as in shaving, brushing teeth or exposure to breeze or wind, or movement of these parts in chewing, talking, or yawning cause paroxysms of pains. Patients can pinpoint these areas and will constantly avoid any stimulation of them. Sensory or motor loss in the distribution of the fifth nerve is not uncommon.

PHYSICAL EXAMINATION

The main goal of physical examination is to identify abnormalities that suggest other diagnoses. A careful examination of the head and neck focused on the mouth, teeth, ears, and temporomandibular joint, with an emphasis on the neurological examination of the trigeminal nerve including corneal reflex, should be performed. The majority of patients with classical trigeminal neuralgia have a normal neurological examination but sensory abnormalities of trigeminal nerve area are not uncommon. However, symptomatic trigeminal neuralgia or other underlying causes should be investigated if the examination shows sensory abnormalities associated with abnormal corneal reflex or other cranial nerve abnormalities.

DIAGNOSIS

The diagnosis of trigeminal neuralgia should be considered for all patients with unilateral facial pain. It must be distinguished from other forms of facial and cephalic neuralgia and pain arising from diseases of jaw, teeth, or sinuses. Symptomatic trigeminal neuralgia is always secondary to another disorder: multiple sclerosis (may be bilateral), aneurysm of the basilar artery, or tumor (acoustic neuroma, trigeminal neuroma, or meningioma) in the cerebellopontine angle. Age younger than 40 years, the presence of trigeminal sensory deficits, or bilateral involvement of the trigeminal nerves may increase the risk of symptomatic trigeminal neuralgia.

DIAGNOSTIC TESTS

Laboratory studies generally are not helpful in patients with typical symptoms of trigeminal neuralgia. The American Academy of Neurology and the European Federation of Neurological Societies (AAN-EFSN) concluded that routine head imaging of MRI identifies structural causes in up to 15% of patients and may be considered useful in patients with trigeminal neuralgia. Trigeminal reflex testing including blink reflex could distinguish classical from symptomatic trigeminal neuralgia with a sensitivity of 96% and a specificity of 93%.

TREATMENT

The initial treatment of choice for classical trigeminal neuralgia is pharmacological therapy. Patients who have minimal response to medical therapy, who relapse with medical therapy, or who cannot tolerate medications should be considered for surgical treatment. Referral for a surgical consultation seems reasonable in patients with trigeminal neuralgia refractory to medical therapy for early surgical intervention (Fig. 23.1.1).

The two drugs to consider as the first-line therapy are carbamazepine (200–1200 mg/day) and oxcarbazepine (600–1800 mg/day). Baclofen and lamotrigine are possibly effective for controlling pain in patients with classical trigeminal neuralgia. Clonazepam, gabapentin, phenytoin, tizanidine, topical capsaicin, and valproate are used but their efficacy is not established. Some evidence supports add-on therapy with lamotrigine or a switch to baclofen. The effect of other drugs commonly used in neuropathic pain is unknown. There are no published studies directly comparing polytherapy with monotherapy.

Surgical interventions are targeted to portions of the trigeminal nerve distal to the Gasserian ganglion by peripheral techniques (cryotherapy, neurectomies, alcohol injection, phenol injection, peripheral acupuncture, radiofrequency, and thermocoagulation), targeting the Gasserian ganglion by percutaneous Gasserian ganglion techniques (rhizotomies), the trigeminal nerve root by gamma knife radiosurgery, and vessels by posterior fossa vascular decompression techniques. Percutaneous procedures on the Gasserian ganglion, gamma knife, and microvascular decompression are possibly effective in the treatment of trigeminal neuralgia. The use

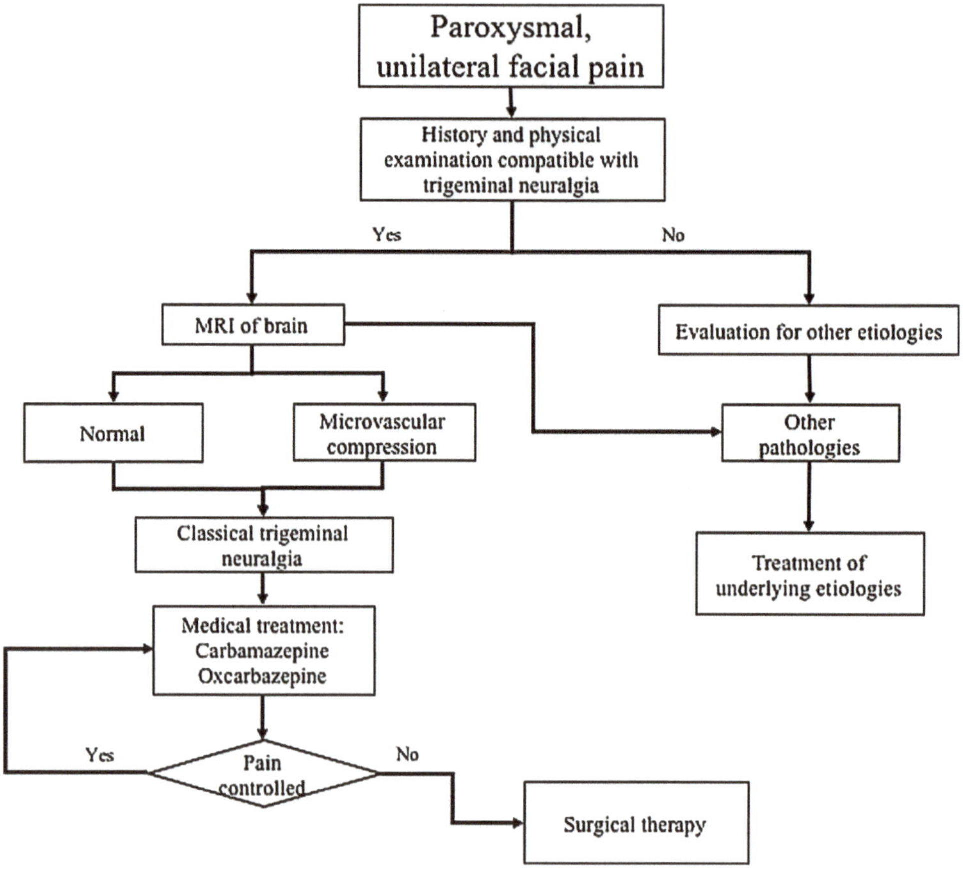

FIGURE 23.1.1 Management algorithm for trigeminal neuralgia.

of microvascular decompression has increased recently. Patients undergoing microvascular decompression have a longer duration of pain control than patients undergoing other surgical interventions.

Further Reading

[1] Cruccu G, et al. AAN-EFNS guidelines on trigeminal neuralgia management. Eur J Neurol 2008;15:1013–28.

[2] Headache Classification Subcommittee of the International Headache Society. The international classification of headache disorders: 2nd edition. Cephalalgia 2004;24(Suppl. 1):9–160.

[3] Maarbjerg S, Gozalov A, Olesen J, Bendtsen L. Trigeminal neuralgia – a prospective systematic study of clinical characteristics in 158 patients. Headache 2014;54:1574–82.

[4] Reddy GD, Viswanathan A. Trigeminal and glossopharyngeal neuralgia. Neurol Clin 2014;32:539–52.

CHAPTER

24

Neurogenesis as a Factor in the Functional Plasticity of the Nervous System

A. Alfonso

University of Illinois at Chicago, Chicago, IL, United States

INTRODUCTION

The birth/generation, differentiation, and integration of neurons and glia into circuits are essential for the establishment of a functional nervous system. In organisms with a nervous system, whether invertebrates or vertebrates, neurons and glia are generated in embryonic development from ectodermal derivatives and as a result of differential but coordinated expression of transcription factors in temporally and spatially regulated domains. External cues are also known to have a significant influence on the fate of the embryonic neural stem cell or neuron progenitor cell descendants. Both intrinsic and -extrinsic factors continue to exert a role on fate-choice as development proceeds; their relative contribution is an active area of research but will not be discussed further as it is beyond the scope of this chapter.

Embryonic neurogenesis results in the formation of circuits that mediate behavior and can be modified as the organism encounters environmental stimuli, a process generally attributed to the plasticity of the system. Neural plasticity has been defined as either morphological change in brain areas as a result of changes in dendrite density in response to a stimulus, or alterations in neuronal networks via changes in connectivity and birth of neurons. In the field of adult neurogenesis, it has been suggested that newly generated neurons in the adult brain can serve as a mechanism to increase the plasticity of the circuit level and complement plasticity at the molecular and cellular levels. In practice, therefore, in those instances where there is post-embryonic, postnatal, or adult neurogenesis, the newly formed neurons have to integrate into existing circuits to function appropriately and without detrimental effects. This chapter gives an overview of the process by which the nervous system is formed and reviews some of the evidence for neurogenesis in adult (meaning mature organisms) rodents and humans in the context of the olfactory system. It also highlights work on how and whether adult neurogenesis influence the plasticity of the behaviors mediated by the olfactory system of mammals including humans.

Research in the field of adult neurogenesis in vertebrates is voluminous (vast). A comprehensive survey of the field requires more depth and space than what is allocated in the context of this

Conn's Translational Neuroscience
http://dx.doi.org/10.1016/B978-0-12-802381-5.00039-7

document. Thus, in most instances original reports are not cited to substantiate a claim, my apologies to the many researchers whose names are not mentioned. However, the reader is encouraged to look at the citations and the references therein as well as a set of recent, excellent, in-depth reviews on adult mammalian neurogenesis in the hippocampus, the birth and integration of adult born neurons into the dentate gyrus (DG) circuitry, and their role(s) in memory and pattern separation to learn more about this important area of inquiry.

EMBRYONIC NEUROGENESIS, HOW DOES IT ALL START?

Neurogenesis and gliogenesis are multistep processes and consist of the proliferation of multipotent neuroectodermal precursors (a.k.a. neural stem cells), generation and commitment of progenitors (neural vs. glial), and acquisition of unique and distinct cellular phenotypes displaying cell type-specific features (terminal differentiation). The neurons and glia present in the adult organism are primarily generated during embryonic development via distinct and well-orchestrated events. In vertebrates, neurogenesis precedes gliogenesis during embryonic development but postembryonic or postnatal generation of neurons and glia is common in most organisms studied. That said in most organisms the number of multipotent neural stem cells and neural precursors decreases with age and is very limited in the adult stage. In the case of vertebrates, the neural stem cells assume many different identities and become restricted in their differentiation potential as development proceeds or their local environment changes. They include neuroepithelial cells, radial glial cells, intermediate rapidly amplifying progenitor cells, and short neural precursors (Fig. 24.1).

Progenitor cell descendants exit the cell cycle, and commit to producing different types of neurons and glia. For example, neuronal lineages give rise to excitatory or inhibitory motor neurons and interneurons, as well as neurons with distinct sensory identities. Glial lineages, on the other hand, give rise to astrocytes, oligodendrocytes, microglia, or Schwan cell identities depending on whether we are considering progenitor descendants in the central or peripheral nervous system of vertebrates or invertebrates. Once the neural descendants undergo terminal differentiation—dendrite growth, axonal pathfinding, synapse formation and maturation, and cell survival—they integrate into circuits, and this holds true for invertebrates and vertebrates.

ADULT NEUROGENESIS, WHERE DOES IT TAKE PLACE?

Seminal evidence that neurogenesis takes place in adult vertebrates was collected during the 1960s by Altman and colleagues. Their data demonstrated proliferation of cells in the DG of the hippocampus and the olfactory bulb (OB) of adult rats. Almost five decades later, there is no doubt that adult neurogenesis takes place in rodents and other mammals including humans. In rodents, adult neurogenesis occurs and is robust in the DG of the hippocampus and the OB, whereas in adult humans there is detectable and substantial neurogenesis in the DG but no evidence for neurogenesis in the OB. It is also generally accepted that the precursors for the adult neurons that are generated in the DG of the hippocampus come from the subgranular zone of the hippocampus (in close proximity) and those generated in the OB are located in the subventricular zone (SVZ) of the lateral ventricles of the forebrain (far away).

Further characterization of the location and developmental potential of the SVZ precursors in postnatal and adult rodents has revealed that they are primarily located along the walls of the lateral ventricles but are also found in other positions of the ventricle walls and the rostral migratory stream (RMS). Moreover, they are not homogeneous in the types of neurons they can

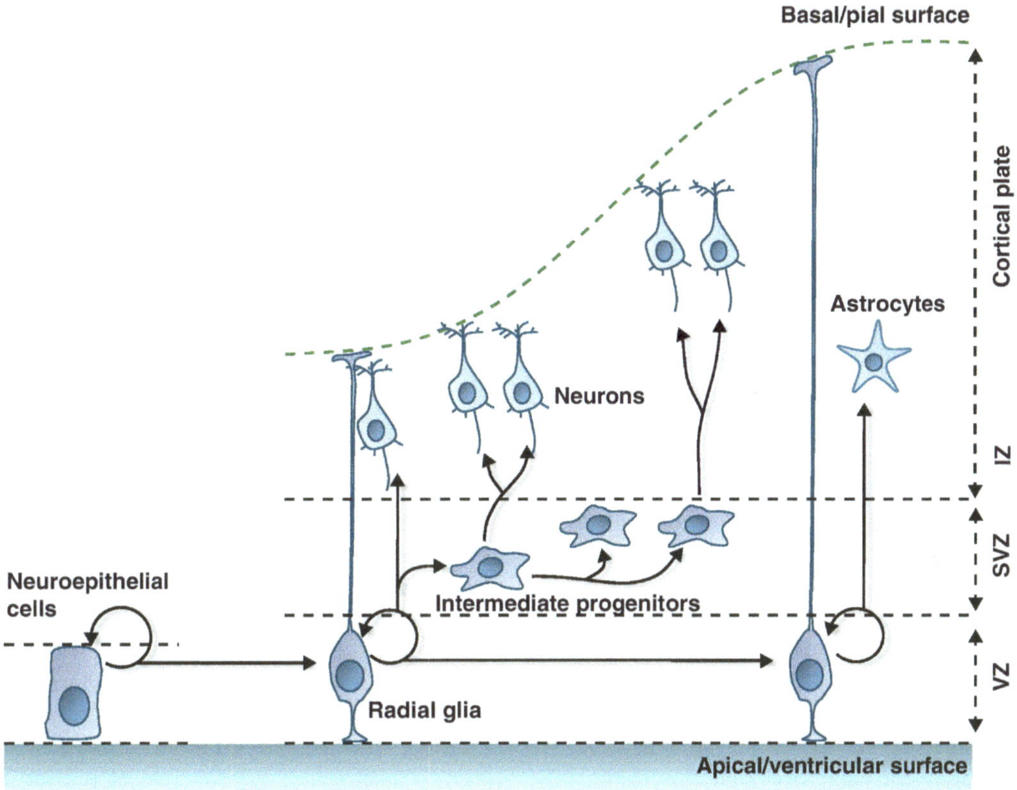

FIGURE 24.1 **Cortical neurogenesis during mouse embryonic development.** Several types of cortical progenitors and their modes of division toward neurons and glial cells: neuroepithelial cells, radial glial cells, intermediate progenitors, with their specific location in ventricular (VZ) and subventricular (SVZ) and intermediate (IZ) zones. *From Tiberi L, Vanderhaeghen P, van den Ameele J. Cortical neurogenesis and morphogens: diversity of cues, sources and functions. Curr Opin Cell Biol April 2012; 24(2):269–76.*

produce. The physical environment in which these precursors are located and where postnatal and adult neurogenesis occurs is called the neurogenic niche. These places (environments) contain and support the neural progenitor cells and their descendants, but also include blood vessels and astroglial cells that promote their survival, proliferation, and maturation.

Back in 1992, when Reynolds and Weiss showed that they could isolate cells from the "periventricular area of the mammalian forebrain" and that these cells had both self-renewing and multipotent capabilities, features of a stem cell population, it was not widely accepted that the mammalian brain could generate new neurons in adulthood. In their work they isolated cells from the adult mouse striatum and cultured them in vitro in the presence of epidermal growth factor (EGF). The clumps of cells they obtained through proliferation, called neurospheres, contained cells displaying differentiation features of both neurons and astrocytes. Since then many reports have been published that corroborate the presence of self-renewing neural precursors in adult rodents as well as humans. There is also evidence that adult-born neurons have unique features that distinguish them from embryonically generated cells including increased excitability and plasticity, as well as differences in their response to regulatory signals. The finding that, in contrast to rodents, there is no detectable adult neurogenesis in the OB of humans, and that instead neurogenesis takes place in the adult human striatum is surprising and tantalizing but awaits confirmation. In fact, the assertion that there is neurogenesis in the human striatum is controversial and has been challenged as a recent study using differential gene expression and bromodeoxyuridine (BrdU) incorporation finds no evidence it happens in adult humans or rhesus monkey brains. Regardless of whether this assertion holds true, given the abundant evidence for continued neurogenesis throughout life in the OB of rodents, and the apparent absence of neurogenesis in the adult human OB, an important question to address is whether there is plasticity in the function of the human olfactory system as it is strongly suggested there is in the rodent and even invertebrates like the nematode *Caenorhabditis elegans*. First, let us first briefly review the organization of the olfactory system in mammals.

STRUCTURAL COMPONENTS OF THE OLFACTORY SYSTEM IN MAMMALS

The main olfactory system is composed of three stations, the main olfactory epithelium (OE), the main OB, and the olfactory-processing centers in the cortex (OC). It is a sensory system that displays plasticity at multiple levels, including molecular, cellular, and circuit (structural) levels. Detection of chemicals involved in innate behavioral responses are thought to be mediated by the accessory OE (vomeronasal organ) and accessory OB in reptiles and mammals but it is thought these structures are nonfunctional in humans so we will not discuss them further.

The main olfactory epithelium (MOE) is the first station in the relay of olfactory information toward higher areas of decoding in the central nervous system. The MOE is a component of the peripheral nervous system and its neurons have a direct connection with the central nervous system. The MOE is populated by ciliated neurons, glia, support cells, glandular cells, and stem cells. The ciliated neurons are olfactory sensory neurons (OSNs). In rodents, these cells are regenerated throughout life and there are two classes of stem cells that can produce them; each with a unique morphology and location [called horizontal basal cells (HBCs) and positioned below the globose basal cells (GBCs)]. In humans there is no direct evidence that OSNs are continually replaced in the organism under normal physiological conditions. However, studies of postmortem and living tissue biopsies show that the human OE also has stem cells. The presumed human OE stem cells, however, have a homogeneous appearance and are round (globose) in shape. The inference that adult-born OSN are also generated in humans is likely to result from the fact that the immature OSN axon marker GAP43 can be detected in the expected location of OB biopsies from individuals 39 to 89 years old. Unfortunately, with the current experimental approaches, it will be hard to confirm this inference, given the limitations of working with human subjects.

The OSNs detect water-soluble and airborne odorants via 7-pass G-protein-coupled olfactory receptors (ORs) and trace-amine-associated receptors (TAARs), respectively. Mice and rats have over 1000 ORs, whereas humans have about 400 functional receptors. Each OSN expresses a single receptor gene in the organism's repertoire and the one expressed appears to be a stochastic decision that is maintained throughout the life of the cell via epigenetic mechanisms. In mice it is thought that ORs have a role in the guidance of the OSN axons toward the cells they synapse with at the next station in the relay, the mitral cells of the OB glomeruli. Rodent ORs have also been implicated in the process of convergence of like-minded OR-expressing ORNs to a single glomerulus. A recent publication from the laboratory of Charles Greer discusses this literature and provides support for an updated model where the authors propose that pathfinding and entering of the ORN axon into the outer neuron layer of the OB is OR independent, but coalescence of ORN axons expressing similar receptors into glomeruli is OR dependent. Undoubtedly, further work in this area will increase our understanding of the role of these receptors in axon guidance and synapse formation.

The main OB is the second and intermediate station in the relay of olfactory information and the processing of odors and odorants. Although there is an OB on each side of the midline in mammals, I refer to the OB as singular throughout this document for the sake of simplicity. The OB, a structure in the central nervous system, is strategically located between the input structure, the OE, and the output centers, the piriform (PIR) cortex and other higher centers of olfactory information decoding. The position and size of the OB in rodents, when compared to humans, is thought to be a strong indicator of the importance of olfactory-based behaviors in rodents but not humans (as discussed later, there are others that argue olfaction is very important sensory system in humans too). In rodents the OB comprises a prominent protuberance at the anterior ventral surface of the brain, whereas in humans it is much smaller, and located inside the brain on the ventral side of the frontal lobes. Per brain weight, the OB is about 300 times bigger in rodents (20%) than in humans (0.064%).

The differences in the anatomical organization of the OB between rodents and humans extend further and are very well described in a paper by Maresh, Greer, and colleagues. As mentioned earlier, the axons of rodent OSNs expressing identical ORs coalesce onto one glomerulus. Moreover, the rodent OB has a stereotypic and reproducible laminar organization that results in the segregation of cell types and as a result easy-to-identify structures. Furthermore, rodent glomeruli are even spaced, spherical, and display a lateral and medial symmetry within each OB. Given the organization of the rodent glomeruli, and the coalescence of OSN axons expressing the same OR on the ipsilateral side to a glomerulus, each rodent OR is represented by two well-defined glomeruli in each OB. In contrast, the human OB is less structured and there is more variability among individual humans. The analysis by Maresh and colleagues revealed that although most human glomeruli were even spaced and spherical, some were clustered and most importantly their volumes varied widely. Their most unexpected observation, however, was that the number of glomeruli ranged from about 3000 to 9000 with a mean around 5600 glomeruli in each human OB. The observation that humans have about five times more glomeruli than mice, but only about a third as many OR proteins is surprising and suggests that the representation of each OR is mapped onto more than 10 glomeruli. The meaning of this observation in terms of how olfactory information decoding is accomplished in humans is not fully understood but ongoing research in this area should provide insights about its significance.

Even though the OB is smaller and there are fewer ORs in humans, it has been argued that human olfaction is nevertheless a very important sensory modality because humans can discriminate single molecules in highly similar odors (pattern separation) and there is evidence of strong interactions among olfaction, memory, and emotional or behavioral responses, among other factors. Lastly, since anosmia is common in individuals with Alzheimer's or Parkinson's before they present full-fledged neurodegenerative diseases, one could consider the cells in the OE and the OB as serving the role of the "canary in the mine," an indicator of their susceptibility to environmental stressors. Moreover, it supports the argument that there is some adult neurogenesis in the olfactory system in humans but this process is impaired at the inception of neurodegenerative diseases such as Alzheimer's and Parkinson's.

Regardless of its size the mammalian OB is composed of glutamatergic, excitatory, projection neurons (mitral and tufted cells) and two classes of interneurons (excitatory and inhibitory) that arise sequentially in time. The excitatory interneurons (juxtaglomerular cells) arise first during embryogenesis. The inhibitory interneurons (periglomerular cells and granule cells) arise later in development. Moreover, granule cells are continually generated throughout life and account for 90% of the adult-born neurons. Periglomerular cells use GABA and dopamine as a neurotransmitter, whereas the granule cells only use GABA. Interestingly, and in contrast to other regions in the central nervous system, the OB has an overwhelming preponderance of inhibitory to excitatory synapses and this functional organization has been suggested to underlie the ability of the system to refine odor representations. The chapter written by Gheusi and Lledo (2014) is an outstanding source of primary information devoted to explaining the synaptic organization and synaptic mechanisms that underlie the function of the OB.

As mentioned earlier, OB inhibitory granule cells are generated throughout the entire life of the animal. The rodent precursors to these cells arise in the SVZ and migrate to the OB where they differentiate. The migration of the cells is along a scaffold produced by astroglial cells and the endothelial cells of the blood capillaries and is named the RMS. The RMS connects the cells from their site of origin (birth) to their final destination, the OB. Although there are known similarities in the anatomical features of the RMS in vertebrates, there are also differences between those features in human (primates) and rodents, including the size and shape of the RMS. For example, the RMS is smaller and more contorted in the human, and there are fewer neuronal precursors detected in the human RMS. It has been suggested that the observed differences are likely to result from the differences in the size, structure, and complexity of the brains of rodents and humans.

The last steps in the relay of olfactory information take place in several parts of the ventral forebrain, a region collectively called the olfactory cortex, which includes the anterior olfactory nuclei (AON), PIR cortex, olfactory tubercles (OTU), entorhinal (ENT) cortex, and amygdala (AMY) (Fig. 24.2). The olfactory cortex receives input from the OB (afferents) but also sends connections to the OB (top-down inputs). Within the OB the mitral cells have been shown to project to nearly all areas of the olfactory cortex, whereas the tufted cells primarily make contact with the anterior parts of the olfactory cortex. There is also evidence suggesting that the top-down connections are important for odor discrimination, food intake, and the survival of newly born neurons in the OB. How the perception of odors is processed in the higher centers of the brain is not fully understood but it is currently a question addressed by several laboratories.

IS THERE ADULT NEUROGENESIS IN THE OLFACTORY BULB OF HUMANS?

The first evidence that adult neurogenesis occurs in humans was provided by Eriksson, Gage, and colleagues in their 1998 report that cells in the DG of the hippocampus of patients treated with BrdU for health reasons unrelated to neurological diseases incorporate this DNA precursor and express neuronal-specific proteins. Subsequently others showed that cells with the ability to proliferate and give rise to neurons and glia populated the human hippocampus by in vitro cultivation of surgical postmortem specimens. Since it is not ethical to inject humans with radiolabeled DNA precursors in order to examine the time course and location of cell proliferation, the ability to determine the number and position of newly generated cells in the human brain has lagged behind that in other organisms.

A breakthrough in the assessment of whether neurogenesis takes place in humans throughout their lifespan came from work by Frisén and colleagues; their study took advantage of the

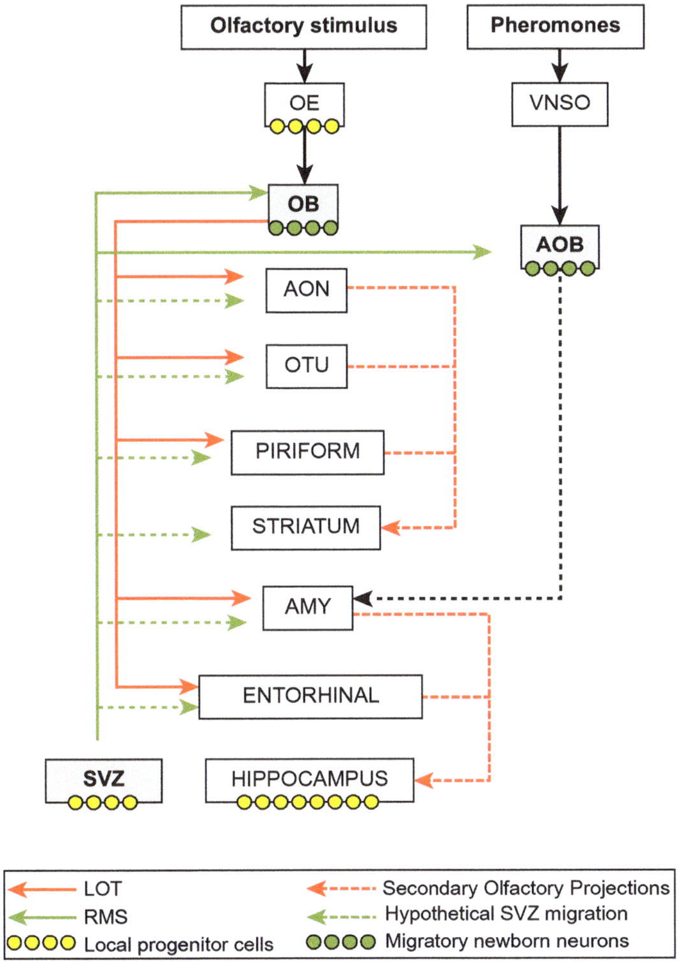

FIGURE 24.2 **Schematic illustration of the connections of the olfactory system and neurogenic pathways.** The olfactory epithelium (OE) contains olfactory receptors which detect olfactory stimulus and send axons through the olfactory nerve to the olfactory bulb (OB). The OE is in constant replacement of neurons through neurogenic precursors, as represented by *yellow circles* in the OE. In addition, pheromones are detected by vomeronasal organ (VNSO) that is connected through the vomeronasal nerve to the accessory olfactory bulb (AOB), from where information is sent mainly to amygdala (AMY). Both OB and AOB receive new neurons (*green circles*) from the rostral migratory stream (RMS; *solid green line*), which originates in the subventricular zone (SVZ) of the lateral ventricle (LV). The lateral olfactory tract (LOT; *solid red line*) projects primary olfactory information from the OB to many regions, such as the anterior olfactory nuclei (AON), piriform (PIR) cortex, olfactory tubercles (OTU), entorhinal cortex (ENT), and AMY. Olfactory information is also relayed (*dashed red lines*) to the striatum and hippocampus. All of the structures that receive direct or indirect olfactory input also exhibit neurogenesis in adulthood. Hypothetically, the newborn neurons in most of these structures migrated from the LV (*dashed green lines*), but some of them also appear to have a local population of progenitor cells. In the hippocampus, new neurons are generated in the subgranular zone of the dentate gyrus (*yellow circles*). Many internuclei connections, such as from the PIR cortex to hippocampus and AMY, are not represented in the scheme. *From Arisi GM, Foresti ML, Mukherjee S, Shapiro LA. The role of olfactory stimulus in adult mammalian neurogenesis. Behav Brain Res February 14, 2012;227(2):356–62.*

presence of ^{14}C in the cellular DNA of individuals that were born or lived throughout the period when atomic bombs were being exploded above ground. They isolated nuclei from OB and cerebellar cortex, sorted them as neuronal or nonneuronal according to antigenic markers, and then measured ^{14}C in their DNA. Retrospective birth-dating of neuronal and nonneuronal cells from postmortem samples of individuals ranging from 21 to 78 years of age allowed them to infer that there is very little (negligible) neurogenesis in the OB of humans but detectable neurogenesis in the human striatum. These observations are surprising as adult neurogenesis in the rodent OB is robust and adult hippocampal neurogenesis is detectable in both rodents and humans. Just as important is the observation that the extent of adult neurogenesis in the DG is also very different between species. In humans about one in three cells are generated throughout a lifetime, whereas in mice it is only one in ten. In the future, it will be important to determine whether the interpretation of the ^{14}C in DNA retrospective birth-dating findings is accurate and can be reproduced, it will be also important to determine how the differential location of adult neurogenesis in humans versus rodents relates to or influences the plasticity of the olfactory system in humans.

ADULT NEUROGENESIS, DOES IT IMPACT BEHAVIOR?

As mentioned at the beginning of this chapter, there is abundant evidence that adult mammalian neurogenesis in the hippocampus occurs and the integration of new-born neurons into the DG circuitry has a role in memory and what has been called pattern separation. Pattern separation has been defined as "the process by which similar images or experiences are transformed into distinct and nonoverlapping neural representations." In the context of the olfactory system, Sakamoto and colleagues have provided evidence, and recently a great summary of what is known about the role of newly generated neurons in rodent olfaction. They suggest, based on their own work, that adult neurogenesis in the OB also has a role in pattern separation in the OB. They also state that although much less is known when compared to what happens in the hippocampus, newly generated neurons are important in olfaction-related plastic activities in the OB and they are necessary to "rewrite acquired odor memory and modify the value of odor-associated memory."

To address the role of adult neurogenesis in the rodent OB researchers have used antimitotic drugs in adult mice to block cell proliferation in the RMS and find an impaired ability of the treated individuals to detect and remember odors. Others have used irradiation and chemical ablation of newborn neurons, as well as conditional genetic models and genetically labeling newborn neurons to study their migration, differentiation, and integration into mature circuitry, all while correlating these manipulations with behavioral changes. Unfortunately, not all studies performed and experimental approaches used to inhibit adult neurogenesis in the OB have resulted in congruent or consistent results. Gheusi, Lledo, Sakamoto, and others have suggested that perhaps the reason for the discrepancies arise from the fact that some of the manipulations affect both neurogenesis in the DG of the hippocampus and the OB, and as such there might be confounding effects. The advent of targeted interventions to only affect generation of adult neurons in the OB should help clarify their role in olfactory behavior and help understand how they contribute to the adult organization of the system.

SUMMARY

In summary, there is now substantial evidence that neurons generated throughout adult neurogenesis may have many and unique functional roles depending on the ages of the animal and the newly born neurons, as well as where they are generated. Moreover, the breadth of the findings suggests that adult-born neurons act along with those born during the embryonic or early postnatal period.

Further Reading

[1] Abrous DN, Wojtowicz JM. Interaction between Neurogenesis and Hippocampal Memory System: New Vistas. Cold Spring Harb Perspect Biol June 01, 2015;7(6).

[2] Curtis MA, Monzo HJ, Faull RL. The rostral migratory stream and olfactory system: smell, disease and slippery cells. Prog Brain Res 2009;175(0):33–42.

[3] Ernst A, Frisén J. Adult neurogenesis in humans – common and unique traits in mammals. PLoS Biol 2015;13(1):e1002045.

[4] Gheusi G, Lledo PM. Adult neurogenesis in the olfactory system shapes odor memory and perception. Prog Brain Res 2014;208:157–75.

[5] Jessberger S, Gage FH. Adult neurogenesis: bridging the gap between mice and humans. Trends Cell Biol October 2014;24(10):558–63.

[6] Lepousez G, Nissant A, Lledo P. Adult neurogenesis and the future of the rejuvenating brain circuits. Neuron April 22, 2015;86(2):387–401.

[7] Lim DA, Alvarez-Buylla A. Adult neural stem cells stake their ground. Trends Neurosci October 2014;37(10): 563–71.

[8] Sakamoto M, Kageyama R, Imayoshi I. The functional significance of newly born neurons integrated into olfactory bulb circuits. Front Neurosci May 06, 2014;8:121.

CHAPTER

25

Axonal Degeneration and Regeneration in the Peripheral and Central Nervous Systems

P.J. Reier[1], *L.V. Zholudeva*[2], *M.A. Lane*[2]

[1]University of Florida, Gainesville, FL, United States; [2]Drexel University, Philadelphia, PA, United States

INTRODUCTION

The devastating disabilities resulting from neural insults, especially to the CNS, have been viewed for many years as being irreversible, but advances in basic and translational neurobiology have begun to temper centuries of pessimism. Novel therapeutic strategies to promote functional recovery are being identified which can lead to significant neurological improvements. To appreciate future bench-to-bedside translational possibilities for improving the quality of life after central and peripheral nervous system (CNS and PNS, respectively) injuries, neurodegenerative disease, and stroke, it becomes increasingly important to establish a basic understanding of principles underlying neuronal injury and disease. The following chapter is intended to fill a void in many graduate and medical neuroscience curricula by providing an introduction to neural responses to trauma and differences in PNS versus CNS pathologies that affect axonal regrowth and functional outcomes.

AXONAL AND NEURONAL CELL BODY RESPONSES TO NERVE INJURY AND DISEASE IN THE PERIPHERAL AND CENTRAL NERVOUS SYSTEMS

Axotomy and Anterograde Axonal Degeneration

The term "axotomy" refers to an interruption of axonal continuity due to during injury or disease and thereby divided into two cytoplasmic compartments. The part of the axon remaining in continuity with the cell body is referred to as the proximal nerve stump, while the axonal region isolated from the neuronal soma is the distal segment. One of the major consequences of axotomy affecting neurons associated with either the CNS or the PNS is a progressive demise of the distal axonal segment. This phenomenon is referred to as "Wallerian" or "anterograde" degeneration (Figs. 25.1 and 25.2).

Conn's Translational Neuroscience
http://dx.doi.org/10.1016/B978-0-12-802381-5.00040-3

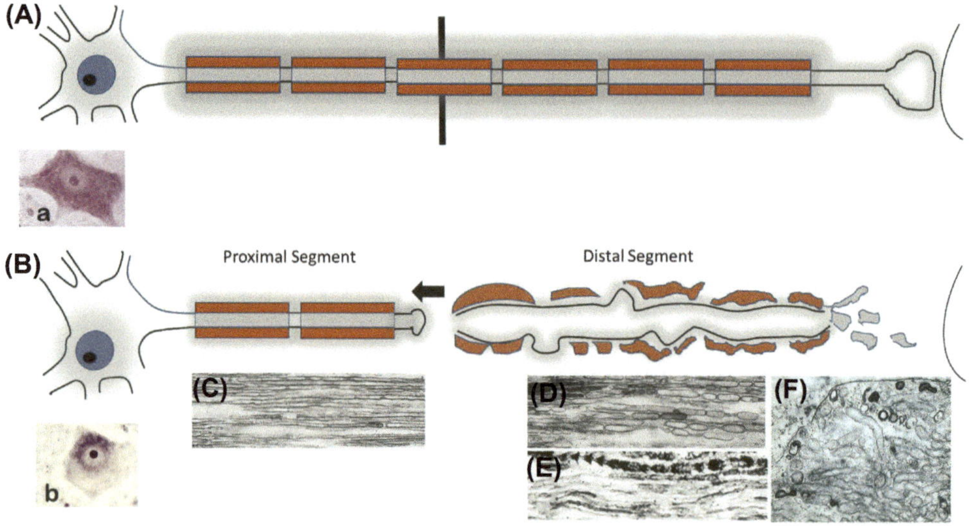

FIGURE 25.1 (A) Transection or compression of a peripheral nerve (*line*) results in the axon being divided into two compartments. (B) The axon in continuity with the neuronal cell body is the proximal (nerve) segment. The axonal region separated from the cell body is the distal (nerve) segment which undergoes subsequent Wallerian degeneration involving overall axonal breakdown and disruption of surrounding myelin sheaths. Proximal dieback is indicated by the *arrow* (see text for details). (C) Longitudinal section of a normal peripheral nerve. (D) Early distal degeneration following nerve injury. (E) Advancing Wallerian degeneration. (F) An electron micrograph of an axon undergoing Wallerian degeneration showing an accumulation of organelles, mostly mitochondria, at various stages of breakdown. In some cases, an overt cell body response to axotomy is seen referred to as chromatolysis (b) characterized in part by the loss of Nissl substance (a), displacement of the nucleus, and cell body swelling (b).

Cytological features of Wallerian degeneration (WD) include axonal swelling accompanied by abnormal accumulations of axonal organelles (Fig. 25.2A–C). Dissolution of cytoskeletal components (eg, microtubules and neurofilaments) ensues followed by axonal collapse often leaving hollowed out and highly irregular myelin profiles (Fig. 25.2A and B). WD has been shown to entail a biphasic process. This involves an initial latent interval of 36–44-h post-axotomy during which time distal axon segments remain intact and conduct action potentials. This interval is then followed by an abrupt and total axonal fragmentation phase reaching totality even as quickly as 1 h from onset (Fig. 25.2D).

A considerable degree of axonal breakdown is caused by an unregulated influx of calcium (Ca^{2+}), which leads to the activation of calcium-dependent proteolytic enzymes (eg, calpains) that promote degradation of the cytoskeleton and membrane proteins (Fig. 25.3). Evidence indicates that interruption of axonal transport may contribute to primary triggering events within the distal stump. Nevertheless, Ca^{2+}-related breakdown of the axonal cytoskeleton and concurrent disruption of the integrity of the axolemma contribute most directly to axonal fragmentation and collapse. While this scenario conveys the impression of WD being a relatively passive process, a large body of evidence has demonstrated a more complex pathophysiology being involved, as described in "Wallerian Degeneration: New Insights" section.

WD also entails a secondary breakdown of the myelin sheath surrounding degenerating axons. Such myelin pathology accompanying axonal disintegration implies an intimate dependency of myelin-producing Schwann cells and

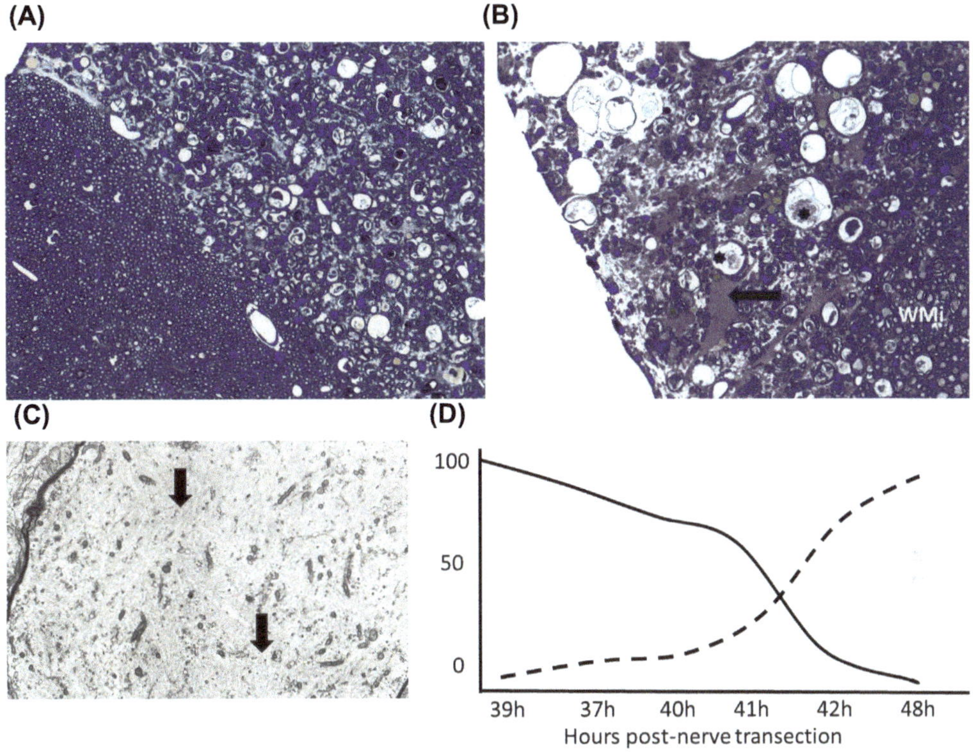

FIGURE 25.2 (A) An image from an experimental spinal cord injury showing a region of intact white matter (left half of image) adjacent to a region of injured white matter undergoing Wallerian degeneration (right half of image). (B) Higher magnification view of Wallerian degeneration in the CNS showing various stages of myelin breakdown and presence of macrophages engulfing myelin debris. Examples of advanced axonal fragmentation are seen (*) in addition to astroglial hypertrophy (*arrow*). (C) Early axonal degeneration as seen at the ultrastructural level. The axon is swollen and its surrounding myelin sheath is attenuated. The region is characterized by mitochondria within a dense cytoskeletal matrix (*arrows*). (D) A graph illustrating the biphasic nature of Wallerian degeneration after peripheral nerve injury. The *solid line* represents intact axons and the *dashed line* represents fragmented axons. Initially, there is a latent interval following axotomy during which axons remain intact and still exhibit axonal transport capacity and can conduct nerve impulses. This interval is then followed by a catastrophic loss of intact fibers and dramatic increase in axonal fragmentation. *The graph is based upon data from Beirowski B, Adalbert R, Wagner D, et al. The progressive nature of Wallerian degeneration in wild-type and slow Wallerian degeneration (WldS) nerves. BMC Neurosci 2005;6:6.*

oligodendrocytes (in the PNS and CNS, respectively) on axonal integrity. This principle is a corollary of development, during which the axon dictates whether myelin is formed and the ultimate thickness of the sheath to be established. Peripheral nerve injury can thus lead to the anterograde degeneration of large populations of axons and their surrounding myelin sheath in the distal nerve stump (ie, secondary demyelination). It is important to appreciate, however, that WD *only* applies to the axon and its contents. The distal nerve stump itself does not degenerate. Vascular and connective tissue components of the nerve remain along with cohorts of orphaned Schwann cells (see section: Regenerative Responses After Peripheral Axotomy).

It is also worth noting that in diseases such as multiple sclerosis, primary demyelination in the CNS (ie, myelin breakdown around intact axons) has been associated with subsequent

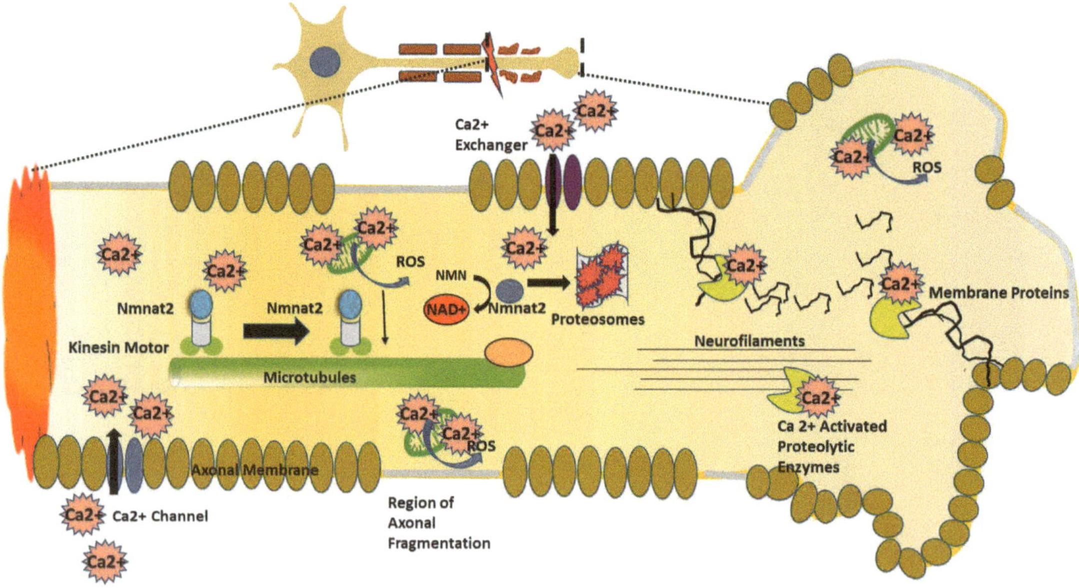

FIGURE 25.3 A potential mechanism of axonal degeneration. Normally, an intra-axonal balanced turn-over exists between Nmnat2 (an axonal survival signal), which is delivered to the axon by anterograde transport (*large arrow*), and its degradation by proteasomes (*small arrow*). Enzymatic activity of Nmnat2 may potentially regulate intra-axonal Ca^{2+} levels and thereby prevent activation of Ca^{2+}-dependent protease activity. Impaired axonal transport resulting from nerve injury impairs delivery of Nmnat2 (*down arrow*) leading to diminished NAD^{+} levels and increasing intra-axonal levels of Ca^{2+} via influx through Ca^{2+} channels and exchangers. Consequently, Ca^{2+}-dependent proteases (eg, calpains) are activated leading to the breakdown of membrane proteins and the axonal cytoskeleton. Elevated Ca^{2+} levels also promote loss of mitochondrial membrane potential and increased production of reactive oxygen species (ROS). Ultimately, these events contribute to axonal fragmentation. *Adapted from Wang JT, Medress ZA, Barres BA. Axon degeneration: molecular mechanisms of a self-destruction pathway. J Cell Biol 2012;196(1):7–18.*

axonal damage leading to WD, thereby illustrating axonal dependence on trophic support from surrounding oligodendrocytes. Other lines of evidence suggest, however, that chronic inflammation, independent of primary demyelination, contributes to axonal injury. Therefore, some common pathobiology (ie, inflammation) exists between axonal damage in relation to disease and trauma.

Lastly, anterograde axonal degeneration leads to the loss of presynaptic terminals (eg, terminal degeneration). One of the early and most comprehensive electrophysiological and ultrastructural investigations performed revealed that nerve transection is followed, after a delay of some days, by structural disintegration of the axon terminals together with failure of transmission and stoppage of the spontaneous end-plate activity. Later studies showed a persistence of end-plate function for 8–10h following, for example, transection of the phrenic nerve in the neck; however, by 20h all neuromuscular transmission had abruptly ceased. Interestingly, the length of the distal stump appeared to correlate with the timing of neuromuscular transmission failure. Structural disruption of the nerve terminal closely coincides with the loss of physiological activity. Fragmentation of the terminals is accompanied by Schwann cells extending cytoplasmic processes into the postsynaptic cleft to engulf remnants of the nerve terminal.

Wallerian Degeneration: New Insights

The progressive degeneration of the distal segment has been thought to reflect the fact that the metabolic maintenance and integrity of the axon is primarily dependent on synthetic activity in the cell body and anterograde axonal transport. Thus once separated from the neuronal cell body, the distal axon segment does not have sufficient inherent biosynthetic capacity to maintain itself. While this conclusion may still hold some qualified validity, a strikingly different perspective emerged with the serendipitous discovery of the WD slow (Wld^S) mutant mouse. Unlike after nerve injury in wild-type mice and other species where the distal segments of severed axons (both CNS and PNS) begin degenerating within 3day postinjury, the distal axon segments in Wld^S mutants remain intact for weeks in both the CNS and the PNS and are still able to conduct action potentials before undergoing WD.

The Wld^S mutation is the resulting protein of a chimeric gene product of an ubiquitination factor and the full sequence of a NAD^+ synthetic enzyme (nicotinamide mononucleotide adenylyltransferase 1, NMNAT1), localized on chromosome four in mice. In contrast to the normally nuclear NMNAT1, the resulting Wld^S protein has the ability to localize in the axoplasm and in axonal organelles such as the mitochondria and phagosomes. This results in the distribution of NMNAT1 throughout the axon and enables its neuroprotective functions in the distal part of the axon after axotomy. Studies have shown that activity of an endogenous NMNAT isotype within the axon (NMNAT2) is essential for axonal survival and rapidly degrades within 4h after nerve injury, while Wld^S activity persists. This sustained NMNAT activity in the axon via Wld^S allows for the promotion of axonal survival via NAD^+-related mechanisms. One hypothesis for such a mechanism involves the role of NAD^+ in modulation of ion channel opening in organelles that have the capacity of sequestering Ca^{2+}, namely endoplasmic reticulum (ER), Golgi, and mitochondria. Thus it has been suggested that the local increase in NMNAT activity via Wld^S and the resulting increase in NAD^+ levels leads to the decrease in intra-axonal Ca^{2+} concentrations either by enhancing Ca^{2+} buffering capacity of mitochondria or by decreasing intracellular release by axoplasmic Ca^{2+} storage sites. In the absence of Wld^S, an opposite scenario unfolds after axotomy (Fig. 25.3).

Although the mechanisms of Wld^S action are far from being completely understood, the genetic evidence to date has had tremendous impact on our understanding of the basis of axonal destructive processes. Such evidence has significant implications for future treatments of various neurological disorders in which WD or WD-like events play significant roles. It is now clear that WD involves active axonal death-signaling events that could possibly be modulated via therapeutics based on gene editing. Fruitful development of future therapeutic approaches will clearly benefit from further identification of axonal death-signaling molecules and learning how to inhibit their action.

Axonal Dieback and Wallerian Degeneration

Histopathological features closely resembling WD in transected nerves have also been associated with a large group of disorders of the PNS and CNS collectively referred to as "axonal dying back" neuropathies (or axonopathies). Such disorders have a spectrum of genetic or neurotoxic origins ranging from peripheral nerve diseases to traumatic brain injury and even Alzheimer's disease. Laboratory investigations have indicated that dieback often, but not always, begins at neuromuscular junctions or axonal presynaptic terminals and then appears to proceed in a retrograde direction. More extensive studies, however, have shown that small foci of axonal pathology (ie, localized axonal swellings or spheroids) can appear at various sites along axons and are characterized by amyloid protein (APP) accumulation. Axonal degeneration then progresses anterogradely from each site (Fig. 25.4) resulting in microtubule disassembly, fragmentation of the axonal cytoskeleton, and subsequent disruption of axonal transport. In many cases, dieback is limited in distance and more proximal portions of affected axons remain intact. The evidence for common mechanistic pathways in both axonal dieback and WD was demonstrated in mouse models of peripheral neuropathies cross bred with Wld^S mice. In such cases, dieback pathology was significantly delayed by Wld^S, which likewise offered protection to genetic and toxic insults in the CNS. Other axon destructive pathways also may be involved in dieback axonopathies, since there are disease models in which the Wld^S protein is deemed ineffective. Aging and species-wide differences may likewise impact Wld^S efficacy. Overall these and other lines of multidisciplinary investigation have shed new light on a fundamental pathological process, namely WD that may have much greater significance to a large range of neurological conditions by virtue of several convergent pathways involving deficient axonal transport, mitochondrial dysfunction, and increased intracellular concentrations of Ca^{2+} (Fig. 25.3).

Proximal Axonal Segment Following Axotomy

Leakage of axoplasm from the proximal segment is another immediate aftermath of

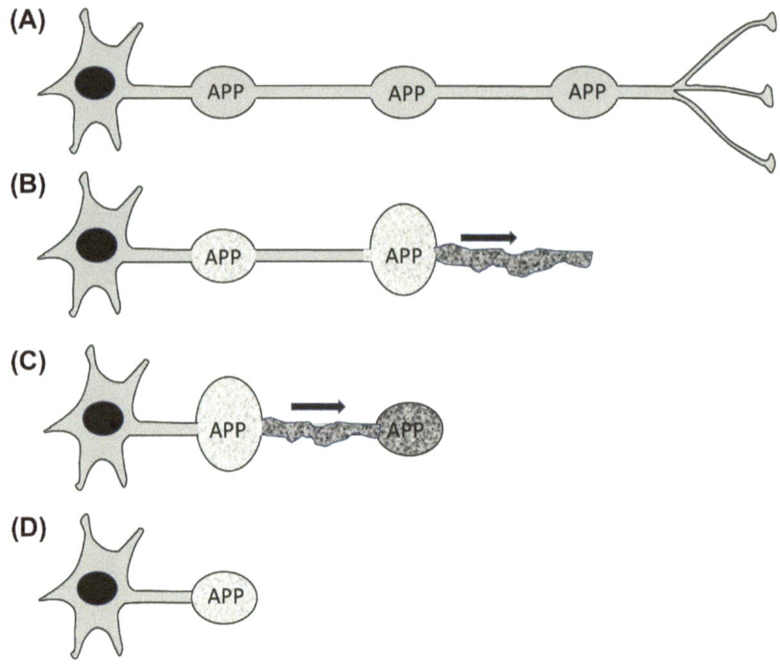

FIGURE 25.4 Wallerian degeneration in disease. (A) Impaired anterograde axonal transport has been associated with many neurodegenerative conditions and may be characterized by accumulation of amyloid precursor protein in axonal swellings or spheroids. Axonal varicosities can be multifocal along the length of an axon and further characterized by a markedly disorganized cytoskeleton (Fig. 25.2C). (B–D) As the swellings continue to enlarge, axonal transport is increasingly compromised and Wallerian degeneration (*arrows*) is triggered without necessarily involving axonal transection or crush. Ultimately, only an end bulb remains at the end of the proximal stump. *Adapted from Coleman M. Axon degeneration mechanisms: commonality amid diversity. Nat Rev Neurosci 2005;6(11):889–98.*

axotomy that can persist until healing occurs by resealing of the axonal membrane. In addition, the cut end of the proximal segment can undergo degenerative retraction from the original site of injury—a phenomenon referred to as "proximal (segment) dieback" (Fig. 25.1B) that has been reported to be greater in the CNS than in the PNS. Such proximal axonal breakdown is usually limited in distance (up to the preceding node of Ranvier), unless the lesion occurs close to the cell body, in which case proximal dieback can advance to the soma and thus result in retrograde neuronal degeneration (see later).

Axotomy Induces Responses at the Level of the Neuronal Cell Body

The effects of axotomy also can be exhibited at the level of the cell soma in the form of metabolic, molecular, cytological, and neurophysiological changes. The cell body response depends on many variables (eg, Table 25.1), and it cannot be generalized how any given neuronal population will respond to axotomy. Although many possibilities have been proposed for ways axonal damage can trigger changes in the cell soma, it is likely that multiple, temporally related signals are involved. Massive neuronal depolarization due to influxes of calcium and sodium ions at the site of axotomy is one of the more immediate and plausible signals of injury. Such pronounced ionic feedback could, in turn, activate a large variety of molecular responses at nuclear and cytoplasmic levels of the neuron. For example, some attention has been drawn to the glycoprotein 130 family of cytokines, which by acting through a common signal receptor (gp130) can trigger changes in neuronal gene expression after axotomy.

TABLE 25.1 Variables Related to Cell Body Responses to Axotomy

- The distance of the lesion from the cell body
- The age of the subject at the time of injury
- Species differences
- The nature and severity of the lesion
- Gender
- The extent of axonal collateralization between the site of axotomy and the cell body
- The functional type of cells involved
- The PNS or CNS location of the axotomized neuronal populations

A classic cytological example of a retrograde, cell body response to axotomy is chromatolysis. Characteristic features include formation of cytoplasmic vacuoles, nucleolar enlargement, displacement of the nucleus from a central to an eccentric location in the neuronal cell body, and swelling of the cell body (Fig. 25.1A and B). A prominent aspect of chromatolysis observed under the light microscope is the dissolution of Nissl substance (ie, rough ER). At the ultrastructural level, this is accompanied by a concomitant increase in the density of free polyribosomes, the appearance of which is consistent with elevated RNA metabolism and protein synthesis more typically seen in young neurons. It should be emphasized that chromatolysis is not a universal neuronal response, but one that is exhibited only by certain neuronal populations and subject to many variables. While certain features of chromatolysis are consistent with imminent cell death (eg, nuclear changes, cytoplasmic vacuolation, and cellular swelling), the appearance of polyribosomal clusters, seen in normally developing neurons, is more closely reflective of anabolic responses supporting regeneration.

Fate of Axotomized Neurons in the Peripheral and Central Nervous Systems

Cell death following peripheral nerve injury is dependent upon several variables, the most prominent of which include the neuronal cell type(s) affected, proximity of the lesion to the cell body, type of injury, and age-related factors (Table 25.1). For example, large dorsal root ganglion cells and some peripheral and spinal preganglionic neurons are more susceptible to peripheral nerve injuries than motoneurons. The latter are especially vulnerable after spinal root avulsions which can occur with spinal cord injuries due to large tensile forces on nerve roots. Such root avulsions interrupt motor axons and also damage motoneuron collaterals within the spinal cord. Otherwise motoneurons and neurons located completely within the PNS will often survive and shift their metabolism toward the biosynthesis of cell survival and repair molecules rather than constituents associated with normal neuronal function. An example is neurotransmitter synthesis which is strikingly decreased after nerve injury, whereas production of certain cytoskeletal and growth-related proteins is up-regulated.

Cell death in the CNS has been extensively discussed elsewhere in this text (Koliatsos, Chapter 21) relative to traumatic brain injury, but necrosis and apoptosis are not limited to that neurological condition. The present discussion is directed more specifically to outcomes of axotomy without primary cell body damage. In some instances, axotomized neurons survive but show a significant reduction in cell body size (ie, atrophy) and limited, if any, anabolic responses to axotomy (Fig. 25.5). This state of "cellular atrophy" may be associated with overall neuronal dysfunction. In other cases, axonal damage may cause some neurons to undergo regressive changes that could cascade into molecular events directed at neuronal death (eg, programmed cell death or apoptosis). Thus some neurons that are unable to survive axonal damage succumb to what is referred to as retrograde cell death (Fig. 25.5). As a general rule, lesions occurring close to the cell body of origin usually result in a greater chance of cell death.

Cell death also can occur in neuronal populations that are synaptically associated with damaged neurons (ie, "transneuronal cell death", see Chapter 21). For instance, axotomy, with or without ensuing cell death, can lead to a phenomenon referred to as "anterograde (orthograde) transneuronal" or "trans-synaptic degeneration." As a result of WD, postsynaptic neurons become at least partially denervated. A significant reduction in presynaptic input can then precipitate an eventual degeneration of the target cell. Conversely, neurons that synapse with axotomized cells can die by virtue of "retrograde transneuronal" or "retrograde trans-synaptic degeneration" (Fig. 25.5). An important corollary of these transneuronal degenerative responses to axotomy is that similar principles apply to many neurodegenerative diseases, as well as traumatic brain injury (see Chapter 21). For instance, the loss of certain neuronal populations that are the primary targets of a particular neurological disorder could lead to more widespread cell degeneration elsewhere in the nervous system as a result of deafferentation (eg, orthograde transneuronal degeneration) or the loss of postsynaptic targets (eg, retrograde transneuronal degeneration). One illustration of this point is Huntington's disease in which pathology is not solely restricted to the striatum but also extends to neuronal loss in other parts of the basal ganglia, as well as in cortical regions during progression of the disease. Such global transneuronal cell death also explains the spectrum and progression of associated behavioral and cognitive consequences of many neurodegenerative diseases.

REGENERATIVE RESPONSES AFTER PERIPHERAL AXOTOMY

As mentioned earlier, neurons located exclusively within the PNS or those that send axons from the CNS into peripheral nerves (eg, spinal or cranial motoneurons) are fairly resilient after axonal injury and have the innate capacity to support robust axonal regrowth (ie, axonal regeneration), even for long distances. For example, axotomy can up-regulate intracellular cyclic AMP expression that then sets in motion

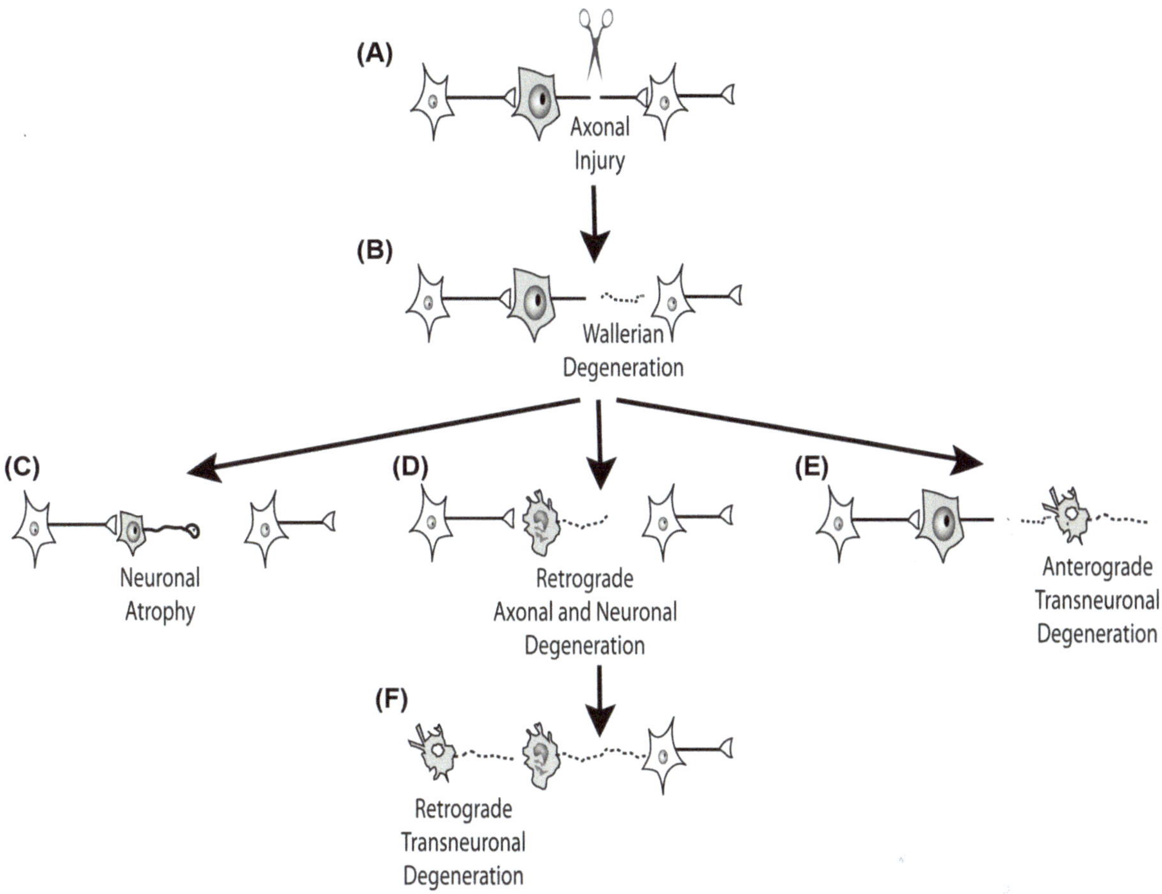

FIGURE 25.5 Several basic neuronal responses to injury are illustrated in this diagram (see main text). This schematic is comprised of three interconnected neurons. Following axonal injury to the second neuron in the series, the neuron cell body becomes enlarged (A) and the distal part of the axon undergoes Wallerian degeneration (B). If the cell body survives the axonal injury, it may still undergo neuronal atrophy (C). Alternatively, it may undergo retrograde axonal degeneration (of proximal axon stump) and neuronal degeneration (D). Secondary neuronal loss may subsequently occur via anterograde (E) and/or retrograde (F) "transneuronal" degeneration.

other molecular pathways leading to the transcription of several growth-promoting genes. In some cases, these responses can be enhanced by a "conditioning lesion" in which the extensive nerve injury is preceded by an initial axotomy with the aim of up-regulating growth-promoting genes that should enhance axonal regrowth.

The onset of axonal regrowth does not occur immediately after axotomy, and the length of delay depends on variables such as the severity of the injury or the proximity of the lesion to the cell body. This has other implications given that the regenerative potential can be time limited in the best of circumstances. Another important consideration is that robust axonal regeneration is not necessarily synonymous with functional recovery. Instead, the degree and quality of functional return is highly dependent on the ability of axons to reconnect with their proper targets (ie, reinnervation).

Regeneration Is Initiated by the Formation of Growth Cones That Possess Selective Preferences for Certain Tissue Constituents

The preserved end of the proximal nerve segment undergoes several morphological changes that lead to the formation of multiple sprouts from each damaged axon. Appearing at the end of each sprout is an amoeboid-like structure referred to as the growth cone (Fig. 25.6) that can be best visualized in neuronal tissue culture preparations. Growth cones are highly specialized motile structures that move in a crawling fashion by extending and retracting cytoplasmic extensions (eg, filopodia) or flattened veil-like projections (eg, lamellipodia). Proximal nerve sprouts increase in length as the growth cones advance, and their formation represents the earliest overt phase of axonal regeneration.

In addition to their mobility, growth cones have the remarkable capacity to sense and respond to cues in the surrounding tissue milieu. They express a range of receptors with high affinities for molecules that either attract, repulse, arrest, or cause growth cone collapse (Fig. 25.6). Using these receptors, filopodia attach to favorable or growth-promoting and permissive substrates such as laminin. Alternatively, if filopodia encounter adverse tissue environments, their initial attachments are unstable, they retract, and the growth cone is arrested in its advance (ie, growth cone collapse) or turns in the direction of those filopodia that have found more favorable substrates. Growth cones also are attracted by certain neurotropic molecules and are repulsed by others, with calcium playing a major role in defining the direction of growth cone movement.

Thus the mechanisms of growth cone motility are complex and in several respects parallel those that underlie motility seen in other cell types. Briefly stated, the movement of growth cones has been associated with specific cytoskeletal constituents—most notably actin and myosin II. Actin is intimately involved in the extension of filopodia, whereas myosin II underlies traction-based movements. Guidance cues

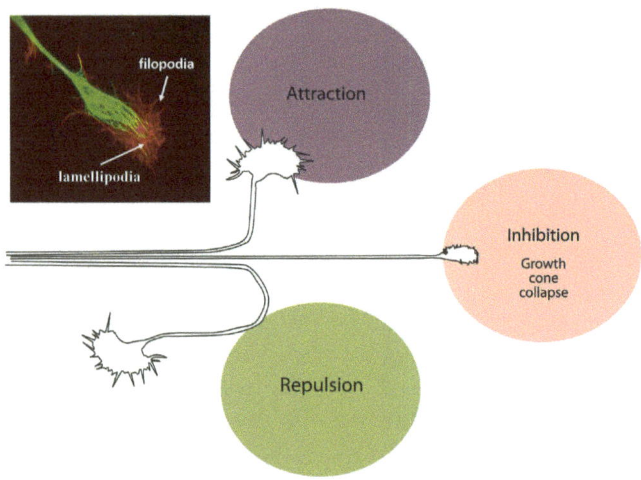

FIGURE 25.6 Growing axons are guided by a range of attractive, repulsive, and inhibitory molecules. During development extending lamellipodia and filopodia interact with attractive and repulsive molecular cues by pulling the growing axon toward or away, respectively. In contrast, increase in inhibitory molecules that occurs as a result of CNS injury causes the growth cone to collapse. Inset shows a growth cone from a neuron in culture [BIII-tubulin (green); Actin (red)]. *Image used for inset kindly provided by Drs. Anand Rao and Peter Baas (Drexel University, USA).*

seem to influence growth cone trajectories by modulating the rate and location of actin polymerization via intracellular signaling mechanisms that appear to be mediated by transient elevations of intracellular calcium.

Nature of the Cellular Environment at the Site of Nerve Damage Can Influence the Ultimate Success of Regeneration by Affecting Growth Cone Behavior

As growth cones emerge from the proximal nerve stumps in the PNS, they first encounter the cellular and molecular microenvironment of the lesion site which can vary considerably depending on whether nerve damage was caused by severe compression or transection. The setting following a nerve crush stands in stark contrast with a gap between the proximal and distal nerve stumps growth cones must encounter after a nerve transection. An extended lesion site usually presents a very unfavorable environment for directed fiber outgrowth and results in a slow, "staggered" elongation of axons through the lesion. Proliferating cells from the proximal and distal stump endoneurium and epineurium, as well as other surrounding nonneural tissues, infiltrate the wound site, and form a tortuous terrain. In some cases, scarring can be so dense that the nerve ends become blindly embedded in a collagenous fibrotic tissue environment. In other cases, axons can wander and may never make contact with the distal nerve stump. The chances of regenerating fibers reaching the distal nerve stump become progressively less with increasing lesion gap distance, thereby negating the potential for any useful regrowth beyond the injury. Rodent data indicate that wound gaps greater than 6mm lead to negligible distal regeneration and functional recovery.

Scarring at the wound site can likewise promote development of highly disorganized, axonal terminal enlargements at the cut edge of the proximal nerve stump. Such swollen structures often contribute to a tangled meshwork of profusely branched regenerated nerve fibers, Schwann cells, fibroblasts, and collagenous matrices. Collectively this tissue matrix constitutes a posttraumatic neuroma, which can severely impede further regeneration, and even more importantly, be a source of painful stimuli. The latter is due to the close approximation of fine-caliber fibers, which enables the spread of electrical current from one axon to another.

Schwann Cells Provide a Permissive Substrate Essential for Axonal Elongation and Are Responsible for Myelination of Regenerated Axons in the Peripheral Nervous System

Within the first 24h after injury, Schwann cells in the distal nerve stump begin proliferating and reach a peak of mitotic activity by about 3days with cell divisions progressively decreasing over the following 2–3days. Dividing Schwann cells and their progeny become organized into characteristic linear arrays that are virtually aligned parallel to the initial trajectory of the axons they had surrounded. In addition to what occurs in the distal nerve stump, Schwann cells at the separated ends of the distal and proximal stumps often infiltrate the lesion gap and, in the best cases, restore Schwann cell continuity between the proximal and distal cut ends. The extent to which migrating Schwann cells from the two separated ends meet depends on the length of the gap and degree of scarring. In cases of nerve crush or compression, the distribution of Schwann cells at the injury site is virtually identical to that seen more distal to the lesion since the continuity of involved peripheral nerves is not interrupted.

Unless otherwise hindered by adverse conditions in the lesion, regenerating fibers reach the distal stump by as little as 3–5days after injury where they encounter the terrain of linearly oriented Schwann cells, which have long been considered to have a significant role in PNS regeneration. Clusters of Schwann cells are enclosed within a common basal lamina. Growth cones entering the distal stump extend between the basal lamina and Schwann cell plasma membrane. This is mainly because of their affinity for laminin, which is a major constituent of the Schwann cell's basal lamina. The Schwann cell plasma membrane also has growth-promoting and cell adhesion molecules.

In general, the distal stump shows dramatic changes in gene expression involving an up-regulation of neurotrophic factors, neural adhesion molecules, cytokines, and other factors and their related receptors. Many of these responses have been shown to be related to Schwann cells. However, the favorable substrate which Schwann cells establish does not last indefinitely and will gradually be reduced when there is limited or protracted regeneration into the distal stump due to extensive scarring at the lesion or long gaps between separated nerve ends. Thereafter, the distal nerve stumps become increasingly collagenous and lose many growth-promoting qualities.

Inflammatory Responses to Nerve Injury and Regeneration

Another important cellular response to nerve injury is inflammation and associated removal of cellular debris. The rate at which this occurs can influence the extent of regeneration. In the PNS, myelin debris resulting from WD is scavenged primarily by activated macrophages that hone-in on the degenerating nerve stump within 2–3 days after injury. In addition to debris removal, macrophages play other significant roles in wound healing by producing growth factors and certain cytokines with neural growth-promoting properties. For these and other reasons, macrophage responses during WD in the PNS have been implicated in tissue remodeling, differentiation of Schwann cells, and the establishment of an environment that can support axonal regrowth.

Regeneration and Remyelination

As regeneration progresses distally, Schwann cells start remyelinating axons in a manner analogous to events occurring during development, and there is a progressive increase in the number of myelinated axons in the distal stump. Other functionally relevant morphological differences can be seen between regenerated versus normal axons. In contrast with intact, adult axons, the myelin internodes (eg myelinated segments between successive nodes of Ranvier) are shorter and undergo persistent remodeling for a year or more. In addition, regenerated axons are usually of much smaller diameter with thinner myelin sheath than the parent axon and may persist as such throughout the course of regeneration.

Functional Recovery After Peripheral Nerve Injury Is Dependent on the Rate of Axonal Elongation and Distance of the Lesion From the Original Targets

The total time for regeneration to be completed is based on (a) an initial proximal healing time and dieback, (b) when the earliest growth cones navigate through the lesion and enter the distal stump, and (c) the distance that regenerating axons need to grow to reach their target. A related consideration is that not all cells regenerate at the same time. The rate of axonal regeneration is influenced by a variety of factors including the onset of regrowth, age, the location and type of injury, and the period that the distal stump is denervated before the first regenerating fibers arrive. As a rule, nerve regeneration progresses 3–4 mm/day after crush and about 2–3 mm/day after transection injury. In humans, it has been commonly held that axonal regrowth proceeds at a relatively constant rate of 1 mm/day or 1 in./month, but it has been shown that this rate diminishes over time. Given the slow progress of regeneration, persistence of viable targets becomes highly dependent upon how far regenerating axons must grow. Denervated muscles eventually become atrophic due to the loss of neurotrophic support and functional disuse. If the injury is far from the target then even under optimal conditions the chance of reinnervating viable muscles becomes increasingly less likely.

Although distal Schwann cell tubes provide a compatible cellular highway for axonal regrowth, this matrix does not guarantee that fibers will reach appropriate target sites. Even at a macroscopic level, the characteristic multifascicular nature of a normal peripheral nerve is readily apparent. These perineurial fascicles do not maintain a uniform appearance; geometry even along short lengths of a peripheral nerve presents a complex interconnecting landscape (Fig. 25.7). Thus any disruption of the continuity of this internal architecture by injury can readily cause regenerating axons to become misdirected as they enter the distal stump. These features may partially explain the potentially different functional outcomes of crush versus transection injuries.

Misdirection within the distal stump, however, is not the only possible contributor to suboptimal functional outcomes during regeneration. Inappropriate reinnervation can even occur at the level of the end organ itself. For example, motor axons may randomly reinnervate muscle fibers within a muscle that were not the original postsynaptic targets. Peripheral nerve injury also results in neural circuitry changes within the spinal cord which can lead to altered sensory and motor functions independent of circumstances in the regenerating nerve.

A nerve crush/compression injury yields the most favorable setting for unimpeded axonal regeneration and end organ reinnervation because the integrity of surrounding epi- and perineurial connective tissue sheaths of the nerve is preserved. Moreover, after nerve

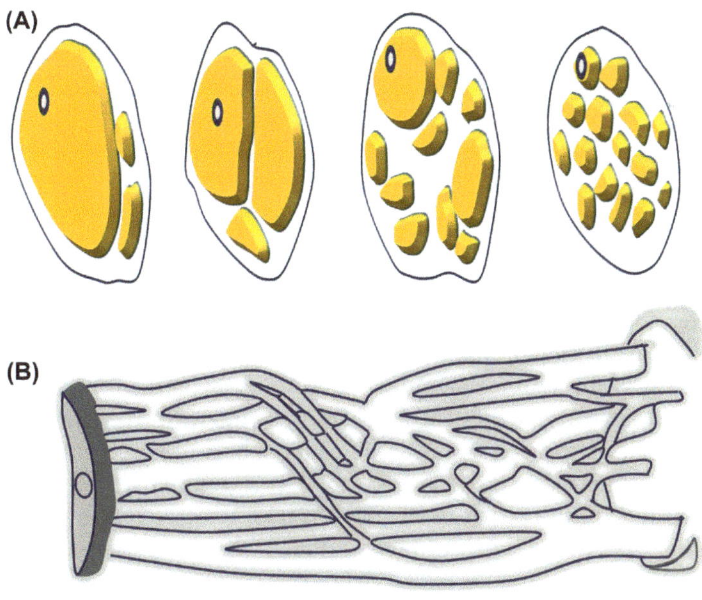

FIGURE 25.7 (A) Tracings of perineurial fascicles over a short distance of a peripheral nerve reveal considerable heterogeneity in fascicle sizes and distributions. (B) Three-dimensional reconstruction of perineurial fascicles demonstrate a complex and interconnecting architecture that can lead to misdirection during regeneration. *Adapted from Sunderland S. Nerves and nerve injuries. 2nd ed. Edinburgh: Churchill-Livingstone; 1978.*

compression perineurial fascicular geometry is not radically perturbed, thereby minimizing scar tissue in the lesion site and preserving the internal geometry of the nerve. In cases of nerve transection, one surgical procedure is to suture the proximal and distal nerve ends together in such a way that opposing proximal and distal perineurial fascicles are matched as much as possible. Unfortunately, even this approach has its limitations due to scarring at the site of repair and poor realignment of the fascicles. From the combined perspective of regeneration rate, distance, end-organ atrophy, and internal nerve geometry, the closer the lesion is to the original sites of innervation, the greater the probability of achieving appropriate functional outcomes.

Summary of Post-axotomy Responses in the Peripheral Nervous System

This general review of peripheral nerve axonal degeneration and regeneration has highlighted several basic aspects of nerve injury and regrowth that will serve as a basis for later discussions of regeneration in the injured CNS. As shown in Table 25.2, the essential prerequisites for optimal regeneration fall under two main headings—intrinsic and extrinsic regulation. The former refers to properties inherent to the neuron itself, whereas the latter addresses molecular and cellular influences from the environment surrounding the injured neuron. Under optimal conditions, regeneration in the PNS holds promise for useful functional outcomes because neurons can often muster sufficient metabolic capacity to initiate and sustain active regrowth through various alterations in neuronal gene expression. Furthermore, the growth cones of the newly formed neurites are exposed to a cellular and molecular milieu that, in the absence of severe scar formation and other limiting factors, is favorably disposed to extensive axonal elongation. However, the speed of regrowth and successful targeting often represent major limitations to optimal functional returns.

TABLE 25.2 Conditions Favoring Optimal Axonal Regeneration

Intrinsic Neuronal Attributes

- Neurons must have the metabolic/synthetic capacity to initiate and maintain regrowth of damaged axons over long distances.

Extrinsic Conditions

- The presence of a cellular microenvironment that is compatible (actively or passively) with growth cone mobility and sustained axonal elongation.
- The cues should be present in the cellular terrain that can enable guidance of regenerating axons toward appropriate postsynaptic targets.
- Regenerating axons should be responsive to environmental cues and ultimately establish appropriate functional connections.

REGENERATIVE FAILURE IN THE CENTRAL NERVOUS SYSTEM

As in the case of peripheral nerve injury, CNS trauma and certain disease conditions entail axotomy and subsequent WD. While PNS axonal regrowth can be robust, very contrasting circumstances exist in the CNS. Frequently cited are observations of the renowned Spanish neuroscientist Santiago Ramón y Cajal whose research showed that nerve fibers in the adult mammalian CNS display an initial outgrowth that may extend at best for limited distances. Thereafter, regeneration comes to a halt, and the newly formed sprouts from injured axons die-back and become stationary terminal enlargements. The transient regrowth of axons within the CNS has led to the long-held view of "abortive regeneration" which anchored the view in clinical and basic science that CNS injuries were totally irreversible.

New perspectives began to evolve in the early 1980s, with experiments involving peripheral nerve grafts to regions of the CNS, including primary visual pathways. In one set of experiments, investigators introduced an end of an isolated peripheral nerve segment into the spinal

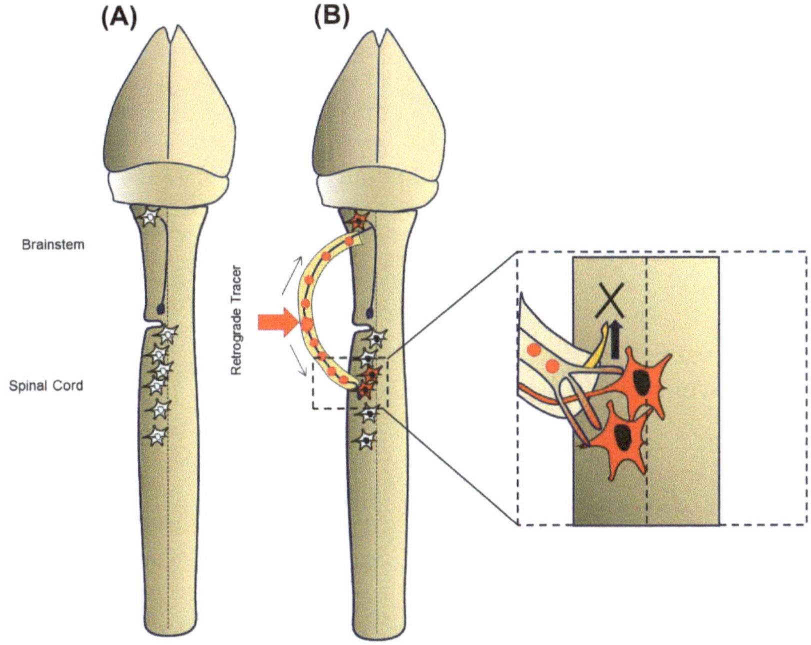

FIGURE 25.8 (A) Illustration of an experimental spinal injury which interrupts continuity of a population of descending bulbospinal axons. The injured axons terminate in arrested end bulbs at the lesion site. (B) A peripheral nerve bridge formed between the lower medulla and upper cervical spinal cord and upper lumbar spinal cord. Histological analysis subsequently showed axons extending through such grafts. To determine the source of those axons, the bridge was cut and a retrograde tracer was introduced. The tracer was then transported to the neurons of origin (*red circles*), which were located in the brainstem and spinal cord (red neurons). While growing for long distances (about 30 mm) within the peripheral nerve bridge, the axons extended only for short distances upon reentering the CNS (inset). This classic experiment *(David S, Aguayo AJ. Axonal elongation into peripheral nervous system "bridges" after central nervous system injury in adult rats. Science 1981;214(4523):931–3.)* illustrated the intrinsic capacity of CNS neurons to initiate and sustain axonal growth over substantial distances when the CNS cellular microenvironment is replaced by that of the PNS.

cord of a rat and the other end into the medulla (Fig. 25.8). An axon-free bridge (recall that WD occurs in this preparation), consisting of Schwann cells and other peripheral nerve tissue elements was thus made that extended several millimeters from end to end. Subsequent neuroanatomical evaluations showed that neurons intrinsic to the CNS could extend fibers into these bridges, and traverse their entire length. In other studies, grafts of purified Schwann cells have been made with comparable results, thereby further underscoring the important role Schwann cells play in promoting axonal growth not only in the PNS, but the CNS as well.

Such PNS–CNS nerve-grafting experiments also revealed how unfavorable the CNS milieu can be to regeneration. When centrally derived axons growing through PNS grafts exit at the opposite end and reestablish contact with CNS tissue, they invariably stop elongation (Fig. 25.8). Therefore, an important corollary to the demonstration of regenerative potential in the injured CNS with PNS grafts is that such a capacity can only be seen when the CNS tissue environment is replaced by a cellular microenvironment that is permissive to axonal elongation.

While CNS neurons have more intrinsic growth potential than previously recognized, the need for finding ways to enhance this growth capacity remains as not all neuronal populations exhibit that potential. Studies of experimental spinal cord injury (SCI) have shown that long-tract axons, such as corticospinal tract fibers, in particular, are especially refractory to regrowth after injury and lack the pro-regenerative molecular programs seen in embryonic, peripheral, and CNS neurons of certain lower vertebrates.

Extrinsic Factors Governing Axonal Regeneration in the Central Nervous System

For many years, researchers held the notion that a major limitation to regeneration in the CNS was the absence of Schwann-like cells that could facilitate axonal regrowth. In addition, several other nonpermissive cellular and molecular environmental conditions have now been identified that currently center mostly on the inhibitory effects of astroglial scars. Inadequate inflammatory responses, coupled with protracted debris removal, also have been implicated. Axonal regenerative failure in the CNS is incredibly complex with no absolute rules. Accordingly, the lack of axonal regrowth is not exclusively due to the presence of any one nonpermissive tissue environments, but can be coupled with modest growth responses exhibited by different neuronal populations after injury.

Cellular and Molecular Inhibition of Axonal Outgrowth and Neuroplasticity in the Central Nervous System

Astroglial Impediments to Axonal Elongation

Loss of neurons and/or their processes triggers an astroglial response in which glial cells increase in size, as well as in thickness and number of their cytoplasmic processes. This overall astrocytic response is referred to as gliosis or astrogliosis which, alone or in conjunction with other cells, form glial or connective tissue-glial scars. In many ways, astroglial scarring is a form of wound healing, and experiments have shown that if the CNS is rendered deficient in astrocytes, the degree of secondary tissue damage is much greater. While in many respects gliosis may be a way to restore physiological homeostasis within the CNS and provide a protective barrier to harmful inflammatory processes, it also has a downside since considerable evidence exists showing that in conjunction with other cells, glial scarring represents a significant physical and chemical impediment to regeneration.

In the case of laceration injuries causing the interior of the CNS to be breached (eg, by stab wound, gunshot or bone fragments entering the CNS), astrocytes respond by reinstating a glial limiting membrane along the exposed CNS neural parenchyma apposing non-CNS tissues (eg, fibroblasts, meninges, and collagen) (Fig. 25.9). There are several stages in the evolution of a glial scar that is triggered partly by the transforming growth factor β (TNFβ) bound to fibrinogen and other molecules that contribute to the complex meshwork architecture of an astrocytic scar. The process eventually culminates in the formation of a type of one or more layered glial capsule or partition which walls off the interior of CNS tissue from non-CNS elements in the lesion (Figs. 25.9 and 25.10A, B). This form of astroglial activity thus serves to repartition the CNS compartment from non-CNS tissue elements after injury.

Astrocytic reactivity occurs along damaged white matter tracts and in gray matter (Fig. 25.10B–E). In effect, astrocytes function to fill in postinjury tissue voids. In degenerated white matter, a dense astrocytic meshwork is formed over time that most notably consists of enlarged and intertwining hypertrophic astroglial processes, which at the ultrastructural level contain large accumulations of glial intermediate filaments (Fig. 25.11D). Similarly, astrocytes exhibit considerable reactivity in regions of gray matter after fiber and axon terminal

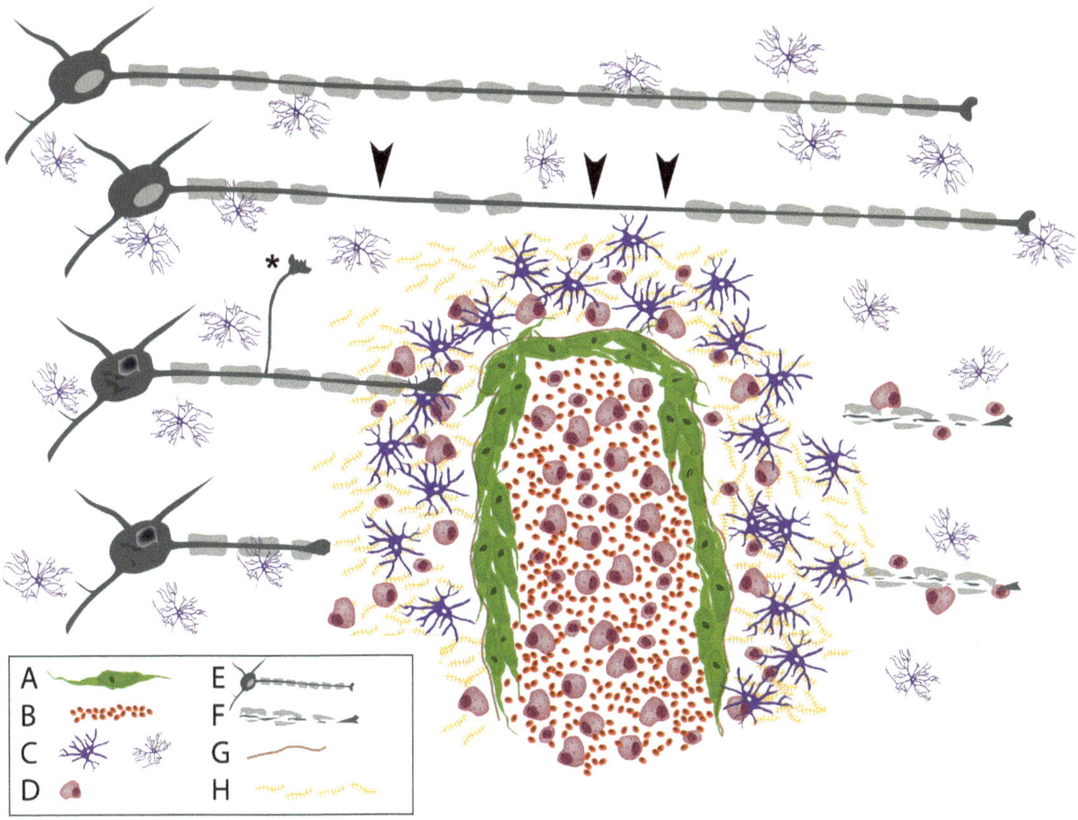

FIGURE 25.9 Traumatic injury within the CNS results in a cellular and molecular "wound healing" response that limits infiltration of peripheral cells (eg, red and white blood cells, B) into the nervous system, but also limits its ability for neuronal (E) repair. Fibroblasts (A) and glial (astrocytic and microglial) cells form a scar at the lesion site. A basal lamina (G) develops around fibroblasts and astrocytes (C). Glial cells become reactive (enlarged cell body with shortened processes, C left) and secrete pro-inflammatory cytokines. Cells comprising this scar also secrete chondroitin sulfate proteoglycans (CSPG's, H) that limit axonal growth and repair. Breakdown of oligodendrocytes as a result of either primary (*arrow heads*) or secondary demyelination leads to an increase in additional growth inhibitory molecules (eg, myelin-associated glycoprotein, Nogo) that can trigger axonal growth cone collapse. Macrophages (D) migrate to the injury site and the surrounding host tissue, phagocytosing debris and contributing to the ongoing inflammatory cascade. While injured axons can undergo Wallerian (F) and proximal degeneration, spontaneous axonal sprouting can also be seen (*).

degeneration and in response to neuronal loss, but this usually does not develop into the extensive three-dimensional astroglial scars seen in lesions and in degenerated white matter.

There are several examples of reactive astrocytes having inhibitory effects on growing axons. One example is that of a crushed or cut dorsal root axons which regenerate and advance vigorously within the PNS compartment of an injured spinal sensory root. As they reach the spinal cord–dorsal root entry zone (DREZ), these axons terminate either as large bulbous enlargements, assume random growth patterns, or are deflected back toward the spinal ganglion of origin. The DREZ, also known as the PNS–CNS transition zone (Fig. 25.11A), is characterized by a progressive shift from a peripheral to a CNS tissue terrain. Injury

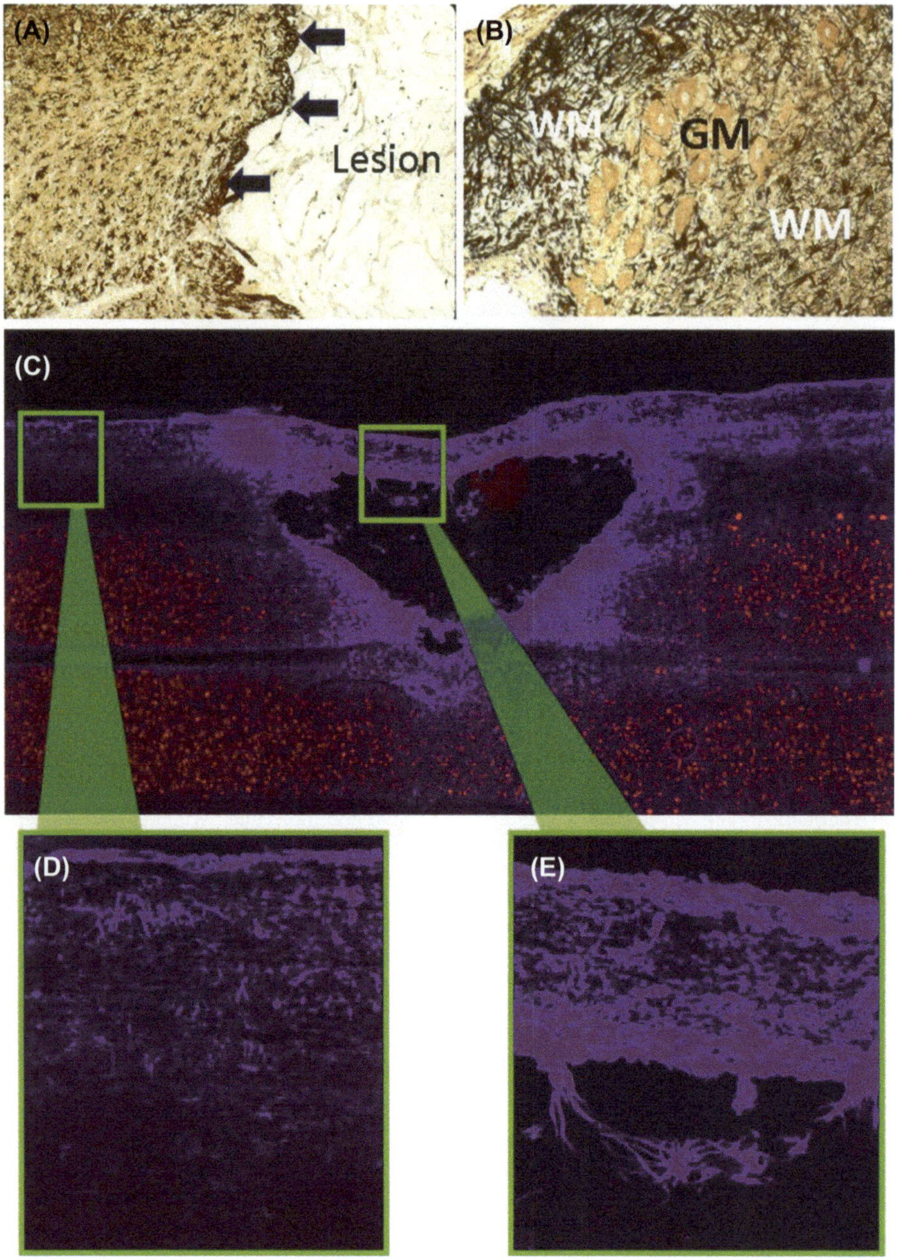

FIGURE 25.10 Photomicrographs of the traumatically injured spinal cord showing labeling for astrocytes [glial fibrillary acidic protein (GFAP), (A, B) brown, (C–E) blue and neurons (C–E) red]. Close-up images from (C) are shown in (D) and (E). Astrocytes become activated and accumulate at the border of the injury site (A (*arrows*), C, and inset in E). Note the distribution of activated astrocytes in both white (WM) (A, C–E) and gray matter (GM) regions (B, C).

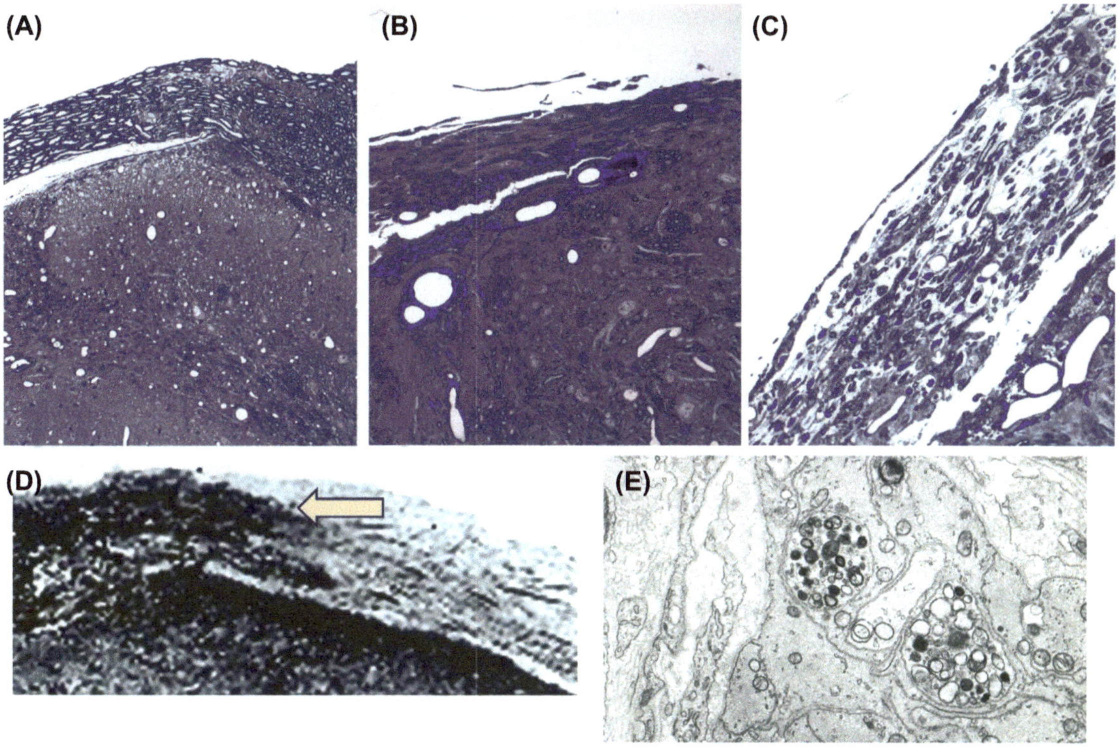

FIGURE 25.11 (A) The PNS–CNS transition or dorsal root entry zone (DREZ) (*arrow*) as seen immediately above the superficial dorsal horn of a normal rat spinal cord. (B) Injury to the dorsal root results in Wallerian degeneration and the formation of an astroglial "dome" at the DREZ (*arrow*). (C) Axonal regeneration ensues (*arrows* denote Schwann cell myelinated axons) but most terminate at the astroglial dome (D, *arrow*), where they end blindly, turn back on themselves, or degenerate (E) within a convoluted astroglial matrix extending peripherally from the DREZ.

to a dorsal root results in WD extending back into the spinal cord (Fig. 25.11B). Regeneration subsequently ensues (Fig. 25.11C), however, as axons approach the PNS–CNS transition zone, they encounter a convoluted astroglial matrix, with astrocytic processes extending out into the PNS environment. Axonal growth is thereafter very limited resulting in fibers either ending blindly or being deflected away from the DREZ. The altered axonal growth pattern at this interface appears to be associated with the presence of astrocytic processes that are interspersed with Schwann cells and endoneurial connective tissue (comprised of fibroblasts and collagen) that do not become confluent with the CNS tissue domain. Axons can also be found engulfed by astroglial processes and exhibiting degenerative features (Fig. 25.11D).

Because of the complex network of enlarged and intertwined cytoplasmic processes that characterize astroglial scars, one is left with the impression that reactive astrocytes form physical obstacles to growing axons. However, specific biosynthetic properties of these cells also seem to have bearing on glial inhibition of axonal elongation. Several well-defined chemorepellent or growth-inhibitory molecules, some of which are major extracellular matrix (ECM) constituents, have been identified in the CNS with some being produced by reactive astrocytes.

Chondroitin sulfate proteoglycans (CSPGs), which are a complex class of macromolecules, are of particular interest since a number of laboratories have shown that CSPGs are potent inhibitors of axonal growth in tissue culture and *in vivo* both in the CNS and in the PNS. CSPGs consist of a protein core with chondroitin sulfate glycosaminoglycan side-chains. Chondroitin sulfate interactions with proteins in the ECM are responsible for regulating a diverse array of cellular activities. CSPG expression is up-regulated after CNS injury and is especially prominent in regions of glial scar formation.

Cleaving of the glycosaminoglycan side chains with the bacterial enzyme chondroitinase ABC (ChABC) can enhance axonal regeneration and fiber sprouting in SCI and other CNS injury/disease rodent models, as well as in peripheral nerve injuries. Efficacy of ChABC has been extended to cat SCI models in which accelerated recovery of skilled locomotor behavior was observed. Functional benefits of ChABC treatment can be markedly amplified by combining it with rehabilitation.

Other Cellular Constituents of the Glial Scar

Impeded growth of axons is also seen in the presence of neuroglial-2-proteoglycan (NG2)-producing oligodendrocyte precursor cells that reside at the lesion site within the realm of the glial scar. In fact, arrested, dystrophic axon tips are more often associated with NG2 cells than astrocytes. Axons within the CNS exhibit long-distance retraction (or dieback) from the site of injury, and macrophages that infiltrate the lesion and evolving glial scar directly induce this process. NG2 cells, which express CSPGs, actively divide and become interspersed with macrophages. Experimental data indicate that the NG2 cells counteract the dieback-promoting effects of astrocytes and thereby stabilize arrested terminal end bulbs of severed axons for as long as years after injury. Thus the beneficial effect of NG2 cells is counterproductive to potential forward growth of proximal axonal tips. It is unlikely that NG2 cells are the sole cells responsible for glial scar-related inhibition of axonal growth. Other studies of SCI, for example, have shown population of the lesion by a variety of inflammatory cells, as well as activation of normally quiescent ependymal cells around the central canal which ultimately begin displaying a reactive astrocyte phenotype. Fibroblast-like cells also populate glial scars and are derived from surrounding meninges or vasculature in the vicinity of the lesion. The presence of such non-CNS cells provides an additional stimulus for astrocytes to establish new boundaries between CNS and non-CNS cellular environments which can contribute to axonal entrapment.

CSPGs and Neuroplasticity

It is now well established that the CNS possesses considerable neuroplasticity potential. For the purposes of this chapter, the term "neuroplasticity" can be simply defined as persistent neuronal changes after an experience or event (such as CNS trauma). Neuroplasticity can be exhibited in many different ways ranging from anatomical, physiological, pharmacological, and/or molecular changes in neural circuitries above and/or below the level of injury. From a neuroanatomical perspective, injured and uninjured, axons can sprout new fibers that can then grow short distances and establish new synaptic connections that may result in behavioral changes. In that regard, the CNS has the capacity to undergo significant beneficial re-wiring in various circumstances with new connections contributing to either partially restored functions or the acquisition of adaptive or compensatory behaviors.

CSPGs are a major ECM component throughout the normal adult CNS with some proteoglycan family members being intimately involved in formation of a structure produced by neurons and astrocytes called perineuronal net (PNN). The PNN encapsulates synapses as the associated CSPGs are believed to stabilize synaptic contacts during development and after injury. In doing so, CSPGs associated with PNNs are believed to restrict plasticity. It follows that disruption of CSPGs with ChABC should enhance plasticity which experimental evidence relative to various CNS regions (eg, spinal and visual) has demonstrated to be the case.

White Matter–Associated Inhibition

Among other axonal elongation inhibitors, proteins associated with myelin and myelin-producing oligodendrocytes contribute to growth cone collapse and overall regenerative failure in the CNS. One of the group of myelin-associated proteins that can impede axonal outgrowth is referred to as Nogo-A, receptors for which are present on growth cones of regenerating axons. Receptor activation by this inhibitory protein results in activation of molecular pathways within the growth cone that result in depolymerization of cytoskeletal constituents associated with growth cone mobility and calcium influx, both of which can cause growth cone collapse or arrested movement. Various experiments have shown that either an antibody directed at Nogo-A or the application of antagonists to the Nogo receptor itself can enhance axonal sprouting and regeneration with associated behavioral improvements.

While white matter inhibition of axonal growth may still be involved, regeneration and behavioral recovery have not been significantly improved even when the three major myelin proteins—Nogo, myelin-associated glycoprotein, and oligodendrocyte myelin protein—which are thought to limit axonal elongation in the CNS are deleted in triple-mutant mice. Clinical trials with Nogo antibody also failed to demonstrate significant effects that could be attributed to axonal regeneration. Myelin inhibition, on the other hand, appears to play a much greater role in limiting short-distance axonal sprouting which can be an integral aspect of neuroplasticity, as discussed later.

Microglia, Macrophages, and Inflammation After Central Nervous System Injury

Microglia have been estimated to represent as much as 20% of the CNS cellular population. Unlike astro- and oligodendrocytes, these cells are not neuroectodermal derivatives but instead are derived from the yolk sac during embryogenesis and ultimately migrate into the developing CNS in concert with the early establishment of a functional circulatory system. After CNS injury and in disease, microglia are rapidly activated, undergoing morphological transformations from ramified resting cells to amoeboid-activated cells (Fig. 25.12) that migrate toward a lesion site and become involved in scar formation. Overall microglia are part of an inflammatory

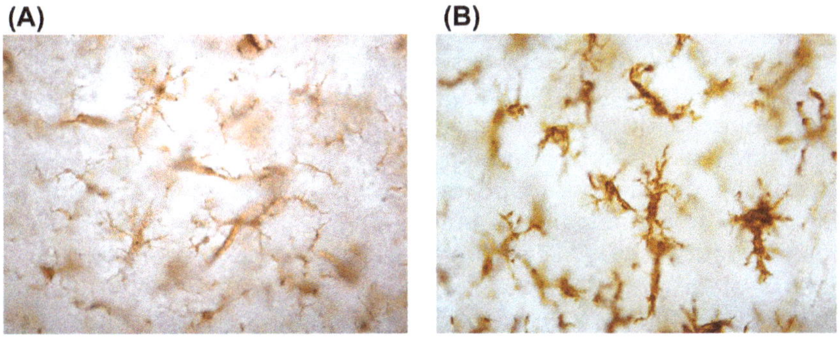

FIGURE 25.12 (A) Resting, ramified (eg, process bearing) microglia are shown as seen in the normal human CNS. (B) Microglial activation ensues rapidly following most CNS perturbations. Under most pathological conditions, these cells increase in number and exhibit hyper-ramification. *Micrographs courtesy of Dr. Wolfgang J. Streit, University of Florida.*

response dominated by macrophages derived from microglia directly or blood monocytes (ie, monocyte-derived macrophages). Microglia within a lesion become activated and phagocytic, much like activated macrophages.

Pro-inflammatory cytokines and other molecules differentially polarize macrophages into M1- (pro-inflammatory) or M2 (antiinflammatory)-activated cell subtypes that can play roles in either tissue repair or damage postinjury, respectively. It has been demonstrated that as in the case of astrocytes, the initial microglial/macrophage response to injury is directed toward wound healing, suggesting early predominance of the M2 phenotype and subsequent conversion to M1. Genetic or pharmacological inhibition of microglial activation results in amplified lesion histopathology and increasingly impaired functional outcomes in injury models. On the other hand, the accumulation of macrophages at lesion sites may have significant ramifications related to axonal growth. In fact, macrophages have been shown to be potent inducers of dystrophic axonal dieback, whereas macrophage depletion reduces such retraction. General inhibition of matrix metalloproteinases (MMP) and specific inhibition of MMP-9 prevents macrophage-induced axonal retraction. The growth-inhibitory effect has been proposed to be a function of the M1 microglial/macrophage subtype which expresses a variety of repulsive guidance molecules upon activation. While our understanding of the complex and contradictory nature of the inflammatory response to injury and disease are far from being complete, the effects of microglial/macrophage populations appears to be dependent on the activation state of the cells and the relative ratio of M1/M2 cells at any given time and place after injury or the onset of disease. Importantly, what is already known has drawn attention to new windows of novel therapeutic opportunity use cell-based and other therapeutic strategies.

CHAPTER OVERVIEW AND THERAPEUTIC PERSPECTIVES

As noted in the introduction to this chapter, the primary goal is to provide fundamental insights into the dynamics associated with axonal injury in the PNS and CNS. In doing so, many experimental details have been intentionally omitted in order to keep emphasis on the basic principles which can provide a framework for advanced literature searches for those interested in pursuing neural injury and repair in more depth.

Over the past several decades, there has been steady progress in determining molecular mechanisms that are responsible for the cellular dynamics triggered by axonal injury and factors that affect regenerative potential in the PNS and CNS. WD can no longer be considered a passive event, but one that involves a complex molecular-genetic signature distinct from neuronal apoptotic mechanisms. Neurons within or projecting axons to the PNS are capable of mounting a vigorous growth response after axotomy with Schwann cells and inflammatory responses to peripheral nerve injury playing a significant role establishing favorable conditions for axonal elongation.

While repairing an injured nervous system still represents a formidable hurdle, advances over the last three decades have identified many therapeutic possibilities for restoring useful functions and improving the quality of life for affected individuals. In the PNS, the issue is not of whether or not injured axons can regenerate, but more a matter of accuracy and overcoming undesirable neuroplastic changes in the CNS that ensue peripheral nerve injury. In addition, a variety of biomaterials and scaffolding designs have been identified to improve axonal growth into and through sites of injury. Whether such scaffolding approaches can alone be exploited to improve the directionality and rate of axonal elongation beyond the lesion site remains to be seen. The application of hormones, electrical stimulation and other strategies, including ChABC treatments in conjunction with scaffolding approaches holds considerable promise for improving functional regeneration in the PNS.

The challenges of neuronal injury and axonal regrowth are much greater in the CNS and must also take into account the likely need to replace neurons which die as a consequence of injury or disease. Neurons with cells and axonal and dendritic processes entirely restricted to the CNS do not possess the same degree of regenerative capacity as neurons in the PNS or those neurons in the brain stem and spinal cord, which extend axons into the PNS. The problem appears even more daunting when considering the inhibitory cellular and molecular milieu of the CNS. Advances are being made, however, with the emergence of genome-wide expression profiling and high-throughput screening strategies to enable discovery of novel regulators of axon growth. The identification of CSPGs and other inhibitory molecules in the CNS has likewise provided new therapeutic options as exemplified by ChABC. Studies of the CNS inflammatory response to injury have also provided new insights into the complex cellular environment of the glial scar and the diverse nature of macrophages and their respective roles in tissue repair and regenerative failure.

Meanwhile, numerous demonstrations of the neuroplastic potential of the CNS have provided instructive examples where even short-distance axonal growth can lead to the establishment of new and functionally beneficial neuronal connections. While long-range growth of injured axons, as in the case of SCI, remains the ideal objective, it is likely that more immediate benefits can be derived by enhancing restorative or compensatory plasticity. It also has become clear that no single therapeutic approach will provide optimal results. Rather, a carefully orchestrated combination of treatments, including rehabilitation, will most likely yield the most favorable outcomes. Further progress in neural repair research, based upon the basic principles presented in this chapter will undoubtedly continue to reveal exciting prospects for improving clinical outcomes following a range of functionally devastating disorders.

Further Reading

[1] Bradbury EJ, Carter LM. Manipulating the glial scar: chondroitinase ABC as a therapy for spinal cord injury. Brain Res Bull 2011;84(4–5):306–16.

[2] Stoll G, Muller HW. Nerve injury, axonal degeneration and neural regeneration: basic insights. Brain Pathol 1999;9(2):313–25.

[3] Coleman MP, Freeman MR. Wallerian degeneration, Wld(S), and Nmnat. Annu Rev Neurosci 2010;33:245–67.

[4] Hu X, Leak RK, Shi Y, et al. Microglial and macrophage polarization—new prospects for brain repair. Nat Rev Neurol 2015;11(1):56–64.

[5] Silver J, Miller JH. Regeneration beyond the glial scar. Nat Rev Neurosci 2004;5(2):146–56.

[6] Hur EM, Saijilafu, Zhou FQ. Growing the growth cone: remodeling the cytoskeleton to promote axon regeneration. Trends Neurosci 2012;35(3):164–74.

[7] Silver J, Schwab ME, Popovich PG. Central nervous system regenerative failure: role of oligodendrocytes, astrocytes, and microglia. Cold Spring Harb Perspect Biol 2015;7(3):a020602.

[8] Gordon T, English AW. Strategies to promote peripheral nerve regeneration: electrical stimulation and/or exercise. Eur J Neurosci 2016;43(3):336–50.

CHAPTER

25.1

Translational Correlation: Sarcopenia

B. Hwang, J.C. DeToledo

Texas Tech University Health Sciences Center, Lubbock, TX, United States

SARCOPENIA

Age-related reduction of muscle mass is the predominant change of body composition in healthy elderly people. Sarcopenia is the term used to describe the loss of muscle tissue as a natural part of the aging process. In physically inactive people, after the age of 30, there is a loss of about 3–5% of muscle mass per decade and a parallel decline in muscle strength. Men and women are equally affected; it often begins in the fourth decade with a prevalence that increases from ±5% in the sixth decade to ±10% by the eight decade of life. Although sarcopenia is mostly seen in physically inactive individuals, it is also evident in individuals who remain physically active throughout their lives suggesting that physical inactivity is not the only contributing factor to sarcopenia.

Although neurons are the longest living cells in the body, the number of spinal cord motor neurons continuously declines with age and this is considered an irreversible process. Human nerve cells have a predetermined life span and the loss of muscle fibers begins with the loss of motor neurons. Motor neurons die with age resulting in a denervation of the muscle fibers within the motor unit. This denervation causes the muscle fibers to atrophy and eventually die, leading to a decrease in muscle mass. When a motor neuron dies, an adjacent motor neuron may reinnervate the muscle fibers, preventing atrophy. These motor neurons tend to have less precise control of movements, less force production, and slowing of muscle mechanics, and this may help explain the loss of balance and speed of movement with age.

Further Reading

[1] Hwang B, Lim JY, Lee J, Choi NK, Ahn YO, Park BJ. Prevalence rate and associated factors of sarcopenic obesity in Korean elderly population. J Korean Med Sci July 2012;27(7):748–55.

Conn's Translational Neuroscience
http://dx.doi.org/10.1016/B978-0-12-802381-5.00041-5

CHAPTER

26

Audition

J.K. Bizley

University College London, London, United Kingdom

The auditory system provides our only long-range panoramic sensory system and is fundamental to human communication. Hearing loss (HL) has profound socioeconomic consequences: with an increasingly aged population HL can be a debilitating condition that leads to social isolation.

WHAT IS SOUND?

Sounds are emitted by or from people, objects, and events. But what is a sound? Sound is a pressure wave, a vibration in the air. Just like water, air is a fluid, comprised of loosely packed molecules. When an object vibrates—for example, imagine a drum skin moving after being struck—air molecules are displaced causing those next to the drum skin to be compressed. As the drum skin moves backward a pocket of air is stretched. These alternating pockets of compressed and stretched air form the sound wave that travels through the environment and may, ultimately, be detected by the ears. The way in which an object vibrates determines the sound that it emits: for example, a large bell being struck will emit relatively low-frequency oscillations, whereas if we strike a glass, the air around it vibrates more rapidly causing higher-frequency sound waves to be emitted. In turn we perceive the bell as making a low pitch sound, whereas the glass makes a higher pitch note. A pure tone is a sound comprised of only a single frequency component: a sine wave that can be defined according to its frequency and amplitude. The frequency, that is, how many times a second it vibrates, will determine its perceived pitch with higher-frequency sounds having a higher pitch. The amplitude will determine the perceived loudness. However, real-world sounds are usually comprised of many different sound frequencies and the relationship between the frequency components present in a sound and how that sound is perceived is not necessarily straightforward. Nevertheless, using the Fourier transform, we can describe any sound according to the collection of frequencies that it is comprised of and, as we shall see shortly, the way in which the auditory system encodes sound is, to a first approximation, a spectral analysis analogous to a Fourier transform.

THE EAR

The ear is exquisitely designed to detect sound waves. The role of the outer and middle ear (Fig. 26.1A) is to detect, transmit, and amplify sound waves in the air and transmit these to the fluid

Conn's Translational Neuroscience
http://dx.doi.org/10.1016/B978-0-12-802381-5.00042-7

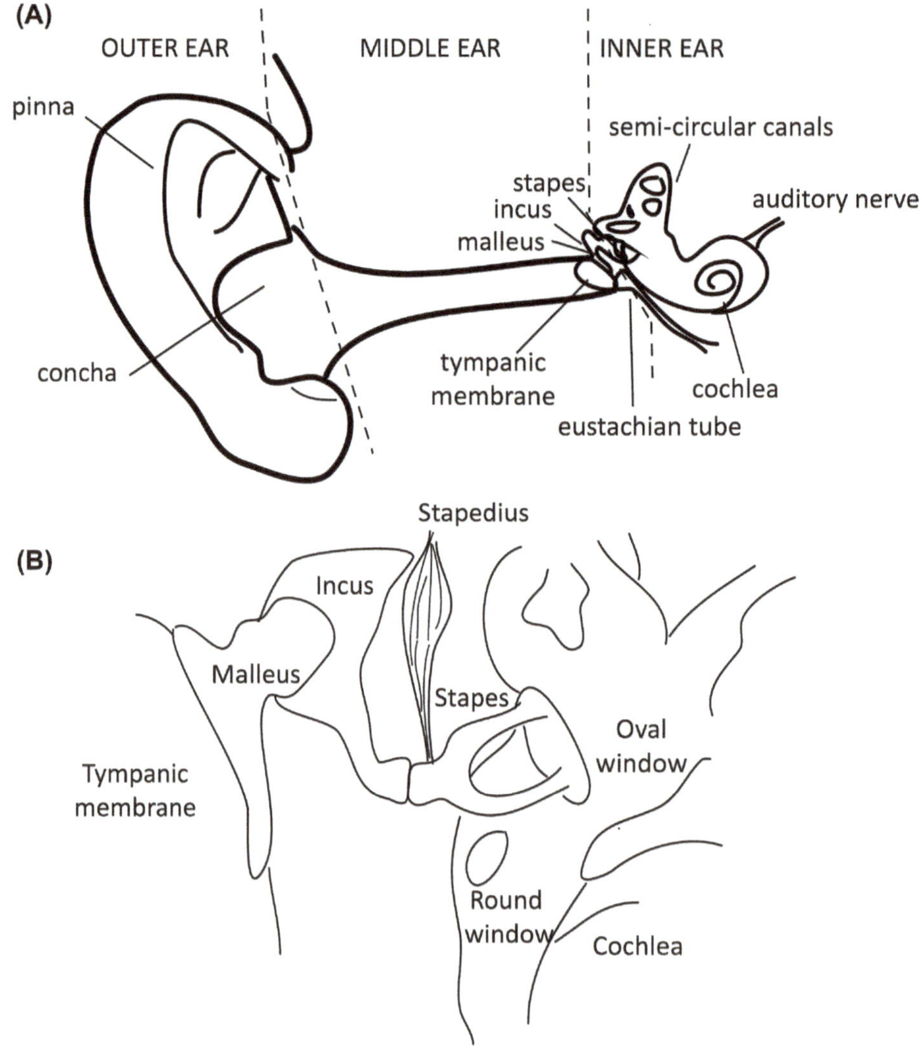

FIGURE 26.1 **The ear.** (A) Anatomy of the ear. Shown are the major structures that comprise the outer, middle, and inner ear. (B) Ossicles. The *malleus, incus,* and *stapes* transmit the vibrations of the tympanic membrane to the inner ear via the contact between the stapes footplate and the oval window. The *stapedius* muscle can act to dampen vibrations providing protection against loud sounds for the delicate machinery of the inner ear.

of the inner ear. The role of the outer ear is to "collect" sound by funneling it toward the ear canal. The pinnae also act as filters adding information about the direction of the incoming sound source relative to the head (see section: How do we Localize a Sound?) The middle ear contains the eardrum, or tympanic membrane. Sound waves in the air cause the tympanic membrane to vibrate. It in turn stimulates the ossicles, a small chain of bones called the malleus, incus, and stapes, thus transmitting the vibration from the tympanic membrane to the oval window of the cochlear. Such a complicated structure is necessary to overcome the impedance mismatch between the air and the fluid of the cochlear: water has a much higher impedance than air and passive transmission would result in a 30dB (99.9% loss) attenuation of the incoming sound wave. The physical structure of the middle ear facilitates this: the eardrum has a much larger surface area than the stapes footplate, which makes contact with the oval window. Much like hitting the head of a nail with a hammer, by focusing the force of the tympanic membrane vibration on a much smaller area, the force is increased. Secondly, the arm of the malleus is longer than that of the incus ensuring that it acts as a lever to further amplify vibrations. Lastly, the way in which the tympanic membrane is deflected and buckles in response to sound amplifies vibrations. The combination of the funneling effect of the outer ear, and the amplifying effects in the middle ear, means that some sound frequencies are boosted more effectively than others, with sounds between roughly 300Hz and 2–3kHz being most efficiently transferred through the middle ear. This transfer function leads to our audiogram displaying a similar shape with our sensitivity being the greatest to sounds in this frequency range and sounds from 20 to 20,000Hz being audible. The gain applied by the middle ear ossicles is adaptive—that is, it can be decreased by activating the stapedius muscle, which acts to limit the movement of the stapes (Fig. 26.1B). This is one of the mechanisms that protect the ear from exposure to very loud sounds.

THE COCHLEA

The inner ear houses the cochlea, along with the semicircular canals, and vestibular organs. The cochlea is a coiled structure, in humans about 1cm wide and about 5mm in length from base to apex. It is connected to the middle ear via the contact between the footplate of the stapes and the oval window. The cochlear comprises three compartments: the *scala tympani*, the *scala media*, and the *scala vestibuli* (Fig. 26.2A). Vibrations are transmitted to the cochlear via the oval window in the *scala vestibuli* and exit via the round window in the *scala tympani*. Both the *scala tympani* and *scala vestibuli* contain perilymph, a fluid containing low potassium ion concentration. In contrast, the endolymph within the *scala media* has a high potassium concentration. Recall that the extracellular fluid usually has a high sodium ion concentration and a low potassium concentration. The high potassium concentration in the *scala media* must be actively maintained and this is achieved via the richly vascularized *stria vascularis*. As we shall see shortly, unlike other sensory receptor cells and most neurons where depolarization is triggered by the inward movement of sodium or calcium ions, the flow of potassium ions down their electrochemical gradient into the cell triggers depolarization in the sensory receptor cells of the cochlea. Dividing the *scala media* from the *scala tympani* is the basilar membrane on which the Organ of Corti sits.

The structure of the basilar membrane changes from base to apex of the cochlear with the basal end being both narrower and stiffer (Fig. 26.2C). Consequently, the way in which the basilar membrane vibrates depends on its location: the basal end preferentially vibrates to high-frequency sounds and the apical end to low-frequency sounds. Thus, vibrations form what Von Békésy termed a "traveling wave" that peaks at a certain location on the basilar membrane. This frequency tuning occurs as a result of the physical structure of the basilar membrane, and can be observed in cadavers. However, in living animals, frequency tuning is further enhanced by active processes described in more detail further.

The organ of Corti (Fig. 26.3) is the sensory receptor organ responsible for transmitting information about sound to the brain. Within it are three rows of outer hair cells (OHCs) and one row of inner hair cells (IHCs) as well as various supporting cells (Fig. 26.3). The hair cells are so called due to the bundles of stereocilia that

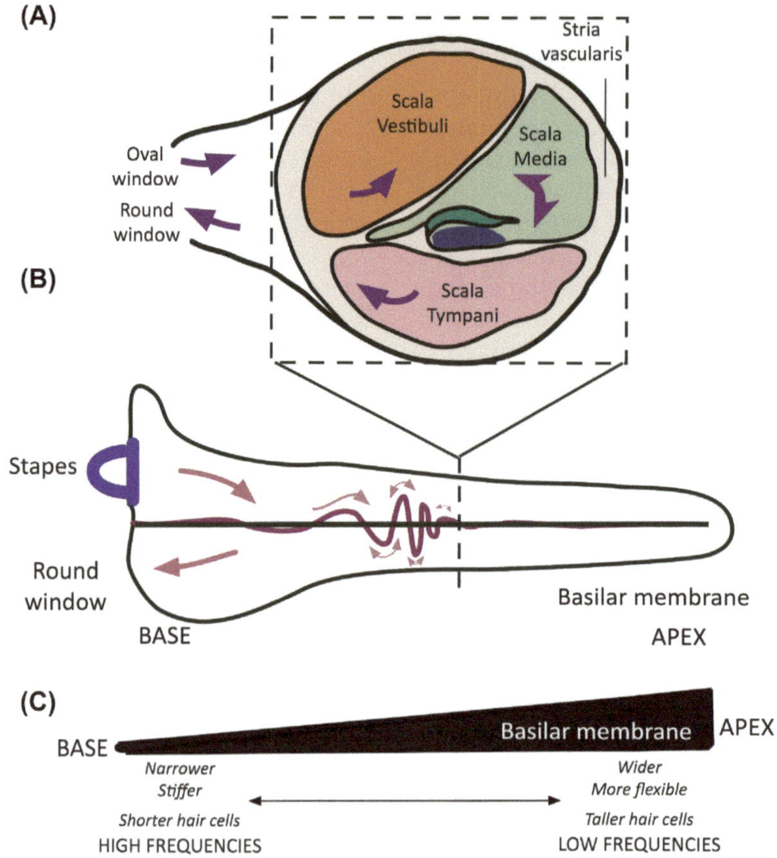

FIGURE 26.2 **The cochlea.** (A) Cross section through the cochlea. The inner ear is composed of three compartments—the *scala vestibule*, *scala media*, and *scala tympani*. Vibrations are introduced at the oval window and exit via the round window. The basilar membrane, where the sensory hair cells are located, sits at the base of the *scala media*. (B) Lateral view of the cochlear showing the formation of a traveling wave. (C) Mechanical properties of the basilar membrane. The basilar membrane varies in both thickness and stiffness. These physical properties effectively tune the membrane to vibrate optimally at specific sound frequencies such that the basal end vibrates in response to high frequencies, while the apical end vibrates to low frequencies.

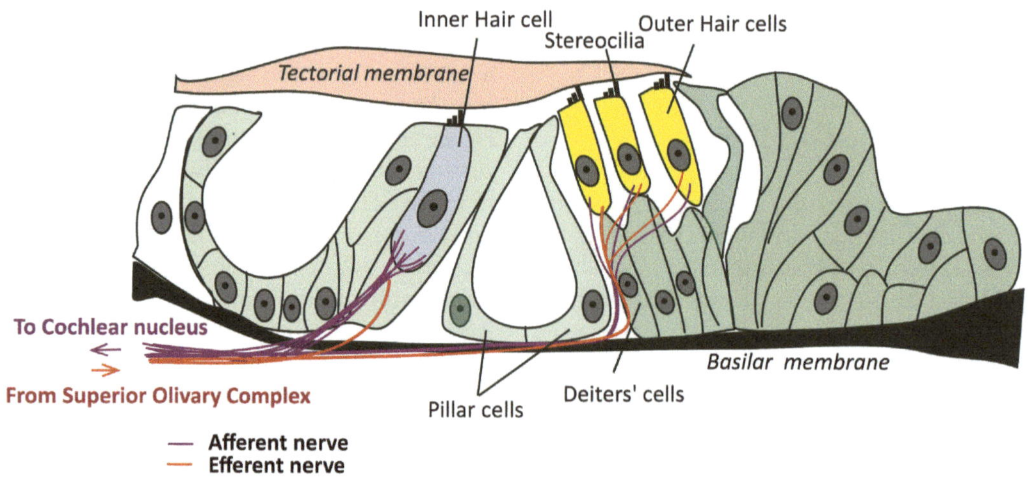

FIGURE 26.3 **The organ of Corti.** A schematic depicting a cross-section through the basilar membrane showing, in addition to the various classes of supporting cells, the location of the three rows of outer hair cells and the inner hair cells, including their afferent (purple) and efferent (red) innervation.

sit on their apical surface. The tectorial membrane sits on top of the organ of Corti and the OHC stereocilia make contact with it. Vibrations of the basilar membrane cause deflection of the hair cells. Tip-links between rows of stereocilia attach to an elastic gating spring, which means that as the stereocilia bend, ion channels on the stereocilia are opened. Because the ion channels on the stereocilia are physically gated, they can respond extremely rapidly to vibrations of the basilar membrane. This rapidity is in contrast to, for example, the slower second messenger signaling cascade that allows light-sensitive cells in the retina to convert light signals into depolarization. The opening of ion channels on the stereocilia causes an influx of potassium ions into the hair cell, which flow down their electrochemical gradient into the stereocilia. Both IHCs and OHCs respond to deflection by depolarizing. However, the consequence of depolarization in both cell types is rather different. In the IHC, this causes a change in membrane potential from −70 to −60 mV which is sufficient to trigger the opening voltage-gated calcium channels on the hair cell body, which in turn cause neurotransmitter release at the base of the hair cell where the hair cell synapse is located. The OHC responds to a change in voltage in a rather different way: the cell membrane of the OHC is filled with a protein called *prestin* that changes shape in the presence of a change in voltage. It is sufficiently densely packed within the cell membrane that a change in voltage triggers a change in length of the OHC. What is the consequence of this change in length? The electro-motility exhibited by the OHC has a twofold effect: it amplifies the vibrations at the basilar membrane and causes a sharpening of the traveling wave. Consequently, the OHCs are crucial to both our sensitivity to very quiet sounds and our ability to resolve small differences in sound frequency. The OHCs are vulnerable to damage by noise and ototoxic drugs.

The role of the OHCs in amplifying quiet sounds results in the healthy cochlear actually *generating* sounds: the ability to evoke an otoacoustic emission (OAE) forms the basis of a simple noninvasive test for screening newborn infant hearing. Distortion product OAEs (DPOAEs) result when two tones (f_1 and f_2) are played to the ear and can be recorded via a microphone at the entrance to the ear canal. This results in OAEs at $2f_1-f_2$ (used for clinical testing) and at the difference tone f_2-f_1. Thus by simply combining a microphone and a headphone, in order to both play and record sound, it is possible to rapidly and noninvasively assess cochlear function.

THE AUDITORY NERVE

Two types of afferent auditory nerve fibers (ANFs) innervate the cochlear and transmit information to the central nervous system: Type I fibers target IHCs and are most numerous making up 90–95% of the total number of fibers. Each Type I fiber innervates only a single IHC, with up to 30 fibers innervating each IHC. Type I fibers are thick, myelinated bipolar cells whose cell bodies reside in the spiral ganglion. The remaining fibers are Type II fibers, and these innervate OHCs. They are thin, unmyelinated monopolar cells and can innervate up to 10 OHCs (Fig. 26.3).

Recordings from ANFs demonstrate that each is tuned to a particular sound frequency (Fig. 26.4A). The properties of a sound-responsive neuron are usually characterized by measuring a "frequency-response area (FRA)." This involves presenting pure tone stimuli at different combinations of frequency and sound levels and quantifying the spiking response elicited in a given neuron by each frequency–level combination. Typical ANFs from a healthy cochlear have "V shaped," FRAs at low sound levels they respond to a restricted range of sound frequencies. As sound levels are increased, the width of the tuning curve broadens to encompass more frequencies. The characteristic frequency of an ANF is the frequency of the lowest intensity sound that elicits an increase in spiking

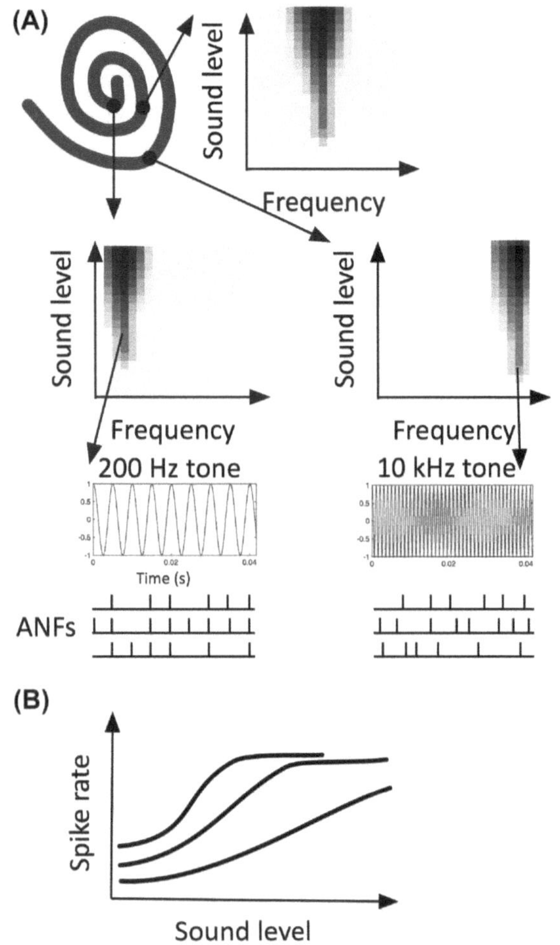

FIGURE 26.4 **Auditory nerve fiber.** (A) ANF responses at three points on the cochlear (positions marked in the top left panel). Apical ANFs (bottom left) exhibit low-frequency tuning as shown in the corresponding frequency-response area (FRA) which shows the spiking response (grayscale, *darker colors* indicate a stronger response) for a range of pure tones presented at different frequencies and sound levels. More basal ANFs have progressively higher-frequency responses. Below the high- and low-frequency FRAs are the example spiking responses of three ANFs in response to a pure tone at the neurons best frequency. The low-frequency ANFs exhibit phase-locking—that is, their discharges are locked to the phase of the sound wave. In contrast, in the high-frequency fiber the spikes are not precisely timed. (B) Three classes of ANFs have been described according to their spontaneous rate. Low, medium, and high spontaneousrate fibers are activated by high-, medium-, and low-intensity sounds, respectively. High spontaneousrate fibers saturate at relatively low sound intensities, whereas low spontaneousrate fibers are able to respond to an increase in sound intensity with an increase in firing rate across a wider range of sound levels.

above baseline. ANF responses in a damaged cochlear are likely to have an elevated threshold and display broader tuning. Since each ANF contacts only one IHC, the position of the IHC on the basilar membrane is the key determinant of the characteristic frequency of ANF (Fig. 26.4A). The spike discharge of an ANF can convey information not simply in its rate, but also in its pattern. For low sound frequencies, the discharge pattern of the ANF reflects the waveform of the sound—that is, spikes are elicited only once per period of the sound wave. This is a consequence of the way in which the IHCs depolarize and hyperpolarize in response to sound—as the stereocilia are periodically deflected by the basilar membrane vibrations, the ion channels on the stereocilia are also periodically opened and closed once per stimulus cycle. This periodic alternation between hyperpolarization and depolarization will, in turn, lead to periodic neurotransmitter release and spike patterns in the ANF. Above a critical frequency the ion-channel kinetics are no longer able to follow the dynamics of the stereocilia and preserve the temporal relationship between the stimulus waveform and the spiking output. Instead, a steady train of spikes are elicited. While the upper limit of phase-locking has not been measured in humans, in other mammals it declines above 1–2 kHz and is absent for frequencies above 5 kHz. It is not necessary for an individual ANF to fire on every cycle of the stimulus waveform, only that it fires at a fixed point during the waveform period: accurate phase-locking can then be created across a population of ANFs (Fig. 26.4A, bottom panel).

ANFs vary in their spontaneous rate (Fig. 26.4B), as well as their sensitivity and dynamic range. High spontaneous–rate fibers have a low threshold, requiring very little acoustic input to increase their spike rate above baseline. However, they rapidly saturate at relatively modest sound levels. Medium and low spontaneous–rate fibers have lower spontaneous rates and require moderate and high sound levels, respectively, to raise their firing rates above their spontaneous baseline. Medium-rate fibers saturate at high sound levels, while low spontaneous–rate fibers are capable of encoding increases in sound level as an increase in spike rate at high sound levels.

One of the greatest challenges faced by the auditory system is maintaining sensitivity to sound pressure changes over the 10^{10} range that we typically experience. Sound levels are measured in decibels (dB), a logarithmic unit with an increase in sound level of 10 dB corresponding to a 10 times increase in power. The threshold for hearing is around 0 dB and we experience sounds in excess of 100 dB, although sounds of 95 dB and above can damage our hearing. Our ears have several protective mechanisms to prevent damage by very loud sounds such as contracting the stapedius muscle in the inner ear and inactivating the OHC gain. We are able to encode sounds over such a large range due to the dynamic nature of the gain provided by the OHCs, in addition to the variety of ANF fiber types that allow sensitivity to be maintained throughout a large intensity range.

CENTRAL AUDITORY PROCESSING: COCHLEAR NUCLEUS

ANFs send their axons via the VIII cranial nerve to the cochlear nucleus (CN, Fig. 26.5). The CN comprises three major divisions: the dorsal

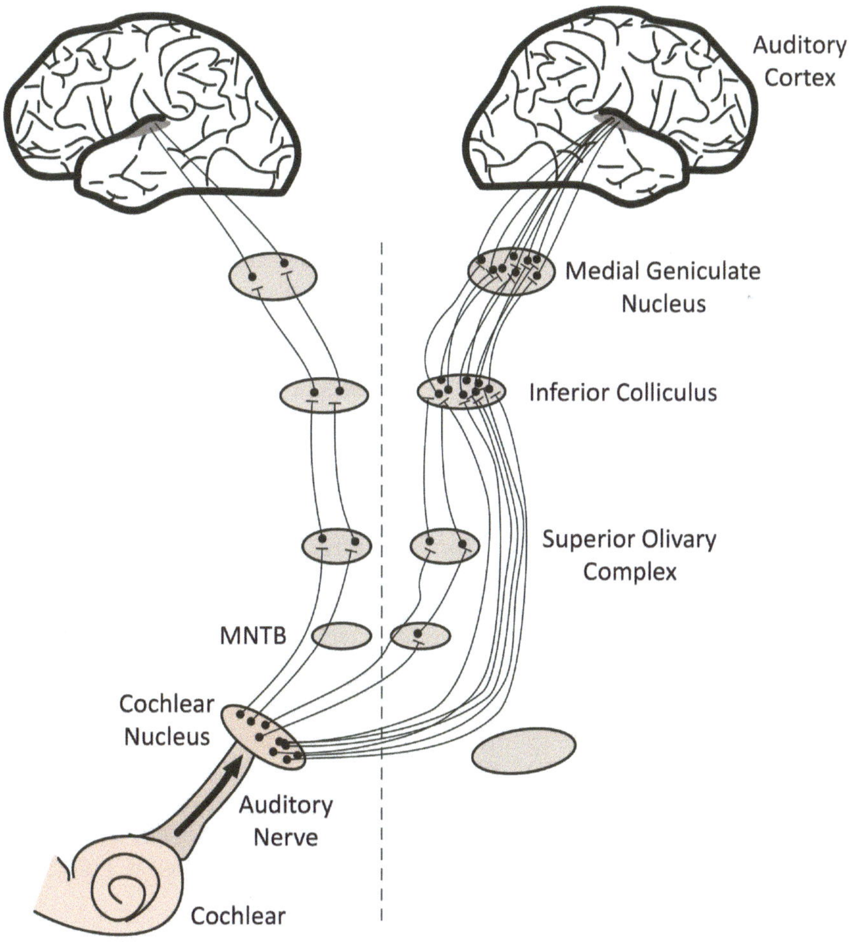

FIGURE 26.5 **Central auditory pathways.** Schematic showing the major ascending pathways originating from the left cochlear.

(DCN), the anteroventral (AVCN), and posteroventral (PVCN). Thus, even at the earliest stage of the auditory brain, signals are processed in parallel as the auditory nerve branches to form a posterior and an anterior branch. The posterior branch innervates the DCN and PVCN, while the anterior branch innervates the AVCN. The CN preserves the tonotopic organization of the cochlea as the projections from the cochlear are topographically organized. This tonotopic organization, whereby neighboring neurons share a common characteristic frequency, and the frequency tuning of neurons varies systematically from high to low from one end of the nucleus to the other, is a characteristic that is preserved throughout the central auditory system, all the way to auditory cortex.

Within the cochlear nuclei different cell types exist, each with a characteristic morphology and a distinct physiological response to sounds. Within the AVCN there are two major cell types: globular bushy cells and spherical bushy cells. Spherical bushy cells are innervated by a few (1–4) ANFs, which form giant synaptic terminals known as the end-bulbs of Held. At low sound levels, activating one end-bulb of held is sufficient to cause spiking in the spherical bushy cell ensuring that the spherical bushy cell can follow the firing of ANFs. As sound levels increase, inhibitory inputs reduce the sensitivity of the spherical bushy cells so that a larger number of inputs are required to drive the cell to spike. Consequently, the temporal precision of the spherical bushy cell is in fact greater than that of its individual inputs. Spherical bushy cells project to the medial and lateral superior olivary nucleus (MSO and LSO). Compared to spherical busy cells, globular bushy cells are innervated by a greater number of inputs (4–40), each of which form a smaller end-bulb of Held. The globular bushy cells project to the medial nucleus of the trapezoid body (MNTB) on the contralateral side of the brain. The MNTB is an inhibitory nucleus that projects to the LSO and together with the MSO and LSO plays an important role in extracting information about where in space a sound originates from (see section: How do we Localize a Sound?). In summary therefore, the AVCN preserves and enhances the temporal precision with which auditory information is encoded. Frequency-response areas resemble those of ANFs being narrow and V shaped.

Cells in the PVCN are influenced by a broader range of sound frequencies than those in the AVCN. Their frequency-response areas are, correspondingly, broader than those in the ANFs. Major cell types here include stellate cells and octopus cells. Octopus cells in particular integrate information across a wide frequency range, but fire only at the onset of a sound. Therefore, they are able to represent the temporal fluctuations of a broadband stimulus with great fidelity locking to stimuli at up to 800 Hz. They are innervated by a large number of ANFs (60+) and require many of these inputs to be driven simultaneously in order to fire. The stellate cells have narrow excitatory areas and wider inhibitory ones and send collaterals widely within the cochlear nuclei. Octopus cells are also known as chopper cells, as they often fire at regularly spaced intervals in response to a sustained stimulus. The rate of the chopping pattern is intrinsic to the cell rather than dictated by the frequency of the sound.

Like cells within the PVCN, cells in the DCN have response properties that are considerably more complex than those observed in ANFs. The DCN contains several different cell types, many of which have complex frequency-response areas that can include inhibitory areas. For example, type IV cells are excited only by very specific frequency–level combinations and are thought to be responsible for detecting spectral notches which underlie spectral cue processing for monaural sound localization and in particular for localization in the vertical plane.

As noted before, the AVCN projects to the MSO, LSO, and MNTB, which, in turn, project to the inferior colliculus (IC) which is located in the midbrain. Information from the MSO is relayed ipsilaterally via the lateral lemniscus to the midbrain, while the LSO transmits information contralaterally to the midbrain. The DCN and PVCN project both directly to the IC and also additionally send branches to the dorsal and lateral nucleus of the lateral lemniscus. These in turn also project to the IC (Fig. 26.5).

THE INFERIOR COLLICULUS

The IC is a near obligatory relay for ascending auditory information (Fig. 26.5). The primary, or lemniscal, auditory pathway runs through the central nucleus of the IC (ICCN). Here, information is once again represented in a tonotopic manner, with bands of cells sharing common characteristic frequencies. Information from the many different brain stem sources outlined earlier are brought together within an isofrequency laminae. Cells within the ICCN typically display sharp frequency tuning, but within an isofrequency laminae cells vary in their tuning to other stimulus features—notably in their binaural response properties and responses to amplitude modulation.

In addition to the primary ascending pathway through the ICCN, a nonlemniscal route through the IC runs through the dorsal cortex and exterior nucleus of the IC. Cells in the nonlemniscal IC tend to be more broadly tuned than those in the ICCN and show more rapidly habituating responses. The external nucleus is innervated by the DCN, whereas the dorsal cortex is innervated by the contralateral IC and is also targeted by descending inputs from the auditory cortex.

THE MEDIAL GENICULATE NUCLEUS

Like the IC, the medial geniculate nucleus (MGN) also comprises of multiple subdivisions. The ventral nucleus forms part of the lemniscal pathway and is the main ascending route for information transfer from the ICCN to the primary auditory cortex. The ventral division is tonotopically organized with bands of similarly tuned cells forming isofrequency laminae. Cells in the ventral division have sharp frequency tuning and show sustained responses to sounds.

The medial and dorsal divisions also convey information to auditory cortex. The medial division has a wide range of auditory, multisensory, and vestibular inputs as well as inputs from the ICCN. The medial division projects widely to both primary and nonprimary auditory cortex. Cells here are broadly tuned, often having multiple peaks in their frequency-response area and responds to multimodal stimuli. The dorsal division is principally innervated by the dorsal division of the IC, as well as receiving inputs from somatosensory nuclei. It projects predominantly to secondary auditory cortical areas. Cells within the dorsal division have short latencies and sharply tuned responses.

THE AUDITORY CORTEX

The auditory cortex comprises multiple sound responsive fields that are hierarchically organized. The "core" auditory area includes primary auditory cortex and areas within the core are characterized by being predominantly innervated by the ventral division of the MGN. Cells in the core areas have relatively short latencies and have well-defined frequency-response areas. Responses of cells have low thresholds and are well tuned to a restricted range of pure tones. While frequency-response areas are typically a little broader than those in the midbrain and thalamus, like stations earlier in the auditory pathway, these areas are tonotopically organized. Secondary, or belt, auditory areas are predominantly innervated by the core auditory areas, in addition to contributions from the auditory thalamus (mainly medial and dorsal divisions of the MGN). Secondary areas may be tonotopically organized and cells may show frequency tuning in response to pure tones or narrowband noise. Parabelt areas represent the next level in the auditory hierarchy. They are principally innervated by belt areas and project to auditory association areas concerned with memory or decision-making. Cells here are unlikely to respond to pure tones and prefer complex stimuli such as vocalizations. In nonhuman primates, anatomical tract tracing experiments delineate two projection pathways emanating from core auditory cortex and traveling either caudally via the posterior auditory fields to dorsolateral prefrontal cortex (PFC) or rostrally via the anterior belt to the ventrolateral PFC. These pathways are hypothesized to have distinct behavioral roles mirroring the "what/where" division proposed in visual cortex. The anterior processing pathway is thought to be key for extracting sound identity, while the caudal pathway functions to encode a sound source's location). Alternative models challenge this clear division of function, or, instead, propose a division of function between perception and action ("what/how").

DESCENDING CONNECTIONS IN THE AUDITORY SYSTEM

As with other sensory systems, feedback connections equal or outnumber feed-forward connections between cortex and thalamus. Descending connections from auditory cortex target both the IC and auditory thalamus and are thought to be crucial for active listening. Cortico-collicular connections have been shown to play a key role in auditory plasticity and appear crucial for some sorts of learning, and studies that inactivate auditory cortex while recording neuronal responses in the IC or MGB show that the cortex exerts a profound influence on the response properties of many neurons. In turn, the IC sends descending connections to the superior olivary complex and cochlear nucleus.

The olivo-cochlear efferent system runs from the olivary complex directly to the hair cells in the cochlea. The cochlear efferent system can be divided into the medial and lateral systems. Lateral olivo-cochlear (LOC) fibers originate from small neurons in and around the lateral superior olive (LSO) and terminate on the dendrites of afferent ANFs underneath the IHCs. They are therefore well placed to change the firing rates of ANFs. Indeed, stimulation of the lateral system has been shown to shift the rate-level functions of ANFs toward higher sound levels, thereby increasing the dynamic range of the ANFs. Medial olivo-cochlear fibers originate from the relatively large neurons around the MSO and terminate directly on the cell bodies of the OHC. Activation inhibits the extent of the mechanical amplification provided by the OHCs at the basilar membrane. By controlling the response magnitude of the OHCs the medial system is therefore well placed to modify and control the gain of the cochlear amplifier.

HOW DO WE LOCALIZE A SOUND?

As we have seen earlier, the cochlea encodes sound frequency. Since the frequency content of a sound offers little information about its location in space, the spatial origin of a sound source must be computed by the brain using various cues. These cues fall into two types: first, binaural cues that are dependent on comparing the signal available at the two ears, and second, monaural cues that are available from one ear alone. Binaural cues result from the separation of the ears on the head, which facilitate a comparison of the sound signal at the two ears. A sound source to the right of the head will be louder in the right ear than the left, and will arrive sooner in the right

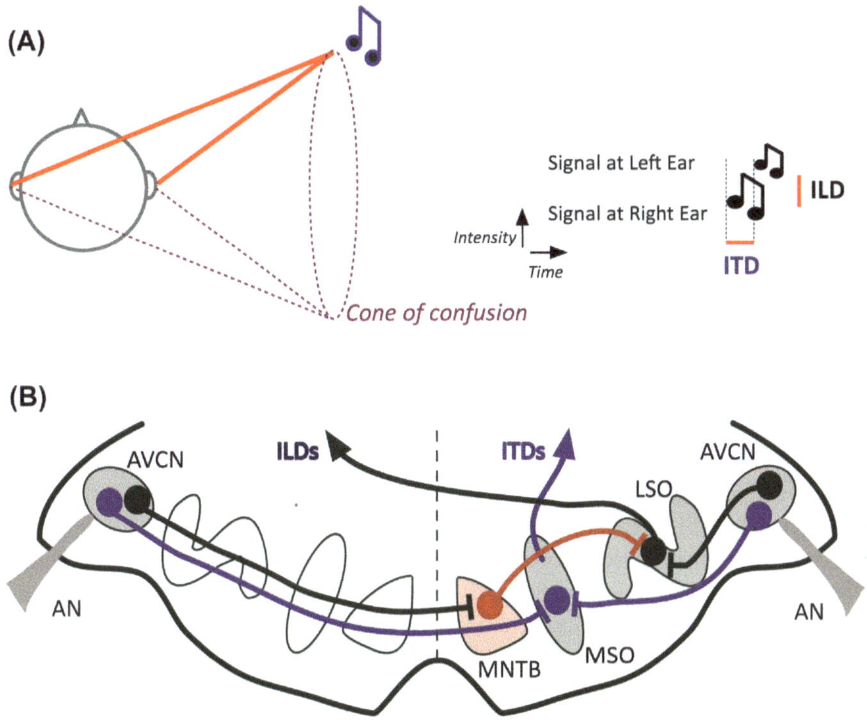

FIGURE 26.6 **Sound localization cues and pathways.** (A) Sounds to the right of the head will arrive earlier and be louder in the right (near) ear than in the left (far) ear resulting in two types of binaural localization cues: interaural timing and interaural intensity differences. The cone of confusion describes the ambiguity that results because any given binaural cue does not uniquely specify a single location in space. (B) Major pathways involved in the extraction of ITDs (shown in blue) and ILDs (shown in black). Centers that send excitatory outputs are shown in gray, those that send inhibitory outputs are shown in red.

ear than in the left (Fig. 26.6A). A difference in sound level between the ears (interaural level difference, ILD) is a consequence of the shadowing effect of the head and is most effective for high-frequency sounds. Low-frequency sounds elicit less-effective ILDs as their longer wavelength means they can effectively bend around the head. In contrast, ITD cues are most effective for low-frequency sounds; the encoding of ITDs requires neural responses to be temporally precise. For an ongoing sound, there will be a difference in phase between the two ears—an interaural phase difference (IPD), the magnitude of which is a product of the azimuthal location of the source and the frequency of the sound. Thus, neuronal responses must also be capable of transmitting the phase of a sound signal faithfully. Once the IPD exceeds 180 degrees it becomes ambiguous placing an upper frequency limit on the utility of time/phase cues for localization. This upper limit is determined by the maximum range of ITDs experienced, which in turn is decided by the size of the head, with larger heads yielding larger ITDs. This, coupled with the upper limits to phase-locking, leads to the *duplex theory* of sound localization which proposes that we use ITDs for low-frequency sounds and ILDs for high-frequency sounds. In addition to these binaural cues, monaural spectral cues also play an important role in sound localization—in particular in the midline, where no interaural cues are available, and for resolving the front–back ambiguities that result from the cone for confusion (Fig. 26.6A). Finally, monaural spectral cues result from the filtering effect of the pinnae and outer ear, which impose characteristic location-dependent spectral features on the incoming sound source. These cues are particularly helpful for resolving sounds in the midline, where no binaural cues are available, and for sounds that fall on the cone of confusion. However, in order to utilize such cues the listener must be able to disambiguate the spectral characteristics introduced by the outer ear from those present in the initial sound source.

Binaural cues are extracted in parallel in the midbrain by neurons in the superior olivary complex (Fig. 26.6B). The LSO receives direct excitatory input from the ipsilateral AVCN and indirect inhibitory input from the contralateral AVCN via the medial nucleus of the trapezoid body (MNTB). Thus, a large response will be driven in the LSO by a sound which is louder in the ipsilateral ear. In contrast, in LSO contralateral to the sound source inhibition from the MNTB will be attenuating the LSO response. A sound presented in the midline will elicit equivalent responses in both LSOs. The medial superior olive is responsible for ITD encoding. Listeners are exquisitely sensitive to timing differences—threshold ITD detection is 10 μs. In order to perceive such incredibly small timing differences, it is crucial that timing information is accurately and precisely encoded. Neurons in the MSO receive direct excitatory input from both the ipsilateral AVCN and the contralateral AVCN. Two theories of how ITDs might be encoded have been proposed. The first, proposed by Jeffress in 1948, suggests that the excitatory inputs from the contralateral ear act as a series of delay lines and that cells in the MSO act as coincidence detectors such that a cell fires when it receives coincident input from the two ears, with each cell being tuned to a particular delay and the neural population representing the range of delays that a listener might experience. Such a scheme operates in the nucleus laminaris of the barn owl where an orderly map of interaural delay is formed. More recently, an alternative model has been proposed for the mammalian MSO based on the observation that in addition to excitatory input from both AVCNs, MSO cells also receive glycinergic inhibitory inputs from both ears from the LNTB and MNTB, both of which show adaptations for fast, high-fidelity synaptic transmission. Monaural spectral cues are thought to be first encoded by type IV neurons in the DCN and relayed from here to the IC.

Within the IC sound localization cues are processed largely according to their frequency. Therefore, the excitatory connections from the MSO target low-frequency areas of the IC and are largely separate from those that originate in the LSO, which target high-frequency areas. Within the high-frequency areas of the IC there is convergence of the projections from the LSO with those from the DCN. However, inhibitory connections from the LSO target similar areas to excitatory MSO connections offering the potential to integrate localization cues. Neurons in the IC may also extract information about stimulus motion or dynamic localization cues. Given the early extraction of sound localization cues in the auditory midbrain, it is tempting to imagine that auditory cortex is not necessary for sound localization. Nevertheless, lesion experiments in animal models, as well as human stroke patients, consistently demonstrate the requirement of an intact auditory cortex for sound localization behavior. It seems likely that auditory cortex is necessary to integrate the various localization cues available as any one alone provides a potentially ambiguous representation of space. How auditory cortex represents space remains an unresolved question: spatial tuning in cortical neurons is typically broad and contralateral. Although higher auditory areas show increased spatial tuning, it seems likely that some sort of population coding is necessary to achieve the resolution that humans and other animals are capable of behaviorally. In an analogy with the dorsal and ventral divisions of visual "where" and "what" processing, a spatial processing pathway has been proposed to exist within the brain. Within such a scheme a series of areas emanating from primary auditory cortex and terminating in PFC are increasingly specialized for extracting the spatial features of a sound source.

SOUND IDENTIFICATION

A sound can be described according to five key perceptual attributes: its intensity, duration, location in space, pitch, and timbre. Of these, two are crucial for identifying a sound source—namely pitch and timbre. The American National Standards Institute (ANSI, 1994) defines pitch as "that auditory attribute of sound according to which sounds can be ordered on a scale from low to high." Most people have an intuitive understanding for the percept of pitch, which underlies melody perception, conveys meaning in tonal languages such as Mandarin Chinese, emotional content in nontonal languages, and crucially, enables us to track the speech of one talker in a mixture of voices (Fig. 26.7A). Timbre is defined as the attribute of a complex sound that allows a listener to distinguish that two sounds having the same loudness, pitch, and duration are dissimilar. More intuitively, it is the quality of a sound that is sometimes described as "color" and allows a listener to distinguish two musical instruments that are playing the same note, or two different vowel sounds spoken with the same voice pitch. Both the onset dynamics of a sound and its spectral envelope determine the perceived timbre (Fig. 26.8A). For a harmonic sound such as that generated by the vocal chords or a musical instrument the distribution of energy across harmonics is crucial to determining the resulting timbre.

It is pitch that allows us to perceive melody, but what physical sound attributes determines our percept of pitch? For a pure tone (ie, a sound of a single frequency) the perceived pitch is determined by the frequency (Fig. 26.7B and C). However, most natural sounds are not pure tones but instead are composed of harmonically related frequency components—that is, pure tones spaced at integer multiples of the lowest, or fundamental, frequency (F0, eg, a vowel sound such as Fig. 26.7H). The perceived pitch of such sounds is at the F0. However, if the fundamental frequency is removed, the sound maintains the same pitch. Indeed, it is possible to generate a number of complex sounds that have the same pitch but that differ considerably in their frequency structure (Fig. 26.7D–I). Instead, periodicity is thought to be the major determinant of pitch perception. A periodic sound is one that repeats at regular intervals and it is the number of repetitions per second that determine the perceived pitch. Sounds with a long period will repeat fewer times a second and have a lower pitch than sounds that repeat more rapidly. Musical pitch (in western music) is based upon the octave—one octave is a doubling in F0, and each octave is divided into 12 semitones corresponding to 1/12th octave steps (Fig. 26.7A). Notes that differ by one full octave are said to differ in their pitch height, but to share the same chroma, and have the same nomenclature. The musical scale is anchored to "middle A" which corresponds to an F0 of 440 Hz.

As noted earlier, both temporal and spectral features can determine perceived timbre and the same features that determine what timbre a type of musical instrument has, in particular the energy distribution across harmonics (or spectral envelope) and the speed with which a note is generated and subsequently decays. For example, the guitar and piano both generate notes with a very rapid nature of attack, that decay rapidly, and have very little energy in the higher harmonics (Fig. 26.8A). In contrast, a wind instrument, such as the oboe (Fig. 26.8B), has a much more consistent distribution of energy across time, and has energy across many more harmonics. These characteristic differences allow us to recognize and identify the many musical instruments in an orchestra.

Both pitch and, in particular, timbre are crucial for spoken word and speaker identification.

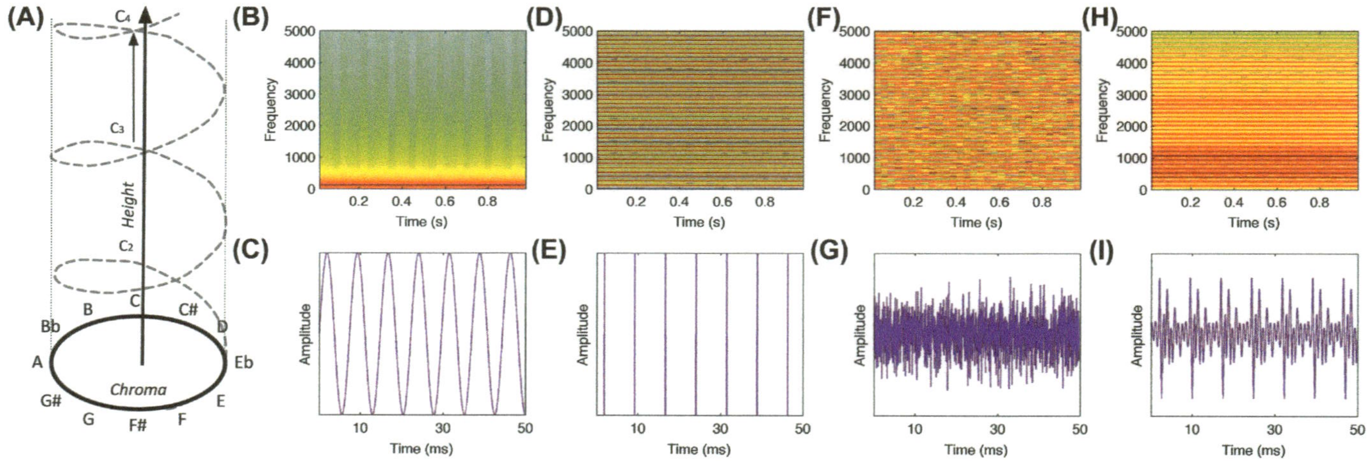

FIGURE 26.7 **Pitch.** (A) Pitch is a helical percept having both a chroma and height. In western music, sounds separated by an octave share a common chroma, but differ in height. (B–I) Spectrograms (top row) and amplitude waveforms (bottom row) of four sounds that share the same pitch (F0 = 135 Hz). (B,C) A pure tone. (D,E) A clicktrain with a repetition rate of 135 Hz. (F,G) Iterated ripple noise. (H,I) An artificial vowel with an F0 of 135 Hz. Despite having very different spectral characteristics, all sounds share a common periodicity and therefore elicit the same pitch.

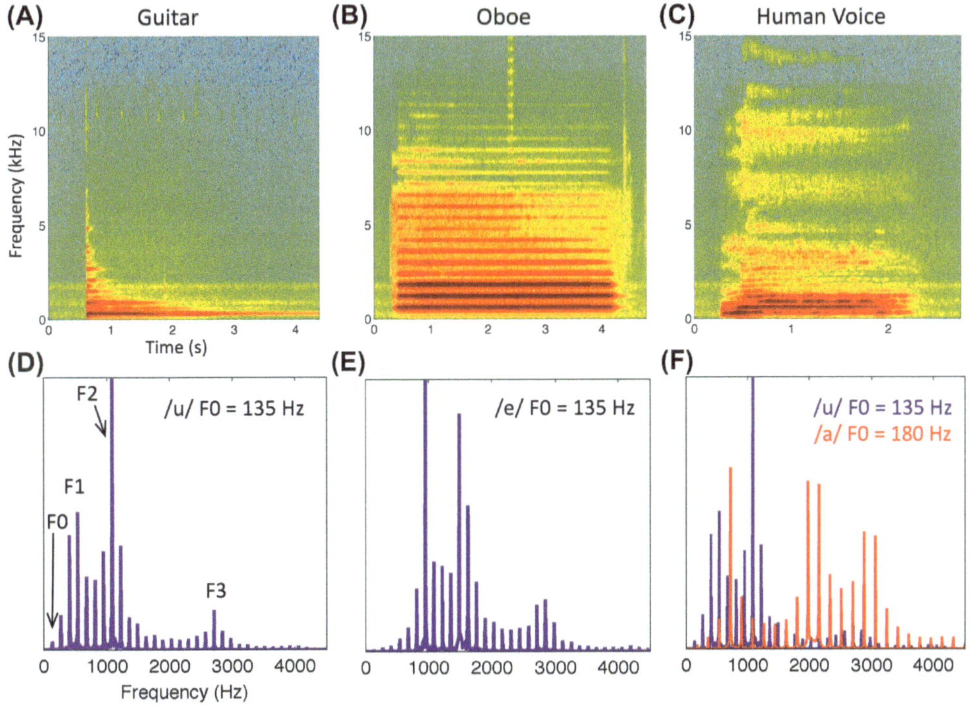

FIGURE 26.8 **Sound identification.** (A–C) Spectrograms of a guitar, oboe, and human voice playing/singing the same note. All three sounds share a harmonic structure with the same underlying periodicity (and, therefore, share a common fundamental frequency and pitch), but each has a characteristic timbre due to the differences in energy distribution across harmonics and the temporal profile of the sound offset and decay. (D–F) Frequency spectra of two artificial vowels /u/ and /e/ (D,E) generated with the same pitch (F0 = 130 Hz). If these two vowels were spoken separately, they would be easily identified. However, if they were spoken simultaneously, a single vowel with an ambiguous identity would be perceived. In (F) the frequency spectra of the vowel /e/ and /a/ are overlaid. Because /a/ (in red, F0 = 180 Hz) is spoken with a higher pitch than the /e/ in this example, these sounds are heard as two distinct vowels, each of which can be identified.

Voiced sounds, such as vowels and voiced consonants (eg, /b/, /g/, and /d/), are produced by forcing air through the vocal chords to generate a glottal pulse train. The periodicity of the resulting sound will determine its pitch. As the sound passes through the vocal tract, it is filtered by the mouth, lips, and tongue, which impose characteristic peaks, or formants, onto the sounds energy spectrum and allow us to generate a rich repertoire of sounds. The frequency of these formants, in particular of the first and second formants, is the key determinant for vowel identity (Fig. 26.8D–F). The vocalizations emitted by one talker will have a characteristic pitch and timbre depending on their age, size, and gender. Thus, children and females tend to have higher pitched voices than males, and, due to their shorter vocal tracts, generate resonances with more closely spaced formants, thus providing listeners with important information about the individual who is speaking. During speech, humans can shape the glottal pulse train to vary the location of the formants and hence generate different vowel sounds and voiced consonants.

At the level of the auditory nerve, the cues necessary for extracting pitch and timbre are represented both in the temporal discharge patterns of low-frequency fibers and in the pattern of activity across the tonotopic array. The discharge pattern of ANFs accurately represents the spectrum of a vowel sound, although temporal information can provide a more robust estimate of a spectrum of a vowel when background noise is present and even when a vowel is whispered rather than voiced. Both place and temporal information will also convey information about the periodicity of a sound source providing the auditory brain with an accurate estimate of the pitch of a sound. Nevertheless, for both pitch and timbre, integration across multiple sound frequencies is necessary in order to estimate the spectral envelope or underlying periodicity. For timbre, integration across the tonotopic array is necessary to estimate the spectral envelope in a manner that is robust to changes in the underlying spectral details or periodicity. Neural responses in the brain stem and midbrain are sensitive to the modulation rate of a stimulus but do not have responses that represent the pitch of a sound regardless of its level or timbre. As we ascend the auditory system neural frequency tuning becomes broader offering the potential to integrate information across a wider range of frequencies. The representation of pitch and timbre at the level of the auditory cortex remains controversial. However, humans with damage to auditory cortex (in particular, the right auditory cortex) are impaired at using spectral and temporal cues for timbre discrimination, suggesting auditory cortex plays an important role. Information about timbre is broadly distributed across multiple auditory fields, but evidence from human functional imaging suggests that higher cortical fields might be specialized for extracting spectral envelope. Human patients with damage to primary auditory cortex are also impaired on pitch discrimination tasks. Functional imaging studies in humans provide evidence for a specialized pitch processing area in the auditory cortex humans, and electrophysiological investigations in nonhuman primates provide evidence for an area of auditory cortex that is selective for the fundamental frequency of a sound. However, other studies in both humans and animals provide evidence that pitch sensitivity is broadly distributed throughout multiple auditory cortical fields and that neurons throughout auditory cortex are informative about the pitch of a sound.

SPEECH

Communication is arguably the most important function of the human auditory system. Understanding spoken language requires that speech sounds are accurately encoded so that the acoustics of the signal can be translated into perceptual features and these perceptual features can be translated into meaning. To communicate, the brain must also be able to convert meaningful sound patterns into motor actions to be able to create speech sounds using the vocal apparatus. A ventral auditory stream is thought to underlie the ability to extract meaning from speech, whereas a left-dominant dorsal auditory stream is thought to translate acoustic-based representations into articulatory representations. The division of speech processing into sensory and motor representations arose from Wernicke–Lichtheim model of language which accounted for the two major forms of speech difficulties—those with production, known as Broca's aphasia, and those of comprehension, known as Wernicke's aphasia. These two categories of communication deficit were localized to two areas beyond the classical auditory cortex. The clinical observations that led to these two categories of deficit followed case studies of patients with lesion damage (usually as a consequence of a stroke) to the areas in the left hemisphere. Wada tests, developed by Wada and Rasmussen in 1960, administer an anesthetic via the carotid vein to either the left or right cerebral cortex, and have determined that 90% of all right-handed patients and about 75% of left-handed patients display classic left-handed dominance for speech processing. Right dominance is relatively rare (1–2%) with the remainder of patients showing either mixed or bilateral representations where both or either hemisphere, respectively, can support speech processing. More detailed mapping of the acoustic, semantic, and motor abilities required for speech is provided during intraoperative recordings of patients undergoing brain surgery for intractable epilepsy. These patients undergo testing during surgery to ensure that their speech functions will not be compromised when brain tissue is removed to prevent future seizures. Such recordings demonstrate a hierarchical organization of speech processing whereby a larger number of increasingly higher order cortical fields are involved as the task complexity increases from acoustic to phonetic to semantic processing.

Animal studies provide insight into the neuronal representation of speech sounds at the single neuron level. At the level of primary auditory cortex the acoustical structure of speech is well represented and speech sounds can be reconstructed from the responses of populations of primary auditory cortical neurons. Speech sounds activate early auditory areas where the acoustical features of sound signals are extracted. Portions of the superior temporal sulcus (STS) are important for representing and processing phonologic information; neural activity here is often selective for speech signals over other complex acoustic stimuli. Functional imaging studies demonstrate that the STS is activated by language tasks that require both perception and production of speech and during the active maintenance of phonological information. Areas in the temporal lobe are thought to be specialized for converting phonological representations into semantic ones. Damage to posterior temporal lobe regions frequently cause auditory comprehension deficits that are thought to be post-phonemic in nature. Sensorimotor integration is crucial for accurate speech perception and production. It is perhaps unsurprising therefore that there are several areas that seem to be involved in both perception and production—for example, the postero-medial planum temporale area is an auditory area but seems to be important in articulation, whereas both real and imagined speech can activate premotor areas known to be important in the production of speech.

AUDITORY SCENE ANALYSIS

One of the greatest challenges that a listener faces is the ability to pick out the voice of one talker amidst a noisy background of competing voices, such as when trying to listen to conversation in a bar or restaurant. The so-called "cocktail party" problem, a term coined by Cherry in the 1950s, is a specific example of a more general challenge that the auditory system faces—that of separating out a sound scene into its component sources. Unlike the visual system, where light from different objects falls on discrete neighboring patches of the retina and where visual objects occlude one another, sounds are effectively transparent and therefore the sound wave that arrives at the ear is the sum of all of the sounds in the environment. The representation of the resulting mixture will be mixed on the cochlea and within the auditory nerve. The challenge faced by the auditory brain therefore is to appropriately separate and group sound elements in order to estimate the sources that produced them. Various cues are available to the brain to achieve this aim. These can be loosely grouped into "sequential" cues—that is, cues that are available because sounds unfold over time, and "simultaneous" cues—those that allow two simultaneous sounds to be perceptually grouped. We are quite good at ignoring interrupting sounds—such as a car horn or someone coughing. This ability can be revealed with the "continuity illusion." In this instance, if a tone with a short gap is presented in silence, listeners accurately hear the gap in the tone. However, if that gap is filled with a relatively loud noise burst, listeners report hearing the tone to be continuous—the brain apparently fills in the missing information. Another example commonly used in research is the "ABA" streaming paradigm. Here high- and low-frequency tones are presented in alternation. When the frequency difference is small, and the presentation rate slow, listeners hear a single stream of alternating tones. However, if the frequency difference is made larger, and/or the presentation rate more rapid, the percept changes so that listeners hear two separate streams of high and low tones. Simultaneous grouping features, as their name suggests, allow us to separate out two instantaneous sounds. There are various cues available to achieve this but the main ones are that sound elements originating from a common source share a common onset time, the same harmonicity (pitch), and a common spatial location. Sound elements from a single source will tend to start and stop at the same time and are likely to increase and decrease in loudness together. Many natural sounds are harmonically related and it is a difference in pitch that allows us to resolve two competing talkers' voices (eg, two vowels as in Fig. 26.8F). Spatial cues also influence grouping—although they are a relatively weak cue if pitted against pitch, for example. The process of auditory scene segregation is multifaceted and likely occurs throughout the auditory system, although the auditory cortex is thought to play a key role. Various models exist for how auditory cortex might achieve auditory scene analysis, most of which propose that in order for two sources to be perceived as separate two discrete neuronal populations must be activated. However, the physiological mechanisms remain controversial and precisely how this happens is poorly understood.

HEARING LOSS

Hearing loss is defined according to pure tone hearing threshold measurements. Hearing thresholds are expressed on the hearing-level scale (dB HL) and severity is classified according to the average hearing level for the thresholds obtained at 500, 1000, 2000, and 4000 Hz. Deafness corresponds to a profound (>91 dB HL) hearing loss. Hearing problems can also occur in the absence of elevated pure tone thresholds and can be the consequences of either peripheral or central impairment. Central auditory processing disorders can result in difficulty in performing complex listening tasks such as sound localization or listening to speech in noise. "Hidden" hearing loss is characterized by hearing problems despite normal pure tone thresholds—while audiometrically normal, these individuals display anomalies in the auditory brain stem response elicited by sounds. Here we consider peripheral hearing loss (HL), that is, that occurring due to dysfunction at the ear or cochlear.

Peripheral HL can be categorized according to whether it occurs as a consequence of conductive HL or sensorineural HL. A conductive HL impairs transmission of sound through the external and/or middle ear, whereas sensorineural HL is a consequence of a deficiency in the cochlea or auditory nerve. Conductive HL is the most common form of problem to affect children in the form of *otitis media* ("glue ear") and is both treatable and reversible. Sensorineural HL is the most common form of deafness to affect adults and can arise as a result of damage to the OHCs, IHCs, stria vascularis, or auditory nerve. Due to the role that OHCs play in boosting the basilar membrane response to quiet sounds and in sharpening frequency resolution, damage to OHCs results in a loss of sensitivity. This in turn leads to elevated detection thresholds, and a loss of the compressive growth function of the basilar membrane that therefore limits the range of sound levels that can be encoded. IHCs are responsible for converting basilar membrane motion into activity in the auditory nerve. Unsurprisingly, therefore, loss of IHCs results in an elevation of the compound action potential threshold (a gross measure of the response to sound), although individual ANFs can display normal thresholds and frequency tuning. Damage to the auditory nerve can occur due to degeneration or demyelination. Lastly, the *stria vascularis* is responsible for maintaining the endocochlear potential by maintaining the high potassium ion concentration in the endolymph. Deficiency in the function of the *stria vascularis* will therefore impact upon both OHCs and IHCs. Sensorineural HL can be hereditary (accounting for 60% of childhood HL at birth) or acquired as a result of ototoxic drugs (eg, aminoglycoside antibiotics), hypoxia, noise exposure, or as a consequence of aging.

Age-related HL is characterized by progressive loss of sensitivity to high frequencies. Age-related HL is accompanied by changes in the *stria vascularis*, hair cells, and by degeneration in the auditory nerve. Whether these changes are a natural consequence of aging or the cumulative effect of a lifetime's exposure to noise is unclear. The consequences of such HL are elevated thresholds, impaired frequency selectivity, and a decrement in temporal coding within the auditory system. Perceptually, this leads to a decrease in sound localization ability, a decline in the ability to use spatial cues for "unmasking," or separating a signal such as speech from a masking noise on the basis of its spatial location, and general difficulties in listening to speech in noise. Sensorineural HL also leads to loudness recruitment whereby quiet sounds are inaudible but more intense sounds are just as loud as for normal hearing listeners.

REHABILITATION

The most common form of treatment for HL is a hearing aid. These comprise a microphone, amplifier, and speaker, and restore audibility by increasing the sound level available within the ear. Most hearing aids contain signal processing devices that attempt to address some of the perceptual deficits that still result from hearing aid usage. For example, amplitude compression is commonly included to deal with the problem of loudness recruitment (whereby audible sounds grow in loudness very rapidly with only small changes in sound intensity), and other signal processing strategies exist to attempt to improve the signal-to-noise ratio of speech signals, often by implementing directional microphones into the hearing aid.

Cochlear implants (CIs) are used to treat profound sensorineural HL and directly electrically stimulate the auditory nerve. CIs are comprised of a microphone and signal processor situated on the skull behind the ear, an implanted receiver, and an array of electrodes that are surgically inserted into the cochlear. Sound signals are processed to break them into frequency bands with different channels then being used to stimulated electrodes, each of which is situated at a different location on the cochlear. Health care providers in many countries now routinely recommend that profoundly deaf children be implanted bilaterally.

The electrode array of a CI typically only covers the first of the two and a half turns that make up the human cochlear and therefore is sited predominantly at the high-frequency end of the basilar membrane, typically from about 20 to 1.2 kHz. A further limitation is that, even without any spread of current between neighboring electrodes, the frequency resolution of a CI cannot be better than the number of electrode contacts on the array which currently is a maximum of 22—contrast this to the 3000 IHCs that the healthy cochlear has available for encoding sound frequency. Nevertheless, a CI is highly successful at encoding information sufficient for speech comprehension (at least in quiet). The speech processor on the device filters acoustic signals into discrete frequency bands and takes the filter outputs as the basis of the signal used to stimulate each electrode. Typically, electrodes are stimulated with a series of high-frequency (1 kHz or more) pulses which effectively remove any temporal information that would usually be available about the fundamental frequency (pitch) of a sound. Given this lack of temporal information, and the position of the electrode contact within the cochlear, it is therefore unsurprising that most implant users have relatively poor pitch perception. For speakers of Indo-European languages, this means that while the CI is tremendously helpful for processing speech, it does not convey musical pitch. However, for speakers of tonal languages such as Mandarin, the limited perception of pitch provided by a CI also limits their speech recognition.

Further Reading

[1] Bendor D, Wang X. Cortical representations of pitch in monkeys and humans. Curr Opin Neurobiol 2006;16(4):391–9.

[2] Bizley JK, Cohen YE. The what, where and how of auditory-object perception. Nat Rev Neurosci 2013;14(10):693–707.

[3] Grothe B, Pecka M, McAlpine D. Mechanisms of sound localization in mammals. Physiol Rev 2010;90(3):983–1012.

[4] Hickok G, Poeppel D. Neural basis of speech perception. Handb Clin Neurol 2015;129:149–60.

[5] Hopkins K. Deafness in cochlear and auditory nerve disorders. Handb Clin Neurol 2015;129:479–94.

[6] Kazmierczak P, Muller U. Sensing sound: molecules that orchestrate mechanotransduction by hair cells. Trends Neurosci 2012;35(4):220–9.

[7] Rauschecker JP, Scott SK. Maps and streams in the auditory cortex: nonhuman primates illuminate human speech processing. Nat Neurosci 2009;12(6):718–24.

[8] Schnupp JW, Nelken I, King AJ. Auditory neuroscience: making sense of sound. Boston: MIT Press; 2012.

CHAPTER

27

Body Clocks in Health and Disease

I.N. Karatsoreos[1], *R. Silver*[2,3]

[1]Washington State University, Pullman, WA, United States; [2]Barnard College, New York, NY, United States; [3]Columbia University, New York, NY, United States

INTRODUCTION TO CIRCADIAN CLOCKS

Overview of Circadian Timing

A frustration in the search for mechanisms and treatments for brain-based disorders and diseases is the complex and seemingly overlapping nature of symptoms and causes associated with clinically defined states of health and illness. The circadian timing system, the subject of this chapter, is one of the system level mechanisms that impact an immense array of responses that either antecede or accompany diseases that claim the attention of neuroscientists. These include disturbed sleep/wake cycles and mood regulation, inappropriate food intake and metabolic factors that regulate fat and glucose metabolism and energy expenditure, and physiological responses including stress responses and the accompanying activation of hypothalamic–pituitary axes regulating hormone secretions. A growing body of evidence indicates that circadian "clocks" are located in most cells of the body and that they modulate functions that control emotional, cognitive, and metabolic processes, thereby pointing to mechanisms that can link observed comorbidity in psychiatric and metabolic disruption. Additionally, polymorphisms in genes associated with the cell-based circadian clocks have been associated with sleep disorders and obesity.

Circadian rhythms can best be understood in the context in which they evolved. Living organisms on our planet have internal timekeeping systems, termed circadian "clocks" that enable anticipation of upcoming events for optimal timing of behavior, metabolism, and physiology with respect to predictable changes in the environment. At the level of a cell, system, and whole organism, each function of the body must find its appropriate temporal niche, as it is not possible to do everything all at once. While correlated with daily changes in day and night caused by the rotation of the earth about its axis, circadian rhythms are endogenously organized and are not dependent on timing signals from the environment. That is, even if all time cues are removed, such as occurs in deep underground caves or in outer space, circadian rhythms continue with a period of about 24h.

One tremendous advantage in the analysis of circadian timing mechanisms is that research

Conn's Translational Neuroscience
http://dx.doi.org/10.1016/B978-0-12-802381-5.00043-9

has advanced to the point that we can monitor time within cells, and/or tissues, or organs or in the behaviors of the whole animal. This is based on our ability to probe the molecular basis of clocks inside cells. The availability of animal models is also a fundamental feature enabling understanding of mechanisms as the molecular elements of cellular clocks are similar in humans and nonhumans, including rodents. Knowledge of molecular circadian clock mechanisms is key to determining the role of clocks in health, dysfunction, and disease.

The aim of this chapter is to provide an overview of the relationships between circadian clocks of the body and the ways in which they can serve to promote health and can cause dysfunction. Some of the hypotheses and research presented in this chapter are derived from very strong experimental evidence, based on the work of many labs, over many years, and confirmed in both animal and human studies. In other cases, we are closer to the frontiers of research and hypotheses and experimental evidence are more tentative. In both cases, we indicate the strength of evidence and direct the reader to relevant in-depth literature reviews.

Defining Circadian Rhythms

Circadian (circa = about, diem = day) rhythms are produced by a biological clock system within the body; they can be distinguished from daily rhythms that are the result of environmental pressures and perturbations. There are several commonly used experimental strategies for distinguishing between endogenously organized and exogenously driven changes (Box 27.1). Furthermore, measures of period, phase, amplitude, and precision are used to assess the status of circadian rhythms. The parameters used for the analysis of circadian periodicity and plasticity include the process of entrainment, determination of period length in free-running conditions, determination of circadian periodicity in response to light disruption (eg, jet lag studies), and evaluation of clock plasticity in non-24-h conditions (termed T-cycles). Studying the properties of circadian (such as their phase, amplitude, and period length) in response to photic perturbation can be particularly useful to help understand how humans respond to jet lag, night shifts, rotating work shifts, or other transient or chronic disruption of environmental surroundings.

Molecular Basis of Circadian Clocks

Circadian rhythms are generated by a clock-like timing mechanism found in virtually all cells of the body. Restated, the fundamental unit of circadian timekeeping is the individual cell. Most generally, the cellular clock is composed of interlocked molecular loops within which the activity of "clock" genes is suppressed by their own protein products. More specifically, circadian clocks in mammals have a highly conserved molecular mechanism based on self-regulating transcriptional–translational feedback loops that underlie rhythmic expression of core "clock" genes (see Fig. 27.1). This consists of the transcriptional activators BMAL1 and CLOCK, which bind to E-box enhancers and activate the transcription of the *Per* and *Cry* repressor gene families. These repressors then feedback to inhibit BMAL1/CLOCK activity, thereby inhibiting their own expression. Each molecular component in the core clock loop is represented by multiple paralogs (eg, *Bmal1*, *Bmal2*; *Per1*, *Per2*, *Per3*; *Cry1*, *Cry2*), which provides the potential for functional redundancy and cell-type specificity. Auxiliary elements such as casein kinase 1/ε (CK1/ε) and glycogen synthase kinase 3, and ubiquitin ligases such as F-box/LRR-repeat protein 3 (FBXL3) fine-tune the period length and amplitude of the molecular circadian clock. One cycle of activation and repression of gene expression takes about 24 h, as befits a daily clock.

BOX 27.1

GLOSSARY OF TERMINOLOGY IN CIRCADIAN RESEARCH

Amplitude. The difference between the peak value and trough value in an oscillation.

Arrhythmia versus Desynchronization. Arrhythmia involves the absence of rhythmicity in a process that is normally rhythmic. Desynchronization refers to the loss of synchrony between a rhythm and its zeitgeber (external desynchronization) or between two rhythms within an organism (internal desynchronization).

Circadian. Circadian events are those having a cycle of about 24h and include the requirement of being endogenously generated. This is determined by the ability to free-run, with a period of about 24h, in constant conditions.

Circadian time (CT) versus Zeitgeber time (ZT). Circadian time is a standard unit of time based on the endogenous free-running period of a rhythm. Zeitgeber time is a unit of time based on the period of a zeitgeber, such as the 12:12 light:dark cycle. By convention, in free-running, constant conditions, the onset of activity of day-active organisms is circadian time zero (CT0) and the onset of activity of night-active organisms is CT12. In a light:dark cycle, for diurnal organisms ZT0 (lights on) is the time of activity onset and ZT12 (lights off) defines activity onset for nocturnal animals.

Clock gene. A gene that is an essential element in the cellular–molecular mechanism of circadian rhythmicity.

Entrain/entrainment. The synchronization of a self-sustained endogenous circadian rhythm by a forcing signal (the zeitgeber). In steady entrainment, the period of the self-sustaining oscillation conforms to that of the zeitgeber, and they maintain a stable phase relationship.

Food- versus light-entrainable oscillator. The SCN is a light-entrainable oscillator. A food-entrainable oscillator is a putative circadian pacemaker, locus unknown, that can be entrained by regularly scheduled food restriction but not by a light–dark cycle.

Free-run/free-running. A self-sustaining oscillation or rhythm that is retained in the absence of effective zeitgebers or other environmental agents that may affect the period of the oscillation.

Masking. Disruption or elimination of the expression of an overt rhythm caused by an external agent, without a direct effect on the period or phase of a pacemaker.

Pacemaker. An entity that can generate endogenous rhythmicity and impose this rhythmicity on other entities. The SCN is a pacemaker. Note that while a pacemaker is an oscillator, not all oscillators are pacemakers.

Phase/phase angle/phase shift. The relative angular displacement between a periodic quantity and a reference angle. A phase shift is a displacement (in time) of an oscillation.

Precision. The daily stability of the clock over successive days. Under free-running conditions, a highly precise clock will time the onset of behavioral activity or physiological processes to the same time of day each day, while a less-precise clock may encounter drift from day-to-day, sometimes starting early, other times starting late.

Subjective day/subjective night. In free-running, constant photic conditions (where day and night are not signaled by light–dark cycles), the segment of a circadian cycle that corresponds to the day is the subjective day (active time for the diurnal animal and inactive time for the nocturnal animal). The segment of a circadian cycle that corresponds to the night is the subjective night (inactive time for the diurnal animal and active time for the nocturnal animal).

Zeitgeber/zeitgeber time. A synchronizing signal capable of resetting a pacemaker or synchronizing a self-sustaining oscillation.

When core clock genes are mutated, the consequence is altered period, amplitude, or robustness of circadian cycles. These can be tracked in behavioral measures of locomotor activity and sleep–wake patterns in animals or humans, or in altered rhythms of electrical activity or clock gene expression in SCN cells, or in tissues as a whole, in vivo or in vitro. Mutations of auxiliary elements that modulate the core clock genes also affect these parameters. For example, hypophosphorylation of *Per2* because of a mutation of the phospho-acceptor site of human PER2 protein, seen in subjects suffering from familial advanced sleep phase syndrome (FASPS), shortens circadian period by 2–4h. Early nighttime sleep onset is accompanied by early occurrence of many responses including sleep onset and wakening, minimum core body temperature, and evening onset of melatonin secretion.

Clock-Controlled Genes

Circadian clocks have a broad impact throughout the body, because downstream genes are rhythmically activated by core clock genes. More specifically, in addition to the core clock genes such as *Per and Cry*, the CLOCK/BMAL complex also drives rhythmic expression of numerous output genes, termed clock-controlled genes. Examples include the control of metabolism by rhythmic nuclear receptors, the modulation of the immune response, and diurnal rhythms of detoxification via *Alas1*. The clock-controlled genes encode proteins that are involved in tissue-specific regulatory and metabolic activities. Genome-wide transcription analysis suggests that clock-controlled genes have only about 10% overlap from tissue to tissue, and have different peak phase distributions in various organs. Circadian influences have been widely documented and include the immune and metabolic systems, sleep–awake cycles, memory consolidation, blood pressure, body temperature, cell division and proliferation, hormone secretion, apoptosis, and senescence. Tissue-specific differences in rhythm generation may be attributable to variations of the core clock genes across tissues, tissue-specific rhythmicity in transcription factors and co-factors, or sensitivity to systemic cues originating in hormone secretions, immune system responses, sympathetic innervation, body temperature, and activity rhythms. Understanding the complexity of tissue-specific circadian expression patterns and how to use them to optimize administration of therapeutic drugs is currently a major research area.

The Master Clock Lies in the Suprachiasmatic Nucleus

There is substantial evidence that a "master" clock, located in the suprachiasmatic nucleus (SCN) of the hypothalamus, guides the temporal structure of a hierarchical system of molecular–cellular clocks located throughout the body, acting to set the phase of rhythms throughout the body (Fig. 27.1). The SCN lies above the optic chiasm, and has about 10,000 cells on each side of the third ventricle (in rodents). When these cells are ablated, animals lose circadian rhythmicity, and never recover. However, rhythmic activity/rest cycles can be restored by brain tissue transplants containing the SCN, and the genotype of the grafted SCN determines the period of the restored rhythm. Thus, if the donor tissue has a period of 20h, while the host originally had a period of 24h, the restored rhythm is strictly that of the donor. The lesion and transplant data together constitute strong evidence that the SCN is the sole origin of the circadian signal that supports circadian oscillation and determines the phase of the behavioral and physiological responses.

Output Signals of the Suprachiasmatic Nucleus

The SCN master clock, directly or indirectly, is the source of signals that synchronize peripheral

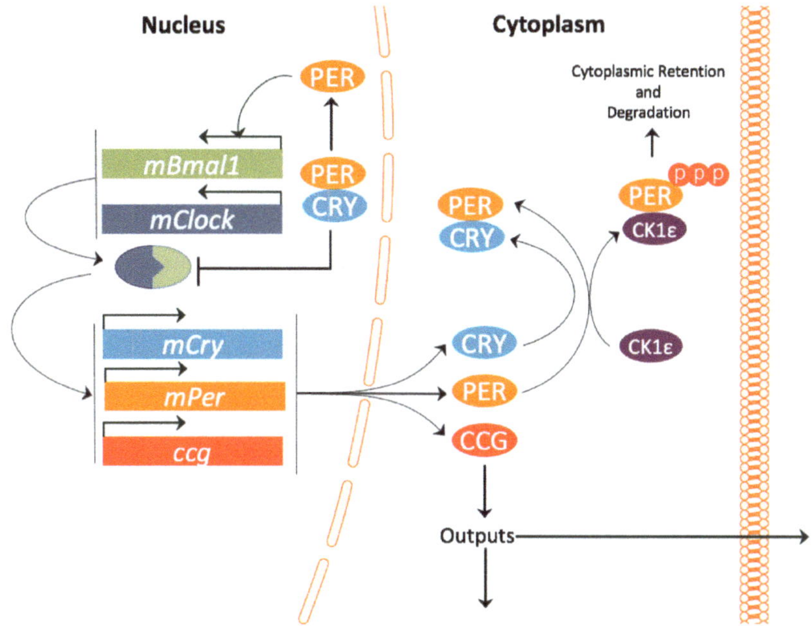

FIGURE 27.1 A simplified illustration of the core circadian clock mechanism depicting the transcriptional–translational feedback loop in mammals. At the core of this loop there is, in the nucleus, a BMAL1/CLOCK protein complex that binds to the promoters of the *Per* and *Cry* family of genes, as well as to clock-controlled genes (*ccg*). This results in increased transcriptional activity of these targets. In the cytoplasm, the PER/CRY proteins dimerize and translocate back to the nucleus to inhibit BMAL1/CLOCK and block their own transcription. In parallel, kinases such as casein kinase 1 epsilon (CK1e) phosphorylate PER protein in the cytosol, targeting them for degradation. This provides additional time delays enabling the molecular clock to cycle with a period of about 24h.

clocks which themselves contain oscillators. This yields a system in which some aspects of mammalian physiology are controlled directly by central signals and others by peripheral clocks. The SCN can signal the rest of the body by many different routes. First, there are monosynaptic and multisynaptic projections that reach the central and peripheral nervous systems. Secondly there is an important role for peptidergic and hormonal signals that circulate locally or via systemic circulation. Finally, the SCN also acts indirectly, via behavioral effects on food intake and temperature, to entrain peripheral clocks. The SCN and peripheral clocks respond to different signals. For example, the SCN does not respond to food-related or thermal cues. Glucocorticoids of the adrenal gland are secreted rhythmically (orchestrated by the SCN), and their receptors are expressed in most peripheral tissues, but not in SCN neurons. Furthermore, in vitro studies indicate a myriad of other pathways, involving signaling by kinases and calcium that can shift peripheral clocks.

Extra-Suprachiasmatic Nucleus Clocks in Central Nervous System and Periphery

The discovery of clock genes in extra-SCN brain regions and in the periphery is of crucial importance in understanding the importance of temporal order in both health and disease (Fig. 27.2). Clocks have been found in many brain regions, including the prefrontal cortex, hippocampus, and amygdala, where they are thought

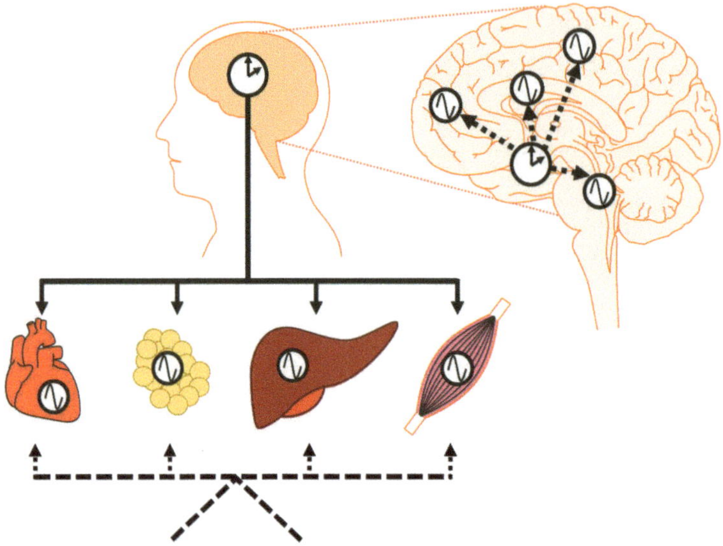

FIGURE 27.2 The master circadian clock is located in the suprachiasmatic nucleus (SCN) of the hypothalamus. The phase of the SCN is adjusted to geophysical time every day by light signals detected in the retina. The phase of clocks in extra-SCN brain regions (such as the cortex, hippocampus, and amygdala) are set by both neural and diffusible signals from the SCN. Clocks in peripheral tissues, such as the heart, adipose, liver, and skeletal muscles, are synchronized by the SCN through a variety of systemic signals, including metabolites, cytokines, body temperature, and peripheral nervous system signals. In addition, nutrients from food, as well as changes in body temperature, can help to synchronize these peripheral clocks. In most tissues, the SCN is necessary for synchronization of individual cellular oscillators, though in some tissues (eg, olfactory bulbs) individual oscillators remain coupled even in the absence of systemic signals from the SCN (see text for further details).

to help regulate homeostatic, cognitive, and emotional function. Rhythmic gene expression has also been found in several hypothalamic nuclei, including the bed nucleus, paraventricular nucleus, and arcuate nucleus. These rhythms are thought to influence numerous homeostatic functions from feeding to reproduction. Rhythmic gene expression is also found in the reward circuitry, with known effects on diverse motivated behaviors. A similar pattern is observed in peripheral organs, including the heart, adipose tissue, liver, and skeletal muscle, where rhythmic clock gene expression is clearly observed. The mechanisms by which these local clocks modulate downstream function still remain a mystery, but current data show that they modulate a vast number of genes. Therefore, their dysregulation in the brain and peripheral tissues is thought to provoke a wide range of pathologies. The relatively familiar temporal dysregulations evident in jet lag and shift work, are produced by differential rates of adjustment of cellular clock and their downstream targets in different tissues. Pathologies such as metabolic syndrome, immune dysfunction, and cancer are additional conditions that accompany and perhaps result from conflict between peripheral and central clocks. While true that nearly all extra-SCN oscillators dampen following upon the removal of the synchronizing influence of the SCN, the olfactory bulb shows sustained rhythmicity that is independent of the SCN. Thus there exist both SCN-dependent and SCN-independent oscillators.

Much evidence supports the hypothesis that cellular physiology is tightly regulated in time, measured by regularly recurrent daily oscillations of gene expression in a tissue-specific metabolic program. Tissue-specific oscillators have unique molecular components, so that the synchronizing factors to which they respond, and their cellular outputs, can be regulated locally. As discussed later, these local tissue–specific molecular factors, especially in extra-SCN regions of the brain, are not well understood.

Suprachiasmatic Nucleus Clocks Differ From Other Tissue Clocks

Insight into how the cells of the brain and body subserve the functions necessary for healthy survival can be understood in the beautifully plastic hierarchical organization of the circadian timing system. As noted, the SCN does not respond to temperature fluctuations. Thus, inverting environmental day–night temperature changes will inverse circadian gene expression in peripheral cells and non-SCN brain regions, but not in the SCN itself. The insensitivity to temperature changes disappears in dispersed SCN cells, indicating that it is a property of its network organization and not attributable to its individual cells. Its network properties enable the SCN to continue to oscillate for many months in vitro, again proving that this is an intrinsic oscillation—not requiring any external input. Interestingly, circadian clock genes and proteins harvested from neural and nonneural tissues, including liver, lung, kidney, SCN, and some brain regions outside the SCN, also continue to oscillate when they are placed in culture. However, in non-SCN tissues, these oscillations tend to damp after several days, likely because their individual cells eventually go out of phase with each other. The circadian oscillations can be resynchronized and thus restarted for another series of days by chemical signals such as those produced by changing the culture medium.

These findings evoke the question of how clock cells of the SCN differ from those in the rest of the body. Part of the answer lies in the complement of clock genes and in the nature of communications among cells in various tissues. While "core clock genes" are present in most tissues, the homologous genes can assume different tissue-specific functions. For example, there is a functional overlap between CLOCK and its homolog NPAS2. However, knockout of CLOCK protein is compensated by the presence of NPAS2 in the SCN, and *Clock* mutant mice are behaviorally rhythmic. In contrast, the deletion of CLOCK produces arrythmicity of circadian oscillators in peripheral tissue explants.

A unique aspect of the SCN is not only the network organization of its neurons but also their access to signals from the external and internal environment. Mammalian tissue clocks, unlike those of other species such as flies, are not intrinsically light sensitive and so, in vivo, depend completely on the SCN for entrainment to the solar day. In mammals, photic input from the retina travels along the retinohypothalamic tract and reaches a subset of the oscillating SCN clock cells via a monosynaptic connection. These directly retinorecipient neurons synchronize the SCN to the external solar day and the SCN then communicates temporal information to peripheral clocks via endocrine and systemic cues. Importantly, the same cells that are directly retinorecipient also are sensitive to temporal information derived from internal cues provided from other parts of the body. As an example, retinorecipient SCN cells in males also contain androgen receptors. Removing the endogenous source of androgens (from testes), significantly alters the function of the SCN, and causes a lengthening of the circadian period measured by behavioral locomotor activity. These are direct, local effects of androgens on the SCN, and do not require actions at other neural sites or on muscle. This finding is consistent with several studies demonstrating that the concentration of androgens reaching the SCN via the general circulation modulates the period of activity rhythms (eg, the higher the concentration of androgens, the shorter the circadian period). Remarkably, both the pattern of androgen receptor expression and androgenic effects on circadian behavior show a clear sexual dimorphism, and this may be key to understanding changes in period and sleep onset times that occur with age (see section: Temporal Order and Phase-Setting Signals in a Life Span Perspective). It is not only gonadal hormones that are peripheral signals that can alter clock function, but also feeding-related and metabolic signals also have very strong effects on circadian timing as well, and are discussed later in the context of obesity and metabolic disease.

Light Sets Suprachiasmatic Nucleus Phase

Because light from the daily day–night cycle is the most salient phase-setting cue, the retinal input to the SCN is considered one of its most important unique features. The endogenous rhythm of the SCN is about 24 h; in humans it tends to be somewhat longer than 24 h, while in mice it tends to be somewhat shorter. To remain synchronized to the daily day–night cycle, the SCN must be reset each day. To do this, the SCN clock shifts its phase in response to light cues during its dark phase. Information about light in the environment reaches the SCN through specialized neurons in the eye called intrinsically photosensitive retinal ganglion cells. In mammals, this is the only system whereby photic information reaches the brain.

Photic Versus Visual Afferents to the Central Nervous System

The intrinsically photosensitive retinal ganglion cells serve the SCN, but do not support vision. Even in mice that are completely blind visually (due to the lack of rods and cones), information from intrinsically photosensitive retinal ganglion cells can synchronize the SCN, and in the absence of these cells photic cues detected by melanopsin or classical photo pigments cannot reach the SCN. Specifically, genetic elimination of melanopsin-containing ganglion cells completely abolishes the ability of mice to set their brain clock. On the other hand, these mice possess normal visual abilities, proving that visual function and resetting the phase of the brain's clock constitute two separate afferent systems. Similar mechanisms occur in people: If the retinal input to the SCN is intact, visually blind individuals can continue to synchronize to the daily light:dark cycle, though they may be totally unaware of the presence or absence of light in their environment. Thus, by virtue of its network of interconnected oscillators, its ability in situ to both synchronize to photic cues of the external environment and to generate endogenous signals within the body is what differentiates the SCN clock from oscillators in other tissues.

COHERENT VERSUS DISRUPTED TEMPORAL ORDER

In modern industrialized societies, the advent of electric lighting has added substantial environmental perturbation to the circadian timing system. In many ways, today our internal clocks are constantly under pressure from changes in light cycles, altered activity patterns, and new social norms. The American Medical Association, in recognition of serious adverse consequences, has adopted a policy statement on the adverse health effects of nighttime lighting, acknowledging a direct link between disrupted circadian rhythms and cell cycle regulation, DNA-damage responses, and metabolism. In recognition of the evidence that nurses who work at night have an increased risk of developing breast cancer, the Danish government, since 2009, pays compensation to women who developed breast cancer after working at night. The mechanisms whereby such environmental disruption produces vast effects on the body can be understood in the context of circadian temporal organization.

Temporal Order and Phase-Setting Signals in a Life Span Perspective

The temporal order of the body is normally synchronized such that sleep, wake, feeding, and other functions are anticipated, occur at appropriate times, and do not interfere with each other. To understand optimal temporal structure, it is important to take a life span perspective. A systematic change in circadian rhythms occurs over a lifetime, measured by changes in rhythm precision, amplitude, and phase. New born infants are nearly arrhythmic, with irregularly occurring episodes of sleep and wakefulness over a day. Over the course of many months, there is a consolidation of sleep–wake cycles, and less prominent circadian responses such as temperature and heart rate. The timing of sleep onset, for example, is progressively later from early childhood to young adulthood, and then becomes gradually earlier into old age. Along with sleep, there are changes with age in the amplitude of hormone secretion (eg, melatonin), body temperature, and other physiological parameters. These age-related phase changes of our circadian rhythms can have severe practical social consequences.

People can be characterized as having different sleep types ("chronotypes") ranging from very early active "larks" to very late active "owls." Extreme larks and owls may not even overlap in their sleep–wake cycles. Furthermore, on average, the chronotype changes significantly during the lifetime of an individual, and timing of sleep is markedly affected by the schedule requirements of work and school. On average mid-sleep time for children, on nonschool days, is about 3:30 am and by early adulthood it is about 5 am. Mid-sleep time then gradually advances to earlier than 3 am in 75–80 year olds. If we assume that the average sleep duration is about 8 h, this means that the normal wake-up time is roughly 7:30 am for elementary school children and 9 am for college students. For these young people, the phase of the internal biological clock is not well adjusted to the socially imposed phase. This has given rise to the idea that in these conditions, young people suffer from "social jet lag," which appears to contribute to poorer performance, and possibly to obesity.

Many of us become familiar with the temporal order of the body only when it is disrupted, as occurs during jet lag. The problems associated with jet lag as a result of travel to a new time zone and changes in the environmental light–dark cycle provide clues to the importance of temporal structure. The disruptions consequent to jet lag are well understood. As we know, human sleep occurs almost entirely at night, during the dark, and not surprisingly, light is a major factor in setting the timing of human sleep. Light at night resets the phase of circadian rhythms, but the ability of the SCN clock to adjust to daily photic resetting signals is limited. In humans tested under controlled laboratory conditions, the circadian clock can maximally advance its phase by about 2 h/day and can delay its phase by about 3 h/day under optimal conditions. That is why it takes several days to adjust to a new time zone after long-distance travel. During the time of adjustment, we suffer from jet lag, since our internal timing system is in conflict with geophysical time. This manifests itself by perturbed sleep–wake cycles, indigestion problems, urination during the night, fatigue, and disrupted cognitive function.

Sleep–Wake Cycles

The rhythm most obvious to us is the daily alternation between wake and sleep. An intuitively pleasing conceptualization of the circadian contribution to the timing of sleep was captured decades ago in the two-process model of sleep—one of the most widely examined hypotheses

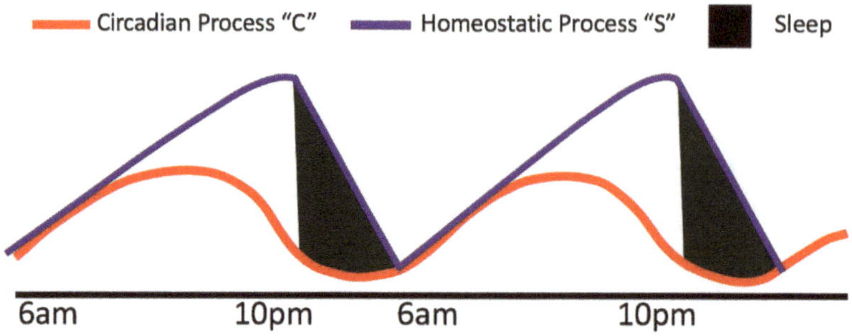

FIGURE 27.3 The two-process model of sleep posits that time spent awake drives a homeostatic need for sleep (Process "S"), represented at the physiological level by buildup of sleep-promoting substances such as adenosine among others. In parallel, the circadian drive (Process "C") promotes wake during what is supposed to be the active phase (eg, daytime in humans), which rhythmically wanes later in the evening. When sleep pressure is high (peak of Process "S") and circadian drive is low (trough of Process "C"), sleep is initiated. Sleep pressure is dissipated through the night, and the circadian drive for wake begins its rhythmic increase early in the morning. This interactive two-process system ensures that sleep need and sleep timing are properly balanced.

regarding sleep regulation. The key element of this model is the proposed interaction of two constituent processes—the sleep–wake homeostatic process, termed "S" and the circadian process, termed "C," to generate the timing of sleep and waking (Fig. 27.3). Process "S" is driven by the buildup of sleep-inducing substances that increase while you are awake. Once you begin to sleep, the levels of these substances decline throughout the sleep period, and will begin to build up once again upon awakening. Potential sleep substances are many, and they include adenosine, some cytokines, and likely many as-yet-unknown compounds. Process "C" on the other hand is driven by the circadian clock, and SCN signals act to stimulate waking. Thus, under normal conditions, wake drive is lowest in the late evening, when process "S" sleep drive is reaching a peak. It is at this point that sleep is induced—at the intersect of these two processes. As sleep need diminishes, and the circadian drive for wake increases once again in the early morning, the sleep bout is terminated and awakening follows. Restated, if you stay awake for 24 h, then as time progresses, you start to feel alert again as the circadian input signals alertness, even though sleep pressure–derived homeostatic process S continues to increase.

Of course, many additional external factors can influence sleep, including work requirements, alarm clocks, ambient temperature, naps, stress, exercise, meal times, and types of food eaten. There is some evidence that foods containing tryptophan (bananas, milk) or carbohydrates (bread) promote sleep, while those containing tyramine (ham, avocado, nuts, red wine) promote night time wakefulness. Also, psychoactive drugs can act either as sedatives or as stimulants to enhance or reduce sleep and advance or delay sleep onset. A multitude of effects modulate circadian timing of sleep, and several industries have arisen to advocate for "sleep hygiene." In the following sections, we explore some of the consequences associated with disruption of temporal order.

METABOLISM AND OBESITY: PERIPHERAL CLOCKS CONTRIBUTE TO METABOLIC DISORDERS

Mechanistic Links Between Circadian and Metabolic Processes

Environmental perturbations can be disruptive, leading to a search for the mechanisms by which they perturb circadian timing and thereby impact metabolic processes. The search for potential mechanisms at the molecular level entails examination of the genes and molecules that provide links between the circadian clock and metabolic processes. It is very clear that the circadian system and metabolic systems are intimately linked. Disruption of the circadian clock hierarchy is observed in obesity and metabolic syndrome, but it is yet to be established whether disruption of the circadian clock is a *symptom*, or a *cause*, of metabolic dysregulation. One way to probe the causal connection between circadian disruption and metabolic disorders is to disrupt clock genes and observe the effects on metabolic function. A significant finding in this regard was a series of studies in which whole-body *Clock* mutant mice were used to probe changes in metabolism. In *Clock* mice, the circadian clock is altered from birth in both central and peripheral tissues. These mice show marked development of obesity, which is a result of decreased metabolic rate and reorganization of feeding patterns. That is, *Clock* mice eat more during the light period than do WT mice. This whole animal model clearly shows that genetic disruption of clock function in the whole animal can lead to metabolic dysregulation. However, it does not provide adequate resolution to determine the precise mechanisms by which this occurs.

Because obesity and metabolic syndrome are associated with defects in glucose homeostasis, this is a logical area of investigation to investigate effects of mutations in circadian clock genes. The first question becomes: how are glucose and insulin altered by genetic mutations of the clock? In humans, polymorphisms in both *Clock* and *Bmal1* have been implicated in the development of Type II diabetes. In animal studies, whole body mutants of *Clock* and *Bmal1* demonstrate impaired glucose tolerance and blunted insulin insensitivity, and mice with mutated *Per2* show increased plasma insulin as well as impaired liver gluconeogenesis. Similarly, knocking out both *Cry1* and *Cry2* causes a hyperglycemic response to acute feeding and impaired glucose tolerance.

How can altering clock gene expression lead to such marked changes in systems controlling these complex metabolic processes? If the molecular clocks are disrupted throughout the body, the central and peripheral clocks no longer function in concert, and this temporal disorganization contributes to metabolic dysregulation, weight gain, and eventually obesity. This can be understood by considering CCG's. Peripheral tissues show rhythms not only in the canonical "clock" genes, but also in thousands of other genes governing nearly all cellular processes. It has been estimated that between 10 and 20% of genes in the liver are rhythmic, and about 15–20% of hepatic proteins show circadian rhythms. Modern bioinformatics techniques, linking gene transcript levels to protein levels and metabolic outputs of biochemical pathways have revealed the intricate temporal relationship in the liver. Taken together, it appears that clock function in peripheral organs is very important for overall function of these tissues.

Today, tools are available for probing the role of specific peripheral clocks using genetic targeting techniques that can disrupt clock genes in specific organs, including some essential for normal glucose homeostasis. Using this clever approach, the specific role of circadian clock genes in particular organs can be interrogated. As an example, it is now known that genetically knocking out the core clock genes *Clock* and *Bmal1* specifically in pancreatic beta cells (leaving them unaltered in other tissues) results in a diabetic phenotype, including very high levels of plasma insulin, and impaired glucose homeostasis.

Circadian Rhythms and Metabolic Function: A Two-Way Relationship

On the one hand, disrupted circadian clocks contribute to metabolic disorder, but it is also the case that this is a "two-way street," and metabolic factors—particularly those involved in the development of obesity and/or diabetes—can also modulate circadian and sleep systems. As an example, if food is presented to a nocturnal animal during the day, the timing of oscillations in peripheral clocks will be inverted, though the clock gene oscillations in the SCN are not affected. Another example, drawn from genetic mutant animals, the *Clock* double mutant mouse can entrain to food, even when its circadian locomotor behavior in the absence of food cues is arrhythmic. The quality of nutrition can also change eating behavior and affect the peripheral clock. A high-fat diet produces obesity and a damped feeding rhythm in mice so that they eat more food during the day and similarly reduces rhythmic gene expression in the liver. Furthermore, when food is withheld for 24h, the number of oscillating hepatic transcripts is reduced to only a few hundred transcripts while controls who consume the same amount of food as ad libitum-fed counterparts show robust rhythm in 5000 transcripts. Importantly, high-fat diet-induced obesity can be prevented by maintaining a strict feeding–fasting rhythm without reducing caloric intake. Similarly, the timing of food availability can "rescue" arrhythmic phenotypes in genetic clock mutants. For instance, Cry1/Cry2 double mutant mice, which completely lack a functional circadian clock, have erratic eating patterns, and lack rhythms in the liver. However, when these genetically modified mice are subject to a feeding–fasting regimen, several hepatic transcripts and metabolites assume daily rhythms characteristic of normal temporal structure.

The evidence is strong that the master circadian clock in the SCN can be entrained by light, while the phase of oscillators of other tissues, such as the liver, can be reset by cues derived from the timing of food intake. However, while the effects of food-derived signals are very well established, the mechanism(s) of entrainment by food is unclear. That said, the independent effects of light and food on SCN and peripheral clocks are likely one mechanism whereby poor temporal structure can arise. Circadian rhythm research in peripheral organs has revealed the importance of eating patterns on peripheral gene expression patterns, uncovered genetic and nutrition factors that modulate eating pattern, and demonstrated that eating patterns can counteract some obesogenic effects.

In humans and nonhuman animals, obesity is linked to disrupted sleep and circadian rhythms. This is likely due to a variety of factors, including clinical disorders such as sleep-disordered breathing, which can lead to sleep disruption and are more prevalent in obese or overweight individuals. Animal studies also indicate that obesity can lead to altered circadian and sleep parameters, including a change in the amount of time spent in wake or sleep. Furthermore, genetically obese animals, such as leptin-deficient mice and Zucker rats also show disruption in the amount of sleep and sleep timing. Taken together, these findings suggest that there is likely a bidirectional relationship between obesity, sleep, and circadian disruption.

Environmental Disruption and Metabolic Dysregulation

Thus far, we have discussed the proximate mechanistic links between circadian timing and metabolic function in terms of genetic mechanisms. However, clock gene mutations and tissue-specific knockouts do not directly address the ultimate causes of circadian disruption–induced metabolic dysregulation. In humans, genetic disruptions are likely not a significant driver of the links between obesity and circadian disruption. Instead, it is more likely that our disrupted circadian environments are the primary drivers. Epidemiologically, the "obesity epidemic" seems to be correlated with a gradual decrease in sleep time and quality. This correlational relationship does not directly imply causation, but it is intriguing. In more focused studies, obvious circadian disruption is found in shift workers (eg, airline crews, truck drivers, medical doctors, law enforcement, and the military). Even in the general public, individuals are exposed to light late into the night (eg, TVs, tablets, smart phones), and have work schedules that may conflict with their endogenous chronotype (ie, circadian preference). This phenomenon has come to be known as "social jet lag," which describes the changes in sleep patterns from work to nonwork days (eg, holidays and weekends). This is significant, as individuals who report poor sleep show increased incidences of diabetes and risk factors for the development of cardiovascular disease, and social jet lag is associated with a higher BMI. In human laboratory studies, misaligned sleep can lead to plasma glucose levels approaching those of diabetics. However, the physiological and cellular mechanisms translating *environmental* circadian disruption to increased weight gain and metabolic syndrome remain unknown. This gap limits our understanding of the mechanisms by which environmental circadian disruption leads to metabolic dysregulation. To explore the mechanisms behind these effects, it is necessary to apply animal models in which we can manipulate the circadian environment (ie, the light–dark cycle) and determine how this affects metabolism.

Understanding the mechanisms whereby the SCN synchronizes peripheral clocks and is able to affect downstream pathways, such as glucose and lipid metabolism, has been a major focus for researchers in this area over the last decade. Since circadian rhythms are highly conserved among species, the findings from animal studies translate well to understanding the interconnections among the environment, circadian rhythms, sleep, and metabolic function in humans. For instance, altering light–dark cycles of rodents is an elegant and simple method to probe how circadian disruption affects myriad physiological systems. Inappropriate lighting (eg, light at night) has been used to explore some of these effects. Mouse studies show that exposure to dim light at night causes body weight increases, but no increase in food consumption. Similarly, using nonstandard lighting periods in mice (eg, 20h days, 10h light, 10h dark cycles) causes weight gain, elevates plasma, leptin insulin, and triglycerides, and an increased insulin:glucose ratio. These are all indications of the metabolic syndrome, and represent a prediabetic or diabetic state. Both of these cases (light at night and 20h days) do not cause an increase in overall food consumption, but instead lead to a reorganization of feeding behavior. This suggests that disruption of these feeding rhythms may be a crucial component for development of circadian-induced obesity.

CIRCADIAN DISRUPTION AND COGNITIVE FUNCTION

Circadian rhythms are a remarkable feat of biological complexity. Perhaps more remarkable is the fact that in most cases we are generally oblivious to their daily operation until we engage in a round of working night shifts, or try to function while severely jet lagged. After these acute circadian disruptions, changes in mood, affect, or cognitive function are commonly reported. The neurobehavioral dysregulation caused by circadian disruption has had some stark consequences, as several of the most notorious industrial accidents in the past few decades, including the Bhopal disaster in India, the Chernobyl nuclear accident in Ukraine, and the Exxon Valdez oil spill in Alaska occurred during the night. It is thought that several factors, including fatigue, interacted in each of these cases to cause or exacerbate the chain of events that lead to catastrophe. Therefore, the evidence is suggestive that disrupted circadian clocks and sleep cycles, coupled with high cognitive workload, could lead to significant performance deficits.

Circadian Effects on Learning and Cognition

While we may be extremely interested in the consequences of disrupted clocks in human performance, most human studies do not allow for mechanistic or causal experiments to be undertaken and basic research in animals can prove helpful. Many studies have demonstrated that circadian rhythms are important for normal learning and cognitive function. Even in simple invertebrates, such as the sea slug *Aplysia*, circadian effects on learning can be observed. In mammals, ablation of the SCN clock affects learning and memory. Hippocampal-mediated learning and memory has been particularly well studied, with clear circadian effects on spatial navigation tasks and object localization tasks, for instance. It is also known that the circadian clock influences hippocampal long-term potentiation (LTP), and aspects of the intracellular MAPK signaling pathway in hippocampal cells, which is essential for plastic changes that occur during learning, are modulated by circadian time.

Modeling Circadian Disruption by Unusual Light Conditions

Not only the evolutionary ubiquity of circadian clocks, but also their role in learning and memory (both simple and complex), underlie the significance of these phenomena and their importance in an organism's survival. Thus, modeling the effects of circadian disruption on learning, memory, and cognitive function is crucial. One such method involves using unique or "exotic" lighting conditions to drive circadian disruption. For instance, using a sequence of light pulses and a phase shift to cause arrhythmia in hamsters also impairs hippocampal learning and memory. Repeated jet lag models, where light cycles are rapidly shifted forward and back have also been helpful in this regard. As an example, chronic "jet lag" by repeatedly phase shifting the light–dark cycle results in learning and memory deficits that are accompanied by reductions in hippocampal neurogenesis. Exotic light–dark cycles can also provide important information, as chronically housing mice in shortened 20h days leads to atrophy of neurons in the prefrontal cortex, and significant degradation of cognitive flexibility. Though the mechanisms are still unknown, accumulating evidence supports a role for circadian disruption as a causative contributor to neurocognitive deficits.

CIRCADIAN DISRUPTION AND PSYCHIATRIC DISORDERS

In many psychiatric disorders, a common comorbidity is disruption in the circadian sleep–wake cycle. By many reports, this is also one of the most significant quality of life issues. What remains unclear is whether disrupted circadian rhythms are a *symptom* or a *cause* of psychiatric disease. While it is unlikely that circadian rhythm disruption could cause de novo development of psychiatric disorders, much research suggests that disrupted clocks could shift probability of their occurrence in already vulnerable individuals.

Depression

Depression is a normal behavior in some instances, but it becomes pathological when it interferes with life, making it impossible for an individual to carry out daily activities. In addition to the classic symptoms of flattened affect and anhedonia, depressed individuals also report sleep and circadian abnormalities, including both insomnia (inability to sleep) and hypersomnia (excessive sleep and sleepiness). In addition to these behavioral outputs, new evidence also shows that circadian clock gene polymorphisms are common in major depressive disorder, and that intensity of major depressive symptoms are correlated with circadian misalignment. Specifically, having a circadian pacemaker more delayed relative to sleep onset is associated with more severe depressive states. These types of data show us that disrupted clocks are associated with depressive disorders, but we still need to explore if this is a causal connection. Such studies are difficult in humans, but there are some populations that can help us unravel this mystery. For instance, shift workers often suffer from mood disturbances, including increased risk for depression. Again, animal studies can provide some additional information about these relationships. At the level of the molecular clock, mice with mutations in several clock genes (*Per*, *Cry*, *Bmal*) show increases in depressive-like behaviors. Conversely, manipulations that are known to cause increased depressive-like behaviors, such as different forms of chronic stress, can lead to abnormal rhythms of clock genes in brain areas associated with depressive-like behaviors, including the amygdala and hippocampus. At the circuit level, as mentioned earlier, chronic circadian phase shifting leads to decreased hippocampal neurogenesis, and reduced neurogenesis is associated with depressive-like behaviors in many animal models. In addition, atrophy of prefrontal cortical neurons that occurs in chronic housing in 20h days is similar in extent and magnitude to decreases observed in chronic stress, which is also related to increased susceptibility to depressive-like phenotypes in animals. Looking at the entire landscape of findings, it seems clear that circadian disruption can contribute to depression, but unraveling symptomology and etiology is extremely difficult. Though it is clear that there is strong relationship between circadian disruption and depression, the effects are likely bidirectional.

Mania and Bipolar Disorder

A seemingly opposite behavioral state to depression is mania, and there are conditions such as bipolar disorder that involve rapid cycling between these two states. Connections between circadian rhythms, mania and bipolar disorder are an active area of research. It is well known that in humans manic episodes result in significant alterations of sleep patterns. In concert with the behavioral sleep abnormalities, several normally rhythmic physiological functions, such as body temperature and plasma hormone levels, are blunted during these episodes. Here again, use of the well-characterized molecular clock genes can help to assess causal connections. For example, mutations in the core clock gene *Clock* can lead to mania-like behaviors. Site-specific knockdown of *Clock* in the ventral tegmental area of mice can also induce manic-like behaviors, including increased locomotor behaviors, and altered responses in novel environments. In addition to effects of clock gene mutations on manic behavior, there are interesting examples in the experimental literature indicating that treatments known to reduce manic symptoms can affect clock mechanisms. Treating hamsters with lithium chloride (one of the most common treatments of bipolar disorder) causes a lengthening in the circadian period. Follow-up work in several species further indicates that lithium modifies many intracellular signaling cascades, including glycogen synthase kinase-3beta, which is an important link to cellular circadian mechanisms. Specifically, lithium indirectly inhibits GSK3β which then results in increased transcription of *Per2*. Thus, a pharmacological treatment that helps control symptoms of mania and bipolar disorder can have direct effects on circadian processes at the behavioral and molecular levels. Therefore, there is very strong evidence that disruption of the circadian clock is a component of some manic or bipolar phenotypes. Furthermore, modification of the molecular clock can recapitulate many of these phenotypes, and treatments that reduce symptoms can directly interact with the molecular clockworks.

Schizophrenia

Over the past decade, links between normal cognitive function, depression, and mania and disrupted circadian timing are becoming clearer. However, schizophrenia has been a more difficult psychiatric condition to probe, both in terms of potential mechanisms and findings in clinical populations. Obviously, a contributing issue to this dearth of solid evidence is that the causes of schizophrenia are unknown, and the fact that the disorder is likely a consequence of both environmental and genetic factors. Even with this added complication, there is some evidence of links between circadian timing and schizophrenia. First, schizophrenic patients show clear fragmentation of rest–activity cycles. Second, sleep problems are also evident, with people suffering from schizophrenia showing both sleep onset and sleep maintenance insomnia. Finally, melatonin, a key hormonal regulator of sleep in many mammals including humans, shows a disrupted rhythm in schizophrenia sufferers. Taken together, the evidence points to disruption of multiple systems that are normally under tight circadian control. There is a genetic contribution to development of schizophrenia, and excellent data exist on the genetic makeup of several human populations. In one such study, polymorphisms in the vasoactive intestinal polypeptide 2 receptor (*VIP2r*) are associated with increased risk for development of schizophrenia. That these polymorphisms are in *VIP2r* is particularly interesting because VIP and its receptors (including *VIP2r*) are very important in the network functioning of the SCN clock. This represents an intriguing link between a human polymorphism associated with schizophrenia risk and a noncore "clock gene." Animal models of schizophrenia are also being actively explored for potential circadian and sleep abnormalities. For instance, the "blind drunk" (Bdr) mouse model of schizophrenia-like symptoms show fragmented circadian rhythms, as well as a phase-advanced period under entrained conditions. So while the links between circadian rhythms and schizophrenia are not as clear as in depression or manic behaviors, there is nonetheless compelling evidence that alteration of circadian timing systems may increase the risk of developing schizophrenia.

SUMMARY AND CONCLUSIONS

Disrupted circadian clocks are observed in many psychiatric conditions including depression, mania, and schizophrenia. While it may be easy to dismiss these changes as merely symptoms of the disorders, or even consequences of their treatments, it is also intriguing to think about the role that circadian rhythms may play in development of such disorders (Fig. 27.4). As mentioned at the outset of this section, it is highly unlikely that circadian disruption is the sole factor causing these disorders to develop, especially since they are usually polygenic, and are influenced by experiential and environmental exposures. However, it is likely that disruption of the clock somehow alters an individual's risk. This is an active area of research, enabled by the current focus on "personalized" treatment protocols that are sensitive to differences among individuals. Thus, while it is still unknown if disruption of circadian rhythms is a symptom or an etiology of psychiatric disorders, they clearly play an important interactive role, and understanding how they interact with other known mechanisms may provide additional insight in the causes of mental disorders.

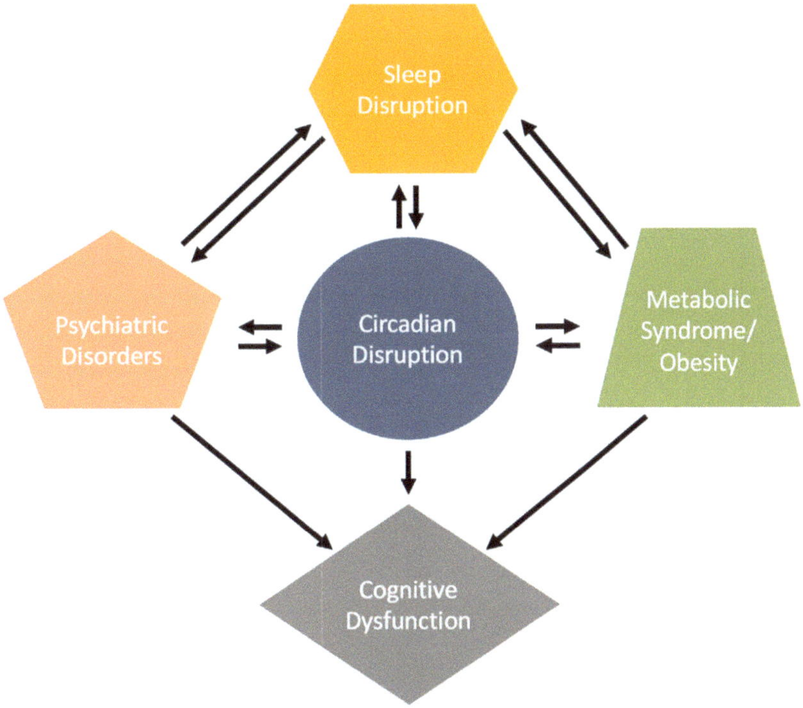

FIGURE 27.4 Summary of the interplay between circadian disruption and sleep disruption, and the subsequent effects on mental and physical health. Note some relationships are bidirectional, while other relationships seem more unidirectional. There is substantial evidence that circadian and sleep disruption can have wide-ranging effects on both mental and physical well-being.

Further Reading

[1] Takahashi JS, Hong H-K, Ko CH, McDearmon EL. The genetics of mammalian circadian order and disorder: implications for physiology and disease. Nat Rev Genet 2008;9(10):764–75. http://dx.doi.org/10.1038/nrg2430.

[2] Roenneberg T, Allebrandt KV, Merrow M, Vetter C. Social jetlag and obesity. Curr Biol 2012;22(10):939–43. http://dx.doi.org/10.1016/j.cub.2012.03.038.

[3] Wallach T, Kramer A. Chemical chronobiology: toward drugs manipulating time. FEBS Lett 2015;589(14):1530–8. http://dx.doi.org/10.1016/j.febslet.2015.04.059.

[4] Antle MC, Silver R. Circadian insights into motivated behavior. Curr Top Behav Neurosci 2015. http://dx.doi.org/10.1007/7854_2015_384. [chapter 384].

[5] Barandas R, Landgraf D, McCarthy MJ, Welsh DK. Circadian clocks as modulators of metabolic comorbidity in psychiatric disorders. Curr Psychiatry Rep 2015;17(12):98. http://dx.doi.org/10.1007/s11920-015-0637-2.

[6] Karatsoreos IN, Silver R. Minireview: the neuroendocrinology of the suprachiasmatic nucleus as a conductor of body time in mammals. Endocrinology 2007;148(12):5640–7. http://dx.doi.org/10.1210/en.2007-1083.

[7] Borbely AA. A two process model of sleep regulation. Hum Neurobiol 1982;1(3):195–204.

[8] Peek CB, Ramsey KM, Marcheva B, Bass J. Nutrient sensing and the circadian clock. Trends Endocrinol Metab 2012;23(7):312–8. http://dx.doi.org/10.1016/j.tem.2012.02.003.

[9] Van Cauter E, Spiegel K, Tasali E, Leproult R. Metabolic consequences of sleep and sleep loss. Sleep Med 2008;9(Suppl. 1):S23–8. http://dx.doi.org/10.1016/S1389-9457(08)70013-3.

[10] Gerstner JR. On the evolution of memory: a time for clocks. Front Mol Neurosci 2012;5:23. http://dx.doi.org/10.3389/fnmol.2012.00023.

[11] Edgar N, McClung CA. Major depressive disorder: a loss of circadian synchrony? BioEssays September 2013. http://dx.doi.org/10.1002/bies.201300086.

CHAPTER

27.1

Translational Correlation: Sleep Disorders

S. Gorantla[1], J.V. Rundo[1], J.C. DeToledo[2]

[1]Cleveland Clinic Foundation, Cleveland, OH, United States; [2]Texas Tech University Health Sciences Center, Lubbock, TX, United States

The average total sleep in a healthy adult varies from 7.5 to 8h, but sleep requirements vary widely from person to person (4–10h). American Academy of Sleep Medicine and Sleep Research Society consensus statement recommended 7–9h of sleep on a regular basis to promote optimal physical and mental health in adults. The International Classification of Sleep Disorders, 3rd edition (ICSD-3) lists more than 70 well-defined sleep disorders. The prevalence of sleep disorders in the adult population is estimated to be about 25–30%. Sleep disorders may cause or aggravate many systemic illnesses and may effect quality of life.

The key anatomical regions in the brain that regulate the sleep–wake cycle include the brain stem, hypothalamus, thalamus, and basal forebrain. Sleep disorders due to lesions in those key anatomical regions are not very common in clinical practice. Monoamines (serotonin, dopamine, and norepinephrine) and acetylcholine are the main neurotransmitters that facilitate sleep. In contrast to anatomic problems, which are rare, altered neurochemical transmission is a common etiology for sleep disorders.

To complement the basic neuroscience of sleep from the previous chapter, the next session is review of some of the most common sleep disorders.

SLEEP-RELATED BREATHING DISORDERS

Sleep-related breathing disorders is a broad term given to describe syndromes with intermittent episodes of apnea, hypopnea, or respiratory effort-related arousals during sleep. Sleep-related breathing disorders are grouped into obstructive sleep apnea (OSA) disorders, central sleep apnea (CSA) disorders, and sleep related hypoventilation disorders based on pathophysiology and polysomnography (PSG) findings.

OSA is the most common sleep-related breathing disorder. OSA is characterized by partial or complete collapse of the respiratory tract during sleep in susceptible individuals, resulting in hypopnea or apnea, respectively. REM sleep often exacerbates respiratory events (apneas and hypopneas) in OSA patients due to

Conn's Translational Neuroscience
http://dx.doi.org/10.1016/B978-0-12-802381-5.00044-0

marked reduction in smooth muscle tone in the upper airway. Supine head position, sedatives, and alcohol consumption also worsen the incidence of airway collapse. Increasing age, male gender, body mass index (BMI) ≥ 30, oropharyngeal crowding, and specific craniofacial features (short mandible, teeth position) are other risk factors for OSA.

The common manifestations of OSA include loud snoring, recurrent nocturnal awakenings, daytime fatigue, memory and concentration issues, dry mouth or sore throat due to mouth breathing, and excessive daytime sleepiness (EDS) concentration issues. Long-term consequences of hypoxia and hypercapnia due to OSA include increased risk of hypertension, right heart failure, pulmonary hypertension, cardiac arrhythmias, stroke, and cognitive impairment. Continuous positive airway pressure (CPAP) is the main modality of treatment, which keeps the airway patent and prevents collapse during sleep. Other treatment options include mandibular advancement devices, expiratory positive airway pressure devices, oral pressure therapy devices, positional therapy, and surgical options. These treatment options are generally less effective than CPAP, however, they may be more tolerable than CPAP for some patients.

CSA, complex sleep apnea syndrome (CompSAS), and sleep-related hypoventilation due to neuromuscular or chest wall deformities are less commonly encountered sleep-related breathing disorders. CSA is characterized by episodes of apnea or hypopneas due to lack of respiratory effort secondary to the loss of ventilator drive in the lower brain stem. Increased responsiveness of the respiratory center to CO_2 and unstable ventilatory control system (high loop gain) is believed to cause CSA. Lesions in the lower brain stem and upper cervical spine, including infarcts, demyelination, and tumors can result in CSA. Cheyne–Stokes breathing pattern, characterized by alternating periods of hyperventilation and hypoventilation, is often seen in patients with CSA. The treatment for CSA varies depending on the etiology. Adaptive servo ventilation (ASV), a device with the ability to adjust to ventilatory requirements, is usually the preferred treatment for CSA, except for patients with systolic congestive heart failure (ejection fraction less than 45%). Sleep-related hypoventilation disorders are characterized by impaired ventilation during sleep resulting in increased CO_2 level in blood. Obesity hypoventilation syndrome (OHS) is a well known form of hypoventilation disorders due to obesity (BMI > 30) manifested by daytime hypercapnea in addition to nocturnal hypercapnea and often coexists with OSA. Weight loss and PAP therapy usually benefit patients with OHS.

SLEEP-RELATED MOVEMENT DISORDERS

Restless legs syndrome (RLS), also known as Willis–Ekbom disease, is characterized by an irresistible urge to move the limbs, usually the legs, which is frequently associated with an uncomfortable sensation. The uncomfortable sensation occurs predominantly or exclusively at night, while inactive, and is temporarily relieved by moving the limbs. While symptoms are usually bilateral, and affect the legs, some patients report symptoms only in one extremity. Symptoms can progress to the arms and trunk over time.

RLS is associated with sleep onset and maintenance insomnia, with frequent awakenings due to periodic limb movements (PLMs). Iron deficiency (low ferritin levels), end-stage renal disease requiring dialysis and pregnancy are some of the conditions associated with RLS. Iron supplementation is recommended in patients with iron deficiency. Dopamine agonists (pramipexole, ropinirole, and rotigotine patch) and calcium channel alpha 2-delta ligands (gabapentin and pregabalin) are considered first-line treatment options for persistent symptoms of RLS.

Periodic limb movement disorder (PLMD) is a condition that is described as repetitive limb movements that occur during sleep and cause sleep disruption. The limb movements usually involve the lower extremities, consisting of extension of the big toe and flexion of the ankle, the knee, and the hip. Patients with PLMD do not have an irresistible urge to move the limbs, as seen in patients with RLS.

PARASOMNIAS

Abnormal behaviors related to sleep are broadly termed as parasomnias. Parasomnias can be divided into REM-related and NREM-related parasomnias. REM sleep behavior disorder (RBD) is a REM parasomnia that is characterized by dream enactment behaviors. RBD is possibly due to failure of atonia circuitry resulting in preserved ability to move and enact dreams. Patients with RBD can injure themselves and their bed partners due to violent movements at night. Patients with idiopathic RBD are at increased risk (80%) of developing alpha synucleinopathies (Parkinson disease, multiple system atrophy and dementia with Lewy bodies) in future. Medications, particularly antidepressants (selective serotonin receptor inhibitors and tricyclic antidepressants) and narcolepsy are common culprits of RBD in young adults. Clonazepam and melatonin are typically used to treat idiopathic RBD.

NREM-related parasomnias include sleep terrors, sleep walking, sleep-related eating disorder, and confusional arousals. They are frequently associated with hypnotic sedatives, especially benzodiazepine receptor agonists (zolpidem), alcohol, sleep deprivation, stress, illness, and coexisting OSA, especially in adults. Management of underlying OSA with PAP therapy and removal of causative agent often improves NREM-related parasomnias.

INSOMNIA

Insomnia can be defined as difficulty with initiating or maintaining sleep when patient is provided with adequate opportunity to sleep. These patients typically experience impaired daytime functioning due to night-time sleep deficit. Psychiatric comorbidities, coexisting primary sleep disorders (OSA, RLS, and circadian rhythm disorders), medications, and several medical conditions are associated with insomnia. Insomnia can be acute or chronic. Short-term or acute insomnia is often triggered by pain or other stressor and typically lasts less than 3 months. Chronic insomnia can be defined as having symptoms of insomnia lasting at least 3 times per week for at least 3 months. Cognitive and behavioral interventions (sleep hygiene education, stimulus control, sleep restriction therapy and relaxation techniques) and pharmacological therapies (benzodiazepines, non-benzodiazepine sedatives, melatonin agonists, and antidepressants) are the two main therapeutic options in patients with insomnia.

HYPERSOMNIA

Hypersomnia is defined as disabling EDS and inability to maintain wakefulness during the daytime. Hypersomnia can be secondary to OSA, circadian rhythm disorders, and depression among other causes. Hypersomnia in the absence of earlier mentioned disorders is referred to as primary hypersomnia of central origin. Narcolepsy, idiopathic hypersomnia, and recurrent hypersomnia (Kleine–Levin syndrome and menstrual-related hypersomnia) are different types of hypersomnias of central origin.

Narcolepsy is REM-related sleep disorder characterized by EDS with or without cataplexy, hypnogogic hallucinations, and sleep paralysis. Narcolepsy results from deficiency of hypocretin-1 (orexin A) in the dorsolateral hypothalamus. Hypocretin inhibits REM sleep and facilitates wakefulness. It prevents inappropriate sleep–wake and REM–NREM sleep transitions. Loss of hypocretin results in intrusion of REM sleep-related phenomena (EDS, hypnogogic hallucinations, and sleep paralysis) into wakefulness. Hypocretin neurons project to the histaminergic tuberomammillary nucleus, the ventrolateral preoptic nucleus, cholinergic pedunculopontine nucleus, basal forebrain, the brain stem monoaminergic system (raphe nucleus, locus ceruleus, ventro-tegmental area, and periaqueductal grey matter). Stimulation of orexin neurons in those monoaminergic areas promotes wakefulness, inhibits REM sleep, and REM sleep–related phenomena (cataplexy, hypnogogic hallucinations, and sleep paralysis). EDS in narcolepsy can also be due to the loss of the inhibitory effect on the sleep-promoting rostral hypothalamus (preoptic area). Cataplexy due to hypocretin deficiency is a consequence of reduced activity of monoaminergic motor excitatory systems in the brain stem or reduced activity of glycine-mediated motor inhibitory systems in the brain stem. Genetic factors play an important role in narcolepsy. About 85–90% of patients with narcolepsy with cataplexy will test positive for HLA-DQB1*0602. An autoimmune hypothesis was proposed but has not been confirmed. Infarcts, vascular malformations, tumors, and other lesions involving posterior hypothalamus and midbrain can cause secondary narcolepsy due to the damage to hypocretin-secreting neurons.

Evaluation of narcolepsy starts with a comprehensive sleep history and epworth sleepiness scale (ESS) to estimate subjective daytime sleepiness. An overnight PSG to rule out other causes of hypersomnia (OSA), followed by a multiple sleep latency test (MSLT) are required to diagnose narcolepsy. The MSLT is a diagnostic tool to objectively measure sleepiness and to determine if the patient is entering REM sleep rapidly. A mean sleep latency (MSL) <8 min along with two sleep onset rapid eye movement (SOREM) periods without any other confounding factors (known sleep deprivation or untreated OSA) is diagnostic for narcolepsy.

Stimulants (amphetamines) and wake-promoting agents (modafinil, armodafinil) are conventional therapeutic options to treat EDS in patients with narcolepsy. Antidepressants can be used to treat cataplexy due to their ability to suppress REM sleep. Sodium oxybate is another medication approved for the treatment of narcolepsy. This medication is effective in treating both EDS and cataplexy.

Idiopathic hypersomnia is less common, and it is a diagnosis of exclusion. MSLT in idiopathic hypersomnia demonstrates reduced mean sleep latency (MSL < 8 min) and 1 or no SOREM periods. Stimulants (amphetamines) and wake-promoting agents (modafinil, armodafinil) are helpful in treating idiopathic hypersomnia. The remaining central origin hypersomnias are very rare.

CIRCADIAN RHYTHM DISORDERS

Circadian rhythm disorders are characterized by chronic (at least 1 month) patterns of sleep disturbances due to either misalignment of endogenous circadian rhythm and sleep–wake cycle or disruption of circadian timing system. About 3% of the adult population is estimated to be suffering from circadian rhythm disorders. There are six well recognized circadian rhythm disorders: (1) Delayed sleep phase disorder. (2) Advanced sleep phase disorder. (3) Irregular sleep–wake rhythm disorder. (4) Non-24-h sleep–wake disorder. (5) Jet-lag. (6) Shift work disorder. Poor sleep hygiene and shift work are frequently associated with delayed sleep phase disorder (DSPD), in which the internal biological clock is delayed with respect to external day-night cycle.

The suprachiasmatic nucleus (SCN) acts as a pacemaker that synchronizes the external environment with the internal biological clock. The SCN has an inhibitory effect on the paraventricular nucleus (PVN). The PVN activates the sympathetic output through superior cervical ganglion and stimulates melatonin secretion from the pineal gland. Light activates the SCN which inhibits the PVN from activating the sympathetic output to pineal gland, resulting in decreased secretion of melatonin.

During night, lack of input from retinohypothalamic tract leads to decreased inhibition of PVN, thus stimulating the production of melatonin from the pineal gland. Non-24-h sleep–wake disorder is commonly seen in blind people due to impaired stimulation of SCN through the retinohypothalamic tract.

A comprehensive sleep history, sleep diary, and/or actigraphy (device that detects movement and estimates wake/sleep patterns) are used to diagnose circadian rhythm disorders. Timed bright light exposure, chronotherapy (progressive delay of sleep time) and exogenous melatonin administration are the therapeutic options utilized in clinical setting to synchronize environmental and internal endogenous rhythms.

OTHER INTERESTING CONDITIONS ASSOCIATED WITH DISTURBED SLEEP

Human African trypanosomiasis (HAT) or Sleeping Sickness is a parasitic disease caused by protozoan parasite *Trypanosoma brucei* (*Trypanosoma gambiense* and *Trypanosoma rhodesiense*). It is mostly confined to sub-Saharan Africa and is transmitted by tsetse fly. The late meningoencephalitic stage with the invasion of central nervous system (CNS) parenchyma is characterized by a sleep disorder. It has a predilection to affect circumventricular organs outside the blood–brain barrier including the choroid plexus, neurohypophysis, pineal gland, arcuate nucleus, and regions of posterior hypothalamus and area postrema in the brain stem. The sleep disorder in HAT consists of excessive daytime sleepiness, intense urges to sleep during daytime and insomnia at night. PSG in the late stages of Trypanosomiasis with sleep disorder often shows SOREM, when REM sleep occurs in the first 20 min of sleep onset. Normally, the first REM period begins 90–110 min after sleep onset. Circadian rhythm dysfunction is also associated with HAT.

Serology is the initial step in diagnosis HAT. Antibody detection against *Trypanosoma* is the first step followed, by parasite identification under microscope to confirm the diagnosis. Cerebrospinal fluid (CSF) analysis is used to stage the disease, as the parasite can be found in the CSF in late stages of the disease. Treatment is based on the species of *Trypanosoma* and the stage of the disease. Early *T. gambiense* is treated with pentamidine and late stage with eflornithine. Early stage of *T. rhodesiense* is treated with suramin and late stage with melarsoprol.

Further Reading

[1] Paiva T, Attarian H. Obstructive sleep apnea and other sleep-related syndromes. Handb Clin Neurol 2014; 119:251–71.

[2] Merlino G, Gigli GL. Sleep-related movement disorders. Neurol Sci June 2012;33(3):491–513.

[3] Peever J, Luppi PH, Montplaisir J. Breakdown in REM sleep circuitry underlies REM sleep behavior disorder. Trends Neurosci May 2014;37(5):279–88.

[4] Howell MJ. Parasomnias: an updated review. Neurotherapeutics October 2012;9(4):753–75.

[5] Foldvary-Schaefer N, Alsheikhtaha Z. Complex nocturnal behaviors: nocturnal seizures and parasomnias. Continuum (Minneap Minn) February 2013;19(1 Sleep Disorders):104–31.

[6] Sutton EL. Insomnia. Med Clin North Am May 2014; 98(3):565–81.

[7] Kim MJ, Lee JH, Duffy JF. Circadian rhythm sleep disorders. J Clin Outcomes Manag November 1, 2013; 20(11):513–28.

[8] Lejon V, Bentivoglio M, Franco JR. Human African trypanosomiasis. Handb Clin Neurol 2013;114:169–81.

CHAPTER

28

The Experience of Consciousness

P.T. Morgan

Yale University, New Haven, CT, United States

Ignorance is Bliss. ***Thomas Gray (1742).***

There is a wisdom that is woe; but there is a woe that is madness. ***Herman Melville (1851).***

INTRODUCTION

Progress in research related to cognition and awareness has accelerated due to the application of modern observational techniques in neurobiology to evolving principles of psychology. However, history suggests that during periods of scientific advancement (such as the ongoing revolution) there is a general optimism that problems once considered impenetrable may crumble under the siege of modern technology and new ways of thinking. Although largely supported by the remarkable progress in science made during such periods, the notion that all questions will eventually succumb to science is fallacious when the questions of interest are not scientific ones. Of particular interest in this regard is the desire to "understand" how the experience of consciousness, or sentience, arises.

Many aspects of brain function are measureable already, and many more will be with the continued application of new technologies. Features of consciousness that fit into this category have been described by David Chalmers as the "easy" problem of consciousness. Examples of this type of brain function include the ability to discriminate stimuli, to enter distinct states like wakefulness and sleep, and to report mental states including response to stimuli and even metacognition (the awareness of being aware), among others. In each case, the neural correlates of these functions, states, and behaviors are measureable, and testable predictions with regard to all these are possible.

In contrast, the "hard" problem of consciousness asks how the experience of consciousness, or "phenomenal consciousness," arises. This problem presupposes that (à la Thomas Nagel) there is "something it is like" to be conscious, and comes from an experience of being wherein visual stimuli are "seen," auditory stimuli are "heard," pain is "felt," pleasure is "enjoyed," thoughts are "thought," and so on. These instances of conscious experience were termed "qualia" by Clarence Irving Lewis in the first half of the 20th century. A commonly invoked example of a "quale" (the singular of *qualia*) is the experience of seeing the "redness" of the color red. However, qualia equally include the feeling of joy, pleasure, or anguish, the manifold sensations of a re-experienced memory, or the experience of some new construct of sensations in a day- or night-dream. Asking questions

Conn's Translational Neuroscience
http://dx.doi.org/10.1016/B978-0-12-802381-5.00045-2

about qualia, or about the experience of consciousness generally, is problematic, and for this reason has been termed the "hard" problem of consciousness.

Central to the "hard" problem of consciousness is that we have the syntax to describe the problem, the philosophy and motivation to ponder it, and no way to test the problem scientifically. The inability to test the problem scientifically has not, however, stopped a steady onslaught of scientific-sounding hypotheses or work-arounds that sincerely aim to answer the question. In this chapter, a description of the problem of measurement of phenomenal consciousness, a discussion of what science does and does not do, a simple model that aims to support an intuitive appreciation of the fundamental problem with the problem itself, as well as a discussion of a relatively recent effort to address the problem and other thoughts are presented. A rationale for why the problem is not more generally recognized as insoluble is also discussed, and any impact related to not being able to answer this question scientifically is considered.

THE PROBLEM WITH MEASUREMENT

Statement of the problem:

Brain activity is measureable. A person's report of their mental state, and behaviors related to their mental state, is measureable. Whether or not the person next to you has a *conscious experience*, however, is not measureable.

Although the reality of this statement may be intuitively obvious, there are nevertheless points worth discussing with regard to the immeasurability that may make it more so. First, this is not a claim that the person next to you lacks a conscious experience, only that it is impossible to measure its presence or the quality of the experience itself. For example, assuming that others have a conscious experience, there is no manner in which the experience of it can be directly compared to yours. The "redness" of red that you experience may appear to someone else like what you experience as "green," or be experienced as something else entirely. Though the neural correlates of "red" color recognition may be measureable and consistent, and observing those correlates allows for the prediction that the observer will report seeing "red," the experience of the "redness" in the observer is *known* only to them (Box 28.1), and *assumed* (albeit in most cases quite deeply) by others.

If anything, it is likely the depth of this assumption about the conscious experience in others that obstructs an intuitive grasp of the immeasurability of the conscious experience. It makes sense that this assumption would be deeply held in a creature that has the capacity to consider the alternative. In particular, a deep assumption related to the conscious experience in others being like one's own likely provides an advantage in natural selection: the ability to predict others' behaviors

> **BOX 28.1 The Problem With Qualia**
>
> The sensation of colour cannot be accounted for by the physicist's objective picture of light-waves. Could the physiologist account for it, if he had fuller knowledge than he has of the processes in the retina and the nervous processes set up by them in the optical nerve bundles and in the brain? I do not think so. We could at best attain to an objective knowledge of what nerve fibres are excited and in what proportion, perhaps even to know exactly the processes they produce in certain brain cells … but even such intimate knowledge would not tell us anything about the sensation of colour … the same physiological processes might conceivably result in a sensation of sweet taste, or anything else … The same holds for other sensations…
>
> *Erwin Schrödinger, from* Mind and Matter

based on a reflection of one's own thoughts and how they predict one's behavior is undoubtedly powerful, and significant doubts (or even disinterest) about the conscious experience in others could have dire consequences for creatures that largely depend on social structure to survive. The depth of this assumption notwithstanding, it can also be assumed, no matter how unlikely it may seem, that the person next to you does not have a conscious experience: there is no way to prove otherwise. This so-called "philosophical zombie" could protest to the contrary, but how would you know? A computer, and perhaps even a parrot, could protest to the contrary, and perhaps in some cases as convincingly, except for the prejudice in the listener who is aware it is a computer or a parrot. As any proposed measurement of neural correlates of conscious experience necessarily relies on the subject's description of the conscious experience to presumably make that correlation, problems ensue. First, the subject's attestation of conscious experience may be inaccurate; second, regardless of whether the attestation is accurate, the observation that is being correlated to the neural activity is the attestation and not the conscious experience itself; third, an inability to make such an attestation cannot be proof that there is no experience.

Another way of looking at this problem is through the consideration of artificial intelligence, where discussing a machine "thinking" is akin to attributing a conscious experience to the machine. Alan Turing proposed in 1950 his famous test to determine whether a machine "thinks," or at least that is how it is popularly described. Actually, he disregarded any notion that he was answering that question:

> The original question, 'Can machines think?' I believe to be too meaningless to deserve discussion … the question, 'Can machines think?' should be replaced by 'Are there imaginable digital computers which would do well in the imitation game?'

Turing's imitation game is nothing more than a test to measure the likeness of a computer to a human in how well it engages in conversation (but it could equally be some other observable human behavior). The measure of success is how well actual humans can or cannot distinguish the computer from a real human. So having discarded the immeasurable—whether a computer can "think"—he proposed a grossly measurable quantity, like *how many of us how often and with what consistency cannot tell the human from the computer?*

Rather than measure whether a computer becomes sentient, or experiences thoughts or consciousness, the test measures whether others would believe that it does. This interpretation of the test follows from the following logic: humans tend to believe that other humans experience consciousness; if a human believes it is or could be conversing with another human, it would tend to believe it is conversing with something that experiences consciousness. In this case, the presumption of conscious experience in others leads naturally to the presumption of conscious experience in a computer—a computer whose identity was erstwhile hidden but perhaps no longer:

> Nevertheless I believe that at the end of the century the … general educated opinion will have altered so much that one will be able to speak of machines thinking without expecting to be contradicted. ***Turing***

Turing never claimed to have a test for whether a machine "thinks," but rather because it is "meaningless" to discuss such a test (being that the conscious experience is immeasurable), we are left to consider only whether something can appear to be human-like in this regard. Consistent with this limitation, we cannot also rule out that computers, or animals, or anything else lacks the conscious experience. Rather, the projection of our own conscious experience onto others may lead us to project the same experience onto whatever appears to behave as we do, as the limits of our observation do not permit us to test scientifically whether conscious experience arises in the *other*. Although Turing's prophecy may have predicted a date too soon, there is no scientific hindrance to its ultimate fulfillment, as only presuppositions about nonhuman computers must change for this to occur. Understandably, the bias against nonhuman computers experiencing consciousness will likely diminish further as nonhuman computers are built or recognized to be more like human computers in their power, capacity, and architecture (and perhaps ability to make strong eye contact). Nevertheless, because of the inability to observe the conscious experience, we will be no closer to a "formula" for how the conscious experience arises or does not.

THE PROBLEM WITH SCIENCE

The immeasurability of the conscious experience should be enough to dissuade scientific endeavors to "explain" how it arises. Nevertheless, such propositions regularly generate substantial interest. Unfortunately, this interest is piqued by couching explanations in terms that give the air of scientific credibility. Although the parts of these explanations that pertain specifically to the hard problem are inherently nonscientific (and the authors tend to conflate the easy and hard problem when referring to measurements that relate to the easy problem), it may nonetheless be instructive to examine the fallacies within the construction of the explanations.

One such fallacy is the fallacy of science governing nature rather than nature governing science. Explanations in this category tend to rely on what Chalmers calls the "extra ingredient." The extra ingredient is something that seems as mysterious as consciousness itself, except that fortunately whoever is explaining it seems to understand it better than, or in a different way from others. Quantum mechanics is a typical candidate, as the uncertainty inherent in quantum mechanical predictions seems well suited to match the "difficulty" around explaining the conscious experience (eg, the Orchestrated Objective Reduction model championed by Roger Penrose and Stuart Hameroff). A fallacy of all such explanations is that they ascribe to the quantum mechanical model of nature *nature* itself, rather than accept that quantum mechanics or any other scientific model of nature is just a predictive model. With the great advancements in scientific discovery and the immense power such advancements bring, it is perhaps not surprising how easy it is to lose sight of what science does. Science predicts. Science does not "explain" nature any more than that; science is not equivalent to nature. Science models nature, and the best scientific models predict well. Quantum mechanics, for example, does not govern nature, but predicts nature with some level of accuracy and precision and to a certain degree of probability (and also happens to predict that there are absolute limits to our predictions). It does not prove that what happens next is not exactly determined by the entire state of nature now (or that it is), only that we cannot predict it better, and perhaps will never be able to do so.

There is no "problem" with science in this regard. The only problem is that the better the model, the more developed the theory, and the greater the breadth of observations that are predicted well, the more we are wont to lose sight of what science does and does not do. Schrödinger addressed this issue in lectures at Trinity College, Dublin, in 1943 (Box 28.2).

Schrödinger asks us to compare the models of science (the theories) to what we experience as nature (the sensual). In nature, we sense electric force, for example, through observation. With scientific observations of electric force, we create models of the behavior of "charged" particles that predict how those particles will move. No scientific models explain why there is electric force, or how it arises. Any such argument would necessarily be circular, like explaining gravitational force in terms of curvature in space-time (ie, it may be a broader predictive model, but still does not explain *how* mass induces the "curvature" any more than *how* mass generates "gravitational force"). Similarly, with regard to explaining how consciousness

BOX 28.2 The Apotheosis of Theory

Scientific theories serve to facilitate the survey of our observations and experimental findings. Every scientist knows how difficult it is to remember a moderately extended group of facts, before at least some primitive theoretical picture about them has been shaped. It is therefore small wonder, and by no means to be blamed on the authors of original papers or of text-books, that after a reasonably coherent theory has been formed, they do not describe the bare facts they have found or wish to convey to the reader, but clothe them in the terminology of that theory or theories. **This procedure,** while very useful for our remembering the fact in a well-ordered pattern, **tends to obliterate the distinction between the actual observations and the theory arisen from them. And since the former always are of some sensual quality, theories are easily thought to account for sensual qualities; which, of course, they never do.**

arises, any explanation that summons a scientific model (like quantum mechanics) to explain how the phenomenon arises without the benefit of measurement is scientifically meaningless, as the model is not nature, and has no *scientific* function outside of the measurable predictions it makes.

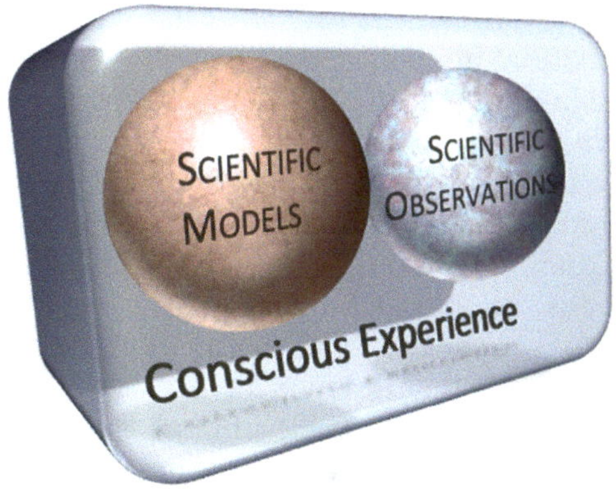

FIGURE 28.1 Scientific models exist only within the conscious experience and rely on observations that exist only within the conscious experience. Observations of an external conscious experience are impossible, and models that claim to predict the existence of conscious experience are necessarily circular, existing only within a subset of the predictor's conscious experience itself, and outside of science.

THE PROBLEM WITH THE PROBLEM: INCOMPLETENESS

Science does not explain the origin of phenomena, nor does *science* acknowledge predictions for what is intrinsically immeasurable. In the particular case of the conscious experience, an additional (or perhaps the fundamental) problem is that scientific models and scientific observations seem to exist for us only within the experience of consciousness itself (Fig. 28.1). This would of course be true for any model or any "observation," whether or not the conscious experience recognizes them as scientific. However, it is perhaps enough that scientific models and observations are contained within the conscious experience for them to be incapable of encapsulating how the conscious experience arises. Although this may also be intuitively obvious, it is nevertheless intriguing that (within the experience of consciousness itself) we have developed the language to "prove" that the experience of consciousness cannot explain itself. To explore this further, it may be helpful to return to Turing and the related example of the computing machine and its limits:

> There are a number of results of mathematical logic which can be used to show that there are limitations to the powers of discrete-state machines. The best known of these results is known as Gödel's theorem* (1931) and shows that in any sufficiently powerful logical system, statements can be formulated which can neither be proved nor disproved within the system, unless possibly the system itself is inconsistent. *[*the incompleteness theorem]*

Turing mentions this argument against the notion of machines "thinking" because others had noted that the logical incompleteness of a discrete-state machine (digital computer) prevents it from being able to answer all questions (ie, there are statements that the machine can make that cannot be proven or disproven by the machine). Many, including Kurt Gödel, Roger Penrose, and J.R. Lucas, have argued that humans are not subject to the same limitations as computers, because there are identified computational tasks that humans can do and computers cannot. Gödel himself, however, admitted that his argument for humans is no more than conjecture, and others (eg, Douglas Hofstadter, author of *Gödel, Escher, Bach*) have effectively argued that humans are generally subject to the same limitations as computing machines.

What links the arguments that humans are somehow not limited by incompleteness are assumptions that reflect a bias in favor of human thinking, like Gödel finding it *implausible* that humans would not be able to decide whether solutions exist for certain equations, or Penrose *believing* that humans have a *special* understanding of the "truth." Although such beliefs may resonate with what we desire, application of such beliefs imparts circularity to any logic based upon them, and of course they are not testable scientifically. Harkening back to Turing:

> … it has only been stated, without any sort of proof, that no such limitations apply to the human intellect … Whenever one of these machines is asked the appropriate critical question, and gives a definite answer, we know that this answer must be wrong, and this gives us a certain feeling of superiority. Is this feeling illusory? It is no doubt quite genuine, but … We too often give wrong answers to questions ourselves to be justified in being very pleased at such evidence of fallibility on the part of the machines … In short, then, there might be men cleverer than any given machine, but then again there might be other machines cleverer again, and so on.

So to the degree that logic generated within the conscious experience can be applied to a theory of how the conscious experience arises (also generated within the conscious experience), it would find the theory either incomplete (containing statements that are neither provable nor falsifiable) or inconsistent (containing statements that contradict one another). Accepting that inconsistent scientific theories are pared down to remove those inconsistencies, any scientific theory that purports to encapsulate how consciousness arises would be found to be incomplete. Of course, the inability to measure the experience of consciousness perhaps makes the point moot, yet it is nevertheless striking that logic—another product of the conscious experience—would nonetheless recognize the incompleteness of ideas that ostensibly explain the generation of the conscious experience.

The argument in this section, however, makes an important assumption that is also held deeply. For the incompleteness theorem, or its equivalent when applied as presented before, to apply at all requires that the experience of consciousness somehow be responsible for, or at least intrinsically connected to, the generation of logic, models, theorems, etc. The deeply held assumption is that our conscious experience does just that—that it has creative power and is not just a "picture" painted by projections of brain activity onto some purely receptive window of experience (Fig. 28.2). Although the assumption—it may appear akin to "free-will"—is not testable scientifically because of the immeasurability of the conscious experience, the alternative possibility leaves us (perhaps frighteningly) less able to describe and test how the conscious experience arises. The alternative suggests that whatever underlies the experience of consciousness is distinct from generative brain activity: the conscious experience may report in some way some of that activity, but it has no influence on it. From this viewpoint, we (ie, our conscious experiences) are not even doing the thinking. It may be worth noting that such a view does not require the absence of "free-will" in our entire selves, it only removes the experience of

FIGURE 28.2 An untestable, highly schematic, alternative model of the conscious experience. Vectors within the sphere represent brain activity (eg, associated with sensation, planning, memory, etc.). Projections of that activity onto a "window of consciousness" (represented by the surface of the sphere) stimulate the conscious experience. Only portions of the activity are projected, some or most parts remain fully "unconscious," represented by the cube. In this model, information flows within the sphere and to the sphere (ie, the projections), but does not necessarily flow from the surface of the sphere back in. This latter possibility is difficult to reconcile with the deep assumption that our conscious experience dictates at least some of our behavior, and requires that the "conscious experience" is somehow "known" to the generative brain activity it receives, perhaps through passive reflection.

consciousness (ie, what we perceive to be ourselves) from the decision-making, or thinking, or everything except the experience itself.

The wholly unsatisfying but no more or less refutable, alternative viewpoint taken in the preceding paragraph is considered for the sake of completeness and as an example of how our ability to think philosophically deflects our attention from scientific voids. In the absence of an ability to measure something, philosophy enters the vacuum. This section, "the problem with the problem: incompleteness" is an example of this phenomenon. We are able to consider whether the "mind" "surpasses" a finite machine. If it does not, we see that it is limited by incompleteness; if it does, then we are left to accept the "impossibility of physicochemical explanations of human reason [Gödel]." By these arguments, we may be satisfied that either way, we are left unable to account scientifically for how consciousness arises as already discussed in this section. However, from a scientific standpoint, these discussions are also meaningless, as they are irrelevant to measurement and testable prediction. From a philosophical standpoint, these arguments occur on the level of meta-consciousness (ie, consciousness is the experience, we are only discussing it) so are at best fundamentally circular with regard to explaining the experience—our ability to discuss the experience of consciousness relies on the experience of consciousness (or worse, doesn't).

A CASE STUDY

Despite the many angles from which our inability to answer scientifically the question, "How does the conscious experience arise?" may be viewed, a steady stream of efforts to explain just that continues. One laudable effort has been spearheaded by Giulio Tononi, whose integrated information theory (IIT) states, "an experience is identical to a conceptual structure that is maximally irreducible intrinsically." IIT seeks to go beyond the "easy" problems of consciousness and provide a "theory of consciousness – one that says what experience is and what type of physical systems can have it ... [that] provides a principled account of both the quantity and the quality of an individual experience (a quale), and a calculus to evaluate whether or not a particular physical system is conscious and of what." Recognizing problems with earlier efforts to do the same, Tononi asserts that the issue is that earlier efforts "start from the brain and ask how it could possibly give rise to experience." To solve the problem, rather, Tononi suggests to "start from consciousness itself, by identifying its essential properties, and then ask what kinds of physical mechanisms could possibly account for them." Tononi outlines several properties of consciousness, and proposes a mathematical model that predicts when conscious experience is occurring and in what form.

Of course, in "starting from consciousness itself," Tononi betrays the circularity inherent in any such argument. This is evident from passages like:

> If someone is awake and acts meaningfully, we have little doubt he is conscious. If he speaks, and especially if he can answer questions about what he is conscious of, we are fully confident. In the laboratory, the ability to report one's experiences has become the gold standard for judging the presence of consciousness ... One can also ask subjects how confident they are in their judgments ... These kinds of meta-cognitive and confidence reports can also be obtained from trained monkeys and other animals, with so many similarities to our own reports that there is little doubt as to the presence of consciousness.

In this passage, the premise is that the experience of consciousness is detectable because *we know it when we see it*. If that were true, a theory like IIT, or any other theory, could be tested for when it matches and does not match with the experience of consciousness as detected or not. However, like other theories, IIT conflates the hard problem with the easy problem, and cites evidence and makes predictions that pertain only to the easy, behaviorally measurable problems of consciousness. But Tononi also recognizes the problems inherent in this simplistic approach of measuring consciousness behaviorally to obtain insight into whether the conscious experience is present:

> ... behaviour can be misleading: a person may walk and speak in her sleep, yet it is quite dubious whether she is experiencing anything. Or a person can be asleep, immobile, silent, and unresponsive, yet she may be dreaming – vividly conscious of an imaginary environment. In such cases, reportability can be used as retrospective evidence of consciousness, by waking up the sleeper to obtain a 'dream report.' However, reportability, too, can be problematic.

Although the problem with reportability as a measure of the conscious experience was discussed previously in this chapter, considering a particular issue with reportability may be helpful in demonstrating the inherent error in conflating the easy and hard problems of consciousness. The particular issue is the confounding of the conscious experience with memory and vice versa. For example, a person in the midst of an alcoholic blackout—the time period during which the person may be interacting normally if drunkenly with others—would be gauged by IIT standards to be experiencing consciousness (ie, based on how the person was acting we would have little doubt as to the presence of consciousness). Similarly, simple cognitive testing in laboratory participants imbibing alcohol does not predict those who will and those who will not later experience a blackout. Yet some persons will experience a blackout and believe that their timeline of consciousness either skipped an entire period or at least jumped here and there with some parts so lost that they report feeling like they were not experienced at all.

This type of problem, where the experience of consciousness is apparently altered after the fact by an issue with memory, leads to paradoxes. In one such paradox, an alpine skier is enjoying a particularly daring run, but falls and tumbles down the mountain. In the experience of the skier, the experience of consciousness jumps from time point A, just before losing balance, to time point D, when the skier finds himself/herself tumbling down the mountain (Fig. 28.3). Although the skier did not hit their head until time point C, some moments after losing balance, the timeline of their conscious experience nevertheless jumped instantaneously from A to D, as is often reported in similar cases. So what of the set

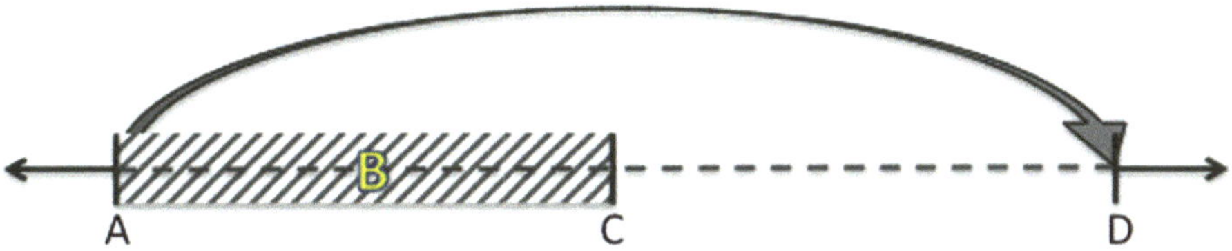

FIGURE 28.3 The confound of memory on the experience of consciousness. A time-line of consciousness is shown. At some point C, an insult occurs that causes the experience of consciousness to jump from before (point A) to after (point D). Reflecting on the experience of consciousness during any time B between A and C paradoxically protects from such an insult, or else the time-line of consciousness would have already jumped to D.

of points B between A and C? Was there an experience of consciousness during that time? The observers on the ski-lift would have no reason not to think so based on the behavior of the skier, fighting to stay upright, and yet the skier could not report it as such. So who are we to believe: the person experiencing the consciousness or the observers of that person? Well, the observers can only assume the conscious experience in another and are biased by their own experience (as previously discussed), and the person does not appear to be a reliable reporter.

The promised paradox generated by this example is that at any moment, someone could claim that, being they are presently experiencing consciousness, they are not in a time period of points B preceding some traumatic event at C that would induce them to have jumped their time-line of consciousness from A to D, and hence are safe from such a traumatic event (or would have jumped in their time-line of consciousness already). In this case, self-reflection on the experience of consciousness is miraculously protective from some memory-impairing event that would have caused the loss of that period of consciousness, no matter how traumatic. The point of this discussion is not only to exemplify the immeasurability of the experience of consciousness by others, but more so to demonstrate the fallibility of the measurement of consciousness in one's self. Even if one could reliably report states when one is experiencing consciousness in the moment (and when someone is experiencing something is just as immeasurable as the experience itself: by the "alternative" model in the preceding section, the experience of consciousness could be delayed substantially from the events that stimulated it), it is clear that one cannot reliably report states during which one is (was) not experiencing consciousness even in hindsight, because of the confound of memory on the experience. Regardless, relatively commonplace examples of dissociation between one's own experience of consciousness and others' perception of it highlight the intrinsic difference between the experience of consciousness and the measurement of awareness and other "easy" problems of consciousness by others.

This problem is not particularly worrisome, however, if a theory like IIT were only to seek an explanation for how measureable brain function works. Science allows for imperfection while still making excellent predictions regarding observable outcomes. Recognizing the problem with immeasurability, Tononi nonetheless goes further:

> Besides accounting for experimental data, ultimately a theory of consciousness should also shed some light on the seemingly ineffable qualitative properties of phenomenology … What makes a color different from a smell? … Viewing an experience as a shape in qualia space implies that features of experience that seem impossible to account for in neural terms – like the redness of red … should instead be accounted for in mathematical terms.

In this statement, Tononi appears to admit the "impossibility of physicochemical explanations" for the experience of consciousness, but he nevertheless offers a mathematical model that purports to describe the experience of consciousness. In this regard, he appears to fall into the previously discussed fallacy of supposing that science, or mathematics, governs nature (ie, a mathematical model cannot account for how a phenomenon arises). Although the mathematical model of IIT may *seem* to Tononi to be like what the experience of consciousness is to him, it is not scientifically testable in this regard, and no matter how well it may predict measurable qualities of brain function, that brings it no closer to predicting the immeasurable.

Another way of looking at the problem with IIT (and other theories) is to consider the *perspective* of the theory. IIT most explicitly constructs its theory of consciousness on axioms that reflect features of how we humans experience consciousness as confounded by our immediate and more distant memory of it. Beyond the previously discussed issues, this leaves out completely any other possible type of conscious experience. Is it possible to rule out a collective experience of consciousness in hymenoptera? Could consciousness not arise in a computer that was different both in experience and underlying structure from our own? Could a person under general anesthesia still have an experience of consciousness, but in a fundamentally different way, yet not be able to recall the experience? There are no scientific grounds to rule out such possibilities, but again, only prejudices or lack thereof that govern whether we view such contentions as possible. As it is, the only conscious experience we can (maybe) be sure of is the one we are experiencing in the moment.

WHAT DRIVES THE SEARCH? WHAT IS THE IMPACT OF INSOLUBILITY?

So if there is no possibility of a scientific explanation for how consciousness arises, why the decades of commitment and lengthy tomes providing incredibly detailed theories for just that? From a psychological perspective, the most likely answer seems to be related to the existential nature of the question. The experience of consciousness is our existence. Although we may be distracted from its contemplation by basic needs, other drives, conflict, pleasure, stress, or joy, and we may be inured from its difficulties by faith in religion or other spiritual or philosophical positions, the drive to seek answers for the question of existence is undeniable. In seeking to explain the conscious experience, we are perhaps tempted because it seems so close in so many ways to what is measurable, or we lack (or suppress) the intuition that tells us it is unanswerable like existence itself. By keeping ourselves ignorant of the insolubility of the problem, we retain hope and perhaps happiness in our pursuit of it. In our pursuit of it, we distract ourselves from contemplating its insolubility and the existential crises that may arise from such and related contemplations.

Fortunately, there is no scientific consequence of insolubility. Because the problem of how consciousness arises is not scientifically answerable, it is also scientifically irrelevant. However, that is not to say that it has been irrelevant to science. The temptation of a great problem—the greatest of all problems—has fueled research and thinking on the problem, with remarkable progress in the answerable questions related to cognitive function. It is clear that much of this work has been inspired by the belief that the great problem is solvable, and some of this work has seemed to come close to the great problem itself, fueling more research with testable hypotheses and the potential for making valuable predictions about basic brain function and dysfunction. In this regard, the great problem has captured and promoted imagination—different from but perhaps not entirely unlike the effect on the budding geometrician of being told that trisecting an arbitrary angle with only a compass and unmarked straightedge is impossible.

Further Reading

[1] Chalmers D. Facing up to the problem of consciousness. J Conscious Stud 1995;2(3):200–19.
[2] Nagel T. What is it like to be a bat? Philos Rev 1974;4:435–50.
[3] Lewis CI. Mind and the world order: Outline of a theory of knowledge. New York: Charles Scribner's Sons; 1929.
[4] Schrödinger E. Mind and matter (1958). In: What is life? With mind and matter and autobiographical sketches. Cambridge: Cambridge University Press; 1992.
[5] Schrödinger E. What is life? The physical aspects of the living cell. 1944. Available at: http://whatislife.stanford.edu/LoCo_files/What-is-Life.pdf.
[6] Turing AM. Computing machinery and intelligence. Mind 1950;49:433–60.
[7] Tononi G, Koch C. Consciousness: here, there and everywhere? Philos Trans R Soc Lond B Biol Sci 2015;370(1668). pii: 20140167.
[8] Tononi G. Integrated information theory of consciousness: an updated account. Arch Ital Biol 2012;150:290–326.
[9] English poetry I: from Chaucer to Gray, vol. XL. The Harvard Classics. New York: P.F. Collier & Son, 1910–14; Bartleby.com. 2001. www.bartleby.com/40/286.html.
[10] Melville H, Moby-Dick; or, the whale. London: Constable & Co., 1922; Bartleby.com, 2013. www.bartleby.com/91/96.html.

CHAPTER

29

The Emotional Brain

L. Pessoa

University of Maryland, College Park, MD, United States

INTRODUCTION

In his seminal contribution, in 1878, Broca described the *great limbic lobe* comprising the interior surface of the cortex and a set of related structures. These structures were later part of the first major circuit theory of emotion, namely, the circuit of Papez. By the middle of the 20th century, ideas by Papez and others were further refined and extended by MacLean into the concept of the "limbic system," according to which emotion depends on a well-defined and delimited set of brain regions. The "emotional brain" was suggested to be responsible for both response systems and feeling states/processes, given that emotion was characterized by a combination of changes in physiological arousal, behavioral expression, and, importantly, subjective experience.

This chapter describes several components of the *emotional brain*. As outlined in the exposition in the following sections, the concept of a "limbic system" has given way to a modern understanding of emotion as highly integrated with other brain systems. It is thus preferable to avoid the term "limbic system" altogether, as advocated for some time by many leading emotion researchers.

REGIONS OF THE EMOTIONAL BRAIN

In this section, several important regions of the emotional brain are discussed. Although the exposition is centered on particular regions, they do not work in isolation. Instead, they participate in large, distributed neural networks whose constituent regions, collectively, carry out functions and behaviors of interest. Therefore, although the exposition emphasizes a region's participation in emotional functions, to understand the emotional brain it is critical to characterize large-scale circuits within which regions are embedded (further discussed in sections From Regions to Circuits and How Are Emotions Organized in the Brain?). In this sense, there is no one, separately identifiable "emotional brain," but instead regions that play important roles in emotion (Fig. 29.1).

Amygdala

Anatomically, the amygdala is a complex structure containing more than a dozen richly interconnected nuclei. However, based on several types of information, including connectivity and distribution of neurotransmitters, some have questioned whether this collection of nuclei

Conn's Translational Neuroscience
http://dx.doi.org/10.1016/B978-0-12-802381-5.00046-4

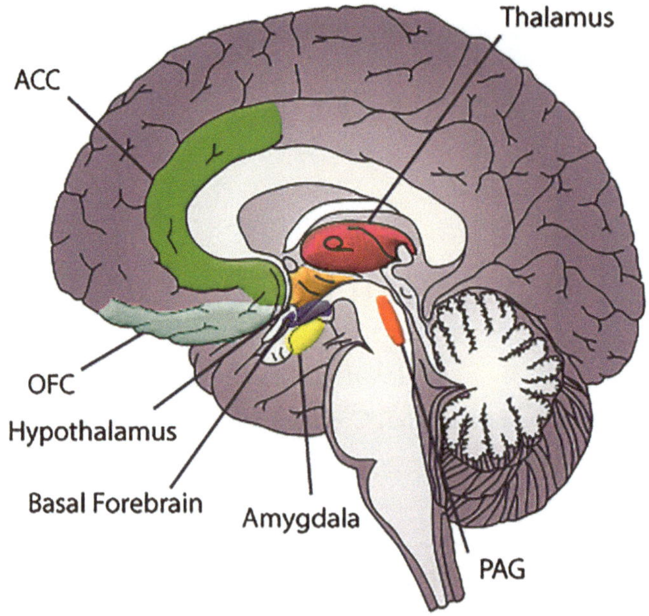

FIGURE 29.1 Emotional brain. Several brain structures participate in emotional mechanisms in important ways. Some of the regions discussed in the chapter are shown schematically here. *PAG*, periaqueductal gray.

should be considered a functional–anatomical unit. Indeed, the title of a provocative paper by leading researchers was "What is the amygdala?" Nevertheless, it is useful to consider at least two subunits of this structure: the "lateral amygdala" (at times called "basolateral amygdala") involving the lateral, basal, and accessory basal nuclei, and the "central amygdala" involving the central nucleus (Fig. 29.2). These nuclei are typically further subdivided into subnuclei, such as the medial and lateral divisions of the central amygdala.

Structurally, the lateral amygdala exhibits a cytoarchitecture that is closer to that seen in cortex, which is organized in a layered manner. The majority of lateral amygdala neurons are glutamatergic (excitatory) spiny projection neurons and about 20% of its neurons are GABAergic (inhibitory) interneurons. The central amygdala exhibits a more simplified cytoarchitecture with incipient lamination, and is a striatum-like structure that is composed of GABAergic medium spiny neurons. Furthermore, In terms of embryological origin, the amygdala consists of distinct groups of nuclei with distinct precursors linked to the lateral and central amygdala (among others).

The two amygdala subdivisions display connectivity patterns that are quite distinct from each other (Fig. 29.3). The lateral amygdala receives substantial sensory information from cortex and is richly connected with temporal, cingulate, insular, and prefrontal cortex; connections with parietal cortex also exist but are more modest. In contrast, the central amygdala has extensive interconnections with the hypothalamus and brain stem nuclei (including sites in the midbrain, pons, and medulla) that are involved in behavioral, autonomic, and neuroendocrine responses; cortically, the central amygdala is interconnected with the medial prefrontal cortex, orbitofrontal cortex, and insula.

The amygdala is a key brain structure for aversive memory acquisition, a notion that is

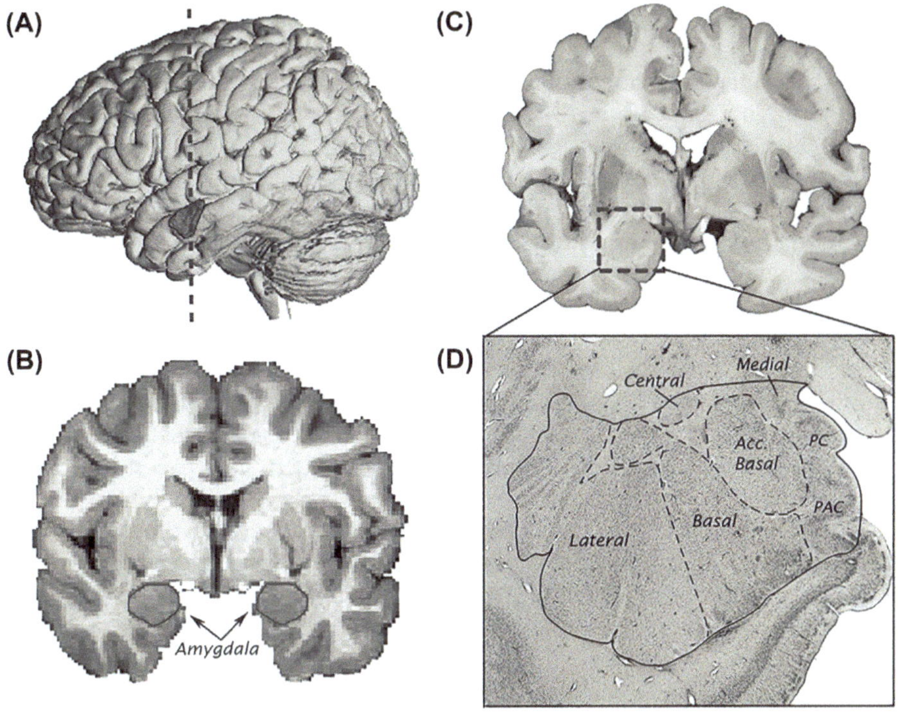

FIGURE 29.2 Human amygdala. (A) Lateral view of an MRI scan with the approximate location of the amygdala outlined on the lateral surface; line represents location of coronal section in panel B. (B) Coronal section showing position of left and right amygdala. (C) Coronal section with box around left amygdala. (D) Enlarged, Nissl-stained section of left amygdala nuclei from panel C. *Acc*, accessory; *PAC*, periamygdaloid cortex; *PC*, posterior cortical. *Reproduced with permission from Schumann CM, Bauman MD, Amaral DG. Abnormal structure or function of the amygdala is a common component of neurodevelopmental disorders. Neuropsychologia 2011;49:745–759.*

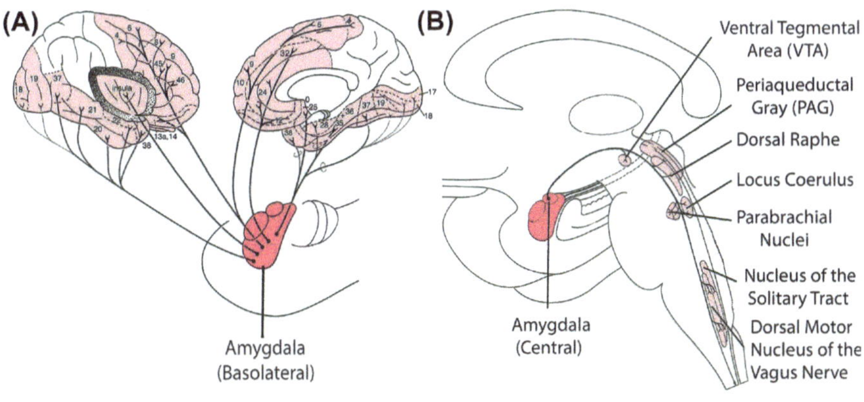

FIGURE 29.3 Amygdala connectivity. (A) Output connections from the lateral amygdala. Connections to parts of the parietal cortex, including the intraparietal sulcus, are not shown. Numbers refer to approximate locations of Brodmann areas. (B) Output connections of the central amygdala. Connections to cortex are not shown. *(A) Reproduced with permission from Nieuwenhuys R, Voogd J, van Huijzen C. The human central nervous system. 4th ed. New York: Springer-Verlag; 2008. (B) Adapted with permission from Nieuwenhuys R, Voogd J, van Huijzen C. The human central nervous system. 4th ed. New York: Springer-Verlag; 2008.*

supported by studies using multiple experimental paradigms and measures of conditioned responses (Box 29.1). Note that the term "aversive conditioning" is used here instead of "fear conditioning" because fear is a subjective mental state not accessible in nonhuman animals (unfortunately, many researchers continue to use the term "fear" when studying rodents or other animals).

Aversive conditioning employs techniques of classical conditioning (also called Pavlovian conditioning). In aversive conditioning, the subject is exposed to a neutral conditioned stimulus (CS), such as a tone or a light, which typically co-terminates with an aversive unconditioned stimulus (US), such as a foot-shock. With training, the CS (tone/light) acquires aversive properties and, when subsequently presented alone, elicits a conditioned response. In rodents, responses include freezing behavior (technically, a nonresponse), alterations in autonomic nervous system activity, release of stress hormones, analgesia, and facilitation of reflexes. Subsequently, conditioned responses can be suppressed if the CS is repeatedly presented alone, a phenomenon called "extinction" (see section: Interactions Between the Amygdala and Medial Prefrontal Cortex).

The lateral amygdala receives multiple sensory signals, including auditory, visual, and somatosensory stimuli that convey information about both conditioned and unconditioned stimuli. Such convergence positions the lateral amygdala to be able to form associations between the two. Evidence supporting the idea that the lateral amygdala is critical for aversive learning comes from multiple sources. First, lesion or inactivation of the amygdala during

BOX 29.1

STUDYING "FEAR" AND "ANXIETY"

It is productive to operationalize "fear" as *acute threat* and "anxiety" as *potential threat*, such as in the Research Domain Criteria (abbreviated RDoC) initiative of the National Institute of Mental Health. Acute threat is linked to imminent and discrete threat, such as when a CS signals the possibility of a shock. Potential threat is linked to contexts in which harm may be experienced, such as when entering a room in which a shock may be delivered. More generally, acute and potential threat can be distinguished according to the degree of certainty regarding the timing and form of the threat. Behaviorally, cues that indicate an immediate threat give rise to "fearful" defensive behaviors, whereas more diffuse, distal, or unpredictable threat cues produce "anxious" risk assessment behavior. In the literature, there is widespread inconsistency in how terms such as "fear," "anxiety," "stress," "distress," "worry," and so on are employed, suggesting that, if these terms are used at all, clear operationalizations are needed. In nonhuman animals, terms such as "fear" and "anxiety" are preferably avoided given that they reflect subjective mental states.

Aversive conditioning constitutes the central experimental paradigm to study acute threat. In this paradigm (see main text), an initially neutral stimulus evokes a negative state through association with an aversive event. Potential threat is studied, among others, via exposure to threatening contexts (such as a brightly lit space for rodents, or a dark room for humans) and via context conditioning (where unpredictable aversive events are paired with a certain context or place). Past research tended to emphasize the role of particular brain areas in generating states linked to acute or potential threat. However, mounting evidence indicates that these states rely on neuronal circuits that involve multiple brain regions.

learning prevents associations from taking place (but typically these lesions, though they might include the lateral amygdala, are not limited to this sector). Second, responses elicited by the CS in the lateral amygdala are modified after pairing with the US. Specifically, during aversive conditioning, cells in the lateral amygdala of rats show plasticity during exposure to a CS. Cells that are responsive to the CS prior to conditioning increase their firing after it. Third, recent studies employing modern techniques based on optogenetics indicate that activity-dependent plasticity in the lateral amygdala is necessary for aversive learning. For example, aversive conditioning can be inactivated through optogenetic depression of sensory inputs to the lateral amygdala, and later reactivated by optogenetic activation of these inputs.

The central amygdala is a major participant in the generation of conditioned responses—at times referred to as "emotional responses." This role is supported by projections to multiple autonomic and skeletomotor regions involved in aversive conditioned responses, including the following: the periaqueductal gray (PAG), which participates in freezing behavior that is characterized by the absence of overt behavior; the lateral hypothalamus, which participates in sympathetic activation; and caudal parts of the reticular formation in the pons, which participate in the potentiation of the startle reflex. The central amygdala also projects to multiple brain stem nuclei that originate neurotransmitter systems with widespread outputs to the forebrain, including the following: the dopaminergic ventral tegmental area and adjacent substantia nigra pars compacta; the noradrenergic locus coeruleus; the serotonergic raphe nuclei; and the cholinergic basal forebrain nuclei. In all, several conditioned responses depend on the central amygdala and its projections to these regions.

One way to summarize some of the functions of the lateral and central amygdala during aversive conditioning is as follows. The lateral amygdala establishes associations, thus allowing the CS to access the motivational value of its particular US. This information can then be used by the central amygdala and its hypothalamic, midbrain, and other brain stem targets, thus giving rise to "emotional" responses such as freezing, fear-potentiated startle, and modulation of arousal and attention.

This serial scheme of amygdala function in which the lateral amygdala passes on information to the central amygdala is a valuable heuristic and research tool. However, it captures only a specific mode of amygdala functioning. This is because the central amygdala receives direct inputs from cortical and subcortical structures too (cortical input includes the medial prefrontal cortex and orbitofrontal cortex), which contribute to the acquisition of Pavlovian associations.

The Bed Nucleus of the Stria Terminalis and the Extended Amygdala

The strial terminalis is a major fiber pathway of the amygdala that courses upward and loops around medially in the direction of the hypothalamus. Among others, it carries fibers from the amygdala to the thalamus, hypothalamus, and septum; it also carries fibers projecting from these areas back to the amygdala. Whereas the stria terminalis fiber bundle is relatively easy to visualize in three dimensions, it is much harder when viewed via standard cross-sectional anatomical planes. Along the stria terminalis, it is possible to identify several small masses of cells, which together comprise the bed nucleus of the stria terminalis (BNST; Fig. 29.4).

Broadly speaking, the BNST can be subdivided into lateral and medial parts. The lateral BNST ($BNST_{LAT}$) is closely associated with the central amygdala, and together they have been suggested to be part of a central "extended amygdala." The $BNST_{LAT}$ projects to the hypothalamus and brain stem, as does the central nucleus of the amygdala (as reviewed previously). Indeed, the central amygdala and the $BNST_{LAT}$ have similar anatomical connectivity patterns. Furthermore, neurons in the central amygdala and the $BNST_{LAT}$ share similar

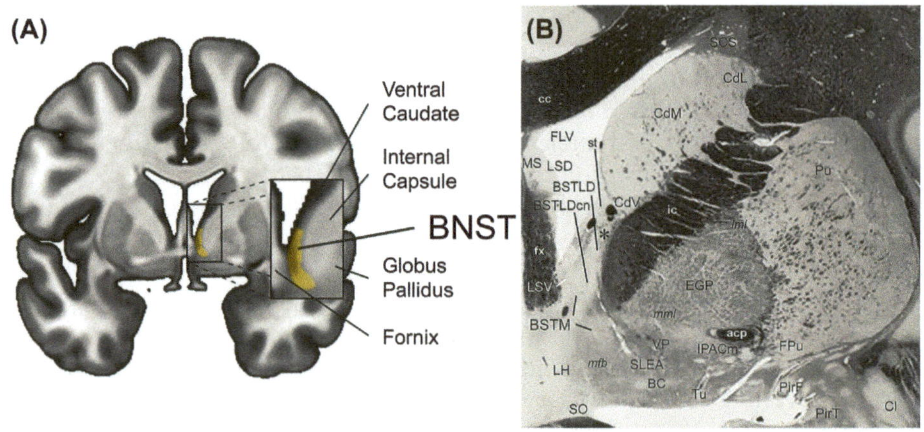

FIGURE 29.4 Bed nucleus of the stria terminalis (BNST). (A) Approximate location of the BNST in the human brain. (B) Coronal myelin sections showing cell masses of the BNST. Abbreviations for the structures of main interest: *BSTLD*, laterodorsal part of the BNST; *BSTLDcn*, central subdivision of the BSTLD; *BSTM*, medial part of the BNST; *CdL*, lateral aspect of the caudate; *CdM*, medial aspect of the caudate; *CdV*, ventral aspect of the caudate; *ic*, internal capsule; *LH*, lateral hypothalamus; *Pu*, Putamen; *st*, stria terminalis. *(B) Reproduced with permission from Yilmazer-Hanke DM. Amygdala. In: Mai JK, Paxinos G, editors. The human nervous system. 3rd ed. Amsterdam: Elsevier; 2012. p. 759–834.*

morphological and neurochemical characteristics (they are mostly GABAergic as found in striatal brain regions).

Functionally, the BNST_{LAT} is important for maintaining prolonged threat preparedness that is associated with the processing of potential threat (see Box 29.1). Consider the paradigm of light-enhanced startle. Rats produce a startle reflex when presented with brief noise bursts. The amplitude of this acoustic startle can be measured during different contexts. In one study, during the first phase, rats were tested in the dark. During the second phase, rats were tested again in the dark or, alternatively, in the presence of bright light (which is highly aversive to rats). Exposure to bright light for 5–20 min led to a significant increase in acoustic startle amplitude going from the dark to the light (light-enhanced startle) compared with startle amplitude when the rats remained in the dark. Critically, inactivation of the BNST (but not the central amygdala) eliminates the light-enhanced startle.

An influential framework suggests that the amygdala is more closely associated with acute threat ("fear") and the BNST is more closely associated with potential threat ("anxiety"). Recent research, including with novel techniques such as optogenetics, is consistent with this general idea but also reveals important ways in which amygdala and BNST circuits participate in multiple aspects of threat processing. For example, activation/deactivation of projection neurons from the lateral to the central amygdala can be used to influence the time mice will spend on open fields (a behavior that is typically actively avoided by mice), a form of potential threat. Thus, the lateral and central amygdala are not only involved in the processing of acute threat (the latter is revealed by standard conditioning paradigms).

Hypothalamus

The hypothalamus contains a number of nuclei and, as its name implies, is located just below the thalamus (and thus just above the brain stem) (Fig. 29.1; see also Fig. 29.4). The importance of the hypothalamus in emotion was established in the first decades of the 20th century, as highlighted by the work of Bard and Cannon, who showed that coordinated emotional expressions were compromised when the hypothalamus was excised. The involvement of the hypothalamus in emotion and other basic functions was fittingly summarized in 1929 by Cushing (considered by many to be the father of modern neurosurgery) in the following way: "Here in this well-concealed spot, almost to be covered with a thumb nail, lies the very mainspring of primitive existence–vegetative, emotional, reproductive."

Since the 1920s and 1930s, our knowledge of hypothalamic function has been greatly extended and refined, and current understanding concurs with the earlier notion that the hypothalamus is involved in several important basic operations. For example, it participates in many complex homeostatic mechanisms and contributes to neuroendocrine outputs. Indeed, hypothalamic mechanisms contribute to a wealth of processes: circadian rhythms, wakefulness and sleep, stress responses, temperature regulation, food intake, thirst, sexual behaviors, and defensive behaviors. In all these cases, the hypothalamus works in concert with a multitude of other sites, several of which are located in the brain stem and spinal cord.

In broad terms, the hypothalamus has been subdivided into several parts: rostrocaudally into the preoptic, anterior, medial (also called tuberal), and posterior regions, and mediolaterally into periventricular, medial, and lateral zones. Within those sectors, multiple "nuclei" or "areas" have been described. However, the hypothalamus poses considerable challenges for systematic study because compact groupings of neuronal elements are not always seen in some "areas." In fact, even the overall borders of the hypothalamus are ill defined (eg, caudally the hypothalamus transitions gradually into the midbrain tegmentum).

Historically, the hypothalamus has been conceptualized in terms of "descending" systems, a view that is conveyed by its designation as the "head ganglion" of the autonomic nervous system. Indeed, the hypothalamus has robust connections that synapse at multiple brain stem sites, including the PAG, midbrain reticular formation, dorsal raphe, parabrachial nucleus, locus coeruleus, nucleus of the solitary tract, as well as sites in the medulla and spinal cord. Notably, several of these sites have "ascending" projections that reach the hypothalamus too.

The stress response is among the many functions in which the hypothalamus participates; in this case as part of the hypothalamic–pituitary–adrenal (often abbreviated HPA) axis. Neurons in the paraventricular nucleus of the hypothalamus, which synthesize corticotropin-releasing hormone (CRH), integrate stress-relevant signals from multiple brain regions and launch the neuroendocrine response to stress. CRH stimulates the release of the adrenocorticotropic hormone from the pituitary, which in turn drives the release of glucocorticoids from the adrenal glands (sitting atop the kidneys). These "stress hormones," such as corticosterone and cortisol, have a wide range of effects, including increasing blood pressure and glucose levels, and suppressing inflammatory and immune responses.

The hypothalamus is also implicated in defensive behavior. In the cat, for example, the behavioral components of this reaction include ear retraction, hissing, growling, and striking with the forepaw; autonomic responses include pupil dilation, piloerection, accelerated heart rate, and elevated blood pressure. Classic work by Hess and many others in the 1920s–50s showed that a defense response can be elicited by electrical stimulation of brain areas in and around the hypothalamus. In the literature, specific defensive behaviors are at times linked to specific "centers." For example, in the rat, a hypothalamic "attack area" has been described, and includes parts of the lateral hypothalamic area and the ventromedial hypothalamic nucleus. Whereas characterizations such as this have conferred a certain degree of specificity to

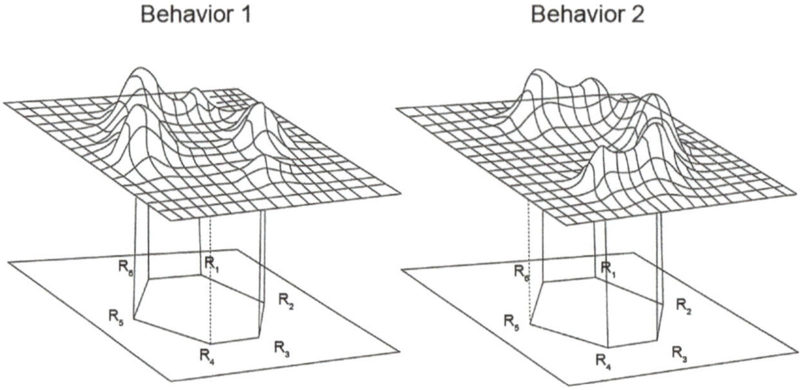

FIGURE 29.5 Hypothalamus and structure–function relationship. Hypothalamic regions (R_1, …, R_6) are involved in many functions (or behaviors), and functions rely on many hypothalamic regions. The three-dimensional landscape represents levels of activity across space. The patterns of activity indicate that multiple regions of the hypothalamus are engaged during a certain function/behavior. In this manner, the function is associated with a pattern of activity across the hypothalamus (and elsewhere in the brain), and not activation of a specific hypothalamic nucleus. *Adapted with permission from Newman SW. The medial extended amygdala in male reproductive behavior. A node in the mammalian social behavior network. Ann N Y Acad Sci 1999;877:242–257.*

hypothalamic areas, it is important to emphasize that there is no one-to-one mapping between area and function. Instead, hypothalamic areas are involved in many functions (eg, stress responses, temperature regulation, food intake, and defensive behavior), and at the same time functions rely on many hypothalamic areas (Fig. 29.5). Most importantly, the functions in which the hypothalamus participates are determined by neuronal networks that extend well beyond this structure.

Periaqueductal Gray

The brain is bathed in a colorless fluid also found in the spine called the cerebrospinal fluid which, among other things, provides buoyancy and protection to the brain. The PAG (also called central gray) surrounds the cerebral aqueduct containing cerebrospinal fluid (hence the name) in the rostral brain stem (Figs. 29.1 and 29.6). Cortically, the PAG receives input from several areas along medial prefrontal cortex, including cingulate areas and dorsomedial sectors, as well as orbitofrontal cortex, insula, and temporal cortex. Subcortically, the PAG receives input from a variety of structures, including amygdala, hypothalamus, dorsal raphe, parabrachial nuclei, pontine and medullary reticular formation, and spinal cord. Many of these regions receive reciprocal projections from the PAG.

Cytoarchitectonic features and physiological responses indicate that PAG subregions are organized more or less like columns that are aligned in parallel to the long axis of the brain stem (vertically in humans; horizontally in rodents) (Fig. 29.6). At least two columns help organize defensive behaviors (although recent data also reveal local integration of inputs, that is, within columns). Excitation of neurons in the "active" column generates behaviors such as facing and backing away, or a full-blown flight reaction, reactions that are very similar to natural ones produced by a rat or cat when threatened or attacked. Excitation of neurons in the "passive" column generates an entirely different reaction, namely, the cessation of ongoing activity and profound hyporeactivity, with the animal neither orienting nor responding to its environment. This type of freezing behavior is similar to

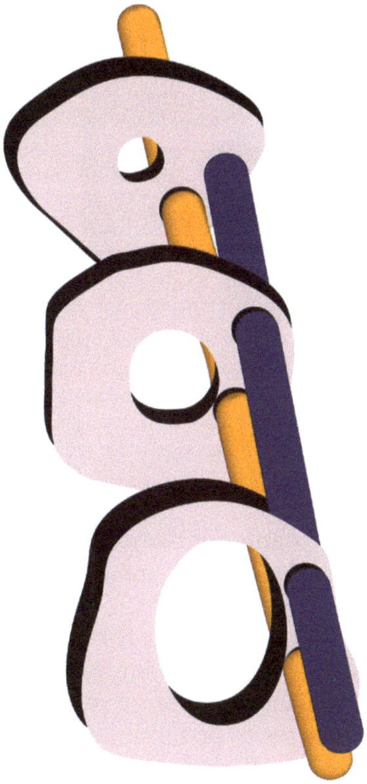

FIGURE 29.6 Periaqueductal gray (PAG). The PAG is organized as elongated columns of neurons. *Figure idea based on Figure 1 of Bandler R, Shipley MT. Columnar organization in the midbrain periaqueductal gray: modules for emotional expression? Trends Neurosci 1994;17:379–389.*

that of an animal that has incurred an injury, or after defeat in a social encounter. An important feature of the behaviors elicited by PAG excitation is that they are coordinated actions; that is, they are not simply isolated reactions (such as a knee-jerk reflex), but full-blown behaviors.

The PAG plays an important role in nociception (pain signaling). The PAG receives direct (and indirect) nociceptive input from the dorsal horn of the spinal cord, which is sensitive to nociceptive information from the viscera, skin, and other organs. In addition, the PAG is reciprocally interconnected with the parabrachial nucleus and the nucleus of the solitary tract. The latter fulfills a key role in the processing of visceral information and, like the parabrachial nucleus, provides an interface between autonomic and sensory systems. Activation and behavioral data indicate that PAG columns are important during responses to different classes of pain. Interestingly, the PAG is involved in analgesia too. Early studies in rats that used simple withdrawal reflexes as a measure of pain showed that stimulation of the PAG inhibited responses to noxious stimulation. Subsequent experiments demonstrated antinociceptive effects in humans too. For example, electrical stimulation of the PAG produced clinically significant pain relief.

Cingulate Cortex

The cingulate cortex surrounds the entire extent of the corpus callosum. The anterior sector of the cingulate gyrus (Fig. 29.7A) is involved in a broad range of processes, including willed action, executive function, and emotion. A remarkable property of this cortical tissue is that its "descending" projections are probably more extensive than that of any other cortical region and includes major connections to autonomic regulatory structures; notably to lateral hypothalamus, PAG, parabrachial nucleus, and the nucleus of the solitary tract. This connectivity is consistent with studies that have documented effects of cingulate electrical stimulation on virtually all autonomic processes, as well as many endocrine mechanisms.

Conversely, the cingulate cortex is the recipient of ascending signals from the brain stem. The most notable of these is perhaps the one from the nucleus of the solitary tract, the major visceral sensory cell group in the brain. Several nociceptive circuits also reach anterior and mid-cingulate areas indirectly via the thalamus. Thus, the cingulate cortex participates in more than "motor" (ie, output) autonomic functions.

The anterior part of the cingulate cortex is one of the most intensely studied areas in human neuroimaging, and is activated by a wide range of experimental paradigms. Based on early

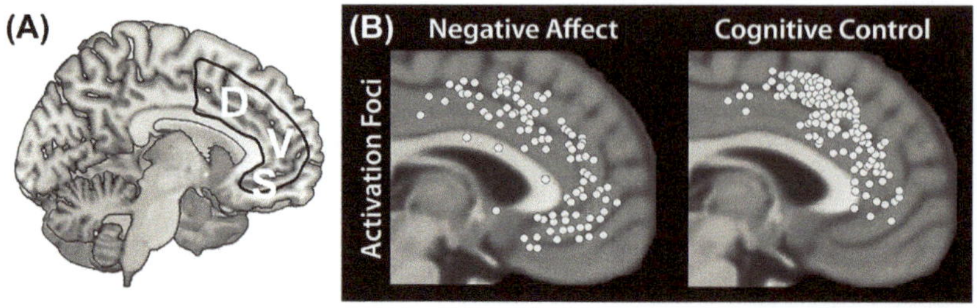

FIGURE 29.7 Anterior sector of the cingulate cortex. (A) Medial surface of cortex with the anterior sector of the medial prefrontal cortex outlined in black. General locations of dorsal (D) and ventral (V) parts of the medial prefrontal cortex are indicated. Anatomically, the cingulate cortex is delimited dorsally by the cingulate sulcus, although cortex dorsal to the sulcus is often referred to as "cingulate cortex." S, subgenual anterior cingulate cortex. (B) Cognition and emotion in the medial frontal cortex. Foci of activation across studies of negative affect and cognitive control, demonstrating extensive overlap between the two. *(B) Kindly provided by Alex Shackman and adapted with permission from Shackman AJ, Salomons TV, Slagter HA, Fox AS, Winter JJ, Davidson RJ. The integration of negative affect, pain, and cognitive control in the cingulate cortex. Nat Rev Neurosci 2011;12:154–167.*

imaging studies, it was proposed that this anterior region could be subdivided into a rostral "affective division" and a caudal "cognitive division," sectors putatively sensitive to emotional and cognitive manipulations, respectively. Recent meta-analyses of neuroimaging studies that take into account hundreds of studies have revealed, instead, considerable overlap of sites in medial prefrontal cortex engaged during negative affect and cognitive control (Fig. 29.7B). Activation sites of negative affect studies encompass a broad expanse of medial prefrontal cortex, as do those of cognitive control studies, although the latter exhibit a discernible concentration of sites in more dorsal regions. Indeed, studies of cognitive control observed activation sites in dorsomedial prefrontal cortex more often than they did in ventromedial prefrontal cortex. Notably, however, in studies of negative affect, activation of dorsal and ventral prefrontal cortex was equally likely. These findings thus undermine the notion of a strict segregation between emotion and cognition in medial prefrontal cortex.

The participation of the anterior cingulate cortex in emotion has important clinical implications. Chronic deep brain stimulation of the white matter in the vicinity of the subgenual cingulate cortex (Fig. 29.7A) produces considerable remission of symptoms in at least some treatment-resistant depression patients. A current hypothesis is that treatment efficacy is mediated via cascading effects of the initial subgenual stimulation on a distributed network of brain regions, such as the amygdala, hypothalamus, striatum, and orbitofrontal cortex.

In rodents, parts of the medial PFC, the so-called "prelimbic" and "infralimbic" cortices, that have important roles in emotion are discussed in section "Interactions Between the Amygdala and Medial Prefrontal Cortex."

Orbitofrontal Cortex

Orbitofrontal cortex is the portion of prefrontal cortex that sits just above the orbits of the eyes and extends posteriorly several centimeters to form the frontal base of the brain (Fig. 29.8). Based on anatomical connectivity information, it has been suggested that the orbitofrontal cortex can be subdivided into *orbital* and *medial* components, or networks. The orbital network receives wide-ranging sensory information and appears to integrate it, particularly in relation to the assessment of food and reward. The medial network is distinctively and heavily connected with regions of the medial wall of the brain, including those of cingulate cortex and surrounding areas. In contrast to the orbital network, the medial

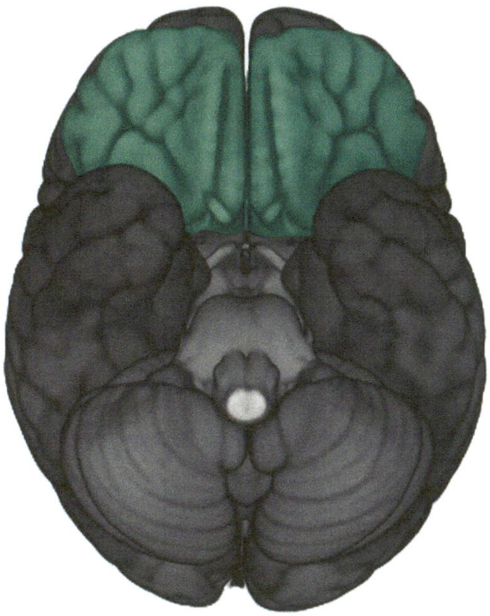

FIGURE 29.8 Orbitofrontal cortex. Ventral surface of the brain showing the approximate location of the orbitofrontal cortex.

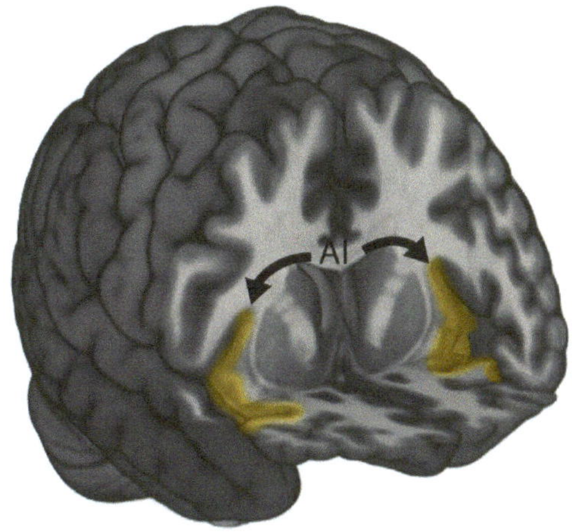

FIGURE 29.9 Anterior insula. The insula is a territory of cortex folded within the lateral sulcus between the frontal and temporal lobes, and is not visible from the lateral surface of the brain. The approximate location of the anterior insula (AI) is shown.

network receives few sensory inputs. Importantly, the medial orbitofrontal cortex projects to the hypothalamus and other areas involved in visceral processing. In fact, via the hypothalamus, the descending medial orbitofrontal influence has the potential to extend as far as autonomic centers in the spinal cord. In contrast, there are relatively few projections to the hypothalamus from the orbital network.

Anterior Insula

Insular cortex is the part of cortex folded within the lateral sulcus (also called the "Sylvian fissure") between the frontal and temporal lobes—it is thus not visible from the lateral surface of the brain (cortex covering it is referred to as the "operculum," meaning "lid") (Fig. 29.9). The anterior insula is strongly involved in the processing of bodily signals as it contains representations of the internal state of the body. This representation involves signals from many body parts, including the viscera, signals reporting temperature, pain, itches, muscular sensations, touch, and other bodily sensations. Both sympathetic and parasympathetic bodily signals are at first conveyed to the posterior insula, and then to the anterior insula via the mid-insula. These properties have led researchers to suggest that the anterior insula participates in the representation of "feelings" from the body.

In all, whereas an output or motor role is often emphasized for the cingulate cortex, an input or sensory role is frequently emphasized for the insula. But as discussed earlier, the cingulate cortex also receives visceral, ascending signals from important autonomic nuclei. In the case of the insula, descending connections to the hypothalamus and brain stem (for instance, the parabrachial nucleus) also exist. Thus, in both cingulate cortex and the insula, bidirectional circuits are present, although they are likely asymmetrical in their efficacies (stronger role for output for the cingulate and stronger role for input for the insula).

Interestingly, in human neuroimaging studies, researchers have found that the anterior insula is engaged by a remarkable array of tasks spanning perception, cognition, and emotion. Indeed, the anterior insula has been suggested to be among the most functionally diverse regions of the brain.

FROM REGIONS TO CIRCUITS

Earlier we reviewed some of the regions that participate in emotional processing. But more broadly, how should we understand the brain basis of complex mental functions such as emotion? Many accounts of mental function emphasize the importance of particular brain regions. In the case of emotion, the hypothalamus, for example, was considered an emotion "center" for several decades. Likewise, the amygdala has played a major role in models of the emotional brain in the past few decades. However, it is increasingly recognized that functions arise via the coordinated action of multiple brain regions that participate in distributed circuits. In this section, particular examples of this type of "distributed processing" are illustrated in the context of emotional processing.

Large-Scale Amygdala Networks

A remarkable property of the amygdala is its massive interconnectivity; as many as 1000 separate cortical and subcortical pathways may exist. The connectivity is all the more notable given that it involves all cortical lobes (though connections to parietal cortex are sparser) as well as subcortex (see Fig. 29.3). Combined, these properties indicate that the amygdala is an extensively interconnected *hub*, namely, a region that exhibits a high degree of anatomical connectivity (Fig. 29.10). Indeed, computational analysis of databases that collate anatomical connectivity suggests that several amygdala nuclei (including the lateral nucleus and the accessory basal nucleus) should be considered as part of a "core" brain circuit, all of whose regions have extremely high connectivity (see section: Principle 1: Massive Anatomical Connectivity).

The pattern of connectivity between the amygdala and prefrontal cortex is of particular interest given the latter's role in cognitive functions. In addition to substantial connections between the amygdala and both medial and orbital aspects of prefrontal cortex, bidirectional connectivity is present along the lateral surface too, although these connections are relatively weak. More generally, to understand how amygdala signals are potentially "broadcast" to all sectors of prefrontal cortex, it is important to consider prefrontal connectivity itself. In one study, although the amygdala was estimated to be directly connected to about 40% of prefrontal regions, about 90% of prefrontal cortex was shown to be capable of receiving amygdala signals after a single additional connection within prefrontal cortex. This "one-step" property shows that amygdala signals are not necessarily confined to orbital and medial prefrontal territories, both of which are directly connected with the amygdala. Furthermore, this architecture indicates how amygdala signals are able to influence responses in lateral prefrontal cortex, even though direct connections to this part of cortex are relatively weak.

Let us consider the potential implications of the amygdala's connectivity on its influence on signal processing in the brain. Even a simplified view of its anatomical connectivity shows that, minimally, it belongs to three networks—in addition to prefrontal networks—as discussed in the previous paragraph. The first is a "visual network," as the amygdala receives fibers from anterior parts of temporal cortex. Indeed, cells in the amygdala respond to visual stimuli, including faces, with response latencies a little longer than cells in anterior visual cortex (a region that responds to abstract visual stimuli such as specific shapes, as well as faces). The amygdala, by its turn, influences visual processing via a set of projections that reach most of ventral occipitotemporal cortex (including primary visual

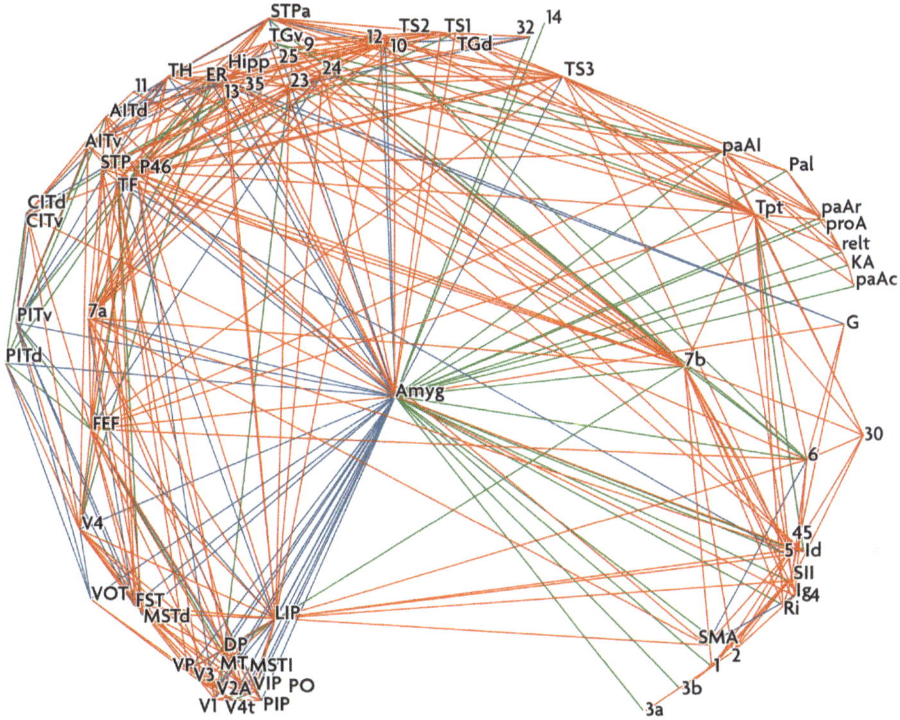

FIGURE 29.10 Brain connectivity graph. The amygdala (Amyg) is highly connected brain region. Note that the central position of the amygdala in the figure does not mean that it is the most connected (or "central") region in the brain; other highly connected regions, including the hippocampus (Hipp) and entorhinal cortex (ER), could have been displayed at the center. *Reproduced with permission from Pessoa L. On the relationship between emotion and cognition. Nat Rev Neurosci 2008;9:148–158 and adapted with permission from Young MP, Scannell JW, Burns GA, Blakemore C. Analysis of connectivity: Neural systems in the cerebral cortex. Rev Neurosci 1994;5:227–249.*

cortex). Thus, for instance, responses to emotional faces are stronger than those to neutral ones in occipito-temporal cortex.

The second is the well-known "autonomic network," as evidenced by connectivity with subcortical structures such as the hypothalamus and PAG. Via this network, the amygdala participates in many complex autonomic mechanisms, such as enhancing bodily arousal.

The third is a "value network," as evidenced by its connectivity with orbitofrontal cortex, medial prefrontal cortex, and striatum. The orbitofrontal cortex, for example, is important for determining the relative value of the current "state," and orbitofrontal responses differentiate events, even those removed from actual reward delivery, if they provide information about the likelihood of a future reward.

In total, the amygdala affiliates with different sets of regions, and thus participates in several networks, in a highly flexible and context-dependent manner. Other notable aspects of amygdala connectivity include interactions between the amygdala and the basal forebrain that are important for attentional functions (see section: Basal Forebrain), as well as how substantial projections from the amygdala to visual cortex influence competition mechanisms in visual cortex, thus contributing to attention as well.

Interactions Between the Amygdala and Medial Prefrontal Cortex

As described, the pairing of an otherwise neutral CS with an aversive US leads to aversive learning; subsequent presentation of the CS alone generates several of the properties of the US, including freezing behavior and other physiological responses that prepare the organism to handle the impending danger. However, upon repeated presentations of the CS without accompanying USs, the acquired responses dissipate. For example, freezing behavior to a tone that was previously paired with shock decreases.

This phenomenon, which is called *extinction*, does not simply amount to forgetting, but instead appears to involve new learning that inhibits or overrides past learning. In other words, extinction is not a passive forgetting process, but an active learning one. In the rat, a ventral portion of the medial prefrontal cortex is critically involved in extinction. Animals with lesions to this so-called infralimbic region of medial prefrontal cortex show resistance to extinction, that is, they show resistance to unlearning the original CS-to-US association.

The role of prefrontal cortex in extinction resonates with the long-held idea that this region helps regulate behavior, for instance, as espoused by the 19th century British neurologist Hughlings Jackson who proposed that "higher" centers exert inhibitory control over "lower" brain regions. However, the picture of higher regions inhibiting lower ones is too impoverished. Indeed, recent studies implicate a region of the medial prefrontal cortex called the prelimbic cortex in "fear" *excitation* (this region is immediately dorsal to the infralimbic cortex). Importantly, the prelimbic cortex appears to be critical for fear expression. Prelimbic neurons exhibit sustained increases in activity lasting tens of seconds that parallel the time course of aversive behaviors. In all, the prelimbic and infralimbic cortex appear to integrate multiple types of inputs with emotional significance to regulate the extinction and expression of behaviors associated with aversive stimuli. Finally, evidence for contributions of different parts of the medial prefrontal cortex in extinction and response expression has also been garnered in humans. Like in rodents, more dorsal (and posterior in humans) parts are involved in response expression (and appraisal/evaluation of emotional stimuli) and ventral (and anterior in humans) parts are involved in response extinction.

Hypothalamic Circuits

However important the hypothalamus may be for descending influences, a significant recent insight is that the mammalian cerebral cortex and the hypothalamus share massive bidirectional connections. In the rat, the best-studied mammal species, there are four major routes from the hypothalamus to the cerebral cortex (Fig. 29.11). These include a robust direct projection to all parts of the cortical mantle, and three indirect routes by way of the thalamus, basal brain nuclei (specifically, basal forebrain and amygdala), and brain stem. In all, the hypothalamus appears to be the largest source of nonthalamic direct input to the cortex. In the rat, some notable targets of hypothalamic input include regions in medial prefrontal cortex (infralimbic, prelimbic, and anterior cingulate cortex) and insular cortex. Interestingly, less prominent projections of the hypothalamus to lateral prefrontal cortex and even to primary sensory areas have been found. The connections between prefrontal cortex and the hypothalamus have also been investigated in nonhuman primates, where they were found to closely resemble those observed in rats. Notably, the hypothalamus has widespread projections to all sectors of the prefrontal cortex.

In sum, whereas the hypothalamus is involved in a host of "basic" control functions, it is part of an extensive bidirectional connective system with cortex and many other subcortical structures in a manner that allows for integration of

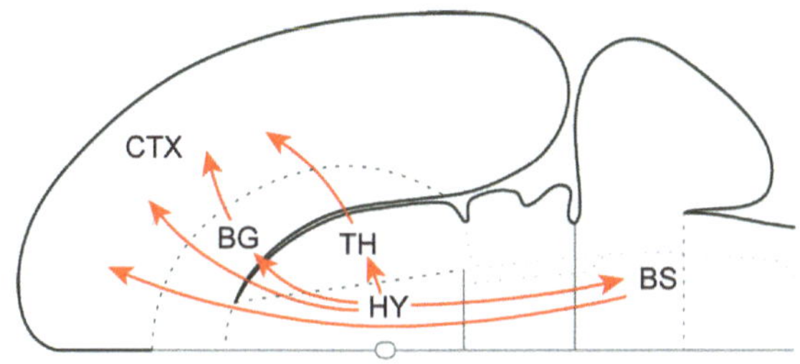

FIGURE 29.11 Hypothalamic ascending connectivity. Summary of the four major pathways from the hypothalamus to cerebral cortex schematized on a flattened representation of the rat brain. The basal ganglia here refer to the basal forebrain and the amygdala complex. Note that one of the indirect connections first descends to the brain stem. *BG*, basal ganglia; *BS*, brain stem; *CTX*, cortex; *HY*, hypothalamus; *TH*, thalamus. *Reproduced with permission from Risold PY, Thompson RH, Swanson LW. The structural organization of connections between hypothalamus and cerebral cortex. Brain Res Brain Res Rev 1997;24:197–254.*

wide-ranging signals. Critically, the hypothalamus is linked to other structures that have themselves broad connectivity, including the basal forebrain and the amygdala, further expanding its potential for influencing information processing. These insights thus resonate with the observation by Morgane that the hypothalamus is "only a single link in a larger multicircuit complex rather than ... an entity unto itself controlling any behavior as some form of command center" (p. 7). This view stands in stark contrast to the traditional viewpoint that the hypothalamus functions as a "head ganglion" that functions as an overall controller region.

Basal Forebrain

The basal forebrain comprises a heterogeneous set of structures close to the medial and ventral surfaces of the cerebral hemispheres (Fig. 29.12). These include the magnocellular basal nucleus, which contains the basal nucleus of Meynert and other cell groups. The basal forebrain originates ascending cholinergic (and GABAergic) projections that innervate extensively throughout the entire cortex. Major projections reach visual, parietal, cingulate, frontal (including lateral and orbital regions), and insular cortex (Mesulam, 2000); substantial connections have also been found to both the hippocampus and amygdala (Mesulam et al., 1992).

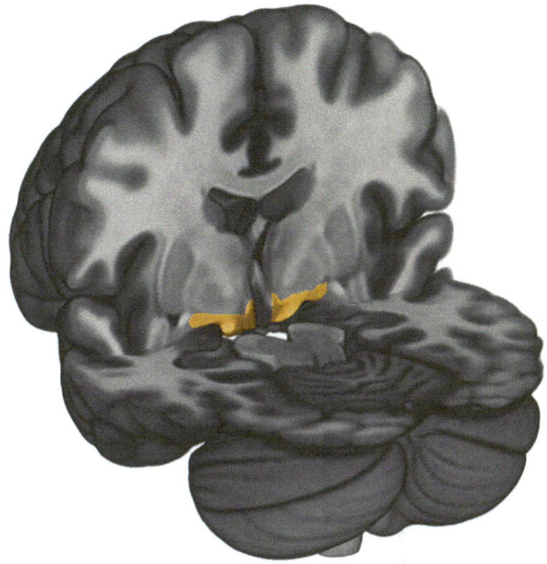

FIGURE 29.12 Basal forebrain. The basal forebrain comprises heterogeneous structures close to the medial and ventral surfaces of the brain. The basal forebrain originates ascending cholinergic (and GABAergic) projections that innervate extensively throughout the entire cortex. The approximate location of the basal forebrain is shown.

Acetylcholine released onto cortical neurons facilitates their response. Studies suggest that directly applying acetylcholine or stimulating the basal forebrain can increase the signal-to-noise ratio of neuronal responses and sharpen response tuning curves. More broadly, given its widespread connections, the basal forebrain is in a favorable position to influence multiple sites across the brain, including sensory regions which play a central role in responding to environmental stimuli. Overall, these topographically distributed effects result in increased vigilance, alertness, or attention.

Emotional content can influence signal processing across the brain via the basal forebrain because of the inputs that this structure receives. Connections from cortex originate from sites in orbitofrontal, temporal, insular, and cingulate cortex, all of which are important for emotion. Major subcortical projections to the basal forebrain originate from the amygdala (in particular, the central amygdala), hypothalamus, and various brain stem cell groups, including midbrain dopaminergic cells and the locus coeruleus. In all, given its overall connectivity pattern, the basal forebrain is in a position to influence processing across the brain based on multiple types of signals, including those central to emotion.

Lateral Prefrontal Cortex

The role of lateral prefrontal cortex in a range of cognitive operations is well documented. Indeed, this region is perhaps the paradigmatic "cognitive" brain region. Yet, the connectivity pattern of lateral prefrontal cortex is quite extensive, indicating that its functions should be conceptualized more broadly. For example, the dorsolateral prefrontal cortex is not only an important hub, but is also strongly connected with other regions that are topologically central themselves, which considerably expands the reach of the dorsolateral prefrontal cortex in influencing, and being influenced by, other regions. Being robustly bidirectionally connected to medial and orbital prefrontal areas, as well as to the insula, the lateral prefrontal cortex has access to afferent signals from the body and, at the same time, the region can influence signals that reach the body. Furthermore, as previously discussed, the amygdala is not only directly connected to a substantial set of prefrontal regions, but nearly all of prefrontal cortex is capable of receiving amygdala signals after a single additional connection within prefrontal cortex itself (what was referred to as the "one-step" property). Overall, the connectivity pattern of the lateral prefrontal cortex allows it to play a significant role in the integration of "cognitive" and "emotional" signals.

Distributed Circuits and the Evolution of the Emotional Brain

From an evolutionary perspective, it is sometimes stated that cognition can be considered "the tip of the iceberg," often associated with "newer" parts of the cortex, while the "rest of the iceberg" involves emotional processes that frequently rely on "old" subcortical brain regions. Furthermore, so the story goes, the "old" emotional system functions in a largely autonomous manner.

The brain has what can be considered old systems. For example, anatomical and behavioral data suggest that ray-finned fishes and land vertebrates probably share an ancestor that already possessed an amygdala. And the amygdala of the ancestral amniote (which includes reptiles, birds, and mammals) is believed to have possessed an amygdala with so-called pallial and subpallial components, namely the ancestral form of the basolateral and central amygdala highlighted in this chapter. However, "old" regions do not exist within a "layered" architecture, with newer structures simply added on top of old ones, such that the ones on top control the ones at the bottom (Fig. 29.13). The examples discussed in the preceding sections, including the amygdala and the hypothalamus, illustrate that the old/new conceptualization that is central to many versions of the emotional brain, does not take into account the extensive anatomical connectivity of brain circuits.

Recent studies focusing on the amygdala further inform the understanding of the evolution of a region that is important for emotion. For example, the lateral, basal, and accessory basal subregions of the amygdala are considerably more "developed" in monkeys than in rats (based on morphological characteristics, such as cell counts and the volume of subregions). One suggestion is that the differences between rats and monkeys are linked to the degree of connectivity of these nuclei with other brain structures, in line with the framework of correlated

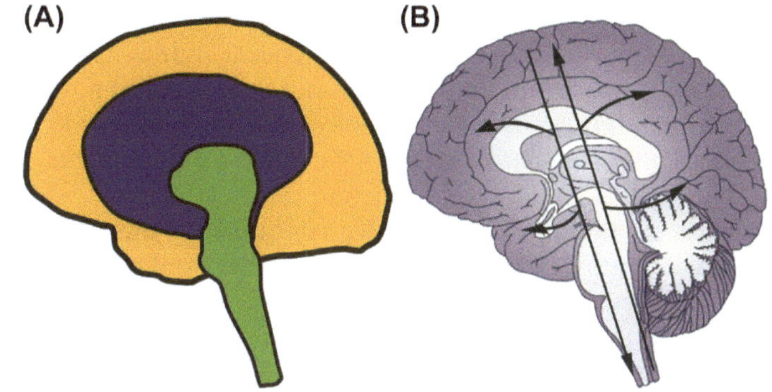

FIGURE 29.13 Brain evolution. (A) Brain evolution conceptualized as an additive process, with newer structures being added on top of older ones, and exerting control over them. Although this view was espoused by some comparative neuroanatomists in the early 20th century, and popularized well into later parts of the century (most notably in the idea of the "triune brain"), it does not reflect current knowledge of comparative neuroanatomy. Importantly, it does not take into account the extensive anatomical connectivity of brain circuits in general, and between cortical and subcortical regions in particular, as schematically shown in part (B).

evolution between components of functional systems. In this view, the lateral, basal, and accessory basal nuclei are more developed in primates than in rodents, paralleling the greater development of the cortical areas with which these nuclei are interconnected in primates. In other words, the relative development of these amygdala nuclei is influenced by their interconnections with other brain structures. Furthermore, such correlated evolution would be responsible for a higher convergence and integration of information in the primate amygdala.

HOW ARE EMOTIONS ORGANIZED IN THE BRAIN?

In the late 1940s, MacLean proposed the idea of a "limbic system" to explain how emotion is organized in the brain. Although the term is probably one of the most widely used in neuroscience, the concept has proven too unwieldy and unstable to be scientifically useful. Importantly, agreement regarding the brain regions that belong to this system has never been attained. Because of this, the term is often used in a circular fashion to indicate the "emotional brain" and, as some have pointed out, the term "limbic system" substitutes naming for understanding. Thus, the designation "emotional brain" is preferable, and the term "limbic system" should be abandoned.

Nevertheless, two concepts related to early notions of the "limbic system" are worth considering. One concept is the broad organizing principle that along a medial–lateral axis of the brain, a gradient of affective/motivational sensitivity is observed. By and large, medial brain structures (along the inner edge or limbus of the brain) are more sensitive to affective and motivational variables, while more lateral brain regions are less so. But the sensitivity is a matter of degree because all brain regions are susceptible to emotional and motivational influences, to some degree. There are important exceptions to this broad principle. For example, the anterior insula, which participates in important emotion-related functions, lies more laterally (although it is covered by the operculum of the frontal and parietal cortices).

Another concept is the broad principle that parts of the brain that are more sensitive to emotional variables are less structured. Many brain regions important for emotional processing are subcortical, and subcortex is cytoarchitecturally poorly structured. Cortex varies in degree of lamination, from three main layers to six or more well-demarcated layers. *Agranular* cortex contains three layers (it lacks a conspicuous layer IV), *dysgranular* cortex has four layers, and *granular* cortex contains six layers. Because cortex along Broca's "great limbic lobe" (ie, the medial edge of the hemispheres) is mostly *not* granular, it follows from its location and the medial–lateral gradient noted previously that less individuated cortex is relatively more involved in emotional processing.

Next, three other principles of the emotional brain are described.

Principle 1: Massive Anatomical Connectivity

Computational analysis of anatomical connectivity demonstrates that both cortical and subcortical brain regions are massively interconnected (Fig. 29.14). Massive connectivity is not limited to specific sectors of the brain but instead includes all of them: occipital, temporal, parietal, and frontal lobes, as well as cingulate, insula, thalamus, and basal brain regions (subcortical nuclei at the base of the forebrain, including the amygdala and basal ganglia).

Related analyses have described a tightly integrated "core circuit" spanning all of the brain sectors mentioned earlier. This core circuit is topologically central, that is, strongly connected to all other regions of the core and the rest of the brain. The core circuit was proposed to have several important properties: (1) it is a subnetwork that is far more tightly integrated than the overall brain network; (2) information likely spreads more swiftly within the core than through the overall brain network; and (3) the overall brain network communicates with itself mainly through the core.

Although the property of massive anatomical connectivity is a principle that applies to the entire brain, the organization of emotion in the brain must be understood from the standpoint of this general architectural property. Furthermore, it is noteworthy that many regions implicated in emotion were proposed to be part of the brain's core circuit. Finally, the property of extensive anatomical connectivity is further illustrated, for example, by the discussion of the amygdala and hypothalamus in section "From Regions to Circuits."

Principle 2: Anatomical Connectivity *to* and *From* the Body

Because emotion is closely linked to bodily states, brain circuits that involve signals to and from the body are important for emotion.

To illustrate the to-the-body part, consider the hypothalamus. Classical studies showed that "decortication" (ie, removal of cortex) alone did not abolish emotional expression, but that decortication plus ablation of the hypothalamus has a much more substantial effect on this expression. Current understanding of hypothalamic function indeed reveals that this structure is involved in important body-related operations. As described in section "Hypothalamus," hypothalamic mechanisms contribute to a wealth of processes, including circadian rhythms, wakefulness and sleep, stress responses, temperature regulation, food intake, thirst, sexual behaviors, and defensive behaviors. To carry out these functions, the hypothalamus works in concert with a multitude of other sites, several of which are located in the brain stem and spinal cord.

Cortical regions with substantial body-related signals include the medial orbitofrontal cortex, the cingulate gyrus, and the insula—illustrating the from-the-body part. Consider, for example, the insula: the physiological condition of the entire body is conveyed to the posterior insular cortex, including individually mapped and distinct feeling-related signals from the body. These signals are also present in the mid-insula and the anterior insula, which integrate body-related signals with activity that is associated

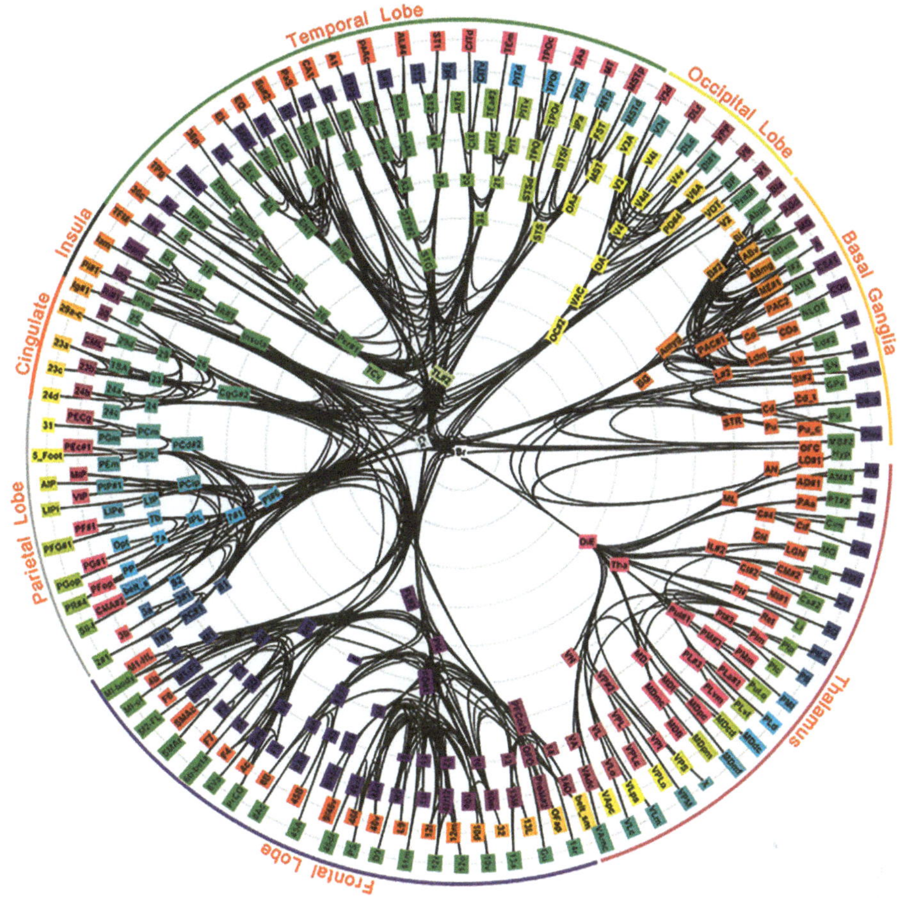

FIGURE 29.14 Computational analysis of whole-brain anatomical connectivity. The analysis considered existing data of an extensive set of cortical and subcortical regions (indicated via *colored rectangles*) spanning major brain sectors. The basal ganglia refer to nuclei at the base of the forebrain, including the amygdala. *Reproduced with permission from Modha DS, Singh R. Network architecture of the long-distance pathways in the macaque brain. Proc Natl Acad Sci USA 2010;107:13485–13490.*

with emotionally salient stimuli (eg, via sensory regions, the temporal pole, and the amygdala). Thus, the insula, especially its posterior part, can be considered interoceptive cortex, much like parts of parietal cortex are somatosensory cortex, for example.

The combination of principles 1 and 2 reveals that brain regions that influence the body, and conversely brain regions that are influenced by the body, extend well beyond those with more direct anatomical pathways to and from the body.

PRINCIPLE 3: HIGH FUNCTIONAL CONNECTIVITY WITH MOST OF THE BRAIN

Understanding how regions contribute to brain function requires characterizing how they are "functionally connected," in addition, of course, to how they are anatomically connected. *Functional connectivity* is a concept that was devised to characterize how neurons interact and was defined as the "coherence" among the

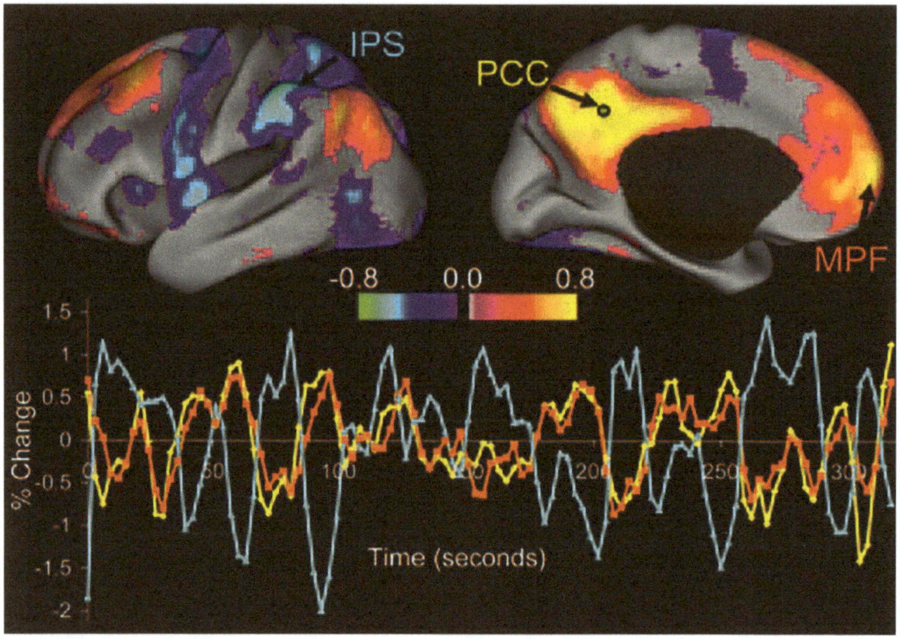

FIGURE 29.15 Functional connectivity. Maps show correlations between a seed region in the posterior cingulate cortex (PCC) and all other voxels in the brain for a single subject during resting fixation (both positively and negatively correlated regions are shown). Thus, all colored voxels are functionally connected with the PCC. The bottom plot shows the time course for a single run for the seed region (PCC, yellow), a region positively correlated with this seed region in the medial prefrontal cortex (MPF, orange), and a region negatively correlated with the seed region in the intraparietal sulcus (IPS, blue). *Reproduced with permission from Fox MD, Snyder AZ, Vincent JL, Corbetta M, Van Essen DC, Raichle ME. The human brain is intrinsically organized into dynamic, anticorrelated functional networks. Proc Natl Acad Sci USA 2005;102:9673–9678.*

activity of different neurons. Functional connectivity thus answers the following question: how coordinated is the activity of two brain regions that may or may not be directly anatomically connected (Fig. 29.15)?

Whereas it is natural to anticipate a functional association between brain regions that are directly connected, the relationship between structural and functional connectivity is not always a simple one. A striking example of structure–function *dissociation* is illustrated by an unusual population of adults without the corpus callosum, a structure that provides the major communication pathway between the two hemispheres. Although starkly different structurally relative to controls, individuals without the callosum exhibited very similar patterns of *functional* connectivity (as measured during rest with functional MRI). Thus, largely normal coordinated activity can emerge in brains with dramatically altered structural connectivity. In all, the functional organization of the brain is driven by factors that go beyond direct structural connectivity.

Therefore, to understand the organization of emotion in the brain, in addition to characterizing the brain's structural anatomy, it is necessary to characterize the functional relationships between brain regions. Importantly, anatomical architectural features support the efficient communication of information even when strong direct structural connections are not present, and support functional interactions that will vary based on context. This is illustrated, for example, by the "one-step" property of amygdala–prefrontal cortex connectivity discussed previously (amygdala signals reach nearly all prefrontal regions via a single connectivity step within prefrontal cortex; section "From Regions to Circuits"). This anatomical property allows the amygdala to engage in functional interactions with, for instance, lateral prefrontal regions that are not supported by strong direct anatomical connections.

CONCLUSIONS

The search to understand the emotional brain is as active today as it ever was. With time, the list of areas that have been proposed to play important roles in emotion has grown from having just a few members (hypothalamus, amygdala) to a large list. Notably, this list is not "fixed," but some members are included and highlighted by some authors but not others. This is not capricious; instead, it is because brain regions do not implement a single function (say, the amygdala is responsible for just aversive learning), but participate in many functions (the amygdala is involved in attention and decision making, among other functions). Therefore, the choice is one of emphasis. For example, the hippocampus is centrally important for memory processes. However, although it was not discussed in this chapter, it also participates in emotional processing, and has been particularly investigated in its contributions to anxiety.

Another reason the list of emotional brain regions is not fixed is because the boundaries of the emotional brain are ill defined. As reviewed in the present chapter, the brain is characterized by an architecture that exhibits massive interconnectivity. This connectivity implies that signals in a given structure can readily influence other parts of the brain in a few synapses—in many cases in as few as one or two synapses. Importantly, brain regions do not implement functions in isolation. Instead, they are part of distributed circuits that involve nonlocal connections that bring regions, which are anatomically far from one another, together into functional circuits.

In all, to understand how emotions are organized requires understanding general principles of information processing in the brain (as a whole). Thus, to understand the emotional brain, it is critical to understand the sensory, motor, and cognitive brain. In this way, it is also possible to characterize and understand the full ramifications of emotional psychopathologies, such as the anxiety disorders and depression, whose impact on mental function is not restricted to emotion, but necessarily encompasses additional dimensions, most notably cognition.

Further Reading

[1] Bains JS, Cusulin JIW, Inoue W. Stress-related synaptic plasticity in the hypothalamus. Nat Rev Neurosci 2015;16:377–88.
[2] Bandler R, Shipley MT. Columnar organization in the midbrain periaqueductal gray: modules for emotional expression? Trends Neurosci 1994;17:379–89.
[3] Craig AD. How do you feel—now? The anterior insula and human awareness. Nat Rev Neurosci 2009;10:59–70.
[4] Davis M, Walker DL, Miles L, Grillon C. Phasic vs sustained fear in rats and humans: role of the extended amygdala in fear vs anxiety. Neuropsychopharmacol 2010;35:105–35.
[5] Fields H. State-dependent opioid control of pain. Nat Rev Neurosci 2004;5:565–75.
[6] Grupe DW, Nitschke JB. Uncertainty and anticipation in anxiety: an integrated neurobiological and psychological perspective. Nat Rev Neurosci 2013;14:488–501.
[7] Heimer L, Van Hoesen GW, Trimble M, Zahm DS. Anatomy of neuropsychiatry. Amsterdam: Academic Press; 2008.
[8] Johansen-Berg H, Gutman DA, Behrens TE, Matthews PM, Rushworth MF, Katz E, et al. Anatomical connectivity of the subgenual cingulate region targeted with deep brain stimulation for treatment-resistant depression. Cereb Cort 2008;18:1374–83.
[9] LeDoux JE. Emotion circuits in the brain. Annu Rev Neurosci 2000;23:155–84.
[10] Martínez-García F, Novejarque A, Lanuza E. Evolution of the amygdala in vertebrates. In: Kaas JH, editor. Evolution of nervous systems, vol. 2. Oxford: Academic Press; 2007. p. 255–334.

[11] Mesulam MM. Principles of behavioral and cognitive neurology. Oxford University Press; 2000.
[12] Mesulam M, Hersh LB, Mash DC, Geula C. Differential cholinergic innervation within functional subdivisions of the human cerebral cortex: a choline acetyltransferase study. J Comp Neurol 1992;318(3):316–28.
[13] Millan MJ. Descending control of pain. Prog Neurobiol 2002;66:355–474.
[14] Nieuwenhuys R, Voogd J, van Huijzen C. The human central nervous system. 4th ed. New York: Springer-Verlag; 2008.
[15] Morgane PJ. Historical and modern concepts of hypothalamic organization and function. In: Morgane PJ, Panksepp J, editors. Handbook of the hypothalamus, vol. 1. New York: Marcel Dekker Inc.; 1979. p. 1–64.
[16] Pessoa L. The cognitive-emotional brain: from interactions to integration. Cambridge: MIT Press; 2013.
[17] Pessoa L. Understanding brain networks and brain organization. Phys Life Rev 2014;11:400–35.
[18] Sotres-Bayon F, Quirk GJ. Prefrontal control of fear: more than just extinction. Curr Opin Neurobiol 2010;20:231–5.
[19] Tovote P, Fadok JP, Lüthi A. Neuronal circuits for fear and anxiety. Nat Rev Neurosci 2015;16:317–31.
[20] Tovote P, Esposito MS, Botta P, Chaudun F, Fadok JP, Markovic M, et al. Midbrain circuits for defensive behaviour. Nature 2016;534:206–12.
[21] Vogt BA, editor. Cingulate neurobiology and disease. Oxford: Oxford University Press; 2008.

CHAPTER

29.1

Translational Correlation: Limbic Encephalitis

C.-K. Wu

Texas Tech University Health Sciences Center, Lubbock, TX, United States

The limbic system is a good starting point for studying the relationship between brain and mind. Our evolving understanding of the concept of the limbic system is a reminder of the difficulties that researchers have encountered in the study of this extraordinary neural network. The very existence of the limbic system was denied and disputed as an entity for decades. Paul Broca in 19th century provided some of the earliest anatomoclinical evidence of the role of the limbic system in memory function. Those initial observations by Broca have expanded considerably, particularly in three areas. First, the key components of the limbic system—the hippocampal formation and amygdala—have been shown to have much more extensive and more complex connectivity than it was initially suspected. This connectivity seems to serve as a link between various brain regions and provides the anatomic substrate for many cognitive functions and emotions. Second, detailed case studies and experiments have clarified the role that some specific regions of the limbic system play in the mechanisms subserving memory function and expression of emotions. Third, advanced neuroimaging tools can now detect specific changes in the limbic systems at the early stage of various disease states. This latter development has enabled clinicians to study these disorders *in vivo* without having to wait to learn from postmortem material as Broca did. There are many good reviews of the various disease states affecting the limbic system and we will limit our discussion to an interesting and often devastating condition that primarily involves the limbic system, limbic encephalitis (LE).

LIMBIC ENCEPHALITIS: OVERVIEW

LE usually affects hippocampi and amygdalae of the cerebral hemispheres. Common etiologies of LE include infections, paraneoplastic syndromes, and autoimmune mechanisms. Herpes simplex virus type-1 (HSV-1)—remains a common cause for LE in clinical practice and often affects otherwise healthy individuals. Before the advent of effective antiviral treatments, untreated herpes encephalitis (HE) was associated with high mortality and severe, disabling cognitive morbidity. Another increasingly common cause of LE are those related to neoplasms. An interesting feature in these cases is that the underlying

Conn's Translational Neuroscience
http://dx.doi.org/10.1016/B978-0-12-802381-5.00047-6

mechanism of cancer-related LE is not direct invasion by the cancer. Rather, certain cancers appear to be able to trigger the host immunity with the intention to target the cancer cells but instead the host immunity targets the limbic system—the so-called "paraneoplastic" mechanism. Another group of immune-mediated LE is unrelated to any underlying cancer and appears to be triggered by different immune mechanisms. The agents in these cases appear to be specific antibodies against neuronal surface markers and synapses, the autoimmunity-mediated LE.

Infectious Encephalitis of the Limbic System: Herpes Encephalitis

The most common and well-established infection involving the major brain regions of limbic system is caused by invasion of HSV-1. The mechanism by which the virus enters the central nervous system (CNS) remains unknown. It has been postulated that HSV-1 invades the olfactory bulb and stays inactive for a period of time. This virus would then be activated and migrate along the olfactory tract and spread into the major regions of the limbic system, namely entorhinal cortex, amygdalae, and hippocampi within the medial portion of bilateral temporal lobes of the human brain. At the acute stage, the HE patients present fever, confusion, and olfactory hallucination. More severe cases are complicated by seizures and falling into a comatose state. Before the advent of effective antiviral treatment, as many as 70% of HE patients would die. Survivors of HE suffered from severe amnesia (memory dysfunction). This unique complication of HE provided a valuable model for memory research in humans. In the past four decades, some HE survivors greatly contributed to our knowledge of the role of these limbic structures in the medial temporal lobe in mechanisms of the human memory and emotions.

Some of the very interesting clinical features observed in HE survivors are outlined as follows. HE results in the destruction of hippocampi and amygdalae of both cerebral hemispheres. Researchers learned that the memory impairment was so severe in these patients due to bilateral nature of the damage. This observation was subsequently confirmed in laboratory animals. In humans, the extensive damage to the limbic systems of the temporal lobes by HE results in an inability to encode new information (anterograde amnesia), to consolidate new memories, and also to retrieve the previously learned materials (retrograde amnesia). It is of interest the fact that severely amnesic HE survivors can still have surprisingly normal performance on neuropsychological batteries for intelligence and immediate memory. These observations show that mechanisms subserving long-term memory, immediate memory, and intelligence are located in different parts of the human brain. HE survivors have a severely compromised ability to recognize facial expression for emotions, particularly fear and sadness, a deficit that has been attributed to the destruction of amygdalae. Finally, not surprisingly, another key component of the limbic system—olfactory function—is always damaged by the HSV-1 infection. This damage leads to olfactory hallucination and seizure as well as inability to identify smells.

Paraneoplastic Encephalitis of the Limbic System

The discovery that paraneoplastic limbic encephalitis (PNLE) could be associated with certain cancers was a surprise and a puzzle to clinicians and investigators. In 1960s, there were several reports of cognitive and behavior changes in patients with lung cancer. Instead of finding the metastatic tumor to limbic structures, the investigators were surprised to find chronic changes, suggesting of encephalitis and the term "limbic encephalitis" (LE), was coined to describe this condition. Although this condition is not exclusively associated with lung malignancies, small cell lung cancer (SCLC) has been reported in most cases. The terminology—"paraneoplastic"—is rightly used to label this cancer-associated condition. PNLE is rare among patients with cancer and it can affect not only the CNS but also the peripheral nervous system (PNS). The diagnosis of PNLE can only be confirmed by either biopsy or postmortem brain examination.

The clinical onset of PNLE is often subtle but it later progresses rapidly. The course has been labeled as "subacute"—over several weeks. There are two common scenarios:

The first scenario is an oncological patient who develops symptoms of memory impairment and behavior changes in the absence of CNS metastasis and whose symptoms cannot be explained by chemotherapy or radiation therapy. SCLC is the most common malignancy associated with PNLE.

The second scenario is a patient who presents with cognitive and behavioral changes in the absence of known malignancy. Symptoms include memory loss, confusion, and personality changes with or without seizures. MRI and cerebrospinal fluid (CSF) examination show evidence of encephalitis in the limbic structures, but serology is negative for infectious causes of LE, such as HSV-1 infection. Until proven otherwise, these patients are considered to have an occult malignancy.

Even though the cognitive and behavioral changes of PNLE may be similar to those of HE, the clinical course of PNLE tends to be progressive and relentless. Memory deficits—anterograde amnesia and retrograde amnesia—are commonly reported by large series of PNLE cases. Behavioral disturbances include acute confusion, disorientation, hallucinations, and changes in mood and personality.

The autoantibody most commonly associated with SCLC related to PNLE is an antineuronal nuclear antibody (ANNA)—called "anti-Hu" antibody. Anti-Hu antibody can be detected immunohistochemically both in the neurons and in the malignant cells of the SCLC. The anti-Hu antibody appears to cause an overproduction of T-cell lymphocytes directed against the neurons in the limbic structures.

Autoimmune Encephalitis of the Limbic System

A number of previously unidentified autoantibodies have been discovered with the study of newly recognized paraneoplastic syndromes. Even though many patients with autoimmune LE have an underlying malignancy, there is a subgroup of these patients in whom no malignancy is detectable. Furthermore, in some others, the antibodies do not seem to be related to any type of known cancer cells. These are defined as autoimmune LE.

The majority of the autoantibodies target either the cell surface or synaptic proteins of neurons. The best described autoantibody associated with autoimmune (nonparaneoplastic) LE appeared to target at the voltage-gated potassium channels (VGKC). Later investigations showed the actual target was the "leucine-rich glioma-inactivated 1" (LGI1) protein. Thus the former VGKC encephalopathy was renamed as "LGI1 limbic encephalitis" in 2010. Extensive search for underlying cancers has proven futile in most patients with LGI1 LE. These patients commonly exhibit clinical features suggesting rapidly progressive dementia, a feature that is sometimes misdiagnosed as Prion disease. Patients with LGI1 LE often have low serum sodium levels and often develop faciobrachial dystonic seizures in the early stages of the disease. Immunotherapy treatment for LGI1 LE is associated with favorable outcomes as demonstrated by a study of neuropsychological measurements before and after the treatment.

Finally, autoimmune LE can rarely be associated with anti-Hu antibody. The anti-Hu LE in adults has usually a poor prognosis and does not respond well to immunotherapy. Similarly, the anti-Hu-associated LE in children is always related to acute behavioral changes that result in permanent cognitive impairment despite immunotherapies. These children often develop epileptic seizures that do not respond well to anticonvulsion treatments.

DIAGNOSIS OF LIMBIC ENCEPHALITIS

Although the clinical hallmarks of damaged limbic system are deranged memory and emotions, the initial presentation of LE of different etiologies is often an episode of "acute confusion" or "rapidly changing mental status" in a period of days or weeks. The picture is further complicated by the fact that some patients in the early stages of HSE LE may not have fever. For the clinician, the most important step in a case of suspected LE is to explore the possibility of an HSV-1 infection. Therefore, all patients with suspected LE should have a lumbar puncture to obtain the CSF. The demonstration of HSV can be done rapidly by polymerase chain reaction (PCR) of HSV DNA from CSF samples. The CT or MRI of the brain can show necrotic lesions of the medial temporal limbic structures and that helps to confirm the diagnosis.

As of paraneoplastic versus autoimmune causes of LE, the history of an underlying cancer is always valuable for differential diagnosis. More than 50% of paraneoplastic LE patients have SCLC. Other cancers that often associated with paraneoplastic LE include breast cancer, ovary cancer, testicular germ-cell tumor, thymoma, and Hodgkin's disease. CT scans are notoriously insensitive in these cases, and MRI usually is the most useful tool that shows the abnormal signal changes of early stage of paraneoplastic and autoimmune LE in limbic structures of the medial temporal lobe (Fig. 29.1.1).

Once the diagnosis a paraneoplastic or autoimmune LE is suspected, the next step is to search for the presence of specific autoantibodies in these cases. It is important to look for elevated titers of these autoantibodies both in the serum and in the CSF, as this provides the most reliable evidence paraneoplastic or autoimmune etiology. In rare instances, anti-LGI1 antibody can only be detected in serum with regard to autoimmune LE. Cases of LE confirmed with postmortem studies who had negative serum and CSF antibodies are a reminder of the limitations of our existing neurodiagnostic methods. Detailed history of strong clinical suspicion complemented by neuroimaging studies remains invaluable for the diagnosis of many cases of LE.

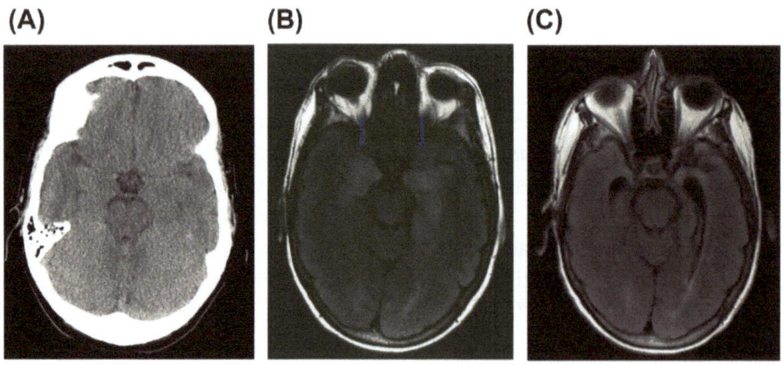

FIGURE 29.1.1 (A) CT at onset does not show any abnormality of medial temporal lobes. (B) MRI at onset shows the abnormal signals (*arrows*) involving medial temporal lobes. (C) After 4 months, the MRI shows obvious destruction of hippocampi and amygdalae in bilateral medial temporal lobes.

THERAPIES FOR LIMBIC ENCEPHALITIS

Acyclovir, a guanosine analog is an effective treatment for HSV-1 LE. Because the HSV-1 infection progresses rapidly, antiviral treatment with acyclovir should be initiated early to prevent more irreparable damage. The timely use of acyclovir dramatically reduces the mortality in HSE LE (from 70% down to 19%), but the residual cognitive and behavior morbidity in these cases remain substantial.

Even though the paraneoplastic and autoimmune LE share a common target, the limbic system, there are different etiologies and prognosis. The immunotherapies to treat paraneoplastic LE or autoimmune LE include steroids, plasma exchange therapy, IVIG infusion, and immunosuppressive drugs. The protocols of using these therapies for either paraneoplastic or autoimmune LE are not well established. The reason is that prognosis in paraneoplastic LE is much less favorable compared with that of autoimmune LE.

Further Reading

[1] Shaw C-M, Alvord Jr EC. Neuropathology of the limbic system. In: Tien RD, editor. The limbic system: normal anatomy and pathology. Neuroimaging clinics of North America, vol. 1 No. 1. Philadelphia: W.B. Saunders Company; February 1997. p. 101–42.

[2] Corsellis JA, Goldberg GJ, Norton AR. 'Limbic encephalitis' and its association with carcinoma. Brain 1968;91:481–96.

[3] Cermak LS. The encoding capacity of a patient with amnesia due to encephalitis. Neuropsychologia 1976;14:311–26.

[4] Insausti R, Annese J, Amaral DG, Squire LR. Human amnesia and medial temporal lobe illuminated by neuropsychological and neurohistological findings for patient E.P. PNAS 2013;110(21):E1953–62.

[5] DeAngelis LM, Posner JB. Paraneoplastic syndromes. In: DeAngelis LM, Posner JB, editors. Neurologic complications of cancer. New York: Oxford University Press; 2009. p. 577–617.

[6] Graus F, Dalmau J. Paraneoplastic neurological syndromes. Curr Opin Neurol 2012;25:795–801.

[7] Armangue T, Leypoldt F, Dalmau J. Autoimmune encephalitis as differential diagnosis of infectious encephalitis. Curr Opin Neurol 2014;27:361–8.

[8] Yoo JY, Hirsch LJ. Limbic encephalitis associated with anti-voltage-gated potassium channel complex antibodies mimicking Creutzfeldt-Jakob disease. JAMA Neurol 2014;71(1):79–82.

CHAPTER

30

Stress, Mood, and Pathways to Depression

K.R.R. Krishnan

Rush University, Chicago, IL, United States

INTRODUCTION

Understanding the neurobiology of behavioral disorders is at its infancy. Deciphering the nature and origin and biology of these disorders has been extraordinarily complex and fraught with problems, much of which is exemplified by the difficulty in creating models that bridge the many facets and dimensions of what it is in essence human nature.

In this chapter we focus primarily on one aspect, namely depression. The state of feeling sad is a normal human emotion. It is when that state affects function that one considers it to be abnormal. It can affect function when it is persistent, when it affects interaction with others and as we now are beginning to discern even medical outcomes.

The state of feeling depressed can arise as a consequence of experiencing an adverse situation or event. When that happens, the sadness is usually transient and disappears over a short period of time. If on the other hand depression becomes persistent, it can evoke additional behavioral responses such as withdrawal from others and the environment, sleep disturbance, either sleeping too much or insomnia, appetite changes—again could be overeating or loss of appetite, fatigue, loss of interest, anhedonia, and many psychological symptoms such as feeling helpless, hopeless, worthless, guilty, and ultimately feeling suicidal. Many of these features can be modeled to a limited extent under conditions of stress where the individual is deprived of options to adapt that is "learned helplessness."

Depression and depressive disorders are indeed very common and can afflict as many as 15–20% of the population. Obviously, a common phenomenon depression is heterogeneous and multifactorial in origin and hence biology.

Theories of Biology of Depression

There are many theories regarding the biological basis of depression. The first is based on the early observations that the medications used to treat depression change monoamines. The monoamine norepinephrine was the initial focus. The theory was bolstered by the finding that a drug used to treat hypertension, namely reserpine depleted norepinephrine, and that many patients who received reserpine became profoundly depressed and even suicidal. Many of the early antidepressants were linked to

Conn's Translational Neuroscience
http://dx.doi.org/10.1016/B978-0-12-802381-5.00048-8

EVIDENCE FOR NOREPINEPHRINE IN DEPRESSION

1. Depleting norepinephrine leads to depressive symptoms in patients on antidepressants.
2. Decreased norepinephrine transporter binding in locus coeruleus of depressed patients.
3. Polymorphism of norepinephrine transporter 182c is linked to risk for depression.
4. Norepinephrine transporter inhibitors that increase norepinephrine in the synapse treat depression.

Moret C, Briley M. The importance of norepinephrine in depression. Neuropsychiatr Dis Treat 2011;7(Suppl. 1):9–13. http://dx.doi.org/10.2147/NDT.S19619 [Epub 2011 May 31].

EVIDENCE FOR SEROTONIN IN DEPRESSION

1. Depleting tryptophan, a precursor for serotonin, in patients on antidepressants can induce depressive symptoms.
2. Serotonin transporter (5HTT) inhibitors that increase serotonin in the synapse treat depression.
3. Decreased serotonin metabolites in the spinal fluid of depressed patients.
4. 5HTT polymorphism related to risk of depression.
5. Altered serotonin receptors in the brains of depressed patients.

Delgado PL, Price LH, Miller HL, Salomon RM, Aghajanian GK, Heninger GR, et al. Serotonin and the neurobiology of depression. Effects of tryptophan depletion in drug-free depressed patients. Arch Gen Psychiatry November 1994; 51(11):865–74.

monoamines. The two monoamines that have been linked are mostly norepinephrine and serotonin.

What must be kept in mind is that although there is correlative evidence for the role of monoamines in the pathophysiology of depression, by no means is this conclusive and/or the sole mechanism that underpins depression.

When individuals become severely depressed, there is usually activation of the hypothalamus–pituitary–adrenal (HPA) axis. Whether the disturbance is the cause or consequence of depression is debatable, but changes in this axis are common in severe depression and have been implicated as a possible mechanism underlying depression.

The mechanisms underlying neural plasticity involve neurotrophins, including BDNF and nerve growth factor (NGF) [4–6]. Manifestation of these neurotrophins and their receptors increases during maximal neuronal differentiation and synaptogenesis. During development, NGF and BDNF are localized to the hippocampus and prefrontal cortex (PFC) [7,8]. Stimuli, such as stressful events, affect neurotrophin expression [5,6]. Neurotrophic (involved in neuroplasticity) factors especially brain-derived neurotrophic factor (BDNF) has been implicated in the pathophysiology of depression. BDNF is rich in the brain, and is also found in the periphery including blood. Stress decreases BDNF expression in the hippocampus, and this decrease can be mitigated by treatment with antidepressants.

Clinical studies show BDNF is reduced in untreated depressed patients, and antidepressant treatment increases BDNF.

EVIDENCE FOR HYPOTHALAMUS–PITUITARY ADRENAL AXIS DYSFUNCTION IN DEPRESSION

1. Urinary free cortisol is increased in depression and returns to normal after resolution of depression.
2. Dexamethasone, a glucocorticoid antagonist, does not suppress cortisol to the same extent in severely depressed patients compared to normal.
3. Adrenal gland is enlarged in depressed patients when compared with normal individuals.
4. Pituitary gland is enlarged in depressed patients, especially in those with elevated HPA function.

Dinan TG, Scott LV. Anatomy of melancholia: focus on hypothalamic-pituitary-adrenal axis overactivity and the role of vasopressin. J Anat September 2005; 207(3):259–64 [Review].

BRAIN-DERIVED NEUROTROPHIC FACTOR AND DEPRESSION

1. BDNF in hippocampus is reduced by stress in animal studies.
2. BDNF is increased by antidepressants in human blood.
3. Depression leads to reduced BDNF in blood.
4. BDNF modulates neurogenesis in the brain and in modifying neuroplasticity and that is believed to be the presumed mechanism.
5. Decreased peripheral BDNF levels and selected *BDNF* gene variants are associated with the risk of developing mood disorders in individuals with early trauma [9,10].
6. The combination of BDNF Met allele carrier status and exposure to trauma early in life predicted reduced gray matter in the hippocampus, lateral PFC and greater depression.
7. The *BDNF* gene is a risk locus for depression [11,12] and with antidepressant response [11,13].

Duman RS. Pathophysiology of depression: the concept of synaptic plasticity. Eur Psychiatry 2002;17(Suppl. 3):306–31 [14].

Vulnerability and Development of Depression

What puts an individual at risk? The cognitive state characterized by a negative bias in information processing (affecting attention, memory, perception, and decision making) and the liability to experience negative affect in response to stress are the features that **increases** vulnerability to the development of depression.

Paths to Vulnerability State

There are many vulnerability paths that give rise to this state and to the development of depression [15–17]. These paths have been discerned from a plethora of studies that have been longitudinal in nature [15–17]. These paths lead to the cognitive state that **increases** vulnerability to the development of depression. These pathways are not uniquely specific to depression vulnerability and can also lead to other psychological and behavioral expressions including anxiety disorders. Understanding this state is critical to our understanding the neural substrates of risk.

The two dimensions of this state are negative bias and sensitivity to stress. Studies in depressed populations indicate a robust relation between negative bias and negative cognitions and the occurrence of depression. For example, normal individuals are positively biased in their perception of information [18], whereas when one is sad, greater attention to negative words is paid [19].

When depressed, individuals view gloomy scenes for a lengthier period of time than non-depressed individuals [20]. They also remember emotionally negative information to a greater extent [21,22]. The scale of negative cognitive bias exhibited in subjects under stress is a predictor for a new onset of a depressive episode up to 2 years later [23]. Negative interpretations of emotionally indistinct stimuli can predict future depressive reaction to stress [24]. The behavioral dimensions to this internal state, namely stress sensitivity and negative bias manifests clinically as low self esteem coupled with anxiety. The two features, stress sensitivity and negative bias often get labeled as neuroticism. Both negative bias and stress sensitivity are key elements of neuroticism, which is a part of the internalizing pathway and a recognized risk for major depression [25–27]. This personality trait moderates the interaction between stressful daily events and subsequent negative affect [28] and predicts responsiveness to negative mood induction [29]. This relationship is supported by a diary study that captured individuals' negative cognitions in response to naturalistic stressors and confirmed a strong association between neuroticism and a negative cognitive style that predicted daily depressive symptoms [30].

This negativity bias is accompanied by a difficulty in disengaging attention from negatively valenced emotional stimuli [19], which results in a pathway through which negative ideation can develop into worry and rumination [31], where attention is repetitively focused on one's negative mood and its potential causes or implications [32]. This process keeps negative cognitions, memories, and interpretations current [32], which results in a passive dwelling on negative life events [33] and may further enhance the negativity bias in information processing [34]. This predisposes individuals toward developing a depressive episode, and the depressed state may also strengthen negative bias and rumination (Fig. 30.1).

Depression is also characterized by increased stress sensitivity, or the tendency to experience negative emotional states such as being sad, worried, irritable, stressed, or overwhelmed [35] in response to minor stressors in daily life [36]. A longitudinal population study found that daily

FIGURE 30.1 Factors that lead to vulnerability for depression.

life stress sensitivity predicted depression onset [37]. Increased stress sensitivity and resultant negative affect represent vulnerability to and a behavioral expression of a risk for depression [23,36]. Furthermore, the strength of the association between daily-life stress sensitivity and depression depended upon both genetic and environmental factors.

Importantly, negative bias and stress sensitivity are linked. This is seen in both community and clinical samples. Negative processing bias predicts both greater negative affect [28] and stress sensitivity as exemplified by cortisol responses to stressful events [38]. This profile is consistent with previous findings for community samples, as well as for diagnosed depression and anxiety [39]. The data is quite clear that negative bias and the increased sensitivity to stress interact and induce a vulnerability to depression characterized by a tendency to have negative cognitions, negative outlook, and a predilection to see the negative elements in a situation and prone to high reactivity to stress.

So if a person is vulnerable and sensitive to stress, then the occurrence of stress can indeed lead to the development of depression. This can begin even before birth, prenatal stress, and extend into childhood events with troubled family environment, childhood abuse, and death or loss of parents or their divorce.

Over time, other experiences of the individual divorce, lifetime trauma, become common stressors that can put the individual at risk. Stress early in life can make the individual more susceptible to additional stress later in life and therefore add to internal factors. Adversity effects can be more if the individual does not have support systems that can help mitigate the stress.

Besides neuroticism factors and adversity, if the individual, for whatever reason, engages in drug and alcohol abuse or bad conduct, then that can aggravate and extenuate vulnerability to depression.

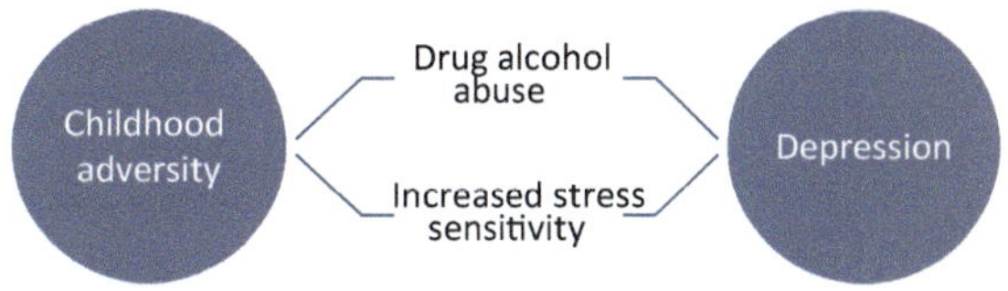

FIGURE 30.2 Mediating factors linking childhood adversity to depression.

Finally, many medical conditions have been associated with risk for depression ranging from B12 deficiency to cancer and vascular disease [17].

Even though the construct outlined earlier makes it appear that the paths are distinct, they are not. Negative memory bias and rumination are associated with vulnerability to depression, and rumination itself increases the frequency of negative memories [40,41]. Depressed individuals exhibit negative interpretations of emotionally ambiguous stimuli [42]. The paths are interlinked in important ways.

Adversity in childhood leads to increased sensitivity to stress, which in turn leads to more experience of stress and can also lead to truancy in childhood, and alcohol and drug abuse [43,44] (Fig. 30.2).

Genetic factors influence and moderate all these pathways, even though we do not know all the specific genes involved [45].

Stressful events experienced during prenatal and early postnatal life influence vulnerability to depression throughout the lifespan [46]. Childhood abuse has been repeatedly related to increased neuroticism in adulthood [27,47,48]. The effects of early life stress is affected by the presence or absence of effective support mechanisms—"buffers," such as social or family support in childhood (Fig. 30.3).

A cognitive frame work that helps understand how these pathways leads to depression is based on the work of Beck and other cognitive theorists [49] as well as that of Abramson [50] and others who have focused on other dimensions. Beck's cognitive theory [49] focuses on negative bias and outlooks that sieve and impact how we perceive societal and emotional

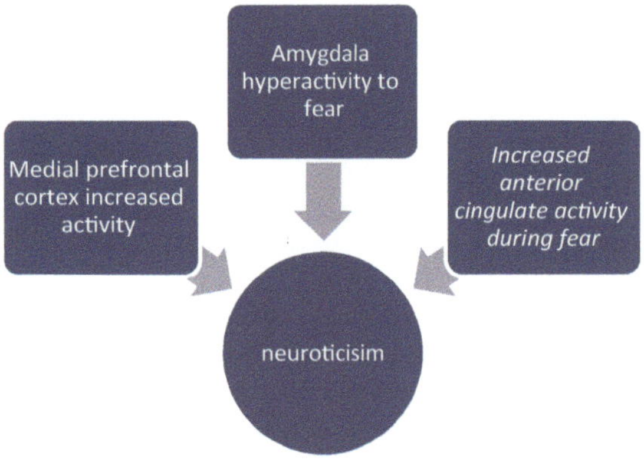

FIGURE 30.3 Neural substrates for neuroticism.

information. Abramson's hopelessness theory [50] underscores one element that of negative extrapolations about the basis of events [18]. This negative extrapolation is an example of negative bias.

NEUROBIOLOGICAL BASIS OF DEPRESSION PATHWAYS

The Structural and Functional Localization of Neuroticism

Negativity bias is intermediated primarily through structural and functional differences in the PFC and amygdala. Trait neuroticism is linked with increased gray matter concentration in the right amygdala [51] and with amygdala hyperreactivity reaction to facial expressions of fear [52]. Greater negativity bias is associated with hypersensitivity to fear cues, increased amygdala activity, and differences in amygdala volume. Individuals with a higher self-rated negativity bias exhibited enhanced activation in fear circuitry, including the amygdala, medial prefrontal cortex (mPFC), and ACC40. For conscious fear, negativity bias was linked to activation of the dorsal mPFC and ACC, while nonconscious fear elicited activation of the ventral-mPFC [53].

Amygdala reactivity also predicts stress effects. Healthy individuals with heightened amygdala reactivity to environmental stresses are at more risk to show more anxious symptoms after exposure [54]. This intrinsic baseline in amygdala reactivity accounts for almost 50% of the variance in the development of depressive symptoms following stress exposure [54]. Thus amygdala hyperreactivity could represent a functional endophenotype for vulnerability to depression [55]. The mechanism by which these develop is uncertain, although there are several possibilities.

Higher neuroticism scores are moreover linked with greater continued mPFC activity throughout blocks of sad facial expressions, but not fearful or happy expressions [56].

These mechanisms probably include Neuronal plasticity, remodeling, and epigenetic mechanisms. The factors in conjunction with functional genetic polymorphisms affecting these systems result in altered plasticity, development, and function of frontotemporal circuitry. Clinically, these changes present as an increased negative cognitive bias and increased stress sensitivity, which increases susceptibility to developing MDD in response to new stressors.

Plasticity in neural circuits involves the strengthening or weakening of existing synapses or formation and elimination of synapses. Axonal boutons turn over in a cell type–specific mode, while dendritic spines develop and decay, and these dynamics are regulated by experience and activity [57]. For example, acute or chronic stressors refashion synapses and dendritic splitting in the medial amygdala and basolateral amygdala (BLA), with vigorous dendritic development in pyramidal and stellate BLA neurons [58–60]. Such changes go with increased anxiety [58,59] and heightening in fear conditioning [61]. When chronic stress stirred dendritic growth in the BLA, it led to larger anxiety-like behavior [59,62].

Chronic stress also produces contemporaneous neuronal remodeling in the PFC and hippocampus [63]. The effects of stress on amygdala morphology are patently unlike those seen in the hippocampus and the PFC. Unlike in the amygdala, there is dendrite reduction and a decrease of spine density in mPFC neurons with shrinkage of dendrites in hippocampal CA3 pyramidal neurons and dentate gyrus granule neurons.

These brain regions are interconnected; it is likely that the structural remodeling in one region influences function and maybe structure in other brain regions. It is likely that genetic factors and stress alter the structure and function of the amygdala, hippocampus, and PFC in a manner similar to what was observed with chronic stress. The evidence that this is indeed the case has not yet been proven in humans (Fig. 30.4).

Another possible way in which brain plasticity is altered is through epigenetic changes. Epigenetic changes are adjustments in gene function without changes in the original nucleotide sequence [64,65]. Epigenetic mechanisms involve alteration of the nucleosome, comprised of histone octamers. The modification of chromatin structure allows the regulation of transcription through controlling access to the regulatory regions of the genome. Such epigenetic changes include DNA methylation of CpG dinucleotides and acetylation and methylation at the N-terminal tails of histones. Methylation of gene promoter regions is connected to transcriptional repression, while histone methylation and demethylation enhance or repress transcription depending upon the histone modification [66]. Environmental stressors can lead to long-lasting changes in behavior through changes in epigenetic regulation [67].

We now deliberate the role of plasticity and epigenetic changes in neurochemical systems implicated in depression, including the HPA axis system, serotonin, and neurotrophic factors.

HPA Axis

Early environmental factors create a long-lasting HPA reactivity to stress that can be conveyed epigenetically to subsequent generations and may be displayed at both the behavioral and biochemical levels. Epigenetic alterations function as a potential tool by which childhood adversity creates persistent changes in gene expression that may directly or indirectly lead to the intergenerational transmission of traits from mother to daughter in both humans and in nonhuman primates [65,68].

Manipulating mother–infant interactions in animals can induce epigenetic changes in the offspring that effect long-term changes on behavior [69,70]. In rhesus macaques, when infants are cross-fostered from nonabusive mothers to abusive mothers, and conversely from abusive mothers to nonabusive mothers, the infants

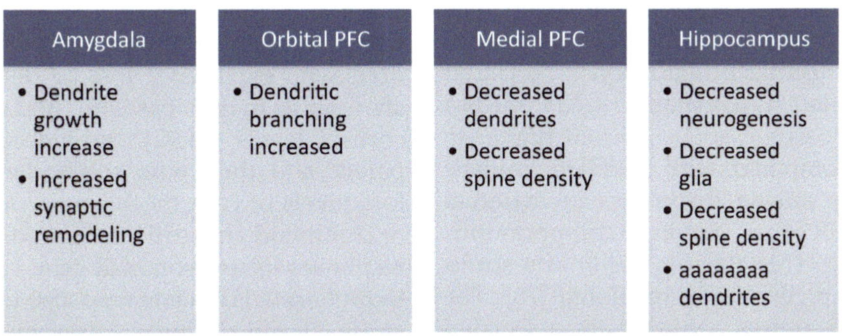

FIGURE 30.4 Neurostructural changes in brain regions linked to depression vulnerability.

show patterns of abuse in adulthood similar to their adoptive mothers, suggesting transmission of behavioral traits [71]. In rats, individual differences in maternal care are also transmitted from mother to the female offspring [72].

Postnatal handling increases the frequency of pup-licking by rat mothers [73], and there are differences in HPA axis responsiveness between the offspring of lactating rats who engage in high and low levels of pup-licking [73]. The changes in pup-licking and HPA axis responsiveness are connected to different histone acetylation and transcription factor binding to the GR promoter with ensuing increased GR mRNA [73,74]. Thus, developmental stress may result in persistently increased activity of one or more central corticotrophin-releasing hormone (CRH) systems and with sensitization of the HPA axis to stress and subsequent alteration in behavior.

Besides epigenetic factors, the persistent effects of handling in rodents can also arise from differentiation of neurons involved in the stress response [75]. Deprivation of mice and animals can change the behavior of animals. Both deprivation and handling lead to deviations in subpopulations of GABAergic neurons in brain areas involved in the regulation of the HPA axis and stress responses [68,76]. This change in the subpopulation may underpin the persistent behavioral expression after the deprivation. In addition, glucocorticoid receptor (GR) function is altered by handling.

Handled rodents show increased GR expression in the hippocampus, which is implicated in glucocorticoid feedback regulation [77]. Under basal conditions, hypothalamic CRH mRNA and protein levels are about 2.5-fold higher in nonhandled compared with handled animals [70]. Naturally arising disparities in maternal care regulate GR expression in the hippocampus of rat offspring. This effect is stable into adulthood, and is epigenetically regulated [78]. The alterations in maternal behavior are associated with changes in GR expression in the offspring, including differences in the expression of the exon 17-containing mRNA transcripts [74]. DNA methylation of the GR17 promoter is greater in the offspring of low-interactive than in those of high-interactive (licking/grooming) mothers [79]. These differences in DNA methylation emerge during the first week of life. Besides the HPA axis, vasopressin has been linked to the transmission of behavioral traits. Repeated episodes of protracted maternal separation in mice alter DNA methylation of the arginine vasopressin gene in the hypothalamus and increase HPA responses to stress [80].

Human Studies

Heredity has a significant effect on basal free cortisol [81]. A link between neuroticism and HPA axis reactivity has been observed [82]. There is evidence for increased CRH activity in patients with both depression and with anxiety disorders [83], and human studies demonstrate a genetic contribution of the HPA axis to anxiety and depression [84].

Early childhood adversity impacts stress reactivity in adulthood [85]. There are significant correlations between childhood attachment styles and salivary cortisol responses to acute stress during adulthood [86–88] in response to conflict situations [89]. For example, in a study of Finnish World War II survivors, separation from parents was connected with lifelong transformed responsiveness of the HPA axis to stress that was persistent more than 60 years later [90]. Other studies have also documented that children who experience loss or death of parents show shifts in both basal [91] and stress-induced cortisol levels [86,92]. Individuals who lost a parent and then who subsequently described low levels of care by the surviving parent [93] or childhood abuse [94] exhibited raised cortisol responses to stressors. GR gene variants appear to moderate HPA axis response to psychosocial stress elicited during the Trier Social Stress Test (TSST) [95].

Such associations are influenced by heredity, as stated in a study on the effects of childhood familial misfortune on salivary cortisol stress reactions to unaccustomed situations [96].

The HPA axis is controlled by the hippocampus, and childhood abuse and neglect has been associated with smaller hippocampal volumes in depressed adults [97]. This decreased hippocampal volume is grasped as evidence for impaired central HPA regulation, and presumably depression [98].

Prenatal stress can lead to a lasting impression on HPA axis response to stress [99,100]. This link between maternal stress and infant emotional development is probably facilitated by cortisol levels in the amniotic fluid [101–103]. Low birth weight could be a result of fetal stress often attributed to increased cortisol responsiveness to stress [104].

Postmortem examination of the hippocampus in suicide victims show reduced levels of GR mRNA and increased cytosine methylation of the exon 1F neuron-specific GR (NR3C1) promoter [105].

Role of Serotonin

The effect of serotonin-related gene variation manifests during ontogeny with the developmental formation of the stress-responsive neural system. During development, changes in serotonin regulation can create long-lasting variations in corticolimbic systems that mediate stress response. Serotonin regulates synaptic development through 5-HT1A receptors [106], and mice missing the 5HT1A receptor gene display increased anxiety [107], and repression of 5-HT1A receptors in the first 4 weeks of age creates adult mice with increased anxious behavior [108].

There is accruing evidence that variation in serotonin receptors can mediate stress-related phenotypes [109–112].

Stressors regulate both 5HTT and tryptophan hydroxylase 2 (Tph2) mRNA expression. Mice with an inactivated TPH2 gene exhibit anxiety- and depressive-like behaviors [113,114], macaques deprived of maternal rearing show heightened cortisol response to stress during adulthood, if they are stressed.

Studies examining genetic influences on serotonergic function, negativity, and stress reactivity have primarily focused upon the *5-HTTLPR* short variant [115]. *5-HTTLPR* short variant has been linked to depression [115–117], and neuroticism [116,118,119]. *5-HTTLPR* short allele carriers who had childhood trauma show greater HPA axis reactivity to stressors [120].

Short allele carriers show less functional coupling between the amygdala and the ACC when processing angry and fearful faces, and this could be the biological underpinning of the link to neuroticism [121,122].

5HTT mRNA transcription is also regulated by epigenetic mechanisms [123]. Changes in expression of BDNF and NGF during development can stimulate a cascade of events that can alter maturation of these regions, and could predispose individuals to altered stress responsiveness during adulthood [69]. Conversely, neurotrophins are sensitive to manipulations of mother–infant interactions in animal studies [124–127]. Maternal stress leads to hypermethylation of the exon IV *BDNF* promoter and decreased hippocampal BDNF levels and concordantly drops the frequency of pup-licking [128]. Maternal deprivation in rats increases *BDNF* gene expression in the PFC and hippocampus and long-term reduction in BDNF expression in the PFC and impaired HPA axis responses to chronic swim-stress in adulthood [126]. The *BDNF* Met-ELS interaction also predicted elevated neuroticism [129].

Neurobiological Model for Vulnerability Paths to Depression

This body of literature gives a neurobiological basis and model of how vulnerability factors and paths make the individual susceptible to

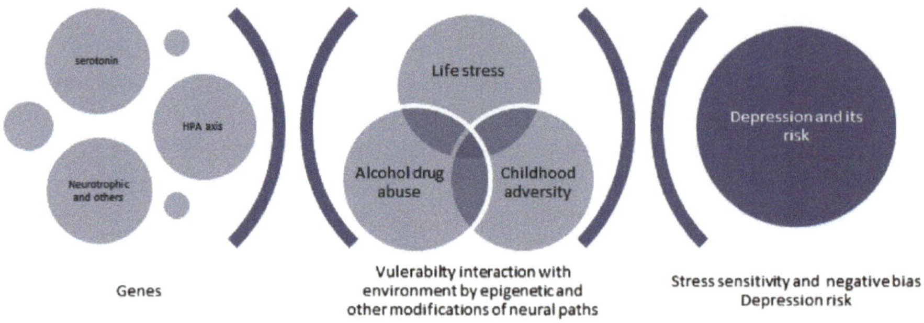

FIGURE 30.5 Interactions between genes environment leading to depression risk.

depression. The vulnerability factors in combination with functional genetic polymorphisms affecting these systems effect changed plasticity, development, and function of frontotemporal circuitry, presenting as increased negative cognitive bias and increased stress sensitivity, which increases vulnerability to developing depression in response to new stressors.

Early life stress and subsequent epigenetic modification can alter frontotemporal function and induce vulnerability. Furthermore, there are likely resiliency factors tempering the relationship between neuroticism and depression, as not all neurotic individuals become depressed.

While we focused on the role of serotonin, BDNF, and the HPA axis, other systems are also likely to be involved in the pathogenesis of depression.

Childhood **adversity** is a substantial predisposing risk factor for depression and anxiety in adulthood, but that does not mean progression is inevitable, nor is it specific to depression as childhood trauma can also lead to anxiety, and impulsivity (Fig. 30.5).

SUMMARY

This review brings together a diverse literature and emphasizes important elements underlying the neural processes related to depression and the vulnerability features, negativity bias and stress sensitivity. Animal and human studies have converged to identify important roles of the hippocampus, amygdala, and PFC in mediating or enhancing the effects of these vulnerability factors and the development of depression. The mechanisms that transduce this vulnerability to depression involve multiple neurotransmitter systems and the HPA axis. Cellular, epigenetic, and neuroplasticity changes contribute to transduction of these functional changes into the onset and maintenance of depressive states. Such influences may be further mediated and modified or even magnified by functional genetic variations, which independently affect functions within these systems. Altered serotonin, HPA axis, and neurotrophin influences brain development and function, resulting in a negative bias in information processing and an increased liability to experience negative affect in response to stress. These cognitive features in turn increase vulnerability to the development of depression.

Further Reading

[1] Moret C, Briley M. The importance of norepinephrine in depression. Neuropsychiatr Dis Treat 2011;7(Suppl. 1):9–13. http://dx.doi.org/10.2147/NDT.S19619. [Epub 2011 May 31].

[2] Delgado PL, Price LH, Miller HL, Salomon RM, Aghajanian GK, Heninger GR, et al. Serotonin and the neurobiology of depression. Effects of tryptophan depletion in drug-free depressed patients. Arch Gen Psychiatry November 1994;51(11):865–74.

[3] Dinan TG, Scott LV. Anatomy of melancholia: focus on hypothalamic-pituitary-adrenal axis overactivity and the role of vasopressin. J Anat September 2005;207(3):259–64. [Review].

[4] Duman RS, Heninger GR, Nestler EJ. A molecular and cellular theory of depression. Arch Gen Psychiatry 1997;54(7):597–606.

[5] Smith MA, Makino S, Kvetnansky R, Post RM. Stress and glucocorticoids affect the expression of brain-derived neurotrophic factor and neurotrophin-3 mRNAs in the hippocampus. J Neurosci 1995;15(3 Pt 1): 1768–77.

[6] Thoenen H. Neurotrophins and neuronal plasticity. Science 1995;270(5236):593–8.

[7] Large TH, Bodary SC, Clegg DO, Weskamp G, Otten U, Reichardt LF. Nerve growth factor gene expression in the developing rat brain. Science 1986;234(4774):352–5.

[8] Yan Q, Rosenfeld RD, Matheson CR, Hawkins N, Lopez OT, Bennett L, et al. Expression of brain-derived neurotrophic factor protein in the adult rat central nervous system. Neuroscience 1997;78(2):431–48.

[9] Aguilera M, Arias B, Wichers M, Barrantes-Vidal N, Moya J, Villa H, et al. Early adversity and 5-HTT/ BDNF genes: new evidence of gene-environment interactions on depressive symptoms in a general population. Psychol Med 2009;39(9):1425–32.

[10] Kaufman J, Yang BZ, Douglas-Palumberi H, Grasso D, Lipschitz D, Houshyar S, et al. Brain-derived neurotrophic factor-5-HTTLPR gene interactions and environmental modifiers of depression in children. Biol Psychiatry 2006;59(8):673–80.

[11] Licinio J, Dong C, Wong ML. Novel sequence variations in the brain-derived neurotrophic factor gene and association with major depression and antidepressant treatment response. Arch Gen Psychiatry 2009;66(5):488–97.

[12] Sklar P, Gabriel SB, McInnis MG, Bennett P, Lim YM, Tsan G, et al. Family-based association study of 76 candidate genes in bipolar disorder: BDNF is a potential risk locus. Brain-derived neutrophic factor. Mol Psychiatry 2002;7(6):579–93.

[13] Taylor WD, McQuoid DR, Ashley-Koch A, MacFall JR, Bridgers J, Krishnan RR, et al. BDNF Val66Met genotype and 6-month remission rates in late-life depression. Pharmacogenomics J 2011;11(2):146–54.

[14] Duman RS. Pathophysiology of depression: the concept of synaptic plasticity. Eur Psychiatry 2002;17(Suppl. 3):306–31.

[15] Kendler KS, Gardner CO, Prescott CA. Toward a comprehensive developmental model for major depression in women. Am J Psychiatry 2002;159(7):1133–45.

[16] Kendler KS, Gardner CO, Prescott CA. Toward a comprehensive developmental model for major depression in men. Am J Psychiatry 2006;163(1):115–24.

[17] Krishnan KR, Taylor WD, McQuoid DR, MacFall JR, Payne ME, Provenzale JM, et al. Clinical characteristics of magnetic resonance imaging-defined subcortical ischemic depression. Biol Psychiatry 2004;55(4):390–7.

[18] Hirsch CR, Mathews A. Impaired positive inferential bias in social phobia. J Abnorm Psychol 2000;109(4):705–12.

[19] Bradley BP, Mogg K, Lee SC. Attentional biases for negative information in induced and naturally occurring dysphoria. Behav Res Ther 1997;35(10):911–27.

[20] Eizenman M, Yu LH, Grupp L, Eizenman E, Ellenbogen M, Gemar M, et al. A naturalistic visual scanning approach to assess selective attention in major depressive disorder. Psychiatry Res 2003;118(2):117–28.

[21] Gilboa-Schechtman E, Erhard-Weiss D, Jeczemien P. Interpersonal deficits meet cognitive biases: memory for facial expressions in depressed and anxious men and women. Psychiatry Res 2002;113(3):279–93.

[22] Ridout N, Noreen A, Johal J. Memory for emotional faces in naturally occurring dysphoria and induced sadness. Behav Res Ther 2009;47(10):851–60.

[23] Rude SS, Valdez CR, Odom S, Ebrahimi A. Negative cognitive biases predict subsequent depression. Cogn Ther Res 2003;27(4):415–29.

[24] Pury CLS. Information-processing predictors of emotional response to stress. Cogn Emot 2002;16(5):667–83.

[25] Kendler KS, Kuhn J, Prescott CA. The interrelationship of neuroticism, sex, and stressful life events in the prediction of episodes of major depression. Am J Psychiatry 2004;161(4):631–6.

[26] Chan SW, Goodwin GM, Harmer CJ. Highly neurotic never-depressed students have negative biases in information processing. Psychol Med 2007;37(9):1281–91.

[27] Williams LM, Gatt JM, Schofield PR, Olivieri G, Peduto A, Gordon E. 'Negativity bias' in risk for depression and anxiety: brain-body fear circuitry correlates, 5-HTT-LPR and early life stress. NeuroImage 2009;47(3):804–14.

[28] Longua J, Dehart T, Tennen H, Armeli S. Personality moderates the interaction between positive and negative daily events predicting negative affect and stress. J Res Pers 2009;43(4):547–55.

[29] Larsen RJ, Ketelaar T. Personality and susceptibility to positive and negative emotional states. J Pers Soc Psychol 1991;61(1):132–40.

[30] Hankin BL, Fraley RC, Abela JR. Daily depression and cognitions about stress: evidence for a traitlike depressogenic cognitive style and the prediction of depressive symptoms in a prospective daily diary study. J Pers Soc Psychol 2005;88(4):673–85.

[31] Lyubomirsky S, Nolen-Hoeksema S. Effects of self-focused rumination on negative thinking and interpersonal problem solving. J Pers Soc Psychol 1995;69(1): 176–90.

[32] Nolen-Hoeksema S. Sex differences in unipolar depression: evidence and theory. Psychol Bull 1987;101(2):259–82.

[33] Nolen-Hoeksema S. Responses to depression and their effects on the duration of depressive episodes. J Abnorm Psychol 1991;100(4):569–82.

[34] Joormann J, Gotlib IH. Updating the contents of working memory in depression: interference from irrelevant negative material. J Abnorm Psychol 2008;117(1):182–92.

[35] Watson D, Clark LA, Tellegen A. Development and validation of brief measures of positive and negative affect: the PANAS scales. J Pers Soc Psychol 1988;54(6):1063–70.

[36] Wichers M, Myin-Germeys I, Jacobs N, Peeters F, Kenis G, Derom C, et al. Genetic risk of depression and stress-induced negative affect in daily life. Br J Psychiatry 2007;191:218–23.

[37] Wichers M, Geschwind N, Jacobs N, Kenis G, Peeters F, Derom C, et al. Transition from stress sensitivity to a depressive state: longitudinal twin study. Br J Psychiatry 2009;195(6):498–503.

[38] Fox E, Cahill S, Zougkou K. Preconscious processing biases predict emotional reactivity to stress. Biol Psychiatry 2010;67(4):371–7.

[39] Bienvenu OJ, Samuels JF, Costa PT, Reti IM, Eaton WW, Nestadt G. Anxiety and depressive disorders and the five-factor model of personality: a higher- and lower-order personality trait investigation in a community sample. Depress Anxiety 2004;20(2):92–7.

[40] Lyubomirsky S, Caldwell ND, Nolen-Hoeksema S. Effects of ruminative and distracting responses to depressed mood on retrieval of autobiographical memories. J Pers Soc Psychol 1998;75(1):166–77.

[41] Nolen-Hoeksema S. The role of rumination in depressive disorders and mixed anxiety/depressive symptoms. J Abnorm Psychol 2000;109(3):504–11.

[42] Lawson C, MacLeod C, Hammond G. Interpretation revealed in the blink of an eye: depressive bias in the resolution of ambiguity. J Abnorm Psychol 2002;111(2):321–8.

[43] Rosenman S, Rodgers B. Childhood adversity and adult personality. Aust NZ J Psychiatry 2006;40(5):482–90.

[44] Molnar BE, Buka SL, Kessler RC. Child sexual abuse and subsequent psychopathology: results from the National Comorbidity Survey. Am J Public Health 2001;91(5):753–60.

[45] Foley DL, Eaves LJ, Wormley B, Silberg JL, Maes HH, Kuhn J, et al. Childhood adversity, monoamine oxidase a genotype, and risk for conduct disorder. Arch Gen Psychiatry 2004;61(7):738–44.

[46] Heim C, Nemeroff CB. The role of childhood trauma in the neurobiology of mood and anxiety disorders: preclinical and clinical studies. Biol Psychiatry 2001;49(12):1023–39.

[47] Gamble SA, Talbot NL, Duberstein PR, Conner KR, Franus N, Beckman AM, et al. Childhood sexual abuse and depressive symptom severity: the role of neuroticism. J Nerv Ment Dis 2006;194(5):382–5.

[48] McFarlane A, Clark CR, Bryant RA, Williams LM, Niaura R, Paul RH, et al. The impact of early life stress on psychophysiological, personality and behavioral measures in 740 non-clinical subjects. J Integr Neurosci 2005;4(1):27–40.

[49] Beck TT. Cognitive models of depression. J Cogn Psychother 1987;1:5–37.

[50] Abramson LY, Alloy LB, Metalsky GI. Hopelessness depression – a theory-based subtype of depression. Psychol Rev 1989;96(2):358–72.

[51] Omura K, Todd Constable R, Canli T. Amygdala gray matter concentration is associated with extraversion and neuroticism. Neuroreport 2005;16(17):1905–8.

[52] Chan SW, Norbury R, Goodwin GM, Harmer CJ. Risk for depression and neural responses to fearful facial expressions of emotion. Br J Psychiatry 2009;194(2):139–45.

[53] Williams LM, Liddell BJ, Kemp AH, Bryant RA, Meares RA, Peduto AS, et al. Amygdala-prefrontal dissociation of subliminal and supraliminal fear. Hum Brain Mapp 2006;27(8):652–61.

[54] Admon R, Lubin G, Stern O, Rosenberg K, Sela L, Ben-Ami H, et al. Human vulnerability to stress depends on amygdala's predisposition and hippocampal plasticity. Proc Natl Acad Sci USA 2009;106(33):14120–5.

[55] Hariri AR, Drabant EM, Munoz KE, Kolachana BS, Mattay VS, Egan MF, et al. A susceptibility gene for affective disorders and the response of the human amygdala. Arch Gen Psychiatry 2005;62(2):146–52.

[56] Haas BW, Constable RT, Canli T. Stop the sadness: neuroticism is associated with sustained medial prefrontal cortex response to emotional facial expressions. NeuroImage 2008;42(1):385–92.

[57] Holtmaat A, Svoboda K. Experience-dependent structural synaptic plasticity in the mammalian brain. Nat Rev Neurosci 2009;10(9):647–58.

[58] Vyas A, Jadhav S, Chattarji S. Prolonged behavioral stress enhances synaptic connectivity in the basolateral amygdala. Neuroscience 2006;143(2):387–93.

[59] Vyas A, Mitra R, Shankaranarayana Rao BS, Chattarji S. Chronic stress induces contrasting patterns of dendritic remodeling in hippocampal and amygdaloid neurons. J Neurosci 2002;22(15):6810–8.

[60] Vyas A, Pillai AG, Chattarji S. Recovery after chronic stress fails to reverse amygdaloid neuronal hypertrophy and enhanced anxiety-like behavior. Neuroscience 2004;128(4):667–73.

[61] Conrad CD, LeDoux JE, Magarinos AM, McEwen BS. Repeated restraint stress facilitates fear conditioning independently of causing hippocampal CA3 dendritic atrophy. Behav Neurosci 1999;113(5):902–13.

[62] Vyas A, Chattarji S. Modulation of different states of anxiety-like behavior by chronic stress. Behav Neurosci 2004;118(6):1450–4.

[63] McEwen BS, Chattarji S. Neuroendocrinology of stress. In: Lajtha A, Blaustein JD, editors. Handbook of neurochemistry and molecular neurobiology. New York (NY): Springer; 2007. p. 571–94.

[64] Bird A. Perceptions of epigenetics. Nature 2007;447(7143):96–8.

[65] Zhang TY, Meaney MJ. Epigenetics and the environmental regulation of the genome and its function. Annu Rev Psychol 2010;61:439–66. C431–433.

[66] Szyf M, McGowan P, Meaney MJ. The social environment and the epigenome. Environ Mol Mutagen 2008;49(1):46–60.

[67] Bale TL, Baram TZ, Brown AS, Goldstein JM, Insel TR, McCarthy MM, et al. Early life programming and neurodevelopmental disorders. Biol Psychiatry 2010;68(4):314–9.

[68] Meaney MJ. Maternal care, gene expression, and the transmission of individual differences in stress reactivity across generations. Annu Rev Neurosci 2001;24:1161–92.

[69] Cirulli F, Berry A, Alleva E. Early disruption of the mother–infant relationship: effects on brain plasticity and implications for psychopathology. Neurosci Biobehav Rev 2003;27(1–2):73–82.

[70] Plotsky PM, Meaney MJ. Early, postnatal experience alters hypothalamic corticotropin-releasing factor (CRF) mRNA, median eminence CRF content and stress-induced release in adult rats. Brain Res Mol Brain Res 1993;18(3):195–200.

[71] Maestripieri D, Higley JD, Lindell SG, Newman TK, McCormack KM, Sanchez MM. Early maternal rejection affects the development of monoaminergic systems and adult abusive parenting in rhesus macaques (*Macaca mulatta*). Behav Neurosci 2006;120(5):1017–24.

[72] Francis D, Diorio J, Liu D, Meaney MJ. Nongenomic transmission across generations of maternal behavior and stress responses in the rat. Science 1999;286(5442):1155–8.

[73] Meaney MJ, Szyf M. Environmental programming of stress responses through DNA methylation: life at the interface between a dynamic environment and a fixed genome. Dialogues Clin Neurosci 2005;7(2):103–23.

[74] Szyf M, Weaver IC, Champagne FA, Diorio J, Meaney MJ. Maternal programming of steroid receptor expression and phenotype through DNA methylation in the rat. Front Neuroendocrinol 2005;26(3–4):139–62.

[75] Meaney MJ, Diorio J, Francis D, Widdowson J, LaPlante P, Caldji C, et al. Early environmental regulation of forebrain glucocorticoid receptor gene expression: implications for adrenocortical responses to stress. Dev Neurosci 1996;18(1–2):49–72.

[76] Giachino C, Canalia N, Capone F, Fasolo A, Alleva E, Riva MA, et al. Maternal deprivation and early handling affect density of calcium binding protein-containing neurons in selected brain regions and emotional behavior in periadolescent rats. Neuroscience 2007;145(2):568–78.

[77] Meaney MJ, Aitken DH, Viau V, Sharma S, Sarrieau A. Neonatal handling alters adrenocortical negative feedback sensitivity and hippocampal type II glucocorticoid receptor binding in the rat. Neuroendocrinology 1989;50(5):597–604.

[78] Liu D, Diorio J, Tannenbaum B, Caldji C, Francis D, Freedman A, et al. Maternal care, hippocampal glucocorticoid receptors, and hypothalamic-pituitary-adrenal responses to stress. Science 1997;277(5332):1659–62.

[79] Weaver IC, Cervoni N, Champagne FA, D'Alessio AC, Sharma S, Seckl JR, et al. Epigenetic programming by maternal behavior. Nat Neurosci 2004;7(8):847–54.

[80] Murgatroyd C, Patchev AV, Wu Y, Micale V, Bockmuhl Y, Fischer D, et al. Dynamic DNA methylation programs persistent adverse effects of early-life stress. Nat Neurosci 2009;12(12):1559–66.

[81] Bartels M, Van den Berg M, Sluyter F, Boomsma DI, de Geus EJ. Heritability of cortisol levels: review and simultaneous analysis of twin studies. Psychoneuroendocrinology 2003;28(2):121–37.

[82] Zobel A, Barkow K, Schulze-Rauschenbach S, Von Widdern O, Metten M, Pfeiffer U, et al. High neuroticism and depressive temperament are associated with dysfunctional regulation of the hypothalamic-pituitary-adrenocortical system in healthy volunteers. Acta Psychiatr Scand 2004;109(5):392–9.

[83] Lopez-Duran NL, Kovacs M, George CJ. Hypothalamic-pituitary-adrenal axis dysregulation in depressed children and adolescents: a meta-analysis. Psychoneuroendocrinology 2009;34(9):1272–83.

[84] Leonardo ED, Hen R. Genetics of affective and anxiety disorders. Annu Rev Psychol 2006;57:117–37.

[85] Entringer S, Kumsta R, Hellhammer DH, Wadhwa PD, Wust S. Prenatal exposure to maternal psychosocial stress and HPA axis regulation in young adults. Horm Behav 2009;55(2):292–8.

[86] Luecken LJ. Childhood attachment and loss experiences affect adult cardiovascular and cortisol function. Psychosom Med 1998;60(6):765–72.

[87] Luecken LJ. Attachment and loss experiences during childhood are associated with adult hostility, depression, and social support. J Psychosom Res 2000;49(1):85–91.

[88] Quirin M, Pruessner JC, Kuhl J. HPA system regulation and adult attachment anxiety: individual differences in reactive and awakening cortisol. Psychoneuroendocrinology 2008;33(5):581–90.

[89] Powers SI, Gunlicks M, Laurent H, Balaban S, Bent E, Sayer A. Differential effects of subtypes of trauma symptoms on couples' hypothalamus-pituitary-adrenal (HPA) axis reactivity and recovery in response to interpersonal stress. Ann NY Acad Sci 2006;1071:430–3.

[90] Pesonen AK, Raikkonen K, Feldt K, Heinonen K, Osmond C, Phillips DI, et al. Childhood separation experience predicts HPA axis hormonal responses in late adulthood: a natural experiment of World War II. Psychoneuroendocrinology 2010;35(5):758–67.

[91] Nicolson NA. Childhood parental loss and cortisol levels in adult men. Psychoneuroendocrinology 2004;29(8):1012–8.

[92] Tyrka AR, Wier L, Price LH, Ross N, Anderson GM, Wilkinson CW, et al. Childhood parental loss and adult hypothalamic-pituitary-adrenal function. Biol Psychiatry 2008;63(12):1147–54.

[93] Luecken LJ, Rodriguez AP, Appelhans BM. Cardiovascular stress responses in young adulthood associated with family-of-origin relationship experiences. Psychosom Med 2005;67(4):514–21.

[94] Luecken LJ, Appelhans BM. Early parental loss and salivary cortisol in young adulthood: the moderating role of family environment. Dev Psychopathol 2006;18(1):295–308.

[95] Wust S, Van Rossum EF, Federenko IS, Koper JW, Kumsta R, Hellhammer DH. Common polymorphisms in the glucocorticoid receptor gene are associated with adrenocortical responses to psychosocial stress. J Clin Endocrinol Metab 2004;89(2):565–73.

[96] Ouellet-Morin I, Boivin M, Dionne G, Lupien SJ, Arseneault L, Barr RG, et al. Variations in heritability of cortisol reactivity to stress as a function of early familial adversity among 19-month-old twins. Arch Gen Psychiatry 2008;65(2):211–8.

[97] Frodl T, Reinhold E, Koutsouleris N, Reiser M, Meisenzahl EM. Interaction of childhood stress with hippocampus and prefrontal cortex volume reduction in major depression. J Psychiatr Res 2010;44(13):799–807.

[98] Bremner JD, Randall P, Vermetten E, Staib L, Bronen RA, Mazure C, et al. Magnetic resonance imaging-based measurement of hippocampal volume in posttraumatic stress disorder related to childhood physical and sexual abuse – a preliminary report. Biol Psychiatry 1997;41(1):23–32.

[99] Huizink AC, Mulder EJ, Buitelaar JK. Prenatal stress and risk for psychopathology: specific effects or induction of general susceptibility? Psychol Bull 2004;130(1):115–42.

[100] Weinstock M. The potential influence of maternal stress hormones on development and mental health of the offspring. Brain Behav Immun 2005;19(4):296–308.

[101] Bergman K, Sarkar P, Glover V, O'Connor TG. Maternal prenatal cortisol and infant cognitive development: moderation by infant-mother attachment. Biol Psychiatry 2010;67(11):1026–32.

[102] Davis EP, Glynn LM, Schetter CD, Hobel C, Chicz-Demet A, Sandman CA. Prenatal exposure to maternal depression and cortisol influences infant temperament. J Am Acad Child Adolesc Psychiatry 2007;46(6):737–46.

[103] de Weerth C, van Hees Y, Buitelaar JK. Prenatal maternal cortisol levels and infant behavior during the first 5 months. Early Hum Dev 2003;74(2):139–51.

[104] Wust S, Entringer S, Federenko IS, Schlotz W, Hellhammer DH. Birth weight is associated with salivary cortisol responses to psychosocial stress in adult life. Psychoneuroendocrinology 2005;30(6):591–8.

[105] McGowan PO, Sasaki A, D'Alessio AC, Dymov S, Labonte B, Szyf M, et al. Epigenetic regulation of the glucocorticoid receptor in human brain associates with childhood abuse. Nat Neurosci 2009;12(3):342–8.

[106] Gaspar P, Cases O, Maroteaux L. The developmental role of serotonin: news from mouse molecular genetics. Nat Rev Neurosci 2003;4(12):1002–12.

[107] Savitz J, Lucki I, Drevets WC. 5-HT(1A) receptor function in major depressive disorder. Prog Neurobiol 2009;88(1):17–31.

[108] Gross C, Zhuang X, Stark K, Ramboz S, Oosting R, Kirby L, et al. Serotonin1A receptor acts during development to establish normal anxiety-like behaviour in the adult. Nature 2002;416(6879):396–400.

[109] Celada P, Puig MV, Casanovas JM, Guillazo G, Artigas F. Control of dorsal raphe serotonergic neurons by the medial prefrontal cortex: Involvement of serotonin-1A, GABA(A), and glutamate receptors. J Neurosci 2001;21(24):9917–29.

[110] Jankowski MP, Sesack SR. Prefrontal cortical projections to the rat dorsal raphe nucleus: ultrastructural features and associations with serotonin and gamma-aminobutyric acid neurons. J Comp Neurol 2004;468(4):518–29.

[111] Christianson JP, Ragole T, Amat J, Greenwood BN, Strong PV, Paul ED, et al. 5-hydroxytryptamine 2C receptors in the basolateral amygdala are involved in the expression of anxiety after uncontrollable traumatic stress. Biol Psychiatry 2010;67(4):339–45.

[112] Heisler LK, Pronchuk N, Nonogaki K, Zhou L, Raber J, Tung L, et al. Serotonin activates the hypothalamic-pituitary-adrenal axis via serotonin 2C receptor stimulation. J Neurosci 2007;27(26):6956–64.

[113] Beaulieu JM, Zhang X, Rodriguiz RM, Sotnikova TD, Cools MJ, Wetsel WC, et al. Role of GSK3 beta in behavioral abnormalities induced by serotonin deficiency. Proc Natl Acad Sci USA 2008;105(4):1333–8.

[114] Gardner KL, Hale MW, Oldfield S, Lightman SL, Plotsky PM, Lowry CA. Adverse experience during early life and adulthood interact to elevate tph2 mRNA expression in serotonergic neurons within the dorsal raphe nucleus. Neuroscience 2009;163(4):991–1001.

[115] Caspi A, Sugden K, Moffitt TE, Taylor A, Craig IW, Harrington H, et al. Influence of life stress on depression: moderation by a polymorphism in the 5-HTT gene. Science 2003;301(5631):386–9.
[116] Munafo MR, Freimer NB, Ng W, Ophoff R, Veijola J, Miettunen J, et al. 5-HTTLPR genotype and anxiety-related personality traits: a meta-analysis and new data. Am J Med Genet B Neuropsychiatr Genet 2009;150B(2):271–81.
[117] Risch N, Herrell R, Lehner T, Liang KY, Eaves L, Hoh J, et al. Interaction between the serotonin transporter gene (5-HTTLPR), stressful life events, and risk of depression: a meta-analysis. J Am Med Assoc 2009;301(23):2462–71.
[118] Schinka JA, Busch RM, Robichaux-Keene N. A meta-analysis of the association between the serotonin transporter gene polymorphism (5-HTTLPR) and trait anxiety. Mol Psychiatry 2004;9(2):197–202.
[119] Sen S, Burmeister M, Ghosh D. Meta-analysis of the association between a serotonin transporter promoter polymorphism (5-HTTLPR) and anxiety-related personality traits. Am J Med Genet B Neuropsychiatr Genet 2004;127B(1):85–9.
[120] Way BM, Taylor SE. The serotonin transporter promoter polymorphism is associated with cortisol response to psychosocial stress. Biol Psychiatry 2010;67(5):487–92.
[121] Pezawas L, Meyer-Lindenberg A, Drabant EM, Verchinski BA, Munoz KE, Kolachana BS, et al. 5-HTTLPR polymorphism impacts human cingulate-amygdala interactions: a genetic susceptibility mechanism for depression. Nat Neurosci 2005;8(6):828–34.
[122] Frokjaer VG, Vinberg M, Erritzoe D, Baare W, Holst KK, Mortensen EL, et al. Familial risk for mood disorder and the personality risk factor, neuroticism, interact in their association with frontolimbic serotonin 2A receptor binding. Neuropsychopharmacology 2010;35(5):1129–37.
[123] Philibert RA, Sandhu H, Hollenbeck N, Gunter T, Adams W, Madan A. The relationship of 5HTT (SLC6A4) methylation and genotype on mRNA expression and liability to major depression and alcohol dependence in subjects from the Iowa Adoption Studies. Am J Med Genet B Neuropsychiatr Genet 2008;147B(5):543–9.
[124] Branchi I, D'Andrea I, Fiore M, Di Fausto V, Aloe L, Alleva E. Early social enrichment shapes social behavior and nerve growth factor and brain-derived neurotrophic factor levels in the adult mouse brain. Biol Psychiatry 2006;60(7):690–6.
[125] Cirulli F, Micera A, Alleva E, Aloe L. Early maternal separation increases NGF expression in the developing rat hippocampus. Pharmacol Biochem Behav 1998;59(4):853–8.
[126] Roceri M, Cirulli F, Pessina C, Peretto P, Racagni G, Riva MA. Postnatal repeated maternal deprivation produces age-dependent changes of brain-derived neurotrophic factor expression in selected rat brain regions. Biol Psychiatry 2004;55(7):708–14.
[127] Liu D, Diorio J, Day JC, Francis DD, Meaney MJ. Maternal care, hippocampal synaptogenesis and cognitive development in rats. Nat Neurosci 2000;3(8):799–806.
[128] Roth TL, Lubin FD, Funk AJ, Sweatt JD. Lasting epigenetic influence of early-life adversity on the BDNF gene. Biol Psychiatry 2009;65(9):760–9.
[129] Gatt JM, Nemeroff CB, Dobson-Stone C, Paul RH, Bryant RA, Schofield PR, et al. Interactions between BDNF Val66Met polymorphism and early life stress predict brain and arousal pathways to syndromal depression and anxiety. Mol Psychiatry 2009;14(7):681–95.
[130] Barr CS, Newman TK, Becker ML, Parker CC, Champoux M, Lesch KP, et al. The utility of the non-human primate; model for studying gene by environment interactions in behavioral research. Genes Brain Behav 2003;2(6):336–40.
[131] Shannon C, Schwandt ML, Champoux M, Shoaf SE, Suomi SJ, Linnoila M, et al. Maternal absence and stability of individual differences in CSF 5-HIAA concentrations in rhesus monkey infants. Am J Psychiatry 2005;162(9):1658–64.
[132] Maekawa T, Kim S, Nakai D, Makino C, Takagi T, Ogura H, et al. Social isolation stress induces ATF-7 phosphorylation and impairs silencing of the 5-HT 5B receptor gene. Eur Mol Biol Organ J 2010;29(1):196–208.
[133] Pham TM, Soderstrom S, Henriksson BG, Mohammed AH. Effects of neonatal stimulation on later cognitive function and hippocampal nerve growth factor. Behav Brain Res 1997;86(1):113–20.
[134] Cirulli F, Alleva E, Antonelli A, Aloe L. NGF expression in the developing rat brain: effects of maternal separation. Brain Res Dev Brain Res 2000;123(2):129–34.
[135] Aloe L, Bracci-Laudiero L, Alleva E, Lambiase A, Micera A, Tirassa P. Emotional stress induced by parachute jumping enhances blood nerve growth factor levels and the distribution of nerve growth factor receptors in lymphocytes. Proc Natl Acad Sci USA 1994;91(22):10440–4.
[136] Hadjiconstantinou M, McGuire L, Duchemin AM, Laskowski B, Kiecolt-Glaser J, Glaser R. Changes in plasma nerve growth factor levels in older adults associated with chronic stress. J Neuroimmunol 2001;116(1):102–6.
[137] Lippmann M, Bress A, Nemeroff CB, Plotsky PM, Monteggia LM. Long-term behavioural and molecular alterations associated with maternal separation in rats. Eur J Neurosci 2007;25(10):3091–8.
[138] Tsankova NM, Berton O, Renthal W, Kumar A, Neve RL, Nestler EJ. Sustained hippocampal chromatin regulation in a mouse model of depression and antidepressant action. Nat Neurosci 2006;9(4):519–25.

[139] Wichers M, Kenis G, Jacobs N, Mengelers R, Derom C, Vlietinck R, et al. The BDNF Val(66)Met x 5-HTTLPR x child adversity interaction and depressive symptoms: an attempt at replication. Am J Med Genet B Neuropsychiatr Genet 2008;147B(1):120–3.

[140] Miller AH, Maletic V, Raison CL. Inflammation and its discontents: the role of cytokines in the pathophysiology of major depression. Biol Psychiatry 2009;65(9):732–41.

[141] Carpenter LL, Gawuga CE, Tyrka AR, Lee JK, Anderson GM, Price LH. Association between plasma IL-6 response to acute stress and early-life adversity in healthy adults. Neuropsychopharmacology 2010;35(13):2617–23.

[142] Uddin M, Koenen KC, Aiello AE, Wildman DE, De los Santos R, Galea S. Epigenetic and inflammatory marker profiles associated with depression in a community-based epidemiologic sample. Psychol Med 2011;41(5):997–1007.

[143] Backdahl L, Bushell A, Beck S. Inflammatory signalling as mediator of epigenetic modulation in tissue-specific chronic inflammation. Int J Biochem Cell Biol 2009;41(1):176–84.

[144] Valinluck V, Sowers LC. Endogenous cytosine damage products alter the site selectivity of human DNA maintenance methyltransferase DNMT1. Cancer Res 2007;67(3):946–50.

CHAPTER

31

Language Processing, Development and Evolution

M. Catani, M.S. Dawson

King's College London, London, United Kingdom

INTRODUCTION

Complex language has been fundamental to the evolution of our species and sets us apart from every other animal on the planet. As humans we combine a vocabulary of thousands of words with grammatical rules to articulate thoughts and emotions, solicit information, express negation, and create rich metaphors. Development of this skill begins early in our lives—with most human infants showing impressive perceptive and language-learning abilities—before continuing throughout adolescence and adulthood. In contrast, while our closest living relatives, chimpanzees and bonobos, share some of our prelinguistic social skills, their auditory memory, grammatical capacities, and pragmatic abilities are limited or nonexistent. Some of these unique faculties are likely the result of specific anatomical differences, which, as humans, we have acquired during evolution and continue to acquire during brain development. In the human species, modern neuroimaging techniques have identified specific brain regions and connecting white matter that implement a system for social communication and language. Preliminary findings indicate that some of these structures are shared with other nonhuman primates, yet others are unique to humans.

In this chapter, we explore the anatomy of language processing and provide insights from infant development and animal comparison studies. We use this knowledge to then explore a model for the acquisition of higher social, linguistic and pragmatic competence, and apply it to help us understand some of the more common disorders that affect the ability to develop social communication in childhood and to use language in adult life.

Those unfamiliar with technical terms relating to language, cortical anatomy, and white matter anatomy may find Tables 31.1 and 31.2 a useful starting point.

LANGUAGE CORTEX

For a long time, our understanding of language anatomy relied on cortical maps, which were first identified by the pattern of cortical folding and later through cytoarchitectonics

Conn's Translational Neuroscience
http://dx.doi.org/10.1016/B978-0-12-802381-5.00049-X

TABLE 31.1 Major Components of the Structure of Language

Language Term	Definition	Example
Semantics	Word meaning—the literal meaning of a word (or phrase)	Semantic processing—encoding a word's meaning and relating it to similar words with similar meanings
Syntax	Sentence meaning—the rules by which a sentence is formed (eg, grammar)	Syntactic processing—encoding the arrangement of words in a sentence and the subsequent meaning
Lexicon	Vocabulary—includes visual form, phonetic form, semantic form, etc.	Lexical decision—classifying stimuli as words or nonwords
Phonetics	Classification of speech sounds—sounds that can change the meaning of a word but have no meaning themselves	Phonological processing—the ability to hear a word, break it down into discrete sounds, and associate each sound with a letter/s
Pragmatics	Context meaning—includes metaphor, irony, indirect meaning, etc.	Pragmatic integration—the ability to understand the indirect meaning of a word or phrase

TABLE 31.2 Anatomical Terms Related to Language

Anatomical Term	Definition	Example
CORTICAL ANATOMY		
Center	19th century term used to define small parts of the cortex dedicated to words storage	Broca's center (storage of spoken words), Wernicke's center (storage of heard words), reading center (storage of written words)
Area	A discrete cortical region defined with cytoarchitectonic or functional methods	Brodmann areas, Broca's area (BA 44 and 45), Wernicke's area (BA 22), visual word form area
Territory or region	Large cortical regions defined according to diffusion tractography	Broca's territory (defined according to the cortical projections of the arcuate fasciculus to posterior middle and inferior frontal gyrus and inferior precentral gyrus); Wernicke's territory (arcuate projections to posterior regions of the superior and middle temporal gyrus); Geschwind's territory (arcuate projections to the inferior parietal lobe)
WHITE MATTER ANATOMY		
Fibers	Term used for myelinated or unmyelinated axons identified using anatomical methods	U-shaped fibers (connecting adjacent gyri)
Bundle or segment	Term that indicates smaller groups of fibers	Anterior, posterior, and long segment of the arcuate fasciculus
Fascicle or fasciculus	Large white matter tracts composed of several bundles	Arcuate fasciculus (defined according to his arching form), uncinate fasciculus (hook-like tract connecting anterior temporal and ventral frontal regions)
Tract or pathway	Similar to fasciculus. It can indicate connections identified with either anatomical or tractography methods	Frontal aslant tract

(the postmortem analysis of gray matter's cellular distribution). However, the advent of probabilistic cytoarchitectonic maps and receptor architectonic analysis (the study of receptor distribution in gray matter) revealed cortical regions to be extremely complex, more than previously thought. Whether these anatomically defined areas have a corresponding functional role in language is still debated. Fig. 31.1 shows the principal cortical and subcortical regions considered to have a prominent role in communication in humans, along with possible equivalent regions in the monkey brain.

Human Adults

Classical Cortical Regions

Language development in humans is traditionally considered to occur as a direct consequence of specialization of two areas in the left hemisphere: Broca's area in the frontal lobe allowing word production and Wernicke's area in the temporal lobe allowing auditory comprehension.

Broca's region corresponds to the most posterior part of the inferior frontal gyrus (IFG), namely the pars opercularis and pars triangularis. This region is largely occupied by Brodmann areas (BA) 44 and 45 (Fig. 31.1A), although modern analysis with receptor architectonics has further subdivided these two areas to add an extra layer of complexity. As such, the current definition of Broca's region includes a mosaic of areas whose functional specialization has not been detailed. Nevertheless, lesion and functional activation studies suggest a functional gradient within Broca's area in the left hemisphere of most humans. Posterior Broca's area (BA 44) is preferentially activated by tasks involving syntactic processing and lexical decision, whereas more anterior regions (BA 45) activate for semantic tasks. BA 44 is also activated bilaterally in response to personal and observed movement implying the existence of mirror neurons.

Wernicke's region has historically been more difficult to define. Upon discovery, it occupied the posterior superior temporal gyrus (STG), but with time this has expanded to include posterior middle temporal gyrus (MTG). These gyri incorporate BA 21, 22, 42, and part of BA 37 (Fig. 31.1A). Functional imaging studies show that the STG responds to distinct phonetic features of spoken language, whereas posterior MTG and ITG map semantic and lexical representations.

Other Cortical Regions

In addition to these core regions, imaging studies propose other areas are relevant to language and speech. Broca's and Wernicke's regions are linked to areas involved in working memory, semantic knowledge, and social cognition by Geschwind's region in the inferior parietal lobule (IPL). Geschwind's region incorporates BA 39 in the angular gyrus and BA 40 in the supramarginal gyrus (Fig. 31.1A) Some researchers have split these areas further, with five subdivisions within BA 40 and two subdivisions within BA 39. Of note, are similarities in receptor architecture between the three most anterior subdivisions of BA 40 and BA 44 in Broca's region. The similarities are not limited to architecture because the IPL, like Broca's region, also contains mirror neurons that activate during observation of both intended and nonintended actions.

Broca's region relies on further cortex for the planning, initiation, and monitoring of speech and mentalizing. These include the mesial aspect of the superior frontal gyrus (SFG) extending from BA 6 (eg, anterior supplementary motor area, SMA and pre-SMA) to part of BA 8, BA 9, BA 32 (dorsomedial prefrontal cortex, DMPFC) and BA 24 and BA 33 in the cingulate gyrus (Fig. 31.1C). These areas allow distinguishing our own actions from those of others and are involved in mentalizing, such as the representation of our own mental state and those of others. They are also directly involved in speech production.

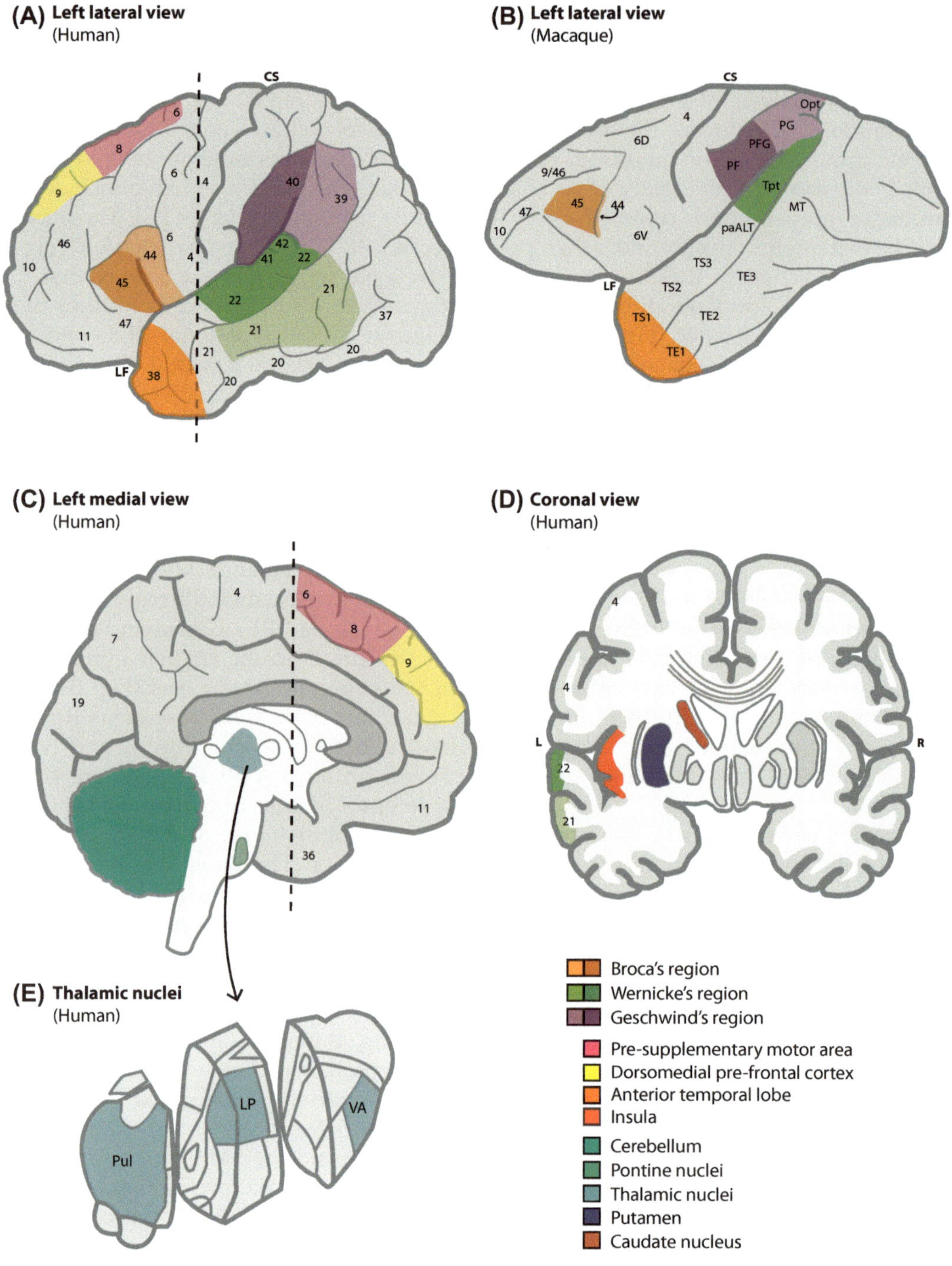

FIGURE 31.1 Left hemisphere human cortical and subcortical regions known to play a role in language, compared with macaque homologs. (A) The core human cortical regions are numbered based on Brodmann's areas (BA). Broca's area corresponds to BA 44 and 45 and may extend into BA 6. Wernicke's area encompasses anterior BA 22 and 41/42, and extends into BA 21. Geschwind's territory incorporates much of BA 39 and 40. (B) Macaque cortical regions are based on a combination of functional maps, as per Schmahmann and Pandya (2006). Broca's area homolog is areas 45 and 44 (in the arcuate sulcus). Wernicke's area homolog is area Tpt, which is equivalent to the human anterior superior temporal gyrus. Early evidence suggests Geschwind's area homolog is areas PF, PFG, PG, and Opt. (C) and (D) Additional regions with a role in language include the presupplementary motor area and dorsomedial prefrontal cortex, as well as subcortical structures such as the thalamus striatum (putamen and caudate) and cerebellum. (E) The main nuclei of the thalamus that connect to the core language cortex: Ventral anterior (VA), Lateral posterior (LP), Pulvinar (Pul). *CS*, Central sulcus; *LF*, Lateral fissure. *Dotted line* represents the approximate location of the coronal view.

The anterior insula is directly connected to Broca's region and activates during the production of language. In addition, the processing of written word relies on its own specialized area in the occipitotemporal region and has been suitably named the visual word form area (VWFA).

Subcortical Regions

Cortical language regions form complex networks with subcortical structures, including the basal ganglia, thalamus, and cerebellum. These corticosubcortical networks are often segregated (Fig. 31.1C–E).

In the anterior thalamus, the ventral anterior nucleus connects with Broca's region and pre-SMA; damage to this area or its connections are associated with stuttering and reduced spontaneous speech. Moving more posteriorly, the lateral posterior nucleus connects with Geschwind's region and lesions are commonly associated with reduced comprehension. The main cortical afferents to Wernicke's region come from the pulvinar (Fig. 31.1E).

The cerebellum connects to the language regions via the thalamus (efferent connections) and the pontine nuclei (afferent connections) (Fig. 31.1C). Telegraphic speech and even mutism have been reported after lesions to parts of the cerebellum.

An additional important relay station is the basal ganglia. Involuntary language production can be induced by stimulation of the striatum (caudate and putamen) (Fig. 31.1D). Lesions, in contrast, can lead to nonfluent speech.

Nonhuman Primates

Given how vital Broca's and Wernicke's regions are to complex language, it is perhaps surprising that neither is unique to humans. Homologs exist in the brains of apes, chimpanzees and macaques based, for example, on cytoarchitectonics or shared nonlinguistic functions. However, there is some structural, and much functional, divergence.

Human Broca's region (BA 44 and 45) is larger in the left hemisphere than the right, a lateralization not seen in other primates. It also has wider mini-columns compared to the macaque homolog (area 44 or F5. Fig. 31.1B), yet such differences are not seen in other somatosensory or motor regions.

Differences are also reported between Wernicke's region and the macaque (area Tpt) and chimpanzee homologs (Fig. 31.1B). In these primates, the visual area MT lies within the superior temporal sulcus (STS), but the equivalent area in humans lies more posterior. So while the entire length of the human STS responds to species-specific calls, in primates the MT processes visual motion.

In the macaque, the IPL is much smaller than the human equivalent and splits into four subregions (PF, PFG, PG, and Opt). Whether these are homologs to those of the Geschwind's territory in humans remains unclear, although studies in connectivity analysis propose a shared anteroposterior subdivision (PF, PFG, and PG). Like humans, the IPL in macaques also contains mirror neurons that fire during the observations and execution of motor actions. Nevertheless, unlike humans, macaques have very poor auditory and verbal working memory, functions that both rely on posterior temporal and inferior parietal regions.

LANGUAGE WHITE MATTER

Language cortex is connected through a complex system of association, projection, and commissural tracts, some of which are left lateralized. The structure and function of this white matter is more controversial than the regions it connects. Whereas similar techniques have been used to study the cortex of all primates, the method of white matter investigation has varied by species. For a long time, the most reliable way to study the terminations of long-range connections was by injecting radioactively labeled tracers directly into the brain, a method known as axonal tracing.

For ethical reasons this method cannot be used in humans or chimpanzees. As such, the most detailed knowledge available on white matter came from macaques, which was directly translated to the human brain. This has been considered a valid approach given the limited ability of classical "blunt dissections" used to identify the course of large white matter pathways in postmortem human brains.

Diffusion tensor imaging (DTI) is a noninvasive technique now commonly used to map the organization of human white matter pathways. While the method has problems—generating fibers that are artifacts or only partially showing the anatomy of existing tracts—it is the only alternative to axonal tracing in living humans. Fig. 31.2 shows the principal tracts thought to be involved in human language, along with possible equivalents in the monkey brain.

Human Adults

Classical Model

The classical model of language connectivity was first proposed by Wernicke–Lichteim (1885) and expanded much later by Geschwind (1970). It proposed Broca's and Wernicke's regions to be directly connected by the arcuate fasciculus, a dorsal perisylvian association tract that arches around the lateral sulcus (Fig. 31.2A). Originally described by blunt fiber dissection this tract is now known to be left lateralized in the adult human brain when studied with tractography. The tract provides a fast and efficient way to transfer information between auditory temporal regions and motor frontal regions. A combined DTI and functional imaging study has shown that our ability to learn new words relies on the strength of functional connectivity and microstructural properties of this direct connection. It also plays a role in word repetition and syntactic processing.

Dual-Network Model

The classical model is now thought an oversimplification with the majority of language models based on a dual stream network: a dorsal (phonological) route for mapping sound into words and a ventral (semantic) route for mapping sound into meaning.

When compared to the classical model, the direct connection of the arcuate fasciculus is analogous to the dorsal phonological pathway (Fig. 31.2B). The possible tracts involved in the ventral pathway are more controversial, with some tightly packed through the extreme capsule region making them difficult to distinguish with low-resolution DTI.

An indirect ventral pathway, the inferior longitudinal fasciculus, connects the posterior occipital and posterior temporal regions to the anterior temporal lobe. It plays a role in object recognition, linking the objects to their lexical labels and reading. From the anterior temporal lobe, the uncinate fasciculus projects into the inferior frontal gyrus, including Broca's region. The uncinate fasciculus is severely damaged in patients with the semantic variant of primary progressive aphasia, implying a role in lexical retrieval, semantic associations, and naming.

The inferior fronto-occipital fasciculus offers a direct connection between the occipital cortex and the frontal lobe. It is theorized that this fiber bundle plays a role in reading and other aspects of language (eg, semantic), although this is yet to be demonstrated.

The extreme capsule (temporal to frontal lobes) and middle longitudinal fasciculus (occipital to temporal lobes) are possible additional tracts in the ventral stream. However, many DTI studies in humans are unable to distinguish the extreme capsule from the inferior fronto-occipital fasciculus. What role these tracts perform in human language is yet to be established.

Extended Network

One criticism of the dual-stream network is that it ignores connections to the IPL (Geschwind's territory) and DMPFC, regions that underlie aspects of communication at the interface of language and social cognition.

With this in mind, DTI has shown the classical arcuate fasciculus (dorsal route) to be more

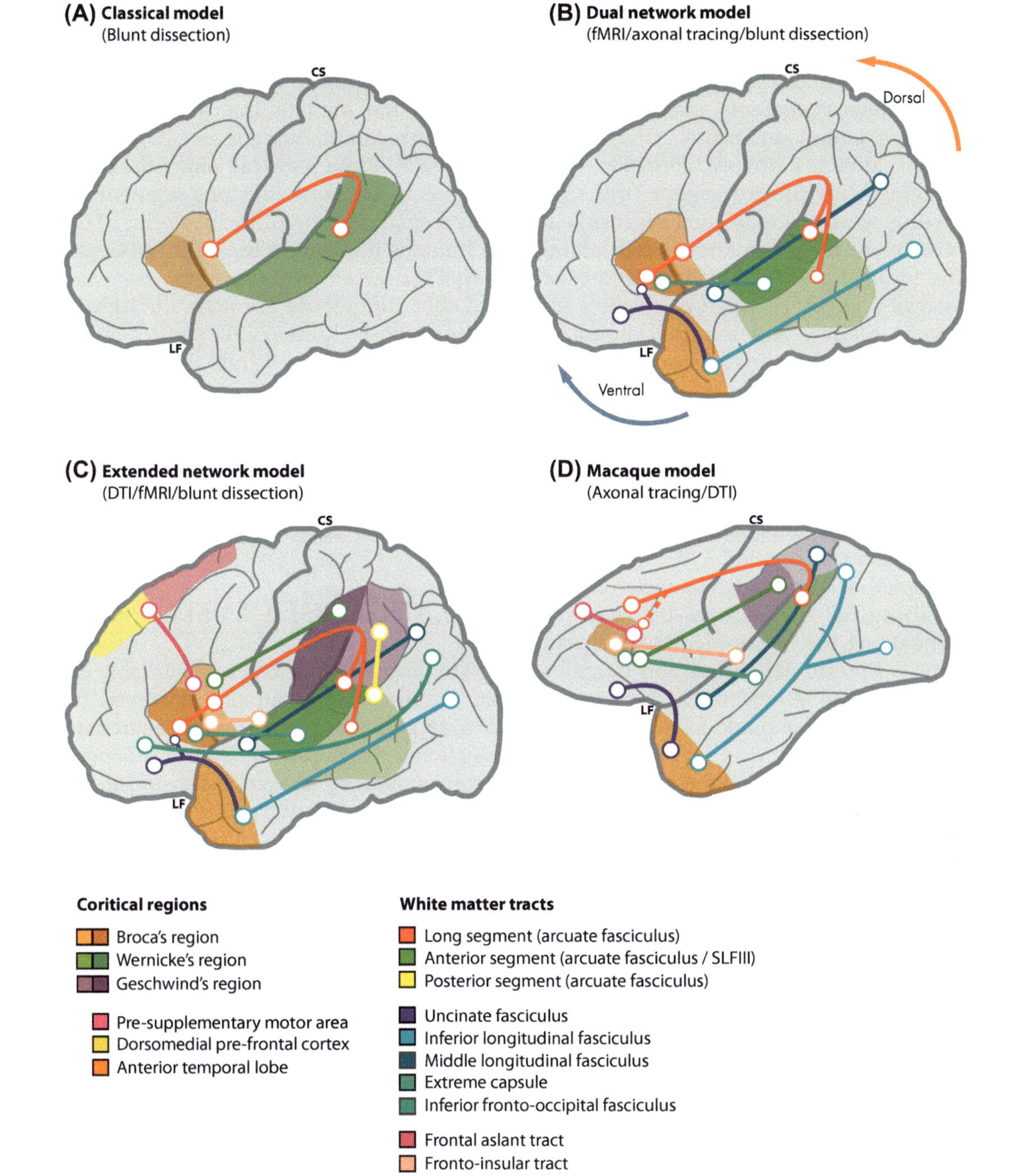

FIGURE 31.2 Networks for language and communication in humans and the macaque. (A) The arcuate fasciculus linking Broca's to Wernicke's area. (B) Functional magnetic resonance imaging (fMRI) (in humans) and axonal racing (in macaques) have expanded this network into a dual stream model with a dorsal route (red) and ventral route (blues). (C) Diffusion tensor imaging (DTI) and fMRI studies extended the model further to incorporate connections to Geschwind's territory via the anterior and posterior segments of the arcuate fasciculus, and to the presupplementary motor area via the frontal aslant tract. (D) Most knowledge of macaque white matter comes from axonal tracing studies. Indirect comparison of humans and nonhuman primates suggests extensive similarities but also remarkable divergences in the connections of their respective communication networks. Note that termination points are simplified; additional terminations exist for the tracts. *CS*, Central sulcus; *LF*, Lateral fissure.

complex than previously thought by splitting it into three different segments: one medial and direct, and two lateral and indirect (Fig. 31.2C). The medial, direct pathway connects Wernicke's region with Broca's region (the arcuate fasciculus *sensu stricto* or long segment). In adult humans this white matter is left lateralized. The indirect pathways consist of a bilateral posterior segment between Wernicke's and Geschwind's regions and a right lateralized anterior segment that links Geschwind's to Broca's regions (also called the superior longitudinal fasciculus III). While the long segment remains important for word repetition, the posterior and anterior segments are employed in verbal working memory, phonological and semantic processing and the interpretation of context (pragmatics).

The frontal aslant tract connects Broca's region with medial frontal areas that facilitate speech, including the presupplementary motor region, medial prefrontal cortex, and cingulate cortex. Patients with lesions to this tract present with a nonfluent/agrammatic form of primary progressive aphasia.

In addition to these tracts, the insula is connected to the frontal operculum via a system of short U-shaped fibers called frontoinsular tracts. Direct connection to Broca's via this tract provides the ability to modulate speech output with emotional and visceral information based on our internal state.

Nonhuman Primates

Comparisons between human and primate white matter have revealed extensive similarities but also remarkable divergences in human and primate communications networks (Fig. 31.2D).

Similarities exist in the anatomy of the anterior segment of the arcuate fasciculus (superior longitudinal fasciculus III), uncinate fasciculus, inferior longitudinal fasciculus and frontal aslant tract.

However, probabilistic DTI of the long segment of the arcuate fasciculus in macaques, chimpanzees, and humans reveals a pronounced expansion of the pathway along the phylogeny of scale. In macaques the long segment connects only a small portion of the posterior STG, while in humans it connects the STG, MTG, and possibly ITG.

Other differences are seen through indirect comparison of human DTI studies and axonal tracing in macaques. While the inferior fronto-occipital fasciculus connects frontal to occipital regions in humans, the analogous extreme capsule of macaques connects only the frontal to temporal regions. No posterior segment of the arcuate fasciculus between MTG and angular gyrus has been reconstructed in macaques.

LANGUAGE DEVELOPMENT

The human brain has a remarkable capacity for change throughout life; however, the first two decades are pivotal for its anatomical and functional development. During this period the brain undergoes a large expansion in volume, neurogenesis, synaptic pruning, and white matter myelination. These changes are driven by a mixture of genetic and environmental factors. However, how these anatomical changes link to the development of increasingly complex communicative and linguistic abilities is still unknown.

Cortical Development

Cortical maturation is region specific with different areas developing at different times and rates. Imaging studies suggest that motor and visual cortices show fast myelination during the first year of life, but this then slows during the second year. Importantly for language, frontal and temporal cortices, as well as the angular gyri, are slow to myelinate during the first year but display rapid growth during year two.

However, imaging studies support the belief that Broca's and Wernicke' regions still exist to some extent at birth. The inferior frontal and temporal regions respond to speech processing as early as 2 days old. In contrast to adults, this processing may have a preponderance in the right temporal auditory cortex. Indeed, the left STS may lag behind the right in maturation in the early stages of life, with dorsal regions of the gyrus also more mature than ventral ones. Cortical maturation is slower still in Geschwind's region, with the angular gyrus one of the latest regions to fully myelinate in the human brain.

White Matter Development

While Broca's and Wernicke's regions may show some activation during speech processing in newborn infants, evidence from functional imaging and DTI suggest they are not strongly connected. The long segment of the arcuate fasciculus is still myelinating at 7 years of age (Fig. 31.3). Some researchers have proposed that maturation of the arcuate fasciculus is synchronized with maturation of posterior STS and IFG. A different connection from Broca's region, the frontal aslant tract, matures earlier during the first years of life.

The lateralization of the arcuate fasciculus is a dynamic process beginning early in life and adolescence. The long and anterior segments are lateralized left and right, respectively, before adolescence and this remains stable into early adulthood (Fig. 31.4). However, the posterior segment is right lateralized in childhood but becomes progressively bilateral during adolescence. This change appears to be driven by a reduction in volume in the right hemisphere and, compared to the other

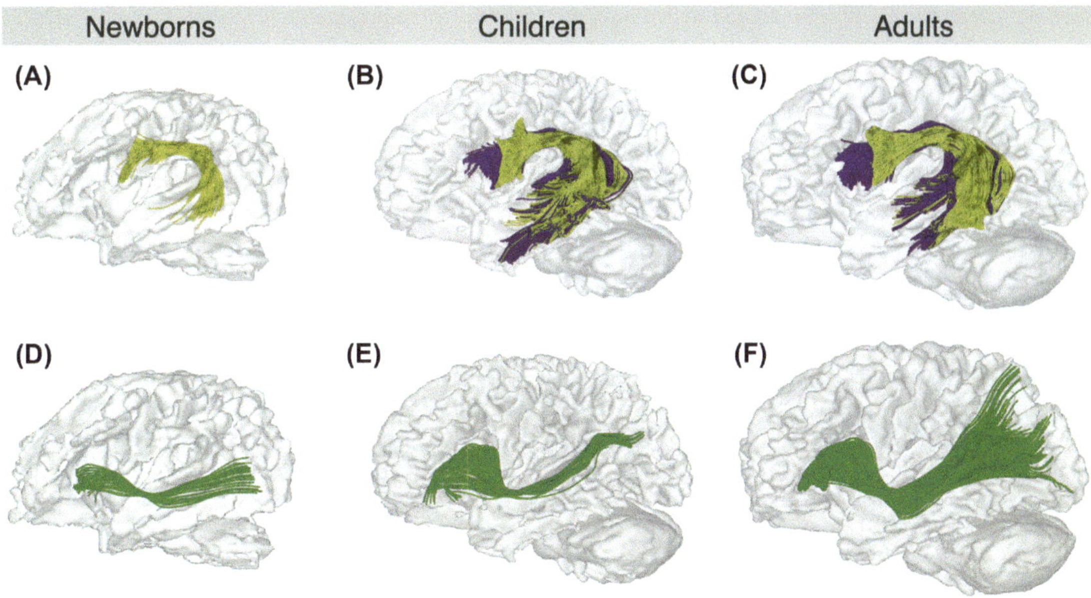

FIGURE 31.3 Maturational changes in the arcuate fasciculus (A–C) and inferior fronto-occipital (D–F) in newborn infants, 7-year-old children, and adults. In the dorsal pathway, the most prominent connection of the arcuate fasciculus projects to the inferior premotor cortex (yellow). By 7 years the connection between the IFG and temporal regions is more visible (blue). In the ventral pathway, the inferior fronto-occipital fasciculus (green) already exists in newborns and continue to mature later in life. *Modified with permission from Brauer J, Anwander A, Perani D, Friederici AD. Dorsal and ventral pathways in language development. Brain Lang 2013;127:289–95.*

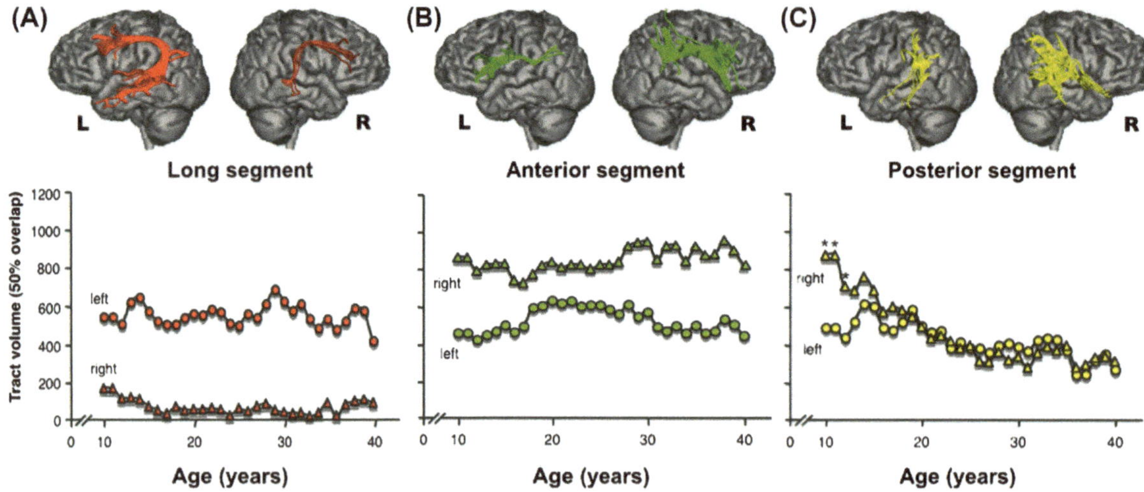

FIGURE 31.4 Age-related differences in the volume of the long, anterior, and posterior segments of the arcuate fasciculus. Group volumes for all ages are indicated by *triangles* for the right segments and *circles* for the left segments. (A) The volume of the long segment is left lateralized in 10-year-old subjects and remains left lateralized throughout adolescence and early adulthood. (B) The anterior indirect segment is right lateralized in 10-year-old subjects and remains right lateralized throughout adolescence and early adulthood. (C) The posterior segment shows right lateralization before adolescence (age 9–11), but left and right differences diminish progressively throughout adolescence and adulthood. *Modified with permission from Budisavljevic S, et al. Age-related differences and heritability of the Perisylvian language networks. J Neurosci. 2015;35(37):12625-34.*

tracts, more influenced by specific environmental factors.

The immaturity of the dorsal pathway early in life may mean infants are likely to connect the frontotemporal network via the tracts of the ventral route. This route has consistently been shown to be more mature during the initial months of life than many other tracts involved in language. These frontotemporal tracts are then slow to fully mature, with the inferior longitudinal fasciculus and inferior fronto-occipital fasciculus still maturing in adolescence and uncinate fasciculus as late as early adulthood.

Functional Development

Like the anatomy, human communication and language skills develop throughout time. While some children suffer some discontinuity due to language impairments (see section: Disorders Affecting Language in Childhood, Adolescence, and Adult Life) most follow a similar rate of progression. Specific language abilities develop during the first 2 to 3 years of life: phonological processing during the first few months, semantic processing by 12 months and syntactic processing at around 30 months.

Humans often display other social communication skills before evidence of language acquisition. Two-day-old newborns are sensitive to visual cues with purposeful movements suggesting that a primitive form of sensory–motor association is already present. Then, during the first year of life, infants begin to obtain a "social sense." Within months newborns direct special attention to humans over other objects, as well as displaying the ability to attribute beliefs to others. Older infants even show pragmatic functions such as requesting by pointing.

Nonhuman primates share some of these prelinguistic communicative abilities. Great apes are capable of identifying other members of their species as communicative agents. They also employ intentional communication, such as gestures that aim to attract attention or initiate an action. Linguistically, there is evidence that chimpanzees can understand some spoken words and learn a limited vocabulary via sign language, but have no capacity for the higher aspects of semantic and syntactic processing or pragmatics.

SOCIAL COMMUNICATION AND LANGUAGE EVOLUTION AND DEVELOPMENT MODEL

The anatomical and functional knowledge discussed so far can be integrated with observations from animal behavior and neuropragmatics to create models of communicative behavior. One such approach is the social communication and language evolution and development (SCALED) model (Fig. 31.5). This five-tiered model provides an anatomical account for the evolution and development of social communication and language. It consists of at least eight white matter connections that are deemed crucial for the development and evolution of five increasing complex levels of communication: from the understanding and ability to engage in social interaction, to the attainment of higher social, linguistic, and pragmatic competence. The individual levels are explained in more detail in the following paragraphs, but in summary: From a developmental point of view, levels 1 and 2 are attained within the first year of life, levels 3 and 4 during infancy and level 5 by adolescence. From an evolutionary point of view, levels 1, 2, and partly 3 are shared between humans and nonhuman primates, while the higher communicative levels 4 and 5 rely on uniquely human connections.

Level 1: A Frontoparietal Network for Informative Actions

The anterior segment of the human arcuate fasciculus links the mirror neurons of posterior Broca's region to anterior Geschwind's region. This network myelinates early in infant humans and is present in nonhuman primates via the corresponding third branch of the superior longitudinal fasciculus.

The frontoparietal mirror neuron network maps the sensory representations of sounds and actions of others into internal motor representations helping to implement a cognitive mechanism of action understating. This network represents a first step in the evolution and development of communication where the observer recognizes his/her own ability to reproduce the actions of others and so engage in social interaction.

Level 2: A Frontal Aslant Network for Communicative Intentions

The frontal aslant tract connects Broca's region to the DMPFC. This network myelinates early in infant humans and the presence of the frontal aslant tract in macaques suggests a similar network in nonhuman primates.

The frontal aslant network is involved in both the lower and higher aspects of mentalizing, allowing the observer to understand what the sender intends. In infants and nonhuman primates, this may represent itself as an imperative gesture such as pointing. This network is the second step in the evolution and development of social communication where the observer recognizes that they can communicate a message and that others intend to communicate back.

Level 3: Anterior Temporal Networks for Lexical and Semantic Processing

The ventral network in humans comprises the inferior longitudinal fasciculus, inferior fronto-occipital fasciculus, uncinate fasciculus,

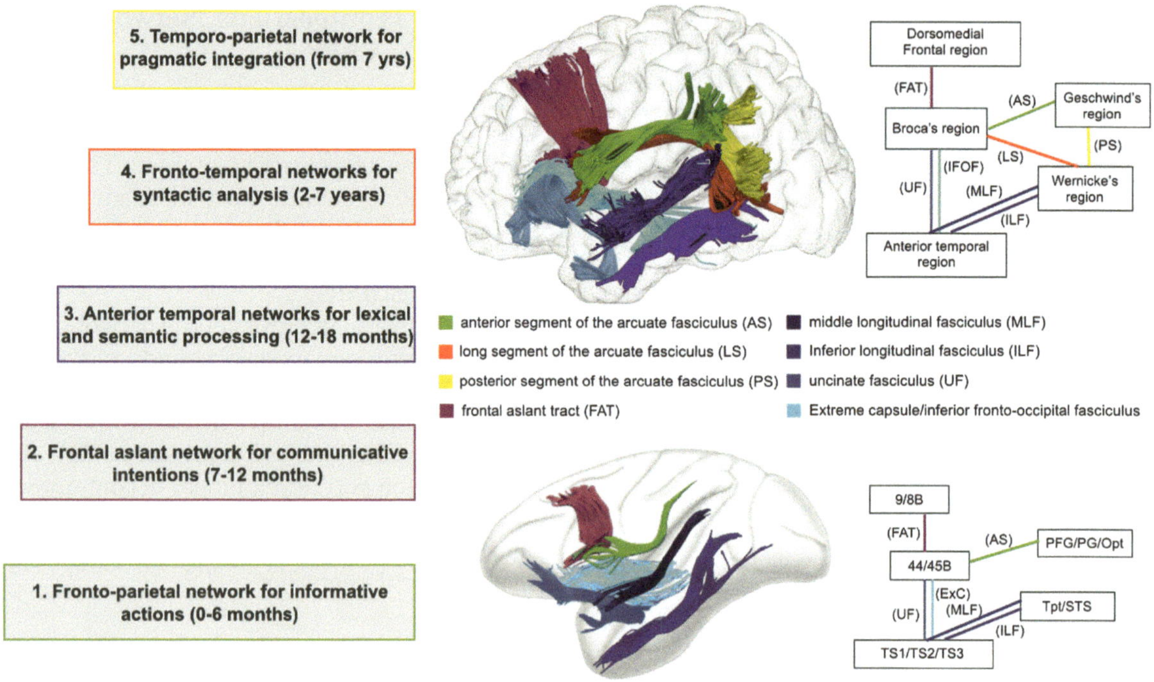

FIGURE 31.5 The social communication and language evolution and development (SCALED) model. Left: Major networks and functions for language and social communication in a hierarchical scale, reflecting developmental and evolutionary trajectories. Center: The colors of the boxes correspond to the tractography reconstruction of the tracts for the human (upper) and macaque monkey brain (lower). Right: The diagrams show the brain regions connected by each network. Level 1: in humans, the anterior segment of the arcuate fasciculus (AS; green) connects posterior cortex of Broca's region to anterior cortex of Geschwind's region (supramarginal gyrus), implementing a system for the recognition and production of informative actions. In the macaque monkey brain, this tract corresponds to the third branch of the superior longitudinal fasciculus (SLF III). Level 2: the frontal aslant tract (FAT; purple) connects Broca's region to the dorsomedial frontal cortex, allowing the processing of communicative intentions. This tract is present in the macaque monkey brain. Level 3: the ventral network for lexical and semantic processing is composed of several tracts. The anterior portion of the middle longitudinal fasciculus (MLF; dark blue) and inferior longitudinal fasciculus (ILF; like blue) connect Wernicke's region to the anterior temporal region, whereas the uncinate fasciculus (UF; azure) connects the anterior temporal region to Broca's region. Some fibers of the inferior fronto-occipital fasciculus (IFOF; cyan) (corresponding to extreme capsule in monkey anatomy; ExC) connect posterior temporal regions (including perhaps Wernicke's region in humans) to the frontal lobe. Significant differences exist between the human and macaque brain in the anatomy of these ventral pathways. Level 4: the long segment of the arcuate fasciculus (LS; red) connects Broca's and Wernicke's region, supporting a frontotemporal network for syntactic analysis and other functions. In the monkey, the portion of the long segment projecting to the middle temporal lobe is absent. Level 5: the posterior segment of the arcuate fasciculus (PS; yellow) links Wernicke's region to Geschwind's region (primarily to the angular gyrus), forming a temporoparietal network for pragmatic integration at the highest level of communication. In the monkey brain, these connections are poorly developed. *Taken from Catani M, Bambini V. Curr Opin Neurobiol 2014;28:165–71.*

and the anterior portion of the middle longitudinal fasciculus. This network develops early in human infants. A similar, but less developed route, exists in monkeys, where the inferior longitudinal fasciculus connects occipital rather than temporal regions.

In humans, this anterior temporal network is involved in lexical and semantic processing. Development of this route in 2-year old children correlates with the "vocabulary spurt," while nonhuman primates show a limited ability to learn vocabulary through the use of sign language. This network is the third step in the evolution and development of language where sound is mapped into meaning.

Level 4: A Frontotemporal Network for Syntactic Analysis

As discussed, the long segment of the arcuate fasciculus in humans directly connects Broca's regions with all aspects of Wernicke's region. Unlike the tracks of the previous three levels, the long segment takes longer to fully myelinate in infants. In macaques, the tract is much smaller and connects only the posterior STG.

This frontotemporal network plays an important role in lexical processing (repetition and learning), syntax (production and comprehension), as well the integration of the two. Nonhuman primates have limited abilities in both word production and grammatical sense. This network represents a fourth step, and an evolutionary leap, in the development of social communication where words are arranged into comprehensible sentences.

Level 5: A Temporoparietal Network for Pragmatic Integration

The posterior segment of the arcuate fasciculus is unique to humans and connects Wernicke's and Geschwind's areas. This temporoparietal network is one of the slowest regions to mature in the language network and undergoes a shift in lateralization during adolescence.

This network plays a role in complex sentence comprehension and higher semantic processing and acts as a hub for multisensory integration, especially visual–auditory, allowing the observer to attain pragmatic abilities. While nonhuman primates are able to recognize the actions of others (level 2), they have not attained this fifth step in the evolution of social communication. The ability to communicate in flexible and complex ways that take into account the beliefs and emotional states of others is attained by humans in childhood.

DISORDERS AFFECTING LANGUAGE IN CHILDHOOD, ADOLESCENCE, AND ADULT LIFE

As discussed in section "Language Development," the complex ongoing formation and remodeling of cortical regions and their white matter connections during human development is a prerequisite for normal communicative abilities acquired in childhood, adolescence, and adult life. Alterations to this process during childhood may be responsible for a delay in cognitive development and lead to the emergence of neurodevelopmental disorders such as autism conditions, dyslexia, and other conditions affecting language abilities. In addition, differences in the development of certain anatomical features (eg, bilateral vs. lateralized language pathways) could explain resilience to disorders affecting people in adolescence (eg, schizophrenia) or later life (eg, aphasia).

The hierarchical nature of the SCALED model is supported by studies into these neurodevelopmental disorders. In autism spectrum disorder (ASD) and schizophrenia, for example, language and social abilities are affected at different levels.

Autism Spectrum Condition

ASD is a relatively common neurodevelopmental disorder affecting 1% of children. It is typically diagnosed early in life, with atypical language and social development frequent features and important aspects of a symptom triad: repetitive and stereotypic behavior, impaired language communication, and deficits in social reciprocity. Such a deficit in communication has made the human language network an important target for research in autistic children.

At one end of the spectrum, children with low functioning autism show impairments in the recognition of action and the intentions of others to communicate (levels 1 and 2 in the SCALED model). Such children show disruption in networks involved in Theory of Mind, including those connecting the DMPFC.

There are also deficits in the development of normal linguistic skills (levels 3 and 4). Tractography studies on ASD show altered white matter connections in some tracts of the language network. The volume of the arcuate fasciculus is also significantly less left lateralized in children with ASD compared with neurotypical children. Differences in anatomy of the uncinate fasciculus and inferior longitudinal fasciculus between children with autism spectrum condition and healthy controls have also been reported.

At the other end of spectrum, children with Asperger's syndrome display no language delay but present with impairments in higher communication such as pragmatics (level 5). For example, children with Asperger's are commonly less accurate in detecting the intent behind a speaker's remark, especially in relation to irony, than a typically developing child.

Schizophrenia

Schizophrenia is less common and presents later in development than autism spectrum condition. Symptoms typically occur in late adolescence or early adulthood and include delusions, auditory hallucinations, abnormalities of the emotional sphere (eg, lack of motivation and reduced social interaction) and thought disorder.

A common aspect of these symptoms is impairment to pragmatics and social cognition skills for which the IPL plays a vital role (level 5). Indeed, bilateral reductions in mean fractional anisotropy (a measure that reflects white matter density and myelination) has been reported in the long and posterior segments of the arcuate fasciculus. Such reductions are highest in those patients reporting auditory verbal hallucinations. This abnormal anatomy suggests a higher vulnerability to specific anatomical connections and supports the theory that schizophrenia may be the result of a disconnection between frontal, parietal, and temporal language networks late in development.

Acknowledgment

Marco Catani is the recipient of the Welcome Trust Investigator Award 103759/Z/14/Z.

Further Reading

[1] Brauer J, Anwander A, Perani D, Friederici AD. Dorsal and ventral pathways in language development. Brain Lang 2013;127:289–95.
[2] Catani M, Bambini V. A model for social communication and language evolution and development (SCALED). Curr Opin Neurobiol 2014;28:165–71.
[3] Friederici AD. The cortical language circuit: from auditory perception to sentence comprehension. Trends Cogn Sci 2012;16:262–8.
[4] Frith CD, Frith U. Mechanisms of social cognition. Annu Rev Psychol 2012;63:287–313.
[5] Price CJ. A review and synthesis of the first 20 years of PET and fMRI studies of heard speech, spoken language and reading. Neuroimage 2012;62:816–47.
[6] Thiebaut de Schotten M, Dell'Acqua F, Valabregue R, Catani M. Monkey to human comparative anatomy of the frontal lobe association tracts. Cortex 2012;48:82–96.

CHAPTER

32

Memory

M.L.R. Meister, E.A. Buffalo
University of Washington School of Medicine, Seattle, WA, United States

INTRODUCTION

Multiple memory systems exist within the brain. The brain structures needed to remember how to ride a bicycle, for example, are distinct from the ones needed to remember the story of the first time riding a bicycle. This chapter will review these separate types of memory and their corresponding neural basis.

An important caveat to keep in mind about separate memory systems is that "separate" within the brain is a reductionist term. Because the brain is a single organ within which these regions are connected, separate memory systems interact to function in concert during daily life. For example, what a person is currently attending to with their "working memory" system may be encoded strongly in the brain because of an intense emotional drive from the "emotional memory system" to forge a memory. Additionally, while sensory and motor areas of the brain are not addressed here, their activity is certainly linked to memory. For example, sensory input from visual areas may initially drive, as the neural substrate of visual experience, separate memory systems for the creation of a memory trace distributed in distinct memory systems and even within the visual cortex itself. Similarly, activity in visual areas can stimulate recall by activating distributed neural ensembles across the brain and then constitute the perceptual experience of the visual aspect of that memory during recall. In this way, noticing a periwinkle color may activate the visual cortex so that it cues a memory of previously seeing a view of periwinkle-colored mountains. In sum, although separate memory systems are introduced here, along with experiments aimed at isolating the function of separate brain areas, coordinated activity is the norm in a working brain during daily life.

The significance of memory is enormous, as it is arguably memory which chiefly defines who we are as individuals, allowing our individual histories to guide our current actions and beliefs. Disorders that affect memory such as Alzheimer's disease are often terrifying in that memory loss is accompanied by loss of one's sense of self through remembered experiences and loss of meaning of the world. As ongoing research probes these memory systems, there is growing potential that new findings will advance our ability to treat memory impairments suffered by people with conditions such as Alzheimer's disease, traumatic brain injury, and posttraumatic stress disorder (PTSD).

Before describing the separate memory systems, the first section of this chapter will review a fundamental mechanism of memory, in which experience alters the structure of the brain.

Conn's Translational Neuroscience
http://dx.doi.org/10.1016/B978-0-12-802381-5.00050-6

SYNAPTIC PLASTICITY

In order for experience to become a memory, it must cause a physical change in the brain. The connections between neurons in the brain, called synapses, can be structurally modified by experience. Experience, which equates to patterns of neuronal activity, can cause new synapses to appear and old synapses to disappear and can also change a synapse's structure so that the connection becomes a stronger or a weaker connection. These synaptic changes, known as *synaptic plasticity*, are a physical manifestation of memory in the brain.

Repetitive electrical stimulation in brain tissue slides can produce stronger (or weaker) connections between neurons. This technique allows for the study of how the stimulation triggers molecular-signaling pathways that lead to altering synaptic structure and hence altered synaptic function. This electrical stimulation used in the laboratory is an artificial substitute for the stimulation neurons naturally receive within a living organism. Within a living organism, stimulation from the environment is turned into action potentials by peripheral sensory neurons and transmitted to the brain. Within the brain, the pattern of activity may be strong enough to produce synaptic change and thereby be encoded as a memory. A general rule is that simultaneously active neurons achieve the conditions that strengthen the synaptic connection between them through changes in each neuron, captured by the maxim that neurons which "fire together, wire together."

Activity-induced synaptic changes show a wide range of stability and temporal dynamics, which appear to correspond to the range of memory shown by behavior. Activation can trigger a molecular cascade that changes the strength of a synapse for a only few minutes, or more intense activation could set off a molecular cascade that, due to new protein synthesis, takes hours or days to strengthen a synapse. This stronger activation may even cause entirely new synapses to grow that can be observed weeks later. Similarly, the range of time laboratory animals retain a memory can reflect this time scale. Most tellingly, when the molecular cascades causing these synaptic structural changes are blocked within an organism, memory is also blocked.

A change in the strength of a synapse between two neurons translates to what neural activity will happen the next time the presynaptic neuron is activated. For example, consider the case in which the activity triggers a cascade that causes the number of receptor proteins on the surface of the postsynaptic neuron to increase (Fig. 32.1). Due to this change in number of receptors, the release of neurotransmitters into the synapse from the presynaptic cell will cause a bigger electrical change on the postsynaptic cell than they previously did. In other words, the activity of the presynaptic neuron is now more coupled to the activity of the postsynaptic neuron because the synapse changed structure.

If we extrapolate from our understanding of changing synaptic strength to the resulting change in multi-neuronal activity patterns, then one can appreciate how synaptic strength dictates which neurons will be activated together during memory recall because of their strong connections. The physical manifestation of separate memories is generally understood as activation of unique groups of neurons, or *neural ensembles*. The strong synaptic connections between neurons of an ensemble cause those neurons to be activated together. Theoretically, this means that if only some of the neurons of a neural ensemble are stimulated, then the rest of the ensemble will also be stimulated due to strong connections. This theoretical phenomenon, which could allow a single cue to activate an entire memory because of a connected ensemble, is called *pattern completion*. For example, the smell of coffee could be a single cue that evokes a vivid, detailed memory of the experience of walking into a coffee shop earlier that day.

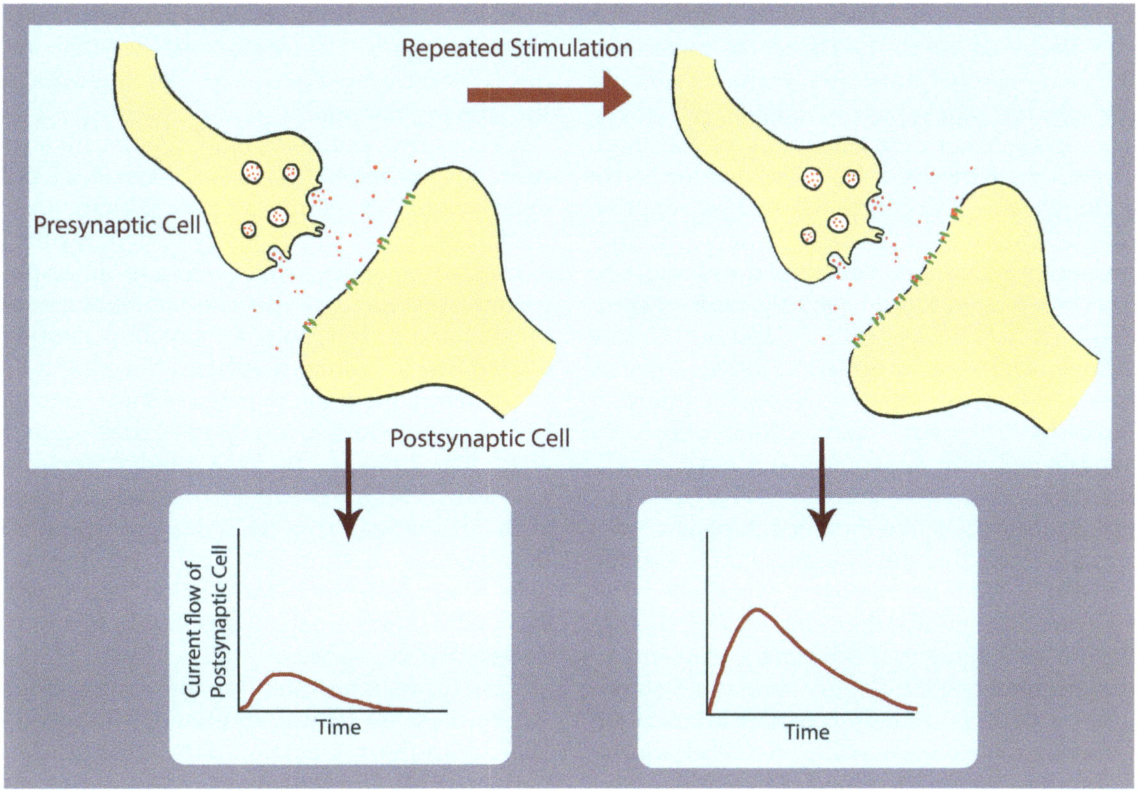

FIGURE 32.1 Synaptic plasticity is observed when repeated stimulation causes a change in synaptic structure that makes a postsynaptic cell respond differently to the presynaptic cell. The left panel shows the initial structure of the synapse, where neurotransmitters (orange) are released into the synapse to stimulate receptors (green) on the postsynaptic cell. The graph below shows the amplitude of current flow observed in the postsynaptic cell in response to this release of neurotransmitter. After repeated stimulation causes a change in the number of receptor proteins on the postsynaptic cell (right top panel), then the same size stimulation causes a larger current flow (right bottom panel) in the postsynaptic cell in response to the release of neurotransmitters. *Figure by S. Schleufer.*

Upon recall of a memory, a subset of the neurons that were originally active during the formation of that memory are again activated. If synaptic connections between neurons representing a memory were to change, then that would potentially modify the memory. For example, synaptic pruning may weaken connections so that part of the neural ensemble is not activated upon recall with the rest of the ensemble. Consequently, the portion of the memory represented by that missing ensemble activity is not experienced during recall.

The strength of a memory does not necessarily entail a larger number of strong connections between neurons, but instead *specific* strong connections that represent a certain memory. At the cellular level, a strong memory rehearsed over and over again would not be expected to correspond to a specific neural ensemble with dramatic new growths that continue to occur, but rather one where only the appropriate synapses are strengthened while irrelevant synapses are pruned.

The human brain is estimated to have hundreds of trillions of synapses, and so the capacity to encode unique information with a unique set of strongly connected neurons is by no means small. However, this does not mean that the brain keeps an everlasting store of memories. Like their neuronal substrates, memories are not only lost, but also subject to change, which can create an altered or even false memory. Although it is perhaps alarming to hear that memories can be altered, this also can be viewed as adaptive because it allows integration of new information, and useful reorganization of the memory so it serves a person better when it is next recalled. Forgetting is also arguably adaptive, because it allows irrelevant information to leave neural resources that can now store useful information instead. If neural connections were only strengthened with experience and never weakened or lost, such a system might challenge this important flexibility of memory. A person with eternally strengthening synapses could lose the ability to (1) focus on important aspects of memory instead of unimportant details, (2) advance wisdom by linking and separating elements of different memories, (3) apply new understanding to past events, and (4) store new information in memory. The human ability to forget, diminish, or alter memory is especially clear as adaptive for those who suffer a traumatic event and can avoid being debilitated by a strong, unfading memory for that traumatic experience.

DECLARATIVE MEMORY

The memory for facts and events, or in other words, memory for which we can verbally declare its contents, is referred to as *declarative memory* or *explicit memory*. This type of memory encompasses both memory for episodes that occur in one's life, known as *episodic memory*, and memory for facts, called *semantic memory*. Episodic memory enables a person to tell a personal story about past personal events, including what one had for breakfast, whether one stopped at the post office this morning, as well as memories of important life events like a marriage or the birth of a child. Semantic memory includes memory for information such as "Madrid is the capital of Spain," "Lincoln was the 16th president of the United States," or "the Blue whale is the largest cetacean."

Declarative memory relies on brain structures within medial temporal lobe (Fig. 32.2), most notably the hippocampus. Damage to the medial temporal lobe memory system results in *anterograde amnesia*, which refers to an impairment in forming new declarative memories. In most cases, individuals with medial temporal lobe damage retain memories formed many years before the brain damage but are not capable of remembering any events that occurred after the damage. Because medial temporal lobe structures are the first to degrade in people with Alzheimer's disease, anterograde amnesia develops as its most prominent, early symptom. Diagnostic tests for Alzheimer's disease especially target a patient's ability to form new, declarative memories.

The appreciation that medial temporal lobe structures were critical for memory occurred in 1957, with the publication of the case study of patient H.M. Due to severely debilitating epileptic seizures, the medical decision was made to perform a surgery to remove medial temporal lobe structures in both hemispheres in an attempt to destroy the brain area producing the seizures. Although the surgery was successful in alleviating the seizures, it left H.M. with a profound and persistent memory impairment. He was unable to form any new declarative memories. For example, Dr. Brenda Milner, who first met H.M. after his surgery and then studied his behavior for many years thereafter, needed to reintroduce herself to him every single time they met. The experiments Dr. Milner conducted on H.M. marked the beginning of the modern era of the neuropsychology of memory and led to several key insights about the neurobiology of memory. First, memory was dissociable from other cognitive functions, including intelligence and attention. H.M. still could converse intelligently and

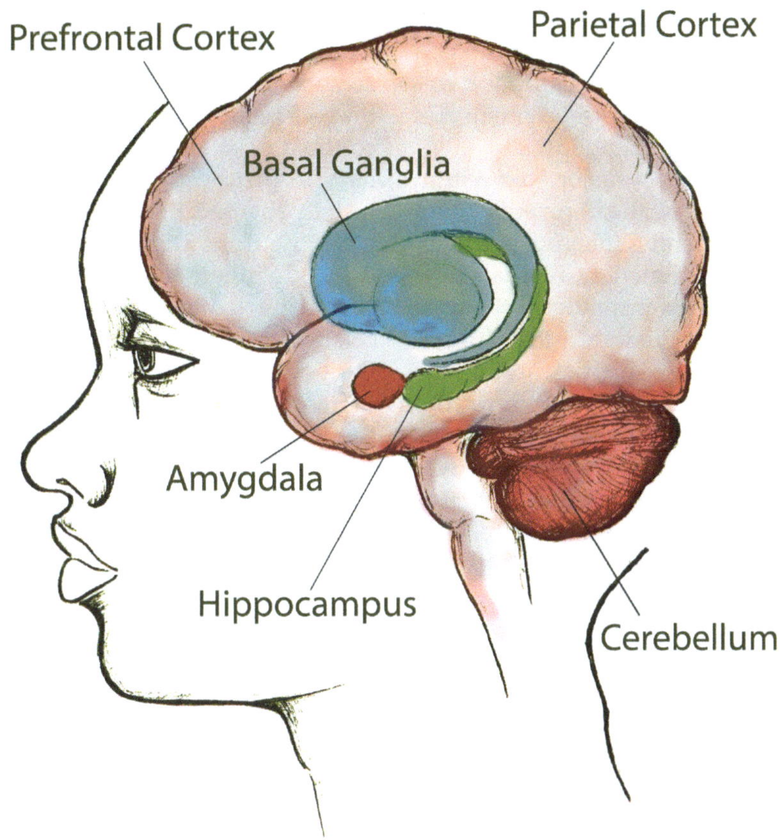

FIGURE 32.2 Key brain structures for separate memory systems are labeled. *Figure by S. Schleufer.*

even remember things for as long as 15 min if he continuously rehearsed the memoranda and maintained his focus upon them. Second, not all forms of memory were impaired following medial temporal lobe damage. For example, H.M. demonstrated normal learning in forms of perceptual and motor skills: he showed clearly improved performance through experience with the difficult perceptual task of mirror-drawing star-shaped figures, while he had no explicit memory of ever having done the task before.

Impaired memory for events that occurred before the brain damage, a condition called *retrograde amnesia,* also occurs after damage to the hippocampal memory system, but to a lesser extent because of the *consolidation* of declarative memories. The process of consolidation entails a transfer of the memory from the hippocampus to the cortical areas distributed throughout the brain. It is thought the hippocampus gradually makes a memory independent of the hippocampus (achieving consolidation) by serving as a "hub" that reactivates, upon recall, the disparate cortical areas that subserve the distinct perceptual aspects memory. This simultaneous reactivation causes an increase in the synaptic strength between the ensemble of reactivated cortical neurons so that eventually, triggering the memory's cortical ensemble can occur completely through activation from a part of that ensemble, thus bypassing the need for the hippocampus's coactivation. Regardless of whether this is the exact mechanism behind consolidation, it is a firmly established fact that hippocampal damage impairs recall for recently formed memories that are still hippocampal dependent.

Synaptic plasticity plays a major role within the hippocampus in ways that are still being investigated. One overarching theoretical framework for hippocampal function is that new memories must be encoded instantly, at the time the event occurs, which requires fast formation of strong, but unique synaptic connections to create a unique neural memory trace. The process of creating and activating a memory's unique neural pattern without overlap activation of other memory traces is referred to as *pattern separation*. Conversely, in order to recall a memory from a partial cue, the neural substrate of the memory needs to be capable of re-creating the same, original ensemble activation from only partial activation, referred to as *pattern completion*. These two types of hypothetical activation within the hippocampal circuitry impose a complexity for synaptic weighting. It means that the hippocampus needs simultaneously to allow the creation of unique ensemble activity and the re-creation of old ensemble activity. Currently, different regions within the hippocampus with distinct wiring patterns are thought to perform these opposite processes. Such a wiring scheme can allow for positive feedback loops in which a neuron that fires once would, as a consequence of that initial firing, be further activated. This type of self-exciting activity in a hippocampal wiring scheme is thought to be supportive of pattern completion and is also considered to be one reason why this brain area is commonly the source of seizures.

The activity of individual hippocampal cells also reflects the content of declarative memory. One well-known example of this is "the Jennifer Anniston neuron," named this way by the popular media because scientists recording the activity of medial temporal lobe neurons in an awake, human surgical patient identified a neuron that fired an action potential every time a picture of celebrity Jennifer Anniston was presented. Unlike prototypical sensory neurons that respond to image features, such as local contrast or shape, this neuron responded similarly to pictures with very different image features, provided they were of Jennifer Anniston. Other neurons with different preferred images were identified (Fig. 32.3), with some neurons preferring distinct individuals, including both celebrities and non-celebrities, as well as entities beyond individuals, like a neuron that fired in response to a specific television show (eg, the Simpsons neuron). Referred to as "concept neurons," these types of cells are considered to reflect a high level of cognitive processing and can even activate before the subject utters the concept in a free recall task (eg, answering the question "What did you see?"). However, the name "concept-neuron" is a somewhat misleading title because any given stimulus will activate an ensemble of specific cells, rather than a single cell alone. In other words, specific ensembles of neurons within the medial temporal lobe memory system selectively signal memory recall and formation.

A vigorous area of current research focuses on the spatial representation of the environment by neurons in the medial temporal lobe memory system. Most well known are *place cells*, which are found predominantly in the hippocampus and fire selectively for a certain location as a rodent locomotes within its environment (Fig. 32.4A). *Grid cells* found predominantly in the entorhinal cortex, just adjacent to the hippocampus, fire selectively for periodic locations that form a grid pattern if one is looking down on the environment from a bird's-eye view (Fig. 32.4B). Drs. John O'Keefe, May-Britt Moser, and Edvard Moser were jointly awarded the 2014 Nobel Prize in Physiology or Medicine for the discovery of these cells, hypothesized to be "the brain's GPS." It is still unknown exactly how these spatial signals might be used by the medial temporal

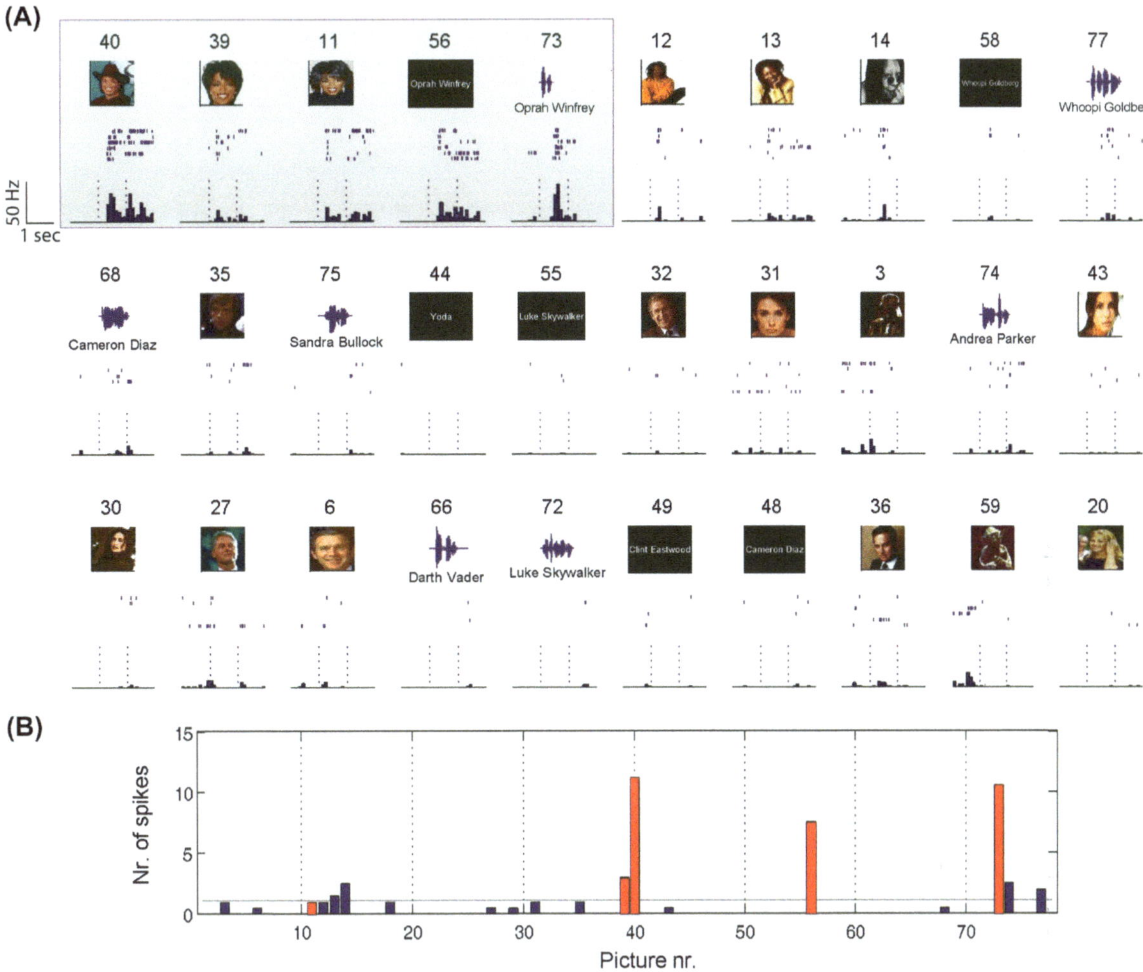

FIGURE 32.3 A hippocampal "concept neuron" selective for Oprah Winfrey fires in response to different pictures of her, her name written on screen, and her name being spoken. Neurosurgery for epilepsy allowed researchers to record the activity of this hippocampal neuron from the brain of a patient, while the patient was presented with the pictures of various people. The written and spoken names of people were also used as stimuli. (A) Examples of different stimuli are shown along with the neuron's response to each stimulus. The top left box highlights the heightened neural response for Oprah Winfrey stimuli. Stimulus number is indicated on top, and the neural response to multiple trials of a particular stimulus are shown below each stimulus. The neural response is shown in two ways below each stimulus. The *dashed vertical lines* indicate the start and end of a 1-s visual presentation of a stimulus, unless it was an auditory stimulus, which varied in length. Immediately below each stimulus, action potentials are indicated as *blue tick marks* across time for multiple trials (each row is a trial). The second way the neural response is shown, in the bottom plot under each stimulus, the height of a bar indicates the average firing rate for the neuron in that time point across multiple trials. The rate can be estimated from the legend, which shows the height for a rate of 50 times per second (50 Hz). (B) Each bar represents the response of the neuron to a stimulus. The stimulus number is indicated on the horizontal axis and the height of the bar indicates the median number of spikes across trials for that stimulus. Oprah Winfrey stimuli are colored red. *Figure from Quian Quiroga R, Kraskov A, Koch C, Fried I. Explicit encoding of multimodal percepts by single neurons in the human brain. Curr Biol 2009;19:1308–313.*

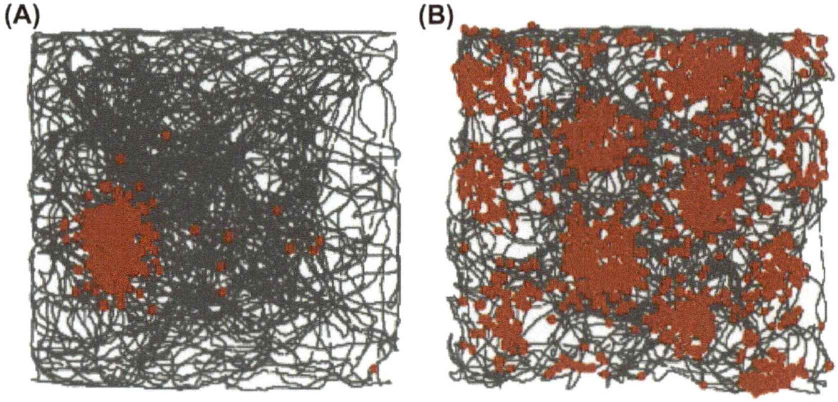

FIGURE 32.4 Neurons in the medial temporal lobe memory system exhibit spatially specific firing. From the perspective of someone looking down into an enclosure containing a rat, the *gray line* indicates the path of the rat, and each *red dot* indicates the location of the rat when the neuron fired an action potential. (A) A place cell in the hippocampus fires action potentials only when the rat is in one particular area of space. (B) A grid cell in the entorhinal cortex, which is intensively connected with the hippocampus, fires when the rat is located within multiple, periodic locations that form a grid pattern. *Figure from Derdikman D, Moser E. A manifold of spatial maps in the brain. Trends Cog Sci 2010;14:561–69.*

lobe memory system. One prominent theory suggests that these representations provide the spatial dimension of a multidimensional memory, which is consistent with demonstrations of neural representations in this region of not only space but also time, providing another dynamic component of memory for an episode. How these extremely intriguing neural representations of space relate to memory in the hippocampal memory system is a topic of ongoing research.

EMOTIONAL MEMORY

The powerful link between emotion and memory becomes intuitive when considering one's strongest memories, which are most likely the most emotional memories. The impact emotions have on memory is clearly advantageous for survival because the emotional aspect of a memory allows us to attach personal relevance to experiences. For example, if a person had a great weekend trip with a friend, the emotion attached to that experience would probably power the formation of a lasting memory, and also can guide behavior to readily accept if the friend ever extended another trip invitation. Also, the emotional quality of the memory may cause happiness upon seeing that friend, and approach behavior to that social partner. In this way, the great trip with the friend is more emotionally charged and important for guiding future behavior than a nonemotionally charged commute trip that is quickly forgotten. Processing within the amygdala is important for this emotional guidance of learning and memory. Without the amygdala, rewarding or punishing stimuli are not selectively highlighted in memory and thus not subsequently recognized for their emotional significance. Anatomical structure reflective of this role can be found in the large number of connections between the amygdala and the nearby hippocampus, an area critical for creating episodic memories.

Although the amygdala plays a major role in a range of emotions, its role in fear learning has been most intensively investigated through classical conditioning experiments in rodents. In these experiments, an aversive stimulus, such as a foot shock, is paired with a neutral

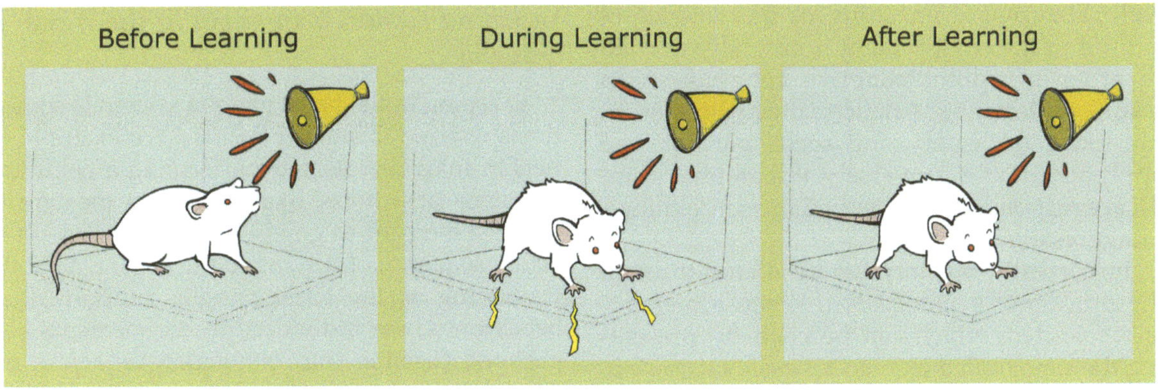

FIGURE 32.5 A rat learns to show fear to the sound of a tone. Initially, the rat has no reaction to the sound of a tone (left). Then, after repeated pairing of the tone with a mild electric shock to the feet delivered through the floor (middle), the rat learns to show the fear response of freezing to the sound of the tone even without the delivery of the shock (right). This learning depends on the amygdala. *Figure by S. Schleufer.*

stimulus, such as a tone (Fig. 32.5). Repeated pairing produces fear behavior (freezing) when presented with the tone in future trials. If the amygdala is lesioned, rodents do not learn to associate the aversive stimulus with the tone and do not demonstrate fearful behavior at the sound of the tone. Neural activity within the amygdala also tracks this fear learning, with neuronal firing reflecting both acquisition of the learned tone–shock association and extinction (which occurs when the tone is repeatedly presented without a shock). During acquisition, neurons begin to fire at the start of the tone that predicts the shock, and conversely, cease this activity during extinction when the rat has learned to no longer freeze to the tone. Unsurprisingly, synaptic plasticity has also been shown to occur within the amygdala during this learning.

With available technology that allows for more selective manipulation of brain anatomy, more delicate experimental work probing amygdala function has exposed substructural function.

Activation or inactivation of specific neuronal pathways between certain areas of the amygdala is feasible, and such methods can reveal more precisely how the different parts of amygdala circuitry contribute to specific emotional behavior, like the scaling up or down of anxious behavior. Stimulation of appropriate input pathways also allows researchers to substitute stimulation for real-world stimuli like the tone or the shock. Because such stimulation can lead to similar learned associations as evident through freezing behavior, the experimentalists arguably accomplished artificial implantation of a fear memory. Additionally, inhibiting or exciting only certain cell types can prevent fear learning, or only alleviate behavioral expression of it. Finally, selectively abolishing only the neurons that were active immediately after training effectively abolishes fear learning, suggesting that those specific neurons constituted the memory trace.

Neurons in the amygdala are valence selective. This means that neuronal-firing activity represents the valence of the outcome (eg, reward or punishment) with which a cue is paired and not the sensory features of the cue itself. Neurons that fire in response to a fear cue will not respond to a reward cue, and vice versa. For example, if the sensory cues for reward and punishment were switched, then through learning, amygdala neurons would switch the sensory cue they responded to as well. This neural encoding of the valence of stimuli in the environment is critically needed in order to exhibit adaptive behavior. As a result of learning valence, rewarding stimuli, like food, safety, and social partners are approached, and conversely, threatening stimuli are responded to with vigilance, avoidance, or aggression.

Importantly, memory can be altered through the process of *reconsolidation*. When a memory is recollected, which can be cued by presenting the tone from a learned tone–shock association, the memory is labile and can be altered by concurrent stimuli at the time of recollection. In both rodents and humans, drugs that block protein synthesis administered during this time-window interfere with the memory so that it is not reconsolidated. This time-window for interfering with memory has been exploited therapeutically with patients who suffer from PTSD who repeatedly relive a traumatic event because of a strong, debilitating fear memory. In therapy, the traumatic memory is recalled to render it vulnerable to alteration, and then, it is hopefully remodeled by the current, safe, therapeutic environment in which it is recalled, thus diminishing its original, debilitating, emotional quality.

MOTOR MEMORY

Another type of memory is memory for how to perform actions, referred to in popular culture as "muscle memory." This type of memory, along with other types of memory that are not declarative memories for facts or events, is *implicit memory*. Different types of implicit motor memory are associated with two different brain areas: the basal ganglia and the cerebellum. The *basal ganglia* exert control over learning of action sequences so that even a complex set of action steps can become automated, and the *cerebellum* contributes to using error signals to learn appropriate fine movement adjustments.

Learning Action Sequences in the Basal Ganglia

Learning in the basal ganglia system is sometimes referred to as *habit learning*. For example, a person may find herself performing a complex sequence of actions, like driving to the office, when she meant to go elsewhere—because that action sequence of driving to the office is a habit. Processing in the basal ganglia supports this behavior, where individual motor elements are grouped together into one action sequence, or "chunk," that is then performed with decreasing variation as learning progresses. Another widely understood example of this is when a person first learns to write, each stroke of each letter requires a conscious effort. After practice and chunking, the action sequence becomes stable, and a person can quickly perform the sequence of strokes that correctly constitutes a letter effortlessly, with little variation. As individual action steps become chunked into action sequences, the time between each step of the action sequence tends to decrease, and the action sequence becomes more stable and stereotyped. At the same time this occurs, basal ganglia neurons begin to fire action potentials at the beginning and ends of the entire action sequence, corresponding to the chunk boundaries observed in behavior. Also, as one would expect with a honing of synaptic connections, the variability of when neurons fire relative to each other decreases during learning in both the basal ganglia and the motor cortex, and neurons within the basal ganglia show an increase in synaptic efficacy. If proteins needed for synaptic plasticity are blocked from acting in the basal ganglia neurons that project to other brain structures, then behavioral evidence of chunking action sequences does not occur. In other words, synaptic plasticity between the basal ganglia and the cortex is critical for action-sequence learning.

Learned action sequences become similar to the innate "hard-wired" behaviors, we are born with due to our genetic memory, which dictated our initial neural connections (eg, the neural connections critical for breathing). After basal-ganglia-assisted learning has remodeled the synaptic connections between neurons, an action sequence can be produced much like a reflex in response to a stimulus, as a highly reproducible action below the level of conscious awareness. As a result of this learning, an author can focus on the content of the writing more than the complex action sequence of writing itself.

One way to conceptualize the neural basis of action sequence learning supported by basal ganglia is that it allows us to develop "habits." Habits are extremely important to survival because they greatly improve the ease with which we navigate our environments by making behaviors that we need to perform accurate, stable, and less thought-consuming. People with Parkinson's disease, which results from the degeneration of certain cells within the basal ganglia, do not enjoy such ease of action performance, and are impaired at initiating and terminating voluntary action sequences.

At first, however, to learn how to do something might require trying it different ways, with lots of variability. Interestingly, the basal ganglia is also important for producing this salutary diversity in action or exploratory behavior; a bird learning to mimic a tutor's song first needs the basal ganglia for injecting variability into the song. Until the right sequence of notes is achieved, this injected variability is important for improving performance. If a major basal ganglion output region was lesioned during learning, then the song would "freeze" in its imperfect state, and never complete the learning process.

Adjusting Movement in the Cerebellum

Individuals with cerebellar dysfunction show impairments in learning from movement errors. For example, people with cerebellar damage might over- or undershoot when reaching for a target despite experience in a particular reach task. Specifically, it is thought that the cerebellum performs a learning process, referred to as *adaptation*, in which input error signals are used to calibrate movements by making fine adjustments to movement amplitude and timing.

This type of learning manifests clearly in a much-studied experimental paradigm of eyelid response conditioning, in which an animal learns a very specifically timed response to a sensory stimulus (Fig. 32.6). In this paradigm, an animal is conditioned over repeated trials to associate a tone or a light with an aversive puff of air to the eye that, when delivered alone, causes the animal to close its eyelid. Eventually, the eyelid closure becomes a learned, conditioned response, in which the animal closes its eyelid in response to the tone alone, and does so just the moment in time before the air puff occurs. The cerebellum is necessary in order for this learning to occur; lesions or reversible inactivation of the cerebellum ipsilateral to the conditioned eye block this form of learning.

Within the cerebellum, the cellular circuitry has also been probed during this paradigm to reveal alterations that cooccur with the behavioral expression of learning. Such alterations include changes in synaptic structure. Synapses of various types have been documented as increasing or decreasing in number, increasing in size, and decreasing in number of receptor proteins. Drugs that block protein synthesis, which is necessary for synaptic remodeling, also impair eyelid response learning. These findings suggest that cerebellar motor learning is accomplished through the calibrating of synaptic weights in order to output the appropriate timing and amplitude of movements. Additionally, cerebellar activity reflecting the learning emerges over the course of training; cerebellar neurons change their activity and increase firing in response to the conditioned stimulus (the tone or light) at the same time as behavioral evidence of motor learning emerges.

Although the cerebellum's role in motor memory is by far the best understood, evidence

FIGURE 32.6 A rabbit learns to close its eye in response to the sound of a tone. On every trial, a tone is paired with a puff of air delivered to the eye at the end of the tone. Before learning (top), the rabbit closes its eye after the puff of air is delivered. After repeated trials (bottom), the rabbit learns to close its eye in response to the tone alone, before the puff of air is delivered. This learning requires the cerebellum. *Figure by S. Schleufer.*

has also accumulated suggesting that the cerebellum is perhaps involved in cognitive and emotional perceptual processing. Because the cellular architecture of the cerebellum is highly repetitive, cerebellar tissue is consequently hypothesized to perform similar neural processing on the various inputs it receives across its expanse, including inputs from motor areas as well as areas involved in higher order cognition or emotional processing. This suggests the advances made in understanding the cerebellar processes in the context of motor memory can translate in the near future to nonmotor cerebellar domains.

WORKING MEMORY

Working memory sustains the content of immediate conscious awareness and is thought to be primarily dependent upon the integrity of the prefrontal and posterior parietal cortex (PPC).

It is different from the other types of memory discussed so far in this chapter foremost because it is not known to necessarily translate to synaptic plasticity. Likely related directly to this point, working memory has a very limited capacity over time on the scale of seconds, as well as a limited capacity on load size equivalent to attention. The capacity limits of this type of "online" memory, where one can manipulate information, can be understood as the inability to consider the optimal seating arrangement (without writing it down) for a dinner party of 15 people. Because the load size limit for working memory is in the range of four to seven items, 15 far exceeds that working-memory limit. Another example of working memory is the ability to contribute to a conversation by knowing what was just said, by combining each previously uttered word with the next one in the sequence in order to comprehend meaning of a sentence, and likewise combine the meaning of a sentence with the meaning of previously uttered sentences, and then considering how to respond to it even though a conversational partner's speech finished several seconds ago. Working memory is flexible in that it allows manipulation of its contents, like in the dinner party seating example, and also as a general property of "thought" itself; the process of thinking is one in which a person can access relevant pieces of sensory information and pieces of information from long term memory, and evaluate them together in order to achieve a current goal, like figuring out the answer to "Should I buy this book?"

The content of working memory is fleeting unless it's kept in working memory by sustaining focus on it (like repeating a phone number to oneself), or it is transformed into a more lasting form of memory (requiring some form of synaptic plasticity). The transformation of information from working memory into a more lasting form is facilitated by processing the information by using it (like evaluating whether one agrees with it, or telling someone else about it). Such "tricks" are common in the effective teaching strategy of "active learning" that rebuffs passive lecture formats and instead requires students to use introduced course material within the class as frequently as possible by, for example, routine discussions with a partner to answer posed questions. Such elaborative processing diminishes the potential for losing introduced information from working memory on the route implied by the expression "in one ear and out the other."

Working memory limits can be measured quantitatively in experiments that test an individual's time, load, and accessibility capacity for memoranda. A common format of such tests is to present people with several nonlogically connected items to remember (eg, several nonsense strings of letters like "czq" and "pfa") and then prevent them from rehearsing the information, usually by asking them to perform a distracting cognitive task like a performing a math problem. Then, the number of items that they recall is used as a quantity reflecting load limit of working memory. The time limit of working memory is evident in the exponentially decreasing number of items recalled as the time gap grows (on the scale of seconds) between the presentation of the memoranda and the recall test. The accessibility limit of working memory can be gauged by the reaction time on the scale of hundreds of milliseconds for answering a question on the contents of working memory. This metric of reaction time grows (indicating decreased accessibility) along with the size of the items held in working memory.

The neural substrate of this working memory in the brain is thought to be located in the prefrontal cortex (PFC) and PPC. Human fMRI studies show that the level of activation in these two brain areas correlates with the quantitative properties of an individual's working memory capacity. Specifically, the activation of these brain regions increases during the imposed time delay between presentation of memoranda and a recall test. Their activity also scales up with increasing number of memoranda, the length of time of the delay, and the accuracy of recall performance.

The massive flow of sensory information coming at us through our sensory systems cannot be simultaneously processed with the limited size of our working memory; instead, working memory content is guided by attentional mechanisms. It is clear from experimental work that attention is directed at a prioritized portion of the environment, while distracting information is filtered out. For example, when looking for someone in a crowd, eye movements are chiefly directed to the places in the visual field where a person is located, and target people with specific features like a certain height or hair color. Neural activity in PPC and PFC appear to correlate with the focus of attention. The PPC is hypothesized to be a "priority map" for visual space because PPC neurons that individually represent different parts of the visual field fire more when attention is deployed to a visual location within a neuron's *response field*. For example, a flashed stimulus draws attention, and causes a PPC (as well as PFC) neuron with a response field at the location of the flashed stimulus to increase its activity. Similarly, in a visual search task in which a monkey is rewarded for visually fixating an upside down "T" in an array of similar-looking distractors, a PPC neuron will fire more when the target falls into the neuron's response field. In general, whenever attention is deployed to a PPC neuron's response field, the neuron will increase its firing. This neural activity therefore echoes the spatial map of attention for salient objects in the visual environment of the subject, a seemingly useful neural representation for guiding behavior.

Most compellingly, neurons in the PPC and PFC show *persistent memory activity* for the duration of attentional deployment to a neuron's response field (Fig. 32.7). Although many visual cortical neurons show heightened firing rate for spatial deployment of attention to the neuron's response field and can to some extent also represent an attentional map, the PPC and PFC are unique. The PPC and PFC putatively reflect working memory because neurons within those structures exhibit persistent activity, even when a remembered stimulus is visually absent. In a commonly studied paradigm for evoking this activity, a trained monkey looks at a central fixation point on a computer monitor until it disappears. During the time, the monkey is looking at the central fixation point, a stimulus is flashed somewhere on the screen. The disappearance of the fixation point cues the monkey to make an eye movement to the remembered location of the flashed visual stimulus in order to receive a juice reward (Fig. 32.7). By varying the delay between the flash of the visual stimulus and the "go signal" (disappearance of the central fixation point), it becomes apparent that neurons in the PPC and PFC show sustained, heightened firing for the duration of the memory delay period, corresponding precisely to the duration of the working memory of the stimulus. In addition, the neurons are spatially selective so that this activity only occurs for a memory delay when the monkey is planning to make an eye movement to a location within the neuron's response field. This persistent activity of neurons firing during the time period in which the target location must be "kept in mind" suggests that this activity is the neural correlate of the remembered stimulus.

Causal evidence for a behavioral role of persistent activity in PFC and PPC is found in the behavior of experimental animals with lesions in these areas and patients with damage to these regions. PFC lesions prevent animals from learning a conditioned eyelid response to a tone (Fig. 32.6), if there is a time gap between the end of the tone and delivery of the air puff to the eye. This is because the PFC lesion prevents transmission of a persistent activity

FIGURE 32.7 A neuron in the parietal cortex of a monkey shows persistent memory activity in a task where working memory is used to direct an eye movement. On each trial of this task, a trained monkey looks at a central fixation point displayed on a screen until it disappears, which is his cue to make an eye movement in order to receive a rewarding drop of juice (not shown). At the beginning of each trial (left panel), the monkey looks at the central fixation point. The *dotted circle* indicates the location of the response field of the neuron whose activity is being observed. While the monkey is looking at the central fixation point, a target is briefly flashed in a peripheral location. The monkey continues to look at the central fixation point for a few seconds until the fixation point vanishes. The monkey then looks at the location where the target was flashed earlier. The monkey receives a juice reward for looking at the correct location. A parietal neuron shows sustained activity for the entire duration of the delay in which the monkey is remembering the location of the flashed target, suggesting that its activity represents the monkey's working memory. Not shown are examples of trials where the remembered target is presented outside of the neuron's response field, when the neuron would not be active. *Figure by S. Schleufer.*

signal from the PFC to the cerebellum which bridges the temporal gap in animals and supports successful learning of a conditioned response despite the time gap. In primates, lesions, and inactivations of PFC and PPC impair performance on attention-demanding tasks, whereas microstimulation can enhance performance. In humans, patients with damage in the PFC and PPC display attentional deficits. One interesting type of attentional deficit shown in humans with PPC damage is spatial neglect, where despite having completely intact sensory cortices, everything on the side contralateral to the brain damage is ignored. Most famously representative of this impairment, spatial neglect patients will draw a clock by drawing a semicircle with only half the numbers of the clock face, and report that everything looks correct. Another intriguing example more directly illustrative of this attentional deficit within working memory but *from* a longer-lasting memory is from reports from Italian spatial neglect patients. The patients were asked to imagine they were standing in a certain location of a famous square in Milan and report what they saw. The patients reported items on only one side of the square, corresponding to the visual field processed by noninjured hemisphere. When asked to again report what they see, but to do so imagining that they were standing in a location directly across the square from the original location, the previously ignored half became the reported half, and vice versa. The spatial attention limits within "the mind's eye" of their conscious awareness thus reflected the damage within the PPC.

Although the mechanism that produces working memory is not completely understood, one prevalent theory is that the PFC and PPC sustain activity for a stimulus by sending reverberatory activity through the reciprocal connections between the two brain structures. Progress on understanding how working memory arises has potential for informing treatment of attentional disorders where the attentional spotlight seems degraded, such as schizophrenia.

Further Reading

[1] Bailey C, Kandel E, Harris K. Structural components of synaptic plasticity and memory. Cold Spring Harb Perspect Biol 2015;7:a021758.

[2] Freeman J. Cerebellar learning mechanisms. Brain Res 2015;1621:260–9.

[3] Ikkai A, Curtis C. Common neural mechanisms supporting spatial working memory, attention and motor intention. Neuropsychologia 2011;29:1428–34.

[4] Janak P, Kay T. From circuits to behaviour in the amygdala. Nature 2015;517:284–92.

[5] Jin X, Costa R. Shaping action sequences in basal ganglia circuits. Curr Opin Neurobiol 2015;33:188–96.

[6] Manns J, Buffalo E. Learning and memory: brain systems. In: Squire L, Darwin B, Bloom F, Du Lac S, Ghosh A, Spitzer N, editors. Fundamental neuroscience. 4th ed. Waltham (MA): Academic Press; 2012. p. 1029–51.

[7] Scoville WB, Milner B. Loss of recent memory after bilateral hippocampal lesions. J Neurol Neurosurg Psychiatry 1957;20:11–21.

[8] Smith E, Kosslyn S. Working memory. In: Cognitive psychology: mind and brain. 1st ed. Upper Saddle River (NJ): Pearson; 2007. p. 239–79.

[9] Therrien A, Bastian A. Cerebellar damage impairs internal predictions for sensory and motor function. Curr Opin Neurobiol 2014;33:127–33.

CHAPTER

32.1

Translational Correlation: Alzheimer's Disease

C.-K. Wu

University of California Irvine, School of Medicine, Orange, CA, United States

INTRODUCTION

In order to understand higher cortical functions of the human brain, behavioral neurologists and cognitive neuroscientists employ several approaches. The animal models only provide a reductive method of analyzing cortical functions at their simplistic levels. Historically, observations on patients with specific brain lesions show the evidence that the human intellect attributes, including attention, memory, language, motor action, and higher cortical perceptions are localized in cerebral cortices. The modern neuroimaging tools, such as functional MRI, enable researchers to formulate plausible cognitive processes of individual higher cortical functions. Often a brain disease can provide a unique window that fascinates clinicians as well as researchers in their views of how higher cortical functions interact with the environment for survival.

More than 100 years ago, Dr. Alois Alzheimer described not only a vivid account of a patient with the disease of the cerebral cortex that later bears his name (Alzheimer's disease, AD), but also linked progressively failing higher cortical functions with specific pathological signatures of cerebral cortices. Since mid-1970s, research on AD has stimulated enthusiasm for exploration into the mechanism of higher cortical functions. In this chapter, major discoveries about AD are highlighted. Most importantly, the fact that AD can affect higher cortical functions is discussed.

ALZHEIMER'S DISEASE: OVERVIEW

In 2015, the Alzheimer Association reported more than five million patients with AD demand a vast amount of care provided at homes and facilities. By 2050, this number will be tripled as the majority of population ages. The most common risk factor of AD is aging. Infrequently, a positive family history of autosomal dominant hereditary dementia posts an increased risk of AD.

Due to the aging processes, many alterations occur in neurons and glial cells as well as in penetrating arteries (arteriosclerosis). Studies on the aged human brain in 60s showed cell loss as the major feature. By definition, the diagnosis of

Conn's Translational Neuroscience
http://dx.doi.org/10.1016/B978-0-12-802381-5.00051-8

dementia means declined intellectual functions. Dementia in the elderly population was called "senile dementia," as the aging-related changes (mainly cell loss and arteriosclerosis) were believed to be the cause of this type of dementia. By contrast, because the first patient reported by Dr. Alzheimer showed dementia in her 50s, AD was originally recognized as a kind of young-onset dementia, so-called "presenile dementia." By 1980, the neuropathological studies noted the hallmarks of the brain examination between "senile dementia" and "presenile dementia" were essentially the same. The terms—senile dementia and presenile dementia—were discarded; they have been called AD universally ever since.

While majority of AD occurs in the elderly—older than 65 of age, AD can occur in the younger individuals (in early 60s or 50s)—so-called young-onset AD. Rarely a certain kind of young-onset AD actually is an autosomal dominant hereditary disease. At the present time, three chromosomes, 1, 14, and 21, are found to have specific mutations that can cause AD. The gene products of these mutation lead to the discovery of amyloid precursor protein (APP at chromosome 21), presenile 1 (PSEN1 at chromosome 14), and presenilin 2 (PSEN2 at chromosome 1). Regarding the major aging changes that can lead to AD, studies demonstrate a wide variety of biochemical and genetic mechanisms, including oxidative stress theory and excitotoxicity. The basal forebrain cholinergic neurons which reside in the nucleus basalis of Meynert have shown neuronal loss in the early stage of AD. In the aging model and Alzheimer brains, researchers have found the aged-related loss of an intracellular calcium-binding protein (calbindin) in the basal forebrain cholinergic neuron in part can provide an explanation of the aging process that can been linked with the cell death in AD.

Moreover, the biochemical and molecular research of AD brains discover the protein structures of pathological hallmarks of AD—senile (amyloid or neuritic) plaques and neurofibrillary tangles, which were originally described by Dr. Alzheimer in 1907. Amyloid plaques are made of by peptides of 1–42 amino acids, named β-amyloid (1–42) peptides (Aβ). These peptides are primarily formed in the extracellular spaces by cleavage from a large membrane protein, called amyloid precursor protein. Their cellular function remains uncertain. Aggregation of Aβ becomes oligmers and even further insoluble filaments; research demonstrates abnormal deposition of Aβ-aggregated proteins can cause neuron death. Processes of dead neurons (dendrites and axons) form "neurites" in the neurophil around the core-aggregated amyloid proteins. Under the microscope with histological staining method, a dense core of amyloid surrounded by break-down neurites are recognized as "neuritic plagues"; this is the classic evidence of AD in the human brain. Numbers of neuritic plaques counted and recorded in the brain regions determine the severity of AD according to the neuropathological criteria.

As of neurofibrillary tangles, these are intracellular aggregated proteins. Under the electronic microscope, these proteins appear like pair-helical filaments. Biochemical analysis has verified that the microtubule-associated protein—tau (τ) protein—is the key component of tangles. For unknown reasons, these τ proteins lose the control of the proper phosphorylation sites and turn into so-called "hyperphosphorylated" state. Therefore, the abnormally hyperphosphorylated τ proteins no longer can function as a transporter protein attached to microtubules within the cell body and dendrites and axon of the neuron. They aggregate together to form longer and larger fibrils and eventually fill the cell body to form "tangles," demonstrated by histochemical staining methods and seen by the light microscope. After the neurons die, the intracellular tangles remain to become the tombstone of dead neurons. As a result, counts of neurofibrillary tangles under the microscope in the brain regions represent numbers of neuron death. These numbers are also used to diagnose AD and to show the severity of AD in neuropathological examination.

In the neurochemical studies, the AD brains display early depletion of the basal forebrain cholinergic neurons. The nuclei of other neurotransmitter systems, such as serotonin and norepinephrine systems, are also affected by Alzheimer pathology. At the present time, only drugs that can enhance cholinergic transmission have established therapeutic effectiveness in treating AD.

For the past three decades, neuropsychological research has discovered several phenotypes of AD. Based on the clinical–pathological correlation studies, about 80% of AD patients begin with memory impairment in the early stage of the disease course. The early stage of AD brains shows amyloid deposition, tangle formation as well as extensive loss of neurons in the medial temporal–limbic system structures. The rest of AD patients, instead of memory impairment, exhibit other phenotypes of clinical presentation—frontal/executive dysfunction, language dysfunction, and visuo-perceptual/visuo-spatial dysfunction. Since 2014, the International Working Group (IWG) recognizes all the phenotypes of AD and applies the diagnostic biomarkers to support the clinical diagnosis of AD. All the phenotypes of AD are discussed in the following sections:

TYPICAL ALZHEIMER'S DISEASE: AMNESIA PHENOTYPE

This type is the most common presentation of AD. Most patients experience progressive memory impairment in the early stage of the disease. They particularly note difficulty learning and remembering new materials. Even if they can make efforts to learn new words or details, they experience rapid forgetting. In other words, they cannot retain the newly formed memory materials. The early changes in the brain are related to limbic structures of the medial temporal lobe, including hippocampal formation and amygdala. Eventually, the Alzheimer pathology spreads to other cerebral cortices of frontal, temporal, and parietal lobes. In the typical MRI of AD, the medial temporal structures show significant atrophy; the degree of atrophy in temporal lobe/limbic structures is outstanding, compared with other cortices (Fig. 32.1.1A).

ATYPICAL ALZHEIMER'S DISEASE

In this category, there are three variants (phenotypes) of AD. According to the clinical-pathological studies, about 15–20% of AD patients present these unique syndromes in the beginning stage. These patients are often younger than 65 years of age and commonly receive wrong diagnoses before the reliable biomarkers of AD are used for diagnosis. Their behavioral changes are associated with specific cognitive difficulties.

Frontal/Behavioral Variant of Alzheimer's Disease

For several decades, the frontal/behavioral variant of AD has been misdiagnosed as the frontotemporal dementia. These patients do not have memory impairment. They are not aware of social norms and display bizarre behaviors with poor social judgment. For example, they ask inappropriate questions to strangers in public. The neuropsychological examination demonstrates dysexecutive syndrome. The brain-imaging study often shows outstanding atrophy of bilateral prefrontal portions of frontal lobes (Fig. 32.1.1B).

Logopenia/Language Variant of Alzheimer's Disease

The second variant of AD is regarding the progressive language impairment. In the most

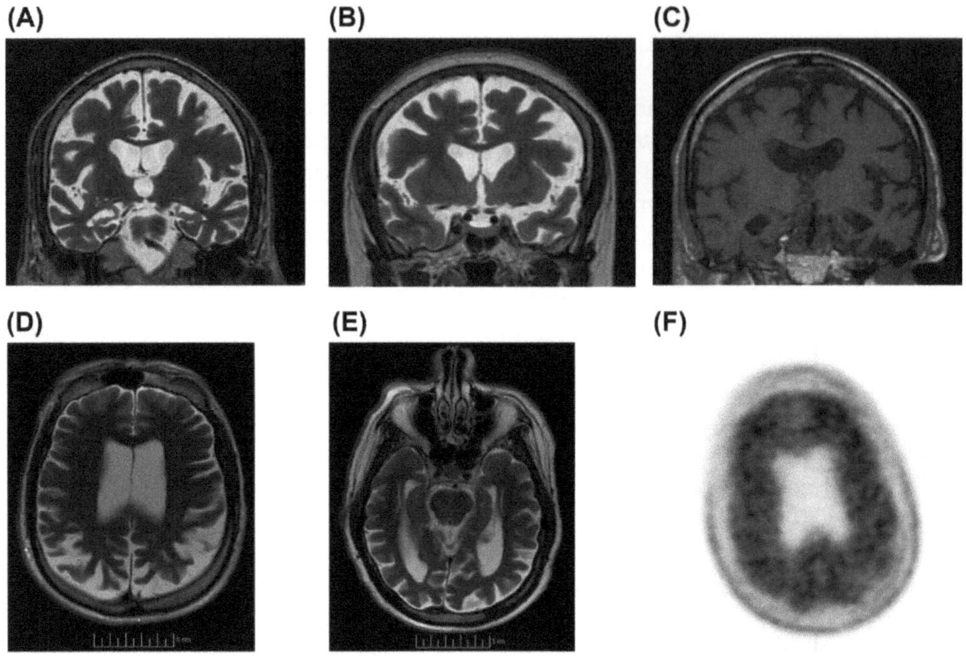

FIGURE 32.1.1 (A) Typical Alzheimer's disease. (B) Atypical-frontal variant of Alzheimer's disease. (C) Atypical-logopenia variant of Alzheimer's disease. (D) Atypical-posterior variant of Alzheimer's disease (biparietal). (E) Atypical-posterior variant of Alzheimer's disease (occipitotemporal). (F) Amyloid PET image of frequent cortical neuritic plaques.

human brains, the function of left hemisphere is dominant in language function. For unknown reason, Alzheimer pathology can begin to occur in the left perisylvian brain regions that are crucial for language function. The brain MRI shows asymmetric atrophy of the left hemisphere (Fig. 32.1.1C). The patients develop aphasia with difficulty in speaking, comprehension, repetition, writing, and reading. Eventually, they cannot comprehend conversations and suffer from communication barriers. Their memory is intact in the early phase of the disease.

Posterior Variant of Alzheimer's Disease

The posterior variant of AD in fact has two different subtypes: a bi-parietal variant (predominant visuo-spatial dysfunction) and an occipitotemporal variant (predominant visuoperceptual dysfunction).

The bi-parietal variant occurs when the Alzheimer pathology first develops in the association cortices of bilateral parietal lobes (Fig. 32.1.1D). The patient is disabled to orient the eyes to search objects in the three-dimensional space. The depth perception is impaired; the right–left distinction is blurred. The patient shows difficulty looking for the correct directions and has trouble dressing (dressing apraxia) and paying attention to objects. Moreover, other common deficits are inabilities to calculate (acalculia) and to perform tasks with tools under verbal commands (apraxia).

The cardinal clinical manifestation of the occipitotemporal variant includes inability to recognize words (alexia), objects (agnosia), and faces (prosopagnosia). The Alzheimer pathology begins to affect the association cortices of posterior temporal lobes and occipital lobes (Fig. 32.1.1E).

DIAGNOSTIC TOOLS OF ALZHEIMER'S DISEASE

To diagnose AD, first the clinician usually needs to determine whether the patient has fulfilled the diagnostic criteria of dementia. Dementia, by definition, indicates a significantly progressive decline in intellectual functions that incapacitates an individual so that he or she no longer can perform daily living skills. The diagnosis of dementia is made based on the clinical symptoms and signs. The next step for the clinician to do is to search for the underlying causes of dementia, reversible causes versus irreversible causes. AD is one of the irreversible causes of dementia. Therefore, a general medical workup, including blood tests for anemia, hypothyroidism, and vitamin B12 deficiency, should be ordered to determine the treatable, reversible causes of dementia. The neuroimaging tests, such as noncontact CT of the head or brain MRI, would be needed to rule out structural lesions, such as cerebral infarction. Afterward, when the diagnostic tests do not find the explanatory causes of dementia, the clinician concludes that the patient's clinical features likely meet the diagnostic criteria of AD, as there are no other causes of dementia that can be established. According to the clinic-pathological studies, the clinical diagnosis of AD can be confirmed by the postmortem autopsy of the brain in majority of patients (more than 80% of cases).

Since mid-1990s, two diagnostic approaches ensure the clinician to diagnose AD with more certainty. One is to use the measurements of amyloid-β (1-42) peptides and τ protein in the cerebrospinal fluid (CSF); the other is to develop the radiopharmaceutical tracers to label amyloid-β or τ shown by positron emission tomography (PET). In AD, the CSF tests typically show dramatically reduced levels of amyloid-β (1–42) and abnormally increased levels of τ. The PET scan of AD demonstrates frequent cortical deposition of neuritic amyloid plaques (Fig. 32.1.1F). Compared with the gold standard diagnostic tool—pathological examination of the AD brain, these two diagnostic tools raise the accuracy of AD diagnosis up to more than 90% of AD cases.

THERAPEUTIC STRATEGIES FOR ALZHEIMER'S DISEASE

At the present time, no cure is available for AD and the disease relentlessly progresses over the years so that AD patients are all in need of supportive care for their daily living. Several therapeutic strategies have developed since 2000s, aiming to alleviate the symptoms and progression of AD. One approach is to solve the abnormal accumulation of amyloid-β (1–42) peptides and τ protein in the AD brain at the earliest stage of disease course—so-called "the disease-modifying drugs." Another approach is to deal with the disturbed cholinergic system in the AD brain. As the basal forebrain cholinergic neurons are depleted by AD, the cholinergic transmission to the cortical regions is disrupted. The drugs, cholinesterase inhibitors, can provide modest effectiveness in cognitive symptoms (memory and attention) for typical AD. However, for the clinical symptoms of atypical AD, like language, visuo-perceptual, and executive dysfunction, cholinesterase inhibitors are not beneficial. Currently, three cholinesterase inhibitors (donepezil, galantamine, and rivastigmine) are routinely prescribed for AD. Finally, memantine has been approved by FDA for its mechanism of NMDA receptor inhibition in the brain that can seemingly delay the mechanism of cell death in AD. Memantine is indicated for the advanced stage of AD.

Almost all AD patients experience behavioral and psychiatric symptoms of dementia (BPSD). The common symptoms are depression, hallucination, delusion, suspiciousness, wandering, agitation, and aggressive behavior. Despite cognitive symptoms of AD, these behavioral issues are the primary causes of caregiver burden.

Although clinicians have tried to apply psychoactive medicines to treat BPSD since 2000s, all currently available medicines fail to show consistent efficacy so that none has gained the FDA approval for BPSD of AD. One reason has been that AD affects multiple psychopharmacological targets through the entire course of disease with progressive loss of neurons in different brain centers of neurotransmitter systems. It is nearly impossible to concurrently treat the behavioral symptoms and cognitive deficits of AD by targeting any one neurotransmitter system.

Further Reading

[1] Hodges JR. Alzheimer's centennial legacy: origins, landmarks and the current status of knowledge concerning cognitive aspects. Brain 2006;129:2811–22.
[2] Wu C-K, Thal L, Pizzo D, Hansen L, Masliah E, Geula C. Apoptotic signals within the basal forebrain cholinergic neurons in Alzheimer's disease. Exp Neurol 2005;195:484–96.
[3] Galton CJ, Patterson K, Xuereb JH, Hodges JR. Atypical and typical presentation of Alzheimer's disease: a clinical, neuropsychological, neuroimaging and pathological study of 13 cases. Brain 2000;123:484–98.
[4] Gefen T, Gasho K, Rademaker A, Lalehzari M, Weintraub S, Rogalski E, et al. Clinically concordant variations of Alzheimer pathology in aphasic versus amnestic dementia. Brain 2012;135:1554–65.
[5] Ossenkoppele R, Pijnenburg YAL, Perry DC, Cohn-Sheehy BI, Scheltens NME, Vogel JW, et al. The behavioral/dysexecutive variant of Alzheimer's disease: clinical, neuroimaging and pathological features. Brain Adv July 2, 2015:1–18.
[6] Dubois B, Feldman HH, Jacova C, Hampel H, Molinuevo JL, Blennow K, et al. Advancing research diagnostic criteria for Alzheimer's disease: the IWG-2 criteria. Lancet Neurol 2014;13:614–29.
[7] Crutch SJ, Lehmann M, Schott JM, Rabinovici GD, Rossor MN, Fox NC. Posterior cortical atrophy. Lancet Neurol 2012;11:170–8.
[8] Clark CM, Pontecorvo MJ, Beach TG, Bedell BJ, Coleman RE, Doraiswamy PM, et al. Cerebral PET with Florbetapir compared with neuropathology at autopsy for detection of neuritic amyloid-beta plaques: a prospective cohort study. Lancet Neurol 2012;11:669–78.

CHAPTER

33

Neurobiology of Drugs of Abuse

N. Dafny, G.C. Rosenfeld

University of Texas McGovern Medical School Houston, TX, United States

INTRODUCTION

Drug abuse, now considered a brain disorder, refers to the use of a drug in ways that are not approved for therapy but rather for the sole purpose of causing pleasure, euphoria, an alteration in perception, or to prevent symptoms of withdrawal. A national survey (2010) on drug use and health estimated that about 22 million Americans 12 years of age and older have a substance abuse or drug-dependent disorder.

Addiction, resulting from the use of psychostimulant, also referred to as psychological dependence, is defined as a compulsive, relapsing, craving, and drug-seeking and drug-taking behavior despite negative consequences of drug consumption. Psychostimulants bind to specific receptors in well-defined brain areas known as the brain reward centers. These centers are involved in cognitive function, learning and memory, attention, reasoning, impulse control, and pleasure. The reward, or good feeling, including euphoria, induced by psychostimulant consumption encourages repeated drug use due to modulation of neurotransmitters in the brain reward areas, as does the avoidance of discomfort that occurs when drug use is terminated. Users move from voluntary drug use to compulsive drug use, primarily to maintain homeostasis. They become addicted despite the negative consequences.

Dependence, also known as physical dependence, refers to the physical and behavioral signs that occur, such as the withdrawal syndrome. The withdrawal syndrome occurs when a drug used continuously is no longer available to the user. In addition to occurring with continuous use of certain drugs taken for nonmedical purposes, it may also occur with continuous drug use for the treatment of a legitimate medical condition that may or may not lead to drug abuse and addiction. Dependence may also occur with the use of nonpsychoactive drugs such as bronchodilators.

The following are often associated with the transition from addiction to dependence.

Craving is the overwhelming desire to reuse a drug, whether abused or not, that can occur spontaneously or can be induced by "triggers," for example, emotional or situational cues, such as depression or anxiety, certain individuals, and joyful social situations that have been previously associated with the user's use of the drug.

Tolerance is the body's reversible adaptation to the repeated use of an addictive drug that results in a reduction in many of its initial physical and behavioral effects. The degree of tolerance may

Conn's Translational Neuroscience
http://dx.doi.org/10.1016/B978-0-12-802381-5.00052-X

vary for different effects of the same drug. It has been hypothesized that repeated consumption of psychostimulants desensitizes their receptors to protect them from over stimulation. Thus an increase in drug dose is necessary to achieve the usual initial effects. Pharmacodynamic tolerance, which may be caused by a variety of mechanisms, represents the cellular responses to the repeated use of a drug. Metabolic tolerance represents the increased synthesis of enzymes, usually hepatic, that are responsible for the metabolism of the drug.

Withdrawal is a constellation of unpleasant symptoms and behavioral manifestations that result from drug cessation. The withdrawal syndrome, also referred to as abstinent withdrawal, represents the body's reaction to the reduction in drug dose or its discontinued use. The symptoms of withdrawal, which are often the opposite of a drug's usual effects and are somewhat drug specific, may promote continued drug use to reverse the unpleasant feelings. The symptoms may include irritability, headaches, anxiety, stomach distress, diarrhea, tremors, sweating, difficulty with cognition, and possibly seizures. Symptoms typically peak after 36 h and can last for even longer periods of time. Following administration of a drug antagonist withdrawal may develop explosively and is referred to as precipitated withdrawal.

Neurocircuitry

As mentioned before, drug abuse is a CNS disorder. To understand the changes elicited by repetitive consumption of a psychostimulant, the brain sites where the drug binds to exert its effect needs to be identified. As a general principle, all psychostimulants bind to the mesocorticolimbic circuit.

The brain is divided into several distinct regions, each involved in performing different functions. The main brain centers participating in the effects stemming from drug abuse are connected to each other and is known collectively as the reward or the motive center; this center is responsible for human motivation. Drugs of abuse elevate mainly the brain catecholamines (CAs) in this center that controls physiological and behavioral function, as well as modulating other neurotransmitters.

All substances of abuse share a common neurocircuitry known as the brain reward or motive system. This brain circuitry is targeted by addictive drugs that mediate pleasure and that strengthen behaviors associated with natural reinforcers, such as food, water, sex, music, and electronic games. The motive circuit includes the following brain structures: ventral tegmental area (VTA), nucleus accumbens (NAc), prefrontal cortex (PFC), hippocampus (Hipp), amygdala (Amyg), caudate nucleus (CN), locus coeruleus (LC), and the raphe nuclei (RN).

In general, drugs of abuse bind to dopamine (DA), norepinephrine (NE), and 5-hydroxytryptamine (5HT; also known as serotonin) transporters. This binding results in an elevation of DA, NE, and 5HT in the synaptic cleft of the brain nuclei belonging to the reward circuit. Their elevation activates the reward/pleasure centers. The balance between the dopaminergic, noradrenergic, and serotonergic systems is essential for normal behavioral activity. Brain nuclei in the brain stem are the main sources of DA, NE, and 5HT, respectively. There is strong evidence showing that drugs of abuse activate the mesocorticolimbic circuit and its projections sites, which suggest that there are interactions between the brain stem, limbic system, motor system, and the cortex. The following are the main brain sites belonging to the motive circuit.

BRAIN SITES

Ventral Tegmental Area

The VTA is comprised of a group of neurons located around the midline of the midbrain floor and contains mainly neurons that produce DA.

The VTA–DA neurons participate in drug addiction, behavioral disorders, cognition, motivation, and locomotor activity. The activity of dopaminergic neurons in the VTA is linked to reward prediction. Most drugs of abuse modulate this system. Moreover, the VTA–DA neurons are essential participants in the rewarding effect of psychostimulants. The VTA is the anatomical site critical to the development of behavioral reward and behavioral sensitization following repetitive drug administration. For example, it has been demonstrated that local amphetamine microinjection into the VTA initiates behavioral sensitization, while VTA lesions result in failure to express behavioral sensitization to chronic drug administration. VTA–DA neurons have reciprocal projections to and from the NAc, Amyg, cingulate gyrus, dorsal raphe, PFC, basal ganglia (BG), and olfactory bulb. Drugs of abuse activate VTA–DA neurons that release DA in the mesocorticolimbic system that in turn affects many different cell populations through VTA circuitry that regulate motivated behaviors. Studies have shown that glutamatergic neurons from the PFC regulate the VTA–DA neurons and that activation of the PFC by psychostimulants participates in eliciting drug dependence.

Nucleus Accumbens

The NAc is a region in the basal forebrain rostral part located below the corpus striatum. Some call this structure the ventral striatum. It is comprised of two parts; the core and shell. The NAc, referred to as the pleasure center, plays a role in motivation and cognitive processing of aversion. It has a significant role in response to drugs of abuse, for example, their reward and reinforcing effect and the translation of an emotional stimulus into behaviors. This anatomical site is critical to the expression of behavioral sensitization resulting from repetitive drug exposure. VTA–DA neurons project to NAc. The NAc acts as a functional interface between the limbic system and the motor system. It has reciprocal innervations with the VTA, PFC, Amyg, Hipp, basal ganglia, and the reticular formation. The NAc influences the planning and inhibition of behavior through its projection to PFC, brainstem and basal ganglia. The same projection also mediates the rewarding effects of pleasurable stimuli from psychostimulant drugs. The NAc neurons are activated by the drugs that create the rewarding experiences felt during exercise, sex, food, and music. The NAc informs the subject what is pleasurable. Local injection of a psychostimulant within the NAc elicits behavioral sensitization, while lesioning of NAc prevents the development of behavioral sensitization to repetitive psychostimulant exposure [5,8]. Moreover, chronic psychostimulant administration also upregulates the NAc cAMP pathway and the transcription factor delta Fos B. Delta Fos B in NAc is critical for the reinforcing effects of music and sex as well as drugs of abuse. All these reports indicate the critical role of the NAc in the response to drugs of abuse and suggest that the NAc is a pleasure center. The NAc and the VTA are the central components of the circuitry underlying reward and the memory of reward.

Prefrontal Cortex

The PFC is involved in those characteristics that distinguish humans from other animals, such as self-awareness, the capacity for complex planning, problem solving, learning and memory, executive functions, personality expression, decision making, and modulation of social behavior. The PFC controls intense emotions and impulses and suppresses emotional urges such as sex, food, and drugs. The PFC uses working memory to guide behavior and attention. It has interconnections with different cortical, diencephalic, and brain stem structures. The PFC integrates sensory information from both the outside world and the body. Dysfunction of the PFC results in many behavioral disorders such as schizophrenia, bipolar disease, attention-deficit/hyperactivity disorder (ADHD), and more. The PFC has reciprocal connections with other cortical, subcortical, and brain stem sites, especially with brain regions involved with attention, action, and cognition. The reciprocal projection from PFC to VTA and Amyg are involved in the expression of the behavioral responses produced by chronic exposure to drugs of abuse. Bilateral electrical, 6-hydroxydopamine (6-OH DA), and ibotonic acid lesions of the PFC modulate the chronic effects of psychostimulants and prevent the development of psychostimulant-induced behavioral sensitization [6], indicating the important role of the PFC in the response to these drugs.

Hippocampus

The Hipp is a major component of the *limbic system* and plays important roles in memory formation and the processing of novel and contextual information. The Hipp dentate gyrus contains neuronal stem progenitor cells (PNIPCs) capable of the generation of new neurons, that is, neurogenesis. Most drugs of abuse, such as opiates, psychostimulants, nicotine, and alcohol, suppress hippocampal neurogenesis. This suggests the important role of hippocampal neurogenesis in the initiation, maintenance, and treatment of drug addiction. Acute psychostimulant application enhances hippocampal long-term potentiation (LTP) and long-term depression (LTD). Repetitive drugs of abuse induce prolonged impairment of memory in rats. All together, these results suggest that drugs of abuse modify the function of Hipp. Furthermore, the Hipp receives a significant DA input from the VTA that acts to modulate Hipp plasticity and, subsequently, learning and memory.

Amygdala

The Amyg forms part of the limbic system and plays a key role in the processing of emotions, learning, memory, reward, attention, and arousal. All are related to processes involved with drug abuse-induced addiction. Emotional reactivity is one of the most disabling symptoms associated with behavioral disorders. The Amyg is involved in various cues associated with the consumption of drugs of abuse and plays a key role in the initial learning of that association. Moreover, inactivation of the Amyg prevents cues from inducing reinstatement of drug self-administration. The Amyg is known to mediate cognitive and emotional processes such as fear and pleasure. Chronic exposure to drugs of abuse has been shown to disrupt the pathways that projects from the Amyg to the PFC. Moreover, neuroimaging of the Amyg has revealed abnormal activation in people who abuse drugs. Of particular importance is the basolateral Amyg (BLA) in permitting emotional regulation and decision-making functions facilitated by the medial PFC (mPFC). The Amyg participates in the response to stress. Stress, defined as a threat to the maintenance of hemostatic balance, is one of the major factors in drug abuse, since it contributes to relapse. The mesocorticolimbic DA circuit, including the Amyg, is implicated in the processing of drug-related reward to mediate the hedonic aspects of rewarding stimuli. Animal models exposed to repeated psychostimulants, such as amphetamine or cocaine, and other drugs of abuse demonstrated modulation of the Amyg DA system. DA exerts neuromodulatory control over excitatory and inhibitory transmission in Amyg–PFC pathway. DA D1 receptor activity suppresses Amyg-evoked excitatory response in PFC neurons. Conversely, DA D2 (but not D1) activity attenuates Amyg-evoked inhibition. Thus mesocorticolimbic DA may exert modulatory control over the balance of excitatory and inhibitory transmission in the Amyg–PFC pathway.

Caudate Nucleus

The CN receives information from all parts of the cortex, thalamus, limbic system, hypothalamus, brain stem nuclei, and the sensory sites. The CN processes information then relays information back to the cortex and the mesocorticolimbic structure. The CN is implicated in voluntary movement, learning, memory, sleep, pain, social behavior and is affected by exposure to drugs of abuse. The CN is one of the CNS sites that regulate locomotor functions. It was shown that selective bilateral lesions to the CN alter the chronic effect of drug of abuse [1]. Drug of abuse administration results in an increased DA release in the CN and increased locomotion in rodents. Acute and chronic administration of drugs of abuse elicit different dose-dependent increases in cFOS immunoreactivity and ΔFos B transcription factor in the CN [3,7]. In human studies, it was demonstrated that psychostimulants accumulate mostly in the CN where they bind to the DA transporters. Reciprocal innervations between the CN, the VTA, NAc, Hipp, and the Amyg are the core motor and limbic circuit interactions and the circuits affected by psychostimulant exposure.

Locus Coeruleus

This brain stem area is the sole source of NE release to different brain areas. NE is a neurotransmitter involved in arousal, cognitive function, memory, the sleep–wake cycle, attention, emotion, and inhibition of stress via the opiate system. Noradrenergic neurons that originate in the LC regulate sensory network processing via NE release with the VTA, brain stem nuclei, cerebellum, thalamus, hypothalamus, Hipp, Amyg, basal ganglia, cortex, and spinal cord [2]. It is thought that NE plays an important role in arousal-dependent and attention-induced modulation of sensory signal processing and activation of LC neurons. Drugs of abuse enhance noradrenergic transmission by blocking NE reuptake via its transporter to increase extracellular NE concentration [4]. This increase in NE regulates the waking state [2]. The LC–NE system is one of the major systems implicated in stress. Drugs of abuse, and opiates in particular (mainly kappa), interact at LC–NE afferents to alter behavior. Treatments for drug abuse elicit withdrawal symptoms that affect the LC–NE system. Thus the role of LC and NE in drug abuse therapy is significant.

Raphe Nuclei

The RN is a collective term of several nerve cell groups in and along the median plane of the medulla oblongata. These nuclei include neurons that synthesize 5HT. Their axons carry this neurotransmitter to subcortical and cortical areas such as the hypothalamus, limbic structure, brain stem area, and spinal cord. The RN synthesizes and releases 5HT and is affected by drugs of abuse. 5-HT plays a role in many brain processes like mood, emotion, aggression and sleep and is implicated in the pathophysiology of anxiety disorders as well as in memory, appetite, pain, and temperature regulation. Serotonergic RN and their forebrain projections are targets of drugs of abuse, mainly alcohol, opioids, psychostimulants, MDMA (3,4-methylenedioxymethamphetamine, commonly known as ecstasy), and others. The serotoninergic system ascends from the RN to the PFC and has been implicated in the etiology of several neuropsychiatric disorders, including obsessive–compulsive disorder, schizophrenia, ADHD, and the cognitive sequelae of certain drugs of abuse. Psychostimulants modulate the serotonergic system. This modulation alters the balance of the catecholaminergic system and is in part responsible for the therapeutic effects of psychostimulants on some behavioral disorders. A relationship has been reported between low levels of 5-HT and the impulsive and violent behavior of the drug addict.

There are complex reciprocal interconnections between the CNS sites described earlier that result in the integration of the internal and the external stimuli, including those resulting from consumption of drugs of abuse that modulate the subjects' behaviors.

GENETICS

Comparing monozygotic twins with dizygotic twins shows that drug abuse, such as alcoholism, is hereditary, suggesting that genetic composition is a risk for addiction. In addition, the environment of drug use in family and friends is a strong factor to becoming a drug user. Although the neurobiological basis of addiction is speculated to be a genetic disorder, the specific, likely multiple, genes involved are not well established.

NEUROCHEMISTRY

Neurons communicate with each other via chemicals that they synthesize and release into the synaptic cleft. These chemicals are known as neurotransmitters, peptides, or neuromodulators. These substances are endogenous chemicals that are synthesized and released from nerve cells. Neurotransmitters released from presynaptic terminals bind to postsynaptic receptors to initiate an electrical impulse that transmits information from one nerve cell to another. There are about 50 different neurotransmitters. Neurotransmitters can exert either excitation and/or inhibition depending on which receptors they bind. The most common neurotransmitters are acetylcholine (ACh), DA, NE, 5-HT, and gamma aminobutyric acid (GABA). Each neurotransmitter can directly or indirectly influence neurons in specific brain sites that affect behavior. Several neurotransmitters have a direct effect on arousal and alertness, learning, memory, mood, motor activity, and the modulation of the brain reward/pleasure center. Neurons that produce DA are called dopaminergic neurons, those that produce NE are called adrenergic neurons, those that produce 5HT are called serotonergic neurons, and those that produce peptide are called peptidergic neurons. These neurotransmitter systems are affected by drugs of abuse and have an important function in the development of addiction to drugs, and to food, smoking, sex, gambling, and more. The chemicals abused by humans are structurally diverse and produce different behavioral effects in the users. All of these drugs share common features in their ability to modulate the brain reward/pleasure center. From an evolutionary perspective, these brain reward sites have ensured survival from generation to generation. Drugs of abuse exert alteration of the brain reward neurotransmitters directly, or by altering other neurotransmitters or neuropeptides that exert a modulatory influence on this neuronal circuit. The most common drugs of abuse stimulate the release of CAs which are responsible for their rewarding and psychomotor effects.

CA are neurotransmitters involved in drug addiction. Major drugs of abuse increase CA release and levels directly or indirectly. An excess level of CA elicit feelings of pleasure and also promote addiction. This increase in CA levels due to drug exposure changes the set point of homeostasis. The brain seeks to maintain the new DA levels constant to keep the new set point. If the drug is not provided, the subject experiences withdrawal. To relieve this unpleasant feeling of withdrawal, the subject needs to be readministered the drug. This craving is powerful and will control the person's life.

For example, a psychostimulant, such as amphetamine, causes release of CA that bind to DA, NE, and 5HT transporters to prevent their reuptake to the presynaptic terminals. DA, NE, and 5HT accumulate in the synapse and thus prolong their effect on their postsynaptic receptors. Other drugs have indirect effects by altering the activity of other neurotransmitters or neuromodulators that affect the CA system.

GABA and glutamate are the brain's most plentiful neurotransmitters. Over half of all brain synapses contain glutamate and 30–40% use GABA. Drugs of abuse that cause an increase in GABA or a decrease in glutamate in the synaptic cleft result in depression. Drugs of abuse that alter GABA and/or glutamate levels are tranquilizers or stimulants depending on the dose and frequency consumed. Alcohol drinking results in decreased glutamate and increased GABA levels, phencyclidine (PCP) causes an increase in mainly glutamate, caffeine causes an increase in glutamate and a decrease in GABA, and tranquilizers increase GABA levels. The changes in these neurotransmitters are the underlying causes of addiction and dependence.

Drugs of abuse are administered intravenously, intraperitoneally, orally, and by inhalation through the nose or lungs. Inhaled drugs have an effect within seconds. Capillary-rich mucous membranes of the nose rapidly absorb the drug into the blood stream. Inhaled drugs are absorbed directly into arterial blood, thus bypassing the venous system on their way to the brain to exert their effects. The effects following intravenous drug injection occur in 1 to 3 min, whereas the effects following intraperitoneal administration are slower to develop. A drug's effect following oral administration takes even longer to develop since the drug must be first absorbed from the gastrointestinal system and traverse the liver before entering the general circulation to exert its effects on the brain. The route of drug administration and the speed with which the drugs reach the brain determine to a large degree their abuse potential.

DRUG OF ABUSE

Substance abuse includes the following:

Alcohol and barbiturates: They are classified as CNS depressants. They cause drowsiness and reduce anxiety.

Opioids: The opioids are powerful analgesics, and very addictive. They include morphine, meperidine (Demerol), hydromorphone (Dilaudid), oxycodone, heroin, and opium which can cause excitement, euphoria, drowsiness, and sedation.

Psychostimulants: The drugs in this class stimulate the CNS and include cocaine, amphetamine, dextroamphetamine, methamphetamine (speed), and methylphenidate (Ritalin).

Benzodiazepines: These drugs are classified as sedative hypnotics and include chloral hydrate, diazepam (Valium), alprazolam (Xanax), lorazepam (Ativan), and paraldehyde. Their use can lead to dependence.

Hallucinogens: LSD, mescaline, phencyclidine (PCP/angel dust), and psilocybin (mushrooms) can cause hallucinations and psychological dependence.

Others: Tetrahydrocannabinol (THC), marijuana (cannabis), hashish, smoking (nicotine), gambling, sex, food, caffeine, and electronic games all elicit addiction.

Drug Changed Brain and Morphology

Repeated administration of drugs of abuse rewires brain circuits. For example, postmortem magnetic resonance imaging (MRI), positron emission tomography (PET), and computed tomography (CT) scans show that alcoholics exhibit smaller brains compared to nonalcoholics of the same age and gender. Many chronic users of alcohol exhibit cognitive impairment. Alcoholism is the second leading cause of dementia. Methamphetamine (speed) damages DA-producing neurons and triggers apoptosis. In general, repetitive use of addictive drugs alters the function of the brain by one or more of the following: excess production and activity of neurotransmitters, depletion and decreased activity of neurotransmitters, inhibition of neurotransmitter binding to their transporter and receptors, modulation of protein synthesis, and changes in neuronal morphology and plasticity [7]. In addition, studies indicate that repetitive consumption of abused drugs modulates the production of transcription factors such as the cAMP response element–binding protein

(CREB) and the transcription factor of the Fos family including a truncated product of the Fos B gene, the delta Fos B. Delta Fos B regulates protein synthesis which results in morphological neuroplasticity [3,7]. These transcription factors participate in regulating protein synthesis that regulates neuronal plasticity and modulates neuronal morphology including alterations of dendrite branching and arborization, dendritic spine morphology and density [9].

Further Reading

[1] Claussen CM, Chong SL, Dafny N. Selective bilateral lesion to caudate nucleus modulates the acute and chronic methylphenidate effect. Pharmacol Biochem Behav 2012;101:208–16.

[2] Drouin C, Page M, Waterhouse B. Methylphenidate enhances noradrenergic transmission and suppresses mid- and long-latency sensory responses in the primary somatosensory cortex of awake rats. J Neurophysiol 2006;96:622–32.

[3] Kim Y, Teylan MA, Baron M, Sands A, Nairn AC, Greengard P. Methylphenidate-induced dendritic spine formation and DeltaFosB expression in nucleus accumbens. Proc Natl Acad Sci USA 2009;106(8):2915–20.

[4] Kuczenski R, Segal DS. Effects of methylphenidate on extracellular dopamine, serotonin, and norepinephrine: comparison with amphetamine. J Neurochem 1997;68:2032–7.

[5] Kuczenski R, Segal DS. Locomotor effects of acute and repeated threshold doses of amphetamine and methylphenidate: relative roles of dopamine and norepinephrine. J Pharmacol Exp Ther 2001;296:876–83.

[6] Li Y, Wolf ME. Ibotenic acid lesions of the prefrontal cortex do not prevent the expression of behavioral sensitization to amphetamine. Behav Brain Res 1999;84:285–9.

[7] Nestler EJ. Review. Transcriptional mechanisms of addiction: role of DeltaFosB23. Philos Trans R Soc Lond B Biol Sci 2008;363:3245–55.

[8] Podet A, Lee MJ, Swann AC, Dafny N. Nucleus accumbens lesion modulates the effect of methylphenidate. Brain Res Bull 2010;82:293–301.

[9] Russo SJ, Dietz DM, Dumitriu D, Morrison JH, Malenka RC, Nestler EJ. The addicted synapse: mechanisms of synaptic and structural plasticity in nucleus accumbens. Trends Neurosci 2010;33(6):267–76.

Index

'*Note*: Page numbers followed by "f" indicate figures, "t" indicate tables and "b" indicate boxes.'

D

F

G

H

J

K

L

M

N

O

Q

R

W

X

Z

Kolibri 360 GmbH
Ferdinand-Jühlke-Straße 7
99095 Erfurt, Deutschland
produktsicherheit@kolibri360.de